Andrew Streitwieser, Jr.
Clayton H. Heathcock

Introduction to Organic Chemistry

SECOND EDITION

Solutions Manual and Study Guide

prepared by

Paul A. Bartlett
University of California, Berkeley

Macmillan Publishing Co., Inc.
NEW YORK
Collier Macmillan Publishers
LONDON

Printed in the United States of America

Macmillan Publishing Co., Inc.
866 Third Avenue, New York, New York 10022

Collier Macmillan Canada, Ltd.

ISBN: 0-02-418060-2

Printing: 6 7 8 Year: 3 4 5 6 7 8

ISBN 0-02-418060-2

ACKNOWLEDGMENT

To those who deciphered my handwriting, typed the manuscript, and drew the 8,791 structures we owe a great deal of thanks: Mary Bergstrom, Mary Browne, Nancy Monroe, Cordelle Yoder, and especially Wendy Zukas, whose incomparable sense of style and unparalleled accuracy are a continual source of delight.

CONTENTS

NOTES ON STUDYING

If you ask a group of college students which course is the most difficult, they will probably say "organic chemistry." This is intimidating to say the least, especially if you are about to start the course. There is no denying that a course in organic chemistry contains a great deal of material, including many names and facts and an extensive framework of basic principles. However, many students discover the logical foundation to the subject and thoroughly enjoy learning it. It requires work, but so do many other skills worth knowing and doing.

Learning organic chemistry requires logical reasoning as well as memorization. The importance of memorization is great, because the framework of organic chemistry rests on a foundation of facts, but you will find that memorization alone is not going to be enough. (If it were, organic chemistry would not have the reputation it has. Sheer memorization is easy--dull perhaps, but basically easy.) To be able to discern a general principle from a collection of facts (reason deductively) and then to be able to predict the behavior of an unknown system (reason inductively) is really what is required to learn organic chemistry beyond the level of simple memorization. The subject can be compared to the game of chess: the chemical behavior of each class of compounds is the counterpart of the moves that different chess pieces are allowed to make. Each chess game becomes intricate and unique, but it always develops within the logical framework of the rules. Similarly, although the range and variety of chemical reactions is often bewildering, they too arise from a finite set of logical principles. We derive a great deal of satisfaction from organic chemistry, as do many of our friends from the game of chess. Hopefully we can help you learn and enjoy the game of organic chemistry as well.

Learning organic chemistry is not easy: it requires you to commit both time and intellectual effort. Learning anything requires reinforcement through repetition, especially repetition in different contexts. In a typical chemistry course repetition is provided in several different ways in order to involve the whole brain. Among these are:

Lectures. The lecture is often an underrated or misunderstood part of the learning process, but it shows you what aspects of the subject the instructor feels are most important, which reactions or mechanisms should receive emphasis, and what should be considered background information. Moreover, the lecturer will present the material in a different way than the text, provide different examples and thereby complement the text. Finally, you listen to a lecturer, and involve a different part of your brain than you do reading.

Lecture Notes. Because the material that the lecturer presents both orally and on the blackboard is clearly important, it is necessary to recall it while studying. Many students in this electronic age use tape recorders for this purpose, but we feel that these are only a poor supplement to taking notes. Unless you videotape the lecture, you lose the blackboard material. Moreover, you can review a 50-minute lecture more effectively and quickly from notes than you can by spending 50 minutes listening to it again. All of which is to say that lecture notes are important. They help you focus your attention on the lecturer and the material (an important aid on Monday mornings if you or the lecturer is not at peak performance...), they force you to process the information (providing another form of repetition), and they make you use a different brain function as well.

Taking good notes does not end with the lecture. You should transcribe the rough notes into a more organized, legible, and complete form as soon as possible after the lecture. This further processing of information (more repetition...) should be done while the lecture is still fresh in your mind and before the short term memory has started to fade.

Textbook Study. As pointed out above, learning organic chemistry requires an intellectual commitment on top of the time commitment of studying. Mere reading is not study: study is work and requires self-discipline. It requires reading with concentration and thought. For instance, you should constantly ask yourself questions and formulate examples on your own as you proceed through the text. In studying a new reaction, run through specific examples in your mind, using those provided in the text as a guide. Interspersed throughout the text are exercises that ask you to do just this. You will find additional specific suggestions on what to look for and how to learn new reactions in this Study Guide.

Most reactions fall into a very few categories such as proton transfer, displacement, addition, elimination, etc. In studying each reaction, see how these categories apply, and see how the mechanism of that reaction compares with similar ones of that class. The most important thing you should remember about a reaction is that it exists; that is, that a given functional group transformation or interconversion can be accomplished. It is secondarily useful, although still important, to know the overall reaction conditions and reagents required and any structural limitations involved. It is generally sufficient to know these aspects in general terms, such as the need for heating or cooling the reaction mixture, having acid or base present, etc., rather than memorizing a host of specific details.

Many students find it useful to construct "flash cards" for learning reactions. If you use this technique, be sure to make your own. Remember that the process of writing out the flash cards helps you to learn what's on them.

Problem Sets. Except for the first and last chapters, each chapter in the text is followed by a number of problems. Working these problems is the most important way to learn organic chemistry. Working a problem forces you to think, points out the things you don't know, and is good practice for exams. When you review your notes, or go back over the text, it is only human nature to say to yourself, "I know that reaction....I've seen that mechanism before....I'll remember that tomorrow morning...." But your professor is not going to be very sympathetic if you write, "I understood that yesterday" as an answer to an exam question. Working problems is really the only way to probe your mind to find out if you know something when you don't have it right in front of you. Multistep syntheses for example provide an excellent test of knowledge. You have to know each reaction, but that alone is not enough; you have to be able to put them together in the proper sequence.

It should be obvious that you have to work the problems on your own; that is, without thumbing back through the text to find the answer, or worse, simply looking it up in this Study Guide. You should make a serious attempt to work a problem before you look for the answer; if you honestly can't, then use the text and your notes to help you. Only when you are completely stymied by a problem, or you have solved it, should you look up the answer in this Study Guide. It will be a valuable check to see if you got the right answer, and it shows you how to solve the problem if you didn't. If you look the answer up right away, you'll just say to yourself, "Oh, I knew that," or "I'll remember that tomorrow" (when you probably didn't and won't).

If the answer given in the Study Guide doesn't make sense to you, check back through the text and your notes again. If you still are confused, by all means see your professor and/or teaching assistant. That is what they are paid for, and most organic chemistry classes are too large for the instructor to come to you. Most instructors like to teach, and they will be pleased if you use their office hours judiciously.

Laboratory. Most introductory organic chemistry courses have a laboratory associated with them. The primary purpose of the laboratory part of the course is for you to learn the experimental operations of organic chemistry and to gain facility in the associated physical manipulations. It takes actual practice to set up an experimental apparatus and carry out the preparation of a compound, just as it takes dexterity and practice to play the piano or make a sauce béarnaise. Moreover, your laboratory experience provides still another reinforcement route to learning organic chemistry. Organic compounds are stuff that melt, boil, smell, have colors and crystalline form; modern organic chemistry has many abstract theories and principles, but they apply to real physical substances. This character of organic compounds is emphasized in the text, but will be reinforced by your laboratory work.

Further Suggestions. Finally, but most importantly, you should study frequently and in continuous pieces. If you limit your study to "cramming" for examinations, you will not learn organic chemistry. There is a great deal of structural heirarchy in the science, and you should understand each level thoroughly before you go on to the next one.

WHAT THIS STUDY GUIDE CONTAINS

The Study Guide is organized in chapters which parallel those of the text. Each chapter contains the following sections:

A. Outline of the chapter in the text, with keywords (bold-faced terms), important ideas introduced, a brief equation illustrating each reaction discussed in the chapter, and where appropriate the major points to keep in mind for these reactions.

B. A discussion section which: 1) provides an overview of the chapter and a look at the important reactions from a different point of view, and 2) gives specific hints on how to approach certain topics and how to solve certain types of problems.

C. Answers to the exercises which are interspersed throughout the chapter in the text.

D. Answers and explanations to the problems at the end of the chapter in the text.

E. Supplementary problems to provide you with additional practice. These problems tend to be a little harder than those in the text, and they often incorporate ideas from previous chapters as well. This can be valuable from the point of view of review, of course.

F. Answers to the supplementary problems. Although these are very near the questions themselves, don't look until you've worked the problems!

The Study Guide concludes with two Appendices:

Appendix I is a glossary of terms, and contains definitions and comments on the vocabulary and concepts which are important in organic chemistry. It is arranged alphabetically and is cross-referenced both among the terms themselves and to the section of the text where the terms are introduced.

Appendix II is a summary of Functional Group Preparations, listing essentially every reaction in the text. It is organized according to the type of product formed, and it gives the page(s) where specific examples can be found.

We hope that this Study Guide is useful to you in learning organic chemistry. Even more importantly, we hope that it helps you to appreciate the beauty and excitement of this area of science, and that you will be one of the many students who enjoy the subject. We would appreciate any comments or suggestions you might have for this Study Guide; we apologize for any mistakes and will be grateful if you bring them to our attention.

2. ELECTRONIC STRUCTURE AND BONDING

2.A Chapter Outline and Important Terms Introduced

2.1 Periodic Table (important elements in organic chemistry)

2.2 Lewis Structures (where the electrons are)

octet configuration
valence electrons
ionization potential and electron affinity
electropositive and electronegative
Lewis structures
Kekulé structures
covalent bonds
formal charges

2.3 Geometric Structure (bond angles and bond lengths)

single bond
double bond

2.4 Resonance Structures (when a single Lewis structure is inadequate)

resonance hybrids
move electrons, not nuclei
importance of octet configuration
minimize charge separation

2.5 Atomic Orbitals (electrons as waves)

Heisenberg Uncertainty Principle
wave function
atomic orbital
probability function
electron density distribution
quantum number
node
shape of orbitals

2.6 Electronic Structure of Atoms (1s, 2s, 2p, etc.)

2.7 Bonds and Overlap (how orbitals are combined)

reinforcement and interference
covalent bond
molecular orbital
bonding and antibonding orbitals
bent bonds

2.8 Hybrid Orbitals and Bonds (the electronic structure of molecules)

hybrid orbital
hybridization (sp, sp^2, sp^3)
bond angle

2.B Important Concepts and Hints

The topics of electronic structure and bonding are presented from two different points of view in Chapter 2. The first is concerned with the accounting of electrons, predicting how many covalent bonds an atom will make, and assigning formal charges. All of this falls under the heading of *Lewis*, or "electron dot", structures. Being familiar with this accounting procedure is important, because as you progress in organic chemistry, you will be expected to keep track mentally of the information which Lewis structures provide, although you will write the simpler Kekulé structures primarily. Another important concept in the accounting of electrons and bonds is that of *resonance*, which enables you to understand the structure and reactivity of a molecule which does not conform to any single Lewis structure.

In parallel with an understanding of Lewis structures and resonance, you will also need to be familiar with the concept of an *orbital*, and many of the topics pertaining to it. Because modern organic chemists describe almost all reaction mechanisms in terms of the orbitals involved, you should understand

1) what an orbital is

2) why it can be filled or empty

3) how a covalent bond is formed from the overlap of two atomic orbitals

4) what *molecular* and *bonding* orbitals are

5) what *hybridized* orbitals are, how and why they are formed, and what their shape and orientation are

and 6) how the geometry of a molecule depends on this hybridization

We don't exaggerate when we say that these concepts are the foundation of the modern approach to understanding organic chemistry. You would be wise, therefore, to understand them from the start.

HINTS: For determining the hybridization of an atom, a convenient approach is to count up the number of things bonded to that atom. By "things" we mean other atoms and unshared electron pairs. For example, methane, ammonia, and water all have four things bonded to the central atom;

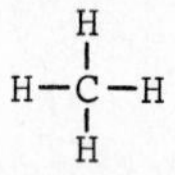

$$H-\overset{\overset{\textstyle H}{|}}{\underset{\underset{\textstyle H}{|}}{C}}-H \qquad\qquad H-\overset{\overset{\textstyle H}{|}}{\underset{\underset{\textstyle \cdot\cdot}{|}}{N}}-H \qquad\qquad H-\overset{\cdot\cdot}{\underset{\cdot\cdot}{O}}-H$$

$$\qquad methane \qquad\qquad\qquad ammonia \qquad\qquad\qquad water$$

they are all, therefore, sp^3-hybridized. The carbon in methyl cation, the carbon and oxygen in formaldehyde, the nitrogens in diimide, and the boron in borane are all bonded to only three things and are sp^2-hybridized. The carbons and boron in these examples are bonded to three atoms only, although in formaldehyde the carbon still has four bonds. Each nitrogen in diimide is bonded to a hydrogen, another nitrogen, and a lone pair. The oxygen in

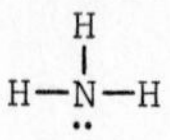

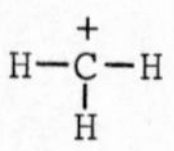

$$H-\overset{\overset{\textstyle +}{|}}{\underset{\underset{\textstyle H}{|}}{C}}-H \qquad\qquad \overset{\textstyle H}{\underset{\textstyle H}{>}}C=\overset{\cdot\cdot}{\underset{\cdot\cdot}{O}} \qquad\qquad H-\overset{\cdot\cdot}{N}=\overset{}{\underset{\cdot\cdot}{N}}-H \qquad\qquad H-B\overset{\textstyle H}{\underset{\textstyle H}{<}}$$

$$\quad methyl\ cation \qquad\quad formaldehyde \qquad\qquad diimide \qquad\qquad borane$$

formaldehyde is bonded to a carbon and two lone pairs. Using these criteria, it is clear why all the non-hydrogen atoms in the structures below are sp-hybridized:

$$H-Be-H\ , \qquad H-C\equiv N\ , \qquad :C\equiv C-H\ , \qquad \overset{+}{:}O\equiv N:$$

The only difficulty arises in attempting to predict the hybridization of the halogens, which by the method outlined above should be sp^3-hybridized. However, their hybridization is unimportant, because for our purposes the significance of evaluating hybridization lies in the prediction of bond angles, and the halogens are ordinarily monovalent. Nonetheless, in some instances (e.g., the fluorine in HF), they are considered to be unhybridized (see Section 8.1 later on in the text).

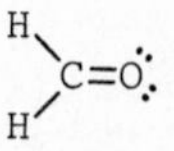

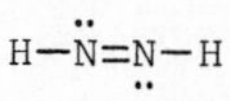

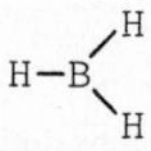

2.C Answers to Exercises

(2.2) (a) chloride ion, :C̈l:⁻ (e) ozone, ⁻:Ö:Ö⁺::Ö

(b) water, H:Ö:H (f) ammonia, H:N̈:H with H below N

(c) hydroxide ion, H:Ö:⁻ (g) cyanogen chloride, :C̈l:C:::N:

(d) hypochlorite ion, :C̈l:Ö:⁻ (h) nitric oxide, :N::Ȯ

(NOTE: with an odd number of valence electrons, one of the atoms must accept an unfilled octet.)

(2.3) Because the Lewis structure for NO indicates a N=O double bond, the bond distance expected is approx. 1.15 Å. The actual distance is 1.15 Å.

Resonance Structures Hybrid

(2.4) nitric acid:

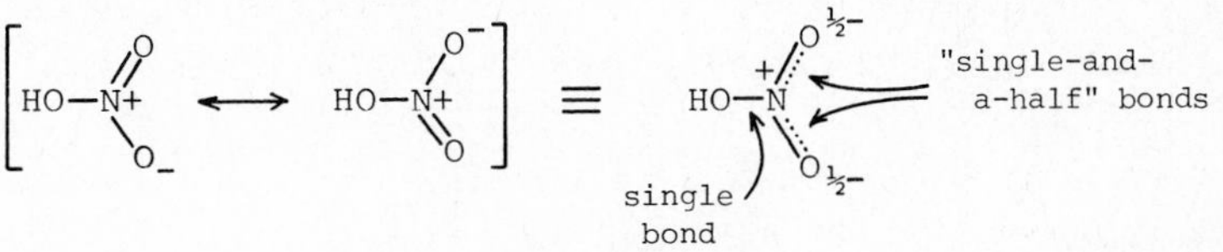

nitrate ion:

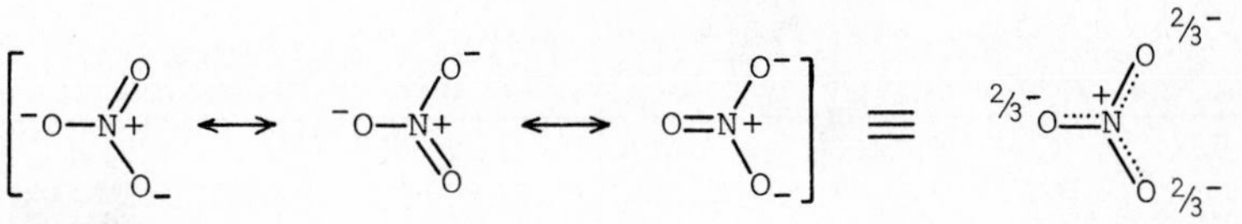

each N-O bond is a "single-and-a-third"

2.D Answers to Explanations for Problems

1(a)

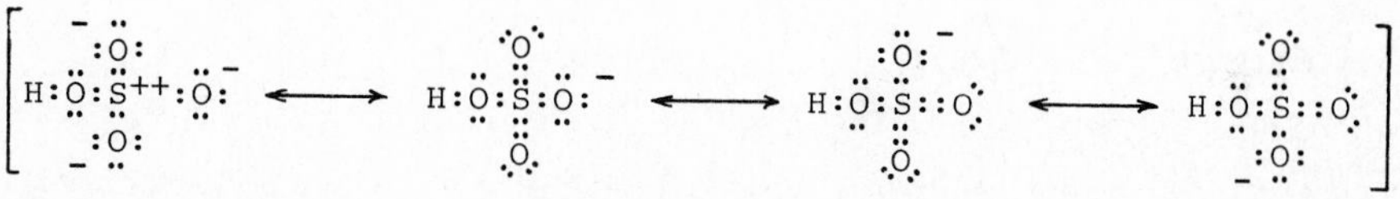

See the Lewis structure of the sulfate ion given in Section 2.2 of the text, and the discussion of resonance hybrids with expanded valence shells in Section 2.4. Note the equivalence of the three resonance structures at the right above.

1(b) H:N̈:H⁻ Nitrogen contributes five valence electrons, each hydrogen contributes one, and the negative charge represents one more for a total of eight.

(c) [:Ö::N̈:Ö:⁻ ⟷ ⁻:Ö:N̈::Ö:]

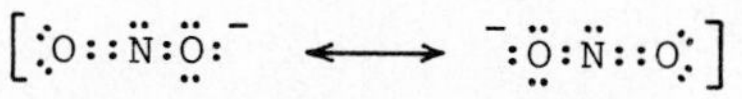

Nitrogen contributes five valence electrons, the two oxygens contribute twelve, and the negative charge adds one for a total of eighteen.

Incorrect:

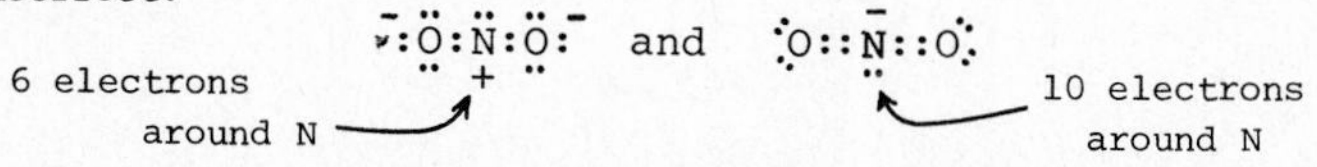

⁻:Ö:N̈⁺:Ö:⁻ and :Ö::N̄::Ö:

6 electrons around N 10 electrons around N

These structures are wrong because nitrogen should not have less or more than an octet.

(d) :Ö::N̈:Ö:N̈::Ö: (e) ⁻:N::N⁺::Ö: (f) ⁻:Ö:N̈:H with H below N

(g) 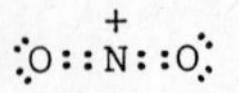 Compare with nitrite ion (ONO^-, 1(c)), which has two more electrons.

(h) 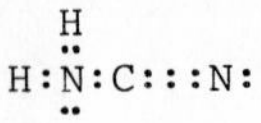

(i) 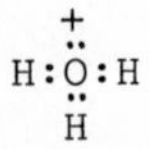 (j) 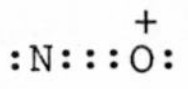 Note the NO triple bond. Compare with N_2, in Section 2.2 of the text.

(k) 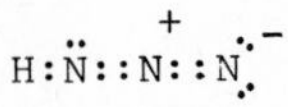

(ℓ) 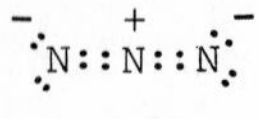 compare with carbon dioxide (OCO, Section 2.2 of the text).
NOTE: An alternative Lewis structure with filled octets is possible, but it is less important than the one above because one nitrogen must have two formal charges: 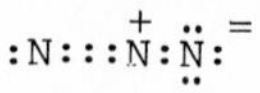

(m) 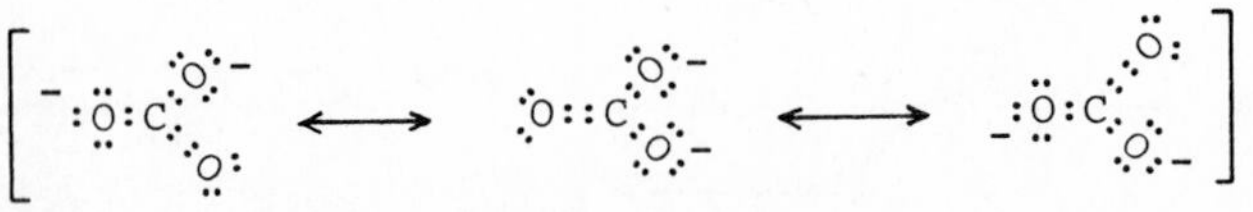

(n) 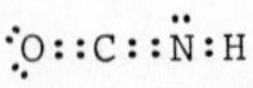

2(a) 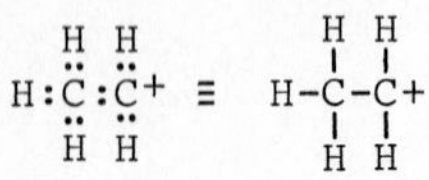 (b) 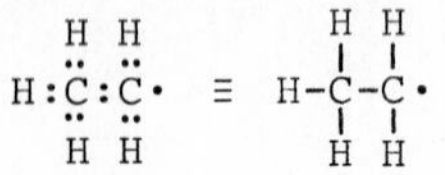 (c) 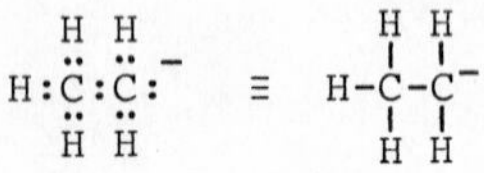

Note the change in charge on carbon as it goes from six (2a) to seven (2b) to eight (2c) valence electrons.

(d) 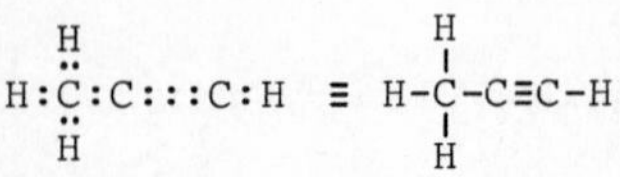 (e) 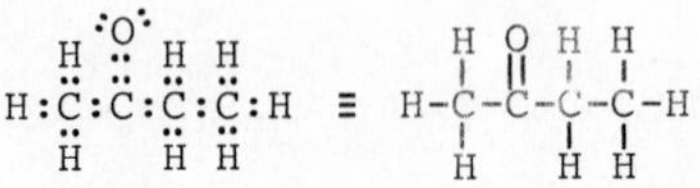

(f) 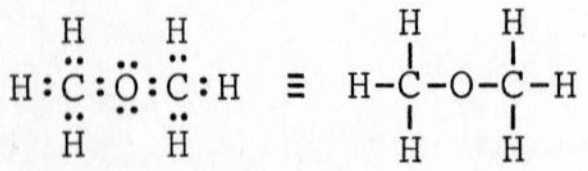 (g) 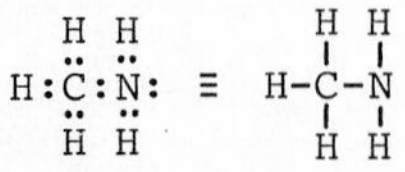 Compare with ethyl anion (2c).

(h) 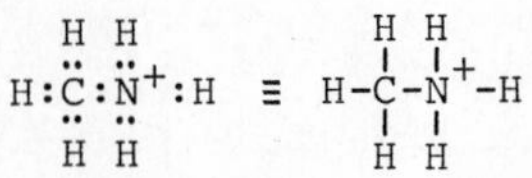 (i) 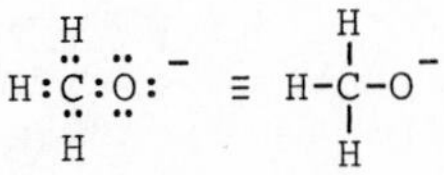

(j) 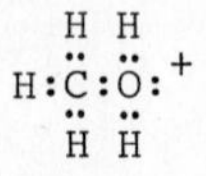 Compare the electronic arrangement in the series: ethyl anion (2c), methylamine (2g), and methyl oxonium ion (2j).

(k) 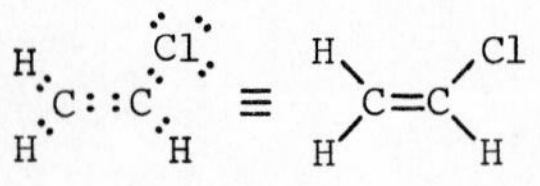

(ℓ) 

Note the non-equivalent resonance structures. The one on the right is less important than the other because the carbon has only six electrons around it. Compare with the nitrosonium ion (NO^+, 1j).

NOTE: for many applications it is convenient to use lines to indicate electron-pair bonds and dots to show lone pairs; for instance:

2(c) 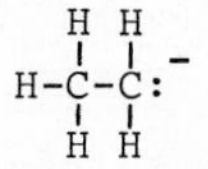

2(e) 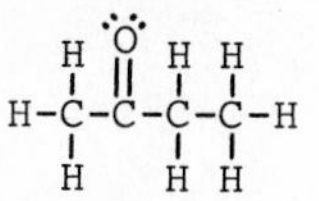

2(j) 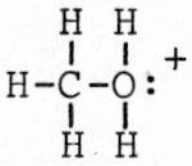

3(a) 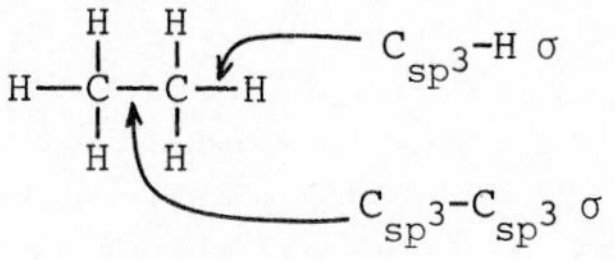

(b) 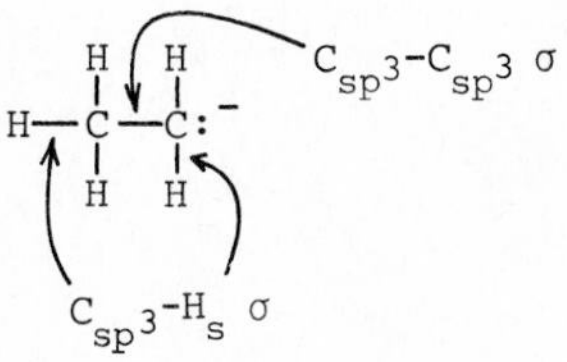

Each carbon has four things bonded to it, and therefore is sp^3-hybridized.

(c) 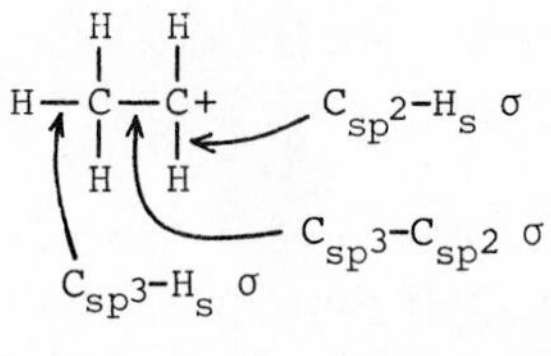

The carbon with only three things bonded to it is planar, sp^2-hybridized.

3(d) 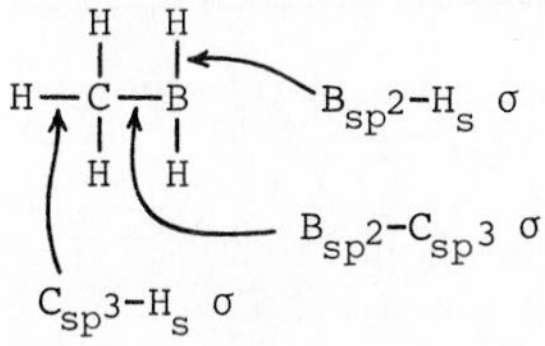

Methylborane has the same arrangement of electrons as ethyl cation (3c).

(e) 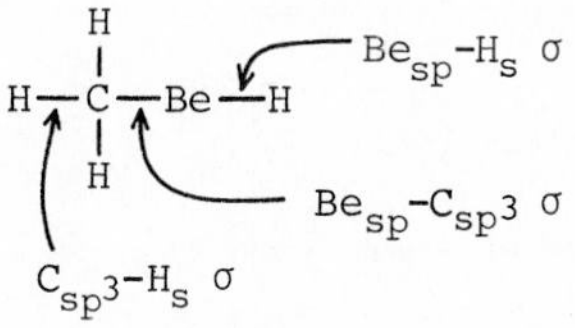

Beryllium, bonded to only two other atoms, is linear, sp-hybridized.

(f) 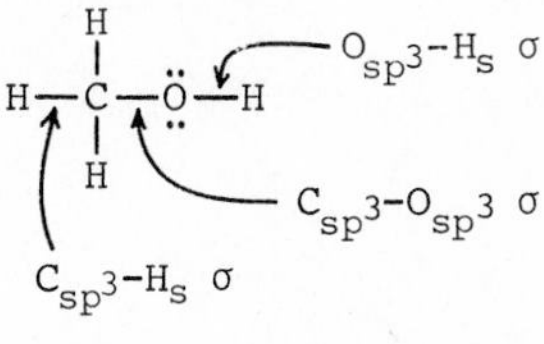

Oxygen, with four things bonded to it (a carbon, a hydrogen, and two lone electron pairs), is approximately sp^3-hybridized.

4. Resonance structures involve a shift in the position of electrons only. They do not involve changes in the positions of the atomic nuclei. Therefore (b), (c), (f), (g), (j), and (k) are not pairs of resonance structures because one or more atoms change position.

5(a) 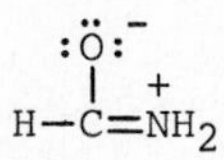 is less important because charge separation leads to higher energy.

(b) 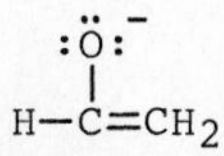 is more important (lower energy) because the negative charge is on a more electronegative atom (oxygen rather than carbon).

(c) $\overset{..}{\underset{..}{O}}:^{-}$ is more important for the same reason given in 5(b): oxygen is more
     H—C≡NH   electronegative than nitrogen.

(d) $\overset{:O:}{\underset{||}{CH_2=CH-CH}}$ is the most important because of the absence of charge separation.

    $^{-}:CH_2-CH=\overset{:O:^{+}}{\underset{|}{CH}}$ is the least important (highest energy) because an electronegative
atom (oxygen) is forced to accept only six valence electrons, while
a much less electronegative atom (carbon) must carry a negative
charge.

(e)   These two structures are exactly equivalent, and therefore contribute equally to the
resonance hybrid.

(f)   These two structures are also equivalent.

(g) $CH_3-\overset{+}{N}≡C-O^{-}$ is more important than $CH_3-N-C≡O^{+}$ because oxygen is more
electronegative than nitrogen. Neither resonance structure is as important as
$CH_3-N=C=\overset{..}{O}:$,   which has no charge separation.

5(h)   O=O=O is of much higher energy (is less important) because the central oxygen
atom has ten valence electrons around it, as can be seen by drawing the full Lewis
dot structure: $\overset{.}{\underset{.}{O}}::\overset{.}{\underset{.}{O}}::\overset{.}{\underset{.}{O}}.$ (three oxygens × 6 = 18 valence electrons)

    NOTE that no octet structure can be written for ozone which does not involve
charge separation.

(i) $\overset{.}{N}=O$ is more important because it has no charge separation and $^{-}N=O\overset{.}{}^{+}$ does.

6.   Na· ⟶ Na^{+} + e^{-} *requires* 118.0 kcal/mole
   Cl· + e^{-} ⟶ Cl^{-} *liberates* 83.3 kcal/mole

   Na· + Cl· ⟶ Na^{+} + Cl^{-} requires 34.7 kcal/mole energy *input*

Electrostatic energy is proportional to $\dfrac{1}{r^2}$; therefore if a positive and negative
charge separated by 1Å have an energy of 330 kcal/mole, Na^{+} and Cl^{-} at a distance of
2.36 Å will have an energy of roughly $\dfrac{1}{(2.36)^2}$ × 330 = 59.3 kcal/mole. This is more
than sufficient to overcome the unfavorable ionization energy calculated above.

2.E Supplementary Problems

S1. (a) Write Lewis structures for BH_4^{-}, CH_4, and NH_4^{+}. Do you expect
the structures of these molecules to be similar or different?

(b) Do the same for AlH_4^{-} and GaH_4^{-}. How should these structures
compare to BH_4^{-}? to PH_4^{+}?

(c) Write the Lewis structures of BH_3 and NH_3. Do these molecules
have similar structures?

S2. Write Lewis structures for the following molecules.

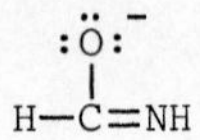

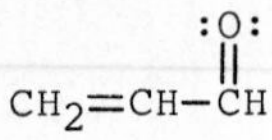

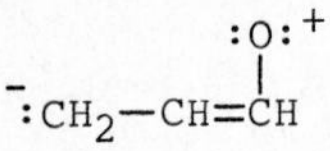

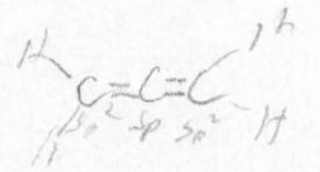

(a) CH_3F (c) ClNO   (e) H_2CCCH_2

(b) H_2O_2 (d) CH_3CHNH (f) H_3BNH_3 (assign formal
                                                              charges)

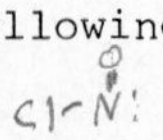

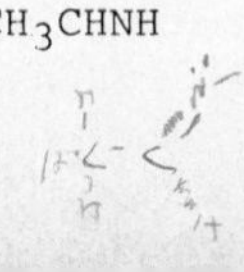

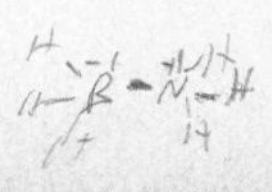

S3. For each of the compounds in problem #S2 above, show the hybridization of all atoms except hydrogen or halogens.

S4. Write out the Lewis structures and corresponding Kekulé structures for at least two resonances structures of each of the following compounds. Circle the one which represents the most important contributor to the structure of each compound, and justify your choice.

(a) $[CH_2CHNH_2]$ (b) $[CH_2NO]^-$ (c) $[CH_2NCH_2]^+$ (d) $[CH_2\overset{O}{C}_{OCH_3}]^-$

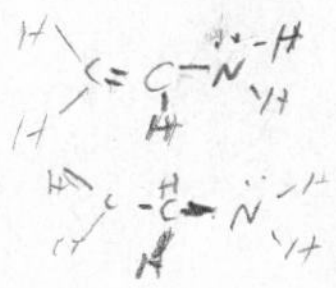

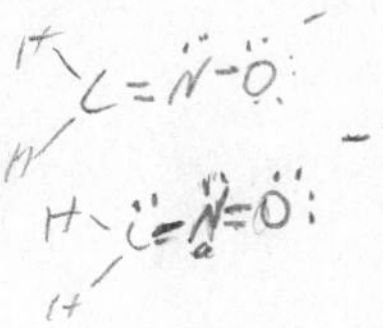

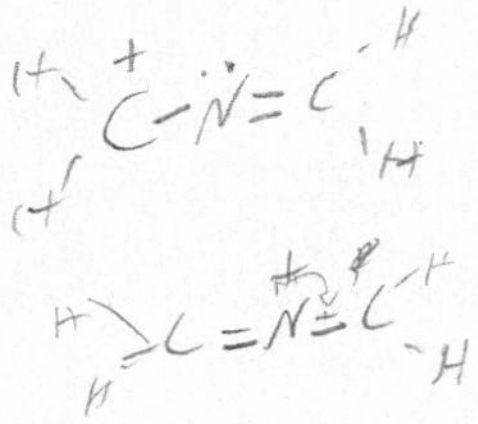

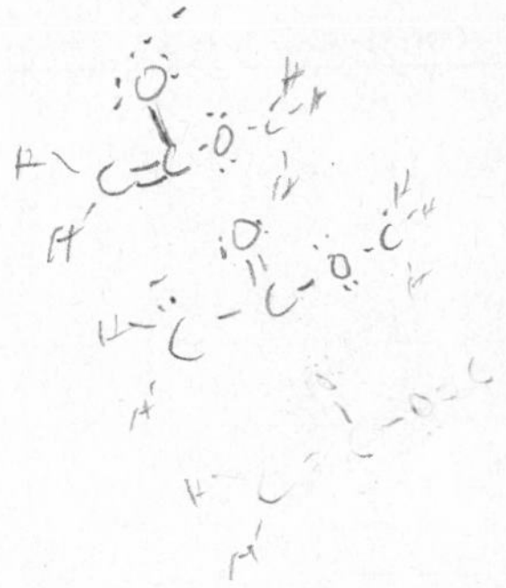

2.F Answers to Supplementary Problems

S1. (a)

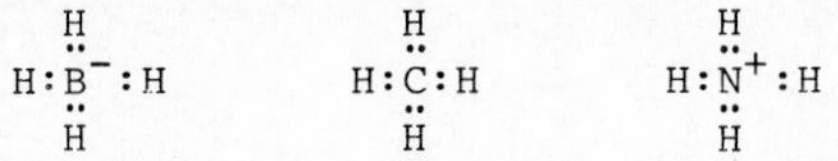

Since the electronic structure is the same for all three molecules (they are isoelectronic), the structures will be similar: they are tetrahedral with sp^3-hybridized central atoms.

(b)

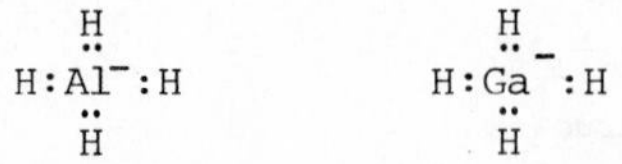

Aluminum and gallium are in the same column of the periodic table as boron, hence their valence shells are similar. BH_4^-, AlH_4^-, and GaH_4^- are all tetrahedral in shape (sp^3-hybridized), although the molecules get bigger on going from B → Al → Ga. PH_4^+ is isoelectronic with AlH_4^- and is tetrahedral as well.

(c)

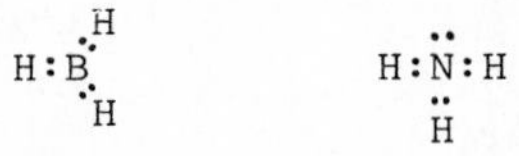

BH_3 has six valence electrons, the boron is sp^2-hybridized and the molecule is therefore planar; NH_3 has eight valence electrons, the nitrogen is sp^3-hybridized, and the molecule is therefore pyramidal.

S2,3.

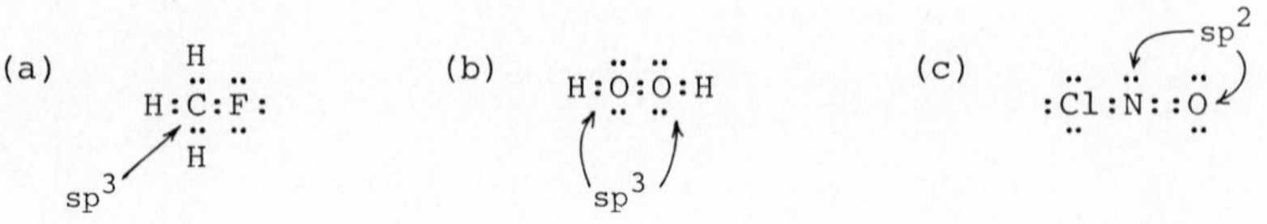

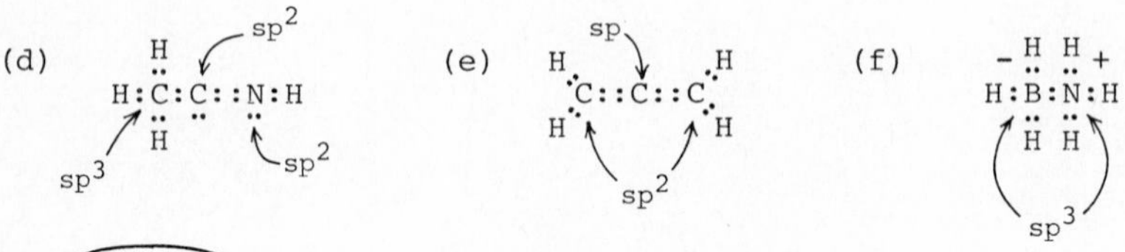

S4. (a)

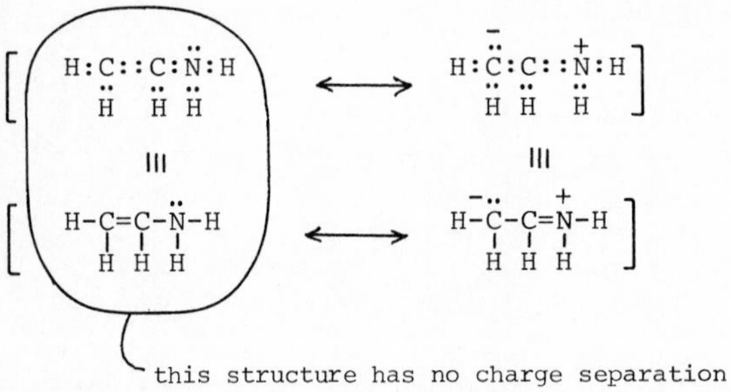

this structure has no charge separation

(b)

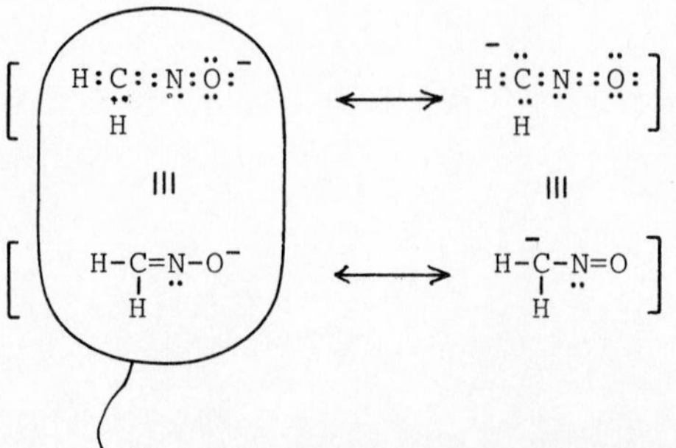

in this structure, the more electronegative atom carries the negative charge

(c)

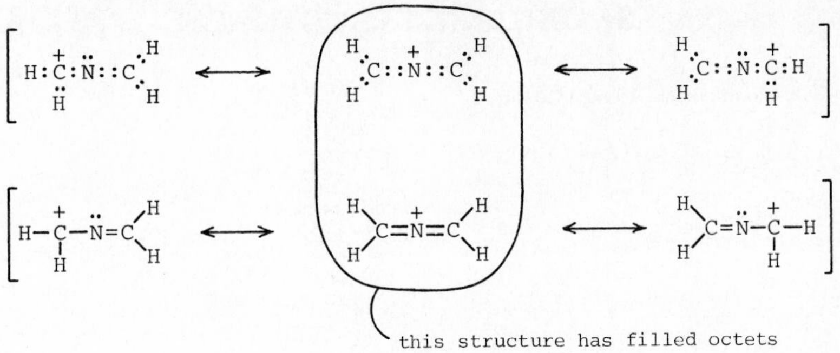

this structure has filled octets

(d)

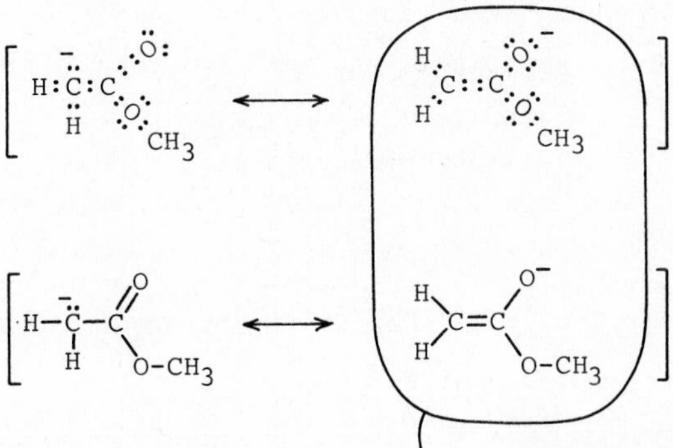

in this structure, the more electronegative element carries the negative charge

3. ORGANIC STRUCTURES

3.A Chapter Outline and Important Terms Introduced

3.1 Introduction (historical)

isomerism

structural formulas

condensed formulas

3.2 The Shape of Molecules (their three-dimensionality)

molecular models

3.3 Functional Groups (commonly-encountered subunits)

hydrocarbons (alkanes, alkenes, alkynes, cyclo-)

alcohols

ethers

aldehydes

ketones

carboxylic acids

(all other functional groups)

3.4 The Determination of Organic Structure (how to figure out the formula)

characterization

combustion analysis

empirical formula

molecular formula

spectroscopy

3.5 n-Alkanes, the Simplest Organic Compounds (the backbone of organic structure)

homologous series

saturated and unsaturated hydrocarbons

3.6 Systematic Nomenclature (naming compounds)

IUPAC system

isomers

common names (iso-, neo-)

chemical literature

3.B Important Concepts and Hints

Chapter 3 continues to discuss structure, but emphasizes the molecular level as opposed to the electronic aspects which are presented in Chapter 2. A good way to organize your thoughts on this topic is to consider the chart on the following page, in which each level down the scheme corresponds to a more specific or detailed description of the structure of a molecule. You should know the difference between empirical and molecular formulas, and

$$CH_3-CH_2-CH_2-\underset{\underset{OH}{|}}{CH_2}$$

A

$$CH_3-CH_2-\underset{\underset{OH}{|}}{CH}-CH_3$$

B

$$CH_3-CH_2-O-CH_2-CH_3$$

C

$$CH_3-CH_2-\underset{\underset{OH}{|}}{CH}-CH_3$$

D

$$CH_3-CH_2 \diagdown \atop CH-OH \atop CH_3 \diagup$$

E

$$CH_3-\underset{\underset{OH}{|}}{CH}-CH_2-CH_3$$

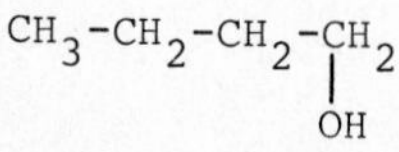

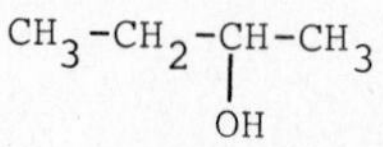

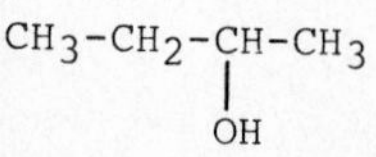

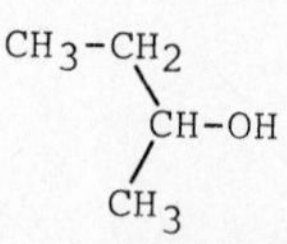

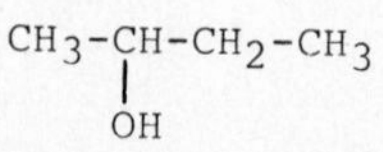

F

understand the concept of *isomerism*. For instance, you should be able to draw all the structural isomers of a compound, given its molecular formula. In this regard, you should also be able to tell that compounds A, B, and C above are structural isomers of each other, while compounds D, E, and F are not.

Another important concept is the idea of functional groups, subunits of molecular structure which are larger than single atoms. These are important not only in considering structural isomerism, but (as you will discover as the course progresses) also in understanding and predicting the reactions a compound will undergo.

After discussing the various levels of molecular structure, how this information can be obtained for a compound is presented. In this chapter you should learn how to determine empirical and molecular formulas, and you should understand the concept of functional group tests. The more detailed methods of studying a molecule's structure are presented in the chapters which discuss spectroscopy (Chapters 9, 14, 17, and 21).

In addition to molecular and structural formulas, we also have to have names for compounds, so that we can discuss them verbally, index them, etc. Although nomenclature seems like a sideline to reactions when studying chemistry, it is important to learn it at the outset of the course so you will know what is being discussed the rest of the time. The systematic nomenclature (IUPAC) is the most important; it does become complex (as the molecules to be named become complex), but it is systematic and the step-by-step rules to follow are easily remembered. The simpler aspects of common nomenclature are useful too, just as clichés and contractions are necessary in our everyday language.

Level	Example
Empirical Formula	$(C_3H_6O)_n$ (this excludes CH_4O, C_3H_8O, etc.)
Molecular Formula	$C_6H_{12}O_2$ (as opposed to C_3H_6O or $C_9H_{18}O_3$, etc.)
Structural Formula	

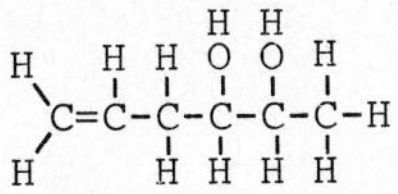

at this level, one can rule out molecules having the same molecular formula (isomers) but with different topology; i.e., with the atoms connected in a different order, such as

$$CH_3-O-CH_2-CH_2-\underset{\underset{OH}{|}}{CH}-CH=CH_2 \quad \text{or} \quad CH_2=CH-\underset{\underset{OH}{|}}{CH}-CH_2-\underset{\underset{OH}{|}}{CH}-CH_3$$

| Stereochemical Formula | |

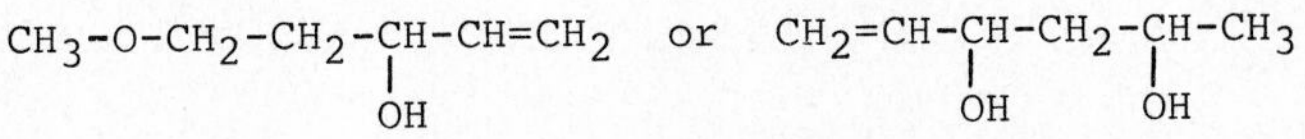

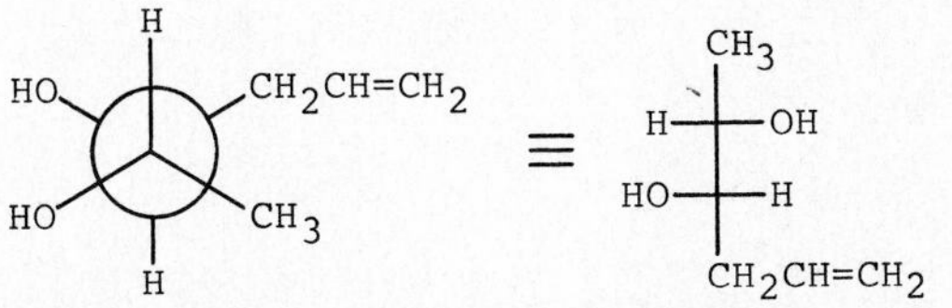

NOTE: the concept of stereochemistry is not discussed in the text until Chapter 7: STEREOISOMERISM)

HINTS: *BUY A SET OF MODELS!* As pointed out in Section 3.2, molecules are three-dimensional and much of organic structure and chemical reactions cannot be understood without thinking in three dimensions. You may not like to admit it, but most people spend their lives on the two-dimensional surface of the earth and have little experience in thinking in three dimensions. Using models is the best way for you to really get an idea of what the two-dimensional illustrations in the text or on the blackboard represent in three dimensions. Other aspects of structure, such as rotation around single bonds (and therefore the floppiness of big molecules), are easily understood using models, too.

Unless the three-dimensional aspects of structure are specifically indicated (for instance, as will be introduced in Chapters 5 and 7), structural formulas only show "connectedness", i.e., what atom is bonded to what atom. A common trap that students fall into arises from a misunderstanding of this point. Although the three drawings below look different, they all represent the same molecule:

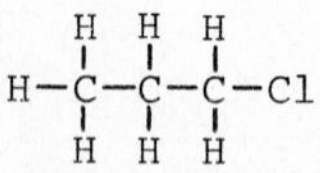

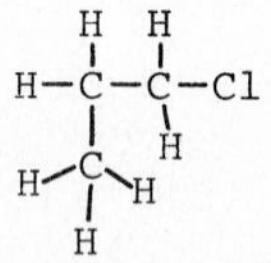

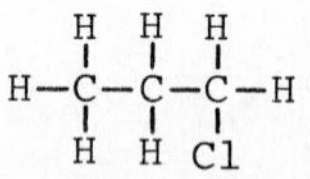

Similarly, the three drawings below represent the same molecule, because of the ability of groups joined by single bonds to rotate relative to each other. The best way to convince yourself of this may be to make models.....

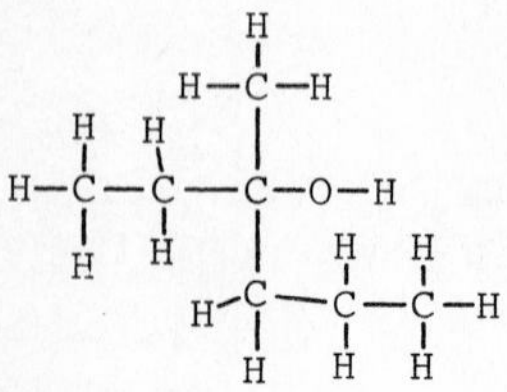

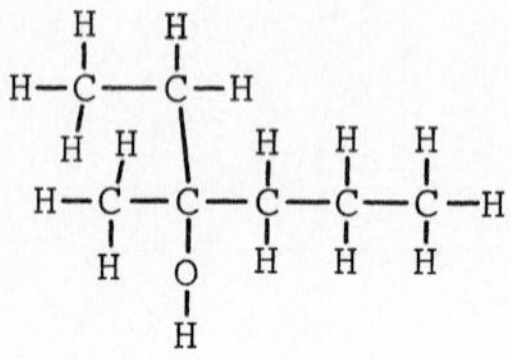

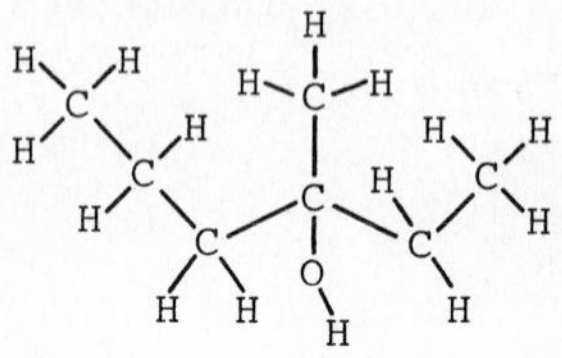

3.C Answers to Exercises

(3.1) C_4H_{10}: 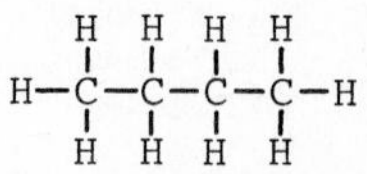 ≡ $CH_3CH_2CH_2CH_3$; 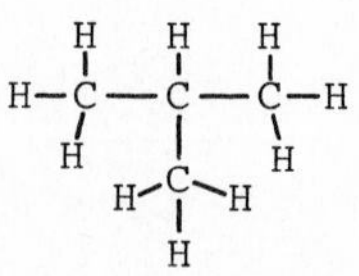 ≡ $(CH_3)_3CH$

C_5H_{12}: 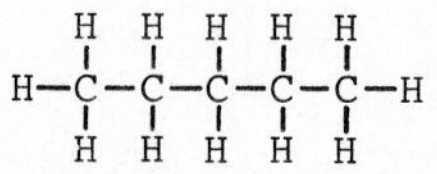 ≡ $CH_3CH_2CH_2CH_2CH_3$;

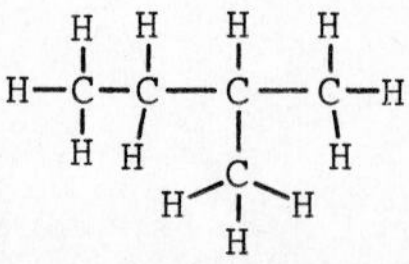 ≡ $CH_3CH_2CH(CH_3)_2$;

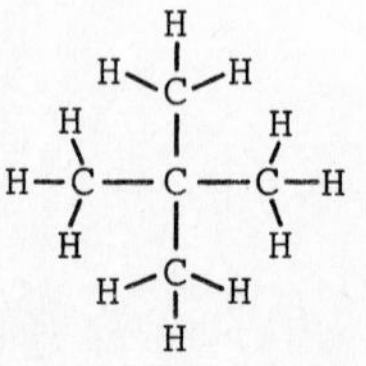 ≡ $(CH_3)_4C$

(3.4) The molecular weight of $C_6H_{12}O$ is $(6 \times 12) + 12 + 16 = 100$, therefore 3.74 mg of $C_6H_{12}O$ represents 3.74×10^{-5} mole. According to the equation for combustion, this will produce

$$C_6H_{12}O + 8\tfrac{1}{2}\,O_2 \longrightarrow 6\,CO_2 + 6\,H_2O$$

$6 \times 3.74 \times 10^{-5}$ mole $= 2.24 \times 10^{-4}$ mole of CO_2. Multiply this by 44 g/mole to find that 9.87 mg of CO_2 are produced.

$6 \times 3.74 \times 10^{-5}$ mole $= 2.24 \times 10^{-4}$ mole of H_2O. Multiply this by 18 g/mole to find that 4.04 mg of H_2O are produced.

3.D Answers and Explanations for Problems

1. (a) "70.4% C, 13.9% H" means: of 100 g of the compound, 70.4 g is carbon, 13.9 g is hydrogen, and the rest (15.7 g) is assumed to be oxygen. A 100-g sample would then contain 70.4/12 = 5.87 moles of carbon, 13.9/1 = 13.9 moles of hydrogen, and 15.7/16 = 0.98 moles of oxygen. This empirical formula of $C_{5.87}H_{13.9}O_{0.98}$ is clearly $C_6H_{14}O$ in integral values. Because this formula represents a fully saturated compound, no multiple of it is possible (see problem #6).

	Wght ratio	Mole ratio	Empirical formula		
(b)	92.1% C	92.1/12 = 7.68	$C_{7.68}$	$= (CH)_n$	(Benzene is C_6H_6)
	7.9% H	7.9/1 = 7.9	$H_{7.9}$		
(c)	71.6% C	71.6/12 = 5.97	$C_{5.97}$	divide by the smallest number →	$(C_4H_5N)_n$
	7.5% H	7.5/1 = 7.5	$H_{7.5}$		
	20.9% N	20.9/14 = 1.49	$N_{1.49}$		(Pyrrole is C_4H_5N)
(d)	71.6% C	71.6/12 = 5.97	$C_{5.97}$	divide by the smallest number →	$(C_{17}H_{19}NO_3)_n$
	6.7% H	6.7/1 = 6.7	$H_{6.7}$		
	4.9% N	4.9/14 = 0.35	$N_{0.35}$		(Morphine is $C_{17}H_{19}NO_3$)
	(16.8% O)	16.8/16 = 1.05	$O_{1.05}$	← the weight not accounted for is assumed to be oxygen	
(e)	74.1% C	74.1/12 = 6.18	$C_{6.18}$	divide by the smallest number →	$(C_{10}H_{12}NO)_n$
	7.5% H	7.5/1 = 7.5	$H_{7.5}$		
	8.6% N	8.6/14 = 0.614	$N_{0.614}$		(Quinine is $C_{20}H_{24}N_2O_2$)
	(9.8% O)	9.8/16 = 0.613	$O_{0.613}$	← the weight not accounted for is assumed to be oxygen	

	Wght ratio	Mole ratio	Empirical formula			
(f)	47.4% C	3.95	divide by the smallest number →	$C_{2.8}H_{1.85}Cl$ →	take multiples until all subscripts are integers → (× 5) →	$(C_{14}H_9Cl_5)_n$
	2.6% H	2.6				
	50.0% Cl	1.41				(DDT is $C_{14}H_9Cl_5$)
(g)	38.4% C	3.2				
	4.9% H	4.9	$(C_2H_3Cl)_n$		(Vinyl chloride is C_2H_3Cl)	
	56.7% Cl	1.6				

(h)

23.4% C		1.95		
1.4% H		1.4		
65.3% I	65.3/127 =	0.514	$(C_{15}H_{11}I_4NO_4)_n$	(Thyroxine is $C_{15}H_{11}I_4NO_4$)
1.8% N		0.13		
(8.1% O)		0.51		

2. (a) 0.0132 g of camphor gives 0.0382 g of CO_2, which is equivalent to 8.68×10^{-4} mole (0.0382 divided by 44). Therefore 0.0132 g of camphor contains 8.68×10^{-4} mole = 0.0104 g of carbon.

0.0132 g of camphor gives 0.0126 g of H_2O, which is equivalent to 7.0×10^{-4} mole (0.0126 divided by 18). Therefore 0.0132 g of camphor contains 14×10^{-4} mole = 0.0014 g of hydrogen.

0.0104 g of carbon plus 0.0014 g of hydrogen leaves 0.0014 g which is unaccounted for from the 0.0132-g sample of camphor; this is assumed to be oxygen (0.0014 divided by 16 = 0.875×10^{-4} mole).

mole ratio in camphor: $C_{8.68}H_{14}O_{0.875}$ —| divide by the smallest number |→ $(C_{10}H_{16}O)_n$

(Camphor is $C_{10}H_{16}O$)

(b) 1.56 mg of sex attractant:

3.73 mg = 0.0848 mmole of CO_2
0.0848 mmole of carbon = 1.018 mg

1.22 mg = 0.0678 mmole of H_2O
0.1356 mmole of hydrogen = 0.136 mg

1.154 mg accounted for

The remaining 0.41 mg is assumed to be oxygen = 0.0256 mmole

$C_{0.0848}H_{0.136}O_{0.0256}$ —| divide by 0.0256 |→ $C_{3.31}H_{5.31}O$ $\xrightarrow{\times 3}$ $(C_{10}H_{16}O_3)_n$

(the molecular formula for this compound is $C_{10}H_{16}O_3$)

(c) 2.16 mg of benzo[a]pyrene:

7.5 mg = 0.17 mmole of CO_2
0.17 mmole of carbon = 2.05 mg

0.92 mg = 0.051 mmole of H_2O
0.102 mmole of hydrogen = 0.10 mg

2.15 mg (this accounts for all of the benzo[a]pyrene)

$C_{0.17}H_{0.102}$ —| divide by 0.102 |→ $C_{1.66}H$ $\xrightarrow{\times 3}$ $(C_5H_3)_n$

(benzo[a]pyrene is $C_{20}H_{12}$)

3. 2.03 mg of sample:

4.44 mg = 0.101 mmole of CO_2
0.101 mmole of carbon = 1.21 mg = 59.7% C

0.91 mg = 0.051 mmole of H_2O
0.102 mmole of hydrogen = 0.10 mg = 5.0% H

5.31 mg of sample gives X mmole of Cl⁻ in solution. Because one mmole of AgNO$_3$ is required for each mmole of Cl⁻, X is equal to 4.80×0.0110 = 0.0528 mmole of Cl. This is equivalent to 1.87 mg of Cl (0.0528×35.5) in 5.31 mg of sample = 35.2% Cl.

$$
\begin{array}{ll}
\text{59.7\% C} & \text{59.7/12 = 4.98} \\
\text{5.0\% H} & \text{5.0/1 = 5.0} \\
\text{35.2\% Cl} & \text{35.2/35.5 = 0.99}
\end{array}
\qquad \text{C}_5\text{H}_5\text{Cl}
$$

4. $CH_3CH_2CH_2CH_2CH_2Br$ 1-bromopentane

$CH_3CH_2CH_2CHBrCH_3$ 2-bromopentane (this is the same as $CH_3CHBrCH_2CH_2CH_3$; also, 4-bromopentane is just an incorrect name for 2-bromopentane)

$CH_3CH_2CHBrCH_2CH_3$ 3-bromopentane

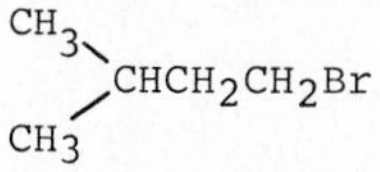

$\begin{array}{c} CH_3 \\ \\ CH_3 \end{array}$ CHCH$_2$CH$_2$Br   1-bromo-3-methylbutane

$\begin{array}{c} CH_3 \\ \\ CH_3 \end{array}$ CHCHBrCH$_3$ 2-bromo-3-methylbutane

$\begin{array}{c} CH_3 \\ \\ CH_3 \end{array}$ CBrCH$_2$CH$_3$ 2-bromo-2-methylbutane
(note that the direction of numbering has changed; 3-bromo-3-methylbutane is an incorrect name)

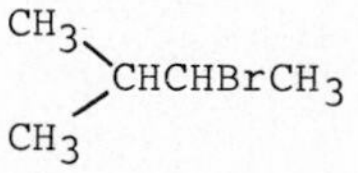

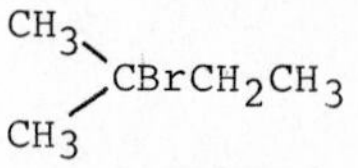

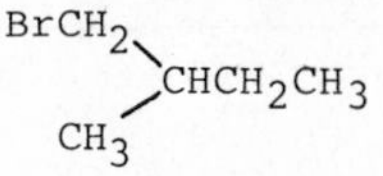

$\begin{array}{c} BrCH_2 \\ \\ CH_3 \end{array}$ CHCH$_2$CH$_3$ 1-bromo-2-methylbutane
(this is the same as $\begin{array}{c} CH_3 \\ \\ BrCH_2 \end{array}$ CHCH$_2$CH$_3$)

$CH_3 - \overset{\overset{\displaystyle CH_3}{|}}{\underset{\underset{\displaystyle CH_3}{|}}{C}} - CH_2Br$ 1-bromo-2,2-dimethylpropane

5. $CH_3CH_2CH_2CH_2OH$ $CH_3OCH_2CH_2CH_3$

$CH_3CH_2CHOHCH_3$ $CH_3OCH(CH_3)_2$

$(CH_3)_2CHCH_2OH$ $CH_3CH_2OCH_2CH_3$

$(CH_3)_3COH$

6.   The formula C_5H_{12} represents a fully saturated hydrocarbon; if the molecular formula were a multiple of this, there would be too many hydrogens to go around.

7.   [no answer]

8.   (a) alkene   (b) ether   (c) alcohol   (d) alkyl halide
(e) carboxylic acid   (f) ketone   (g) aldehyde   (h) disulfide
(i) thiol   (j) sulfide   (k) aromatic ring   (l) primary amine
(m) alkyne   (n) organometallic

9.   (a) alcohol, alkene   (b) ester, alcohol, alkene   (c) tertiary amine, ester, aromatic ring, primary amine   (d) sulfide, carboxylic acid, amide (two), ether, aromatic ring   (e) ketone   (f) sulfide, primary amine, carboxylic acid   (g) alcohol (two), aromatic ring, ether,

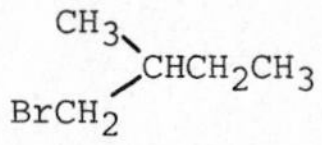

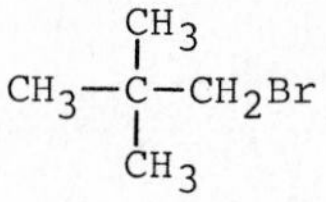

alkene, tertiary amine (h) organometallic (i) alkyl halide (two), sulfide

One of the values of problems #8 and #9 is to train you to look at all of the functional groups in a molecule. A common mistake for students (and experienced chemists, too, sometimes!) is to focus on only one part of a molecule, and to overlook functional groups in the other part of the molecule that will be affected by a reaction one is planning.

10. (a) 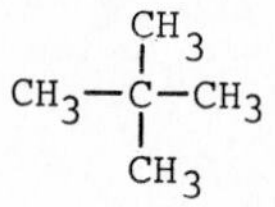 (b) 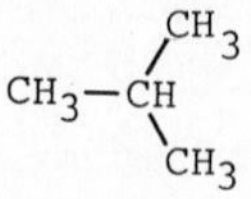 (c) 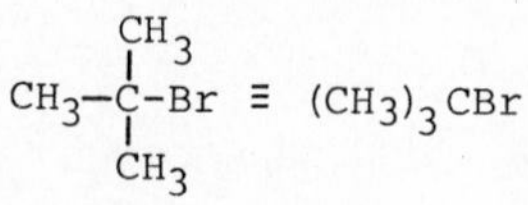

(d) 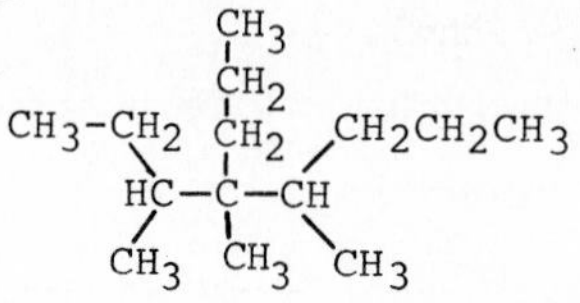 (e) 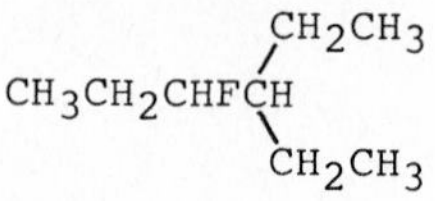

(f) 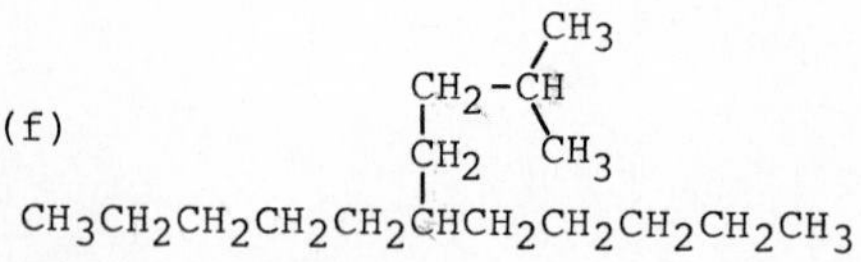 (g) 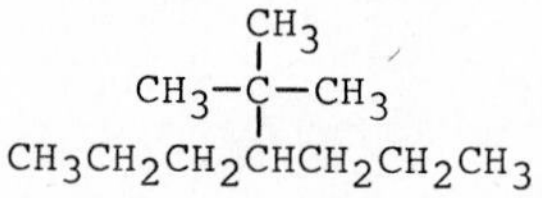

(h) 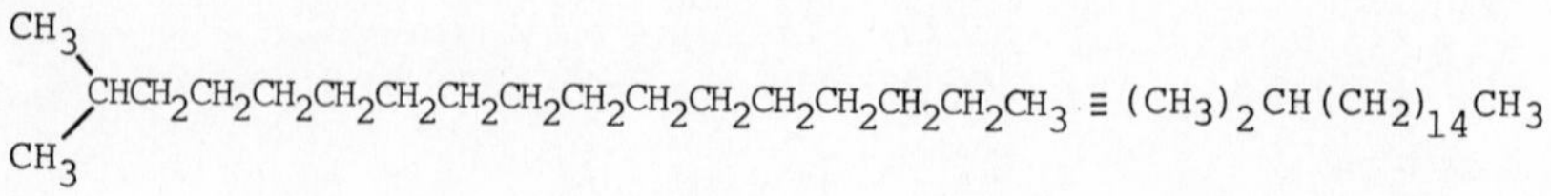

(i) 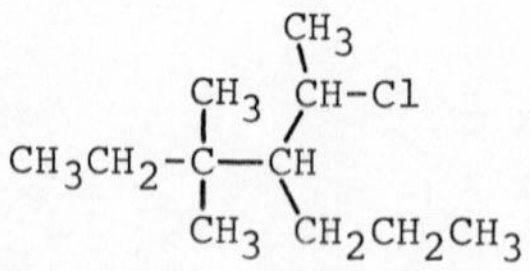 (j) $(CH_3)_2CHCH_2I$ (k) 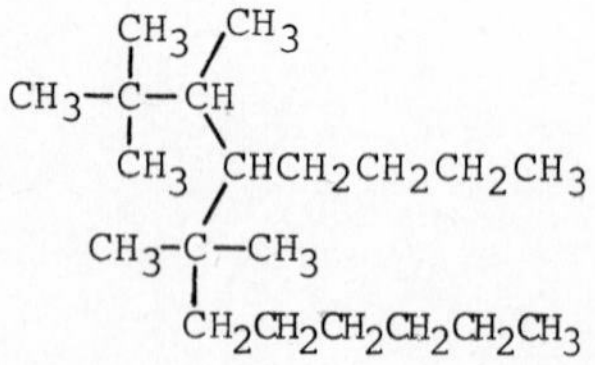

11. (a) 2,5-dimethylhexane
(b) 3-ethyl-5,5,7-trimethylnonane
(c) 4-ethyl-3-methylheptane
(d) 1-bromo-4-chloro-2-methylpentane
(e) 5-ethyl-4-iodo-2,2-dimethyloctane
(f) 3,6-diethyl-2,6-dimethyloctane
(g) 7-(4,4-dimethylhexyl)-3,3,11,11-tetramethyltridecane
(h) 3,3-diethylpentane
(i) 3-ethyl-4-methylhexane
(j) 4-ethyl-3,3-dimethylhexane

12. (a) Where on the heptane backbone is the methyl attached? This has to be specified in a complete name.

(b) "4-Methylhexane" should be numbered from the other end to give 3-methylhexane.

(c) The longest chain in "4-propylhexane" has seven carbons; it should be named 4-ethylheptane.

(d) When choosing between chains of equal length, the one which has more substituents should be chosen: 3-ethyl-2,5,5-trimethyloctane.

(e) Alkyl and halo substituents should be listed in alphabetical order: 3-chloro-4-methylhexane.

(f) The prefix "di-" does not count in alphabetizing: 3-ethyl-2,2-dimethylpentane

(g) Numbered from the wrong end: 3,4,5,7-tetramethylnonane

(h) A position must be specified for every substituent, even if the position is the same: 2,2-dimethylpropane.

3.E Supplementary Problems

S1. From the analytical values for each compound, derive its empirical formula.

(a) Cecropia moth juvenile hormone: 73.6% C, 10.1% H

(b) Valium: 67.5% C, 4.6% H, 12.5% Cl, 9.8% N

(c) Nicotine: 74.0% C, 8.7% H, 17.3% N

(d) Sarin (a nerve gas): 34.3% C, 7.2% H, 13.6% F, 22.1% P

S2. The following compounds were shown to contain only carbon, hydrogen, oxygen, and (if indicated) nitrogen. Calculate the empirical formula for each case.

(a) Combustion of 5.63 mg of aspirin gave 12.39 mg of CO_2 and 2.27 mg of H_2O

(b) Combustion of 1.87 mg of vitamin E gave 5.55 mg of CO_2 and 1.97 mg of H_2O

(c) Combustion of 2.79 mg of caffeine gave 5.06 mg of CO_2, 1.30 mg of H_2O, and 0.80 mg of N_2

(d) Combustion of 1.07 mg of epinephrine (adrenaline) gave 2.31 mg of CO_2, 0.69 mg of H_2O, and 0.08 mg of N_2

S3. On oxidation of sulfur-containing compounds, the sulfur is oxidized to sulfate, which can be determined by conversion to the very insoluble barium salt $BaSO_4$. Combustion of 3.27 mg of saccharin gave 5.50 mg of CO_2, 0.81 mg of H_2O, and 0.25 mg of N_2. Oxidation of a 6.73 mg sample of saccharin and conversion of the sulfate to the barium salt gave 8.59 mg of $BaSO_4$. What is the empirical formula for saccharin?

S4. (a) What is the percent elemental composition of vitamin C ($C_6H_8O_6$)?

(b) How much CO_2 and H_2O do you expect to obtain on combustion of a 3.97 mg sample of vitamin C?

S5. Write out condensed formulas and IUPAC names for all of the isomers of each of the following formulas.

(a) $C_3H_5Cl_3$

(b) C_6H_{12}

(c) C_4H_8ClI

S6. Write condensed structural formulas for each of the following compounds.

(a) 2,5,5-trimethylheptane

(b) 1-bromo-3-ethylpentane

(c) neopentyl bromide

(d) 4,4-di(2,2-dimethylpropyl)-2,2,6,6-tetramethylheptane

(e) 1,2-dichloro-1,1,2,2-tetrafluoroethane ("Freon 114")

(f) 4-(1,1-dimethylpropyl)-2,2,3-trimethyloctane

S7. Give the IUPAC name for each of the following compounds.

(a)
$$CH_3-\underset{\underset{\displaystyle H_3C}{|}}{\overset{}{CH}}-\underset{\underset{\displaystyle CH_3}{|}}{\overset{\overset{\displaystyle CH_3}{|}}{C}}-CH_3$$

(d) $((CH_3)_3C)_2CHCH_3$

(b) $(CH_3CH_2)_2CHCH_2CH_2CH_2Cl$

(e) $(CH_3)CHCH_2\underset{\underset{\displaystyle CH_3}{|}}{CH}CCl F_2$

(c)
$$\underset{\displaystyle BrCH_2CH_2CH_2}{\overset{\displaystyle CH_3CH_2CH_2}{\diagdown}}\overset{}{\underset{\diagup}{C}}{}_{CH_3}-CH_2CH_2CH_3$$

(f)
$$(CH_3)_2CHCH_2\diagdown\underset{\displaystyle CH}{}\diagup CH_2CHCH_2CH_3$$
$$\overset{}{\underset{\displaystyle CH_2}{|}}\qquad CH(CH_3)_2$$
$$(CH_3)_2\overset{}{\underset{|}{C}}$$
$$CH_3CH_2\underset{\underset{\displaystyle CH_3}{|}}{CH}CH_2$$

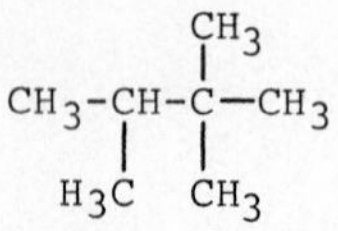

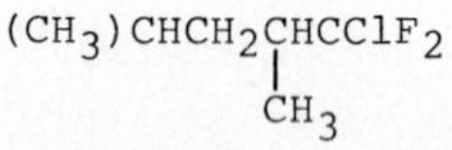

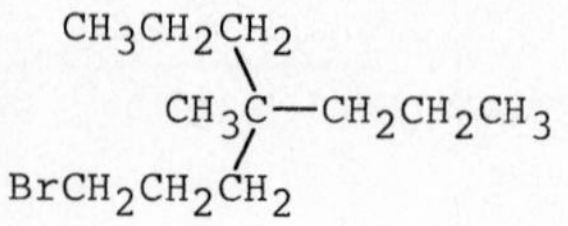

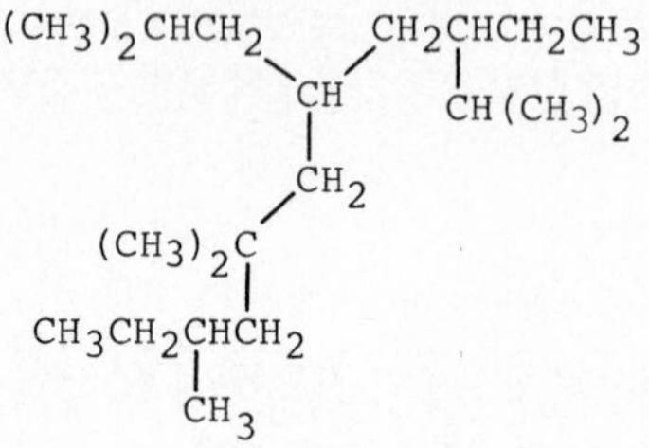

3.F Answers to Supplementary Problems

S1. (a) $C_{6.13}H_{10.1}O_{1.02} = (C_6H_{10}O)_n$ (Cecropia juvenile hormone is $C_{18}H_{30}O_3$)

(b) $C_{5.63}H_{4.6}Cl_{0.35}N_{0.70}O_{0.35} = (C_{16}H_{13}ClN_2O)_n$ (Valium is $C_{16}H_{13}ClN_2O$)

(c) $C_{6.17}H_{8.7}N_{1.24} = (C_5H_7N)_n$ (Nicotine is $C_{10}H_{14}N_2$)

(d) $C_{2.86}H_{7.2}F_{0.72}O_{1.43}P_{0.71} = (C_4H_{10}FO_2P)_n$ (Sarin is $C_4H_{10}FO_2P$)

S2. (a) $\frac{12.39}{44} \times \frac{12}{5.63} \times 100 = 60.0\%\ C$ $\frac{2.27}{18} \times \frac{2}{5.63} \times 100 = 4.5\%\ H$

$100\% - 60.0\% - 4.5\% = 35.5\%\ O$

$C_{60/12}H_{4.5/1}O_{35.5/16} = C_5H_{4.5}O_{2.2} = C_{2.27}H_2O = (C_9H_8O_4)_n$ (Aspirin is $C_9H_8O_4$)

(b) 80.9% C, 11.7% H $100\% - 80.9\% - 11.7\% = 7.4\%\ O$

$C_{6.74}H_{11.7}O_{0.46} = C_{14.6}H_{25.4}O = (C_{25}H_{50}O_2)_n$ (Vitamin E is $C_{25}H_{50}O_2$)

(c) 49.5% C, 5.2% H, 28.7% N $100\% - 49.5\% - 5.2\% - 28.7\% = 16.6\%\ O$

$C_{4.13}H_{5.2}N_{2.05}O_{1.04} = (C_4H_5N_2O)_n$ (Caffeine is $C_8H_{10}N_4O_2$)

(d) 58.9% C, 7.2% H, 7.5% N $100\% - 58.9\% - 7.2\% - 7.5\% = 26.4\%\ O$

$C_{4.9}H_{7.2}N_{0.54}O_{1.65} = C_{9.07}H_{13.33}NO_{3.05} = (C_9H_{13}NO_3)_n$ (Epinephrine is $C_9H_{13}NO_3$)

S3. 45.9% C, 2.75% H, 7.6% N $\frac{8.59}{233.3} \times \frac{32}{6.73} \times 100 = 17.5\%\ S$

$100\% - 45.9\% - 2.75\% - 7.6\% - 17.5\% = 26.25\%\ O$

$C_{3.8}H_{2.75}N_{0.54}O_{1.64}S_{0.55} = (C_7H_5NO_3S)_n$ (Saccharin is $C_7H_5NO_3S$)

S4. (a) $MW = (6 \times 12) + (8 \times 1) + (6 \times 16) = 176$

$\frac{6 \times 12}{176} \times 100 = 40.9\%\ C$ $\frac{8 \times 1}{176} \times 100 = 4.5\%\ H$ $\frac{6 \times 16}{176} \times 100 = 54.5\%\ O$

(b) $\frac{0.409 \times 3.97}{12} \times 44 = 5.95$ mg of CO_2 $\frac{0.045 \times 3.97}{2} \times 18 = 1.61$ mg of H_2O

S5. (a)

$CH_3CH_2CCl_3$	1,1,1-trichloropropane
$CH_3CHClCHCl_2$	1,1,2-trichloropropane
$CH_2ClCH_2CHCl_2$	1,1,3-trichloropropane
$CH_3CCl_2CH_2Cl$	1,2,2-trichloropropane
$CH_2ClCHClCH_2Cl$	1,2,3-trichloropropane

(*NOTE:* $CH_2ClCCl_2CH_3$ is 1,2,2-trichloropropane)

(b)

$CH_3(CH_2)_4CH_3$	hexane
$(CH_3)_2CH(CH_2)_2CH_3$	2-methylpentane
$CH_3CH_2CH(CH_3)CH_2CH_3$	3-methylpentane
$(CH_3)_3CCH_2CH_3$	2,2-dimethylbutane
$(CH_3)_2CHCH(CH_3)_2$	2,3-dimethylbutane

(c)

$CH_3CH_2CH_2CHClI$	1-chloro-1-iodobutane
$CH_3CH_2CHClCH_2I$	2-chloro-1-iodobutane
$CH_3CHClCH_2CH_2I$	3-chloro-1-iodobutane
$ClCH_2CH_2CH_2CH_2I$	1-chloro-4-iodobutane
$CH_3CH_2CHICH_2Cl$	1-chloro-2-iodobutane
$CH_3CHICH_2CH_2Cl$	1-chloro-3-iodobutane
$CH_3CH_2CClICH_3$	2-chloro-2-iodobutane
$CH_3CHClCHICH_3$	2-chloro-3-iodobutane
$(CH_3)_2CHCHClI$	1-chloro-1-iodo-2-methylpropane
$(CH_3)_2CClCH_2I$	2-chloro-1-iodo-2-methylpropane
$(CH_3)_2CICH_2Cl$	1-chloro-2-iodo-2-methylpropane
$ClCH_2\underset{\underset{CH_3}{\mid}}{C}HCH_2I$	1-chloro-3-iodo-2-methylpropane

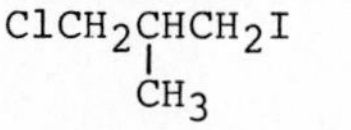

S6. (a) $(CH_3)_2CHCH_2CH_2C(CH_3)_2CH_2CH_3$

(e) $F_2ClCCClF_2$

(b) $BrCH_2CH_2CH(CH_2CH_3)_2$

(c) $(CH_3)_2CCH_2Br$

(f) $(CH_3)_3C\underset{\underset{CH_3-C-CH_3}{\mid}}{\overset{\overset{CH_3}{\mid}}{C}}HCH(CH_2)_3CH_3$
$\underset{CH_2CH_3}{\mid}$

(d) $(CH_3)_3CCH_2-\overset{\overset{(CH_3)_3C}{\mid}}{\underset{\underset{CH_2C(CH_3)_3}{\mid}}{\overset{\overset{CH_2}{\mid}}{C}}}-CH_2C(CH_3)_3$

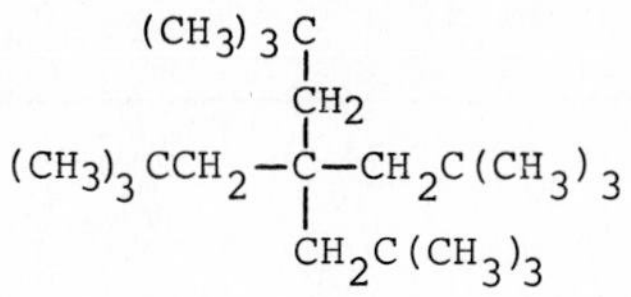

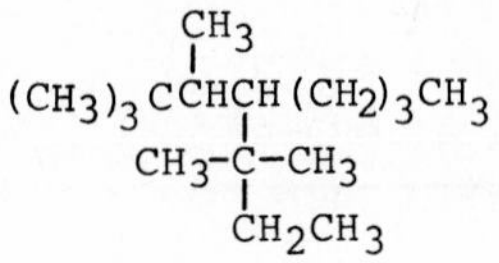

S7. (a)   2,2,3-trimethylbutane

(b)   1-chloro-4-ethylhexane

(c)   1-bromo-4-methyl-4-propylheptane

(d)   2,2,3,4,4-pentamethylpentane

(e)   1-chloro-1,1-difluoro-2,4-dimethylpentane

(f)   3-ethyl-2,7,7,9-tetramethyl-5-(2-methylpropyl)undecane

4. ORGANIC REACTIONS

4.A Chapter Outline and Important Terms Introduced

4.1 Introduction (distinction between equilibrium and rate)

mechanism

4.2 An Example of an Organic Reaction: Equilibria (how far it goes)

"go to completion"

Gibbs Standard Free Energy change, $\Delta G°$

$\Delta G° = -RT \ln K$

driving force

enthalpy, $\Delta H°$

entropy, $\Delta S°$

exothermic and endothermic

4.3 Reaction Kinetics (how fast it goes)

energy barrier

activation energy (enthalpy), $\Delta H^{\ddagger}$

rate constant

first- and second-order reactions

pseudo first-order reactions

4.4 Reaction Profiles and Mechanism (graphs of energy changes in a reaction mechanism)

transition state

maximum energy

activation energy

reaction coordinate

theory of absolute rates ($\Delta G^{\ddagger} = -RT \ln k$ + constant)

reaction mechanism

rate-determining step

reaction intermediate

multi-step reactions

4.5 Acidity and Basicity (an important review -- back to the basics)

solvation

dissociation constant

pK_a

dependence of pK_a on structure

4.B Important Concepts and Hints

Whereas Chapters 2 and 3 present the underlying principles of structure, Chapter 4 discusses the principles of reactivity. Two quite separate concepts are those of *equilibrium* (a measure of how completely a reaction proceeds) and *rate* (a measure of how fast a reaction proceeds). Although they are independent, both depend on potential energy differences which organic chemists like to display pictorially with "reaction profile" or "reaction coordinate" diagrams, such as the simple one sketched below. These diagrams

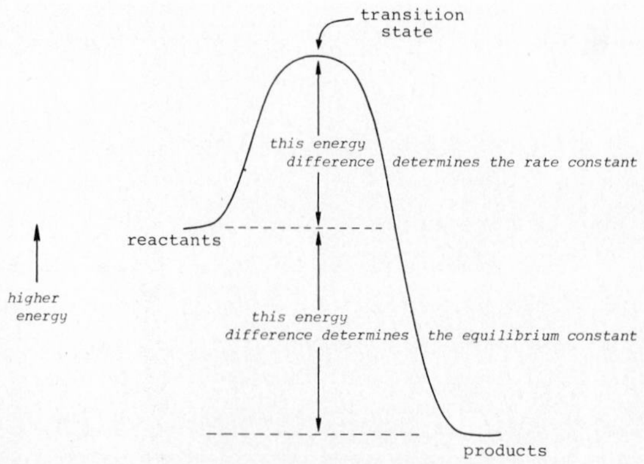

are seldom used in a quantitative sense, but they frequently provide a picture of the energy changes associated with a reaction.

A helpful analogy can be drawn between molecules passing over the energy barrier (through the transition state) leading to product, and water passing over a physical barrier and running downhill. The higher the barrier, the more vigorously the water must be agitated before very much splashes over the edge. This is analogous to raising the temperature of a reaction mixture until enough molecules have the kinetic energy necessary to overcome the activation energy $\Delta G^{\ddagger}$. The energy released as the water runs downhill, on the other hand, depends only on the difference in height between the upper and lower reservoirs, just as the energy released in a reaction depends only on the overall free energy of reaction, $\Delta G°$.

Reaction coordinate diagrams can be useful for an intuitive understanding of the energetics of reaction kinetics and equilibria. However, you should recognize their qualitative nature as well as some pitfalls which you may encounter when using them. Because they really refer only to standard states (approx. 1M for liquid-phase reactions), they cannot be rigorously applied to reactions in which relative rates and equilibria depend on concentration. For example, see Problem #S9 in this Chapter of the Study Guide.

Equilibrium: the equilibrium constant, $K = \frac{[\text{product(s)}]}{[\text{reactant(s)}]}$, is a measure of how far a reaction will proceed before there is no further change in the concentration of reactants and products. It depends on the potential energy difference between the reactants and products, as expressed by the equation $\Delta G° = -RT \ln K$. This is a useful equation to remember, as it can be applied in many situations. Also useful to remember is a particular consequence of this equation: at normal temperatures ($T \cong 300° K$), every 1.37 kcal $mole^{-1}$ change in $\Delta G°$ corresponds to a ten-fold change in the equilibrium constant K (and vice versa).

Rate: the rate of a reaction depends on a constant which is characteristic of each reaction (the rate constant, k), and on the concentration of the reactants. In this regard, be sure you understand the difference between *rate* and *rate constant*, and first- and second-order reactions. The rate constant, k, depends on the activation energy for the reaction, $\Delta G^{\ddagger}$, as shown by the equation:

$$k = \text{constant} \times e^{-\Delta G^{\ddagger}/RT},$$

which rearranges to: $\Delta G^{\ddagger} = -RT \ln k + \text{another constant}$.

This equation is similar to the one which relates the standard free energy of reaction, $\Delta G°$, and the equilibrium constant, K, so the same "1.37 kcal $mole^{-1}$ $\equiv$ a factor of 10" rule-of-thumb applies. Although these equations are similar, don't get $\Delta G^{\ddagger}$ and $\Delta G°$ mixed up.

Chapter 4 also discusses acid/base equilibria fairly thoroughly, although you may feel that the topic belongs more to General Chemistry courses. In fact, the fundamental aspects of acid/base equilibria will be referred to repeatedly throughout your course in organic chemistry. These concepts will be useful for explaining why reactions take place, what mechanisms they proceed by, and so on. You should be familiar with the idea that any molecule which donates a proton in the course of a reaction is an "acid", and any molecule which accepts a proton is a "base". All of the equilibria below are acid/base reactions. Note that some molecules can act either as acids or bases, depending on the situation.

Acids		Bases		Bases		Acids
HCl	+	NH_3	$\rightleftharpoons$	Cl^-	+	NH_4^+
HCl	+	H_2O	$\rightleftharpoons$	Cl^-	+	H_3O^+
H_2O	+	CH_3O^-	$\rightleftharpoons$	OH^-	+	CH_3OH
H_2SO_4	+	$O{=}CH_2$	$\rightleftharpoons$	HSO_4^-	+	$H{-}\overset{+}{O}{=}CH_2$
NH_4^+	+	OH^-	$\rightleftharpoons$	NH_3	+	H_2O
NH_3	+	CH_3^-	$\rightleftharpoons$	NH_2^-	+	CH_4

The most useful measure of how easily a compound gives up a proton is its pK_a (frequently called simply "pK"). The easiest way to remember what the numbers mean is the following: in aqueous solution, when the pH equals the pK_a of a compound, it is half-ionized (i.e., half of it is protonated and half of it is unprotonated). Weaker acids have higher pK_a's and therefore require higher pH's before they are half-ionized, and vice versa for stronger acids. Obviously, this relationship is only realistic for pK_a's in the range attainable in water (0-14), but the device is still useful for remembering what pK_a's outside this range mean; for instance, a negative pK_a means that the compound is a very strong acid. On the other hand, a pK_a of 35 means that the compound is a very weak acid, but that the deprotonated form (called the "conjugate base") is a very strong base. Each pK_a unit represents a factor of 10 in equilibrium because it is a logarithmic scale. Remember: a factor of 10 ≡ 1.37 kcal $mole^{-1}$ in energy difference.

N.B. Later in your course in organic chemistry you may hear: "the pK_a of ammonia is 9". This is a careless statement, although a common one; what is intended is "the pK_a of the ammonium ion (NH_4^+) is 9". Although we often think of both the acidity of a protonated compound and the basicity of the unprotonated form in terms of pK_a, we should always be aware that pK_a refers to the ability of the protonated species to give up its proton; i.e., for the compound to function as an acid. The pK_a of ammonia ($NH_3 \rightleftharpoons NH_2^- + H^+$) is actually 35.

If the whole concept of pH, and of acids and bases in general, has receded from your memory since you took introductory chemistry, you should definitely go back and review this subject . . . . .

4.C Answers to Exercises

(4.2) $\Delta G° = \Delta H° - T\Delta S°$

(a) At 27° C (300° K): $\Delta H° = -10$ kcal $mole^{-1}$

$T = 300°$ K

$\Delta S° = -22$ e.u. $= -22$ cal deg^{-1} $mole^{-1}$

Don't overlook the difference between cal and kcal!

$$\therefore \Delta G° = -10{,}000 - 300\cdot(-22) = -3400 \text{ cal mole}^{-1} = -3.4 \text{ kcal mole}^{-1}$$

$\Delta G° = -RT \ln K$

$\Delta G° = -3.4$ kcal $mole^{-1}$

$R = 1.987$ cal deg^{-1} $mole^{-1}$
(cal, not kcal)

$T = 300°$ K

$$\therefore \ln K = \frac{-3400}{-1.987 \times 300} = 5.70$$

and $K = e^{5.7} = 300$ at 300° K

(b) At 227°C (500° K): $\Delta G° = -10{,}000 - 500\cdot(-22) = +1000$ cal mole^{-1} $= +1.0$ kcal mole^{-1}

$$\ln K = \frac{1000}{-1.987 \times 500} = -1.01, \text{ and } K = e^{-1.01} = 0.36$$

(4.3) rate = $k[OH^-][CH_3Cl]$

$k = 6 \times 10^{-6}\ M^{-1}\ sec^{-1}$

(a) $k \times 0.1\,\underline{M} \times 1.0\,\underline{M} = 6 \times 10^{-7}\ \underline{M}\ sec^{-1}$

(b) $k \times 0.1\,\underline{M} \times 0.1\,\underline{M} = 6 \times 10^{-8}\ \underline{M}\ sec^{-1}$

(c) $k \times 0.01\,\underline{M} \times 0.01\,\underline{M} = 6 \times 10^{-10}\ \underline{M}\ sec^{-1}$

(4.4) Rate-determining step is A → B:

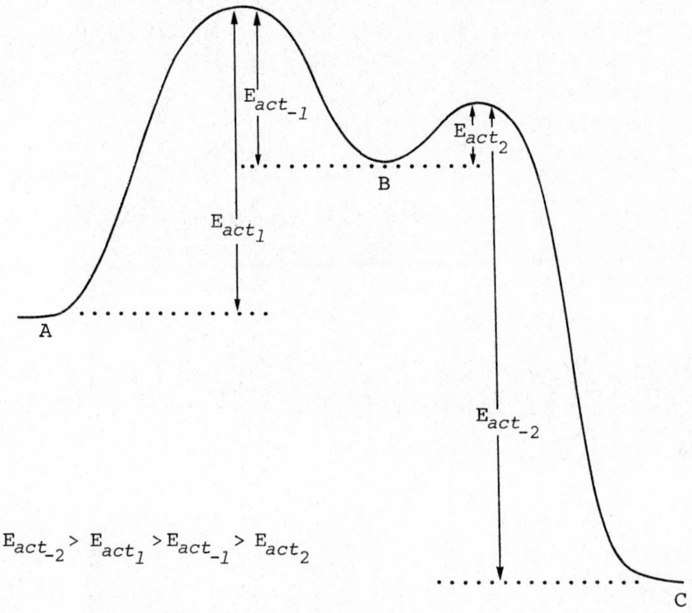

$E_{act_{-2}} > E_{act_1} > E_{act_{-1}} > E_{act_2}$

The rate of B → C is the faster one. For a two-step reaction A ⇌ B → C, in which B is an unstable intermediate (i.e., A → B is endothermic), the rate of A → B can become faster than B → C only if the first transition state is lower than the second (i.e., if $k_{-1} > k_2$).

(4.5) $[HA] \rightleftharpoons [H^+] + [A^-]$, $K = \dfrac{[H^+][A^-]}{[HA]}$. $pK_a = -\log K$ and $K = 10^{-pK_a}$

$$[H^+] \cong [A^-] = 1 - [HA] \quad \text{and} \quad K = \frac{(1 - [HA])^2}{[HA]}$$

<u>For HI and HCl</u>: when K is very large, [HA] ≅ 0 and $[H^+] \cong 1\,\underline{M}$; pH = 0

<u>For HF</u>: $K = 10^{-3.2} = 6.3 \times 10^{-4}$. In this case, K is small, $[HF] \cong 1\,\underline{M}$ and

$$\frac{[H^+][F^-]}{1} \cong 6.3 \times 10^{-4}; \quad [H^+] = [F^-] = \sqrt{6.3 \times 10^{-4}} = 2.5 \times 10^{-2}\ \underline{M} \text{ and pH} = 1.6$$

For acetic acid: $K = 10^{-4.76} = 1.74 \times 10^{-5}$. As for HF, very little of HA dissociates: $[HA] \cong 1\underline{M}$; $\frac{[H^+][A^-]}{1} \cong 1.74 \times 10^{-5}$; $[H^+] = \sqrt{1.74 \times 10^{-5}}$

$= 4.17 \times 10^{-3}\,\underline{M}$

and pH = 2.4

For H_2S: $K = 10^{-7}$. Again, very little of H_2S ionizes: $[H_2S] \cong 1\,\underline{M}$ and $[H^+] = [HS^-] = \sqrt{10^{-7}}$

$= 3.2 \times 10^{-4}\,\underline{M}$ and pH = 3.5

4.D Answers and Explanations for Problems

1. (a)

$\Delta H° = 7.3$ kcal mole^{-1}

T = 298° K

$\Delta S° = 0.3$ cal deg^{-1} mole^{-1}

Note difference between kcal and cal

$$\Delta G° = \Delta H° - T\Delta S° = 7300 - (298 \times 0.3)$$
$$= 7211 \text{ cal mole}^{-1} = 7.21 \text{ kcal mole}^{-1}$$

(b)

$\Delta G° = 7210$ cal mole^{-1}

R = 1.987 cal deg^{-1} mole^{-1} *(cal, not kcal)*

T = 298° K

$\Delta G° = -RT \ln K$, so that $\ln K = \frac{7210}{-1.987 \times 298} = -12.2$

and $K = e^{-12.2} = 5 \times 10^{-6}$

(c) No; the reaction would actually "go to completion" in the opposite direction.

2. (a) $\Delta G° = \Delta H° - T\Delta S°$

$= 22,200 - (298 \times 33.5)$

$= 12,200$ cal mole^{-1} = 12.2 kcal mole^{-1}

The equilibrium lies far to the left (a positive $\Delta G°$ indicates an unfavorable reaction in the direction written).

(b) $\Delta G° = \Delta H° - T\Delta S°$

At 800° K, $\Delta G° = 22,200 - (800 \times 33.5)$

$= -4,600$ cal mole^{-1} = -4.6 kcal mole^{-1}

At this temperature, the equilibrium lies to the right.

(c) The contribution of $\Delta H°$ to $\Delta G°$ is unaffected by temperature, but the contribution of $\Delta S°$ depends directly on temperature because of the $-T\Delta S°$ term. Therefore, at higher temperatures the entropy term $\Delta S°$ becomes more important.

3. (a) At room temperature (300° K), compare $\Delta G°$ for K and for 10K:

$\Delta G°_A = -RT \ln 10K$ $\quad \Delta G°_B = -RT \ln K$

$$\Delta G°_A - \Delta G°_B = -RT(\ln 10K - \ln K) = -RT \ln 10$$
$$= -1.987 \times 300 \times 2.303$$
$$= -1370 \text{ cal mole}^{-1} = -1.37 \text{ kcal mole}^{-1}$$

For a factor of 100, $\Delta G°_A - \Delta G°_B = -RT \ln 100 = -2.75$ kcal mole^{-1}

(b) A factor of 10 in K equals a change in $\Delta G°$ of 1.37 kcal $mole^{-1}$ at room temperature. This could arise from a change in $\Delta H°$ of 1.37 kcal $mole^{-1}$ if $\Delta S°$ remained constant, or a change in $\Delta S°$ of $-\frac{1370}{298} = -4.6$ e.u., if $\Delta H°$ remained constant.

4. (a) $k = 6\times10^{-6}\ \underline{M}^{-1}\ sec^{-1}$

$[OH^-] = 0.10\ \underline{M}$

$[CHCl_3] = 0.05\ \underline{M}$

$$\begin{aligned} rate &= k[OH^-][CHCl_3] \\ &= 6\times10^{-6}\ \underline{M}^{-1}\ sec^{-1} \times 0.10\ \underline{M} \times 0.05\ \underline{M} \\ &= 3.0\times10^{-8}\ \underline{M}\ sec^{-1} \text{ at the start of reaction} \end{aligned}$$

(b) and (c) Each 10% of reaction means a change in $[CHCl_3]$ of 0.005 $\underline{M}$.

% Reaction	$k, \underline{M}^{-1} sec^{-1}$		$[OH^-], \underline{M}$		$[CHCl_3], \underline{M}$		Rate, $\underline{M}\ sec^{-1}$	Time for 10% reaction
0	6×10^{-6}	×	0.10	×	0.050	=	3.0×10^{-8}	
								1.67×10^{5} sec
10	6×10^{-6}	×	0.095	×	0.045	=	2.6×10^{-8}	
								1.92×10^{5} sec
20	6×10^{-6}	×	0.090	×	0.040	=	2.2×10^{-8}	
								2.27×10^{5} sec
30	6×10^{-6}	×	0.085	×	0.035	=	1.8×10^{-8}	
								2.78×10^{5} sec
40	6×10^{-6}	×	0.080	×	0.030	=	1.4×10^{-8}	
								3.57×10^{5} sec
50	6×10^{-6}	×	0.075	×	0.025	=	1.1×10^{-8}	

Total time = 1.22×10^{6} sec
= 339 hours

[Using calculus instead of this approximate method, the correct length of time is calculated to be 375 hours. See problem #14 in this Chapter.]

5. (a) The entropy is negative because more order is introduced into the system (less freedom of motion) when two molecules combine to give one.

(b) $\Delta G° = \Delta H° - T\Delta S°$

NOTE!

$$= -15.5 \text{ kcal mole}^{-1} - (298 \text{ deg} \times -31.3 \text{ cal deg}^{-1} \text{ mole}^{-1})$$

$$= -6.17 \text{ kcal mole}^{-1}$$

(c) $\Delta G° = -RT \ln K$, so that $\ln K = \dfrac{-6170 \text{ cal mole}^{-1}}{-1.987 \text{ cal deg}^{-1} \text{mole}^{-1}\ \ 298 \text{ deg}} = 10.4$

and $K = e^{10.4} = 3.29\times10^{4}$

$$K = \frac{[C_2H_5Cl]}{[HCl](C_2H_4]} = \frac{P_{C_2H_5Cl}}{P_{HCl}\times P_{C_2H_4}}$$

Because the reaction essentially goes to completion, $P_{C_2H_5Cl}$ at equilibrium = 1 atm (1 atm HCl + 1 atm C_2H_4 $\rightleftharpoons$ 1 atm C_2H_5Cl). Furthermore, because $P_{HCl} = P_{C_2H_4}$ at the start of reaction, and one molecule of HCl is consumed for every molecule of C_2H_4, P_{HCl} will always equal $P_{C_2H_4}$.

$$K = \frac{1}{(P_{HCl})^2} = 3.29\times10^{4} \Longrightarrow P_{HCl} = P_{C_2H_4} = 5.5\times10^{-3} \text{ atm}$$

(d) For all three components to be present in equal amounts, i.e., at equal pressures of each component: $K = \dfrac{P}{P\cdot P} = \dfrac{1}{P} = 3.29\times10^{4} \Longrightarrow P = 3\times10^{-5}$ atm.

With each component at this pressure, the total pressure will be 9×10^{-5} atm.

6. (a)

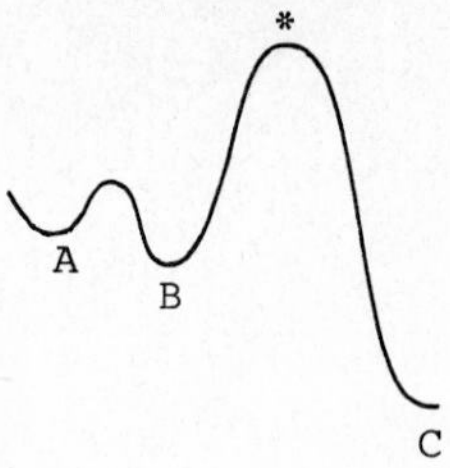

or

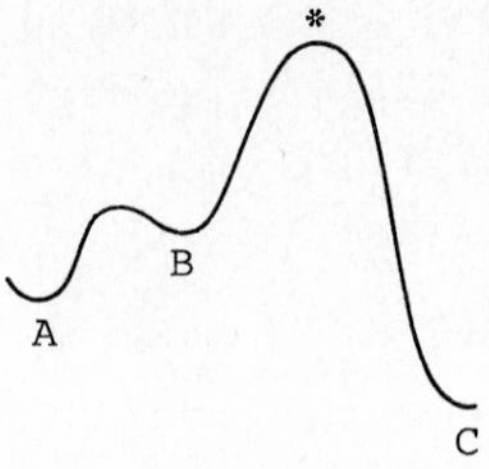

(b)

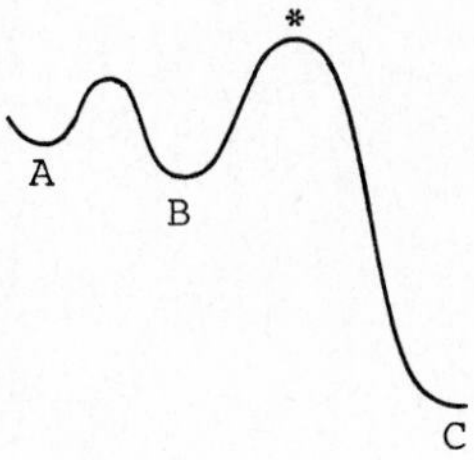

or

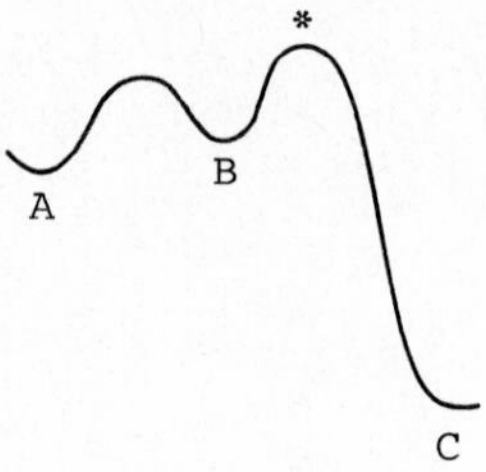

(c)

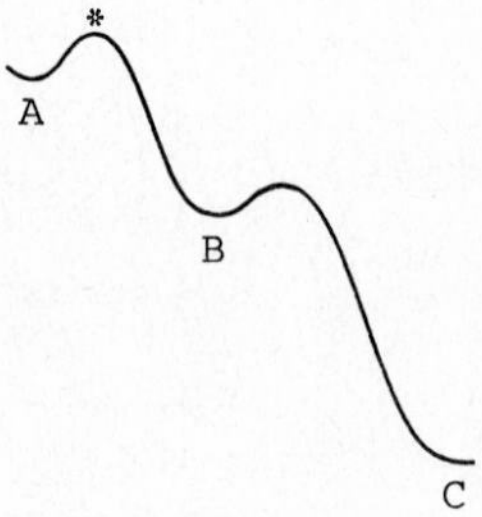

(d)

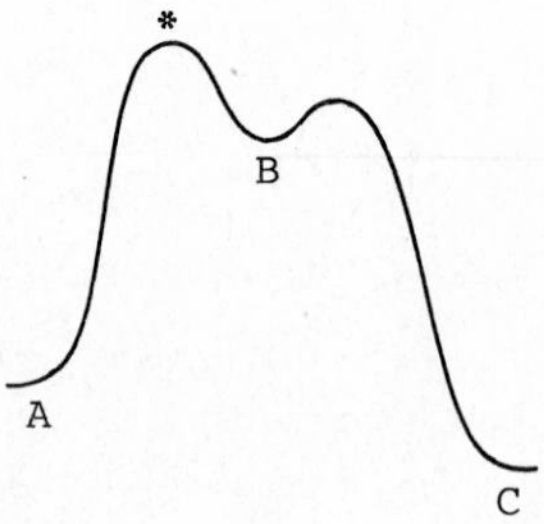

In each case, an asterisk (*) indicates the rate-determining transition state

7. (a) Endothermic. An "uphill" reaction requires you to put energy <u>in</u>.

(b)

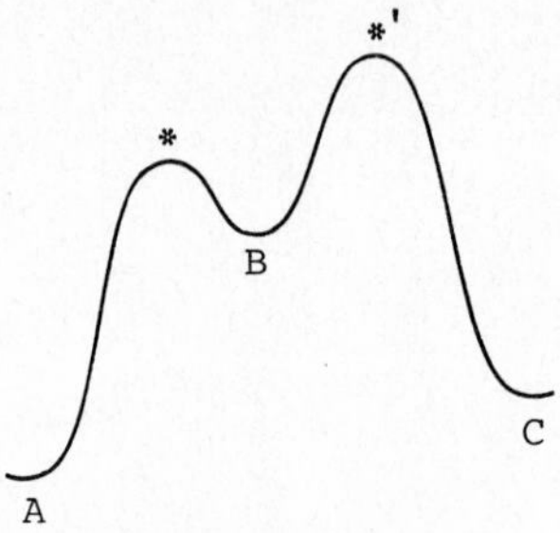

*, *' = transition states
*' = rate-determining transition state

(c) $k_2 > k_3 > k_1 > k_4$

(d) A is most stable (e) B is least stable

8.

(a)

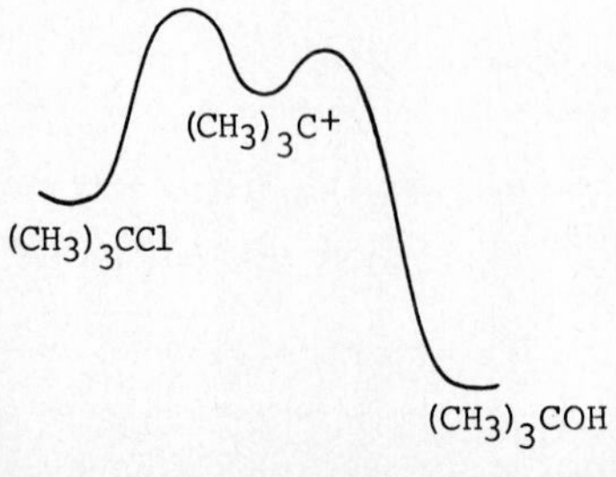

(b) endothermic; uphill to the carbocation

(c) exothermic; downhill overall

(d) the first step, because that is the highest barrier to cross

9. Because NH_2^- is such a strong base, the equilibrium with water lies completely to the right: $H_2O + NH_2^- \rightleftharpoons HO^- + NH_3$, and $[OH^-]$ will equal 0.1 M.

$$[H^+][OH^-] = 10^{-14} \Rightarrow [H^+] = 10^{-13} \text{ and pH} = 13.$$

This result is essentially the same as adding 0.1 M of NaOH itself.

PH_3 is expected to be a stronger acid because the P-H bonds are weaker than N-H bonds (P is below N in the periodic table).

10. Hydrogen is a very, very weak acid (i.e., hydride ion is a very strong base).

11. (a) and (g) The presence of one negative charge makes it more difficult to generate another.

(b), (d), (f), and (h) The more electronegative oxygen substituents present, the better the negative charge can be stabilized.

(c) The bonds become weaker farther down the periodic table.

(d) Nitrite ion is also stabilized by resonance relative to the hydroxylamine ion.

(e) Because it lies farther to the right in the periodic table, sulfur is more electronegative than phosphorus.

12. At 50% reaction, $[A] = \frac{1}{2}[A]_{initial}$ and $[B] = \frac{1}{2}[B]_{initial}$

$$\text{rate} = k\,\frac{1}{2}[A]_{initial} \times \frac{1}{2}[B]_{initial} = \frac{1}{4}\text{(initial rate)}$$

13. (a) $[H_2O]$ in water is $\dfrac{1000 \text{ g liter}^{-1}}{18 \text{ g mole}^{-1}} = 55\ \underline{M}$

(b)

$$k_1[CH_3Cl] = k_2[CH_3Cl][H_2O]$$
$$k_1 = k_2[H_2O] \Rightarrow k_2 = \frac{3\times10^{-10}\text{ sec}^{-1}}{55\ \underline{M}} = 5.45\times10^{-12}\ \underline{M}^{-1}\text{ sec}^{-1}$$

The rate constant for the reaction with OH^- is $6\times10^{-6}\,M^{-1}\text{ sec}^{-1}$, i.e., much faster.

(c)

$$\frac{0.69}{3\times10^{-10}\text{ sec}^{-1}} = 2.3\times10^{9}\text{ sec} = 73\text{ years}$$

14. $\text{rate} = k[OH^-][CH_3Cl]$ Let $r = [OH^-]_{reacted} = [CH_3Cl]_{reacted}$

At time t, $[OH^-] = 0.10 - r$ and $[CH_3Cl] = 0.05 - r$

$$\frac{dr}{dt} = k[0.10 - r][0.05 - r] \quad \text{so that} \quad \frac{dr}{[0.10-r][0.05-r]} = k\,dt$$

In a table of integrals, we find:

$$\int \frac{dx}{(a+bx)(c+dx)} = \frac{1}{ad-bc}\ln\frac{c+dx}{a+bx}$$

For our case, $a = 0.10$, $c = 0.05$, and $b = d = -1$

$$\frac{1}{-0.05}\ln\left(\frac{0.05-r}{0.10-r}\right)\Bigg]_{r=0}^{r=0.025} = kt$$

$$8.1 = kt = 6\times10^{-6}\,t$$

so that $t = 1.35\times10^{6}$ sec = 375 hours

[compare this solution with the approximate one worked out for problem #4]

4.E Supplementary Problems

S1. (a) At 25 °C, the equilibrium constant K for the addition of water to ethylene in the gas phase is 23.1 $\underline{M}^{-1}$. Calculate ΔG° for this reaction at 25 °C.

$$CH_2{=}CH_2 + H_2O \rightleftharpoons CH_3CH_2OH \qquad K = 23.1\ \underline{M}^{-1} \text{ at } 25\ °C$$

(b) At 400 °K, the equilibrium constant is 0.213 $\underline{M}^{-1}$. Calculate ΔH° and ΔS° for the reaction.

(c) Why does the equilibrium constant decrease at higher temperature?

S2. The rate constant k for the reaction: $CH_3CH_2Br + CH_3CH_2O^- \longrightarrow CH_3CH_2OCH_2CH_3 + Br^-$ in ethanol (CH_3CH_2OH) solvent at 25 °C is $7.6 \times 10^{-5}\ \underline{M}^{-1}\ sec^{-1}$, and the rate equation is: rate = $k[CH_3CH_2Br][CH_3CH_2O^-]$.

(a) If 0.05 mole of CH_3CH_2Br and 0.05 mole of $CH_3CH_2O^-\ Na^+$ are dissolved in 250 ml of ethanol, to a first approximation how long will it take before 10% of the CH_3CH_2Br has reacted?

(b) If the same quantities are dissolved in one liter of ethanol?

(c) If 0.05 mole of CH_3CH_2Br is dissolved in one liter of 0.2 $\underline{M}$ $CH_3CH_2O^-\ Na^+$ in ethanol?

S3. The rate constant k' for the reaction

$$(CH_3)_3CBr + CH_3CH_2O^- \longrightarrow (CH_3)_3COCH_2CH_3 + Br^-$$

in ethanol solvent at 25 °C is: $5 \times 10^{-4}\ \underline{M}^{-1}\ sec^{-1}$, and the rate equation is: rate = $k'[(CH_3)_3CBr]$.

(a) If 0.05 mole of $(CH_3)_3CBr$ and 0.05 mole of $CH_3CH_2O^-\ Na^+$ are dissolved in 250 ml of ethanol, to a first approximation how long will it take before 10% of the $(CH_3)_3CBr$ has reacted?

(b) If the same quantities are dissolved in one liter of ethanol?

(c) If 0.05 mole of $(CH_3)_3CBr$ is dissolved in one liter of 0.2 $\underline{M}$ $CH_3CH_2O^-\ Na^+$ in ethanol?

S4. (a) When 0.05 mole of CH_3CH_2I and 0.05 mole of $CH_3CH_2O^-\ Na^+$ are dissolved in 250 ml of ethanol solvent at 25 °C, it requires 65 minutes before 10% of the starting material has reacted according to the following equation:

$$CH_3CH_2I + CH_3CH_2O^-\ Na^+ \longrightarrow CH_3CH_2OCH_2CH_3 + Na^+\ I^-$$

What is the approximate rate of the reaction under these conditions?

(b) If the same amounts of starting material are dissolved in 500 ml of the solvent at 25 °C, 130 minutes are required before 10% reaction has occurred. What is the rate of reaction this time?

(c) What is the form of the rate equation? Is it a first- or second-order reaction?

(d) What is the rate constant?

S5. Construct a reaction profile diagram for the following reaction sequence:

$$A \underset{k_{-1}}{\overset{k_1}{\rightleftharpoons}} B \underset{k_{-2}}{\overset{k_2}{\rightleftharpoons}} C \underset{k_{-3}}{\overset{k_3}{\rightleftharpoons}} D$$

$k_1 = 3 \times 10^{-3}\ sec^{-1}$ $\quad k_2 = 10^{-1}\ sec^{-1}$ $\quad k_3 = 6 \times 10^{5}\ sec^{-1}$

$k_{-1} = 4 \times 10^{3}\ sec^{-1}$ $\quad k_{-2} = 10^{-3}\ sec^{-1}$ $\quad k_{-3} = 10^{-5}\ sec^{-1}$

(b) Which compound has the highest potential energy?

(c) Which compound reacts the fastest?

(d) What is the equilibrium constant $K = \frac{[D]}{[A]}$?

(e) What is the rate-limiting step in the conversion of A to D?

(f) Is the conversion of A to B endothermic or exothermic?

(g) What is the most exothermic step?

S6. For each of the following pairs, choose the compound with the higher pK_a.

(a) HCl , H_2S (b) H_2O , HF (c) $HO{-}\overset{O^-}{\underset{OH}{P^+}}{-}OH$, $HO{-}\overset{O^-}{\underset{OH}{P^+}}{-}O^-$

(d) H_2O , NH_3 (e) H_3O^+ , H_2O (f) H_3O^+ , NH_4^+ (g) HOOH , H_2O

S7. (a) Why can there be no stronger base in water than hydroxide ion?

(b) What is the strongest acid possible in water?

S8. From the pK_a's given below, calculate the pH of a solution obtained when one mole of each substance is dissolved in one liter of water.

(a) $NH_4^+ Cl^-$ $pK_a = 9.2$ (b) NH_3 (pK_a of $NH_4^+ = 9.2$)

S9. Consider the following reaction sequence:

$$A \underset{k_{-1}}{\overset{k_1}{\rightleftharpoons}} B \text{ , then } B + C \underset{k_{-2}}{\overset{k_2}{\rightleftharpoons}} P$$

$k_1 = 10^{-5}\ sec^{-1}$ $k_2 = 2 \times 10^{-2}\ M^{-1}\ sec^{-1}$

$k_{-1} = 10^{-3}\ sec^{-1}$ $k_{-2} = 10^{-8}\ sec^{-1}$

(a) Draw a reaction coordinate diagram for the overall process $A + C \rightleftharpoons P$.

(b) The rate equation for formation of P is: $rate = \frac{d[P]}{dt} = \frac{k_1 k_2 [A][C]}{k_{-1} + k_2[C]}$

Under standard state conditions ($[A] = [C] = 1\ \underline{M}$), $k_{-1} + k_2[C]$ is approx. equal to $k_2[C]$, and the expression reduces to: $\frac{d[P]}{dt} \cong \frac{k_1 k_2[A][C]}{k_2[C]} = k_1[A]$.

What is the rate-limiting step under these conditions?

(c) What is the approximate form of the rate equation if $[A] = [C] = 0.001\ \underline{M}$? What is now the rate-limiting step? What does this suggest about the limitations of reaction coordinate diagrams?

S10. Consider the following reaction sequence:

$$A \underset{k_{-1}}{\overset{k_1}{\rightleftharpoons}} B + C \text{ , then } B + D \underset{k_{-2}}{\overset{k_2}{\rightleftharpoons}} P$$

$k_1 = 2 \times 10^{-5}\ sec^{-1}$ $k_2 = 3 \times 10^{-2}\ \underline{M}^{-1}\ sec^{-1}$

$k_{-1} = 10^{-2}\ \underline{M}^{-1}\ sec^{-1}$ $k_{-2} = 10^{-8}\ sec^{-1}$

(a) Draw a reaction coordinate diagram for the overall process $A + D \rightleftharpoons C + P$

(b) The complete rate equation for this reaction is:

$$\text{rate} = \frac{d[P]}{dt} = \frac{k_1k_2[A][D]}{k_{-1}[C] + k_2[D]}$$

If [A] = [D] = 0.1 $\underline{M}$ and [C] = 0 at the start of the reaction, what is the approximate form of the rate equation and what is the rate-determining step?

(c) What is the approximate form of the rate equation if [C] = 2 $\underline{M}$? What is now the rate-determining step?

4.F Answers to Supplementary Problems

S1. (a) $\Delta G° = -RT \ln K$; $\Delta G°_{298} = -1.86$ kcal mole^{-1}

(b) To calculate ΔH° and ΔS°, you need ΔG° at two different temperatures: $\Delta G° = \Delta H° - T\Delta S°$. If K = 0.213 at 400° K, $\Delta G°_{400} = +1.23$ kcal mole^{-1}.

$$\Delta G°_{298} = \Delta H° - (298 \times \Delta S°) = -1.86$$
$$\Delta G°_{400} = \Delta H° - (400 \times \Delta S°) = +1.23$$

$$\Delta G°_{298} - \Delta G°_{400} = (-298 + 400)\Delta S° = -3.09 \text{ kcal mole}^{-1};$$
$$\Delta S° = -30.3 \text{ cal deg}^{-1} \text{ mole}^{-1} \text{ (e.u.)}$$

$$\Delta H° = \Delta G° + T\Delta S° = -10.9 \text{ kcal mole}^{-1}$$

(c) The equilibrium constant decreases at higher temperatures because the unfavorable entropy term (TΔS°) becomes more important. The entropy for this reaction is unfavorable (negative) because two molecules are combining to form one.

S2. rate (moles liter^{-1} sec^{-1}) × time (sec) = amount (moles liter^{-1})

$$k[CH_3CH_2Br][CH_3CH_2O^-]t = [CH_3CH_2OCH_2CH_3]$$

(Note that this approximation is valid only for very short reaction times. As soon as $[CH_3CH_2Br]$ and $[CH_3CH_2O^-]$ change, the rate changes too.)

For 10% reaction, $[CH_3CH_2OCH_2CH_3] = 0.1 \times [CH_3CH_2Br]$

and the equation reduces to: $k\,\cancel{[CH_3CH_2Br]}[CH_3CH_2O^-]t \cong 0.1 \times \cancel{[CH_3CH_2Br]}$

For 10% reaction, $t \cong \dfrac{0.1}{k[CH_3CH_2O^-]}$; $k = 7.6 \times 10^{-5}\ \underline{M}^{-1}\ \text{sec}^{-1}$

(a) $[CH_3CH_2O^-] = \dfrac{0.05 \text{ mole}}{0.25 \text{ liter}} = 0.2\ \underline{M} \Rightarrow t = \dfrac{0.1}{7.6 \times 10^{-5} \times 0.2} = 6580$ sec $= 1.83$ hr

(b) $[CH_3CH_2O^-] = \dfrac{0.05 \text{ mole}}{1.0 \text{ liter}} = 0.05\ \underline{M} \Rightarrow t = \dfrac{0.1}{7.6 \times 10^{-5} \times 0.05} = 26{,}300$ sec $= 7.31$ hr

(c) $[CH_3CH_2O^-] = 0.2\ \underline{M} \Rightarrow t = 1.83$ hr (as in part (a))

S3. $k'[(CH_3)_3CBr]t = [(CH_3)_3COCH_2CH_3]$

For 10% reaction, $[(CH_3)_3COCH_2CH_3] = 0.1 \times [(CH_3)_3CCBr]$

and the equation above reduces to: $k'\,[\cancel{(CH_3)_3CBr}]\,t \cong 0.1 \times [\cancel{(CH_3)_3CBr}]$

For 10% reaction, $t \cong \dfrac{0.1}{k'} = \dfrac{0.1}{5 \times 10^{-4}\ \text{sec}^{-1}} = 200\ \text{sec} = 3.3\ \text{min}$

Because the rate of the reaction is *independent* of $[CH_3CH_2O^-]$ (a first-order reaction), the length of time for 10% reaction is the same for all three cases ((a), (b), and (c)). As far as each molecule of $(CH_3)_3CBr$ is concerned, the presence or absence of a nearby $CH_3CH_2O^-$ molecule is unimportant in determining how fast it is going to react.

S4. (a) 10% reaction $= \dfrac{0.005\ \text{mole}}{0.25\ \text{liter}} = 0.02\ \underline{M}$ (change in $[CH_3CH_2I]$)

rate $= \dfrac{0.02\ \underline{M}}{65} = 3.08 \times 10^{-4}\ \underline{M}\ \text{min}^{-1}$ $(5.13 \times 10^{-6}\ \underline{M}\ \text{sec}^{-1})$

(b) 10% reaction $= \dfrac{0.005\ \text{mole}}{0.5\ \text{liter}} = 0.01\ \underline{M}$ (change in $[CH_3CH_2I]$)

rate $= \dfrac{0.01\ \underline{M}}{130} = 7.69 \times 10^{-5}\ \underline{M}\ \text{min}^{-1}$ $(1.28 \times 10^{-6}\ \underline{M}\ \text{sec}^{-1})$

(c) The two likely possibilities are:

rate $= k[CH_3CH_2I][CH_3CH_2O^-]$ <u>OR</u> rate $= k'[CH_3CH_2I]$

Try the first-order equation: case (a) rate $= 5.13 \times 10^{-6}\ \underline{M}\ \text{sec}^{-1} = k'[CH_3CH_2I]$

so that $k' = \dfrac{5.13 \times 10^{-6}}{0.2} = 2.57 \times 10^{-5}\ \text{sec}^{-1}$

INCONSISTENT

case (b) rate $= 1.28 \times 10^{-6}\ \underline{M}\ \text{sec}^{-1} = k'[CH_3CH_2I]$

so that $k' = \dfrac{1.28 \times 10^{-6}}{0.1} = 1.28 \times 10^{-5}\ \text{sec}^{-1}$

Next, try the second-order equation:

case (a) rate $= 5.13 \times 10^{-6}\ \underline{M}\ \text{sec}^{-1} = k[CH_3CH_2I][CH_3CH_2O^-]$

so that $k = \dfrac{5.13 \times 10^{-6}}{0.2 \times 0.2} = 1.28 \times 10^{-4}\ \underline{M}^{-1}\ \text{sec}^{-1}$

CONSISTENT

case (b) rate $= 1.28 \times 10^{-6}\ \underline{M}\ \text{sec}^{-1} = k[CH_3CH_2I][CH_3CH_2O^-]$

so that $k = \dfrac{1.28 \times 10^{-6}}{0.1 \times 0.1} = 1.28 \times 10^{-4}\ \underline{M}^{-1}\ \text{sec}^{-1}$

Therefore, this is a second-order reaction, and the rate equation is:

rate $= k[CH_3CH_2I][CH_3CH_2O^-]$

(d) As calculated in (c), $k = 1.28 \times 10^{-4}\ \underline{M}^{-1}\ \text{sec}^{-1}$

S5. (a)

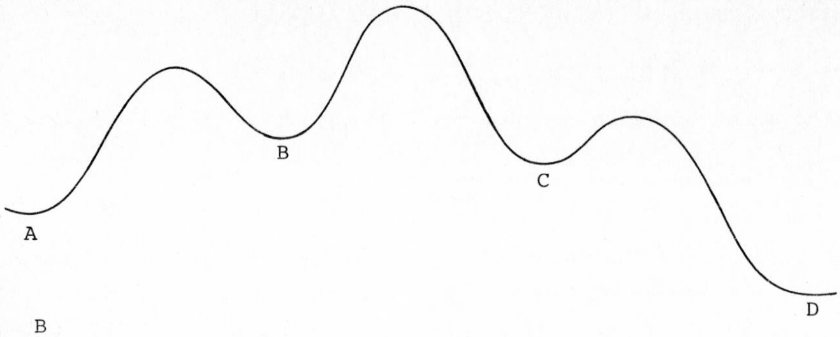

(b) B

(c) C (to go to D), because it has the lowest activation energy and thus the fastest rate constant of any of the reactions.

(d) $K = \frac{[D]}{[A]} = \frac{k_1 \times k_2 \times k_3}{k_{-1} \times k_{-2} \times k_{-3}} = 4.5 \times 10^6$

(e) B to C (f) endothermic (g) C to D

S6. (a) H_2S is less basic (= higher pK_a) because S lies to the left of Cl in the periodic table (it is less electronegative).

(b) H_2O (same reason as for (a))

(c) $H_2PO_4^-$, because one negative charge destabilizes a second.

(d) NH_3 (same reason as for (a))

(e) H_2O, because loss of a proton from a cationic molecule is easier than from a neutral one, other factors being equal.

(f) NH_4^+ (same reason as for (a))

(g) H_2O, because in HOOH, one oxygen acts as an electronegative substituent on the other.

S7.. (a) The equilibrium: $\text{Base}^- + H_2O \rightleftharpoons \text{Base-H} + OH^-$ will always take place. A base stronger than OH^- (pK_a of Base-H > 14) will simply react with water to give OH^-.

(b) For a similar reason, H_3O^+ is the strongest acid possible in water.

S8. (a) $\frac{[NH_3][H^+]}{[NH_4^+]} \cong \frac{[NH_3][H^+]}{1} = 10^{-9.2}$

so that $[H^+] = \sqrt{10^{-9.2}} = 2.51 \times 10^{-5}$

and pH = 4.6

(compare with Exercise at the end of Section 4.5)

(b) $NH_3 + H_2O \rightleftharpoons NH_4^+ + OH^-$ $[NH_3] = 1 - [NH_4^+] = 1 - [OH^-]$

$[NH_4^+] = \frac{10^{-14}}{[H^+]}$ so that $\frac{[NH_3][H^+]}{[NH_4^+]} = \frac{\left(1 - \frac{10^{-14}}{[H^+]}\right)[H^+]}{\frac{10^{-14}}{[H^+]}}$

$10^{14} \times [H^+]^2 - [H^+] - K = 0$

$[H^+] = 2.5 \times 10^{-12}$ and pH = 11.6

S9. (a)

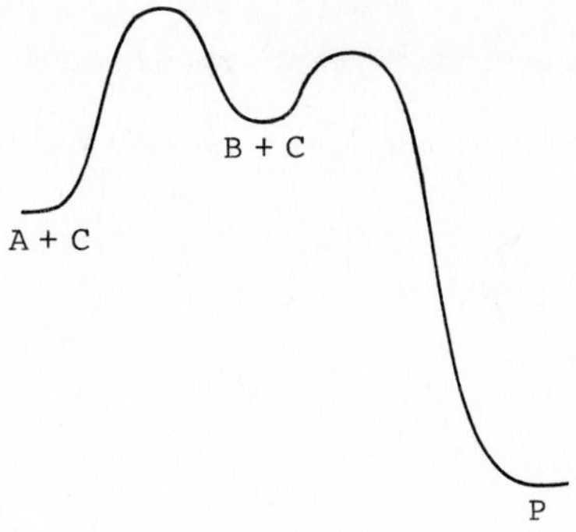

(b) Under standard state conditions, the conversion of A → B is the slow step.

(c) If [C] = 0.001, k_2[C] (2×10^{-5}) is small relative to k_{-1} (10^{-3}) and the rate equation reduces to:

$$\frac{d[P]}{dt} = \frac{k_1 k_2}{k_{-1}}[A][C]$$

The reaction is now second-order and B + C → P is the rate-determining step. Notice how the rate-determining step can change with concentration. The reaction coordinate diagram does not provide a valid picture for concentrations other than standard state, unless all the transformations involved are first-order.

S10. (a)

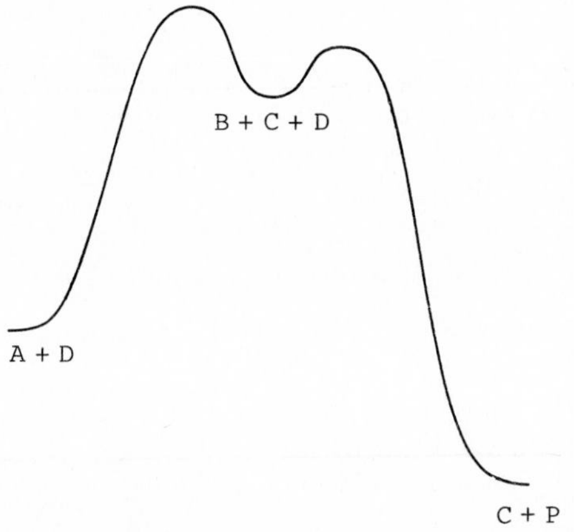

(b) k_{-1}[C] = 0 during the early part of the reaction, and

$$\frac{d[P]}{dt} = \frac{k_1 k_2 [A][D]}{k_2 [D]} = k_1[A]$$

The first step is rate-limiting.

(c) If [C] = 2, then $k_{-1}[C] + k[D] = (10^{-2} \times 2) + (3 \times 10^{-2}) \times 0.1$

$$= 2.3 \times 10^{-2} \cong k_{-1}[C]$$

so that $$\frac{d[P]}{dt} = \frac{k_1 k_2 [A][D]}{k_{-1}[C]}$$

The second step is now rate-determining.

5. ALKANES

5.A Chapter Outline and Important Terms Introduced

5.1 n-Alkanes; Physical Properties (boiling points, melting points, and density)

non-polar forces of attraction — London force, dispersion force

Van der Waals forces — dependence on surface area

5.2 n-Alkanes; Barriers to Rotation (rotation about single bonds)

conformations — Newman projections

eclipsed — anti

staggered — gauche

5.3 Branched Chain Alkanes (more on physical properties, conformations)

conformational "isomers"

5.4 Cycloalkanes (ring structures, physical properties, nomenclature)

5.5 Heats of Formation (how to calculate relative stabilities)

heat of formation, $\Delta H°_f$ — distinction from free energy, $\Delta G°_f$

standard state

5.6 Cycloalkanes: Ring Strain (bending bonds in small rings)

5.7 Cycloalkanes: Conformation (specific shapes of cyclic molecules)

A. Cyclopropane and Cyclobutane
   bent bonds

B. Cyclopentane
   envelope

C. Cyclohexane
   chair ⇌ boat interconversion
   axial   equatorial

D. Larger-Ring Cycloalkanes
   transannular repulsions

5.8 Substituted Cyclohexanes (axial vs. equatorial substituents)

conformational energies — skew boat

5.9 Occurrence of Alkanes (or lack thereof . . . .)

crude oil — natural (living) sources

5.B Important Concepts and Hints

Chapter 5 presents the structural and physical properties of a specific class of organic compounds, the saturated hydrocarbons. Although there are only a few chemical reactions to be discussed for the alkanes (Chapter 6), carbon chains form the backbone of all other organic molecules, and it is therefore important to know what their shapes are, how their conformations differ in three dimensions, and how much more stable one isomer is compared to another. In this connection, the following terms are used over and over, and you should know what they mean: *anti* and *gauche*, *eclipsed* and *staggered*, *ring strain*, *chair conformation*, and *axial* and *equatorial*. Be sure you also understand the concept of *heat of formation* as an indication of relative thermodynamic stability.

By now you should be completely familiar with the relationship between free energy differences and relative amounts ($\Delta G° = -RT \ln K$), and also quite convinced of the value of your molecular model kit.

HINTS: In problems which ask you to calculate energy differences ($\Delta H°$ or $\Delta G°$), the best way to figure out what to subtract from what is:

(1) Write the equation for the reaction, with the products on the right;
(2) Write the thermodynamic value you are interested in comparing under the formula or name of each component, keeping the signs straight;
(3) ADD the values for the products and SUBTRACT the values for the starting materials, again paying attention to the signs.

As an example, calculate $\Delta H°$ for the following reaction:

(1) $CH_2{=}CH_2 \quad + \quad HBr \longrightarrow CH_3CH_2Br$

(2) $\Delta H°_f = \quad 12.5 \quad\quad -8.7 \quad\quad -15.2$

(3) $\Delta H° = +(-15.2) - (12.5 - 8.7) = -19 \text{ kcal mole}^{-1}$

This will give you the thermodynamic value for the overall transformation, with the sign correct for the direction in which you have written the equation. If you write the reaction backward, you change the sign of the value which you calculate.

If you like to remember things visually, the following scheme can also help you keep track of when to add and when to subtract.

(1) Across the page, draw a line representing zero energy.
(2) To determine the energy content of the reactants, go UP for positive values and DOWN for negative values, moving sequentially. For instance, "$CH_2{=}CH_2$ ($\Delta H°_f = +12.5$) + HBr ($\Delta H°_f = -8.7$)" would result in:

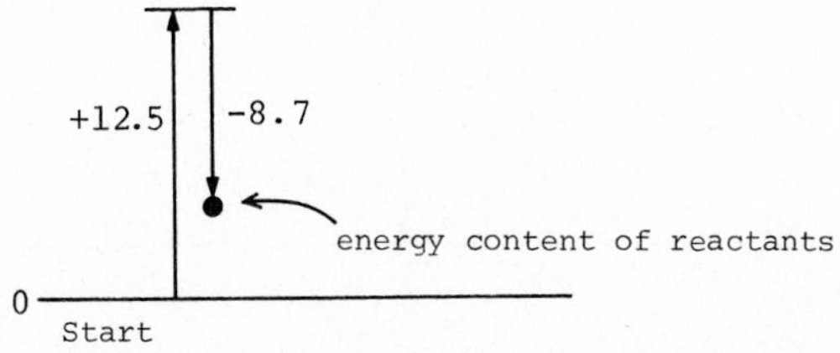

(3) Determine the energy content of the products the same way:

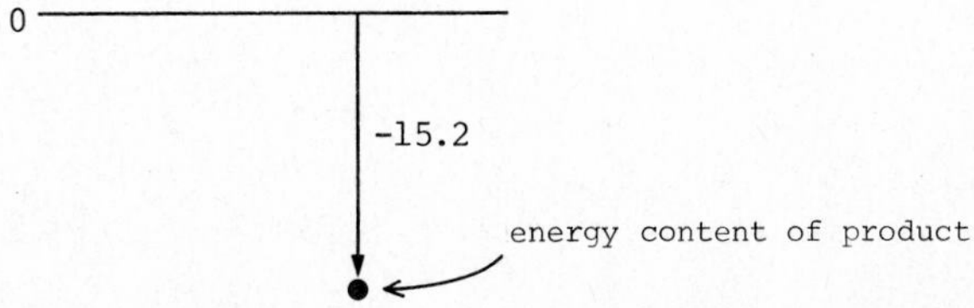

(4) Visually, then, it is easy to see whether going from reactants to products is uphill (endothermic, $\Delta H° > 0$) or downhill (exothermic, $\Delta H° < 0$)

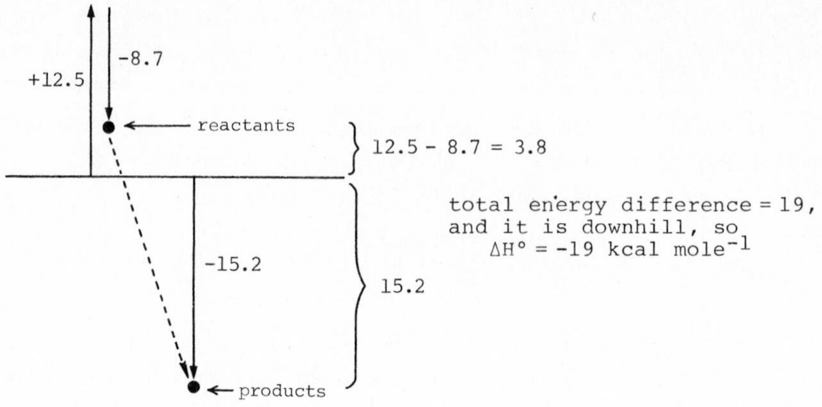

Also, remember that ΔH and ΔG are usually presented as "kcal mole^{-1}", whereas ΔS and the gas constant R are given as "cal deg^{-1} mole^{-1}" (the same as an entropy unit, e.u.). A kcal is 1000 cal.

5.C Answers to Exercises

(5.2) The distance between elipsed hydrogens in ethane is about 2.3 Å, as indicated by molecular models (trigonometry with r(C-H) = 1.10 Å, r(C-C) = 1.54 Å, and tetrahedral angles gives 2.27 Å). The distance between staggered hydrogens in ethane is about 2.5 Å. By contrast, in gauche-butane one pair of 1,4-hydrogens in only 2.0 Å apart. It is this close approach of these two hydrogens that probably accounts for most of the relative instability of the gauche conformation.

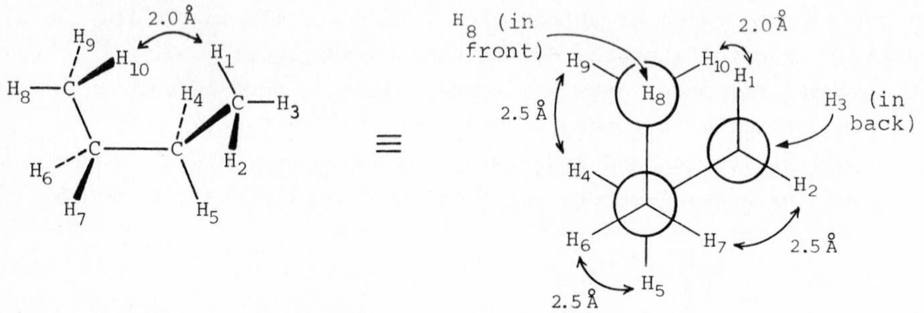

(5.3)

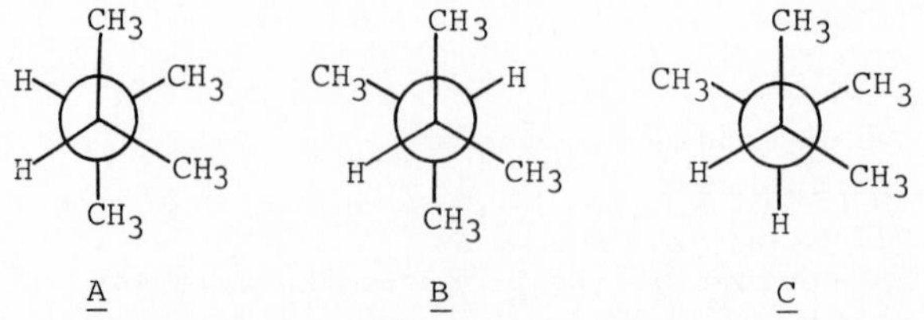

A and C are equivalent and have equal energy

(5.4)

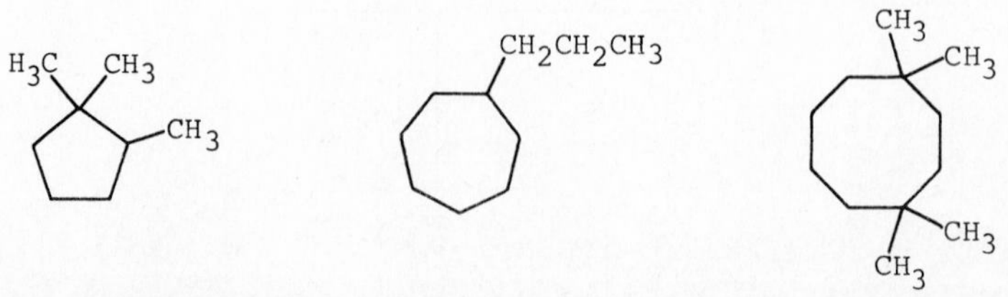

(5.5) n-pentane ⇌ isopentane

$\Delta H^\circ_f =$ −35.1 −36.9 $\Delta H^\circ = -36.9 - (-35.1) = -1.8 \text{ kcal mole}^{-1}$

isopentane ⇌ neopentane

$\Delta H^\circ_f =$ −36.9 −40.3 $\Delta H^\circ = -40.3 - (-36.9) = -3.4 \text{ kcal mole}^{-1}$

(5.7) Although you may think we're trying to teach you art in part (b), consider this point: many exam answers are marked wrong because the structures drawn are incomprehensible to the grader .....

(5.8) *cis*-1,2-dimethylcyclohexane:

There is one axial and one equatorial methyl in each conformation, therefore they are of equal energy.

trans-1,2-dimethylcyclohexane:

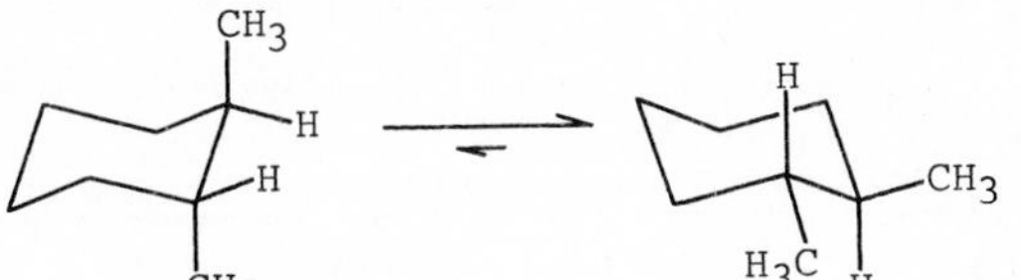

The diaxial conformation is less stable than the equatorial.

5.D Answers and Explanations for Problems

1. (a) 1-ethyl-1-methylcyclohexane
   (b) *trans*-1-isopropyl-3-methylcyclohexane or *trans*-1-methyl-3-(1-methylethyl)cyclohexane
   (c) t-butylcyclodecane or (1,1-dimethylethyl)cyclodecane
   (d) 1,1-dimethylcyclopropane
   (e) isobutylcyclopentane or (2-methylpropyl)cyclopentane
   (f) 1-cyclobutyl-3-methylpentane
   (g) *cis*-1-bromo-3-methylcyclohexane
   (h) *trans*-1-ethyl-2-iodocyclopentane

2.

```
(1)  C-C-C-C-C-C-C        heptane

(2)  C-C-C-C-C-C          2-methylhexane
       |                   (isoheptane)
       C

(3)  C-C-C-C-C-C          3-methylhexane
         |
         C

(4)  C-C-C-C-C
       |   |              2,4-dimethylpentane
       C   C

(5)  C-C-C-C-C            2,3-dimethylpentane
       | |
       C C

         C
         |
(6)  C-C-C-C-C            2,2-dimethylpentane
         |
         C

         C-C
         |
(7)  C-C-C-C-C            3-ethylpentane

         C
         |
(8)  C-C-C-C-C            3,3-dimethylpentane
         |
         C

         C
         |
(9)  C-C-C-C              2,2,3-trimethylbutane
         | |
         C C
```

Note how we systematically search through the substituted hexanes (with methyl substituents), pentanes (with dimethyl and with ethyl substituents), and butanes (trimethyl; methylethyl is one of the others already counted).

3. C_2-C_3 anti

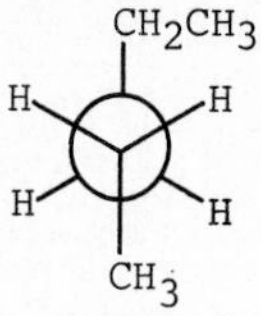

This corresponds to (a) in Figure 5.12

C_2-C_3 gauche

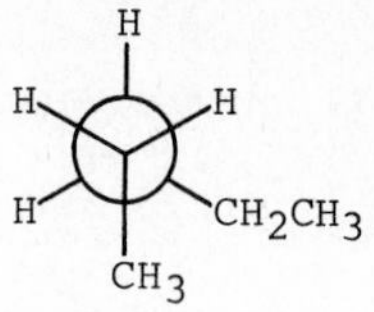

and

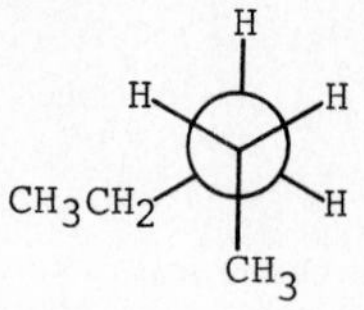

This corresponds to (b) in Figure 5.12

4.

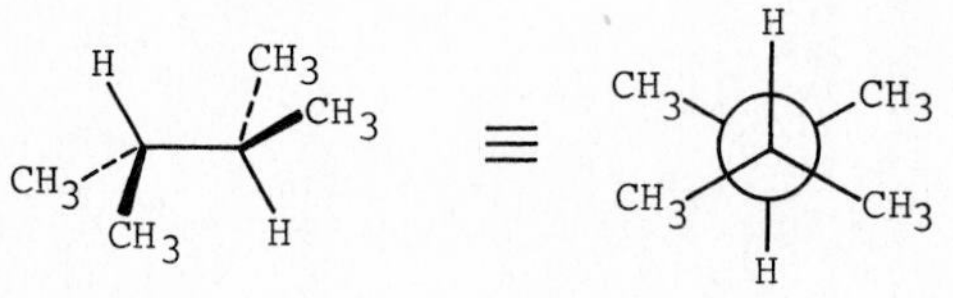

A

Lower energy; only two gauche interactions

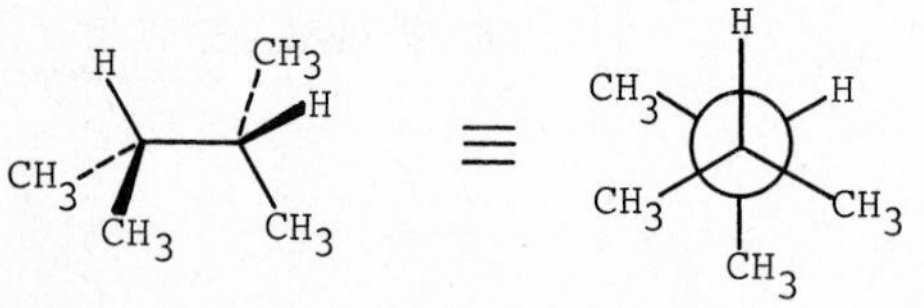

B

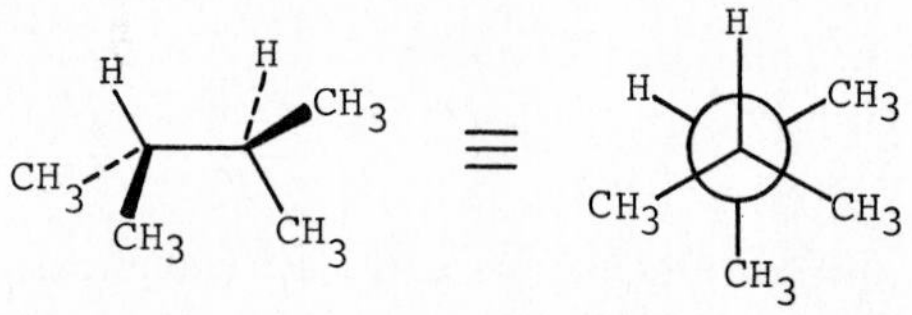

B'

Both B and B' have <u>three</u> gauche interactions, and so are 0.9 kcal $mole^{-1}$ less stable than A.

$$K = \frac{[A]}{[B]} = \frac{[A]}{[B']} \qquad \text{and} \qquad \Delta G° = -RT \ln K$$

$$-0.9 \text{ kcal mole}^{-1} = -1.987 \text{ cal deg}^{-1} \text{ mole}^{-1} \times 298 \text{ deg} \times \ln K$$

K = 4.5 , therefore, the mixture will consist of:

4.5 parts A : 1 part B : 1 part B'

$$\frac{4.5}{6.5} = 69\% \text{ A} \qquad \frac{2.0}{6.5} = 31\% \text{ B + B'}$$

This percentage is almost the same as that of the equilibrium composition of anti (72%) and gauche (28%) conformers of butane, because in each case the two conformers differ by one gauche interaction only.

5. Adamantane is $C_{10}H_{16}$ (mw = 136) and is very symmetrical -- almost spherical. For comparison, 2,2,3,3-tetramethylbutane (C_8H_{18}; mw = 114) has bp 106 °C and mp 100 °C. Adamantane would be expected to have a similar bp (it is even more symmetrical than tetramethylbutane, but a little larger). The symmetrical structure would suggest a high mp. In fact, adamantane melts at 270 °C (in a sealed tube). At atmospheric pressure, adamantane sublimes instead of melting; that is, it goes directly from the solid to the vapor state.

6.

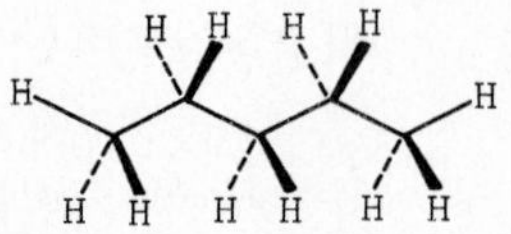

most stable; all anti

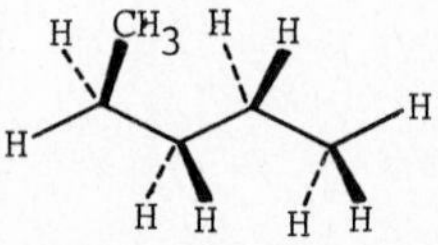

one gauche interaction

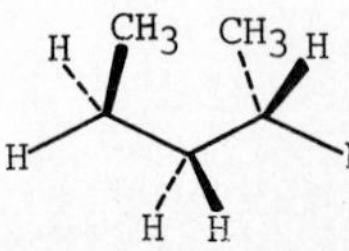

two gauche interactions

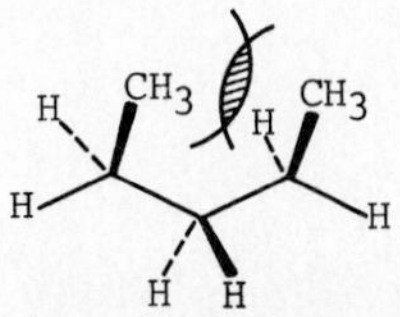

least stable; methyls colliding

7. <u>Isopentane:</u>

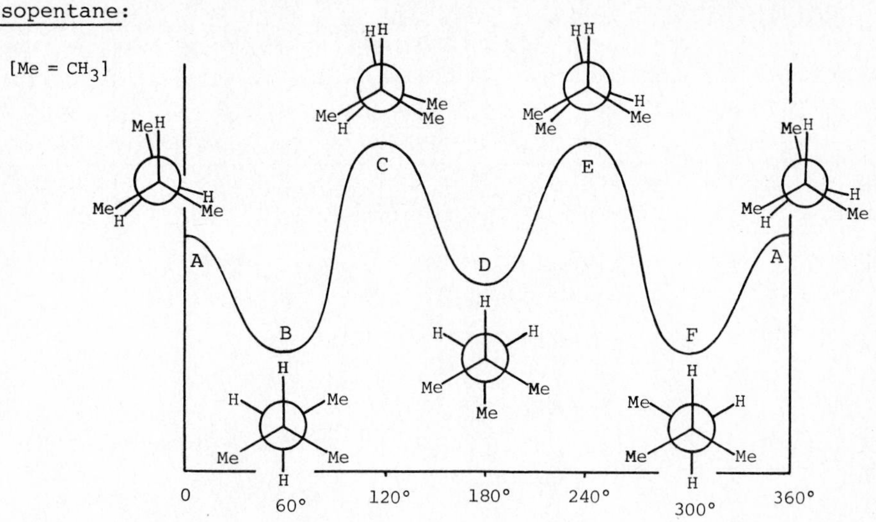

The three energy minima are B, D, and F. Conformations B and F are equivalent and equal in energy; they both have one gauche interaction. Conformation D, with two gauche interactions, is less stable. In rotating about the C_2-C_3 bond from B to F, the molecule passes through an eclipsed conformation (energy maximum) in which there are three CH_3-H interactions. In passing from B to D and also from D back to F, the molecule passes through another type of eclipsed conformation. In these conformations (C and E), there is one CH_3-CH_3 interaction, one CH_3-H interaction, and one H-H interaction. Conformations C and E are equivalent and are of higher energy than conformation A.

<u>2,3-Dimethylbutane:</u> Conformation D, with two gauche interactions, is of lower energy than the other two minima, B and F, which have three gauche interactions each. Of the three energy maxima, C and E each have two CH_3-H interactions and one CH_3-CH_3 interaction. Eclipsed conformation A has two CH_3-CH_3 interactions and one H-H eclipsed interaction.

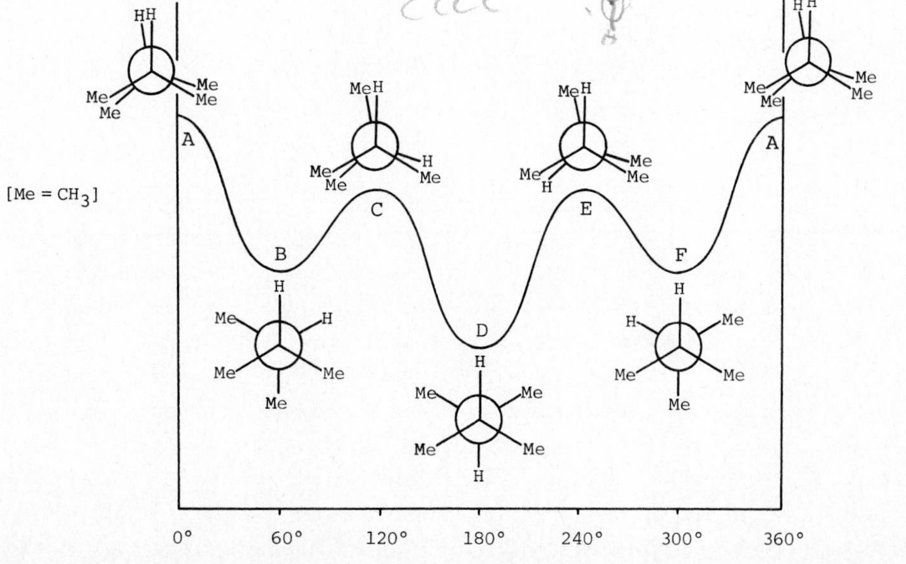

In ethane, the barrier is 3.0 kcal mole^{-1}, or 1.0 kcal mole^{-1} per pair of eclipsed hydrogens. In propane, the barrier is 3.4 kcal mole^{-1}. Thus, the CH_3-H contribution is approximately 1.4 kcal mole^{-1}. In butane, the eclipsed conformation C (Fig. 5.11) is 3.8 kcal mole^{-1} above D (1.0 + 1.4 + 1.4 = 3.8), and the eclipsed conformation A is about 4.5 kcal mole^{-1} above D. Since this conformation has two H-H interactions (2.0 kcal mole^{-1}) and one CH_3-CH_3 interaction, we may estimate the CH_3-CH_3 interaction as: 4.5 - 2.0 ≅ 2.5 kcal mole^{-1}.

Returning to 2,3-dimethylbutane, we may now estimate the heights of the barriers corresponding to conformations A, C, and E as:

A: 1.0 + 2.5 + 2.5 = 6.0 kcal mole^{-1}
C: 1.4 + 1.4 + 2.5 = 5.3 kcal mole^{-1}
E: 1.4 + 1.4 + 2.5 = 5.3 kcal mole^{-1}

2,2,3,3-Tetramethylbutane:

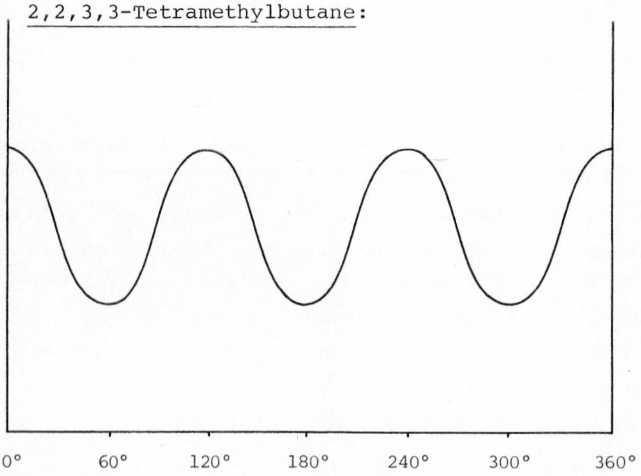

Maxima:

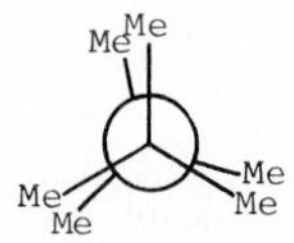

Minima:

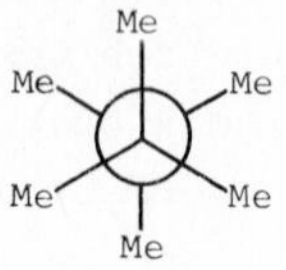

All minima are equal, and all maxima are equal.

8. (a) $CH_2{=}CH_2 + H_2 \longrightarrow CH_3CH_3$

ΔH°_f = +12.5 (0) -20.2 ΔH° = -20.2 - 12.5 = -32.7 kcal mole^{-1}

(b) $CH_2{=}CH_2 + HCl \longrightarrow CH_3CH_2Cl$

ΔH°_f = +12.5 -22.1 -26.1 ΔH° = -26.1 - (12.5 - 22.1) = -16.5 kcal mole^{-1}

(c) $CH_2{=}CH_2 + H_2O \longrightarrow CH_3CH_2OH$

ΔH°_f = +12.5 -57.8 -56.2 ΔH° = -56.2 - (12.5 - 57.8) = -10.9 kcal mole^{-1}

These calculations indicate that the enthalpy term (ΔH°) of the free energy change ΔG° is favorable (negative), but without knowing the entropy term (ΔS°) you can't say very much about the equilibrium constant. See, for example, Supplementary Problem #S1, Chapter 4, which is concerned with the change in the position of equilibrium (c) above with temperature. However, you could predict that the entropy change for each of these reactions would be about the same (two molecules ⟶ one molecule), and because of the enthalpy differences between (a), (b), and (c), the equilibrium constant for reaction (a) would be greater than that for (b), and (b) greater than (c).

The enthalpy change of the overall reaction bears no particular relationship to the energy of activation, which is what governs the rate of reaction.

9. 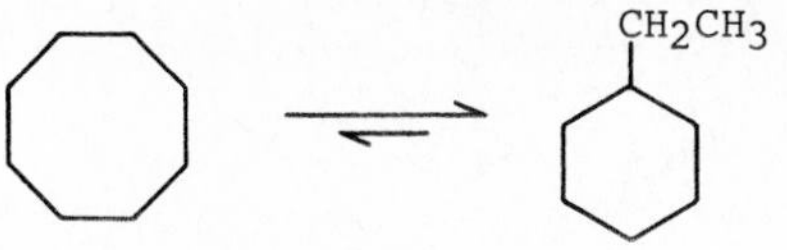

$$K = \frac{[\text{ethylcyclohexane}]}{[\text{cyclooctane}]}$$

$\Delta H^\circ_f =$ -29.7   -41.0 $\Delta H^\circ = -11.3$ kcal mole^{-1}

If $\Delta H^\circ = \Delta G^\circ = -RT \ln K$, then $K = 1.9 \times 10^8$

The fact that the equilibrium is even more in favor of ethylcyclohexane results from the favorable contribution from entropy to the free energy of this isomerization ($\Delta G^\circ = \Delta H^\circ - T\Delta S^\circ$). There is more freedom of motion in ethylcyclohexane than in cyclooctane (the ethyl group can spin relative to the cyclohexane ring, for instance); this results in a positive ΔS° and therefore a negative contribution to ΔG°.

10.

(a) 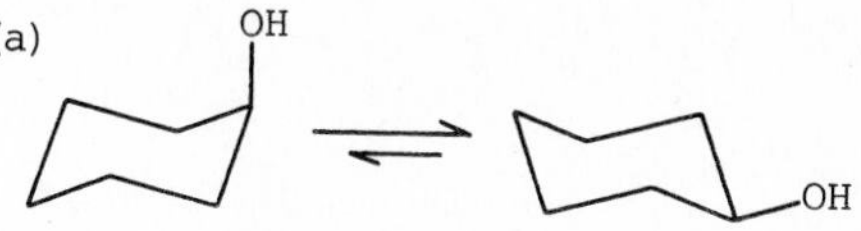

(b) 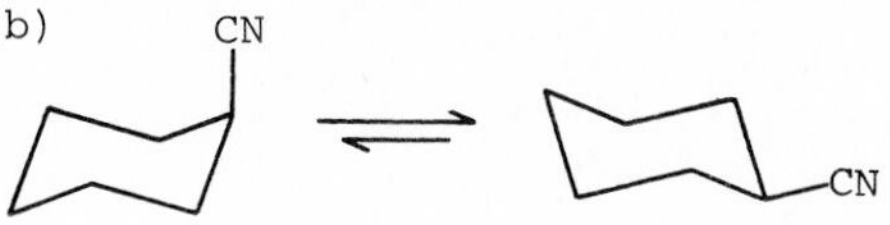

(c) 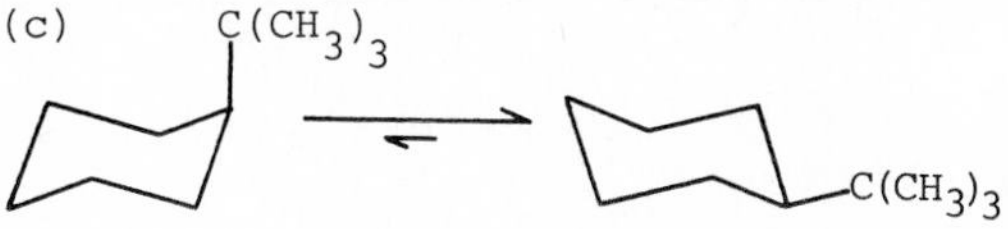

(d) 

	ΔG° (kcal mole^{-1})	ln K	K	Percent Equatorial
(a)	-1.0	1.69	5.42	84.4
(b)	-0.2	0.338	1.40	58.3
(c)	-5 to -6	8.44 to 10.1	4.6×10^3 to 2.5×10^4	99.98 to 99.996
(d)	-0.25	0.422	1.53	60.5

11. (a) Four H-H eclipsing; eight H-H almost eclipsing

(b) Equatorial methylcyclohexane has no gauche interactions: 

Axial methylcyclohexane has two: 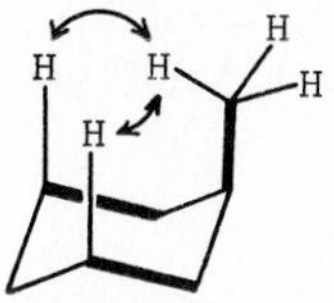

In butane, the gauche conformations are less stable than the anti conformation by 0.9 kcal mole^{-1}. Two gauche interactions therefore account quite neatly for the 1.7 kcal mole^{-1} difference in stability between axial and equatorial methylcyclohexane.

NOTE: a "gauche interaction" is actually a steric interaction of non-bonded hydrogens (see the answer to Exercise (5.2)):

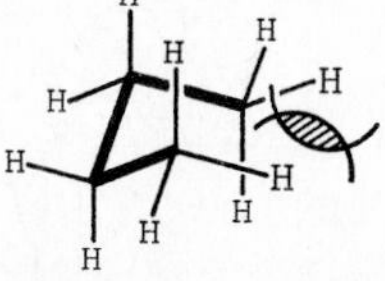

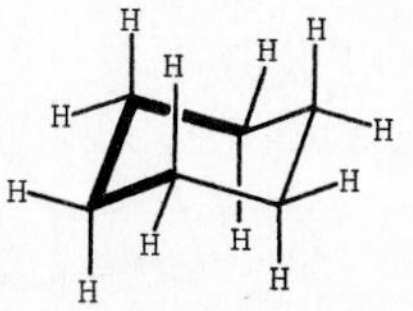

At first glance, it may appear that cyclohexane itself has a number of such gauche relationships of carbon-carbon bonds, as highlighted in the drawing at the far right. However, closer inspection shows that the hydrogen-hydrogen interference of an actual gauche interaction is not present in cyclohexane.

12. (a)

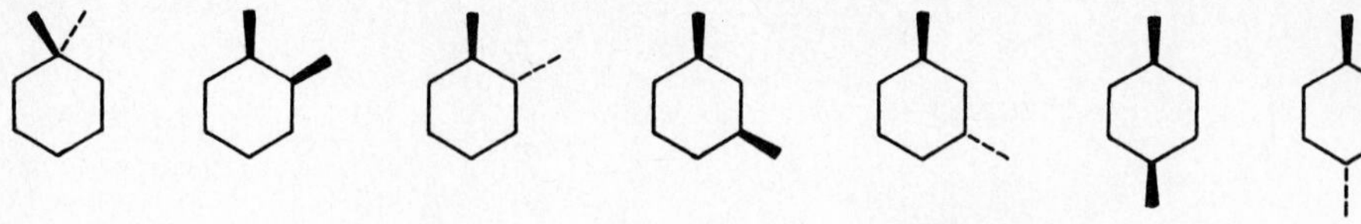

(b)

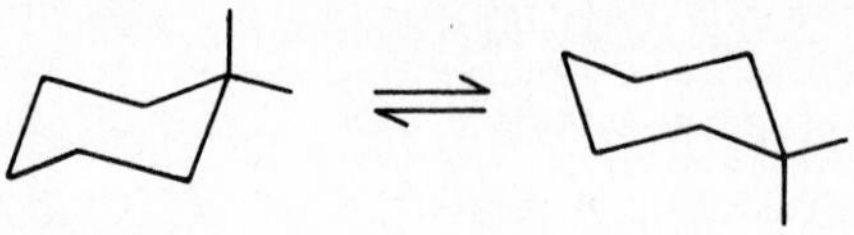

one axial, one equatorial in each, so that $\Delta G° = 0$

(c)

In all three pairs, each conformation has one axial and one equatorial, and $\Delta G° = 0$

(d)

two axial methyls (2×1.7 kcal mole^{-1})

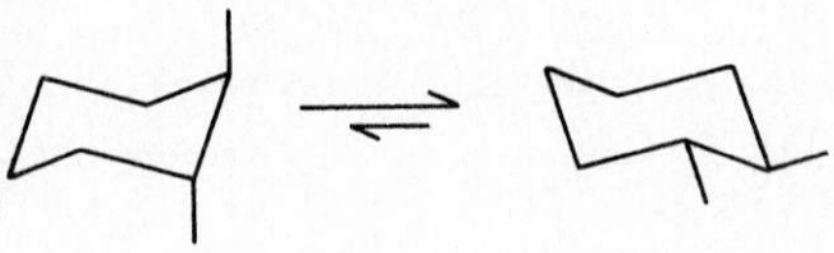

one gauche interaction (0.9 kcal mole^{-1})

$$\Delta G° \cong 0.9 - (2 \times 1.7) = -2.5 \text{ kcal mole}^{-1}$$

two axial methyls (2×1.7 kcal mole^{-1})

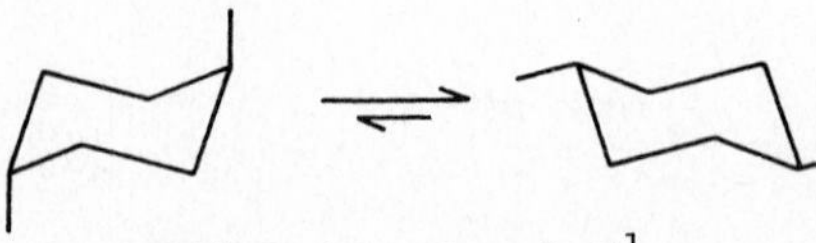

no interactions

$$\Delta G° \cong -3.4 \text{ kcal mole}^{-1}$$

(e)

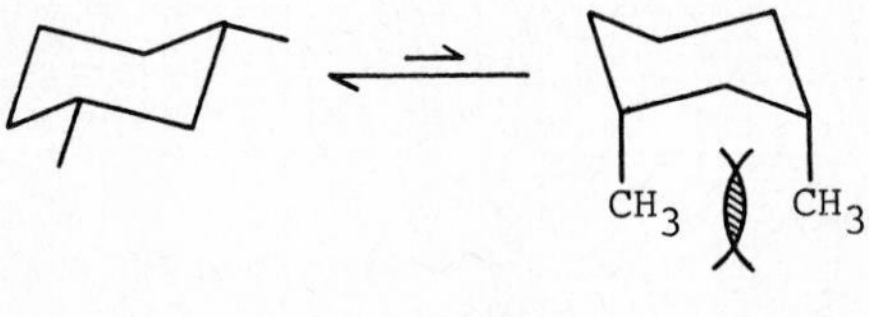

cis-1,3-"diaxial" interaction of two methyl groups

(f) "Gauche-butane" interactions are shown for the methyl groups:

1,1:

(two)

cis-1,2:

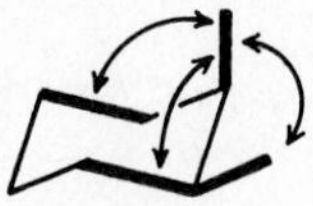

(three)

trans-1,2:

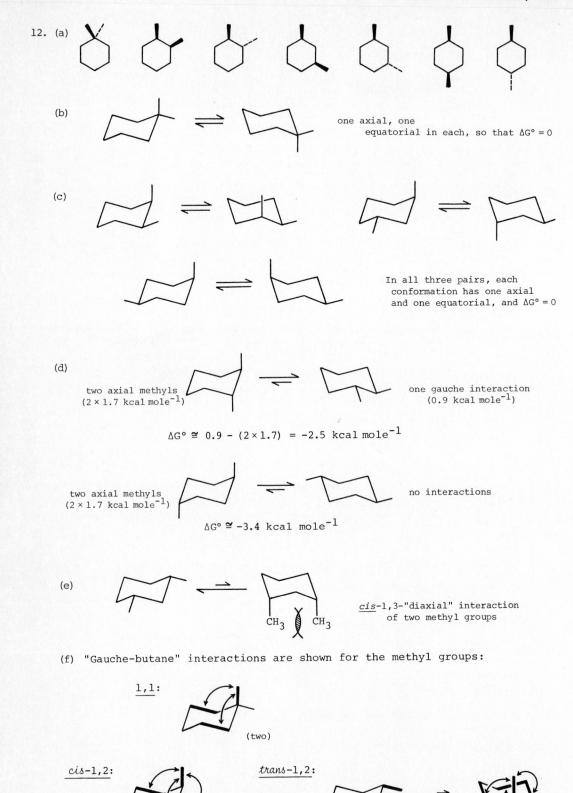

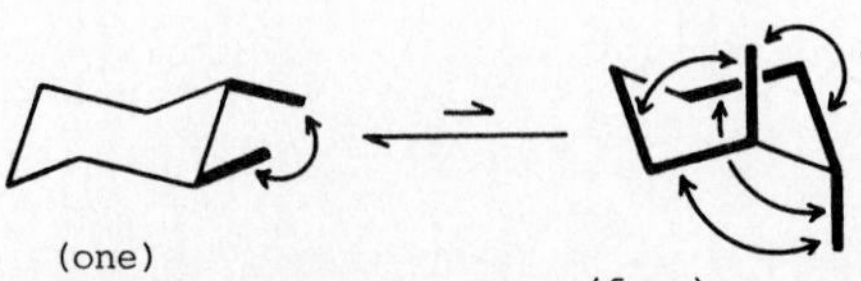

(one) (four)

cis-1,3:

(none) (four, plus 1,3-diaxial)

trans-1,3:

(two)

cis-1,4:

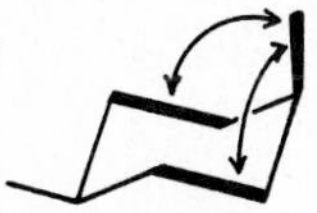

(two)

trans-1,4:

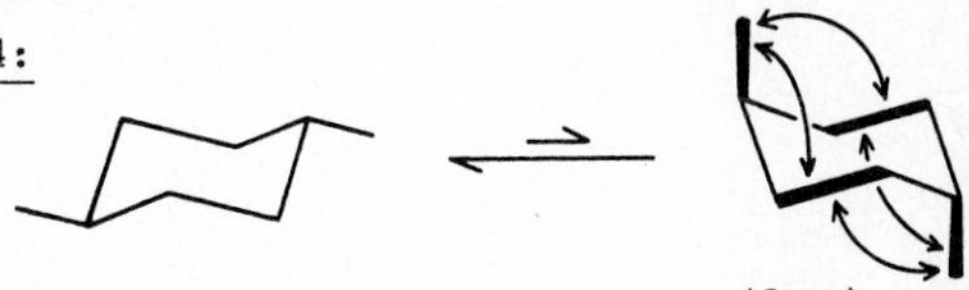

(none) (four)

13.

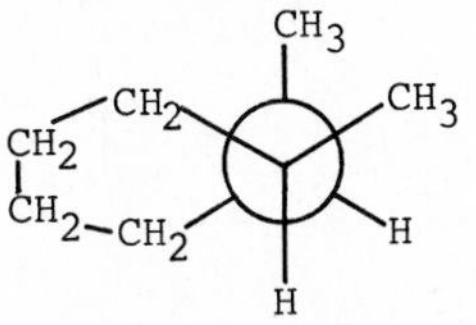

14. $\Delta G° = \Delta H° - T\Delta S°$

$-0.89 = -2.05 - (298 \times \Delta S°)$, so that $\Delta S° = -3.89$ cal deg^{-1} $mole^{-1}$

A negative entropy change indicates more "restriction of freedom". In this instance, it arises because there are fewer different conformations possible for isobutane than for butane.

$\Delta G° = -RT \ln K = -0.89$ kcal $mole^{-1}$; therefore $K = 4.5$

15. pentane $\rightleftharpoons$ isopentane

$\Delta G° = -1.54$ kcal $mole^{-1}$ and $K_1 = \dfrac{[\text{isopentane}]}{[\text{pentane}]} = 13.5$

isopentane $\rightleftharpoons$ neopentane

$\Delta G° = -0.10$ kcal $mole^{-1}$ and $K_2 = \dfrac{[\text{neopentane}]}{[\text{isopentane}]} = 1.2$

Ratio of pentane/isopentane/neopentane = 1 : 13.5 : (13.5 × 1.2), which is equivalent to 3%/44%/53%.

16. Compare:

	ethane		propane		butane		pentane
$\Delta H°_f$ =	-20.2		-24.8		-30.4		-35.1
difference:		-4.6		-5.6		-4.7	

average = -5.0 kcal $mole^{-1}$ per $-CH_2-$ group

ethane: two CH_3- ; $\Delta H°_f = -20.2$, which is equivalent to -10.1 per CH_3-

propane: two CH_3- + one $-CH_2-$ (-5.0) $\Delta H°_f = -24.8 \Rightarrow$ -9.9 per CH_3-

butane: two CH_3- + two $-CH_2-$ (-10.0) $\Delta H°_f = -30.4 \Rightarrow$ -10.2 per CH_3-

pentane: two CH_3- + three $-CH_2-$ (-15.0) $\Delta H°_f = -35.1 \Rightarrow$ -10.1 per CH_3-

average = -10.1 kcal $mole^{-1}$ per CH_3- group

isobutane: three CH_3- (-30.3) + one $-\overset{|}{C}H-$ $\Delta H°_f = -32.4 \Rightarrow$ -2.1 per $-\overset{|}{C}H-$

isopentane: three CH_3- (-30.3) + one $-CH_2-$ (-5.0) + one $-\overset{|}{C}H-$ $\Delta H°_f = -36.9 \Rightarrow$ -1.6 per $-\overset{|}{C}H-$

average = -1.9 kcal $mole^{-1}$ per $-\overset{|}{C}H-$

neopentane: four CH_3- (-40.4) + one $-\underset{|}{\overset{|}{C}}-$ $\Delta H°_f = -40.3 \Rightarrow$ +0.1 per $-\underset{|}{\overset{|}{C}}-$

PREDICTIONS

		ΔH°_f (kcal $mole^{-1}$) Estimated	Found
hexane:	two CH_3- + four $-CH_2$- (-20.2) (-20.0)	-40.2	-39.9
2-methylpentane: 3-methylpentane:	three CH_3- + two $-CH_2$- + one $-\overset{|}{C}H-$ (-30.3) (-10.0) (-1.9)	-42.2	-41.8 -41.1
2,2-dimethylbutane:	four CH_3- + one $-CH_2$- + one $-\overset{|}{\underset{|}{C}}-$ (-40.4) (-5.0) (+0.1)	-45.3	-44.3
2,3-dimethylbutane:	four CH_3- + two $-\overset{|}{C}H-$ (-40.4) (-3.8)	-44.2	-42.6
nonane:	two CH_3- + seven $-CH_2$- (-20.2) (-35.0)	-55.2	-54.7
2,2,4,4-tetramethylpentane:	six CH_3- + one $-CH_2$- + two $-\overset{|}{\underset{|}{C}}-$ (-60.6) (-5.0) (+0.2)	-65.4	-57.8

The steric strain of

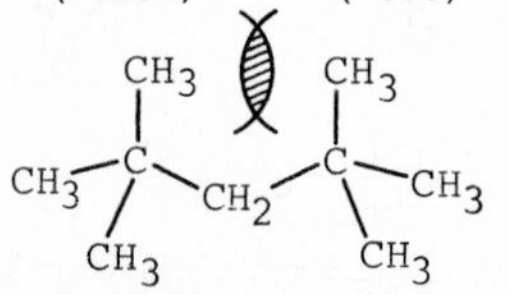

makes it difficult to put the molecule together; i.e., less energy than estimated is released on forming this compound from its elements.

NOTE: from a more extensive comparison of hydrocarbon heats of formation, the ΔH°_f incremental values below have been calculated:

CH_3-	-10.12 kcal $mole^{-1}$
$-CH_2$-	-4.93
$-\overset{|}{C}H-$	-1.09
$-\overset{|}{\underset{|}{C}}-$	+0.80

17. Boiling points and melting points have nothing to do with relative thermodynamic stability. Mp and bp are only indications of the stability of the solid vs. liquid and liquid vs. gaseous states of a molecule, whereas thermodynamic stability is in relation to possible reactions or decompositions. For example, hexahydro-1,3,5-trinitro-1,3,5-triazine (cyclonite or RDX) is a high-melting solid (mp = 204 °C) which on decomposition releases over 300 kcal $mole^{-1}$. (This substance is used as a high explosive.)

5.E Supplementary Problems

S1. Write the structures for all of the isomers of C_5H_{10} which have only C-C single bonds, and provide IUPAC names for each structure.

S2. Write the structures of each of the following compounds:

(a) *trans*-1-ethyl-2-propylcyclopentane

(b) 1,1,4-tribromocyclohexane

(c) 3-chloro-1,1-dicyclopropylcycloheptane

(d) *trans*-1-(3-chloropropyl)-4-t-butylcyclohexane

(e) cyclotetradecane

S3. Using the heats of formation given in Appendix I, calculate or estimate $\Delta H°$ for the following reactions:

(a)  + H_2 ⟶ $CH_3CH_2CH_3$

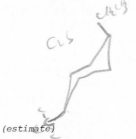

(b)  + H_2 ⟶ $CH_3CH_2CH_2CH_3$

(c)  ⇌ 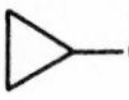

(estimate)

(d) CH_3CH_3 + HCl ⟶ CH_4 + CH_3Cl

(e)  + HCl ⟶ $CH_3CH_2CH_2Cl$

(f)  + HCl ⟶ $CH_3(CH_2)_4CH_2Cl$ *(estimate)*

S4. From the data provided in Appendix I, calculate $\Delta H°$ for the following isomerizations:

$CH_3CH_2CH_2CH_2CH_3$ ⇌ $CH_3CH_2CH(CH_3)_2$

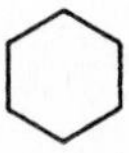 ⇌ 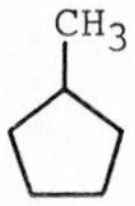

How do you account for the fact that increased branching is favorable in one case and not the other?

S5. (a) Draw the most stable chair conformation of *trans*-1-t-butyl-4-ethylcyclohexane. Construct a potential energy diagram for rotation about the C-C bond between the cyclohexane ring and the ethyl group, with estimations of the energy differences between the different conformations.

(b) Draw the most stable conformation of *cis*-1-t-butyl-4-ethylcyclohexane. How do you expect the potential energy diagram for rotation of the ethyl group in this isomer to differ from the one you drew for part (a)?

(c) Calculate the equilibrium constant for the isomerization of the *cis* isomer to the *trans* isomer at 25 °C (use Table 5.6).

S6. Make a model of t-butylcyclobutane. Which conformation do you expect would be the most stable? Do the same for t-butylcyclopentane.

S7. Predict the difference in heat of formation, $\Delta H°_f$, for each pair of isomers drawn below. Compare your predictions with the data given in Appendix I.

(a) 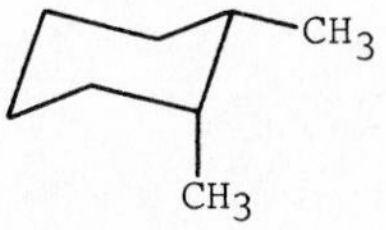

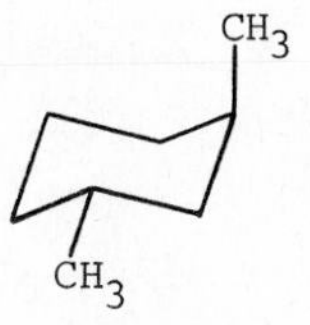

(b) 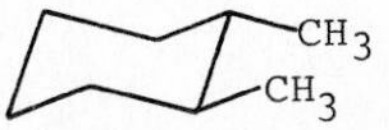

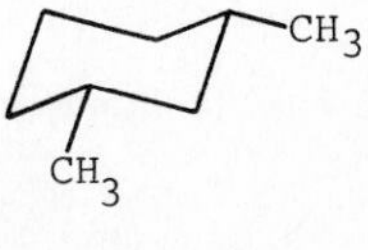

(c) 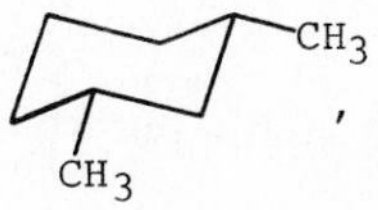

, 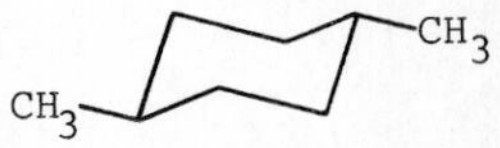

5.F Answers to Supplementary Problems

S1.

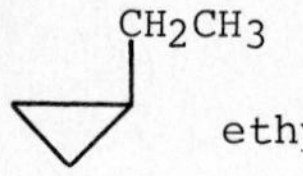

ethylcyclopropane

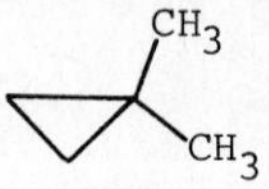

1,1-dimethylcyclopropane

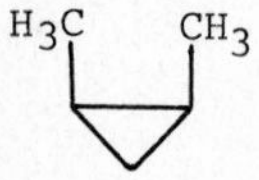

cis-1,2-dimethylcyclopropane

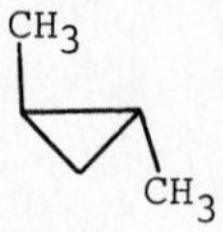

trans-1,2-dimethylcyclopropane

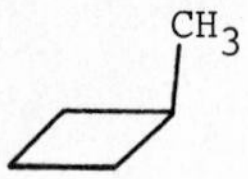

methylcyclobutane

cyclopentane

S2. (a)

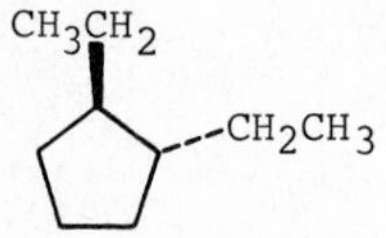

(c)

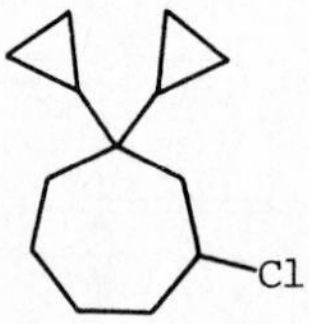

(d)

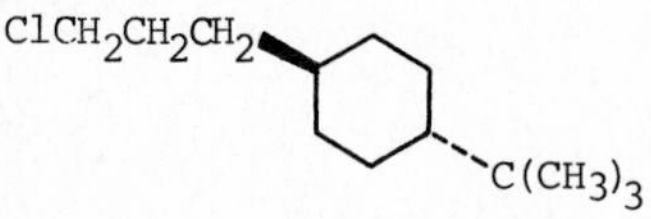

(b)

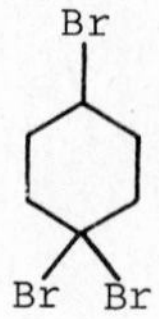

(e)

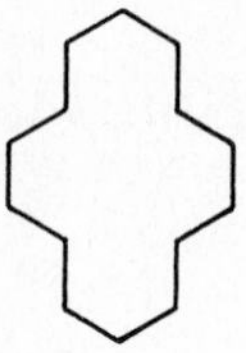

S3. ΔH°_f (products) − ΔH°_f (starting materials) = ΔH° of the reaction

<u>*NOTE:*</u> ΔH°_f for $H_2 = 0$

(a) $-24.8 - 12.7 = -37.5$ kcal mole^{-1}

(b) $-30.4 - 6.8 = -37.2$ kcal mole^{-1}

(c) To calculate the change in ΔH°_f caused by the addition of a methyl group:

ΔH°_f(isobutane) − ΔH°_f(propane) = −7.6 kcal mole^{-1}
ΔH°_f(methylcyclopentane) − ΔH°_f(cyclopentane) = −6.9 kcal mole^{-1}
ΔH°_f(methylcyclohexane) − ΔH°_f(cyclohexane) = −7.5 kcal mole^{-1}
} Average is −7.3

Using this calculated average, the estimated ΔH°_f(methylcyclopropane) is $12.7 - 7.3 = 5.4$ kcal mole^{-1}, and ΔH° for the conversion cyclobutane ⟶ methylcyclopropane is $5.4 - 6.8 = -1.4$ kcal mole^{-1}.

(d) $-17.9 - 20.6 - (-20.2) - (-22.1) = +3.8$ kcal mole^{-1} (an endothermic reaction)

(e) $-31.0 - 12.7 - (-22.1) = -21.6$ kcal mole^{-1}

(f) Estimate that $\Delta H^\circ_f(CH_3(CH_2)_4CH_2Cl)$ is approx. $\Delta H^\circ_f(CH_3CH_2CH_2Cl) + 3\,\Delta H^\circ_f(-CH_2-)$:

= −46.0 kcal mole^{-1} (see problem #16 in this Chapter)

ΔH° for the reaction $= -46.0 - (-29.5) - (-22.1) = +5.6$ kcal mole^{-1}

S4. pentane ⇌ 2-methylbutane

$H^\circ = -36.9 - (-35.1) = -1.8$ kcal mole^{-1}

cyclohexane $\rightleftharpoons$ methylcyclopentane

$$\Delta H° = -25.3 - (-29.5) = +4.2 \text{ kcal mole}^{-1}$$

There is some strain in cyclopentane (6 kcal $mole^{-1}$, in fact), and none in cyclohexane, and this effect overcomes the small difference in $\Delta H°_f$ attributed to branching.

S5. Whenever possible, a t-butyl group occupies an equatorial position.

(a)

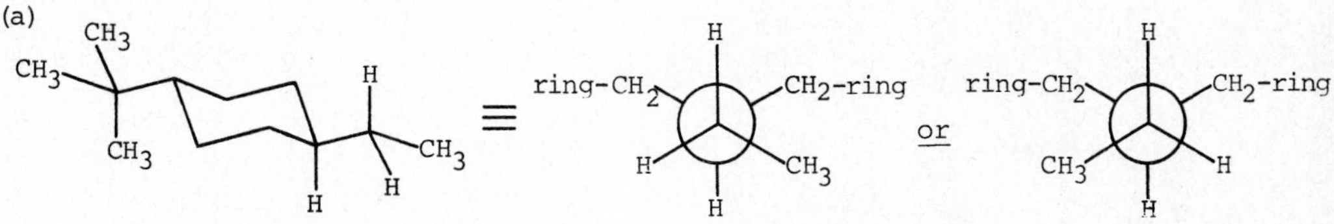

[Me = CH_3]

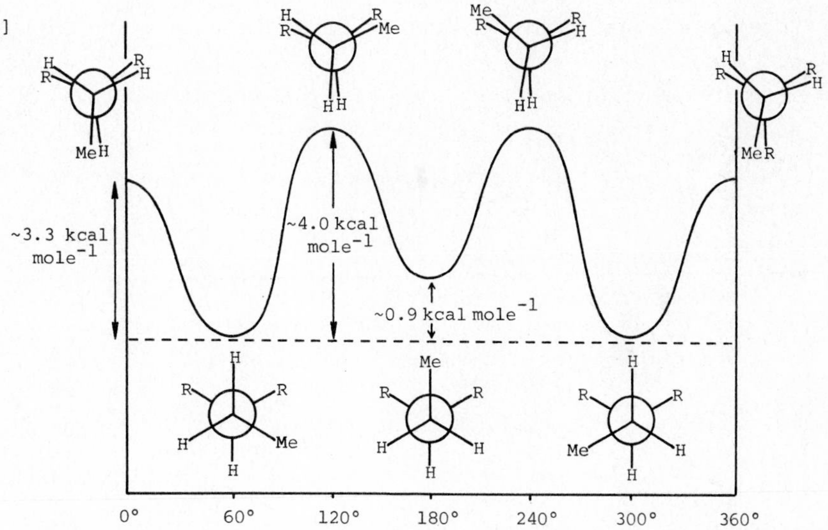

This diagram is qualitatively the same as that for 2-methylbutane (isopentane). [See problem #7.]

(b)

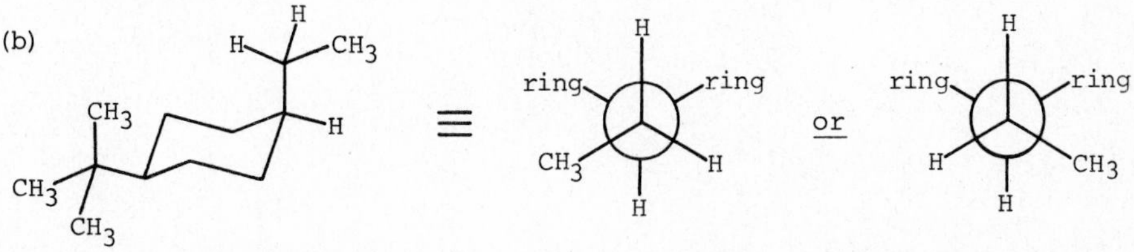

The conformations in which the methyl group is rotated over the cyclohexane ring will be much higher in energy because of steric interaction with the axial hydrogens:

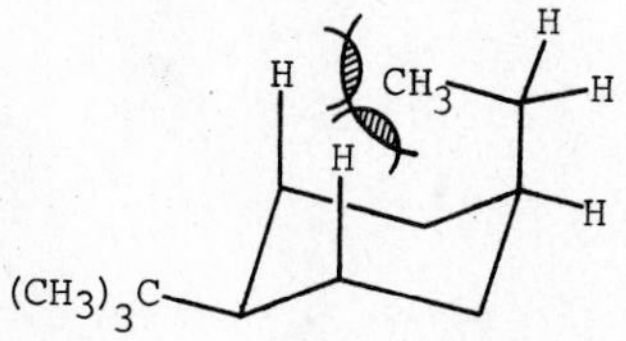

(c) The difference between *cis*- and *trans*-1-t-butyl-4-ethylcyclohexane is the same as that between axial and equatorial ethylcyclohexane, for which

$$\Delta G° = -1.8 \text{ kcal mole}^{-1} = -RT \ln \frac{[\text{trans}]}{[\text{cis}]} \quad ; \quad \text{so that } K = 21$$

S6.

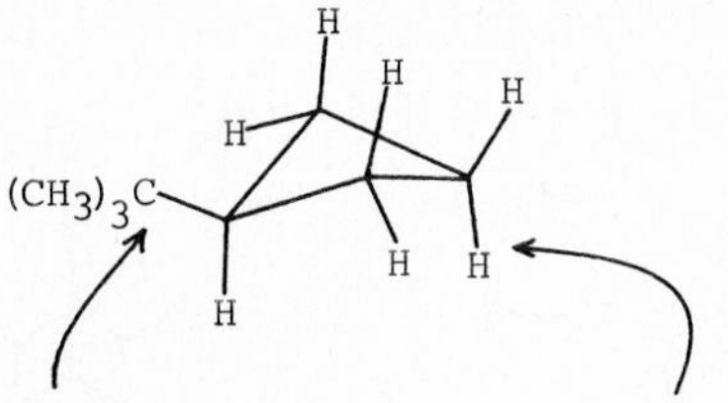

"equatorial" better than "axial"

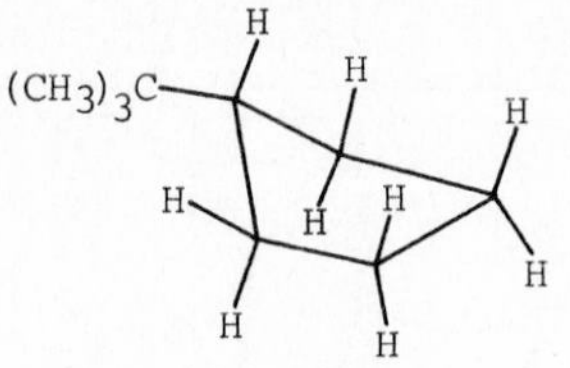

when t-butyl is attached to the "flap" of the "envelope", it is in the least crowded position

S7. (a) Both *cis*-1,2- and *trans*-1,3-dimethylcyclohexane have one axial and one equatorial methyl group, but the *cis*-1,2- isomer has an additional gauche interaction which would be expected to destabilize it by about 0.9 kcal mole^{-1}. In fact, the *trans*-1,3- isomer is 1.1 kcal mole^{-1} more stable.

(b) Similarly, *trans*-1,2-dimethylcyclohexane has an extra gauche interaction that the *cis*-1,3- isomer does not have. Again, the *cis*-1,3- isomer is actually more stable by 1.1 kcal mole^{-1}.

(c) There are no obvious differences between *cis*-1,3- and *trans*-1,4-dimethylcyclohexane as far as steric interactions are concerned, and it is not surprising to observe that they have the same $\Delta H°_f$.

6. REACTIONS OF ALKANES

6.A Chapter Outline, Important Terms Introduced, and Reactions Discussed

6.1 Bond Dissociation Energies (what it takes to break a bond)

vibrational energy levels

zero point energy

free radicals

stability of radicals: tertiary > secondary > primary > methyl

6.2 Pyrolysis of Alkanes: Cracking (reactions of alkanes at high temperature)

disproportionation:

$$CH_3CH_2CH_2CH_2CH_3 \longrightarrow CH_3CH_3 + CH_2{=}CHCH_3$$

6.3 Halogenation of Alkanes (the first thorough analysis of a reaction mechanism)

A. Chlorination

$$RH + Cl_2 \longrightarrow RCl + HCl \quad \textit{(can use } SO_2Cl_2\textit{, too)}$$

homolysis vs. heterolysis

chain reaction

initiation, propagation, and termination steps

principle of microscopic reversibility

relative reactivity

statistical factor

B. Halogenation with Other Halogens

selectivity of bromination

6.4 Combustion of Alkanes (how to determine relative stabilities)

heat of combustion

alkylperoxy radical

6.5 Average Bond Energies (to generalize between compounds)

heat of atomization

6.B Important Concepts

In Chapter 6 you are introduced for the first time to a detailed description of a reaction: free-radical halogenation. Although many industrial processes and other important reactions involve free-radical chemistry, such reactions usually receive less attention in introductory organic courses than do reactions which proceed by ionic mechanisms (involving charged intermediates). Nevertheless, free-radical halogenation provides you with the opportunity to analyze closely the various aspects of chemical reactivity. Many students approach each new reaction as another group of facts to be memorized. As we said in the introduction, memory is important, but it is not enough by itself. You should approach each new reaction with a series of questions, such as: "What sort of functional group transformation does it accomplish? What is the mechanism? What is the generality of the reaction? What are the stereochemical features to keep in mind? What limitations does it have?" These are general questions which will help you to understand the chemistry involved and fit the reaction into your chemical knowledge. Only after you have asked these questions should you ask: "What is unusual or unique about the reaction?" During the process of answering the first questions, you will probably have found the answer to the last one.

One of the essential facts to keep in mind is that all of organic chemistry makes logical sense. For free-radical halogenation, the details of each step are discussed thoroughly, and it is pointed out how the mechanism makes sense in light of the basic principles discussed in previous chapters. Many of the subsequent reactions presented in the text will be discussed in as much detail as this one, many more in less detail, but they are all sensible. Don't pretend that organic reactions are magic; ask yourself questions about the reactions you see, and if you can't make sense of them, ask your teachers.

6.C Answers to Exercises

$\Delta H°$ (kcal mole^{-1})

(6.1) $CH_4 \longrightarrow \cdot CH_3 + H\cdot$

$\Delta H°_f = -17.9 \quad +34 \quad +52$ $\qquad 34 + 52 - (-17.9) = 104$

NOTE: Be aware of significant figures. A sum cannot be more accurate than the least accurate of the numbers that went into it. $\Delta H°_f$ for the radicals are known less precisely than for methane. The overall enthalpy change cannot be determined more precisely than $\Delta H°_f$ for the radicals: $34 + 52 - (-17.9) = 103.9$ is therefore not a correct answer.

$CH_3CH_3 \longrightarrow 2\ \cdot CH_3$
$\Delta H°_f = -20.2 \quad 2 \times 34$ $\qquad 108$

$CH_3CH_3 \longrightarrow CH_3CH_2\cdot + H\cdot$
$-20.2 \quad 26 \quad 52$ $\qquad 98$

$CH_3CH_2CH_3 \longrightarrow (CH_3)_2CH\cdot + H\cdot$
$-24.8 \quad 17.5 \quad 52$ $\qquad 94$

$(CH_3)_3CH \longrightarrow (CH_3)_3C\cdot + H\cdot$
$-32.4 \quad 6.7 \quad 52$ $\qquad 91$

$\Delta H°$ (kcal mole^{-1})

$(CH_3)_4C \longrightarrow (CH_3)_3C\cdot + CH_3\cdot$
$-40.3 \quad 6.7 \quad 34$ $\qquad 81$

$(CH_3)_3CH \longrightarrow (CH_3)_2CHCH_2\cdot + H\cdot$
$-32.4 \quad 13 \quad 52$ $\qquad 97$

(6.3A) *Initiation:* $Cl_2 \longrightarrow 2\ Cl\cdot$ $\qquad 58$

Propagation:

$Cl\cdot + CH_3CH_3 \longrightarrow HCl + CH_3CH_2\cdot$
$\Delta H°_f = 29 \quad -20.2 \quad -22.1 \quad 26$ $\qquad -5$

$CH_3CH_2\cdot + Cl_2 \longrightarrow CH_3CH_2Cl + Cl\cdot$
$26 \quad 0 \quad -26.1 \quad 29$ $\qquad -23$

Termination:

$2\ CH_3CH_2\cdot \longrightarrow CH_3CH_2CH_2CH_3$
$\Delta H°_f = 2 \times 26 \quad -30.4$ $\qquad -82$

$2\ Cl\cdot \longrightarrow Cl_2$ $\qquad -58$

$CH_3CH_2\cdot + Cl\cdot \longrightarrow CH_3CH_2Cl$
$26 \quad 29 \quad -26.1$ $\qquad -81$

$\Delta H°$ for the overall reaction:

$CH_3CH_3 + Cl_2 \longrightarrow CH_3CH_2Cl + HCl$
$\Delta H°_f = -20.2 \quad 0 \quad -26.1 \quad -22.1$ $\qquad \Delta H° = -28.0$ kcal mole^{-1}

Note that the initiation and termination steps are not part of the overall reaction.

(6.3B) $\Delta H^\circ_f(\underline{t}\text{-}C_4H_9\cdot) + \Delta H^\circ_f(X\cdot) - DH^\circ(\underline{t}\text{-}C_4H_9\text{-}X) = \Delta H^\circ_f(\underline{t}\text{-}C_4H_9X)$

X = F	7	+	18.9	-	108	=	-82
X = Cl	7	+	28.9	-	79	=	-43
X = Br	7	+	26.7	-	65	=	-31
X = I	7	+	25.5	-	50	=	-18

	$(CH_3)_3CH$ +	X_2 ⟶	$(CH_3)_3CX$ +	HX	ΔH° (kcal $mole^{-1}$)
ΔH°_f, X = F	-32.4	0	-82	-65.0	-115
ΔH°_f, X = Cl	-32.4	0	-43	-22.1	-33
ΔH°_f, X = Br	-32.4	7.4	-31	-8.7	-15
ΔH°_f, X = I	-32.4	14.9	-18	+6.3	+6

NOTE: ΔH°_f for Br_2 and I_2 in the gas phase are not zero because the standard states of these elements are the liquid and solid phases, respectively. It takes energy to transfer them to the gas phase where the comparison between reactants and products is made.

(6.4) butane: $C_4H_{10} + 6\frac{1}{2}\,O_2 \longrightarrow 4\,CO_2 + 5\,H_2O \quad \Delta H^\circ_{comb.} = -634.82$ kcal $mole^{-1}$

isobutane: $C_4H_{10} + 6\frac{1}{2}\,O_2 \longrightarrow 4\,CO_2 + 5\,H_2O \quad \Delta H^\circ_{comb.} = -632.77$ kcal $mole^{-1}$

graphite and hydrogen: $4\,C + 5\,H_2 + 6\frac{1}{2}\,O_2 \longrightarrow 4\,CO_2 + 5\,H_2O$

$\Delta H^\circ_f = 0 \quad 0 \quad 0 \qquad 4 \times (-94.05) + 5 \times (-57.80) = -665.20$ kcal $mole^{-1}$

Therefore, for butane, $\Delta H^\circ_f = -665.20 - (-634.82) = -30.38$ kcal $mole^{-1}$

isobutane, $\Delta H^\circ_f = -665.20 - (-632.77) = -32.43$ kcal $mole^{-1}$

6.D Answers and Explanations for Problems

1. (a) $CH_3CH_2CH_2CH_2CH_3 \xrightarrow{\Delta} CH_4 + CH_3CH_3 + CH_2{=}CH_2 + CH_3CH_2CH_3 + CH_3CH{=}CH_2 + CH_3CH_2CH_2CH_3 + CH_3CH_2CH{=}CH_2$

(b) $CH_3CH_2CH_2CH_2CH_3 \xrightarrow{\Delta} CH_3CH_2CH_2CH_2\cdot + CH_3\cdot$

$\longrightarrow CH_3CH_2CH_2\cdot + \cdot CH_2CH_3$

$2\,CH_3\cdot \longrightarrow CH_3CH_3$

$CH_3\cdot + CH_3CH_2\cdot \longrightarrow CH_3CH_2CH_3$

$\longrightarrow CH_4 + CH_2{=}CH_2$

$2\,CH_3CH_2\cdot \longrightarrow CH_3CH_2CH_2CH_3$

$\longrightarrow CH_3CH_3 + CH_2{=}CH_2$

$CH_3\cdot + CH_3CH_2CH_2\cdot \longrightarrow CH_3CH_2CH_2CH_3$

$\longrightarrow CH_4 + CH_3CH{=}CH_2$

$CH_3CH_2\cdot + CH_3CH_2CH_2\cdot \longrightarrow CH_3(CH_2)_3CH_3$

$\longrightarrow CH_2{=}CH_2 + CH_3CH_2CH_3$

$\longrightarrow CH_3CH_3 + CH_3CH{=}CH_2$

$CH_3\cdot + CH_3CH_2CH_2CH_2\cdot \longrightarrow CH_3(CH_2)_3CH_3$

$\longrightarrow CH_4 + CH_3CH_2CH{=}CH_2$

Note that longer alkanes can also be produced by recombination of $CH_3CH_2CH_2\cdot$ and $CH_3CH_2CH_2CH_2\cdot$ radicals.

(c)

			$\Delta H°$, kcal mole^{-1}
$CH_3CH_2CH_2CH_2CH_3$	⟶	$CH_3CH_2CH_2CH_2\cdot + CH_3\cdot$	+85
	⟶	$CH_3CH_2CH_2\cdot + CH_3CH_2\cdot$	+82
$CH_3\cdot + CH_3\cdot$	⟶	CH_3CH_3	-88
$CH_3\cdot + CH_3CH_2\cdot$	⟶	$CH_3CH_2CH_3$	-85
	⟶	$CH_4 + CH_2{=}CH_2$	-65
$CH_3CH_2\cdot + CH_3CH_2\cdot$	⟶	$CH_3CH_3 + CH_2{=}CH_2$	-60
$CH_3\cdot + CH_3CH_2CH_2\cdot$	⟶	$CH_3CH_2CH_2CH_3$	-85
	⟶	$CH_4 + CH_3CH{=}CH_2$	-68
$CH_3CH_2\cdot + CH_3CH_2CH_2\cdot$	⟶	$CH_2{=}CH_2 + CH_3CH_2CH_3$	-59
	⟶	$CH_3CH_3 + CH_3CH{=}CH_2$	-62
$CH_3\cdot + CH_3CH_2CH_2CH_2\cdot$	⟶	$CH_4 + CH_3CH_2CH{=}CH_2$	-68

OVERALL REACTIONS:

			$\Delta H°$, kcal mole^{-1}
$CH_3CH_2CH_2CH_2CH_3$ ($\Delta H°_f = -35.1$)	↗	$CH_4 + CH_3CH_2CH{=}CH_2$ (-17.9, -0.2)	+17.0
	⟶	$CH_3CH_3 + CH_3CH{=}CH_2$ (-20.2, 4.9)	+19.8
	↘	$CH_3CH_2CH_3 + CH_2{=}CH_2$ (-24.8, 12.5)	+22.8

2. (a) $Br_2 \longrightarrow 2\,Br\cdot \qquad \Delta H° = DH° = 2\,\Delta H°_f(Br\cdot) - \Delta H°_f(Br_2)$

$= 46$ kcal mole^{-1}

(b) $CH_3CH_3 + Br\cdot \longrightarrow CH_3CH_2\cdot + HBr$

To use DH° values, you must break this step into two reactions:

$CH_3CH_3 \longrightarrow CH_3CH_2\cdot + H\cdot \qquad DH° = 98$

$H\cdot + Br\cdot \longrightarrow HBr \qquad -DH° = -87.5$

$CH_3CH_3 + \cancel{H\cdot} + Br\cdot \longrightarrow CH_3CH_2\cdot + \cancel{H\cdot} + HBr$, and $\Delta H° = 11$ kcal mole^{-1}

The use of heats of formation (Appendix I) gives a similar answer:

$CH_3CH_3 + Br\cdot \longrightarrow CH_3CH_2\cdot + HBr$

$\Delta H°_f = \;-20.2 \quad 26.7 \qquad 26 \quad -8.7 \qquad H° = +11$ kcal mole^{-1}

(c) $CH_3CH_2\cdot + Br_2 \longrightarrow CH_3CH_2Br + Br\cdot$

$\Delta H° = DH°(Br\text{-}Br) - DH°(CH_3CH_2\text{-}Br)$

$= 46 - 68 = -22$ kcal mole^{-1}

(d) (b) + (c) $= \Delta H°$ for the overall reaction

$= 11 + (-22) = -11$ kcal mole^{-1}

To check: $Br_2 + CH_3CH_3 \longrightarrow CH_3CH_2Br + HBr$

$\Delta H°_f = \;7.4 \quad -20.2 \qquad -15.2 \quad -8.7 \qquad \Delta H° = -11.1$ kcal mole^{-1}

The values are the same, except for precision. $\Delta H°_f$ for radicals are not known as accurately as for normal compounds *(see explanatory Note accompanying answer to Exercise 6.3B)*

3. The reaction of Br· with ethane is an endothermic reaction, and therefore is likely to be slow, because the activation energy $\Delta H^{\ddagger}$ is $\geq \Delta H° = +11$ kcal mole^{-1}. In contrast, the reaction of ethyl radical with Br_2 is very exother-

mic ($\Delta H^\circ = -22$ kcal mole^{-1}). Because radical hydrogen abstraction has a relatively low $\Delta H^{\ddagger}$ for exothermic reactions, the reaction of $C_2H_5\cdot$ with Br_2 is expected to be fast. Therefore $[Br\cdot] > [C_2H_5\cdot]$.

A good analogy is the scene on a ski slope: on the average, there are as many people coming down the mountain as are going up, but there are more waiting in line at the bottom than waiting at the top.

4. (a) All of the hydrogens in spiropentane are equivalent. Thus, there is only one possible monochlorospiropentane. Furthermore, the dichloro compounds have higher boiling points and can be separated by distillation.

(b) Mechanism:

$$Cl_2 \longrightarrow 2\,Cl\cdot \qquad \text{(initiation)}$$

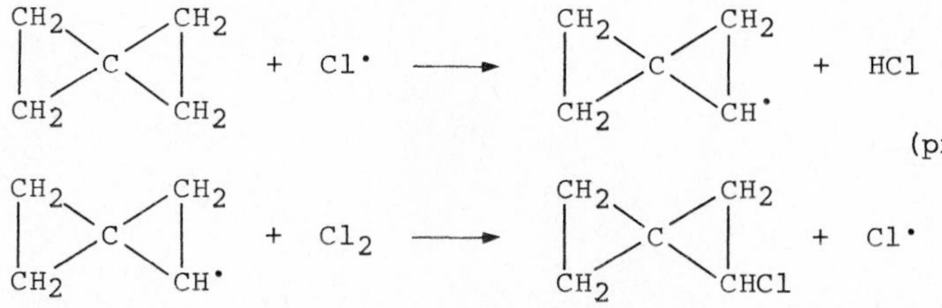

2 R· ⟶ R-R (R· =  or Cl·) (termination)

5. In Section 6.3A, we found the relative reactivities of primary (1°), secondary (2°), and tertiary (3°) hydrogens to be 1 : 4.0 : 5.1 in chlorination reactions.

(a) $CH_3-CH_2-CH_2-CH_3$ (these six hydrogens [of the two CH_3 groups] are equivalent; these four hydrogens [of the two CH_2 groups] are equivalent)

Product	Statistical Factor		Relative Reactivity		Relative Amount	Percent of Mixture
$CH_3CH_2CH_2CH_2Cl$	6	×	1	=	6	6/22 = 27%
$CH_3CH_2CHClCH_3$	4	×	4	=	16	16/22 = 73%
					22	100%

(b) $(CH_3)_2CH-CH_2-CH_3$ (these six hydrogens [of the two CH_3 groups on CH] are equivalent; these three hydrogens [of the terminal CH_3] are different from the other primary ones)

Product						
$ClCH_2(CH_3)CHCH_2CH_3$	6	×	1	=	6	6/22.1 = 27%
$(CH_3)_2CClCH_2CH_3$	1	×	5.1	=	5.1	5.1/22.1 = 23%
$(CH_3)_2CHCHClCH_3$	2	×	4	=	8	8/22.1 = 36%
$(CH_3)_2CHCH_2CH_2Cl$	3	×	1	=	3	3/22.1 = 14%
					22.1	100%

(c)

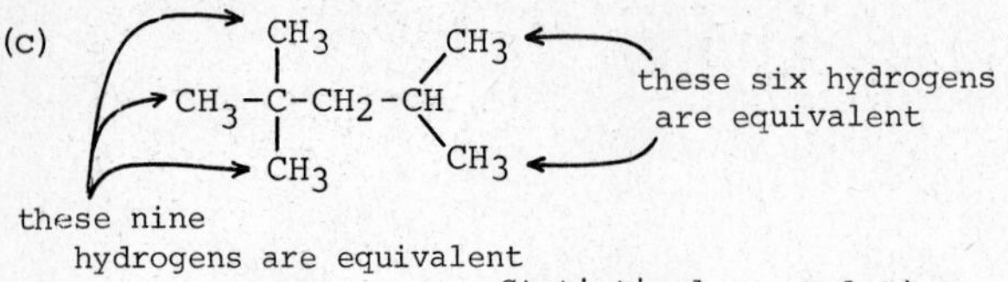

Product	Statistical Factor		Relative Reactivity		Relative Amount	Percent of Mixtur
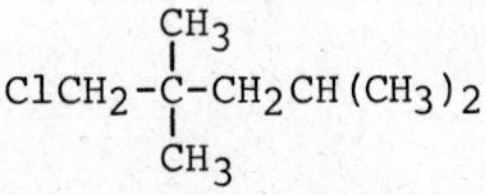 $ClCH_2-C(CH_3)_2-CH_2CH(CH_3)_2$	9	×	1	=	9	9/28.1 = 32%
$(CH_3)_3CCHClCH(CH_3)_2$	2	×	4	=	8	8/28.1 = 28%
$(CH_3)_3CCH_2CCl(CH_3)_2$	1	×	5.1	=	5.1	5.1/28.1 = 18%
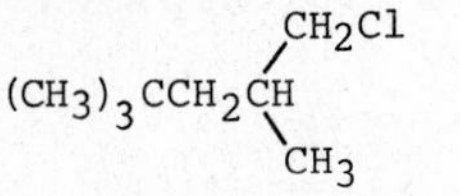 $(CH_3)_3CCH_2CH(CH_2Cl)CH_3$	6	×	1	=	6	6/28.1 = 21%
					28.1	(99%)

(d) $(CH_3)_3CCH(CH_3)_2$

Product						
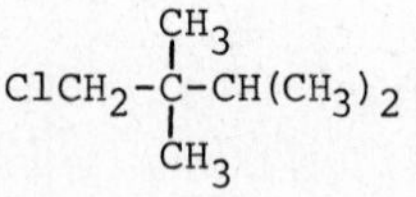 $ClCH_2-C(CH_3)_2-CH(CH_3)_2$	9	×	1	=	9	9/20.1 = 45%
$(CH_3)_3CC(CH_3)_2Cl$	1	×	5.1	=	5.1	5.1/20.1 = 25%
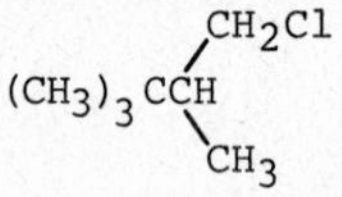 $(CH_3)_3CCH(CH_2Cl)CH_3$	6	×	1	=	6	6/20.1 = 30%
					20.1	100%

(e)

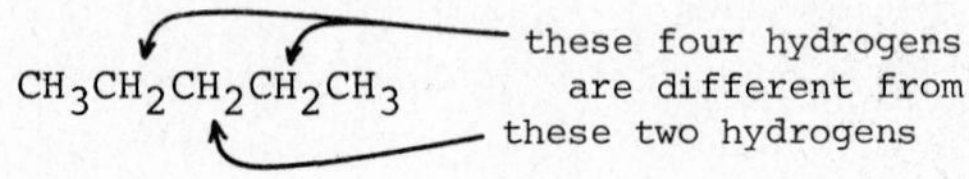

Product						
$CH_3CH_2CH_2CH_2CH_2Cl$	6	×	1	=	6	6/30 = 20%
$CH_3CH_2CH_2CHClCH_3$	4	×	4	=	16	16/30 = 53%
$CH_3CH_2CHClCH_2CH_3$	2	×	4	=	8	8/30 = 27%
					30	100%

Note how chlorination is generally impractical as a synthetic method when the molecule contains non-equivalent hydrogens.

6. (a)

rate(2) = 4 × 220 = 880

$CH_3CH_2CH_2CH_3$

rate(1) = 6 × 1 = 6

total rate = 6 + 880 = 886

%(1) = (6/886) × 100 = 0.7%

%(2) = (880/886) × 100 = 99.3%

(b)

$BrCH_2\overset{CH_3}{C}HCH_2CH_3$	(6/19449) × 100 = 0.03%
$(CH_3)_2CBrCH_2CH_3$	(19000/19449) × 100 = 97.7%
$(CH_3)_2CHCHBrCH_3$	(440/19449) × 100 = 2.3%
$(CH_3)_2CHCH_2CH_2Br$	(3/19449) × 100 = 0.02%

(c)

$BrCH_2\overset{CH_3}{\underset{CH_3}{C}}CH_2CH(CH_3)_2$	(9/19455) × 100 = 0.05%
$(CH_3)_3C\overset{Br}{C}HCH(CH_3)_2$	(440/19455) x 100 = 2.3%
$(CH_3)_3CCH_2CBr(CH_3)_2$	(19000/19455) x 100 = 97.7%
$(CH_3)_3CCH_2\overset{CH_3}{C}HCH_2Br$	(6/19455) × 100 = 0.03%

(d)

$BrCH_2\overset{CH_3}{\underset{CH_3}{C}}CH(CH_3)_2$	0.05%
$(CH_3)_3C\overset{Br}{C}(CH_3)_2$	99.9%
$(CH_3)_3C\overset{CH_3}{C}HCH_2Br$	0.03%

(e)

$CH_3CH_2CH_2CH_2CH_2Br$	0.5%
$CH_3CH_2CH_2CHBrCH_3$	66.4%
$CH_3CH_2CHBrCH_2CH_3$	33.2%

Bromination is a generally more practical reaction than chlorination, especially for the preparation of tertiary bromides.

7\. (a)

$\Delta H°$ (kcal $mole^{-1}$)

$C_2H_6 + Cl_2 \longrightarrow CH_3CH_2Cl + HCl$ −28.0

$\Delta H°_f$ = −20.2, 0 → −26.1, −22.1

$C_2H_6 + Cl_2 \longrightarrow 2\ CH_3Cl$ −21.0

$\Delta H°_f$ = −20.2, 0 → 2 × (−20.6)

(b)

$C_2H_6 + Cl\cdot \longrightarrow CH_3Cl + CH_3\cdot$ +6.9

$\Delta H°_f$ = −20.2, 26.7 → −20.6, 34

$CH_3\cdot + Cl_2 \longrightarrow CH_3Cl + Cl\cdot$ −27.9

$\Delta H°_f$ = 34, 0 → −20.6, 26.7

(c) There is no thermodynamic difficulty with either step; for instance, the first step in the propagation of the bromination of ethane is endothermic by +10.5 kcal $mole^{-1}$ (see problem #2) and yet it proceeds rapidly.

The difficulty with Cl· cleavage of the C-C bond in ethane is one of selectivity: since abstraction of a hydrogen atom is exothermic by -4.9 kcal mole^{-1} (see Exercise 6.3A), that reaction proceeds many orders of magnitude faster, and the alternative C-C bond cleavage never has a chance to occur.

8. $C_4H_{10} \longrightarrow 4\ \cdot\ddot{C}\cdot + 10\ H\cdot$

$\Delta H°_f = -30.4 \qquad 4 \times 170.9 \qquad 10 \times 52.1 \qquad \Delta H°_{atomization} = 1235$ kcal mole^{-1}

From bond energies: three C-C bonds = 3 × 83 = 249

ten C-H bonds = 10 × 99 = 990

1239 kcal mole^{-1}

9.  $+ 4\frac{1}{2}\ O_2 \longrightarrow 3\ CO_2 + 3\ H_2O$

$\Delta H°_f = 12.7 \qquad 0 \qquad 3 \times (-94.05) \qquad 3 \times (-57.80) \qquad \Delta H° = \Delta H°_{combustion} = -468.25$ kcal mole^{-1}

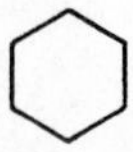 $+ 9\ O_2 \longrightarrow 6\ CO_2 + 6\ H_2O$

$\Delta H°_f = -29.5 \qquad 0 \qquad 6 \times (-94.05) \qquad 6 \times (-57.80) \qquad \Delta H° = -881.6$ kcal mole^{-1}

Since the molecular weight of cyclohexane is twice that of cyclopropane, on a weight basis cyclopropane releases $\frac{2 \times 468}{882}$ = 1.06 times as much energy on combustion as cyclohexane does. If the two fuels cost the same, cyclopropane would be the more economical.

10. (a) $\Delta H°_f(CH_3\cdot) + \Delta H°_f(F\cdot) - DH°(CH_3\text{-}F) = \Delta H°_f(CH_3F)$

$34 + 19 - 109 = -56$ kcal mole^{-1}

$CH_4 + ClF \longrightarrow CH_3Cl + HF \qquad \Delta H° = -55$ kcal mole^{-1}

$\Delta H°_f = -18 \quad -12.2 \qquad\qquad -20.6 \quad -65$

$CH_4 + ClF \longrightarrow CH_3F + HCl \qquad \Delta H° = -48$ kcal mole^{-1}

$\qquad\qquad\qquad\qquad -56 \quad -22.1$

(b) CH_3Cl and HF, because their formation releases the greatest amount of energy.

(c) $CH_3\cdot + ClF \longrightarrow CH_3Cl + F\cdot \qquad \Delta H° = -23.5$ kcal mole^{-1}

$\Delta H°_f = 34 \quad -12.2 \qquad\qquad -20.6 \quad 18.9$

$CH_3\cdot + ClF \longrightarrow CH_3F + Cl\cdot \qquad \Delta H° = -51$ kcal mole^{-1}

$\qquad\qquad\qquad\qquad -56 \quad 26.7$

One would predict from this comparison that CH_3F would be formed faster.

This case provides a good illustration of a situation which is frequently encountered: the product which is formed more quickly (called the product of "kinetic control") is not always the most stable one (which is called the product of "thermodynamic control"). This sort of situation results from the independence of the activation energy ($\Delta G^{\ddagger}$) and the overall energy change ($\Delta G°$) in a reaction.

11.

	Reaction	$\Delta H°$ (kcal mole^{-1})	
	$CH_4 + HNO_3 \longrightarrow CH_3NO_2 + H_2O$		
$\Delta H°_f =$	$-17.9 \quad -32.1 \qquad -17.9 \quad -57.8$	-25.7	
	$CH_4 + \cdot NO_2 \longrightarrow CH_3\cdot + HNO_2$		
	$-17.9 \quad 7.9 \qquad 34 \quad -18.4$	$+25.6$	this step is highly endothermic and is therefore slow; it requires high temperature
	$HNO_2 + HNO_3 \longrightarrow 2\ NO_2\cdot + H_2O$		
	$-18.4 \quad -32.1 \qquad 2 \times 7.9 \quad -57.8$	$+8.5$	
	$CH_3\cdot + \cdot NO_2 \longrightarrow CH_3NO_2$		
	$34 \quad 7.9 \qquad -17.9$	-59.8	

Possible alternatives for $CH_3\cdot$:

$$CH_3\cdot + HNO_3 \longrightarrow CH_3OH + \cdot NO_2 \qquad (34,\ -32.1 \rightarrow -48.1,\ 7.9) \qquad -42$$

$$CH_3\cdot + HNO_3 \longrightarrow CH_3NO_2 + HO\cdot \qquad (34,\ -32.1 \rightarrow -17.9,\ 9.3) \qquad -11$$

Resonance structures for $NO_2\cdot$:

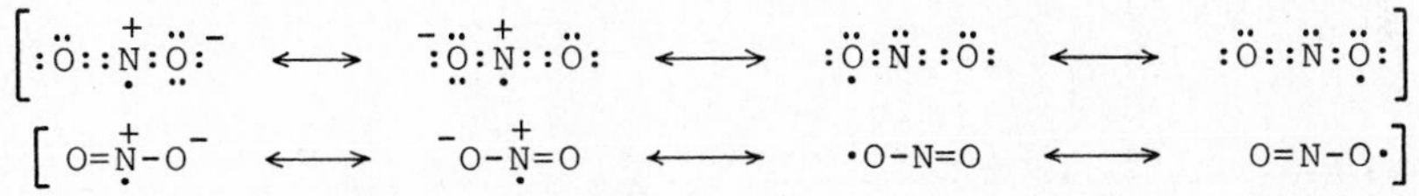

$$CH_3\cdot + NO_2\cdot \longrightarrow CH_3ONO \qquad (34,\ 7.9 \rightarrow -15.8) \qquad \Delta H° = -57.7 \text{ kcal mole}^{-1}$$

$$CH_3\cdot + NO_2\cdot \longrightarrow CH_3NO_2 \qquad (34,\ 7.9 \rightarrow -17.9) \qquad \Delta H° = -59.8 \text{ kcal mole}^{-1}$$

12. At high concentrations of CH_4 and low concentrations of Cl_2, the rate at which chlorine atoms find and react with CH_4 molecules will become faster than the rate at which methyl radicals find and react with Cl_2. At equal concentrations of CH_4 and Cl_2, the reaction of chlorine atoms with CH_4 will be rate-determining. The chlorine atoms will be present at higher concentration (see problem #3). The relative rates of the propagation steps determine which radical is in excess, which in turn determines which termination process is faster. Termination processes involving $Cl\cdot$ will predominate if $Cl\cdot + CH_4$ is the slow step.

13.

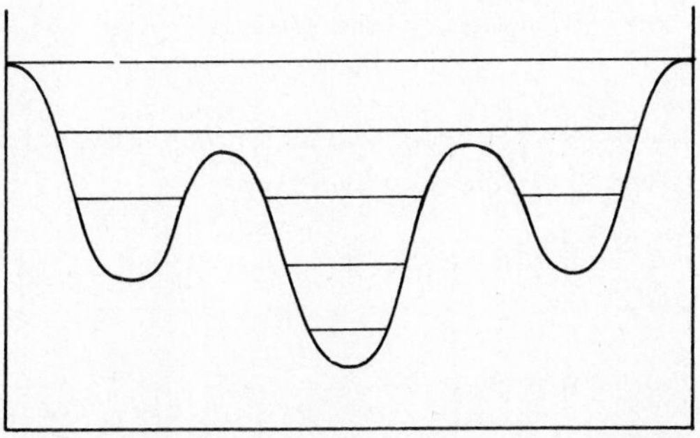

Free rotation at the 5th or 6th quantum level.

6.E Supplementary Problems

S1. Cracking propane to give a mixture of methane and ethylene is an endothermic process (compare with problem #1 in this chapter). However, from the point of view of entropy, splitting one molecule into two is a favorable process, as reflected by the standard entropy change, $\Delta S°$, for the reaction in question:

$$CH_3CH_2CH_3 \longrightarrow CH_4 + CH_2=CH_2 \qquad \Delta S° = +33 \text{ e.u.}$$

(a) Using this number and Appendix I, calculate the temperature at which equilibrium favors the smaller molecules; i.e., $K > 1$.

(b) In a more efficient cracking process, called hydrocracking, hydrogen is mixed with the hydrocarbons to be cracked. What products would you expect from the cracking of propane in the presence of hydrogen?

(c) Calculate $\Delta H°$ for the reaction you have written for part (b).

S2. From the data below and in Table 6.2 and Appendix I, calculate or predict $DH°$ for the weakest bond in each of the compounds listed below.

(a)
$$\begin{array}{c} CH_3 \qquad CH_3 \\ \diagdown \qquad \diagup \\ CH-CH \\ \diagup \qquad \diagdown \\ CH_3 \qquad CH_3 \end{array}$$

(b)
$$\begin{array}{c} CH_3 \;\; CH_3 \\ | \quad\;\; | \\ CH_3-C-C-CH_3 \\ | \quad\;\; | \\ CH_3 \;\; CH_3 \end{array} \quad (\Delta H°_f = -54.0 \text{ kcal mole}^{-1})$$

(c) ⬜

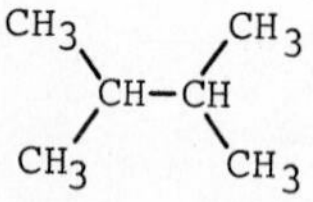

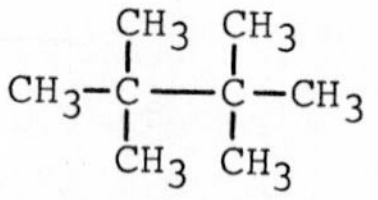

(d) △

S3. Free radical halogenation can also be accomplished using sulfuryl chloride ($Cl-\overset{\overset{O}{\|}}{\underset{\underset{O}{\|}}{S}}-Cl$) via the following propagation steps:

$$R\cdot + ClSO_2Cl \longrightarrow RCl + \cdot SO_2Cl \;\;\Big\}$$
$$\cdot SO_2Cl + RH \longrightarrow R\cdot + HSO_2Cl \;\;\Big\} \quad \textit{Propagation}$$

$$(HSO_2Cl \xrightarrow{\textit{fast}} SO_2 + HCl)$$

(a) From the product compositions below, calculate the selectivity of the $\cdot SO_2Cl$ radical for primary, secondary, and tertiary hydrogens.

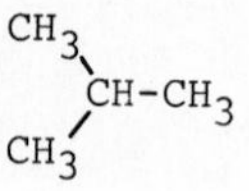

$$\begin{array}{c} CH_3 \\ \diagdown \\ CH-CH_3 \\ \diagup \\ CH_3 \end{array} + ClSO_2Cl \xrightarrow{h\nu} (CH_3)_3CCl + (CH_3)_2CHCH_2Cl$$
$$\qquad\qquad\qquad\qquad\qquad\qquad 31\% \qquad\quad 69\%$$

$$CH_3CH_2CH_2CH_2CH_3 + ClSO_2Cl \xrightarrow{h\nu} CH_3CH_2CH_2CH_2ClCH_3 + CH_3CH_2CH_2CH_2CH_2Cl + CH_3CH_2CHClCH_2CH_3$$
$$\qquad\qquad\qquad\qquad\qquad\qquad\qquad\qquad 48\% \qquad\qquad\qquad 28\% \qquad\qquad\qquad 24\%$$

(b) Predict the relative amounts of monochlorinated products obtained from this reaction with 2,4-dimethylpentane.

S4. On heating with HI, alkyl iodides react to give alkanes and I_2.

(a) Using $CH_3I + HI \longrightarrow CH_4 + I_2$ as an example, calculate $\Delta H°$ for the reaction.

(b) Propose a mechanism for this reaction and justify the steps involved by calculating $\Delta H°$ for each one. (*HINT:* take advantage of the principle of microscopic reversibility.)

S5. One of the mechanisms by which HBr can add to alkenes is illustrated
     below for ethylene:

$$Br\cdot \quad CH_2{=}CH_2 \longrightarrow Br\text{-}CH_2\text{-}CH_2\cdot$$

$$BrCH_2CH_2\cdot \quad H{-}Br \longrightarrow BrCH_2CH_3 + Br\cdot$$

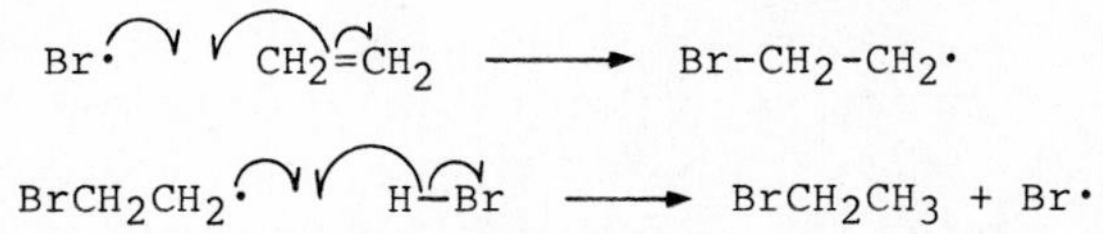

 (a) Using the data in Table 6.2 and in Appendices I and II, estimate $\Delta H°$
      for each of the steps above and for the overall reaction.

 (b) Do the same calculations for the reaction of HCl by the same mech-
      anism. What do you conclude about the relative rates of the two
      reactions?

 (c) Two isomeric products are possible for the addition of HBr to 2-
      methylpropene. Write out the steps involved in the formation of each
      one and estimate $\Delta H°$ for each of them. Which isomer is more stable?
      Which one will be the major product (formed faster)?

S6. (a) Calculate the average bond dissociation energy for CCl_4 using the
          data from Appendix I.

 (b) Calculate the C-H bond dissociation energy for $CHCl_3$ (chloroform).

 (c) Is the reaction $CH_4 + CCl_4 \longrightarrow CH_3Cl + CHCl_3$ likely to take
      place on thermodynamic grounds?

 (d) Is the reaction likely to take place on kinetic grounds? In other words,
      are any of the steps in the mechanism you wrote very endothermic?

6.F Answers to Supplementary Problems

S1. (a) $C_3H_8 \longrightarrow CH_4 + C_2H_4$ $\Delta H° = +19.4$ kcal mole^{-1}

$\Delta S° = +33$ e.u.

K is greater than 1 when $\Delta G°$ is less than zero;

since $\Delta G° = \Delta H° - T\Delta S°$, $19{,}400 - T \times 33 < 0$, so that $T > 588\,°K$

(b) $C_3H_8 + H_2 \longrightarrow CH_4 + C_2H_6$

(c) $\Delta H°_f$: −24.8 0 −17.9 −20.2 $\Delta H° = -13.3$ kcal mole^{-1}

S2. (a) $(CH_3)_2CH\text{-}CH(CH_3)_2 \longrightarrow 2\ (CH_3)_2CH\cdot$

$\Delta H°_f$: −42.6 2×17.5 $\Delta H° = DH° = 77.6$ kcal mole^{-1}

(b) $(CH_3)_3C\text{-}C(CH_3)_3 \longrightarrow 2\ (CH_3)_3C\cdot$

$\Delta H°_f$: −54.0 2×6.7 $\Delta H° = DH° = 67.4$ kcal mole^{-1}

(c) To calculate DH° for  $\longrightarrow \cdot CH_2CH_2CH_2CH_2\cdot$, follow the sequence below:

(1) H_2 +  $\longrightarrow CH_3CH_2CH_2CH_3$ $\Delta H° = -37.2$ kcal mole^{-1}

(2) $CH_3CH_2CH_2CH_3 \longrightarrow 2\,H\cdot + \cdot CH_2CH_2CH_2CH_2\cdot$ $\Delta H° = 2\,DH°$(primary C-H) = +196 kcal mole^{-1}

(3) $2\,H\cdot \longrightarrow H_2$ $\Delta H° = -DH°(H_2) = -104$ kcal mole^{-1}

DH° for the ring opening of cyclobutane is the sum of *(1)*, *(2)* and *(3)*:

$DH° = -37.2 + 196 + (-104) = 55$ kcal mole^{-1}

(d) The same sequence may be applied to  $\longrightarrow \cdot CH_2CH_2CH_2\cdot$; only for step *(1)* is the value for $\Delta H°$ different:

(1)' H_2 +  $\longrightarrow CH_3CH_2CH_3$ $\Delta H° = -37.5$ kcal mole^{-1}

DH° for the ring opening of cyclopropane is calculated to be +55 kcal mole^{-1} also. For your interest, compare this figure with the values you calculate for the cleavage of the cyclopentane and cyclohexane rings.

S3. (a) Reaction of $(CH_3)_2CH$ results in 31% tertiary and 69% primary chlorination.

Relative Amount		Statistical Factor		Relative Reactivity
31%	=	1	×	3° reactivity
69%	=	9	×	1° reactivity

Rel. reactivity: $3°/1° = \dfrac{31}{69/9} = 4$

Reaction of $CH_3CH_2CH_2CH_2CH_3$ results in 48% 2-chloro isomer, 24% 3-chloro isomer, and 28% 1-chloro isomer.

48%	=	4	×	2° reactivity
24%	=	2	×	2° reactivity
28%	=	6	×	1° reactivity

Rel. reactivity: $2°/1° = \dfrac{24/2}{28/6} = 2.6$

Therefore, the ratio of 3° reactivity : 2° reactivity : 1° reactivity = 4 : 2.6 : 1

(b)

$(CH_3)_2CHCH_2CH(CH_3)_2 \longrightarrow (ClCH_2)(CH_3)CHCH_2CH(CH_3)_2 + (CH_3)_2CClCH_2CH(CH_3)_2 + (CH_3)_2CHCHClCH(CH_3)_2$

Statistical factor:	12	2	2
Relative reactivity:	1	4	2.6
Relative amounts:	12 (= 12×1)	8 (= 2×4)	5.2 (= 2×2.6)
Percent of mixture:	48%	32%	20%

S4. (a) $CH_3I + HI \longrightarrow CH_4 + I_2$

ΔH°_f: 3.4   6.3   -17.9   0 $\Delta H^\circ = -27.6$ kcal mole^{-1}

(b) $CH_3{-}I \longrightarrow CH_3\cdot + I\cdot$ $DH^\circ = 56$ kcal mole^{-1}

but many other initiation steps are possible; for example, from traces of peroxide, light, etc.

propagation

$I\cdot + CH_3I \longrightarrow CH_3\cdot + I_2$

ΔH°_f: 25.5   3.4   34   0 $\Delta H^\circ = +5$ kcal mole^{-1}

$CH_3\cdot + HI \longrightarrow CH_4 + I\cdot$

ΔH°_f: 34   6.3   -17.9   25.5 $\Delta H^\circ = -33$ kcal mole^{-1}

S5. (a) $Br\cdot + CH_2{=}CH_2 \longrightarrow BrCH_2CH_2\cdot$

Calculate from the sequence:

			ΔH° (kcal mole^{-1})
(1)	$H\cdot + Br\cdot \longrightarrow HBr$	$\Delta H^\circ = -DH^\circ$(HBr):	-87.5
(2)	$HBr + CH_2{=}CH_2 \longrightarrow CH_3CH_2Br$		-19.0
(3)	$CH_3CH_2Br \longrightarrow H\cdot + \cdot CH_2CH_2Br$	$\Delta H^\circ = DH^\circ$(primary C-H):	98

The sum of *(1)*, *(2)*, and *(3)* = ΔH° for the first step: -8.5

$\cdot CH_2CH_2Br + HBr \longrightarrow CH_3CH_2Br + Br$

Calculate from the sequence:

(4)	$HBr \longrightarrow H\cdot + Br\cdot$	$\Delta H^\circ = DH^\circ$ (HBr):	+87.5
(5)	$\cdot CH_2CH_2Br + H\cdot \longrightarrow CH_3CH_2Br$	$\Delta H^\circ = -DH^\circ$ (primary C-H):	-98

The sum of *(4)* and *(5)* = ΔH° for the second step: -10.5

ΔH° for the overall reaction was calculated as eq. (2) above.

(b) For HCl addition, the following changes in the sequence of equations are made:

		ΔH° (kcal mole^{-1})
(1)'	$\Delta H^\circ = -DH^\circ$(HCl)	-103.2
(2)'	$\Delta H^\circ = \Delta H^\circ_f(C_2H_5Cl) - \Delta H^\circ_f(HCl) - \Delta H^\circ_f(C_2H_4)$	-16.5
(3)'	remains approximately the same	+98

The sum of *(1)'*, *(2)'*, and *(3)'* = ΔH° for the first step ($Cl\cdot + CH_2{=}CH_2 \longrightarrow \cdot CH_2CH_2Cl$): -21.7

(4)'	$\Delta H^\circ = DH^\circ$(HCl)	+103.2
(5)'	remains approximately the same	-98

The sum of *(4)'* and *(5)'* = $\Delta H°$ for the second step

$$(\cdot CH_2CH_2Cl + HCl \longrightarrow CH_3CH_2Cl + Cl\cdot)$$

$$= +5.2 \text{ kcal mole}^{-1}$$

Again, the overall reaction is eq. (2) above: $\Delta H° = -16.5$ kcal mole^{-1}

Because the second step in the chain reaction involving HCl is endothermic ($\Delta H° = +5.2$ kcal mole^{-1}), this reaction proceeds much more slowly than the HBr addition.

(c)

$$(CH_3)_2CBrCH_2\cdot \xleftarrow[\text{Path A}]{Br\cdot} (CH_3)_2C{=}CH_2 \xrightarrow[\text{Path B}]{Br\cdot} (CH_3)_2\dot{C}CH_2Br$$

$$(CH_3)_2CBrCH_2\cdot \xrightarrow{HBr} (CH_3)_2CBrCH_3 + Br \qquad (CH_3)_2\dot{C}CH_2Br \xrightarrow{HBr} (CH_3)_2CHCH_2Br + Br\cdot$$

for Path A:

$$\Delta H°_f((CH_3)_3CBr) = \Delta H°_f(Br\cdot) + \Delta H°_f((CH_3)_3C\cdot) - DH°((CH_3)_3C\text{-}Br)$$

$$= \boxed{-32} \text{ kcal mole}^{-1}$$

			$\Delta H°$ (kcal mole^{-1})
(6)	$H\cdot + Br\cdot \longrightarrow HBr$	$\Delta H° = -DH°(HBr)$:	-87.5
(7)	$(CH_3)_2C{=}CH_2 + HBr \longrightarrow (CH_3)_3CBr$		
	$\Delta H°_f$: -4.3 -8.7 $\boxed{-32}$ *(from above)*		-19
(8)	$(CH_3)_3CBr \longrightarrow (CH_3)_2CBrCH_2\cdot + H\cdot$	$\Delta H° = DH°$(primary C-H):	+98

The sum of *(6)*, *(7)*, and *(8)* = $\Delta H°$ for the first step

$$(Br\cdot + (CH_3)_3C{=}CH_2 \longrightarrow (CH_3)_2BrCH_2\cdot): \quad -8.5$$

$$(CH_3)_2CBrCH_2\cdot + HBr \longrightarrow (CH_3)_3CBr + Br\cdot$$

$$\Delta H° = -DH°(\text{primary C-H}) + DH°(HBr): \quad -10.5$$

for Path B:

$$\Delta H°_f((CH_3)_2CHCH_2Br) = \Delta H°_f(Br\cdot) + \Delta H°_f((CH_3)_2CHCH_2\cdot) - DH°(\text{primary C-Br})$$

$$= \boxed{-28} \text{ kcal mole}^{-1}$$

			$\Delta H°$ (kcal mole^{-1})
(6)	same as for Path A		-87.5
(9)	$(CH_3)_2C{=}CH_2 + HBr \longrightarrow (CH_3)_2CHCH_2Br$		
	$\Delta H°_f$: -4.3 -8.7 $\boxed{-28}$ *(from above)*		-15
(10)	$(CH_3)_2CHCH_2Br \longrightarrow (CH_3)_2\dot{C}CH_2Br + H\cdot$	$\Delta H° =$ DH°(tertiary C-H):	+91

The sum of *(6)*, *(9)*, and *(10)* = $\Delta H°$ for the first step

$$((CH_3)_2C{=}CH_2 + Br\cdot \longrightarrow (CH_3)_2\dot{C}CH_2Br): \quad -11.5$$

$$(CH_3)_2\dot{C}CH_2Br + HBr \longrightarrow (CH_3)_2CHCH_2Br + Br\cdot$$

$\Delta H°$ for the second step = $-DH°$(tertiary C-H) + $DH°$(HBr) : -4.5

t-Butyl bromide ($\Delta H°_f = -32$ kcal mole^{-1}) is more stable than isobutyl bromide ($\Delta H°_f = -28$ kcal mole^{-1}). However, the isobutyl isomer will be formed predominantly because the most favorable reaction of Br• with 2-methylpropene is the one which leads to the tertiary radical ($\Delta H° = -11.5$ kcal mole^{-1}; *Path B*) rather than to the primary radical ($\Delta H° = -8.5$ kcal mole^{-1}; *Path A*).

This is another example of a reaction in which the less stable product is formed more rapidly than the more stable product.

S6. (a) $CCl_4 \longrightarrow \cdot\ddot{C}\cdot + 4\ Cl\cdot$

ΔH°_f: −25.2   170.9   4 × 28.9 $\Delta H^\circ = 311.7$ kcal mole^{-1};

average per C−Cl bond = 78 kcal mole^{-1}

(b) $CHCl_3 \longrightarrow H\cdot + \cdot CCl_3$

Calculate from the sequence: ΔH° (kcal mole^{-1})

(1) $CHCl_3 + Cl\cdot \longrightarrow CCl_4 + H\cdot$

ΔH°_f: −24.6   28.9   −25.2   52.1   22.6

(2) $CCl_4 \longrightarrow \cdot CCl_3 + Cl\cdot$ $\Delta H^\circ = DH^\circ(C\text{-}Cl)$:   78 *(from (a) above)*

The sum of *(1)* and *(2)* = ΔH° for the overall reaction :   101

(c) $CH_4 + CCl_4 \longrightarrow CH_3Cl + CHCl_3$

ΔH°_f: −17.9   −25.2   −20.6   −24.6 $\Delta H^\circ = -2.1$ kcal mole^{-1}

On the basis of a negative overall enthalpy, one would assume that the reaction could take place.

(d) Assuming that an initiation step can occur, the most reasonable propagation steps are:

(I) $\cdot CCl_3 + CH_4 \longrightarrow CHCl_3 + \cdot CH_3$

ΔH°_f: 23.9   −17.9   −24.6   34 $\Delta H^\circ = +3$ kcal mole

($\Delta H^\circ_f(\cdot CCl_3) = \Delta H^\circ_f(CHCl_3) + DH^\circ(Cl_3C\text{-}H) - \Delta H^\circ_f(H\cdot)$ ← *use value calculated for (b) above*

= 23.9 kcal mole^{-1})

(II) $\cdot CH_3 + CCl_4 \longrightarrow CH_3Cl + \cdot CCl_3$

ΔH°_f: 34   −25.2   −20.6   23.9 $\Delta H^\circ = -5.5$ kcal mole^{-1}

Neither of these steps is unacceptably exothermic.

7. STEREOISOMERISM

7.A Chapter Outline and Important Terms Introduced

7.1 Chirality and Enantiomers (mirror images which are different)

chirality
chiral and achiral
non-superimposable
enantiomers
asymmetric

7.2 Physical Properties of Enantiomers: Optical Activity (the only physical difference between enantiomers)

plane polarized light
dextrorotatory and levorotatory
specific rotation, [α]
polarizability

7.3 Nomenclature of Enantiomers: the R-S Convention (describing three dimensions using one dimension)

absolute configuration
sequence rule
higher atomic number
first point of difference

7.4 Racemic Mixtures (chiral molecules, optically inactive mixtures)

racemate
racemic compound
racemization

7.5 Fischer Projections (describing three dimensions in two dimensions)

90° vs. 180° rotation
exchange of a pair of substituents

7.6 Compounds Containing More than One Asymmetric Atom, Diastereomers (stereoisomers with different physical properties)

n asymmetric atoms ⇒ 2^n stereoisomers *(but look out for meso compounds)*

7.7 Stereoisomeric Relationships in Cyclic Compounds (looking for planes and points of symmetry)

symmetry plane

7.8 Chemical Reactions and Stereoisomerism (what happens when an asymmetric center is involved)

achiral intermediates, racemization
enantiomeric transition states
diastereomeric transition states
asymmetric induction

7.B Important Concepts and Hints

Stereoisomerism can be the most exciting aspect of organic chemistry. If you can take a molecule right off a two-dimensional page and see it in three dimensions with your mind's eye, you will have a lot of fun studying Chapter 7 and discovering the details of stereochemistry throughout the course. Stereoisomerism can also be a very difficult and challenging subject, simply because you have to be able to imagine three-dimensional objects when confronted with two-dimensional pictures. This takes practice, which very few people have had (remember, most people have been stuck on the two-dimensional surface of the earth all their lives....). However, it is easy for you to have this practice: USE YOUR MODELS! As you go through the chapter, make models of the structures and compounds described, so you can see a three-dimensional representation alongside the two-dimensional representation in the text. For instance, make a model of (R)-1-bromo-1-chloroethane, and view it from the twelve possible directions depicted for the Fischer projections of Figure 7.9. Work the exercises and problems with models until you are confident of your understanding. Then, go through them again without models so that you can practice and develop your ability to make the mental transition between two and three dimensions.

The most important concepts in Chapter 7 are those of *chirality* and *optical activity*, and how they differ (see the discussion under *Chiral* in the Glossary); how to describe *absolute configurations* using the *R-S convention*; how to draw molecules so as to depict their absolute configuration (*Fischer projections*) and how to manipulate these pictures in your mind and on paper; and the relationships between *enantiomers*, *diastereomers*, and *meso* compounds.

7.C Answers to Exercises

(7.1) Items c, d, e, f, g, h, j, k, and l are clearly chiral; a, b, i, and m are not chiral, unless one takes into account the printing on them. A portrait (n) is probably chiral, unless it is perfectly symmetrical; i.e., a face-on view, hair parted in the middle, arms not folded, etc.

(7.2) $\alpha = [\alpha] \times \ell \times d \qquad V = \pi r^2 \ell, \quad \ell = \dfrac{V}{\pi r^2} = \dfrac{300\ \text{cm}^3}{\pi \times 6.25\ \text{cm}^2} = 15.3\ \text{cm}$

$[\alpha] = 66°, \quad \ell = 15.3\ \text{cm} = 1.53\ \text{dc}, \quad d = 60/300 = 0.2\ \text{g mL}^{-1}$
therefore $\alpha = +20.2°$

(7.3)

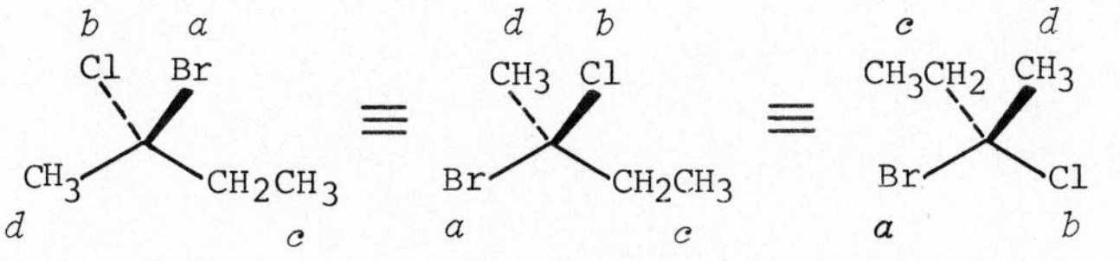

(ranking priority is shown)

(7.3)

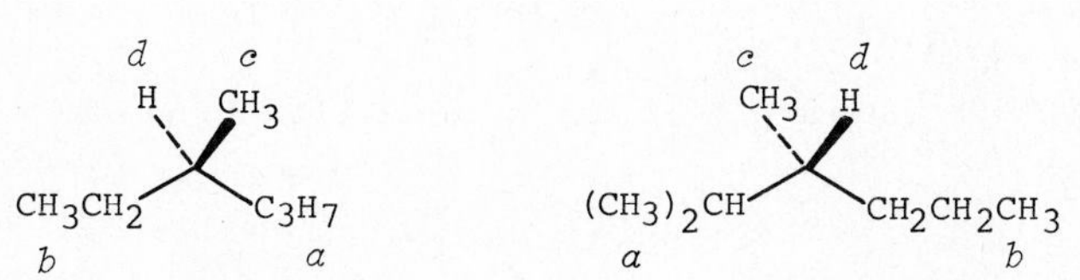

(note the precedence of an isopropyl over an n-propyl group)

(7.5)

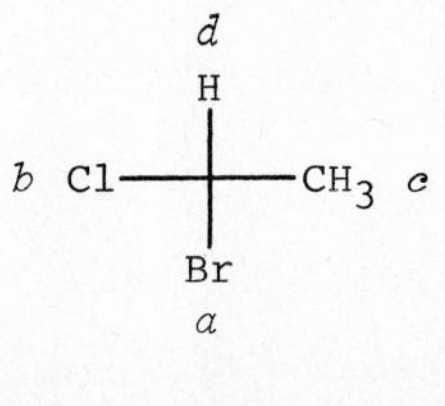

(7.6)

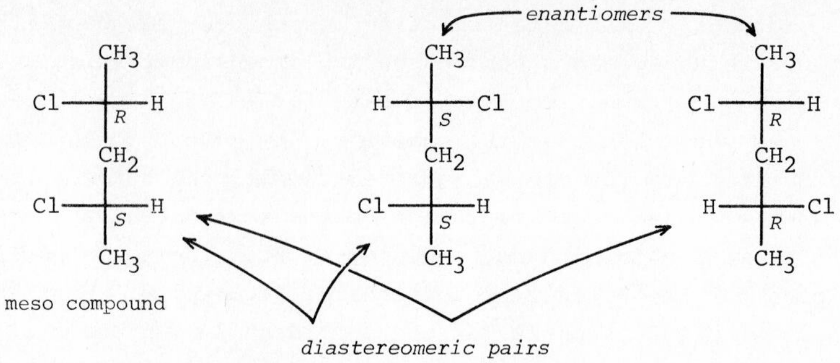

7.D Answers and Explanations for Problems

1. (a) $1\ \underline{M}\ C_5H_{11}Cl = 106.5\ \text{g liter}^{-1} = 0.1065\ \text{g ml}^{-1} = c$

$l = 10\ \text{cm} = 1\ \text{dc}$

$\alpha = +3.64°; \quad [\alpha]_D = \dfrac{\alpha}{l \times c} = \dfrac{3.64}{1 \times 0.1065} = +34.2°$

(b) $c = 0.096\ \text{g ml}^{-1}, \quad l = 0.5\ \text{dc}, \quad \alpha = -1.80°;$

therefore, $[\alpha]_D = -37.5°$

2. (a) and (g), three: RR, SS, and meso (RS)

(b) and (c), four: RR, SS, RS, and SR

(d) and (e), eight: RRR, RRS, RSR, SRR, RSS, SRS, SSR, and SSS

When no possibility of meso compounds exists, as in these cases, the number of possible stereoisomers is equal to 2^n, where n = the number of asymmetric carbons in the molecule.

(f), (h), and (i) are all achiral.

3. (a)

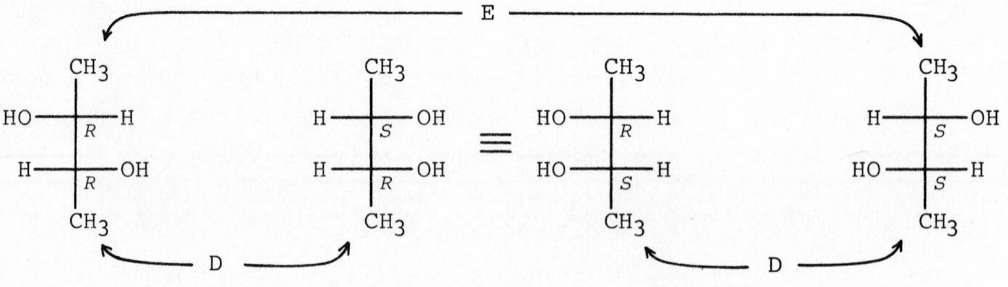

these two structures are equivalent; they represent a meso compound

E = enantiomeric relationship D = diastereomeric relationship

(b)

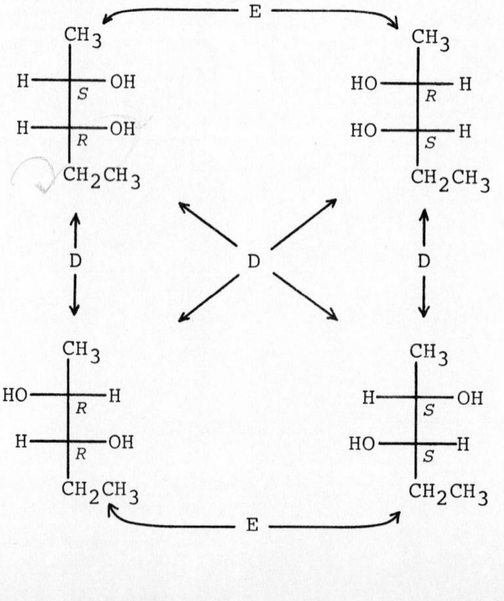

There is no meso isomer because the two ends of the molecules are different.

(c)

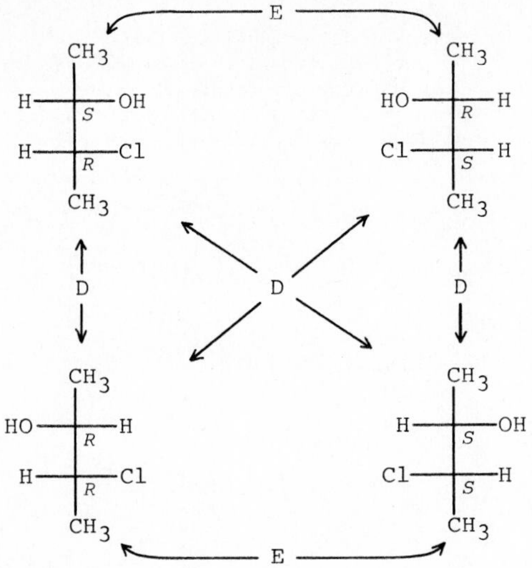

There is no meso isomer because the substituents on C-2 and C-3 are different.

(d)

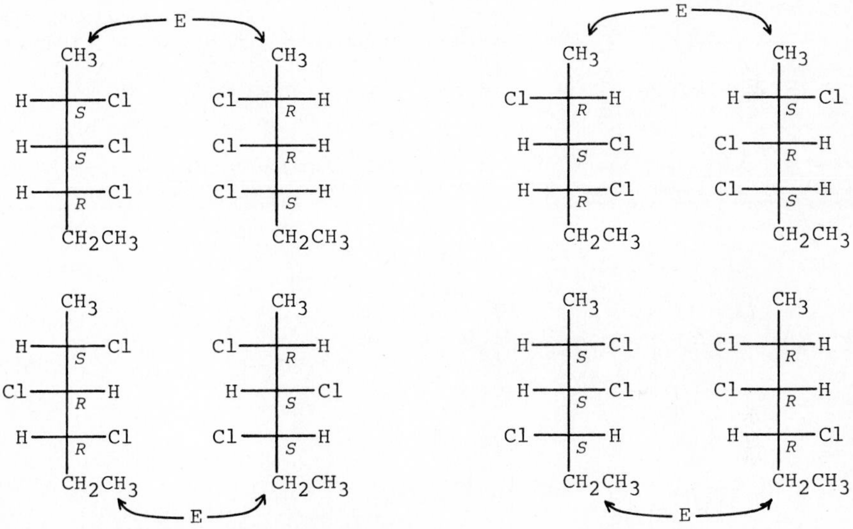

All the pairs not shown as enantiomers are diastereomers.

(e)

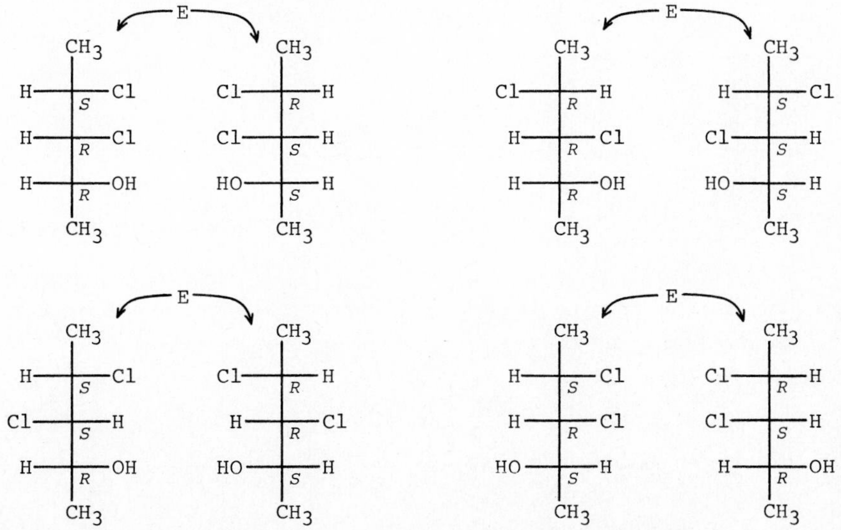

All the pairs not shown as enantiomers are diastereomers.

4. *NOTE:* There are several correct Fischer projections for compounds having one or more chiral centers. Only one or two possibilities are given for each answer below; if the one you have drawn is different, switch pairs of substituents until it matches one of the answers (see Section 7.5). If an even number of switches is required, the Fischer projections are equivalent and your answer is correct; if an odd number of switches are necessary, your answer is wrong.

(a)

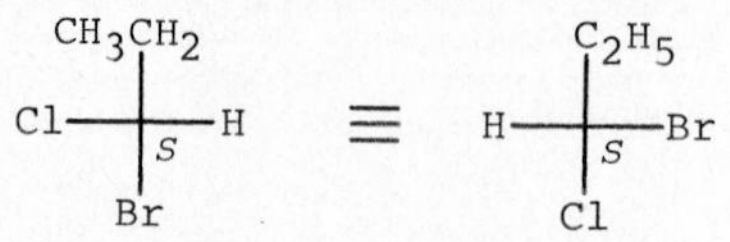

(b)

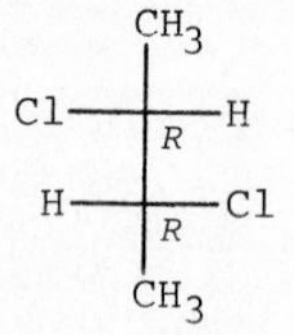

(c)

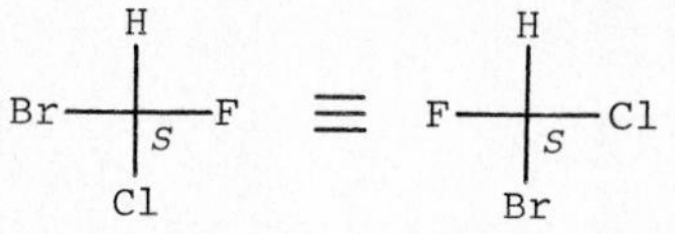

(d)

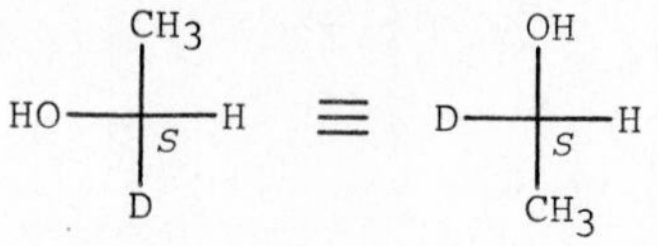

(D has a higher priority than H)

(e)

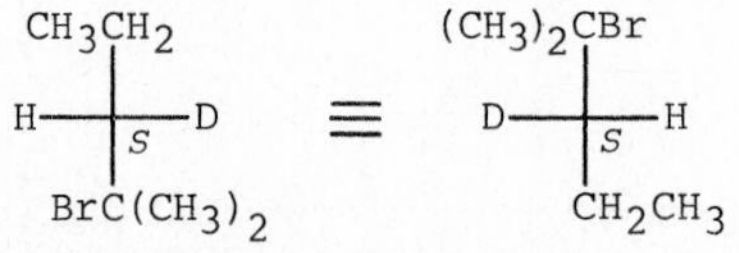

(Although the question shows the bromine-containing carbon with wedged and dotted bonds, it is not a chiral center and should not be displayed in a Fischer projection.)

(f)

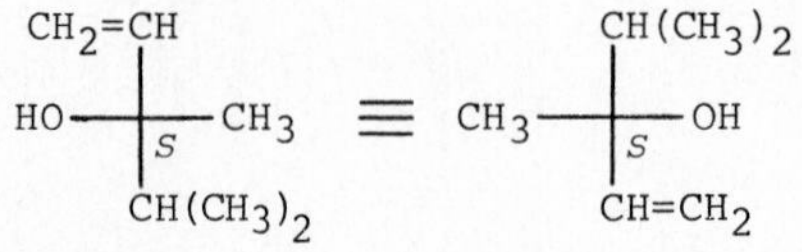

($-CH{=}CH_2$ counts as: 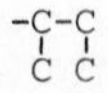 and therefore has priority over 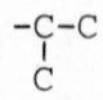 .)

(g)

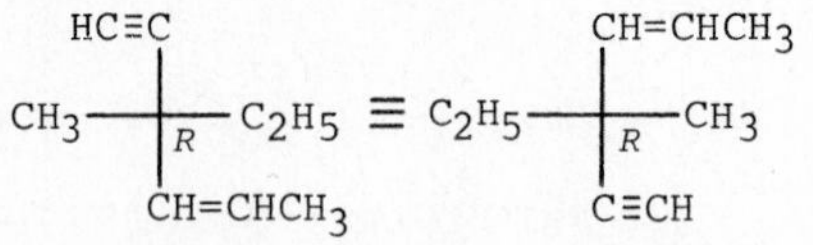

($-C{\equiv}CH$ counts as: 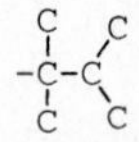 and has priority over $-CH{=}CHCH_3$ which counts as: 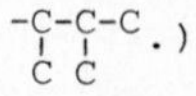 .)

(h)

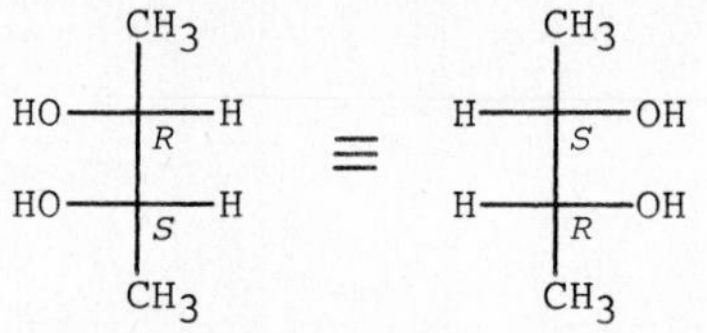

(This is a meso compound)

5. The compound shown is S:

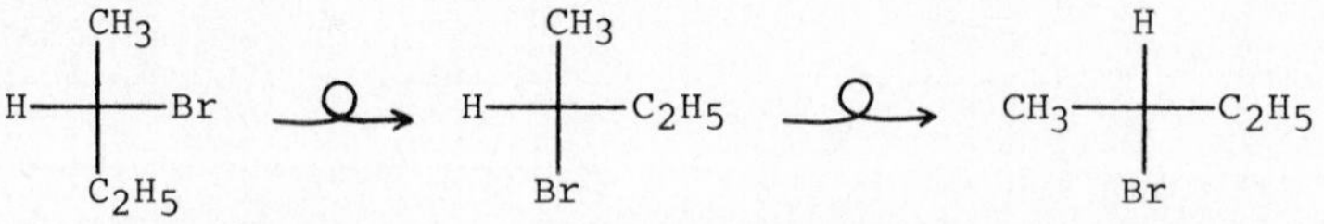

Two pair-switches give the same compound back

c ⟵ b , therefore S

(a) One switch from beginning structure, therefore enantiomeric

(b) Equivalent (see above)

(c)

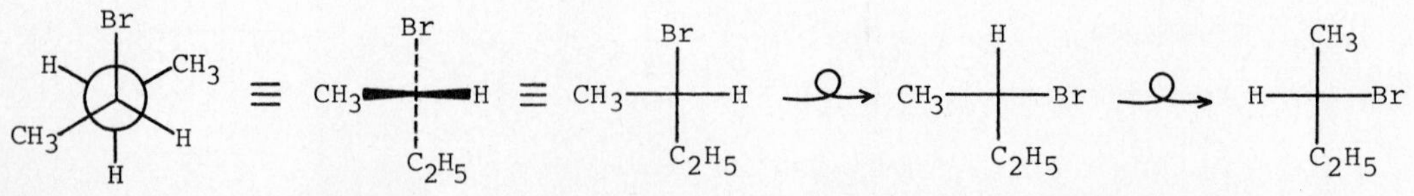

Two pair-switches; therefore equivalent

(d)

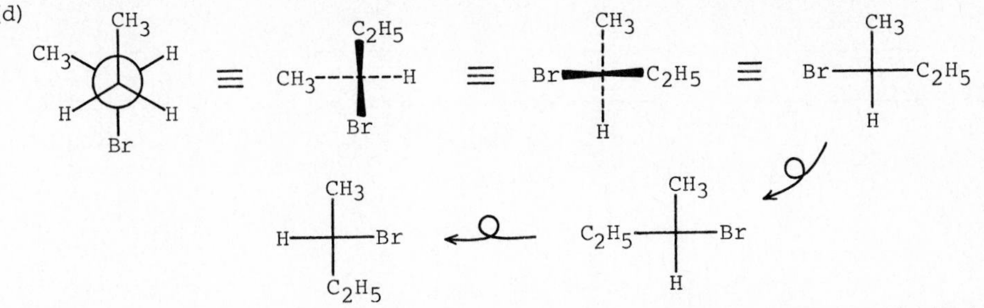

Two pair-switches; therefore equivalent

(e)

[90° rotation]

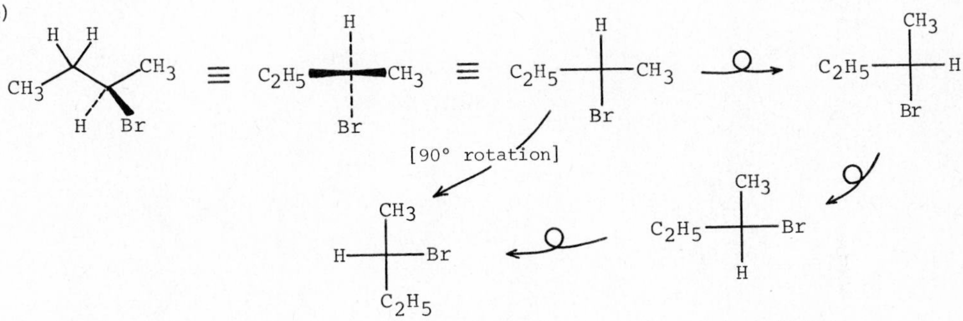

90° rotation or three pair-switches;
therefore enantiomeric

(f)

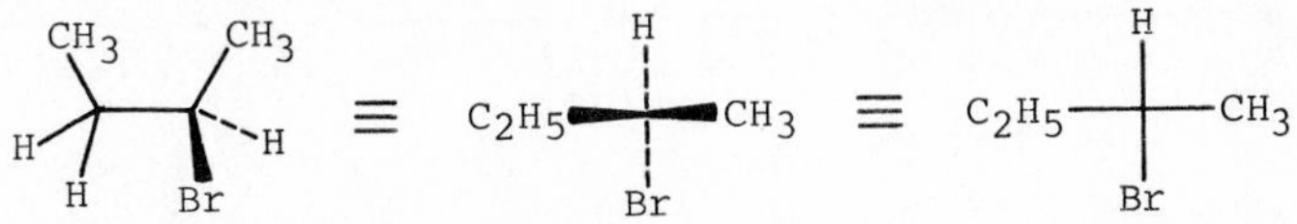

(same as (e),
enantiomeric)

MODELS HELP A LOT IN FIGURING OUT THE ANSWERS TO THIS PROBLEM!

6.

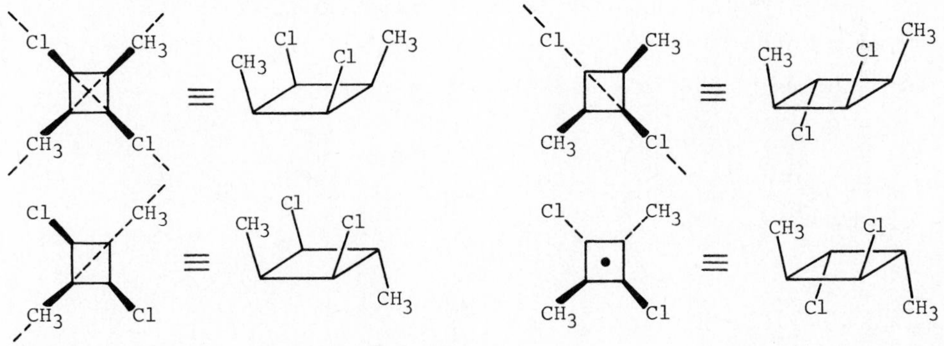

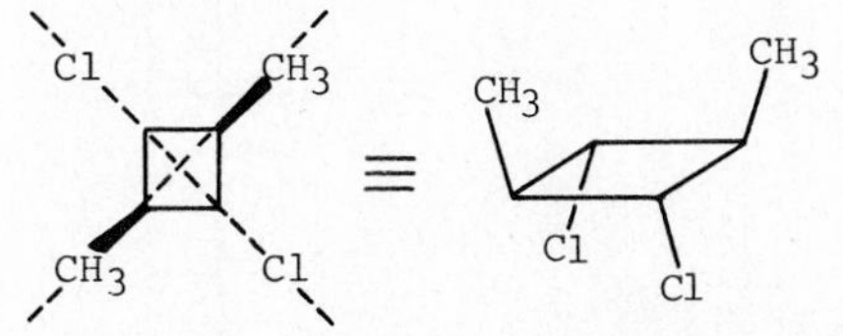

------ = plane of symmetry
● = center of symmetry

<u>None</u> of these stereoisomers is chiral.

7.

$(CH_3CH_2)_2CHCH_2Cl$ achiral

$(CH_3CH_2)_2C\begin{smallmatrix}Cl\\CH_3\end{smallmatrix}$ achiral

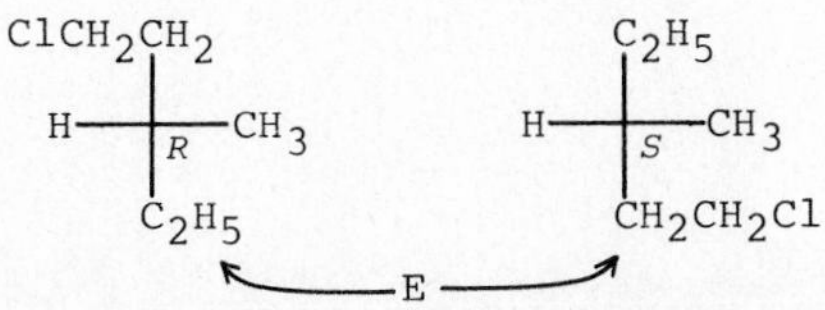

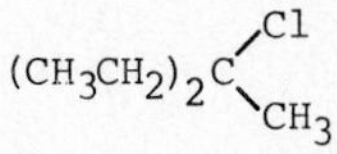

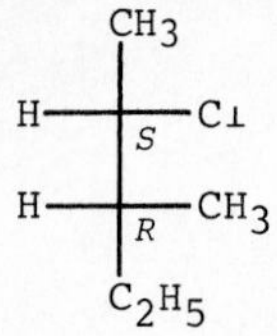

A

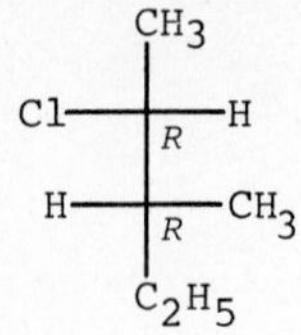

B

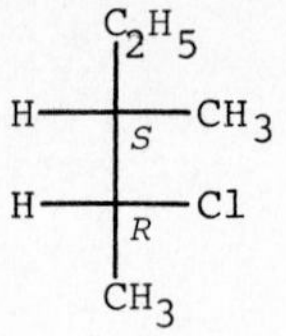

C

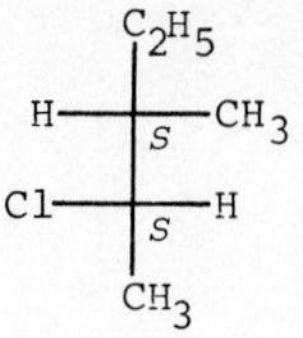

D

Enantiomeric pairs: A,C; B,D Diastereomeric pairs: A,B; A,D; B,C; C,D

8. (a)

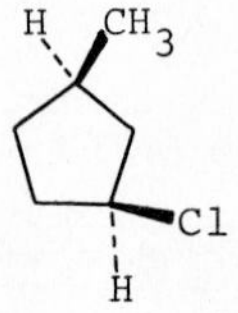

A

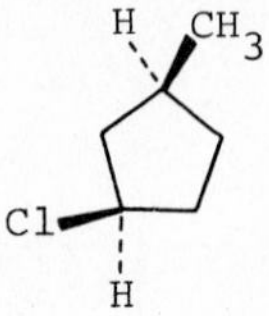

B

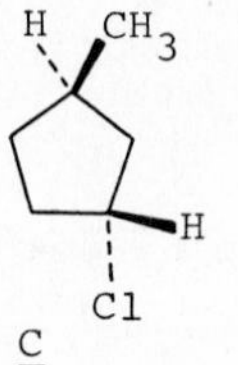

C

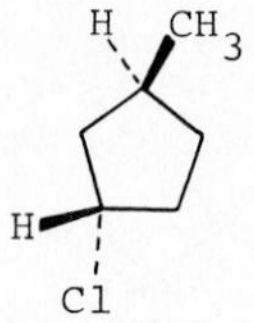

D

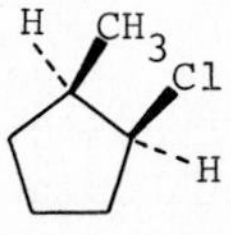

E

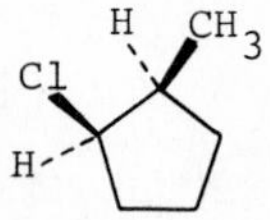

F

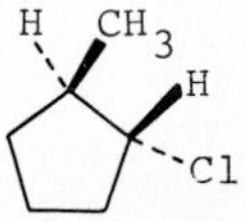

G

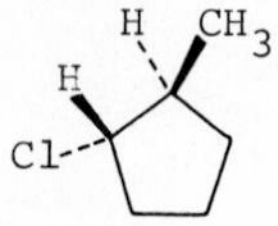

H

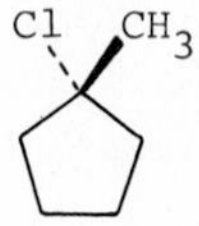

I

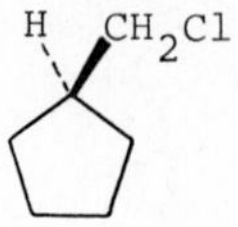

J

(b) The following are pairs of enantiomers: A,B; C,D; E,F; G,H

(c) The following are pairs of diastereomers: A,C; A,D; B,C; B,D; E,G; E,H; F,G; F,H

(d) Isomers I and J are achiral.

9.

10. (a)

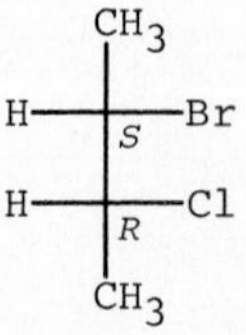

(b)

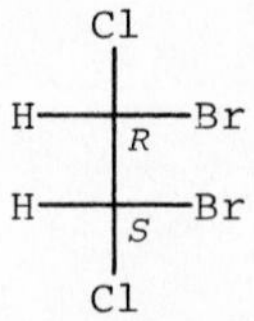

(c)

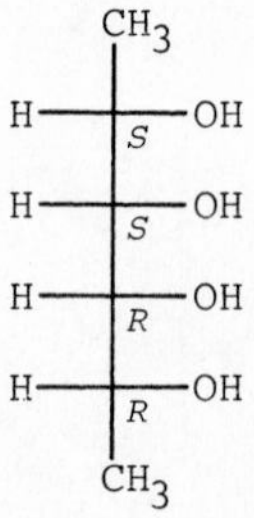

(d)

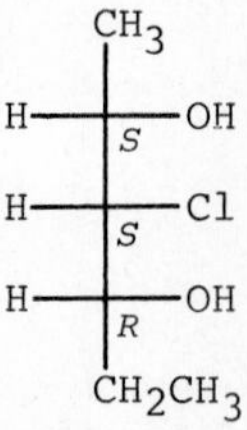

(e)

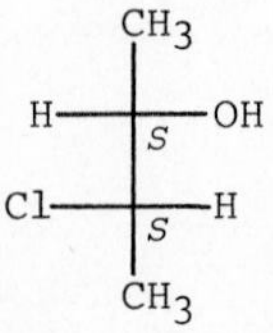

(f)

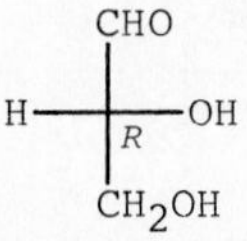

11. (a)

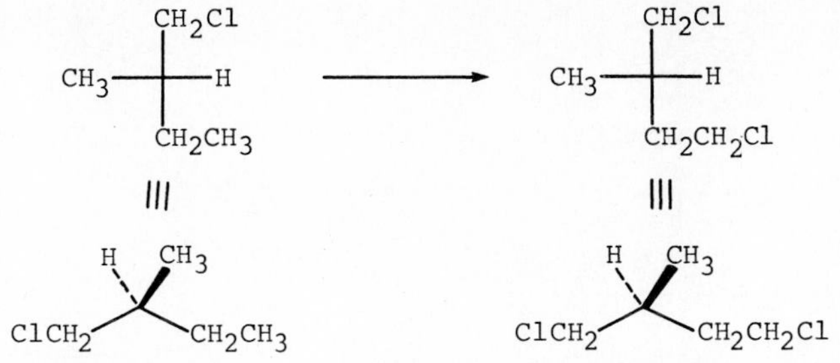

(asymmetric center is unaffected by reaction at the 4-position)

This demonstrates that there is no *a priori* relationship between sign of rotation ((+) or (−)) and configuration (R or S).

(b) $ClCH_2C(CH_3)(Cl)CH_2CH_3$ is produced *via* the $ClCH_2\dot{C}(CH_3)CH_2CH_3$ free radical intermediate. Since the product from this intermediate is racemic, the intermediate itself is probably achiral, either because it is achiral by symmetry (planar radical), or (if the radical is pyramidal) because it inverts much faster than it reacts. Other evidence is required to distinguish between these two alternatives.

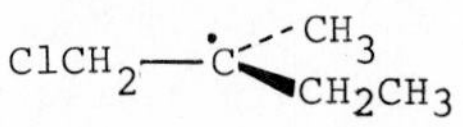

planar free radical (achiral)

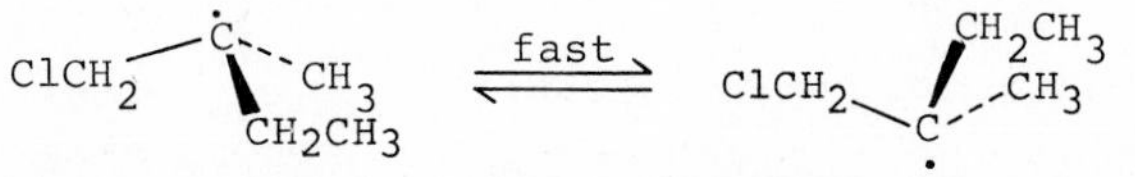

rapidly equilibrating pyramidal free radicals (effectively achiral)

12.

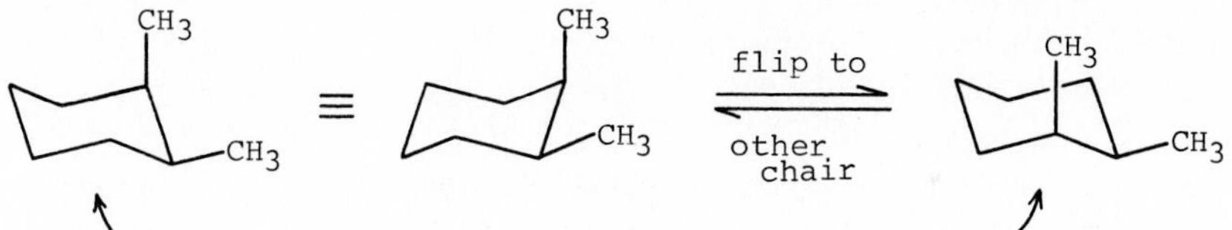

These are mirror images of each other.

That is, the two conformations are enantiomeric. The two enantiomers of *cis*-1,2-dimethylcyclohexane are simply different conformations of the same molecule. Because the chair ⇌ chair interconversion is fast at ordinary temperatures, *cis*-1,2-dimethylcyclohexane behaves, on the time-average, as an achiral, meso compound.

13.

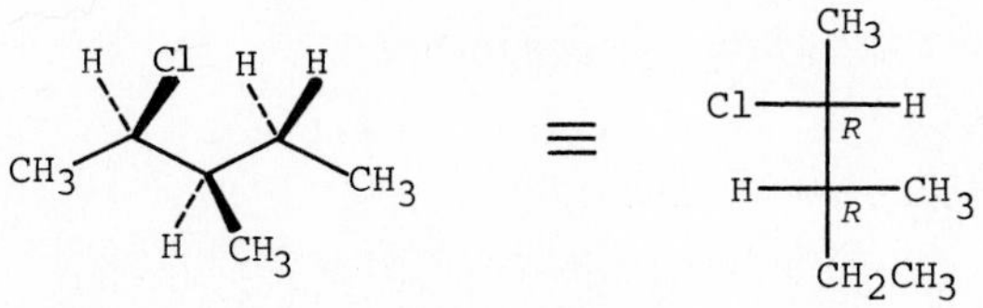

Eight fractions could potentially be seen:

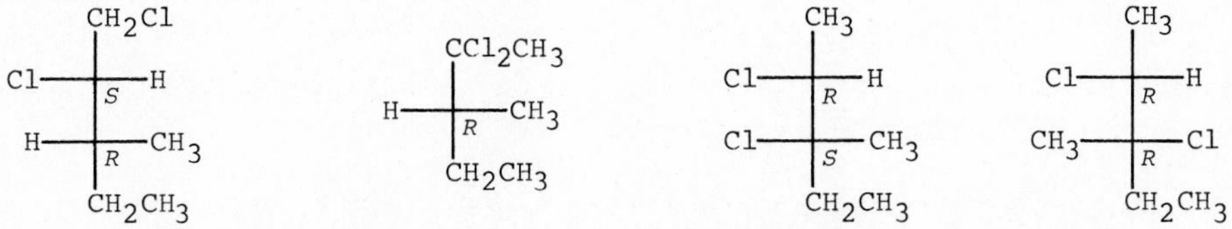

[Remember that the radical can react on either side.]

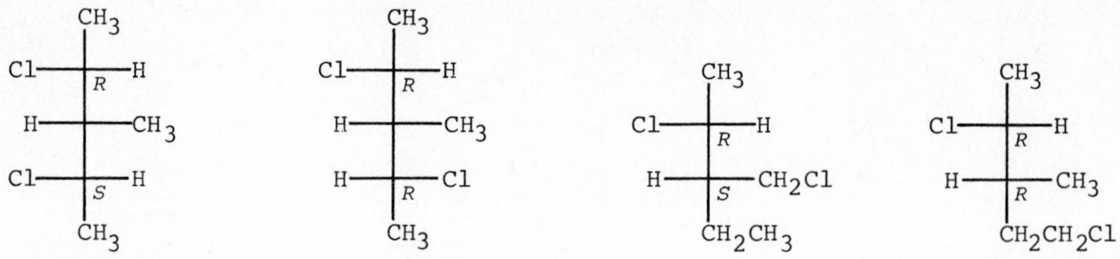

achiral, therefore optically inactive

All except one of the fractions is optically active.

14.

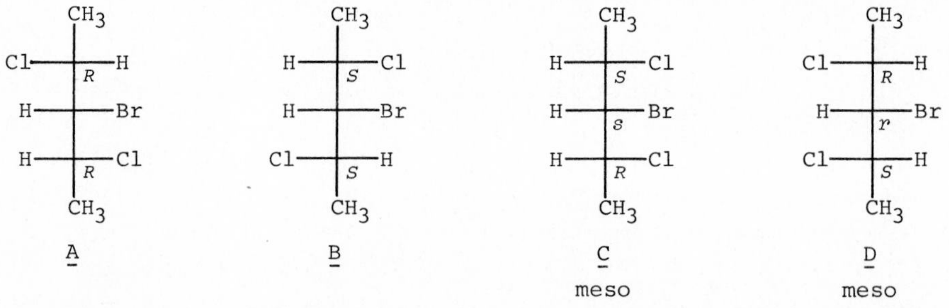

There are only four stereoisomers; any Fischer projection you draw with the bromine on the left can be seen to be equivalent to one of those illustrated above by rotating 180°.

For each of the chiral isomers A and B, no designation for the middle carbon is possible, because it is not a chiral center (two of the substituents are identical, and interchanging the Br and H gives the same compound back again).

For the two achiral isomers C and D, the middle carbon seems to be asymmetric because it is attached to four different groups (H, Br, (R)-$CHClCH_3$, and (S)-$CHClCH_3$). However, it is located on the symmetry plane of the molecule. Such a carbon is said to be "pseudoasymmetric" and its configuration is designated with the lower-case letters *r* and *s*. In C and D, the priority of substituents on the middle carbon is Br > (R)-$CHClCH_3$ > (S)-$CHClCH_3$ > H.

7.E Supplementary Problems

S1. Calculate $[\alpha]_D$ for each of the following compounds:

(a) A 0.13 M solution of strychnine (mw = 334.4) in ethanol in a 10-cm cell gives an observed rotation of -2.26°.

(b) A solution of 3.2 g of common sugar (sucrose, mw = 342.3) in 15 ml of water in a 5-cm cell gives an observed rotation of +7.1°.

S2. Predict the observed rotation (α) for the following solutions. Assume a a cell length of 10 cm for each case.

(a) 3 g of morphine hydrate (mw = 303.3, $[\alpha]_D$ = -132°) in 50 ml of methanol.

(b) Pure (-)-2-chlorobutane (d = 0.87, $[\alpha]_D$ = -8.48°).

S3. (a) Pure (-)-α-pinene has a specific rotation, $[\alpha]_D^{20}$ = -51.3. What is the optical purity of a sample of α-pinene which shows a specific rotation of -35.9° (i.e., how much of each enantiomer is present)?

(b) Predict the specific rotation of a mixture of 30% of (−)-2-bromobutane ($d = 1.254$, $[\alpha]_D^{20} = -23.13°$) and 70% of the (+)-enantiomer.

S4. For each of the following compounds, write a Fischer projection and label each asymmetric center R or S as appropriate.

(a)

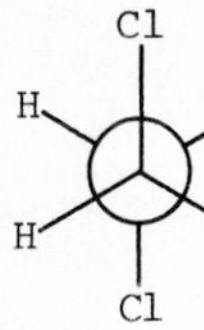

(b)

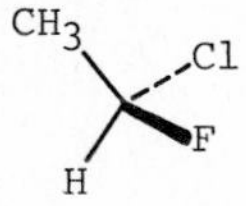

(c)

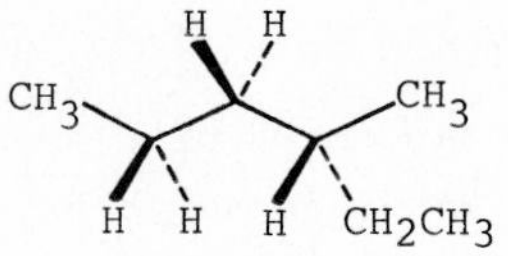

(d)

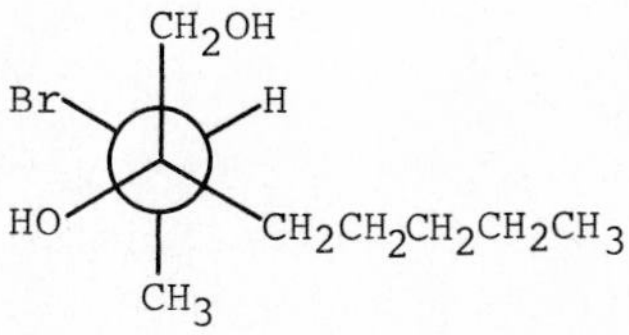

S5. What is the relationship between the molecules of each of the following pairs (i.e., are they the same, enantiomers, diastereomers, structural isomers, etc.)? For each asymmetric center in these molecules, assign R or S as appropriate.

(a)

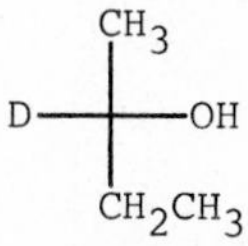

and

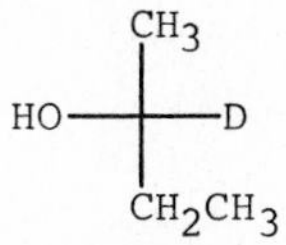

(b)

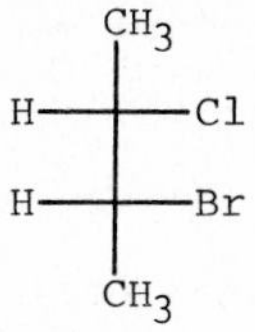

and

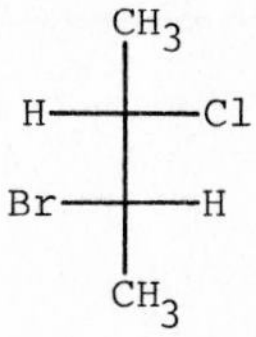

(c)

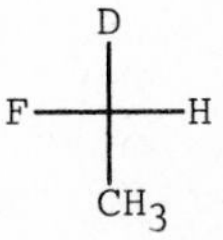

and

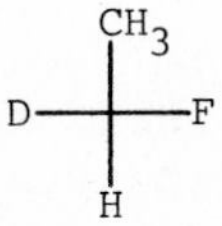

(d)

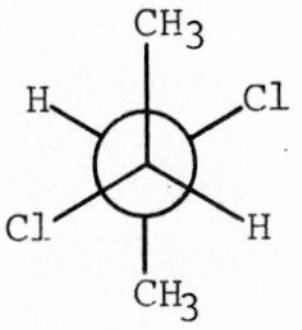

and

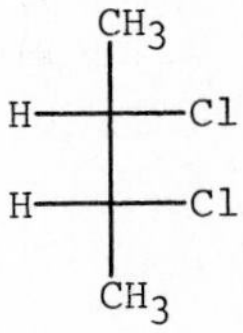

(e)

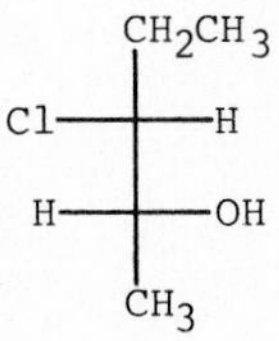

and

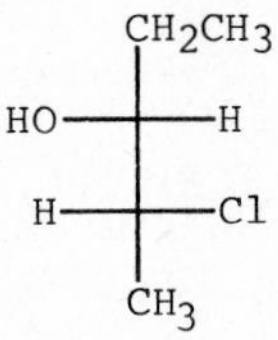

(f)

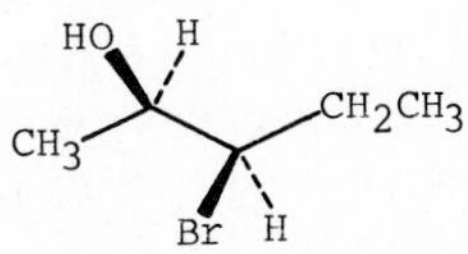

and

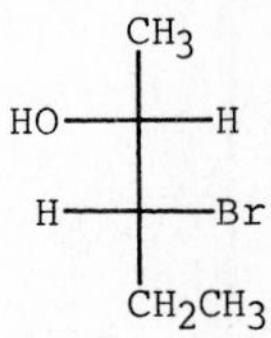

(g)

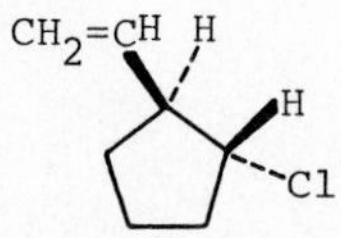

and

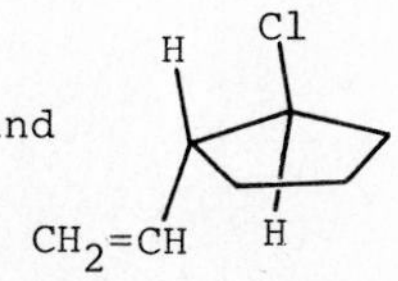

(h)

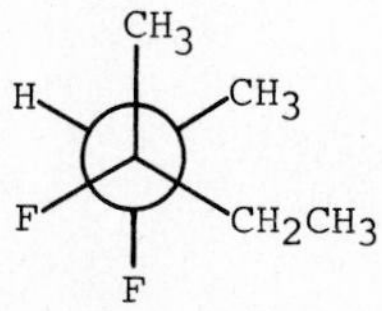

and

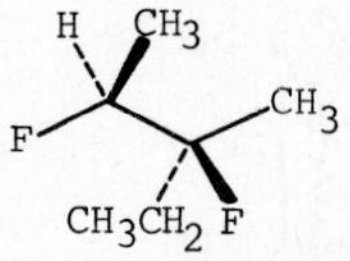

S6. For each of the compounds illustrated below, draw: (1) its enantiomer, (2) a diastereomer, and (3) a Fischer projection.

(a)

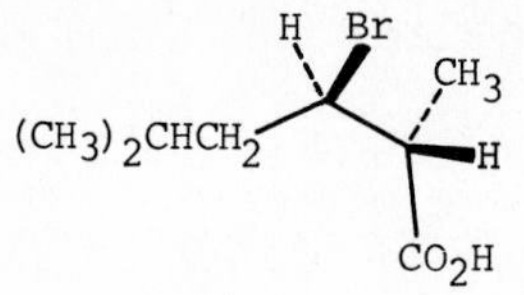

(b)

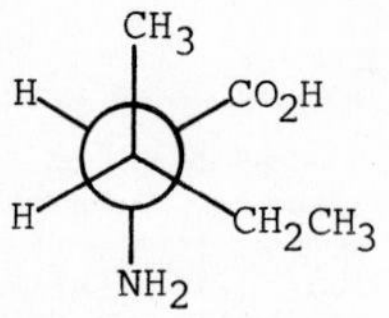

S7. Write structures for all of the isomers of trimethylcyclopentane. Which ones are chiral and which ones are achiral?

S8. Consider the free-radical chlorination of 1,1,4,4-tetramethylcyclohexane with Cl_2.

(a) Write the structures of all of the possible monochloro isomers.

(b) Assign the configuration of each chiral center as you have drawn it.

(c) Predict the relative amounts of each isomer.

7.F Answers to Supplementary Problems

S1. (a) $\frac{0.13 \times 334.4}{1000} = 0.0435 \text{ g ml}^{-1}$; $[\alpha]_D = \frac{-2.26}{1 \times 0.0435} = -52.0°$

(b) $[\alpha]_D = \frac{+7.1}{0.5 \times 3.2/15} = +66.6°$

S2. (a) $\alpha = [\alpha]_D \times \ell \times c$
$= -132 \times 1 \times 3/50 = -7.92°$

(b) $\alpha = -8.48 \times 1 \times 0.87 = -7.38°$

S3. (a) $-35.9°$ is $\frac{35.9}{51.3} = 70\%$ of the rotation expected for the pure enantiomer. Therefore "30%" of the mixture is racemic and "70%" of it is the pure (-)-enantiomer. For the whole mixture, then, 70% + 15% = 85% of it is the (-)-enantiomer and 15% of it is the (+)-enantiomer. The optical purity is 70%; this is frequently also referred to as the "enantiomeric excess".

(b) Consider this mixture to be 60% racemic and 40% (+)-enantiomer:

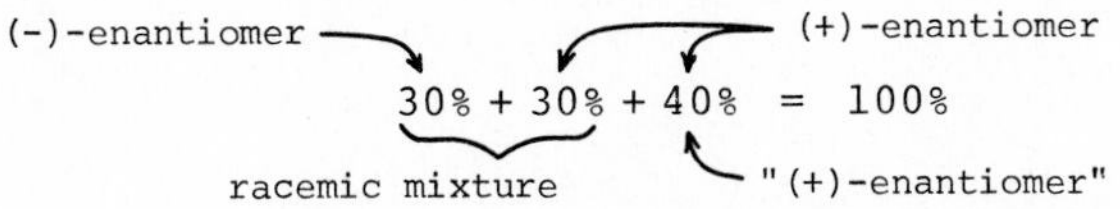

Because the racemic portion shows no rotation, the observed rotation will be 40% of that expected for the pure (+)-enantiomer: $0.4 \times (-23.13) = -9.25°$.

S4. (a)

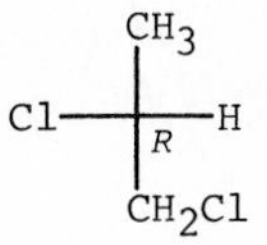

(b)

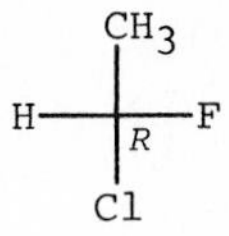

(c)

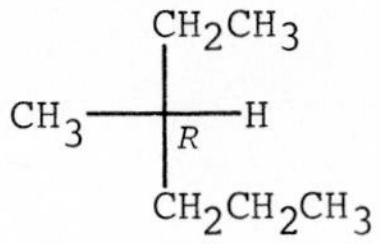

(d)

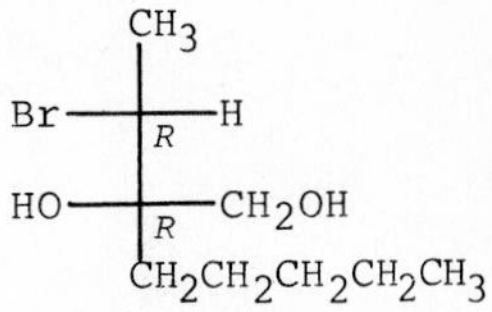

S5. (a), (c), and (g) are enantiomeric pairs; (b) and (h) are diastereomeric pairs; (d) and (f) are the same; (e) is a pair of structural isomers

(a)

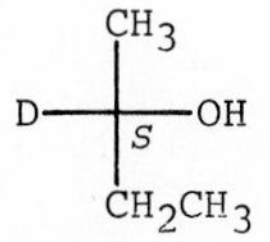

(b)

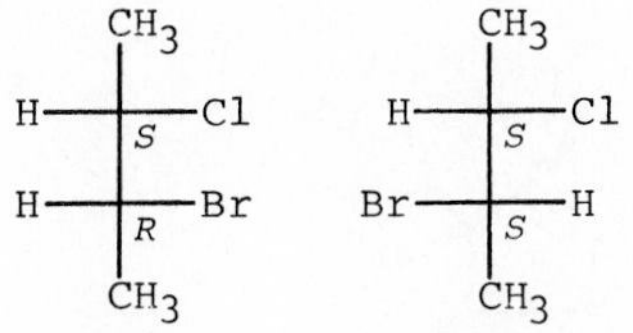

(c)

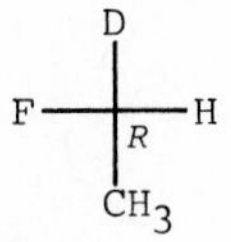

(d)

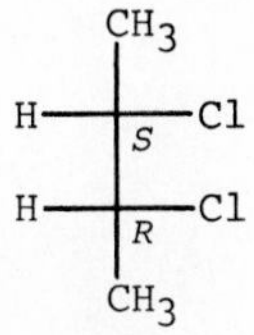

a meso compound

(e)

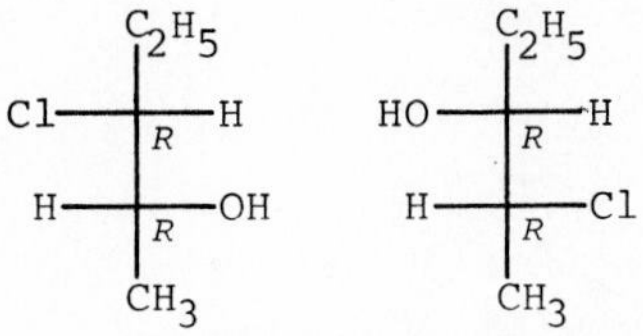

(f)

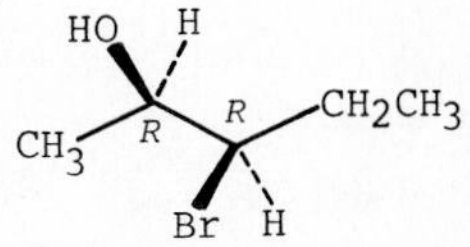

(g)

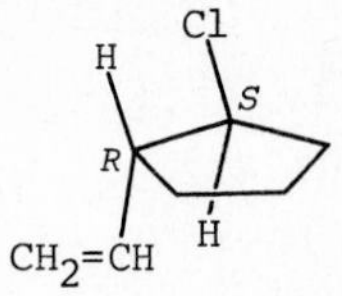

[$-CH=CH_2$ has higher priority than $-CH_2CH_2-$]

(h)

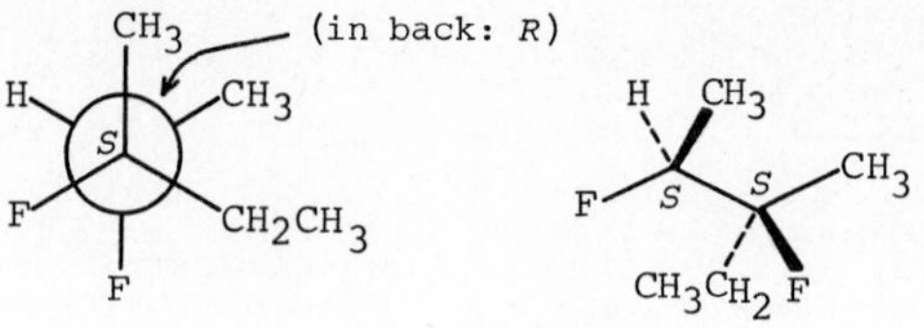

S6. (a)

(1)

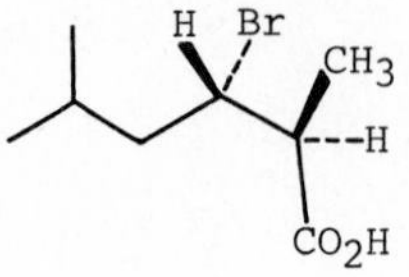

(2)

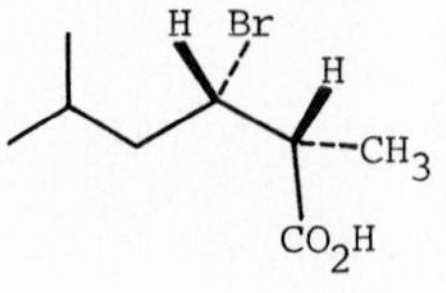

or

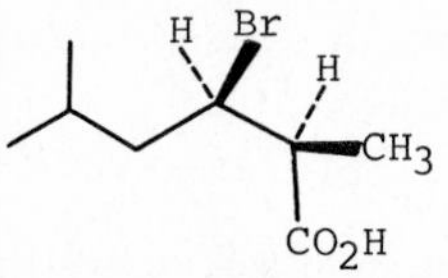

(3)

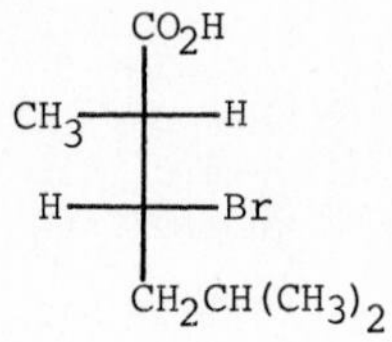

(b)

(1)

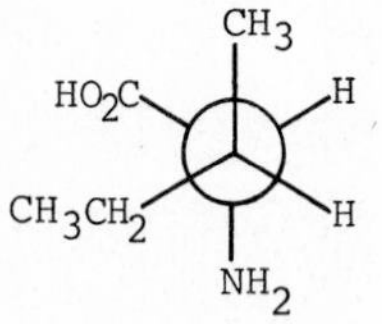

(2)

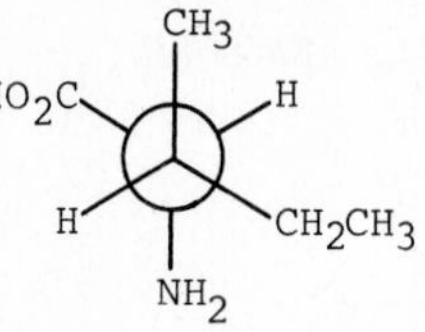

or

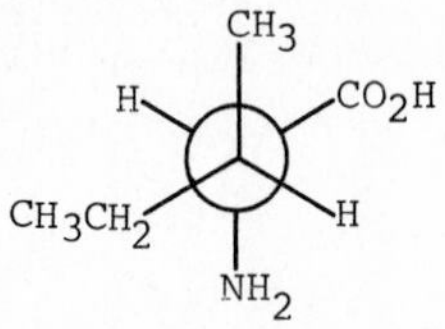

(3)

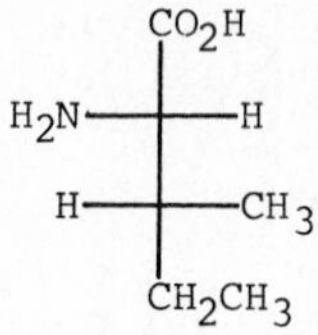

S7. Chiral:

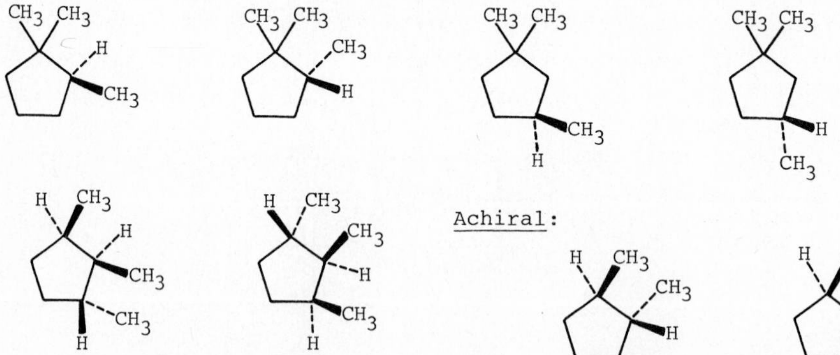

Achiral:

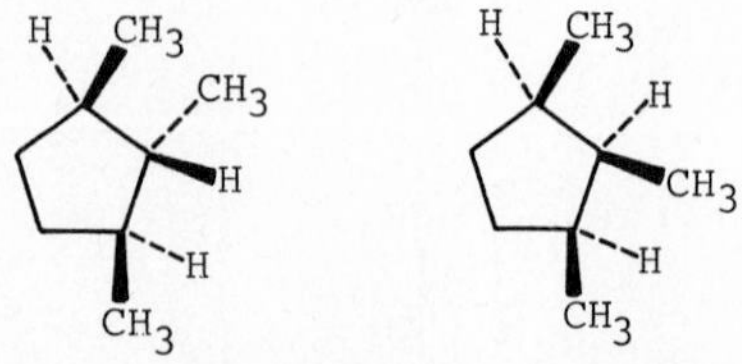

S8.

(a), (b)

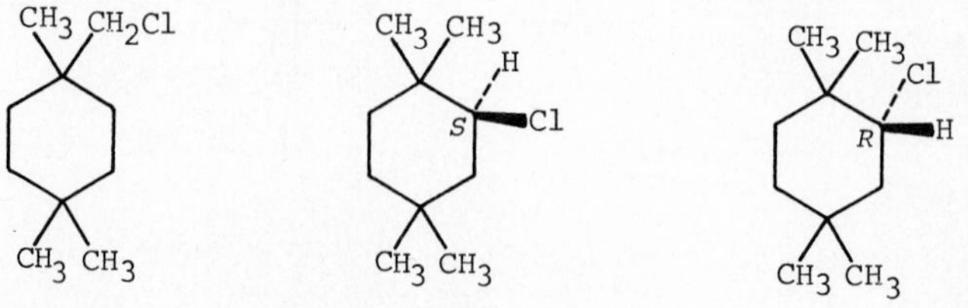

(c) $12 \times 1 = 12$; 27% $4 \times 4 = 16$; 36% $4 \times 4 = 16$; 36%

8. ALKYL HALIDES; NUCLEOPHILIC SUBSTITUTION AND ELIMINATION

8.A Outline of Chapter 8, Important Terms Introduced, and Reactions Discussed

8.1 Structure (size of halogens, bond lengths)

dipole moment — Van der Waals radius

8.2 Physical Properties (boiling points, melting points)

polarizability

8.3 Conformations

barriers to rotation

8.4 Some Uses of Halogenated Hydrocarbons (from dry cleaning to DDT)

8.5 The Displacement Reaction (the first "ionic" reaction discussed in detail)

$$Nu:^{-} \rightarrow R{-}X \longrightarrow Nu{-}R \quad X^{-}$$

second-order kinetics — (not via frontside attack)
bimolecular mechanism

8.6 Stereochemistry of the Displacement Reaction (a detailed look at the mechanism)

backside attack — solvation energy
inversion — sp^2-hybridized transition state

8.7 Generality of the Displacement Reaction (a great variety of nucleophiles and alkyl halides)

nucleophile, nucleophilicity — S_N2
lone pair of electrons — molecularity and kinetic order

8.8 Effect of Substrate Structure on Displacement Reactions

steric hindrance — β-branching
α-branching — "neopentyl-type" systems

8.9 Nucleophilicity and Solvent Effects (factors affecting the nucleophile)

A. Hydroxylic Solvents
   - hydrogen bonding
   - dependence of nucleophilicity on basicity and polarizability
   - ambident nucleophiles

B. Polar Aprotic Solvents
   - lack of hydrogen bonding

8.10 Leaving Groups

not limited to halides — acid catalysis
leaving group ability vs. basicity

8.11 E2 Elimination (a competing reaction)

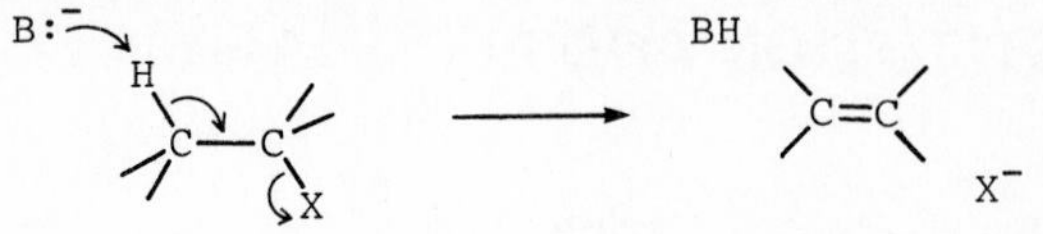

bimolecular elimination   steric hindrance

8.12 S_N1 Reactions: Carbocations

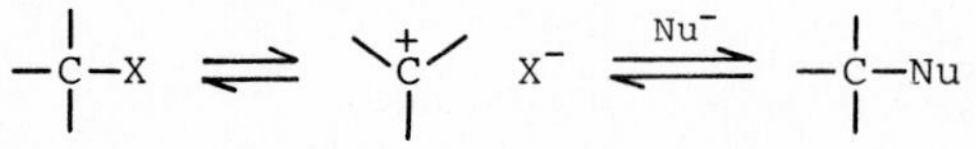

solvolysis reaction

unimolecular, first-order kinetics

carbocation

relative stability of carbocations: $3° > 2° > 1° > CH_3$

hyperconjugation

elimination; E1

8.13 Summation: Elimination vs. Substitution; Unimolecular vs. Bimolecular (useful rules of thumb)

8.14 Ring Systems (special aspects in cyclic molecules)

intramolecular vs. intermolecular   bond angle strain

8.B Important Concepts and Hints

This chapter is your introduction to a detailed analysis of an organic reaction which proceeds via an ionic (heterolytic) mechanism instead of by a free-radical (homolytic) mechanism. The displacement reaction is the primary reaction of interest for alkyl halides and a reaction of major importance in organic chemistry. It is analyzed in detail in Chapter 8 because it provides an opportunity to examine all of the factors which are generally important in understanding any reaction.

In Section 6.B of this Study Guide, we suggested that you approach each new reaction with a series of questions to help you organize your thinking. As an example, the sort of questions you should ask and the sort of answers you should give yourself are illustrated below for displacement reactions.

(1) *Qu:* What functional group transformation does the reaction accomplish?

ANS: In general terms, $Nu^- + R\text{-}X \xrightarrow{\text{solvent}} R\text{-}Nu + X^-$

where Nu = nucleophile, R = alkyl group, and X = leaving group.

(2) *Qu:* What is the mechanism of the reaction?

ANS: For some combinations of reactants, the bimolecular S_N2 mechanism is involved, with simultaneous attack of Nu and departure of X from opposite sides of the carbon atom; for other combinations of reactants, the stepwise S_N1 mechanism is involved, with initial loss of X to give a carbocationic intermediate (slow) and subsequent attack by Nu to give product (fast).

(3) *Qu:* What is the generality of the reaction? That is, what characteristics must the reactants have?

ANS: Nu, the nucleophile, must have a lone pair of electrons (i.e., must be a Lewis base) in order to form a bond to the carbon. *Nucleophilicity* increases with basicity (to the left in the Periodic Table, other factors being equal) and with *polarizability* (down in the Periodic Table, other factors being equal), and helps to determine whether the reaction will proceed via the S_N2 or S_N1 mechanism. Stronger nucleophiles favor S_N2 reactions (other factors being equal).

R, the alkyl group, plays the major role in determining the mechanism (S_N1 or S_N2) and rate of the reaction. Displacement reactions at tertiary carbon occur by the S_N1 mechanism and at primary carbon by the S_N2 mechanism. Both situations reflect the combined influences of steric hindrance and carbocation stability. Substitution at secondary carbon is the gray area, and whether these reactions proceed by the S_N1 or S_N2 mechanism, or both simultaneously, depends on other factors. Steric effects are not limited to substitution on the carbon undergoing reaction (α-*branching*), but are seen at more remote positions as well (β-*branching*, effects of cyclic systems).

X, the leaving group, must be stable when it departs with two electrons (usually as an anion). This is most easily evaluated by considering the pK_a of HX: low pK_a $\Rightarrow$ HX is a strong acid $\Rightarrow$ X^- is a weak base $\Rightarrow$ X^- is stable and a good leaving group. Under most circumstances, the pK_a of HX should be less than 2 or 3 for the reaction to occur at a reasonable rate. ($R'O^-$ [R' = alkyl or H] is essentially <u>never</u> a leaving group in an S_N1 or S_N2 reaction.) The "better" the leaving group is (i.e., the more stable X^- is), the faster the displacement reaction occurs, but the influence is greater on S_N1 reactivity than on S_N2 reactivity.

Solvent plays a role in the way it stabilizes the reactants in comparison to intermediates and transition states. For instance, if a polar, hydroxylic solvent (such as methanol) can form hydrogen bonds to the nucleophile (such as chloride ion) more strongly than it can to the transition state, the S_N2 reaction will be slowed; if a neutral alkyl halide (such as <u>t</u>-butyl bromide) must ionize in order to react (by the S_N1 mechanism), then a polar solvent will speed up the reaction.

(4) *Qu:* What are the stereochemical features to keep in mind?

ANS: <u>S_N2</u> Backside attack and <u>inversion</u> of configuration at the carbon undergoing substitution.

<u>S_N1</u> Planar, carbocation intermediate which the nucleophile can approach from either side. Leads to racemic products if that carbon is the only chiral center in the molecule.

(5) *Qu:* What are the limitations; that is, what possible side reactions should be kept in mind?

ANS: Elimination reactions (E2) compete with substitution when the nucleophile is fairly basic (RO^- [R = alkyl or H], CN^-, RS^-, NH_2^-, etc.) and S_N2

displacement is slowed because of steric hindrance. Elimination competes when the nucleophile is impatient for reaction (basic; i.e., unstable as its anion Nu^-) and doesn't want to wait around for S_N1-type ionization to occur. This is always an important point to keep in mind: very basic nucleophiles give mostly elimination products, except with unhindered primary alkyl halides.

This Question-and-Answer outline for displacement reactions is quite lengthy and detailed for two reasons: first, because it covers a complex topic, and second, because we want to provide you with a comprehensive example. A good exercise for you would be to make a similar outline for free-radical halogenation (Chapter 6). A more skeletal outline of the questions and answers above, using key words to trigger your memory, would be:

(1) Functional Group Transformation

$$Nu^- + R\text{-}X \longrightarrow R\text{-}Nu + X^-$$

(2) Mechanism

S_N2:

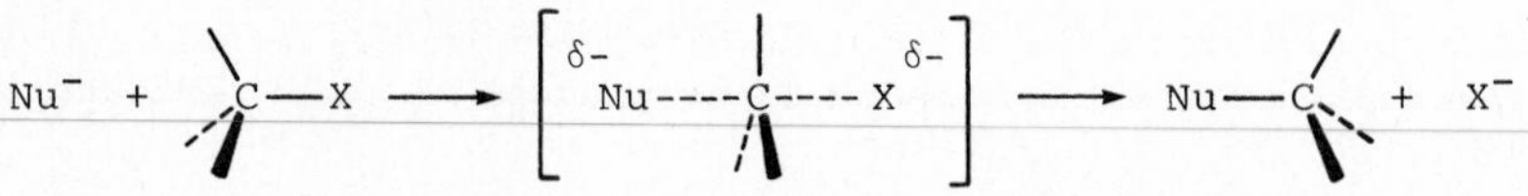

S_N1:

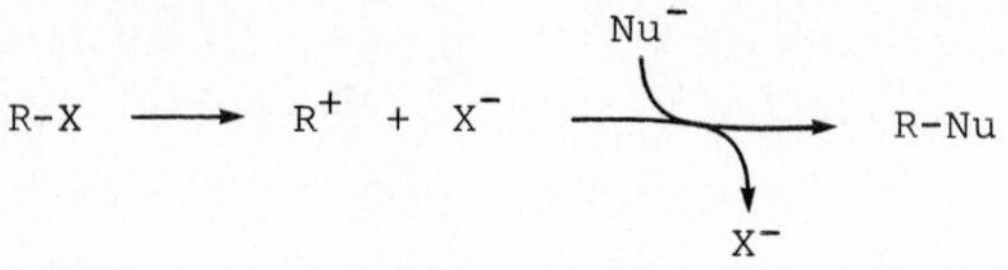

(3) Generalizations

Nu: Lewis base ; more basic, more polarizable $\Rightarrow$ better nucleophile

R: S_N2 $CH_3 > 1° > 2° > 3°$; β-branching slows reaction, too

S_N1 $3° > 2° > 1° >> CH_3$; decision hard only for 2°

X: X^- less basic $\Rightarrow$ better leaving group

Solvent: dipolar solvents speed reactions
H-bonding slows S_N2 by tying up Nu^-

(4) Stereochemistry

S_N2: inversion

S_N1: loss of configuration via planar carbocation

(5) Side Reactions

Elimination (E2) if Nu is basic and/or R-X is sterically hindered

HINTS: Organic chemists love to show what is going where in a reaction mechanism by drawing arrows, and writing these mechanisms often involves a lot of "arrow-pushing". A brief catalog of arrows which you will encounter in organic chemistry follows, as well a description of those used when writing mechanisms.

A single-headed arrow written in an equation (for example, $A + B \longrightarrow C$) is used by organic chemists instead of the = sign common to general chemistry equations (for example, $P_4 + 3\ OH^- + 3\ H_2O = PH_3 + 3\ H_2PO_2^-$). Reversible reactions are written with two single-headed arrows pointing in opposite directions. Often, an idea of the position of equilibrium is given by the relative length of the arrows: $HCN + OH^- \rightleftharpoons CN^- + H_2O$. These double arrows often connect different conformations of the same molecule, for example:

A double-headed arrow is distinct from the equilibration idea of $\rightleftharpoons$, and is specifically used between resonance structures, for example:

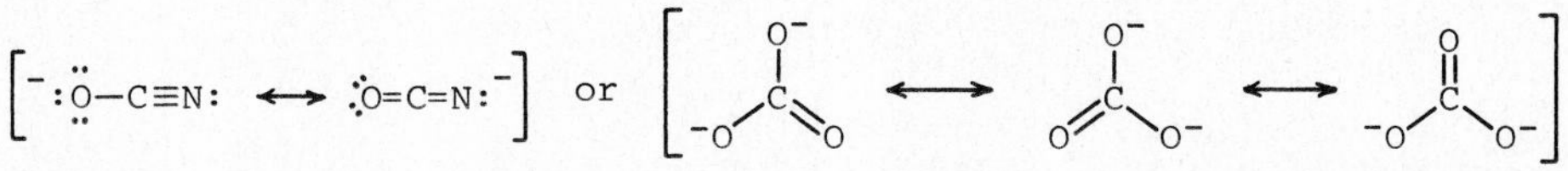

The curved arrows that you see leading molecules around in many depictions of reaction mechanisms represent a pair of electrons. The vast majority of reactions which you will encounter involve heterolytic cleavage of bonds (and their formation) and in essentially every instance the electrons travel in pairs. For instance, in a displacement reaction, the leaving group (X) always departs with the two electrons in the C-X bond, and the two electrons in the C-Nu bond come in with the nucleophile Nu:

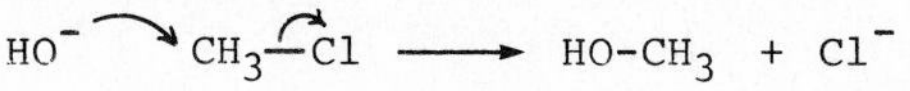

The situation is more complicated for an E2 elimination reaction, because more bonds are being formed and broken, but the arrows help to keep everything organized:

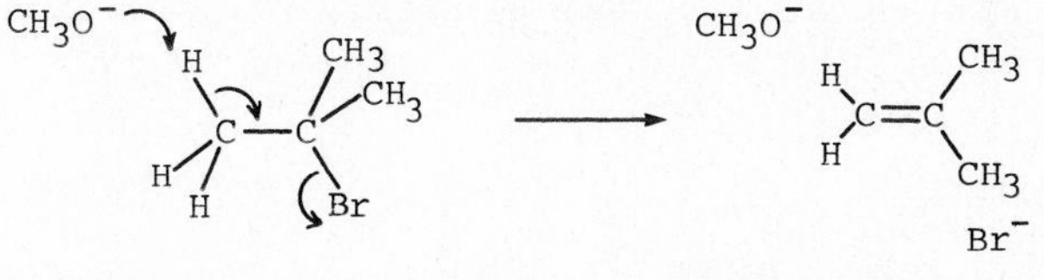

Pushing arrows is valuable because it helps you to keep track of electrons and charges and often prevents you from writing absurd mechanisms. Always bear in mind that an arrow represents the movement of a pair of electrons.

Sometimes, when describing free-radical reactions in which odd electron species are reacting, it is useful to show the movement of a single electron. Organic chemists often use "fish hook" or single-hooked arrow to represent this:

8.C Answers to Exercises

(8.2) (a) t-Butyl bromide has a higher melting point than n-butyl bromide. Little internal freedom of motion needs to be lost on incorporation of the t-butyl bromide molecule into the crystal, whereas the n-butyl bromide molecule loses its freedom to rotate about the C_1-C_2 and C_2-C_3 bonds. Therefore entropy makes it more difficult for n-butyl bromide to crystallize (see Section 5.3).

(b) To estimate boiling point of $CH_3CH_2CH_2CH(CH_3)CH_2CH_2Cl$,

compare: $CH_3CH_2CH_2CH_2CH_3$ (bp 36°) and $CH_3CH_2CH(CH_3)CH_2CH_3$ (bp 63°) $\Delta bp = 27°$

Add this to boiling point of $CH_3(CH_2)_5Cl$: 134.5° + 27° = 161.5° *(est.)*

Because each increment becomes less important as the molecule becomes larger, you would expect the actual bp of 1-chloro-3-methylhexane to be lower than 161.5°.

To estimate boiling point of $CH_3(CH_2)_3CH(C_2H_5)CH_2Cl$,

compare: $CH_3(CH_2)_6CH_3$ (bp 126°) and $CH_3(CH_2)_7Cl$ (bp 182°) $\Delta bp = 56°$ for Cl

Addition of a branching methyl group raises boiling point by about 25°, so to estimate boiling point of 3-(chloromethyl)heptane, add

bp (heptane) + Δbp (branching Me) + Δbp (Cl)
98° + 25° + 56° = 179° *(est.)*

(8.6)

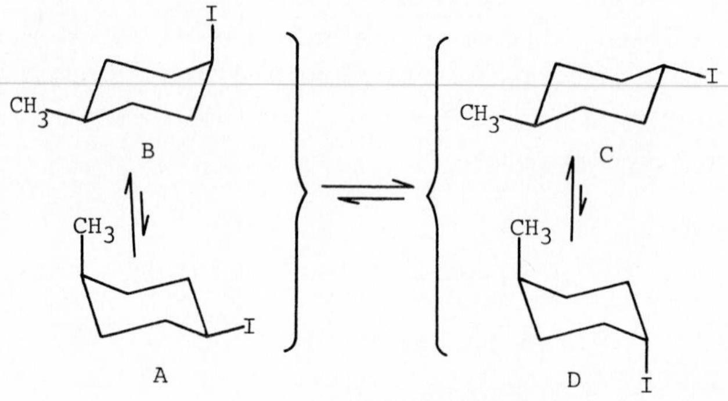

$\Delta G°_f$ differences for isomers:
A- axial methyl: +1.7 kcal mole^{-1}
B- axial iodine: +0.45 kcal mole^{-1}
C- no axial substituents
D- axial methyl and axial iodine: +2.15 kcal mole^{-1}

A:B:C:D = 0.06 : 0.47 : 1 : 0.03

cis (A + B): 34%; *trans* (C + D): 66%

(8.7)

CH_3OH	methanol	$CH_3\overset{+}{O}H_2$	methyloxonium iodide
$CH_3OC_2H_5$	ethyl methyl ether	$CH_3O\text{-}\overset{O}{\overset{\Vert}{C}}CH_3$	methyl acetate
CH_3SH	methanethiol	CH_3ONO_2	methyl nitrate
CH_3SCN	methyl thiocyanate	$(CH_3)_4P^+\ I^-$	tetramethylphosphonium iodide
CH_3CN	methyl cyanide (acetonitrile)	$CH_3\overset{+}{N}(C_2H_5)_3\ I^-$	methyltriethylammonium iodide
CH_3N_3	methyl azide	$CH_3\overset{+}{S}(C_2H_5)_2\ I^-$	methyldiethylsulfonium iodide
$CH_3NH_3^+\ I^-$	methylammonium iodide		

(8.9A) rate = k_2 [CH_3I] [Nu] ; 99.9% reaction means [CH_3I] goes from 0.1 M to 0.0001 M and [Nu] goes from 1.0 M to 0.9 M. To simplify the calculation, you can treat [Nu] as a constant because it changes by only 10% over the course of the reaction (pseudo-first-order conditions); choose [Nu] = 0.95 M.

$$-\frac{d[CH_3I]}{dt} = k_2[CH_3I][Nu]$$

$$-\int_{0.10}^{0.0001} \frac{d[CH_3I]}{[CH_3I]} = k_2[Nu]\int_0^t dt$$

$$-\ln[CH_3I]\Big]_{0.1}^{0.0001} = k_2[Nu]\,t\Big]_0^t$$

$$-(-2.303 - (-9.21)) = 10^{-4} \times 0.95(0-t)$$

$$t = 7.3 \times 10^4 \text{ sec} = 20 \text{ hr}$$

(8.10)

$$CH_3-O-\overset{\overset{O^-}{|}}{\underset{\underset{O_-}{|}}{S}}{}^{++}-O-CH_3 \quad :N(CH_3)_3 \longrightarrow CH_3-O-\overset{\overset{O^-}{|}}{\underset{\underset{O_-}{|}}{S}}{}^{++}-O^- + (CH_3)_4N^+$$

$$\underset{CH_3}{\overset{CH_3}{\underset{}{\diagdown}}}\overset{+}{S}-CH_3 \quad :N(CH_3)_3 \longrightarrow (CH_3)_2S + (CH_3)_4N^+$$

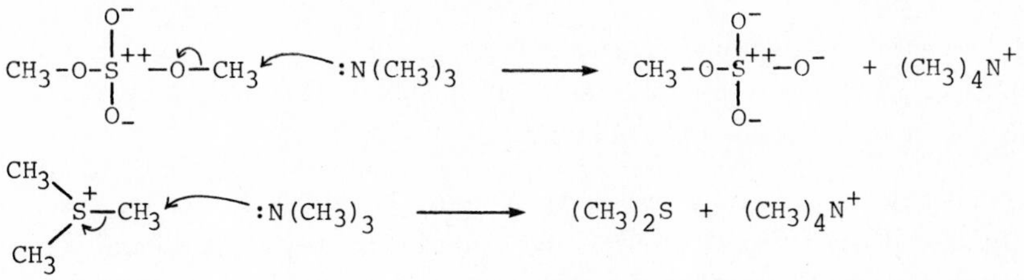

8.D Answers and Explanations to Problems

1.

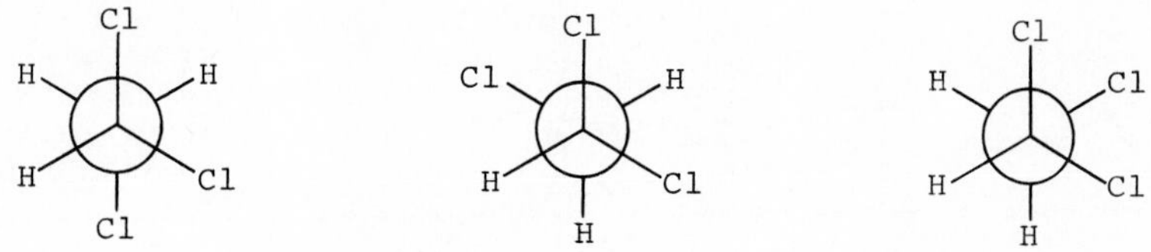

A   B   C

A and B are equivalent and equal in energy, and are more stable than C.
A and B have one gauche and one anti interaction; C has two gauche interactions. Anti is more stable in the gas phase where dipole-dipole interactions are more important. Such interactions are less important in the
liquid phase because they are masked by the dielectric effect of the liquid,
and other interactions can dominate. Interconversion of A and B is accomplished through the eclipsed conformation D, in which there are three H-Cl
interactions. Conversion to C goes through E, in which there is a Cl-Cl
interaction.

D   E

2. Because the C-O bond is not affected in the first step, the intermediate
    mesylate (s-butyl-$^{18}OSO_2CH_3$) still has the R configuration. The departure of
    the ^{18}O with the mesylate group shows that the substitution proceeds with
    cleavage of the C-O bond and by either an S_N2 or S_N1 mechanism (as opposed
    to cleavage of the S-O bond). Because OH$^-$ is a strong nucleophile and the
    alkyl group is not tertiary, the reaction is expected to go by an S_N2 mech-
    anism and inversion of configuration. Therefore the product is (S)-2-butanol.

    Note: there is probably a lot of elimination taking place as a side reaction in this example,
    but the question only concerns the stereochemistry of the 2-butanol that *is* formed.

$$(CH_3)_2CXCH_2CH_3 \xrightarrow{k_x} X^- + (CH_3)_2\overset{+}{C}CH_2CH_3$$

$(CH_3)_2\overset{+}{C}CH_2CH_3 \longrightarrow$

$$CH_2=\overset{\underset{\displaystyle CH_3}{|}}{C}CH_2CH_3$$

$$(CH_3)_2C=CHCH_3$$

$$(CH_3)_2\overset{\underset{\displaystyle \overset{+}{H}OCH_3}{|}}{C}CH_2CH_3 \xrightarrow{-H^+} (CH_3)_2\overset{\underset{\displaystyle OCH_3}{|}}{C}CH_2CH_3$$

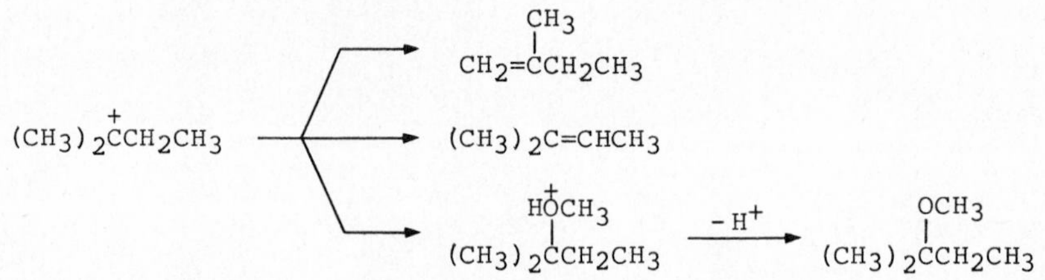

The k_x step is rate-determining and is different in rate for X = Cl, Br, I. The same carbocation is produced from each halide, and gives the same mixture of products.

4. (a) A tertiary halide is sterically hindered and leads to a relatively stable carbocation; water is a weak nucleophile. Both imply an S_N1 substitution mechanism, by way of a planar, <u>achiral</u> carbocation:

either $\cdots$ OH₂

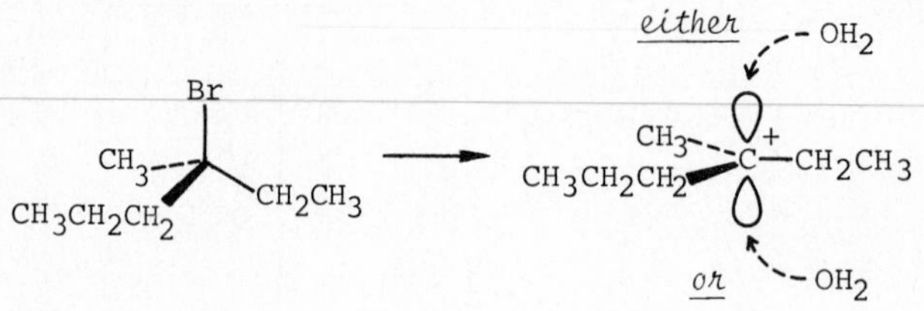

reaction is equally likely on both sides of the molecule, resulting in racemic product

or $\cdots$ OH₂

(b) In this case, the chiral center is not involved in the reaction and is therefore unchanged:

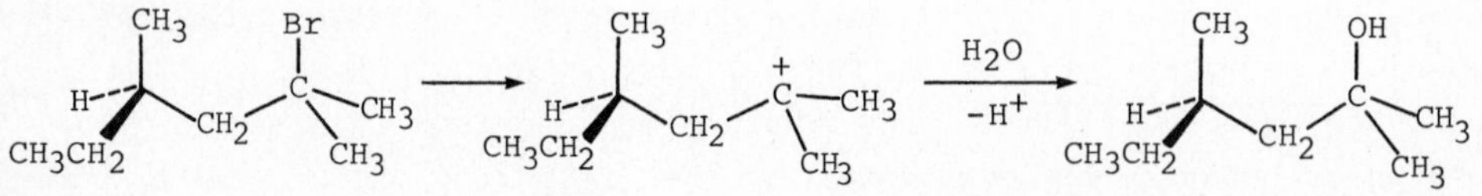

5. (a) In general, the more stable the leaving group is as the free Lewis base, the faster the S_N2 reaction. This is conveniently estimated by considering how basic the leaving group is (or how acidic the protonated compound (the conjugate acid) would be). The leaving group acquires additional electron density in the S_N2 transition state, and its basicity is a good indication of how easily it can accommodate (stabilize) this increased negative charge. For instance in this case, I^- is a weaker base than Cl^- (HI is a stronger acid than HCl; see Appendix IV), and alkyl iodides are generally more reactive in displacement reactions than alkyl chlorides.

(b) Water is a weak nucleophile, therefore the S_N1 mechanism is the most probable; <u>t</u>-butyl bromide reacts faster by an S_N1 mechanism than isopropyl bromide does because the tertiary carbocation is more easily formed than a secondary one.

(c) The methyl branch slows down the S_N2 reaction by steric hindrance; the straight-chain halide reacts faster.

(d) ⁻SH is stabilized by hydrogen-bonding solvation to methanol and its reactivity is reduced; ⁻SH reacts faster in dimethylformamide because this polar aprotic solvent does not form hydrogen bonds to anions.

(e) Reaction with $:NH_3$ is an S_N2 reaction which is faster with the unbranched, less sterically hindered, primary halide.

(f) Sulfur is more nucleophilic than oxygen, in protic solvents such as water; hence ^-SH reacts faster than ^-OH. In this case, the additional polarizability of the third-period atom (sulfur) more than compensates for the higher basicity of ^-OH compared to ^-SH.

(g) Phosphorus is more nucleophilic than nitrogen. In general, third-period atoms are more nucleophilic than their 2nd-period counterparts; hence, trimethylphosphine is more reactive toward methyl bromide than trimethylamine.

(h) The reaction giving CH_3CH_2SCN is faster. SCN^- is an ambident anion and may react on sulfur or nitrogen. Sulfur is more nucleophilic, and the faster reaction occurs there.

(i) Even though the anions are about equal in basicity, sulfur is more nucleophilic because of its greater polarizability.

(j) In a displacement reaction, the carbon atom undergoing substitution changes hybridization from sp^3 (bond angles 109°) to sp^2 (bond angles 120°) when proceeding to the transition state. Ring strain will oppose this "spreading apart" of the bonds to the carbon and will make the reaction more difficult. Therefore cyclobutyl chloride reacts much more slowly than cyclopentyl chloride.

(k) The three-membered and four-membered rings have about the same amount of strain (see Table 5.5). The likelihood that the ends of the chain will find each other to make a three-membered ring is higher than for the ends of a chain which forms a four-membered ring. Therefore  (oxirane, ethylene oxide) is formed faster than 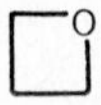 (oxetane).

6. (a) No; ^-CN is a poor leaving group. HCN is a relatively weak acid; recall that there is a good correlation between the acidity of H-Y and the leaving ability of Y^- in displacement reactions.

(b) Slow; F^- is a relatively strong base (HF is a weak acid) and a poor leaving group.

(c) No; ^-OH is a strong base and an exceptionally poor leaving group. The only reaction observed is an acid-base reaction:

$$(CH_3)_3COH + NH_2^- \rightleftharpoons (CH_3)_3CO^- + NH_3$$

(d) Okay; $CH_3OSO_3^-$ is a weak base and a good leaving group. The corresponding acid, CH_3OSO_2OH, is a strong acid, comparable to H_2SO_4.

(e) No; $^-NH_2$ is a very strong base and a perfectly miserable leaving group. NH_3 is a very weak acid.

(f) Okay; I^- is a perfectly good leaving group.

(g) No; ^-OH is a strong base and an exceptionally poor leaving group.

7. (a)

$$N_3^- + CH_3Cl \xrightarrow{\text{(slower)}} CH_3N_3 + Cl^- \quad \text{more basic}$$

$$N_3^- + CH_3I \xrightarrow{\text{(faster)}} CH_3N_3 + I^- \quad \text{less basic}$$

Many other examples could have been cited, for this is the most common situation. If HX is more acidic than HY, X^- is less basic than Y^-, and RX is more reactive than RY.

(b) $CH_3Br + SCN^- \longrightarrow CH_3SCN + CH_3NCS$
(major) *(minor)*

(c) $(CH_3)_3CCl \xrightarrow[\text{acetone}]{H_2O} (CH_3)_3COH$ *(faster)*

$(CH_3)_2CHCl \xrightarrow[\text{acetone}]{H_2O} (CH_3)_2CHOH$ *(slower)*

(d) $CH_3CH_2CH_2I + CH_3S^- \longrightarrow CH_3CH_2CH_2SCH_3$ *(faster)*

$(CH_3)_2CHI + CH_3S^- \longrightarrow (CH_3)_2CHSCH_3$ *(slower)*

Other examples could be chosen among primary halides with branching in the β-position. Recall that relative rates are: $CH_3CH_2CH_2- > (CH_3)_2CHCH_2- > (CH_3)_3CCH_2-$, entirely because of steric hindrance effects.

(e) $CH_3CH_2Cl + SH^- \xrightarrow{C_2H_5OH} CH_3CH_2SH$ *(slower)*

$CH_3CH_2Cl + SH^- \xrightarrow{DMF} CH_3CH_2SH$ *(faster)*

In general, anions are less reactive in S_N2 reactions in hydroxylic solvents (such as alcohol) than in polar aprotic solvents (such as DMF, HMPT, and DMSO). In hydroxylic solvents, hydrogen bonds to the anion need to be broken in order to form the S_N2 transition state.

(f) $CH_3CHBrCH_3 + OH^- \longrightarrow CH_3CH{=}CH_2 + (CH_3)_2CHOH$

$CH_3CHBrCH_3 + SH^- \longrightarrow CH_3CH{=}CH_2 + (CH_3)_2CHSH$

(less $CH_3CH{=}CH_2$ is obtained from the second reaction)

8. (b) (assuming that the carbon atom undergoing the substitution is the only chiral center in the molecule); (d), (e), (g), (h)

9. (a), (c), (f), (h) Note that (h) is true for both S_N1 and S_N2 reactions.

10. For the cyclization reaction, both reactants are part of the same molecule, and the reaction can only have first-order kinetics. The reaction of the amino group of one molecule with the alkyl bromide of another (the *inter*molecular displacement rather than the *intra*molecular or cyclization reaction) has second-order kinetics; hence, the rate of this reaction is reduced much more than the other by reducing the concentration:

$$\frac{inter}{intra} = \frac{k_{inter}\left[H_2N(CH_2)_4Br\right]^2}{k_{intra}\left[H_2N(CH_2)_4Br\right]} = \frac{k_{inter}}{k_{intra}}\left[H_2N(CH_2)_4Br\right]$$

The ratio of the two is concentration-dependent, and the cyclization reaction is favored by "high dilution" methods.

11. (a) SCN^- (b) I^- (c) $P(CH_3)_3$ (d) CH_3S^-

In each case, the nucleophile giving the larger substitution/elimination ratio contains an atom further down the Periodic Table, and is therefore more polarizable. Greater polarizability enhances nucleophilicity more than basicity; that is, polarizability is relatively more important in S_N2 reactions at carbon and is relatively less important in E2 reactions at hydrogen.

12. (a) $\Delta\Delta G° = \Delta G°_A - \Delta G°_B = -RT(\ln K_A - \ln K_B)$; $\ln K = 2.3 \log K = -2.3\,pK$

$\Delta\Delta G° = 2.3\,RT\,\Delta pK = 1.36\,\Delta pK$ (at 25°C) $= 6.1$ kcal mole^{-1}

(b) $\Delta G° = +6.1$ kcal mole^{-1}; $K = 3.5 \times 10^{-5}$

(c) $\dfrac{k_2(CN^-)}{k_2(CH_3CO_2^-)} = \dfrac{k'}{k} = 241$; $\quad \Delta\Delta G^{\ddagger} = RT \ln \dfrac{k'}{k} = 3.2 \text{ kcal mole}^{-1}$

Note that the $\Delta\Delta G^{\ddagger}$ for the S_N2 reaction is only about half that for the proton transfer equilibrium. The equilibrium concerns fully formed anions. In the S_N2 transition state, both entering and leaving groups must share the negative charge, and each has about half (see problem #13).

13. (a) The slope is about 0.5. That is, a difference in acidity of two pK_a units gives a change in rate of one log unit. This suggests that only about half of the negative charge is donated at the transition state, and half remains on the attacking group.

$$Y^- + R\text{-}X \longrightarrow \left[Y^{\frac{1}{2}-} \cdots\cdots R \cdots\cdots X^{\frac{1}{2}-} \right]^{\ddagger} \longrightarrow Y\text{-}R + X^-$$

(b) If one were to plot log k vs. pK for this reaction, the resulting two-point line would have a slope of $\log(40)/3.0 = 1.6/3.0 = 0.53$. In this case, in the transition state, the leaving group has about one-half of a negative charge. Various methods of this sort indicate that the negative charge in the transition state is split roughly equally between entering and leaving groups.

14. $Br(CH_2)_5Br + NaOH \longrightarrow HO(CH_2)_5Br + NaBr$

$\text{(pyran ring)} + NaBr \longleftarrow Na^+ \ ^-O(CH_2)_5Br + H_2O$

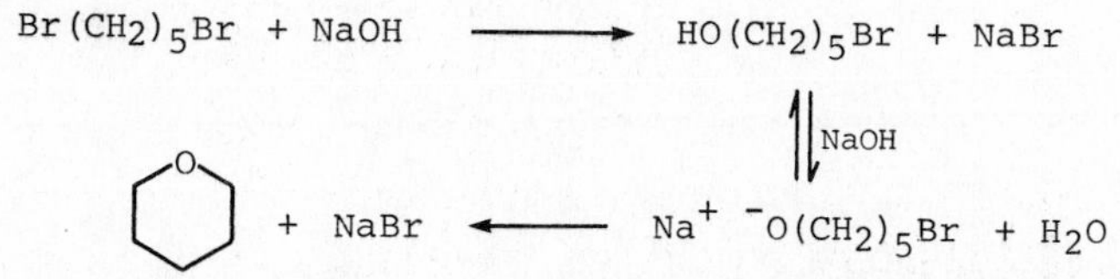

Dilution will minimize the bimolecular side reaction

$$HO(CH_2)_5Br + NaOH \longrightarrow HO(CH_2)_5OH + NaBr$$

but it will not affect the equilibrium

$$HO(CH_2)_5Br + NaOH \rightleftharpoons Na^+ \ ^-O(CH_2)_5Br + H_2O$$

or the cyclization.

To form the ten-membered ring, 1,9-dibromononane is required as substrate, and high dilution is needed to favor ring formation over intermolecular reactions.

15.

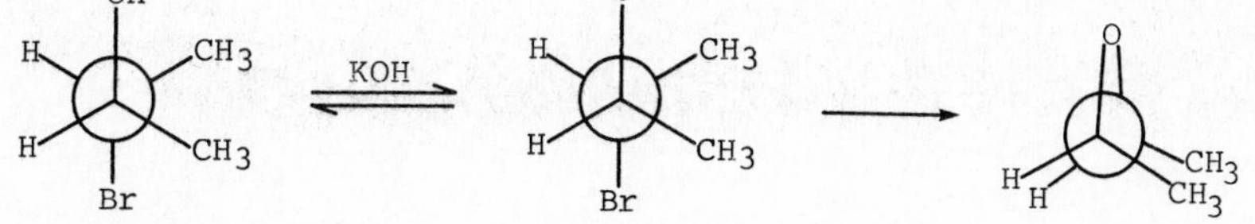

(or the enantiomer)

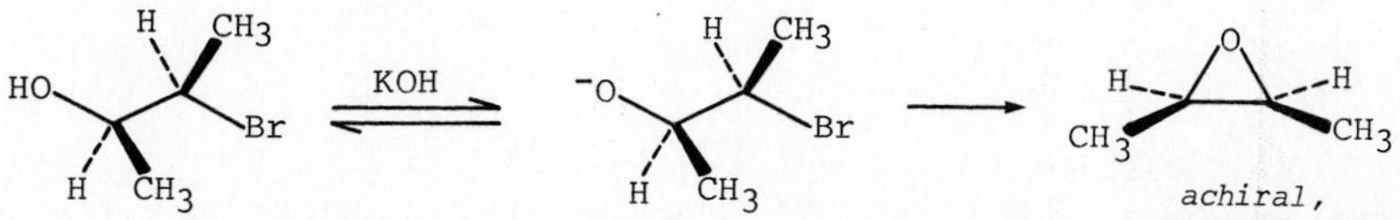

achiral,
optically inactive

Although the cyclization reaction is unimolecular, it still proceeds by the "S_N2" mechanism involving backside attack and inversion of configuration.

Work out the sterochemistry of this reaction with the other *enantiomer* of the starting material and demonstrate that it gives the same product. Then do it for a *diastereomer*, and show that it gives a chiral product.

16.

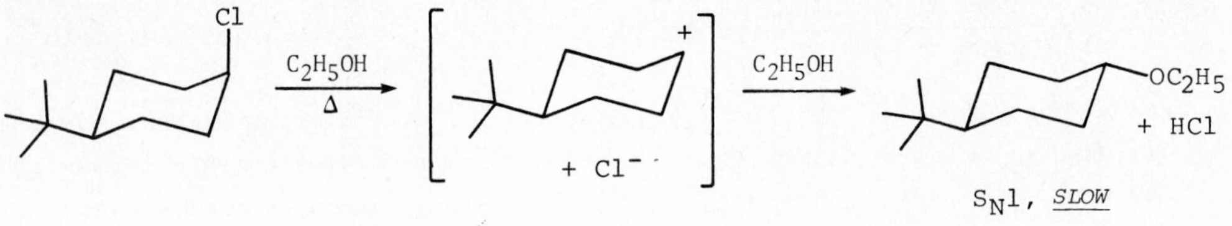

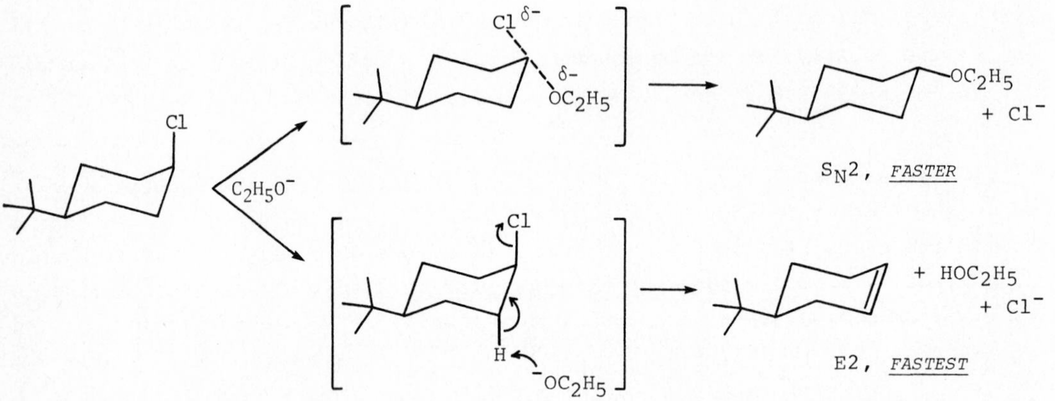

17. (a) $CH_3CH_2CH_2Br + NaSH \xrightarrow{C_2H_5OH} CH_3CH_2CH_2SH + NaBr$

(b) $(CH_3)_2CHCH_2CH_2Cl + KCN \xrightarrow{C_2H_5OH} (CH_3)_2CHCH_2CH_2CN + KCl$

(c) $CH_3CH_2CH_2I + Na^+ \, ^-OCH_3 \xrightarrow{CH_3OH} CH_3CH_2CH_2OCH_3 + NaI$

<u>or</u> $CH_3CH_2CH_2O^- \, ^+Na + CH_3I \xrightarrow{CH_3CH_2CH_2OH} CH_3CH_2CH_2OCH_3 + NaI$

(d) $CH_3CH_2CH_2Br + NaOH \xrightarrow{H_2O} CH_3CH_2CH_2OH + NaBr$

(e) $NaNO_3 + CH_3I \xrightarrow{CH_3OH} CH_3ONO_2 + NaI$

(f) $CH_3CH_2CH_2CH_2Cl + NaN_3 \xrightarrow{CH_3OH} CH_3CH_2CH_2CH_2N_3 + NaCl$

18. (a) Look at models or pictures of chair structures:

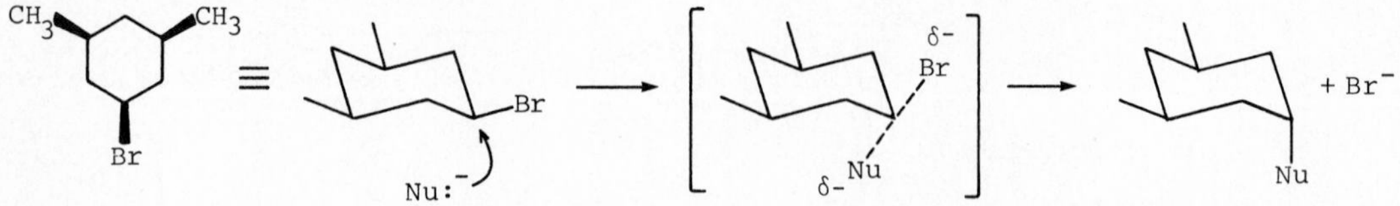

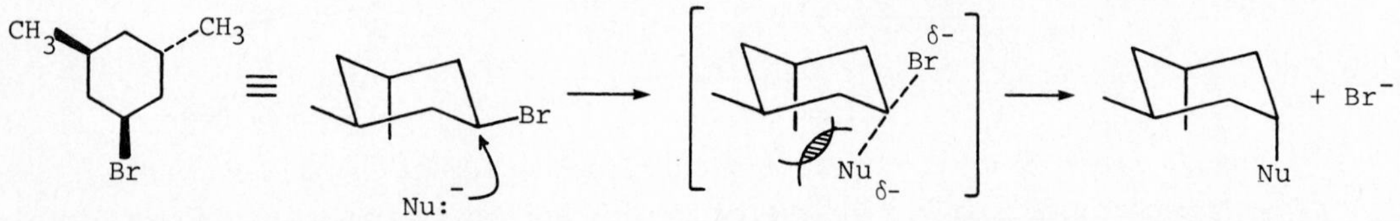

Because of the axial methyl group in isomer D, there are steric interactions which destabilize the transition state for S_N2 displacement.

(b)

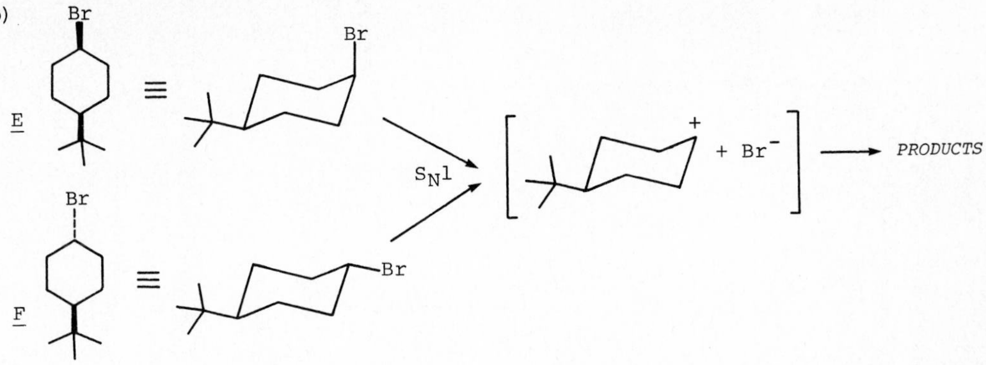

Both E and F lead to the same carbocation intermediate in an S_N1 reaction. However, E is less stable than F to begin with because the bromine is in the axial position (t-butyl is *always* equatorial), so that it doesn't have to go as far "uphill" in order to react. This can be depicted in a reaction profile diagram:

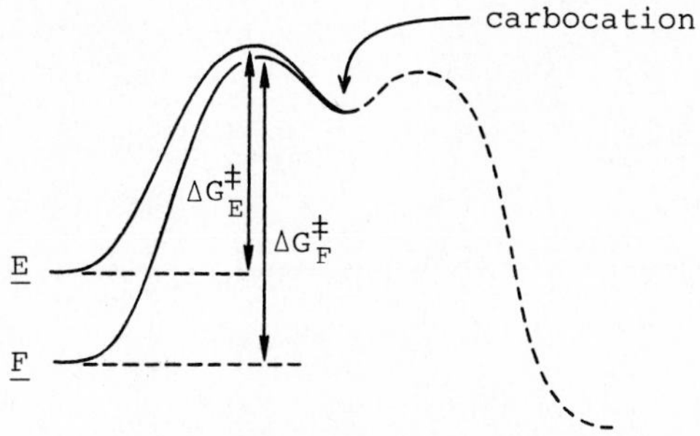

$$\Delta G^{\ddagger}_E < \Delta G^{\ddagger}_F$$

(c) t-Butyl chloride reacts via the S_N1 mechanism involving the intermediate t-butyl cation. The higher the dielectric constant of a solvent, the more it can mask the attraction of a negative charge (the departing Cl^- ion) for a positive charge (the t-butyl cation), and the easier it is for the t-butyl chloride molecule to ionize. The dielectric constant of methanol is higher than that of ethanol, therefore S_N1 reactions proceed faster in methanol.

8.E Supplementary Problems

S1. Predict the major product and mechanism of each of the following reactions:

(a) $(CH_3)_2CHCH_2Cl + NaN_3 \xrightarrow{CH_3OH}$

(b) $CH_3CH_2CH_2I$ +

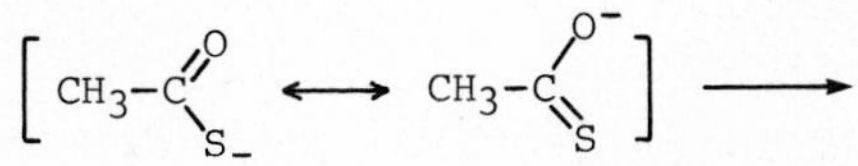

(c) NaCN +

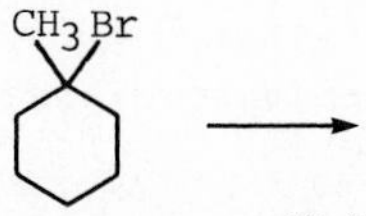

(d) $(CH_3)_3CBr + NaCl \xrightarrow{CH_3OH}$

(e) $(CH_3)_2C(OH)CH_2CH_2CH_2Br + NaNH_2 \longrightarrow$

(f) $CH_3CH_2CH_2Cl$ + $:N(CH_3)_3$ $\longrightarrow$

(g) $CH_3CH_2-\overset{\underset{|}{CH_3}}{\underset{|}{\overset{|}{C}}}-CHClCH_3$ + NaOH $\longrightarrow$

(h)   + CH_3SCH_3 $\longrightarrow$

(i) $BrCH_2CH_2CH_2-\overset{\cdot\cdot}{\underset{\underset{CH_3}{|}}{N}}-CH_2CH_2CH_2CH_2Br$ $\xrightarrow{\text{acetone}}$

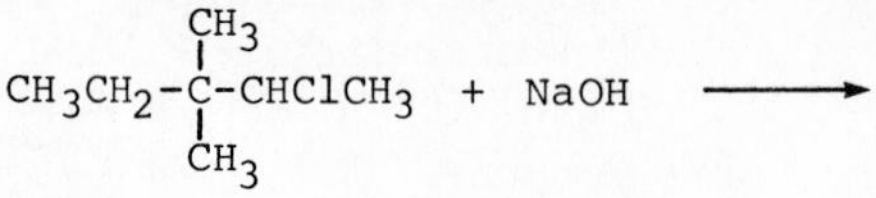

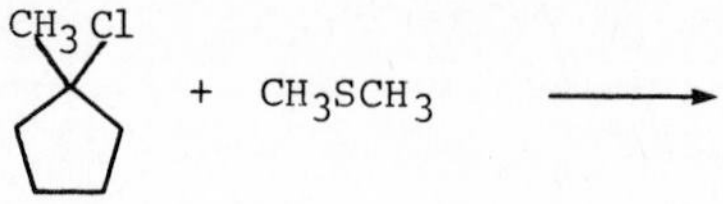

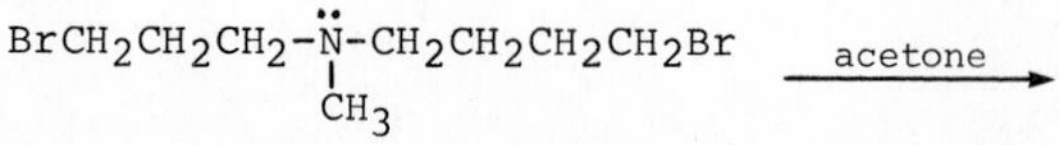

S2. What are the absolute configurations of the products of the following reaction sequences?

(a)

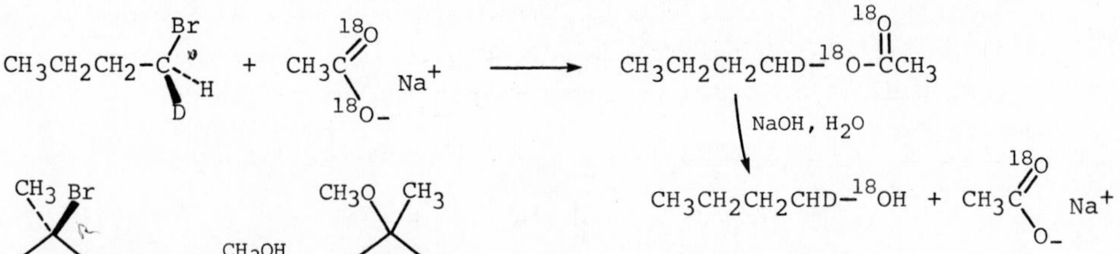

(b)

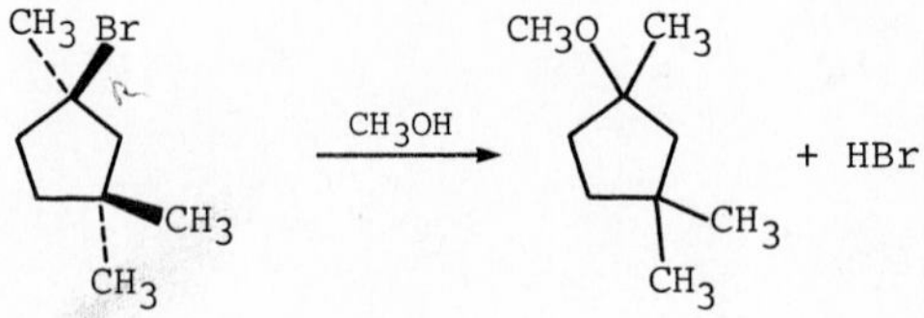

(c)

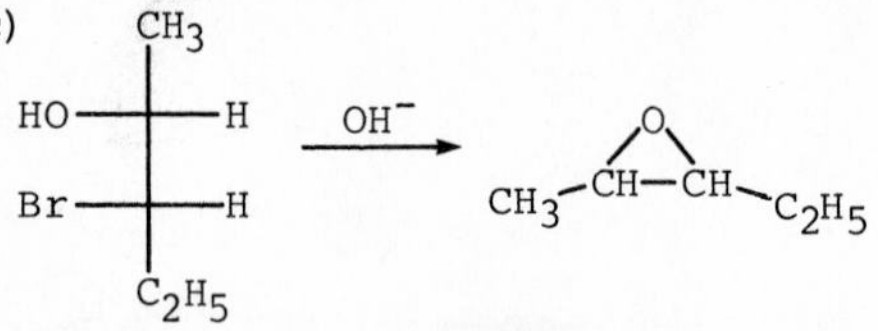

(d)

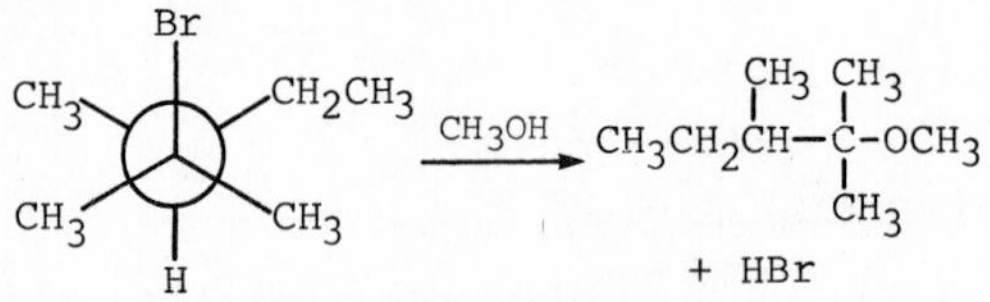

(e)

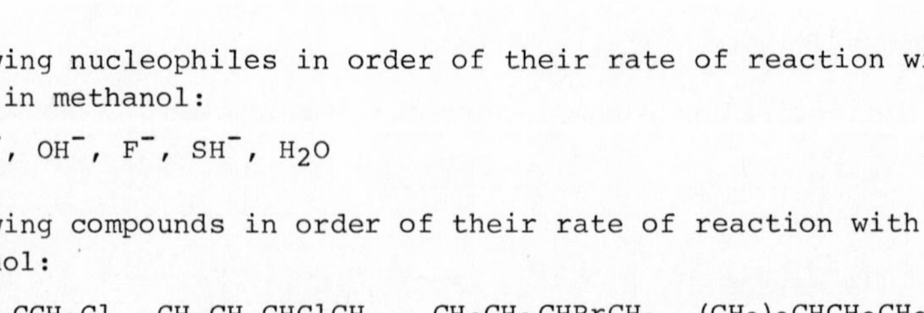

S3. Rank the following nucleophiles in order of their rate of reaction with methyl sulfate in methanol:

$$Cl^-,\ OH^-,\ F^-,\ SH^-,\ H_2O$$

S4. Rank the following compounds in order of their rate of reaction with sodium azide in methanol:

$$CH_3Br,\ (CH_3)_3CCH_2Cl,\ CH_3CH_2CHClCH_3,\ CH_3CH_2CHBrCH_3,\ (CH_3)_2CHCH_2CH_2Br$$

S5. Rank the following compounds in order of their rate of reaction in refluxing (boiling) methanol:

$$CH_3CH_2CHBrCH_3,\ CH_3CH_2CHClCH_3,\ (CH_3)_3CCH_2Br,$$

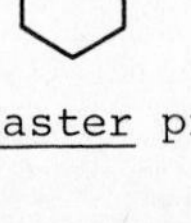

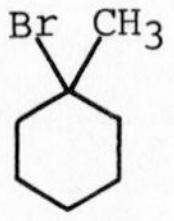

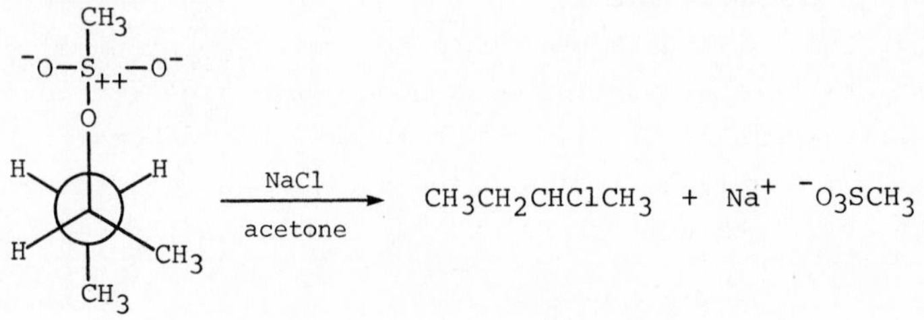

S6. For each of the sets of reactions and conditions below, choose that corresponding to the faster process. Indicate which explanations in the list of statements following are the most pertinent ones for each case. Note that more than one explanation may apply to each set of reactions.

(a) $CH_3CH_2CH_2Br + Y^- \xrightarrow{CH_3OH} CH_3CH_2CH_2Y + Br^-$

$Y^- = CH_3O^-$ or $CH_3CO_2^-$

(b) $(CH_3)_3CCl + CH_3O^- \xrightarrow{CH_3OH} A + Cl^-$

$A = (CH_3)_3COCH_3$ or $(CH_3)_2C=CH_2 + CH_3OH$

(c) $R-Br + HS^- \longrightarrow R-SH + Br^-$

$R = (CH_3)_2CHCH_2CH_2Br$ or $CH_3CH_2\underset{\underset{CH_3}{|}}{C}HCH_2Br$

(d) $Cl^- + CH_3CH_2OSO_2CH_3 \xrightarrow{solvent} CH_3CH_2Cl + CH_3SO_3^-$

$solvent = CH_3\overset{O}{\overset{||}{C}}CH_3$ or CH_3CH_2OH

(e) $CH_3CH_2CHICH_3 + (CH_3)_3Z \longrightarrow CH_3CH_2\underset{\underset{+Z(CH_3)_3}{|}}{C}HCH_3 + I^-$

$Z = N$ or P

(f) $CH_3CH_2-\underset{\underset{CH_3}{|}}{\overset{\overset{CH_3}{|}}{C}}-CHBrCH_3 + CH_3O^- \xrightarrow{CH_3OH} D + Br^-$

$D = CH_3CH_2C(CH_3)_2\underset{\underset{OCH_3}{|}}{C}HCH_3$ or $CH_3CH_2C(CH_3)_2C=CH_2 + CH_3OH$

(g) $H-\overset{\overset{CH_3}{|}}{\underset{\underset{CH_2CH_3}{|}}{\rule{2em}{0.5pt}}}-S^- + CH_3Y \longrightarrow H-\overset{\overset{CH_3}{|}}{\underset{\underset{CH_2CH_3}{|}}{\rule{2em}{0.5pt}}}-SCH_3 + Y^-$

$Y^- = F^-$ or Br^-

(h) $R-Br + CH_3OH \longrightarrow R-OCH_3 + HBr$

R-Br = or

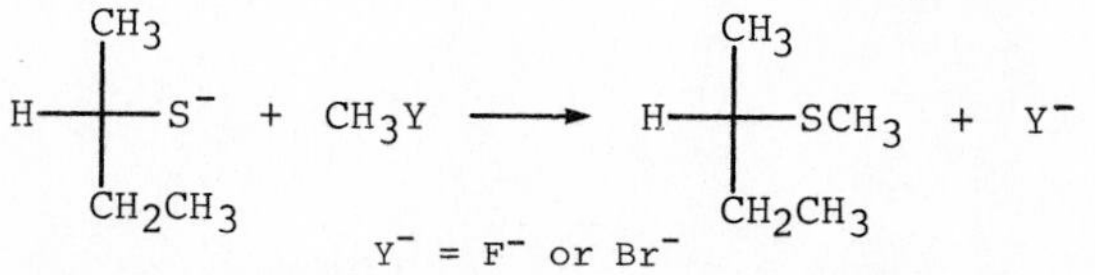

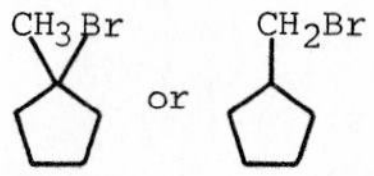

(i) $CH_3CH_2CHBrCH_3 + Y^- \longrightarrow CH_3CH_2CHYCH_3 + Br^-$

$Y^- = N_3^-$ or NH_2^-

EXPLANATIONS

(A) Steric hindrance from α-branching results in slower S_N2 reactions.

(B) Steric hindrance from α-branching has little effect on S_N2 reactions.

(C) Steric hindrance from β-branching results in slower S_N2 reactions.

(D) Steric hindrance from β-branching has little effect on S_N2 reactions.

(E) The order of stability of carbocations is 3° > 2° > 1°.

(F) The order of stability of carbocations is 3° < 2° < 1°.

(G) E2 reactions are generally poor for tertiary systems.

(H) Stronger bases make poorer leaving groups.

(I) The reactivity of anions in protic solvents is diminished by hydrogen bonding.

(J) The reactivity of cations in polar solvents is diminished by interaction with the solvent lone pair electrons.

(K) Very strong bases favor elimination over substitution.

(L) Other factors being equal, stronger bases are generally better nucleophiles.

(M) Other factors being equal, the more polarizable reagent is the better nucleophile.

(N) Reagents with no lone pair electrons are relatively poor nucleophiles.

(O) Chiral molecules are generally more effective nucleophiles.

(P) Neopentyl-type systems are exceptionally slow in S_N2 reactions because of steric hindrance.

8.F Answers to Supplementary Problems

S1. (a) $(CH_3)_2CHCH_2N_3$; S_N2

(b) $CH_3CH_2CH_2S\text{-}\overset{O}{\overset{\|}{C}}CH_3$; S_N2

(c) 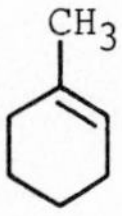 + HCN ; E2

(d) $(CH_3)_3COCH_3$ + HBr ; S_N1

(don't forget that the solvent can react!)

(e)

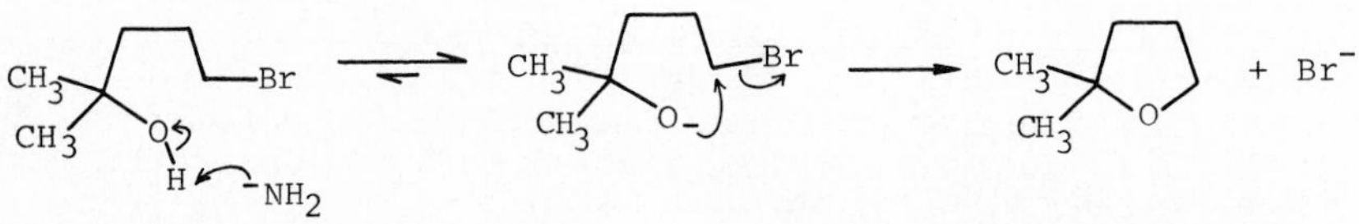

(f) $CH_3CH_2CH_2\overset{+}{N}(CH_3)_3\ Cl^-$; S_N2

(g) $CH_3CH_2C(CH_3)_2CH{=}CH_2$; E2

(too hindered for S_N2, and strong base)

(h) 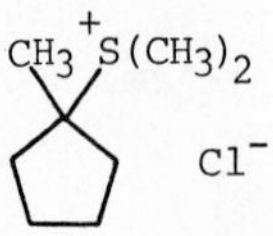 Cl^- ; S_N1

(i)

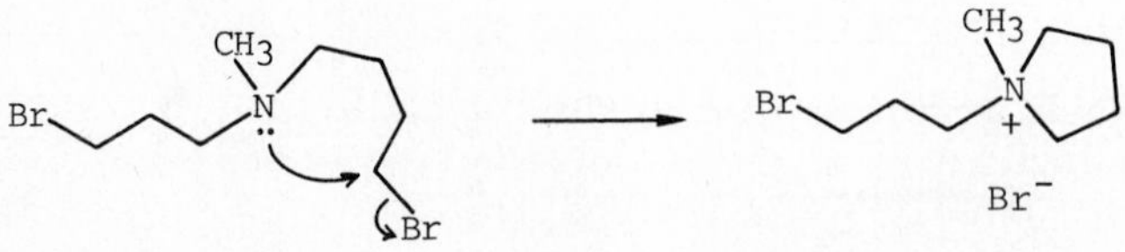

S2. (a) 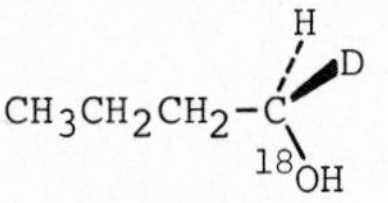 (R) ; the first step is S_N2 with inversion; the second step does not affect the CHD-O bond.

(b) 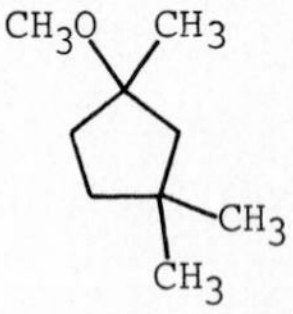 racemic; reaction proceeds via achiral carbocation (S_N1)

(c)

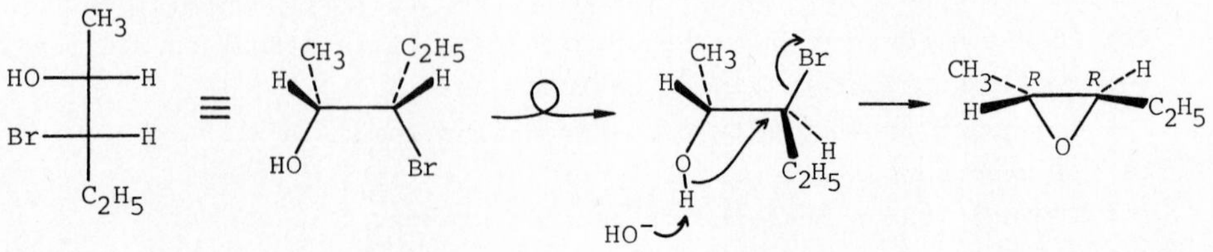

(d) The reaction is S_N1, but does not involve the chiral center:

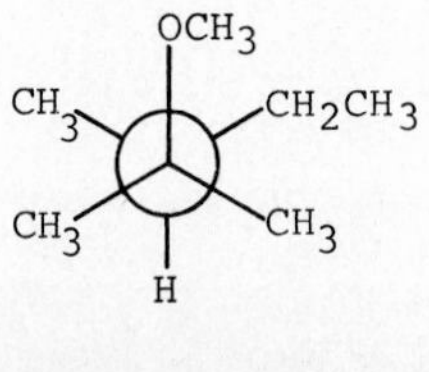
(S)

(e) A mixture of S_N2 and S_N1 reactions, resulting in partial inversion, partial racemization.

Major enantiomer: 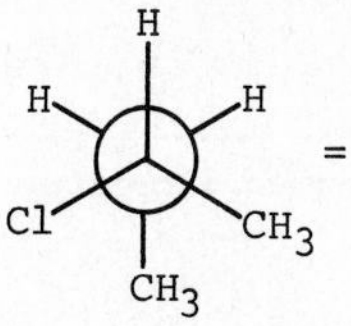

= (S)-2-chlorobutane

S3. $SH^- > OH^- > Cl^- > F^- > H_2O$

S4. <u>S_N2:</u> $CH_3Br > (CH_3)_2CHCH_2CH_2Br > CH_3CH_2CHBrCH_3 > CH_3CH_2CHClCH_3 > (CH_3)_3CCH_2Cl$

S5. <u>S_N1:</u> 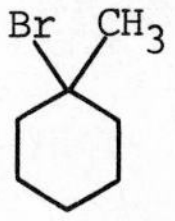

$> CH_3CH_2CHBrCH_3 > CH_3CH_2CHClCH_3 > (CH_3)_3CCH_2Br$

S6. (a) $Y^- = CH_3O^-$; L

(b) $A = (CH_3)_2C{=}CH_2 + CH_3OH$; A (K)

(c) $R = (CH_3)_2CHCH_2CH_2Br$; C

(d) solvent = $CH_3\overset{O}{\overset{\|}{C}}CH_3$; I

(e) Z = P ; M

(f) $D = CH_3CH_2C(CH_3)_2CH{=}CH_2 + CH_3OH$; P

(g) $Y^- = Br^-$; H

(h) RBr = 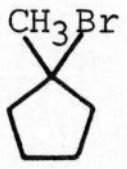

; E

(i) $Y^- = N_3^-$; K

<u>*NOTE:*</u> *"Explanations" B, D, F, and O are <u>false</u> statements.*

9. NUCLEAR MAGNETIC RESONANCE SPECTROSCOPY

9.A Chapter Outline and Important Terms Introduced

9.1 Structure Determination (what does it involve?)

9.2 Introduction to Spectroscopy (the most important methods of structure determination)

- quantization
- energy level differences
- $\Delta E = h\nu$
- microwave spectroscopy
- infrared spectroscopy
- ultraviolet-visible spectroscopy

9.3 Nuclear Magnetic Resonance (the physics behind the technique)

- nuclear spin
- magnetic moment
- α- and β-spins
- spin "flipping"
- magnetogyric ratio
- $\nu = \frac{\gamma H}{2\pi}$

9.4 Chemical Shift (information about the environment of the proton)

- shielding
- diamagnetic shielding
- resonance
- upfield-downfield
- tetramethylsilane (TMS)
- part per million (ppm)
- δ and τ scales
- Heisenberg uncertainty principle
- nmr time scale
- magnetic equivalence

9.5 Relative Peak Areas (information on numbers of protons)

- integration of spectra
- saturation/relaxation

9.6 Spin-Spin Splitting (information on adjacent protons)

- magnetic non-equivalence
- applied vs. effective field
- (singlet), doublet, triplet, (quartet, multiplet, etc.)
- coupling constant J
- binomial pattern
- "first-order" spectra
- conformational effects
- dihedral angle dependence
- Karplus curve

9.7 More Complex Splitting (the real world ...)

- $J_{ab} \neq J_{ac}$
- overlap of patterns
- non-first order spectra

9.8 Solving Spectral Problems (useful generalizations and hints)

9.9 Nmr Spectroscopy of Other Nuclei (carbon magnetic resonance)

- natural abundance
- Fourier transform instrumentation
- proton decoupling and off-resonance decoupling
- symmetry
- chemical shift prediction
- α-, β-, and γ- effects

9.10 Dynamical Systems (slow or fast on the "nmr time scale")

- chair-chair interconversion
- variable temperature nmr

9.B Important Points and Hints

Nmr spectroscopy has become the most important method available to organic chemists for determining the structure of a compound. Chapter 9 presents the physical principles that underly the technique, and describes the spectra observed for alkanes and alkyl halides. As other functional groups are introduced in subsequent chapters, their characteristic nmr resonances will be discussed. Although the facts that make up the topic of nmr form a logical framework, one of the most practical ways to absorb the details is by memorization. While we usually discourage simple memorization as an approach to understanding (it is usually a way to avoid understanding ...), at the very least you should "know" the following facts about nmr:

I. "To the right" in an nmr spectrum is:

   A. more "shielded"

   B. "upfield" (= higher field strength for a given resonant frequency)

   C. "lower frequency" (to reach resonance for a given magnetic field) and vice versa ("to the left" = "deshielded" = "downfield" = "higher frequency")

II. For proton spectra (not cmr), *area* is proportional to *number of protons*

III. Chemical shift depends on substituents:

   A. For proton spectra:

      1. learn the 5 generalizations in Section 9.8 of the text

      2. learn Table 9.2

      3. learn the nmr characteristics of each functional group as they are presented in subsequent chapters

   B. For carbon spectra:

      1. learn some basic shifts (e.g. methane-butane)

      2. learn the substituent effects (α-, β-, γ- effects) (When it comes to solving spectral problems, there is no substitute for knowing right away what chemical shift corresponds to what possibilities.)

IV. As a basic rule, n adjacent protons leads to a splitting of n + 1 peaks. This can easily get more complicated, however.

V. You should also be aware of complications in splitting patterns which arise from:

   A. non-first order spectra

   B. non-equivalent coupling constants

   C. dihedral angle dependence

Nmr spectral problems are usually one of two types: given the structure of the compound, predict the spectrum it would give; or, given the spectral data, figure out the structure of the compound. Solving the first type of question is fairly straightforward:

first: Pick out the non-equivalent nuclei (nuclei can be equivalent either through symmetry alone, or through conformational interconversions and symmetry);

second: Predict the chemical shift for each group;

third: For proton spectra:

(a) assign relative area = number of nuclei in each group,

(b) calculate splitting patterns

For carbon spectra: sometimes you are asked to predict the multiplicity of the off-resonance decoupled spectrum;

finally: Look for complications ...

Solving the second kind of problem is like solving a puzzle. It can be fun and there is a system. It is most complex for proton spectra, where there are the added complications of splitting patterns and relative areas to deal with, so we will point out a systematic way to approach this type of question. Take the following example:

Formula: $C_6H_{12}Br_2$; nmr spectrum: δ, ppm: 0.9 (t, 3H), 1.4 (s, 6H), 1.8 (m, 2H), 4.6 (t, 1H)

1. Write out, underlined, groups corresponding to the relative areas of each resonance (starting with the most probable combinations), including enough unprotonated carbons and other substituents to satisfy the formula:

"3H"	is usually a methyl:	$\underline{CH_3}$-
"6H"	is usually two methyls:	$\underline{CH_3}$-, $\underline{CH_3}$- (but it could be: 3x -$\underline{CH_2}$-)
"2H"	is usually a CH_2:	-$\underline{CH_2}$-
"1H"	is, of course	-$\underline{CH}$- (with one additional bond)
other:		-C- (four bonds), 2 $\underline{Br}$- to satisfy $C_6H_{12}Br_2$

Make sure the number of protons you have written adds up to the number in the molecule; if it doesn't, multiply everything by an integer. (If that doesn't work, then the entire spectrum wasn't given.)

2. Look at the splitting pattern, and include in your part structures the adjacent carbons with an appropriate number of protons, not underlined. Simply circle the multiplets:

(t, 3H):	$\underline{CH_3}$-CH_2-
(s, 6H):	$\underline{CH_3}$-C-, $\underline{CH_3}$-C- (3x -C-$\underline{CH_2}$-C- can be ruled out because there are not enough non-hydrogen bearing substituents to fit)
(m, 2H):	(-$\underline{CH_2}$-) (circled)
(t, 1H):	-$\underline{CH}$-CH_2- or -CH-$\underline{CH}$-CH-
	-C-, 2 $\underline{Br}$

3. Look for some correspondence between the underlined and non-underlined groups.

A. The underlined and circled CH_2 group (3rd set, above) must be the non-underlined CH_2 groups of the 1st and 4th sets:

CH_3-CH_2-CH-

t,3H m,2H t,1H

B.   The underlined, unsubstituted carbon must be the non-underlined, unsubstituted carbons of the 2nd set:

$$CH_3-\overset{\displaystyle |}{\underset{\displaystyle |}{C}}-CH_3$$
$$(s, 6H)$$

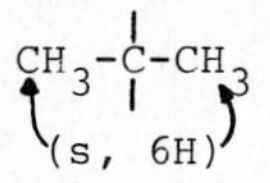

C.   There is only one way to hook these two pieces and two bromine atoms together:

$$CH_3-CH_2-CHBr-\overset{\displaystyle CH_3}{\underset{\displaystyle CH_3}{\overset{|}{\underset{|}{C}}}}-Br$$

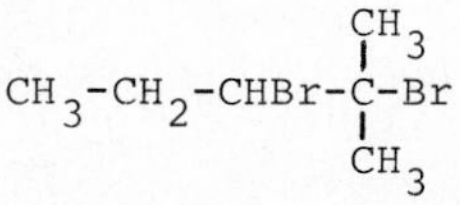

4.   Use the chemical shift information as a check, at least, or as further help if the coupling patterns can't solve the problem (as above):

R_2CH_2 (1.25) with adjacent Br (+0.5) = 1.75

RCH_3 (0.9) with adjacent Br (+0.5) = 1.4

$$CH_3-CH_2-CHBr-CBr(CH_3)_2$$

RCH_3 (0.9)

R_2CHBr (4.1) with adjacent Br (+0.5) = 4.6

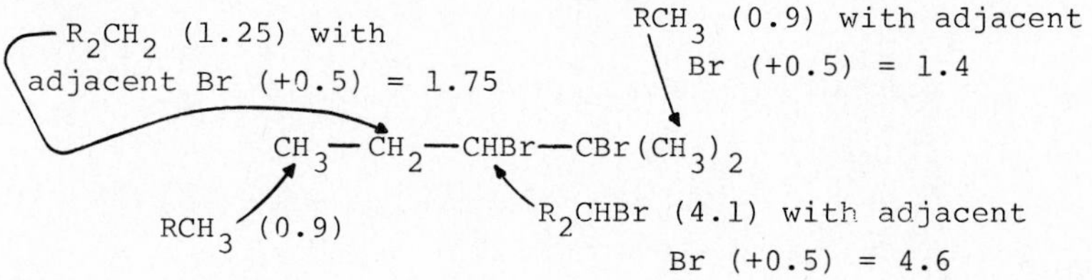

For many problems, you will be able to "see" the structure much more quickly than this stepwise process would suggest, but this sort of methodical approach often avoids confusion in complex cases.

9.C Answers to Exercises

(9.5) $CH_3CO_2CH_3$
         3:3 = 1:1

$$CHCl_2-\overset{\displaystyle CH_3}{\underset{\displaystyle CH_3}{\overset{|}{\underset{|}{C}}}}-CH_2Cl$$
$$1 : 6 : 2$$

rapid chair⇌chair equilibration on the nmr time scale makes all CH_3's equivalent: $\dfrac{12}{8}$ = 3:2
and all CH_2's equivalent:

(9.6) CH_3Cl: singlet; $(CH_3)_4C$: singlet; $CH_3CH_2CH_2CH_3$: triplet;
        CH_3CH_2Br: triplet; CH_3CH_3: singlet; $CH_3CH_2OCH_2CH_3$: triplet;

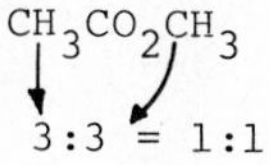

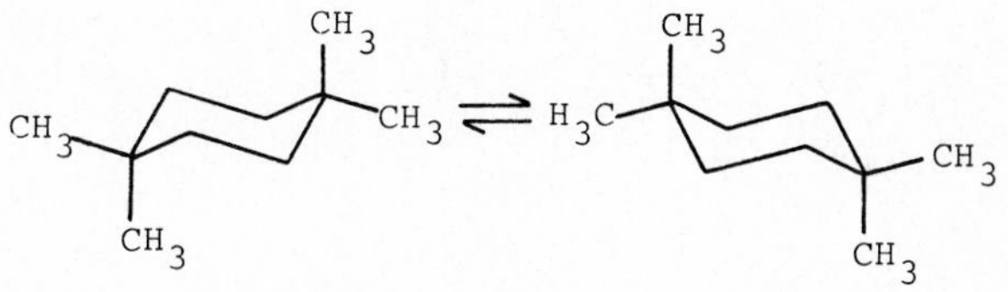

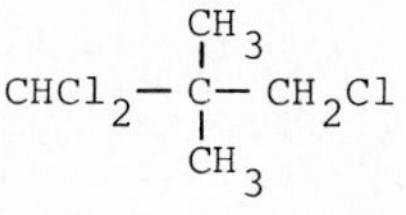

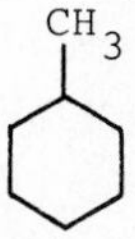

: doublet

(9.9) (a)   δ for (hexane) + increment = predicted δ   (actual δ)

CH_3 CH_3	13.9	9.5(β-CH_3) 23.4	22.4
CH	22.9	9.0(α-CH_3) 31.9	28.4
CH_2	32.0	9.5(β-CH_3) -	
CH_2		2.5(γ-CH_3) 39.0	36.9
CH			
CH_3 CH_3			

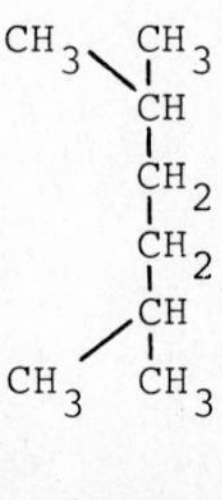

(b)

	δ for (pentane)	+ increments	= predicted δ	(actual δ)
CH_2Cl	13.7	31(α-Cl)	44.7	44.7
CH_2	22.6	11(β-Cl)	33.6	32.9
CH_2	34.5	-5(γ-Cl)	29.5	29.4
CH_2	22.6	--	22.6	22.5
CH_3	13.7	--	13.7	13.9

(c)

	δ for (butane)	+ increments	= predicted δ
CH_2Cl	13.2	31(α-Cl)	44.2
CH_2	25.0	11(β-Cl) - 1(γ-I)	35.0
CH_2	25.0	-5(γ-Cl) + 11(β-I)	31.0
CH_2I	13.2	-6(α-I)	7.2

(d)

	δ for (pentane)	+ increments	= predicted δ
CH_2Br	13.7	20(α-Br)	33.7
CH_2	22.6	11(β-Br) - 2.5(γ-CH_3)	31.1
CH_2	34.5	-3(γ-Br) + 9.5(β-CH_3)	41.0
CH	22.6	9.0(α-CH_3)	31.6
CH_3 CH_3	13.7	9.5(β-CH_3)	23.2

(e)

	δ for (pentane)	+ increments	= predicted δ	(actual δ)
CH_3 CH_3	13.7	9.5(β-CH_3)	23.2	22.7
CH	22.6	9.0(α-CH_3) - 2.5(γ-CH_3)	29.1	25.7
CH_2	34.5	19(2β-CH_3)	53.5	49.0
CH				
CH_3 CH_3				

(9.10) At room temperature you expect a singlet because the resonances from axial and equatorial H's are averaged rapidly by chair ⇌ chair interconversion.

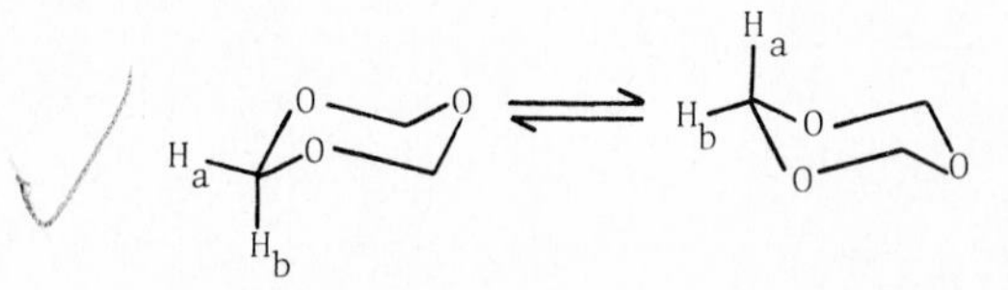

At -100° C, this interconversion is slow ("frozen out"), and separate resonances for axial and equatorial H's will be seen. Because the axial and equatorial H's are not equivalent, they will split each other and a pair of doublets will be observed.

9.D Answers and Explanations to Problems

1. In the nmr spectrum of ethyl bromide, the methyl hydrogens have δ = 1.7 ppm, the methylene hydrogens have δ = 3.3 ppm, and J = 7 Hz. The number of peaks given by the methyl hydrogens is three, with the approximate area ratio 1:2:1. These peaks are separated by 7 Hz.

The number of peaks given by the methylene hydrogens is <u>four</u>, with the approximate area ratio: 1:3:3:1. These peaks are separated by <u>7</u> Hz. The total area of the methyl peaks compared to the methylene peaks is in the ratio <u>3:2</u>. Of these two groups of peaks, the <u>methylene</u> peaks are farther downfield. The chemical shift difference between these peaks of 1.6 ppm corresponds in a 60 MHz instrument to <u>96</u> Hz.

2. (a) Since all peaks are singlets, there are no hydrogens on adjacent carbons. The only possibility is

$$
\begin{array}{c}
CH_3 \\
| \\
BrCH_2CCH_2Br \\
| \\
CH_3
\end{array}
$$

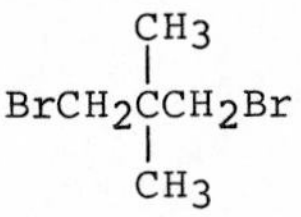

to give only two types of H in a 6:4 ratio.

(b) The triplet and the lower-field quartet are characteristic for CH_3-CH_2; hence,

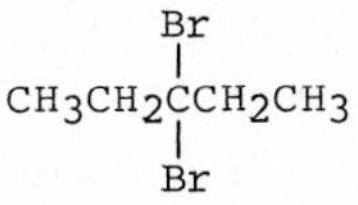

$$
\begin{array}{c}
Br \\
| \\
CH_3CH_2CCH_2CH_3 \\
| \\
Br
\end{array}
$$

(c) δ 0.9 (d,6H) corresponds to $(C\underline{H}_3)_2CH-$;
   δ 1.5 (m,1H) corresponds to the tertiary hydrogen;
   δ 1.85 (t,2H) must be

$$
\begin{array}{ccc}
H & & H \quad\;\; H \\
| & & | \qquad | \\
C\underline{H}_2-C & or & C\llcorner C\underline{H}_2-C \\
| & & \\
H & &
\end{array}
$$

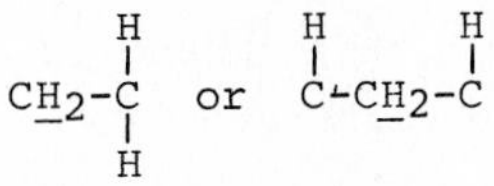

but only the latter fits. Finally, δ 5.3 is so far downfield it must be $-CHBr_2$. Everything fits for $(CH_3)_2CHCH_2CHBr_2$.

(d) δ 1.0 (s,9H) suggests $(CH_3)_3C$;
   δ 5.3 is so far downfield that it suggests $-CHBr_2$.
   Thus the compound must be $(CH_3)_3CCHBr_2$.

(e) δ 1.0 (d,6H) and δ 1.75 (m,1H) as in (c) suggests $(CH_3)_2CH-$;
   both δ 3.95 and δ 4.7 suggest C-Br, therefore

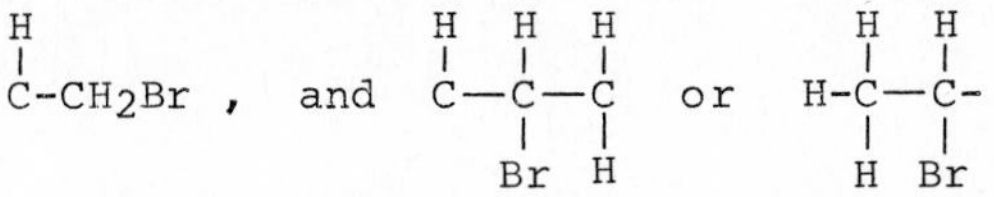

$$
\begin{array}{ccccc}
H & & H \;\; H \;\; H & & H \quad H \\
| & & | \;\; | \;\; | & & | \qquad | \\
C-CH_2Br\;, & and & C-C-C & or & H-C-C- \quad respectively, \\
& & | \quad | & & | \quad | \\
& & Br \;\; H & & H \;\; Br
\end{array}
$$

but only the former fits. The structure is $(CH_3)_2CHCHBrCH_2Br$.

(f) The first thing to notice is that this compound has no methyl groups. $C_5H_{10}Br_2$ can have no rings (replace each Br by H to give C_5H_{12} (C_nH_{2n+2}), an alkane), so it must have $BrCH_2-$ ends. This would fit δ 3.35, and two pairs of H's with the same δ suggests two equivalent $-CH_2Br$ groups. The triplet suggests $-CH_2CH_2Br$, and leads to the structure $BrCH_2CH_2CH_2CH_2CH_2Br$.

$$\delta \; 3.35 \qquad \delta \; 1.85 \qquad \delta \; 1.3$$

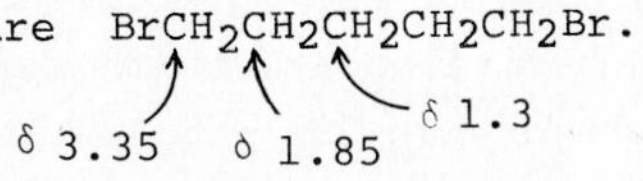

3. A: Two methyls: $CH_3CCl_2CH_3$

B: δ, 1.2(t, 3H) = $C\underline{H}_3\text{-}CH_2$

δ, 1.9(quint., 2H) = $CH_3\text{-}C\underline{H}_2\text{-}CH$ (or $-CH_2\text{-}C\underline{H}_2\text{-}CH_2\text{-}$ but this is inconsistent with the other information)

δ, 5.8(t, 1H) = $-CH_2\text{-}C\underline{H}X_2$

This is all consistent with $CH_3CH_2CHCl_2$.

C: δ, 1.4(d, 3H) = $C\underline{H}_3\text{-}CH$
δ, 3.8(d, 2H) = $-CH\text{-}C\underline{H}_2\text{-}$
δ, 4.3(sext., 1H) = $CH_3\text{-}C\underline{H}_2\text{-}CH_2\text{-}$
} $CH_3CHClCH_2Cl$

D: δ, 2.2(quint., 2H) = $-CH_2\text{-}C\underline{H}_2\text{-}CH_2\text{-}$ (or $CH_3\text{-}C\underline{H}_2\text{-}CH\text{-}$, but this is inconsistent with the other information)

δ, 3.7(t, 4H) = $-C\underline{H}_2\text{-}CH_2\text{-}C\underline{H}_2\text{-}$

The structure must be $ClCH_2CH_2CH_2Cl$.

4. (a) The δ 1.7 ppm doublet, corresponding to six hydrogens, suggests two equivalent $\underline{CH_3}\text{-}\overset{|}{\underset{|}{C}}H$ units. Hence, $CH_3\overset{Br}{\overset{|}{C}}H\text{-}\overset{Br}{\overset{|}{C}}HCH_3$.

The quartet at δ 4.4 ppm is therefore the two equivalent C-H's.

(b) Again, the δ 1.7 ppm doublet suggests CH_3CH-, but now there is only one. Hence, the molecule has the unit: CH_3CHBr-. Since there is no other methyl, the second Br must be at the other end of the chain. Thus the structure is $CH_3CHBrCH_2CH_2Br$.

5. (a)

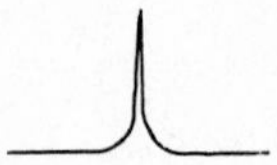

$\delta \approx 2.0\text{-}2.4$

(b) J ≈ 7 Hz

$\delta \approx 3.3$ $\delta \approx 1$

area: 2 3

(c) See problem #4(a).

(d) J ≈ 7 Hz

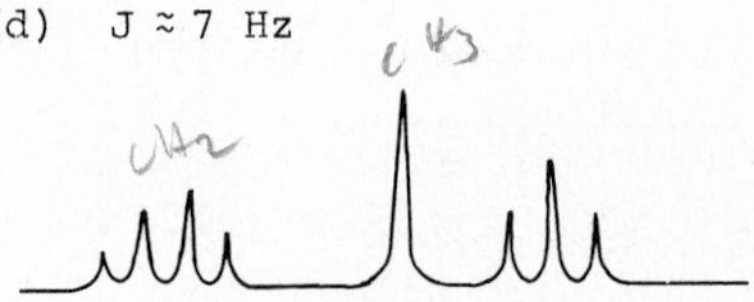

$\delta \approx 3$ $\delta \approx 2$ $\delta \approx 1.2$

area: 2 3 3

(e) See problem #4(b).

6. (a) CH_3CCl_3
(b) Br_2CHCH_2Br
(c) $(CH_3)_3CBr$
(d) $CH_3CBr(CH_2Br)_2$
(e) CH_3CHBr_2
(f) $(CH_3)_2CHBr$
(g) $CH_3CHBrCH_2CH_2Br$
(h) $(CH_3)_2CHCH_2Cl$
(i) $CH_3CH_2CH_2CH_2Cl$
(j) $CH_3CHClCH_2CH_3$

7. $CH_3CHClCHCl_2$

8. The small peaks are due to the $^{13}C\text{-}H$ coupling constant of 209 Hz. They are small because only 1.1% of the molecules have ^{13}C. Together, their intensity is 1.1% of the main peak, which corresponds to the proton resonance in the 98.9% of the $CHCl_3$ molecules having ^{12}C. $^{13}CHCl_3$ has a cmr spectrum which is a doublet with J=209 Hz.

9. (a) $ClCH_2CH_2CH_2CH_2Cl$ (b) $ClCH_2CHClCH_2Cl$
In both cases, note the complexity of the nmr spectra and the simplicity of the cmr spectra.

10. $(CH_3)_2CHCH_2Br$. The nmr spectrum is similar to that for #6(h).

11. The four possible compounds are: $CH_3CH_2CH_2CH_2Cl$, $CH_3CH_2CHClCH_3$, $(CH_3)_2CHCH_2Cl$, $(CH_3)_3CCl$. The cmr spectrum shows *four different* carbons, thus eliminating the last two. The proton-coupled spectrum of n-butyl chloride would show a quartet and three triplets; that of sec-butyl chloride has two quartets, one triplet, and one doublet.

12. On the nmr time scale, the two methyls are averaged by the chair ⇌ chair interconversion:

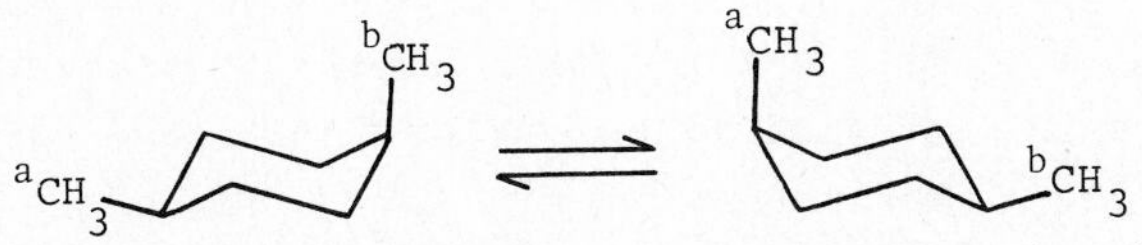

At lower temperatures where this interconversion is slow, separate signals are seen for the two methyls.

13. E and F:

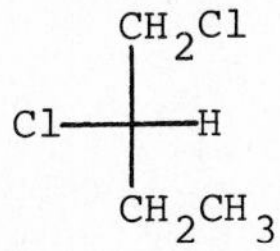

and

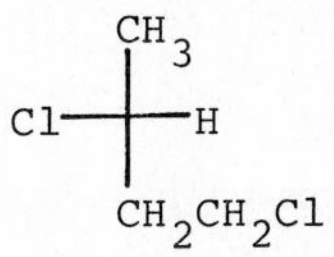

G:

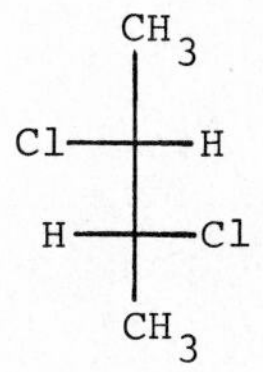

H:

$CH_3CCl_2CH_2CH_3$

I:

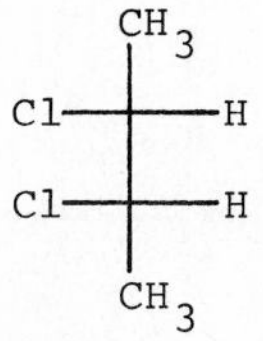

14.

	δ		δ	Δδ	
CH_3	13.2	CH_3	13.9	+0.7	(= δ-effect)
CH_2	25.0	CH_2	19.4	-5.6	(= γ-effect)
CH_2	25.0	CH_2	35.3	+10.3	(= β-effect)
CH_3	13.2	CH_2OH	61.7	+48.5	(= α-effect)

Using the known effects of Cl substitution, the following chemical shifts would be predicted:

$ClCH_2—CH_2—CH_2—CH_2OH$
45 30 30 62

$CH_3—CHCl—CH_2—CH_2OH$
25 50 46 57

$CH_3—CH_2—CHCl—CH_2OH$
9 30 66 61

$(CH_3—CH_2—CH_2—CHClOH)$ (This molecule is unstable)
14 14 46 93

4-Chloro-1-butanol certainly corresponds most closely to the observed values.

15. Three resonances indicate only three different kinds of carbons. The proton coupling indicates that methyl(s), methylene(s), and methine(s) are all present. The only combination to give 5 carbons and 10 hydrogens is:

2 CH_3's, 1 CH_2, and 2 CH's: $CH_3CHBrCH_2CHBrCH_3$.

16. The H's are split by the fluorine spins with $J_{H\text{-}F} = 50$ Hz; two equivalent fluorines give a triplet. Note that in a different region of the spectrum it is possible to determine the fluorine nmr spectrum. This would also show a triplet because of the two H's, and the splitting would have exactly the same value.

17. $CH_3—CH_2—F$

δ 14.6 79.3, each split by ^{19}F. The spectrum is only proton-decoupled.

18. The excess proportion of α is 1×10^{-5} (Sect. 9.5) x 0.01 moles = 1×10^{-7} moles (of α spin converted to β) x 0.0057 cal $mole^{-1}$ (Sect. 9.3) = 5.7×10^{-10} cal.

$$\frac{5.7 \times 10^{-10} \text{ cal}}{1 \text{ cal deg}^{-1} \text{ cc}^{-1}} \times 1 \text{ cc} = 5.7 \times 10^{-10} \text{ deg},$$ a rather small number!

9.E Supplementary Problems

S1. From the molecular formulas and nmr spectral data below, deduce the structure of each compound.

(a) $C_4H_7Cl_3$; δ, 1.4(s, 3H); 4.0(s, 4H)

(b) $C_4H_7Cl_3$; δ, 1.3(d, 3H); 2.1(s, 3H); 4.6(q, 1H)

(c) $C_4H_8Br_2$; δ, 1.0(d, 3H); 2.5(m, 1H); 3.3(d, 4H)

(d) $C_4H_7Br_3$; δ, 1.4(d, 3H); 2.6(t, 2H); 3.6(m, 1H), 5.4(t, 1H)

S2. Sketch the nmr spectra of the following compounds. Be sure to represent the expected δ for each group of peaks, the relative areas, and the splitting patterns.

(a) $ICH_2CH_2CHCl_2$

(b) $(CH_3)_3CCHClCH_2Cl$

(c) $(CH_3)_2CHCHClCH_2Cl$

(d) $(CH_3)_2CHCHCl_2$

(e) $(CH_3CH_2)_2CHCH_2Br$

(f)

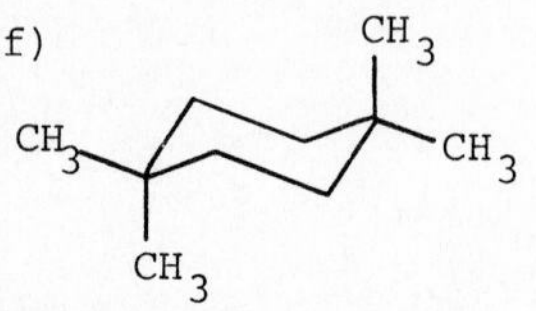

S3. For compounds (a)-(e) in Problem S2, estimate the chemical shifts of all of the resonances in the cmr spectrum.

S4. How many resonances do you expect to see in the cmr spectrum of each of the following compounds at 25° C? At -100° C?

(a)

(b)

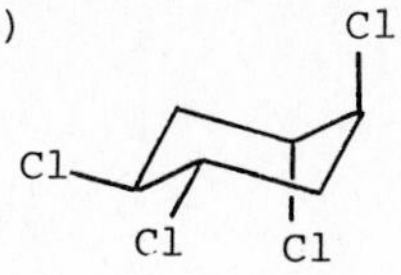

(c)

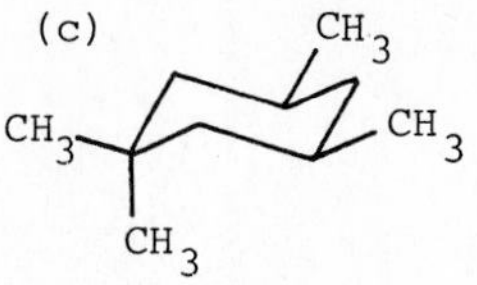

(d)

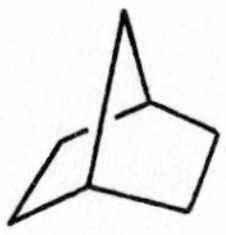

(e)

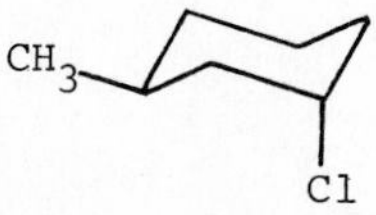

S5. The nmr spectrum of a rapidly (on the nmr time scale) interconverting mixture of conformational isomers is the weighted average of the spectra of the individual conformations. At low temperature, the chemical shifts of H-1 in axial- and equatorial-chlorocyclohexane are as shown below:

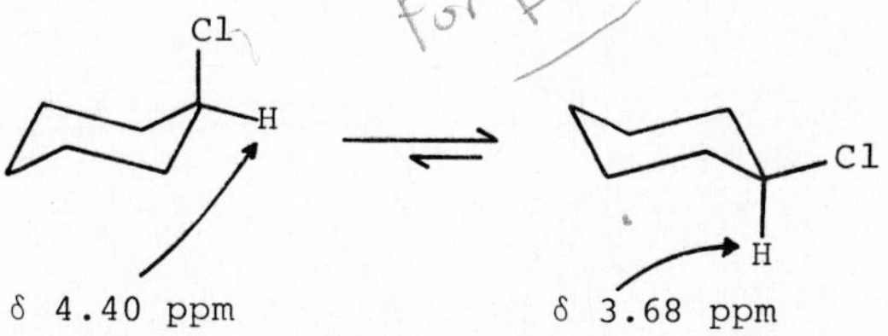

Predict the observed chemical shift of this proton at room temperature.

S6. For 1,1-diphenylpropane, J_{ab} = 7 Hz is in the expected range for acyclic alkanes. In 2-methyl-1,1-diphenylpropane, however, $J_{a'b'}$ is > 10 Hz.

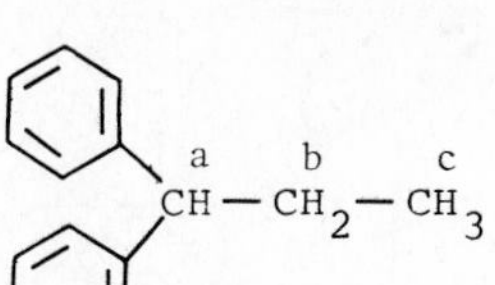

J_{ab} = 7 Hz

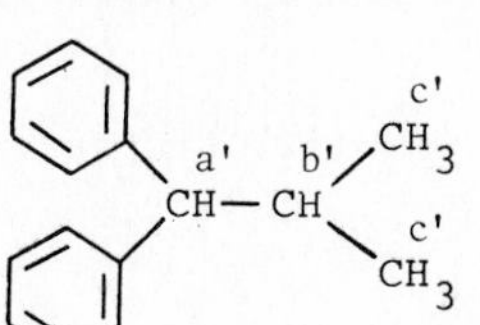

$J_{a'b'}$ = >10 Hz

Using the Karplus curve (Fig. 9.21) and Newman projections, offer an explanation for this increase in the coupling constant J.

9.F Answers to Supplementary Problems

S1. (a) $(ClCH_2)_2CClCH_3$ (b) $CH_3CCl_2CHClCH_3$

(c) $CH_3CH(CH_2Br)_2$ (d) $CH_3CHBrCH_2CHBr_2$

S2. (a)   (b)

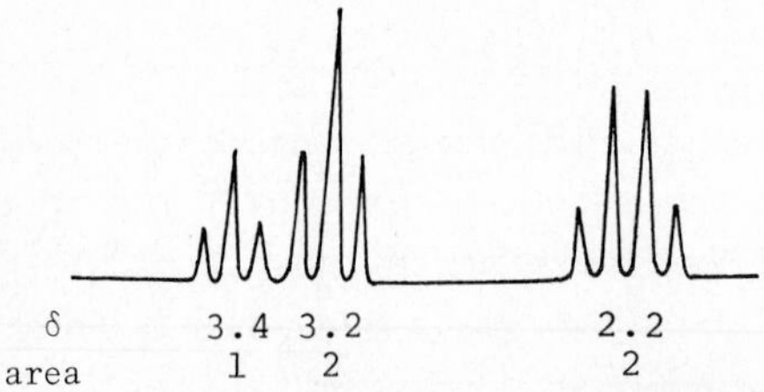

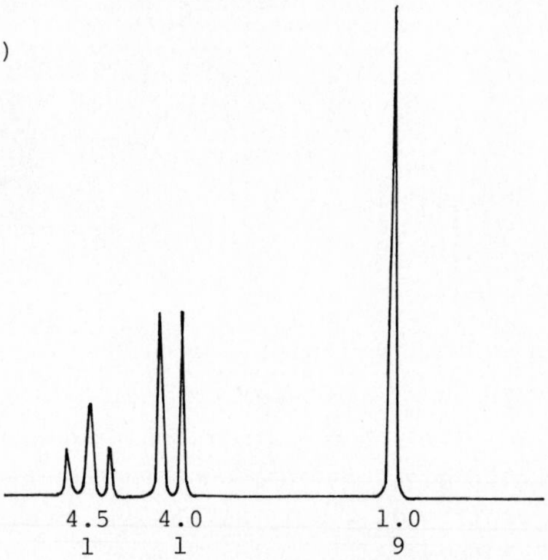

δ	3.4	3.2		2.2
area	1	2		2

δ	4.5	4.0	1.0
area	1	1	9

(c)   (d)

δ	4.5	4.0		2.0	1.6
area	1	2		1	6

δ	5.5	2.5	1.0
area	1	1	6

(e)   (f)

δ	3.3	2.0	1.3	0.9
area	2	1	4	6

	1.3	0.9
	8	12

S3. (a) δ(propane)

$$
\begin{array}{ll}
\underset{|}{CH_2I} & 15.6 \qquad -6(\alpha\text{-}I) - 10(2\gamma\text{-}Cl) \;=\; 0 \\
\underset{|}{CH_2} & 16.1 \qquad +11(\beta\text{-}I) + 22(2\beta\text{-}Cl) \;=\; 49 \\
CHCl_2 & 15.6 \qquad +62(2\alpha\text{-}Cl) - 1(\gamma\text{-}I) \;=\; 77
\end{array}
$$

(b) δ (propane)

$$
\begin{array}{ll}
\underset{|}{(CH_3)_3} & 13.2 \qquad +19(2\beta\text{-}CH_3) - 5(\gamma\text{-}Cl) \hspace{3.5cm}= 27 \\
\underset{|}{C} & 25.0 \qquad +18(2\alpha\text{-}CH_3) + 11(\beta\text{-}Cl) - 5(\gamma\text{-}Cl) \;= 49 \\
\underset{|}{CHCl} & 25.0 \qquad +31(\alpha\text{-}Cl) + 19(2\beta\text{-}CH_3) + 11(\beta\text{-}Cl) \;= 86 \\
CH_2Cl & 13.2 \qquad +31(\alpha\text{-}Cl) + 11(\beta\text{-}Cl) - 5(2\gamma\text{-}CH_3) = 62
\end{array}
$$

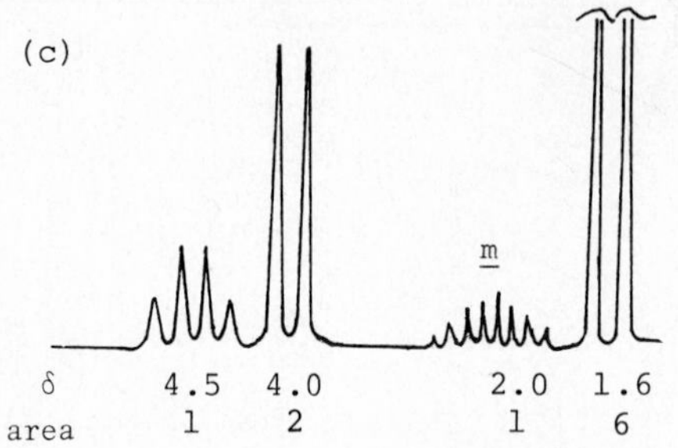

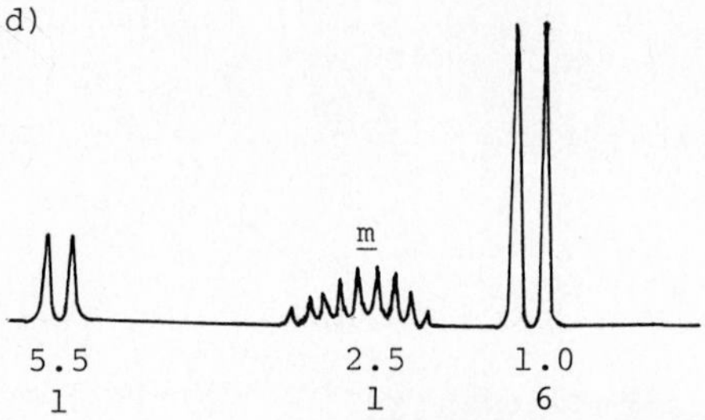

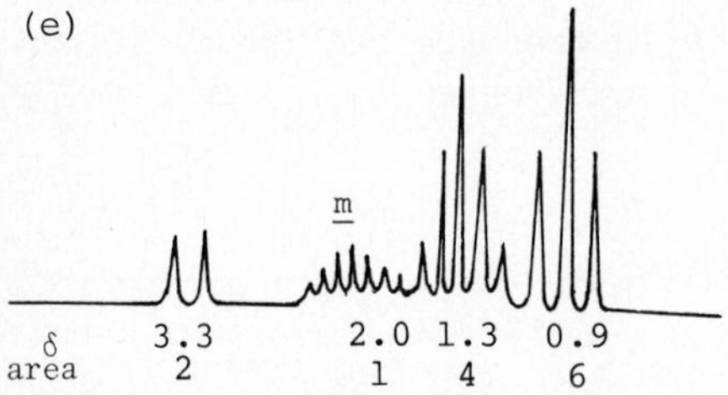

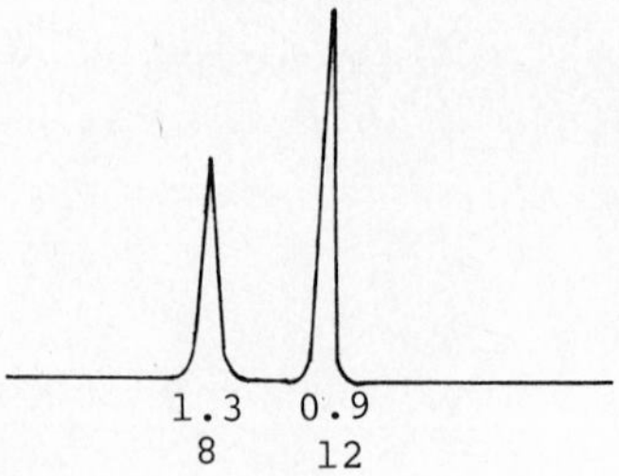

(c)

	δ (butane)		
$(CH_3)_2$	13.2	+9.5(β-CH_3) - 5(γ-Cl)	= 18
CH	25.0	+9.0(α-CH_3) + 11(β-Cl) - 5(γ-Cl)	= 40
CHCl	25.0	+31(α-Cl) + 9.5 (β-CH_3) + 11(β-Cl)	= 77
CH_2Cl	13.2	+31(α-Cl) + 11(β-Cl) - 2.5(γ-CH_3)	= 53

(d)

	δ (propane)		
CH_3 CH_3	15.6	+9.5(β-CH_3) - 10(2γ-Cl)	= 15
CH	16.1	+9.0(α-CH_3) + 22(2β-Cl)	= 47
$CHCl_2$	15.6	+9.5(β-CH_3) + 62(2α-Cl)	= 87

(e)

CH_3 CH_3	13.2	-2.5(γ-CH_3)	= 11
CH_2 CH_2	25.0	+9.5(β-CH_2-) - 2.5(γ-CH_3) - 3(γ-Br)	= 29
CH	25.0	+9.0(α-CH_2-) + 9.5(β-CH_3) + 11(β-Br)	= 55
CH_2Br	13.2	+20(α-Br) + 9.5(β-CH_2-) - 2.5(γ-CH_3)	= 40

S4. (a) 25° C: 4; -100° C: 6 (b) 25° C: 2; -100°C: 3
(c) 25° C: 7; -100° C: 7
(There is too little of the other chair conformation present at equilibrium to be seen.)
(d) 25° C and -100° C: 3 (e) 25° C: 7; -100° C: 14

S5.

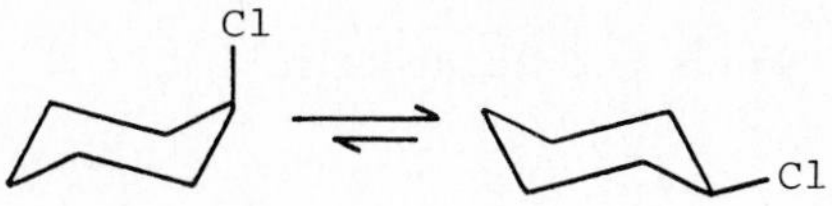

$\Delta G = -0.5$ kcal mole^{-1} = $-RT\ln K$;

$$K = \frac{[\text{equatorial-Cl}]}{[\text{axial-Cl}]} = 2.3$$

relative %: $\frac{1}{3.3} = 30\%$; $\frac{2.3}{3.3} = 70\%$; $\delta_{average} = 0.3\times4.4 + 0.7\times3.7 = 3.9$ ppm

S6. In 1,1-diphenylpropane, the C_2-C_3 bond has two favored staggered conformations which are in rapid equilibrium, thus averaging J_{ab} to the usual 7 Hz value for acyclic hydrocarbons. In the 2-methyl-substituted derivative, only one staggered conformation is favored, with a 180° dihedral angle between $H_{a'}$ and $H_{b'}$.

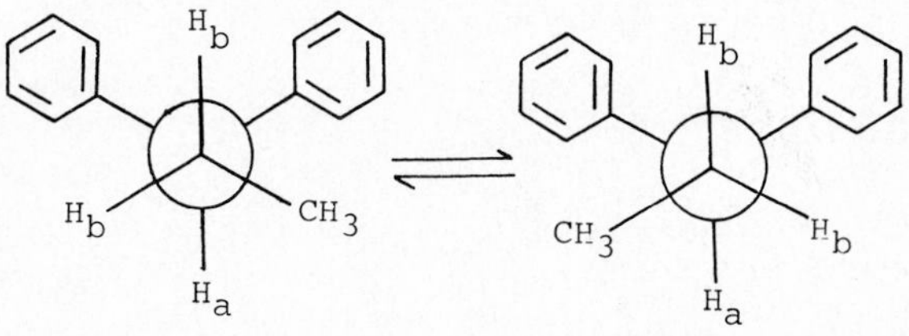

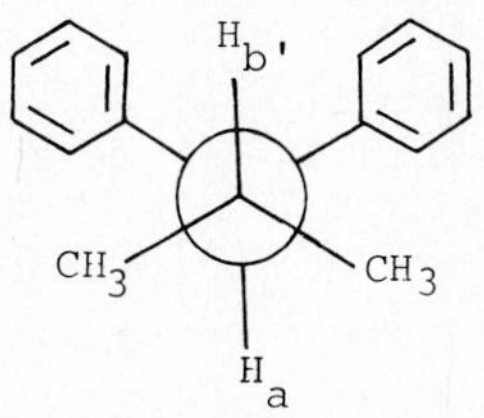

10. ALCOHOLS AND ETHERS

10.A Chapter Outline, Important Terms Introduced, and Reactions Discussed

10.1 Introduction: Structures

alcohols and ethers as functional groups — sp^3-hybridization

10.2 Nomenclature of Alcohols

alkyl alcohol system — carbinol system

phenol (phenyl) — alkanols (IUPAC system)

10.3 Physical Properties (solubility, boiling point, etc.)

dipole moment — dielectric constant

hydrogen bonding

10.4 Acidity of Alcohols: Inductive Effects

reactions with strong bases — inductive effects

effects of electron-withdrawing groups — ion pair

10.5 Nuclear Magnetic Resonance (chemical shifts in nmr and cmr)

proton exchange

10.6 Preparation of Alcohols (functional group interconversions)

1. reduction of aldehydes and ketones
2. addition of organometallics to aldehydes and ketones
3. reduction of carboxylic acids
4. reduction of esters
5. addition of organometallics to esters
6. additions to alkenes

(1–6: *deferred to subsequent chapters*)

A. Preparation from Alkyl Halides

hydrolysis — acetate displacement

$$RX \xrightarrow{H_2O} ROH + HX$$ *(for RX reactive by S_N1 mechanism)*

$$RX \xrightarrow{NaOH} ROH + NaOH$$ *(for RX reactive by S_N2 mechanism, and not E2)*

$$RX \xrightarrow{NaOOCCH_3} ROOCCH_3 \xrightarrow{NaOH} ROH$$ *(for RX reactive by S_N2 mechanism, to avoid E2)*

B. Preparation from Hydrocarbons

autoxidation:

$$RH \xrightarrow[\text{radical reaction}]{O_2} ROOH \xrightarrow{\text{reduction}} ROH + H_2O$$ *(limited utility)*

C. Special Preparations

10.7 Reactions of Alcohols

A. Acidity: Alkoxide Ions (loss of a proton)

$$ROH \rightleftharpoons RO^- + H^+$$

$$ROH + B^- \text{ (strong base)} \rightleftharpoons RO^- + HB$$

B. Alkyloxonium Salts (addition of a proton)

formation of alkyl halides with HX:

$$ROH + HX \longrightarrow RX + H_2O$$ *(via alkyloxonium ion and S_N1 (carbocation rearrangements possible) or S_N2 mechanisms)*

alkylsulfuric acids:

$$ROH + H_2SO_4 \rightleftharpoons ROSO_3H + H_2O$$

C. Formation of Organic Esters (*deferred until Chapter 20*)

D. Formation of Inorganic Esters and Conversion to Alkyl Halides

$$ROH \longrightarrow RX \qquad (with\ SOCl_2,\ PX_3,\ or\ P + I_2)$$

formation of arenesulfonates:

$$ROH + Cl-\overset{\overset{O}{\parallel}}{\underset{\underset{O}{\parallel}}{S}}-Ar \xrightarrow{pyridine} RO_3SAr$$

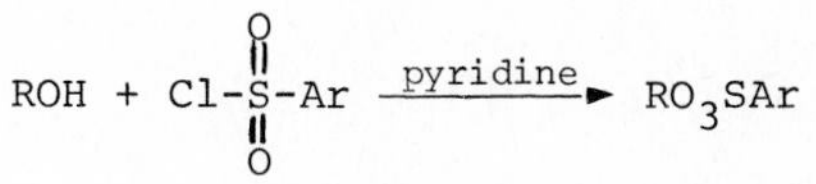

E. Oxidation of Alcohols (loss of two protons and two electrons)

chromic acid   chromate ester
[balancing oxidation/reduction reactions]
nitric acid   catalytic oxidation

$$RCH_2OH \xrightarrow{[O]} [RCHO] \dashrightarrow RCO_2H \quad (hard\ to\ stop\ at\ aldehyde)$$

$$\underset{R}{\overset{R'}{>}}CHOH \xrightarrow{[O]} \underset{R}{\overset{R'}{>}}C=O \qquad ([O] = Na_2Cr_2O_7\ /\ H_2SO_4\ or\ CuO\ /\ \Delta)$$

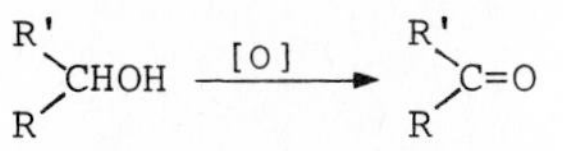

overoxidation:

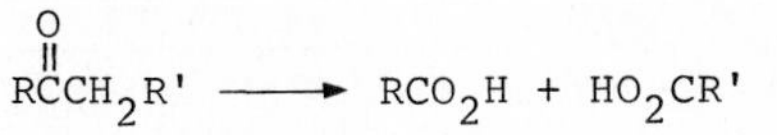

$$R\overset{\overset{O}{\parallel}}{C}CH_2R' \longrightarrow RCO_2H + HO_2CR' \quad (with\ Na_2Cr_2O_7/H_2SO_4\ or\ HNO_3/V_2O_5\\ and\ heat)$$

10.8 Nomenclature of Ethers
      alkyl$_1$ alkyl$_2$ ether   alkoxyalkane

10.9 Physical Properties of Ethers
      THF

10.10 Preparation of Ethers

A. Williamson Ether Synthesis (S$_N$2 displacement)

$$RO^- + R'X \longrightarrow ROR' \qquad (problems\ can\ arise\ from\ E2)$$

B. Reactions of Alcohols with Sulfuric Acid (dehydration)

$$2ROH \xrightarrow{H_2SO_4} ROR' + H_2O \qquad \begin{matrix}(via\ alkylsulfuric\ acid;\ poor\ for\ tertiary\\ R\ or\ for\ making\ unsymmetrical\ ethers)\end{matrix}$$

C. Alkoxymercuration of Alkenes

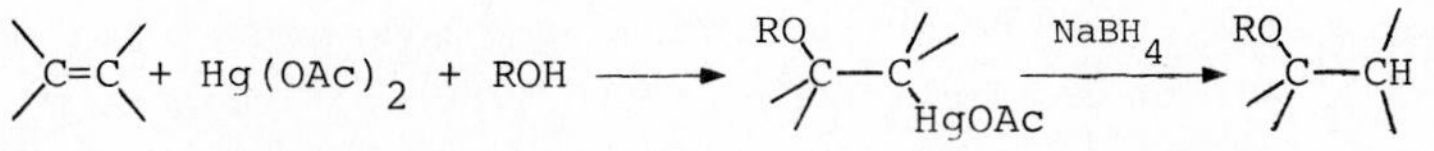

$$\underset{}{>}C=C\underset{}{<} + Hg(OAc)_2 + ROH \longrightarrow \underset{HgOAc}{\overset{RO}{>}C-C<} \xrightarrow{NaBH_4} \overset{RO}{>}C-CH< \quad \begin{matrix}(Markovnikov\\ orientation)\end{matrix}$$

10.11 Reactions of Ethers

A. Reactions with Acids
   protecting group
   acid-catalyzed substitution and elimination:

$$ROR' \overset{HBr}{\underset{H_2SO_4}{\Big<}} \begin{matrix} RBr + HOR' \longrightarrow RBr + R'Br + H_2O \\ \\ Alkenes \end{matrix}$$

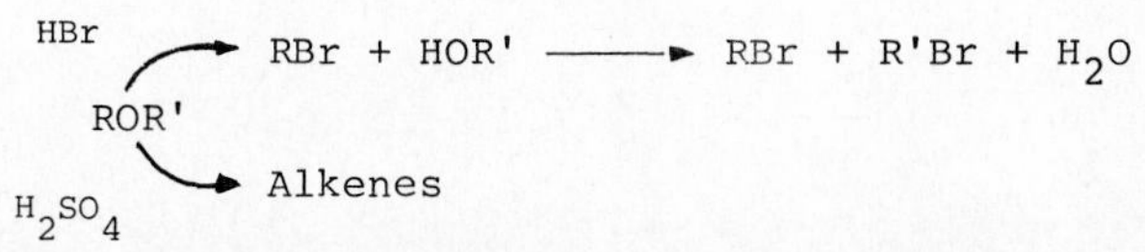

B. Oxidation

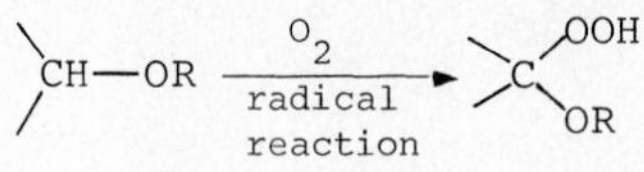

("peroxides")

10.12 Cyclic Ethers

A. Epoxides: Oxiranes

from intramolecular displacement:

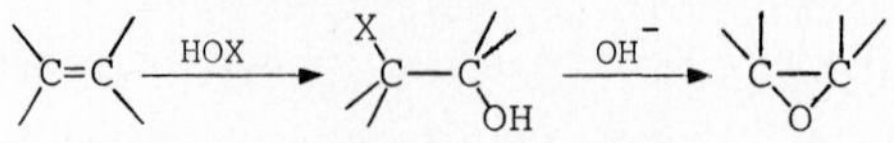

from peracid epoxidation:

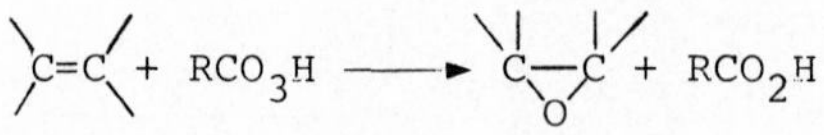

hydrolysis:

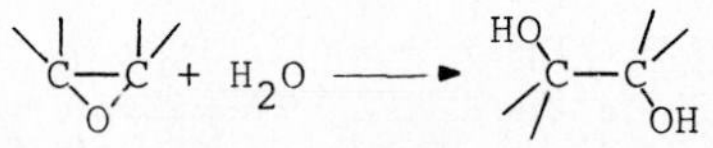

(with acid (S_N1-like) or base (S_N2-like))

B. Higher Cyclic Ethers

oxetanes THF crown ethers

via intramolecular Williamson ether synthesis:

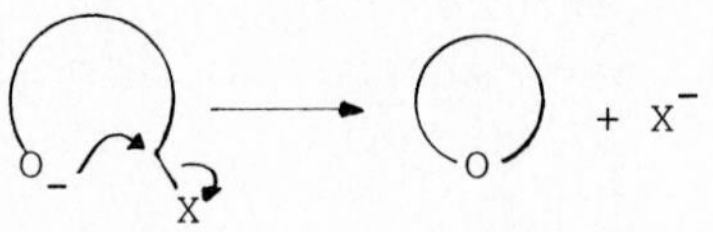

10.B Important Concepts, Hints, and Reactions Discussed

Chapter 10 outlines the chemistry of alcohols and ethers in the way that other functional groups will be discussed throughout the text. First, the nomenclature, physical properties, and spectroscopic properties of molecules which contain the hydroxy group or the ether linkage are presented. Although these topics were introduced in preceeding chapters, they must be expanded upon to show how they apply to each additional functional group.

The bulk of the chapter is devoted to the functional group interconversions in which alcohols and ethers take part. In other words, the ways you can make alcohols and ethers (preparation) and what you can make from them (reactions) are presented. As the course progresses, you will realize that a lot of organic reactions are being discussed (along with a lot of mechanisms). If you learn each reaction by itself, you will soon be overwhelmed by details, at the expense of understanding. You will always need to learn specific facts, but you should try to get an overview and an understanding of the relationship between these facts. As an example, consider the following reactions:

$$(CH_3)_2CHCH_2CH_2OH \xrightarrow{PBr_3} (CH_3)_2CHCH_2CH_2Br$$

$$(CH_3)_2CHCH_2CH_2Br \xrightarrow{NaOH} (CH_3)_2CHCH_2CH_2OH$$

If you approach organic chemistry as a collection of unrelated reactions, you will learn the first reaction twice--as a way to prepare alkyl bromides and as a reaction of alcohols. Similarly, you will learn the second reaction twice --as a way to prepare alcohols and as a reaction of alkyl bromides. On the other hand, if you try to get an overview, you will think of the reactions above as one interconversion of functional groups--alcohol ⇄ alkyl halide. At this point you will have learned the most important aspect of these two reactions--that they exist. Your thinking of functional group interconversions should grow to resemble the diagram below, in a sense, with the connecting arrows representing ways to interconvert the indicated functional groups. With this as a framework, organizing the specific details in your mind becomes easier. At first you will know just a few reactions which interconvert alcohols and alkyl halides; as your knowledge grows you will learn more of them, their limitations and exceptions and side reactions, and so on.

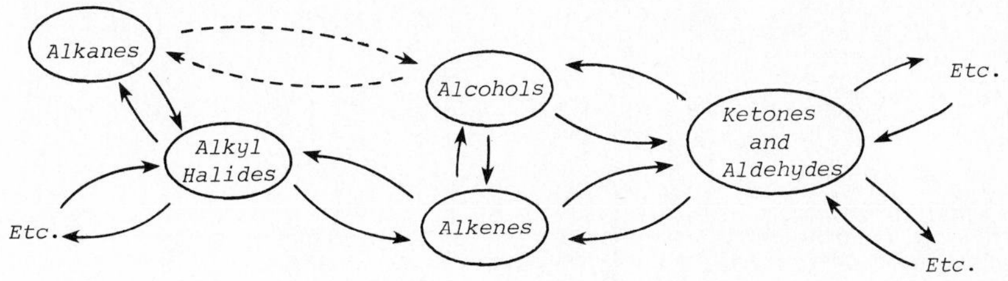

Although an understanding of the framework outlined above is a starting point, you must still know the specific reactions involved. After all, a singer has to know some songs. Think of the functional group interconversions as music and the specific reactions as your repertoire. In this regard, Appendix II (Study Guide) is useful. It is a list of reactions, organized on the basis of functional group preparations and carbon-carbon bond forming reactions. You can try to memorize it if you want to, but it would be wiser to learn a little at a time, referring to it frequently to refresh your memory, using it in the framework suggested above.

10.C Answers to Exercises

(10.2) 2,2,4,6-tetramethyl-4-heptanol 2,3,3-trimethyl-2-butanol

(10.4) (a) $CH_3CHClCH_2OH$ is more acidic because the electronegative substituent is closer to the hydroxy group.

(b) $CH_3OCH_2CH_2OH$ is more acidic because CH_3O is more electron-withdrawing than ethyl.

(c) The dichloro compound is more acidic because it has more electron withdrawing groups than the monochloro compound.

(d) *cis*-3-Chlorocyclohexanol is more acidic than *trans*-4-chlorocyclohexanol for the same reason given in (a) above.

(10.5) (a)

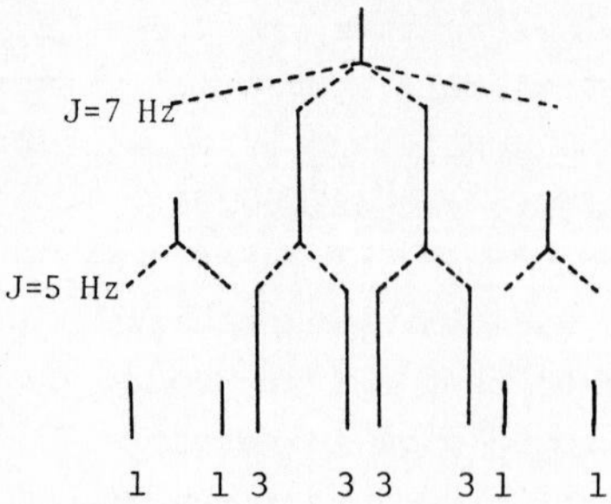

If both J's the same:

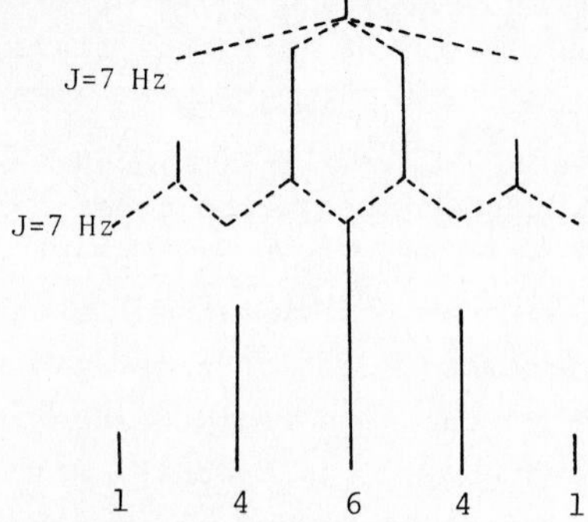

(b) assume CH—CH_3 J=7; CH—OH J=5 for

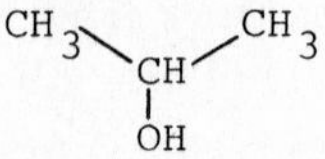

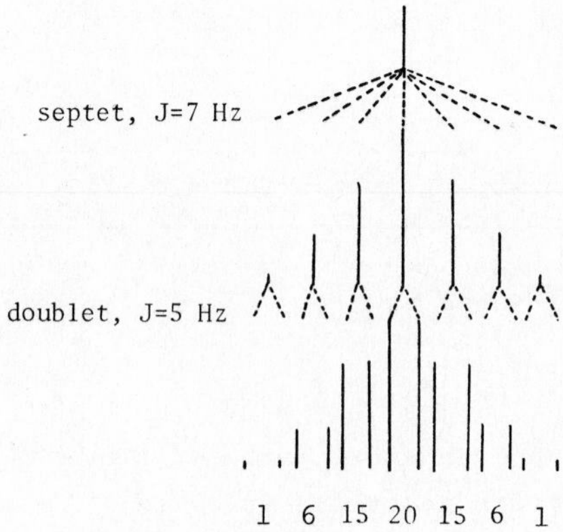

(10.7.B) (a)

$(CH_3)_2CHCH_2CH_2CH_2OH \xrightarrow{SOCl_2} (CH_3)_2CHCH_2CH_2CH_2Cl$ *(other reagents are possible)*

(b)

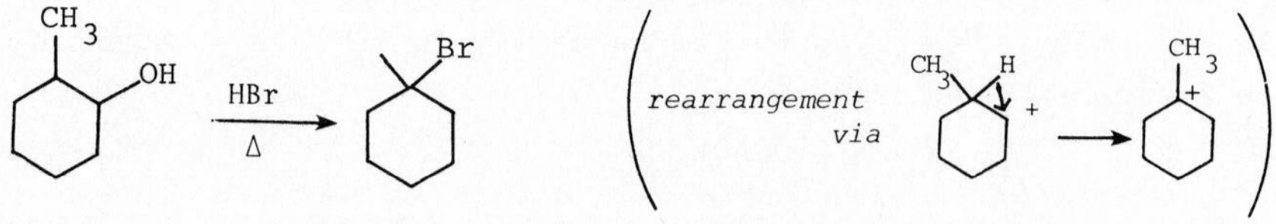

(c)

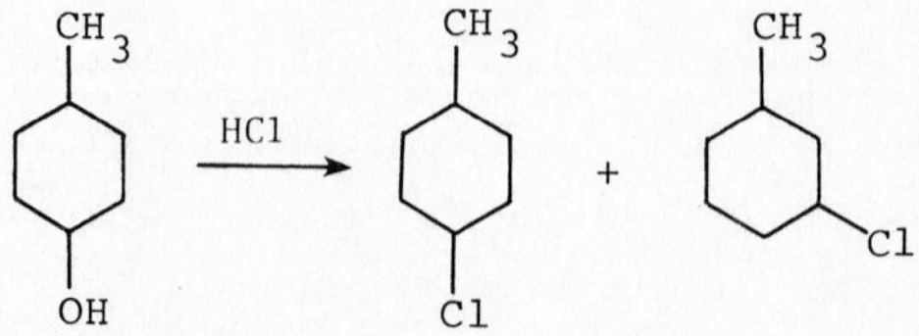

(d)

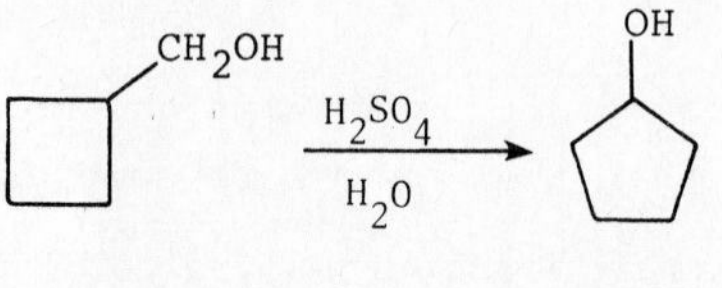

(10.7.D)

(a) $(CH_3CH_2)_3COH \xrightarrow[\text{pyridine}]{SOCl_2} (CH_3CH_2)_3CCl$

(b) $ROH \xrightarrow[\text{pyridine}]{SOCl_2} RCl$ (c) and (d) $ROH \xrightarrow{PBr_3} RBr$

For clean inversion in the conversion of (R)-2-octanol to (S)-2-chlorooctane, $SOCl_2$ in pyridine is the best reagent. Another excellent method is the two-step sequence alcohol → sulfonate ester → halide. One problem to be avoided in forming the halide with clean inversion is the tendency for excess halide ion to racemize an alkyl halide (see section 8.6 in text); this is minimized in the case of the methods cited above because of the high reactivity of the chlorosulfite or sulfonate ester intermediates: they undergo reaction to give (S)-2-chlorooctane much faster than the (S)-2-chlorooctane is racemized by chloride ion.

(10.7.E) (a) $3\ (CH_3CH_2CH_2)_2CHOH + 2\ CrO_3 + 6\ H^+ =$

$3\ (CH_3CH_2CH_2)_2C{=}O + 2\ Cr^{+3} + 6\ H_2O$

(b) $3\ C_4H_9CH_2OH + 4\ H_2CrO_4 + 12\ H^+ = 3\ C_4H_9COOH + 4\ Cr^{+3} + 13\ H_2O$

(c) $CH_3OH + CuO = CH_2O + Cu + H_2O$

(10.10.A)

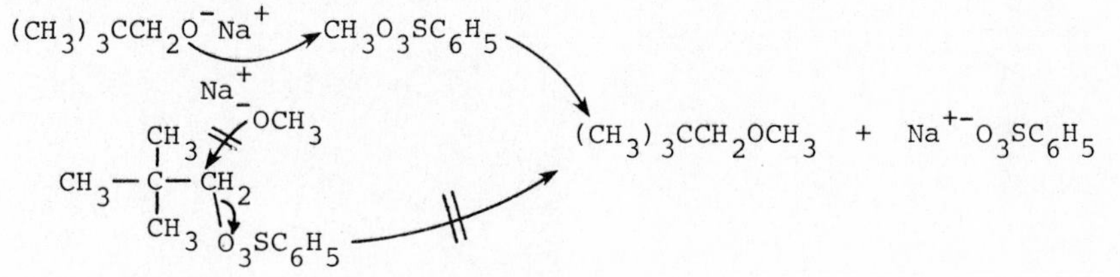

Steric hindrance (β-branching) blocks backside attack of methoxide ion on neopentyl benzenesulfonate.

(10.10.B)

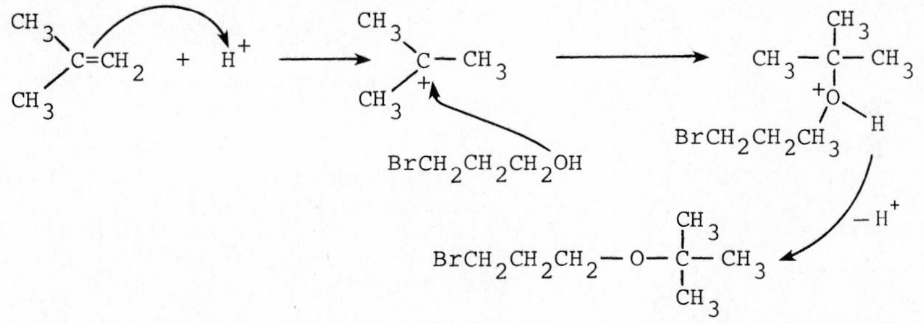

(10.12.A)

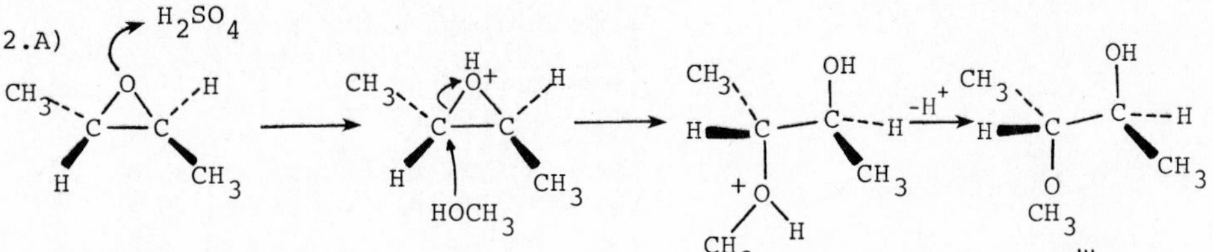

(attack at the other carbon leads to the same compound)

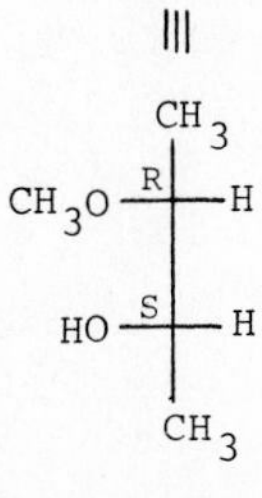

The enantiomeric epoxide leads to the (2R,3S) product.

(2S,3R)-3-methoxy-2-butanol

10.D   Answers and Explanations to Problems

1. (a) $CH_3OCH_2CH(CH_3)_2$

    (b) $(CH_3)_3CCH_2OH$

    (c) $CH_3CH_2CH_2\overset{\overset{\displaystyle CH_3CH_2O}{|}}{\underset{\underset{\displaystyle CH_3}{}}{C}}HCHCH_3$

    (d) $(CH_3)_2CHCH_2\overset{\overset{\displaystyle OH}{|}}{C}HCH_3$

    (e) $(C_6H_5)_3COH$

    (f)

    (g)

    (h) $CH_3CH_2\overset{\overset{\displaystyle OCH_3}{|}}{C}H-\overset{}{C}HCH_2CH_2CH_3$
         $\qquad\qquad\quad \underset{\underset{\displaystyle CH_3}{|}}{\overset{\overset{\displaystyle CH_3-C-CH_3}{}}{}}$

    (i) $CH_3\overset{\overset{\displaystyle CH_3O}{|}}{C}H\overset{\overset{\displaystyle OCH_3}{|}}{C}HCH_3$

    (j) $C_6H_5-\overset{\overset{\displaystyle OH}{|}}{C}HCH_2C_6H_5$

    (k)

    (l)

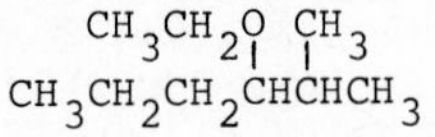

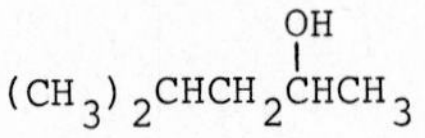

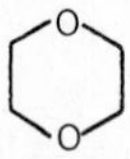

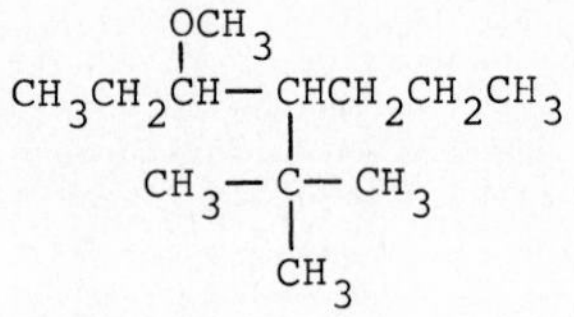

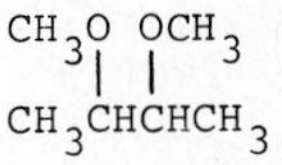

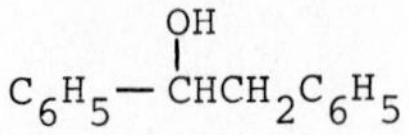

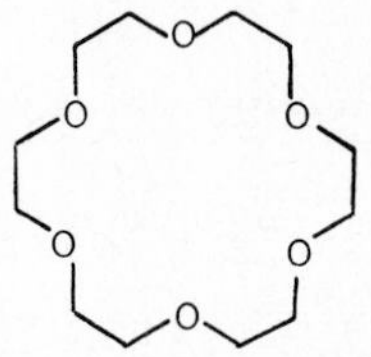

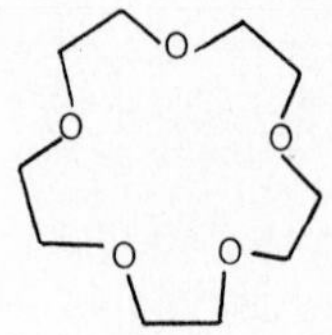

2. (a) 2-methyl-2-propyloxirane

    (b) 2,3,5-trimethyl-3-hexanol

    (c) 2-ethyl-3-methyl-1-pentanol

    (d) 3-methoxy-2-pentanol

    (e) 3,3-diphenylcyclopentanol

    (f) 2,2,4,4-tetramethyloxetane

    (g) 6-chloro-2-ethyl-4-methyl-
         1-hexanol

3. 1) $CH_3OCH_2CH_2CH_2CH_3$ $\qquad$ { 1-methoxybutane
                                      n-butyl methyl ether

    2) $CH_3\overset{\overset{\displaystyle OCH_3}{|}}{C}HCH_2CH_3$ * $\qquad$ { 2-methoxybutane*
                                                                    sec-butyl methyl ether

    3) $CH_3CH_2OCH_2CH_2CH_3$ $\qquad$ { 1-ethoxypropane
                                        ethyl propyl ether

    4) $CH_3CH_2OCH(CH_3)_2$ $\qquad$ { 2-ethoxypropane
                                      ethyl isopropyl ether

    5) $CH_3OCH_2CH(CH_3)_2$ $\qquad$ { 1-methoxy-2-methylpropane
                                      isobutyl methyl ether

    6) $CH_3OC(CH_3)_3$ $\qquad$ { 2-methoxy-2-methylpropane
                                  t-butyl methyl ether

   * chiral; this is the only isomer capable of optical activity

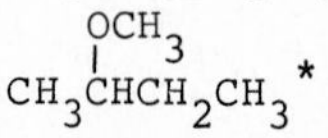

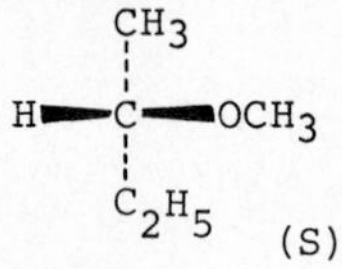

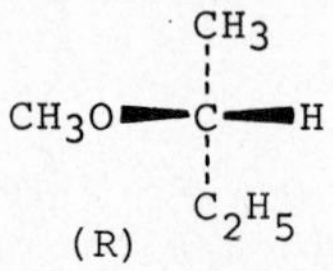

NMR Spectra:

1) $CH_3OCH_2CH_2CH_2CH_3$
a b c d e

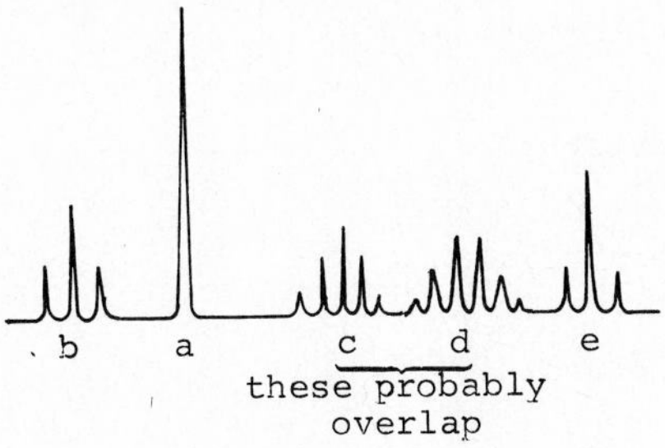

4) $CH_3CH_2OCH(CH_3)_2$
a b c d

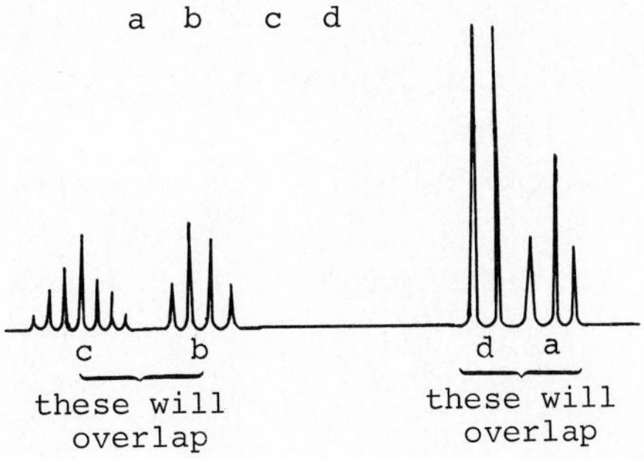

2) $CH_3CH(OCH_3)CH_2CH_3$ (OCH$_3$ = a)
b c d e

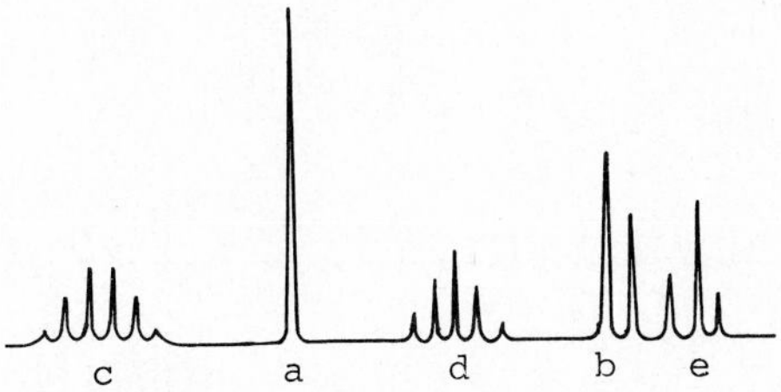

5) $CH_3OCH_2CH(CH_3)_2$
d c b a

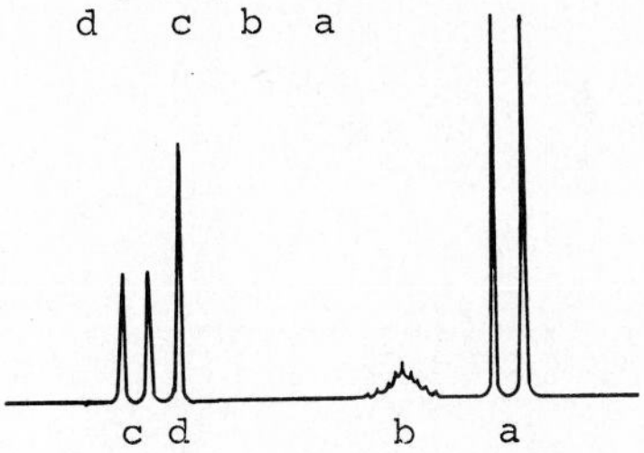

3) $CH_3CH_2OCH_2CH_2CH_3$
a b c d e

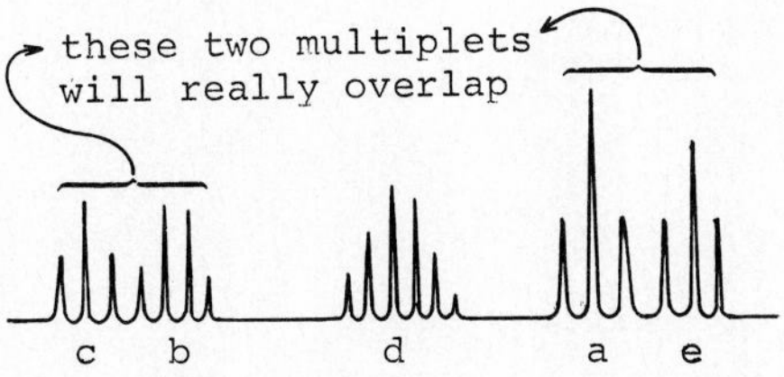

6) $CH_3OC(CH_3)_3$
b a

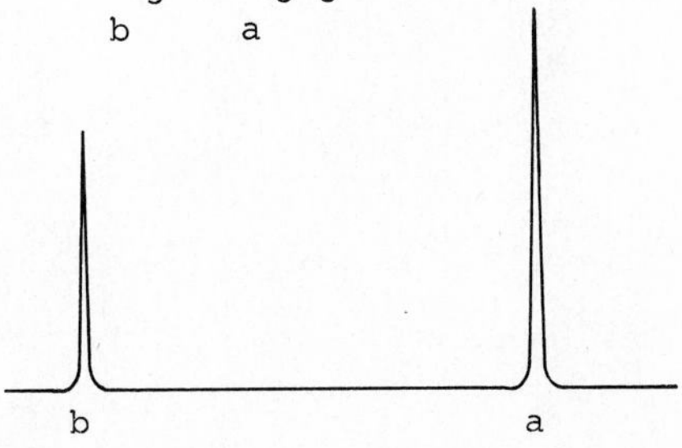

4\. 1) $CH_3CH_2CH_2CH_2CH_2CH_2OH$ 1-hexanol 1°

2) $CH_3CH_2CH_2CH_2CH(OH)CH_3$ 2-hexanol 2°

3) $CH_3CH_2CH_2CH(OH)CH_2CH_3$ 3-hexanol 2°

4) $(CH_3)_2CHCH_2CH_2CH_2OH$ 4-methyl-1-pentanol 1°

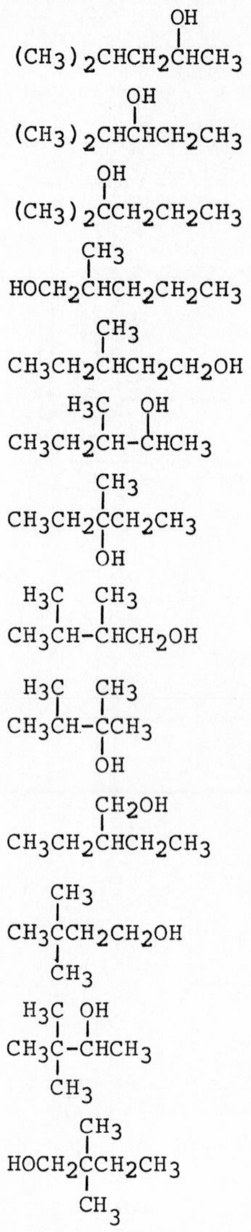

5) $(CH_3)_2CHCH_2\overset{\overset{\displaystyle OH}{|}}{C}HCH_3$ 4-methyl-2-pentanol   2°

6) $(CH_3)_2CH\overset{\overset{\displaystyle OH}{|}}{C}HCH_2CH_3$ 2-methyl-3-pentanol   2°

7) $(CH_3)_2\overset{\overset{\displaystyle OH}{|}}{C}CH_2CH_2CH_3$ 2-methyl-2-pentanol   3°

8) $HOCH_2\overset{\overset{\displaystyle CH_3}{|}}{C}HCH_2CH_2CH_3$ 2-methyl-1-pentanol   1°

9) $CH_3CH_2\overset{\overset{\displaystyle CH_3}{|}}{C}HCH_2CH_2OH$ 3-methyl-1-pentanol   1°

10) $CH_3CH_2\overset{\overset{\displaystyle H_3C}{|}}{C}H-\overset{\overset{\displaystyle OH}{|}}{C}HCH_3$ 3-methyl-2-pentanol   2°

11) $CH_3CH_2\overset{\overset{\displaystyle CH_3}{|}}{\underset{\underset{\displaystyle OH}{|}}{C}}CH_2CH_3$ 3-methyl-3-pentanol   3°

12) $CH_3\overset{\overset{\displaystyle H_3C}{|}}{C}H-\overset{\overset{\displaystyle CH_3}{|}}{C}HCH_2OH$ 2,3-dimethyl-1-butanol   1°

13) $CH_3\overset{\overset{\displaystyle H_3C}{|}}{C}H-\overset{\overset{\displaystyle CH_3}{|}}{\underset{\underset{\displaystyle OH}{|}}{C}}CH_3$ 2,3-dimethyl-2-butanol   3°

14) $CH_3CH_2\overset{\overset{\displaystyle CH_2OH}{|}}{C}HCH_2CH_3$ 2-ethyl-1-butanol   1°

15) $CH_3\overset{\overset{\displaystyle CH_3}{|}}{\underset{\underset{\displaystyle CH_3}{|}}{C}}CH_2CH_2OH$ 3,3-dimethyl-1-butanol   1°

16) $CH_3\overset{\overset{\displaystyle H_3C}{|}}{\underset{\underset{\displaystyle CH_3}{|}}{C}}-\overset{\overset{\displaystyle OH}{|}}{C}HCH_3$ 3,3-dimethyl-2-butanol   2°

17) $HOCH_2\overset{\overset{\displaystyle CH_3}{|}}{\underset{\underset{\displaystyle CH_3}{|}}{C}}CH_2CH_3$ 2,2-dimethyl-1-butanol   1°

5. $CH_3CH_2\overset{\overset{\displaystyle OH}{|}}{C}DCH_3$ + H$^+$ $\rightleftharpoons$ $CH_3CH_2\overset{\overset{\displaystyle +OH_2}{|}}{C}DCH_3$ $\rightleftharpoons$ $CH_3CH_2\overset{+}{C}DCH_3$ $\xrightarrow{Br^-}$ $CH_3CH_2\overset{\overset{\displaystyle Br}{|}}{C}DCH_3$

                 Equilibration of secondary $CH_3\overset{+}{C}HCHDCH_3$ $\xrightarrow{Br^-}$ $CH_3\overset{\overset{\displaystyle Br}{|}}{C}HCHDCH_3$
                 carbocations has occurred.

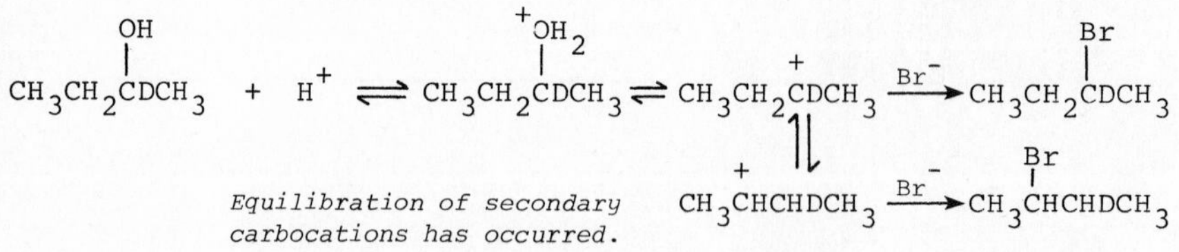

6.   a)   The name is incorrect since it mixes common and systematic nomenclature.

       <u>Correct</u>:   IUPAC: 2-methoxy-2-methylbutane

                  Common: methyl <u>t</u>-pentyl ether, or <u>t</u>-amyl methyl ether

     Amyl is an older common name for pentyl; e.g., <u>n</u>-amyl, isoamyl, <u>tert</u>-amyl. In modern usage, amyl is being replaced increasingly by pentyl, but the name is common in the older literature.

7. (a) $CH_3CH_2CH_2OCH_2CH_2CH_3$ *sulfuric acid-catalyzed ether formation*

(b) $CH_3CH_2CH_2OH + CH_2{=}C(CH_3)_2$ *E2 elimination (remember: 3° halide and strong base)*

(c) $(CH_3)_3COCH_2CH_2CH_3$ *Williamson ether synthesis*

(d) CH_3CH_2CHO *Volatile aldehydes can be isolated under these conditions*

(e)

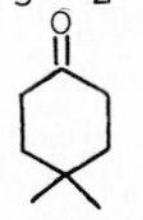

Standard chromic acid ("Jones") oxidation

(f) $CH_3CH_2CH_2I + Na^{+}\,{}^{-}O_3S{-}C_6H_4{-}CH_3$ *S_N2 displacement, sodium tosylate precipitates*

(g) $CH_3CH_2CH_2Br$ *Acid-catalyzed formation of alkyl halide.*

(h) $(CH_3CH_2)_2CBrCH_2CH_2CH_3$ *Reaction proceeds with rearrangement to give the tertiary carbocation as an intermediate*

(i) (R)$-CH_3CHClCH_2CH_3$ *Alkyl chloride formation with inversion of configuration*

(j) $(CH_3CH_2)_2CHO^-K^+ + H_2\uparrow$ *Potassium hydride is a strong base*

(k) 2 $CH_3CH_2CH_2I$

(l) $CH_3CH_2CHBrCH_2CH_3 + CH_3Br$

} *Acid-catalyzed formation of alkyl halide*

8.

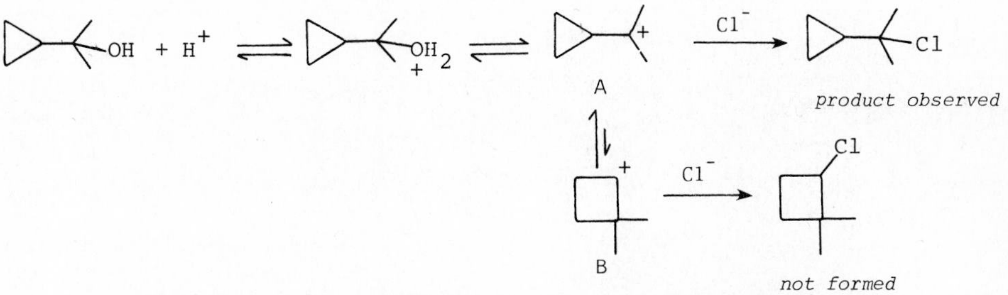

The possible rearrangement, A→B, is not observed. The ring strain of a cyclobutane ring is as high as that of a cyclopropane ring (see Table 5.5 in text), so that the isomerization A→B would involve conversion of a tertiary to a secondary carbocation with no relief of ring strain.

9. (a) and (b): Primary alcohols are best converted to primary alkyl chlorides with $SOCl_2$ in pyridine.

(c) In this case, rearrangement is desired, so a reagent favoring carbocation intermediates should be chosen: $ZnCl_2$ + HCl.

(d) The best way to make alkyl iodides from alcohols is to carry out a displacement on a sulfonate ester intermediate:

$$CH_3CH_2CH_2CH(OH)CH_3 + C_6H_5SO_2Cl \longrightarrow CH_3CH_2CH_2CH(OSO_2C_6H_5)CH_3 \xrightarrow[\text{acetone}]{\text{NaI}} CH_3CH_2CH_2CH(I)CH_3$$

(e) Rearrangement must be avoided: PBr_3 at 0° C.

(f) Tertiary chloride from the alcohol, with no fear of rearrangement: HCl at 0° C.

10. (a) Rearrangement is desired: $ZnCl_2$ + HCl

(b) $(CH_3)_2CH\overset{OH}{\overset{|}{C}}HCH_3 + PBr_3 \xrightarrow{\text{low temp.}} (CH_3)_2CH\overset{Br}{\overset{|}{C}}HCH_3$

(c) $(CH_3CH_2)_3C\text{—}OH + H_2SO_4 \xrightarrow{\Delta} CH_3CH{=}C(CH_2CH_3)_2$

A drop of sulfuric acid or a small amount of sulfonic acid is sufficient. Alternatively, the alcohol can be passed over hot Al_2O_3.

(d) $CH_3CH_2CH_2CH_2OH + SOCl_2 \longrightarrow CH_3CH_2CH_2CH_2Cl \xrightarrow{CN^-} CH_3CH_2CH_2CH_2CN$

(e) $(CH_3)_3CCl + H_2O \xrightarrow{\Delta} (CH_3)_3COH \xrightarrow{K} (CH_3)_3CO^-K^+ \xrightarrow[\Delta]{CH_3I} (CH_3)_3C\text{—}OCH_3$

Note that the reverse procedure of treating $(CH_3)_3CCl$ with $CH_3O^-K^+$ will not work, because the halide is tertiary and E2 elimination would dominate.

(f) $(CH_3)_2CHCH_2OH \xrightarrow[H_2SO_4,\ \Delta]{HBr} (CH_3)_2CHCH_2Br \xrightarrow{CH_3S^-Na^+} (CH_3)_2CHCH_2SCH_3$

11. (a) 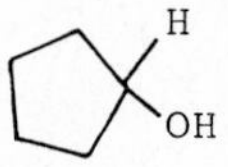 $+ 8\ HNO_3 = HOOC(CH_2)_3COOH + 5\ H_2O + 8\ NO_2$

(b) $ClCH_2CH_2CH_2OH + 4HNO_3 = ClCH_2CH_2COOH + 4\ NO_2 + 3\ H_2O$

(c) $3\ CH_3CHOHCH_3 + 2\ CrO_3 + 6\ H^+ = 3\ CH_3COCH_3 + Cr^{+3} + 6\ H_2O$

(d) 2 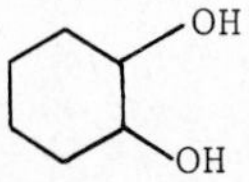 $+ K_2Cr_2O_7 + 8\ H^+ = HO_2C(CH_2)_4CO_2H + 2\ K^+ + 2\ Cr^{+3} + 5\ H_2O$

12. (a) The rearrangement requires a carbocation intermediate. The formation of the high-energy primary carbocation cannot compete with S_N2 displacement on the oxonium ion.

$$(CH_3)_2CHCH_2OH + H^+ \rightleftharpoons (CH_3)_2CHCH_2\overset{+}{O}H_2 \xrightarrow{Br^-} (CH_3)_2CHCH_2Br + OH_2$$

$$(CH_3)_2CHCH_2\overset{+}{O}H_2 \not\longrightarrow (CH_3)_2CHCH_2^{\ +} \longrightarrow (CH_3)_3C^+$$

The secondary carbocation forms far more readily:

$$(CH_3)_2CH\overset{+OH_2}{\overset{|}{C}}HCH_3 \longrightarrow (CH_3)_2CH\overset{+}{C}HCH_3 \longrightarrow (CH_3)_2\overset{+}{C}CH_2CH_3 \xrightarrow{Br^-} (CH_3)_2CBrCH_2CH_3$$

(b) Mixtures of ethers generally result:

$$R'CH_2OH + R''CH_2OH \xrightarrow{H_2SO_4} R'CH_2OCH_2R' + R'CH_2OCH_2R'' + R''CH_2OCH_2R''$$

Either alcohol forms $RCH_2\overset{+}{O}H_2$ and can be displaced by either alcohol. However, t-butyl alcohol readily forms a carbocation,

$$(CH_3)_3COH + H^+ \rightleftharpoons (CH_3)_3C\overset{+}{O}H_2 \longrightarrow (CH_3)_3C^+$$

which reacts more readily with methanol than with the bulkier t-butyl alcohol:

$$(CH_3)_3C^+ + CH_3OH \longrightarrow (CH_3)_3C\overset{+}{O}(H)CH_3 \rightleftharpoons (CH_3)_3COCH_3 + H^+$$

(c) $C_2H_5OC_2H_5 + HI \rightleftharpoons CH_3CH_2\overset{+}{O}(H)CH_2CH_3 + I^-$

$$I^- \; CH_3CH_2\text{—}\overset{+}{O}(H)CH_2CH_3 \longrightarrow CH_3CH_2I + CH_3CH_2OH$$

$$CH_3CH_2OH + HI \rightleftharpoons CH_3CH_2\overset{+}{O}H_2 + I^- \longrightarrow CH_3CH_2I + OH_2$$

13. Spectrum on pg. 273:

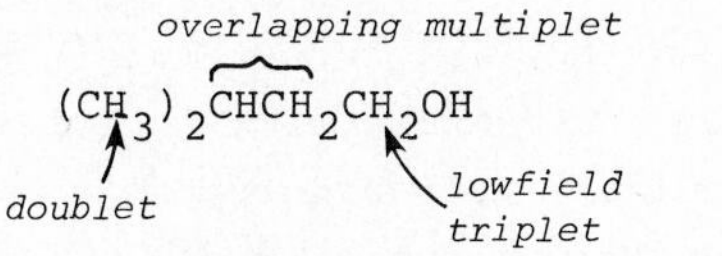

Isopentyl (isoamyl) alcohol. For a discussion of "amyl", see question #6(a).

Spectrum on pg. 274:

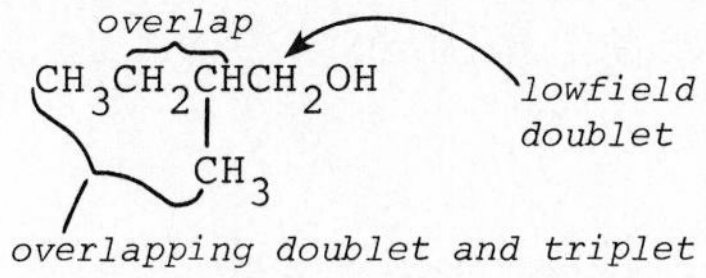

This structure has an asymmetric carbon $(CH_3CH_2\overset{*}{C}H(CH_3)CH_2OH)$ and is chiral. The fermentation product is optically active.

14. $CH_3CH_2OH + Br^- \rightleftharpoons CH_3CH_2Br + OH^-$

ΔH_f°: -56.2   -50.8   -15.2   -32.9 $\Delta H^\circ = +58.9$ kcal mole^{-1}

$CH_3CH_2OH + HBr \rightleftharpoons CH_3CH_2Br + H_2O$

ΔH_f°: -56.2   -8.7   -15.2   -57.8 $\Delta H^\circ = -8.1$ kcal mole^{-1}

Note that the first reaction is highly endothermic; OH^- is a much stronger base than Br^- in the gas phase just as it is in solution. This reaction is also endothermic in solution and is not observed. The second reaction is almost thermoneutral in contrast, and is actually slightly exothermic. Since the entropy change is close to zero (two molecules give two molecules), the equilibrium lies on the right.

The difference between the two reactions can be seen from the following comparison: the O-H bond strength is much greater than the Br-H bond strength, and this difference is greater than for the C-O and C-Br bond strengths (see Appendix III).

15.

$$CH_3CH_2OCH_2CH_2CH_2CH_3 + HBr \rightleftharpoons CH_3CH_2\overset{H}{\overset{|+}{O}}CH_2CH_2CH_2CH_3 + Br^-$$

$$CH_3CH_2\overset{H}{\overset{|+}{O}}CH_2CH_2CH_2CH_3 + Br^- \rightleftharpoons CH_3CH_2Br + HOCH_2CH_2CH_2CH_3$$

$$\rightleftharpoons CH_3CH_2OH + BrCH_2CH_2CH_2CH_3$$

$$(CH_3)_3COCH_2CH_3 + HBr \rightleftharpoons (CH_3)_3C\overset{H}{\underset{+}{O}}CH_2CH_3 + Br^-$$

$$(CH_3)_3C\overset{H}{\underset{+}{O}}CH_2CH_3 \rightleftharpoons (CH_3)_3C^+ + HOCH_2CH_3$$

$$(CH_3)_3C^+ + Br^- \rightleftharpoons (CH_3)_3CBr$$

In the first case, product is formed by S_N2 attack on the protonated ether. Attack can occur at either primary carbon at comparable rates, so both ethyl and n-butyl bromide are produced. In the second case, the ether cleaves by an S_N1 process to give the t-butyl cation, and thus t-butyl bromide is the main product. The tertiary carbocation is sufficiently stable that it is produced readily by concentrated acid, even in the cold. Recall that tertiary alcohols rapidly give the halide with cold concentrated HCl or HBr.

16.

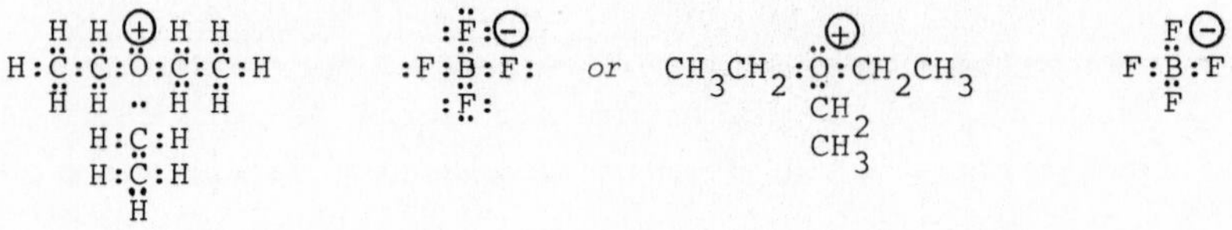

Mechanism of Ethylation

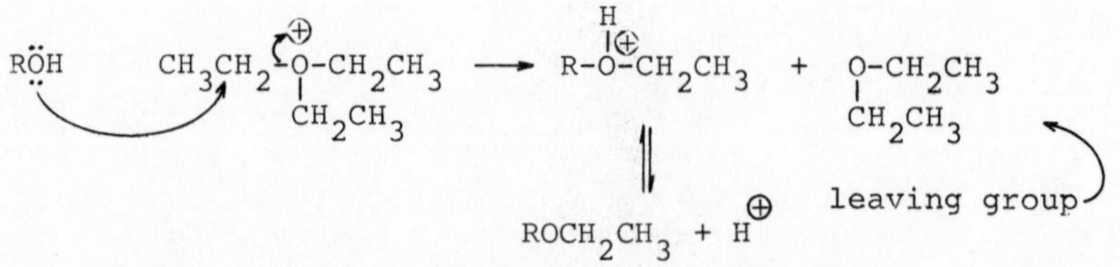

It is a very reactive compound because $(CH_3CH_2)_2O$ is a good leaving group, similar to H_2O.

(the pk_a of $(CH_3CH_2)_2\overset{+}{O}H$ is about -3.6)

17.

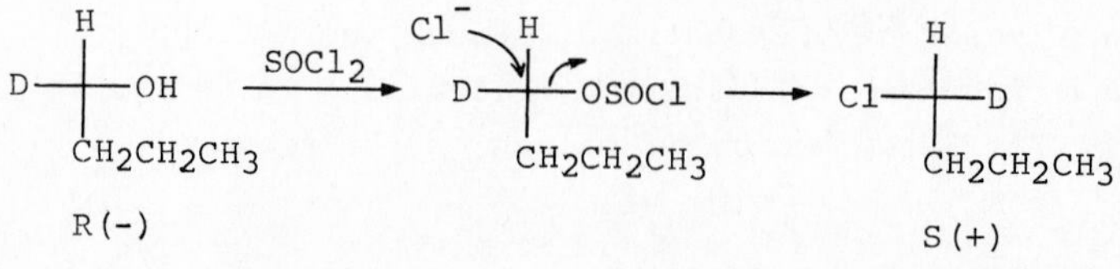

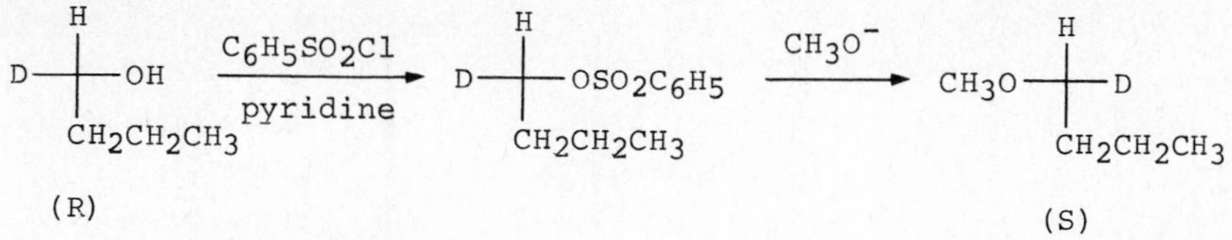

18.

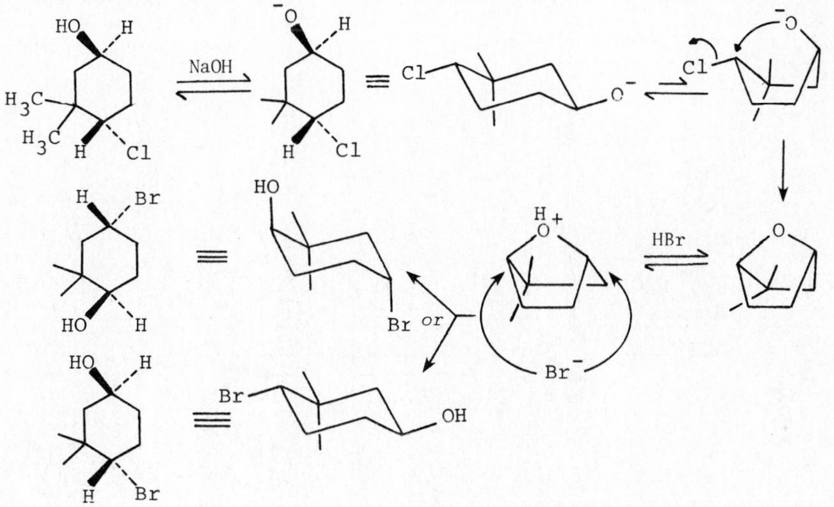

The clues here are the *trans* relationship of the OH and Br in both isomers of the product, and the fact that the Br can end up at either end of the molecule.

19. (a) All resonances are triplets in the proton-coupled spectrum, which means that all carbons are CH_2's:

$$\underset{}{Cl}-\underset{45.4}{CH_2}-\underset{32.9}{CH_2}-\underset{23.7}{CH_2}-\underset{32.1}{CH_2}-\underset{62.2}{CH_2}-OH$$

Carbons 2 and 4 are assigned from the fact that Cl has a larger β-effect (+11 ppm) than OH (+9.5 ppm).

(b) Four resonances for eight carbons means a symmetrical structure:

$$\underset{14.6}{CH_3}-\underset{20.3}{CH_2}-\underset{33.1}{CH_2}-\underset{71.2}{CH_2}-O-\underset{71.2}{CH_2}-\underset{33.1}{CH_2}-\underset{20.3}{CH_2}-\underset{14.6}{CH_3}$$

(c) 75.1 (d) = H—C—O; 35.3(s) = -C-; 25.8(q) = CH_3-, and 18.2(q)= CH_3-
The hint that the δ 25.8 resonance is much more intense than the others gives a clue as to where the other two carbons are: there are 3 equivalent CH_3's which resonate at δ 25.8. This suggests a *t*-butyl group:

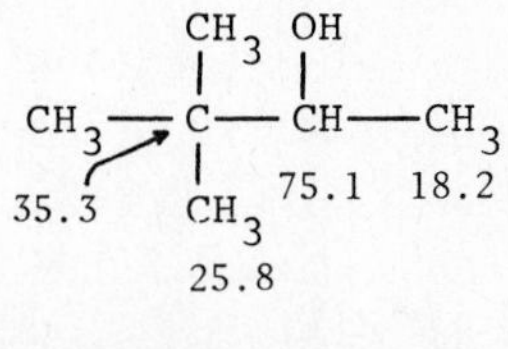

(d)

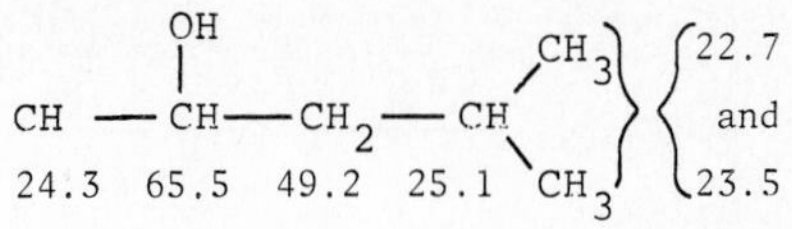

These methyls are non-equivalent because of the chiral center present in the molecule; they are said to be "diastereotopic".

(e) Two resonances suggests a very symmetrical molecule; the formula ($C_nH_{2n}O_m$) indicates that there is one ring or one double bond.

71.1(d) = >CH—O; 33.9(t) = $-CH_2-$:

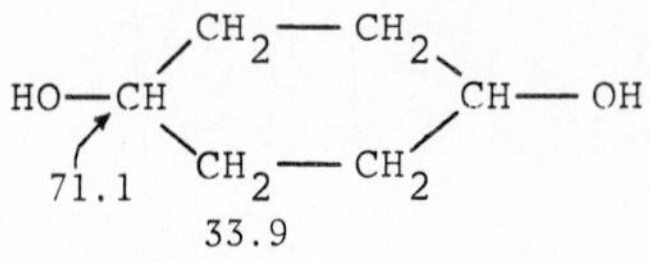

20.

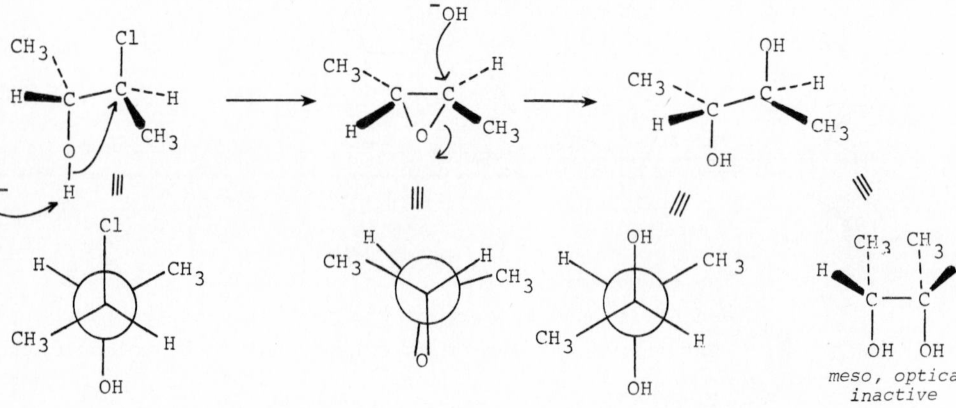

(You will find that you get the same result regardless of which end of the epoxide is attacked in the hydrolysis step)

21.

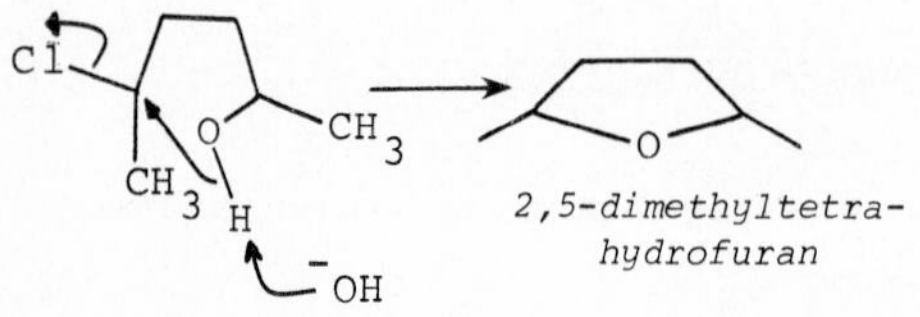

The product could exist as an achiral, *cis* isomer:

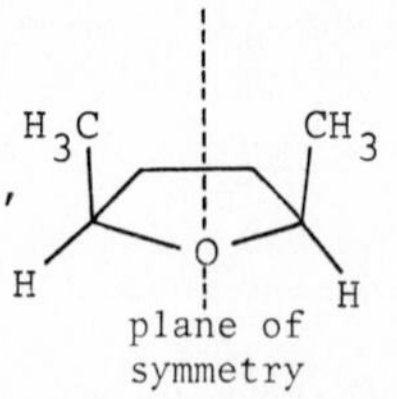

or as a chiral, *trans*, isomer:

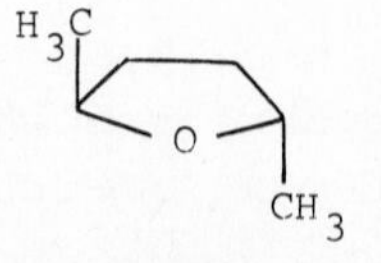

or

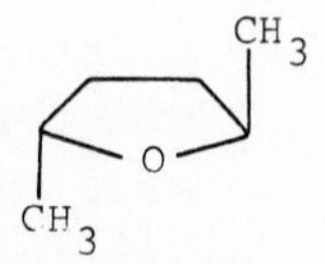

The starting material was *not* racemic (it was optically active), so if the *trans* isomer had been formed it would not have been racemic. However, the product obtained is optically *inactive*, suggesting that the product is the achiral, *meso* isomer: the *cis* compound. This means that the starting material was either

(2R,5R)-5-chloro-2-hexanol or the (2S,5S) isomer, and *not* the (2R,5S) or (2S,5R) isomers:

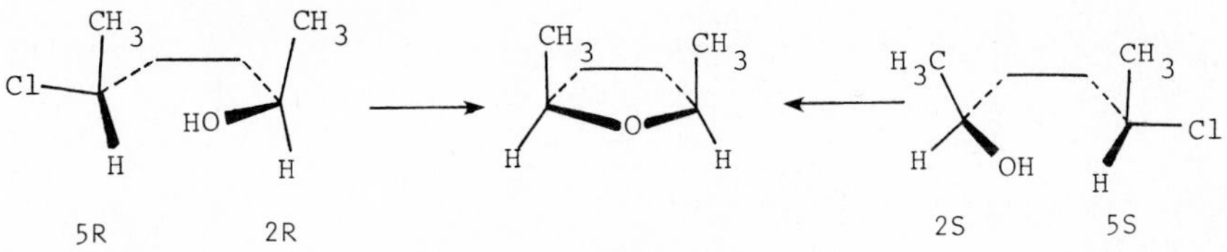

22. The cavity inside 18-crown-6 is just the right size for K^+ to fit in (diameter of K^+ = 2.66 Å); Na^+ is smaller (diameter = 1.96 Å) and it is not coordinated so tightly. Therefore, 18-crown-6 solubilizes potassium salts much more effectively than sodium salts.

23.

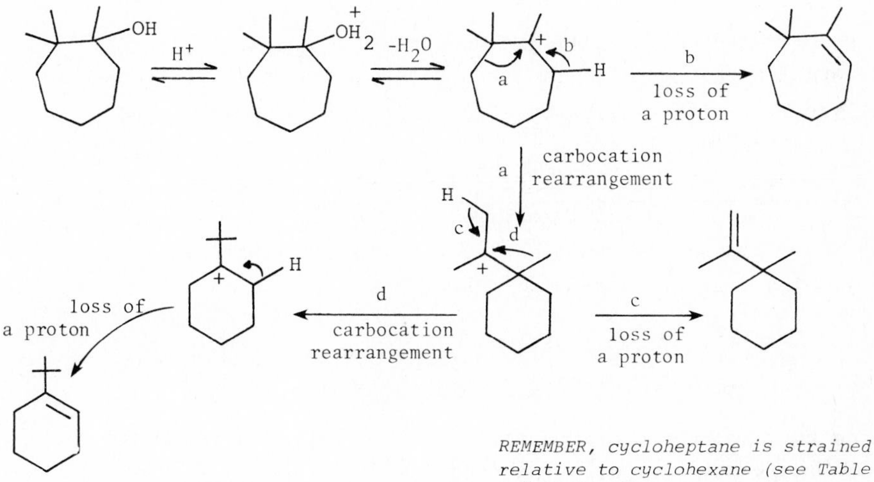

REMEMBER, cycloheptane is strained relative to cyclohexane (see Table 5.5 in text).

(One of the ways to solve mechanism problems is to work both backwards as well as forwards. For instance, 1-*t*-butylcyclohexene must have come from the tertiary carbocation shown above it in the scheme; that in turn must have come from the same carbocation which led to 1-isopropenyl-1-methylcyclohexane, etc.

24.

Charge-dipole:

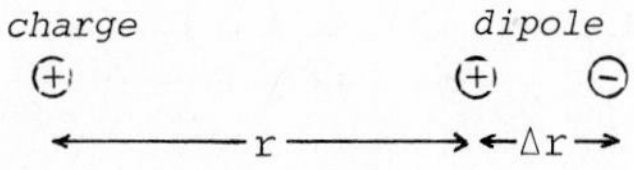

$$E = \frac{1}{r} - \frac{1}{(r + \Delta r)}$$

$$\frac{(r + \Delta r) - r}{r(r + \Delta r)} = \frac{\Delta r}{r(r + \Delta r)} \cong \frac{\Delta r}{r^2}$$

since Δr is much less than r

Dipole-dipole:

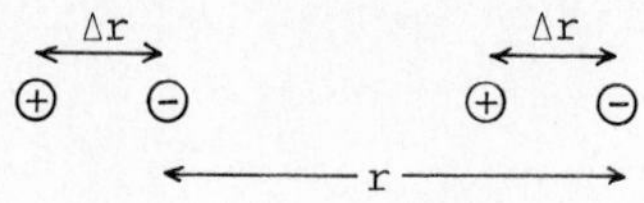

$$E = \frac{1}{r} + \frac{1}{r} - \frac{1}{r+\Delta r} - \frac{1}{r-\Delta r}$$

$$= \frac{2(r+\Delta r)(r-\Delta r) - r(r+\Delta r) - r(r-\Delta r)}{r(r+\Delta r)(r-\Delta r)}$$

$$= \frac{\cancel{2r^2} - 2\Delta r^2 - \cancel{r^2} + \cancel{r\Delta r} - \cancel{r^2} - \cancel{r\Delta r}}{r^3 - r\Delta r^2}$$

$$= \frac{-2\Delta r^2}{r^3 - r\Delta r^2} \cong \frac{-2\Delta r^2}{r^3}$$

(for $\Delta r \ll r$)

25. (a) The best method is to first convert all temperatures into °K and take the reciprocals. A plot of ln P *vs.* 1/T gives a good straight line with a slope of -5.08×10^3. The corresponding slope using log P is -2.21×10^3.

$$\text{Thus, } \Delta H_V = -(-5.08 \times 10^3)(R = 1.986 \text{ cal deg}^{-1} \text{ mole}^{-1})$$
$$= 10.1 \text{ kcal mole}^{-1}$$

Compared to propane, we see that hydrogen-bonding increases the heat of vaporization by about 5.6 kcal mole^{-1}.

(b) $\mu = q.d$ $\quad 1.7 \times 10^{-18}$ esu-cm $= q.\ 0.96 \times 10^{-8}$ cm

$$q = 1.73 \times 10^{-10} \text{ esu}$$

Since the electronic charge is 4.8×10^{-10} esu, q corresponds to 0.36 electronic charges. That is, the dipole moment corresponds to

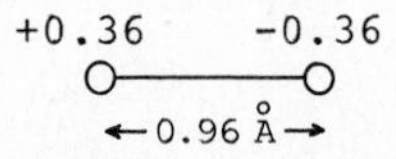

The net electrostatic attraction between two such dipoles is:

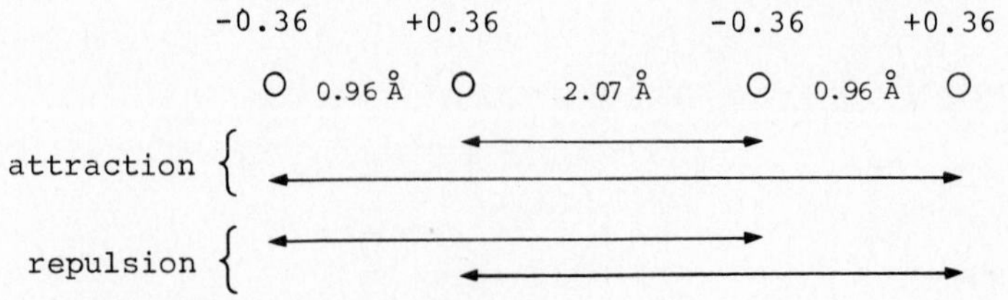

The net attraction is given by:

$$(0.36)^2 \left[-\frac{1}{2.07} - \frac{1}{3.99} + 2 \times \frac{1}{3.03} \right] (332) = -3.2 \text{ kcal mole}^{-1}$$

We see that even this crude model gives a result of the correct order of magnitude.

26.

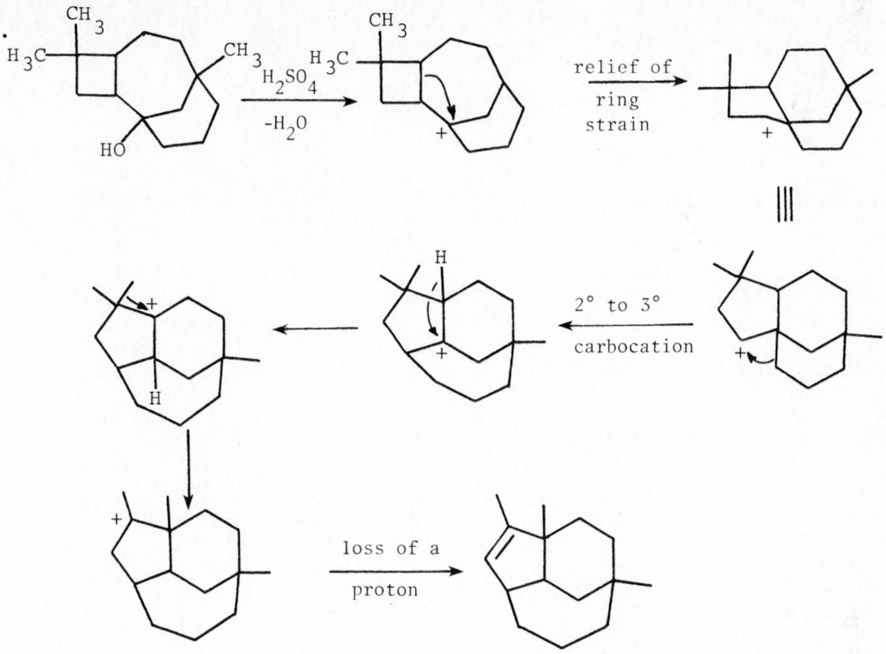

The last two rearrangements are to relieve strain in the tricyclic structure which is hard to see without knowing the stereostructure of the molecule. In a question like this, the problem-solving approach of working backwards, as pointed out in the answer to problem 23, is very useful. For instance, the final product must have come from the last carbocation depicted, which must have arisen from the methyl migration shown, etc.

10.E Supplementary Problems

S1. Provide the structure and IUPAC name for each of the following compounds:

(a) triphenylcarbinol (c) neopentyl alcohol

(b) γ-chloropropyl alcohol (d) diisobutylcarbinol

S2. All of the names below are incorrect. Provide a correct name, either common or IUPAC, for each:

(a) isopropanol (c) β,β-dichloropropanol

(b) 2,2-dimethyl-5-hexanol (d) 2,3-dihydroxybutane

S3. Both the nmr and cmr spectral data are given for each of four isomers of $C_6H_{14}O$. Determine the structure of each one and assign the resonances.

(a) nmr: δ 0.9 (t, 3H), 1.4 (m, 8H), 1.7 (broad s, 1H), 3.6 (t, 3H)
cmr: δ 14.2, 22.8, 25.8, 32.0, 32.8, 61.9

(b) nmr: δ 0.95 (d, 6H), 1.2 (s, 6H), 1.7 (m, 1H), 3.0 (broad s, 1H)
cmr: δ 17.8, 26.6, 39.1, 72.5

(c) nmr: δ 0.9 (s, 9H), 3.1 (s, 3H), 3.3 (s, 2H)

cmr: δ 26.3, 30.6, 50.2, 82.1

(d) nmr: δ 0.9 (s, 9H), 1.5 (t, 3H), 3.0 (broad s, 1H), 3.7 (t, 3H)

cmr: δ 29.7, 29.8, 46.4, 58.9

S4. Give the principal product(s) from each of the following reactions:

(a)

$CH_3CH_2CH_2OH$ $\xrightarrow[\text{pyridine}]{CH_3-\bigcirc-SO_2Cl}$ $\xrightarrow[\text{acetone}]{NaI}$

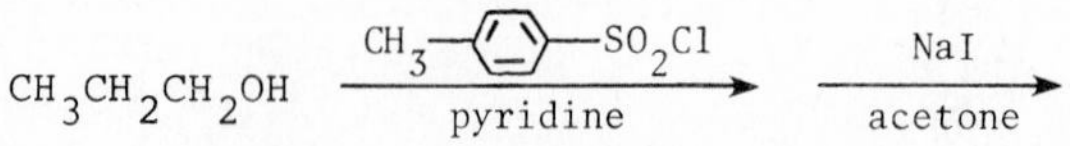

(b) $ICH_2CH_2CH_2CH_2I$ $\xrightarrow{NaOH}$

(c) $(CH_3)_2CHOCH_3$ $\xrightarrow[\text{light}]{O_2}$

(d) $\bigcirc$ $\xrightarrow[\text{peroxide}]{CH_3CO_3H}$ $\xrightarrow{H_3O^+}$

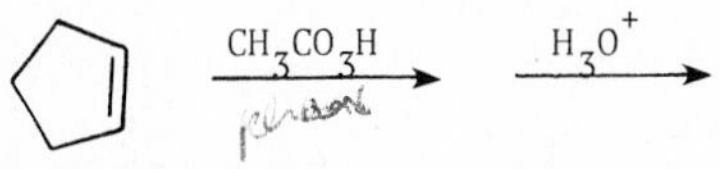

(e) $(CH_3)_3COC(CH_3)_3$ $\xrightarrow[\Delta \text{ (heat)}]{H_2SO_4}$

(f) $CH_3CH_2-\overset{OH}{\underset{D}{\overset{|}{C}}}-H$ $\xrightarrow{PBr_3}$ $\xrightarrow{NaOH}$

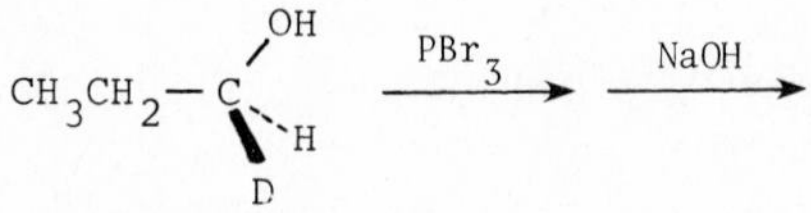

S5. Give the reagents and best conditions for carrying out the following transformations:

(a)

$CH_3-CH_2-\overset{CH_3}{\underset{CH_3}{\overset{|}{\underset{|}{C}}}}-Br$ $\longrightarrow$ $CH_3-CH_2-\overset{CH_3}{\underset{CH_3}{\overset{|}{\underset{|}{C}}}}-OH$

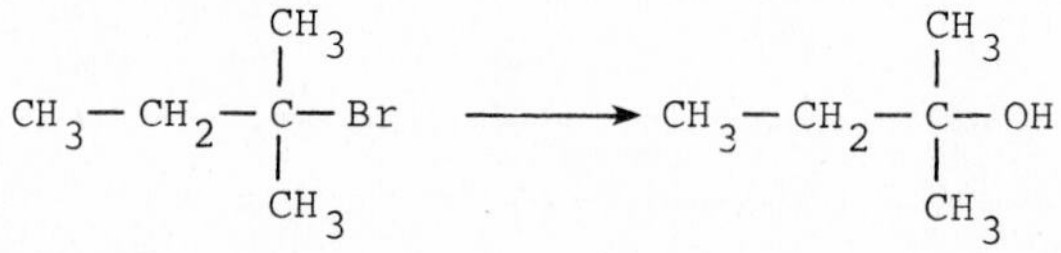

(b) $CH_3-CH_2-CH=CH_2$ $\longrightarrow$ $CH_3-CH_2-CHOH-CH_2OH$

(c)

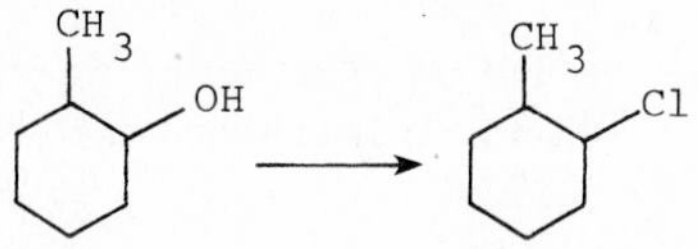

$\bigcirc \overset{CH_3}{-}OH$ $\longrightarrow$ $\bigcirc \overset{CH_3}{-}Cl$

(d)

$\bigcirc-OH$ $\longrightarrow$ $HO_2CCH_2CH_2CH_2CO_2H$

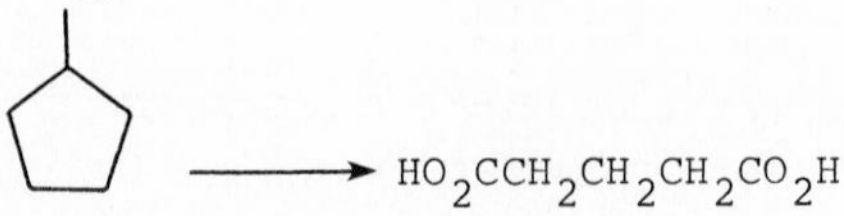

S6. Rank the following compounds in order of decreasing acidity.

(a) CH_3CH_2OH

(b) $CH_3-\overset{CH_3}{\underset{CH_3}{\overset{|}{\underset{|}{C}}}}-OH$

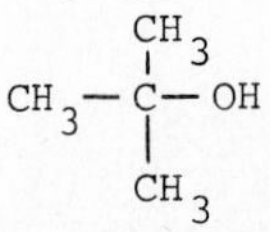

(c) $ClCH_2CH_2OH$

(d) $ClCH_2CH_2CO_2H$

(e) CF_3CH_2OH

(f) CF_3OCH_3

S7. Which of the following reactions give(s) optically active products? Justify your answers.

(a) 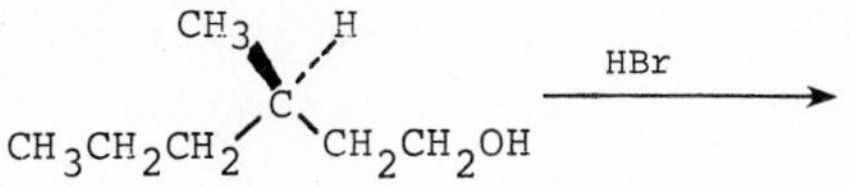

(b) 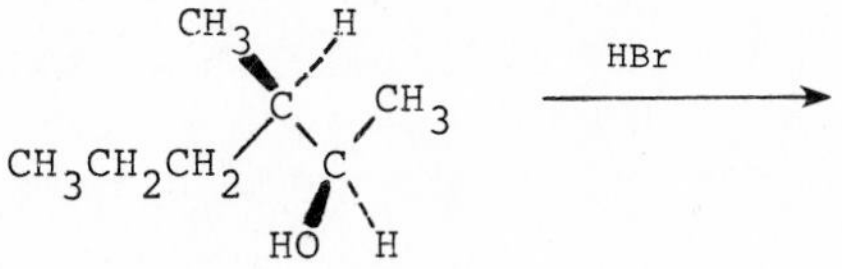

(c) 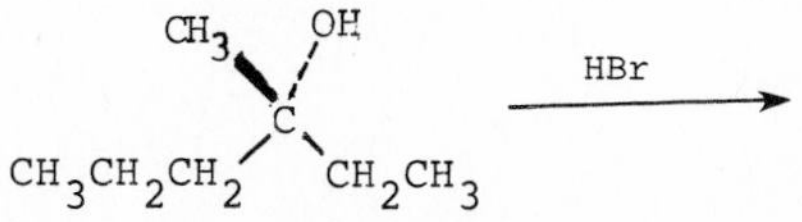

S8. (S)-1-Bromo-2-methylbutan-2-ol is converted to an optically active epoxide with dilute sodium hydroxide as shown below. The epoxide ring can be cleaved either in strong base or in acid to give diol products. What is the difference, if any, between the products formed by the acidic and basic hydrolysis conditions? Write a step-by-step mechanism to explain any differences you expect to see.

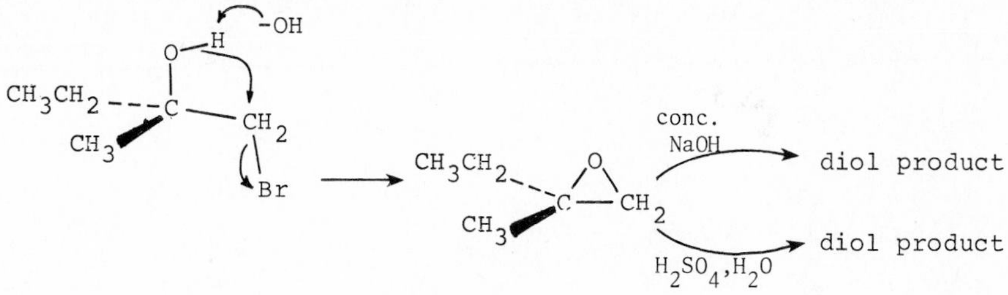

S9. How would you prepare 1-methoxy-1-methylcyclopentane from methylcyclopentane?

10.F Answers to Supplementary Problems

S1. (a) 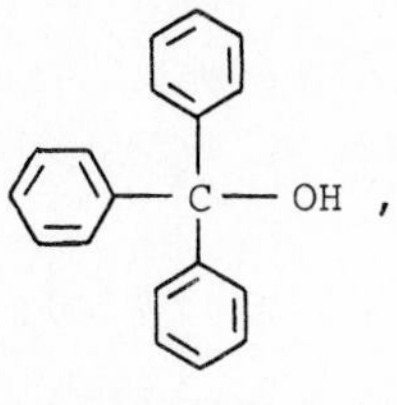 , triphenylmethanol

(b) $ClCH_2CH_2CH_2OH$, 3-chloro-1-propanol

(c) 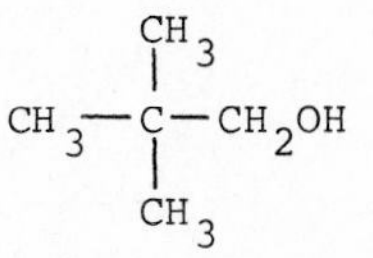 , 2,2-dimethyl-1-propanol

(d) 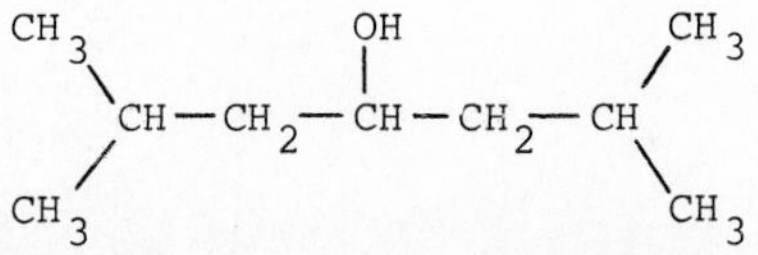 , 2,6-dimethyl-4-heptanol

S2. (a) Mixture of common and IUPAC usage: isopropyl alcohol and 2-propanol are correct.

(b) As an "alkanol", the chain should be numbered to give the lowest number to the hydroxyl group: 5,5-dimethyl-2-hexanol.

(c) Mixture of common and IUPAC usage: β,β-dichloropropyl alcohol or 2,2-dichloro-1-propanol.

(d) Should be named as an alkanediol: 2,3-butanediol.

S3. (a) nmr:

$$\overset{0.9}{CH_3}-\overset{\overbrace{\qquad\qquad 1.4 \qquad\qquad}}{CH_2}-CH_2-CH_2-CH_2-\overset{3.6}{CH_2}-\overset{1.7}{OH}$$

cmr: 14.2   22.8   32.0   25.8   32.8   61.9

(b) nmr: 1.2   3.0   1.7   0.95

$$\overset{\overset{\displaystyle OH}{|}}{(CH_3)C}-CH(CH_3)_3$$

cmr: 26.6   72.5   39.1   17.8

(c) nmr: 0.9   3.3   3.1

$$(CH_3)_3C-\overset{}{CH_2}-O-CH_3$$

cmr: 26.3   30.6   82.1   50.2

(d) nmr: 0.9   1.5   3.7   3.0

$$(CH_3)_3C-CH_2-CH_2-OH$$

cmr: 29.8   29.7   46.4   58.9

S4. (a) $CH_3CH_2CH_2I$

(b)

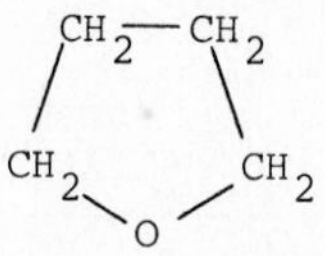

After substitution of one of the iodides with hydroxyl, the second iodide is attacked faster intramolecularly by the alkoxide anion than it is displaced intermolecularly by hydroxide:

$$ICH_2CH_2CH_2CH_2I + NaOH \longrightarrow$$

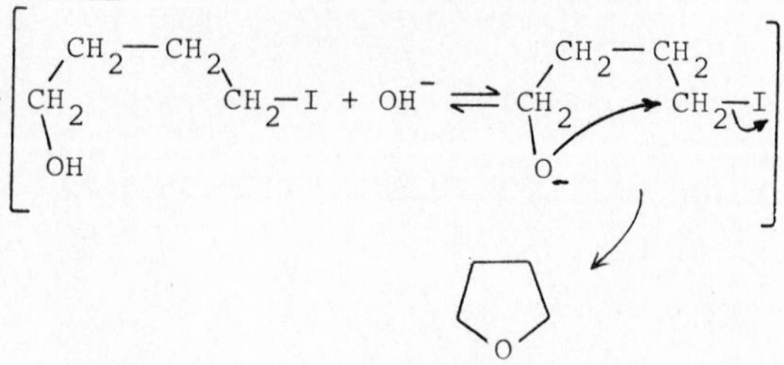

(c)

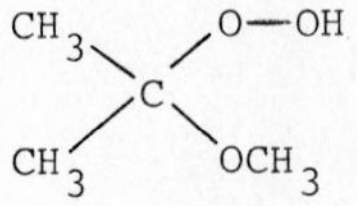

(explosive!)

(d)

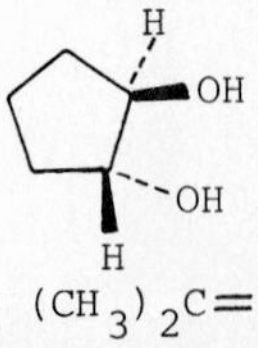

(e) $2\ (CH_3)_2C{=}CH_2$

(f)

$$CH_3CH_2\overset{\displaystyle OH}{\underset{\displaystyle D}{\overset{|}{C}}}\!\!\!-\!H$$

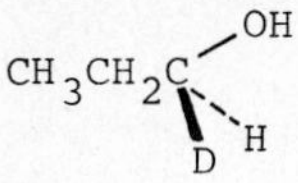

(Two inversions give the same product back again)

(g)

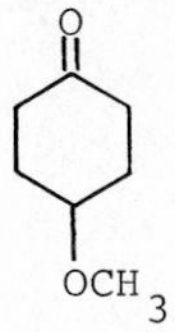

(S5.) (a)

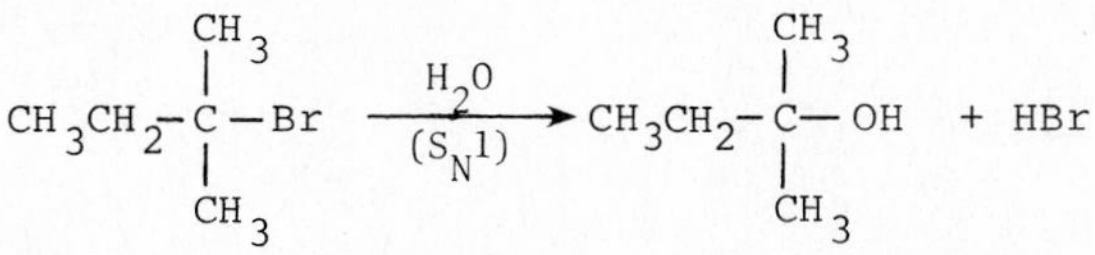

$$CH_3CH_2\overset{\displaystyle CH_3}{\underset{\displaystyle CH_3}{\overset{|}{\underset{|}{C}}}}\!\!-\!Br \xrightarrow[\text{(S}_N1)]{H_2O} CH_3CH_2\overset{\displaystyle CH_3}{\underset{\displaystyle CH_3}{\overset{|}{\underset{|}{C}}}}\!\!-\!OH \ + \ HBr$$

(b)

$$CH_3\!-\!CH_2\!-\!CH\!=\!CH_2 \ + \ CH_3CO_3H \longrightarrow CH_3\!-\!CH_2\!-\!CH\!-\!CH_2 \xrightarrow{H_3O^+}$$

$$CH_3CH_2CHOHCH_2OH$$

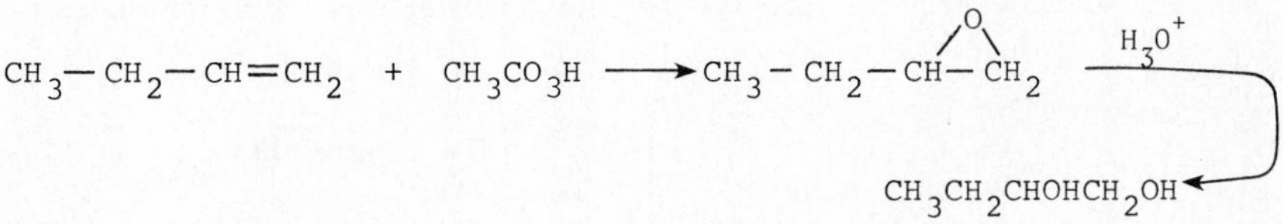

(c)

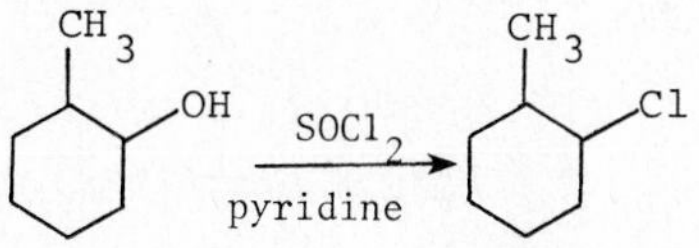

(Carbocation rearrangement must be avoided)

(d)

$$\xrightarrow[H_2SO_4, \ \Delta]{K_2Cr_2O_7} \left[\ \right] \xrightarrow[\text{reaction}]{\text{further}} HO_2CCH_2CH_2CH_2CO_2H$$

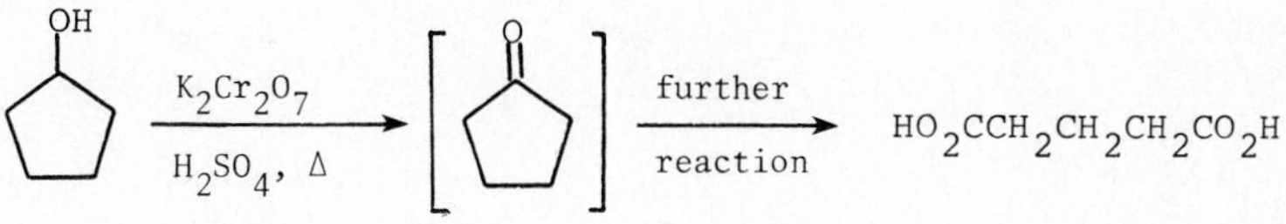

S6. Most acidic: $ClCH_2CH_2CO_2H$ *(it's a carboxylic acid)*

   CF_3CH_2OH

   $ClCH_2CH_2OH$

   CH_3CH_2OH

   $(CH_3)_3COH$

Least acidic: CF_3OCH_3 *(there's no hydrogen which can dissociate)*

S7. (a)

$$\xrightarrow{HBr}$$

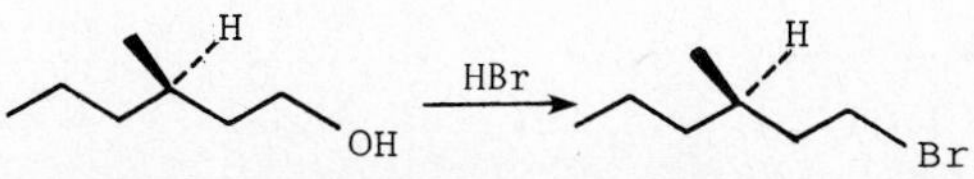

The chiral center is not involved in the reaction so
the product is optically active.

(b)

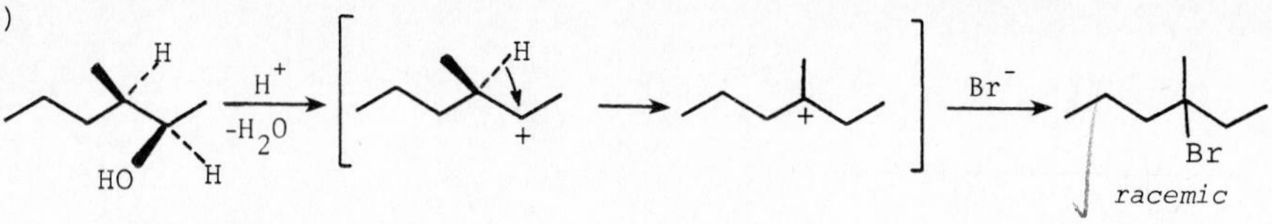

racemic

The reaction involves a rearrangement to give a planar, achiral carbocation intermediate and therefore a racemic product.

(c)

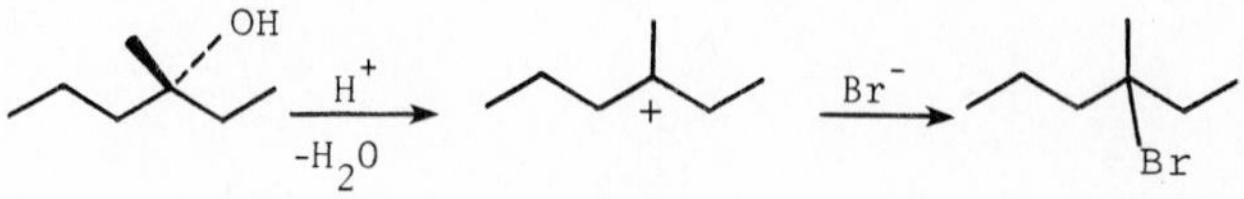

racemic

This reaction involves the same achiral intermediate.

S8. Acidic hydrolysis:

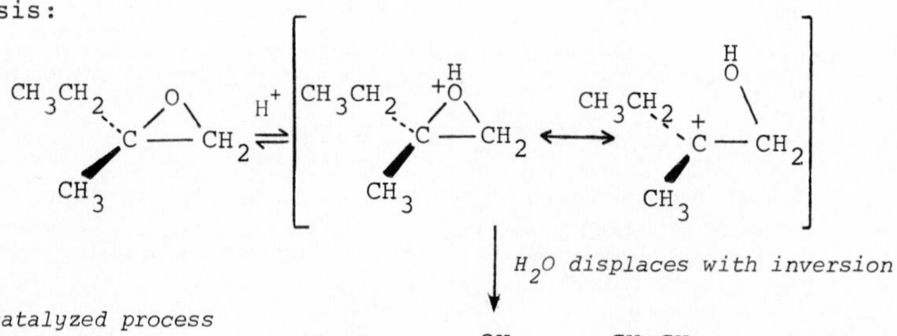

The acid-catalyzed process favors cleavage via a carbocation-like intermediate, and the inversion occurs at the tertiary center.

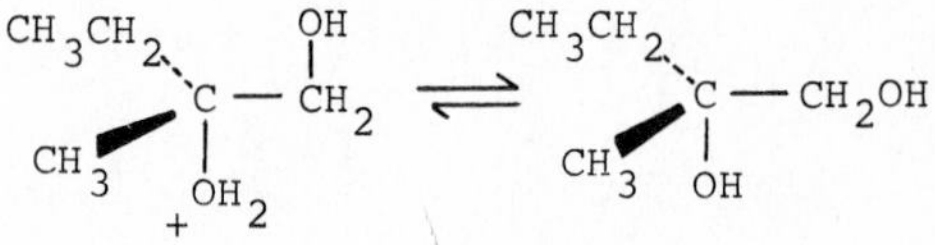

(R)-2-methylbutane-1,2-diol

Basic hydrolysis:

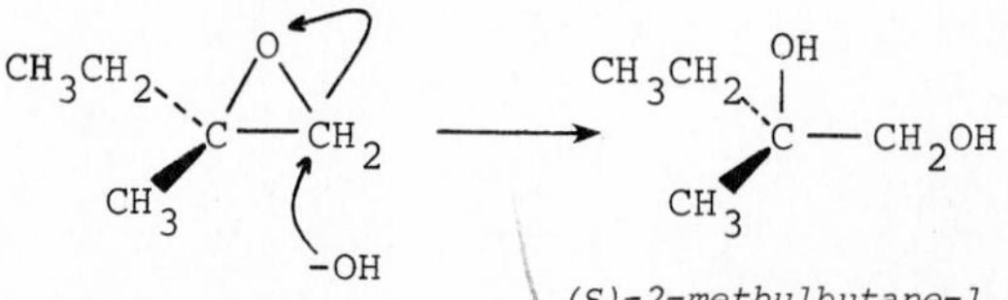

(S)-2-methylbutane-1,2-diol

The mechanism of the basic hydrolysis is like the S_N2 mechanism, and attack and inversion at the less sterically hindered, primary carbon occurs, leading to the enantiomer of the product obtained in acid.

S9.

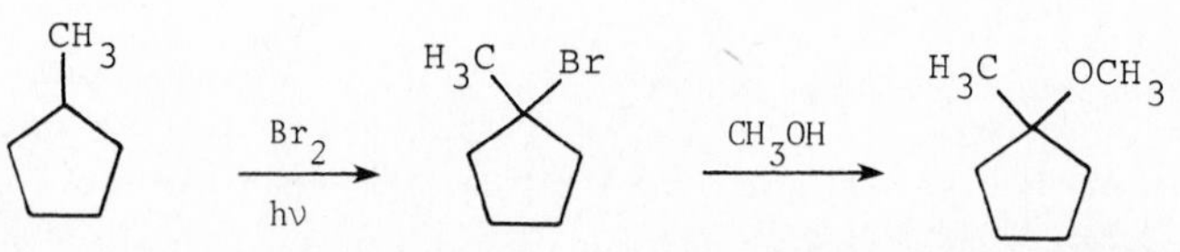

11. ALKENES

11.A Chapter Outline, Important Terms Introduced, and Reactions Discussed

11.1 Electronic Structure

A. Bent-bond Model

B. π Bond Model

σ- and π-bonds | configurational vs. conformational isomers

11.2 Nomenclature of Alkenes

olefin | *cis* and *trans*

alkylenes | *Z* (zusammen) and *E* (entgegen)

11.3 Physical Properties of Alkenes

A. Dipole moments

B. Nmr spectra

anisotropic electron cloud | *cis* and *trans* coupling constants

electron circulation | *cis* and *trans* γ-effects in cmr

geminal

11.4 Relative Stabilities of Alkenes: Heats of Formation

steric hindrance | effects of substitution

11.5 Preparation of Alkenes

A. E2 Bimolecular Elimination of Alkyl Halides (and sulfonates)

rate depends on [RX] and [Base]

isotope effects | rate depends on leaving group

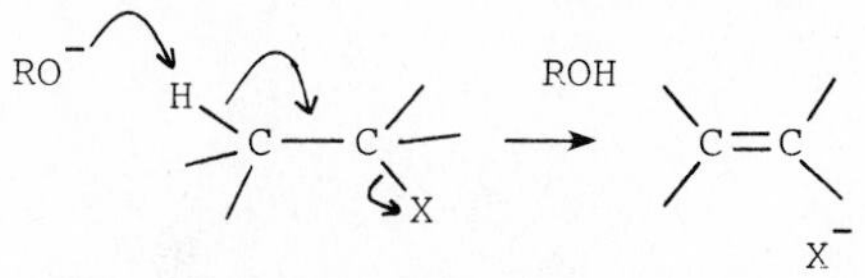

(anti stereochemistry; favors more stable isomer; S_N2 as side reaction for primary RX)

B. Alcohol Elimination

dehydration | alkene isomerization

acid catalyzed:

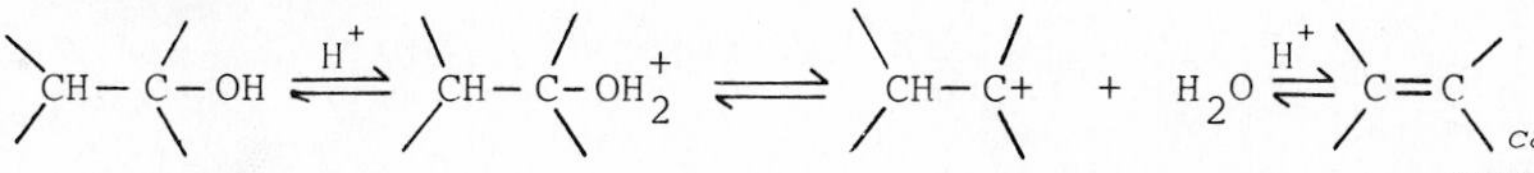

(E1, best for 2° and 3° ROH, carbocation rearrangements possible)

alumina at 350-400°C:

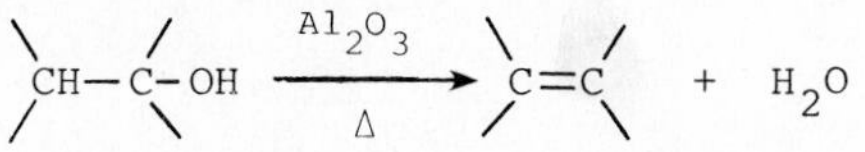

(works for 1° ROH, too)

C. Industrial Preparation of Alkenes

cracking:

$$RCH_2CH_2R' \xrightarrow{700\text{-}900°C} RH + CH_2{=}CHR'$$

11.6 Reactions of Alkenes

A. Catalytic Hydrogenation

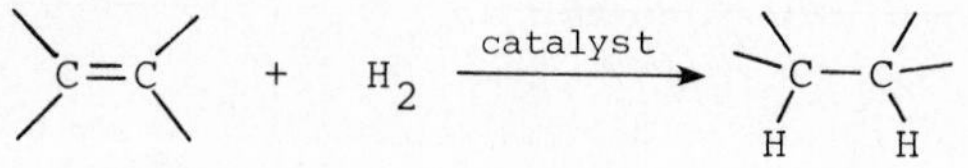

(syn addition; double bond isomerization possible)

B. Electrophilic Additions

1. Addition of HX

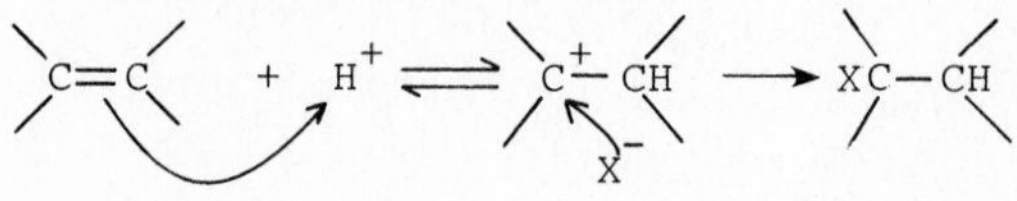

Markovnikov's rule

2. Addition of Water (hydration)

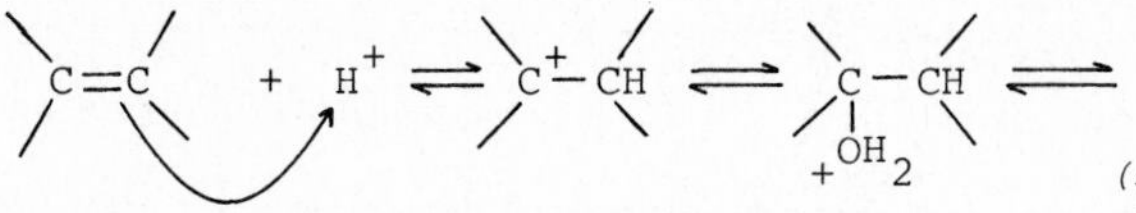

(Markovnikov orientation; reverse of dehydration-- see 11.5.B above)

3. Addition of Halogens

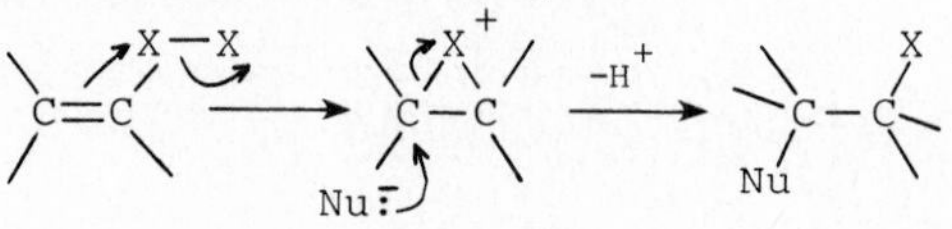

"halonium ion"

(Markovnikov orientation; anti addition, diaxial addition in cyclic systems)

$Nu\colon^- = X^-$ or ROH

4. Addition of Mercuric Acetate ($Hg(OAc)_2$)

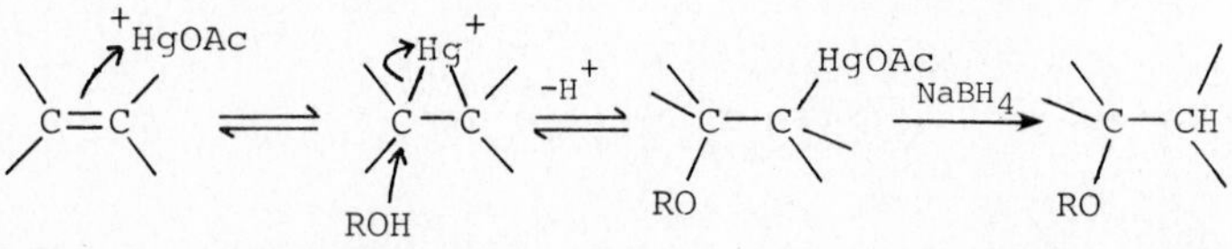

"mercurinium ion"

(Markovnikov orientation; anti addition of AcO-HgOAc)

Unsymmetric ROR

C. Free Radical Additions

HBr or HSR:

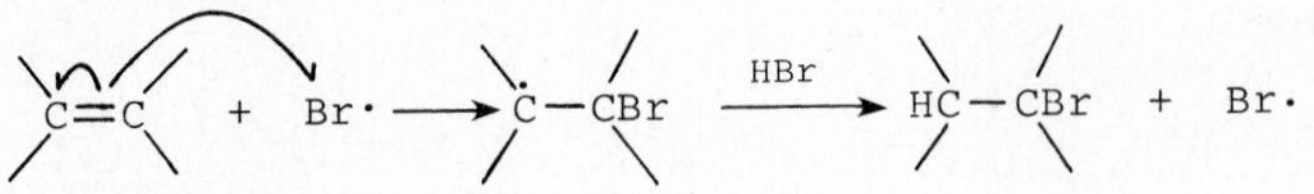

(free radical chain reaction; anti-Markovnikov orientation; not for HCl, HI)

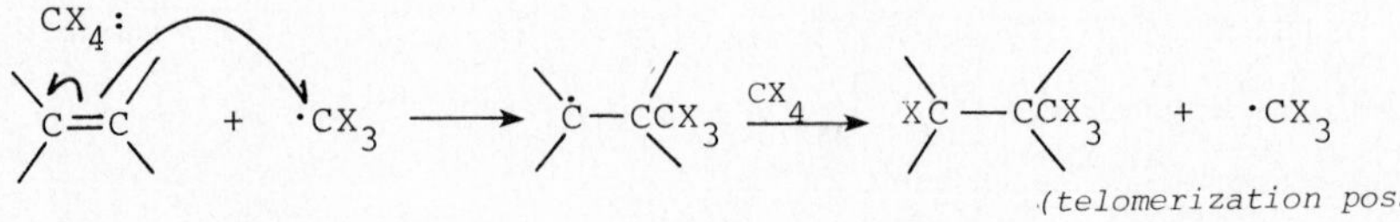

(telomerization possible)

D. Hydroboration

organoboranes

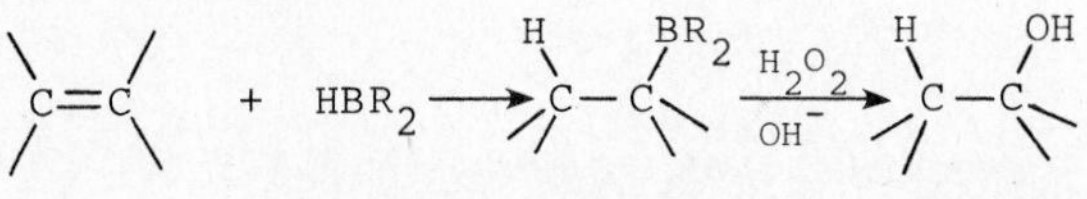

(syn addition; "anti-Markovnikov" orientation)

E.   Oxidation

glycol formation:

syn addition of hydroxyls

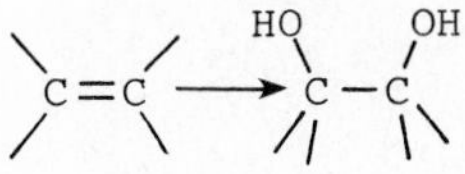

$$\ce{>C=C<} \longrightarrow \underset{\displaystyle \overset{|}{C}-\overset{|}{C}}{\overset{HO \quad OH}{}} \qquad \textit{(cold, dilute KMnO}_4\textit{, or OsO}_4/H_2O_2)$$

anti addition of hydroxyls (see section 10.12.A)

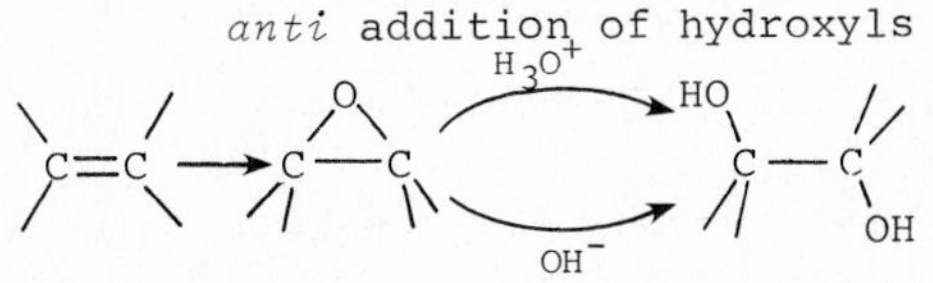

$$\ce{>C=C<} \longrightarrow \underset{O}{\ce{C-C}} \overset{H_3O^+}{\underset{OH^-}{\rightleftarrows}} \underset{OH}{\overset{HO}{\ce{C-C}}}$$

cleavage:

$$\xrightarrow{KMnO_4} \ce{>C=O} + HOOC- \qquad \textit{(does not stop at aldehyde)}$$

$$\ce{>C=C<H}$$

ozonization: O_3

$$\longrightarrow \left[\underset{\displaystyle \overset{|}{C}-O-\overset{|}{\underset{H}{C}}}{\overset{O-O}{}} \right] \xrightarrow[Zn, H_2O]{CH_3SCH_3 \text{ or}} \ce{>C=O} + \underset{H}{O=C<} \qquad \textit{(can stop at aldehyde)}$$

$$\xrightarrow{NaBH_4} \ce{>CHOH} + HOCH_2-$$

epoxidation:

$$\ce{>C=C<} \xrightarrow{RCO_3H} \underset{\displaystyle \overset{|}{C}-\overset{|}{C}}{\overset{O}{}} + RCO_2H \qquad \textit{(see section 10.12.A)}$$

$$\xrightarrow[X_2, H_2O]{} \underset{X}{\overset{OH}{\ce{C-C}}} \xrightarrow{base} \underset{\displaystyle \overset{|}{C}-\overset{|}{C}}{\overset{O}{}} \qquad \textit{(see 11.6.B.3 above, and section 10.12.A)}$$

F.   Addition of Carbenes and Carbenoids: Preparation of Cyclopropanes

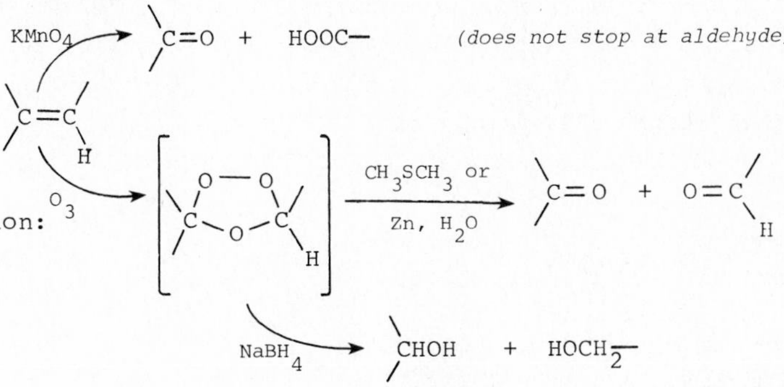

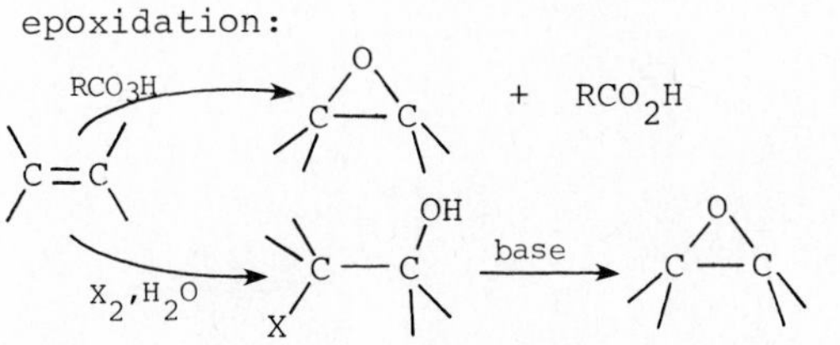

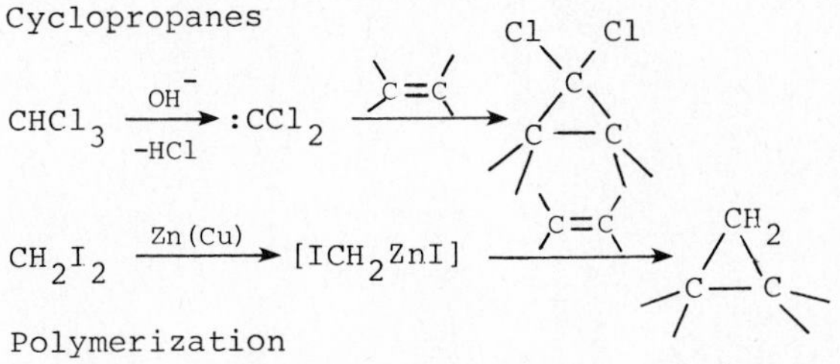

$$CHCl_3 \xrightarrow[-HCl]{OH^-} :CCl_2 \xrightarrow{\ce{>C=C<}} \underset{\displaystyle \overset{|}{C}-\overset{|}{C}}{\overset{Cl \quad Cl}{C}} \qquad \textit{(syn addition)}$$

$$CH_2I_2 \xrightarrow{Zn(Cu)} [ICH_2ZnI] \xrightarrow{\ce{>C=C<}} \underset{\displaystyle \overset{|}{C}-\overset{|}{C}}{\overset{CH_2}{}} \qquad \textit{(syn addition)}$$

G.   Polymerization

monomer, polymer, telomer   dimerization and trimerization

cationic, radical, or anionic polymerizations

$$R* + \ce{>C=C<} \longrightarrow R-\overset{|}{\underset{|}{C}}-\overset{|}{\underset{|}{C}}* \xrightarrow{\ce{>C=C<}} R\left(\overset{|}{\underset{|}{C}}-\overset{|}{\underset{|}{C}}\right)\overset{|}{\underset{|}{C}}-\overset{|}{\underset{|}{C}}* \xrightarrow{termination} R\left(\overset{|}{\underset{|}{C}}-\overset{|}{\underset{|}{C}}\right)R'$$

$$* = +, \cdot, \text{ or } -$$

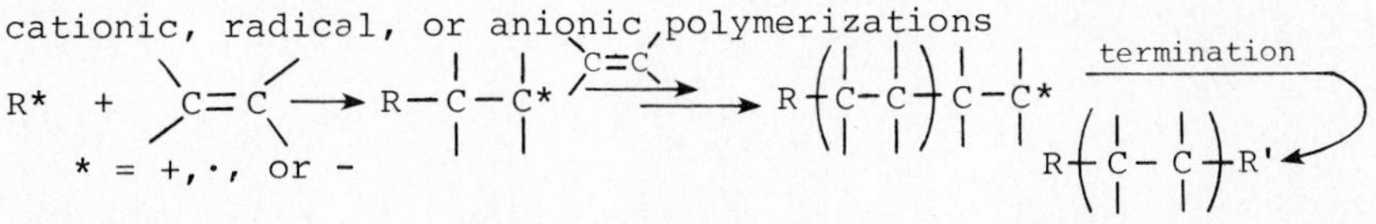

11.B Important Concepts and Hints

The two general methods for formation of carbon-carbon double bonds, E2 elimination of alkyl halides and acid-catalyzed dehydration of alcohols and ethers, have been presented briefly before (sections 8.11, and 10.7.B and 10.11.A). In this chapter they are discussed in greater detail.

The major reactions of alkenes involve addition of a reagent to the two ends of the double bond. This results in rehybridization of the carbons involved, from sp^2 to sp^3, and conversion of the π-bond to two σ-bonds as shown schematically below:

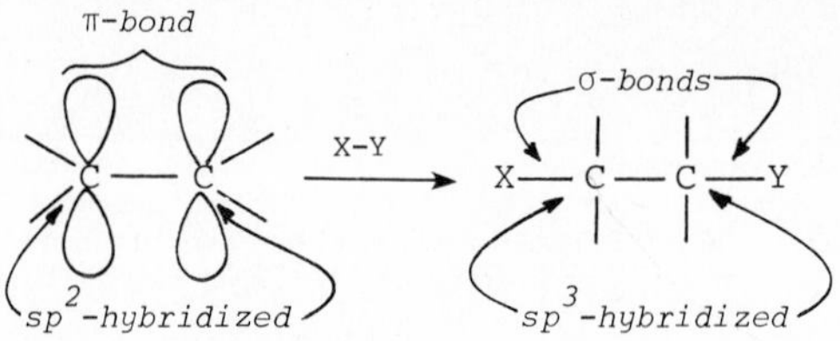

A wide variety of reagents, X-Y, and products are possible, and a number of different mechanisms are observed. The discussion of addition reactions in the text is organized by type of reaction and product. To provide you with a different perspective, we have outlined them below according to the type of *mechanism*, with examples, an indication of the reagents each applies to, and the section in the text where it is discussed.

ONE-STEP ADDITIONS

1.	Hydrogenation	11.6.A
2.	Hydroboration	11.6.D
3.	Glycol formation with $KMnO_4$ or OsO_4/H_2O_2	11.6.E
4.	Ozonide formation	11.6.E
5.	Epoxidation	11.6.E
6.	Carbene addition	11.6.F

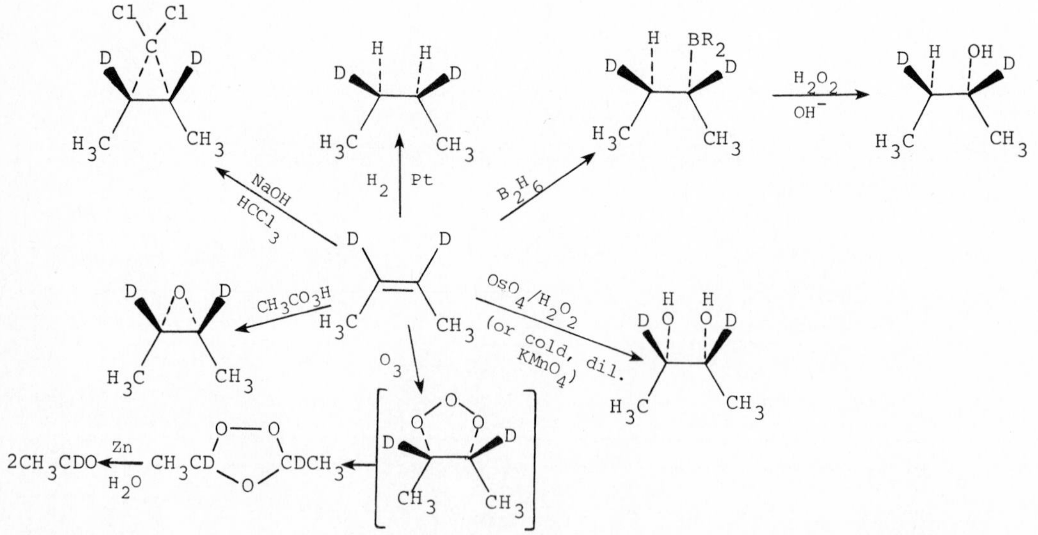

Because they are all one-step additions, the two new σ-bonds are formed with the *syn* relationship, that is, from the same face of the double bond. (You may hear this referred to as "*cis*" addition, but the terms *cis* and *trans* should strictly speaking be applied only to cyclic systems or those in which double bond stereochemistry remains in the product).

Hydroboration is the only case above in which an unsymmetrical addition takes place (i.e. X≠Y). The boron group becomes attached to the sterically

less congested end of the double bond, usually the less substituted end. Note
also that the stereochemistry at that carbon is retained on replacing the C-B
bond with C-OH.

TWO-STEP ADDITIONS

A. Radical chain reactions   11.6.C

$$CH_3-CH{=}CH_2 \quad + \quad Y\cdot \quad \longrightarrow \quad CH_3-\overset{\cdot}{C}H-CH_2Y$$

$$CH_3-\overset{\cdot}{C}H-CH_2Y \quad + \quad X{-}Y \quad \longrightarrow \quad CH_3-\underset{\overset{|}{X}}{C}H-CH_2Y \quad + \quad Y\cdot$$

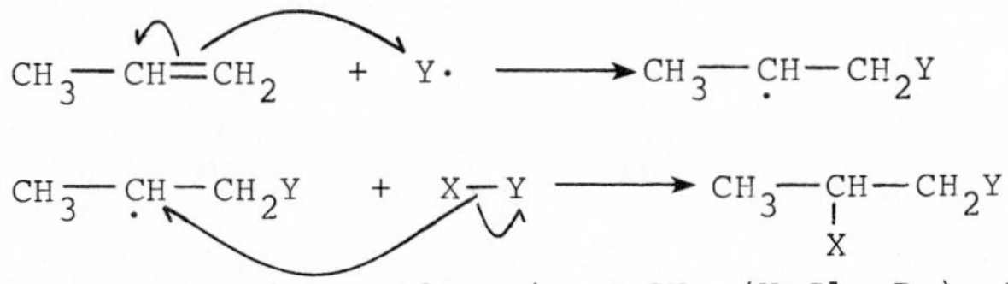

X-Y = H-Br(*not* HCl, HI); X-CX$_3$ (X=Cl, Br); H-SR (R=H, alkyl, etc);

$$\text{or} \quad \overset{\diagdown}{\underset{\diagup}{C}}{=}\overset{\diagup}{\underset{\diagdown}{C}} \quad \text{(polymerization)}$$

The first propagation step is addition of a free radical to one
end of the π-bond to generate an alkyl radical. Note that the radical
is *at the other end* of the original double bond. This reaction occurs
in such a way as to generate the more stable free radical (3°>2°>1°),
so the Y· attacks the least substituted end of the double bond. There
is no stereochemical preference in the second step, so mixtures of
diastereomers are possible. For example:

$$\overset{D}{\underset{CH_3}{}}\diagdown C{=}C\diagup\overset{D}{\underset{CH_3}{}} \quad + \quad HBr \quad \xrightarrow{\text{(peroxides)}} \quad \cdots$$

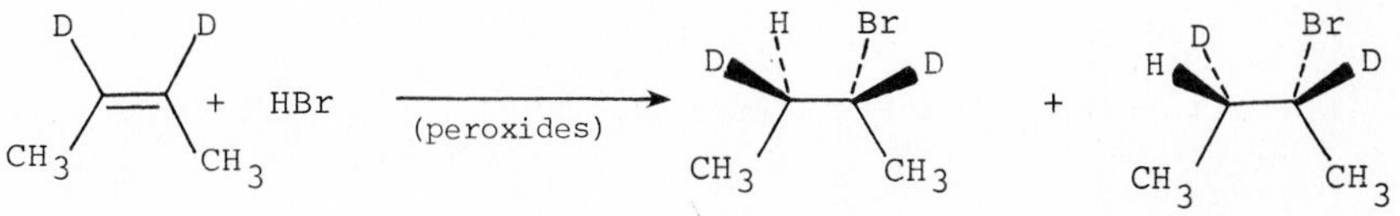

(Each of these is produced in racemic
form; that is, an equal amount of the
enantiomer of each of the above is also
produced.)

B. Electrophilic additions

   1. Acid-catalyzed additions: HX   11.6.B.1

        H$_2$O   11.6.B.2; Polymerization   11.6.G

$$CH_3-CH{=}CH_2 \underset{}{\overset{H^+}{\rightleftharpoons}} CH_3-\overset{+}{C}H-CH_3 \overset{X^-}{\rightleftharpoons} CH_3-CHX-CH_3$$

$$\xrightarrow{H_2O} CH_3-\underset{\overset{|}{\overset{+}{O}H_2}}{C}H-CH_3 \overset{-H^+}{\rightleftharpoons} CH_3CHOHCH_3$$

$$\overset{\diagdown}{\underset{\diagup}{C}}{=}\overset{\diagup}{\underset{\diagdown}{C}} \quad (CH_3)_2CH-(-\overset{|}{\underset{|}{C}}-\overset{|}{\underset{|}{C}}-)_{\overline{n}}$$

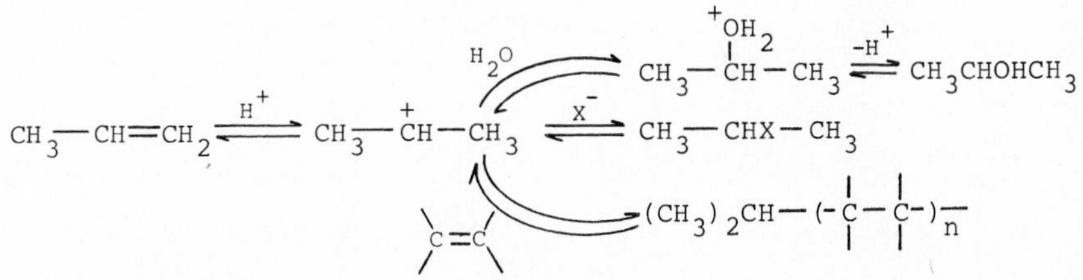

So-called electrophilic additions involve the attack of both
an electrophile and a nucleophile, but the electrophile attacks
first. An isolated double bond is itself weakly nucleophilic, so
its preference for reaction with the electrophile is understandable.
In acid-catalyzed additions, the first species to attack the double
bond is a proton. It forms a C-H bond, using the two electrons
that were in the π-bond and generating a carbocation at the other

end. Because of the greater stability of 3°>2°>1° carbocations, the proton is attached to the less substituted end of the double bond. This generality is the original formulation of Markovnikov's rule. Two other generalities result from the intermediacy of the carbocation: 1) carbocation rearrangements are possible, obviously; and 2) stereochemical preference is usually not seen on attack by the nucleophile. For instance:

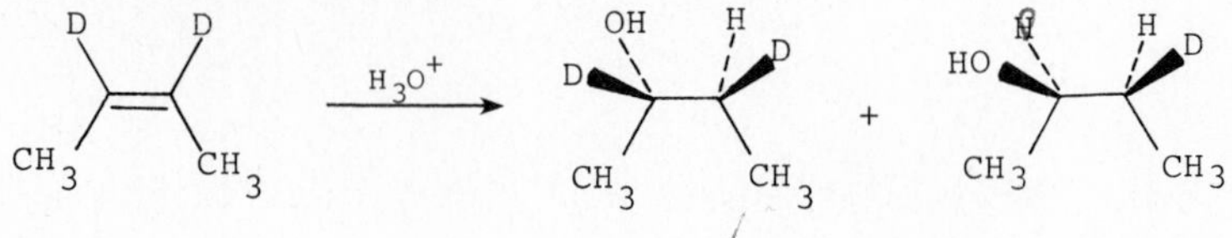

(as racemic mixtures)

Confusion can arise over the use of acid (typically H_2SO_4) to cause *both* the addition of water to a double bond to make an alcohol, and removal of water from an alcohol to make an alkene. Students often ask: how can the same reagent carry out opposing reactions? The situation is easily understood when you realize that there are not two reactions, only one--the equilibration of an alcohol with an alkene plus water, and that the acid is only a catalyst. It is not consumed or formed in the course of the reaction, and it will speed up the reaction in either direction.

$$CH_3-\overset{OH}{\overset{|}{C}}H-CH_3 + H^+ \rightleftharpoons CH_3-\overset{^+OH_2}{\overset{|}{C}}H-CH_3 \rightleftharpoons CH_3-\overset{+}{C}H-CH_3 + H_2O$$

$$\rightleftharpoons CH_3-CH=CH_2 + H_2O + H^+$$

The factors which control the direction that the reaction proceeds are the conditions: in dilute aqueous acid (a *lot* of water around), the equilibrium is driven to the left (as written above), and alcohol is formed from an alkene. Under these conditions an alcohol would not form appreciable amounts of alkene. On the other hand, in concentrated sulfuric acid (60% to 95%, depending on ease of dehydration), especially at higher temperatures, the equilibrium is driven to the right by distillation of the alkene from the reaction mixture and by protonation of the water formed (e.g. $H_2O + H_2SO_4 \rightleftharpoons H_3O^+ + H_2SO_4^-$).

2. Via bridged, cationic intermediates: X_2 11.6.B.3
$Hg(OAc)_2$ 11.6.B.4

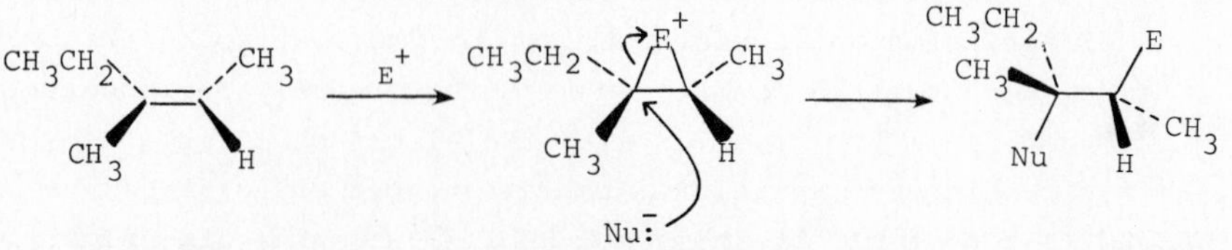

$E^+ = Cl^+, Br^+, I^+$, or ^+HgOAc; $Nu^{:-} = Cl^-, Br^-, I^-, ROH, AcO^-$, etc.

Related reaction: hydrolysis of epoxides (10.12.A)

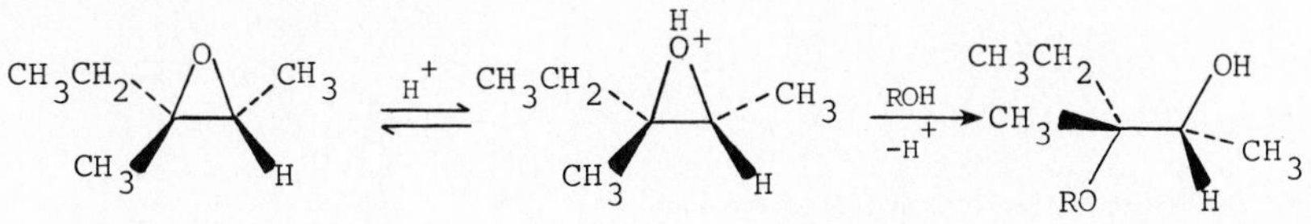

Because of the bridged nature of the cationic intermediate, there are two generalities for reactions of this type: 1) there is a stereochemical preference for *anti* attack by the nucleophile; and 2) Markovnikov's rule is followed in a broader definition: attack of an electrophile occurs in such a way as to form the more stable carbocationic intermediate. You can rationalize this by thinking of the resonance structures possible for the bridged intermediate: the one with a 3° carbocation is more important than the 2° carbocation, and nucleophilic attack occurs faster at that position.

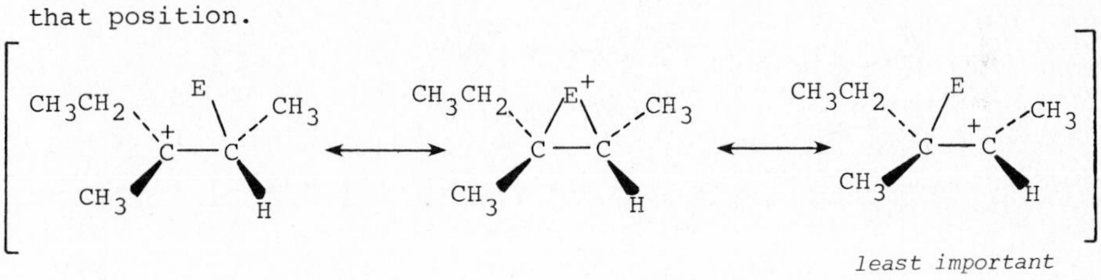

11.C Answers to Exercises

(11.2) (a) *E*-3-methyl-2-pentene
(b) *Z*-1-bromo-2,3,5,5-tetramethyl-2-hexene
(c) *E*-6-chloro-3-octene
(d) 2-(2-bromoethyl)-5-methyl-1-hexene

(11.3) (a)

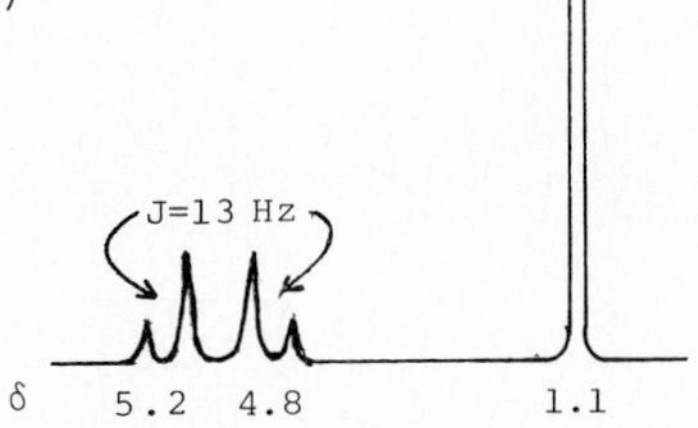

(c)

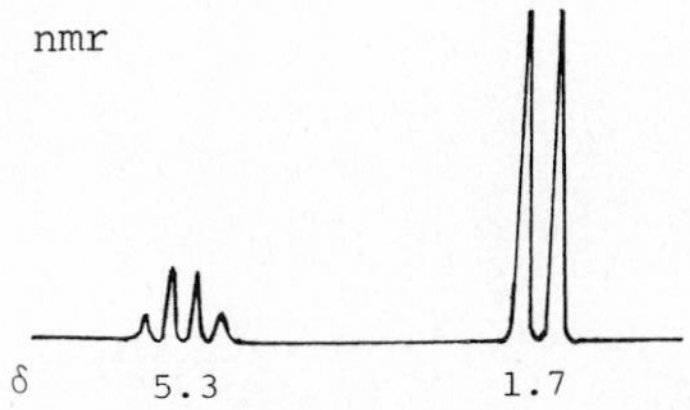

(b)

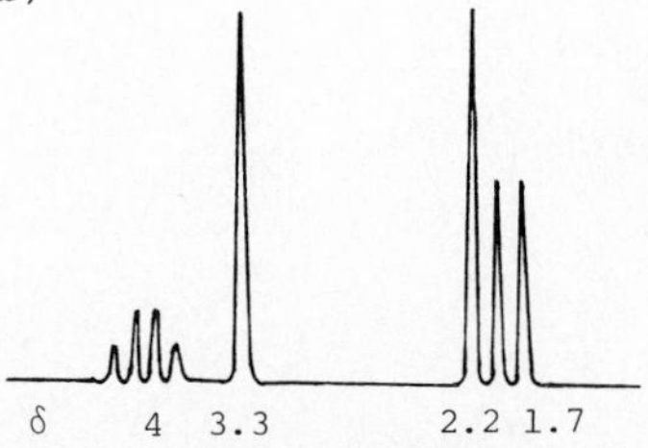

(d)

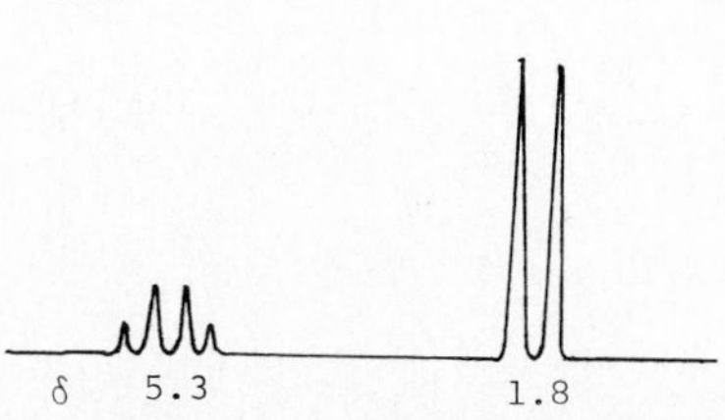

(11.4)   For instance, for $\Delta G° = -2$ kcal mole^{-1}: $\Delta G° = -RT\ln K$

$$\Delta G° = -2000 \text{ cal mole}^{-1} \quad R = 1.987 \text{ cal deg}^{-1} \quad T = 298 °K$$

$$\ln K = 3.38, \quad K = 29.4 = \frac{[B]}{[A]}, \quad \frac{[B]}{[A]+[B]} = \frac{29.4}{1 + 29.4} = 0.967 = 96.7\% \text{ B}$$

trans $\rightleftharpoons$ cis-cyclooctene:

$$\Delta G° = -9.1 \text{ kcal mole}^{-1}; \quad \frac{-9100}{(-1.987)(298)} = \ln K$$

$$K = 4.72 \times 10^6;$$

$$\% \text{ trans at equilibrium} = \frac{1}{(4.72 \times 10^6) + 1} = 2.1 \times 10^{-5}\%$$

(11.5A)

Both of these hydrogens
are anti to the bromine

only one hydrogen is anti to the
bromine, so elimination can only occur on one side

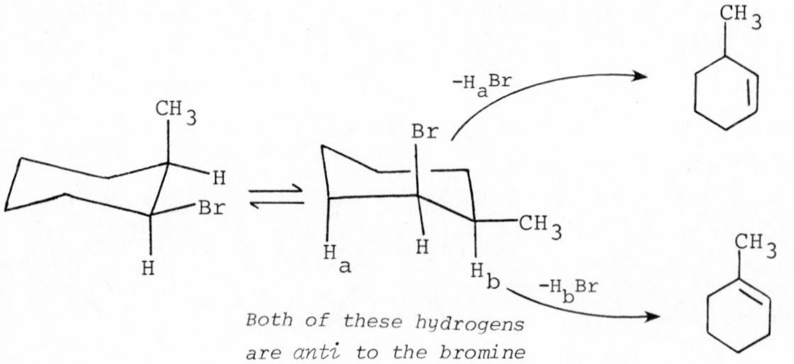

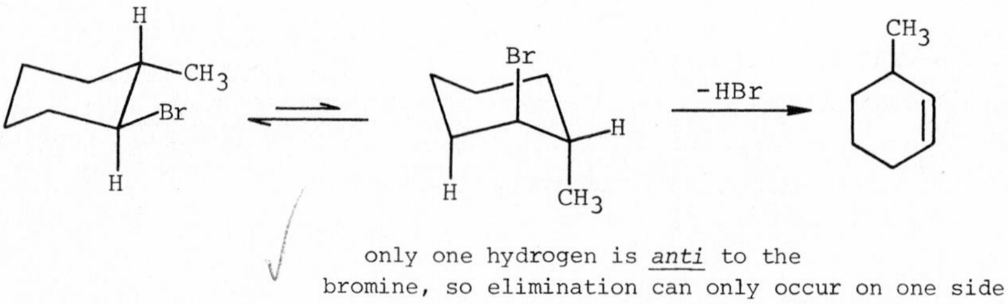

(11.6A)   Palladium as catalyst results in isomerization and loss of chirality
before hydrogenation occurs:

optically active   achiral   racemic

(11.6B.2) $CH_3CH_2CH=CH_2 \rightleftharpoons CH_3CH_2\overset{+}{C}HCH_3 \underset{H_2O}{\rightleftharpoons} CH_3CH_2CHCH_3 \overset{-H^+}{\rightleftharpoons} CH_3CH_2CHCH_3$

$CH_3CH=CHCH_3$

both cis and trans

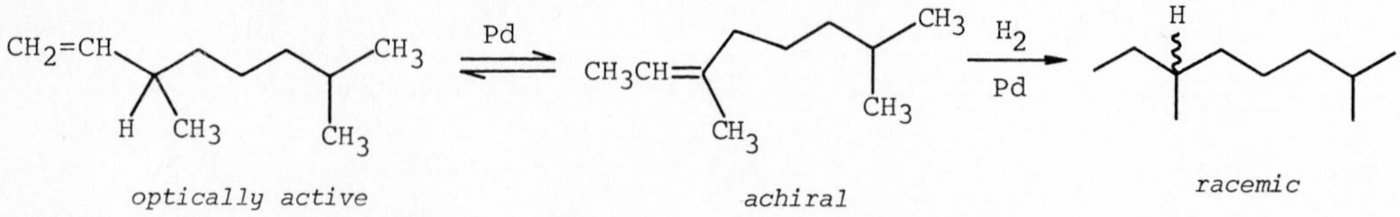

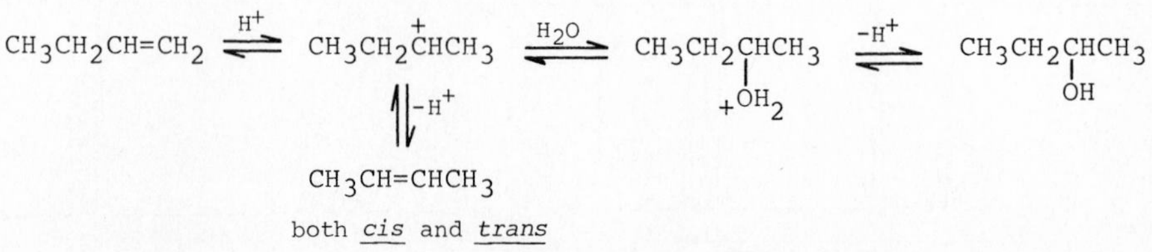

(11.6B.3)

(a) Nu = Br$^-$

(b) Nu = C_2H_5OH

(c) Nu = H_2O

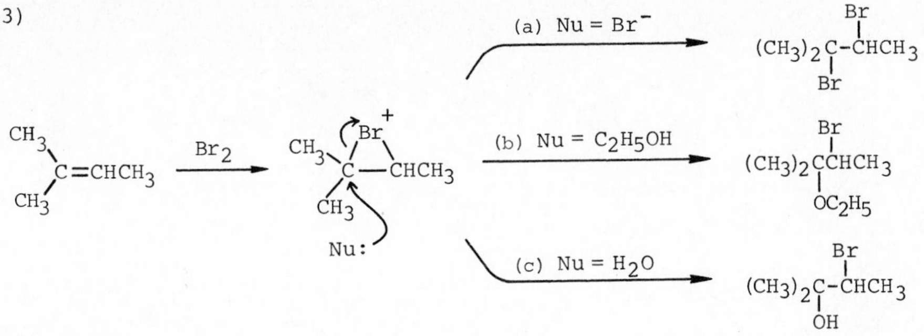

(11.6B.4)

(a) $(CH_3)_2C=CH_2$ + H_2 $\xrightarrow{Pt}$ $(CH_3)_3CH$

(b) $(CH_3)_2C=CH_2$ + HCl $\longrightarrow$ $(CH_3)_3CCl$

(c) $(CH_3)_2C=CH_2$ + H_2O $\xrightarrow{H_2SO_4}$ $(CH_3)_3COH$

(d) $(CH_3)_2C=CH_2$ + Br_2 $\xrightarrow{CCl_4}$ $(CH_3)_2\underset{Br}{\overset{|}{C}}CH_2Br$

(e) $(CH_3)_2C=CH_2$ + Cl_2 + H_2O $\longrightarrow$ $(CH_3)_2\overset{OH}{\overset{|}{C}}CH_2Cl$ + HCl

(f) $\longrightarrow$ $(CH_3)_2\overset{OH}{\overset{|}{C}}CH_2Cl$ + $NaOH$ $\longrightarrow$ $(CH_3)_2C\overset{O}{\overset{\triangle}{—}}CH_2$ + $NaCl$

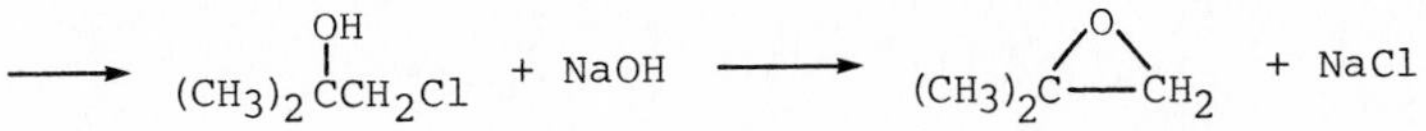

(g) $(CH_3)_2C=CH_2$ + $Hg(OAc)_2$ + H_2O $\longrightarrow$ $(CH_3)_2\overset{OH}{\overset{|}{C}}CH_2HgOAc$ + $HOAc$

$\downarrow$ $NaBH_4$

$(CH_3)_3COH$ + $Hg°$ + borate salts

(11.6D)

(a) $6\ (CH_3)_2CHCH_2CH=CH_2$ + B_2H_6 $\longrightarrow$ $2\ \left[(CH_3)_2CHCH_2CH_2CH_2\right]_3B$

$\xrightarrow{H_2O_2}$ | OH^-

$(CH_3)_2CHCH_2CH_2CH_2OH$ + borate salts

(b)

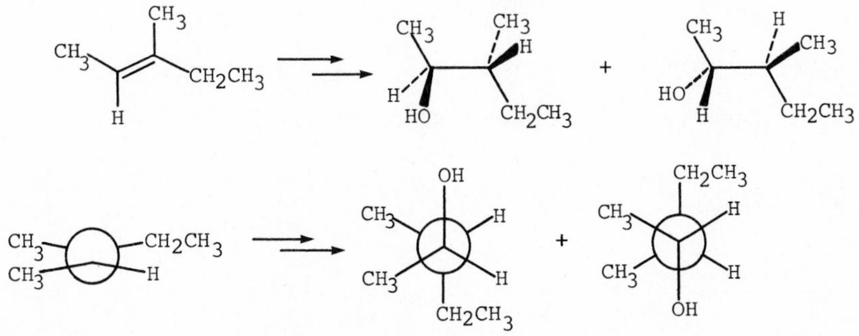

(racemic mixture)

(c)

(racemic)

(11.6E)

(a)

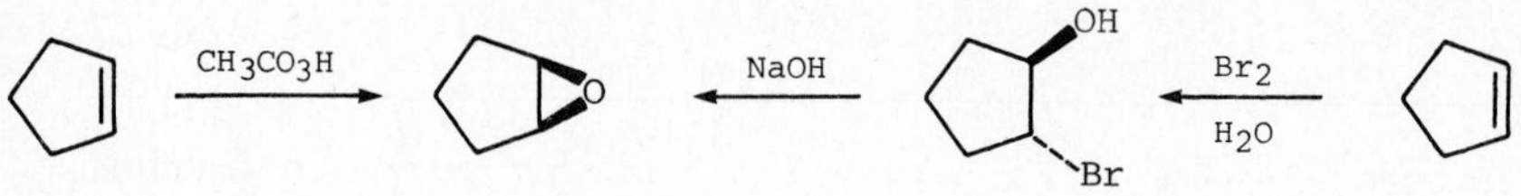

(b)

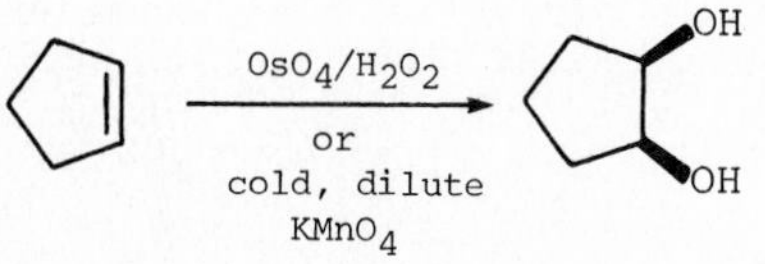

(c)

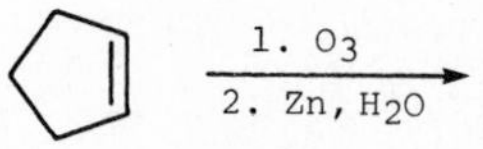

$$\text{cyclopentene} \xrightarrow[\text{2. Zn, H}_2\text{O}]{\text{1. O}_3} OHC(CH_2)_3CHO$$

(d)

$$\text{cyclopentene} \xrightarrow{KMnO_4} HOOC(CH_2)_3COOH$$

(e)

$$\text{cyclopentene} \xrightarrow[\text{2. NaBH}_4]{\text{1. O}_3} HO(CH_2)_5OH$$

(11.6F)

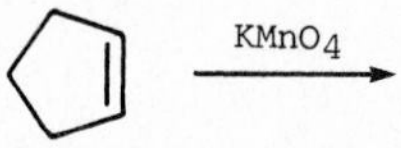

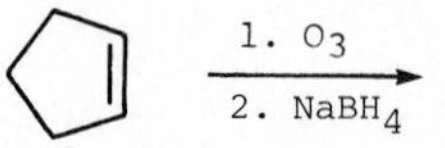

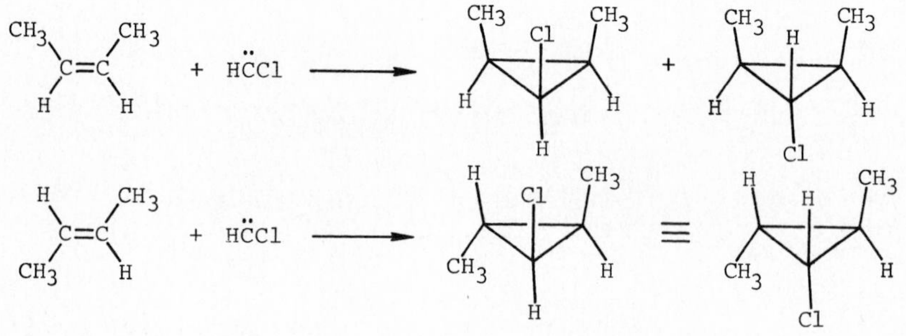

11.D Answers to Problems

1. (a)

$$CH_3CH_2CH_2CH=CH_2 \qquad \text{1-pentene}$$

   cis-2-pentene

   trans-2-pentene

} stereoisomers

$$(CH_3)_2CHCH=CH_2 \qquad \text{3-methyl-1-butene}$$

$$(CH_3)_2C=CHCH_3 \qquad \text{2-methyl-2-butene}$$

$$\underset{\displaystyle CH_2=CCH_2CH_3}{\overset{\displaystyle CH_3}{}} \qquad \text{2-methyl-1-butene}$$

There is only one pair of stereoisomers. None is capable of optical activity.

(b)

   1-methylcyclo-pentene

   (R)-3-methylcyclopentene

   4-methylcyclo-pentene

   (S)-3-methylcyclopentene

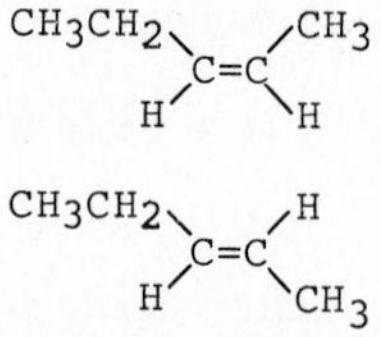

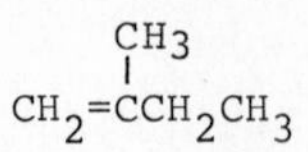

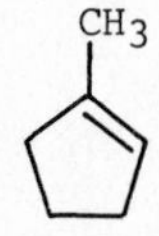

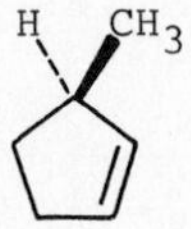

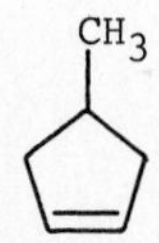

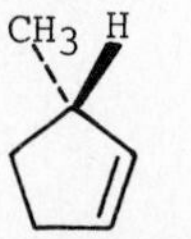

stereoisomers, capable of optical activity

2. (a) 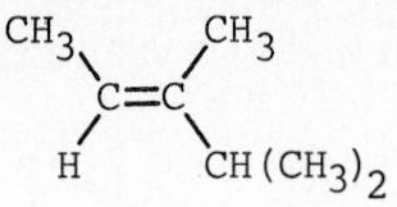

(b) $HOCH_2CH_2CH{=}C(CH_3)_2$

(c) 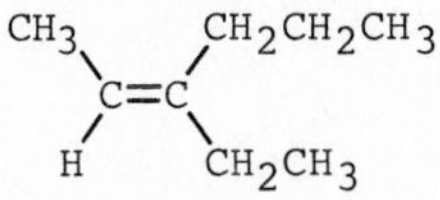

(d) $CH_2{=}CHF$

(e) 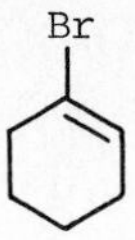

(f) 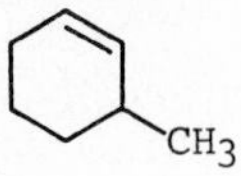

(g) 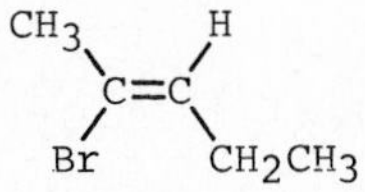

(h) 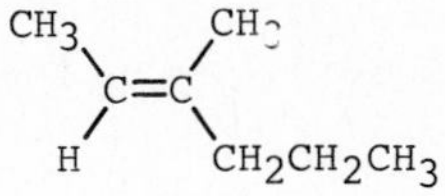

(i) 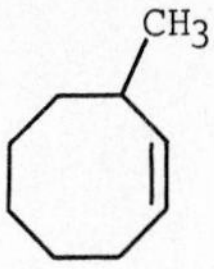

(j) 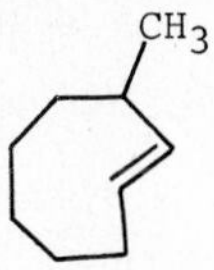

(k) 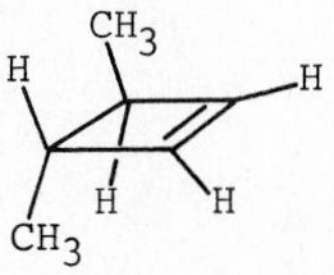

(ℓ) 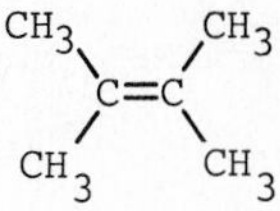

3. (a) 1-methylcyclopentene
   (b) 4-methyl-cis-2-pentene
   (c) 1-butene
   (d) 1-bromo-2-methylpropene
   (e) 1-chlorocyclopentene
   (f) 3-ethyl-2-pentene
   (g) *trans*-3-penten-2-ol

4. (a) 3-methyl-1-hexene
   (b) 1-methylcyclohexene
   (c) (E)-3,4-dimethyl-3-heptene (or 3,4-dimethyl-*trans*-3-heptene)
   (d) (Z)-4-isopropyl-3-methyl-3-heptene (or 4-isopropyl-3-methyl-*trans*-3-heptene)
   (e) (Z)-1-bromo-1-chloro-2-fluoro-2-iodoethene
   (it is difficult to decide whether this should be called *cis* or *trans*, although one possible name is 1-chloro-2-iodo-*trans*-1-bromo-2-fluoroethene.)
   (f) 4-chloro-1-butene
   (g) 4-chloro-2-ethyl-1-butene
   (h) 2-(2-chloroethyl)-1-pentene
   (i) (Z)-1-chloro-3-heptene (or 1-chloro-*cis*-3-heptene)
   (j) 4-penten-1-ol (or pent-4-en-1-ol)
   (k) (E)-hex-4-en-3-ol (or *trans*-hex-4-en-3-ol)
   (ℓ) 2-ethyl-4-methyl-1-hexene

5. (a) $(CH_3CH_2)_3CH$ 3-ethylpentane

(b) 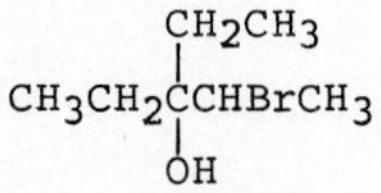 2-bromo-3-ethyl-3-pentanol

(c) 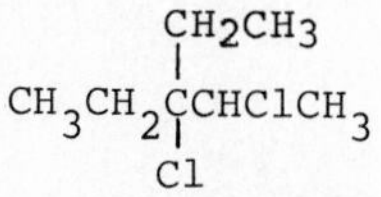 2,3-dichloro-3-ethylpentane

(d) 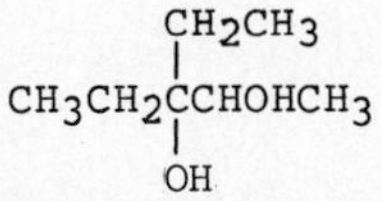 3-ethylpentane-2,3-diol

(e) $CH_3CH_2\overset{\overset{\displaystyle CH_2CH_3}{|}}{\underset{\underset{\displaystyle H}{|}}{C}}CHOHCH_3$ 3-ethyl-2-pentanol

(f) $(CH_3CH_2)_3COH$ 3-ethyl-3-pentanol

(g) $CH_3CH_2COCH_2CH_3$ diethyl ketone or 3-pentanone

      $+ CH_3CHO$ acetaldehyde

(h) $(CH_3CH_2)_3CBr$ 3-bromo-3-ethylpentane

(i) $CH_3CH_2\overset{\overset{\displaystyle CH_2CH_3}{|}}{C}HCHBrCH_3$ 2-bromo-3-ethylpentane

(j) $CH_3CH_2\overset{\overset{\displaystyle CH_2CH_3}{|}}{\underset{\underset{\displaystyle OCH_3}{|}}{C}}CHBrCH_3$ 2-bromo-3-ethyl-3-methoxypentane

(k) $CH_3CH_2\overset{\overset{\displaystyle CH_2CH_3}{|}}{C}\!-\!\underset{\underset{\displaystyle O}{\diagdown\!\diagup}}{C}HCH_3$ 2,2-diethyl-3-methyloxirane

(ℓ) $CH_3CH_2\overset{\overset{\displaystyle CH_2CH_3}{|}}{C}\!-\!CHCH_3$
      $\underset{\displaystyle Br\ Br}{C}$ 1,1-dibromo-2,2-diethyl-3-methylcyclopropane

(m) $CH_3CH_2\overset{\overset{\displaystyle CH_2CH_3}{|}}{C}\!-\!CHCH_3$
      $\underset{\displaystyle CH_2}{}$ 1,1-diethyl-2-methylcyclopropane

6.   <u>Same</u> products from $\underset{H}{\overset{CH_3CH_2}{\diagdown}}C\!=\!C\underset{H}{\overset{CH_2CH_3}{\diagup}}$ and $\underset{H}{\overset{CH_3CH_2}{\diagdown}}C\!=\!C\underset{CH_2CH_3}{\overset{H}{\diagup}}$

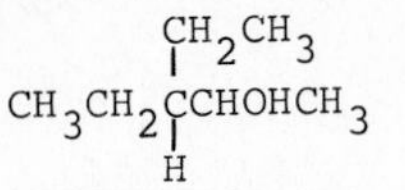

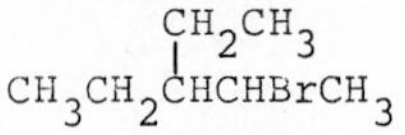

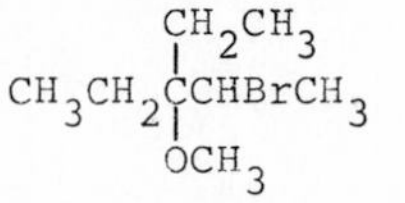

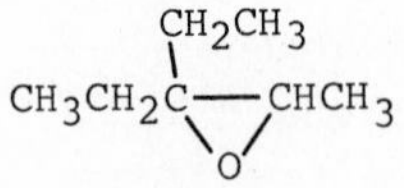

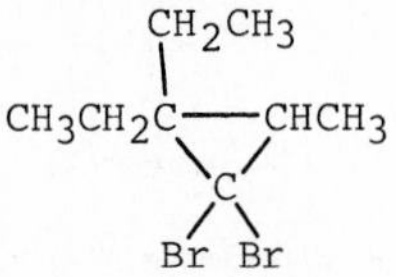

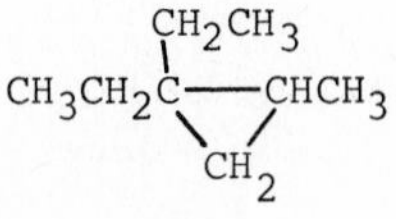

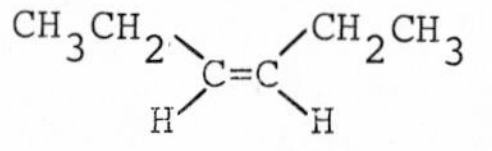

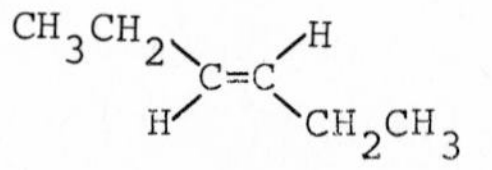

(a)   hexane

(e,f) $CH_3CH_2CHOHCH_2CH_2CH_3$ 3-hexanol

(g) CH_3CH_2CHO propionaldehyde

(h,i) $CH_3CH_2CHBrCH_2CH_2CH_3$ 3-bromohexane

<u>Different</u> products:

(b)

      <u>cis</u>

      (3R,4R)-4-bromo-3-hexanol
      plus an equal amount of the (3S,4S)
                                        enantiomer

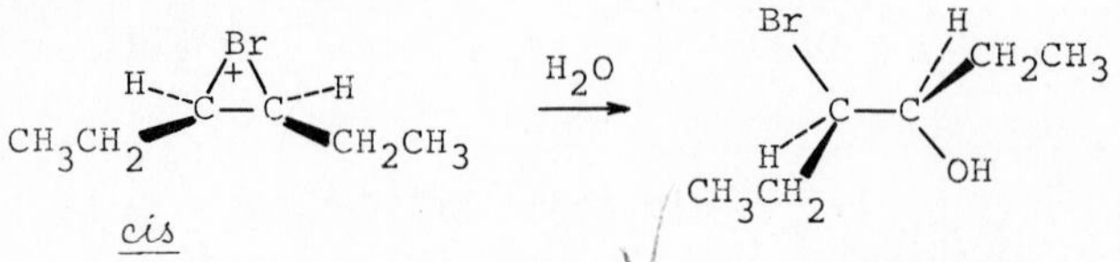

To indicate that this diastereomer is present as a racemic mixture, the designation (3RS,4RS) may be used.

trans-3-hexene gives 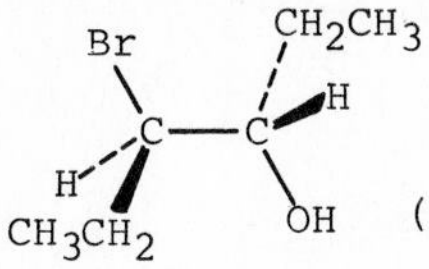 (3RS,4SR)-4-bromo-3-hexanol

(c) *cis* gives (±) (or dl) 3,4-dichlorohexane
trans gives meso-3,4-dichlorohexane

(d) *cis* gives meso-3,4-hexanediol
trans gives (±)-3,4-hexanediol

(j) Same as (b), except for CH_3O instead of OH:
cis gives (3RS,4RS)-3-bromo-4-methoxyhexane
trans gives (3RS,4SR)-3-bromo-4-methoxyhexane

(k) *cis* gives 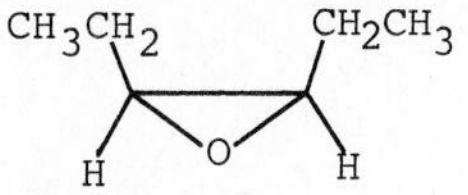 *cis*-2,3-diethyloxirane

trans gives 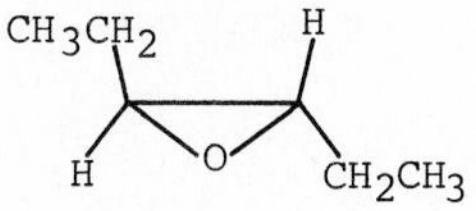 *trans*-2,3-diethyloxirane

(ℓ) *cis* gives 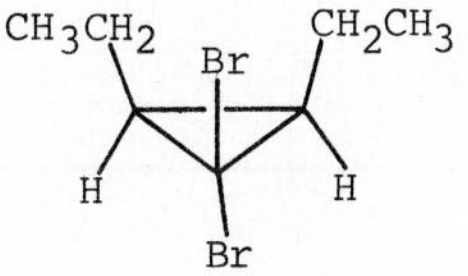 *cis*-1,1-dibromo-2,3-diethylcyclopropane

trans gives 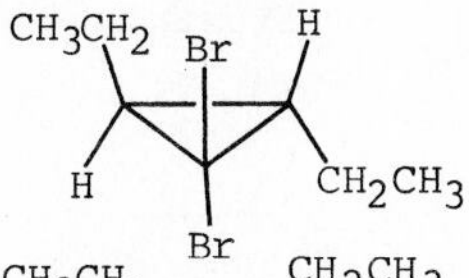 *trans*-1,1-dibromo-2,3-diethylcyclopropane

(m) *cis* gives 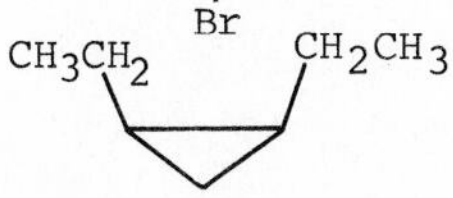 *cis*-1,2-diethylcyclopropane

trans gives 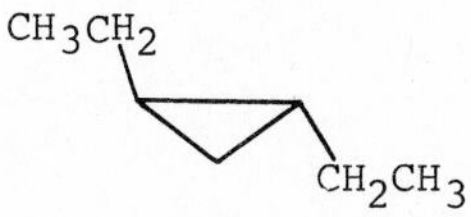 *trans*-1,2-diethylcyclopropane

7. (a) 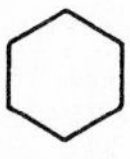 cyclohexane

(h, i) 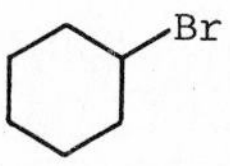 bromocyclohexane

(b) 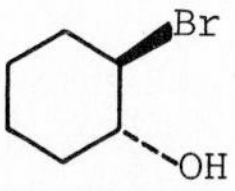 (±)-*trans*-2-bromocyclohexanol

(j) 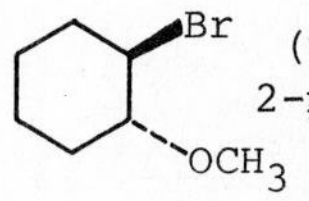 (±)-*trans*-1-bromo-2-methoxycyclohexane

(c) 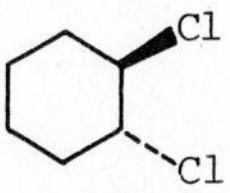 (±)-*trans*-1,2-dichlorocyclohexane

(k) 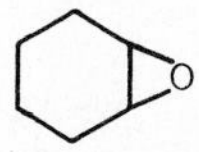 epoxycyclohexane

(d) *cis*-1,2-cyclo-hexanediol

meso

(ℓ) 7,7-dibromo-bicyclo[4.1.0]heptane

(e, f) cyclohexanol

(m) bicyclo[4.1.0]heptane

(g) hexanedial

8. (a) $CH_3CH_2CH_2CH=CH_2$

(b)

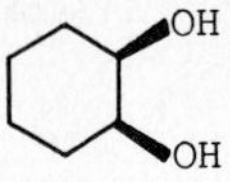

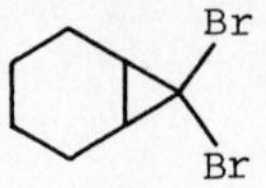

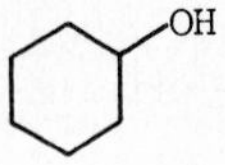

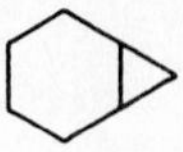

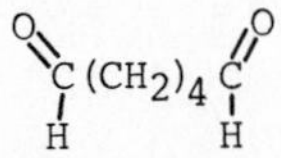

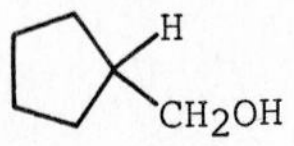

(c) + enantiomer

(d)

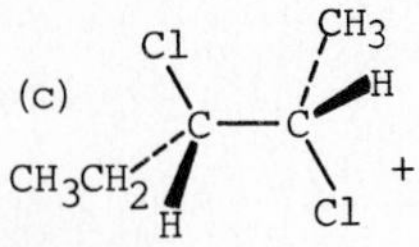

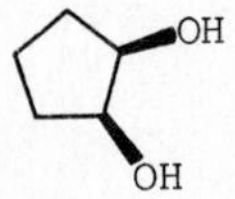

(e)

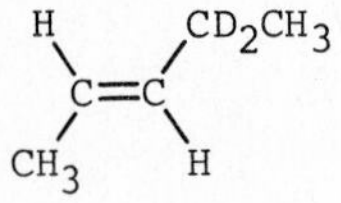

(f)

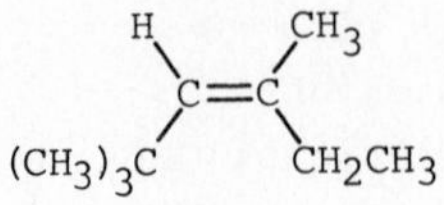

(g)

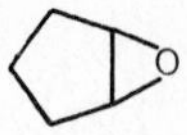

(h)

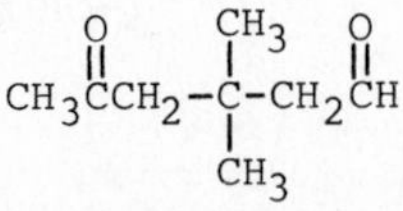

(i)   (t-butyl is always equatorial)

product from *anti*, diaxial addition

9. (a) $CH_3CHBrCH_3$ $\xrightarrow[\substack{or \\ t\text{-}C_4H_9OK/t\text{-}C_4H_9OH \\ \Delta}]{C_2H_5OK/C_2H_5OH}$ $CH_3CH=CH_2$ $\xrightarrow[peroxides]{HBr}$ $CH_3CH_2CH_2Br$

(b) $CH_3CHOHCH_3$ $\xrightarrow[\substack{or \\ H_2SO_4,\Delta}]{Al_2O_3,\Delta}$ $CH_3CH=CH_2$ $\xrightarrow[]{B_2H_6}$ $\xrightarrow[]{\substack{H_2O_2 \\ OH^-}}$ $CH_3CH_2CH_2OH$

(c) $\xrightarrow[h\nu]{Cl_2}$ $\xrightarrow[\Delta]{\substack{C_2H_5ONa, \\ C_2H_5OH}}$ $\xrightarrow[\substack{or \\ H_2O_2/OsO_4}]{\text{cold dil. } KMnO_4}$

(d) $\xrightarrow[]{CH_3CO_3H}$ $\xrightarrow[H_2O]{H_2SO_4}$

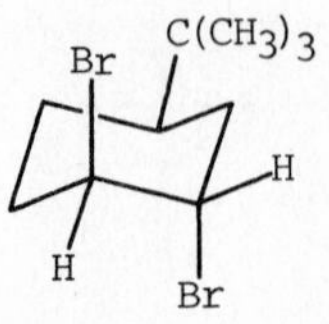

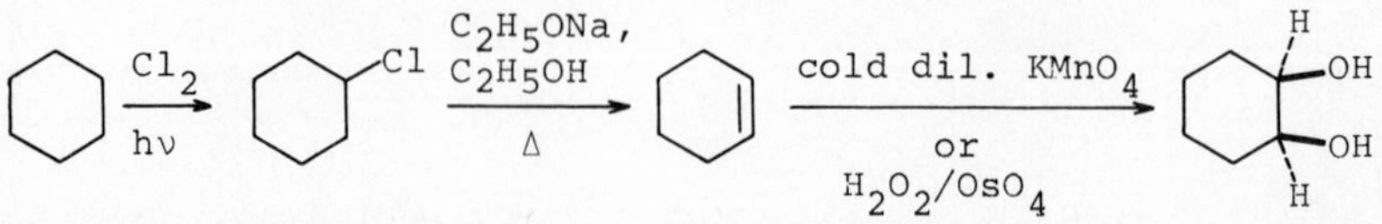

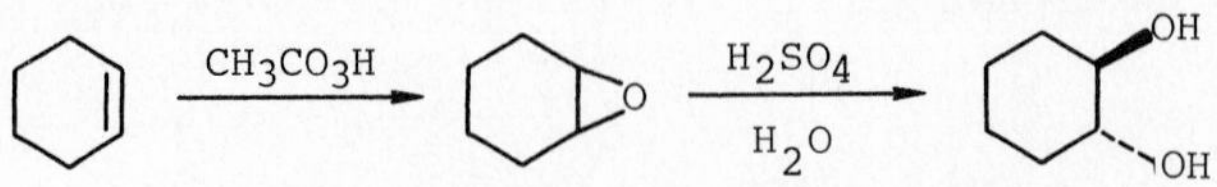

(e) $CH_3CH_2CHBrCH_2Br$ $\xrightarrow[\substack{alc. \\ \Delta}]{Zn}$ $CH_3CH_2CH=CH_2$ $\xrightarrow[\substack{peroxides \\ \Delta}]{CBr_4}$ $CH_3CH_2\overset{\overset{\displaystyle Br}{|}}{C}HCH_2CBr_3$

<u>Note</u>: 2nd step is a radical chain reaction.

$$RCH=CH_2 + \cdot CBr_3 \longrightarrow \cdot RCHCH_2CBr_3$$

$$R\overset{\cdot}{C}HCH_2CBr_3 + CBr_4 \longrightarrow RCHBrCH_2CBr_3 + \overset{\cdot}{C}Br_3$$

(f) $CH_3CH_2\overset{\overset{\displaystyle CH_3}{|}}{C}=CH_2$ $\xrightarrow[CH_3OH]{Hg(OAc)_2}$ $CH_3CH_2\overset{\overset{\displaystyle CH_3}{|}}{\underset{\underset{\displaystyle OCH_3}{|}}{C}}-CH_2HgOAc$ $\xrightarrow[NaOH]{NaBH_4}$ $CH_3CH_2\overset{\overset{\displaystyle CH_3}{|}}{\underset{\underset{\displaystyle OCH_3}{|}}{C}}-CH_3$

or
HCl or HBr/inhibitors

$\longrightarrow$ $CH_3CH_2\overset{\overset{\displaystyle CH_3}{|}}{\underset{\underset{\displaystyle X}{|}}{C}}CH_3$ $\xrightarrow[(S_N1 \ rx)]{CH_3OH}$ $CH_3CH_2\overset{\overset{\displaystyle CH_3}{|}}{\underset{\underset{\displaystyle OCH_3}{|}}{C}}CH_3$

(g) $CH_3CH_2\overset{\overset{\displaystyle CH_3}{|}}{C}=CH_2$ $\xrightarrow[OH^-]{B_2H_6 \quad H_2O_2}$ $CH_3CH_2\overset{\overset{\displaystyle CH_3}{|}}{C}HCH_2OH$ $\xrightarrow{K} \xrightarrow{CH_3I}$ $CH_3CH_2\overset{\overset{\displaystyle CH_3}{|}}{C}HCH_2OCH_3$

(h)

(i)

(j)

(k)

(ℓ) $CH_3CH_2CH_2CH_2CH_2OH$ $\xrightarrow[H_2SO_4]{HBr}$ $CH_3CH_2CH_2CH_2CH_2Br$ $\xrightarrow{t-BuO^-}$

$\downarrow Al_2O_3, \ \Delta$

$CH_3CH_2CH_2CH=CH_2$ $\xrightarrow[\substack{inhibitors \\ (ionic)}]{HBr}$ $CH_3CH_2CH_2\overset{\overset{\displaystyle Br}{|}}{C}HCH_3$

(m) $CH_3CH_2\overset{\overset{\displaystyle CH_3}{|}}{\underset{\underset{\displaystyle D}{|}}{C}}CH_2OH$ $\xrightarrow[\Delta]{Al_2O_3}$ $CH_3CH_2\overset{\overset{\displaystyle CH_3}{|}}{C}=CH_2$ $\xrightarrow[cat.]{H_2}$ $CH_3CH_2CH(CH_3)_2$

(n) $CH_3CH_2CH_2CH(CH_3)_2$ $\xrightarrow[h\nu]{Br_2}$ $CH_3CH_2CH_2\overset{\overset{\displaystyle Br}{|}}{C}(CH_3)_2$ $\xrightarrow{t-BuO^-}$

$CH_3CH_2CH_2\overset{\overset{\displaystyle CH_3}{|}}{C}=CH_2$ $\xrightarrow[OH^-]{B_2H_6 \quad H_2O_2}$ $CH_3CH_2CH_2\overset{\overset{\displaystyle CH_3}{|}}{C}HCH_2OH$

(o)

(p) $CH_3CH_2-\overset{\overset{\displaystyle }{}}{\underset{\underset{\displaystyle Br}{|}}{C}}(CH_3)_2$ $\xrightarrow[HOC(CH_3)_3]{KOC(CH_3)_3}$ $CH_3CH_2-C\overset{\diagup CH_3}{\diagdown CH_2}$ $\xrightarrow[CHCl_3]{KOH}$

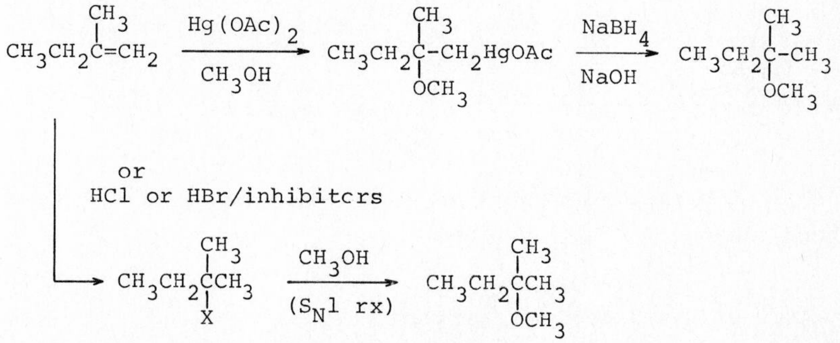

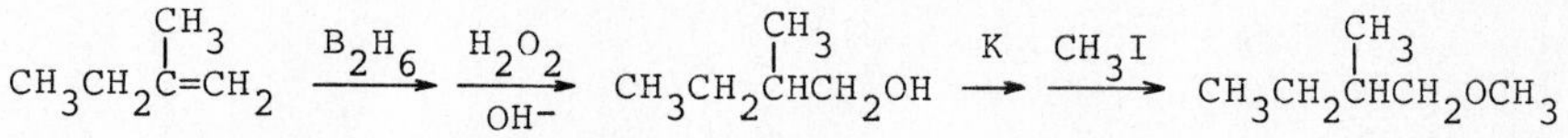

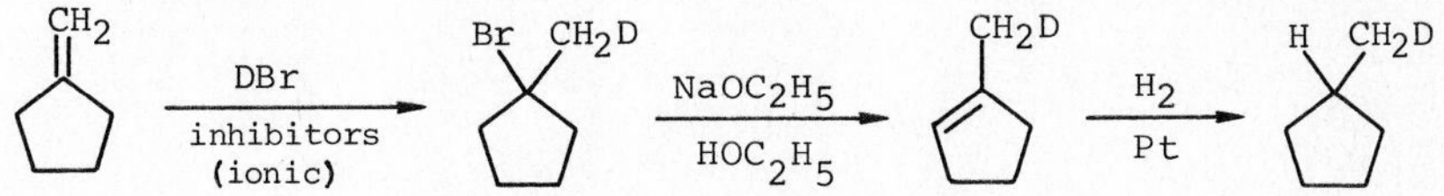

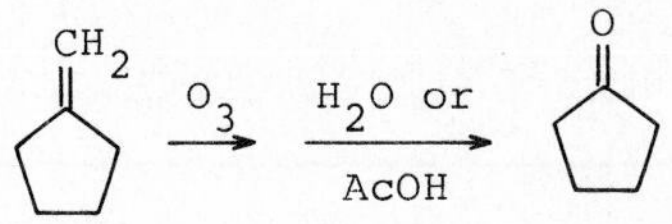

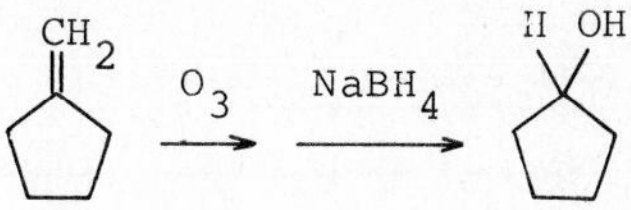

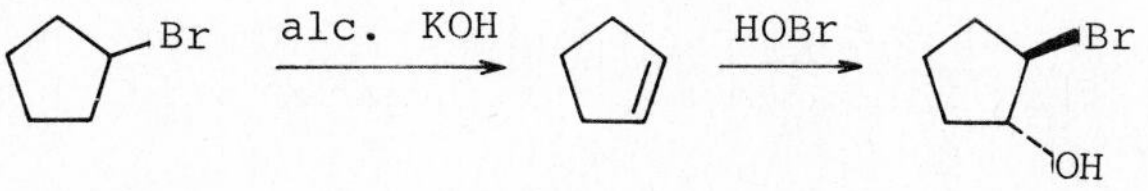

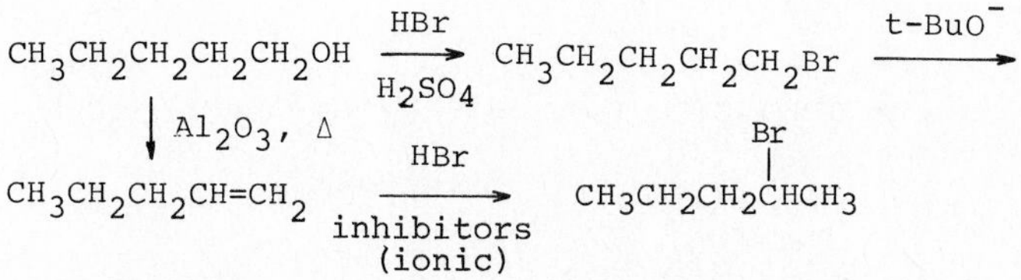

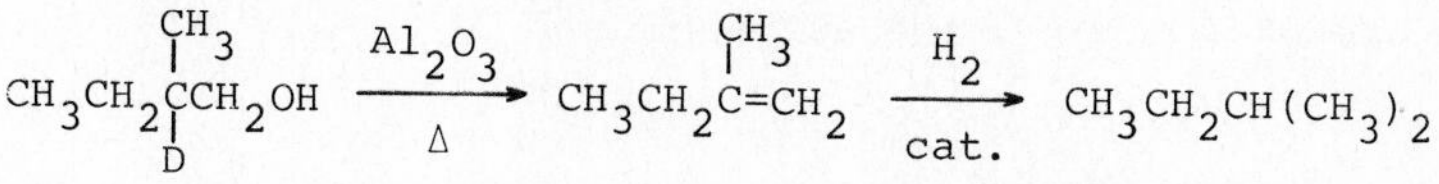

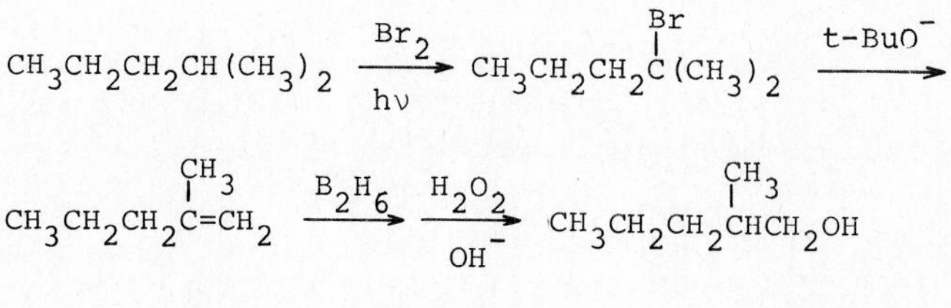

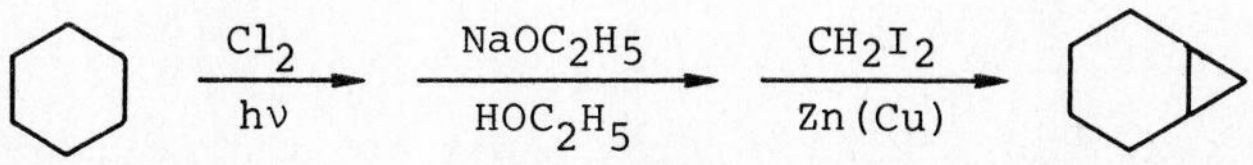

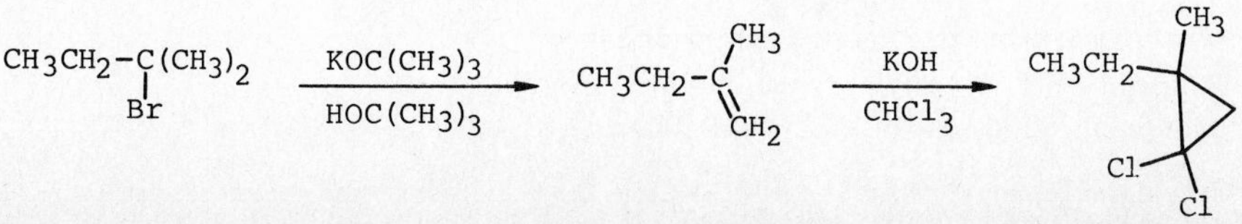

10.

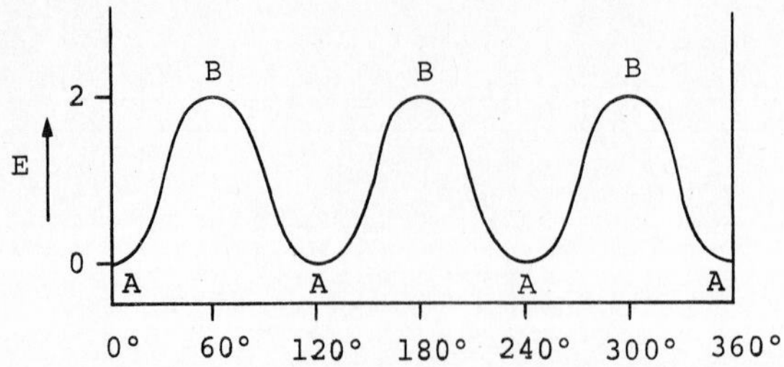

The angle plotted is the dihedral angle between the plane of the double bond and the $C-C_{methyl}-H$ plane.

11.   The structure of 4,4-dimethyl-2-pentene is:

$CH_3-\overset{\overset{\displaystyle CH_3}{|}}{\underset{\underset{\displaystyle CH_3}{|}}{C}}$ $\overset{\text{steric}}{\underset{\text{hindrance}}{}}$ $\underset{H}{\overset{CH_3}{C=C}}\overset{CH_3}{\underset{H}{}}$

$CH_3-\overset{\overset{\displaystyle CH_3}{|}}{\underset{\underset{\displaystyle CH_3}{|}}{C}}$ $\underset{H}{\overset{H}{C=C}}\overset{}{\underset{CH_3}{}}$ (no steric hindrance)

        cis   trans

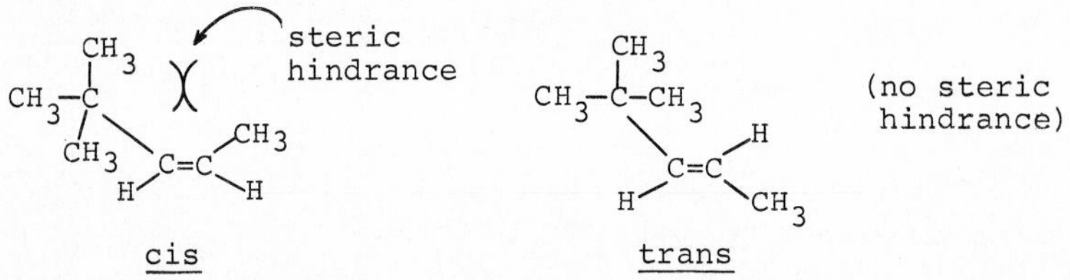

In the cis isomer, the terminal methyl groups are close in space, and the electron clouds interact and repel each other (steric hindrance).

12. $\overset{\overset{\displaystyle CH_3}{|}}{CH_3CCH_2CH_2CH_2CH_2CH_2}$ The tertiary OH groups dehydrate readily
           ↗OH   OH↖   under acid conditions *via* the relatively
                                         stable tert-carbocations.   Primary OH
        tertiary   primary   groups dehydrate much less readily.

$(CH_3)_2\underset{OH}{C}(CH_2)_5OH + H^+ \longrightarrow (CH_3)_2\overset{+}{C}(CH_2)_5OH \longrightarrow (CH_3)_2C=CH(CH_2)_4OH$

                                                    plus some $\overset{\overset{\displaystyle CH_3}{|}}{CH_2=C(CH_2)_5OH}$

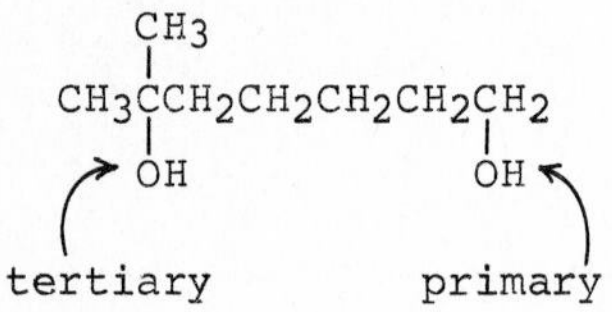

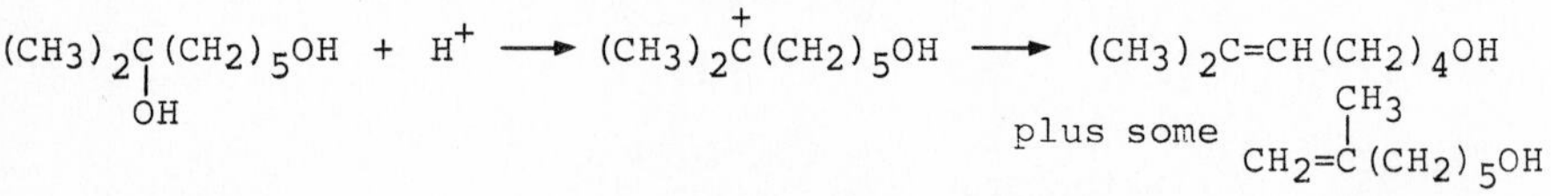

13.   Both isomers are produced *via* the carbocation,

$\overset{CH_3}{\underset{CH_3}{}}\overset{}{\underset{+}{C}}-CH_2C(CH_3)_3$

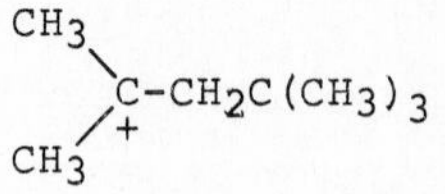

Loss of a primary hydrogen gives $CH_2=\overset{\overset{\displaystyle CH_2}{|}}{C}CH_2C(CH_3)_3$, a disubstituted alkene with little steric hindrance. Loss of a sec-H gives a tri-

$\overset{CH_3}{\underset{CH_3}{}}\overset{}{C=C}\overset{C(CH_3)_3}{\underset{H}{}}$

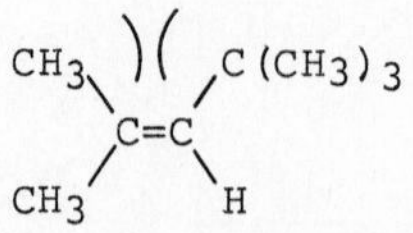

substituted alkene with much steric hindrance between the adjacent methyl and t-butyl. Look at models.

14. 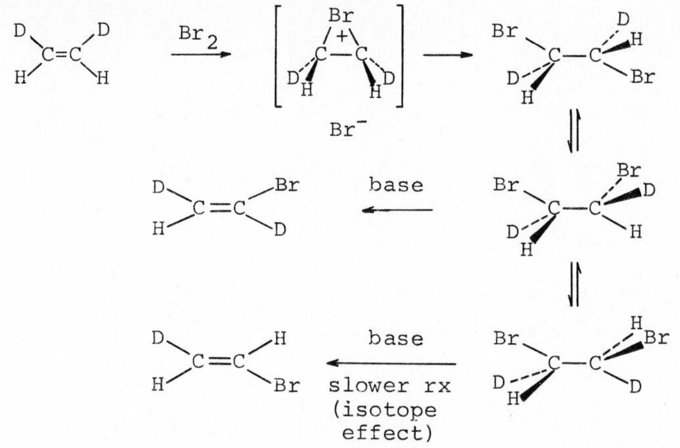

Similarly,

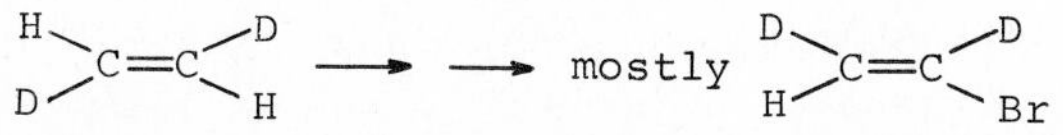

15. Both reactions must have ΔH° that is not too positive; otherwise, $E^{\ddagger}$ is too high and reaction will be slow. For Y = Br, HS, $(CH_3)_3C$, both ΔH°'s are negative (exothermic), and both steps should be facile. For Y = I, the first step (a) has ΔH° = +5 kcal $mole^{-1}$; the activation energy is at least this high, but this reaction could still be possible. For the other compounds one step or the other is no good; that is, one ΔH° is so positive that the reaction has a high activation energy and is slow.

16. 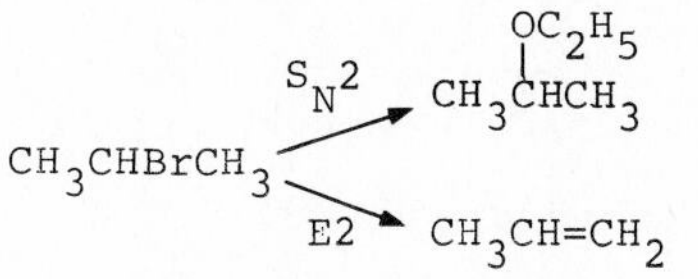

S_N2 reaction is almost unaffected by deuterium. Loss of deuterium in E2 is slower than of hydrogen.

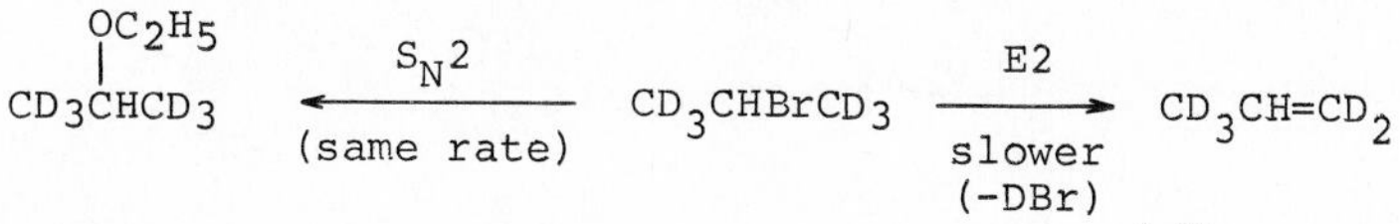

The deuterium isotope effect, k_D/k_H (E2), $= \frac{1/2}{3/1} = 1/6$

17.

Monosubstituted ethylenes	$-\Delta H^\circ_{hydrog}$
propene	29.7
1-butene	30.2
1-pentene	29.8

Disubstituted ethylenes	$-\Delta H^\circ_{hydrog}$
cis-2-butene	28.5
cis-2-pentene	28.1
trans-2-butene	27.4
trans-2-pentene	27.2
2-methylpropene	28.1
2-methyl-1-butene	28.3

Alkyl substitution on a double bond has a stabilizing effect (see Section 11.4), therefore less energy is released on hydrogenation of the alkylethylenes than for ethylene itself. Similarly, *trans* alkenes are more stable than *cis* alkenes, and they have less negative ΔH°_{hydrog}.

18.

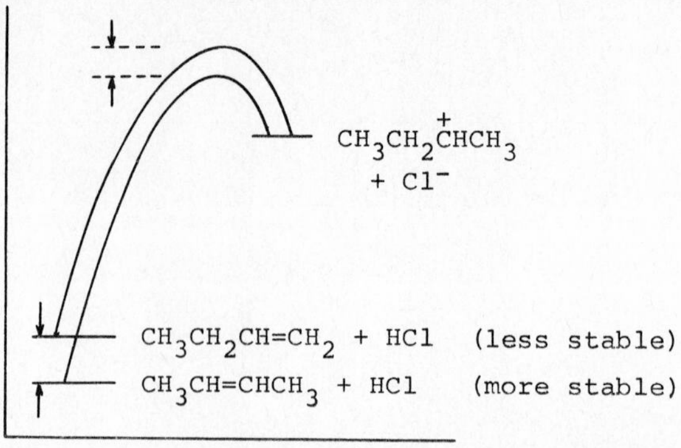

The difference in energies of the two transition states is probably less than the difference in energies of the two reactants. Thus, $\Delta H^{\ddagger}$ for 1-butene is smaller than $\Delta H^{\ddagger}$ for 2-butene, and 1-butene reacts more rapidly. On this basis, cis-2-butene should be more reactive than trans-2-butene. In general, when two isomers give the same intermediate or product *via* comparable transition states, the less stable isomer reacts faster, because it has a smaller energy barrier to overcome.

19. $\Delta H^\circ_f\,(CH_3CH{=}CH_2) = 4.9$ kcal mole^{-1} *(Appendix I)*

$\Delta H^\circ_f\,(\cdot CN) = 99$ kcal mole^{-1} *(Appendix II)*

$\Delta H^\circ_f\,(HCN) = \Delta H^\circ_f\,(H\cdot) + \Delta H^\circ_f\,(\cdot CN) - DH^\circ(H{-}CN)$
$= 52 + 99 - 120 = 31$ kcal mole^{-1}

$\Delta H^\circ_f\,(CH_3CH_2CH_2CN) = \Delta H^\circ_f\,(CH_3CH_2CN) + \Delta H^\circ_f\,(-CH_2-)$
$= 12.1 - 5.0 = 7.1$ kcal mole^{-1} *(see problem #16, Chapter 5)*

$\Delta H^\circ_f\,(CH_3\dot{C}HCH_2CN) = \Delta H^\circ_f\,(CH_3CH_2CH_2CN) + DH^\circ\,(CH_3\underset{H}{C}HCH_2CN) - \Delta H^\circ_f\,(H\cdot)$
$= 7 + 95 - 52 = 50$ kcal mole^{-1}

(a) $CH_3{-}CH{=}CH_2 + HCN \longrightarrow CH_3CH_2CH_2CN$
$\Delta H^\circ_f =$ 4.9, 31, 7 $\qquad \Delta H^\circ_f = -29$ kcal mole^{-1}

(b) STEP 1: $CH_3CH{=}CH_2 + \cdot CN \longrightarrow CH_3\dot{C}HCH_2CN$
$\Delta H^\circ_f =$ 4.9, 99, 50 $\qquad \Delta H^\circ_f = -54$ kcal mole^{-1}

STEP 2: $CH_3\dot{C}HCH_2CN + HCN \longrightarrow CH_3CH_2CH_2CN + \cdot CN$
$\Delta H^\circ_f =$ 50, 31, 7, 99 $\qquad \Delta H^\circ = +25$ kcal mole^{-1}

(c) The overall reaction is exothermic, but the second step is far too endothermic. The intermediate $CH_3\dot{C}HCH_2CN$ radicals would dimerize rather than abstract H from HCN.

20. $CH_3CH_2CH_2\underset{}{\overset{Br}{C}}HCH(CH_3)_2 \xrightarrow{base} CH_3CH_2CH_2CH{=}C(CH_3)_2$ (C) $+ CH_3CH_2CH{=}CHCH(CH_3)_2$ (D + E, *cis + trans*)

C $\downarrow$

$CH_3CH_2CH_2\overset{OH}{C}HCH(CH_3)_2$

F

$$D + E \longrightarrow \underset{F}{CH_3CH_2CH_2\overset{OH}{\overset{|}{C}}HCH(CH_3)_2} + \underset{G}{CH_3CH_2\overset{OH}{\overset{|}{C}}HCH_2CH(CH_3)_2}$$

From the data given, we cannot determine whether D is *cis* or *trans*.

In working this type of "roadmap" problem, it helps to summarize the data in the following manner:

$$C_7H_{15}Br \xrightarrow{\text{base}} \underset{(C_7H_{14})}{C + D + E} \xrightarrow{H_2/\text{cat.}} CH_3\overset{CH_3}{\overset{|}{C}}HCH_2CH_2CH_2CH_3$$

$$C \xrightarrow{B_2H_6} \xrightarrow[OH^-]{H_2O_2} \underset{\text{alcohol}}{F}$$

$$D + E \xrightarrow{B_2H_6} \xrightarrow[OH^-]{H_2O_2} F + G \text{ (isomeric alcohol)}$$

about equal amounts

We can rule out 1-Br and 6-Br (E2 would give a single alkene), 2-Br, (E2 would give two alkenes), 4-Br (E2 would give four alkenes, two *cis-trans* pairs). For 5-Br,

$$(CH_3)_2CHCH_2CH_2\underset{Br}{\underset{|}{C}}HCH_3 \longrightarrow \underset{cis\text{ and }trans}{(CH_3)_2CHCH_2CH{=}CHCH_3} + (CH_3)_2CHCH_2CH_2CH{=}CH_2$$

but with B_2H_6 and H_2O_2/OH^-, 5-methyl-1-hexene would give mainly 5-methyl-1-hexanol, and the 5-methyl-2-hexenes cannot give this alcohol. This leaves only $(CH_3)_2CH\underset{Br}{\underset{|}{C}}HCH_2CH_2CH_3$, which reacts as shown above.

21. (a)

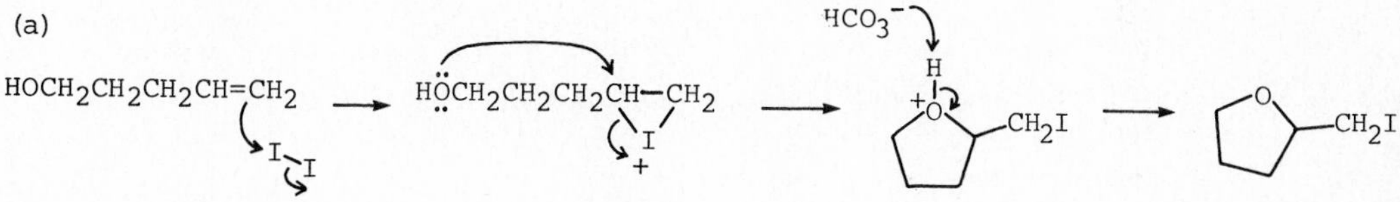

(b)

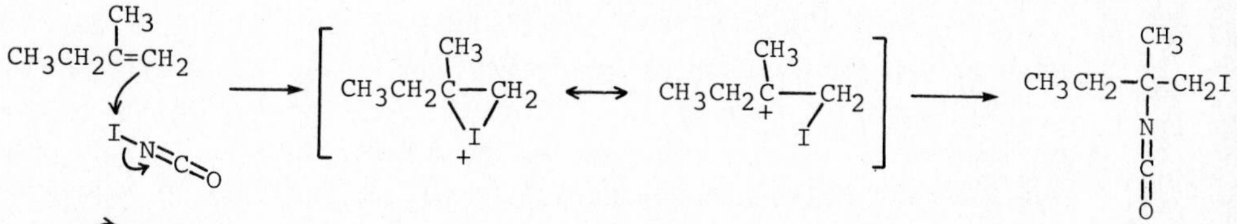

(c)

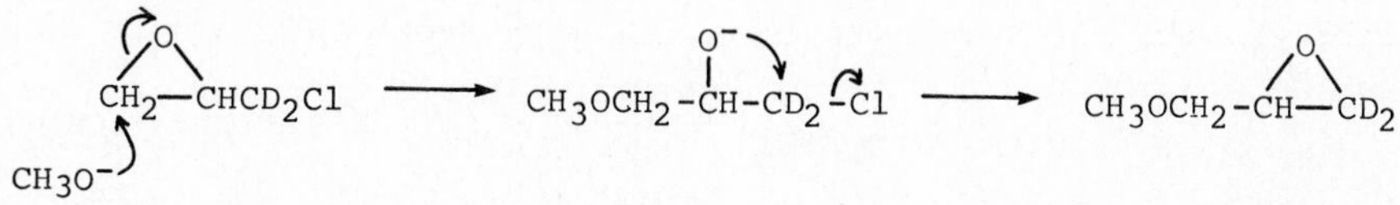

22. Attack from the top side of the molecule (as shown at the right) is sterically hindered, so the incoming reagent approaches from the bottom. With m-chloroperbenzoic acid, the incoming

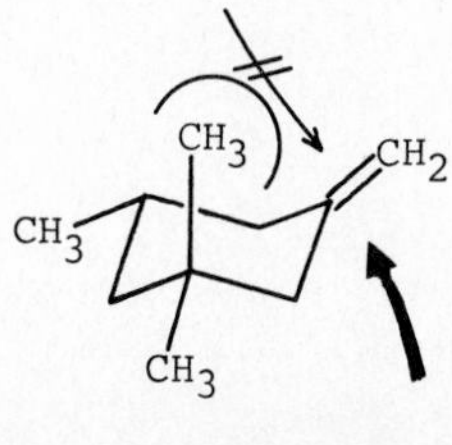

reagent is the peracid and the indicated epoxide
is formed preferentially:

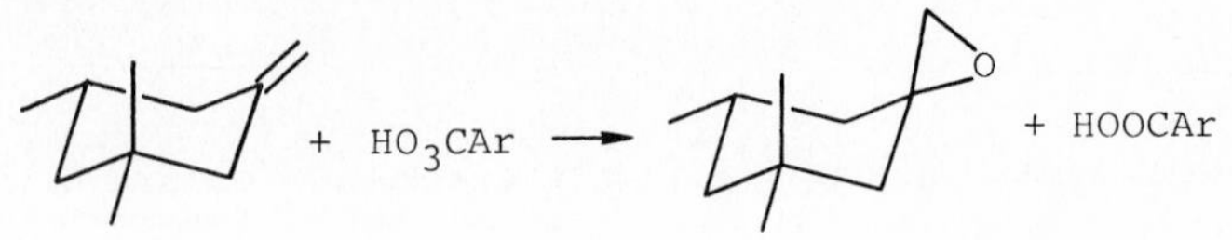

When the epoxide is formed in the two-step procedure via the bromohydrin, it
is the attack of bromine which determines the stereochemistry. Because
addition of water is *anti* to this, the final epoxide has the opposite
configuration:

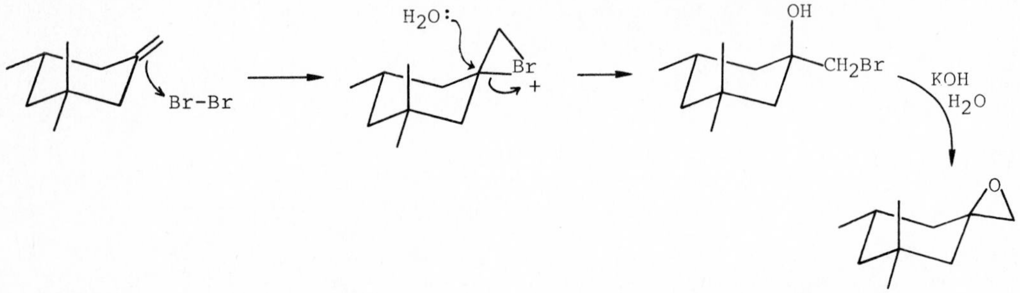

23.

$$\underset{C_{11}H_{24}O}{\underline{H}} \xrightarrow{PBr_3} \underset{C_{11}H_{23}Br}{\underline{I}} \xrightarrow[HOC_2H_5]{KOC_2H_5} \underset{C_{11}H_{22}}{\underline{J}\ (major)} + \underset{C_{11}H_{22}}{\underline{K}\ (minor)}$$

1. O_3
2. $NaBH_4$

$(CH_3)_2CHCH_2OH$ $(CH_3)_2CHCH_2CH_2OH$

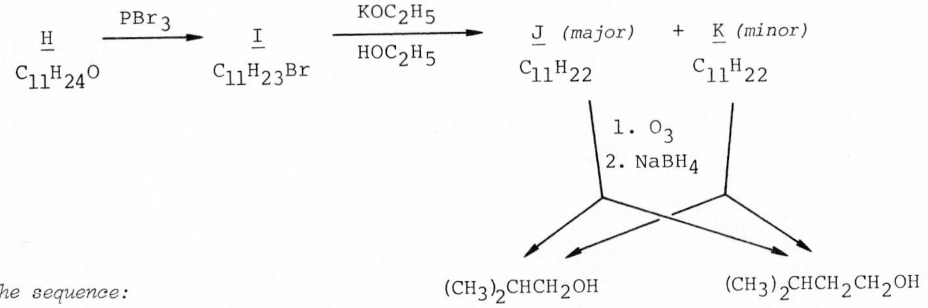

The sequence:
1. O_3 2. $NaBH_4$ cleaves double
bonds and reduces the products
to alcohols, therefore $\underline{J}$ and $\underline{K}$
must be $(CH_3)_2CHCH_2CH=CHCH_2CH_2CH(CH_3)_2$.

$\underline{J}$ must be the *trans* isomer (formed more easily) and $\underline{K}$ the *cis* isomer.
The 2,8-dimethyl-4-nonenes arose from an alkyl bromide by E2 elimination.
Two bromides could have given $\underline{J}$ and $\underline{K}$, but the 4-bromo isomer would have
given 2,8-dimethyl-3-nonene as well. Therefore $\underline{I}$ must be 5-bromo-2,8-
dimethylnonane and $\underline{H}$ must be the corresponding alcohol.

$$(CH_3)_2CHCH_2\overset{\overset{\displaystyle Br}{|}}{C}HCH_2CH_2CH_2CH(CH_3)_2 \xrightarrow[HOC_2H_5]{KOC_2H_5} (CH_3)_2CHCH=CHCH_2CH_2CH_2CH(CH_3)_2$$
$$+ \underline{J}\ and\ \underline{K}$$

$$(CH_3)_2CHCH_2CH_2\overset{\overset{\displaystyle Br}{|}}{C}HCH_2CH_2CH(CH_3)_2 \xrightarrow[HOC_2H_5]{KOC_2H_5} \underline{J}\ and\ \underline{K}\ only$$

$$\underline{I}$$

$$(CH_3)_2CHCH_2CH_2\overset{\overset{\displaystyle OH}{|}}{C}HCH_2CH_2CH(CH_3)_2$$

$$\underline{H}$$

24. 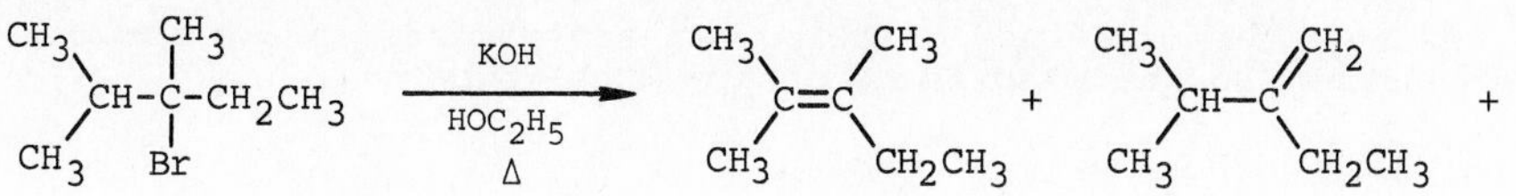

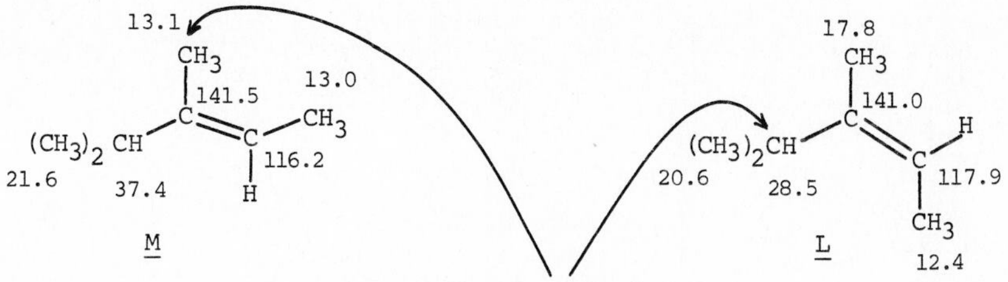

Note how these carbons are shifted upfield when the substituent is *cis*.

25. Compounds N and O: only three kinds of carbons, two of which are sp^2-hybridized, so that the possibilities are:

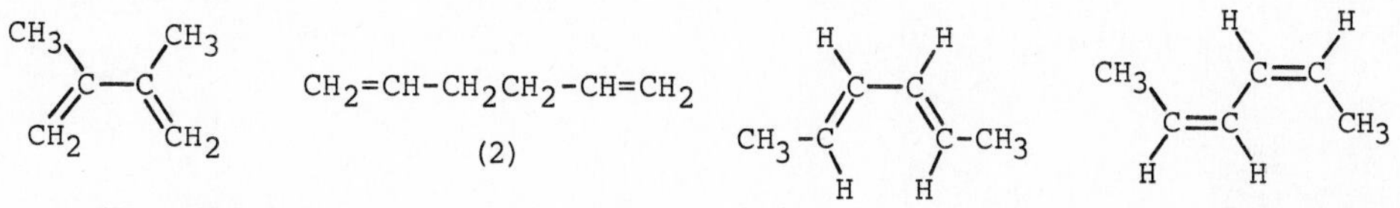

(1) (2) (3) (4)

Structures (1) and (2) can be ruled out because the expected chemical shifts are inconsistent (carbons 2 and 3 of (1) would be expected around 140 ppm, and carbons 1 and 6 of (2) around 113 ppm). In structure (4), because of the *cis* geometry of the double bonds, the resonances for the methyls and central carbons will appear upfield of those in structure (3). Therefore (3) is O and (4) is N.

Compound P is unsymmetrical: 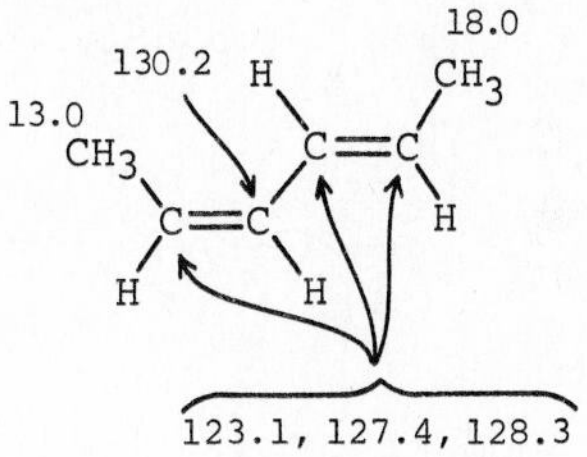

26. 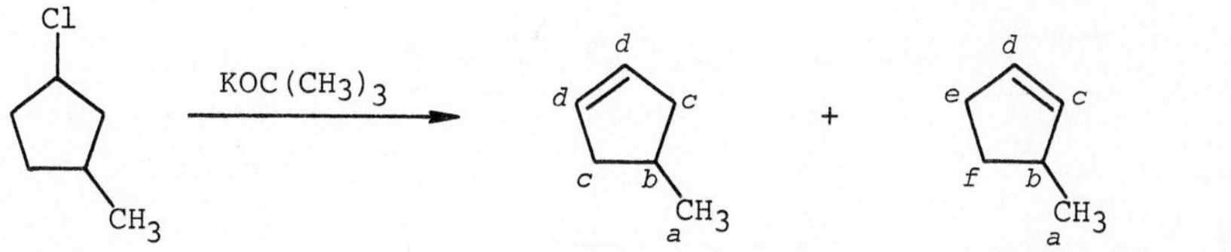

symmetrical structure, only four different kinds of carbons *unsymmetrical, six different resonances*

11.E Supplementary Problems

S1. Write structures for the following compounds:

(a) *cis*-3-methyl-2-heptene

(b) (R)-5-methylhex-4-en-2-ol

(c) (Z)-4-methyl-2-pentene

(d) (E)-5-chloro-3-isopropyl-5-methyl-2-hexene

(e) (Z)-3-bromo-3-hexene

(f) 1-chloro-6-methylcyclohexene

(g) (E)-cyclododecene

S2. Show how to accomplish the following transformations in a practical manner (more than one step is necessary in each case).

(a) $CH_3CH_2CH_2CH_2OH \longrightarrow CH_3CH_2\underset{OH}{CH}CH_3$

(b)

(c) $CH_3CH{=}CH_2 \longrightarrow (CH_3)_2CH\text{-}S\text{-}CH_2CH_2CH_3$

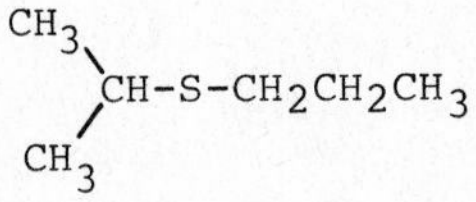

(d)

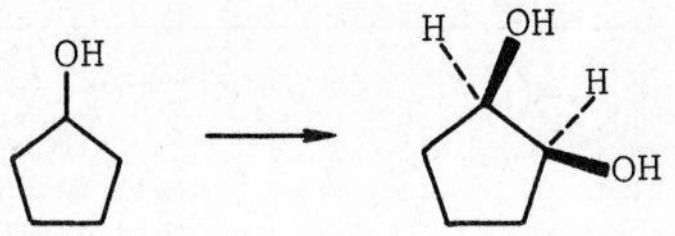

(e)

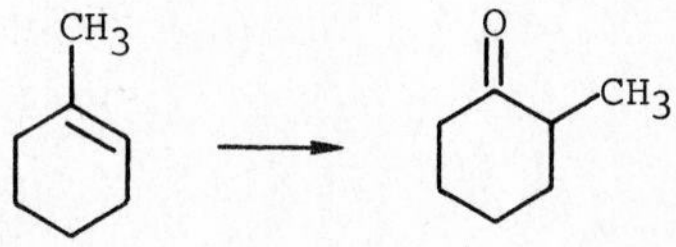

(f)

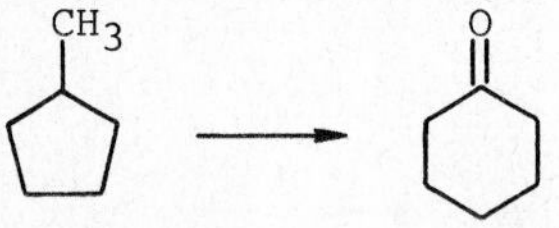

(g)

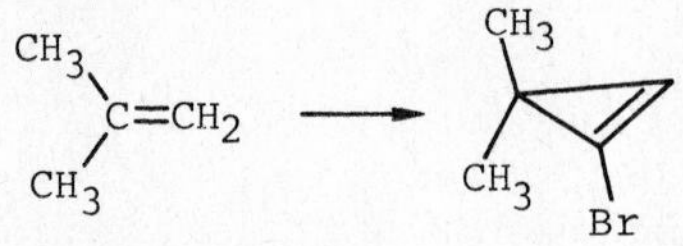

(h)

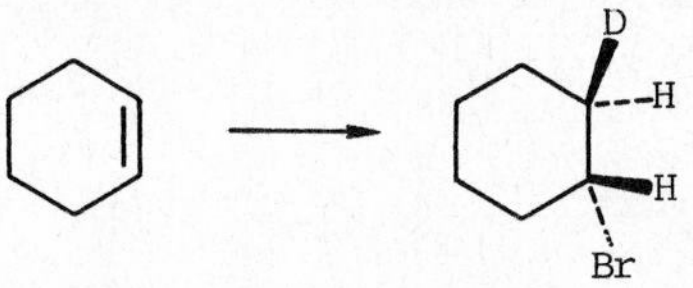

(without any of the <u>*cis*</u> *isomer)*

S3. Predict the product from the reaction of iodinemonochloride (ICl) with 1-methylcyclohexene, and justify your choice by writing a step-by-step mechanism for its formation.

S4. Predict the major products from the following reaction sequences.

(a)

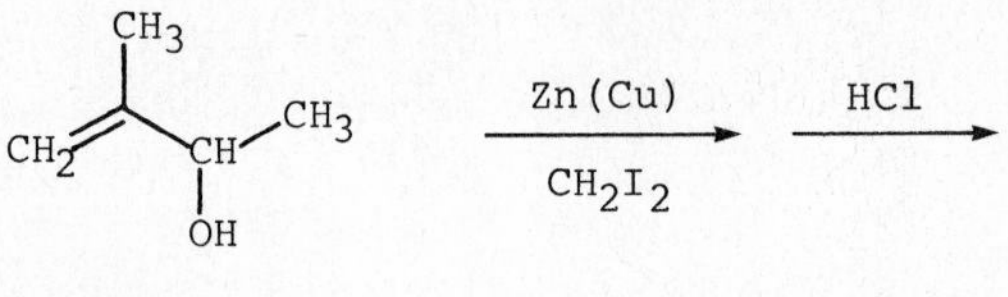

(b)

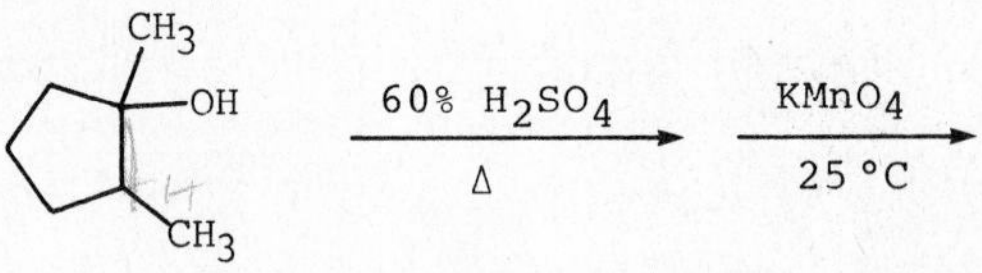

(c) $CH_3CH_2CH{=}CHC(CH_3)_3 \xrightarrow[HOC_2H_5]{Hg(OAc)_2} \xrightarrow{NaBH_4}$

(d) $(CH_3)_2C{=}CHCH_3 \xrightarrow[CH_3OH]{Br_2} \xrightarrow[HOC_2H_5]{KOC_2H_5} \xrightarrow[Pt]{D_2}$

(e) $(CH_3)_2C{=}C(CH_3)_2 + CCl_4 \xrightarrow{h\nu} \xrightarrow[HOC(CH_3)_3]{KOC(CH_3)_3}$

S5. A hydrocarbon (A) of formula C_7H_{12} was treated successively with diborane and alkaline hydrogen peroxide to provide compound B ($C_7H_{14}O$) as the only product. Reaction of B with <u>p</u>-toluenesulfonyl chloride and pyridine, and then with potassium <u>t</u>-butoxide in <u>t</u>-butyl alcohol gave an isomeric hydro-

carbon C (C_7H_{12}). Finally, treatment of C with ozone in methanol, followed
by work-up using sodium borohydride afforded 2-methyl-1,6-hexanediol.
Write complete structures for A, B, and C which are consistent with this
information.

S6.   Write a reasonable mechanism for the following transformation:

$$CH_2=CHCH_2CH_2CH=CH_2 \quad \xrightarrow[\text{H}_2\text{O}]{\text{2 Hg(OAc)}_2} \quad \xrightarrow{\text{NaBH}_4} \quad$$

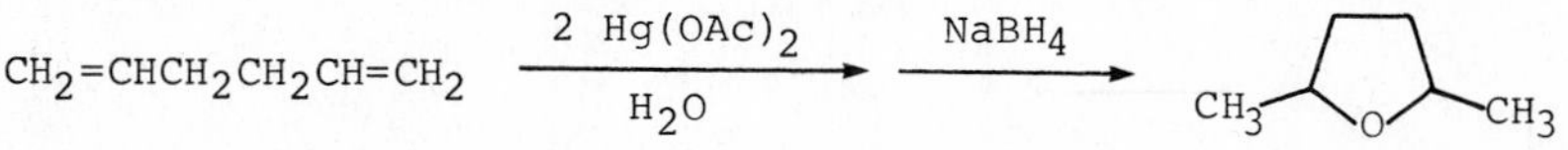

S7.   The bond dissociation energy for H-SH is 90 kcal mole^{-1}. Using this value
and Appendices I and II, determine $\Delta H°$ for each step in the free radical
addition of H_2S to ethylene. Is the proposed reaction feasible by this
mechanism?

S8.   Using Appendices I and II, determine whether the free radical addition of
water to ethylene to give ethanol is feasible thermodynamically.

11.F   Answers to Supplementary Problems

S1.   (a)

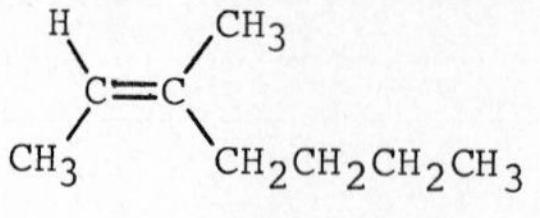

(b)

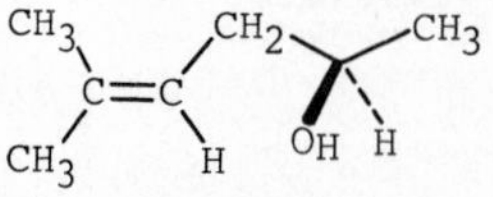

(c)

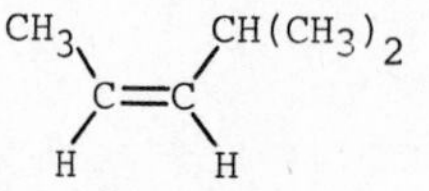

(d)

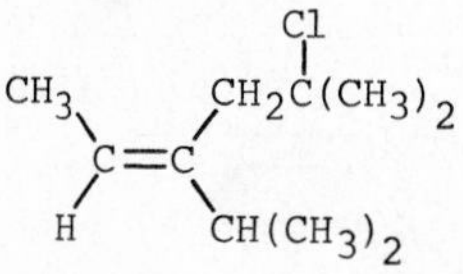

(e)

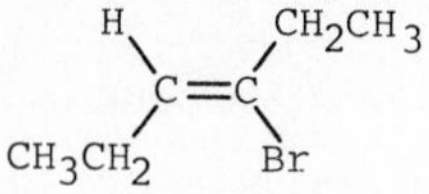

(f)

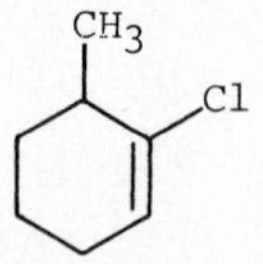

(g)

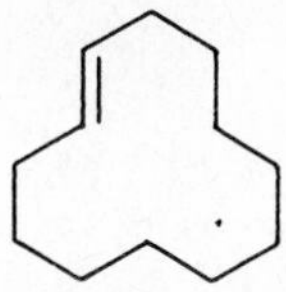

S2.   (a) $CH_3CH_2CH_2CH_2OH$ $\xrightarrow[400°]{Al_2O_3}$ $CH_3CH_2CH=CH_2$ $\xrightarrow[H_2O]{H_2SO_4}$ $CH_3CH_2CHCH_3$ with OH

(b)

$\xrightarrow[\text{(peroxides)}]{HBr}$ $\xrightarrow{NaCN}$

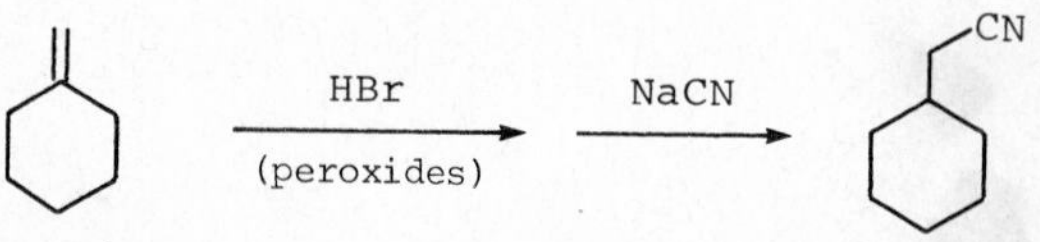

(c) $CH_3CH=CH_2$ $\xrightarrow[\text{2. NaSH}]{\text{1. HCl}}$ CH_3CHCH_3 with SH $\xrightarrow[\substack{100° \\ [O_2]}]{CH_3CH=CH_2}$ $(CH_3)_2CH-S-CH_2CH_2CH_3$

(d)

$\xrightarrow[\Delta]{\text{conc. } H_2SO_4}$ $\xrightarrow[\text{or } OsO_4/H_2O_2]{\text{cold, dilute } KMnO_4}$

(e)

$\xrightarrow{B_2H_6}$ $\xrightarrow[OH^-]{H_2O_2}$ $\xrightarrow[H_2SO_4]{K_2Cr_2O_7}$

(f)

$\xrightarrow[h\nu]{Br_2}$ $\xrightarrow{KOC(CH_3)_2}$ $\xrightarrow{KMnO_4}$

(g) $(CH_3)_2C=CH_2 + HCBr_3$ $\xrightarrow{KOH}$

$\xrightarrow[HOC_2H_5, \Delta]{NaOC_2H_5}$

(h)

$\xrightarrow{B_2D_6}$ $\xrightarrow[OH^-]{H_2O_2}$ $\xrightarrow{PBr_3}$

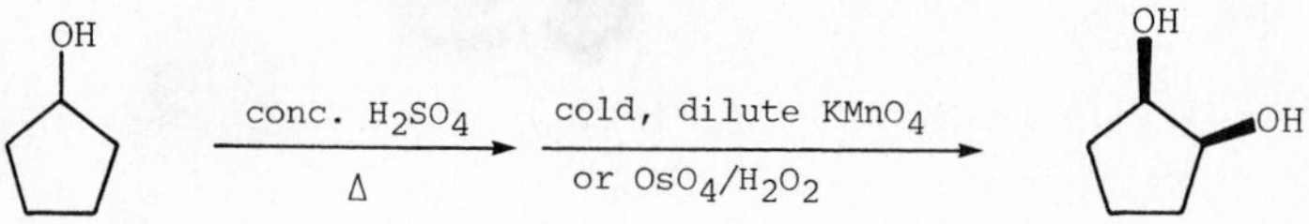

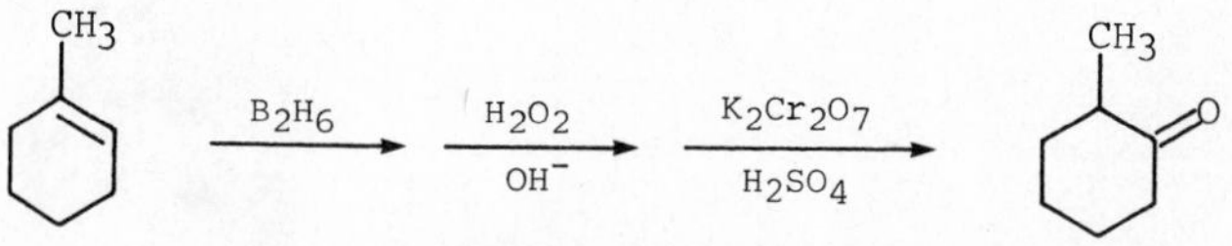

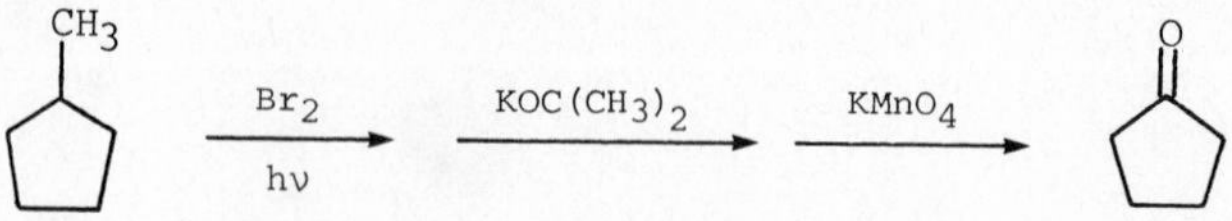

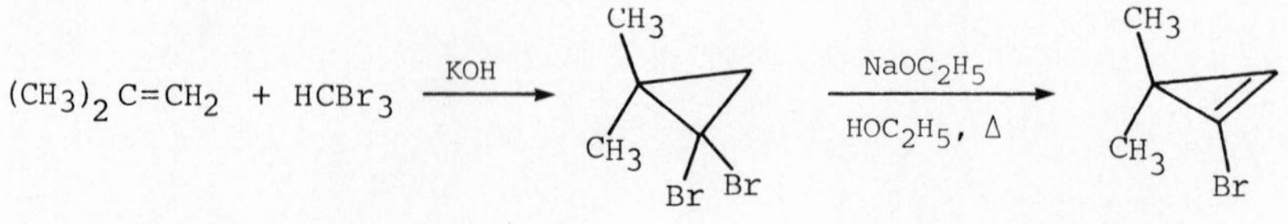

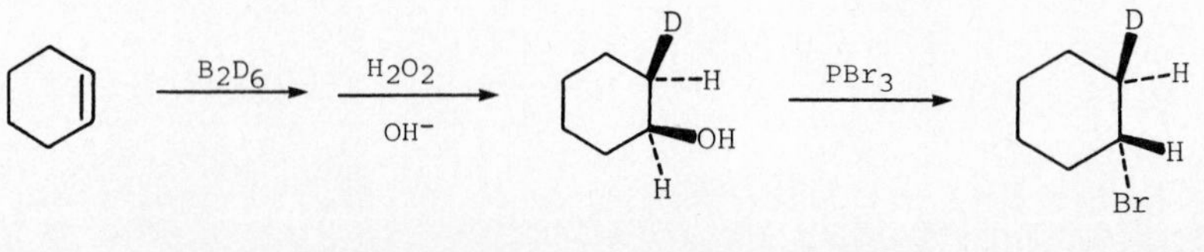

S3.

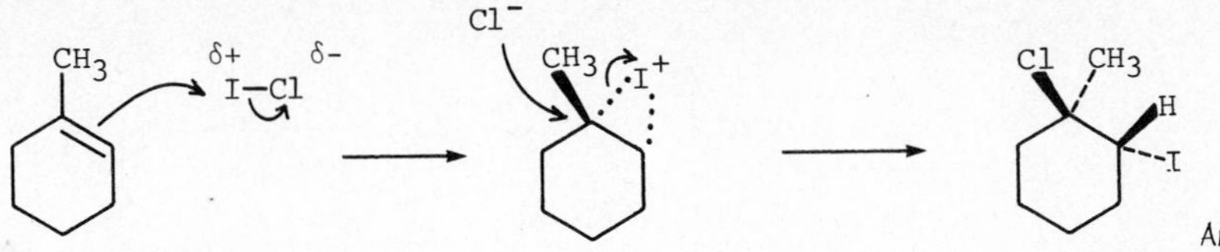

Anti addition,
Markovnikov
orientation

S4. (a)

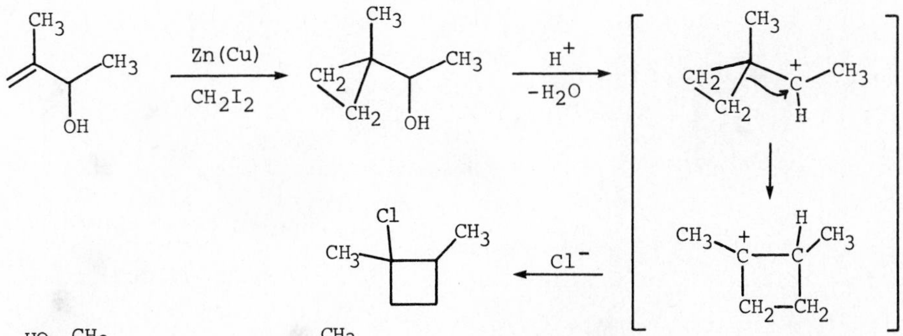

(b)

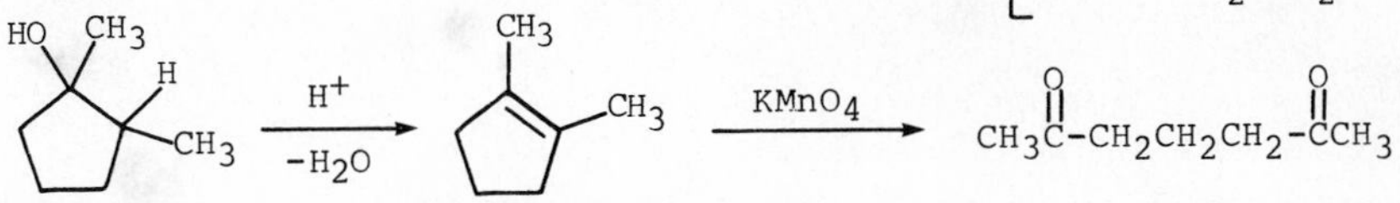

(c)

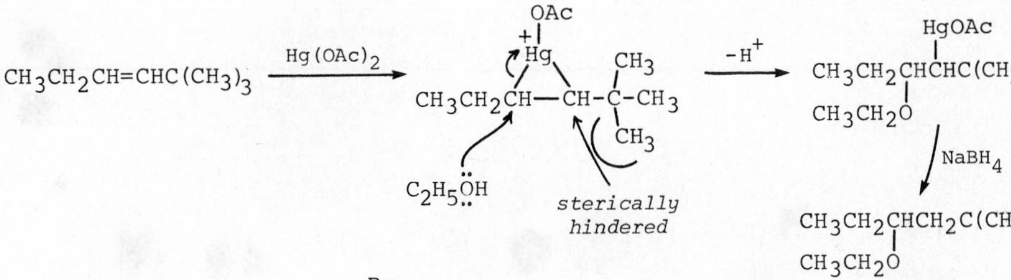

(d)

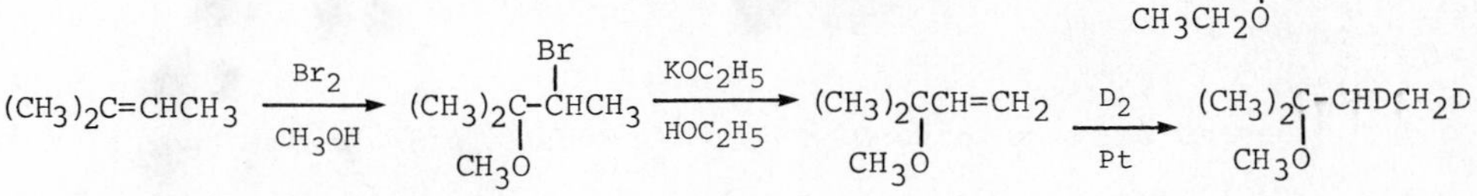

(e)

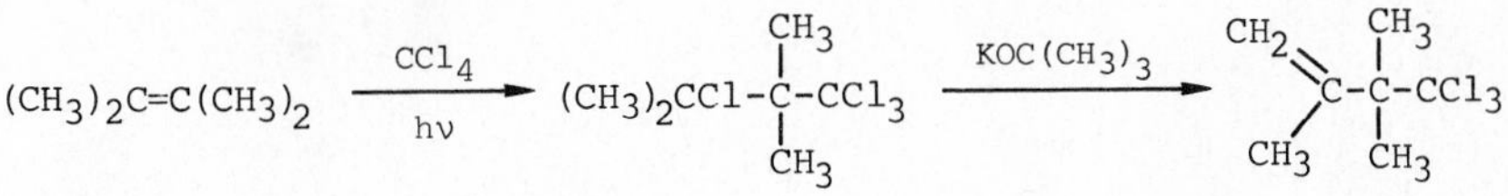

(f)

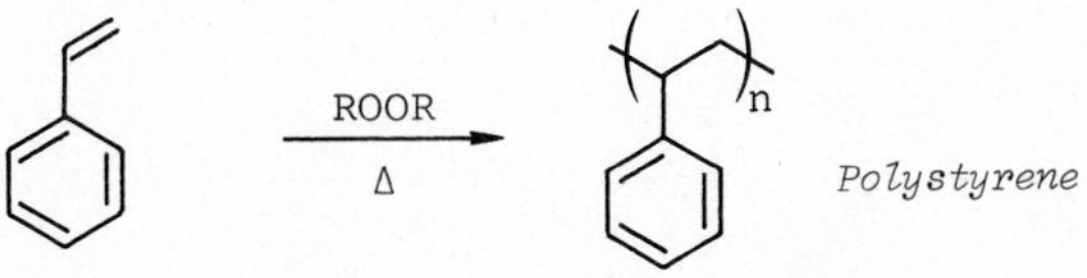

Polystyrene

S5.

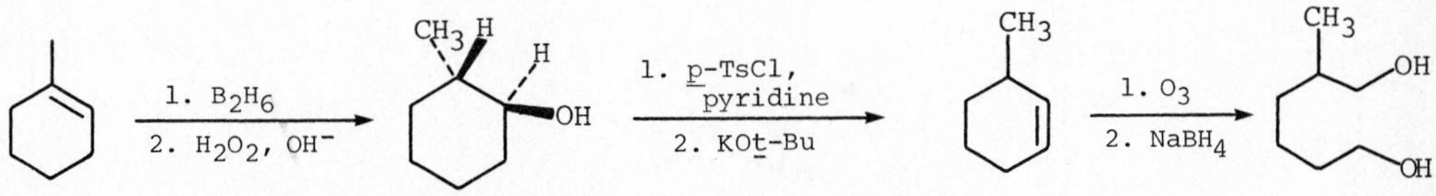

A B C

S6.

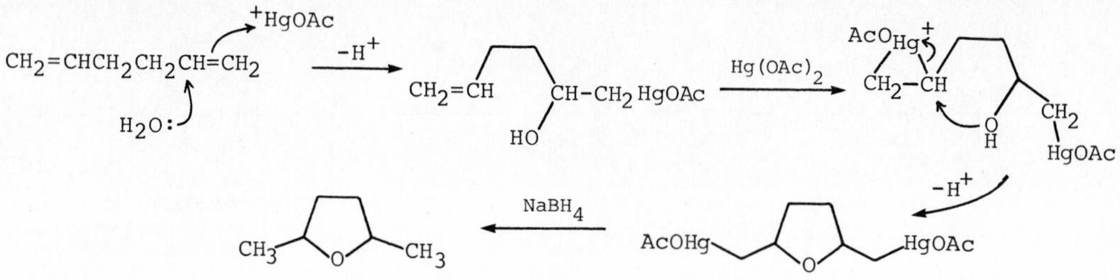

S7. $\Delta H^\circ_f(\cdot SH) = \Delta H^\circ_f(H_2S) + DH^\circ(H\text{-}SH) - \Delta H^\circ_f(H\cdot)$

$= -4.8 + 90 - 52 = 33$ kcal mole^{-1}

$\Delta H^\circ_f(\cdot CH_2CH_2SH) = \Delta H^\circ_f(CH_3CH_2SH) + DH^\circ(\text{primary C-H}) - \Delta H^\circ_f(H\cdot)$

$= -11 + 98 - 52 = 35$ kcal mole^{-1}

$$CH_2{=}CH_2 + \cdot SH \longrightarrow \cdot CH_2CH_2SH$$

$\Delta H^\circ_f =$ 12.5   33   35 $\Delta H^\circ = -11$ kcal mole^{-1}

$$\cdot CH_2CH_2SH + H_2S \longrightarrow CH_3CH_2SH + HS\cdot$$

$\Delta H^\circ_f =$ 35   -4.8   -11   33 $\Delta H^\circ = -8$ kcal mole^{-1}

The proposed reaction is clearly feasible, because each step is exothermic.

S8. $\Delta H^\circ_f(\cdot CH_2CH_2OH) \cong \Delta H^\circ_f(CH_3CH_2OH) + DH^\circ(\text{primary C-H}) - \Delta H^\circ_f(H\cdot)$

$= -56 + 98 - 52 = -10$ kcal mole^{-1}

$$CH_2{=}CH_2 + \cdot OH \longrightarrow \cdot CH_2CH_2OH$$

$\Delta H^\circ_f =$ 12.5   9.3   -10 $\Delta H^\circ = -32$ kcal mole^{-1}

$$\cdot CH_2CH_2OH + H_2O \longrightarrow CH_3CH_2OH + \cdot OH$$

$\Delta H^\circ_f =$ -10   -57.8   -56.2   9.3 $\Delta H^\circ = +21$ kcal mole^{-1}

The second step is too endothermic for this mechanism of hydration to be feasible.

12. ALKYNES

12.A Outline of Chapter 12, Important Terms Introduced, and Reactions Discussed

12.1 Electronic Structure

12.2 Nomenclature

alkyne alkenyne

12.3 Physical Properties

A. Dipole Moments

B. Nuclear Magnetic Resonance

shielding of acetylenic protons long range coupling

12.4 Acidity of Alkynes

12.5 Preparation of Alkynes

A. Acetylene

$$CaC_2 + H_2O \longrightarrow Ca(OH)_2 + HC{\equiv}CH$$

$$CH_4 \xrightarrow{1500°C} HC{\equiv}CH + 3H_2$$

(industrial processes)

B. Elimination Reactions

$$\left.\begin{array}{l} -CX_2{-}CH_2{-} \\ \text{geminal dihalides} \\ -CHX{-}CHX{-} \\ \text{vicinal dihalides} \end{array}\right\} \xrightarrow[\text{(faster)}]{\text{strong base}} -CX{=}CH{-} \xrightarrow[\text{(slower)}]{\text{strong base}} -C{\equiv}C-$$

(can stop after one step)

strong base=RO^- (R=H, alkyl),NH_2^-

base-catalyzed isomerizations:

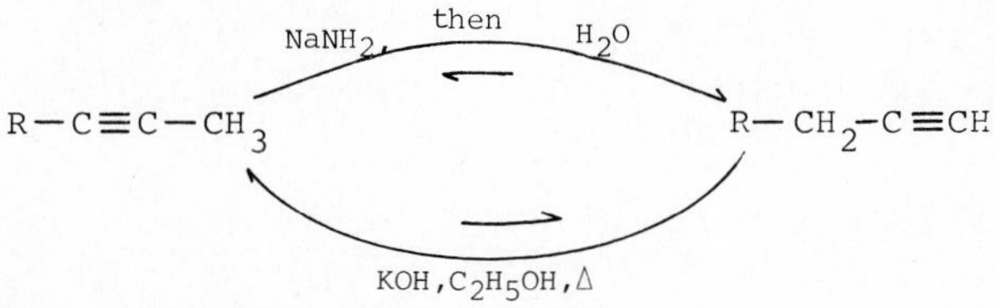

C. Displacement Reactions

alkylation of acetylide anions:

$$R{-}C{\equiv}C{-}H \xrightarrow[-BH]{B^-M^+} R{-}C{\equiv}C\bar{:}\ M^+ \xrightarrow[HMPT]{R'X} R{-}C{\equiv}C{-}R' + MX$$

R = H,alkyl,aryl B^-M^+ = *n*-BuLi,NH_2Na, etc. R' = primary alkyl, unhindered

X = Cl,Br or tosylate

(hindered R' results in E2)

12.6   Reactions of Alkynes

A.   Reduction

Hydrogenation:

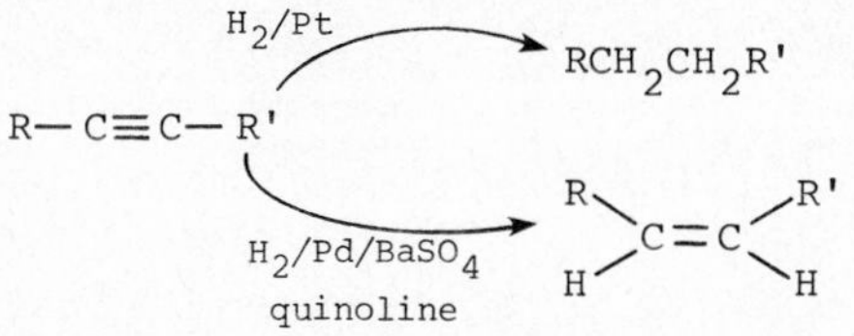

$$R-C\equiv C-R'$$

$$\xrightarrow{H_2/Pt} RCH_2CH_2R'$$ *(with Pt, second step is faster than the first)*

$$\xrightarrow[\text{quinoline}]{H_2/Pd/BaSO_4}$$
$$\underset{H}{\overset{R}{>}}C=C\underset{H}{\overset{R'}{<}}$$ *(Lindlar=poisoned catalyst, stops at alkene; cis stereochemistry)*

Sodium in liquid ammonia:

dissolving metal reduction   radical anion, vinyl radical, vinyl anion intermediates

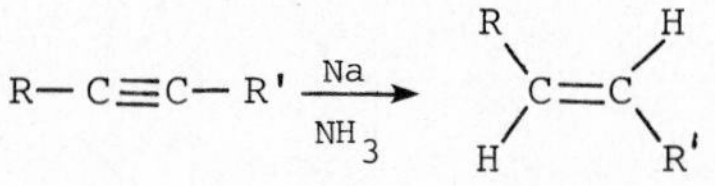

$$R-C\equiv C-R' \xrightarrow[NH_3]{Na} \underset{H}{\overset{R}{>}}C=C\underset{R'}{\overset{H}{<}}$$ *(stops at alkene; trans stereochemistry)*

B.   Electrophilic Additions *(slower than alkenes because vinyl cations are high in energy)*

(can stop at first step, Markovnikov orientation)

$$\underset{X}{\overset{CH_3}{>}}C=CH_2 \quad\xleftarrow{\substack{HX \\ (1\ eq.)}}\quad CH_3-C\equiv C-H \quad\xrightarrow{\substack{HX \\ (2\ eq.)}}\quad CH_3-CX_2-CH_3$$ *(Markovnikov orientation)*

$$\xrightarrow{\substack{\text{dil.}H_2SO_4 \\ Hg^{++}}}$$

$$\underset{}{\overset{O}{\underset{\|}{CH_3CCH_3}}} \xleftarrow{} \left[CH_3-\underset{}{\overset{OH}{\underset{|}{C}}}=CH_2 \right] \quad\text{unstable}$$

$$\xrightarrow{X_2}\ \underset{X}{\overset{CH_3}{>}}C=C\underset{H}{\overset{X}{<}}$$ *(anti addition stereochemistry)*

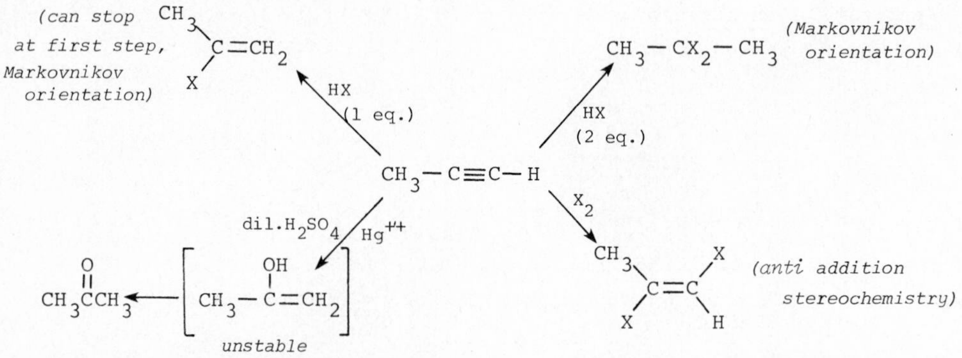

C.   Free Radical Additions

$$CH_3C\equiv CH \xrightarrow[h\nu]{HBr} CH_3CH=CHBr$$ *(anti-Markovnikov)*

D.   Nucleophilic Additions

$$HC\equiv CH \ +\ RO^- \xrightarrow{150°C} ROCH=CH_2$$ *(not a commonly used reaction)*

E.   Hydroboration

vinyl boranes:

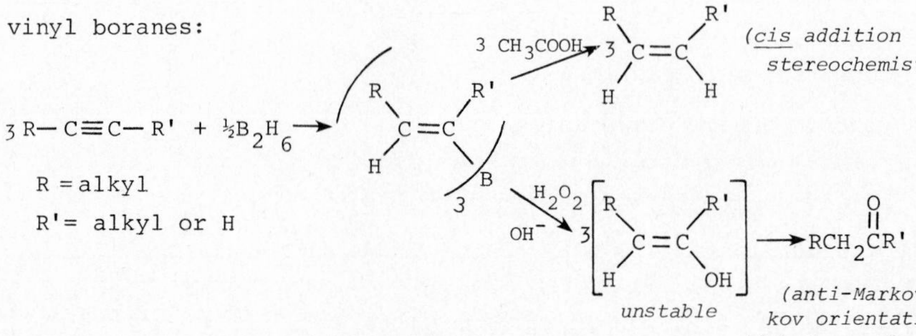

$$3\ R-C\equiv C-R' + \tfrac{1}{2}B_2H_6 \longrightarrow \underset{H}{\overset{R}{>}}C=C\underset{B}{\overset{R'}{<}}\Big]_3$$

R = alkyl
R'= alkyl or H

$$\xrightarrow{3\ CH_3COOH} 3\ \underset{H}{\overset{R}{>}}C=C\underset{H}{\overset{R'}{<}}$$ *(cis addition stereochemistry)*

$$\xrightarrow[OH^-]{H_2O_2} 3\left[\underset{H}{\overset{R}{>}}C=C\underset{OH}{\overset{R'}{<}}\right] \longrightarrow RCH_2\overset{O}{\underset{\|}{C}}R'$$

unstable   *(anti-Markovnikov orientation; aldehyde synthesis)*

F. Oxidation

permanganate oxidation:

$$R-C\equiv C-R' \xrightarrow[\substack{H_2O \\ pH\ 7.5}]{KMnO_4} R-\overset{O}{\overset{\|}{C}}-\overset{O}{\overset{\|}{C}}-R' \xrightarrow[\Delta]{KMnO_4} RCO_2H + R'CO_2H$$

$$(R, R' \neq H)$$

oxidative coupling:

$$2\ R-C\equiv CH \xrightarrow[pyridine]{CuCl_2} R-C\equiv C-C\equiv C-R$$

(Eglinton reaction; for symmetrical diynes)

$$R-C\equiv C-Li + Br-C\equiv C-R' \xrightarrow{CuCl} R-C\equiv C-C\equiv C-R'$$

(Cadiot-Chodkievicz reaction; for unsymmetrical diynes)

$$2\ R-C\equiv CH \xrightarrow[HCl]{CuCl, NH_4Cl} R-CH=CH-C\equiv C-R$$

(Nieuwland enyne synthesis)

12.7 Vinyl Halides (haloalkenes)

12.B Important Concepts and Hints

Chemistry of Alkynes

Because the chemistry of alkynes is so similar to that of alkenes, it is useful to focus on the contrasts. Look particularly at reduction methods: you can hydrogenate alkynes with *syn* delivery of H_2, just as you can in alkene chemistry. But with alkynes there is the added complication of stopping after the first addition or not, depending on the catalyst you use. Sodium in ammonia reduces alkynes (to *trans* alkenes), but doesn't reduce ordinary alkenes.

Electrophilic addition to alkynes is subject to the same orientation and stereochemical effects as alkenes (Markovnikov or anti-Markovnikov, *syn* or *anti* or neither). There are two complications: with alkynes the addition can occur *twice*, and the hydration reactions (Markovnikov-oriented with Hg^{++}, dil. H_2SO_4, or anti-Markovnikov with B_2H_6/H_2O_2, OH^-) do not lead to alcohols but to ketones or aldehydes instead.

$NaNH_2$ vs. Na, NH_3

Much of the chemistry of alkynes involves either sodium amide or sodium in ammonia as reagents, and students get confused over the difference between them. Sodium amide is a salt, comprised of the sodium cation (Na^+) and amide anion (NH_2^-), and it is often used when a very strong base is required ($NH_3 \rightleftharpoons NH_2^- + H^+$; pKa=35). Benzene or liquid ammonia itself are frequently employed as solvents for reactions involving $NaNH_2$.

When sodium metal (Na°) is dissolved in liquid ammonia, it ionizes to give a dark blue solution of sodium cations (Na^+) and solvated electrons (e^-). The solvated electron is a powerful reducing agent; it is for this purpose that "sodium in ammonia" (Na, NH_3) is used.

Organic Synthesis

In the chapter on alkynes you encounter the first reaction useful for formation of carbon-carbon bonds: alkylation of acetylide anions with alkyl halides. Reactions such as this are important because they allow you to build larger molecules, instead of just interchanging functional groups. There are limitations to be sure--the alkyl halide must be primary and unhindered, and there cannot be functional groups which are sensitive to strong base elsewhere in the molecule.

Now that you know a way to build large molecules from small ones, you will encounter a very important type of problem--synthetic problems. These questions ask you to devise a way to make the target compound starting from simpler materials. From the beginning of alchemy to the present day, finding ways to turn simpler (or cheaper) compounds into more complicated (or expensive) ones has been a major pursuit of chemists. When you have learned a greater variety of carbon-carbon bond forming reactions and functional group interconversions, a whole chapter in the text will be devoted to Organic Synthesis (Chapter 16). At this time it is still worthwhile to get you started on synthesis problems the right way by giving you an important hint: work the problems *backwards*.

When you want to go someplace you've never been to before, you find your destination on a map and work your way backward to where you are, before you actually set out. You don't just climb on the first bus that goes by your house, or drive down the first freeway you see, and then decide if it's going to your destination. The same is true for synthesis problems: don't take the starting materials and see what you can turn them into, hoping eventually to bump into the target. You should look at the target and think what its immediate precursor could be. Look for appropriate places for carbon-carbon bond-forming reactions, and work your way backwards to simpler compounds until you get to the starting materials (a lot more will be said on this subject when you get to Chapter 16).

Road-Map Problems

You will also encounter problems which give you an idea what it's like to figure out the structure of a compound by chemical methods. These are called road-map problems (sometimes they seem more like road-block problems...), and are actually just puzzles that need to be solved logically from the clues given. The best way to approach them is to be organized: write down schematically all the information presented in words in the problem. Then, look for the compound about which the most is known relative to the possibilities for it. If the structure of this key compound is actually given, or if you can figure out what it is, work your way outward from that point as logically as you can. The best advice is to be systematic. Supplementary problem S6 and its answer are a good example of this approach.

Degrees of Unsaturation

In solving road-map problems, one particular point is important--deducing "degrees of unsaturation" from the formula. A degree of unsaturation can be a ring or π-bond. Both C=C and C=O double bonds are unsaturations; a C≡C triple bond is two unsaturations. For hydrocarbons and oxygen-containing compounds, the formula tells you directly the number of degrees of unsaturation: for every two hydrogens less than (2 × # carbons + 2) there is one unsaturation. For instance, $C_7H_{10}O_2$ has three degrees of unsaturation, and it could be represented by the following possibilities (and many others).

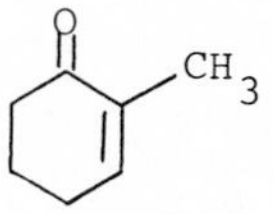

$CH_3C\equiv CCH_2CH_2CH_2\overset{O}{\overset{\|}{C}}H$

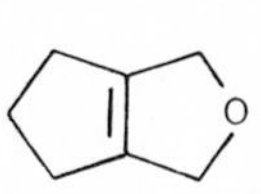

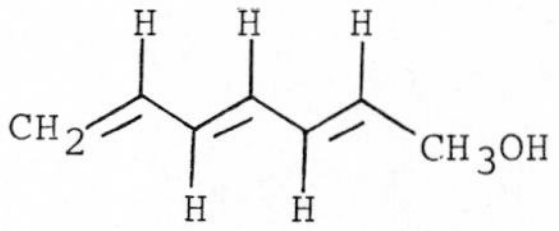

When halogens are present in the molecule, simply count them as hydrogens: $C_3H_5Br_2Cl$ is saturated: $CH_3—CHCl—CHBr_2$ for example. When nitrogens are present, ignore them but also subtract one hydrogen for each nitrogen: $C_5H_{11}N$:

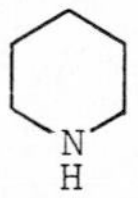

and $CH_3CH_2CH_2CH_2CH=NH$ each have one unsaturation.

You can't distinguish the different types of unsaturation (π-bond or ring) from the formula, but there are often clues given from reactions a compound undergoes. For example, hydrogenation usually removes only C=C or C≡C unsaturation, not rings or C=O. (We say usually, because cyclopropane rings and ketones and aldehydes can be reduced under vigorous hydrogenation conditions).

12.C Answers to Exercises

(12.2)

$HC\equiv CCH_2CH_2CH_2OH$

(R)-4-pentyn-2-ol

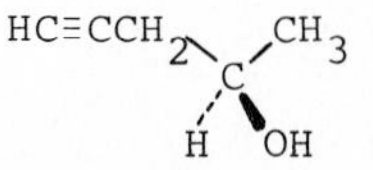

(R)-1-pentyn-3-ol

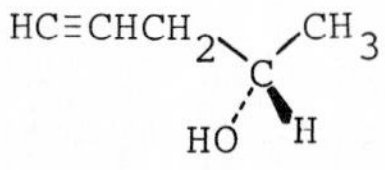

(S)-4-pentyn-2-ol

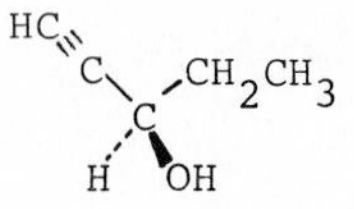

(R)-1-pentyn-3-ol

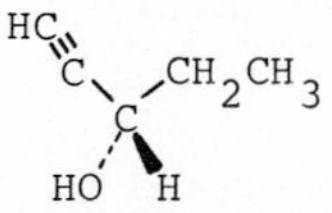

(S)-1-pentyl-3-ol

$HOC\equiv CCH_2CH_2CH_3$

1-pentyn-1-ol

(this molecule is unstable)

$CH_3C\equiv CCH_2CH_2OH$

3-pentyn-1-ol

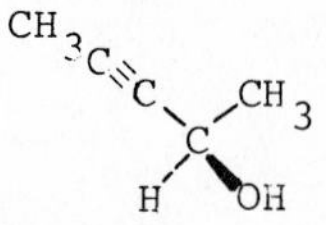

(R)-3-pentyn-2-ol

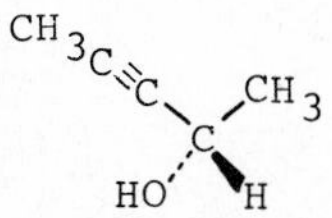

(S)-3-pentyn-2-ol

$HOCH_2C\equiv CCH_2CH_3$

2-pentyn-1-ol

(12.4) 1-Pentyne will form a precipitate with $AgNO_3$ whereas 2-pentyne will not.

(12.5.B) $CH_3CH_2CH{=}CHCH_3 + Br_2 \xrightarrow{CCl_4} CH_3CH_2CHBrCHBrCH_3$

$CH_3CH_2CHBrCHBrCH_3 \xrightarrow{KOH,\ C_2H_5OH,\ \Delta} CH_3CH_2C{\equiv}CCH_3 + 2\ KBr$

$CH_3CH_2C{\equiv}CCH_3 \xrightarrow[150°C]{NaNH_2} CH_3CH_2CH_2C{\equiv}CH$

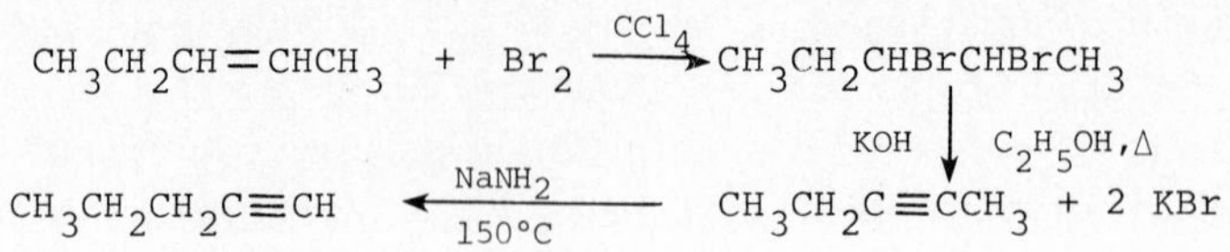

Isomerization of an internal alkyne to the terminal isomer involves the following steps:

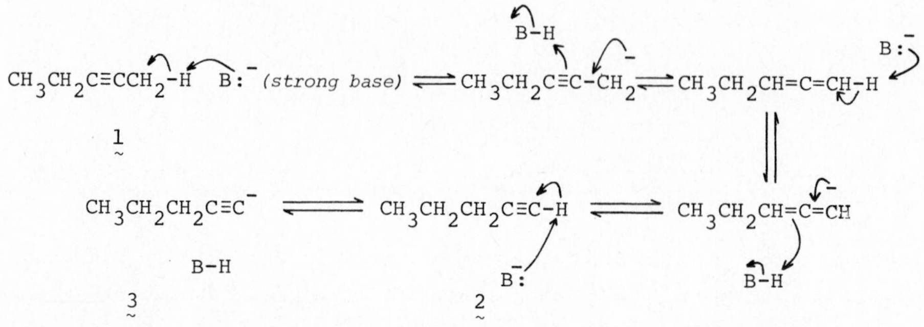

When hydroxide is used as the base (KOH in ethanol, for example), the equilibrium 2 ⇄ 3 favors 2, that is, hydroxide is a weaker base than the acetylide ion. Therefore, the equilibration that takes place is between the internal alkyne 1 and the terminal alkyne 2, and the former (1) predominates because it is more stable.

When amide ion is the base ($NaNH_2$, for example), the equilibrium 2 ⇄ 3 favors 3; that is, amide ion is a stronger base than the acetylide ion. Therefore, the equilibration that takes place favors 3, and when the reaction is worked up the terminal alkyne is obtained.

The ability to obtain either the internal or the terminal alkyne by choosing the appropriate base is a useful aspect of alkyne chemistry.

(12.5.C) The best way to work synthesis problems is backwards:

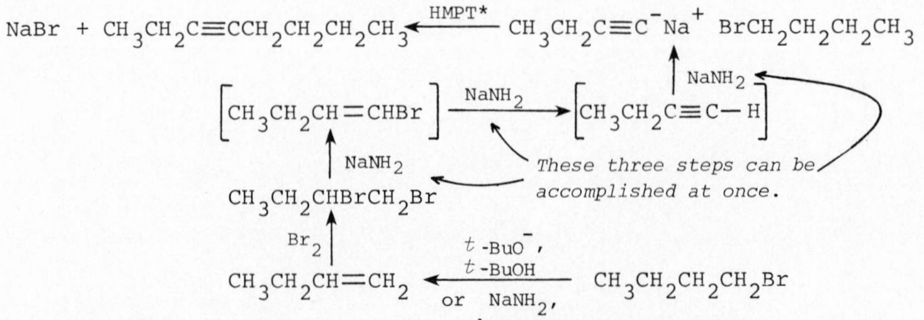

**HMPT=hexamethylphosphorictriamide ($[(CH_3)_2N]_3P{=}O$) is a very polar, aprotic solvent particularly useful for reactions such as this. Unfortunately, it appears to be carcinogenic as well.*

(12.6.A)

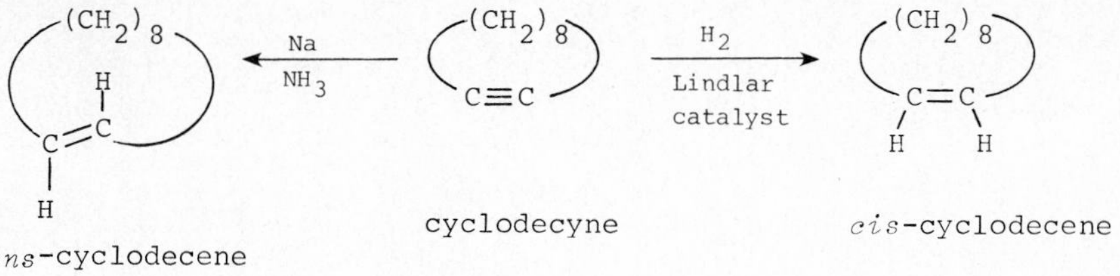

$$\text{trans-cyclodecene} \xleftarrow[\text{NH}_3]{\text{Na}} \text{cyclodecyne} \xrightarrow[\substack{\text{Lindlar} \\ \text{catalyst}}]{\text{H}_2} \text{cis-cyclodecene}$$

(12.6.B)

$$CH_3CH_2C\equiv CH$$

1 eq. HCl → $CH_3CH_2C(Cl)=CH_2$

2 eq. HCl → $CH_3CH_2CCl_2CH_3$

Hg^{++} dil. H_2SO_4 → $CH_3CH_2\overset{O}{\overset{||}{C}}CH_3$

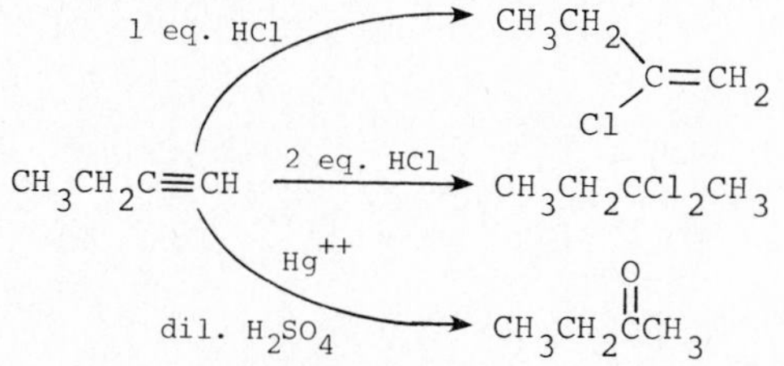

Markovnikov orientation leads to only one product in each case.

$$CH_3C\equiv CCH_3$$

1 eq. HCl → $CH_3C(Cl)=C(CH_3)H$ (and isomer) + $CH_3C(H)=C(Cl)CH_3$

2 eq. HCl → $CH_3CH_2CCl_2CH_3$

Hg^{++} dil. H_2SO_4 → $CH_3CH_2\overset{O}{\overset{||}{C}}CH_3$

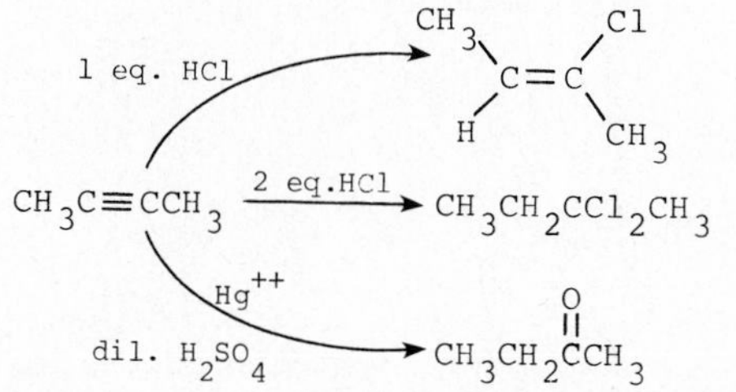

Both ends of the triple bond are the same.

$$CH_3C\equiv CCH_2CH_3$$

1 eq. HCl → $CH_3CH=C(Cl)CH_2CH_3$ *(cis and trans)* + $(CH_3)(Cl)C=CHCH_2CH_3$ *(cis and trans)*

2 eq. HCl → $CH_3CCl_2CH_2CH_2CH_3$ + $CH_3CH_2CCl_2CH_2CH_3$

Hg^{++} dil. H_2SO_4 → $CH_3\overset{O}{\overset{||}{C}}CH_2CH_2CH_3$ + $CH_3CH_2\overset{O}{\overset{||}{C}}CH_2CH_3$

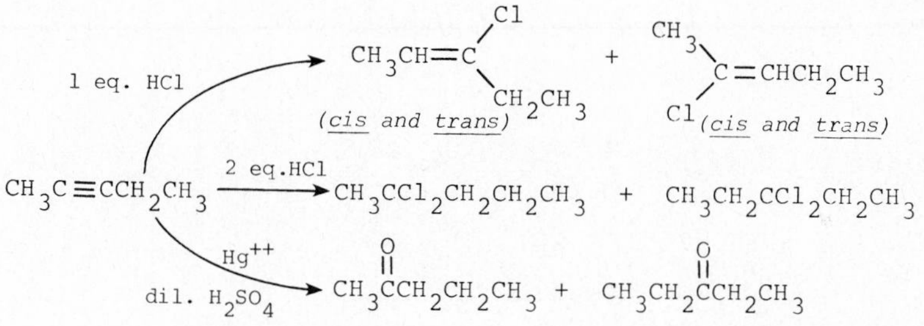

The ends of the triple bond are different, but similar electronically, so mixtures result.

(12.6.E)

$$3\ CH_3CH_2C\equiv CH + \tfrac{1}{2}B_2H_6 \longrightarrow \begin{array}{c} CH_3CH_2 \\ C=C \\ H \qquad B \end{array}_3 \xrightarrow[OH^-]{H_2O_2} 3\left[\begin{array}{c} CH_3CH_2 \\ C=C \\ H \qquad OH \end{array}\right] \longrightarrow 3\ CH_3CH_2CH_2\overset{O}{\overset{||}{C}}H$$

$$3\ CH_3C\equiv CCH_3 + \tfrac{1}{2}B_2H_6 \longrightarrow \begin{array}{c} CH_3 \quad CH_3 \\ C=C \\ H \qquad B \end{array}_3 \xrightarrow[OH^-]{H_2O_2} 3\left[\begin{array}{c} CH_3 \quad CH_3 \\ C=C \\ H \qquad OH \end{array}\right] \longrightarrow CH_3CH_2\overset{O}{\overset{||}{C}}CH_3$$

$$3\ CH_3CH_2C\equiv CCH_3 + \tfrac{1}{2}B_2H_6 \longrightarrow \text{both}$$

$$\begin{array}{c} CH_3CH_2 \quad CH_3 \\ C=C \\ H \qquad B \end{array}_3 \xrightarrow[OH^-]{H_2O_2} CH_3CH_2CH_2\overset{O}{\overset{||}{C}}CH_3$$

+

$$\begin{array}{c} CH_3CH_2 \quad CH_3 \\ C=C \\ B \qquad H \end{array}_3 \xrightarrow[OH^-]{H_2O_2} CH_3CH_2\overset{O}{\overset{||}{C}}CH_2CH_3$$

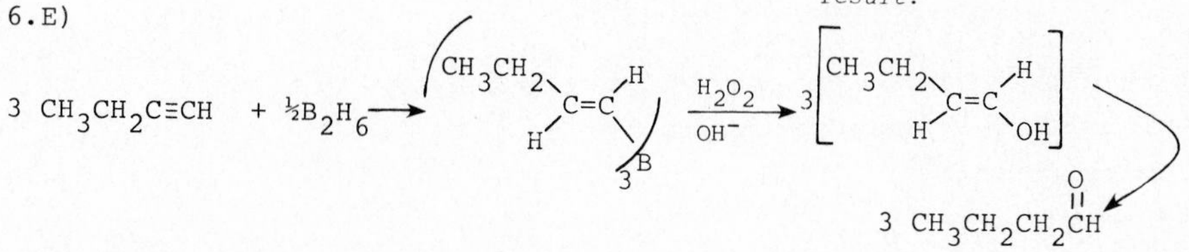

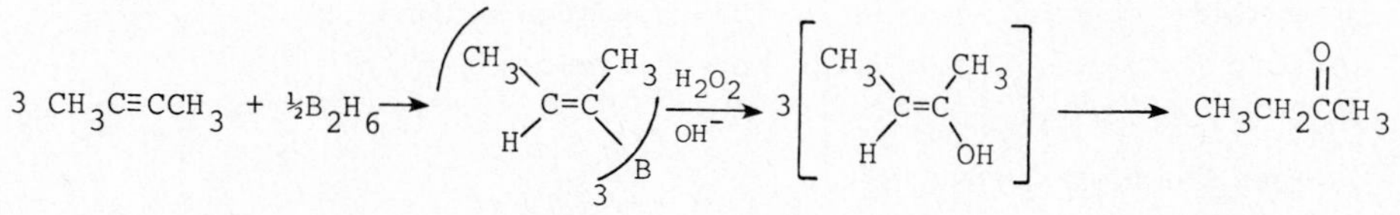

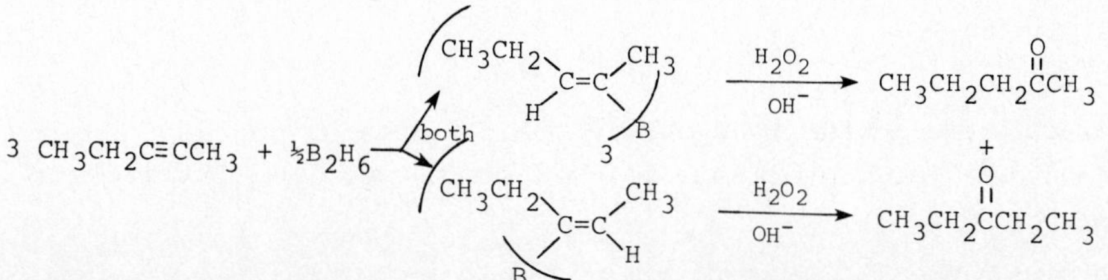

(12.6.F) $CH_2{=}CH{-}C{\equiv}C{-}C{\equiv}C{-}CH{=}CH_2 \xleftarrow[\text{pyridine, } 60^\circ C]{CuCl_2} 2\ CH_2{=}CH{-}C{\equiv}CH \xleftarrow[HCl]{CuCl,\ NH_4Cl} 4\ HC{\equiv}CH$

(12.7)

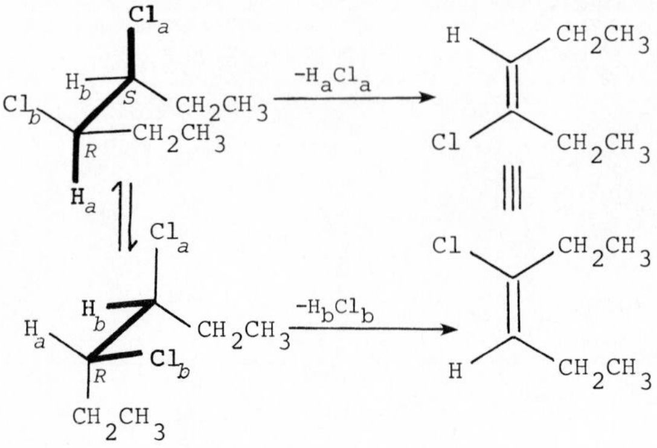

12.D Answers and Explanations to Problems

1. a) $(CH_3)_2CHC{\equiv}CCH_3$

b) 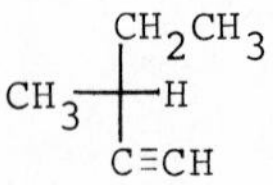

c) $CH_2{=}CHC{\equiv}CH$

d) 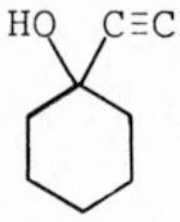

e)  or 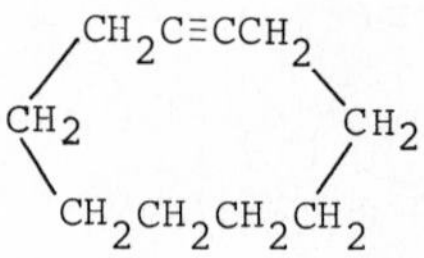

f) $(CH_3)_2CHCH_2C{\equiv}CH$

g) $CH_3OC{\equiv}CH$

h) $CH_3C{\equiv}CC{\equiv}C(CH_2)_6CH_3$

i) $CH_3CH_2C{\equiv}CCH_2OH$

j) $HC{\equiv}CCH(OCH_3)CH_2CH_3$

2. a) 6-bromo-1-hexyne

b) 2,2-dimethyl-3-hexyne

c) iodoethylene or vinyl iodide

d) cyclopropylacetylene or ethynylcyclopropane

e) 4,4-dimethyl-1-phenyl-1-pentyne

f) 1-bromopropyne

g) 3-bromopropyne

Note that the radical $HC{\equiv}CCH_2{-}$ has the common name of propargyl; hence, the bromide has the common name of propargyl bromide

h) (R)-pent-1-en-4-yn-3-ol

i) but-2-yn-1,4-diol

j) 2-methyl-3-hexyne

k) hex-1-en-5-yne

l) (Z)-2-chloro-2-butene or 2-chloro-*trans*-2-butene

m) (Z)-1-bromo-3-methyl-1-butene or *cis*-1-bromo-3-methyl-1-butene

n) 2-chloro-3-hexyne

3. a) $CH_3CH_2COOH + CO_2$
   b) $CH_3CH_2CH_2CH_3$
   c) $CH_3CH_2CBr_2CHBr_2$
   d) N.R.
   e) $CH_3CH_2C{\equiv}CAg\downarrow$
   f) $CH_3CH_2CH_2CHO$
   g) $CH_3CH_2COCH_3$
   h) 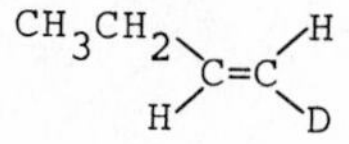
   i) 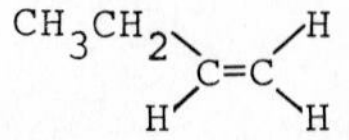
   j) $CH_3CH_2C{\equiv}C\text{-}CH{=}CHCH_2CH_3$
   k) $CH_3CH_2C{\equiv}C\text{-}C{\equiv}CCH_2CH_3$

4. a) CH_3COOH
   b) $CH_3CH_2CH_2CH_3$
   c) $CH_3CBr_2CBr_2CH_3$
   d) N.R.
   e) N.R.
   f) $CH_3CH_2COCH_3$
   g) $CH_3CH_2COCH_3$
   h) 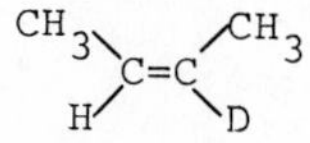
   i) 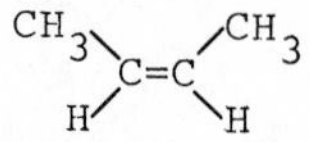
   j) N.R.
   k) N.R.

5. a)

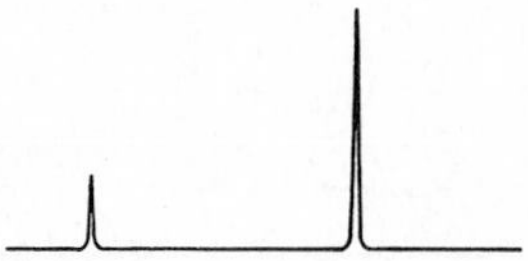

δ ∿4 ∿1
area 1 3

b)

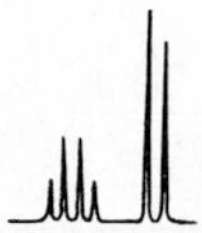

δ ∿2.5 ∿2
area 1 3

These two groups of peaks may overlap to give a more complex pattern looking something like:

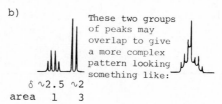

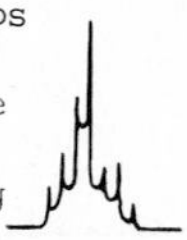

c)

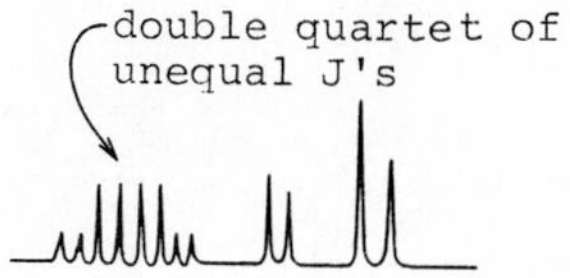

δ ∿4 ∿2 ∿1.5
area 1 1 3

d)

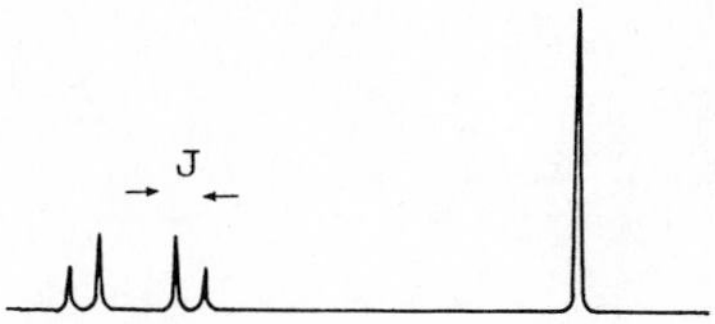

δ ∿7 ∿5 ∿1
area 1 1 9

e) Same as d, except the J value is greater.

6. a) $\xrightarrow[\text{Lindlar}]{H_2} CH_3CH{=}CH_2 \xrightarrow{HBr} (CH_3)_2CHBr$

   b) $\xrightarrow[H_2SO_4]{HgSO_4} CH_3COCH_3$

   c) $\xrightarrow[\text{2. } H_2O_2,\ OH^-]{\text{1. } B_2H_6} CH_3CH_2CH{=}O$

   d) $CH_3CH{=}CH_2$ (from (a)) $\xrightarrow[\text{peroxides}]{HBr} CH_3CH_2CH_2Br + Na^+\ {}^-C{\equiv}CCH_3 \xleftarrow{NaNH_2} HC{\equiv}CCH_3$

   $\downarrow$

   $CH_3CH_2CH_2C{\equiv}CCH_3 \xrightarrow{H_2/Pt}$ hexane

   or $2\ CH_3C{\equiv}CH \xrightarrow[\text{pyridine}]{CuCl_2} CH_3C{\equiv}CC{\equiv}CCH_3 \xrightarrow{H_2/Pt}$ (hexane)

e) $CH_3CH_2CH_2C{\equiv}CCH_3$ (from d) $\xrightarrow[\text{liq. } NH_3]{Na}$ $CH_3CH_2CH_2(H)C{=}C(H)CH_3$ (trans)

f) $CH_3C{\equiv}CH \xrightarrow{HCl\ (1\ mole)} CH_3CCl{=}CH_2$

g) $CH_3C{\equiv}CH \xrightarrow[\text{ether}]{n\text{-}C_4H_9Li} CH_3C{\equiv}C\,Li \xrightarrow{D_2O} CH_3C{\equiv}CD \xrightarrow[0°]{Br_2/CCl_4} CH_3CBr_2CBr_2D$

h) $CH_3C{\equiv}CH \xrightarrow[\text{pyridine, } 60°]{CuCl_2} CH_3C{\equiv}CC{\equiv}CCH_3$

7. a) $CH_3CH_2CH_2CH_3 \xrightarrow[h\nu]{Cl_2} CH_3CH_2CHClCH_3 + CH_3CH_2CH_2CH_2Cl$

$\xrightarrow{\text{alc KOH}}$ $CH_3CH{=}CHCH_3 + CH_3CH_2CH{=}CH_2$

$\xrightarrow{HCl}$ $CH_3CH_2CHClCH_3$

b) $CH_3CH_2CH_2CH{=}CH_2 \xrightarrow{Cl_2} CH_3CH_2CH_2CHClCH_2Cl \xrightarrow[NH_3]{NaNH_2} CH_3CH_2CH_2C{\equiv}C^-Na^+$

$\xrightarrow{H_3O^+} CH_3CH_2CH_2C{\equiv}CH$

c) $CH_3CH_2CH_2Br \xrightarrow[t\text{-BuOH}]{t\text{-BuO}^-} CH_3CH{=}CH_2 \xrightarrow{Br_2} CH_3CHBrCH_2Br \xrightarrow[\Delta]{KOH}$

$CH_3C{\equiv}CH \xrightarrow{NaNH_2} \xrightarrow{CH_3CH_2CH_2Br} CH_3CH_2CH_2C{\equiv}CCH_3$

(d) $HC{\equiv}CH \xrightarrow[\text{Lindlar catalyst}]{H_2} CH_2{=}CH_2 \xrightarrow{HBr} CH_3CH_2Br \xrightarrow{NaC{\equiv}CH} CH_3CH_2C{\equiv}CH$

or $HC{\equiv}CH$ $\xrightarrow{1.\ B_2H_6;\ 2.\ CH_3COOH}$ $CH_2{=}CH_2$

$CH_3CH_2C{\equiv}CH \xrightarrow[\text{Lindlar catalyst } (Pd/BaSO_4,\ \text{quinoline})]{H_2} CH_3CH_2CH{=}CH_2 \xrightarrow{B_2H_6} \xrightarrow[OH^-]{H_2O_2} CH_3CH_2CH_2CH_2OH$

(e) $CH_3CH_2C{\equiv}CH$ *(from (d))* $\xrightarrow{1.\ NaNH_2;\ 2.\ C_2H_5Br} CH_3CH_2C{\equiv}CCH_2CH_3 \xrightarrow{1.\ H_2/\text{Lindlar};\ 2.\ Hg(OAc)_2,\ CH_3OH;\ 3.\ NaBH_4} CH_3CH_2CH(OCH_3)CH_2CH_2CH_3$

(f) $C_6H_5(H)C{=}C(H)C_6H_5$ (trans) $\xrightarrow{Br_2} \xrightarrow{2\ NaNH_2} C_6H_5C{\equiv}CC_6H_5 \xrightarrow[\text{Lindlar}]{H_2} C_6H_5(H)C{=}C(H)C_6H_5$ (cis)

(g) $CH_3CH_2CH_2OH \xrightarrow[\Delta]{H_2SO_4} CH_3CH{=}CH_2 \xrightarrow{1.\ Br_2;\ 2.\ 2\ NaNH_2} CH_3C{\equiv}CH \xrightarrow[Hg^{++}]{H_2SO_4} CH_3\overset{O}{\overset{\|}{C}}CH_3$

(h) $CH_3CH_2C{\equiv}CH \xrightarrow{\underline{n}\text{-BuLi}} CH_3CH_2C{\equiv}CLi \xrightarrow{D_2O} CH_3CH_2C{\equiv}CD \xrightarrow[2.\ CH_3COOH]{1.\ B_2H_6}$ 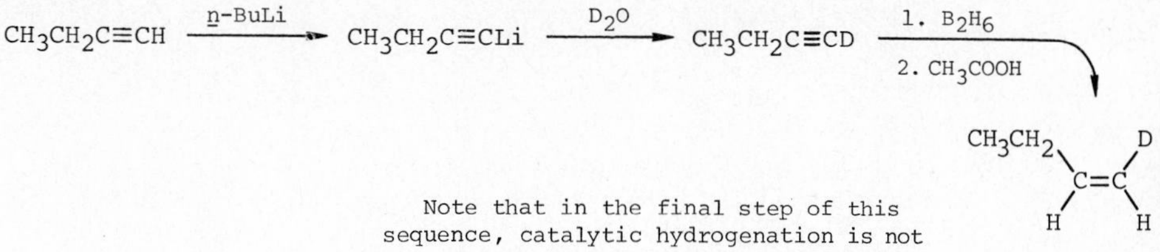

Note that in the final step of this sequence, catalytic hydrogenation is not recommended because of the possibility of some H-D exchange.

(i) $2\ HC{\equiv}CH \xrightarrow[\text{pyridine, } 60\ °C]{CuCl_2} HC{\equiv}C{-}C{\equiv}CH \xrightarrow{H_2/Pt} CH_3CH_2CH_2CH_3$ 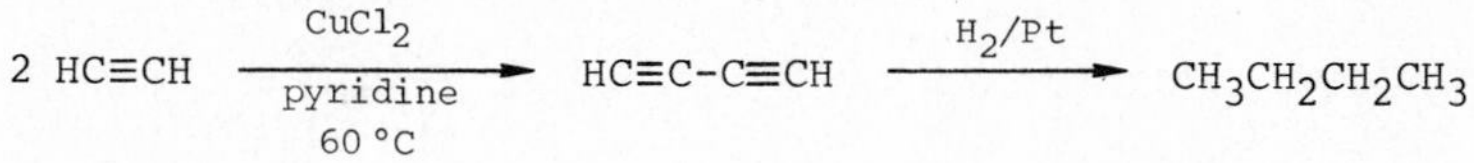

8. (a) $2\ HC{\equiv}C{-}Na + HC(=O)OCH_3 \longrightarrow HC{\equiv}C{-}CH(OH){-}C{\equiv}CH \xrightarrow{H_2/Pt} (CH_3CH_2)_2\,CHOH$ 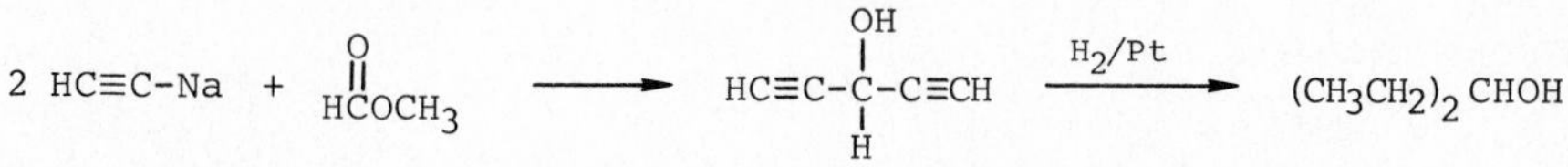

b) $CH_3C{\equiv}CH \xrightarrow[\text{ether, } -80°]{n\text{-BuLi}} \xrightarrow[-20°]{Br_2} CH_3C{\equiv}CBr$ 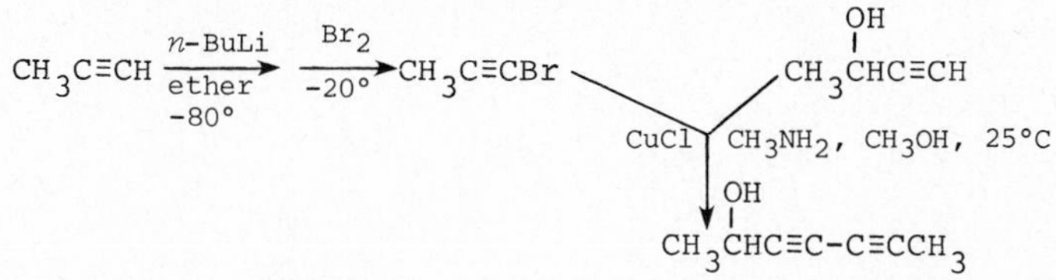

$CH_3C{\equiv}CBr + CH_3CH(OH)C{\equiv}CH \xrightarrow{CuCl,\ CH_3NH_2,\ CH_3OH,\ 25°C} CH_3CH(OH)C{\equiv}C{-}C{\equiv}CCH_3$

c) $HC{\equiv}CH + NaNH_2 \longrightarrow HC{\equiv}CNa \xrightarrow{C_2H_5I} CH_3CH_2C{\equiv}CH \xrightarrow{NaNH_2} CH_3CH_2C{\equiv}CNa \xrightarrow{CH_3CH_2CH_2I} CH_3CH_2C{\equiv}CCH_2CH_2CH_3$ 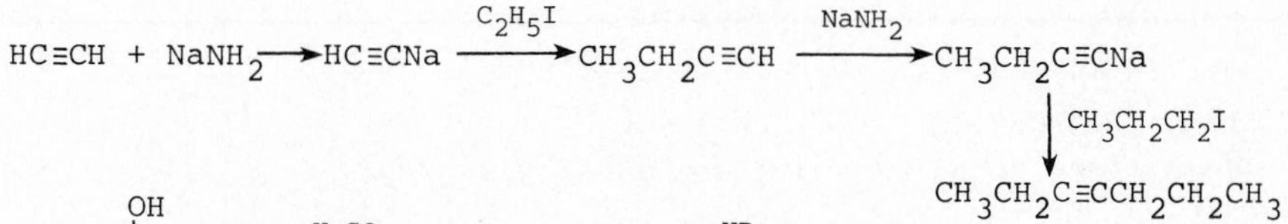

d) $(CH_3)_2C(OH)CH_2CH_3 \xrightarrow[\Delta]{H_2SO_4} (CH_3)_2C{=}CHCH_3 \xrightarrow[\text{peroxide}]{HBr} (CH_3)_2CHCHBrCH_3 \xrightarrow{t\text{-BuO}^-} (CH_3)_2CHCH{=}CH_2$ 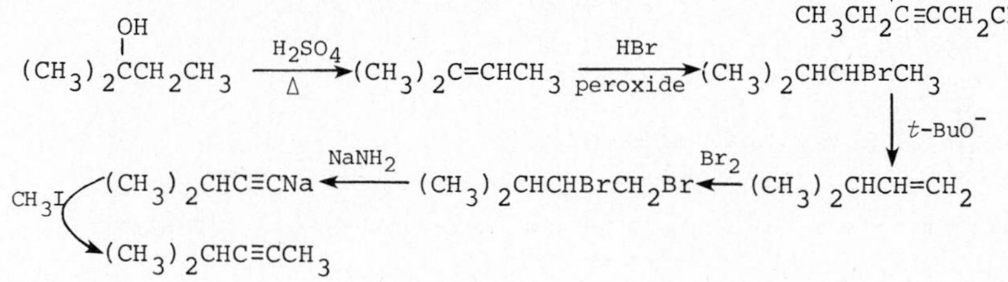

$(CH_3)_2CHCH{=}CH_2 \xrightarrow{Br_2} (CH_3)_2CHCHBrCH_2Br \xrightarrow{NaNH_2} (CH_3)_2CHC{\equiv}CNa \xrightarrow{CH_3I} (CH_3)_2CHC{\equiv}CCH_3$

9. $(CH_3)_2CHCH_2CH_2OH \xrightarrow[\text{or } PBr_3]{HBr/H_2SO_4} (CH_3)_2CHCH_2CH_2Br$ 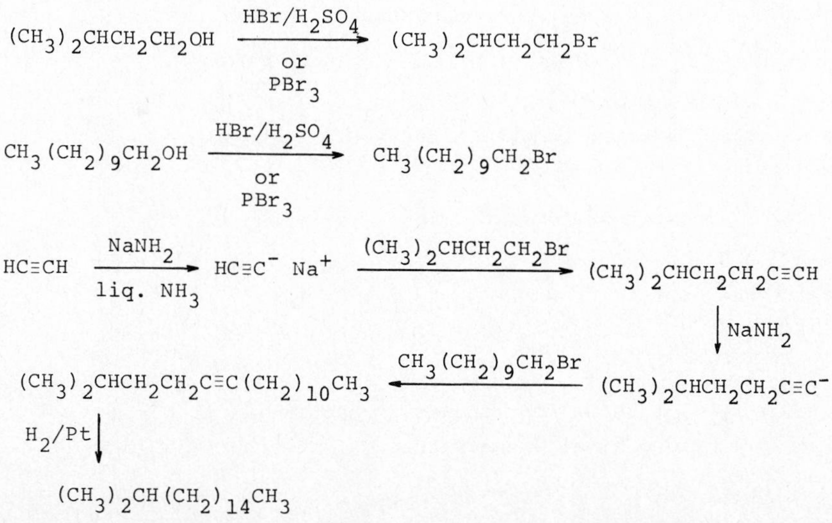

$CH_3(CH_2)_9CH_2OH \xrightarrow[\text{or } PBr_3]{HBr/H_2SO_4} CH_3(CH_2)_9CH_2Br$

$HC{\equiv}CH \xrightarrow[\text{liq. } NH_3]{NaNH_2} HC{\equiv}C^-\ Na^+ \xrightarrow{(CH_3)_2CHCH_2CH_2Br} (CH_3)_2CHCH_2CH_2C{\equiv}CH \xrightarrow{NaNH_2} (CH_3)_2CHCH_2CH_2C{\equiv}C^-$

$(CH_3)_2CHCH_2CH_2C{\equiv}C^- \xrightarrow{CH_3(CH_2)_9CH_2Br} (CH_3)_2CHCH_2CH_2C{\equiv}C(CH_2)_{10}CH_3 \xrightarrow{H_2/Pt} (CH_3)_2CH(CH_2)_{14}CH_3$

10. $CH_3(CH_2)_xCH_2OH \xrightarrow[\substack{\text{or} \\ PBr_3}]{HBr/H_2SO_4} CH_3(CH_2)_xCH_2Br$

$HC\equiv CH \xrightarrow[\text{liq. } NH_3]{NaNH_2} HC\equiv C^- \xrightarrow{CH_3(CH_2)_6CH_2Br} CH_3(CH_2)_6CH_2C\equiv CH \xrightarrow{NaNH_2}$

$\xrightarrow{CH_3(CH_2)_{11}CH_2Br} CH_3(CH_2)_6CH_2C\equiv CCH_2(CH_2)_{11}CH_3 \xrightarrow[\text{quinoline}]{Pd/BaSO_4/H_2}$

$\overset{CH_3(CH_2)_6CH_2}{\underset{H}{\diagdown}} C=C \overset{(CH_2)_{12}CH_3}{\underset{H}{\diagup}}$ or 1. B_2H_6
                                                                     2. CH_3COOH

    cis-9-tricosene

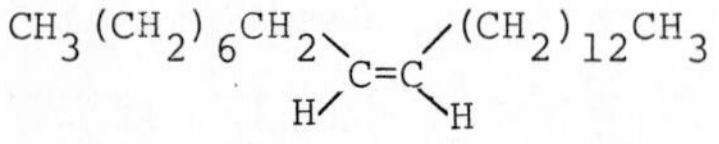

11. a) $K = \dfrac{[R^-][NH_3]}{[RH][NH_2^-]} = \dfrac{[R^-][H^+]}{[RH]} \times \dfrac{[NH_3]}{[H^+][NH_2^-]} = \dfrac{K_{RH}}{K_{NH_3}} = \dfrac{10^{-pka_{RH}}}{10^{-pka_{NH_3}}}$

        for C_2H_6: $K=10^{-15}$

        for C_2H_4: $K=10^{-9}$ for C_2H_2: $K=10^{10}$

   b) Solutions of C_2H_6 or C_2H_4 in $NH_3/NaNH_2$ have so little
       carbanion there is no reaction with CH_3I. $NaNH_2$ will
       react instead to give CH_3NH_2. C_2H_2 is converted
       completely to $HC\equiv C^-$ which can react with CH_3I.

   c) We must consider the following equations:
       $RH = R\cdot + H\cdot$ $\Delta H^\circ = DH^\circ$
       $H\cdot = H^+ + e$ $\Delta H^\circ = $ ionization potential of $H\cdot$, constant
                                          for all RH
       $\underline{R\cdot + e = R^-}$ $\underline{\Delta H^\circ = -(\text{electron affinity of } R\cdot)}$
       $RH = H^+ + R^-$ $= -E.A.$ (Note: positive E.A.
                                          corresponds to negative ΔH°)

                $\Delta H^\circ = DH^\circ + I.P.(H\cdot) - E.A.(R\cdot)$

       The difference in enthalpy, $\Delta H_2^\circ - \Delta H_1^\circ \equiv \Delta\Delta H^\circ$, for
       two different hydrocarbons, R_2H and R_1H, is therefore
       given by:

            $\Delta\Delta H^\circ = [DH_2^\circ - DH_1^\circ] - [E.A.(R_2\cdot) - E.A.(R_1\cdot)]$

       For $R_2H \equiv HC\equiv CH$ and $R_1H \equiv CH_3CH_3$ the negative $\Delta\Delta H^\circ$
       despite the positive $(DH_2^\circ - DH_1^\circ)$ means that
       $[E.A.(HC\equiv C\cdot) - E.A.(CH_3CH_2\cdot)] > [DH^\circ(HC\equiv C-H) - DH^\circ(C_2H_5-H)]$.

            The E.A. of $HC\equiv C\cdot$ corresponds to putting an electron
       in an sp hybrid orbital whereas the E.A. of $C_2H_5\cdot$ involves

putting an electron in an sp^3 hybrid orbital. The greater s-character of the sp-hybrid gives ethynyl radical a high electron affinity. The experimental values of these electron affinities are not known accurately but the available data give E.A.(HC≡C·) - E.A.(C_2H_5·)] = 50 kcal $mole^{-1}$, a value substantially higher than the difference in bond dissociation energies.

In short, although increasing s-character in an orbital increases the stability of a bond involving the orbital, it increases the stability of a lone pair still more. An electron pair in a bond involves 2 orbitals, whereas a lone pair involves a single orbital.

12. a) Overall retention of configuration shows that the vinyl anion is not linear and that protonation of vinyl anion by NH_3 is faster than inversion of the carbanion carbon:

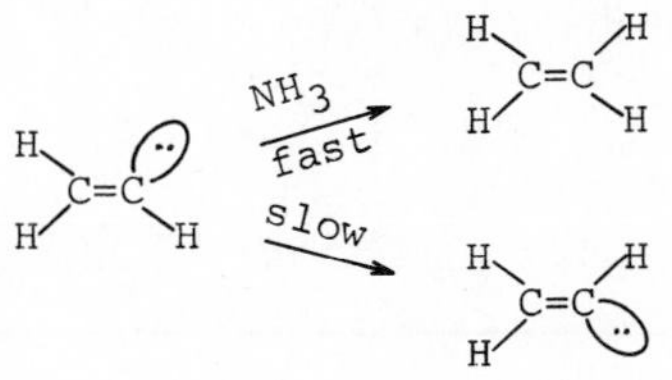

b)

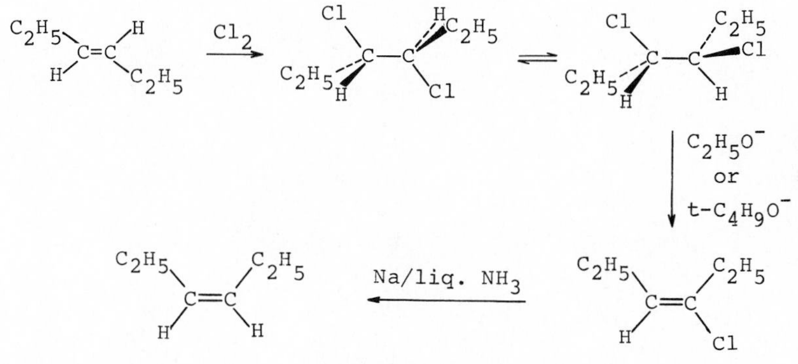

13.

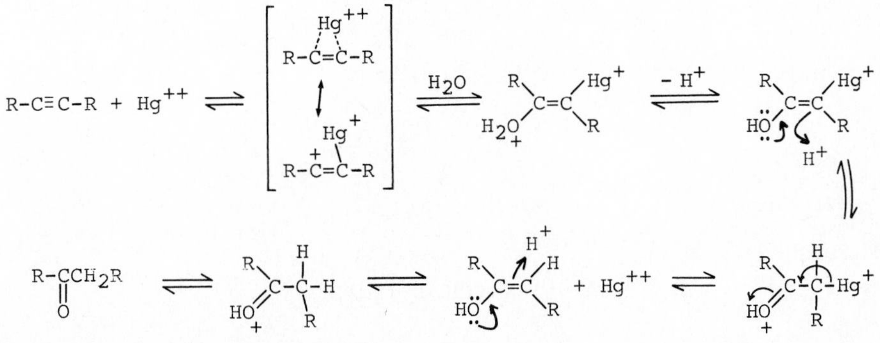

The interconversions of 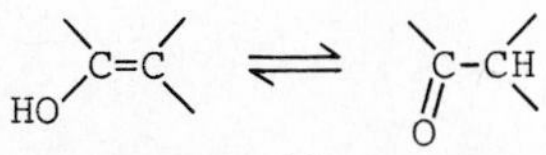 will be discussed in greater detail in Chapter 13.

14. The information in this roadmap problem is summarized as follows:

$$\underset{\substack{C_8H_{18}}}{B} \xleftarrow{H_2/Pt} \underset{\substack{C_8H_{12} \\ \text{optically} \\ \text{active}}}{A} \xrightarrow{H_2/\text{Lindlar}} \underset{\substack{C_8H_{14} \\ \text{optically} \\ \text{active}}}{C}$$

$$A \xrightarrow{Na/NH_3} \underset{\substack{C_8H_{14} \\ \text{optically inactive}}}{D}$$

B, C_8H_{18}, corresponds to C_nH_{2n+2} and must be a saturated hydrocarbon. Thus, A has 3 units of unsaturation (3 C=C or 1 C=C + 1 C≡C). The reactions A → C and A → D show that A has a triple bond. Our part structure is (C=C)(C≡C)C_4H_{12}. From these reductions C has a *cis* double bond and D has a *trans*, yet this difference is sufficient to render one optically inactive. The only rational solution is

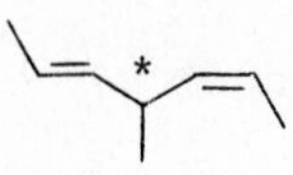

C
opt. act.

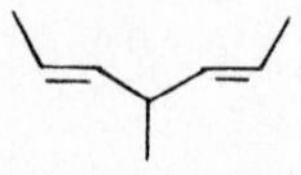

D
opt. inact.

∴ 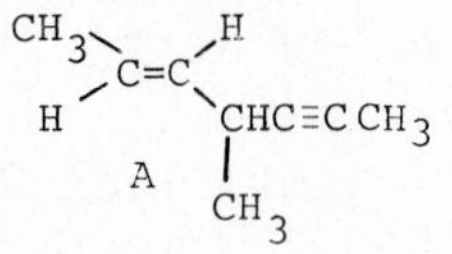

A

$$CH_3CH_2CH_2\overset{\displaystyle CH_3}{\overset{|}{C}}HCH_2CH_2CH_3$$

B

15. $\underset{C_7H_{12}}{E} \xrightarrow[-20°]{HCl} \underset{C_7H_{13}Cl}{F} \xrightarrow{t\text{-}BuO^-} E + \underset{C_7H_{12}}{G} \xrightarrow{O_3}$ 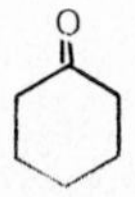 + $CH_2{=}O$

From last reaction G must be 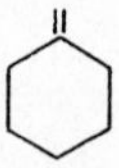 .

Then F must be 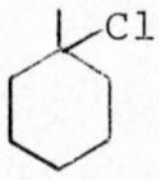 and E is 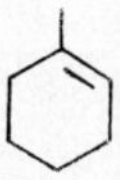 . F cannot be 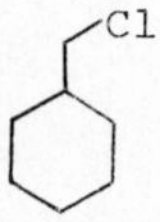 since this chloride cannot be formed by addition of HCl to an alkene.

16. (Br)(H)C=C(H)($CH_2CH_2CH_2Br$) $\xrightarrow{EtO^-}$ (Br)(H)C=C(H)($CH_2CH_2CH_2OEt$)

+

$HC{\equiv}CCH_2CH_2CH_2OEt$

+

$CH_2{=}C(OEt)CH_2CH_2CH_2OEt$

but no $EtOCH{=}CHCH_2CH_2CH_2OEt$

The terminal $-CH_2OEt$ comes from a normal S_N2 reaction on the primary bromide. The terminal HC≡C- group results from E2 elimination of the vinyl bromide. The vinyl ether results from a subsequent reaction of the triple bond:

$$HC{\equiv}CCH_2CH_2CH_2OEt + EtO^- \longrightarrow H\bar{C}{=}C(OEt)CH_2CH_2CH_2OEt \xrightarrow{EtOH} H_2C{=}C(OEt)CH_2CH_2CH_2OEt$$

Ethoxide ion adds to the triple bond to give the primary carbanion, $H\bar{C}{=}$, rather than the less stable secondary carbanion, $={\bar{C}}-C$. This latter mode of addition, if it occurred, would give rise to $EtOCH{=}CHCH_2CH_2CH_2OEt$, but cannot compete with the alternative mode of addition.

17.

	DH°		
CH_3CH_2-F	107	$CH_2{=}CH-F$	?
CH_3CH_2-Cl	81	$CH_2{=}CH-Cl$	88
CH_3CH_2-Br	68	$CH_2-CH-Br$	76
CH_3CH_2-I	53	$CH_2{=}CH-I$	?

For the chloride and bromide, DH° for the vinyl compound is 7-8 kcal $mole^{-1}$ higher than for the ethyl compound. Rough estimates for $CH_2{=}CHF$ and $CH_2{=}CHI$ are 115 and 61 kcal $mole^{-1}$, respectively. The vinyl-halide bond is stronger in part because of increased s character and in part because of delocalization of a halide lone pair electron with the double bond.

$C_{sp^3}-X$
C_2H_5-X
$C_{sp^2}-X$
C_2H_3-X

$$\left[CH_2{=}CH-\ddot{X} \longleftrightarrow \bar{C}H_2-CH{=}\overset{+}{X}\right]$$

This structure contributes a small but significant amount.

18. $CH_3-CH_2-C{\equiv}C-CH_2-CH_3$ $\qquad$ $CH_3-CH_2-CH_2-CH_2-CH_2-CH_3$

15.6 13.2 81.1 $\qquad$ 13.9 22.9 32.0

H
H
C=C
CH_3-CH_2
131.0
CH_2-CH_3
14.3 20.6
H
C
CH_2-CH_3
CH_3-CH_2
C
H
13.9 25.8
131.2

Carbons 2 (and 5) of 3-hexyne experience the same overall shielding effect of the triple bond that an acetylenic proton does (see Section 12.3.B).

12.E Supplementary Problems

S1. Write out the structure corresponding to each of the following names:

(a) diphenylacetylene

(b) 1,3,3-tribromopropyne

(c) sodium acetylide

(d) 1,5,9-cyclododecatriyne

(e) 2,2,7,7-tetramethyl-3,5-octadiyne

(f) ethyne

S2. Give the IUPAC name for each of the following compounds:

(a)

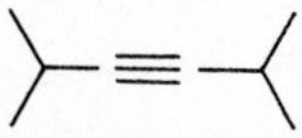

(b)

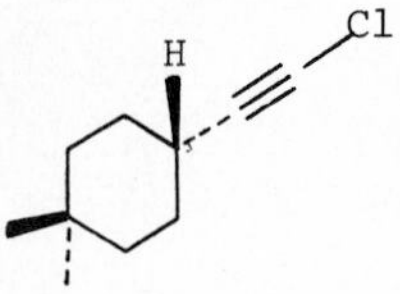

(c)

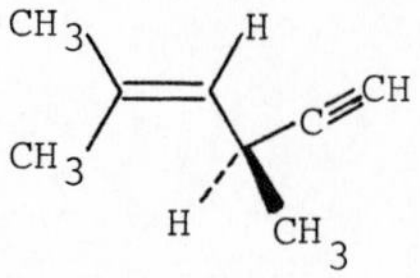

(d)

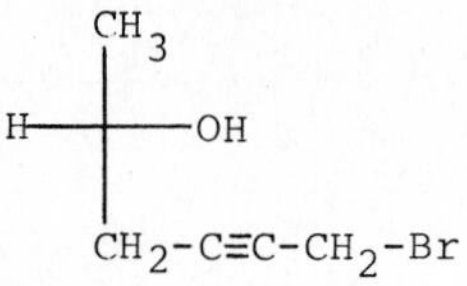

S3. Give the principal product of reaction of compound (c), problem S2 under each of the following conditions:

(a) $\xrightarrow{\text{H}_2/\text{Lindlar catalyst}}$

(b) $\xrightarrow{\text{Na/NH}_3}$

(c) $\xrightarrow{\text{NaNH}_2}$ $\xrightarrow{\text{CH}_3\text{I}}$

(d) $\xrightarrow{\text{NaNH}_2}$

(e) $+ \text{CH}_3\text{C}\equiv\text{CBr}$ $\xrightarrow[\text{CH}_3\text{OH},25°]{\text{CuCl,CH}_3\text{NH}_2}$

(f) $\xrightarrow{\text{HCl (1 eq)}}$

S4. Show how each of the following conversions can be accomplished in good yield. You may use other organic compounds if necessary.

$$(CH_3)_2CH-\overset{\overset{\text{O}}{\|}}{C}-CH_3$$

$$(CH_3)_2CH-CH_2-\overset{\overset{\text{O}}{\|}}{C}-CH_3$$

(a)

(b)

$$(CH_3)_2CH-CH_2-CH=CH_2$$

(e)

(d)

(c)

$$(CH_3)_2CH-\overset{\overset{\text{O}}{\|}}{C}-CH_2-CH_3$$

$$(CH_3)_2CH-CH_2-CH_2-\overset{\overset{\text{O}}{\|}}{C}-H$$

$$(CH_3)_2CH-CH_2-CH_2-\overset{\overset{\text{O}}{\|}}{C}-CH_3$$

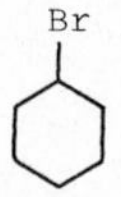

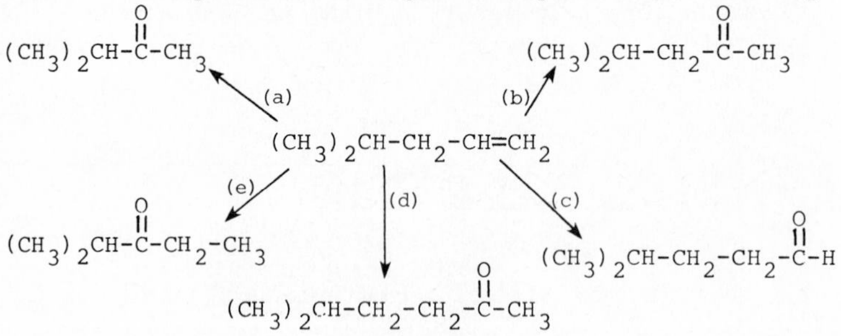

Do you anticipate any problems with the route you've chosen for (e)?

S5. The sex attractant of the galechiid moth *Bryotopha similis* is a derivative of *trans*-9-tetradecen-1-ol, and one of the components of the sex attractant of the butterfly *Lycorea ceresceres* is a derivative of *cis*-11-octadecen-1-ol. Show how to synthesize these two alcohols from 8-bromo-1-octanol and any other compound of four carbons or less. (Hint: protect the hydroxy group during your synthesis by forming the *t*-butyl ether; see Section 10.11.A in the text.)

S6. The nmr spectrum of Compound A shows a doublet at δ 1.0 ppm which has twice the area of the next largest resonance in the spectrum. A reacts with bromine to give B, which in turn reacts with potassium hydroxide in ethanol under mild conditions to give a mixture of C and D. C is the major product. Further reaction of either C or D with KOH in hot ethanol gives E. Combustion of 5.72 mg of E gives 18.41 mg of CO_2 and 6.28 mg of H_2O. Treatment of E with sodium in ammonia gives still another new compound, F. E does not react with alcoholic silver nitrate. However, heating E with sodium amide gives G, which does form a precipitate with $AgNO_3$. What are compounds A-G?

S7. (a) Using Appendix I, calculate the heat of hydrogenation, $\Delta H^\circ_{hydrog.}$ for each step in the hydrogenation of 2-butyne to butane.
(b) Do you expect this to be a good model for calculating $\Delta H^\circ_{hydrog.}$ of cyclodecyne? If not, in which step(s) do you think the biggest difference will be seen?

S8. Show how to synthesize both *cis*- and *trans*-1,2-dichlorocyclododecane stereospecifically from cyclododecane.

S9. Treatment of *cis*-3-hexene with bromine and then KOH in ethanol gives the vinyl halide, Z-bromo-3-hexene. However, when the same sequence of reactions is applied to cyclohexene, no vinyl halide (1-bromocyclohexene) is produced. Instead, 1,3-cyclohexadiene is obtained. Provide an explanation for the difference in behavior of these two alkenes.

12.F Answers to Supplementary Problems

S1. (a) 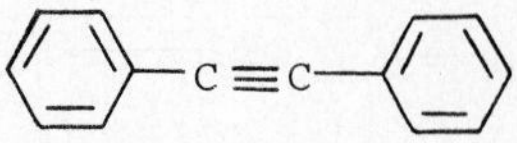

(b) $Br{-}C{\equiv}C{-}CHBr_2$

(c) $HC{\equiv}CNa$

(d) 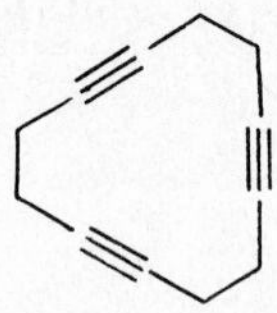

(e) $(CH_3)_3CC{\equiv}C{-}C{\equiv}CC(CH_3)_3$

(f) $HC{\equiv}CH$

S2. (a) 2,5-dimethyl-3-hexyne

(b) *trans*-1-(2-chloroethynyl)-4-bromocyclohexane

(c) (S)-2,4-dimethylhex-2-en-5-yne

(d) (S)-6-bromo-4-hexyn-2-ol

S3. (a) and (b) $(CH_3)_2C{=}CHCH(CH_3)CH{=}CH_2$

(c) $(CH_3)_2C{=}CHCH(CH_3)C{\equiv}CCH_3$

(d) $(CH_3)_2C{=}CHCH(CH_3)C{\equiv}CH$ + 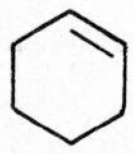 + NaBr *(2° RX undergoes E2)*

(e) $(CH_3)_2C{=}CHCH(CH_3)C{\equiv}C{-}C{\equiv}CCH_3$

(f) $(CH_3)_2CClCH_2CH(CH_3)C{\equiv}CH$ *(electrophilic addition is faster to alkenes than to alkynes)*

S4. (a) $(CH_3)_2CHCH_2CH{=}CH_2 \xrightarrow[2)\ NaBH_4]{1)\ O_3} (CH_3)_2CHCH_2CH_2OH \xrightarrow[\Delta]{Al_2O_3} (CH_3)_2CHCH{=}CH_2 \xrightarrow{Br_2} (CH_3)_2CHCHBrCH_2Br \xrightarrow{NaNH_2} (CH_3)_2CHC{\equiv}CH \xrightarrow[H_2SO_4]{Hg^{++}} (CH_3)_2CHC(=O)CH_3$

(b) $(CH_3)_2CHCH_2CH{=}CH_2 \xrightarrow[2)\ NaNH_2]{1)\ Br_2} (CH_3)_2CHCH_2C{\equiv}CH \xrightarrow[H_2SO_4]{Hg^{++}} (CH_3)_2CHCH_2C(=O)CH_3$

(c) $(CH_3)_2CHCH_2C{\equiv}CH$ *(from (b))* $\xrightarrow{B_2H_6} ((CH_3)_2CHCH_2CH{=}CH)_3B \xrightarrow[OH^-]{H_2O_2} (CH_3)_2CHCH_2CH_2CH{=}O$

(d) $(CH_3)_2CHCH_2C{\equiv}CH$ *(from (b))* $\xrightarrow{NaNH_2} (CH_3)_2CHCH_2C{\equiv}CNa \xrightarrow{CH_3I} (CH_3)_2CHCH_2C{\equiv}CCH_3 \xrightarrow{NaNH_2,\ 150\,°C} (CH_3)_2CHCH_2CH_2C{\equiv}CNa \xrightarrow{H_2O} \xrightarrow[H_2SO_4]{Hg^{++}} (CH_3)_2CHCH_2CH_2C(=O)CH_3$

(e) $(CH_3)_2CHCH_2C{\equiv}CH$ *(from (b))* $\xrightarrow[C_2H_5OH]{KOH} (CH_3)_2CHC{\equiv}CCH_3 \xrightarrow[H_2SO_4]{Hg^{++}} (CH_3)_2CHC(=O)CH_2CH_3 + (CH_3)_2CHCH_2C(=O)CH_3$

The trouble with route (e) is that it will produce a lot of the isomeric ketone as well.

S5.

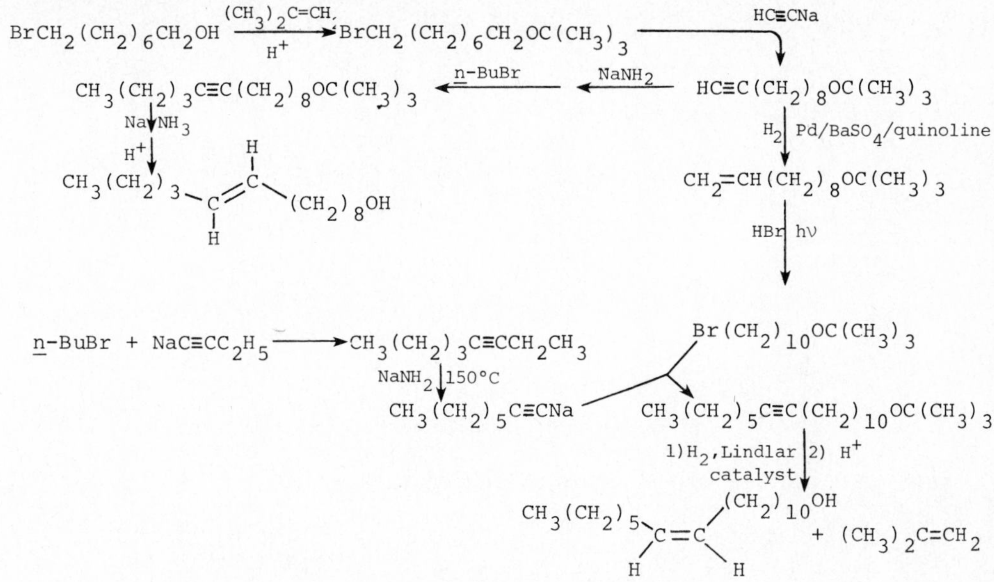

$$BrCH_2(CH_2)_6CH_2OH \xrightarrow[H^+]{(CH_3)_2C=CH_2} BrCH_2(CH_2)_6CH_2OC(CH_3)_3 \xrightarrow{HC\equiv CNa}$$

$$CH_3(CH_2)_3C\equiv C(CH_2)_8OC(CH_3)_3 \xleftarrow[n-BuBr]{} \xleftarrow{NaNH_2} HC\equiv C(CH_2)_8OC(CH_3)_3$$

$$Na \downarrow NH_3$$
$$H^+$$

$$H_2 \Big\downarrow Pd/BaSO_4/quinoline$$

$$CH_2=CH(CH_2)_8OC(CH_3)_3$$

$$CH_3(CH_2)_3\underset{H}{\underset{|}{C}}=\underset{(CH_2)_8OH}{\overset{H}{\overset{|}{C}}}$$

$$HBr \Big\downarrow h\nu$$

$$Br(CH_2)_{10}OC(CH_3)_3$$

$$\underline{n}-BuBr + NaC\equiv CC_2H_5 \longrightarrow CH_3(CH_2)_3C\equiv CCH_2CH_3$$

$$NaNH_2 \Big\downarrow 150°C$$

$$CH_3(CH_2)_5C\equiv CNa \qquad CH_3(CH_2)_5C\equiv C(CH_2)_{10}OC(CH_3)_3$$

$$1)H_2, Lindlar \Big\downarrow 2) H^+$$
$$catalyst$$

$$CH_3(CH_2)_5\underset{H}{\underset{|}{C}}=\underset{H}{\overset{(CH_2)_{10}OH}{\overset{|}{C}}} + (CH_3)_2C=CH_2$$

S6. 5.72 mg of E $\longrightarrow$ 18.41 mg of CO_2 and 6.28 mg of H_2O

(=5.02 mg of C, $\therefore$ 87.8%) (=0.70 mg of H, $\therefore$ 12.2%)

87.8% C and 12.2% H $\Longrightarrow C_{7.32}H_{12.2} = C_6H_{10}$

$$A + Br_2 \xrightarrow[C_2H_5OH]{KOH} B \xrightarrow{KOH} C + D$$

NMR: δ 1.0(d, twice the area of next largest resonance)

$$or$$
$$KOH \Big| C_2H_5OH, \Delta$$

precipitate $\xleftarrow{AgNO_3}$ $\xleftarrow{H_2O}$ $\xleftarrow{NaNH_2}$ E $\xleftarrow[NH_3]{Na}$ F

$$C_6H_{10}$$

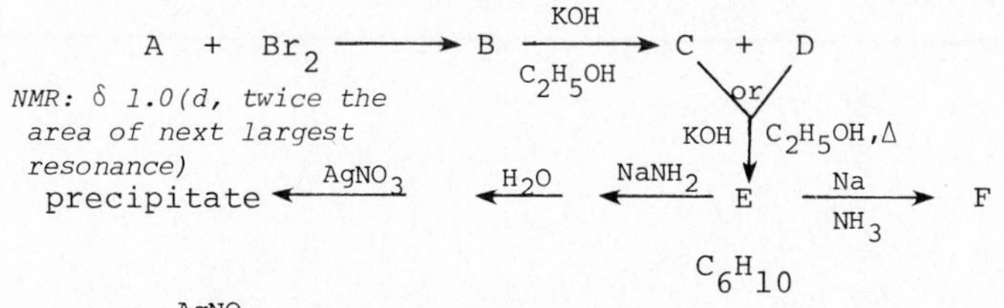

1) G $\xrightarrow{AgNO_3}$ ppt: G is a terminal alkyne

2) E $\xrightarrow{NaNH_2} \xrightarrow{H_2O}$ G : E is an internal alkyne

3) Formula C_6H_{10} for an internal alkyne: $CH_3C\equiv CCH_2CH_2CH_3$,
    $CH_3CH_2C\equiv CCH_2CH_3$, or $(CH_3)_2CHC\equiv CCH_3$

4) A $\xrightarrow[2)KOH, C_2H_5OH]{1) Br_2}$ E: A is an alkene, C_6H_{12}

5) NMR of A shows a doublet at δ 1.0, twice area of next most intense
    resonance:

    A has an isopropyl group, $(CH_3)_2CHCH=CHCH_3$. E is
    therefore $(CH_3)_2CHC\equiv CCH_3$.

6) E $\xrightarrow[NH_3]{Na}$ F: F is

$$(CH_3)_2CH\underset{C}{\underset{|}{}}\overset{H}{\underset{H}{\overset{|}{C}}}=\overset{H}{\underset{CH_3}{\overset{|}{C}}}$$

7) A$\neq$F: A is $(CH_3)_2CH\underset{H}{\overset{}{C}}=\underset{H}{\overset{CH_3}{C}}$

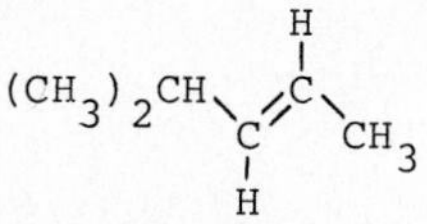

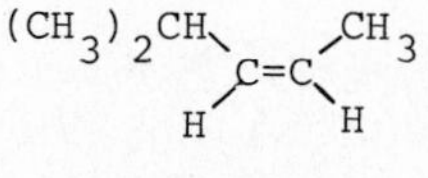

8) A + Br_2 ⟶ B: B is 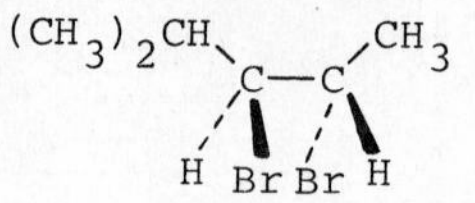

9) B $\xrightarrow[C_2H_5OH]{KOH}$ C + D: C and D are 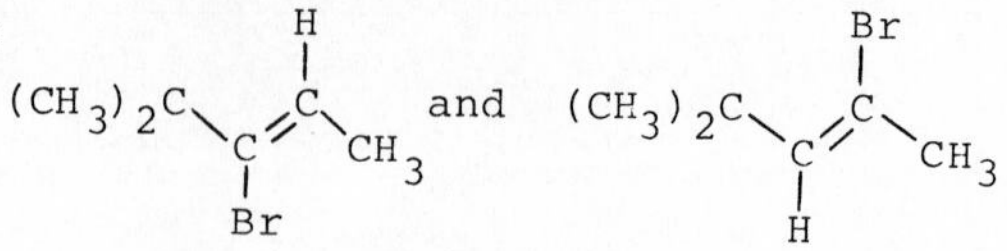

10) C>D: C is 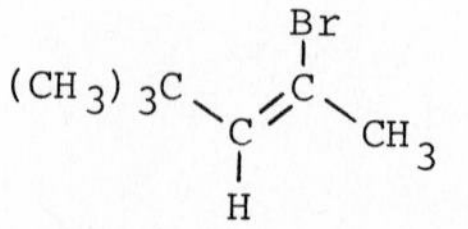 *(less steric hindrance to approach of base in formation of this isomer)*

(A could be $(CH_3)_3CCH_2CH{=}CH_2$ (with B,C and D the corresponding dibromide and vinyl bromides, respectively), if you assume that loss of HBr from C and D to give the terminal alkyne G was followed by isomerization (KOH/hot C_2H_5OH) to the internal alkyne E).

S7. (a) $CH_3C{\equiv}CCH_3 + H_2 \longrightarrow$ *cis*-$CH_3CH{=}CHCH_3$

ΔH_f°: 34.7   0   -1.9 $\Delta H^\circ_{hydrog(1)} = -36.6$ kcal $mole^{-1}$

cis-$CH_3CH{=}CHCH_3 + H_2 \longrightarrow CH_3CH_2CH_2CH_3$

ΔH_f°: -1.9   0   -30.4 $\Delta H^\circ_{hydrog(2)} = -28.5$ kcal $mole^{-1}$

(b) In cyclodecyne there is a lot of strain because the C-C≡C-C group of atoms must bend to fit into the ring structure. This strain will be released in going from cyclodecyne to *cis*-cyclodecene, so $\Delta H^\circ_{hydrog(1)}$ will be more negative than predicted in (a). The second step should be similar to the acyclic model.

S8.

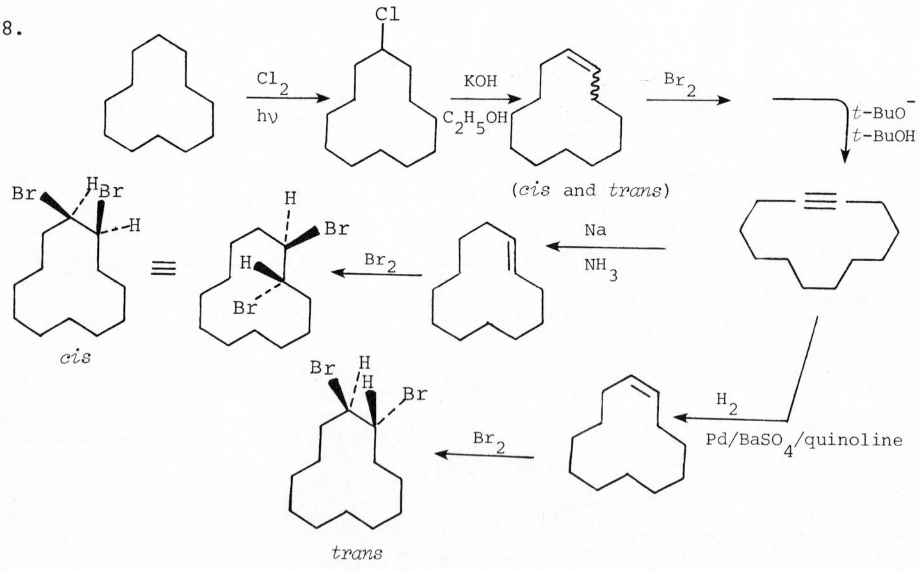

S9. 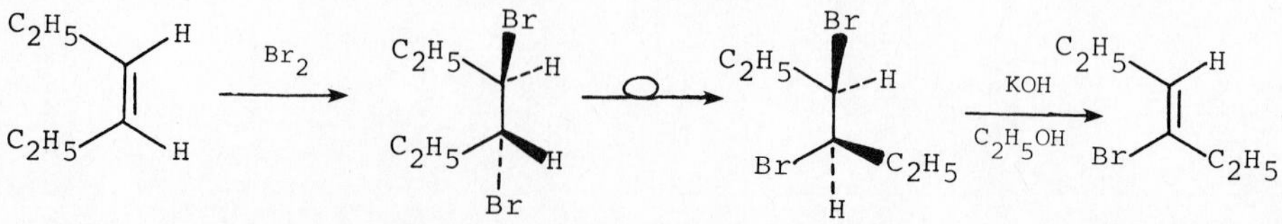

via anti elimination

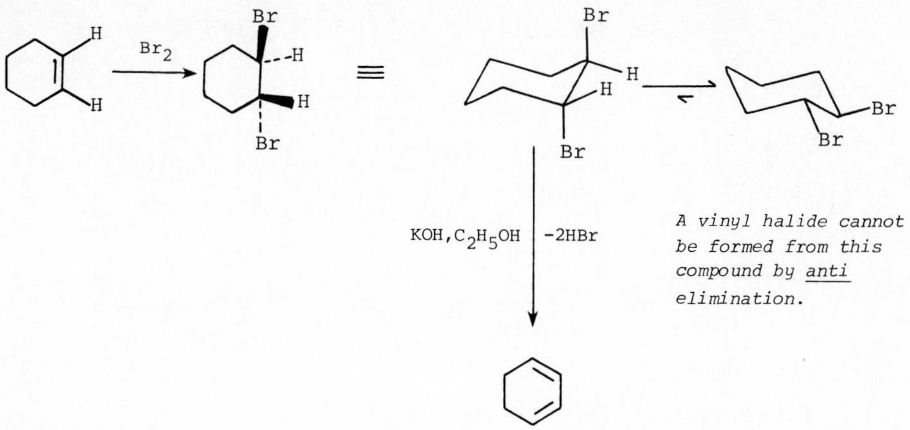
H
H
Br2
Br
H
H
Br
≡
Br
H
H
Br
Br
Br
KOH,C2H5OH
-2HBr
A vinyl halide cannot be formed from this compound by anti elimination.

13. ALDEHYDES AND KETONES

13.A Chapter Outline, Important Terms Introduced, and Reactions Discussed

13.1 Structure

carbonyl group (hybridization, polarization)

13.2 Nomenclature

A. Common Names

"-ic acid" "-aldehyde" $\alpha, \beta, \gamma, \ldots$

B. IUPAC Names

alkanal	alkanone
-carbaldehyde	oxo-
formyl	(numbers, instead of $\alpha, \beta, \gamma, \ldots$)

13.3 Physical Properties

13.4 Nuclear Magnetic Resonance

-CH=O ~9.5 ppm

$-\overset{\displaystyle O}{\overset{\|}{C}}CH_3$ ~2.0 ppm

$R\overset{\displaystyle O}{\overset{\|}{C}}R'$ ~200 ppm

13.5 Synthesis of Aldehydes and Ketones

A. Oxidation of Alcohols (see Section 10.7.E)

Cr^{+6} reagents:

$RCH_2OH \longrightarrow RCH{=}O$ *(use CrO_3/pyridine to avoid overoxidation)*

$R\overset{\displaystyle OH}{\overset{|}{C}}HR' \longrightarrow R\overset{\displaystyle O}{\overset{\|}{C}}R'$ *($Na_2Cr_2O_7/H_2SO_4$ (Jones reagent) is most common)*

B. Oxidation of Alkenes (see Section 12.6.E)

Ozonolysis: line structures

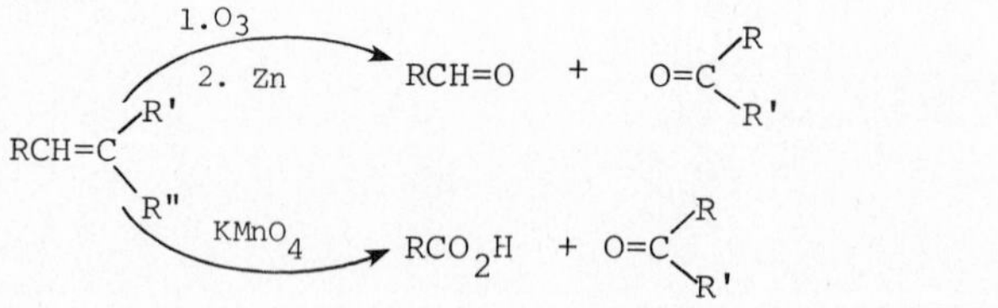

C. Hydration of Alkynes (see Sections 12.6.B and 12.6.E)

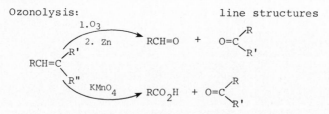

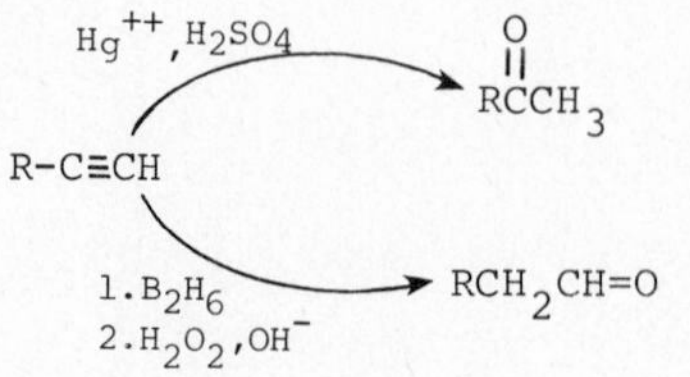

13.6 Enolization (carbonyl compounds as nucleophiles)

A. Keto-Enol Equilibria

enol, enolate ion (resonance structures)

tautomerism:

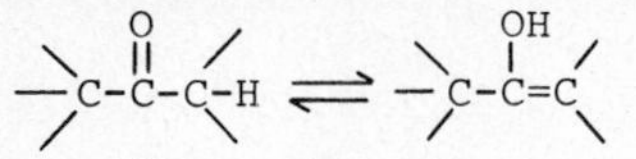

$$\overset{O}{\underset{}{\overset{\|}{-C-C-C-H}}} \rightleftharpoons \overset{OH}{\underset{}{\overset{|}{-C-C=C}}} \quad \textit{(acid- or base-catalyzed)}$$

deuterium exchange:

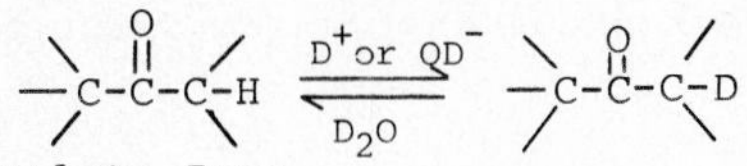

$$\overset{O}{\underset{}{\overset{\|}{-C-C-C-H}}} \underset{D_2O}{\overset{D^+ \text{or } QD^-}{\rightleftharpoons}} \overset{O}{\underset{}{\overset{\|}{-C-C-C-D}}} \quad \textit{(all } \alpha\text{-protons exchangeable)}$$

B. Enolate Ions

(pK_a) amibident anions

lithium diisopropylamide (LDA)   degree of association

Formation and silylation of enolates:

$$\overset{O}{\underset{}{\overset{\|}{-C-C-C-H}}} + Li^+\bar{N}(i\text{-Pr})_2 \longrightarrow \overset{O^-Li^+}{\underset{}{\overset{|}{-C-C=C}}} \overset{(CH_3)_3SiCl}{\longrightarrow} \overset{OSi(CH_3)_3}{\underset{}{\overset{|}{-C-C=C}}}$$

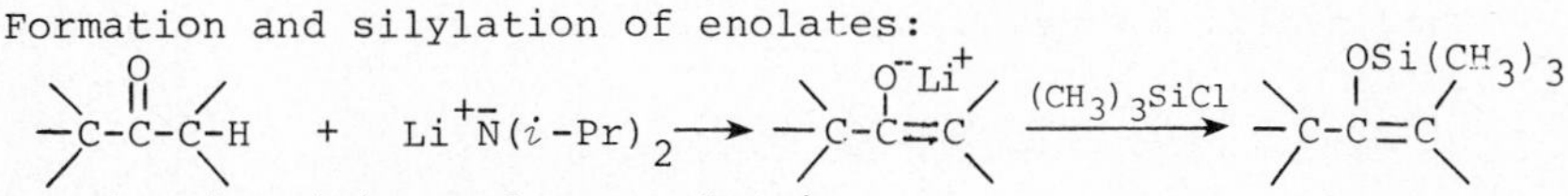

C. Racemization (via enol or enolate)

chiral center α to carbonyl

D. Halogenation

autocatalytic   induction period

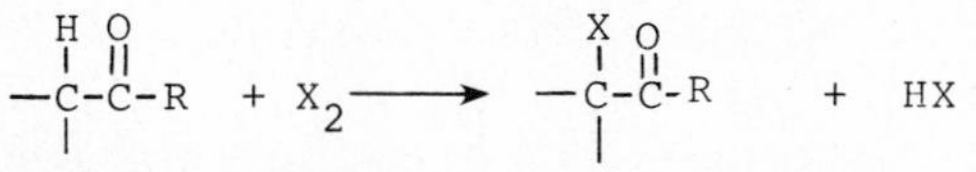

$$\overset{H \ O}{\underset{|}{\overset{| \ \|}{-C-C-R}}} + X_2 \longrightarrow \overset{X \ O}{\underset{|}{\overset{| \ \|}{-C-C-R}}} + HX \quad \textit{(acid- or base-catalyzed)}$$

haloform reaction:

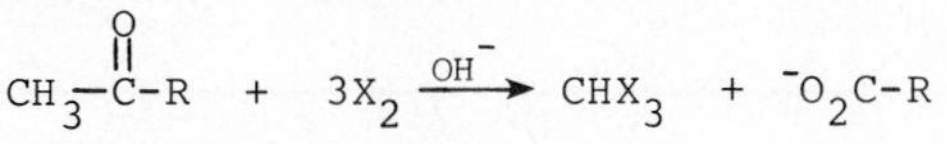

$$\overset{O}{\underset{}{\overset{\|}{CH_3-C-R}}} + 3X_2 \overset{OH^-}{\longrightarrow} CHX_3 + {}^-O_2C-R \quad \textit{(only for methyl ketones)}$$

13.7 Carbonyl Addition Reactions

A. Carbonyl Hydrates: Gem-diols

^{18}O-exchange

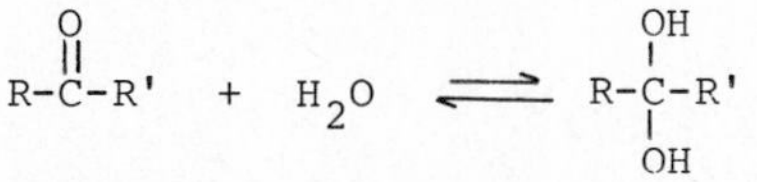

$$\overset{O}{\underset{}{\overset{\|}{R-C-R'}}} + H_2O \rightleftharpoons \overset{OH}{\underset{OH}{\overset{|}{R-C-R'}}} \quad \textit{(acid- or base-catalyzed)}$$

B. Acetals and Ketals

hemiacetal or hemiketal formation:

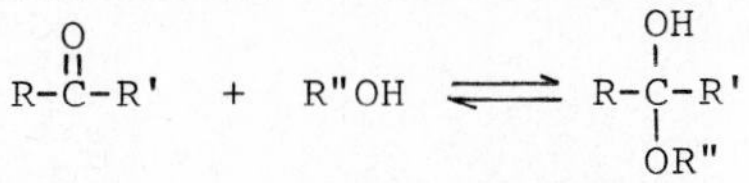

$$\overset{O}{\underset{}{\overset{\|}{R-C-R'}}} + R''OH \rightleftharpoons \overset{OH}{\underset{OR''}{\overset{|}{R-C-R'}}} \quad \textit{(acid- or base-catalyzed)}$$

acetal or ketal formation:

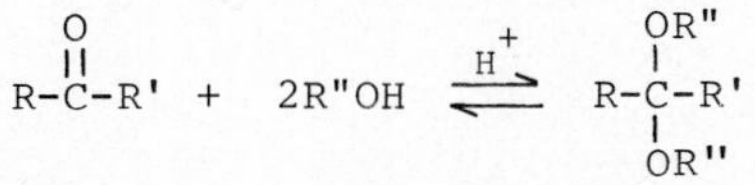

$$\overset{O}{\underset{}{\overset{\|}{R-C-R'}}} + 2R''OH \overset{H^+}{\rightleftharpoons} \overset{OR''}{\underset{OR''}{\overset{|}{R-C-R'}}} \quad \textit{(only acid-catalyzed; ketal stable to base)}$$

azeotropic distillation   protecting group

enol ether formation and hydrolysis:

$$\overset{OR''}{\underset{OR''}{\overset{|}{R-C-CH}}} \overset{\Delta}{\longrightarrow} \overset{OR''}{\underset{}{\overset{|}{R-C=C}}} \overset{H_3O^+}{\longrightarrow} \overset{O}{\underset{}{\overset{\|}{R-C-CH}}}$$

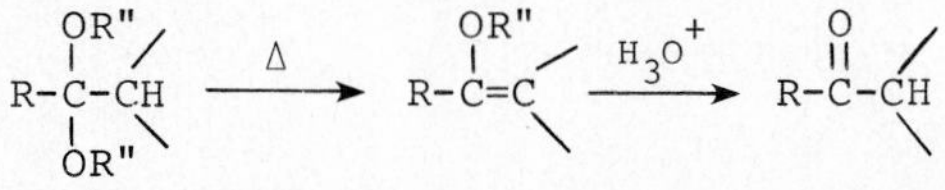

C. Reaction with Derivatives of Ammonia

imine = Schiff base   oximes, hydrazones, azines, phenyl
                                hydrazones, semicarbazones

condensation   hemiaminal = carbinolamine

$$R-\overset{\overset{\displaystyle O}{\|}}{C}-R' \ + \ H_2N-Y \rightleftharpoons R-\overset{\overset{\displaystyle OH}{|}}{\underset{\underset{\displaystyle HNY}{|}}{C}}-R' \ \rightleftharpoons \ R-\overset{\overset{\displaystyle NY}{\|}}{C}-R' \ + \ H_2O$$

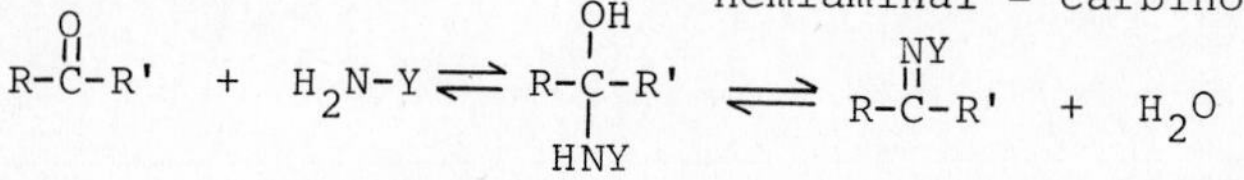

D. Addition of Acetylide Anions   (carbon-carbon forming reaction)

$$R-\overset{\overset{\displaystyle O}{\|}}{C}-R' \ + \ M^{+ \ -}C{\equiv}CR'' \ \longrightarrow \ R-\overset{\overset{\displaystyle O^- M^+}{|}}{\underset{\underset{\displaystyle R'}{|}}{C}}-C{\equiv}C-R'' \ \overset{H_2O}{\longrightarrow} \ R-\overset{\overset{\displaystyle OH}{|}}{\underset{\underset{\displaystyle R'}{|}}{C}}-C{\equiv}C-R''$$

$$M^+=Li,Na,MgBr$$

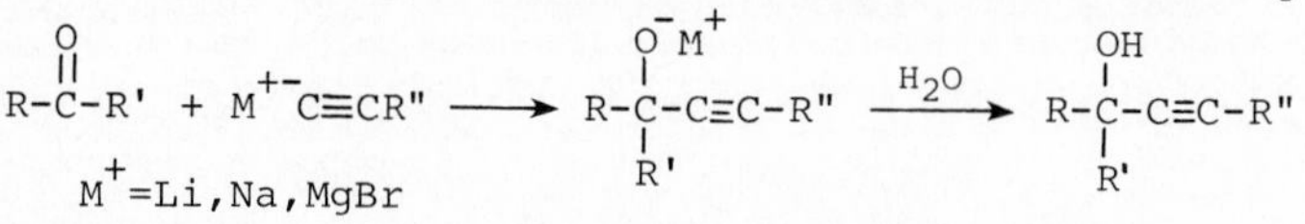

E. Addition of HCN

cyanohydrin formation:

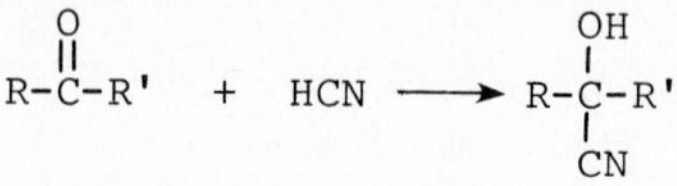

$$R-\overset{\overset{\displaystyle O}{\|}}{C}-R' \ + \ HCN \ \longrightarrow \ R-\overset{\overset{\displaystyle OH}{|}}{\underset{\underset{\displaystyle CN}{|}}{C}}-R' \quad \textit{(reversal by strong base)}$$

F. The Aldol Condensation (an important carbon-carbon bond-forming reaction)

α,β-unsaturated aldehydes

intramolecular aldol condensations   cyclic compounds

mixed aldol condensations using preformed enolate

$$R-\overset{\overset{\displaystyle O}{\|}}{\underset{\underset{\displaystyle R'}{|}}{C}}-CH_2 \ + \ R''-\overset{\overset{\displaystyle O}{\|}}{C}-R''' \longrightarrow R-\overset{\overset{\displaystyle O}{\|}}{C}-\overset{\overset{\displaystyle}{}}{\underset{\underset{\displaystyle R'}{|}}{C}}H-\overset{\overset{\displaystyle OH}{|}}{\underset{\underset{\displaystyle R''}{|}}{C}}-R''' \longrightarrow R-\overset{\overset{\displaystyle O}{\|}}{C}-\overset{}{\underset{\underset{\displaystyle R'}{|}}{C}}=C\overset{\diagup R'''}{\diagdown R''} \ + \ H_2O$$

       (acid- or base-catalyzed; frequently can stop at hydroxyketone
          intermediate)

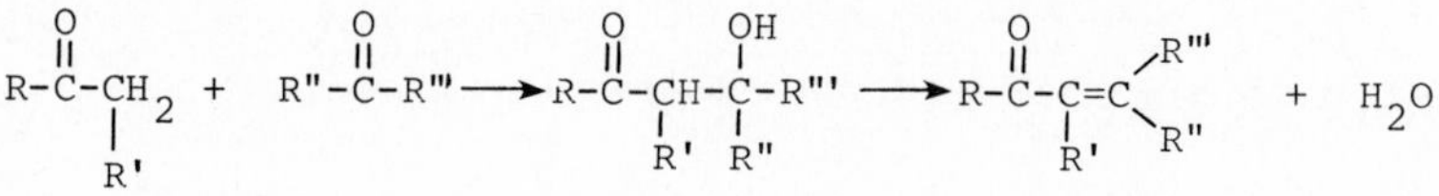

G. Wittig Reaction

phosphonium salts   oxaphosphetane

ylide   betaine

phosphorane

$$R-\overset{\overset{\displaystyle O}{\|}}{C}-R' \ + \ \overset{R''}{\underset{R'''}{\diagup}}C=P(C_6H_5)_3 \ \longrightarrow \ R-\overset{\overset{\displaystyle -O}{|}}{\underset{\underset{\displaystyle R'}{|}}{C}}-\overset{\overset{\displaystyle +P(C_6H_5)_3}{|}}{\underset{\underset{\displaystyle R''}{|}}{C}}-R''' \ \longrightarrow \ \overset{R}{\underset{R'}{\diagup}}C=C\overset{\diagup R''}{\diagdown R'''} \ + \ (C_6H_5)_3P=O$$

          (useful for introduction of C=C bond in
             a specific position)

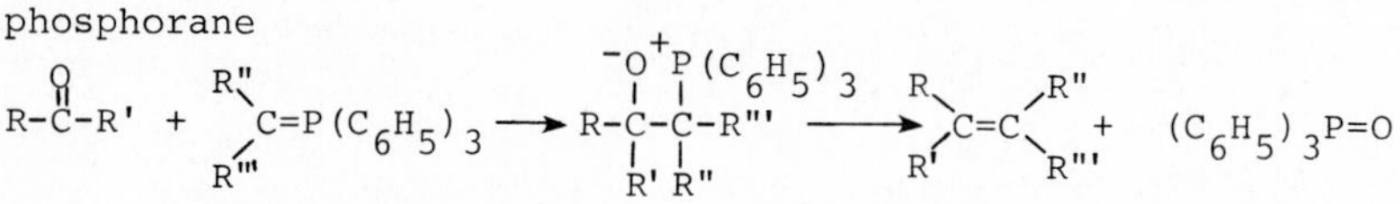

13.8 Oxidation and Reduction

A. Oxidation of Aldehydes and Ketones

autooxidation of aldehydes:

$$RCH=O \ \overset{air}{\longrightarrow} RCO_2H \quad \textit{(and many other oxidizing agents)}$$

Baeyer-Villiger reaction: migratory aptitudes $(H>C_6H_5>3°>2°>$
                                 $1°$ alkyl$>CH_3)$
peroxycarboxylic acids

$$R-\overset{\overset{\displaystyle O}{\|}}{C}-R' \ + \ R''CO_3H \longrightarrow R-\overset{\overset{\displaystyle O}{\|}}{C}-O-R' + R''CO_2H$$

Oxidative cleavage of ketones:

$$RC\overset{\overset{\displaystyle O}{\|}}{}CH_2R' \ \overset{HNO_3}{\underset{V_2O_5}{\longrightarrow}} RCO_2H + HO_2CR' \quad \textit{(hard to control)}$$

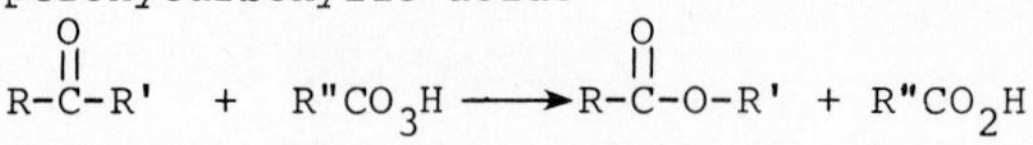

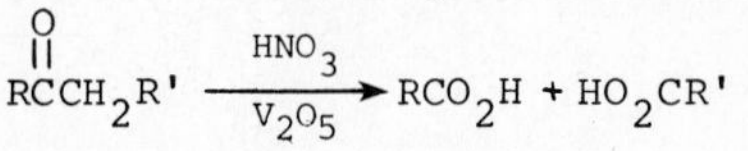

B. Metal Hydride Reduction   *(addition of protons and electrons)*

lithium aluminum hydride ($LiAlH_4$)   sodium borohydride ($NaBH_4$)

$$R-\overset{\overset{\displaystyle O}{\|}}{C}-R' \xrightarrow[\underset{NaBH_4/ethanol}{or}]{LiAlH_4/ether} R-\overset{\overset{\displaystyle OH}{|}}{C}H-R' \quad \text{\textit{(NaBH}}_4 \text{ \textit{more selective than LiAlH}}_4\text{\textit{)}}$$

C. Catalytic Hydrogenation

$$R-\overset{\overset{\displaystyle O}{\|}}{C}-R' \ + \ H_2 \xrightarrow[\underset{Ni}{or}]{Pt} R-\overset{\overset{\displaystyle OH}{|}}{C}H-R' \quad \text{\textit{(harder than hydrogenation of C=C)}}$$

D. Cannizzaro Reaction

disproportionation

$$2\ RCHO \xrightarrow{OH^-} RCO_2H \ + \ RCH_2OH \quad \text{\textit{(non-enolizable aldehyde only)}}$$

crossed Cannizzaro reactions

$$RCHO \ + \ CH_2O \xrightarrow{OH^-} RCH_2OH \ + \ HCO_2H$$

E. Deoxygenation Reactions

Wolff-Kisher reduction:

$$\overset{\overset{\displaystyle O}{\|}}{RCR'} \ + \ H_2NNH_2 \xrightarrow[\underset{\Delta\Delta}{HOCH_2CH_2OH}]{KOH} RCH_2R' \ + \ N_2 \ + \ H_2O \quad \text{\textit{(only for molecules stable}} \atop \text{\textit{to strong base)}}$$

Clemmensen reduction:

$$\overset{\overset{\displaystyle O}{\|}}{RCR'} \xrightarrow[HCl]{Zn} RCH_2R' \quad \text{\textit{(only for molecules stable to strong acid)}}$$

hydrogenolysis of dithioketals:

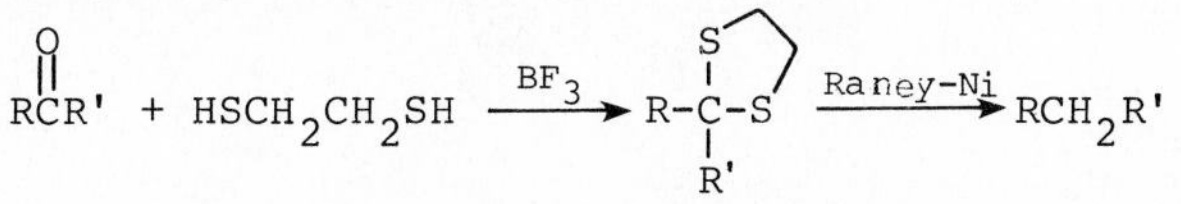

$$\overset{\overset{\displaystyle O}{\|}}{RCR'} \ + \ HSCH_2CH_2SH \xrightarrow{BF_3} R-\overset{\overset{\displaystyle S}{|}}{\underset{\underset{\displaystyle R'}{|}}{C}}-S \xrightarrow{Raney-Ni} RCH_2R' \quad \text{\textit{(for acid- or base-}} \atop \text{\textit{sensitive molecules)}}$$

13.B Important Concepts and Hints

Reactions of Double Bonds: C=O vs. C=C

If you are developing an intuitive understanding of organic chemistry, the chapter on aldehydes and ketones should give you a view of its logical foundations as well as its complexity. Intuition is difficult to teach, but as you study this chapter, try to apply the Question and Answer outline suggested in Section 8.B of this Study Guide. Hopefully, you will not only see similarities among many of the reactions discussed in this chapter, but also recognize analogies between these reactions and those discussed earlier, in the chapter on alkenes for instance. As an illustration of the sort of analogy we want you to see, study the examples below, in which we've juxtaposed reaction mechanisms from carbonyl and alkene chemistry. There are great differences in the conditions under which these reactions occur, and in their rates and the position of equilibrium, but there is a lot of similarity among the mechanisms.

$$CH_3\text{-}C(CH_3)\text{=}CH_2 \; \overset{H^+}{\underset{}{\rightleftharpoons}} \; (CH_3)_2\overset{+}{C}\text{-}CH_3 \; \overset{:OH_2}{\underset{}{\rightleftharpoons}} \; (CH_3)_2C(\overset{+}{O}H_2)\text{-}CH_3 \; \overset{-H^+}{\underset{}{\rightleftharpoons}} \; (CH_3)_2C(OH)\text{-}CH_3$$

$$(CH_3)_2C\text{=}\overset{..}{O}: \; \overset{H^+}{\underset{}{\rightleftharpoons}} \; \left[(CH_3)_2C\text{=}\overset{+}{O}H \;\updownarrow\; (CH_3)_2\overset{+}{C}\text{-}OH \right] \; \overset{H_2O}{\underset{}{\rightleftharpoons}} \; (CH_3)_2C(\overset{+}{O}H_2)(OH) \; \overset{-H^+}{\underset{}{\rightleftharpoons}} \; (CH_3)_2C(OH)_2$$

$$CH_3\text{-}CH\text{=}CH_2 \;+\; HCl \;\rightleftharpoons\; CH_3\text{-}\overset{+}{C}H\text{-}CH_3 \quad \overset{^-Cl}{\longrightarrow} \quad CH_3\text{-}CHCl\text{-}CH_3$$

$$CH_3\text{-}CH\text{=}O \;+\; HCN \;\rightleftharpoons\; \left[CH_3\text{-}CH\text{=}\overset{+}{O}H \;\updownarrow\; CH_3\text{-}\overset{+}{C}H\text{-}OH \right] \; \overset{^-CN}{\rightleftharpoons} \; CH_3\text{-}CH(CN)\text{-}OH$$

$$CH_3\text{-}CH(OH)\text{-}CH_3 \; \overset{H^+}{\rightleftharpoons} \; CH_3\text{-}CH(\overset{+}{O}H_2)\text{-}CH_3 \; \overset{-H_2O}{\rightleftharpoons} \; CH_3\text{-}\overset{+}{C}H\text{-}CH_3 \; \overset{-H^+}{\rightleftharpoons} \; CH_3CH\text{=}CH_2$$

$$CH_3\text{-}CH(OH)\text{-}NH\text{-}NH_2 \; \overset{H^+}{\rightleftharpoons} \; CH_3CH(\overset{+}{O}H_2)\text{-}NH\text{-}NH_2 \; \overset{-H_2O}{\rightleftharpoons} \; \left[CH_3\overset{+}{C}HNH\text{-}NH_2 \;\updownarrow\; CH_3CH\text{=}\overset{+}{N}H\text{-}NH_2 \right] \; \overset{-H^+}{\rightleftharpoons} \; CH_3CH\text{=}N\text{-}NH_2$$

$$CH_3CH_2CH\text{=}CH_2 \; \overset{H^+}{\rightleftharpoons} \; CH_3CH_2\overset{+}{C}HCH_3 \; \overset{-H^+}{\rightleftharpoons} \; CH_3CH\text{=}CHCH_3$$

$$CH_3CH_2CH\text{=}O \; \overset{H^+}{\rightleftharpoons} \; \left[CH_3CH_2CH\text{=}\overset{+}{O}H \;\updownarrow\; CH_3CH_2\overset{+}{C}HOH \right] \; \overset{-H^+}{\rightleftharpoons} \; CH_3CH\text{=}CHOH$$

$$(CH_3)_2C\text{=}CH_2 \;\; \overset{Br\text{-}Br}{\longrightarrow} \;\; \left[CH_3{}_2C(\overset{Br+}{})\text{-}CH_2 \;\updownarrow\; (CH_3)_2\overset{+}{C}\text{-}CH_2Br \right] \; \overset{H_2O}{\underset{-H^+}{\longrightarrow}} \; (CH_3)_2C(OH)\text{-}CH_2Br$$

$$\underset{CH_3}{\overset{HO}{}}C\text{=}CH_2 \;\; \overset{Br\text{-}Br}{\longrightarrow} \;\; \left[HO\text{-}\overset{+}{C}(CH_3)\text{-}CH_2Br \;\updownarrow\; HO\text{-}\overset{+}{C}(CH_3)\text{=}CH_2Br \right] \; \overset{-H^+}{\longrightarrow} \; O\text{=}C(CH_3)\text{-}CH_2Br$$

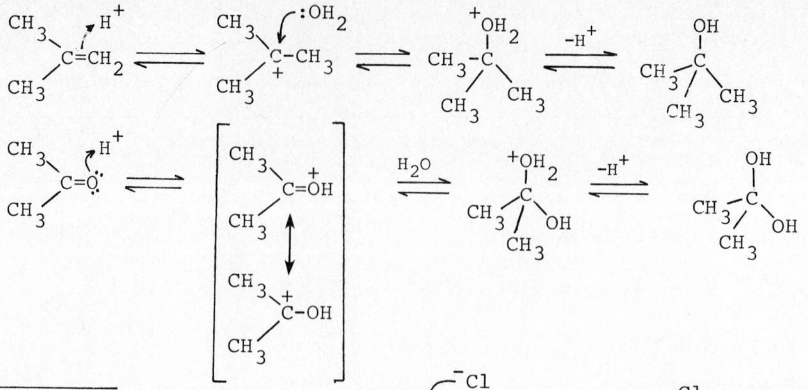

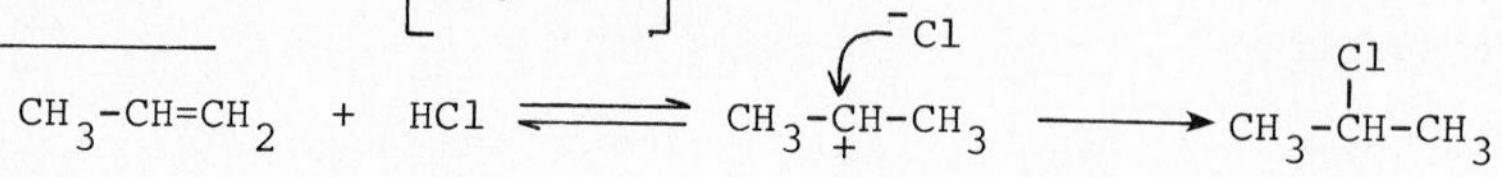

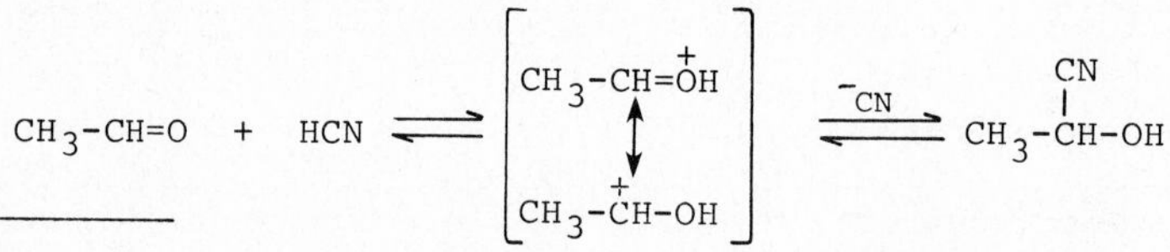

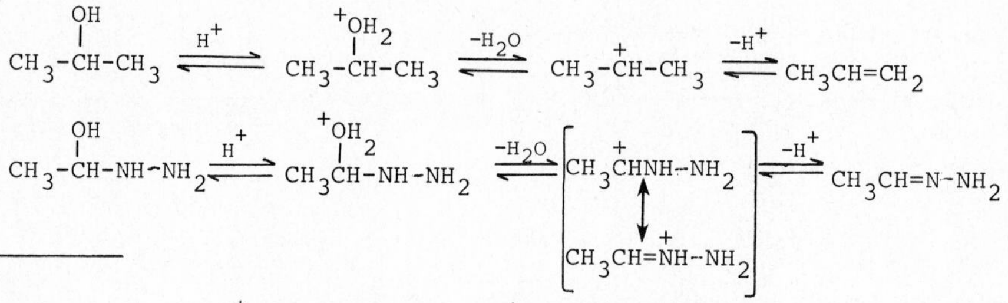

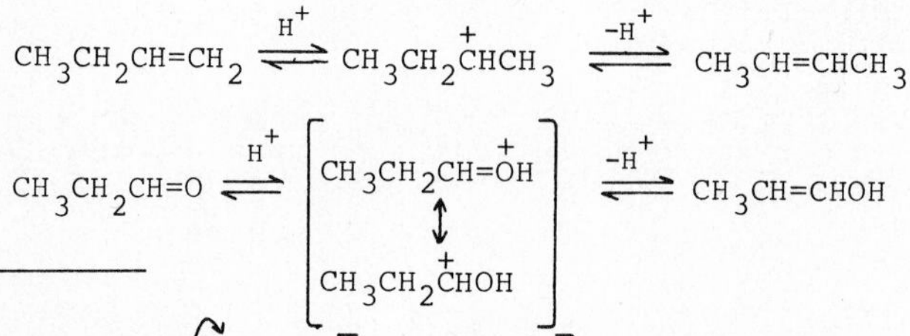

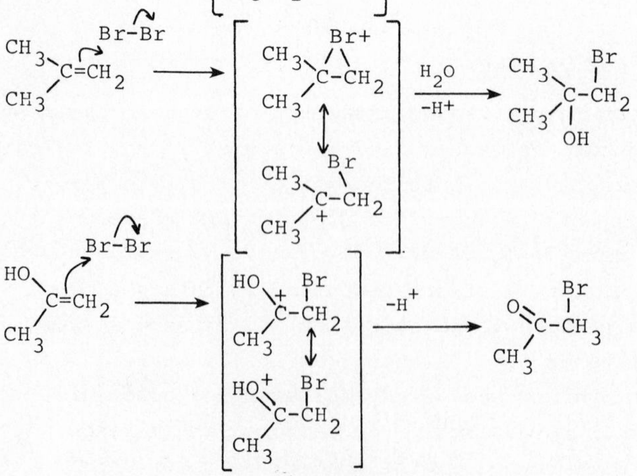

Acid- vs. Base-Catalyzed Reactions

In this chapter you encounter reactions which can be either acid- or base-catalyzed. The equilibration of ketone + alcohol $\rightleftharpoons$ hemiketal,

and the bromination of a ketone are just two examples. Usually, the
mechanisms of the acid- and base-catalyzed transformations are electronically identical; that is, the "electron flow" is the same in both
cases. The mechanisms differ only in the sequence of proton addition
or loss. For instance, compare the two mechanisms below for the bromination of acetone.

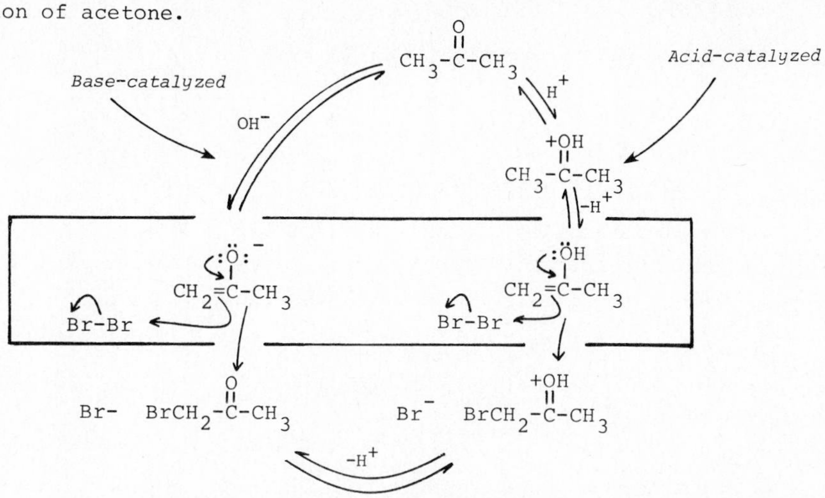

*(Note: Only the major resonance structure of each intermediate
is drawn.)*

Examine in particular the key steps enclosed in the boxes; they differ
only in the number of protons involved. For another comparison, see
the mechanisms of acid- and base-catalyzed hemiacetal formation depicted
in Section 13.7.B in the text. You will encounter many more reactions
which can proceed by either acid- or base-catalysis in the discussion of
carboxylic acid derivatives (Chapter 19).

A few generalizations are useful to keep in mind when you are
studying reactions like these, or trying to recall their mechanisms
during a test:

A. In acid-catalyzed reactions, the protons usually go on the
molecule before other steps occur (nucleophilic attack, loss
of a proton elsewhere (= tautomerization), etc.). Anionic
intermediates are very seldom observed (except counterions
of strong acids like Cl^-, tosylate ion, etc.); enolates and
alkoxide ions $\left(-\overset{O^-}{\underset{|}{\overset{|}{C}}}-\right)$ are not involved as intermediates.

B. In base-catalyzed reactions, proton removal usually preceeds
other steps. Cationic intermediates are rare, unless they
are the protonated forms of good bases (ammonium ions, etc);
oxonium ions and carbocations are not involved as intermediates.

The equations below are examples of incorrect mechanisms:

$$CH_3\overset{O}{\overset{||}{C}}-CH_2-\overset{OH}{\overset{|}{C}}(CH_3)_2 \underset{-H^+}{\overset{H^+}{\rightleftharpoons}} CH_3-\overset{O}{\overset{||}{C}}-CH_2-\overset{+OH_2}{\overset{|}{C}}(CH_3)_2 \underset{-H^+}{\overset{}{\rightleftharpoons}} CH_3-\overset{O^-}{\overset{|}{C}}=CH-\overset{+OH_2}{\overset{|}{C}}(CH_3)_2$$

$$CH_3\overset{O}{\overset{||}{C}}CH=C(CH_3)_2$$

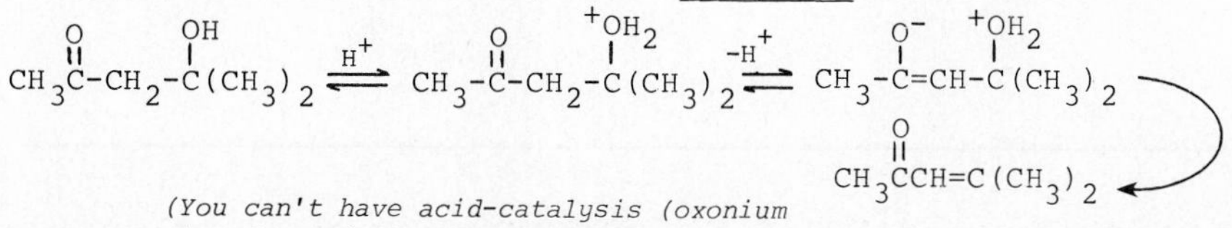

*(You can't have acid-catalysis (oxonium
ion formation) and base-catalysis (enolate
formation) at the same time)*

$$CH_3-\overset{\overset{\displaystyle O}{\|}}{C}-CH_3 \underset{acid}{\overset{H^+}{\rightleftharpoons}} CH_3-\overset{\overset{\displaystyle +OH}{\|}}{C}-CH \overset{CH_3O^-}{\underset{base}{\rightleftharpoons}} CH_3-\overset{\overset{\displaystyle OH}{|}}{\underset{\underset{\displaystyle OCH_3}{|}}{C}}-CH_3$$

$$CH_3-\overset{\overset{\displaystyle O}{\|}}{C}-CH_3 \overset{H^+}{\rightleftharpoons} CH_3-\overset{\overset{\displaystyle +OH}{\|}}{C}-CH_3 \overset{HC\equiv CNa}{\longrightarrow} CH_3-\overset{\overset{\displaystyle OH}{|}}{\underset{\underset{\underset{\underset{\displaystyle H}{|}}{\overset{\displaystyle C}{\|\|}}}{\overset{\displaystyle C}{|}}}{C}}-CH_3$$

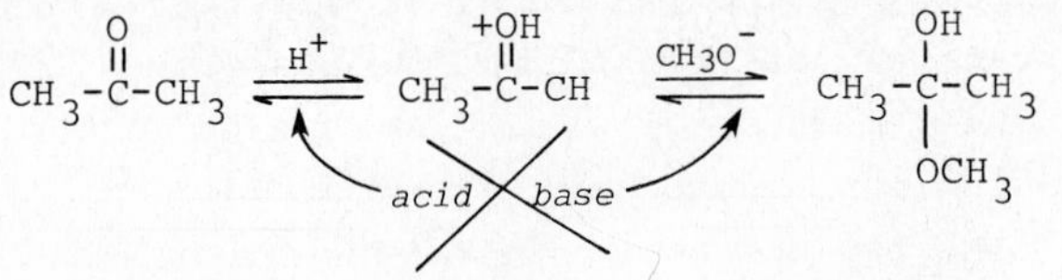

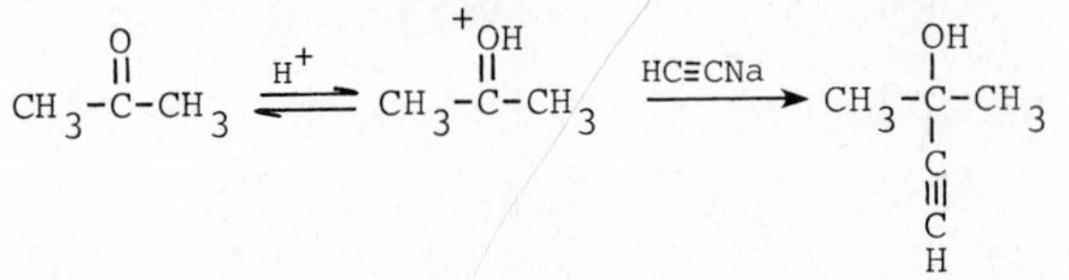

(Acetylide ion (strong base) will react with the acid used to protonate ketone)

Deprotonation vs. Oxidation; Protonation vs. Reduction

When you convert cyclohexanol to cyclohexanone, all you do is remove two protons, right? Wrong! You also have to remove two electrons, which is why an oxidizing agent (electron acceptor) is required, instead

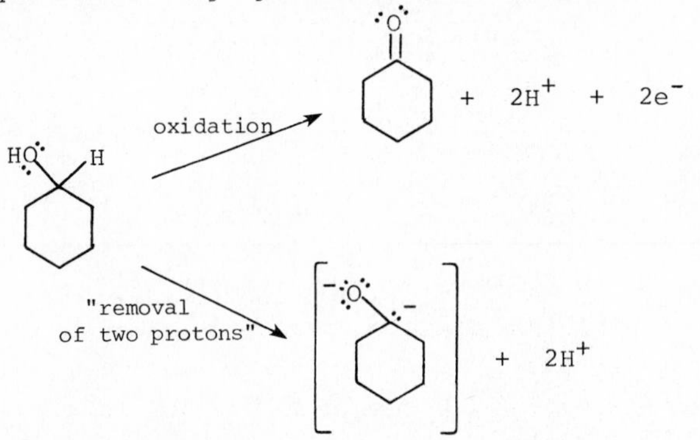

of just a base. Similarly, to convert cyclohexanone back to cyclohex-anol, two electrons need to be supplied by a reducing agent. The simple addition of two protons by a strong acid (even if possible) would not give cyclohexanol:

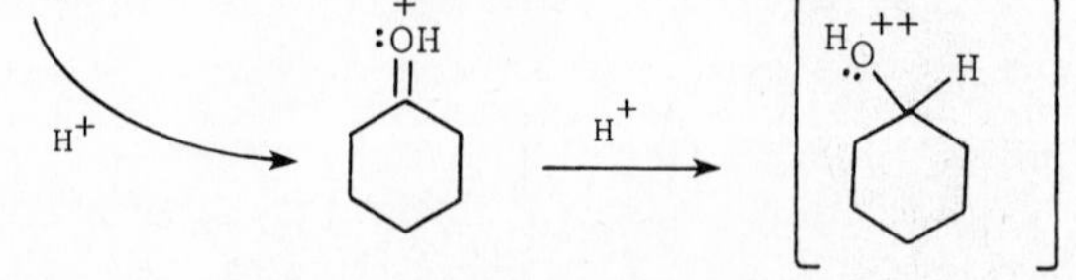

Neither oxidation nor reduction is accomplished with base or acid alone.

13.C Answers to Exercises

(13.2.A)   a) $(CH_3)_2CH\overset{\overset{\displaystyle O}{\|}}{C}CH_2CH_2CH_3$

b) $(CH_3)_2CHCH_2\overset{\overset{\displaystyle O}{\|}}{C}CH_2C(CH_3)_2$

c) $H\overset{\overset{\displaystyle O}{\|}}{C}CH_2CH_2CH_2OCH_3$
    $\quad\quad\alpha\;\;\beta\;\;\gamma$

d) $CH_3CH_2CHBrCH_2\overset{\overset{\displaystyle O}{\|}}{C}CH_3$

e)

f) $CH_3CH_2\overset{\overset{\displaystyle O}{\|}}{C}H\overset{\overset{\displaystyle O}{\|}}{C}CH_2CH_3$
    $\quad\quad\quad\quad CH_3$

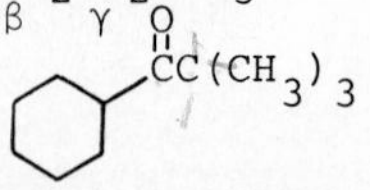

(13.2.B) a) 2-methyl-3-hexanone b) 2,6,6-trimethylheptanone

c) 4-methoxybutanal d) 4-bromo-2-hexanone

e) 1-cyclohexyl-2,2-dimethyl-1-propanone

f) 4-methyl-3-hexanone

(13.5.C)

$CH_3CH_2C{\equiv}CH \xrightarrow{Hg^{++}} CH_3CH_2\overset{O}{\overset{\|}{C}}CH_3 \xleftarrow[\text{or } 1.\ B_2H_6\ 2.H_2O_2,OH^-]{Hg^{++},\ H_2SO_4} CH_3\text{-}C{\equiv}C\text{-}CH_3$

$CH_3CH_2C{\equiv}CH \xrightarrow{1.\ B_2H_6\ \ 2.H_2O_2,OH^-} CH_3CH_2CH_2\overset{O}{\overset{\|}{C}}H$

$CH_3CH_2C{\equiv}CCH_3 \xrightarrow[\text{reagent}]{\text{either}} CH_3CH_2CH_2\overset{O}{\overset{\|}{C}}CH_3$ and $CH_3CH_2\overset{O}{\overset{\|}{C}}CH_2CH_3$

$CH_3CH_2C{\equiv}CCH_2CH_3 \xrightarrow[\text{reagent}]{\text{either}} CH_3CH_2CH_2\overset{O}{\overset{\|}{C}}CH_2CH_3$

(13.6.A) a) $(CH_3)_3C\overset{O}{\overset{\|}{C}}CD(CH_3)_2$ *(one only)*

b) $(CH_3CH_2)_2\overset{O}{\overset{\|}{C}}DCH$ *(one only)*

c) 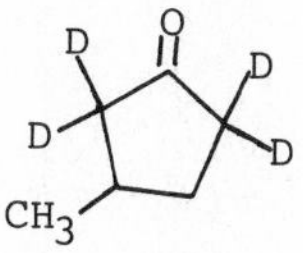 *(four)*

d) $CH_3CH{=}CHCH_2CH_3$ *(none)*

(13.6.B)

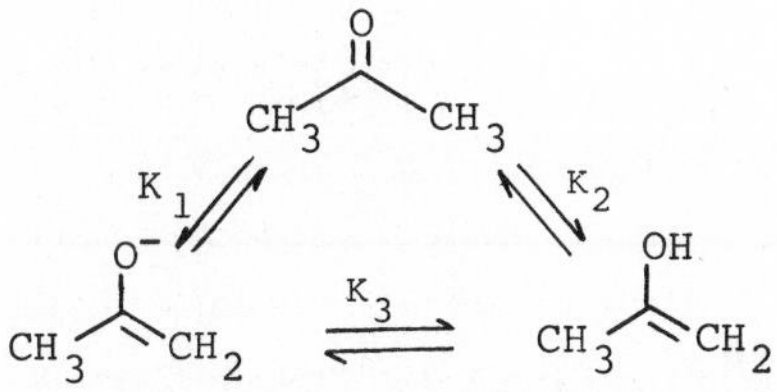

$K_1 = 10^{-pKa\ (acetone)} = 10^{-20}$

$K_2 = \dfrac{[enol]}{[ketone]} = 0.01\% = 10^{-4}$

$K_3 = 10^{-pKa}\ (enol)$

$K_1 = K_2 \cdot K_3;\ K_3 = \dfrac{K_1}{K_2} = \dfrac{10^{-20}}{10^{-4}}$; pKa (enol)=16 (about the same as an alcohol)

The difference in $\Delta H°$ for these two reactions is precisely equal to the difference in $\Delta H_f°$ for the two isomeric forms of acetone. *Since the enol form is much less stable, it is more acidic.*

i.e.

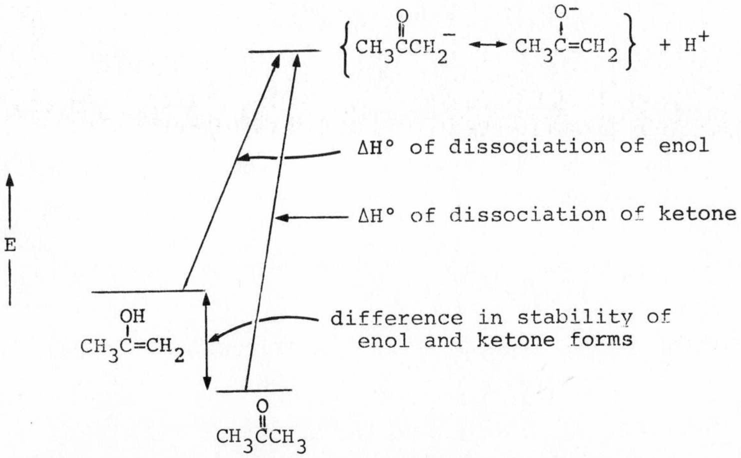

(13.6.C) Only those in which the chiral center is α to the carbonyl: (a) and (b) will racemize, (c) will not.

(13.6.D)

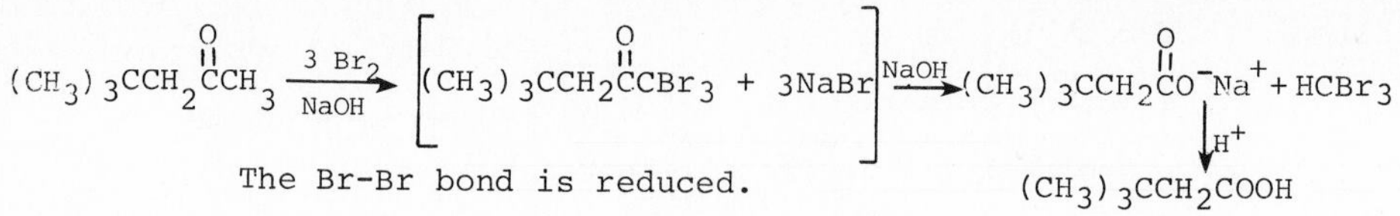

$$(CH_3)_3CCH_2CCH_3 \xrightarrow[\text{NaOH}]{3\ Br_2} \left[(CH_3)_3CCH_2CCBr_3 + 3NaBr \right] \xrightarrow{NaOH} (CH_3)_3CCH_2CO^- Na^+ + HCBr_3$$

The Br-Br bond is reduced.

$$(CH_3)_3CCH_2COOH \xleftarrow{H^+}$$

(13.7.A)   pH 3:

$$CH_3\overset{O}{\underset{\|}{C}}CH_3 \underset{}{\overset{H^+}{\rightleftharpoons}} CH_3\overset{+OH}{\underset{\|}{C}}CH_3 \overset{-H^+}{\rightleftharpoons} CH_3\overset{OH}{\underset{}{C}}CH_2 \quad \text{enolization}$$

$$H_2O \downarrow\!\uparrow$$

$$CH_3\overset{OH}{\underset{+OH_2}{\overset{|}{\underset{|}{C}}}}CH_3 \overset{-H^+}{\rightleftharpoons} CH_3\overset{OH}{\underset{OH}{\overset{|}{\underset{|}{C}}}}CH_3 \quad \text{hydration}$$

pH 11:

$$CH_3\overset{O}{\underset{\|}{C}}CH_3 \overset{OH^-}{\rightleftharpoons} CH_3\overset{O^-}{\underset{}{C}}CH_2 + H_2O \overset{-OH^-}{\rightleftharpoons} CH_3\overset{OH}{\underset{}{C}}CH_2 \quad \text{enolization}$$

$$OH^-, H_2O \downarrow\!\uparrow$$

$$CH_3\overset{O^-}{\underset{OH}{\overset{|}{\underset{|}{C}}}}CH_3 + H_2O \overset{-OH^-}{\rightleftharpoons} CH_3\overset{OH}{\underset{OH}{\overset{|}{\underset{|}{C}}}}CH_3 \quad \text{hydration}$$

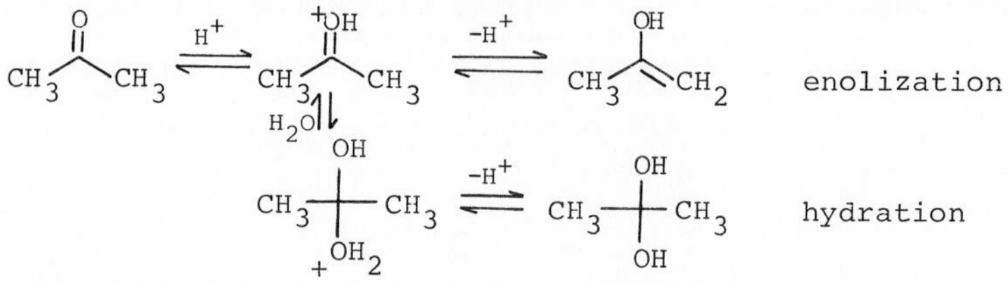

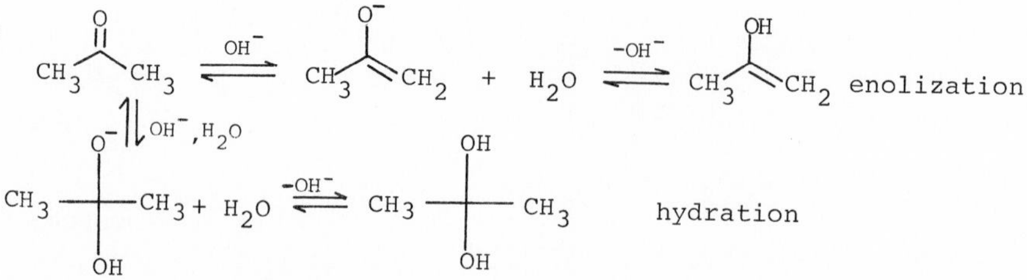

Notice the difference between pH 3 and pH 11: there are no
steps involving OH⁻ or anions under acidic conditions and there
are no steps involving protonated intermediates or cations under
basic conditions. The acid- and base-catalyzed mechanisms differ
only in the sequence of protonation and deprotonation steps.

(13.7.B)   The sequences below would not work:

$$CH_2=CHCH_2CH_2CCH_3 \xrightarrow[\text{2.H}_2O_2,]{1.B_2H_6} \times HOCH_2CH_2CH_2CH_2CH_2CCH_3 \quad \textit{(ketone would be reduced, too)}$$

$$\downarrow \begin{matrix}1.B_2H_6\\2.H_2O_2,OH^-\end{matrix}$$

$$HOCH_2CH_2CH_2CH_2CH_2\overset{OH}{\underset{}{C}}HCH_3 \xrightarrow{Cr^{+6}} \times HOCH_2CH_2CH_2CH_2CH_2CCH_3 \quad \textit{(both alcohols would be oxidized)}$$

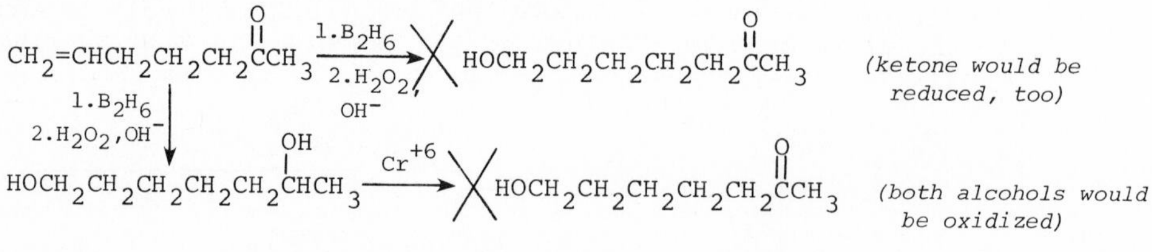

Therefore, the ketone must be protected during the hydroboration/
oxidation step:

$$CH_2=CHCH_2CH_2CCH_3 \xrightarrow[\text{H}_2O]{HOCH_2CH_2OH, H^+} CH_2=CHCH_2CH_2CCH_3 \xrightarrow{1.B_2H_6\ \ 2.H_2O_2,OH^-}$$

$$HOCH_2CH_2CH_2CH_2CH_2CCH_3 \xleftarrow{H_3O^+} HOCH_2CH_2CH_2CH_2CH_2CCH_3$$

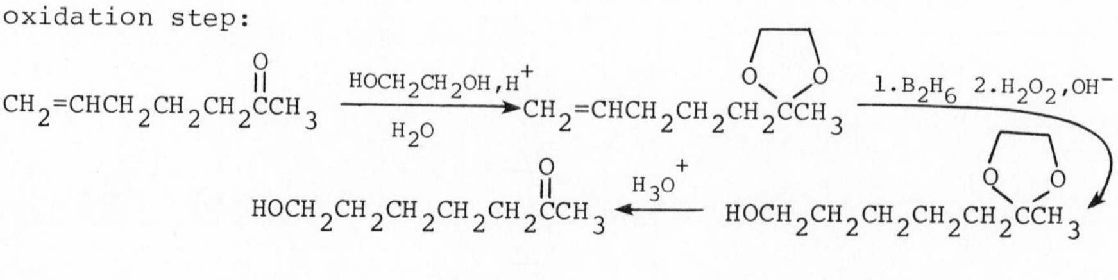

The ketal is stable to the alkaline
hydroboration/oxidation conditions, and is removed by
hydrolysis in aqueous acid.

(13.7.C) Oxime:

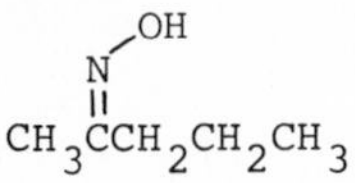

(*cis*)

phenylhydrazone:

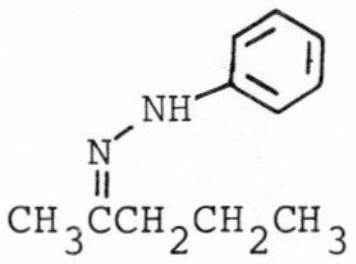

(*cis*)

hydrazone:

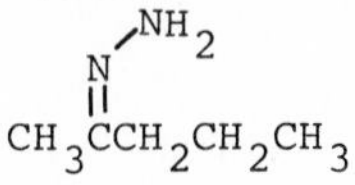

(*cis*)

semicarbazone:

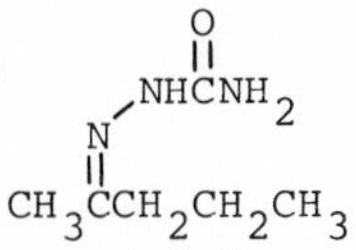

(*cis*)

aniline Schiff base:

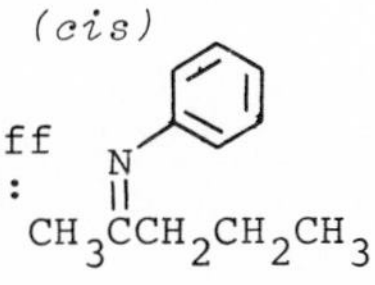

(*cis*)

the *trans* isomers are:

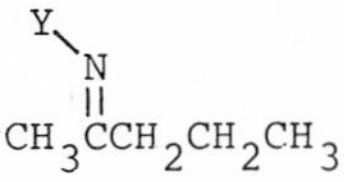

2-pentanone azine exists as 3 stereoisomers:

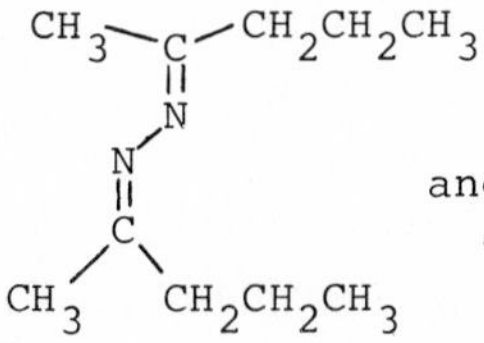

(*cis*, *trans*)

and (*cis*,*cis*) and (*trans*,*trans*)

(13.7.D)

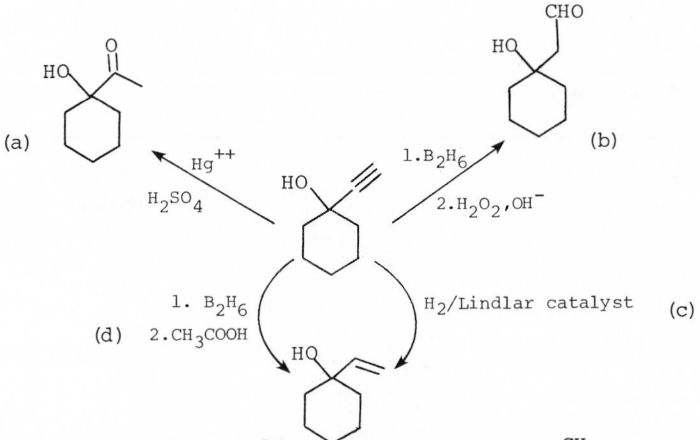

(13.7.F) (a)

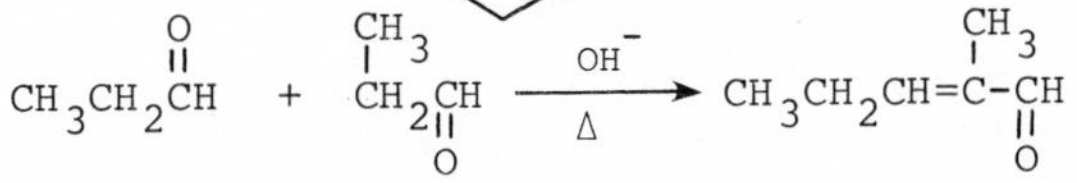

(b)

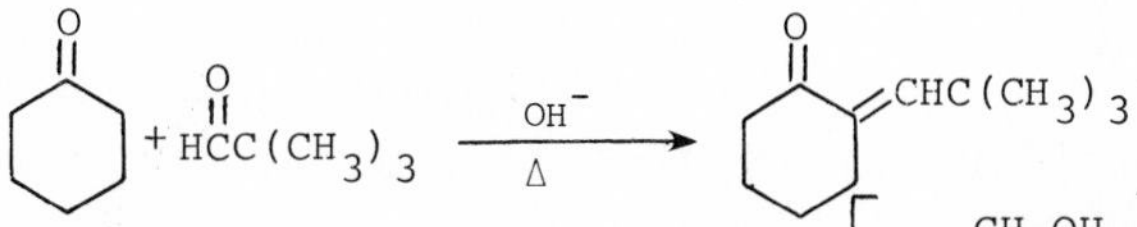

(c)

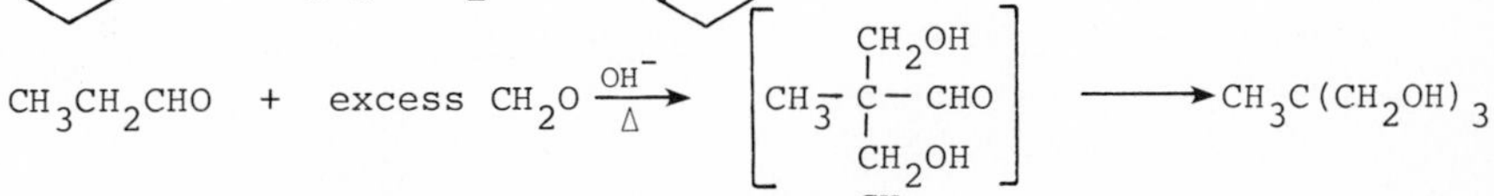

(d)

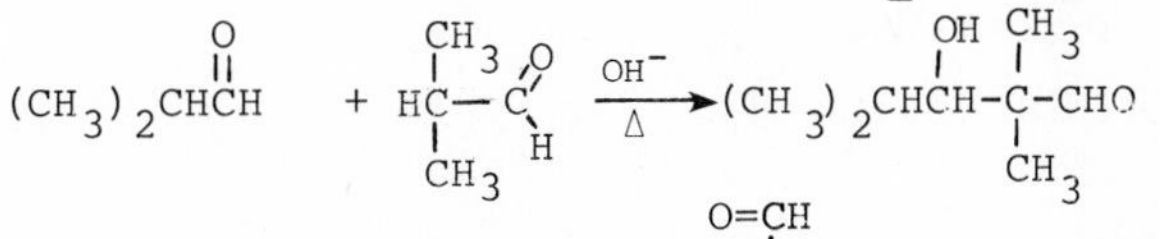

(e) Work backwards:

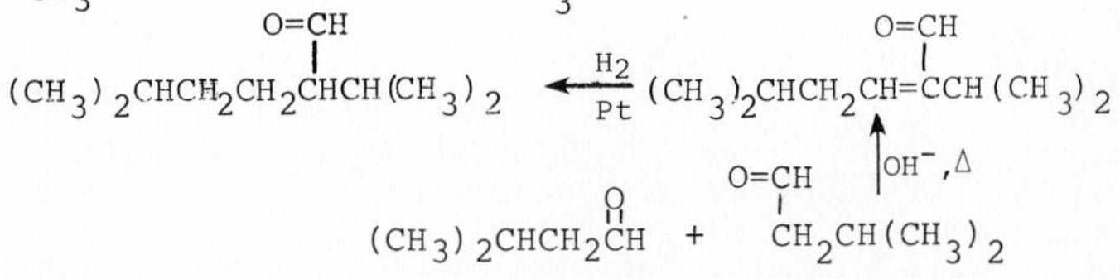

(13.8.A)

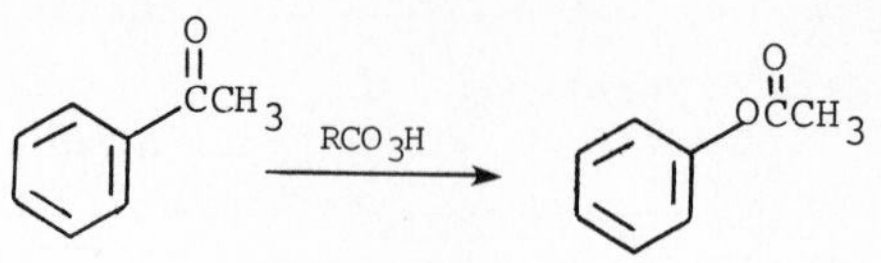

;

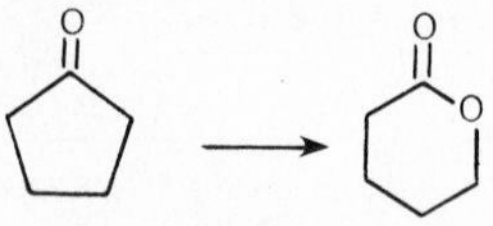

;

$(CH_3)_3\overset{O}{\overset{\|}{C}}CH_3 \longrightarrow (CH_3)_3CO\overset{O}{\overset{\|}{C}}CH_3$;

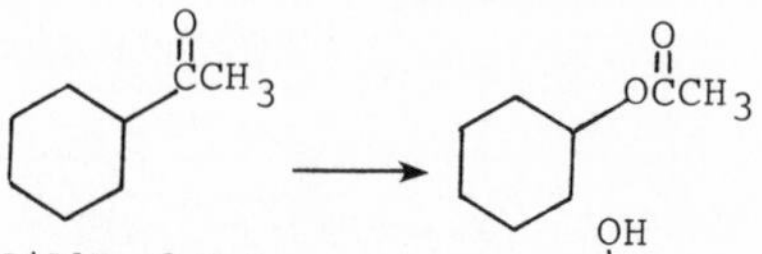

(13.8.B) $CH_3(CH_2)_5C{\equiv}CH \xrightarrow[H_2SO_4]{Hg^{++}} CH_3(CH_2)_5\overset{O}{\overset{\|}{C}}CH_3 \xrightarrow[\text{or } NaBH_4]{1. LiAlH_4\ 2. H_2O} CH_3(CH_2)_5\overset{OH}{\overset{|}{C}}HCH_3$

(13.8.E) (a) Clemmensen reduction (1° Bromide will not survive conditions of Wolff-Kishner or desulfurization reactions).

(b) Desulfurization of dithioacetal (Aldol product will dehydrate in acid or base).

(c) Any of the methods; no sensitive functional groups.

(d) Desulfurization of dithioacetal (Epoxide will be hydrolyzed in acid or base).

13.D Answers and Explanations to Problems

1. a) methyl ethyl ketone; 2-butanone
   b) propyl isopropyl ketone; 2-methyl-3-hexanone
   c) methyl t-butyl ketone; 3,3-dimethyl-2-butanone C gets lowest number
   d) ethyl neopentyl ketone; 5,5-dimethyl-3-hexanone
   e) methyl cyclopentyl ketone; acetylcyclopentane or 1-cyclopentyl-1-ethanone (uncommon)
   f) isobutyraldehyde; 2-methylpropanal
   g) α-bromopropionaldehyde; 2-bromopropanal
   h) β-methoxybutyraldehyde; 3-methoxybutanal
   i) ethyl vinyl ketone; pent-1-en-3-one
   j) methyl cyclopropylmethyl ketone; 1-cyclopropyl-2-propanone

2. a) 4-methylhexanal
   b) 2,6-dimethyl-4-octanone
   c) 4-oxohexanal
   d) 2-ethylcyclopentanone
   e) 4-bromo-2-pentanone
   f) 4-methyl-3-pentenal
   g) 5-phenyl-3-heptanone

3. a) $CH_3\overset{O}{\overset{\|}{C}}CH_2CH(CH_3)_2$
   b) $CH_3CH_2CH(OC_2H_5)_2$
   c) $CH_3\overset{Cl}{\overset{|}{C}}HCH_2CHO$
   d) 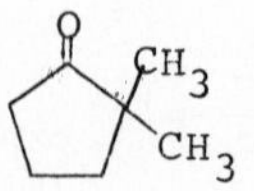
   e) 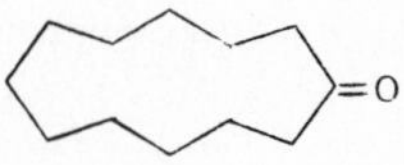
   f) 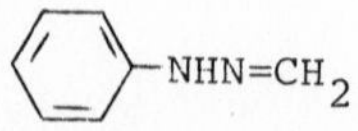
   g) 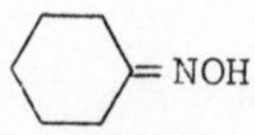
   h) $(CH_3)_2C{=}NNH\overset{O}{\overset{\|}{C}}NH_2$

4.. a) 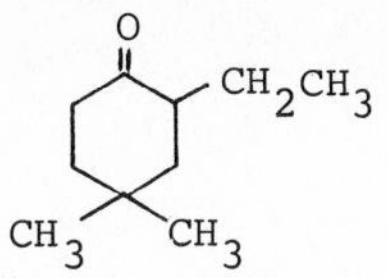

d) 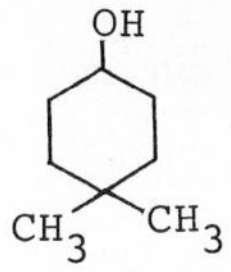

g) 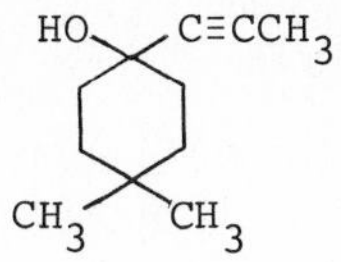

b) 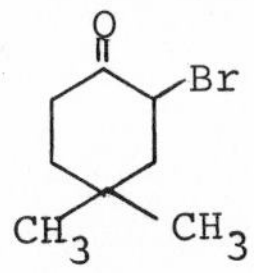

e) 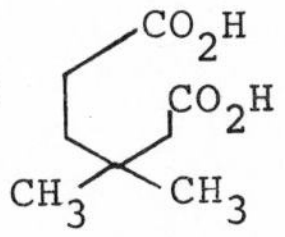

h) 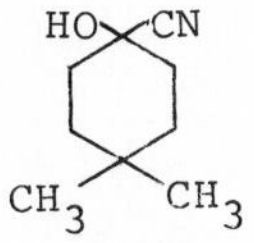

c) 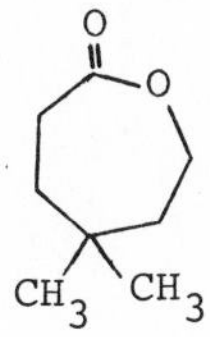

f) 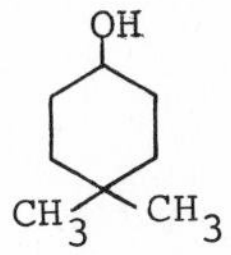

i) 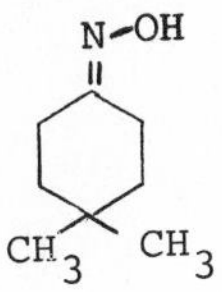

j) 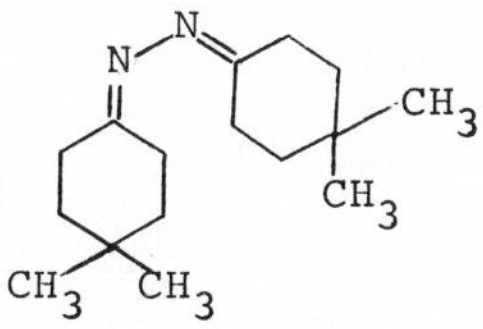

m) 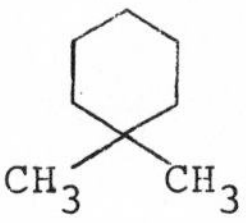

k) 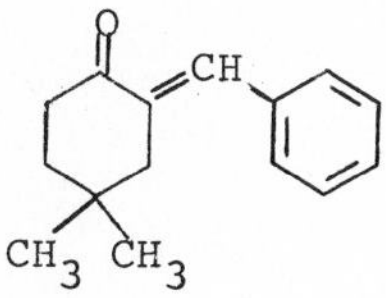

n) 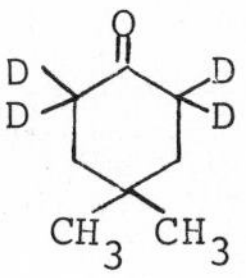

p) 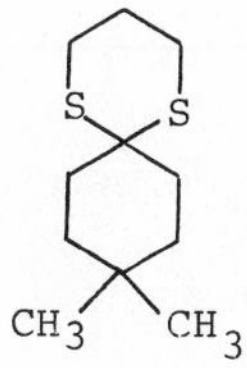

l) 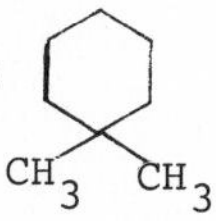

o) 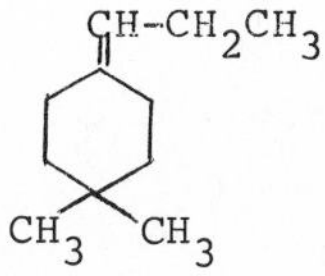

5. a) 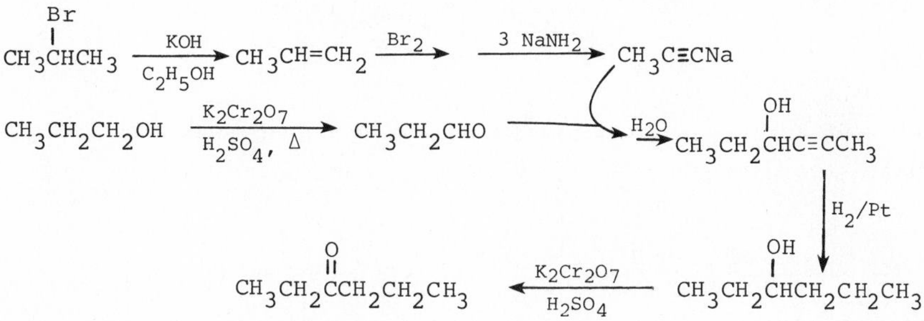

b) 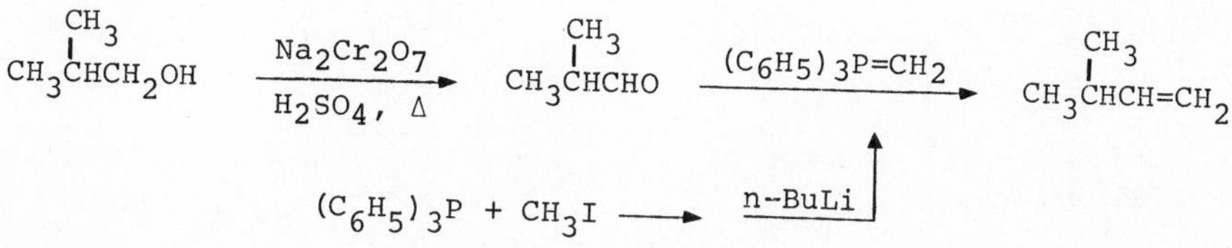

c) $CH_3CH_2CH_2CH_2OH$ $\xrightarrow[H_2SO_4, \ \Delta]{Na_2Cr_2O_7}$ $CH_3CH_2CH_2CHO$ $\xrightarrow[80°]{NaOH, \ H_2O}$

$CH_3CH_2CH_2CH=C-CH_2OH$ $\xleftarrow{LiAlH_4}$ $CH_3CH_2CH_2CH=CCHO$
$\qquad\qquad\quad |$ $\qquad\qquad\qquad\qquad\qquad\qquad\qquad |$
$\qquad\qquad\quad CH_2CH_3$ $\qquad\qquad\qquad\qquad\qquad\qquad CH_2CH_3$

d) $CH_3CH_2CHOHCH_3$ $\xrightarrow[H_2SO_4]{K_2Cr_2O_7}$ $CH_3CH_2\overset{\overset{O}{\|}}{C}CH_3$

$CH_3CH_2CH_2Br$ + $(C_6H_5)_3P:$ $\xrightarrow{n-BuLi}$ $(C_6H_5)_3P=CHCH_2CH_3$ $\rightarrow$ $CH_3CH_2CH=C\overset{CH_2CH_3}{\underset{CH_3}{\big\langle}}$

$\qquad\qquad\qquad\qquad\qquad\qquad\qquad\qquad\qquad\qquad\qquad$ *cis* and *trans*

e) CH_3CH_2OH $\xrightarrow[\Delta\Delta]{H_2SO_4}$ $CH_2=CH_2$ $\xrightarrow{Br_2}$ $\xrightarrow{3 \ NaNH_2}$ $HC\equiv CNa$ + $BrCH_2CH_2CH_2CH_3$

$CH_3(CH_2)_3C\equiv C(CH_2)_3CH_3$ $\xleftarrow[2.n-\atop BuBr]{1.NaNH_2}$ $HC\equiv CCH_2CH_2CH_2CH_3$

$CH_3(CH_2)_3\overset{\overset{O}{\|}}{C}(CH_2)_4CH_3$ $\xleftarrow[H_2SO_4]{Hg^{++}}$ $\xrightarrow{NaBH_4}$ $CH_3(CH_2)_3CHOH(CH_2)_4CH_3$

f) $HC\equiv CNa$ (from (e)) + $BrCH_2CH_2CH_3$ $\rightarrow CH_3CH_2CH_2C\equiv CH$ $\xrightarrow[2.H_2O_2, \atop OH^-]{1.B_2H_6}$ $CH_3CH_2CH_2CH_2CHO$

$\qquad\qquad\qquad\qquad\qquad\qquad\qquad\qquad\qquad\qquad\qquad$ excess CH_2O $\Big\downarrow$ OH^-

$\qquad\qquad\qquad\qquad\qquad\qquad\qquad\qquad\qquad\qquad\qquad$ $CH_3CH_2CH_2C(CH_2OH)_3$

6. a) $CH_3CH_2CH_2CH_2CHO$ $\xrightarrow[\substack{or \\ H_2NNH_2-NaOH \\ (HOCH_2CH_2)_2O \\ \Delta}]{Zn(Hg)-HCl}$ $CH_3CH_2CH_2CH_2CH_3$

b) $CH_3(CH_2)_3CHO$ + $HC\equiv C^-Na^+$ $\xrightarrow{NH_3}$ $\xrightarrow{H^+}$ $CH_3(CH_2)_3\overset{\overset{OH}{|}}{C}HC\equiv CH$

c) $CH_3(CH_2)_3CHO$ + KCN $\xrightarrow{H_2SO_4}$ $CH_3(CH_2)_3\overset{\overset{OH}{|}}{C}HCN$

d) $CH_3(CH_2)_3CHO$ $\xrightarrow{Ag_2O}$ $\xrightarrow{H_3O^+}$ $CH_3CH_2CH_2CH_2CO_2H$

e) $2CH_3(CH_2)_3CHO$ $\xrightarrow{KOH}$ $CH_3CH_2CH_2CH_2CH=\overset{\overset{CHO}{|}}{C}-CH_2CH_2CH_3$

$\qquad\qquad\qquad\qquad\qquad\qquad\qquad\qquad\qquad\qquad$ $\Big\downarrow$ H_2-Pd

$CH_3CH_2CH_2CH_2CH_2\overset{\overset{CH_3}{|}}{C}HCH_2CH_2CH_3$ $\xleftarrow[\substack{(HOCH_2CH_2)_2O \\ (Wolff-Kishner)}]{H_2NNH_2,NaOH,}$ $CH_3CH_2CH_2CH_2CH_2\overset{\overset{CHO}{|}}{C}HCH_2CH_2CH_3$

f) $CH_3CH_2CH_2CH_2CHO$ $\xrightarrow{NaBH_4}$ $\xrightarrow{PBr_3}$ $CH_3(CH_2)_4Br$

g)
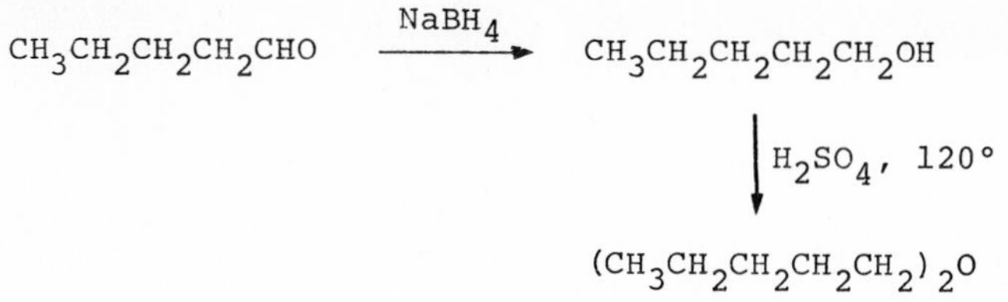
CH3CH2CH2CH2CHO
NaBH4
CH3CH2CH2CH2CH2OH
H2SO4, 120°
(CH3CH2CH2CH2CH2)2O

h)
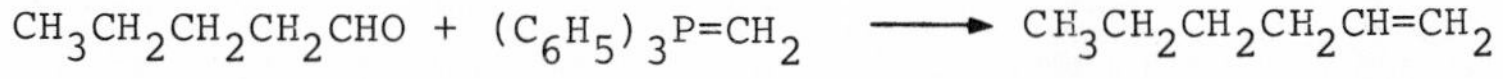
CH3CH2CH2CH2CHO + (C6H5)3P=CH2
CH3CH2CH2CH2CH=CH2

7. a)
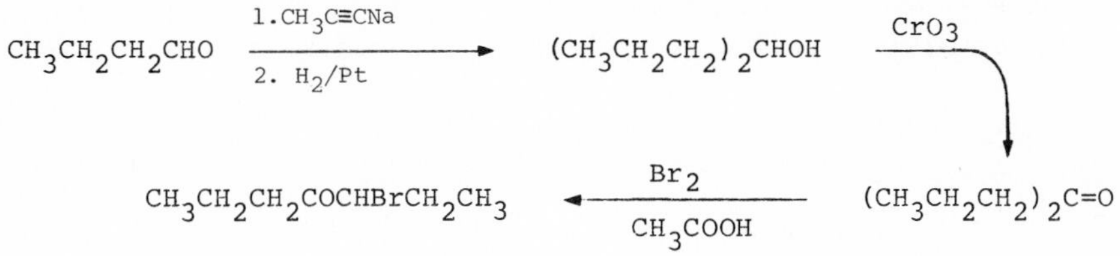
CH3CH2CH2CHO
1. CH3C≡CNa
2. H2/Pt
(CH3CH2CH2)2CHOH
CrO3
(CH3CH2CH2)2C=O
Br2
CH3COOH
CH3CH2CH2COCHBrCH2CH3

b)
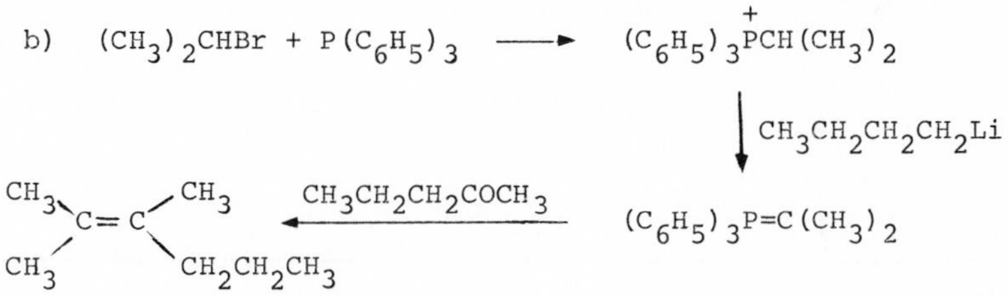
(CH3)2CHBr + P(C6H5)3
(C6H5)3PCH(CH3)2
CH3CH2CH2CH2Li
(C6H5)3P=C(CH3)2
CH3CH2CH2COCH3
CH3
CH3
CH3
CH2CH2CH3

c)
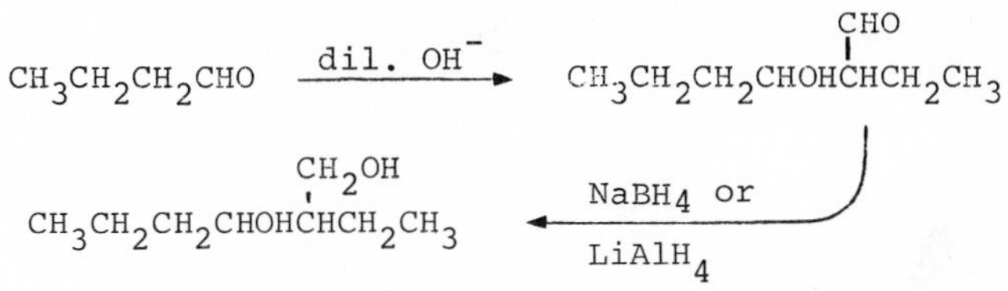
CH3CH2CH2CHO
dil. OH−
CHO
CH3CH2CH2CHOHCHCH2CH3
NaBH4 or
LiAlH4
CH2OH
CH3CH2CH2CHOHCHCH2CH3

d)
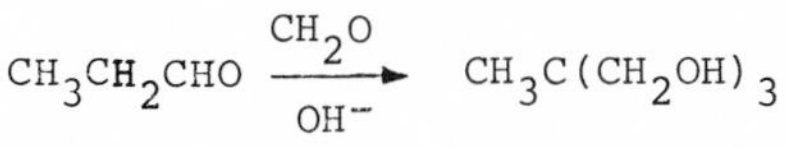
CH3CH2CHO
CH2O
OH−
CH3C(CH2OH)3

e)
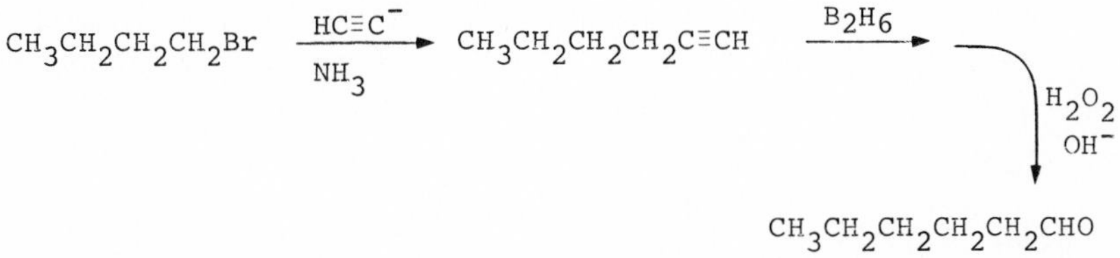
CH3CH2CH2CH2Br
HC≡C−
NH3
CH3CH2CH2CH2C≡CH
B2H6
H2O2
OH−
CH3CH2CH2CH2CH2CHO

f)
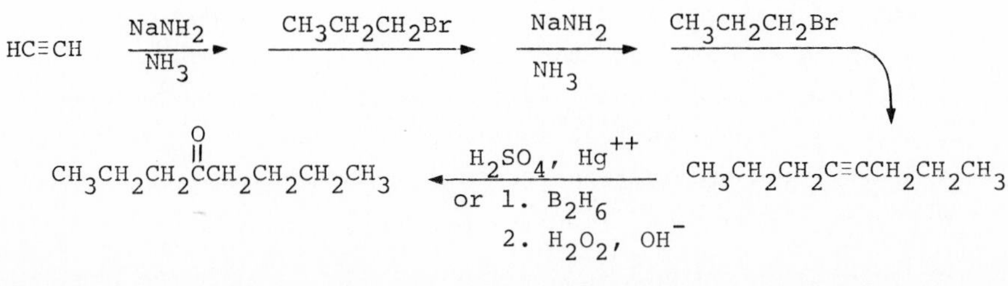
HC≡CH
NaNH2
NH3
CH3CH2CH2Br
NaNH2
NH3
CH3CH2CH2Br
CH3CH2CH2C≡CCH2CH2CH3
H2SO4, Hg++
or 1. B2H6
2. H2O2, OH−
O
CH3CH2CH2CCH2CH2CH2CH3

8. Consider the reaction of a base at the two sites:

reaction on oxygen

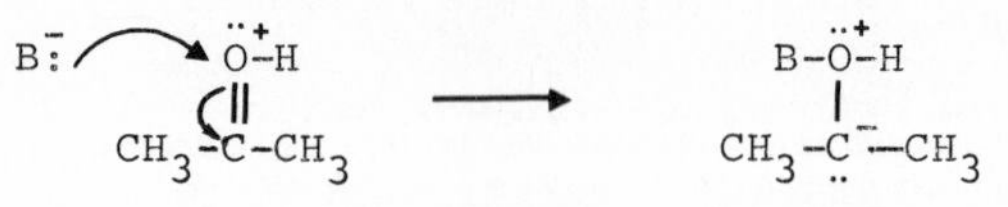

Since the oxygen already has 8 electrons, one pair in the double bond must be displaced onto carbon as the new bond is formed using the electron pair of the base. The resulting intermediate would be a high-energy charge-separated species.

reaction on carbon

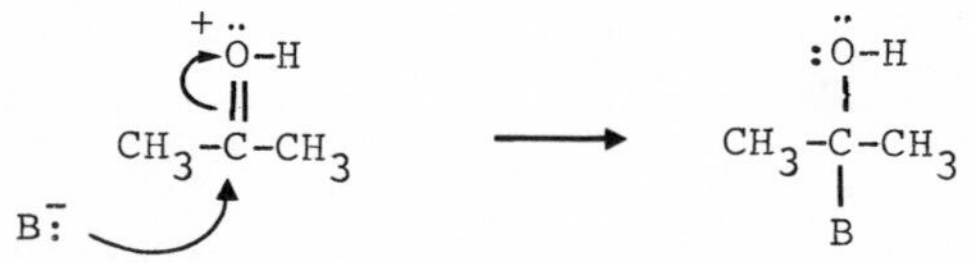

Again, the carbon already has 8 electrons, so one pair in the double bond must be displaced onto oxygen. However, in this case, the resulting product is neutral, and of much lower energy.

This example points up the value of using resonance structures. Of the two contributions to the resonance hybrid

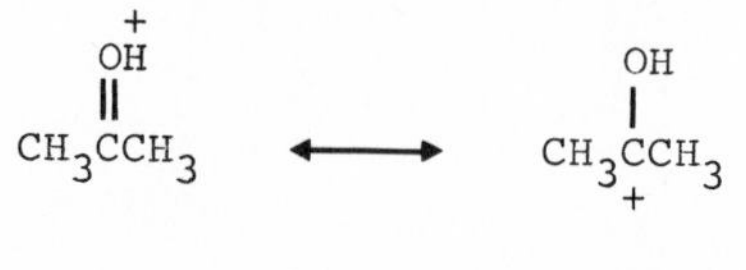

the former is more important since both C and O have octet configurations. However, the latter structure shows that C also has positive character and, to the extent it has positive character, is electron deficient and can react with a base.

9. This example is another case of kinetic vs equilibrium control. The acid-catalyzed reaction involves the more highly substituted enol which is formed faster than its isomer, but 1,3-dibromoacetone is apparently more stable than 1,1-dibromoacetone, perhaps because of steric effects. The important implication in these results is that the acid-catalyzed bromination reaction is reversible.

$$CH_3C=CHBr \xrightarrow{Br_2} CH_3\overset{+OH}{\overset{\|}{C}}CHBr_2 \rightleftharpoons CH_3\overset{O}{\overset{\|}{C}}CHBr_2 + HBr$$

$$+Br^-$$

less stable

$$CH_3\overset{O}{\overset{\|}{C}}CH_2Br + H^+ \rightleftharpoons CH_3\overset{+OH}{\overset{\|}{C}}CH_2Br$$

faster

slower

$$CH_2=\overset{OH}{\overset{|}{C}}CH_2Br \xrightarrow{Br_2} CH_2\overset{Br}{\overset{|}{\underset{}{C}}}\overset{+OH}{\overset{\|}{C}}CH_2Br = BrCH_2\overset{O}{\overset{\|}{C}}CH_2Br$$

$$+Br^- \qquad +HBr$$

more stable

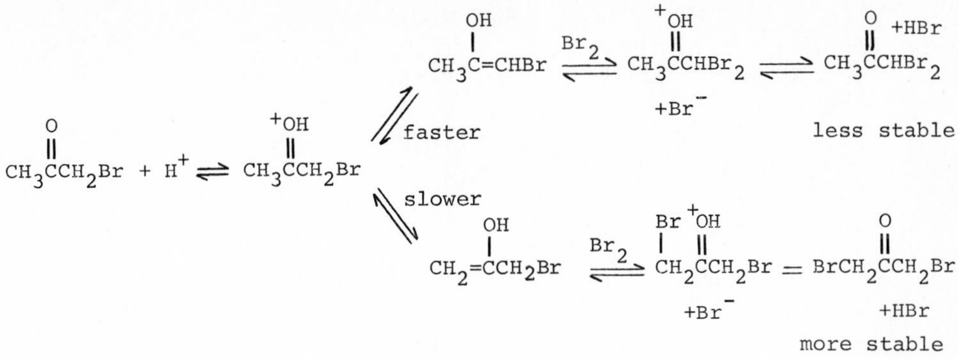

10. When 2-methylcyclohexanone is treated with t-BuOK in t-BuOH, the following equilibria are set up:

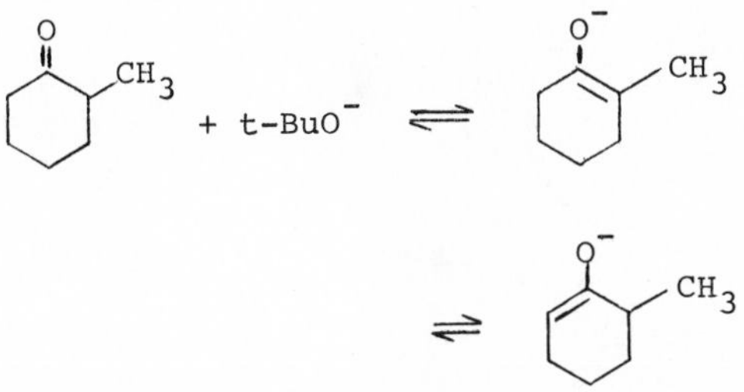

Either of these enolates may react with CH_3I:

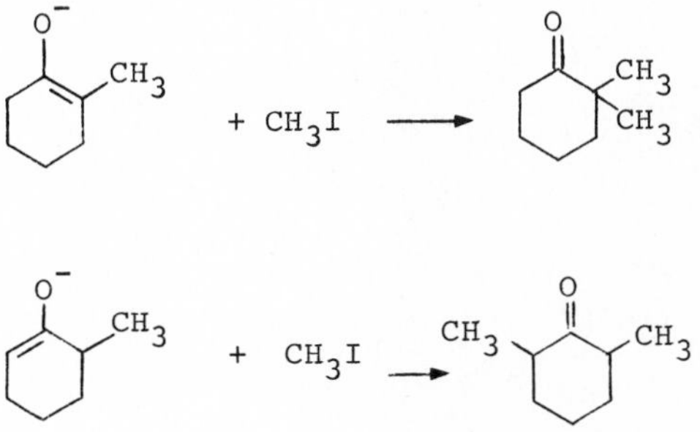

After some product is obtained (above), it can also be enolized and alkylated.

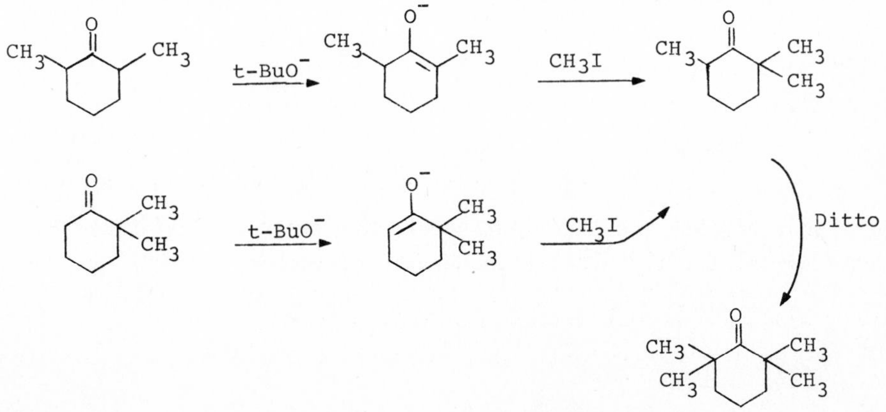

11.

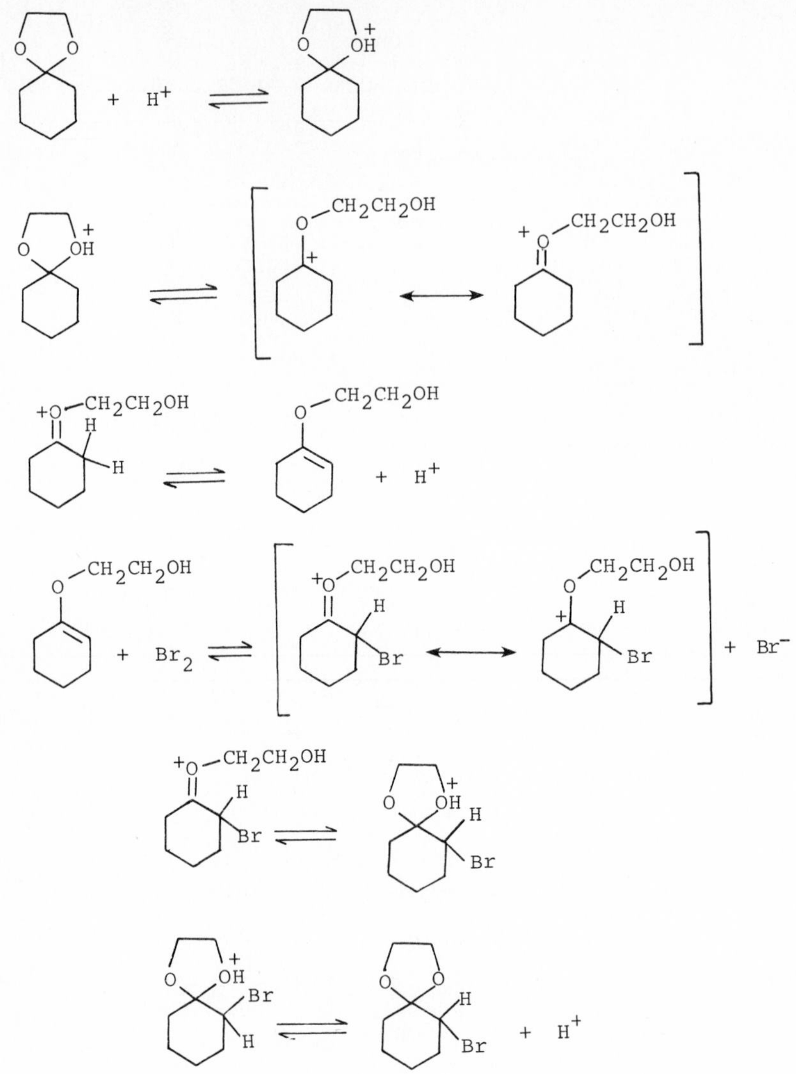

12. (1) $CH_3CH(OEt)OEt + H^+ \rightleftharpoons CH_3CH(\overset{+}{H}OEt)OEt$ (2) $CH_3CH(\overset{+}{H}OEt)OEt \rightleftharpoons CH_3CH{=}\overset{+}{O}Et + EtOH$

(3) $CH_3CH{=}\overset{+}{O}Et + H_2O \rightleftharpoons CH_3CH(\overset{+}{O}H_2)OEt$ (4) $CH_3CH(\overset{+}{O}H_2)OEt \rightleftharpoons CH_3CH(OH)OEt + H^+$

(5) $CH_3CH(OH)OEt + H^+ \rightleftharpoons CH_3CH(OH)\overset{+}{O}(H)Et$ (6) $CH_3CH(OH)\overset{+}{O}(H)Et \rightleftharpoons CH_3CH{=}\overset{+}{O}H + EtOH$

(7) $CH_3CH{=}\overset{+}{O}H \rightleftharpoons CH_3CHO + H^+$

13. Greater than unity. The fluorinated carbonyl group is destabilized because the dipolar resonance structure is less stable.

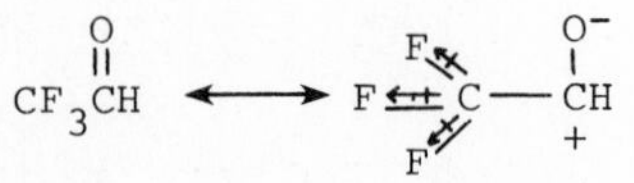

destabilized by electrostatic repulsions between C-F dipoles and C^+.

14. a) a cyclic acetal

b) (1) $HOCH_2CH_2CH_2\overset{O}{\overset{\|}{C}H}$ + H^+ $\rightleftharpoons$ $HOCH_2CH_2CH_2\overset{+OH}{\overset{\|}{C}H}$

   (2) $HOCH_2CH_2CH_2\overset{+OH}{\overset{\|}{C}H}$ $\rightleftharpoons$

   (3)

   (4)

   (5)

   (6)

   (7)

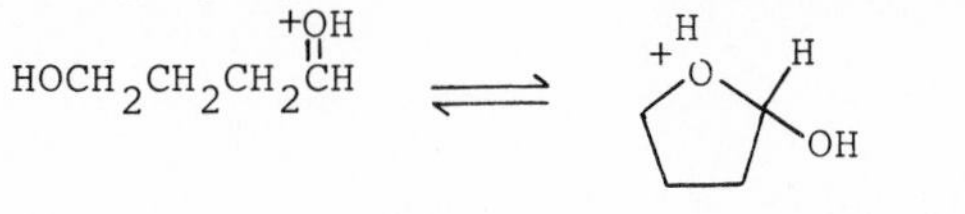

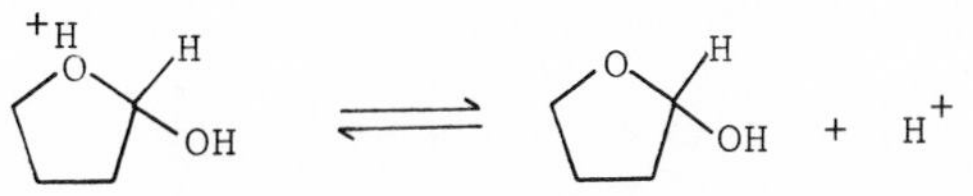

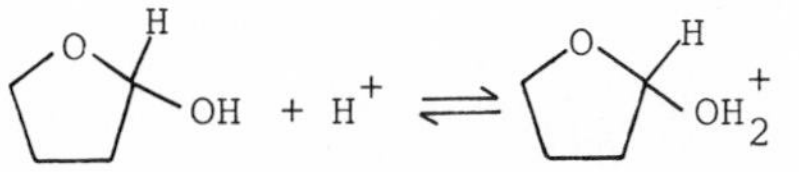

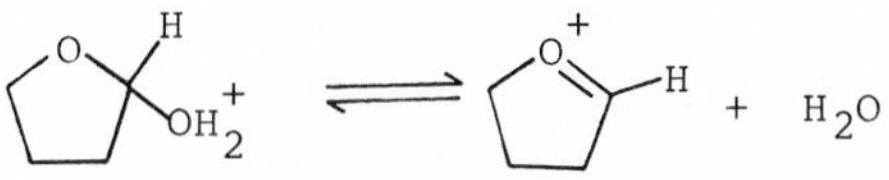

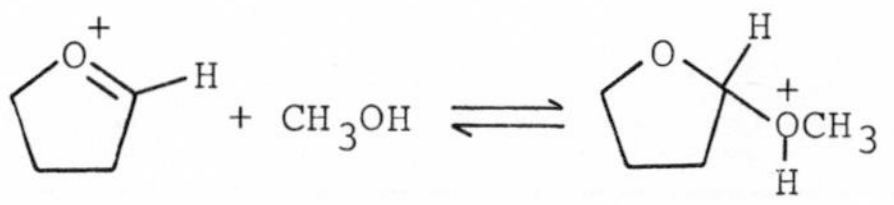

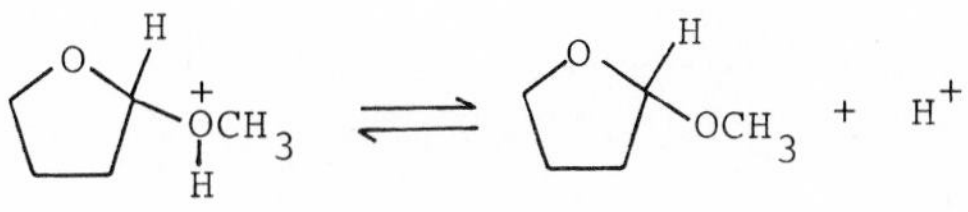

c) A cyclic hemi-acetal

d) See steps (1)-(3) in part b.

15. $CH_3\overset{O}{\overset{\|}{C}H}$ + H^+ $\rightleftharpoons$ $CH_3\overset{+OH}{\overset{\|}{C}H}$ $CH_3\overset{+OH}{\overset{\|}{C}H}$ + CH_3CHO $\rightleftharpoons$ $CH_3\overset{OH}{\underset{H}{C}}-\overset{+}{O}{=}\overset{H}{C}CH_3$

   $CH_3CHOH-\overset{+}{O}{=}\overset{H}{C}CH_3$ + $O{=}\overset{H}{C}CH_3$ $\rightleftharpoons$ $CH_3\overset{OH}{\underset{H}{C}}-O-\overset{H}{\underset{CH_3}{C}}-\overset{+}{O}{=}\overset{H}{C}CH_3$ $\rightleftharpoons$

   The depolymerization of paraldehyde involves this same mechanism, starting from paraldehyde. The acetaldehyde is removed by distillation to displace the equilibrium.

   H^+ +

   paraldehyde

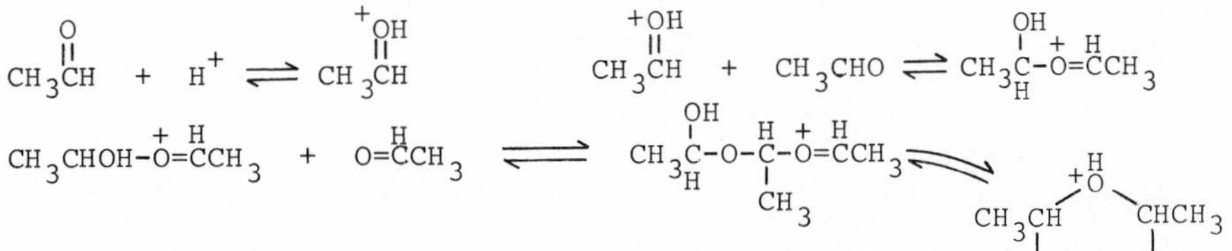

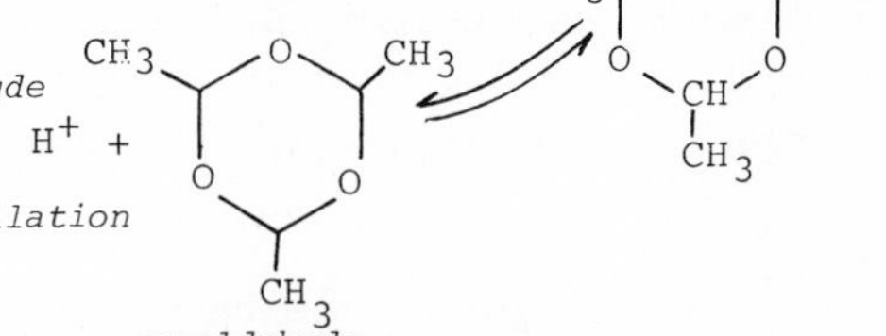

16. (a) $CH_3(CH_2)_8CH_2OH$ $\xrightarrow{PBr_3}$ $CH_3(CH_2)_8CH_2Br$ $\xrightarrow[NH_3]{HC{\equiv}C^-}$ $CH_3(CH_2)_8CH_2C{\equiv}CH$

   $\downarrow H_2$ | Lindlar catalyst

   $CH_3(CH_2)_8CH_2CHO$ $\xleftarrow[AcOH]{Zn}$ $\xleftarrow{O_3}$ $CH_3(CH_2)_8CH_2CH{=}CH_2$

   (b) $CH_3(CH_2)_9CH_2CH_2OH$ $\xrightarrow[\Delta]{Al_2O_3}$ $CH_3(CH_2)_9CH{=}CH_2$ $\xrightarrow{O_3}$ $\xrightarrow[AcOH]{Zn}$

   $CH_3(CH_2)_9CHO$

17.   A $\xrightarrow[\text{H}_2\text{SO}_4]{\text{Na}_2\text{Cr}_2\text{O}_7}$ B $\xrightarrow{\text{NaOD}}$ mw 116

      $C_7H_{16}O$ $C_7H_{14}O$

                       Ag_2O $\downarrow$

                   no reaction

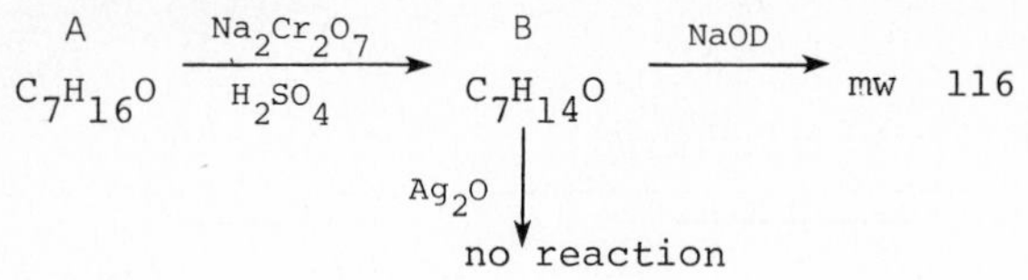

1) A, $C_7H_{16}O$: Formula indicates no rings or double bonds

2) A $\xrightarrow{\text{Cr}^{+6}}$ B: A is an alcohol, B is an aldehyde or ketone

3) B $\xrightarrow{\text{Ag}_2\text{O}}$ N.R.: B is not an aldehyde

4) B, $C_7H_{14}O$: molecular weight of B is 114

5) B $\xrightarrow{\text{NaOD}}$ mw 116: Only 2 hydrogens are exchangeable (α to the carbonyl group of B)

    The only possibilites for this latter case are:

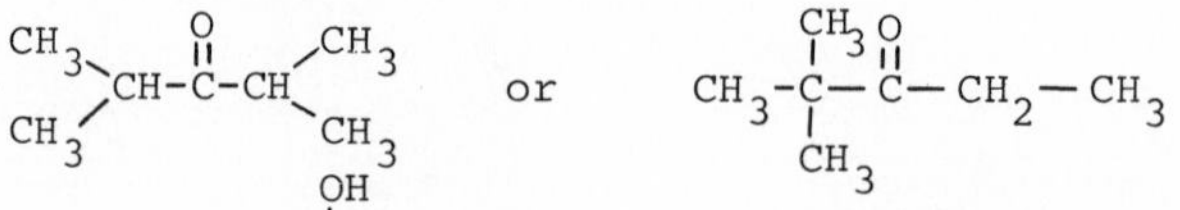

A is therefore $(CH_3)_2CHCHCH_2(CH_3)_2$ or $(CH_3)_3CCHOHCH_2CH_3$. To distinguish between these two possibilities, further information is needed.

18.   C $\xrightarrow{\text{O}_3}$ E $\xrightarrow{\text{H}_2\text{NOH}}$ F

      $C_{12}H_{20}$ $C_6H_{10}O$ $C_6H_{11}NO$

  (optically active)   *(optically*

                 active) $\searrow$ DCl

              NMR: only

  H_2 | Pt   one methyl, as a

             doublet   mw 101

  $\downarrow$

  D_1 and D_2

  $C_{12}H_{22}$

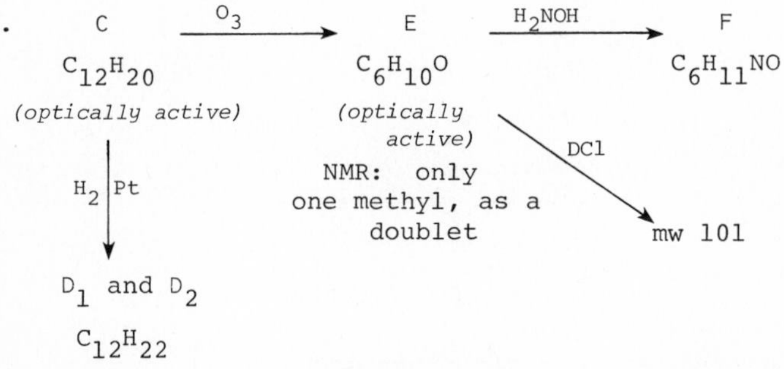

1) C, $C_{12}H_{20}$: Formula indicates 3 "degrees of unsaturation", that is: 3 rings, 2 rings and 1 double bond, 1 ring and 2 double bonds, or 1 ring and 1 triple bond, etc.

2) C $\xrightarrow[\text{Pt}]{\text{H}_2}$ D_1 and D_2, $C_{12}H_{22}$: Hydrogenation adds only 2 H's, so unsaturation is 2 rings and 1 double bond.

3) C $\xrightarrow{\text{O}_3}$ E, $C_6H_{10}O$: C has been cleaved into 2 of the same formula, E, which is either a ketone or an aldehyde.

4) E, $C_6H_{10}O$: Formula indicates mw 98 and 2 unsaturations: 1 C=O and 1 ring

5) E $\xrightarrow{\text{DCl}}$ mw 101: There are 3 exchangeable protons, so E cannot be an aldehyde

6) E, only 1 methyl group, as a doublet in nmr: $\overset{\diagdown}{\diagup}$CH–CH$_3$ is part of the structure.

7) There are many ketones of formula $C_6H_{10}O$ with one ring, but only 1 which satisfies 4), 5), and 6) above:

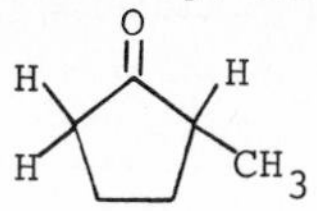

8) E is optically active: it is not a racemic mixture, and the two halves of C must have the same configuration.

9) C could be either:

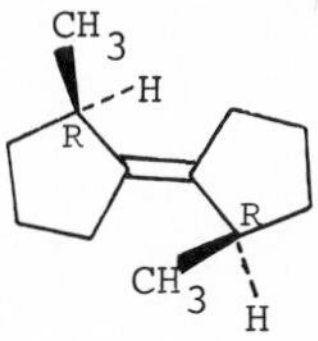

and

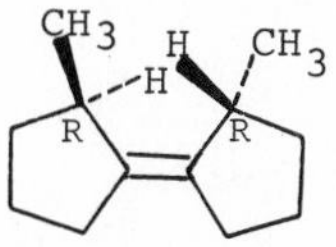

(or the (S,S)-enantiomers)

10) C $\xrightarrow[Pt]{H_2}$ D_1 and D_2: C is not the *cis* isomer in 9); only the *trans* isomer can give two isomers on hydrogenation:

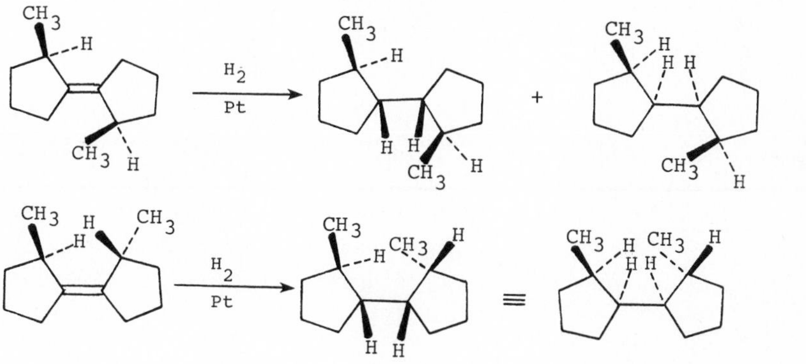

(You can see this easily with models)

11) Finally, F is the oxime:

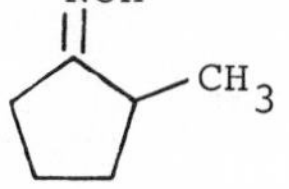

19. G (C_6H_8O) $\xrightarrow[Pd]{H_2}$ H ($C_6H_{10}O$, *(ketone)*) $\xrightarrow{CH_3CO_3H}$ I ($C_6H_{10}O_2$)

H $\xrightarrow{NaOD}$ $C_6H_7D_3O$

I: NMR: only 1 methyl, as a doublet, δ 1.9

1) G, formula, C_6H_8O: 3 degrees of unsaturation.

2) G $\xrightarrow[Pd]{H_2}$ H, $C_6H_{10}O$ *(ketone)*: Only one unsaturation is a C=C; that leaves 1 ring and 1 C=O.

3) H $\xrightarrow{NaOD}$ $C_6H_7D_3O$: 3 hydrogens α to C=O.

4) H $\xrightarrow{CH_3CO_3H}$ I, $C_6H_{10}O_2$: H undergoes the Baeyer-Villiger reaction:

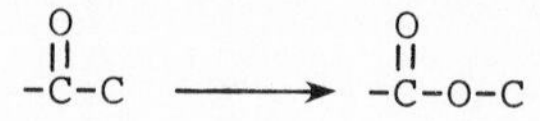

5) I, nmr: one methyl, doublet at δ 1.9:

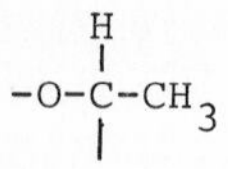

6) 3) & 4): I must be

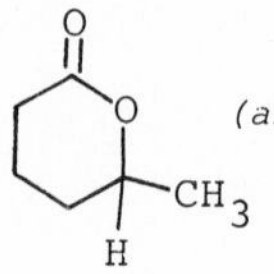

(any smaller ring would require another methyl group)

7) ∴ H must be

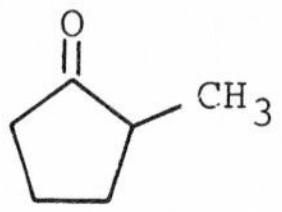

8) G, nmr: methyl is a singlet: G is

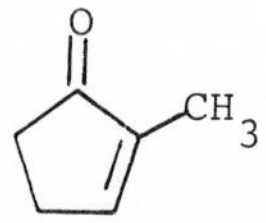

20.

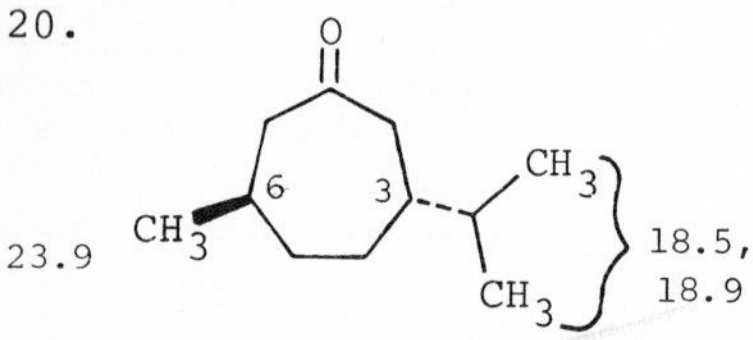

18.5 (q), 18.9 (q), and 23.9 (q) all belong to methyl groups. It is reasonable to assume that the isopropyl methyls are the ones with similar resonances. (Note: the resonances are not identical because the methyls are diastereotopic; see answer to problem 19(d), Chapter 10).

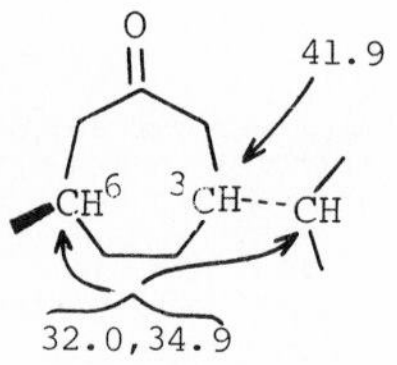

32.0 (d), 34.9 (d), and 41.9 (d) are the CH's. C-3 clearly has the most β-branches and is therefore expected to come the furthest downfield.

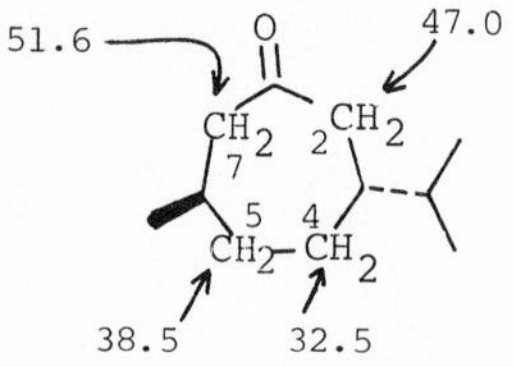

32.5 (t), 38.5 (t), 47.0 (t), and 51.6 (t) are the CH_2's. C-2 and C-7 should be downfield from C-4 and C-5 (presence of electron withdrawing C=O group), therefore they are 47.0 and 51.6. C-2 and C-4 have 2 γ-substitutents more than C-7 and C-5, respectively, because of the isopropyl vs. methyl groups. Therefore they should come ≈5 ppm further upfield.

By comparison, it is clear what the stereochemistry of the 3,6-dimethyl isomer is.

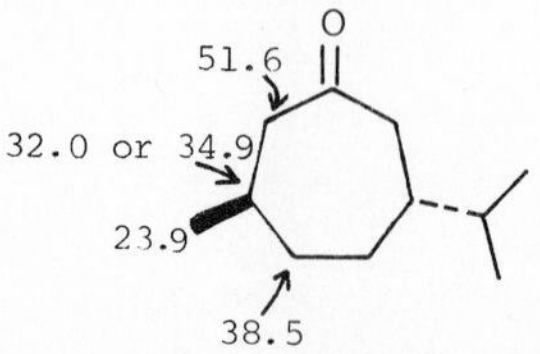

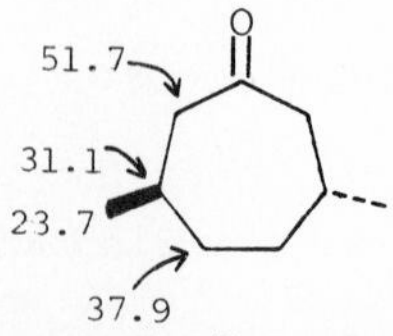

K

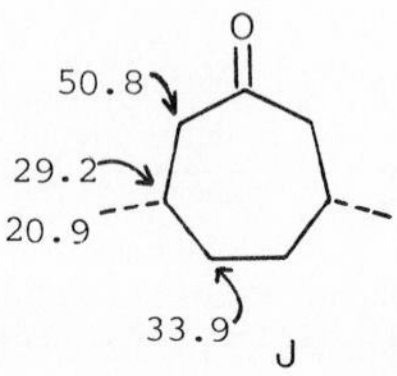

J

21.

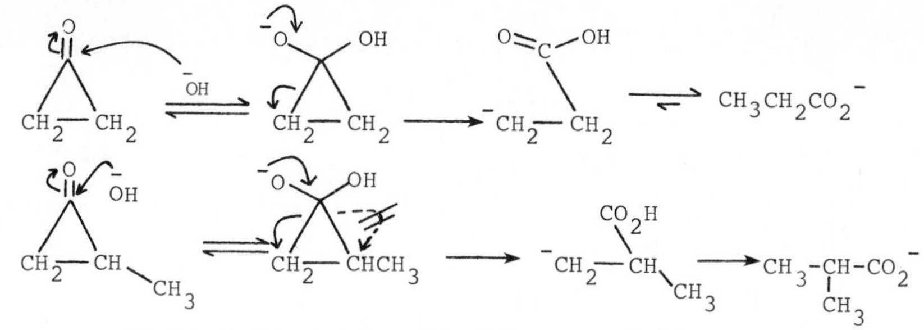

Cleavage occurs to give the more stable,
primary carbanion.

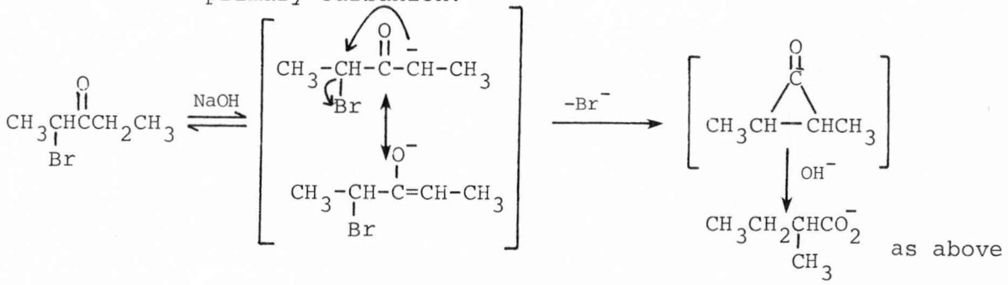

$$CH_3\overset{O}{\underset{}{C}}CHCH_2CH_3 \underset{Br}{} \xrightarrow{NaOH} \left[\begin{array}{c} CH_3-CH-\overset{O}{\underset{}{C}}-CH-CH_3 \\ \underset{Br}{} \\ \updownarrow \\ CH_3-\underset{Br}{CH}-C=CH-CH_3 \\ \underset{O^-}{} \end{array} \right] \xrightarrow{-Br^-}$$

$$\left[\begin{array}{c} \overset{O}{\underset{}{C}} \\ CH_3CH-CHCH_3 \end{array} \right] \xrightarrow{OH^-} CH_3CH_2\underset{CH_3}{CHCO_2^-} \quad \text{as above}$$

22. $CH_3\overset{O}{\underset{}{C}}CH_2CH_2C\equiv CCH_3 \rightleftharpoons$ (hydration of ketone)   (five-membered rings form faster than six)

$CH_3-\underset{OH}{\overset{HO}{C}}-CH_2CH_2C\equiv CCH_3$

$$CH_3\overset{O}{\underset{}{C}}CH_2CH_2\underset{+ Hg+}{\overset{OH}{C}-CHCH_3} \xleftarrow{H^+} CH_3\overset{O}{\underset{}{C}}CH_2CH_2\underset{Hg+}{\overset{OH}{C}}=CCH_3 \xleftarrow{-H^+}$$

$$CH_3\overset{O}{\underset{}{C}}CH_2CH_2\overset{OH}{C}=CHCH_2 \rightleftharpoons CH_3\overset{O}{\underset{}{C}}CH_2CH_2\overset{O}{\underset{}{C}}CH_2CH_3$$

23.

$$CH_2=CHOEt \xrightarrow{H^+} \left[\begin{array}{c} CH_3-\overset{+}{C}H-OEt \\ \updownarrow \\ CH_3-CH=\overset{+}{O}Et \end{array} \right] \xrightarrow{n-BuOH} CH_3-CH\underset{OEt}{\overset{\overset{H}{\underset{}{O}}^+-Bu}{}} \xrightarrow{-H^+} CH_3-CH\underset{OEt}{\overset{OBu}{}}$$

$R,R'=Et, n-Bu$

$$CH_3-CH\underset{OR'}{\overset{OR}{}} \xrightarrow{H^+} CH_3-CH\underset{OR'}{\overset{\overset{H+}{\underset{}{O}}R}{}} \xrightarrow{-ROH} \left[\begin{array}{c} CH_3-\overset{+}{C}H-OR' \\ \updownarrow \\ CH_3-CH=\overset{+}{O}R' \end{array} \right] \xrightarrow{H_2O} CH_3-CH\underset{OR'}{\overset{\overset{+}{O}H_2}{}}$$

$$CH_3CH=O \xrightarrow{-H^+} \left[\begin{array}{c} CH_3\overset{+}{C}HOH \\ \updownarrow \\ CH_3CH=\overset{+}{O}H \end{array} \right] \xrightarrow{-R'OH} CH_3CH\underset{\overset{+}{O}R'}{\overset{OH}{}}\underset{H}{} \xrightarrow{H^+} CH_3-CH\underset{OR'}{\overset{OH}{}}$$

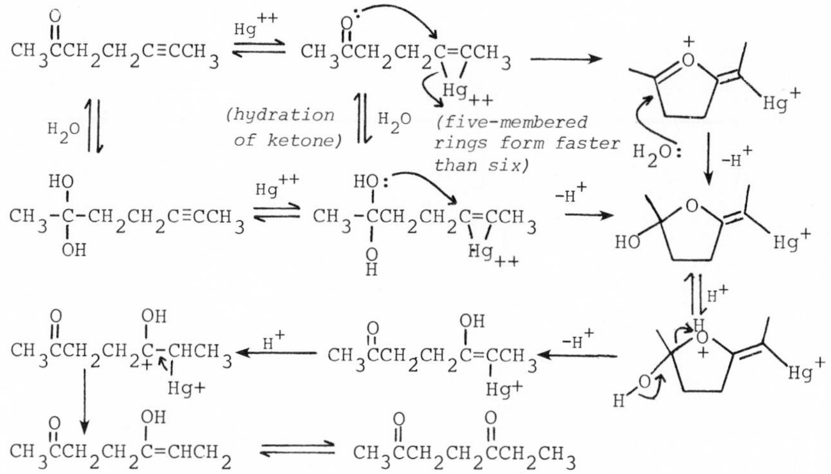

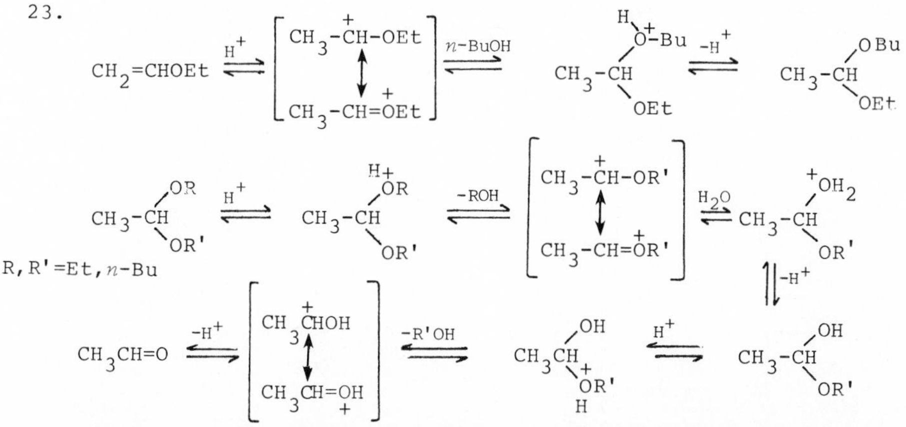

The normal conditions for $LiAlH_4$ or $NaBH_4$ reduction of ketones
to alcohols are alkaline, and the bromohydrin would cyclize to the

epoxide: CH_2—CH~~~~   Protect the alcohol by formation

of the mixed acetal, by using ethyl vinyl ether:

$$HOCH_2\overset{Br}{\underset{}{C}}HCH_2CH_2CH_2\overset{O}{\overset{\|}{C}}CH_3 \xrightarrow[H^+]{CH_2=CHOEt} EtO\overset{CH_3}{\underset{|}{C}}HOCH_2\overset{Br}{\underset{|}{C}}HCH_2CH_2CH_2\overset{O}{\overset{\|}{C}}CH_3 \xrightarrow{NaBH_4}$$

$$HOCH_2\overset{Br}{\underset{|}{C}}HCH_2CH_2CH_2\overset{OH}{\underset{|}{C}}HCH_3 + CH_3CHO + EtOH \xleftarrow{H_3O^+} EtO\overset{CH_3}{\underset{|}{C}}HOCH_2\overset{Br}{\underset{|}{C}}HCH_2CH_2CH_2\overset{OH}{\underset{|}{C}}HCH_3$$

24. a)

5-oxohexanal

b)

2,7-octanedione

c)

25.   The 2.8 Hz coupling corresponds to an equatorial-equatorial or an
        axial-equatorial H-C-C-H coupling; the 11.8 Hz coupling corresponds
        to axial-axial coupling. Since the hydrogen at C-5 is axial,
        the initial bromo compound must have Br axial (C-6 hydrogen
        equatorial). This isomerizes to the more stable product with
        bromine equatorial.

J = 2.8 Hz   J = 10.8 Hz

26. $(CH_3)_2CH\overset{O}{\overset{\|}{C}}CH_3 \xrightarrow[\substack{THF \\ -70°C}]{LDA} \left[(CH_3)_2CH\overset{O^-Li^+}{\overset{|}{C}}=CH_2 \right] \xrightarrow{(CH_3)_3CCHO} \xrightarrow{H_2O} (CH_3)_2CH\overset{O}{\overset{\|}{C}}CH_2\overset{OH}{\overset{|}{C}}HCH(CH_3)_2$

$\qquad$ only

        The amide base is very bulky, and for steric reasons it can't reach
        the tertiary hydrogen at the 3-position as easily as a primary
        hydrogen at the 1-position.

27. $ClCH_2\overset{O}{\overset{\|}{C}}OC_2H_5 \xrightarrow{^-NH_2} \left[\begin{array}{c} Cl\overset{O}{\overset{\|}{C}}H\overset{}{C}OC_2H_5 \\ \updownarrow \\ ClCH=\overset{O^-}{\overset{|}{C}}OC_2H_5 \end{array} \right] \xrightarrow{CH_3\overset{O}{\overset{\|}{C}}CH_3} \left[(CH_3)_2\overset{O^-}{\overset{|}{C}}-\overset{}{C}H\overset{O}{\overset{\|}{C}}OC_2H_5 \right]$

(so far, just
like an aldol
condensation)

$(CH_3)_2\overset{O}{\overset{}{C}}-CH\overset{O}{\overset{\|}{C}}OC_2H_5 + Cl^-$

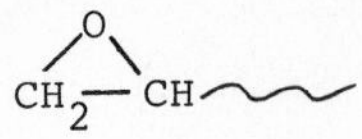

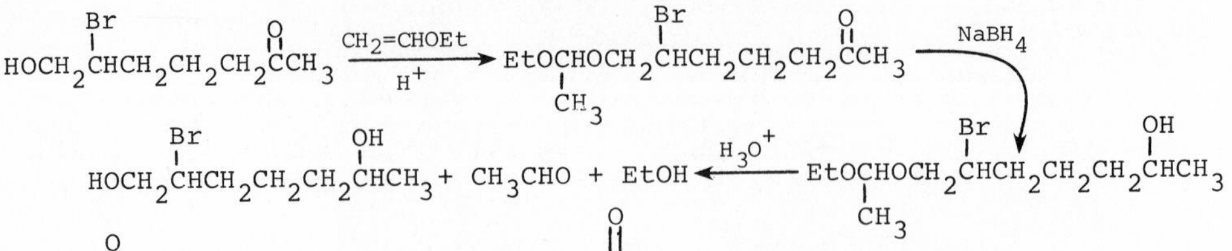

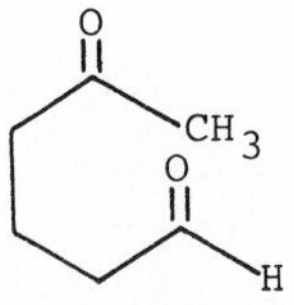

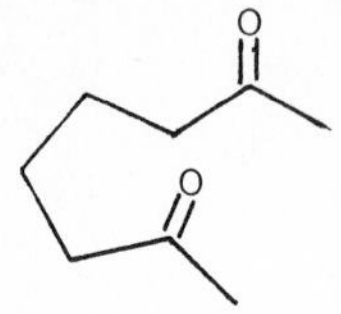

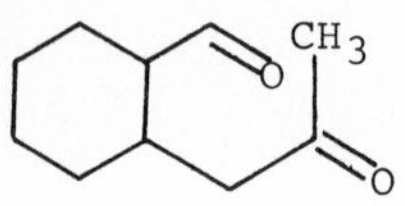

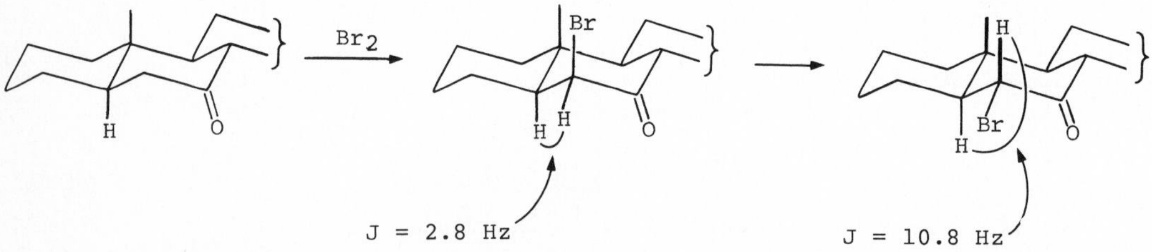

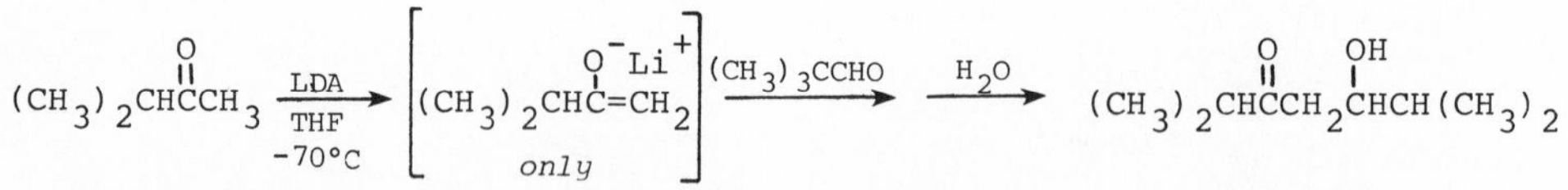

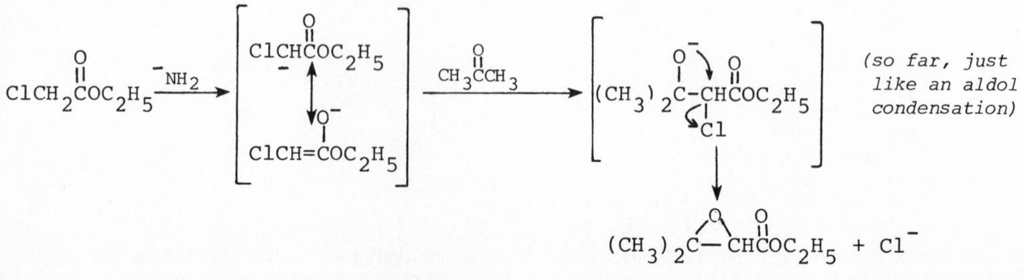

28.

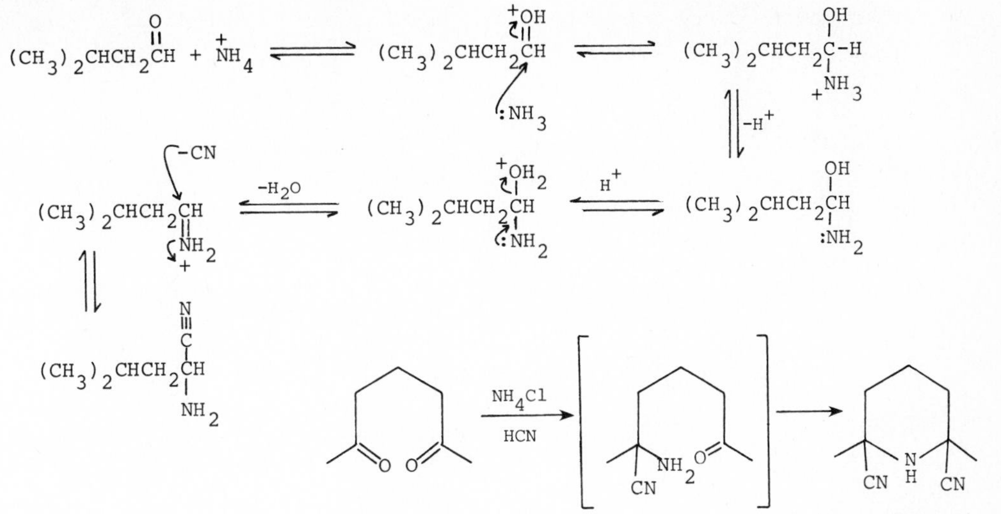

(*cis* and *trans*)

13.E Supplementary Problems

S1. Provide IUPAC names for the following compounds

(a)

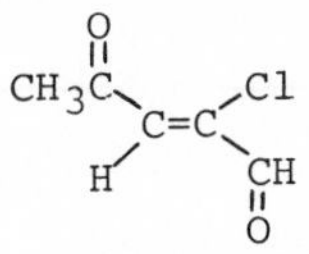

(b)

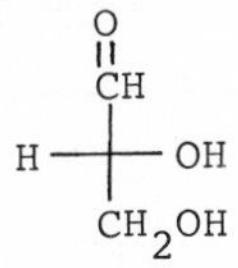

(c)

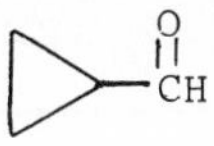

(d)

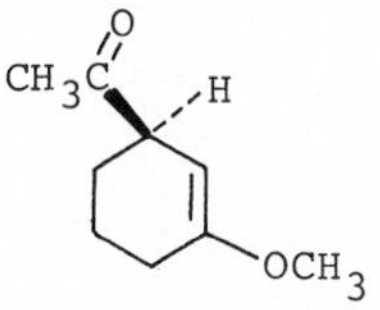

(e)

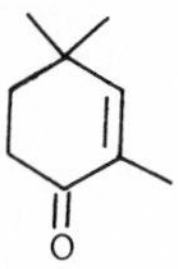

(f)

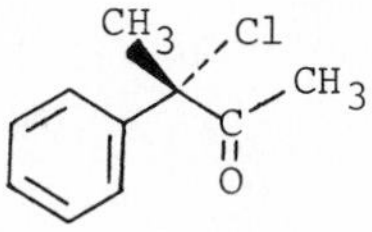

S2. Write the structure of the principal product formed in each of the following reaction sequences.

(a)

(b)

(c) $CH_3CH_2CH_2C{\equiv}CH \xrightarrow[2.\,H_2O_2,\,OH^-]{1.\,B_2H_6} \xrightarrow{HCN}$

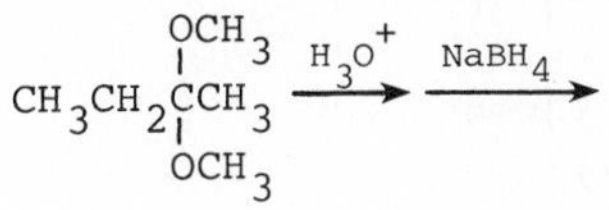

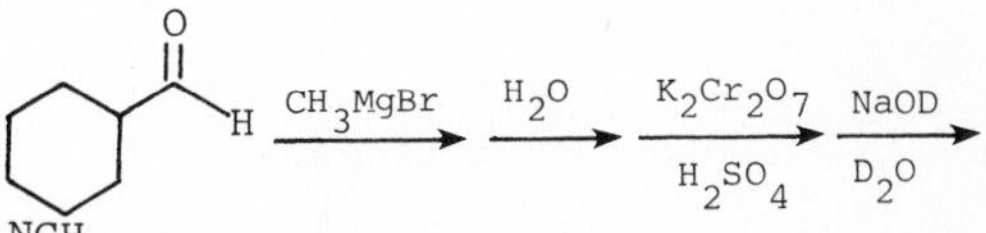

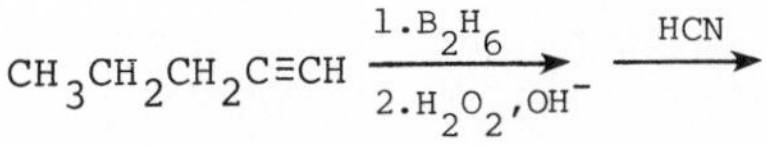

(d)

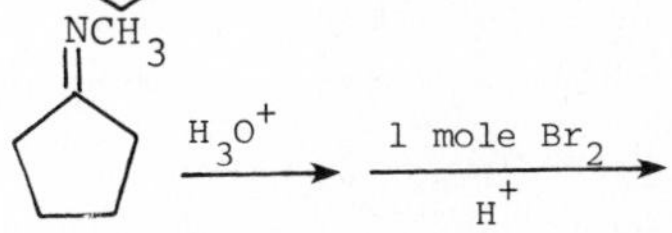

(e)

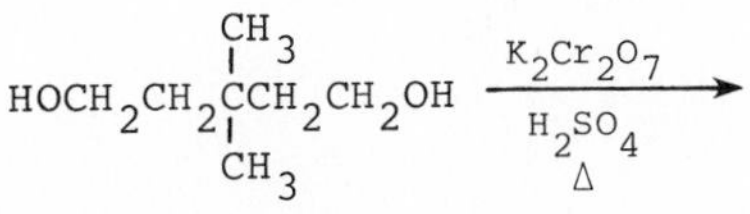

(f)

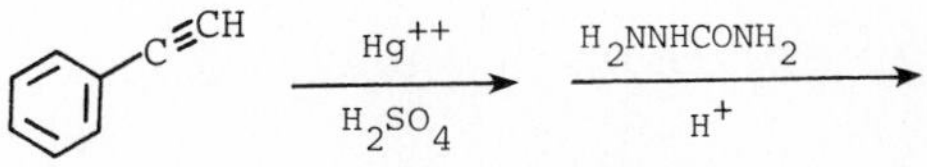

(g)

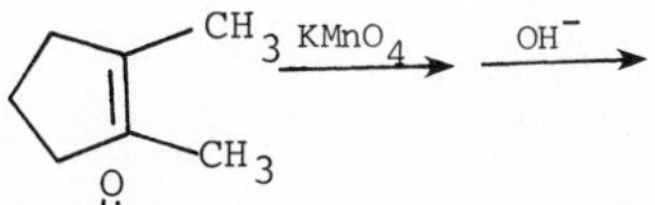

(h)

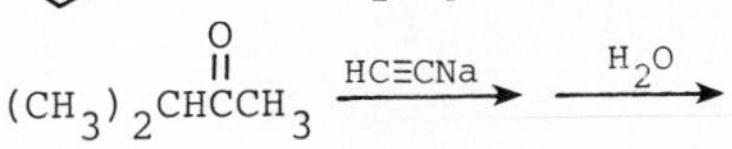

(i)

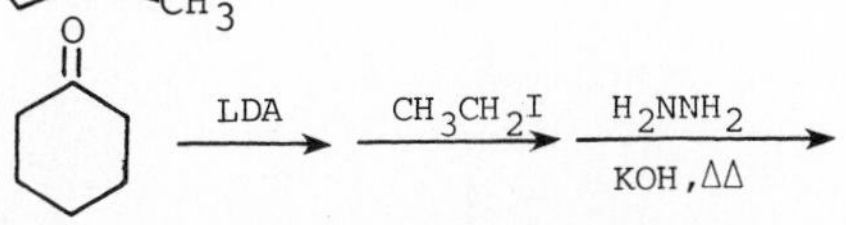

S3. Show how to carry out each of the following transformations. More
than one step is involved in each case.

(a)

$$CH_3CH_2CH_2\overset{\overset{\displaystyle O}{\|}}{C}H \longrightarrow CH_3CH_2CH_2\overset{\overset{\displaystyle NOH}{\|}}{C}CH_2CH_2CH_3 \qquad H\diagdown \diagup CH(CH_3)_2$$

(b)

$$+ \; (CH_3)_2CH\overset{\overset{\displaystyle O}{\|}}{C}H \longrightarrow$$

(c)

$$+ \; H\overset{\overset{\displaystyle O}{\|}}{C}\text{—} \longrightarrow$$

(d) $CH_3CH_2\overset{\overset{\displaystyle O}{\|}}{C}H \longrightarrow HOCH_2\overset{\overset{\displaystyle OH}{|}}{C}H\overset{\underset{\displaystyle CH_3}{|}}{C}HCH_2CH_3$

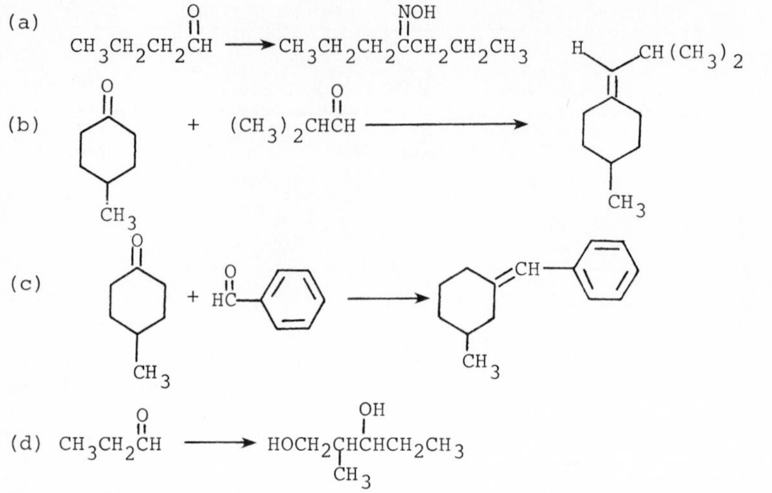

S4. Show how each of the following compounds can be prepared from materials
containing four carbons or less.

(a) $CH_3CH_2CH_2\overset{\underset{\displaystyle CH_3}{|}}{C}H\overset{O}{\underset{O}{\diagup}}CH$

(b) $CH_3CH_2\overset{\overset{\displaystyle OH}{|}}{\underset{\underset{\displaystyle CH_3}{|}}{C}}CH=O$

(c) $CH_3\overset{\overset{\displaystyle OH}{|}}{C}HCH=CHCH(CH_3)_2$

(e) $(CH_3)_2CHCH_2\text{—}\diamondsuit$

S5. On heating *cis*-6,7-epoxy-2-nonanone with a trace of acid, it is converted
to brevicomin, the aggregating pheromone of the female western pine
beetle *Dendroctomus brevicomis:*

$$CH_3\overset{\overset{\displaystyle O}{\|}}{C}CH_2CH_2CH_2 \underset{\underset{\displaystyle H}{|}}{C}\text{—}\underset{\underset{\displaystyle O}{}}{C}\underset{\underset{\displaystyle H}{|}}{\diagup}^{CH_3} \quad \overset{H^+}{\underset{\Delta}{\longrightarrow}}$$

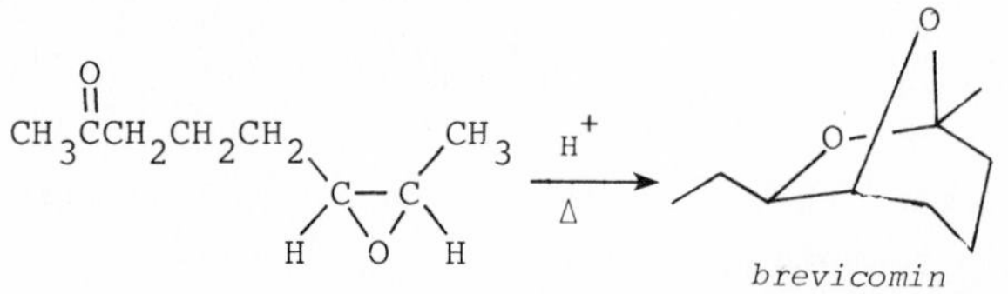

brevicomin

A proposed synthesis of this epoxyketone is shown below. Each
of the proposed steps contains a flaw and will lead to a significant
amount of side products. For each step, point out the flaw or
likely side reaction. Then, suggest a synthesis which avoids these
problems.

$$CH_3C\equiv CNa + BrCH_2CH_2CH_2OH \longrightarrow CH_3C\equiv CCH_2CH_2CH_2OH \overset{Hg^{++}}{\underset{H_2SO_4}{\longrightarrow}}$$

$$(C_6H_5)_3P=CHCH_3 \qquad CH_3\overset{\overset{\displaystyle O}{\|}}{C}CH_2CH_2CH_2CH=O \overset{K_2Cr_2O_7,}{\underset{H_2SO_4,\Delta}{\longleftarrow}} CH_3\overset{\overset{\displaystyle O}{\|}}{C}CH_2CH_2CH_2CH_2OH$$

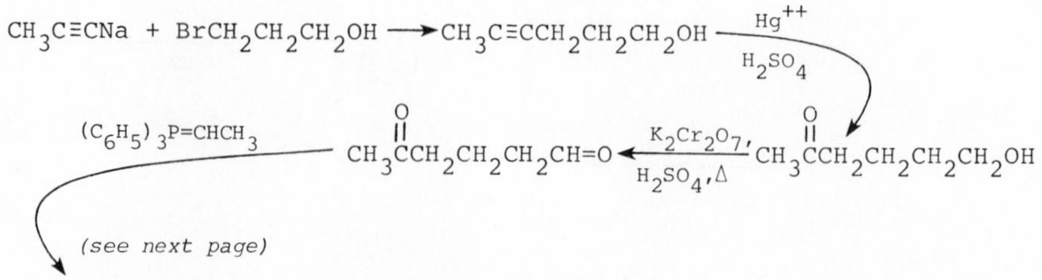

(see next page)

(from last page)

$$CH_3\overset{O}{\overset{\|}{C}}CH_2CH_2CH_2\underset{H}{\overset{}{C}}=\underset{H}{\overset{CH_2CH_3}{C}} \quad \xrightarrow{CH_3\overset{O}{\overset{\|}{C}}OOH} \quad CH_3\overset{O}{\overset{\|}{C}}CH_2CH_2CH_2\underset{H}{\overset{}{C}}\underset{O}{-}\underset{H}{\overset{CH_2CH_3}{C}}$$

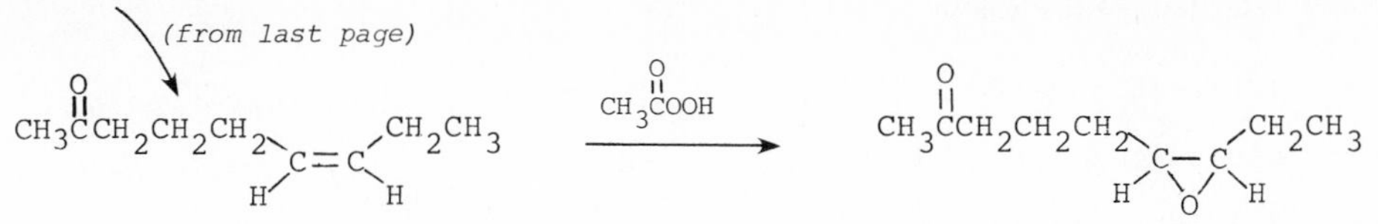

S6. Show all the intermediates involved in the following reaction sequence

$$\xrightarrow{O_3} \quad \xrightarrow[H_2O]{Zn} \quad \xrightarrow[\Delta]{OH^-}$$

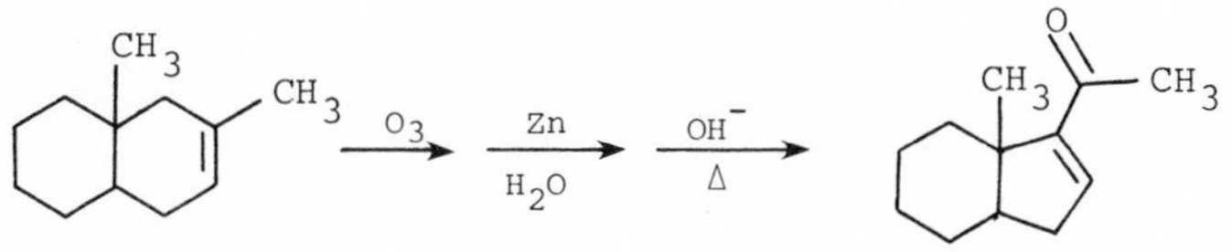

S7. Compound A, of formula C_7H_{14}, reacts with ozone followed by zinc
dust to give B and C. B reacts with semicarbazide to give a crystalline
product of formula $C_3H_7N_3O$. Compound C does not undergo any reaction
with Ag_2O, but in the presence of iodine and base reacts to give a
precipitate of iodoform (CHI_3). The reaction of B with the ylide
formed from ethyltriphenylphosphonium bromide gives two compounds,
A and an isomer D. The cmr spectra of A and D are given below:

> A: δ 12.0, 12.8, 20.2, 22.1, 32.7, 118.3, 134.8 ppm
> D: δ 12.0, 12.6, 14.2, 20.4, 41.2, 117.6, 134.7 ppm

What are A,B,C and D?

S8. Suggest a mechanism for the following conversion:

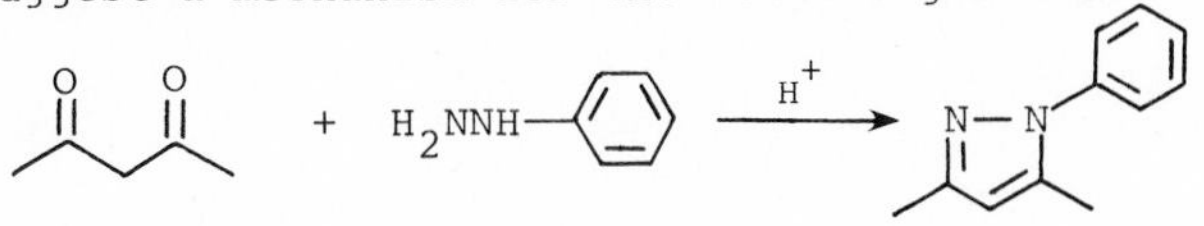

S9. Write a mechanism which accounts for the formation of butyrolactone

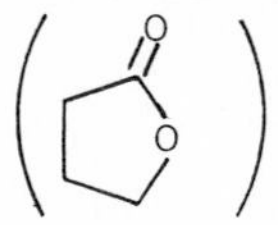

when 1,4-butanediol is treated with CrO_3/pyridine.

13.F Answers to Supplementary Problems

S1. (a) (Z)-2-chloro-4-oxo-2-pentenal   (b) (R)-2,3-dihydroxypropanal

(c) cyclopropanecarbaldehyde   (d) (R)-1-methoxy-3-(1-oxoethyl)cyclo-hexene

(e) 2,4,4-trimethyl-2-cyclohexenone

(f) (S)-3-chloro-3-phenyl-2-butanone

S2.

(a) $\underset{\underset{OH}{|}}{CH_3CH_2CHCH_3}$

(b)

(c) $\underset{\underset{OH}{|}}{CH_3CH_2CH_2CH-C{\equiv}N}$

(d)

(e) $HO_2CCH_2\underset{\underset{CH_3}{|}}{\overset{\overset{CH_3}{|}}{C}}CH_2CO_2H$

(f)

(g)

(h) $(CH_3)_2\underset{\underset{CH_3}{|}}{\overset{\overset{OH}{|}}{C}}HC-C{\equiv}CH$

(i)

S3. (a) $CH_3CH_2\overset{\overset{O}{\|}}{C}H + NaC{\equiv}CCH_3 \longrightarrow \xrightarrow{H_2O} CH_3CH_2\underset{\underset{OH}{|}}{C}HC{\equiv}CCH_3 \xrightarrow[Pt]{H_2} (CH_3CH_2CH_2)_2CHOH$

$\xrightarrow{K_2Cr_2O_7,H_2SO_4}$

$(CH_3CH_2CH_2)_2C{=}NOH \xleftarrow[H^+]{H_2NOH} (CH_3CH_2CH_2)_2C{=}O$

(b) $(CH_3)_2CH\overset{\overset{O}{\|}}{C}H \xrightarrow{NaBH_4} (CH_3)_2CHCH_2OH \xrightarrow{PBr_3} (CH_3)_2CHCH_2Br \xrightarrow{(C_6H_5)_3P}$

$(CH_3)_2CHCH{=}P(C_6H_5)_3 \xleftarrow{n-BuLi} (CH_3)_2CHCH_2\overset{+}{P}(C_6H_5)_3 \\ Br^-$

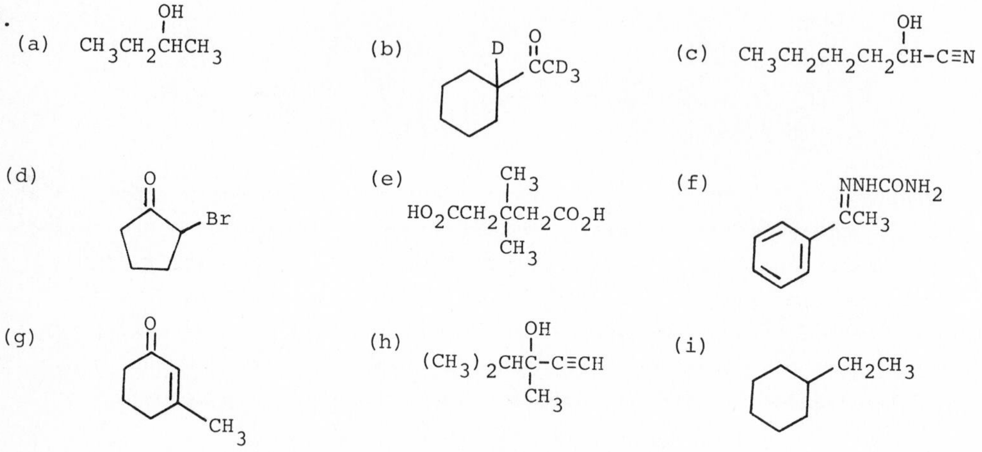

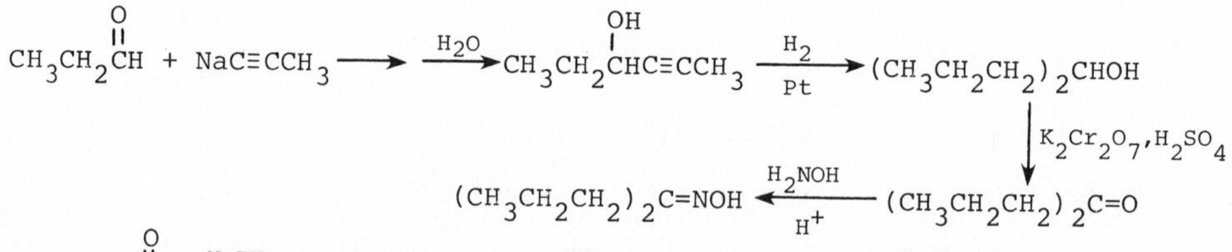

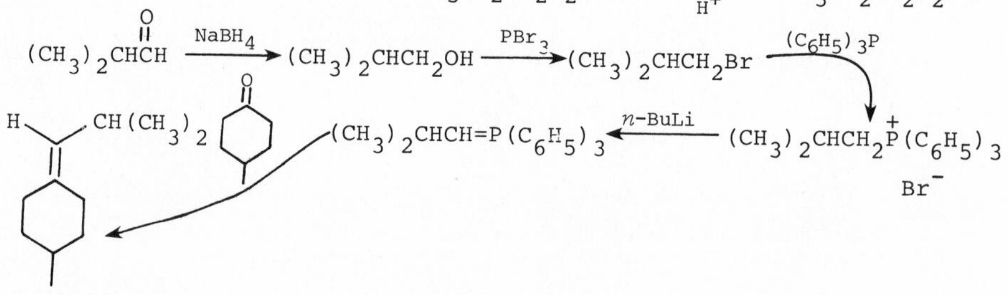

(c)

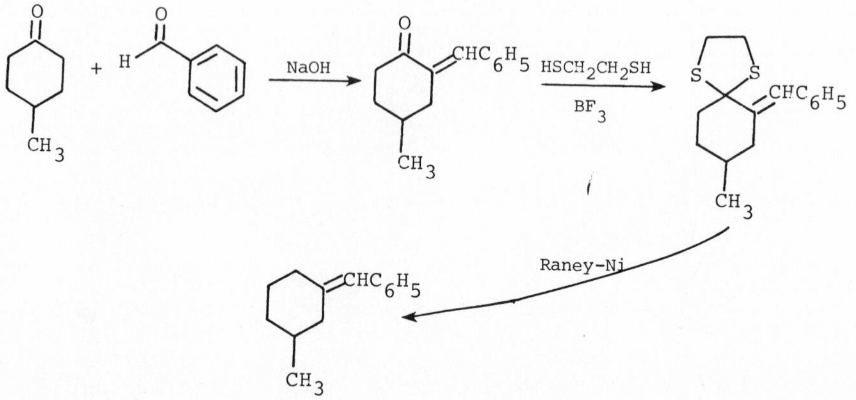

(d) $2CH_3CH_2\overset{\overset{O}{\|}}{C}H \xrightarrow{NaOH} H\overset{\overset{O}{\|}}{C}CH\overset{OH}{\underset{CH_3}{|}}CHCH_2CH_3 \xrightarrow{NaBH_4} HOCH_2\overset{}{C}H\overset{OH}{\underset{CH_3}{|}}CHCH_3$

S4. (a) $2CH_3CH_2\overset{\overset{O}{\|}}{C}H \xrightarrow[\Delta]{NaOH} CH_3CH_2CH=\overset{}{\underset{CH_3}{|}}CCH \xrightarrow[Pt]{H_2} CH_3CH_2CH_2\overset{}{\underset{CH_3}{|}}CH\overset{\overset{O}{\|}}{C}H \xrightarrow[H^+]{HOCH_2CH_2OH}$

$CH_3CH_2CH_2\overset{}{\underset{CH_3}{|}}CHCH\overset{O}{\underset{O}{\diagdown}}$

(b) $CH_3CH_2\overset{\overset{O}{\|}}{C}CH_3 + NaC\equiv CH \longrightarrow \xrightarrow{H_2O} CH_3CH_2\overset{OH}{\underset{CH_3}{|}}CC\equiv CH \xrightarrow[\text{catalyst}]{\underset{Lindlar}{H_2}} CH_3CH_2\overset{OH}{\underset{CH_3}{|}}CCH=CH_2$ $\begin{array}{l}1.O_3 \\ 2.Zn, \\ H_2O\end{array}$

$CH_3CH_2\overset{OH}{\underset{CH_3}{|}}C-CH=O$

(c) $CH_3\overset{\overset{O}{\|}}{C}CH_3 \xrightarrow[\underset{-78°C}{THF}]{LDA} CH_3\overset{OLi}{\underset{}{|}}C=CH_2 \xrightarrow{(CH_3)_2CH\overset{\overset{O}{\|}}{C}H} CH_3\overset{\overset{O}{\|}}{C}CH_2\overset{OH}{\underset{}{|}}CHCH(CH_3)_2 \xrightarrow{H^+}$

$CH_3\overset{OH}{\underset{}{|}}CHCH=CHCH(CH_3)_2 \xleftarrow{NaBH_4} CH_3\overset{\overset{O}{\|}}{C}CH=CHCH(CH_3)_2$

(d) $(CH_3)_2CHCH_2Br + (C_6H_5)_3P: \longrightarrow \xrightarrow{n\text{-BuLi}} (CH_3)_2CHCH=P(C_6H_5)_3$

$(CH_3)_2CHCH_2\diamondsuit \xleftarrow[Pt]{H_2} (CH_3)_2CHCH=\diamondsuit$

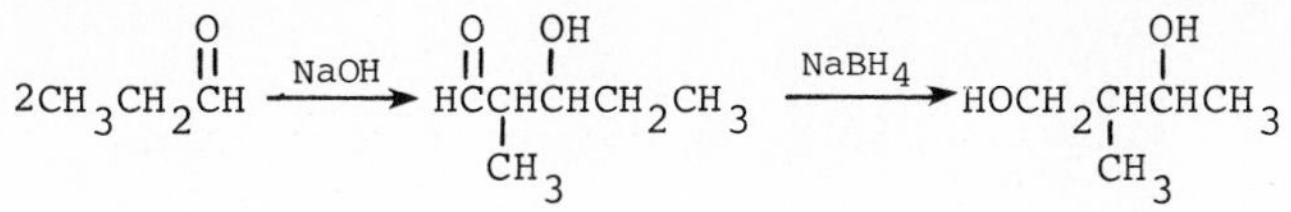

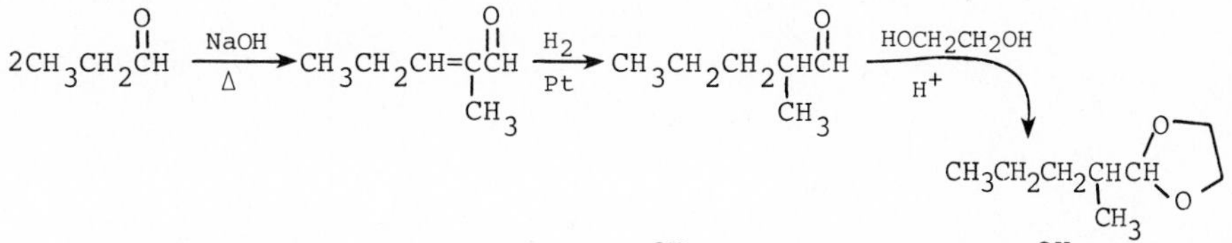

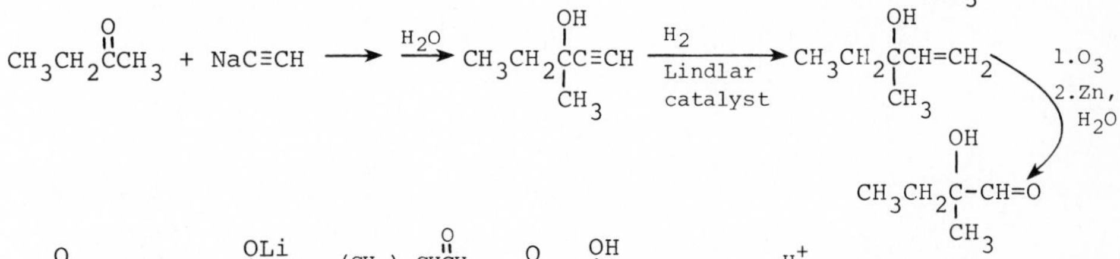

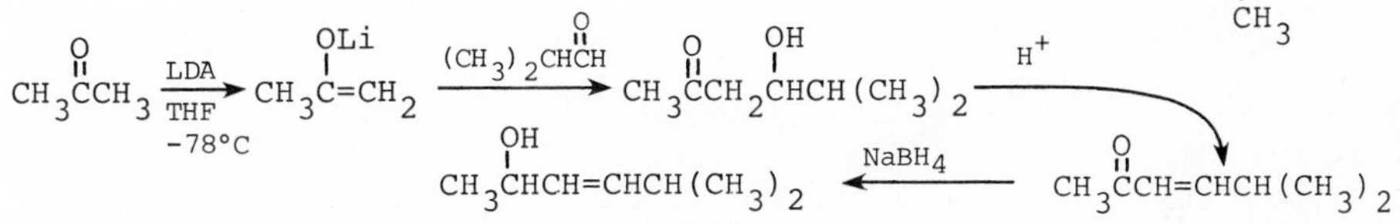

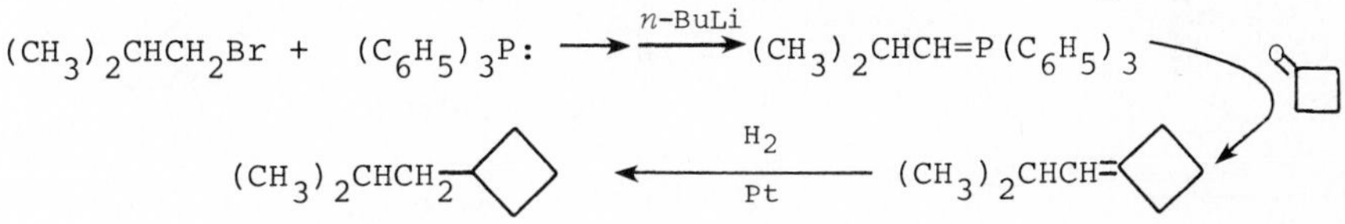

S5. (a) $CH_3C\equiv CNa$ will react with the alcohol: $CH_3C\equiv CNa + ROH \longrightarrow CH_3C\equiv CH$
$+$
RO^-Na^+

(b) Two isomeric ketones can be formed: $CH_3\overset{\overset{O}{\|}}{C}CH_2CH_2CH_2CH_2OH$

and $CH_3CH_2\overset{\overset{O}{\|}}{C}CH_2CH_2CH_2OH$

(c) Aldehyde will be overoxidized, to give acid: $CH_3\overset{\overset{O}{\|}}{C}CH_2CH_2CH_2COOH$

(d) Wittig reagent can react with ketone, too: $CH_3\overset{\overset{CHCH_3}{\|}}{C}CH_2CH_2CH_2$---
and it can also give a mixture of *cis* and *trans* isomers:

$\cdots CH_2CH_2\overset{H}{\underset{H}{\overset{|}{\underset{|}{C}}}}\diagup\overset{}{\underset{}{C}}\diagdown\,^{CH_3}$

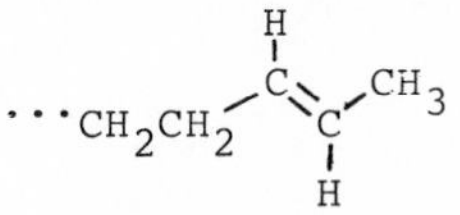

(e) The ketone can undergo Baeyer-Villiger reaction in the presence of a peracid: $CH_3\overset{\overset{O}{\|}}{C}OCH_2CH_2CH_2\cdots$

A synthesis that avoids these problems would make use of protecting groups:

$BrCH_2CH_2CH_2OH + C_2H_5OCH=CH_2 \xrightarrow{H^+} BrCH_2CH_2CH_2OCH\overset{}{\underset{CH_3}{|}}OC_2H_5 \xrightarrow{HC\equiv CNa}$

$\overset{HOCH_2CH_2OH}{\underset{H^+}{}}\overset{}{\underset{}{}}CH_3\overset{\overset{O}{\|}}{C}CH_2CH_2CH_2OH \xleftarrow[H_2SO_4]{Hg^{++}} HC\equiv CCH_2CH_2CH_2OCH\overset{}{\underset{CH_3}{|}}OC_2H_5$

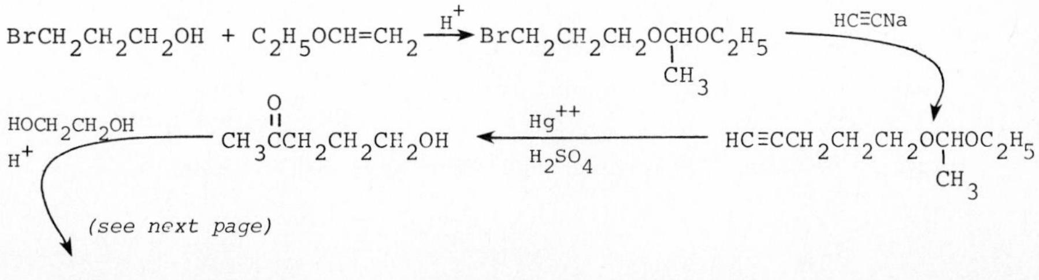

(see next page)

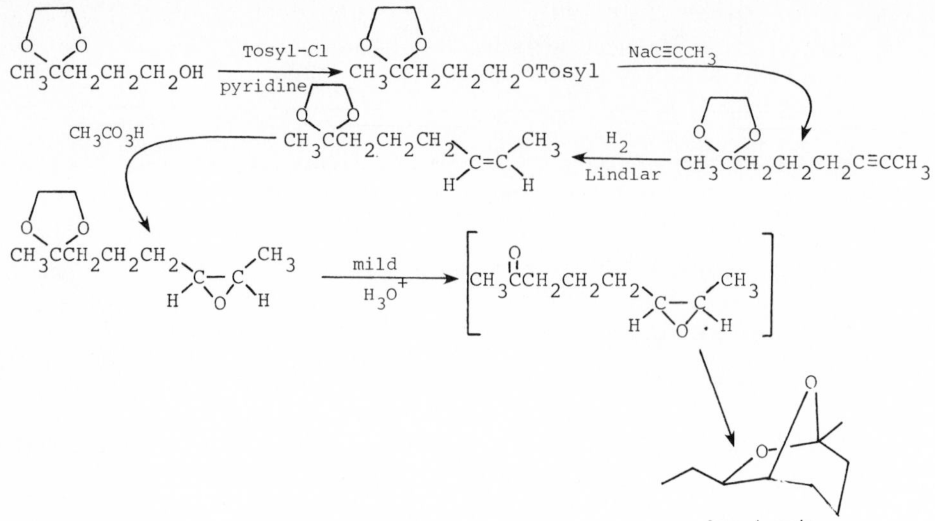

brevicomin

S6.

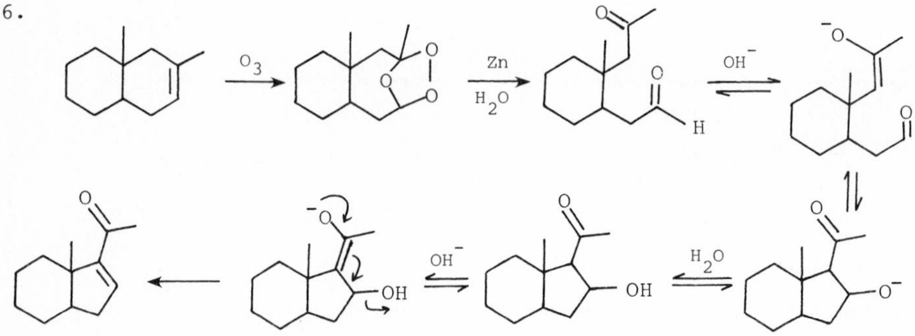

S7.

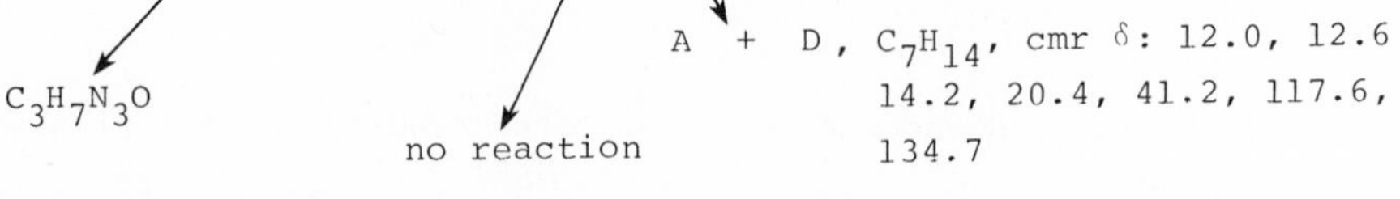

1) A, C_7H_{14}, cmr δ 118.3, 134.8: A is an alkene.

2) A $\xrightarrow[\text{2. Zn}]{\text{1. } O_3}$ B + C; B + $H_2NNHCONH_2$ $\longrightarrow$ $C_3H_7N_3O$ ($CH_3CH{=}NHNHCNH_2$, with C=O): B is acetaldehyde.

3) C $\xrightarrow{Ag_2O}$ n.r.; C $\xrightarrow[I_2]{OH^-}$ HCI_3: C is a methyl ketone, either $CH_3\overset{O}{\overset{\|}{C}}CH_2CH_2CH_3$ or $CH_3\overset{O}{\overset{\|}{C}}CH(CH_3)_2$.

4) cmr of A shows 7 carbons: no isopropyl group present, so C is $CH_3\underset{O}{\underset{\|}{C}}CH_2CH_2CH_3$

5)   A and D are therefore the two diastereomers of 3-methyl-2-hexene:

A   D

(A *cis* substitutent causes an upfield shift in the cmr)

S8.

S9.   HOCH$_2$CH$_2$CH$_2$CH$_2$OH $\xrightarrow[\text{pyridine}]{\text{CrO}_3}$

this is oxidized like
any other 2° alcohol

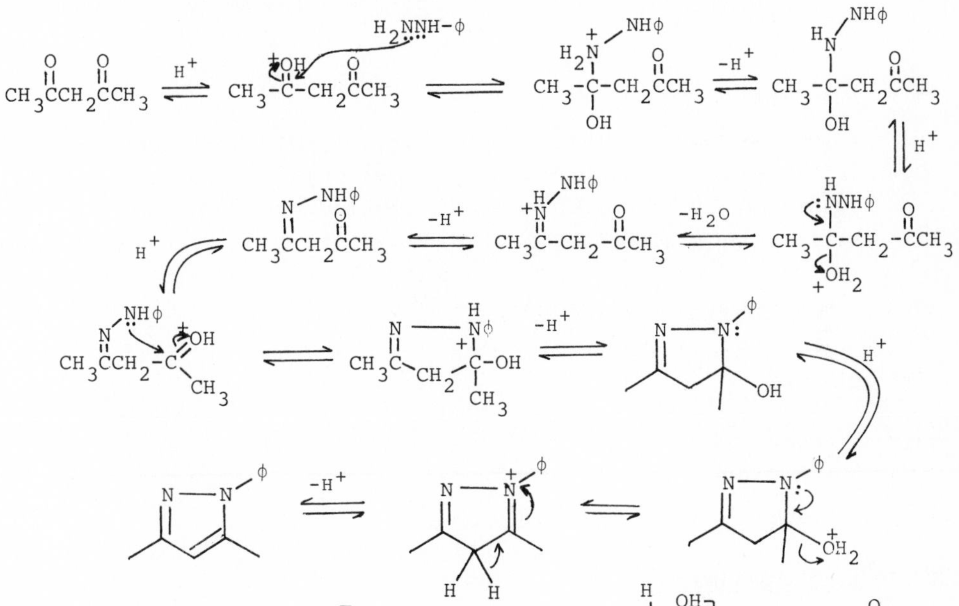

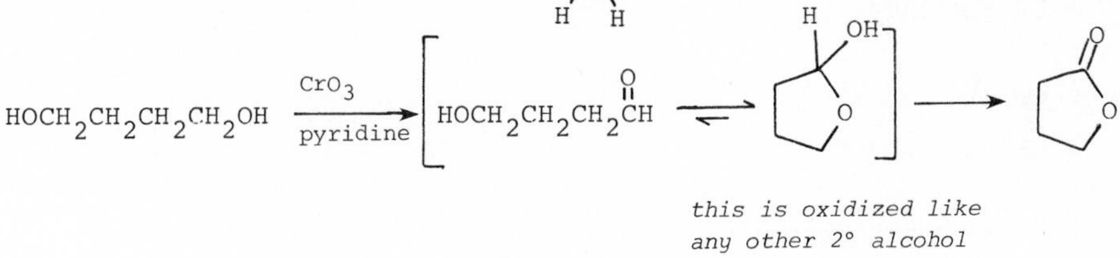

14. INFRARED SPECTROSCOPY

14.A Chapter Outline and Important Terms Introduced

14.1 The Electromagnetic Spectrum

electromagnetic radiation — $E = Nh\nu$, $\nu = \frac{c}{\lambda}$, $\tilde{\nu} = \frac{1}{\lambda}$

spectroscopy

spectrum

middle infrared ($\tilde{\nu}$ = 3333-333 cm^{-1})

14.2 Molecular Vibration

selection rule

infrared active and inactive

fundamental vibrational modes

stretching, bending

degenerate vibrations

overtones

combination bands

harmonic oscillator approximation

14.3 Characteristic Group Vibrations

14.4 Alkanes

14.5 Alkenes

14.6 Alkynes

14.7 Alkyl Halides

14.8 Alcohols and Ethers

effects of hydrogen bonding

14.9 Aldehydes and Ketones

dependence of $\nu_{C=O}$ on structure

14.10 Summary: Principal Functional Group Absorptions (time to start memorizing)

14.11 Instrumentation

single beam or double beam

monochromator

14.B Important Concepts and Hints

What IR Spectroscopy Is Good For:

IR spectroscopy is complementary to nmr because it gives you information primarily about functional groups instead of the carbon skeleton. With this technique you can decide immediately if the compound has a hydroxy group, for example, or a carbonyl, a nitrile, and so on. Often the

IR spectrum can tell you additional details about a functional group: Is the double bond *cis*, *trans*, or a 1,1-disubstituted olefin? Is the carbonyl group in a 5- or 6-membered ring, or part of an acyclic ketone?, etc. In the chapter on carboxylic acid derivatives (Chapter 19) you will learn that the position of the carbonyl stretching frequency can tell you even more.

Generally, the IR spectrum will not tell you much about the hydrocarbon backbone of a molecule. All CH_2 and CH_3 groups, for instance, have the same stretching and bending vibrations, and the fundamental vibrational modes of larger structural units are not readily assignable.

What You Should Learn:

The subject of IR spectroscopy has a logical foundation, but its routine application relies primarily on memory. The characteristic absorption frequencies of the various functional groups in organic molecules can be understood from first principles, as pointed out in the introductory sections of the chapter. However, the day-to-day interpretation of infrared spectra relies on simply knowing what IR band corresponds to what functional group. You will have to do a moderate amount of memorization in order to be able to use IR easily in solving problems. Start with Table 14.2, and then expand your knowledge until you are familiar with all the *highlighted entries* in Table 14.6. In subsequent chapters, the IR characteristics of additional functional groups will be presented, and you should add them to your memory.

What You Should Avoid:

The tendency of most students is to "overinterpret" an IR spectrum, and attempt to assign every band that appears. Although much useful information is contained in an IR spectrum, there are a lot of peaks which cannot be identified with any particular functional group and so are not useful in understanding the structure of a compound. When you look at an IR spectrum, you should first pick out the unambiguous peaks (C=O stretch, O-H stretch, aldehyde C-H, for example), and see what you can do with that information and the rest available in the problem. If necessary, you can return to the spectrum and use it to get finer details.

You should also recognize that average absorptions given for a particular vibrational mode are averages for a wide variety of compounds. As with all averages, individual cases can be very different. For example, you should not assume that all *trans* alkenes will show the C-H out-of-plane bend exactly at 970 cm^{-1}, or that a molecule with a strong band at 1720 cm^{-1} cannot be a 6-membered ring ketone.

14. C Answers to Exercises

(14.9) The C=O stretching motion requires a simultaneous C=C compression, a movement which will be resisted strongly by the C=C bond in comparison to the normal ${C \atop C}\!\!>\!C{=}O$ arrangement. Therefore, the carbonyl stretching frequency of ketene is very high.

14.D   Answers and Explanations to Problems

1. The Hooke's Law equation (eq. 14-1) gives:

$$\tilde{\nu} = 4.120 \sqrt{f \cdot \frac{m_1 + m_2}{m_1 m_2}}$$

(Recall that m_1 and m_2 in eq. 14.1 are the masses of atoms 1 and 2 in grams.) The value 4.120 is obtained as follows:

$$4.120 = \frac{1}{2\pi c} \sqrt{\frac{1}{6.023 \times 10^{-23}} \Bigg/ \frac{1}{(6.023 \times 10^{-23})^2}}$$

so that m_1 and m_2 are expressed as gram-atoms.

For single, double, and triple bonds, the approximate values are:

$$\tilde{\nu}_{single} = 2913 \sqrt{\frac{m_1 + m_2}{m_1 m_2}} \ cm^{-1}$$

$$\tilde{\nu}_{double} = 4120 \sqrt{\frac{m_1 + m_2}{m_1 m_2}} \ cm^{-1}$$

$$\tilde{\nu}_{triple} = 5046 \sqrt{\frac{m_1 + m_2}{m_1 m_2}} \ cm^{-1}$$

a) $3003 \ cm^{-1}$ b) $2185 \ cm^{-1}$ c) $1682 \ cm^{-1}$ d) $2060 \ cm^{-1}$

e) $1112 \ cm^{-1}$ f) $1573 \ cm^{-1}$ g) $1985 \ cm^{-1}$ h) $1074 \ cm^{-1}$

2. Given the value of $3350 \ cm^{-1}$ for 1-octyne, one can calculate a more accurate force constant, f.

$$3350 \ cm^{-1} = 4.120 \sqrt{f \cdot \frac{12+1}{12 \cdot 1}} \ ; \quad f = 6.103 \times 10^5 \ dynes \ cm^{-1}$$

Using this value for the two compounds, we have

$$\tilde{\nu} = 3219 \sqrt{\frac{m_1 + m_2}{m_1 m_2}} \ cm^{-1}$$

$C_6H_{13}C{\equiv}C\text{-}D, \quad \tilde{\nu} = 2459 \ cm^{-1}$ $C_6H_{13}C{\equiv}^{13}C\text{-}H, \quad \tilde{\nu} = 3341 \ cm^{-1}$

3. (a) The intense band at ~965 cm^{-1} suggests $\begin{smallmatrix} R \\ \\ H \end{smallmatrix}\!\!>\!\!C{=}C\!\!<\!\!\begin{smallmatrix} H \\ \\ R \end{smallmatrix}$.

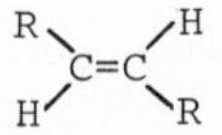

(b) Bands at 3300, 2150, and 630 cm^{-1} conclusively indicate $RC{\equiv}CH$.

(c) Broad band at 3350 cm^{-1} shows OH. Complex absorption peaking at 1050 cm-1 suggests primary.

(d) Bands at 3080, 1825, 1640, 995, and 905 cm^{-1} are conclusive evidence for $RCH{=}CH_2$.

(e) Broad band at 3400 cm^{-1} shows OH, as does absorption at 1020 cm-1 (C-O stretch), probably primary. Sharp band at 2120 cm-1 and band at 650 cm-1 show terminal acetylene. Thus, there are two functional groups, OH and $C{\equiv}CH$.

(f) Strong band at 1720 cm^{-1} (C=O stretch) could be an aldehyde or ketone (acyclic or 6-membered ring), but the absence of bands at 2720 and 2820 shows that it is not an aldehyde.

(g) Strong band at 1725 cm^{-1} (C=O stretch) and bands at 2720 and 2820 cm^{-1} (C-H stretch) are characteristic of the aldehyde group.

4. (a) IR shows $\begin{matrix}R\\R'\end{matrix}\!\!>C{=}CH_2$ clearly. Nmr shows that the two vinyl protons are non-equivalent and that there is a t-butyl group (nine-proton singlet at δ=0.9 ppm). The other five protons occur as two-proton and three-proton singlets. The best structure is:

$$\begin{matrix} & CH_3 & \\ & | & \\ CH_2{=} & C & CH_2C(CH_3)_3 \end{matrix}$$

(b) IR shows OH (3400 cm^{-1}) and some kind of double bond (3100, 1830, 1650 cm^{-1}). Although the C-H out-of-plane bending region is complex, the nmr pattern is clearly due to $R{-}CH{=}CH_2$. In addition, there are two equivalent CH_3's (singlet at δ=1.3 ppm). The OH is the singlet at δ=4.2 ppm.

$$\begin{matrix} & OH & \\ & | & \\ (CH_3)_2 & C & CH{=}CH_2 \end{matrix}$$

(c) IR shows OH (3400 cm^{-1}). Note also the spike at 3300 cm^{-1}, almost hidden by the OH band, the weak band at 2120 cm^{-1}, and the strong band at 670 cm^{-1}, all suggestive of RC≡CH. Also note the singlet at δ=2.5 ppm in the nmr, correct for C≡C-H. Nmr shows two equivalent CH_3's and OH at δ=4.3 ppm.

$$\begin{matrix} & OH & \\ & | & \\ (CH_3)_2 & C & C{\equiv}CH \end{matrix}$$

(d) IR shows OH (3350 cm^{-1}). The nmr spectrum has two low-field resonances, a broad doublet at δ=4.5 ppm and a multiplet at δ=3.6 ppm, each due to one proton. Addition of HCl moves the δ=4.5 ppm and collapses it to a singlet. These two bands are probably:

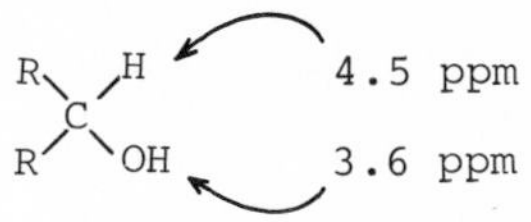
R
H
C
R
OH
4.5 ppm
3.6 ppm

The twelve-proton doublet at δ=0.9 ppm suggests two equivalent isopropyl groups. So far we have

$$-CH(CH_3)_2,\ -CH(CH_3)_2,\ C\begin{matrix}\diagup OH\\ \diagdown H\end{matrix}$$

The most reasonable structure is:

$$\begin{matrix} & OH & \\ & | & \\ (CH_3)_2CHCH_2 & C & HCH_2CH(CH_3)_2 \end{matrix}$$

5. IR shows OH (3500 cm^{-1}) and acyclic or 6-membered ring ketone (1710 cm^{-1}). The absorption at 1100 cm^{-1} suggests that the alcohol is secondary. The cmr spectrum confirms the presence of a carbonyl, and shows that there are two nonequivalent upfield methyls (δ=7.1 and 8.5 ppm) and one which could be the methyl of a methyl ketone (δ=30.7 ppm).

A reasonable structure is:

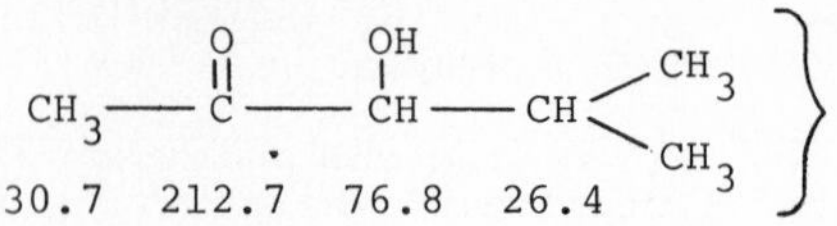

7.1, 8.5 Diastereotopic methyls; see answer to problem 19(a), Chapter 10.

6. Because compounds (a) and (b) are isomeric and because the cmr spectrum of (b) shows 6 carbons, there must be some carbons in (a) which are equivalent.

(a) IR: 3400, 1160 cm^{-1}: tertiary alcohol

cmr: 7.9 ppm: shielded methyl(s)

72.6 ppm: $-\overset{|}{\underset{|}{C}}-O$

This is consistent with:

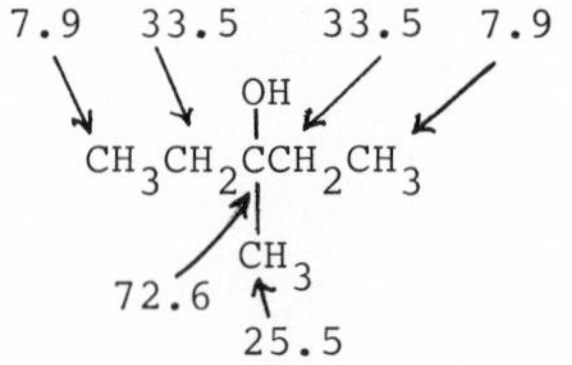

(b) 1) IR: 335 cm^{-1}: alcohol (there are several bands in the 1020-1160 cm^{-1} range and it is not possible to tell if it's a 1°, 2°, or 3° alcohol).

2) Cmr: six resonances: this means that the alcohol cannot be derived from the neohexyl skeleton (C—C(C)(C)—C—C), because symmetry would result in fewer than 6 peaks. For the same reason, the following alcohols can also be ruled out:

$CH_3CH_2CH_2C(CH_3)_2OH$, $HOCH_2CH_2CH(CH_3)_2$,

$HOCH_2CH(CH_2CH_3)_2$, $CH_3C(OH)(CH_2CH_3)_2$, $HO-C(CH_3)_2CH(CH_3)_2$

3) Cmr: no resonance at $\delta<20$: there is no CH_3 at the end of an alkyl chain without a β-substituent. This rules out some more possibilities:

$CH_3CH(OH)CH_2CH_2CH_2CH_3$, $CH_3CH_2CH(OH)CH_2CH_2CH_3$, $CH_3CH_2CH_2CH_2CH_2CH_2OH$,

$HOCH_2CH(CH_3)CH_2CH_2CH_3$, $(CH_3)_2CHCH(OH)CH_2CH_3$, $HOCH_2CH_2CH(CH_3)CH_2CH_3$, $CH_3CH(OH)CH(CH_3)CH_2CH_3$

4) All we have left are:

$CH_3-CH(CH_3)-CH_2-CH(OH)-CH_3$ and $CH_3-CH(CH_3)-CH(CH_3)-CH_2-OH$.

(NOTE: in both of these compounds the two methyls of the isopropyl group are not equivalent because of the chiral center present in the molecule. Therefore they can have different chemical shifts.)

Using Table 9.4 and the α-, β-, and γ-substituent effects of OH (see problem 14, Chapter 9) we predict the following chemical shifts for these two isomers:

	δ (2-methylpentane)	+ substituent effect	= predicted δ	(actual δ)
$(CH_3)_2$	22.5	+0.7(δ-OH)	= 23 (2 resonances)	22.7, 23.5, 24.3
CH	27.8	-5.6(γ-OH)	= 22	
CH_2	41.8	+10.3(β-OH)	= 52	49.2
CHOH	20.7	+48.5(α-OH)	= 69	65.5
CH_3	14.1	+10.3(β-OH)	= 24	25.1

	δ (2,3-dimethylbutane)	+ substituent effect	= predicted δ
$(CH_3)_2$	19.3	+0.7(δ-OH)	= 20 (2 resonances)
CH	34.1	-5.6(γ-OH)	= 28.5
CH	34.1	+10.3(β-OH)	= 44
CH_3	19.3	-5.6(γ-OH)	= 14
CH_2OH	19.3	+48.5(α-OH)	= 68

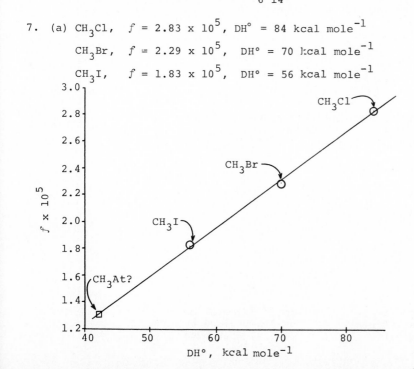

The chemical shifts predicted for 4-methyl-2-pentanol clearly fit the data best.

Notice that the only piece of information gathered from the IR spectrum was that the compound is an alcohol. Nevertheless, that was crucial, because it eliminated from consideration all 16 ethers which have the formula $C_6H_{14}O$.

7. (a) CH_3Cl, $f = 2.83 \times 10^5$, DH° = 84 kcal mole^{-1}

CH_3Br, $f = 2.29 \times 10^5$, DH° = 70 kcal mole^{-1}

CH_3I, $f = 1.83 \times 10^5$, DH° = 56 kcal mole^{-1}

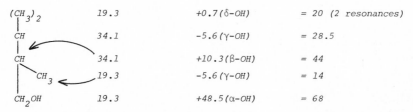

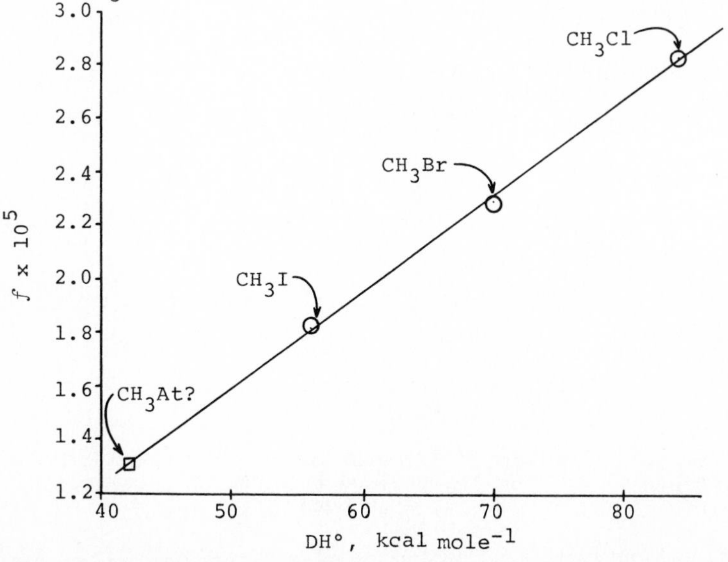

(b) Notice that DH° goes down by 14 kcal mole^{-1} from CH_3Cl to CH_3Br and from CH_3Br to CH_3I. If we project this incremental change to CH_3At, then DH° ≈ 42 kcal mole^{-1} and f from the above graph is ≈ 1.3 x 10^5 dynes cm^{-1}.

$$\tilde{\upsilon} = 4.120\sqrt{1.3 \times 10^5 \frac{12+210}{12\cdot 210}} = 441 \text{ cm}^{-1}$$

(c) For CH_3SH, the Hooke's Law model gives f = 2.56 x 10^5 dynes cm^{-1}. From the graph in part a, this corresponds to DH° = 76.5 kcal mol^{-1}.

8. IR: 3350, 1080 cm^{-1}: -OH
1175 cm^{-1} (s): C-F
cmr: two quartets (each carbon coupled to 3 fluorines)
} $HOCH_2CF_3$

9. The O-O stretch is not infrared active <u>if</u> the molecule exists in a conformation where there is no dipole moment change.

i.e.

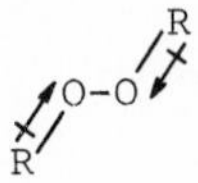

For a gauche conformation, there is a change in dipole moment and the band is active.

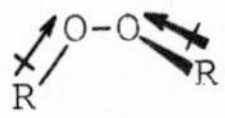

For a 900 cm^{-1} stretch, f = 3.82 x 10^5 dynes cm^{-1}. The value is lower due to the low DH° of O-O bond.

10. $\frac{dE}{dr} = 2f(r-r_0)$ $\qquad \frac{d^2E}{dr^2} = 2f$

The larger the force constant the greater the change of E with r; i.e., the sharper the curvature. But f is also related to the energy difference between vibrational levels. Compare:

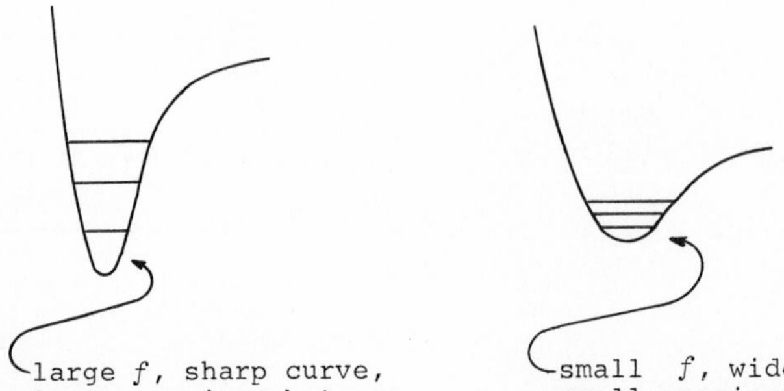

large f, sharp curve, large spacings between levels, strong bond (large DH°)

small f, wide curve, small spacings between levels, weak bond (small DH°)

14.E Supplementary Problems

S1. IR spectral bands and some nmr characteristics are given for four isomers of formula C_6H_{12}. What can you deduce about the structures from the information for each?

(a) IR: 1170, 1374, 1450, 2900 cm^{-1} (all strong); nmr: single resonance.

(b) IR: 960 (s), 1374 (w), 1450, 2940 (s) cm^{-1}; nmr: only ethyl and vinyl protons.

(c) IR: 885, 1370 (w), 1640, 2900, 3000 (w) cm^{-1}; nmr: only ethyl and vinyl protons.

(d) IR: 1450 (s), 2950 (s) cm^{-1}; nmr: single resonance.

S2. What are the structures of the compounds in the sequence below? Assign the resonances.

$$A \xrightarrow{KMnO_4} B + CO_2 \xrightarrow{NaBH_4} C \xrightarrow[\Delta]{H_2SO_4} D \xrightarrow[Pt]{H_2} E$$

IR:	A	B	C	D	E
	886 (s)	1450	1067 (s)	1440	nmr: singlet
	1445	1714 (s)	1450	1645 (vw)	
	1618	2875	2860	2865	
	2860	2950 (s)	2940	3000	
	2940		3300 (s)		
	3082				

S3. How could you use IR spectroscopy to distinguish between the following pairs of isomers?

(a) $HOCH_2CH_2CH_2CH_2\overset{O}{\overset{\|}{C}}H$ and 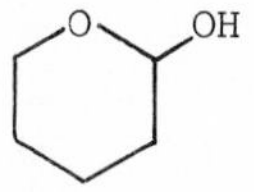

(b) 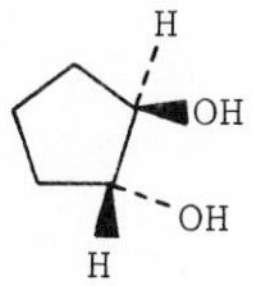 and 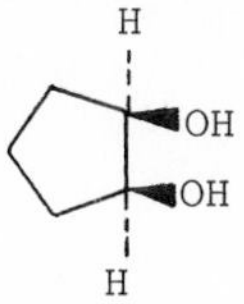

(c)  and 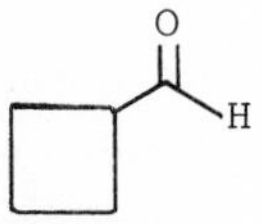

S4. Some IR and nmr spectral properties of five isomers of C_4H_8O are given below. Assign the structures for each isomer.

(a) IR: 1176, 1370, 1718, 3000 cm^{-1}; nmr: singlet (3H) at 2.0 ppm.

(b) IR: 962(s), 1000, 1075, 1666, 2860, 2940, 3300(br) cm^{-1}; nmr: 2 resonances (1H each) between 5 and 6 ppm.

(c) IR: 1724, 2718, 2825, 2878, 2970 cm^{-1}; nmr: triplet (1H) at 9.5 ppm.

(d) IR: 917(s), 1450, 1640(w), 2860, 2940, 3300(br) cm^{-1}; nmr: 3 resonances (1H each) between 5 and 6 ppm, doublet (3H) at 1.0 ppm.

(e) IR: 806, 962, 1030, 1200, 1312, 1612(s), 2950, 3000 cm^{-1}; nmr: indicates presence of ethyl group.

14.F Answers to Supplementary Problems

S1. (a) and (d): nmr: single resonance. The only possibilities are 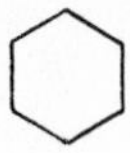

and 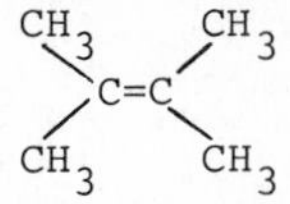. These are hard to assign by IR: the symmetrically-substituted double bond and the lack of vinyl hydrogens mean that no characteristic olefinic absorbances will appear in the spectrum of 2,3-dimethyl-2-butene. However, spectrum (a) has a band at 1374 cm^{-1}, characteristic of $-CH_3$ rocking, indicating that (a) is the olefin. Spectrum (d) therefore corresponds to cyclohexane.

(b) and (c): nmr indicates only ethyl and vinyl hydrogens present. There are three possibilities: $(CH_3CH_2)_2C{=}CH_2$ and *cis*- and *trans*-3-hexene. Spectrum (b) has a strong band at 960 cm^{-1}, consistent with the C-H out-of-plane bending absorption expected for *trans*-3-hexene; (c) has bands at 885 and 1640 cm^{-1}, which are what you would expect to find in the spectrum of 2-ethyl-1-butene (C-H out-of-plane bending and C=C stretching, respectively).

S2.

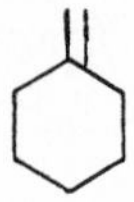
A

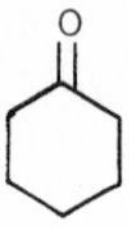

B

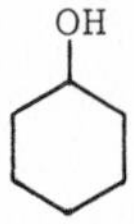

C

D

E

A	B	C	D
886 (C-H bend)	*1450 (CH_2 scissor)*	*1067 (C-O stretch)*	*1440 (CH_2 scissor)*
1445 (CH_2 scissor)	*1714 (C=O stretch)*	*1450 (CH_2 scissor)*	*1645 (C=C stretch)*
1618 (C=C stretch)	*2875,2950 (C-H stretch)*	*2860,2940 (C-H stretch)*	*2865 (alkane C-H)*
2860,2940 (alkane C-H stretch)		*3300 (O-H stretch)*	*3000 (alkene C-H)*
3082 (alkene C-H stretch)			

S3. (a) Open chain form will show strong band due to C=O stretch at 1725 cm^{-1}; cyclic hemiacetal form will not.

(b) *Cis* isomer will show intramolecular hydrogen-bonded O-H stretch (3250-3450 cm^{-1}) at all concentrations; *trans* isomer will show only free O-H stretch (3620-3640 cm^{-1}) at low concentration.

(c) Ketone: C=O stretch at 1745 cm^{-1}, no bands in region 2700-2850 cm^{-1}. Aldehyde: C=O stretch at 1725 cm^{-1}, two medium bands for H-CO stretch at 2720, 2820 cm^{-1}.

S4. (a) $CH_3CH_2\overset{O}{\overset{\|}{C}}CH_3$ (b) 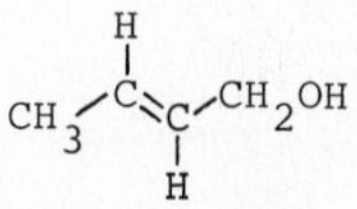 (c) $CH_3CH_2CH_2\overset{O}{\overset{\|}{C}}H$ (d) $CH_3\overset{OH}{\overset{|}{C}}HCH{=}CH_2$

(e) $CH_3CH_2OCH{=}CH_2$

15. ORGANOMETALLIC COMPOUNDS

15.A Outline of Chapter 15, Important Terms Introduced, and Reactions Discussed

15.1 Nomenclature

e.g. butyllithium, ethylmagnesium chloride (without a space between the organic prefix and the name of the metal)

15.2 Structure

electropositive — three-center two-electron bonds

ionic character — Grignard reagents

15.3 Physical Properties

15.4 Preparation of Organometallic Compounds

A. Reaction of an Alkyl Halide with a Metal

Grignard reagent:

$$RX + Mg \xrightarrow[\text{THF}]{\text{ether or}} RMgX \quad (X=Cl,Br,I)$$

alkyllithium:

$$RX + 2Li \xrightarrow{\text{ether}} RLi + LiX \quad (X=Cl,Br,I;\ R\text{-}R \text{ as byproduct})$$

B. Reaction of Organometallic Compounds with Salts

standard reduction potential

$$RM + M'X \rightleftharpoons RM' + MX \quad (E_o \text{ for M more negative than } E_o \text{ for M}')$$

C. Metallation of C-H bonds

e.g.

$$RC{\equiv}CH + MR' \rightleftharpoons RC{\equiv}CM + R'H \quad \text{(acid-base reaction)}$$

(R,R'=alkyl; M=Li,Na,MgX)

D. Special Methods of Preparation

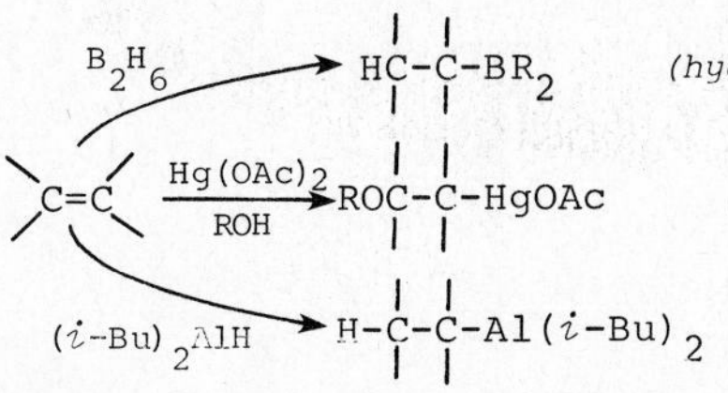

B_2H_6
HC-C-BR_2
$Hg(OAc)_2$
ROH
ROC-C-HgOAc
C=C
$(i\text{-Bu})_2AlH$
H-C-C-Al$(i\text{-Bu})_2$

(hydroboration, see Section 11.6.D)

(oxymercuration, see Section 11.6.B.4)

(hydroalumination with "DIBAL")

15.5 Reactions of Organometallic Compounds

A. Hydrolysis

$$RM + HOR' \longrightarrow RH + MOR'$$

specific deuteration (D instead of H)

B. Reaction with Halogens

$$RM + X_2 \longrightarrow RX + MX \quad \text{(not very useful)}$$

C. Reaction with Oxygen

(need for inert atmosphere over reaction to prevent)

RM + O_2 ⟶ ROOM $\xrightarrow{RM}$ 2 ROM   (can stop at hydroperoxide stage at low temperature)

D. Reaction with Carbonyl Compounds, Carbon Dioxide, and Epoxides

carbon-carbon bond-formation

Grignard reaction:

$$RMgX + R'\overset{O}{\overset{\|}{C}}R'' \longrightarrow R'-\overset{\overset{O^-Mg^{++}X^-}{|}}{\underset{\underset{R}{|}}{C}}-R'' \xrightarrow{H_2O} R'-\overset{\overset{OH}{|}}{\underset{\underset{R}{|}}{C}}-R'' + \text{Mg salts}$$

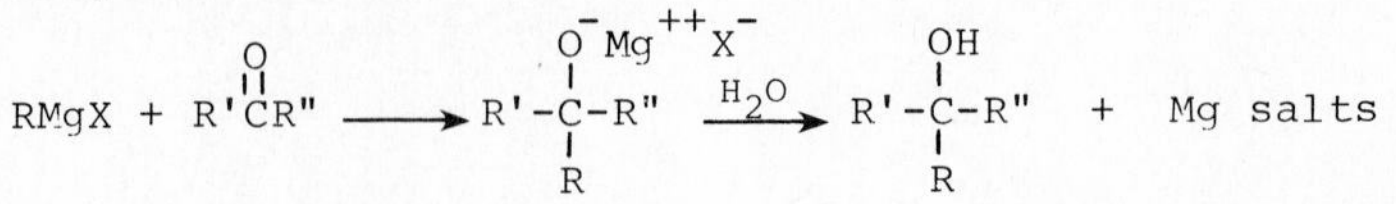

(R=alkyl,aryl; R',R"=H,alkyl,aryl)
(side reactions: enolization and reduction)

carboxylation: formation of carboxylic acids

RM + CO_2 ⟶ RCO_2M $\xrightarrow{H^+}$ RCO_2H (M=MgX,Li,Na)

reaction with ethyl orthoformate: formation of aldehydes

RMgX + $HC(OC_2H_5)_3$ ⟶ $RCH(OC_2H_5)_2$ $\xrightarrow{H_3O^+}$ RCH=O

reaction with epoxides:

$$RM + \overset{O}{\underset{}{C-C}} \longrightarrow R-\overset{|}{\underset{|}{C}}-\overset{|}{\underset{|}{C}}-O^-M^+ \xrightarrow{H_2O} R-\overset{|}{\underset{|}{C}}-\overset{|}{\underset{|}{C}}-OH \quad (R=MgX,Li)$$

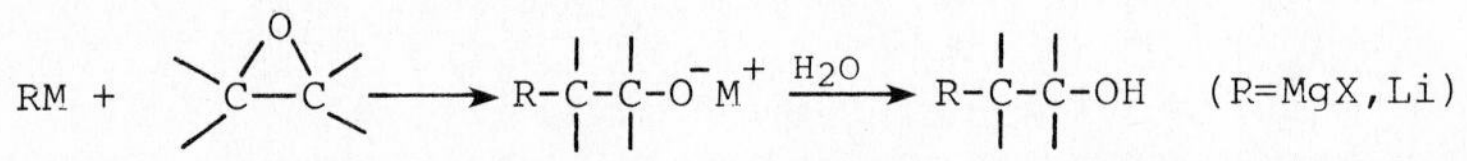

(reacts at less hindered end; rearrangement side reaction with Grignard reagent)

E. Reaction with Other Organometallic Compounds

cuprates

RLi + RCu \
2RLi + CuX / ⟶ R_2CuLi

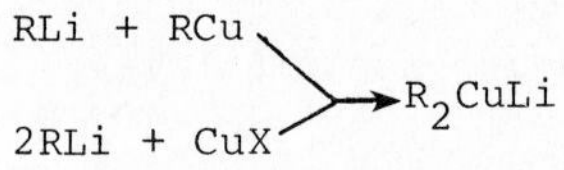

15.6 Transition Metal Organometallic Compounds

16 or 18 electron rule   coordinatively saturated
ligand   counting electrons

types of reactions:

1. Lewis-acid association-dissociation
2. Lewis-base association-dissociation
3. Oxidative addition-reductive elimination
4. Insertion-deinsertion

examples:

hydrogenation of ethylene with $[(C_6H_3)_3P]_3RhHCO$

decarbonylation of $R\overset{O}{\overset{\|}{C}}H$ with $[(C_6H_5)_3P]_3RhCl$ *(Wilkinson's catalyst)*

15.B Important Concepts and Hints

The Grignard Reaction

In this chapter you are introduced to one of the most important carbon-carbon bond-forming reactions: the Grignard reaction. The product of a Grignard reaction is usually an alcohol, which is a useful functional group for further transformations. As you will discover in the course of working the problems in this and subsequent chapters, a wide range of compounds can be made using the Grignard reaction. However, don't fall into the common trap of thinking that the Grignard reaction is the only useful carbon-carbon bond-forming transformation. When you work synthesis problems, you should use your entire repertoire of reactions, including the aldol condensation, Wittig reaction, acetylide alkylation, and so on. The Grignard reaction is versatile, but it's not always the best choice.

One point that often confuses students is the role of the "MgBr" part of the Grignard reagent. It is simply present as a counterion (cation in this case) for the carbanionic reagent or the alkoxide ion product. It plays essentially the same role in the Grignard reaction as the sodium cation plays in reactions of sodium ethoxide ($NaOC_2H_5$), for instance.

The product of addition of a Grignard reagent to a ketone is an ionic compound, the magnesium halide salt of an alcohol, and to obtain the alcohol itself the reaction mixture must be "worked up" by adding water (or, preferably, dilute acid so a messy $Mg(OH)_2$ precipitate doesn't form). Frequently, you will see Grignard reactions written as two-step processes to indicate this point:

$$CH_3\overset{O}{\overset{\|}{C}}CH_3 \xrightarrow{CH_3MgI} \xrightarrow{H_2O} (CH_3)_3COH$$

Just as frequently you will see the second step omitted, with protonation of the alkoxide intermediate simply understood:

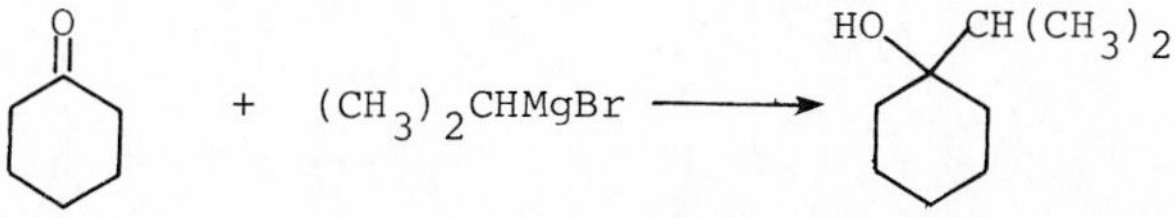

(The same comments apply to reactions with many other organometallic reagents, and to those employing lithium aluminum hydride; that is, as written, the total reaction often incorporates the results of working up the reaction mixture in a normal way).

One other point you should not forget is the strong basicity of carbanions like Grignard reagents: they will react instantly with water, alcohols, and rapidly with amine N-H bonds, in short, with any hydrogen bonded to a hetero-atom. Therefore, any such functional group must be protected in some way if you want to generate a Grignard reagent from a halide elsewhere in the same molecule. Also, if such a functional group is present in the carbonyl compound you expect your Grignard reagent to react with, you will find that the acid-

base reaction of this functional group and your Grignard agent will occur
faster than the desired carbonyl-addition   reaction. If your Grignard
reagent is easy to obtain, you can overcome this problem by using it in
excess, otherwise you will have to protect the sensitive functional group
before the Grignard addition. Some examples of pitfalls you should avoid
are shown below:

$HOCH_2CH_2CH_2CH_2Br \xrightarrow{Mg}$ ✕ $\underline{H}OCH_2CH_2CH_2CH_2MgBr$

$CH_3\overset{O}{\overset{\|}{C}}CH_2CH_2CH_2Cl \xrightarrow{Mg}$ ✕ $CH_3\overset{O}{\overset{\|}{C}}CH_2CH_2CH_2MgCl$

Can't have sensitive function-
ality present in the Grignard
reagent.

+ $HOCH_2CH_2CH_2\overset{O}{\overset{\|}{C}}CH_3 \longrightarrow$ ✕ $\underline{H}OCH_2CH_2CH_2\overset{OMgCl}{\overset{|}{C}}CH_3$

+ $ClMgOCH_2CH_2CH_2\overset{O}{\overset{\|}{C}}CH_3$ instead

(2 + $HOCH_2CH_2CH_2\overset{O}{\overset{\|}{C}}CH_3 \longrightarrow$ + $ClMgOCH_2CH_2CH_2\overset{OMgCl}{\overset{|}{C}}CH_3$

$HOCH_2CH_2CH_2\overset{OH}{\overset{|}{C}}CH_3$ H_2O

(This sequence will work))

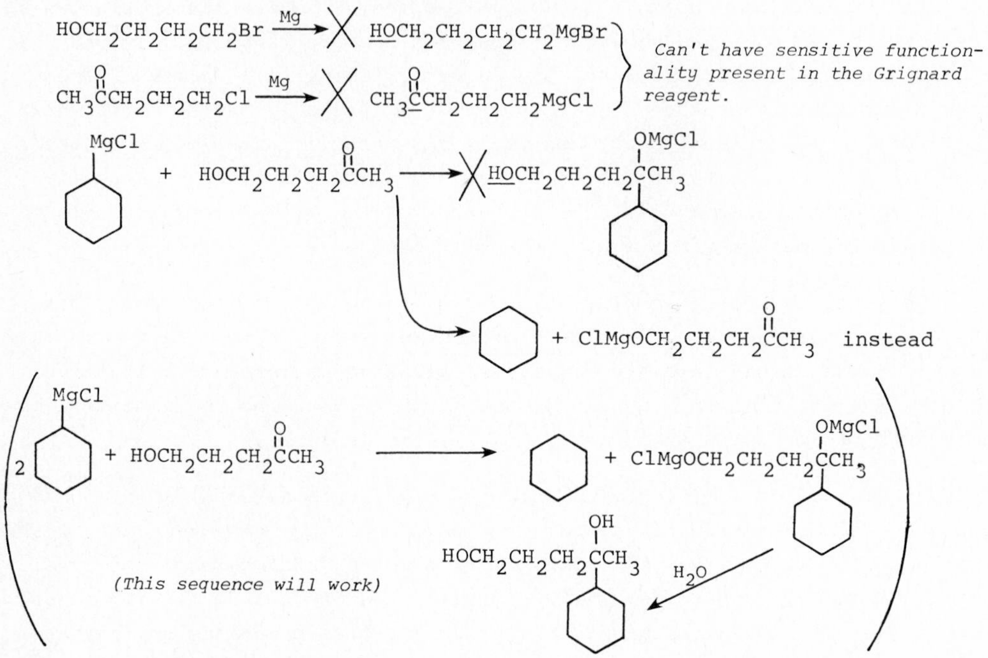

Finally, look out for situations in which a heteroatom will be on a
carbon β to the carbanion in a Grignard reagent. The molecule will undergo
elimination to give an alkene instead of a Grignard reagent. For example:

($CH_3-\overset{OCH_3}{\overset{|}{C}H}-CH_2-Br + Mg \rightarrow \left[CH_3\overset{OCH_3}{C}H-CH_2\overset{-}{:} \quad ^{+}MgBr \right] \rightarrow CH_2CH=CH_2 + CH_3OMgBr$

$\underset{\beta\quad\alpha}{}$

Analogy: $CH_3-\overset{OH}{\overset{|}{C}H}-CH_2-\overset{O}{\overset{\|}{C}}H \xrightarrow{OH^-} \left[\begin{array}{c} CH_3-\overset{OH}{\overset{|}{C}H}-CH-\overset{O}{\overset{\|}{C}}H \\ \updownarrow \quad \overset{..}{} \\ CH_3-\overset{OH}{\overset{|}{C}H}-CH=\overset{O^-}{\overset{|}{C}}H \end{array} \right] \rightarrow CH_3CH=CH\overset{O}{\overset{\|}{C}}H + OH^-)

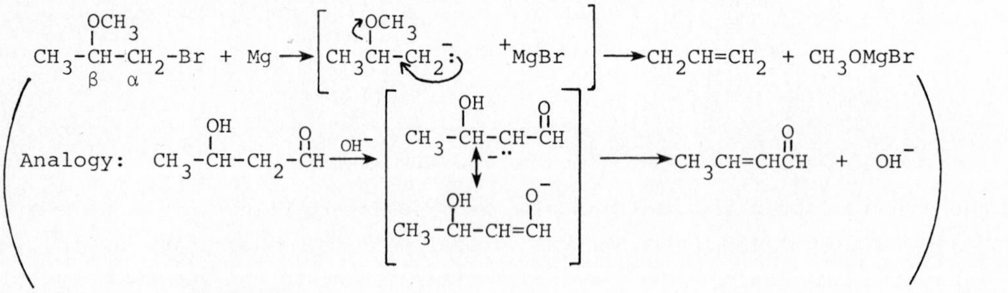

15.C Answers to Exercises

(15.4.A) CH_3MgI methylmagnesium iodide CH_3I methyllithium

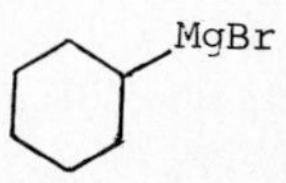

MgBr cyclohexylmagnesium bromide

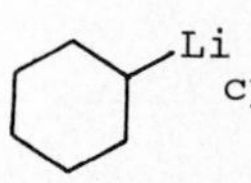

Li cyclohexyllithium

(CH_3)_3CMgCl *t*-butylmagnesium chloride   (CH_3)_3CLi *t*-butyllithium

(15.4.B) (a) $2(CH_3)_3Al + 3\,ZnCl_2 \rightleftharpoons 3\,(CH_3)_2Zn + AlCl_3$

$E_\circ =$ -0.763 -1.66

(b) $2\,(CH_3)_2Hg + SiCl_4 \rightleftharpoons (CH_3)_4Si + 2\,HgCl_2$

$E_\circ =$ -0.840 $+0.854$

(c) $(CH_3)_3Al + 3\,CuBr \rightleftharpoons 3\,CH_3Cu + AlBr_3$

$E_\circ =$ $+0.522$ -1.660

(15.5.A) (a) $CH_3CH_2C{\equiv}CH + CH_3MgBr \longrightarrow CH_3CH_2C{\equiv}CMgBr + CH_4\uparrow \xrightarrow{D_2O} CH_3CH_2C{\equiv}CD$

(b) $(CH_3)_3CH + Br_2 \xrightarrow{h\nu} (CH_3)_3CBr \xrightarrow[\text{ether}]{Mg} (CH_3)_3CMgBr \xrightarrow{D_2O} (CH_3)_3CD$

(c) $(CH_3)_3CBr$ *(from (b))* $\xrightarrow{NaOH} (CH_3)_2C{=}CH_2 \xrightarrow{B_2H_6} [(CH_3)_2CHCH_2]_3B \xrightarrow[\Delta]{CH_3COOD} (CH_3)_2CHCH_2D$

(15.5.D)
(a) (i) $n\text{-}C_4H_9Br \xrightarrow[\text{ether}]{Mg} n\text{-}C_4H_9MgBr \xrightarrow{\text{O}\,/\,CH_2\text{-}CH_2\ (\text{ethylene oxide})} CH_3CH_2CH_2CH_2CH_2CH_2OMgBr \xrightarrow{H_2O}$ 1-hexanol

(ii) $n\text{-}C_4H_9Br \xrightarrow{NaOH} \xrightarrow[H_2SO_4]{Na_2Cr_2O_7} CH_3CH_2CH_2CH(=O) \xrightarrow{CH_3MgBr} \xrightarrow{H_2O} CH_3CH_2CH_2CH(OH)CH_3$

$\xrightarrow{Na_2Cr_2O_7,\ H_2SO_4} CH_3CH_2CH_2C(=O)CH_3 \xrightarrow{CH_3MgBr} \xrightarrow{H_2O}$ 2-methyl-2-pentanol

(iii) $CH_3CH_2CH_2CH(=O)$ *(from (ii)* $\xrightarrow{CH_3CH_2MgBr} \xrightarrow{H_2O}$ 3-hexanol

(iv) $n\text{-}C_4H_9MgBr$ *(from (i))* $\xrightarrow{CH_2{=}O} \xrightarrow{H_2O}$ 1-pentanol

(v) $n\text{-}C_4H_9MgBr$ *(from (i))* $\xrightarrow{CO_2} CH_3CH_2CH_2CH_2CO_2MgBr \xrightarrow{H^+} CH_3CH_2CH_2CH_2COOH$

(b) (i)
$CH_3MgBr + CH_3CH_2C(=O)(CH_2)_5CH_3$
$CH_3CH_2MgBr + CH_3C(=O)(CH_2)_5CH_3$
$CH_3(CH_2)_5MgBr + CH_3C(=O)CH_2CH_3$
$\longrightarrow CH_3CH_2C(OMgBr)(CH_3)(CH_2)_5CH_3 \xrightarrow{H_2O}$ 3-methyl-3-nonanol

(ii)

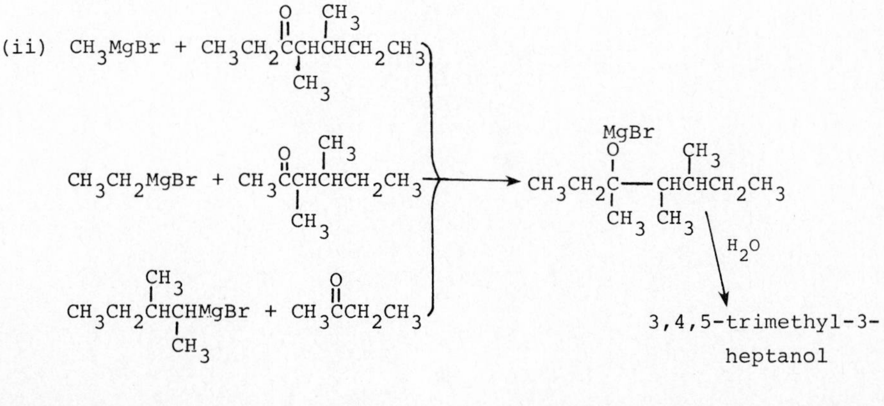

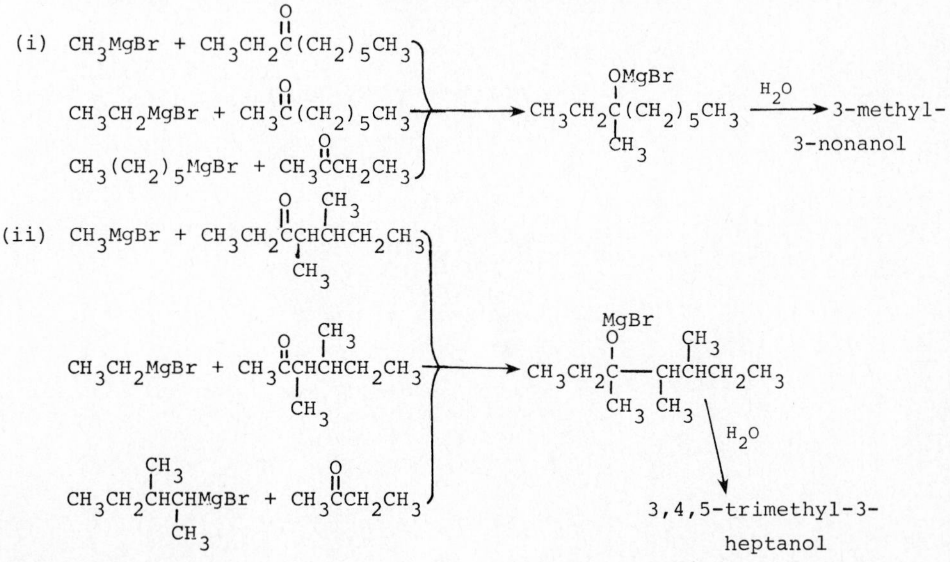

(iii) CH_3CH_2MgBr + $(CH_3)_2CHCH_2\overset{O}{\overset{\|}{C}}CH_2CH_2CH_2CH_3$

$(CH_3)_2CHCH_2MgBr$ + $CH_3CH_2\overset{O}{\overset{\|}{C}}CH_2CH_2CH_2CH_3$

n-C_4H_9MgBr + $(CH_3)_2CHCH_2\overset{O}{\overset{\|}{C}}CH_2CH_3$

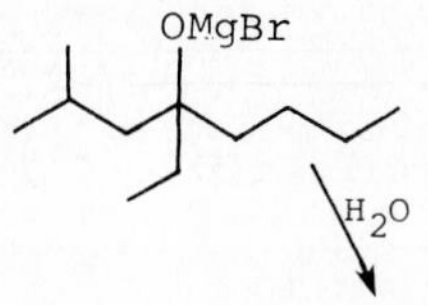

4-ethyl-2-methyl-4-octanol

(15.6) $[(C_6H_5)_3P]_4Ni$

10 electrons on Ni
4×2 = 8 from phosphine ligands
18

$[(C_6H_5)_3P]_4\overset{+}{Ni}H$ Cl^-

10 electrons on Ni
− 1 for + charge
− 1 Ni electron used in Ni-H bond
4×2 = 8 from phosphine ligands
2 the Ni-H bonding pair
18

$^-Mn(CO)_5$

7 electrons on Mn
1 for − charge
5×2 = 10 from CO ligands
18

$CH_3Mn(CO)_5$

7 electrons on Mn
− 1 Mn electron used in Mn-C bond
5×2 = 10 from CO ligands
2 the Mn-C bonding pair
18

$Ni(CO)_4$

10 electrons on Ni
4 × 2 = 8 from CO Ligands
18

$Ni(CO)_3$

10 electrons from Ni
3 × 2 = 6 from CO ligand
16

$(C_6H_5)_3PNi(CO)_3$

10 electrons on Ni
3×2 = 6 from CO ligands
2 from phosphine ligand
18

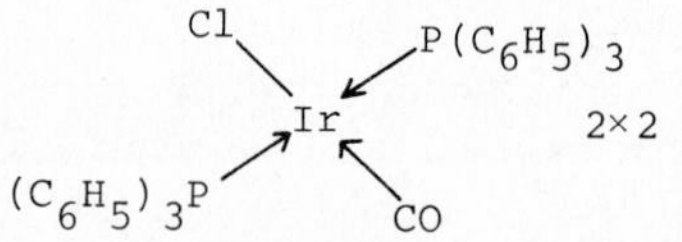

9 electrons on Ir
− 1 Ir electron used in Ir-Cl bond
2×2 = 4 from phosphine ligands
2 from CO ligand
2 from Ir-Cl bond
16

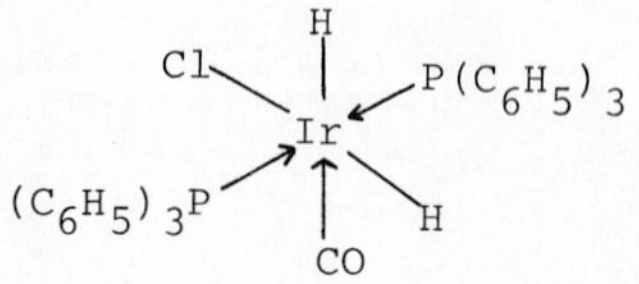

9 electrons on Ir
− 3 Ir electrons used in Ir-Cl and Ir-H bonds
2×2 = 4 from phosphine ligands
2 from CO ligand
2 from Ir-Cl bond
2×2 = 4 from Ir-H bond
18

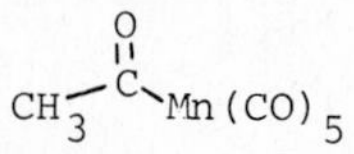

7 electrons on Mn
− 1 Mn electron used in bond to acetyl group
5×2 = 10 from CO ligands
2 from Mn-acyl bond
18

15.D Answers and Explanations to Problems

1. (a) $(CH_3)_3CCH(CH_3)_2 \xrightarrow[h\nu]{Br_2/CCl_4} (CH_3)_3CCBr(CH_3)_2 \xrightarrow{Mg,\ ether} (CH_3)_3CC(MgBr)(CH_3)_2 \xrightarrow{D_2O} (CH_3)_3CCD(CH_3)_2$

(see Sect. 6.3.B and Chapt. 6, problem #6)

(b) $CH_3CH_2CHClCH_3 \xrightarrow[ether]{Mg} CH_3CH_2CH(MgCl)CH_3 \xrightarrow{SnCl_4} (CH_3CH_2CH(CH_3)-)_4Sn$

(c) $(CH_3)_4C \xrightarrow[h\nu]{Cl_2} (CH_3)_3CCH_2Cl \xrightarrow[ether]{Mg} (CH_3)_3CCH_2MgCl$

(d) $CH_3CH_2Cl \xrightarrow[ether]{Mg} CH_3CH_2MgCl \xrightarrow{CdCl_2} (CH_3CH_2)_2Cd + MgCl_2$

(e) $HC{\equiv}CH \xrightarrow{n\text{-}BuLi} HC{\equiv}CLi \xrightarrow{CH_3CH_2CHBrCH_2CH_3} (CH_3CH_2)_2CHC{\equiv}CH \xrightarrow{CH_3MgBr} \xrightarrow{D_2O} (CH_3CH_2)_2CHC{\equiv}CD$

(low yield due to E2)

(f) HC≡CLi *(from (e))* $\xrightarrow{CH_3CH_2Br} CH_3CH_2C{\equiv}CH \xrightarrow{CH_3MgBr} CH_3CH_2C{\equiv}CMgBr \xrightarrow{CH_3COCH_3} CH_3CH_2C{\equiv}CC(OMgBr)(CH_3)_2 \xrightarrow{H_2O} CH_3CH_2C{\equiv}CC(OH)(CH_3)_2$

(g) $(CH_3)_3CCH_2Cl \xrightarrow{Mg} (CH_3)_3CCH_2MgCl \xrightarrow{CH_3CHO} \xrightarrow{H_2O} (CH_3)_3CCH_2CH(OH)CH_3$

(h)

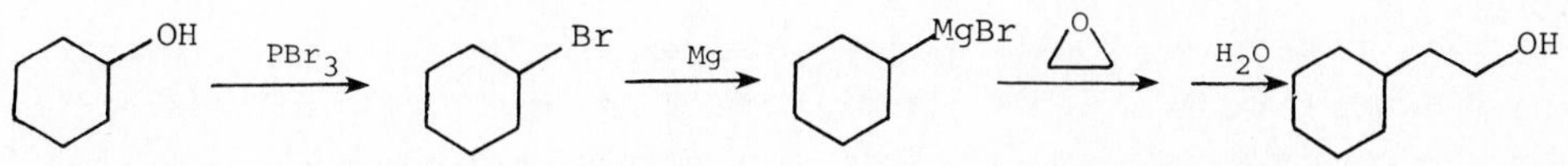

(i)

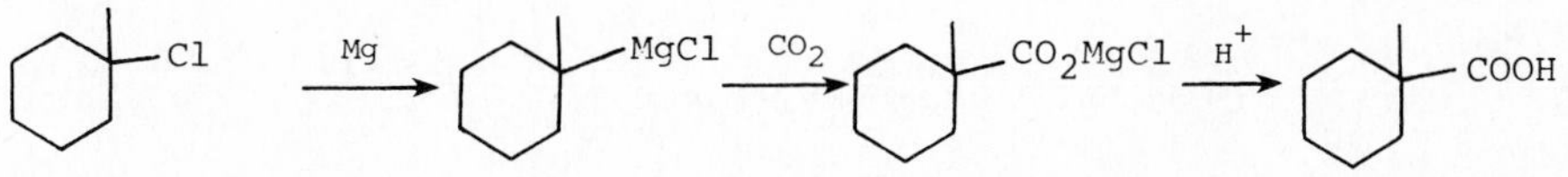

2.

This problem is answered by examination of standard reduction potentials (Table 15.3). The more negative the reduction potential, the more the metal prefers to exist as a cation in an inorganic salt rather than bound to carbon with some covalent bonding.

K > 1: a, c, d   K < 1: b, e

3. (a) For di-*t*-butylberyllium, the *t*-butyl group is too bulky to permit the polymeric structure; instead, it exists as a monomer with sp hybridization:

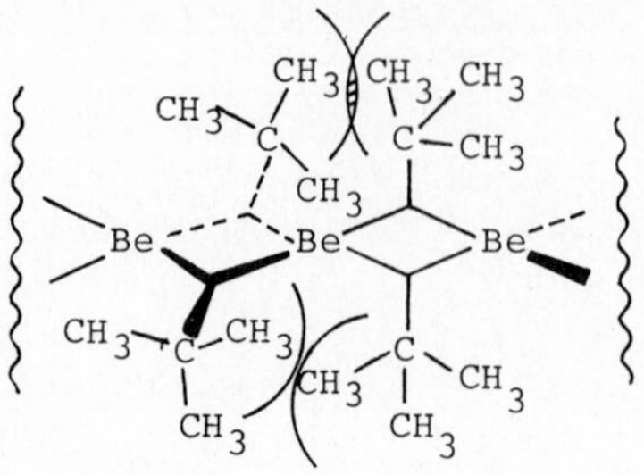

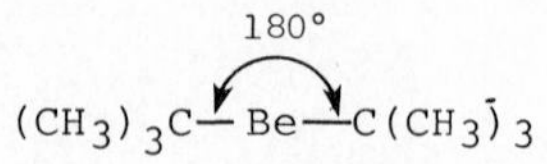

(b) The molecule is linear about Hg: $CH_3-Hg-CH_3$. The bonding is C_{sp^3}-Hg_{sp}. Mercury has two valence electrons; hence, bonds use sp hybrids, much as in beryllium.

4. $(CH_3)_3B + CH_3Li \longrightarrow (CH_3)_4B^- Li^+$. The compound is tetrahedral about boron; that is, C_{sp^3}-B_{sp^3}.

5. The Würtz reaction consists of two steps:

$$RX + 2\ Na \longrightarrow RNa + NaX$$

$$RX + RNa \longrightarrow R\text{-}R + NaX$$

Most simple halides will give an intermediate alkylsodium, but in varying yield. The intermediate R. can disproportionate to give alkane and alkene. This reaction proceeds best with unbranched alkyl halides.

The second step is an S_N2 displacement reaction which gives varying degrees of elimination side reaction (RNa is a strong base: $R\text{-}Na \longleftrightarrow R^- Na^+$). The reaction proceeds very poorly with secondary and tertiary halides, and best with primary halides.

Hence, the Würtz reaction works best for halides of the type RCH_2CH_2X.

$$CH_3CH_2CH_2CH_2Br \xrightarrow{Na} CH_3(CH_2)_6CH_3 \qquad (48\%)$$

Note that even in this relatively favorable case, the yield is only fair.

Obviously, the alkane prepared must be symmetrical. Other examples of alkanes preparable by this reaction are:

$(CH_3)_2CHCH_2CH_2CH_2CH_2CH(CH_3)_2$, $CH_3(CH_2)_8CH_3$, etc.,

but not $CH_3(CH_2)_9CH_3$ or $(CH_3)_2CH(CH_2)_7CH_3$, etc.

6. (a) $CH_3CH_2CH_2MgCl + (CH_3CH_2CH_2)_2C=O \longrightarrow \xrightarrow{H_2O} (CH_3CH_2CH_2)_3COH$

(b) $CH_3CH_2MgBr + H\overset{O}{\overset{\|}{C}}CH_2CH_2CH_3$

or

$CH_3CH_2CH_2MgBr + CH_3CH_2\overset{O}{\overset{\|}{C}}H$

$\xrightarrow{H_2O} CH_3CH_2\overset{OH}{\underset{|}{C}}HCH_2CH_2CH_3$

(c) $2\ CH_3CH_2CH_2MgCl + H\overset{O}{\overset{\|}{C}}CH_2OH \longrightarrow CH_3CH_2CH_2\overset{OMgBr}{\underset{|}{\underset{|}{\underset{H}{C}}}}-CH_2-OMgBr$

$\searrow H_2O$

$CH_3CH_2MgBr + BrCH_2CH=CH_2 \longrightarrow CH_3CH_2CH_2CH=CH_2 \xrightarrow[H_2O_2]{OsO_4} CH_3CH_2CH_2\overset{OH}{\underset{|}{C}}HCH_2OH$

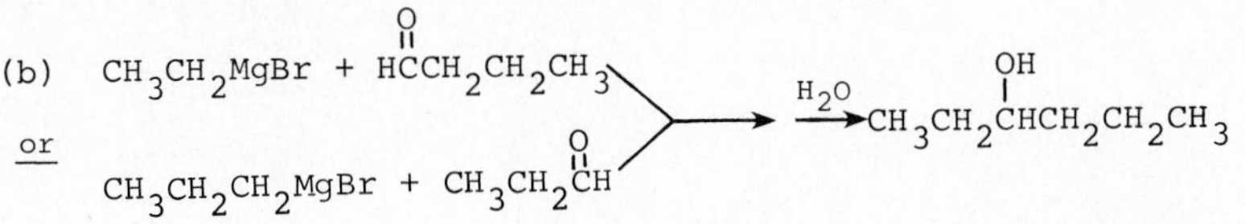

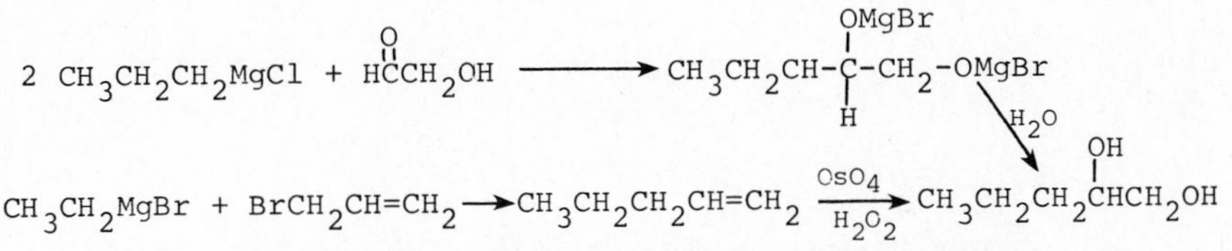

Note: In the first route, two moles of the Grignard reagent are required, because one of them will react with the hydroxy group of the hydroxyaldehyde. For similar reasons, you <u>cannot</u> make Grignard reagents like $CH_3CH_2CH_2\overset{OH}{\underset{|}{C}}HMgBr$ or $BrMgCH_2OH$.

(d) $CH_3CH_2CH_2\overset{CH_3}{\underset{\underset{CH_3}{|}}{\overset{|}{C}}}CH_2MgCl + CH_2=O$

or

$CH_3CH_2CH_2\overset{CH_3}{\underset{\underset{CH_3}{|}}{\overset{|}{C}}}MgCl + CH_2\!-\!CH_2$ (epoxide)

$\xrightarrow{H_2O} CH_3CH_2CH_2\overset{CH_3}{\underset{\underset{CH_3}{|}}{\overset{|}{C}}}CH_2CH_2OH$

(e) $CH_3MgI + CH_3\overset{O}{\overset{\|}{C}}CH_2CH_3$

or

$CH_2CH_2MgCl + CH_3\overset{O}{\overset{\|}{C}}CH_3$

$\xrightarrow{H_2O} (CH_3)_2\overset{OH}{\underset{|}{C}}CH_2CH_3$

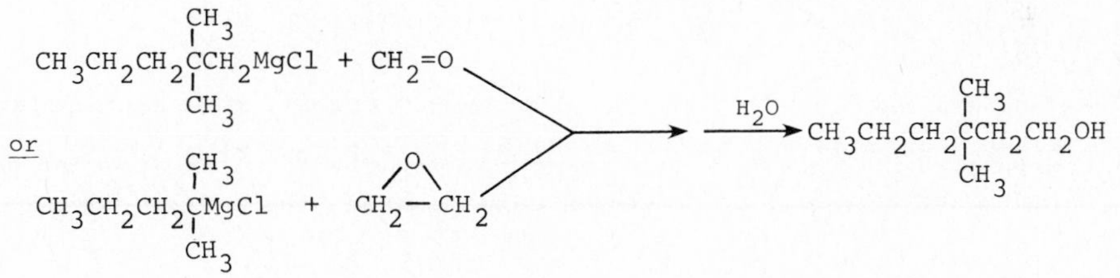

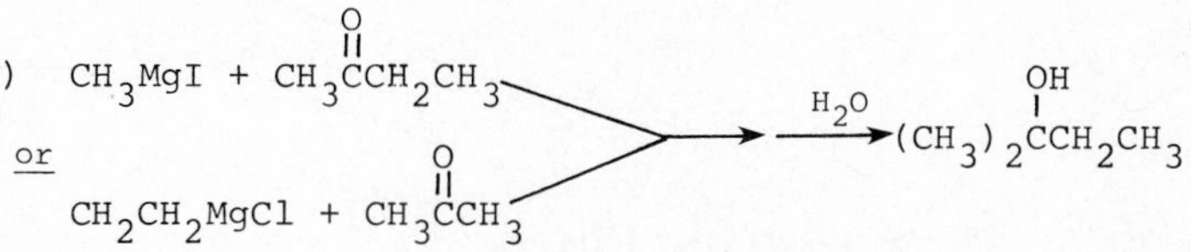

(Throughout this problem, the choice of Cl, Br, or I has been arbitrary.)

7. $CH_3CH_2\!-\!\overset{}{\underset{CH_3CH_2}{C}}\!-\!CH_2$ (epoxide) $\xrightarrow[\text{initial epoxide}]{CH_3CH_2MgBr}$ $\xrightarrow{\text{rearrangement}}$ $\left[(CH_3CH_2)_2CH\overset{O}{\overset{\|}{C}}H \right]$

$\searrow H_2O$

$(CH_3CH_2)_2\overset{OH}{\underset{|}{C}}HCHCH_2CH_3$

direct attack $\downarrow CH_3CH_2Li$

$\downarrow H_2O$

$\overset{OH}{\underset{|}{}}$

$(CH_3CH_2)_2\overset{}{\underset{|}{C}}CH_2CH_2CH_3$

8. $CH_3CH_2\overset{Cl}{\underset{|}{C}}HCH_3 \xrightarrow{Mg} \xrightarrow{\triangle\!\!-\!\!O} \xrightarrow{H_2O} CH_3CH_2\overset{CH_3}{\underset{|}{C}}HCH_2CH_2OH \xrightarrow{PCl_3} CH_3CH_2\overset{CH_3}{\underset{|}{C}}HCH_2CH_2Cl$

$\downarrow Mg$

$CH_3CH_2\overset{CH_3}{\underset{|}{C}}H(CH_2)_4OH \xleftarrow{H_2O} \xleftarrow{\triangle\!\!-\!\!O}$

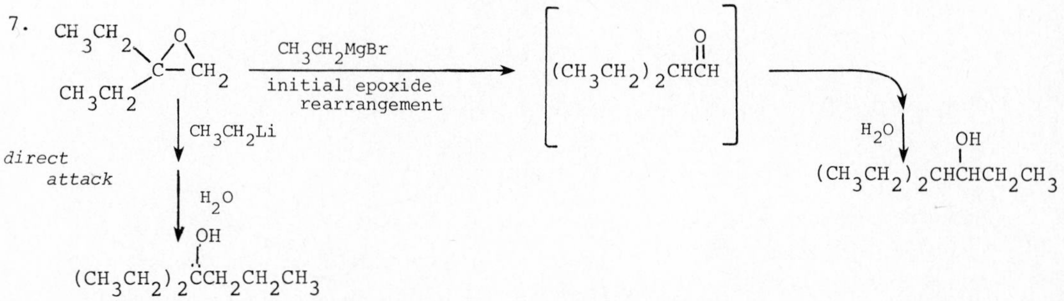

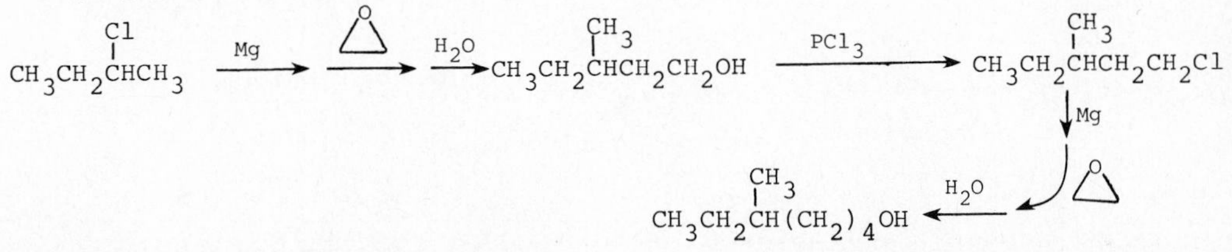

9.

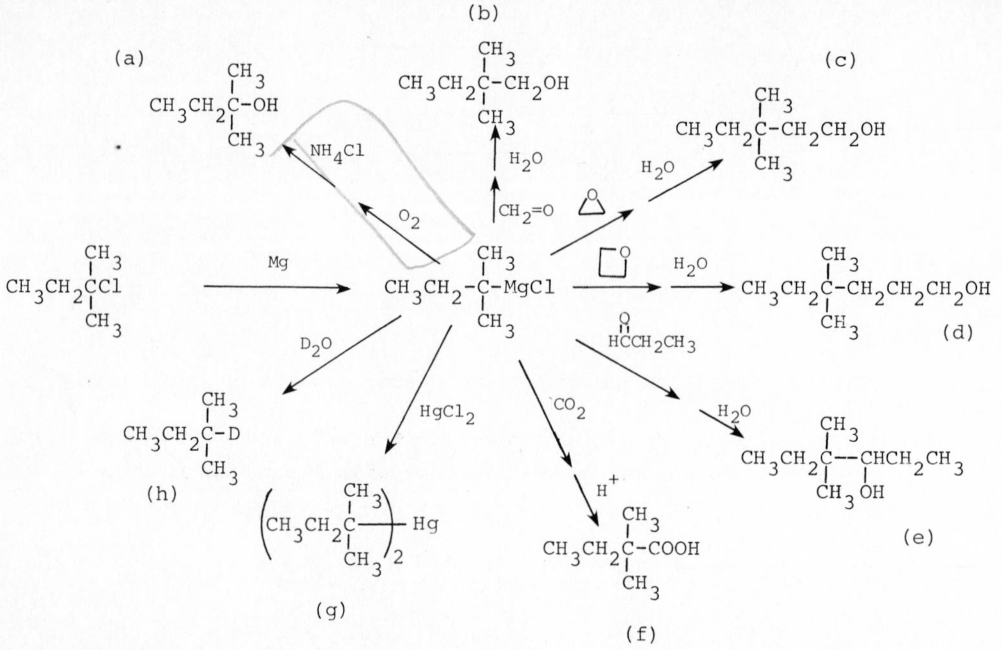

10. (a) To satisfy the 18-electron rule, Cr (6 electrons) needs 6 CO ligands: $Cr(CO)_6$, and Fe (8 electrons) needs 5: $Fe(CO)_5$

(b) $^{-}Mn(CO)_5$ and $^{-}Co(CO)_4$ each have 18 electrons in the valence shell of the metal.

11. The steps involved in this reaction are:

$$Ni(CO)_4 \rightleftharpoons Ni(CO)_3 + CO$$

$$Ni(CO)_3 + (C_6H_5)_3P \rightleftharpoons (C_6H_5)_3PNi(CO)_3$$

The rate expression for formation of product is: $\frac{dP}{dt} = k[Ni(CO)_3][R_3P]$. $Ni(CO)_3$ is in equilibrium with $Ni(CO)_4$, and its concentration is related to that of $Ni(CO)_4$ and CO as shown by the equilibrium constant

$$K = \frac{[Ni(CO)_3][CO]}{[Ni(CO)_4]}.$$

Combining these two expressions gives the following equation:

$$\frac{dP}{dt} = kK\frac{[Ni(CO)_4][R_3P]}{[CO]}$$

and the inverse dependence of rate on CO concentration can be seen. (The concentration of CO in solution is proportional to the pressure of CO in contact with the solution.)

12.

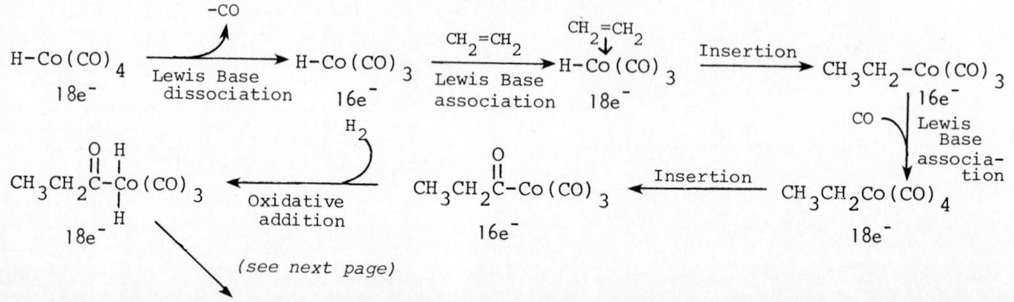

(see next page)

(from last page)

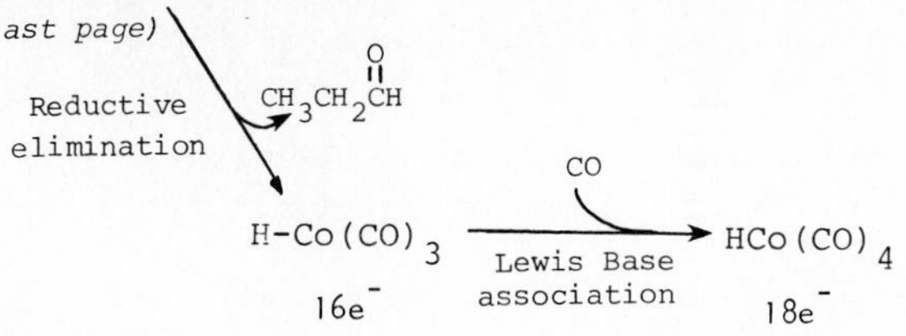

13. $[(C_6H_5)_3P]_3RhCl = L_3RhCl$ = Wilkinson's catalyst

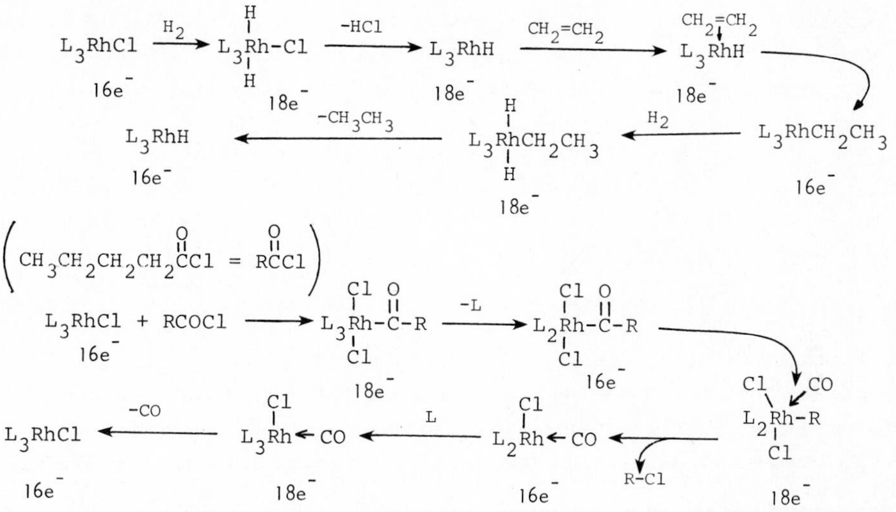

14. (a) If the reaction proceeds entirely through the free radical, one expects a mixture of <u>meso</u>(R,S), (S,S), and (R,R) isomers.

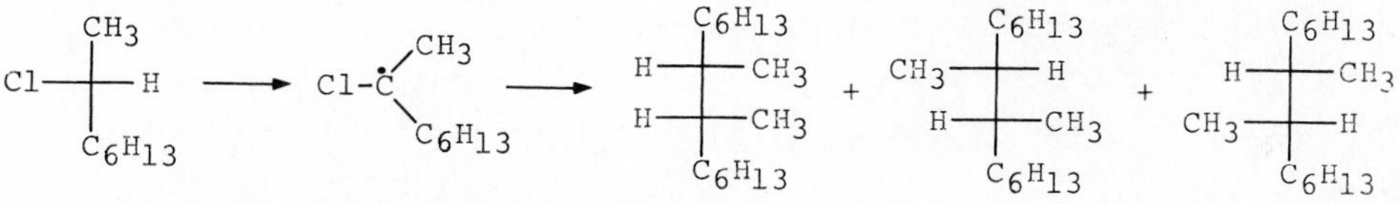

(R,S) (R,R) (S,S)

Since the (S,S) and (R,R) compounds are enantiomers, they should be produced in exactly equal amounts.

(b) The (R,S) isomer is <u>meso</u>, and the (S,S) isomer is optically active.

(c) No. See Sect. 7. 8.

(d) The initial organosodium compound can have either the (R) or the (S) configuration. Displacement by the S_N2 mechanism can give the (R,S) or (S,S) isomer.

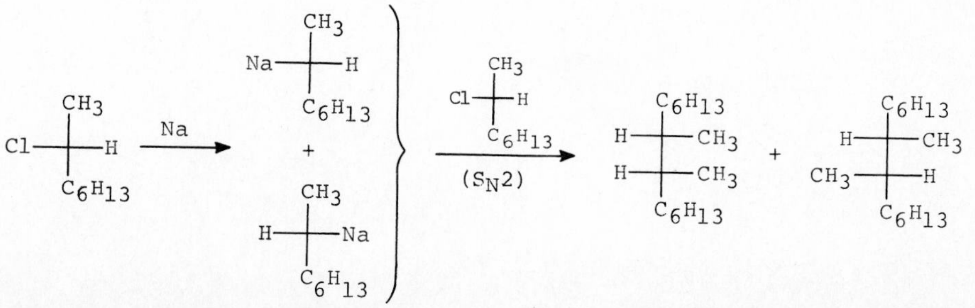

The data are consistent with the S_N2 mechanism (inversion at one 2-octyl unit, racemization at the other), but this does not rigorously prove this mechanism. It does rule out a mechanism proceeding through planar free radicals, since such a mechanism should give some of the (R,R) isomer.

(e) In this case, an intermediate alkylsodium would react *via* an S_N2 mechanism preferentially with n-pentyl iodide rather than with the more hindered neopentyl iodide. Such a mechanism could not give such a large proportion of 2,2,5,5-tetramethylhexane, a product which, according to the S_N2 mechanism, must come from displacement by neopentylsodium on neopentyl iodide. The proportions of product are much more readily explicable on the basis of radical combinations:

$$2\,CH_3(CH_2)_3CH_2\cdot \longrightarrow CH_3(CH_2)_8CH_3$$

$$CH_3(CH_2)_3CH_2\cdot + \cdot CH_2C(CH_3)_3 \longrightarrow CH_3(CH_2)_5C(CH_3)_3$$

$$2(CH_3)_3CCH_2\cdot \longrightarrow (CH_3)_3CCH_2CH_2C(CH_3)_3$$

The Wurtz reaction is clearly a complex reaction whose mechanism depends on the compounds used and the conditions. It is also a heterogeneous reaction, since solid sodium reacts with solutions of alkyl halides. The reaction has important limitations as a synthetic tool.

15. (a) S_N2 displacement reaction by R_3Sn^- as the nucleophilic reagent on an alkyl halide with inversion of configuration.

(b) The overall result of both reactions is regeneration of 2-bromobutane with retention of configuration. Since the first step occurs with inversion of configuration (by analogy with all other known S_N2 reactions), the second reaction must also occur with inversion.

(c) The neopentyl group is highly hindered, and reaction occurs preferentially at the 2-butyl carbon, even though this secondary position is normally more hindered than a straight-chain primary carbon.

15.E Supplementary Problems

S1. Show how to accomplish the following conversions:

(a) 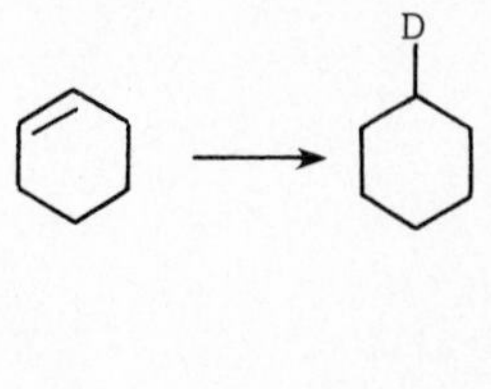

(b) 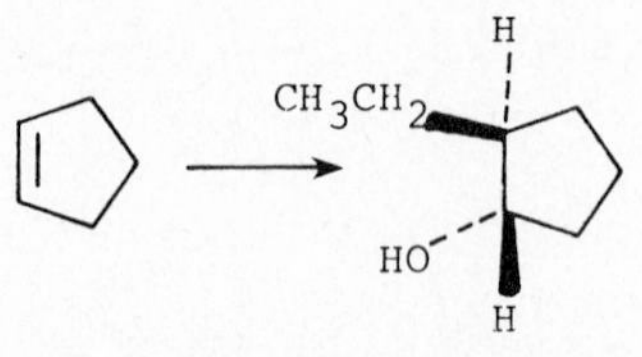

(c) 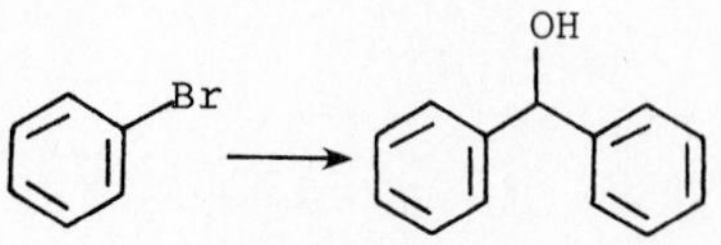

(d) 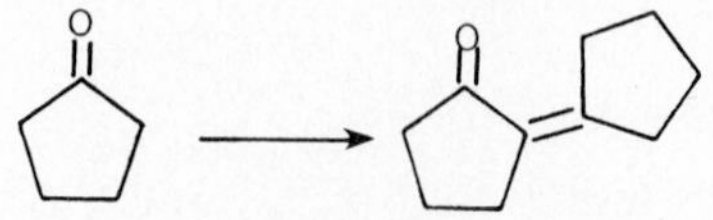

(e)

$$(CH_3)_2CHCH{=}CH_2 \longrightarrow (CH_3)_2CHCH_2CH_2CH_2CH(CH_3)_2$$

(f) $(CH_3)_2CHCH{=}CH_2 \rightarrow (CH_3)_2CHCH_2CH_2CH_2CH_2CH(CH_3)_2$

(g)

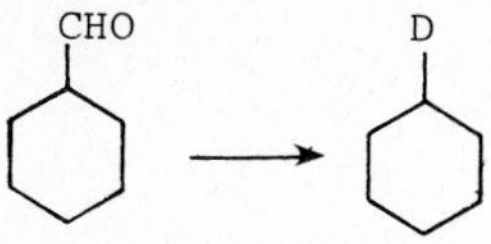

(h) $CH_3CH_2C{\equiv}CH \rightarrow CH_3CH_2\overset{O}{\overset{\|}{C}}CH_3$

S2. What will go wrong with each of the sequences proposed below:

(a)

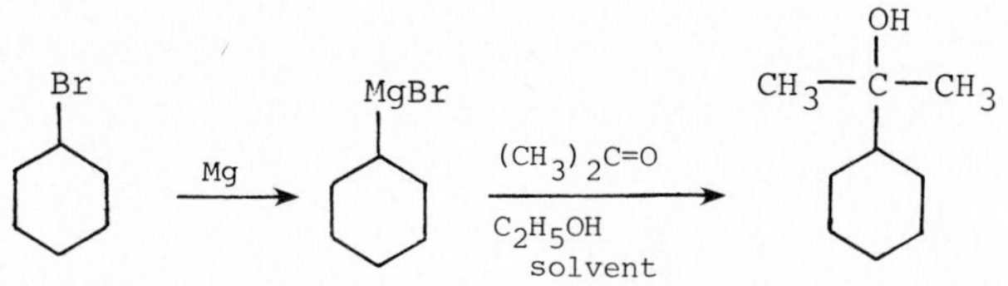

(b) $CH_3CH_2C(CH_3){=}CH_2 \xrightarrow[H_2O]{Br_2} CH_3CH_2C(CH_3)(OH)CH_2Br \xrightarrow{Mg} \xrightarrow{CO_2} \xrightarrow{H^+} CH_3CH_2C(CH_3)(OH)CH_2COOH$

(c) $CH_3CH_2Li + CH_3CH_2\overset{O}{\overset{\|}{C}}CH_2COOH \rightarrow \xrightarrow{H_2O} CH_3CH_2C(OH)(CH_3)CH_2COOH$

(d) $(CH_3)_2C{=}CH_2 \xrightarrow[HOCH_3]{Br_2} (CH_3)_2C(OCH_3)CH_2Br \xrightarrow{Mg} \xrightarrow{CO_2} \xrightarrow{H^+} (CH_3)_2C(OCH_3)CH_2COOH$

(e)

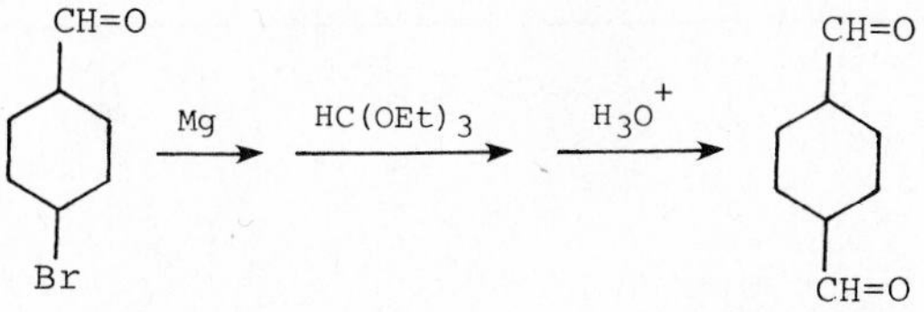

S3. For each of the following compounds, write two different syntheses--one which uses a Grignard reagent (or an alkyllithium) and one which doesn't. For starting materials you may use any compound containing 4 carbons or less.

(a) $(CH_3CH_2)_2C(OH)CH_3$ (b) $CH_3CH_2C(CH_3){=}CH_2$ (c) $CH_3CH_2CH_2CH_2CH_2CH_2OH$

(d) $CH_3CH_2CH_2CH_2\overset{O}{\overset{\|}{C}}CH_3$ (e) $CH_3CH_2CH(Br)CH_2CH_2CH_3$ (f) $CH_3CH_2CH(CH_2OH)CH_2CH_2CH_2CH_3$

S4. Alkyl halides can be converted to ketones using sodium tetracarbonylferrate. A proposed mechanism is shown below. For each intermediate, determine the number of electrons in the valence shell of the iron atom, and classify each step by reaction type (i.e., Lewis acid association, insertion, etc.).

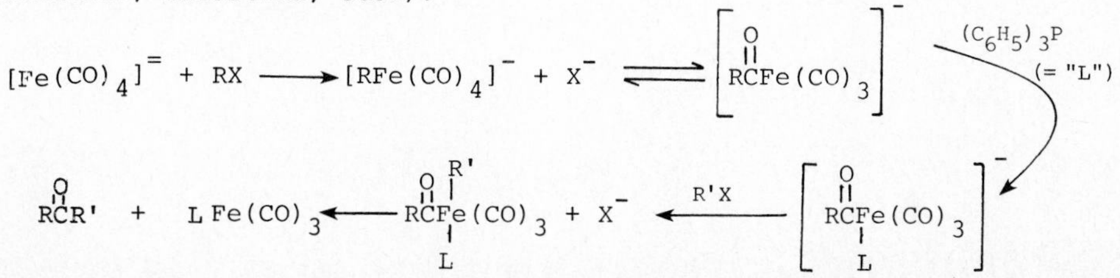

15.F   Answers to Supplementary Problems

S1. (a)

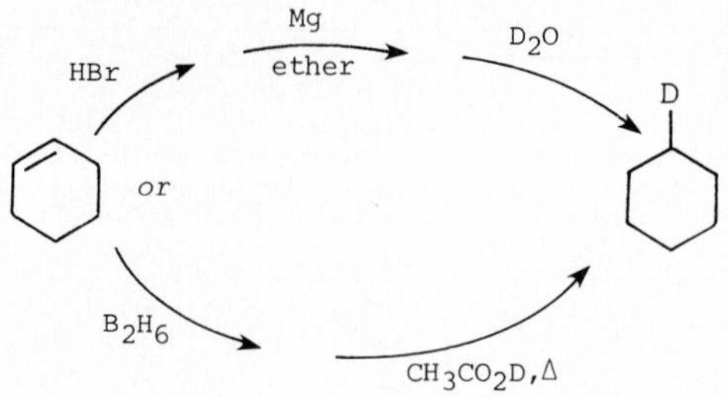

(b)

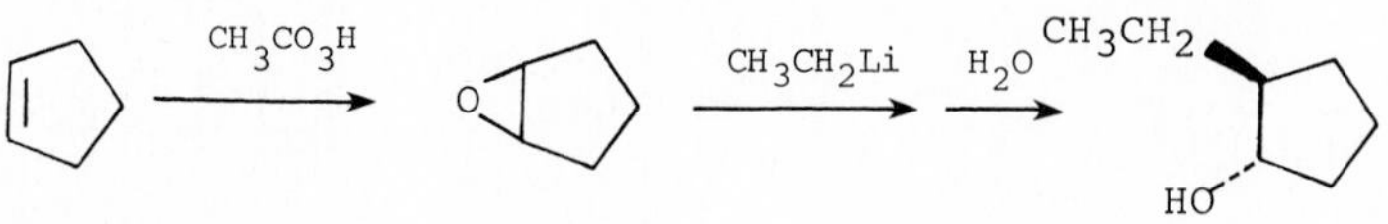

(c)

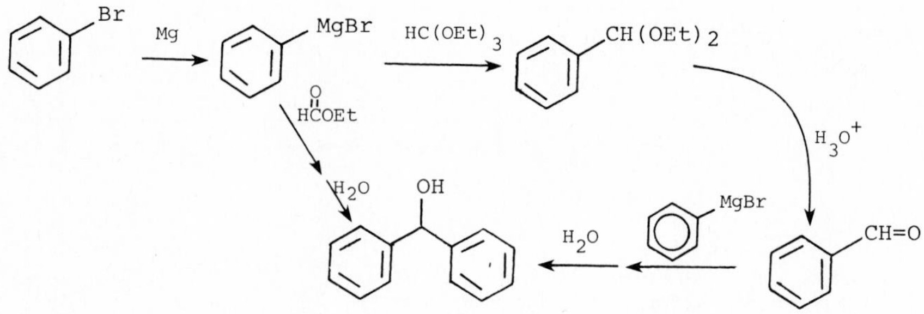

(d)

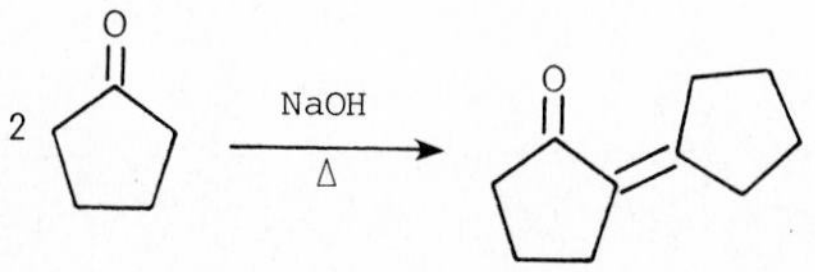

$$2 \quad \xrightarrow[\Delta]{NaOH} \quad \text{(aldol condensation)}$$

(e) $(CH_3)_2CHCH=CH_2 \xrightarrow[\text{peroxides}]{HBr} \xrightarrow{Mg} (CH_3)_2CHCH_2CH_2MgBr \quad + \quad HCCHCH(CH_3)_2 \xrightarrow[\text{2.Zn}]{1.\ O_3}$

$CH_2=CHCH(CH_3)_2$

$(CH_3)_2CH(CH_2)_3CH(CH_3)_2 \xleftarrow[Pt]{H_2} \xleftarrow[\Delta]{H_2SO_4} (CH_3)_2CHCH_2CH_2\overset{OH}{\underset{|}{CH}}CH(CH_3)_2 \xleftarrow{H_2O}$

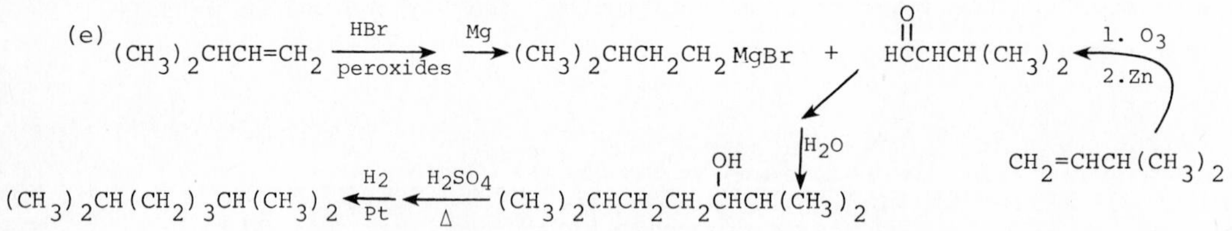

(f) $(CH_3)_2CHCH=CH_2 \xrightarrow[\text{peroxides}]{HBr} (CH_3)_2CHCH_2CH_2Br \xrightarrow{Na} (CH_3)_2CH(CH_2)_4CH(CH_3)_3$

(g)

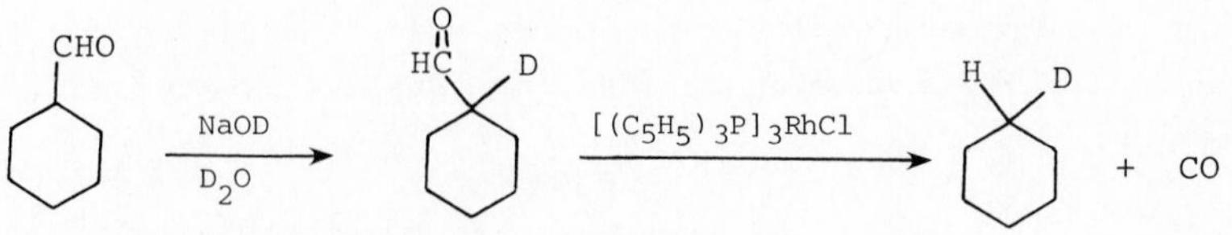

(h) $CH_3CH_2C\equiv CH$ $\xrightarrow[H_2SO_4]{Hg^{++}}$ $CH_3CH_2\overset{\overset{O}{\|}}{C}CH_3$

S2. (a) Grignard reagent will react with ethanol solvent instead of acetone.

(b) Grignard reagent cannot be formed in presence of hydroxyl group:

$$CH_3CH_2\overset{\overset{CH_3}{|}}{\underset{\underset{OH}{|}}{C}}-CH_2-MgBr \quad\longrightarrow\quad CH_3CH_2\overset{\overset{CH_3}{|}}{\underset{\underset{OMgBr}{|}}{C}}CH_3$$

(c) Ethyllithium will react with carboxylic acid instead:

$$RLi + R'COOH \longrightarrow RH + R'COOLi$$

(d) Grignard reagent will undergo elimination as it is formed:

$$"(CH_3)_2\overset{}{\underset{\underset{OCH_3}{|}}{C}}-CH_2-MgBr" \longrightarrow (CH_3)_2C=CH_2 + CH_3OMgBr$$

(e) Grignard reagent will react with itself:

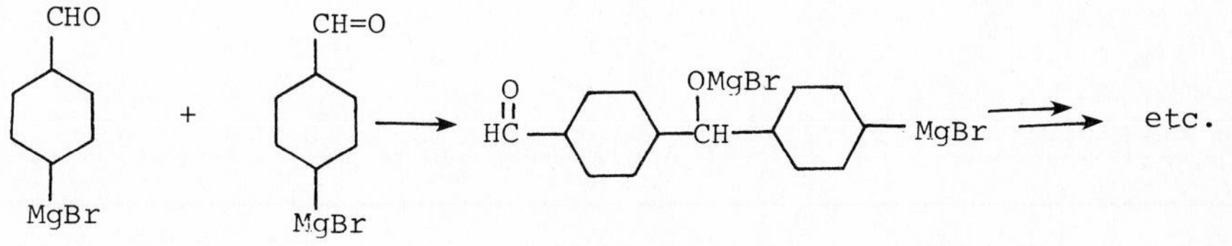

S3. (a) $CH_3CH_2MgBr + CH_3CH_2\overset{\overset{O}{\|}}{C}CH_3 \longrightarrow \xrightarrow{H_2O} (CH_3CH_2)_2\overset{\overset{OH}{|}}{C}CH_3$

$HC\equiv CNa + CH_3CH_2\overset{\overset{O}{\|}}{C}CH_3 \longrightarrow \xrightarrow{H_2O} CH_3CH_2\overset{\overset{OH}{|}}{\underset{\underset{CH_3}{|}}{C}}C\equiv CH \xrightarrow[Pt]{H_2} (CH_3CH_2)_2\overset{\overset{OH}{|}}{C}CH_3$

(b) $CH_3CH_2\overset{\overset{MgBr}{|}}{C}HCH_3 + CH_2=O \longrightarrow \xrightarrow{H_2O} CH_3CH_2\overset{\overset{CH_3}{|}}{C}HCH_2OH \xrightarrow[\Delta]{Al_2O_3} CH_3CH_2\overset{\overset{CH_3}{|}}{C}=CH_2$

(Elimination of tertiary alcohol isomer would give $CH_3CH=C(CH_3)_2$ instead)

$CH_3CH_2\overset{\overset{O}{\|}}{C}CH_3 + (C_6H_5)_3P=CH_2 \longrightarrow CH_3CH_2\overset{\overset{CH_3}{|}}{C}=CH_2 + (C_6H_5)_3P=O$

(c) $CH_3CH_2CH_2CH_2MgBr + CH_2\overset{O}{\overset{/\backslash}{-}}CH_2 \longrightarrow \xrightarrow{H_2O} CH_3(CH_2)_5OH \xleftarrow{NaBH_4}$

$CH_3CH_2CH_2CH_2Br + NaC\equiv CH \longrightarrow CH_3CH_2CH_2CH_2C\equiv CH \xrightarrow[2.\ H_2O_2]{1.\ B_2H_6} CH_3(CH_2)_4\overset{\overset{O}{\|}}{C}H$

(d) $CH_3CH_2CH_2CH_2MgBr + HCCH_3 \longrightarrow \xrightarrow[]{H_2O} \xrightarrow[H_2SO_4]{K_2Cr_2O_7} CH_3(CH_2)_3\overset{\overset{O}{\|}}{C}CH_3$

$CH_3CH_2CH_2CH_2C\equiv CH$ *(from (c))* $\xrightarrow[H_2SO_4]{Hg^{++}}$

(e) $CH_3CH_2MgBr + H\overset{\overset{O}{\|}}{C}CH_2CH_2CH_3 \longrightarrow \xrightarrow{H_2O} \xrightarrow{PBr_3} CH_3CH_2\overset{\overset{Br}{|}}{C}HCH_2CH_2CH_3$

$CH_3CH_2C\equiv CNa + BrCH_2CH_3 \longrightarrow CH_3CH_2C\equiv CCH_2CH_3 \xrightarrow[NH_3]{Na} \xleftarrow{HBr}$

(f) 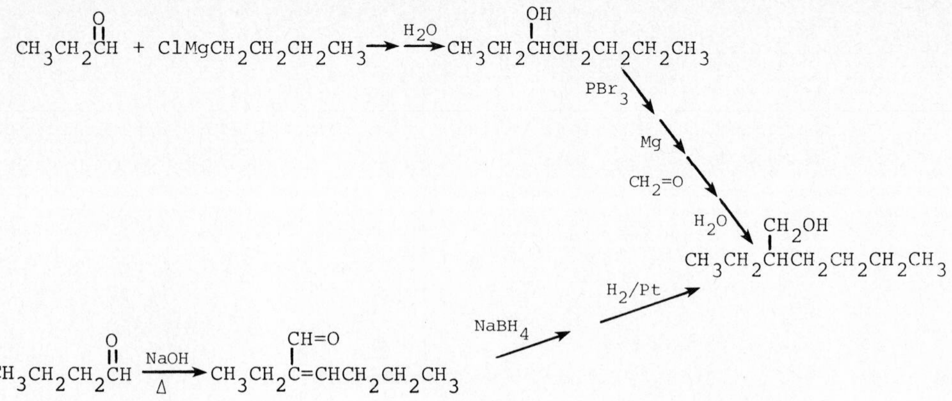

S4. $[Fe(CO)_4]^{=}$ Fe: 8; 4 CO's: 8; $^{=}$charge: 2: Σ =18

↓ RX — Lewis acid association

$[RFe(CO)_4]^-$ Fe: 8; share with R: -1; 4 CO's: 8; R; 2; $^-$charge: 1: Σ =18

⇅ insertion

$[R\text{-}\overset{O}{\overset{\|}{C}}\text{-}Fe(CO)_3]^-$ Fe: 8; share with $R\overset{O}{\overset{\|}{C}}$: -1; 3 CO's: 6; $R\overset{O}{\overset{\|}{C}}$: 2; $^-$charge: 1: Σ =16

↓ $(C_6H_5)_3P$ — Lewis base association

$[R\text{-}\overset{O}{\overset{\|}{C}}\text{-}Fe(CO)_3(\leftarrow L)]^-$ Fe: 8; share with $R\overset{O}{\overset{\|}{C}}$: -1; 3 CO's: 6; L : 2; $R\overset{O}{\overset{\|}{C}}$: 2; $^-$charge: 1: Σ =18

↓ R'X — Lewis acid association

$R\text{-}\overset{O}{\overset{\|}{C}}\text{-}Fe(R')(CO)_6(\leftarrow L)$ Fe: 8; share with $R\overset{O}{\overset{\|}{C}}$,R': -2; 3 CO's: 6; L : 2; $R\overset{O}{\overset{\|}{C}}$,R': 4: Σ =18

↓ reductive elimination

$R\overset{O}{\overset{\|}{C}}R'$ + $L\,FeCO_3$ Fe: 8; 3 CO's: 6; L: 2: Σ =16

16. ORGANIC SYNTHESIS

16.A Chapter Outline and Important Terms Introduced

16.1 Introduction

16.2 Considerations in Synthesis Design

1. Construction of the proper carbon skeleton
2. Placement of desired functional groups in their proper place
3. Control of stereochemistry where relevant

stereospecific

16.3 Planning a Synthesis

16.4 Protecting Groups

16.5 Industrial Syntheses

process development

16.B Important Concepts and Hints

To many organic chemists, the best examples of the elegance and creativity of chemistry are found in synthesis. It is in designing a synthesis of a complex molecule, and carrying it out in the laboratory, that many chemists find their greatest enjoyment. Although you are introduced to this topic with relatively simple molecules as targets, the field has advanced to the extent that we can confidently design syntheses of molecules of bewildering complexity. The successful synthesis of Vitamin B_{12} by the late R. B. Woodward, A. Eschenmoser, and their research groups is a spectacular example of what the chemist can now accomplish.

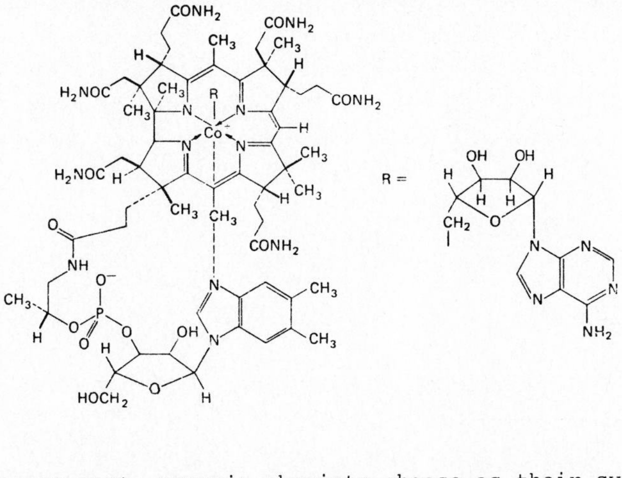

For the most part, organic chemists choose as their synthetic targets compounds isolated from natural sources, or compounds similar to them. Such syntheses are pursued in order to produce compounds of potentially valuable biological activity, to assist in structure determination, to demonstrate the usefulness of a new reaction sequence, or, one suspects in some cases, for the fun of it. The development of efficient syntheses of less complex, commercially important compounds is an equally challenging aspect of organic synthesis. In this regard, elegance and creativity are measured by the simplicity and economy of the processes, their energy efficiency, the absence of byproducts, etc.

Regardless of the type of synthesis you are trying to develop, you must be able to think logically and you must know a broad range of organic reactions. With these two abilities, and practice, you will develop an intuition for organic synthesis and be able to appreciate the creative possibilities. Chapter 16 outlines the logical processes involved in synthesis design; the rest of the text provides you with the background knowledge of reactions (your repertoire). It's up to you to practice synthesis problems and develop the necessary intuition.

The suggestion that you should work synthesis problems backwards was made in Section 12.B. of this Study Guide, and is discussed in more detail in Chapter 16 of the text. An example of the sort of thing that happens when you work a synthesis problem *forwards* is the following scheme, taken from an actual answer to a midterm exam question:

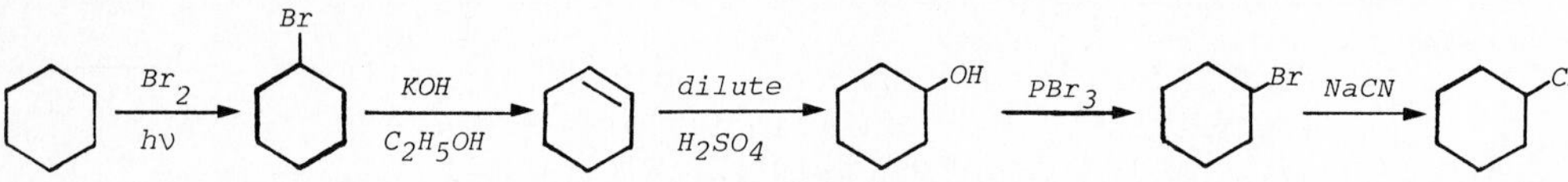

Because the student was thinking "What can I make from this compound?" at each stage, he wasted three steps converting bromocyclohexane into bromocyclohexane!

A more subtle mistake that we all make arises from the usual human tendency to stop looking for something once we think we've found it. Often you will think of a way to carry out a transformation, without evaluating it closely to see whether there are any potential side reactions. Examples of reactions which appear at first glance to accomplish their intended conversion are shown below. Also listed are the side products expected or the incompatibility of functional groups which make these transformations poor choices. You should always look for such problems, and, if you find them, continue to search for another route.

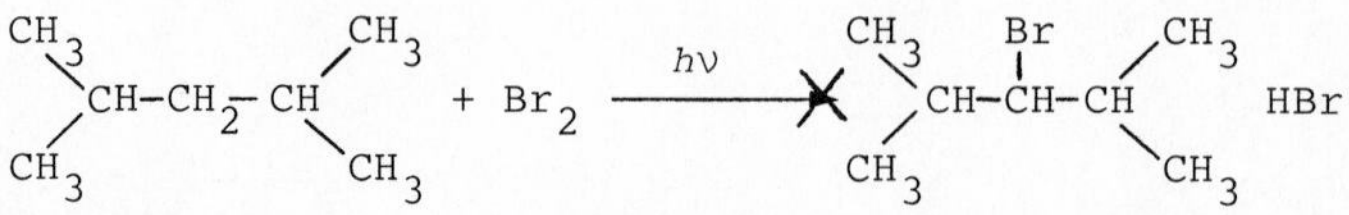

(Bromination is selective for 3° position)

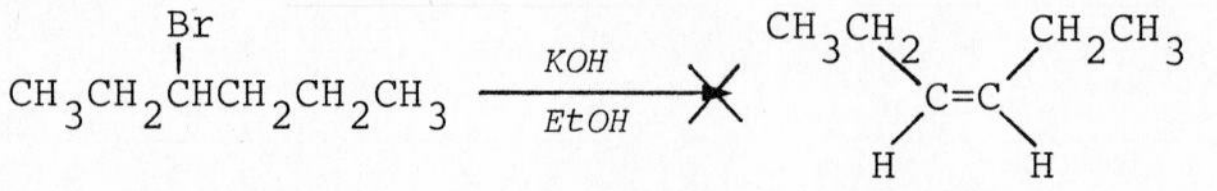

(Three other isomers will be produced, too)

$$CH_3CH_2C{\equiv}CNa + BrCH_2CH_2CH_2OH \not\longrightarrow CH_3CH_2C{\equiv}CCH_2CH_2CH_2OH$$

(Acetylide ion will be protonated by hydroxy proton)

$$2 \ \underset{\text{O}}{\overset{\parallel}{\text{HCCH}_2\text{CH}_2\text{Br}}} \quad \xrightarrow{\text{NaOH}} \quad \cancel{\quad} \ \underset{\text{CH}_2\text{Br}}{\overset{\overset{\text{O} \quad \text{OH}}{\parallel \quad |}}{\text{HCCHCHCH}_2\text{CH}_2\text{Br}}}$$

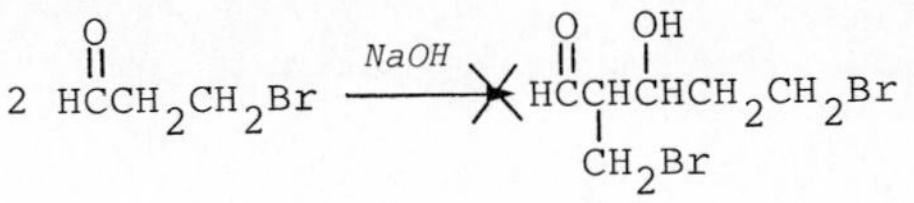

(Bromide will undergo substitution or elimination under the basic conditions)

$$\underset{\text{O}}{\overset{\parallel}{\text{CH}_3\text{CCH}_2\text{CH}_2\text{CH=CH}_2}} \quad \xrightarrow[\text{HBr}]{\text{Br}_2} \quad \cancel{\quad} \ \underset{\text{Br}}{\overset{\overset{\text{O}}{\parallel}}{\text{CH}_3\text{CCHCH}_2\text{CH=CH}_2}}$$

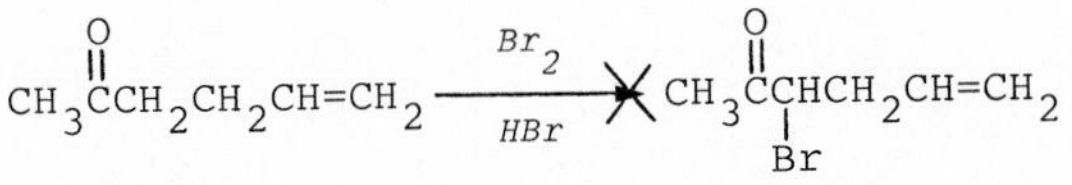

(Double bond will undergo bromine addition)

$$\underset{\text{Br}}{\overset{|}{\text{CH}_2\text{CHCH}_2\text{CH}_2\text{CH}_2\text{Br}}} \quad \xrightarrow{\text{Mg}} \quad \cancel{\quad} \ \underset{\text{Br}}{\overset{|}{\text{CH}_3\text{CHCH}_2\text{CH}_2\text{CH}_2\text{MgBr}}}$$

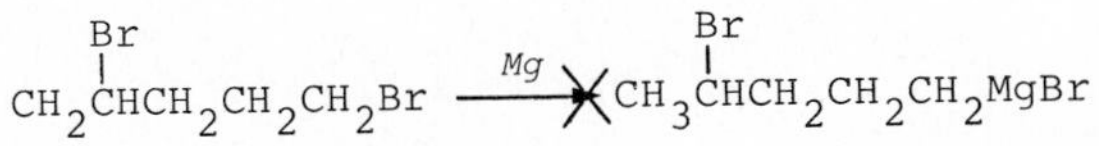

(The other bromide will react, too)

These are only examples to show you the sort of difficulties to look out for.

16.C Answers to Exercises

(16.2) $\text{RY} \xrightarrow{\quad ? \quad} \text{RZ}$

RY \ RZ	RH	RBr	ROH	RCH₂OH	RCHO	RCO₂H	RCN	RCOCH₃
RH	╳	−*	−	−	−	−	−	−
RBr	1	╳	1	1	1	1	1	+
ROH	+	1	╳	+	+	+	+	+
RCH₂OH	+	(−)	(−)	╳	1	1	(−)	+
RCHO	1	(−)	(−)	1	╳	1	(−)	+
RCO₂H	+	(−)	(−)	+	+	╳	(−)	(−)
RCN	+	(−)	(−)	(−)	1	1	╳	(−)
RCOCH₃	+	(−)	+	(−)	(−)	1	−	╳

NOTE: This matrix was constructed based on reactions presented in Chapters 1-16; ( ) indicates entries which would be changed based on reactions introduced in subsequent chapters.

*Although some RH can be converted to RBr by free radical bromination, the reaction is not general because some RH do not react and others give impractical mixtures.

16.D Answers and Explanations to Problems

NOTE: For most synthesis problems, expecially the more complex ones, there is more than one "correct" answer. Two or three routes can be devised which will lead to the target compound. Often one route will stand out above the others because it is shorter, or avoids low yield reactions or isomeric products. Always try to find such a route, because that one is "more correct" as an answer. In this Study Guide, we can't list every conceivable way to make the target compound, and your answer may differ from the one here. Look at our explanations carefully, and learn to evaluate possible syntheses by comparison of yours with ours.

1. (a) $CH_3(CH_2)_5OH$ ← $\xleftarrow{H_2O}$ $CH_3(CH_2)_5OMgBr$ ← $\xleftarrow{CH_2=O}$ $CH_3(CH_2)_4MgBr$ ← $\xleftarrow[ether]{Mg}$ $CH_3(CH_2)_4Br$

$$CH_3(CH_2)_3MgBr \quad \xrightarrow{\triangle}$$

(b) $CH_3(CH_2)_3\overset{\displaystyle OH}{\underset{\displaystyle |}{C}}HCH_3$ ← $\xleftarrow{H_2O}$ $CH_3(CH_2)_3\overset{\displaystyle OMgX}{\underset{\displaystyle |}{C}}HCH_3$

$CH_3(CH_2)_3\overset{O}{\overset{||}{C}}H + CH_3MgI$

$\underline{or}$

$CH_3(CH_2)_3MgBr + H\overset{O}{\overset{||}{C}}CH_3$

(c) $CH_3CH_2CH_2\overset{\displaystyle OH}{\underset{\displaystyle |}{C}}HCH_2CH_3$ ← $\xleftarrow{H_2O}$ $CH_3CH_2CH_2\overset{\displaystyle OMgBr}{\underset{\displaystyle |}{C}}HCH_2CH_3$

$CH_3CH_2CH_2\overset{O}{\overset{||}{C}}H + BrMgCH_2CH_3$

$\underline{or}$

$CH_3CH_2CH_2MgBr + H\overset{O}{\overset{||}{C}}CH_2CH_3$

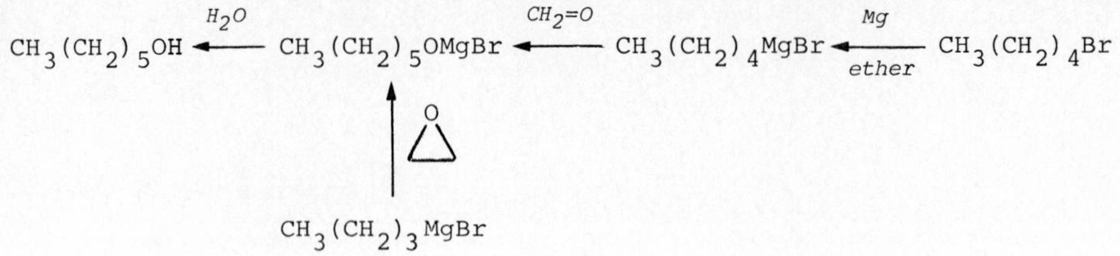

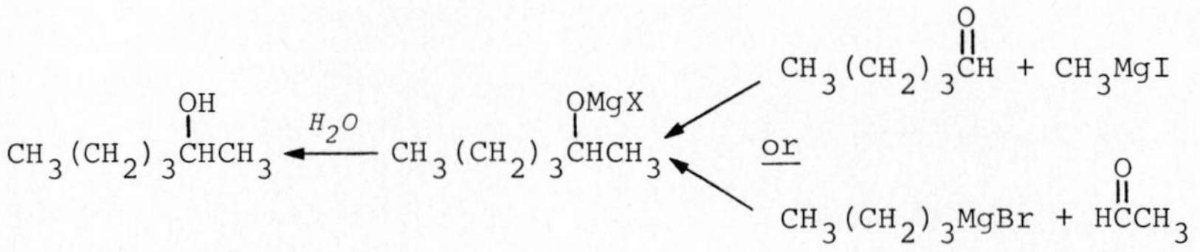

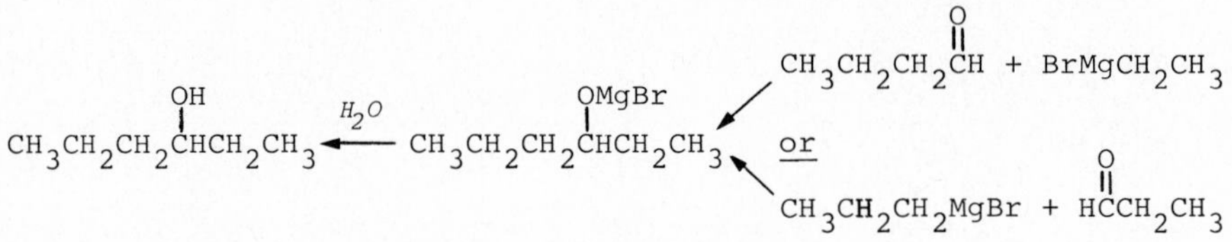

(d) $CH_3(CH_2)_4C{\equiv}CH$ ← $CH_3(CH_2)_4Br + NaC{\equiv}CH$

(e)

$$\underset{\substack{CH_3CH_2 \qquad H \\ \diagdown \qquad \diagup \\ C=C \\ \diagup \qquad \diagdown \\ H \qquad CH_2CH_2CH_3}}{}$$

← $\xleftarrow[NH_3]{Na}$ $CH_3CH_2C{\equiv}CCH_2CH_2CH_3$

$\xleftarrow[]{BrCH_2CH_3 \quad NaNH_2}$ $HC{\equiv}CCH_2CH_2CH_3$

$\underline{or}$

$\xleftarrow[]{BrCH_2CH_2CH_3 \quad NaNH_2}$ $CH_3CH_2C{\equiv}CH$

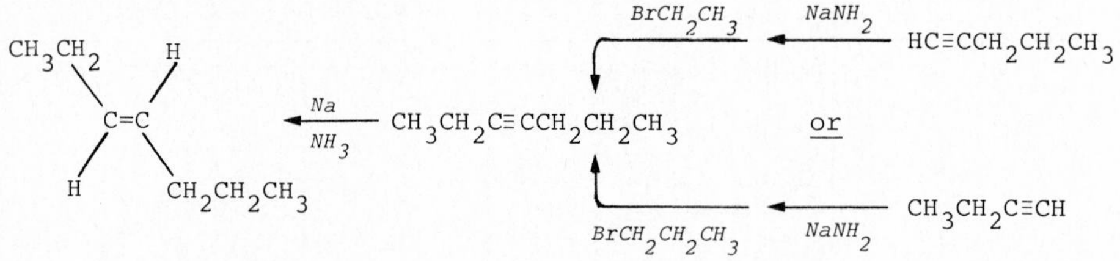

(Any route involving elimination of an alcohol or a halide would give some cis-isomer, too.)

(f) $(CH_3CH_2CH_2CH_2)_2\overset{\displaystyle OH}{\underset{\displaystyle |}{C}}CH_3$ ← $\xleftarrow{H_2O}$ $(CH_3CH_2CH_2CH_2)_2\overset{\displaystyle OMgX}{\underset{\displaystyle |}{C}}CH_3$

$\xrightarrow{CH_3MgI}$ $(CH_3CH_2CH_2CH_2)_2C=O$

$\xleftarrow[]{CH_3CH_2CH_2CH_2MgBr}$ $CH_3CH_2CH_2CH_2\overset{O}{\overset{||}{C}}CH_3$

$\uparrow K_2Cr_2O_7, H_2SO_4$

$(CH_3CH_2CH_2CH_2)_2CHOH$

$\uparrow K_2Cr_2O_7, H_2SO_4$

$CH_3CH_2CH_2CH_2\overset{\displaystyle OH}{\underset{\displaystyle |}{C}}HCH_3$

$\xleftarrow[]{H_2O}$ $\uparrow H_2O$

$\xleftarrow[CH_3CH_2CH_2CH_2MgBr]{H_2O}$ $\xrightarrow[CH_3MgI]{H_2O}$ $\uparrow CH_3\overset{O}{\overset{||}{C}}H$

$CH_3CH_2CH_2CH_2\overset{O}{\overset{||}{C}}H$

$\uparrow Mg$

$CH_3CH_2CH_2CH_2Br$

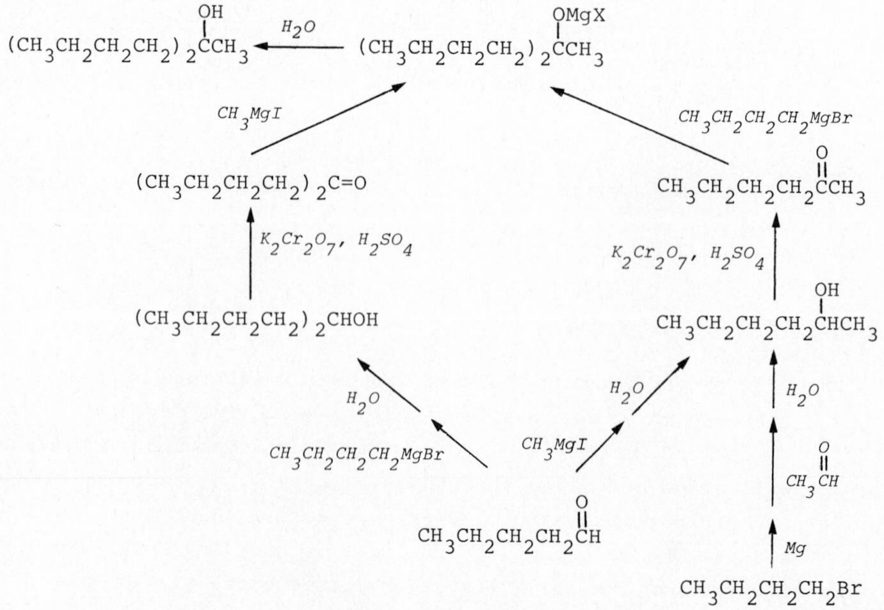

(Note that in many instances a number of similar, equally valid routes are available. In Chapter 19 (Section 19.7.D.) you will find the following reaction, which would be the best way to make 5-methyl-5-nonanol:

$$2 \text{ RMgX} + \text{R'}\overset{\overset{\text{O}}{\|}}{\text{C}}\text{OR''} \longrightarrow \overset{H_2O}{\longrightarrow} \text{R}-\overset{\overset{\text{OH}}{|}}{\underset{\underset{\text{R'}}{|}}{\text{C}}}-\text{R} \quad)$$

(g) $(CH_3)_2\overset{\overset{\text{OH}}{|}}{C}CH_2CH_2CH_2CH(CH_3)_2$

1st decision: which C-C bond to form

$(CH_3)_2\overset{\overset{\text{O}}{\|}}{C}CH_2----CH_2CH_2CH(CH_3)_2$ $\quad$ <u>or</u> $\quad$ $(CH_3)_2\overset{\overset{\text{O}}{\|}}{C}CH_2CH_2----CH_2CH(CH_3)_2$

$\underbrace{\qquad}$ $\underbrace{\qquad\qquad}$ $\qquad$ $\underbrace{\qquad\qquad}$ $\underbrace{\qquad}$

4 carbons$\qquad$*5 carbons*$\qquad\qquad$*5 carbons*$\qquad$*4 carbons*

The first choice is the best, because the functional group in the product is closest to the bond to be formed.

2nd decision: what C-C bond-forming reaction to choose

$(CH_3)_2CH\overset{\overset{\text{O}}{\|}}{C}H + MCH_2CH_2CH(CH_3)_2 \longrightarrow \overset{H_2O}{\longrightarrow} (CH_3)_2CH\overset{\overset{\text{OH}}{|}}{C}HCH_2CH_2CH(CH_3)_2$

This reaction would put the hydroxy group in the wrong place. You could correct this, but the sequence would involve several steps and would therefore be inefficient:

$(CH_3)_2CH\overset{\overset{\text{OH}}{|}}{C}HCH_2CH_2CH(CH_3)_2 \xrightarrow{H_2SO_4} (CH_3)_2CHCH{=}CHCH_2CH(CH_3)_2 + (CH_3)_2C{=}CHCH_2CH_2CH(CH_3)_2$

(minor) $\qquad\qquad\qquad\qquad\qquad$ *(major)*

$\qquad\qquad\qquad\qquad\qquad\qquad\qquad\qquad\qquad\qquad$ dilute H_2SO_4 ↓

$\qquad\qquad\qquad\qquad\qquad\qquad\qquad\qquad\qquad (CH_3)_2\overset{\overset{\text{OH}}{|}}{C}(CH_2)_3CH(CH_3)_2$

Alternatively:

$\underset{CH_3}{\overset{CH_3}{>}}C\overset{\overset{\text{O}}{\triangle}}{-}CH_2 + MCH_2CH_2CH(CH_3)_2 \longrightarrow \overset{H_2O}{\longrightarrow} (CH_3)_2\overset{\overset{\text{OH}}{|}}{C}(CH_2)_3CH(CH_3)_2$

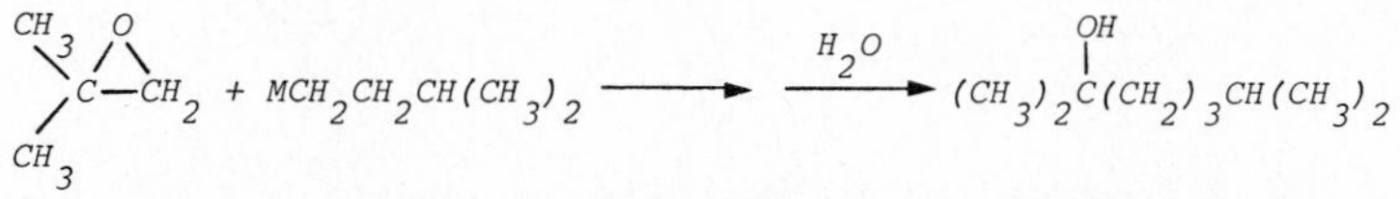

This reaction gives the desired products directly.

Final decision: what M should be: MgX or Li

With an epoxide such as this one, Grignard reagents often induce rearrangement prior to addition (see Section 15.5.D.). The organolithium reaction is therefore the better choice.

$(CH_3)_2\overset{\overset{\text{O}}{\triangle}}{C}-CH_2$

RMgBr ↗ $\left[(CH_3)_2CH\overset{\overset{\text{O}}{\|}}{C}H \right] \longrightarrow \overset{H_2O}{\longrightarrow} (CH_3)_2CH\overset{\overset{\text{OH}}{|}}{C}HCH_2CH_2CH(CH_3)_2$

RLi ↘ $(CH_3)_2\overset{\overset{\text{OLi}}{|}}{C}CH_2CH_2CH_2(CH_3)_2 \xrightarrow{H_2O} (CH_3)_2\overset{\overset{\text{OH}}{|}}{C}(CH_2)_3CH(CH_3)_2$

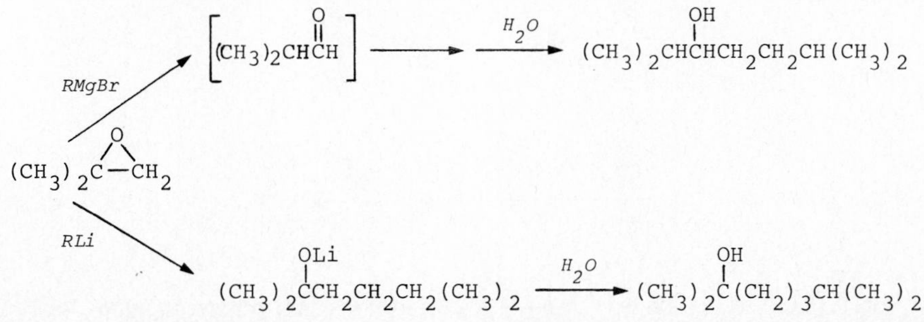

Many other routes can be imagined, such as the multistep sequence below, but the extra steps required make them very inefficient, and therefore poor solutions to the problem.

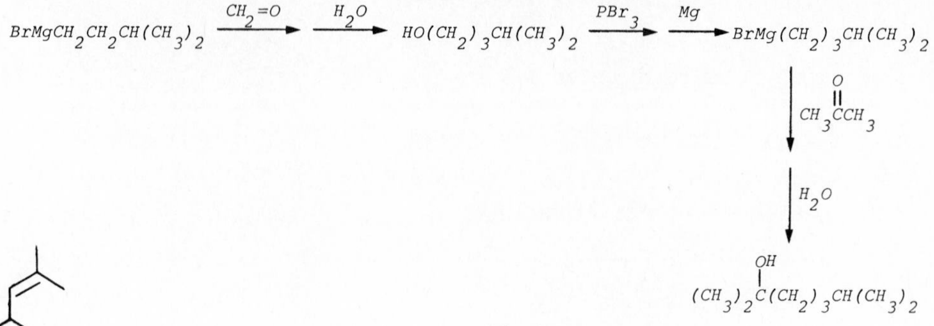

(h) 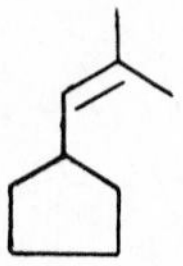

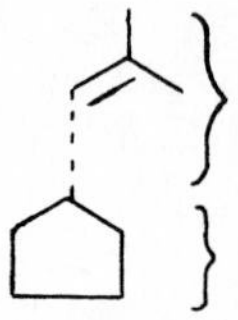

4 carbons

1st decision: which C-C bond to form:

5 carbons

This would be efficient from the point of view of requiring only one C-C bond-forming step. The most obvious way to do this would involve a Grignard reaction and dehydration sequence:

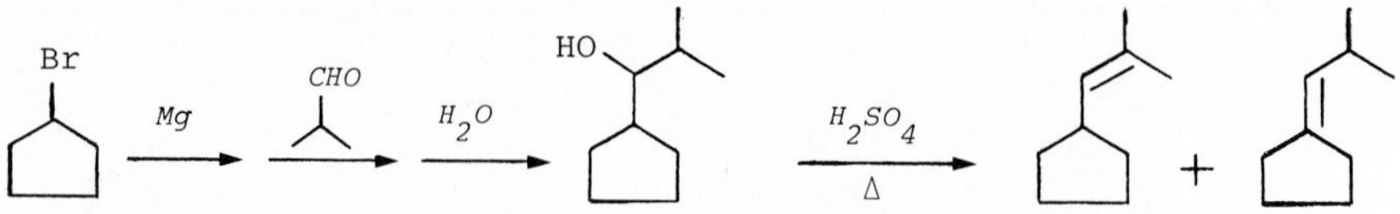

However, this route leads to two double bond isomers (in comparable amounts) which would be hard to separate.

The best general method for introducing a double bond in a specific position is the Wittig reaction. Using this reaction, the last step in the synthesis would be:

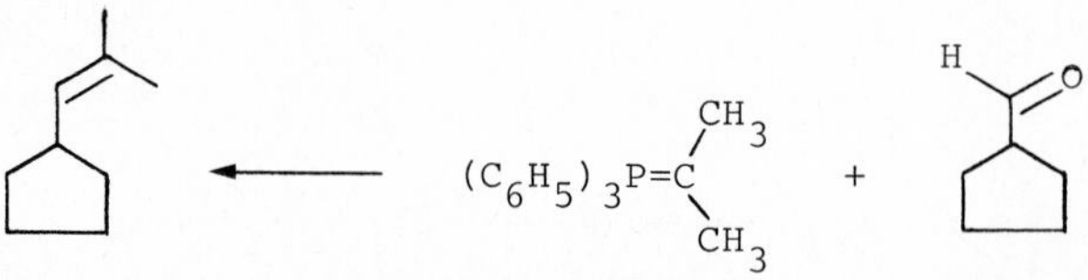

The aldehyde would be synthesized as follows:

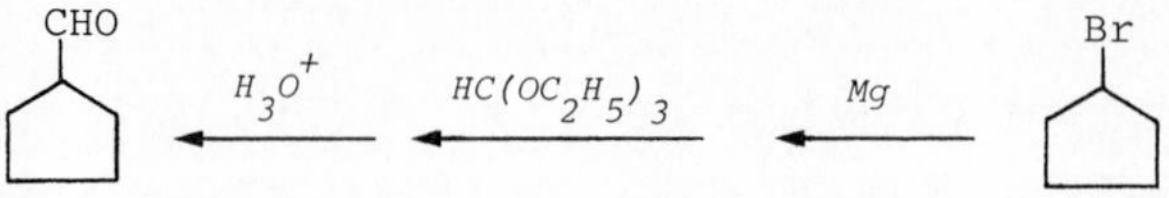

2. (a) 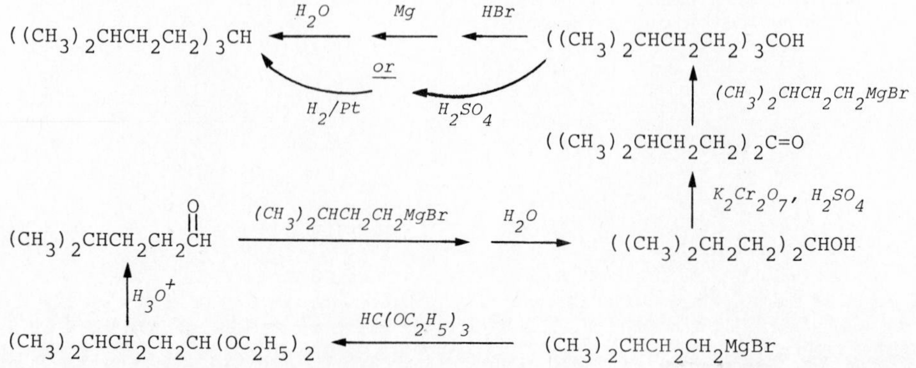

(In Chapter 19 (Section 19.7.D.) you will learn an easier way to do this:

$$3\ RMgBr + R'O\overset{O}{\overset{\|}{C}}OR' \longrightarrow \xrightarrow{H_2O} R_3COH\ )$$

(b)

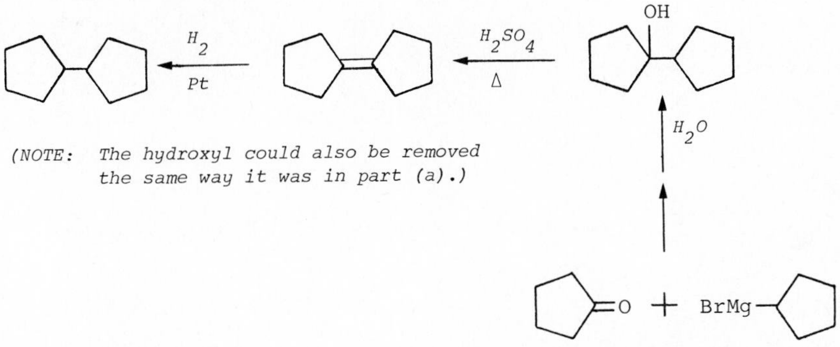

(NOTE: The hydroxyl could also be removed the same way it was in part (a).)

(c) $(CH_3)_2CH(CH_2)_6CH(CH_3)_2$

1st decision:
which C-C bonds to form: $(CH_3)_2CHCH_2CH_2$---CH_2—CH_2---$CH_2CH_2CH(CH_3)_2$

5 carbons   2 carbons   5 carbons

2nd decision: what reaction to use (take the symmetry of the molecule as a hint)

$$(CH_3)_2CHCH_2CH_2Br + NaC{\equiv}CH \longrightarrow (CH_3)_2CHCH_2CH_2C{\equiv}CH \xrightarrow{NaNH_2} \xrightarrow{BrCH_2CH_2CH(CH_3)_2}$$

$$(CH_3)_2CH(CH_2)_6CH(CH_3)_2 \xleftarrow[Pt]{H_2} (CH_3)_2CHCH_2CH_2C{\equiv}CCH_2CH_2CH(CH_3)_2$$

(d) *Don't forget the aldol condensation!*

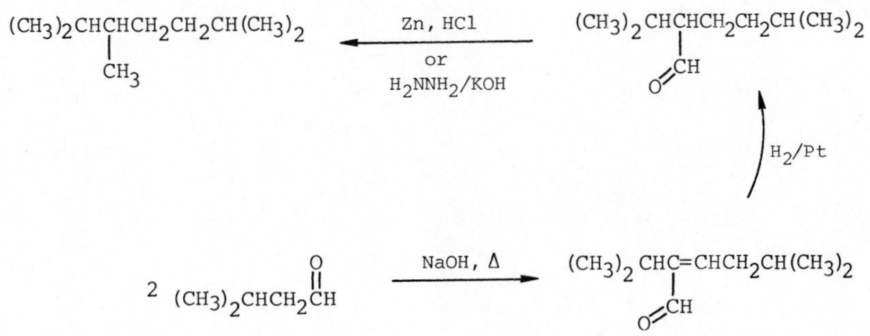

(e)

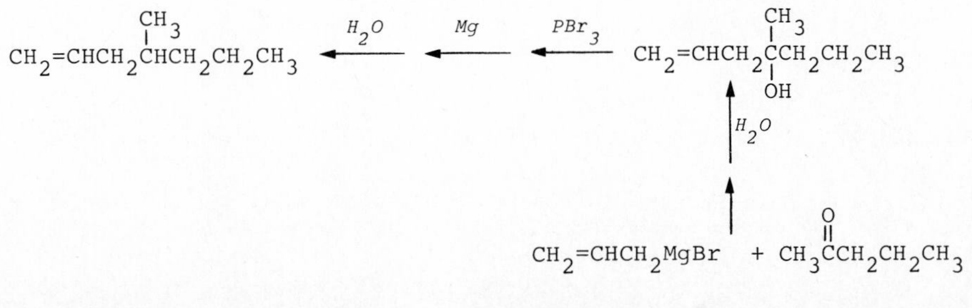

3. (a)

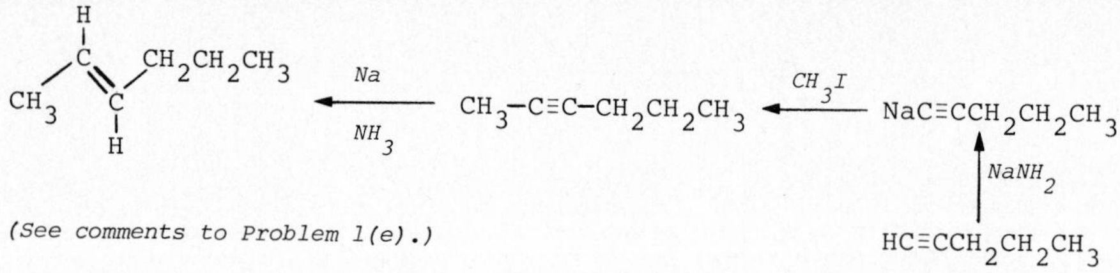

$$\underset{CH_3}{\overset{H}{\underset{|}{C}}}=\underset{H}{\overset{CH_2CH_2CH_3}{\underset{|}{C}}} \quad \xleftarrow[NH_3]{Na} \quad CH_3-C\equiv C-CH_2CH_2CH_3 \quad \xleftarrow{CH_3I} \quad NaC\equiv CCH_2CH_2CH_3$$

$$\uparrow NaNH_2$$

$$HC\equiv CCH_2CH_2CH_3$$

(See comments to Problem 1(e).)

(b)

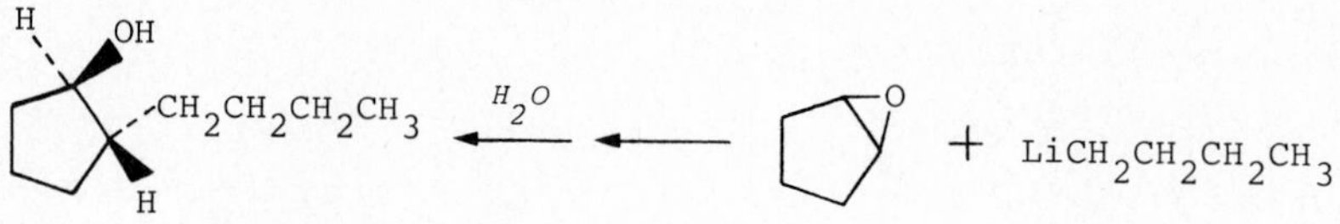

$$\xleftarrow{H_2O} \quad \xleftarrow{} \quad + \quad LiCH_2CH_2CH_2CH_3$$

(via:

$$+ \, Li \quad {}^-R$$

(c)

$$\underset{H}{\overset{CH_3CH_2CH_2}{\underset{}{}}}C=C\underset{H}{\overset{CH_2CH_2CH_3}{}} \quad \xleftarrow[\substack{Lindlar's \\ catalyst}]{H_2} \quad CH_3CH_2CH_2C\equiv CCH_2CH_2CH_3$$

$$\uparrow$$

$$CH_3CH_2CH_2C\equiv CNa \; + \; BrCH_2CH_2CH_3$$

(See comments to Problem 1(e).)

(d)

$$\underset{H\;\;OH}{\overset{OH\;\;H}{\underset{|\;\;\;|}{CH_3CH_2-C-C-CH_2CH_3}}} \quad \xleftarrow[H_2O]{OsO_4} \quad \underset{H}{\overset{H}{\underset{}{}}}\overset{CH_3CH_2}{C=C}\overset{CH_2CH_3}{}$$

$$\uparrow H_3O^+ \qquad\qquad \downarrow Na/NH_3$$

$$CH_3CH_2C\equiv CCH_2CH_3 \quad \xleftarrow{\substack{CH_3CH_2Br \\ + \\ CH_3CH_2C\equiv CNa}}$$

$$\underset{H\;\;O\;\;H}{\overset{CH_3CH_2}{\underset{}{C-C}}}\overset{CH_2CH_3}{} \quad \xleftarrow{CH_3CO_3H} \quad \underset{H}{\overset{CH_3CH_2}{}}C=C\underset{H}{\overset{CH_2CH_3}{}} \quad \xleftarrow[\substack{Lindlar's \\ catalyst}]{H_2}$$

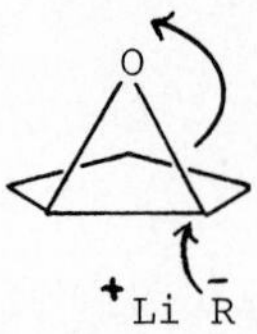

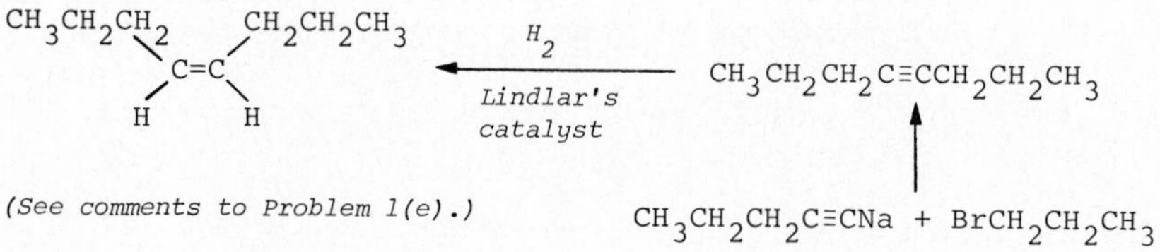

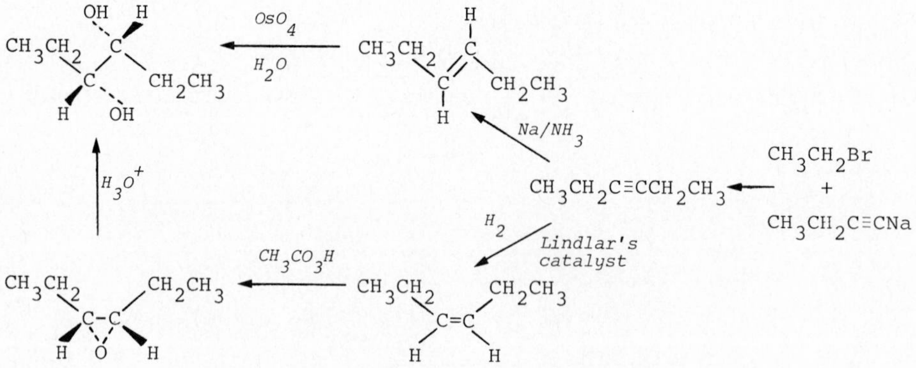

(This is good to work through with models.)

(e)

$$\underset{CH_3\;\;H\;\;CH_3}{\overset{H\;\;OH}{CH_2CH_2CH_2CH_3}} \quad \xleftarrow{H_2O} \quad \underset{CH_3\quad CH_3}{\overset{H\quad O}{}} {}^-H \; + \; LiCH_2CH_2CH_2CH_3$$

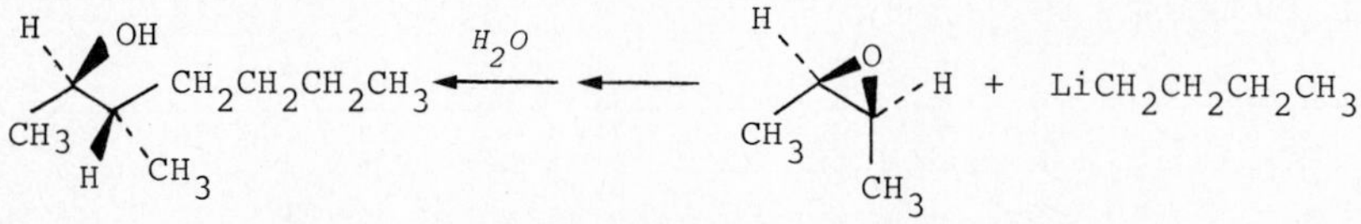

(The trans epoxide would give the (2RS, 3RS) diastereomer.)

(f)

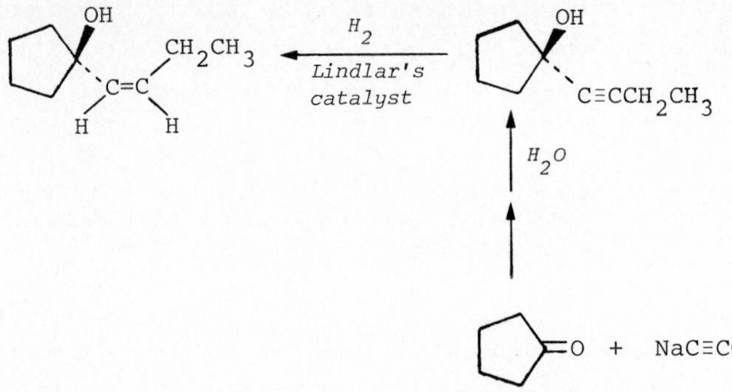

(g)

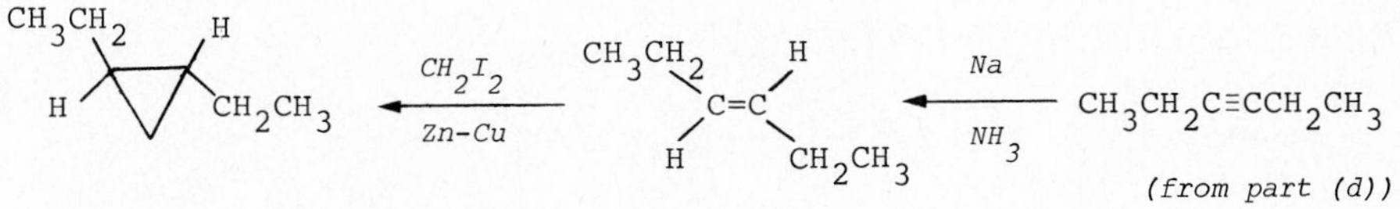

(from part (d))

4. (a)

(S)   (S)   (S)

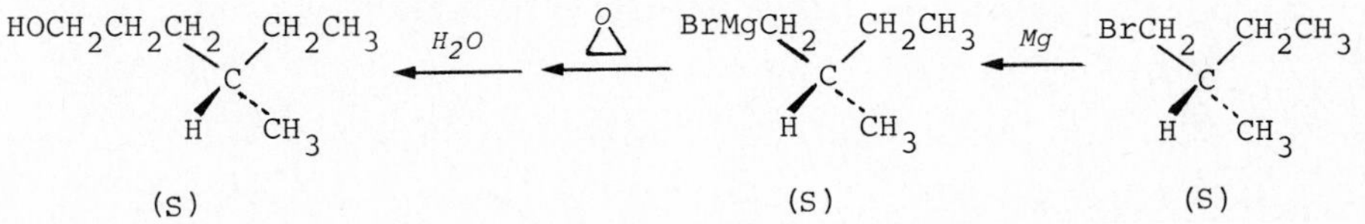

(b)

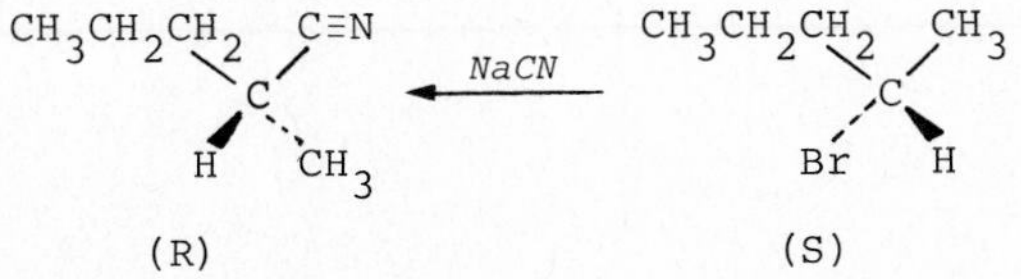

(S_N2 reaction, with inversion)

(R)   (S)

(c)

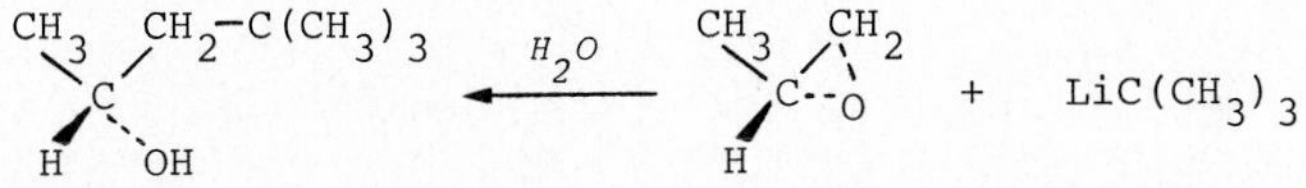

5. (a)

Protect aldehyde to keep it from being oxidized.

(b)

Br /\/\/ CHO $\xrightarrow[H^+]{HOCH_2CH_2OH}$ $\xrightarrow{t-BuO^-K^+}$ $\xrightarrow{H_3O^+}$ /\/\ CHO

Protect aldehyde to prevent condensation.

(c)

$$HOCH_2CH_2CH_2CH_2CH_2\overset{\overset{\textstyle OH}{|}}{C}HCH_3$$

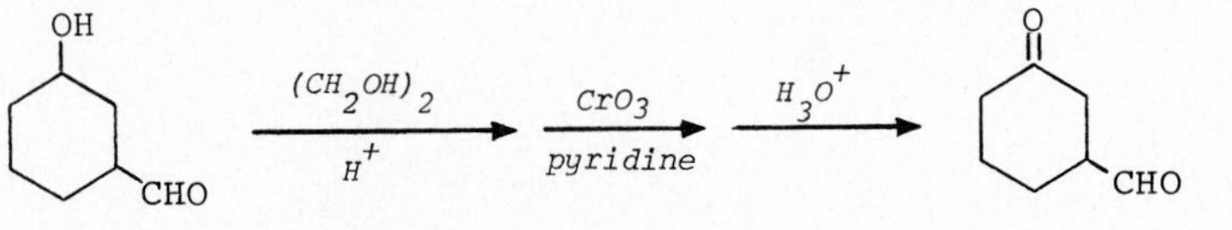

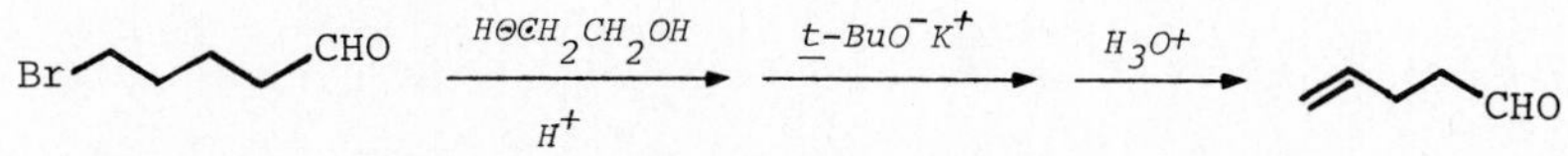

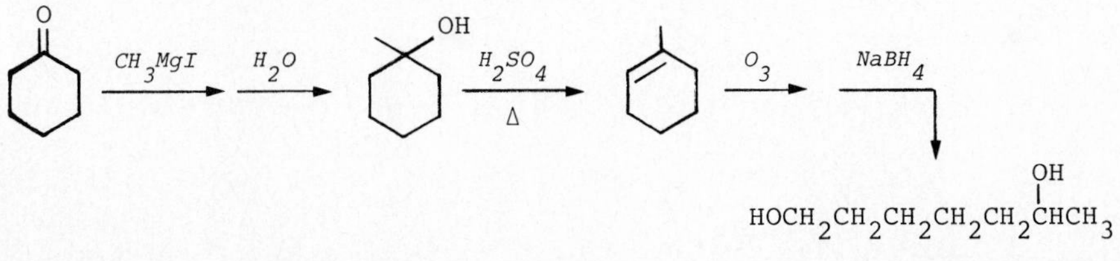

6. (a) Since the problem asks us to synthesize the *cis* isomer of the product, we should avoid sequences which will give a mixture of *cis* and *trans* isomers. One way to prepare a *cis* alkene stereospecifically is by hydrogenation of an alkyne:

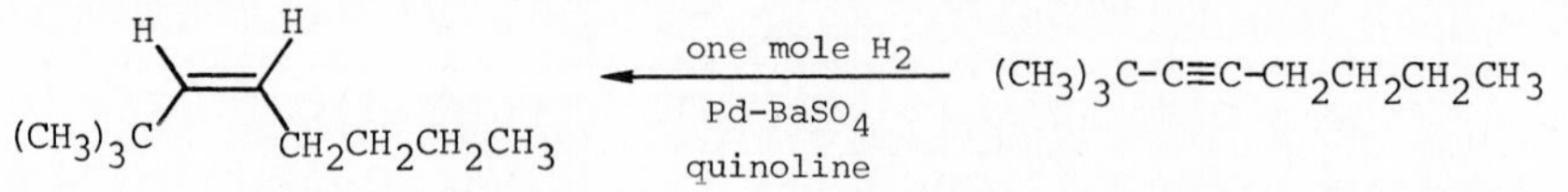

The necessary alkyne has ten carbons and must be built up from smaller fragments. However, one of the alkyl groups attached to the triple bond is tertiary, and cannot be introduced by alkylation; elimination would be the exclusive reaction:

$$(CH_3)_3CBr + Na^+ \ ^-C{\equiv}CR \longrightarrow CH_2{=}C(CH_3)_2 + HC{\equiv}CR + NaBr$$

On the other hand, a mixture of *cis* and *trans* alkenes can be obtained by the Wittig reaction, and then converted to the alkyne:

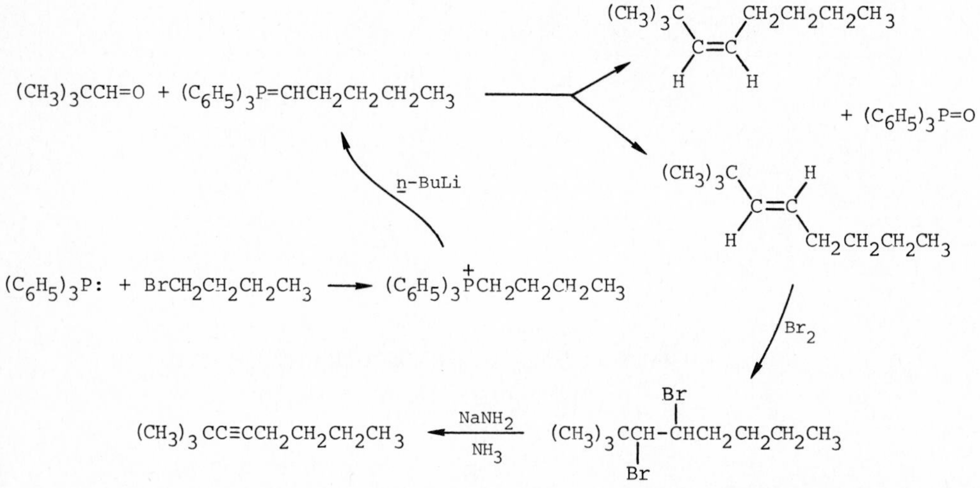

(b) Two possibilities which involve various protecting or masking groups are shown below:

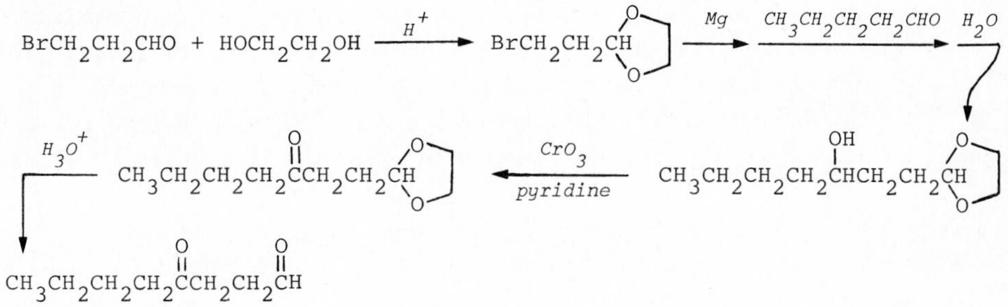

$$BrCH_2CH_2CHO + HOCH_2CH_2OH \xrightarrow{H^+} BrCH_2CH_2CH\begin{pmatrix}O\\O\end{pmatrix} \xrightarrow{Mg} \xrightarrow{CH_3CH_2CH_2CH_2CHO} \xrightarrow{H_2O}$$

$$\xrightarrow{H_3O^+} CH_3CH_2CH_2CH_2\overset{O}{\underset{\|}{C}}CH_2CH_2CH\begin{pmatrix}O\\O\end{pmatrix} \xleftarrow[pyridine]{CrO_3} CH_3CH_2CH_2CH_2\overset{OH}{\underset{|}{CH}}CH_2CH_2CH\begin{pmatrix}O\\O\end{pmatrix}$$

$$CH_3CH_2CH_2CH_2\overset{O}{\underset{\|}{C}}CH_2CH_2\overset{O}{\underset{\|}{CH}}$$

or

$$CH_3CH_2CH_2CH_2MgBr + H\overset{O}{\underset{\|}{C}}CH_2CH_2CH{=}CH_2 \longrightarrow \xrightarrow{H_2O} CH_3CH_2CH_2CH_2\overset{OH}{\underset{|}{CH}}CH_2CH_2CH{=}CH_2$$

$$\xrightarrow{K_2Cr_2O_7} \xrightarrow{H_2SO_4}$$

$$CH_3CH_2CH_2CH_2\overset{O}{\underset{\|}{C}}CH_2CH_2\overset{O}{\underset{\|}{CH}} \xleftarrow[2.\ Zn/H_2O]{1.\ O_3} CH_3CH_2CH_2CH_2\overset{O}{\underset{\|}{C}}CH_2CH_2CH{=}CH_2$$

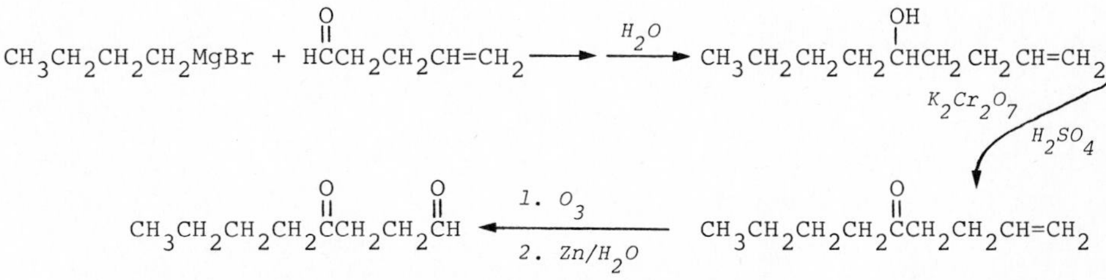

(c) The desired product is a six-carbon keto alcohol. Thus, we must add one carbon at least. Since the material is an alcohol, we may consider a Grignard synthesis. It is a secondary alcohol, so there are two possible combinations:

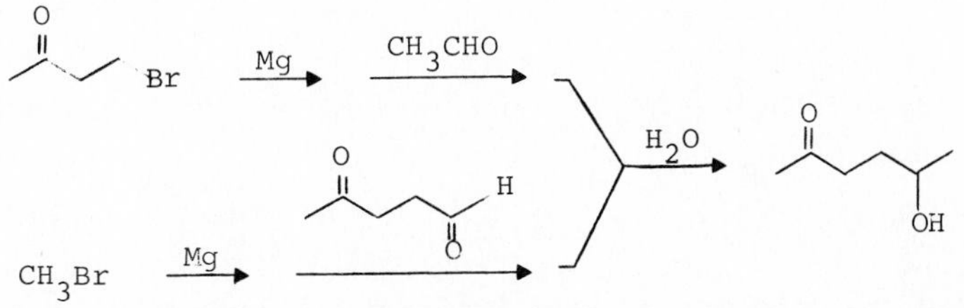

In the latter route, we have a problem of selectivity. The Grignard reagent CH_3MgBr can react with either carbonyl group. Although aldehydes are more reactive than ketones, the Grignard reagent is so reactive it will probably not show much selectivity.

In the first route, we have a different problem: the Grignard reagent *can react with itself*. Polymerization will result. A simple way out of this dilemma is to *protect* the carbonyl group in 4-bromo-2-butanone so that it cannot react with a Grignard reagent. This may be done by converting it into a ketal, which does not react with Grignard reagents.

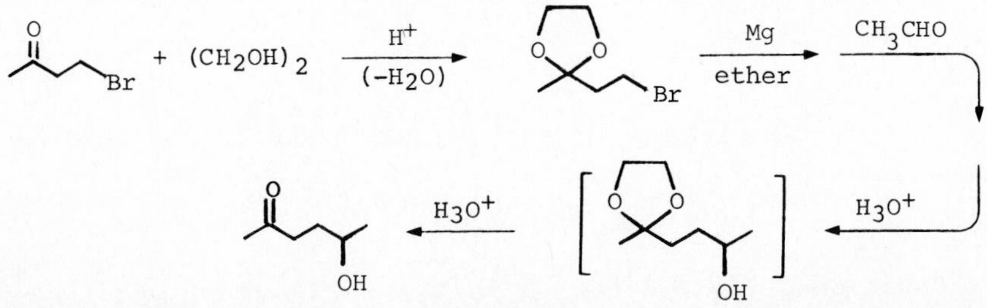

(d) The target: $(CH_3)_2C{=}CH\overset{CH_2CH_3}{\overset{|}{C}}{=}CHCH_2CH_2CH_3$ (no stereochemistry specified). The major challenge is to introduce the double bonds in the correct positions. Routes which involve dehydration of alcohols or elimination of alkyl halides will lead to other isomers in addition to the desired one. The best way to introduce a double bond in a specific position is via the Wittig reaction. With this in mind, you can imagine two carbonyl compounds which would give the desired product:

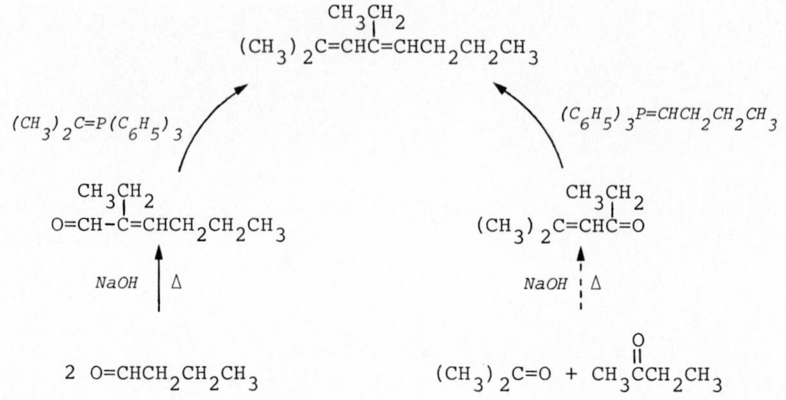

You should recognize that both of these carbonyl compounds are potentially available in one step via the aldol condensation. However, the ketone is clearly a less desirable intermediate because it would require a mixed aldol condensation, leading to many products in addition to the desired one.

(e) $(CH_3)_2C{=}CHCHO + NaC{\equiv}CCH_2CH_3 \longrightarrow \xrightarrow{H_2O} (CH_3)_2C{=}CH\overset{OH}{\overset{|}{C}}HC{\equiv}CCH_2CH_3$

(f) $(CH_3CH_2)_2C{=}O + NaC{\equiv}CCH{=}CH_2 \longrightarrow (CH_3CH_2)_2\overset{OH}{\overset{|}{C}}C{\equiv}CCH{=}CH_2$

(g) The easiest way to form the carbon-carbon bond in this case would be to use a Grignard reaction:

$$\text{"}HOCH_2CH_2CH_2MgBr\text{"} + H\overset{O}{\overset{\|}{C}}CH_2CH(CH_3)_2 \longrightarrow \xrightarrow{H_2O} HOCH_2CH_2CH_2\overset{OH}{\overset{|}{C}}HCH_2CH(CH_3)_2$$

However, this Grignard reagent can't be made, because of the presence of a hydroxy group in the molecule. (A simple proton transfer reaction would destroy the reagent: $HOCH_2CH_2CH_2MgBr \xrightarrow{fast!} BrMgOCH_2CH_2CH_3$.)

The hydroxy group must therefore be protected:

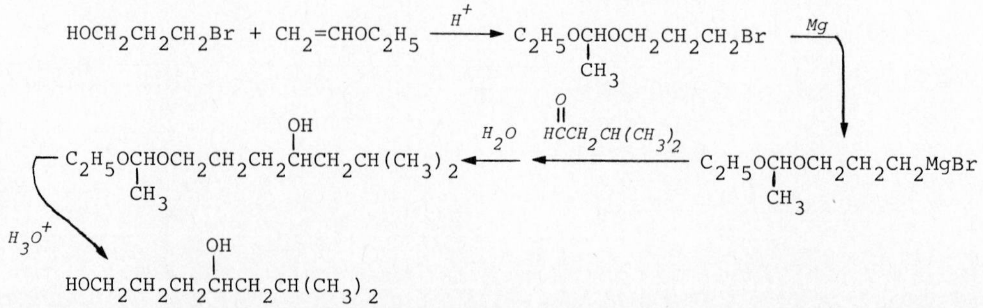

16.E Supplementary Problems

S1. Using monofunctional starting materials of five carbons or less, and any other reagents, outline efficient syntheses of the following compounds.

(a) 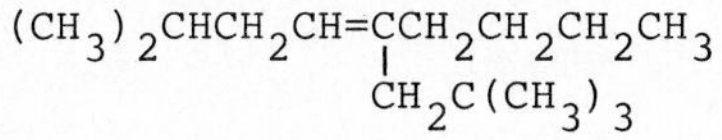

(b) 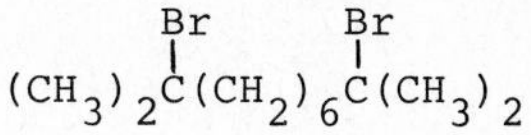

(c) 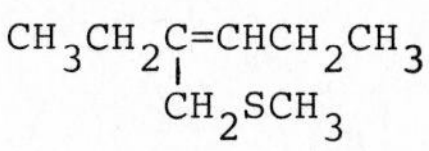

(d) 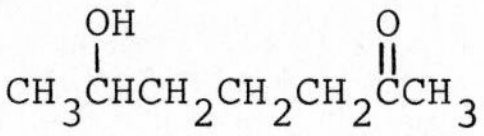

(e) 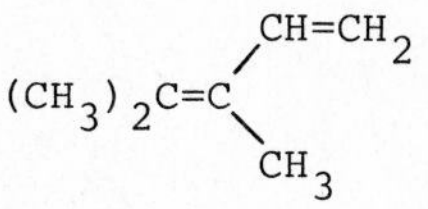

(f) 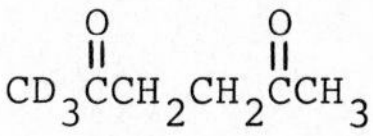

(g) 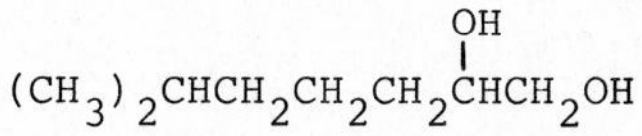

(h) 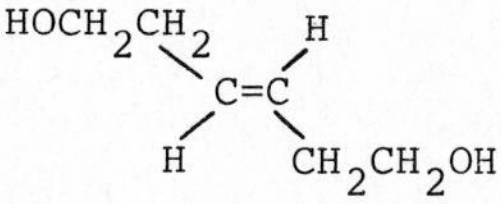

(i) 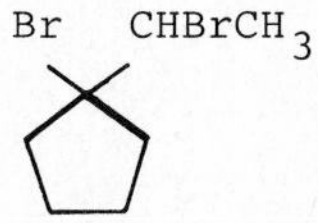

(j) $((CH_3)_2CHCH_2)_2CHCN$

(k) $CH_3C(CH_2Br)_3$

(l) 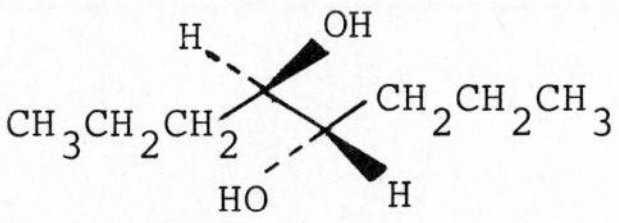

(m) 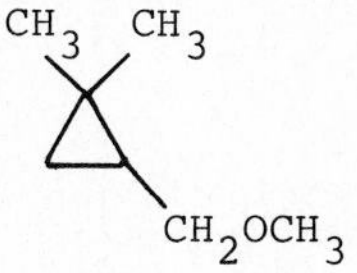

(n) 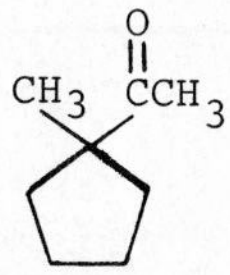

(o) 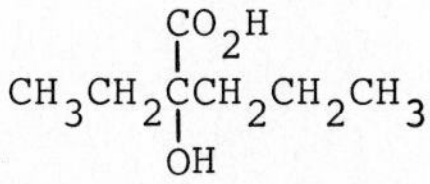

(p) 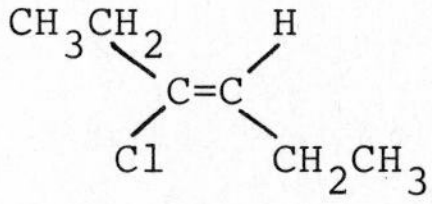

(q) 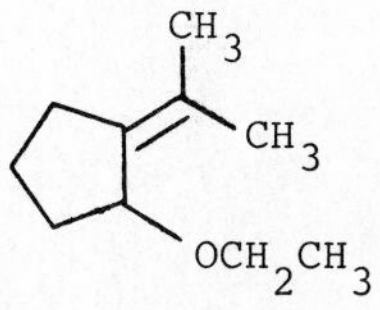

(r) 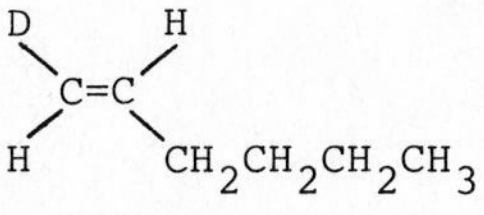

(s) 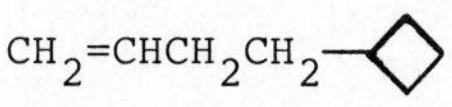

(t) 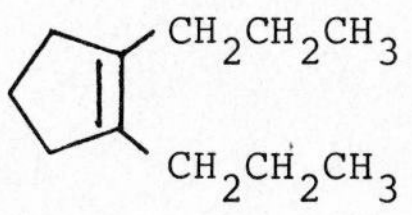

(u) $(CH_3CH_2)_2CHC{\equiv}CCH(CH_2CH_3)_2$

(v) $HON{=}CHCH_2CH_2CH_2CH{=}NOH$

(w) 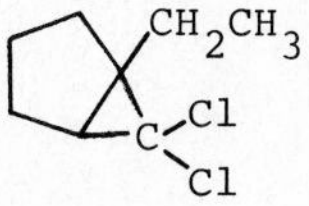

(x) 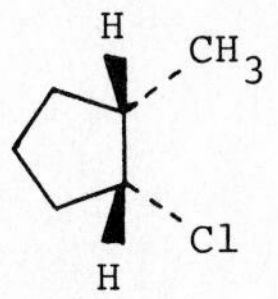

(y) 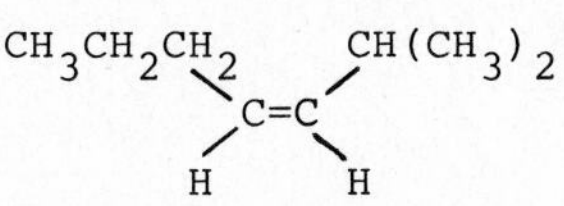

(z) 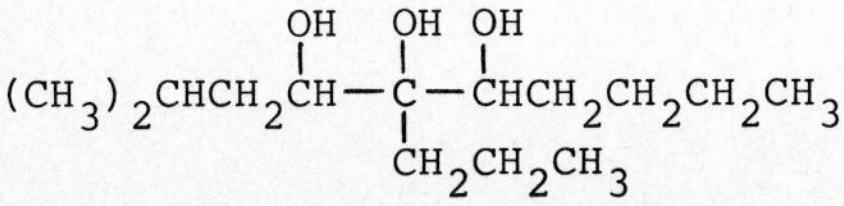

16.F Answers to Supplementary Problems

S1. (a) $(CH_3)_2CHCH_2CH=CCH_2CH_2CH_2CH_3 \longleftarrow (CH_3)_2CHCH_2CH=P(C_6H_5)_3 +$
$$\overset{O}{\overset{\|}{C}}CH_2CH_2CH_2CH_3$$
$$\underset{CH_2C(CH_3)_3}{|} \qquad\qquad\qquad\qquad\qquad\qquad\qquad \underset{CH_2C(CH_3)_3}{|}$$

$\uparrow$ $\underline{n}$-BuLi

$\uparrow$ $(C_6H_5)_3P$

$(CH_3)_2CHCH_2CH_2Br$

$K_2Cr_2O_7 | H_2SO_4$

$$(CH_3)_3CCH_2MgBr + \overset{O}{\overset{\|}{H}}CCH_2CH_2CH_2CH_3 \xrightarrow{H_2O} (CH_3)_3CCH_2\overset{OH}{\overset{|}{C}}HCH_2CH_2CH_2CH_3$$

(b) $(CH_3)_2\overset{Br}{\overset{|}{C}}(CH_2)_6\overset{Br}{\overset{|}{C}}(CH_3)_2 \xleftarrow[h\nu]{2\ Br_2} (CH_3)_2CH(CH_2)_6CH(CH_3)_2 \xleftarrow[Pt]{H_2}$

$\longrightarrow NaC\equiv CNa \xrightarrow{2\ (CH_3)_2CHCH_2CH_2Br} (CH_3)_2CHCH_2CH_2C\equiv CCH_2CH_2CH(CH_3)_2$

$HC\equiv CH + 2\ NaNH_2$

(c) $CH_3CH_2C=CHCH_2CH_3 \xleftarrow{NaSCH_3} CH_3CH_2C=CHCH_2CH_3 \xleftarrow{PBr_3} CH_3CH_2C=CHCH_2CH_2CH_3$
$\underset{CH_2SCH_3}{|} \qquad\qquad\qquad \underset{CH_2Br}{|} \qquad\qquad\qquad \underset{CH_2OH}{|}$

$\uparrow NaBH_4$

$$2\ CH_3CH_2CH_2\overset{O}{\overset{\|}{C}}H \xrightarrow[\Delta]{NaOH} CH_3CH_2C=CHCH_2CH_2CH_3$$
$$\underset{O=CH}{|}$$

(d) $CH_3\overset{OH}{\overset{|}{C}}HCH_2CH_2\overset{O}{\overset{\|}{C}}CH_3 \longleftarrow CH_3\overset{O}{\overset{\triangle}{C}}HCH_2 + CH_2=\overset{OLi}{\overset{|}{C}}CH_3 \xleftarrow[\substack{THF \\ -78°\ C}]{LDA} CH_3\overset{O}{\overset{\|}{C}}CH_3$

(e) $(CH_3)_2C=C\overset{CH=CH_2}{\diagup}_{\diagdown CH_3} \xleftarrow[-H_2O]{H_2SO_4} (CH_3)_2\overset{OH}{\overset{|}{C}}HCCH=CH_2 \xleftarrow{H_2O}$
$\underset{CH_3}{|}$

$\uparrow$

$$(CH_3)_2CH\overset{O}{\overset{\|}{C}}CH_3 + BrMgCH=CH_2$$

(f) $CD_3\overset{O}{\overset{\|}{C}}CH_2CH_2\overset{O}{\overset{\|}{C}}CH_3 \xleftarrow[pyridine]{CrO_3} CD_3\overset{OH}{\overset{|}{C}}HCH_2CH_2\overset{OH}{\overset{|}{C}}HCH_3 \xleftarrow[Pt]{H_2} CD_3\overset{OH}{\overset{|}{C}}HC\equiv C\overset{OH}{\overset{|}{C}}HCH_3$

$H_2O \uparrow$

$$CH_3\overset{O}{\overset{\|}{C}}H \xrightarrow[D_2O]{D^+} CD_3\overset{O}{\overset{\|}{C}}H$$

$H_2O \uparrow$

$$HC\equiv CNa \xrightarrow{\overset{O}{\overset{\|}{H}}CCH_3} HC\equiv C\overset{ONa}{\overset{|}{C}}HCH_3 \xrightarrow{NaNH_2} NaC\equiv C\overset{ONa}{\overset{|}{C}}HCH_3$$

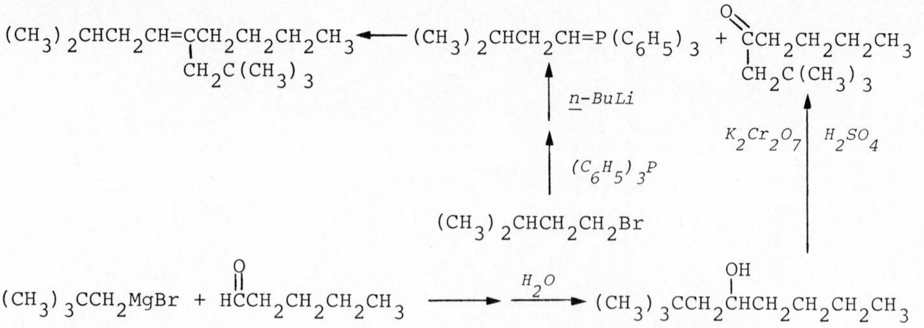

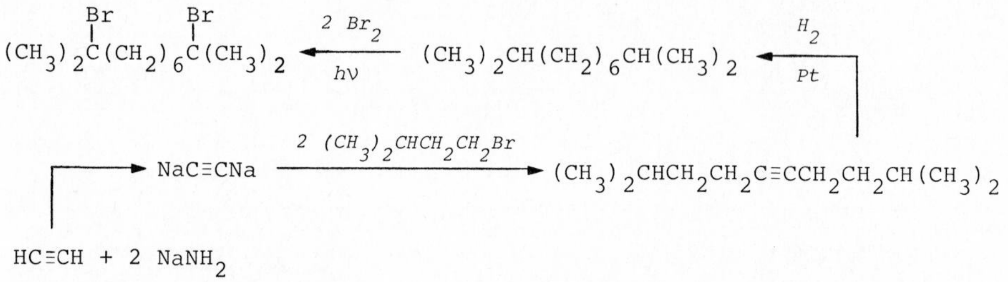

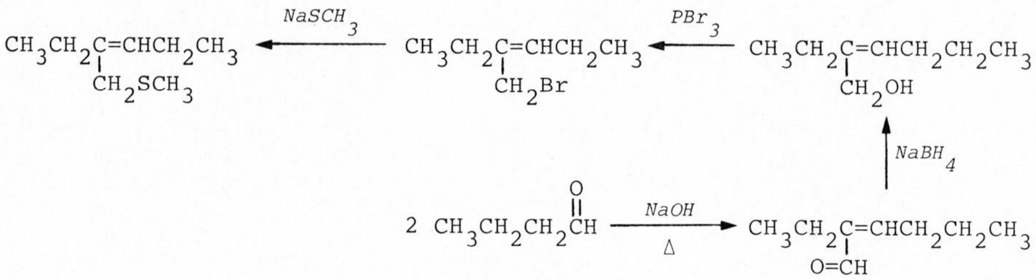

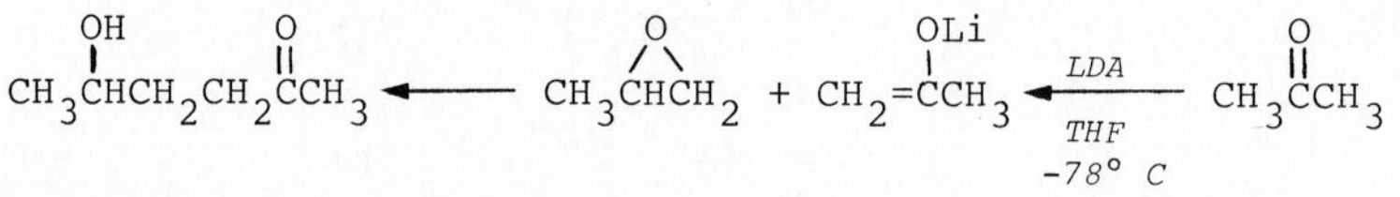

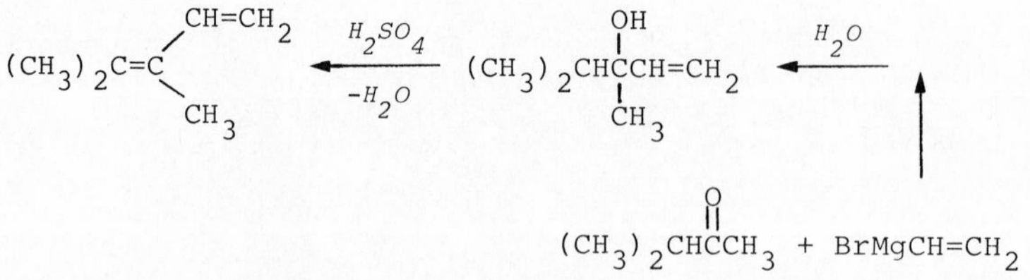

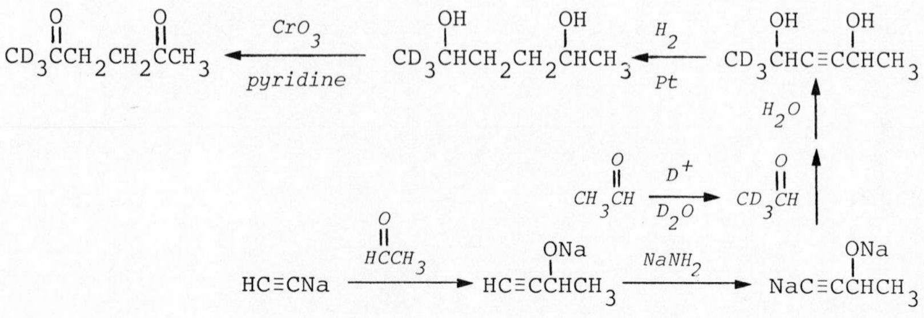

(g)

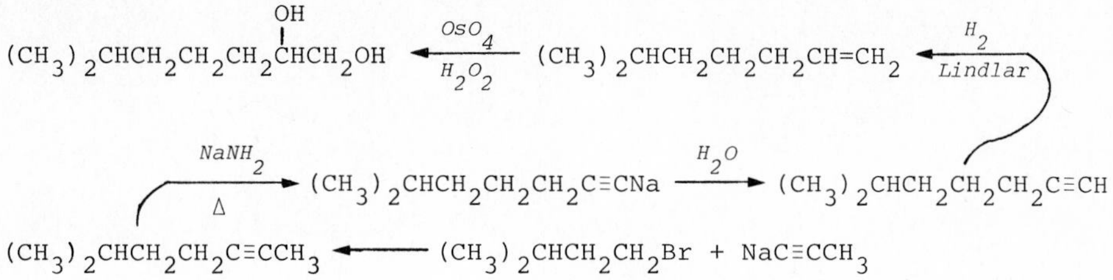
OH
(CH3)2CHCH2CH2CH2CHCH2OH
OsO4
H2O2
(CH3)2CHCH2CH2CH2CH=CH2
H2
Lindlar
NaNH2
Δ
(CH3)2CHCH2CH2CH2C≡CNa
H2O
(CH3)2CHCH2CH2CH2C≡CH
(CH3)2CHCH2CH2C≡CCH3
(CH3)2CHCH2CH2Br + NaC≡CCH3

(h)

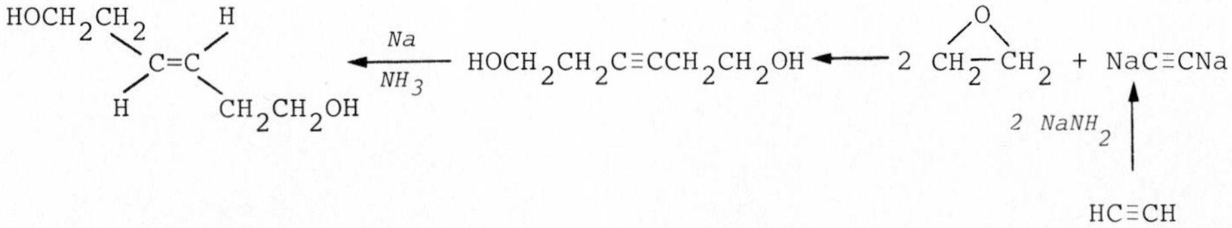
HOCH2CH2
H
C=C
H
CH2CH2OH
Na
NH3
HOCH2CH2C≡CCH2CH2OH
2
O
CH2—CH2
+ NaC≡CNa
2 NaNH2
HC≡CH

(i)

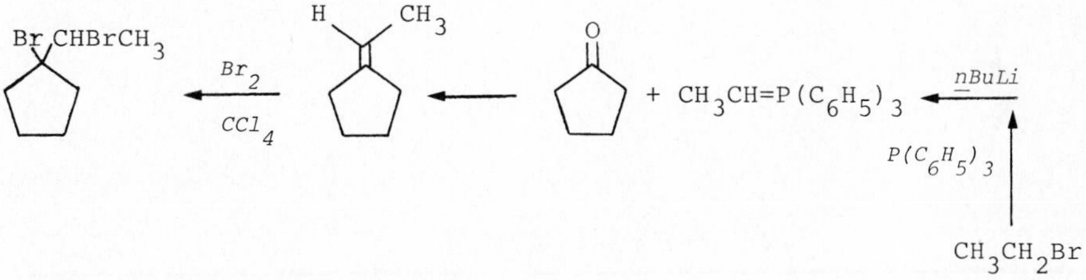
Br CHBrCH3
Br2
CCl4
H
CH3
O
+ CH3CH=P(C6H5)3
nBuLi
P(C6H5)3
CH3CH2Br

(j)

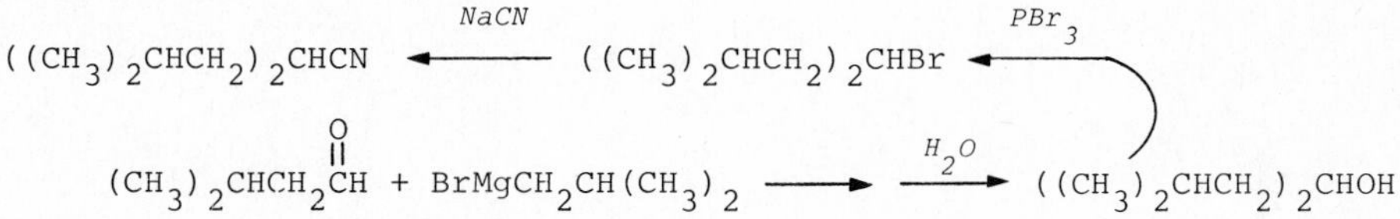
((CH3)2CHCH2)2CHCN
NaCN
((CH3)2CHCH2)2CHBr
PBr3
O
(CH3)2CHCH2CH + BrMgCH2CH(CH3)2
H2O
((CH3)2CHCH2)2CHOH

(k)

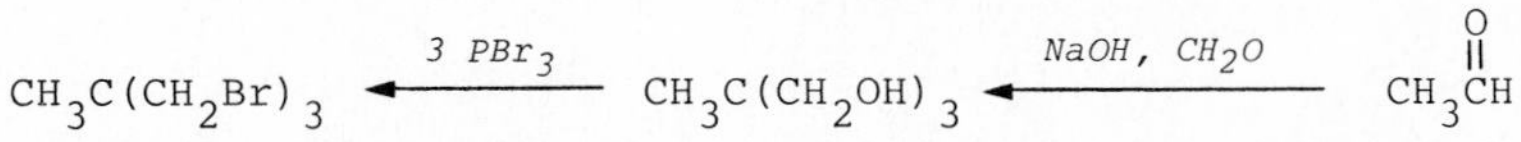
CH3C(CH2Br)3
3 PBr3
CH3C(CH2OH)3
NaOH, CH2O
O
CH3CH

(l)

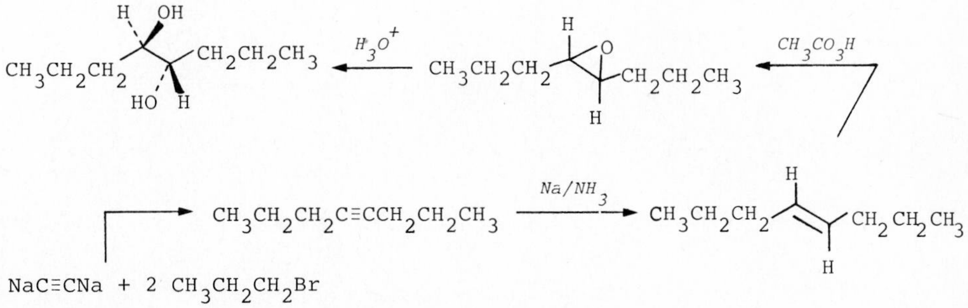
H OH
CH3CH2CH2
CH2CH2CH3
HO H
H3O+
CH3CH2CH2
H
O
CH2CH2CH3
H
CH3CO3H
CH3CH2CH2C≡CCH2CH2CH3
Na/NH3
H
CH3CH2CH2
CH2CH2CH3
H
NaC≡CNa + 2 CH3CH2CH2Br

(m) 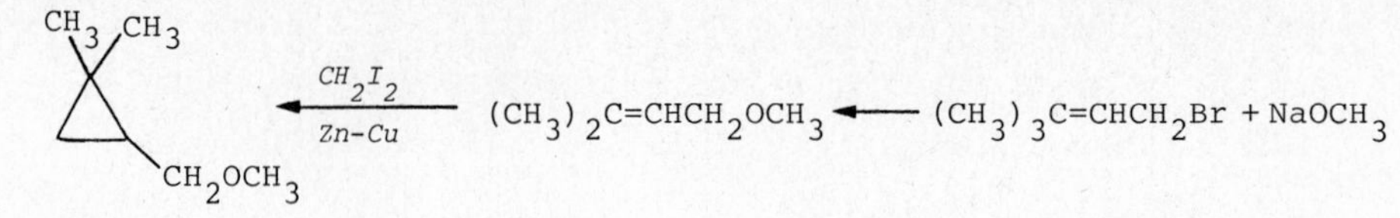
CH3 CH3
CH2OCH3
CH2I2
Zn-Cu
(CH3)2C=CHCH2OCH3
(CH3)3C=CHCH2Br + NaOCH3

(n) 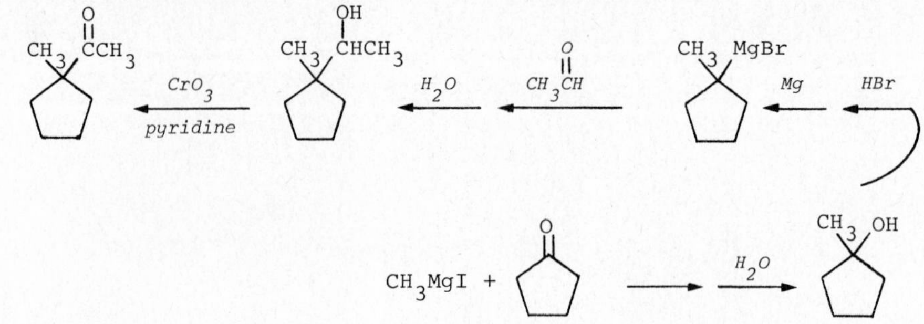
CH3 CCH3
CrO3
pyridine
OH
CH3 CHCH3
H2O
CH3CH
CH3 MgBr
Mg
HBr
CH3MgI +
H2O
CH3 OH

(o) 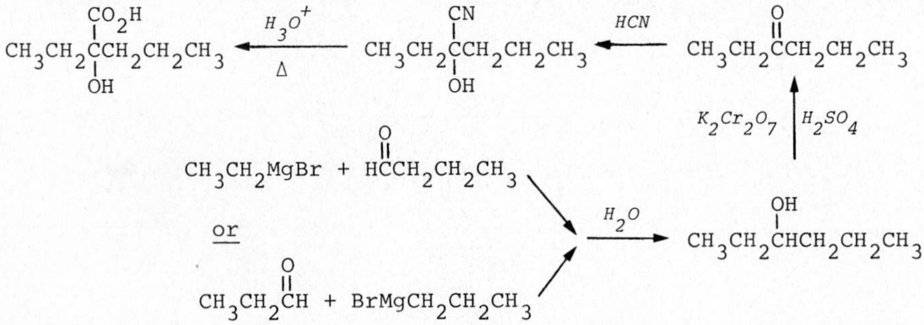
CO2H
CH3CH2CCH2CH2CH3
OH
H3O+
Δ
CN
CH3CH2CCH2CH2CH3
OH
HCN
CH3CH2CCH2CH2CH3
K2Cr2O7
H2SO4
CH3CH2MgBr + HCCH2CH2CH3
or
CH3CH2CH + BrMgCH2CH2CH3
H2O
OH
CH3CH2CHCH2CH2CH3

(p) 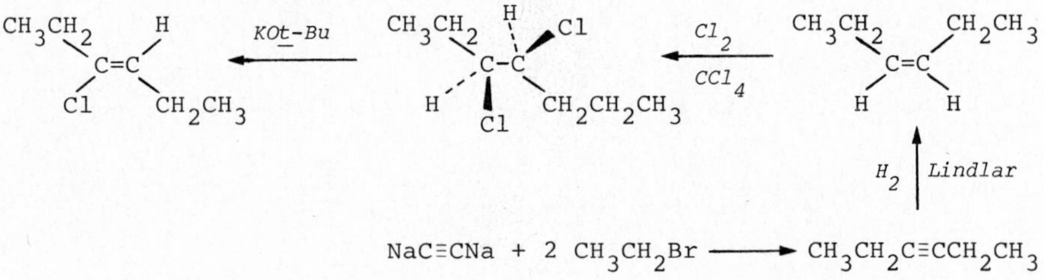
CH3CH2 H
C=C
Cl CH2CH3
KOt-Bu
CH3CH2 H Cl
C-C
H Cl CH2CH2CH3
Cl2
CCl4
CH3CH2 CH2CH3
C=C
H H
H2
Lindlar
NaC≡CNa + 2 CH3CH2Br
CH3CH2C≡CCH2CH3

(q) 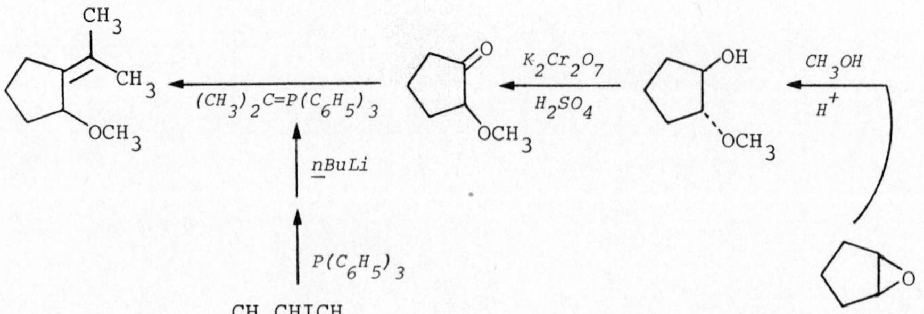
CH3
CH3
OCH3
(CH3)2C=P(C6H5)3
nBuLi
P(C6H5)3
CH3CHICH3
O
OCH3
K2Cr2O7
H2SO4
OH
OCH3
CH3OH
H+
O

(r) 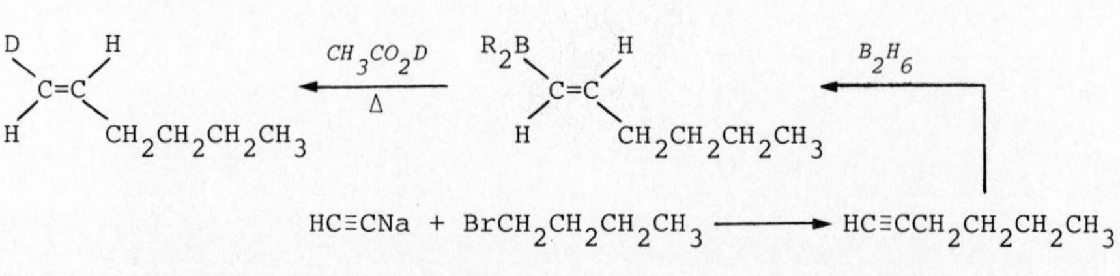
D H
C=C
H CH2CH2CH2CH3
CH3CO2D
Δ
R2B H
C=C
H CH2CH2CH2CH3
B2H6
HC≡CNa + BrCH2CH2CH2CH3
HC≡CCH2CH2CH2CH3

(s)

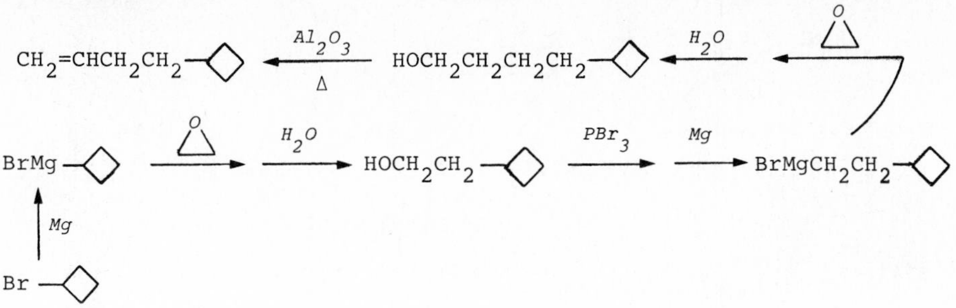
CH_2=CHCH$_2$CH$_2$
Al_2O_3
Δ
HOCH$_2$CH$_2$CH$_2$CH$_2$
H_2O
BrMg
H_2O
HOCH$_2$CH$_2$
PBr_3
Mg
BrMgCH$_2$CH$_2$
Mg
Br

(t)

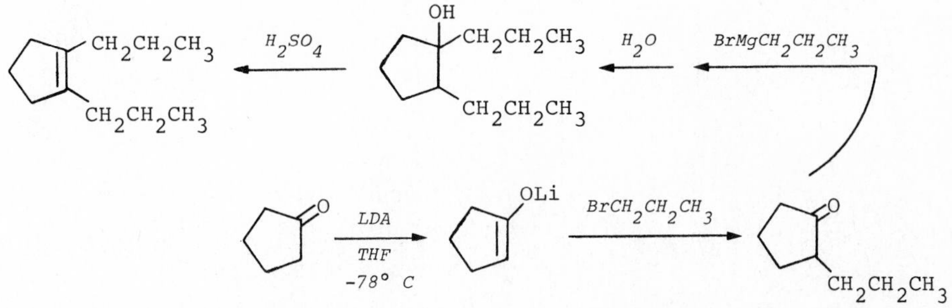
CH$_2$CH$_2$CH$_3$
CH$_2$CH$_2$CH$_3$
H_2SO_4
OH
CH$_2$CH$_2$CH$_3$
CH$_2$CH$_2$CH$_3$
H_2O
BrMgCH$_2$CH$_2$CH$_3$
LDA
THF
-78° C
OLi
BrCH$_2$CH$_2$CH$_3$
CH$_2$CH$_2$CH$_3$

(u)

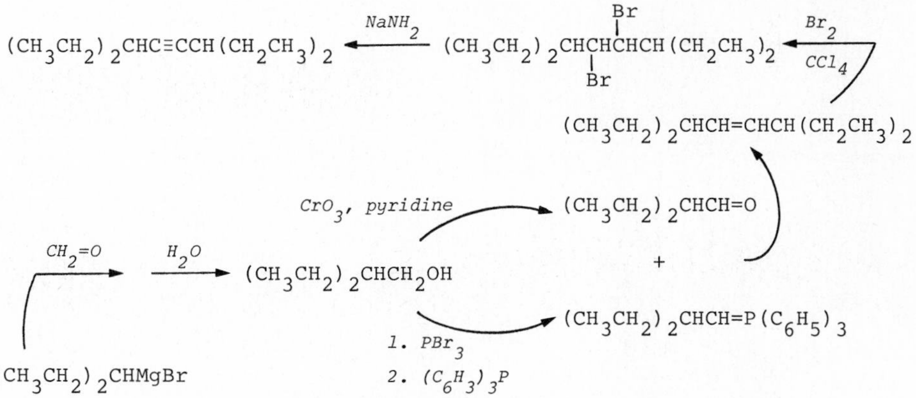
(CH$_3$CH$_2$)$_2$CHC≡CCH(CH$_2$CH$_3$)$_2$
NaNH$_2$
Br
(CH$_3$CH$_2$)$_2$CHCHCHCH(CH$_2$CH$_3$)$_2$
Br
Br$_2$
CCl$_4$
(CH$_3$CH$_2$)$_2$CHCH=CHCH(CH$_2$CH$_3$)$_2$
CrO$_3$, pyridine
(CH$_3$CH$_2$)$_2$CHCH=O
+
CH$_2$=O
H$_2$O
(CH$_3$CH$_2$)$_2$CHCH$_2$OH
(CH$_3$CH$_2$)$_2$CHCH=P(C$_6$H$_5$)$_3$
1. PBr$_3$
2. (C$_6$H$_3$)$_3$P
(CH$_3$CH$_2$)$_2$CHMgBr

(v)

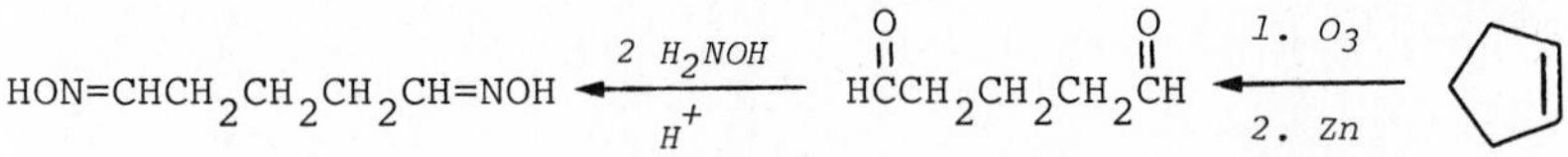
HON=CHCH$_2$CH$_2$CH$_2$CH=NOH
2 H$_2$NOH
H$^+$
HCCH$_2$CH$_2$CH$_2$CH
1. O$_3$
2. Zn

(w)

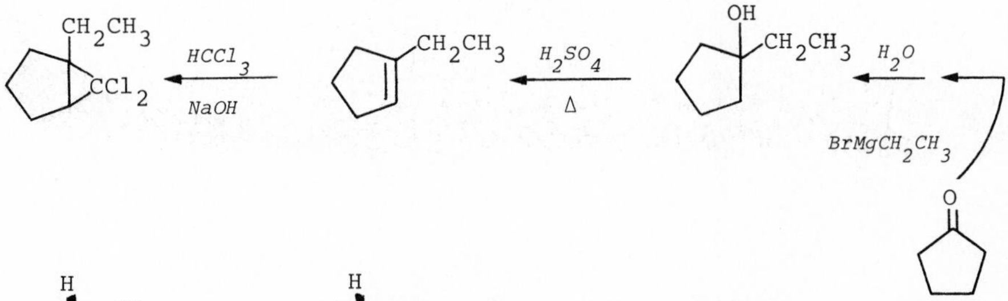
CH$_2$CH$_3$
CCl$_2$
HCCl$_3$
NaOH
CH$_2$CH$_3$
H$_2$SO$_4$
Δ
OH
CH$_2$CH$_3$
H$_2$O
BrMgCH$_2$CH$_3$
O

(x)

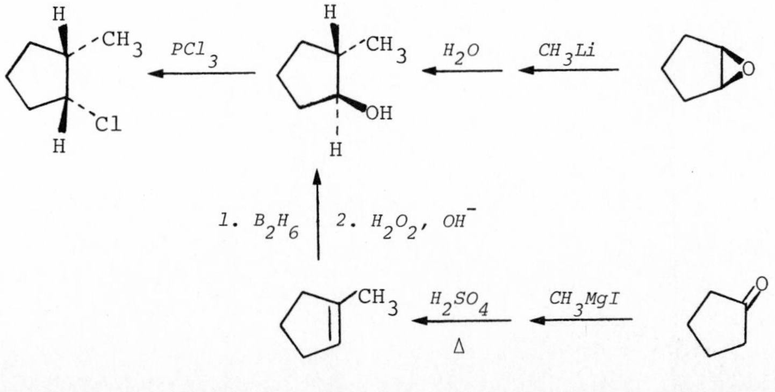
H
CH$_3$
Cl
H
PCl$_3$
H
CH$_3$
OH
H
H$_2$O
CH$_3$Li
O
1. B$_2$H$_6$
2. H$_2$O$_2$, OH$^-$
CH$_3$
H$_2$SO$_4$
Δ
CH$_3$MgI
O

(y) 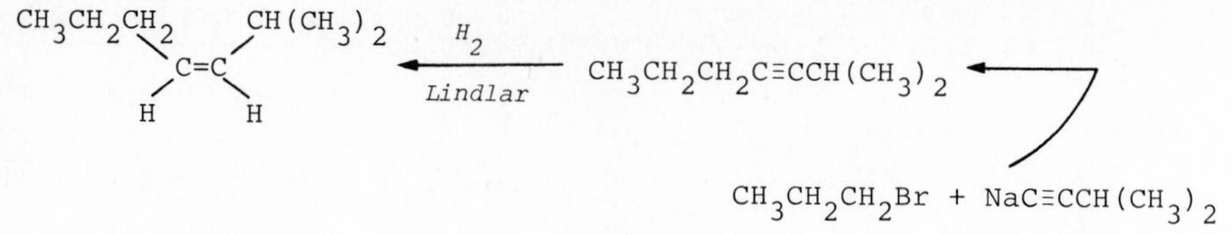

(z) 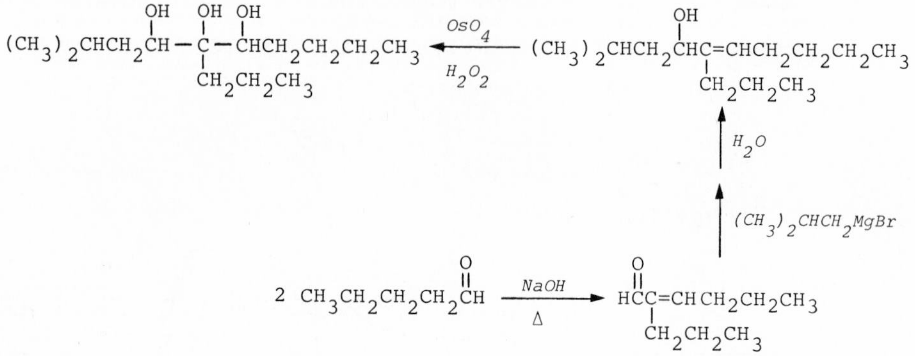

17. MASS SPECTROSCOPY

17.A Chapter Outline and Important Terms Introduced

17.1 Introduction

radical cation — mass spectrum

mass spectrometer

17.2 Instrumentation (the physics behind the technique)

single focussing magnetic deflection — magnetic or electrical scanning

$m/e = H^2r^2/2V$

17.3 The Molecular Ion: Molecular Formula

nominal mass — M+1 peaks

high resolution

17.4 Fragmentation

A. Simple Bond Cleavage
(to give most stable cationic fragment)

B. Two-bond Cleavage, Elimination of a Neutral Molecule
alcohols: loss of water
carbonyl compounds: McLafferty rearrangement

17.B Important Concepts and Hints

How to Interpret a Mass Spectrum ≡ How to Solve Mass Spectral Problems

All the rules listed below apply to molecules containing only C, H, and O. For halogen- or nitrogen-containing compounds, see item #6, entitled "Complications".

1. Decide whether the particle of highest mass/charge ratio (m/e) is the molecular ion (M^+). (*NOTE:* the height of the peak *(intensity)* is something completely different; the tallest peak is often not M^+.)
   (a) If the formula of the molecule is given, you can determine M^+ right away by calculating the molecular weight.
   (b) If m/e for the largest particle is odd, then it's not M^+.
   (c) If the next smaller fragment corresponds to loss of an impossible piece (loss of 7, say, or 22 mass units), then the larger fragment is not M^+. The spectrum of 2-methyl-2-propanol (Figure 17.13) provides an example of both of these generalizations.
   (d) If the particle of highest m/e is not M^+ and no hints or formula were provided, try adding water (to the even-numbered fragments) or alkyl groups (to the odd-numbered fragments) to come up with something reasonable. For instance in the spectrum of 2-methyl-2-propanol, addition of 18 (water) to the next largest fragment and 15 (methyl) to the largest fragment suggest the same molecular ion (m/e 88).

2. Look at the major fragment ions, and decide what pieces have been lost (subtract each m/e from M^+ and/or from a higher m/e).
   (a) Odd-mass pieces are radicals: methyl = 15, ethyl = 29, propyl = 43, butyl = 57, etc.
   (b) Even-mass pieces are neutral molecules: water = 18, ROH = 17 + alkyl; ethylene (from McLafferty rearrangement) = 28; higher alkenes = 28 + 14 for each additional CH_2 group.

3. Before you attempt to interpret cleavage patterns, see what you can deduce about the molecule from other sources. Is there oxygen in the molecule? Is it a ketone or an alcohol? Because you can predict how different classes of compounds will fragment, interpreting an actual cleavage pattern becomes much easier if you know what to expect.

4. Look at even-mass fragments first, the ones that correspond to loss of a neutral molecule.

   (a) Loss of 18 (H_2O) is strong evidence for an alcohol; loss of 17 + alkyl (ROH) is evidence for an ether.

   (b) Loss of 28 ($CH_2{=}CH_2$) or a higher alkene from a carbonyl-containing molecule is the result of McLafferty rearrangement. This means that a carbon γ to the carbonyl group has a hydrogen attached. It also gives you an indication of the type of alkyl group it is (see problem #7, for example).

5. Finally, look at the odd-mass fragments, which correspond to loss of a radical from the molecule. Such cleavages occur to give the most stable cations and (less importantly) most stable radicals.

   (a) Hydrocarbons cleave at branch points so that secondary or tertiary cations can be formed.

   (b) Alcohols, ethers, and carbonyl compounds undergo cleavage adjacent to the functional group (so-called α-cleavage) so that oxonium ions can be formed:

6. *COMPLICATIONS*: molecules which contain several functional groups will obviously produce more complex mass spectra. Cyclic molecules, especially bicyclic molecules, often give fragmentation patterns which are difficult to interpret because more than one bond must be broken to remove a piece.

   (a) Chlorine- and bromine-containing ions show doubled peaks because of the presence of two isotopes for each.

   (b) The presence of an odd number of nitrogen atoms changes statement 1(b) above: molecular ions containing an odd number of nitrogens have odd mass.

17.C Answers to Exercises

(17.3) Use equation 17-5:

(a)
$$\frac{M+1}{M} = \frac{0.01107}{0.98893}(8) + 0.00015\,(16) + 0.00037\,(4)$$
$$= 0.08952 + 0.00240 + 0.00148$$
$$= 0.0934 = 9.3\%$$

Note how ^{13}C dominates M+1.

(b) 13.3% (c) 14.9% (d) 69.0%

Note that the M+1 peak approaches the intensity of the M peak for compounds containing many carbons.

(e) 1.1%
(f) 2.7%

Since ^{127}I and ^{19}F are the only isotopes of these halogens, we only need consider the ^{13}C contribution.

(17.4.A)

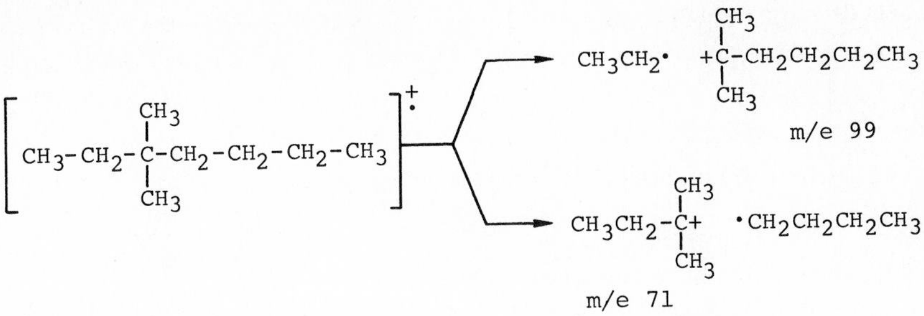

(17.4.B)

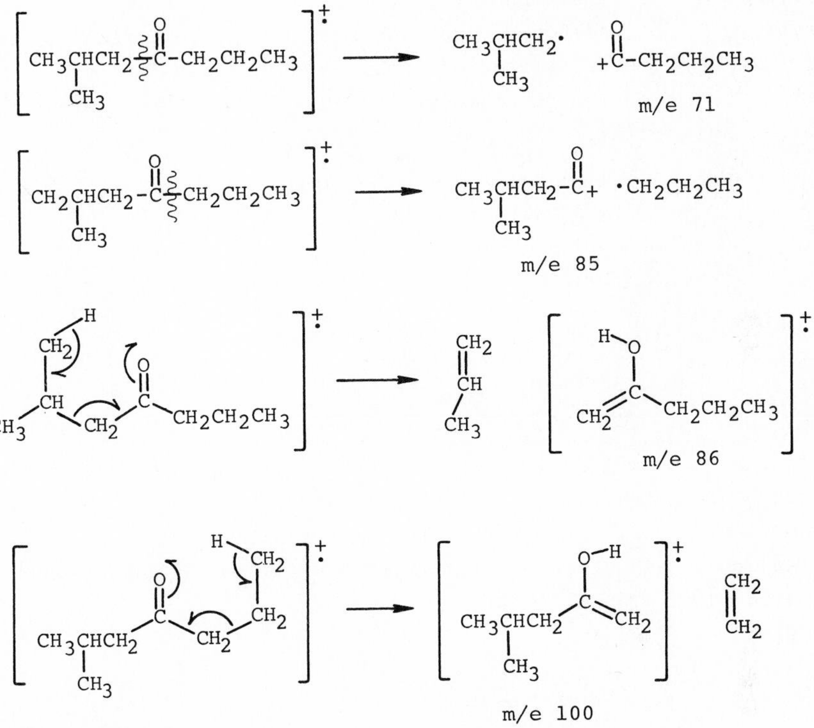

17.D Answers and Explanations for Problems

1. (a) Relative probabilities are: Cl_2^{35} , $(0.7553)^2 = 0.57$; $Cl^{35}Cl^{37}$, $2 \times (0.7553)(0.2447) = 0.37$; Cl_2^{37} , $(0.2447)^2 = 0.06$.

   (b) Predicted $\frac{M+1}{M}$, using equation 17-5:

   $C_{10}H_{18}$: 0.115 ←

   $C_8H_{10}O_2$: 0.092

   $C_8H_{14}N_2$: 0.097

2. Fragmentation of the C_1-C_2 bond produces a cation which is stabilized by the nitrogen non-bonding electrons:

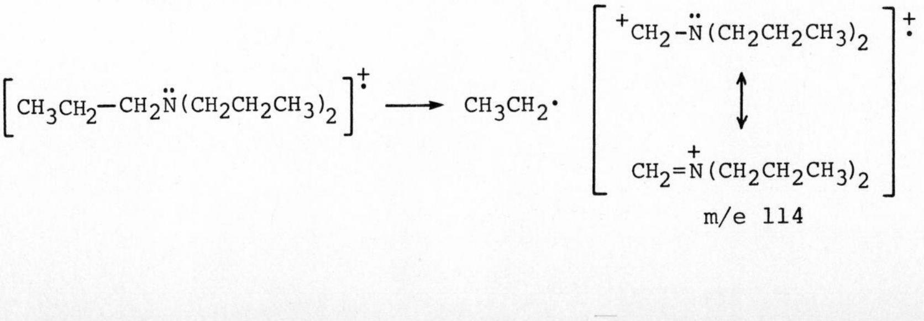

3. M^+ indicates a molecular weight of 128, so possible formulas are $C_{10}H_8$ or C_9H_{20}. But M^+ is small and suggests the saturated hydrocarbon. 113 is M-15 ($M - CH_3$); 43 is $C_3H_7^+$ or $(CH_3)_2CH^+$. Note the virtual absence of 99 ($M - C_2H_5$); hence, no $R-CH_2CH_3$. We deduce:

$$(CH_3)_2CH \overset{\underset{\smile}{}}{-} CH_2 \overset{\underset{\smile}{}}{-} CH_2 \overset{\underset{\smile}{}}{-} CH_2 \overset{\underset{\smile}{}}{-} CH(CH_3)_2$$

$$85 \quad 71 \quad 57 \quad 43$$

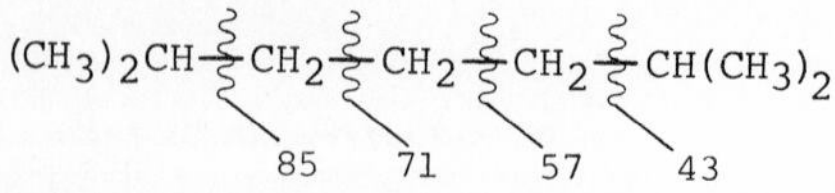

Notice how fragmentations compare with important peaks.

4. Saturated hydrocarbons frequently present problems in mass spectral analyses because of the prevalence of peaks derived from carbocation rearrangements. This example illustrates some of these difficulties. Assignments must be based not just on the presence or absence of given m/e peaks, but on their relative intensities.

$$(CH_3)_2CH \overset{\underset{\smile}{}}{-} CH_2 \overset{\underset{\smile}{}}{-} CH(CH_3)_2$$
$$43 \diagdown \quad \diagdown 43$$

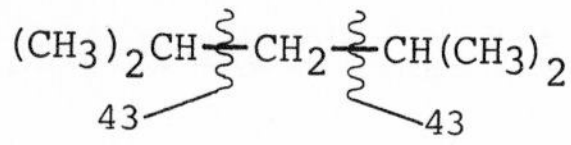

This isomer has two isopropyl ends and is expected to have the largest m/e 43 peak; therefore, c.

$$(CH_3)_3C \overset{\underset{\smile}{}}{-} CH_2CH_2CH_3$$
$$57 \diagup$$

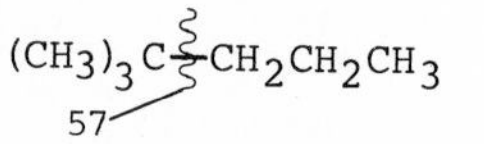

This isomer is the only one with a t-butyl group and is expected to have the largest m/e 57 peak; therefore, b.

$$(CH_3)_2CHCH \overset{\overset{\displaystyle CH_3}{|}}{\underset{\underset{\smile}{}}{}} CH_2CH_3$$
$$71 \diagup$$

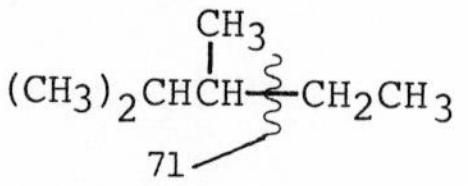

This isomer is expected to have the largest m/e 71 peak; hence, a.

5. The 90 and 92 peaks suggest the presence of Cl; the 55 peak is then R^+ for RCl and corresponds to $C_4H_7^+$, with one element of unsaturation (a saturated ion would be C_nH_{2n+1} or $C_4H_9^+$). The size of 55 suggests that it is an allylic group, $CH_3CH=CHCH_2-$ or $CH_2=CCH_2-$. The ir band at 890 cm^{-1} indicates $CH_2=C$ (Table 14.3); $\quad \overset{|}{CH_3}$

hence, the compound is $\overset{\overset{\displaystyle CH_3}{|}}{CH_2=CCH_2Cl}$.

6. In these spectra, M^+ is 254, corresponding to formulas of $C_{20}H_{14}$, $C_{19}H_{26}$, or $C_{18}H_{38}$. The rather weak M^+ suggests a saturated structure, $C_{18}H_{38}$.

   Acyclic alkanes fragment mainly at branch points, since such fragmentation produces the more stable carbocation. The top spectrum shows extraordinary fragments at m/e 127 and m/e 155, corresponding to loss of 127 (C_9H_{19}) and 99 (C_7H_{15}), respectively. These two fragments add up to $C_{16}H_{34}$, leaving C_2H_4 or $CH_3\overset{\diagup}{CH}$ as the missing piece. This would correspond to n-heptadecane with a single methyl branch at C-8:

m/e 254 $-C_9H_{19}\cdot$ m/e 127

$-C_7H_{15}\cdot$ m/e 155

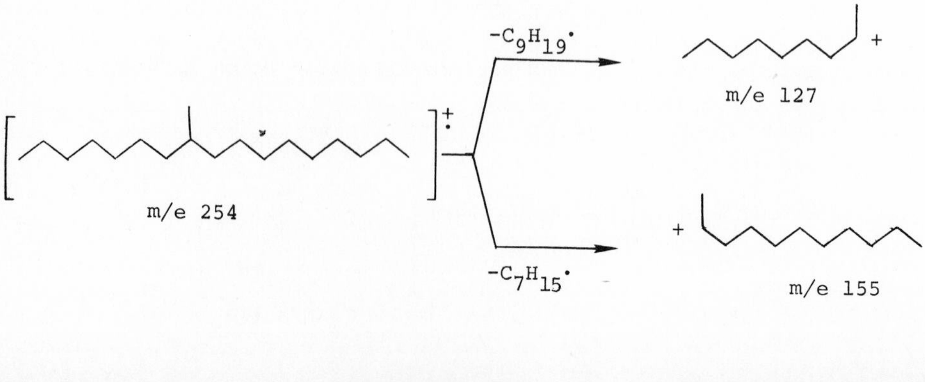

The small M-15 peak, which corresponds to loss of the CH_3 branch, confirms this assignment. Similar logic shows that the bottom spectrum corresponds to 7-methylheptadecane:

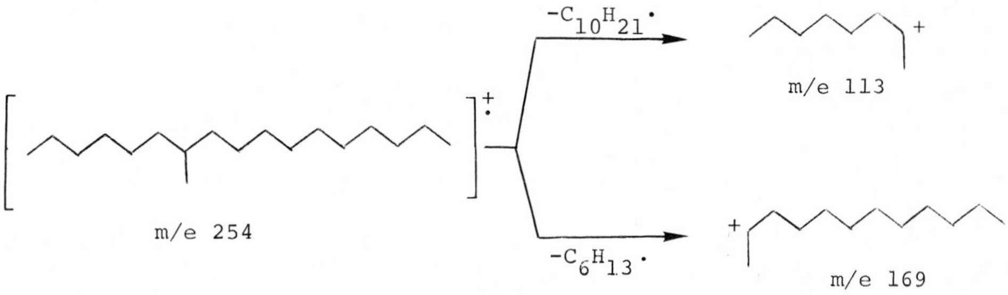

7. The ir band at 1710 cm^{-1} indicates that the compound has a carbonyl group. From the molecular weight of 100, we may deduce that the formula is $C_6H_{12}O$, corresponding to a saturated ketone. Ketones undergo two main types of fragmentation: simple cleavage to produce an oxonium ion, and McLafferty rearrangement. Simple cleavage gives rise to oxonium ions having <u>*ODD*</u> m/e, since an alkyl radical (C_nH_{2n+1}) is lost:

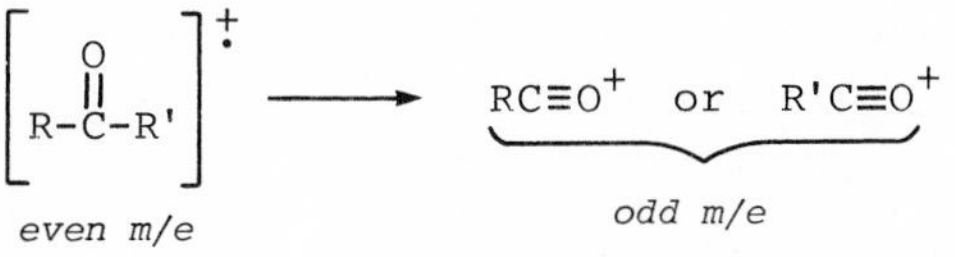

There are three outstanding fragments in the spectrum: m/e 85, m/e 57, and m/e 43, which would correspond to loss of CH_3, C_3H_7, and C_4H_9, respectively. Thus, the ketone must be a methyl butyl ketone:

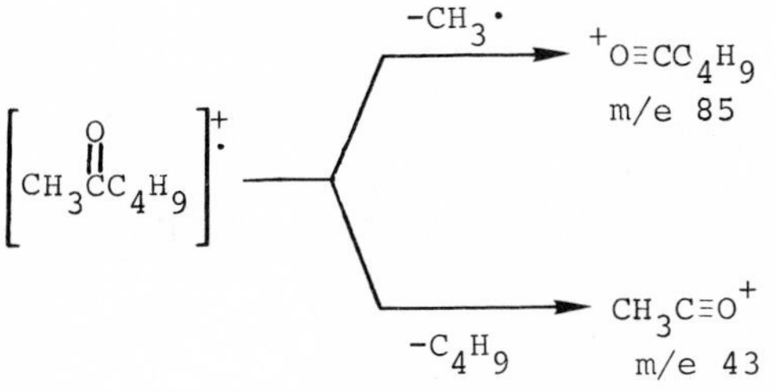

The m/e 57 fragment cannot arise from simple cleavage adjacent to the carbonyl group. (In an acyclic ketone, the two simple cleavage fragments must total M + 28.)

The McLafferty rearrangement ion has <u>*EVEN*</u> m/e, since a neutral molecule is lost. The outstanding fragment with m/e 58 must then be the McLafferty ion (see below). Thus, the ketone could be either $CH_3CH_2CH_2CH_2COCH_3$ or $(CH_3)_2CHCH_2COCH_3$.

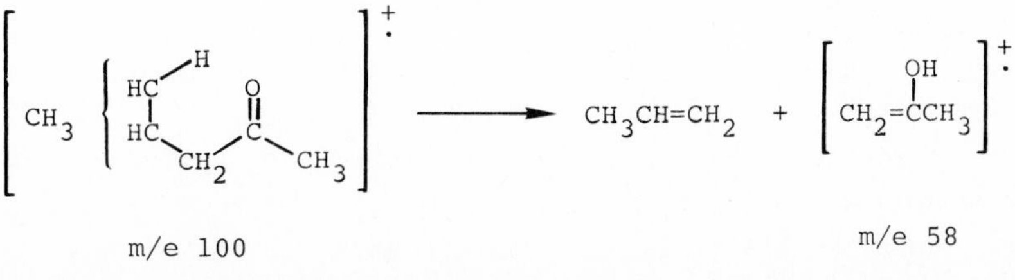

It is difficult to make a conclusive choice between these two possibilities. The McLafferty ion is often the most intense ion in the mass spectrum. In this case it is not, but rather the m/e 43 fragment is. This would tend to suggest methyl isobutyl ketone, since it may give rise to a fragment of m/e 43 in two ways:

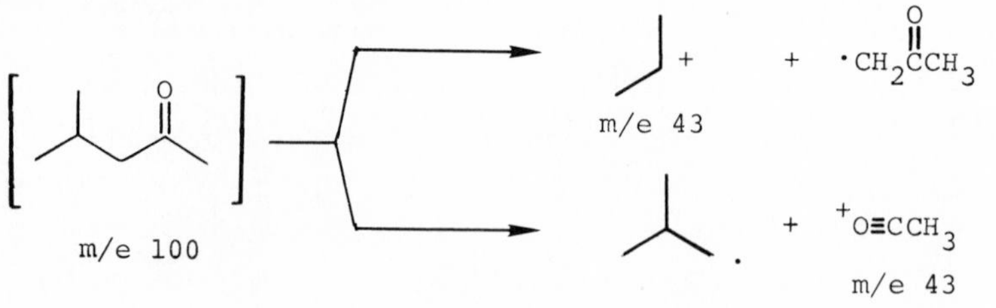

8. As in problem #7, the ir band at 1710 cm^{-1} shows a carbonyl group. From the molecular weight of 114, we deduce the formula $C_7H_{14}O$, corresponding to a saturated ketone. The *ODD* fragments are m/e 71 and m/e 43. Recall that the two simple cleavage fragments must total M+28:

$$\left[R\text{-}\overset{O}{\overset{\|}{C}}\text{-}R' \right]^{+\cdot} \longrightarrow RC{\equiv}O^+ \text{ or } R'C{\equiv}O^+$$

m/e = x m/e = y m/e = z

$$y + z = x + 28$$

In this case, $RC{\equiv}O^+$ <u>*AND*</u> $R'C{\equiv}O^+$ must both have m/e = 71 (71 + 71 = 142 = 114 + 28)

Thus, the ketone is: $C_3H_7\text{-}\overset{O}{\overset{\|}{C}}\text{-}C_3H_7$

A striking feature of this spectrum is the absence of any <u>*EVEN*</u> fragments; there is no McLafferty ion. Thus, the compound must be diisopropyl ketone. The m/e 43 ion must correspond to simple cleavage adjacent to the carbonyl group, with the positive charge residing on the isopropyl fragment:

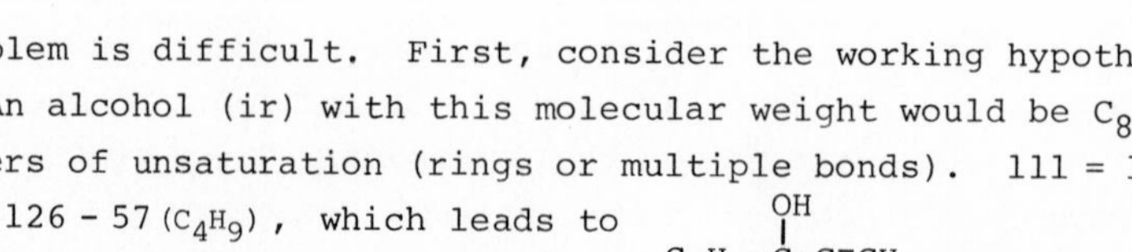

9. This problem is difficult. First, consider the working hypothesis that 126 is M^+. An alcohol (ir) with this molecular weight would be $C_8H_{13}OH$, with two centers of unsaturation (rings or multiple bonds). 111 = 126 - 15 (CH_3) and 69 = 126 - 57 (C_4H_9), which leads to

$$C_4H_9\text{-}\overset{OH}{\underset{CH_3}{C}}\text{-}C{\equiv}CH$$

but then what is the important 87 peak?
And where is the other expected peak from a tertiary alcohol, 126 - 18 (H_2O) = 98? This hypothesis clearly has some loose ends.

Next, consider that 126 is <u>not</u> M^+, especially since alcohols often give no detectable M^+. 126 is then probably M-18, since $RR'C{=}\overset{+}{O}H$ must be an odd number. Thus, M = 144 or $C_9H_{19}OH$ (a saturated alcohol), and 87 = 144 - 57 (C_4H_9).

This gives us a partial structure, $(C_4H_9)RR'COH$, but no other large peak corresponds to 144 - R. Thus, the alcohol must be symmetrical and secondary, $(C_4H_9)_2CHOH$. The 126 peak corresponds to the alkene, $(C_3H_8)C{=}CHC_4H_9$; 111 is probably 126 - CH_3 which indicates branching methyls; however, notice the absence of branching ethyl groups (no 97 = 126 - 29(C_2H_5) peak). Thus, the suggested structure is: $[(CH_3)_2CHCH_2]_2CHOH$.

The 69 peak is left unexplained, but must result from a more deep-seated rearrangement. This is also a loose end, but the second hypothesis explains the data much more completely than the first. Notice how the *absence* of expected peaks can be more significant than the presence of additional m/e peaks.

10. 2-Octanone is $C_8H_{16}O$ and has a molecular weight of 128; 113 = 128 - 15(CH_3), 43 = 128 - 85 (C_6H_{13}):

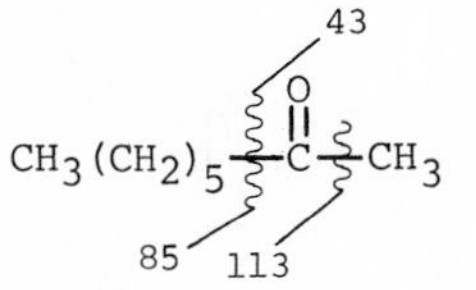

McLafferty rearrangement gives:

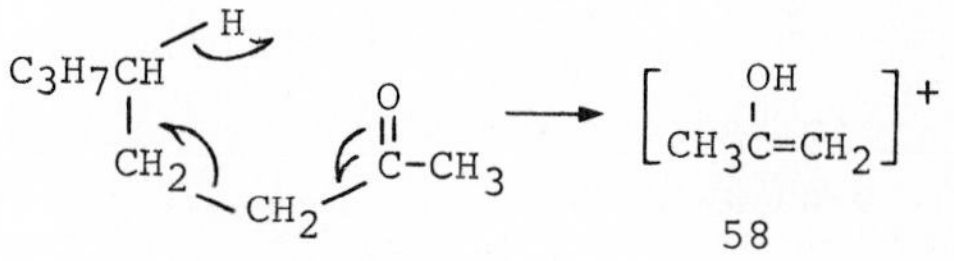

11.

			$\Delta H°$ (kcal mole^{-1})
$[(CH_3)_2CHCH_2CH_3]^+$	⟶	$CH_3^+ + (CH_3)_2CHCH_2^\cdot$	71.7
		$CH_3^\cdot + (CH_3)_2CHCH_2^+$	37.2
		$(CH_3)_2CH^+ + CH_3CH_2^\cdot$	14.2
		$(CH_3)_2CH^\cdot + CH_3CH_2^+$	34.7
		$CH_3^\cdot + CH_3\overset{+}{C}HCH_2CH_3$	24.2
		$CH_3^+ + CH_3\dot{C}HCH_2CH_3$	70.2

$$(CH_3)_2CH^+ > CH_3\overset{+}{C}HCH_2CH_3 > (CH_3)_2CHCH_2^+ > CH_3CH_2^+ \gg CH_3^+$$

17.E Supplementary Problems

S1. Predict the major peaks in the mass spectra of the following compounds:

(a) $CH_3CH_2\underset{}{C}H(CH_3)CH_2CH_2CH(CH_3)_2$ [i.e., $CH_3CH_2CHCH_2CH_2CH(CH_3)_2$ with CH_3 on C-3]

(b) $(CH_3)_2CHCH(OH)CH_2CH_3$

(c) $(CH_3)_3CCH_2CH_2CH_2CH_2OH$

(d) $CH_3\overset{O}{\overset{\|}{C}}CH_2CH_2CH(CH_2CH_3)_2$

(e) $(CH_3)_2CHCH_2OC(CH_3)_3$

(f) 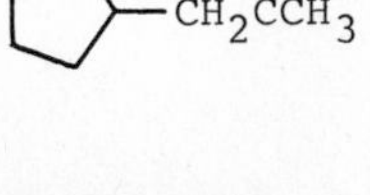

S2. From the following accurate mass measurements, determine the most likely formula for the molecule or fragment.

(a) m/e = 70.0419 (b) m/e = 56.0373 (c) m/e = 81.0861

S3. Deduce the structure of each of the compounds below, and write a mechanism showing the principle fragments in each mass spectrum.

(a) $\mathrm{A} \xrightarrow[\text{2. } H_2O_2,\ OH^-]{\text{1. } B_2H_6} \mathrm{B} + \mathrm{C}$ *Both B and C show a strong band at 1710 cm^{-1} in the ir.*

mass spectrum of B: m/e 114, 86, 85, 57

mass spectrum of C: m/e 114, 99, 58, 43

(b) $\mathrm{D} \xrightarrow{(C_6H_5)_3P{=}CH_2} \mathrm{E}$

$\mathrm{E} \xrightarrow[\text{2. } NaBH_4]{\text{1. } Hg(OAc)_2,\ H_2O} \mathrm{F}$

$\mathrm{E} \xrightarrow[\text{dilute } H_2SO_4]{} \mathrm{G}$

mass spectrum of D: m/e 86, 58, 29
mass spectrum of F: m/e 102, 87, 84, 45
mass spectrum of G: m/e 87, 84, 73

(c) $\mathrm{H} \xrightarrow{C_3H_7MgCl} \xrightarrow[\Delta]{H_2SO_4} \mathrm{I} \xrightarrow[\text{2. } H_2O_2,\ OH^-]{\text{1. } B_2H_6} \mathrm{J} \xrightarrow[H_2SO_4]{K_2Cr_2O_7} \mathrm{K}$

nmr spectrum of H: δ 0.9 (3H,t), 1.2-1.4 (4H,m), 2.1 (2H,dt), 9.5 (1H,t).

mass spectrum of K: m/e 128, 86, 85, 71

17.F Answers to Supplementary Problems

S1. (a)

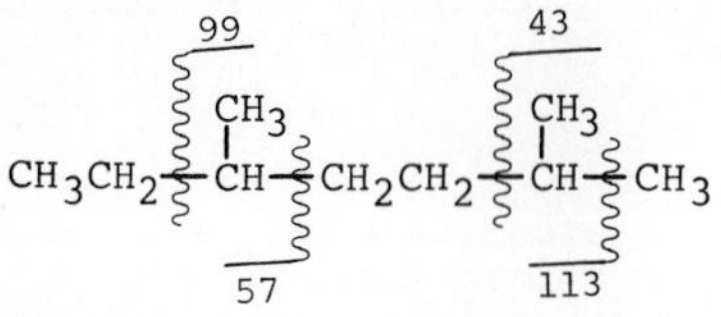

$M^+ = 128$

(b)

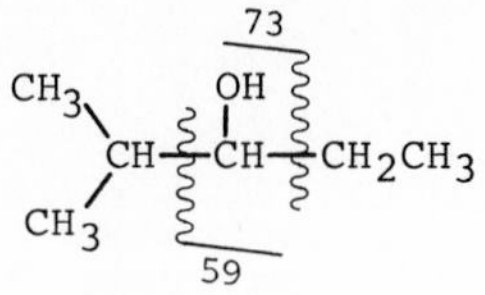

$M^+ = 102$; $M - H_2O = 84$

(c)

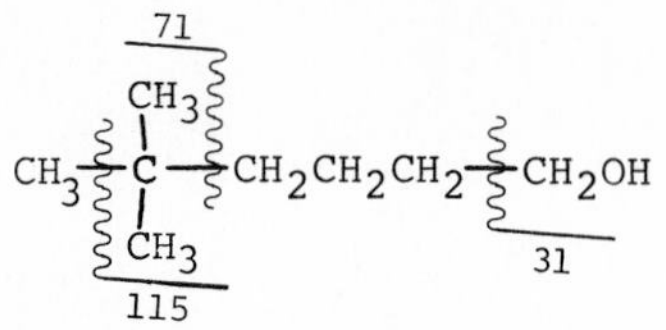

$M^+ = 130$; $M - H_2O = 112$

(d)

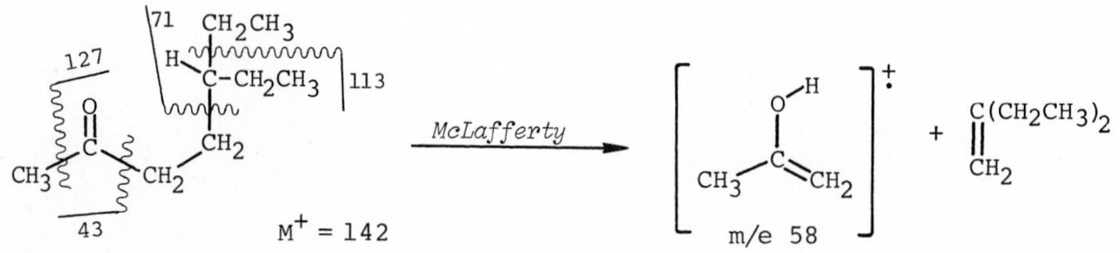

$M^+ = 142$

(e)

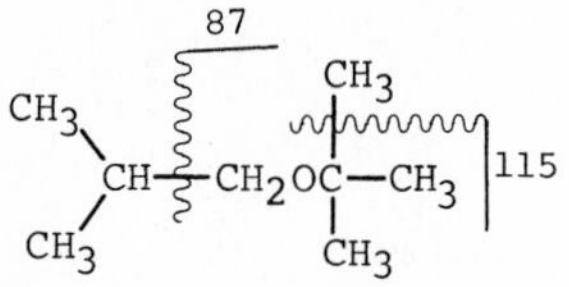

($M^+ = 130$); M - (ROH) = 56

(f)

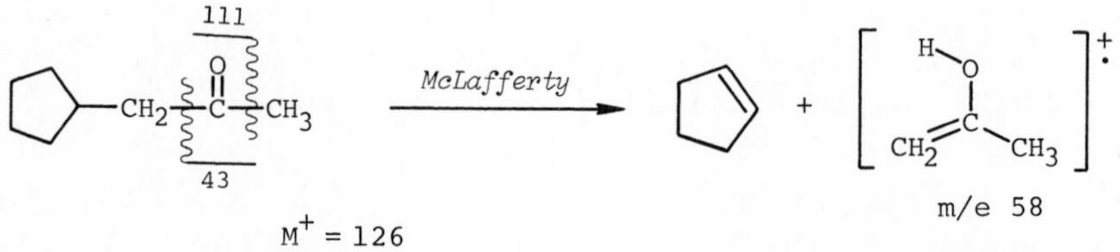

$M^+ = 126$

S2. (a)

Formula	*Calculated Exact Mass*
C_5H_{10}	70.07825
C_4H_6O	70.04186 ←
$C_3H_6N_2$	70.05305
$[C_4H_8N]^{\bullet}$	70.06565

(b)

C_4H_8	56.0626
C_3H_4O	56.0262
$C_2H_4N_2$	56.0374 ←
$[C_3H_6N]^{\bullet}$	56.0500

(c)

Formula	*Calculated Exact Mass*
$[C_5H_{11}]^{\bullet}$	81.08608 ←
$[C_4H_7O]^{\bullet}$	81.04969
$[C_3H_7N_2]^{\bullet}$	81.06088
C_4H_9N	81.07348

S3. (a)

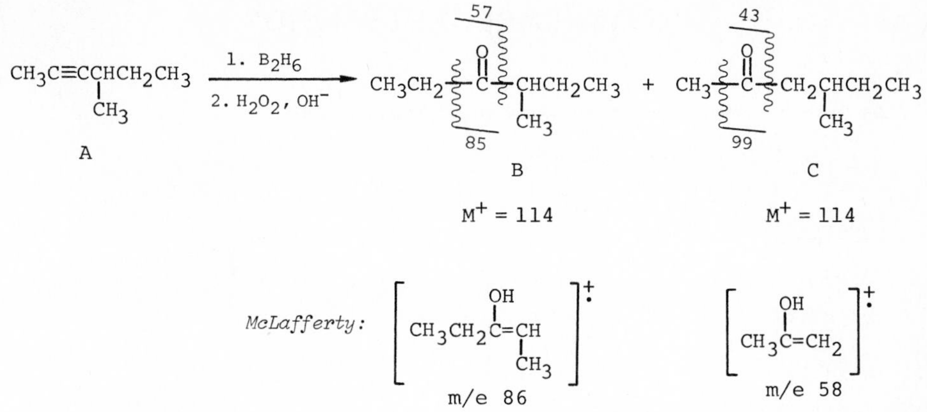
CH3C≡CCHCH2CH3
CH3
A
1. B2H6
2. H2O2, OH−
57
85
CH3CH2C(=O)CHCH2CH3
CH3
B
M+ = 114
43
99
CH3C(=O)CH2CHCH2CH3
CH3
C
M+ = 114
McLafferty:
OH
CH3CH2C=CH
CH3
m/e 86
OH
CH3C=CH2
m/e 58

(b)

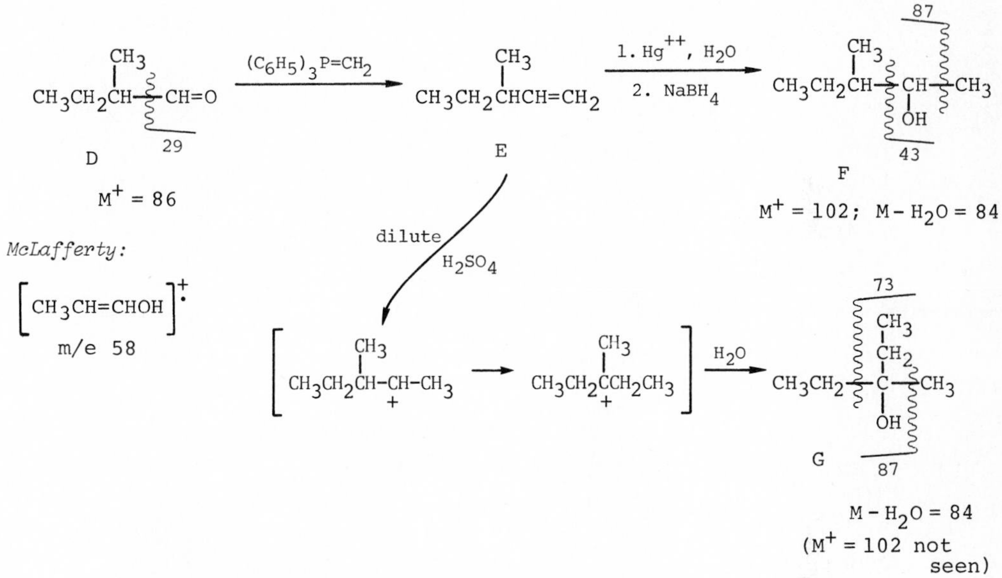
CH3
CH3CH2CH-CH=O
29
D
M+ = 86
(C6H5)3P=CH2
CH3
CH3CH2CHCH=CH2
E
1. Hg++, H2O
2. NaBH4
87
CH3
CH3CH2CH-CH-CH3
OH
43
F
M+ = 102; M - H2O = 84
McLafferty:
CH3CH=CHOH
m/e 58
dilute
H2SO4
CH3
CH3CH2CH-CH-CH3
+
CH3
CH3CH2CCH2CH3
+
H2O
73
CH3
CH2
CH3CH2-C-CH3
OH
G
87
M - H2O = 84
(M+ = 102 not seen)

(c)

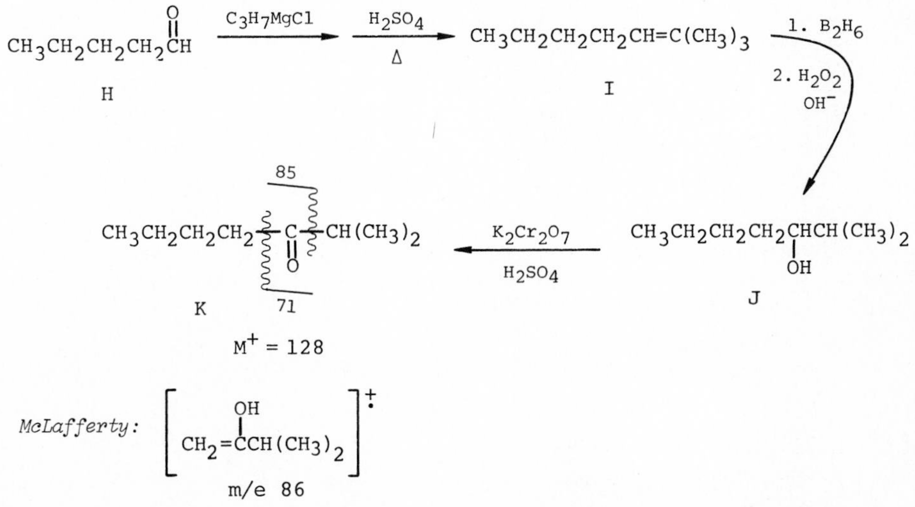
O
CH3CH2CH2CH2CH
H
C3H7MgCl
H2SO4
Δ
CH3CH2CH2CH2CH=C(CH3)3
I
1. B2H6
2. H2O2
OH−
CH3CH2CH2CH2CHCH(CH3)2
OH
J
K2Cr2O7
H2SO4
85
CH3CH2CH2CH2-C-CH(CH3)2
O
71
K
M+ = 128
McLafferty:
OH
CH2=CCH(CH3)2
m/e 86

18. CARBOXYLIC ACIDS

18.A Chapter Outline, Important Terms Introduced, and Reactions Discussed

18.1 <u>Structure</u>

18.2 <u>Nomenclature</u>

-oic acid   1,2,3, ... vs. α, β, γ...

18.3 <u>Physical Properties</u>

18.4 <u>Acidity</u>

A. Ionization
conjugated system

$$R-C\overset{O}{\underset{OH}{\diagup}} \rightleftharpoons \left[R-C\overset{O}{\underset{O^-}{\diagup}} \longleftrightarrow R-C\overset{O^-}{\underset{O}{\diagup}} \right] + H^+$$

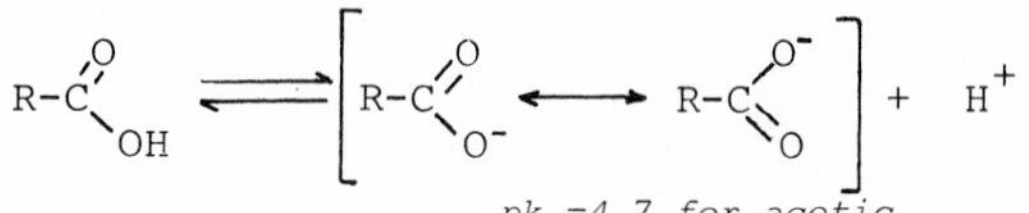

$pk_a = 4.7$ *for acetic*
acid

B. Inductive Effects

C. Salt Formation
-ate

D. Soaps

biodegradable   alkylbenzenesulfonates

18.5 <u>Spectroscopy</u>

A. Nuclear Magnetic Resonance

$\underline{CH_3}COOH$ 2 ppm $CH_3COO\underline{H}$ 10-13 ppm

B. Infrared

C=O   1710-1760 cm^{-1} O-H   2300-3000 cm^{-1}

18.6 <u>Synthesis</u>

A. Hydrolysis of Nitriles

$$R-C \equiv N + H_2O \xrightarrow[\Delta]{H^+ \text{ or } OH^-} RCO_2H + NH_3$$

B. Carbonation of Organometallic Reagents

$$RM + CO_2 \longrightarrow RCO_2^- M^+ \xrightarrow{H^+} RCO_2H \quad M = Li, \ Na, \ MgBr \ (\textit{see Section 15.5})$$

C. Oxidation of Primary Alcohols or Aldehydes

$$RCH_2OH \longrightarrow RCO_2H \quad (\text{using } KMnO_4 \text{ or } K_2Cr_2O_7, \Delta)$$

$$RCHO \longrightarrow RCO_2H \quad (\text{using } KMnO_4, \ HNO_3, \text{ or } Ag_2O)$$

18.7 <u>Reactions</u>

A. Reactions Involving the O-H bond

salt formation:

$$RCO_2H + NaOH \longrightarrow RCO_2^- Na^+ + H_2O$$

esterification with diazomethane:

$$RCO_2H + CH_2N_2 \longrightarrow RCO_2CH_3 + N_2$$

alkylation of salts:

$$RCO_2^- \overset{+}{M} + R'X \longrightarrow RCO_2R' + MX \quad (S_N2 \text{ reaction})$$

B. Reactions Involving the Hydrocarbon Side Chain

Hell-Volhard-Zelinsky reaction:

$$3\ RCH_2CO_2H + PX_3 + 3\ X_2 \longrightarrow 3\ RCH(X)C(=O)X + H_3PO_3 + HBr$$

(X = Cl, Br; acyl halide enolization at α-position)

C. Reactions Occurring at the Carbonyl Center

1. Base-catalyzed Nucleophilic Additions

amide-formation:

$$RCO_2H + NH_3 \longrightarrow RCO_2^- \ ^+NH_4 \xrightarrow{185°C} RC(=O)NH_2 \quad \text{(not commonly used)}$$

lithium aluminum hydride reaction:

$$RCO_2H + LiAlH_4 \longrightarrow \xrightarrow{H^+} RCH_2OH \quad (NaBH_4 \text{ does } \underline{\text{not}} \text{ do this})$$

2. Acid-catalyzed Nucleophilic Additions

esterification:

$$RCO_2H + HOR' \overset{H^+}{\rightleftharpoons} RCO_2R' + H_2O \quad \text{(equilibrium process)}$$

acylium ion formation;

$$RCO_2H + H_2SO_4 \longrightarrow R\text{-}C{\equiv}O^+ \xrightarrow[R''OH]{\text{excess}} RC(=O)OR''$$

(good for hindered, aromatic carboxylic acids)

acyl halide formation:

$$RCO_2H \longrightarrow RC(=O)X + HX \quad (\text{using } XS(=O)X,\ PX_3,\ \text{or } PX_5;\ X = Cl, Br)$$

D. One-Carbon Degradation of Carboxylic Acids

Hunsdiecker reaction:

$$RCO_2H + (Ag_2O \text{ or } HgO) \longrightarrow RCO_2M \xrightarrow{X_2} RX + CO_2 \quad (X = Br \text{ or } I)$$

Kochi reaction:

$$RCO_2H + Pb(O_2CCH_3)_4 + LiCl \longrightarrow RCl$$

18.8 <u>Occurrence of Carboxylic Acids</u>

18.B Important Concepts and Hints

An important, new type of reaction mechanism is quietly introduced in this chapter: the nucleophilic replacement of a substituent on a carbonyl group. The acid-catalyzed formation of an ester from a carboxylic acid and an alcohol is the most important example of this type. The mechanism involves first addition of the nucleophile to the carbon-oxygen double bond, and then elimination of the leaving group to regenerate the carbonyl group. The full scope of this substitution mechanism will become apparent in the following chapter, where it will be discussed at length. For the moment, you should examine closely the comparison of the esterification reaction and the formation, then decomposition of a hemiketal. It's a good analogy, and it helps tie this new mechanism in with one you've seen before.

Aside from that, no other new concepts are introduced in this chapter. The effects of electron-withdrawing substituents on acidity was presented in the chapter on alcohols (section 10.4), and the synthetic reactions which lead to carboxylic acids have been introduced previously. Although the reactions of carboxylic acids are for the most part new, their mechanisms involve the nucleophilic displacement, enolization, or carbonyl-addition mechanisms you have seen before. Consequently there is little else for us to comment on in this section of the Study Guide. This situation will become more common in subsequent chapters of this book, although we will still make (hopefully) helpful comments where appropriate.

18.C Answers to Exercises

(18.2.A) (a) 5-chloropentanoic acid

(b) 5,5-dimethylhexanoic acid

(c) 2-iodo-4-methylhexanoic acid

(d) 2,4-pentadienoic acid (or penta-2,4-dienoic acid)

(18.4.A) $pK_a = -\log K_a = 3.75$

$$\frac{[H^+][HCO_2^-]}{[HCO_2H]} = 1.77 \times 10^{-4} \qquad [H^+] = [HCO_2^-] = x; \qquad [HCO_2H] = 0.1 - x$$

$$\frac{x^2}{0.1 - x} = 1.77 \times 10^{-4}; \quad x = [HCO_2^-] = 4.12 \times 10^{-3}\underline{M}$$

(18.4.B)

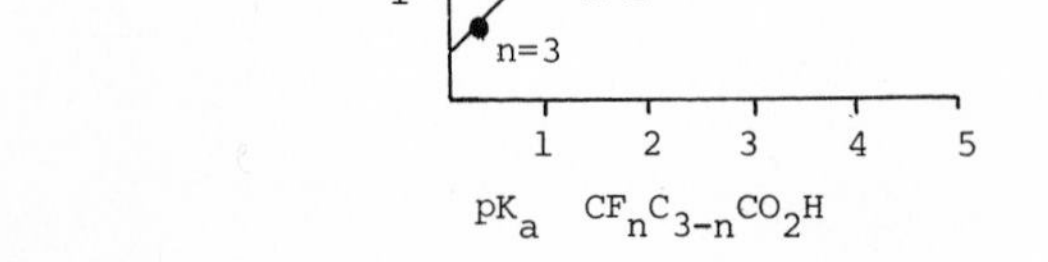

pK_a predicted for F_2CHCO_2H = 0.9

(18.6.C)

(a) $\xrightarrow{Mg}$ $\xrightarrow{CO_2}$ $\xrightarrow{H_3O^+}$ *(NaCN would lead to elimination)*

(b) $\xrightarrow{NaCN}$ $\xrightarrow[\Delta]{H_3O^+}$ *(to make a Grignard reagent would require protection of the hydroxy group)*

(c) $\xrightarrow{B_2H_6}$ $\xrightarrow[OH^-]{H_2O_2}$ $\xrightarrow[H_2SO_4]{K_2Cr_2O_7}$

(18.7.A)

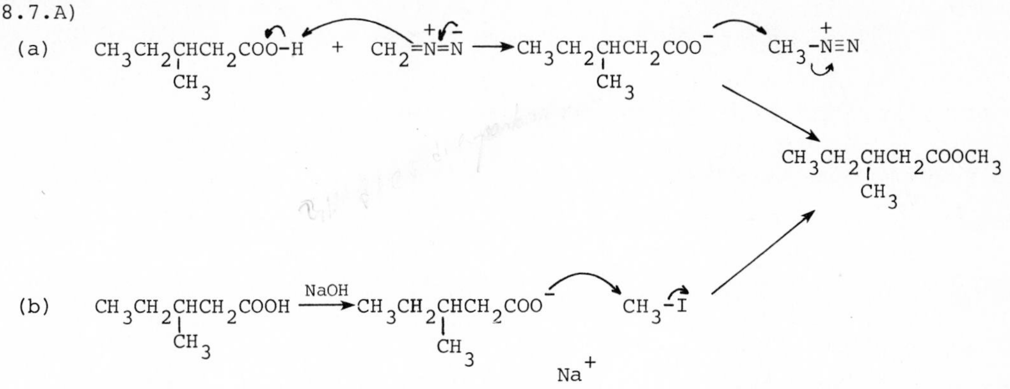

(18.7.B) $CH_3COOH \rightleftharpoons CH_3COO^- + H^+$

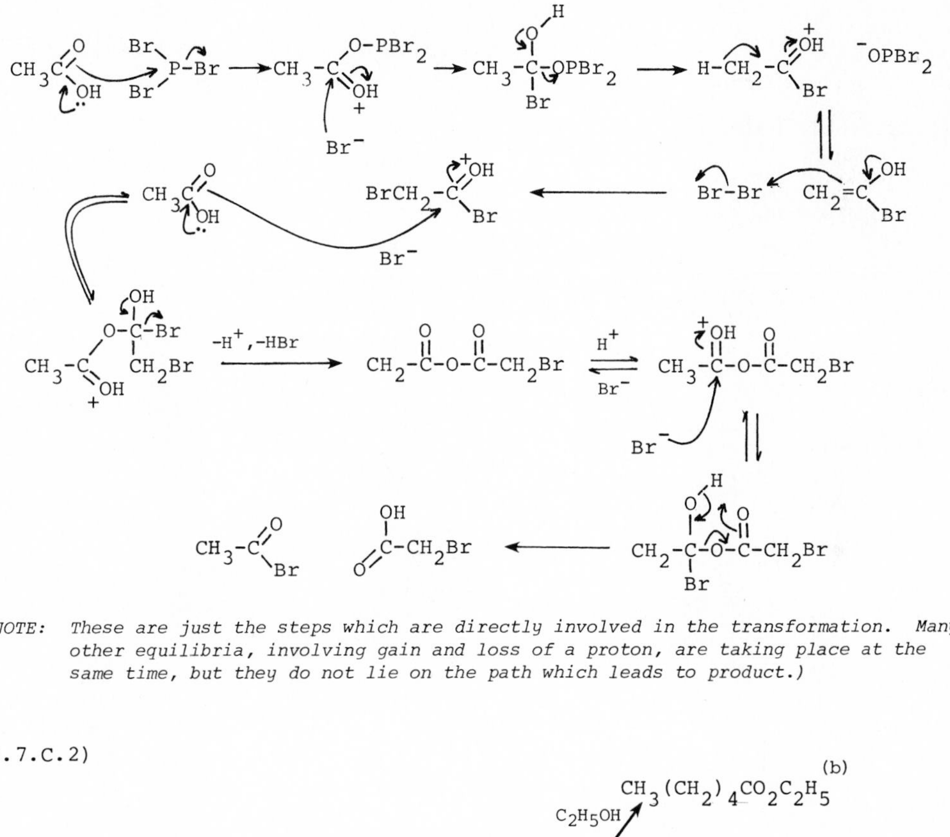

NOTE: These are just the steps which are directly involved in the transformation. Many other equilibria, involving gain and loss of a proton, are taking place at the same time, but they do not lie on the path which leads to product.)

(18.7.C.2)

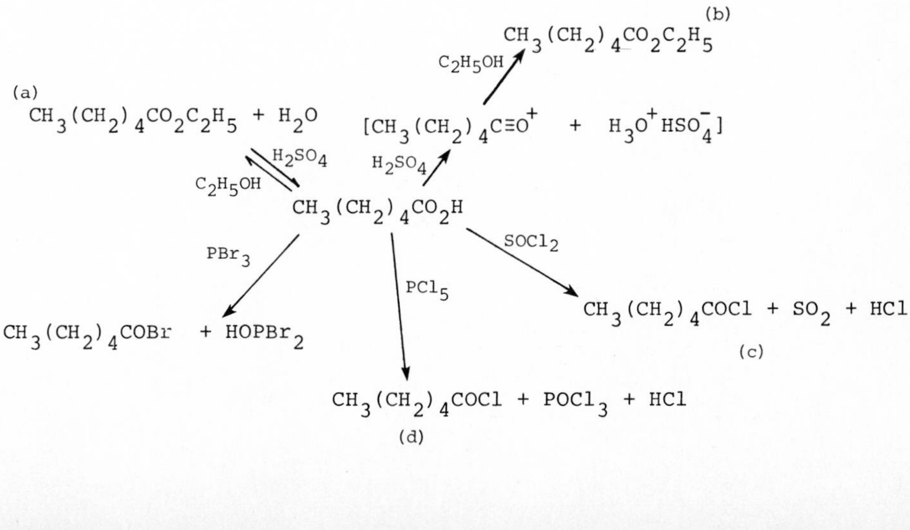

(18.7.D)

(a) $CH_3(CH_2)_4CO_2H \xrightarrow{HgO} (CH_3(CH_2)_4CO_2)_2Hg \xrightarrow{2\ Br_2} 2\ CH_3(CH_2)_4Br + HgBr_2 + CO_2$

(b) $CH_3(CH_2)_4Br + NaCN \longrightarrow CH_3(CH_2)_4C{\equiv}N \xrightarrow[\Delta]{H_3O^+} CH_3(CH_2)_4CO_2H$

18.D Answers and Explanations to Problems

1. IUPAC

a) 3-methylpentanoic acid

b) 2,2-dimethylpropanoic acid
(Note: trivial name is pivalic acid)

c) 4-bromobutanoic acid

d) iodoacetic acid (Note: iodoethanoic acid is a systematic name that is never used in practice)

e) 2-hydroxybutanoic acid

f) decanoic acid

g) cyclobutanecarboxylic acid

h) 3-methoxybutanoic acid

i) 3,4-dimethylpentanoic acid

2. a) 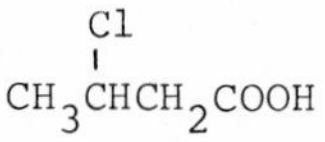 $CH_3CH(Cl)CH_2COOH$

b) $CH_3(CH_2)_4COOH$

c) 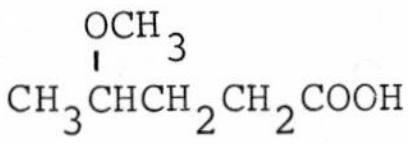 $CH_3CH(OCH_3)CH_2CH_2COOH$

d) 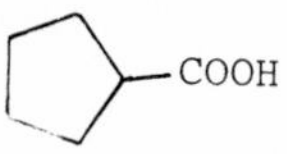

e) 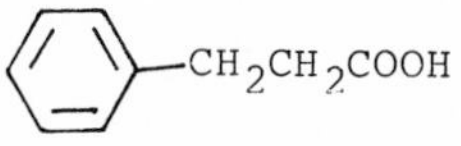

f) 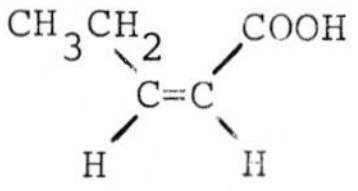

g) 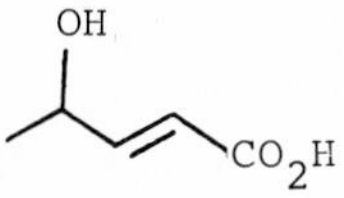

h) 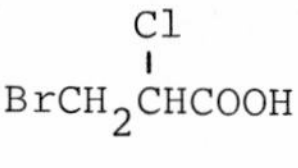 $BrCH_2CH(Cl)COOH$

3. a) 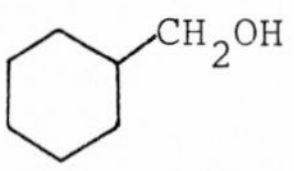

b) 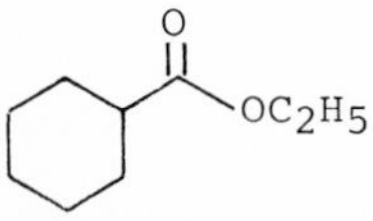

c) 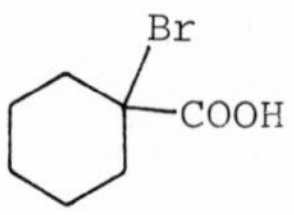

d) 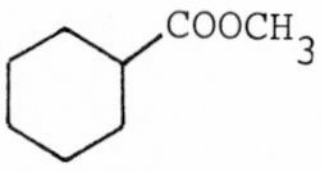

e) 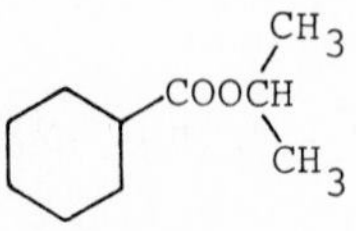

f) 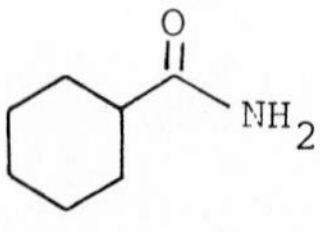

g) 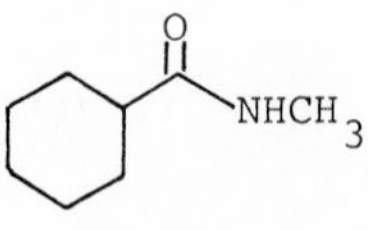

h) 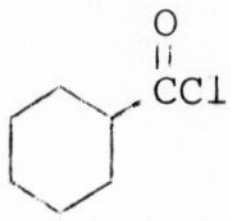

i) 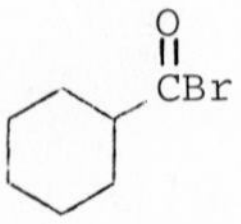

j) NO REACTION

k) 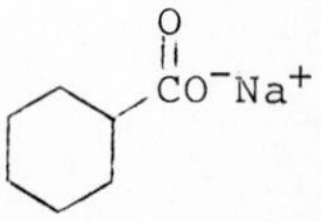

l) 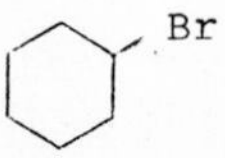

m) 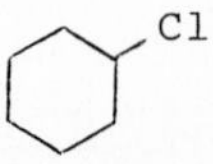

4. a) Grignard (displacement reaction is poor with a tertiary halide).
   b) cyanide (reaction of $BrCH_2CH_2Br$ with Mg $\longrightarrow$ C_2H_4 + $MgBr_2$).
   c) cyanide (the Grignard reagent reacts with carbonyl group).
   d) Grignard (displacement reaction is slow with neopentyl halides).
   e) Both methods work well.
   f) cyanide (Grignard reagent reacts with -OH).

5. a)

$$CH_3-\overset{CH_3}{\underset{CH_3}{\overset{|}{\underset{|}{C}}}}-CH_3 \xrightarrow[h\nu]{Br_2} (CH_3)_3CCH_2Br \xrightarrow[ether]{Mg} (CH_3)_3CCH_2MgBr \xrightarrow{CO_2} \xrightarrow{H^+} (CH_3)_3CCH_2COOH$$

b) $(CH_3)_3CCH_2COOH$ (from a) $\xrightarrow[P]{Br_2} \xrightarrow{H_2O} (CH_3)_3CCHBrCOOH$

c) $(CH_3)_3CCH_2MgBr$ (from a) $\xrightarrow{CH_2O} \xrightarrow{H^+} (CH_3)_3CCH_2CH_2OH$

or $(CH_3)_3CCH_2COOH$ (from a) $\xrightarrow{LiAlH_4} \xrightarrow{H^+} (CH_3)_3CCH_2CH_2OH$

d) $(CH_3)_3CCH_2COOH$ (from a) $\xrightarrow[\Delta]{SOCl_2} (CH_3)_3CCH_2COCl$

6. a) $CH_3CH_2CH_2CHO \xrightarrow{Ag_2O \text{ or } HNO_3} CH_3CH_2CH_2COOH$

b) $CH_3CH_2CH_2CHO \xrightarrow{\text{dil. } OH^-} CH_3CH_2CH_2CH{=}\overset{CHO}{\overset{|}{C}}CH_2CH_3 \xrightarrow{Ag_2O} CH_3CH_2CH_2CH{=}\overset{COOH}{\overset{|}{C}}CH_2CH_3$

c) $CH_3CH_2CH_2CHO \xrightarrow{HCN} CH_3CH_2CH_2\overset{OH}{\overset{|}{C}}HCN \xrightarrow[\Delta]{H+} CH_3CH_2CH_2\overset{OH}{\overset{|}{C}}HCOOH$

d) $CH_3CH_2CH_2CHO \xrightarrow[LiAlH_4]{NaBH_4 \text{ or}} CH_3CH_2CH_2CH_2OH \xrightarrow{PBr_3} CH_3CH_2CH_2CH_2Br \xrightarrow[ether]{Mg} \xrightarrow{CO_2} \xrightarrow{H^+} CH_3CH_2CH_2CH_2COOH$

7. a) $LiAlH_4$

b) $(CH_3)_2CHCH_2COOH \xrightarrow{LiAlH_4} (CH_3)_2CHCH_2CH_2OH \xrightarrow{PBr_3} \xrightarrow{CN^-} \xrightarrow[\Delta]{H^+ \text{ or } OH^-} (CH_3)_2CHCH_2CH_2COOH$

c) $(CH_3)_2CHCH_2CH_2COOH$ (from b) $\xrightarrow{LiAlH_4} (CH_3)_2CHCH_2CH_2CH_2OH \xrightarrow{PBr_3} \xrightarrow[ether]{Mg} \xrightarrow{CH_2O} (CH_3)_2CHCH_2CH_2CH_2CH_2OH \xrightarrow[\Delta]{Al_2O_3} (CH_3)_2CHCH_2CH_2CH{=}CH_2$

$(CH_3)_2CHCH_2CH_2CH{=}CH_2 \xrightarrow{Br_2} (CH_3)_2CHCH_2CH_2CHBrCH_2Br$

d) $(CH_3)_2CHCH_2COOH \xrightarrow[Br_2]{P} (CH_3)_2CHCHBrCOOH \xrightarrow[H^+,\ \Delta]{CH_3OH} (CH_3)_2CHCHBrCOOCH_3 \xrightarrow[CH_3OH]{CH_3ONa} (CH_3)_2C{=}CHCOOCH_3$

e) 1. CH_3Li
   2. H_2O f) NH_3, Δ g) Ag_2O, Br_2

8. a)

  b) $(CH_3)_3CCH=CH_2$ $\xrightarrow{KMnO_4}$ $(CH_3)_3CCOOH$

      or $\xrightarrow{O_3}$ $\xrightarrow[CH_3COOH]{Zn}$ $(CH_3)_3CCHO$ $\xrightarrow{Ag_2O}$

  c) $CH_3COCH_2CH_2C(CH_3)_2Br$ $\xrightarrow[H^+]{HOCH_2CH_2OH}$

  d) CH_3CH_2COOH $\xrightarrow{LiAlH_4}$ $CH_3CH_2CH_2OH$ $\xrightarrow{PBr_3}$ $CH_3CH_2CH_2Br$

     $CH_3CH_2CH_2COOH$ $\xleftarrow[\Delta]{H^+ \text{ or } OH^-}$ $\xleftarrow{CN^-}$

  e) $CH_3CH_2CH_2COOH$ $\xrightarrow[Br_2]{Ag_2O}$ $CH_3CH_2CH_2Br$ $\xrightarrow{OH^-}$ $CH_3CH_2CH_2OH$

     CH_3CH_2COOH $\xleftarrow{HNO_3}$

  f) $CH_3CH_2CH_2Br$ (from e) $\xrightarrow{N_3^-}$ $CH_3CH_2CH_2N_3$

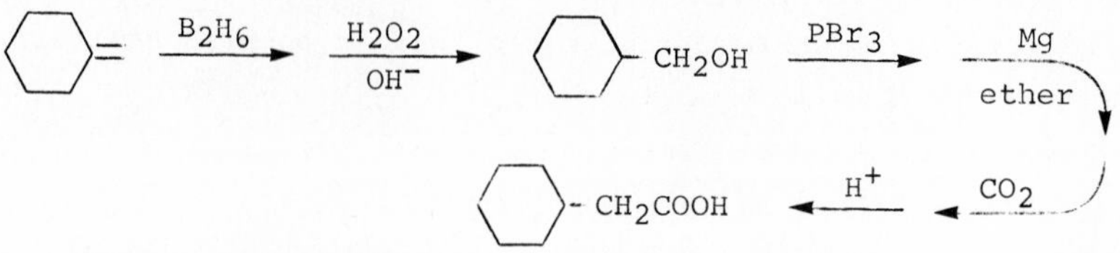

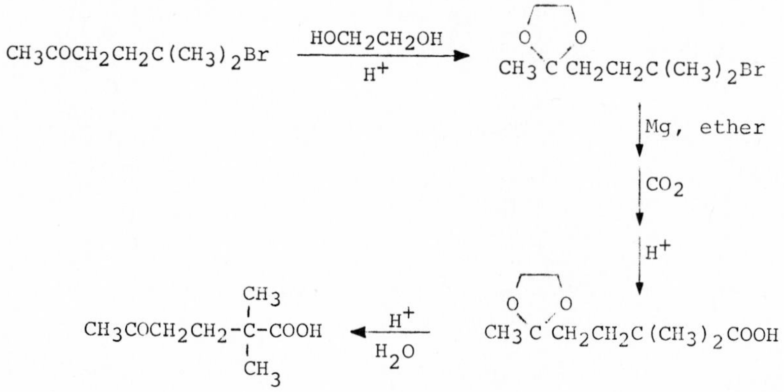

9. The dissociation equilibrium may be written as:

$$HOAc \rightleftharpoons H^+ + AcO^-$$

The equilibrium constant, K, is given by the expression:

$$K = \frac{[H^+][AcO^-]}{[HOAc]} = 1.8 \times 10^{-5}M$$

For each H^+ produced, one AcO^- is also produced. Therefore,

$[H^+] = [AcO^-]$. If we let $x = [H^+] = [AcO^-]$, then [HOAc] = the nominal concentration of HOAc - x.

$$\frac{x^2}{[HOAc]_{nom} - x} = 1.8 \times 10^{-5} \qquad (1)$$

The per cent dissociation will be given by:

$$100 \frac{[H^+]}{[HOAc]}$$

There are two ways to approach this problem. For each value of $[HOAc]_{nom}$, we may solve the quadratic equation (1), or we may simplify by assuming that x is so small that $[HOAc]_{nom} - x \approx [HOAc]_{nom}$.

(a) Applying the simplifying assumption that $0.1 - x \approx 0.1$, we have:

$$\frac{x^2}{0.1} = 1.8 \times 10^{-5}$$

$$x^2 = 1.8 \times 10^{-6}$$

$$x = 1.34 \times 10^{-3}$$

$$\% \text{ dissociation} = \frac{1.34 \times 10^{-3}}{0.1} (100) = 1.34\%$$

(b) $x = 4.24 \times 10^{-4}$; % dissociation = 4.24%

(c) $x = 1.34 \times 10^{-4}$; % dissociation = 13.4%

A more rigorous solution may be obtained by solution of the quadratic,

$$ax^2 + bx + c = 0$$

$$x = \frac{-b \pm \sqrt{b^2 - 4ac}}{2a}$$

Applying this approach to the three concentrations, we find the following % dissociation values:

(a) 1.35% (b) 4.33% (c) 14.35%

Note that the simplfying assumption that $[HOAc]_{nom} \approx (HOAc)_{nom} - [H^+]$ is a useful one, even at very low concentrations. Even at 0.001 M, the approximate value of $[H^+]$ obtained in this way is 90% of the true value of $[H^+]$.

Also note that the *% dissociation increases as the concentration decreases*. For such an equilibrium, *dissociation is a unimolecular reaction*, and *recombination is a bimolecular reaction*.

$$HOAc \underset{\text{bimolecular, } k_2}{\overset{\text{unimolecular, } k_1}{\rightleftharpoons}} H^+ + AcO^-$$

Recall that the equilibrium constant may be viewed as the ratio of the rates of the forward and reverse reactions:

$$K = \frac{\text{rate of forward reaction}}{\text{rate of reverse reaction}} = \frac{k_1\,[HOAc]}{k_2\,[H^+][AcO^-]}$$

Consider a solution of HOAc which is at equilibrium with H^+ and AcO^-. When we dilute the sample by a factor of 10, the rate of dissociation decreases by a factor of 1/10:

$$\text{rate} = k_1 \cdot \frac{[HOAc]}{10}$$

But the rate of recombination decreases by a factor of 1/100:

$$\text{rate} = k_2 \cdot \frac{[H^+]}{10} \cdot \frac{[AcO^-]}{10}$$

Since the forward reaction is now faster than the reverse reaction, more HOAc dissociates, reducing [HOAc] and increasing $[H^+]$ and $[AcO^-]$ until equilibrium is restored.

10 $pK_a = -\log K_a$

a) 4.85 b) 4.20 c) 1.26 d) 0.64 e) 3.68

11. a) $CH_3CH_2O^-$; in acetate ion the negative charge is shared between two oxygens, whereas in ethoxide ion the charge is essentially localized on a single oxygen. Ethoxide ion is therefore relatively less stable and has greater tendency to react with a proton.

b) $CH_3CH_2CH_2CO_2^-$; β-chloropropionate is stabilized by the electron-attracting inductive effect of chlorine.

c) $ClCH_2CH_2CO_2^-$; in α-chloropropionate ion, the chlorine is closer to the center of negative charge and provides greater stabilization.

d) $FCH_2CO_2^-$; two electron-attracting fluorines provide greater stabilization of the anion than does one.

e) $CH_3CH_2CH_2CO_2^-$; the ethynyl group has a somewhat electron-attracting inductive effect compared to an alkyl group, and provides greater stabilization of an anion.

f) $CH_3CO_2^-$; HCl is a stronger acid than acetic acid, hence acetate ion is a stronger base than chloride ion. The reason why HCl is the stronger acid is, however, more difficult to explain. HCl is a stronger acid than CH_3COOH even in the gas phase, but the difference, $\Delta pK = 4\text{-}5$, is smaller than in aqueous solution, $\Delta pK = 11$. This difference means that the solvation energy of Cl^- is greater than that for $CH_3CO_2^-$. The greater acidity of HCl in the gas phase appears to be due in part to the greater bond strength of $CH_3COO\text{-}H$ compared to H-Cl, and in part to the higher electron affinity of $Cl^\bullet$ compared to $CH_3CO_2^\bullet$.

12. $HOOCCH_2CH_2CH(Cl)CH_2\underline{COOH}$

more acidic (Cl is closer).

13. Compare the effect of chlorine substitution on pk_a for the various isomers:

pk_a (butanoic acid) - pk_a (2-chlorobutanoic acid) = 1.90

"   "   "   - pk_a (3-chlorobutanoic acid) = 0.77 } ratio = 2.55

"   "   "   - pk_a (4-chlorobutanoic acid) = 0.30 } ratio = 2.56

pk_a (butanoic acid) - pk_a (4-cyanobutanoic acid) = 0.38, therefore

pk_a (butanoic acid) - pk_a (3-cyanobutanoic acid) is expected to be

   2.56 × 0.38 = 0.97: 4.82 - 0.97 = 3.85 = calculated pk_a of
   3-cyanobutanoic acid. Note that the pk_a of 3-cyanobutanoic acid
   has yet to be determined experimentally.

14.
$$CH_3CH_2\overset{O}{\overset{\|}{C}}OH + H^+ \rightleftharpoons CH_3CH_2\overset{+OH}{\overset{\|}{C}}OH$$

$$CH_3CH_2\overset{+OH}{\overset{\|}{C}}OH + H_2{}^{18}O \rightleftharpoons CH_3CH_2\overset{OH}{\underset{OH}{C}}{-}{}^{18}OH_2{}^+$$

$$CH_3CH_2\overset{OH}{\underset{OH}{C}}{-}{}^{18}OH_2{}^+ \rightleftharpoons CH_3CH_2\overset{OH}{\underset{OH}{C}}{-}{}^{18}OH + H^+ \rightleftharpoons CH_3CH_2\overset{+OH_2}{\underset{OH}{C}}{-}{}^{18}OH$$

$$CH_3CH_2\overset{+OH_2}{\underset{OH}{C}}{-}{}^{18}OH \rightleftharpoons H_2O + CH_3CH_2\overset{}{\underset{OH}{C}}{=}{}^{18}OH^+ \rightleftharpoons H^+ + CH_3CH_2\overset{{}^{18}O}{\overset{\|}{C}}OH$$

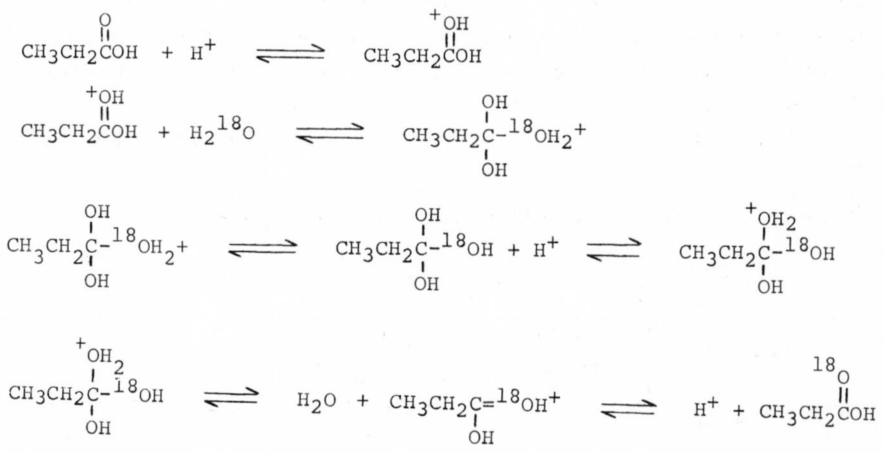

15.

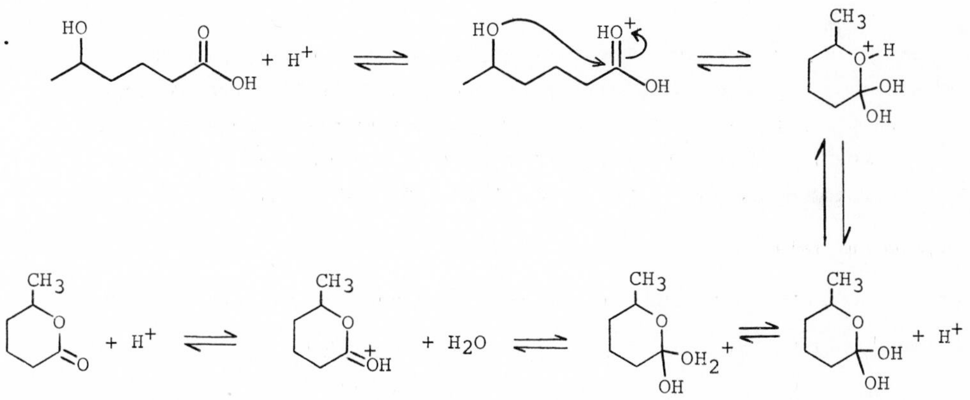

In normal esterification, two reactant molecules (alcohol and acid) give two product molecules (ester and water), with a consequent $\Delta S°$ of about 0. In the present cyclization, one molecule gives two. The additional freedom of motion of the products corresponds to a positive $\Delta S°$ and a larger equilibrium constant.

16.

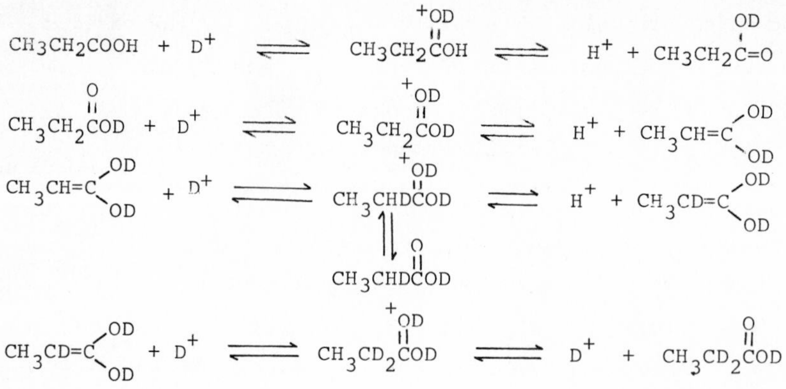

17. ΔH° (gas) = −4.6 kcal $mole^{-1}$. The liquid phase involves solvation energies not present in the ideal gas state. The reactants are solvated more strongly than the products, probably because of hydrogen-bonding, and the esterification reaction is less exothermic.

18. a) The C-Cl dipole stabilizes the carboxylate ion by electrostatic attraction.

b) Trigonometry gives the following geometric results:

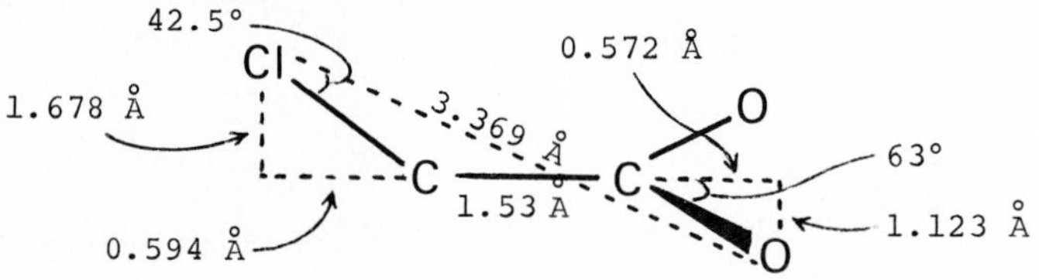

$$E = \frac{(q=0.5)(\mu=1.9)(\cos 42.5^\circ)}{(3.369)^2}(69 \text{ kcal mol}^{-1}) = 4.3 \text{ kcal mol}^{-1}$$

This result is per oxygen. Total stabilization of both oxygens is 8.5 kcal mol^{-1}. The acidity difference between $ClCH_2COOH$ and CH_3COOH in the gas phase corresponds to an energy difference of about 13 kcal mol^{-1}; hence, even this crude electrostatic calculation on one conformer gives a result of the right order of magnitude.

18.E Supplementary Problems

S1. Give the IUPAC name for each of the following compounds:

(a) $BrCH_2CH_2CH_2CO_2H$

(b) 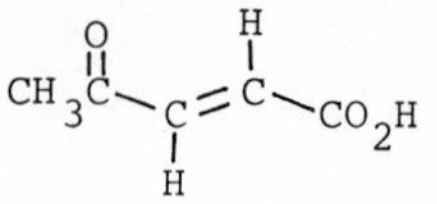

(c) 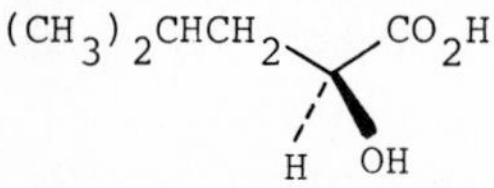

(d) 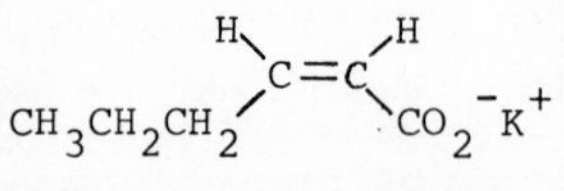

(e) 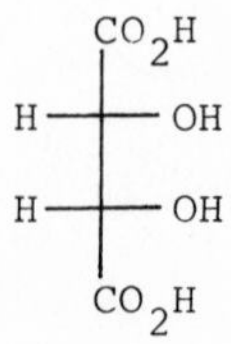

(f) 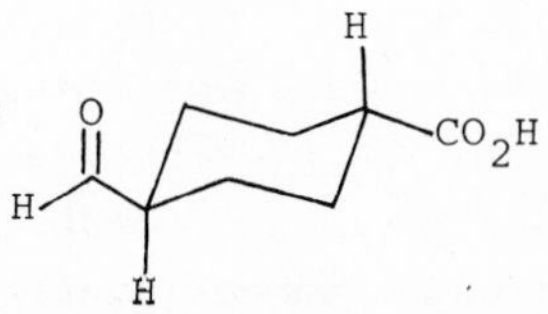

(g) $CH_2=C=CH-CO_2H$

S2. Write out the structure of each of the following compounds:

(a) γ,γ-dibromobutyric acid

(b) 3-chloro-2-hydroxypropanoic acid

(c) sodium 4-hydroxybutanoate

(d) 2-oxopropanoic acid

(e) ammonium valerate

(f) *cis*-cyclopentane-1,2-dicarboxylic acid

S3. What is the major organic product of each of the following reaction sequences?

(a) $(CH_3)_2CHCO_2H \xrightarrow[Cl_2,\ \Delta]{PCl_3} \xrightarrow{H_2O}$

(b) 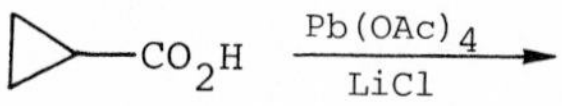

(c) $(CH_3)_3CCH_2C{\equiv}N \xrightarrow[\Delta]{H_3O^+} \xrightarrow{AgO} \xrightarrow{I_2}$

(d) $CH_2{=}CHCH_2CH_2CH_2CO_2H \xrightarrow{LiAlH_4} \xrightarrow[H^+]{CH_3CO_2H}$

(e) $HO_2C{-}C{\equiv}C{-}CO_2H \xrightarrow{CH_2N_2} \xrightarrow{\text{1 mole } Br_2}$

(f) 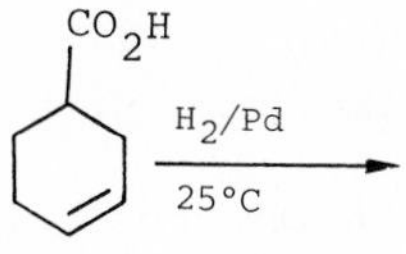

(g) 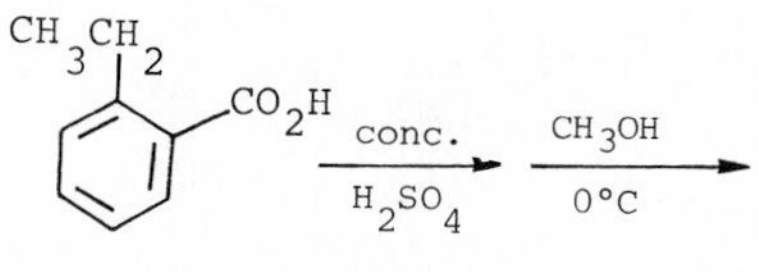

(h) 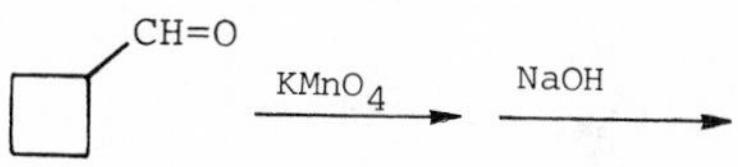

S4. Show how to carry out the following conversions:

(a) $CH_3CH_2CH_2CO_2H \longrightarrow CH_3CH_2CH_2CH_2CH_2OH$

(b) $CH_3CH_2CH_2CO_2H \longrightarrow CH_3CH{=}CHCO_2H$

(c) 

(d) $(CH_3)_2CHCH_2CH{=}CH_2 \longrightarrow (CH_3)_2CHCH_2CH(OH)CO_2H$

(e) 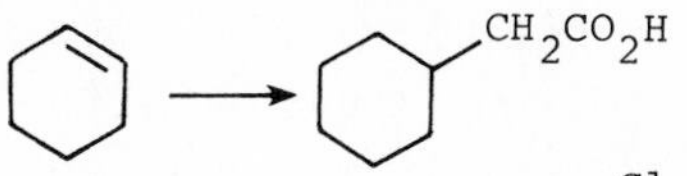

(f) $CH_3CH_2CH_2CH_2Cl \longrightarrow CH_3CH_2CH_2CH(Cl)CO_2CH_3$

(g) $CH_3CH_2CH_2CO_2H \longrightarrow CH_3CH(Br)CH(Br)CO_2CH_3$

(h) 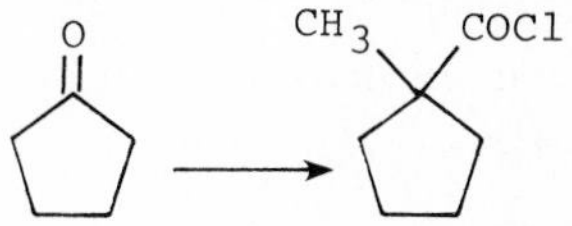

(i) $(CH_3)_3CCH_2C({=}O)CH_3 \longrightarrow (CH_3)_3CCH_2OC({=}O)CH_3$

(j) $(CH_3)_3CCH_2C({=}O)CH_3 \longrightarrow (CH_3)_3CCH_2C({=}O)OCH_3$

S5. Show how to synthesize the following compounds using starting materials of five carbons or less and any other reagents:

(a) $(CH_3)_3CCH_2CO_2CH(CH_3)_2$

(b) $(CH_3)_2CHCH_2CH_2CH_2CH_2CO_2H$

(c) $CH_3CH_2CH_2CH(CO_2CH_3)CH_2CH_2CH_2CH_2CH_3$

(d) 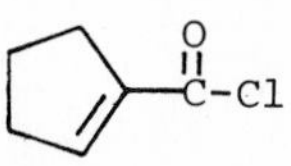

(e) 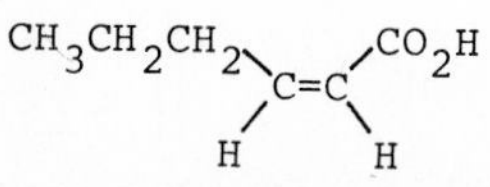

(g) $CH_3CH_2CH(CH_3)CH(OCH_3)C{\equiv}CCO_2CH_3$

(h) 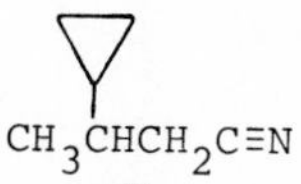

(i) 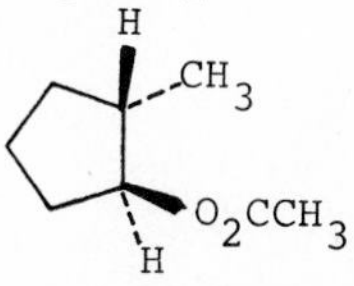

(f) 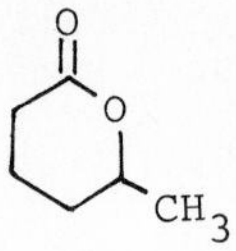

(j) $CH_3CH_2CO_2CH_2CH_2CH_2CH_2O_2CCH_2CH_3$

S6. Calculate the pH when cloroacetic acid is dissolved in water at 1.0 M concentration; at 0.1 M; at 0.001 M.

S7. In a solution which is 1.0 M in dichloroacetic acid and 1.0 M in acetic acid, what are the concentrations of dichloroacetate and acetate ions?

S8. Determine the structure of each of the compounds in the sequence below:

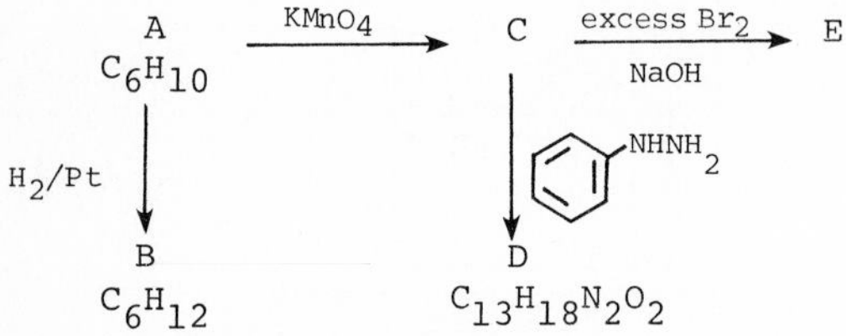

nmr of E: δ 1.3(t, 4), 2.4 (t, 4), 13 (s, 2)

S9. Write a mechanism for the following transformation:

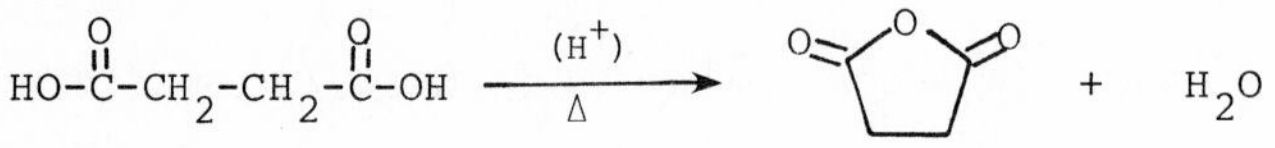

S10. Predict the major product of the following reaction and justify your answer:

$$CH_3CH{=}CHCOOH + Br_2 \xrightarrow[\text{solvent}]{CH_3OH}$$

S11. Rank the following compounds in order of increasing acidity:

F_2CHCO_2H $HOCH_2CH_2CO_2H$ $CH_3CH_2CH_2CH_2OH$ $CH_3CH_2C(CH_3)_2\text{-}CO_2H$ $ClCH_2CO_2H$

$CF_3CO_2CH_3$ CF_3CH_2OH $CH_3CH_2CO_2H$

18.F Answers to Supplementary Problems

S1. (a) 4-bromobutanoic acid
(b) (E)-4-oxo-2-pentanoic acid
(c) (S)-2-hydroxy-4-methylpentanoic acid
(d) potassium (Z)-2-hexenoate
(e) (R,S)-2,3-dihydroxybutanedioic acid (or meso-)
(f) *trans*-4-formylcyclohexanecarboxylic acid
(g) 2,3-butadienoic acid

S2.
(a) $Br_2CHCH_2CH_2CO_2H$
(b) $ClCH_2CH(OH)CO_2H$
(c) $HOCH_2CH_2CH_2CO_2^-\ Na^+$
(d) $CH_3C(=O)CO_2H$
(e) $CH_3CH_2CH_2CH_2CO_2^-\ NH_4^+$
(f) 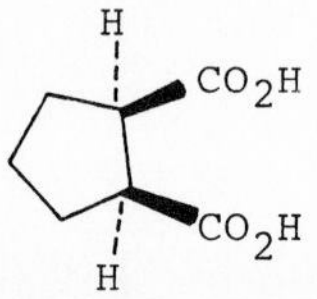

S3.

(a) $(CH_3)_2CClCO_2H$ (Cl drawn on the central C)

(b) 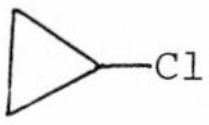

(c) $(CH_3)_3CCH_2I$

(d) $CH_2{=}CHCH_2CH_2CH_2CH_2O_2CCH_3$

(e) 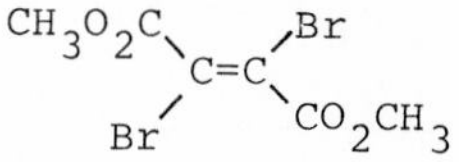

(f) 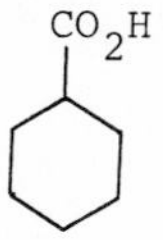

(g) 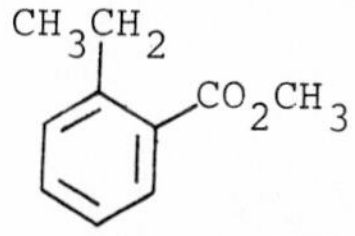

(h) 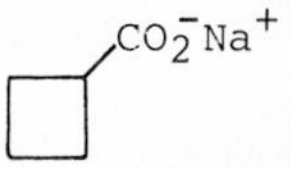

S4. (a) $CH_3CH_2CH_2CO_2H \xrightarrow{LiAlH_4} \xrightarrow{H_2O} CH_3CH_2CH_2CH_2OH \xrightarrow{PBr_3} \xrightarrow{Mg} \xrightarrow{CH_2=O} \xrightarrow{H_2O} CH_3CH_2CH_2CH_2CH_2OH$

(b) $CH_3CH_2CH_2CO_2H \xrightarrow[\Delta]{P + Br_2} \xrightarrow{H_2O} CH_3CH_2CHBrCO_2H \xrightarrow[HOC_2H_5]{NaOC_2H_5} \xrightarrow{H^+} CH_3CH{=}CHCO_2H$

(c) 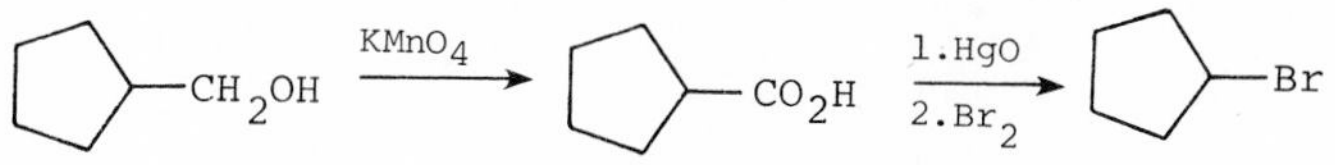

(d) $(CH_3)_2CHCH_2CH{=}CH_2 \xrightarrow[2.\,Zn]{1.\,O_3} (CH_3)_2CHCH_2CHO \xrightarrow{HCN} (CH_3)_2CHCH_2CH(OH)C{\equiv}N \xrightarrow[\Delta]{H_3O^+} (CH_3)_2CHCH_2CH(OH)CO_2H$

(e) 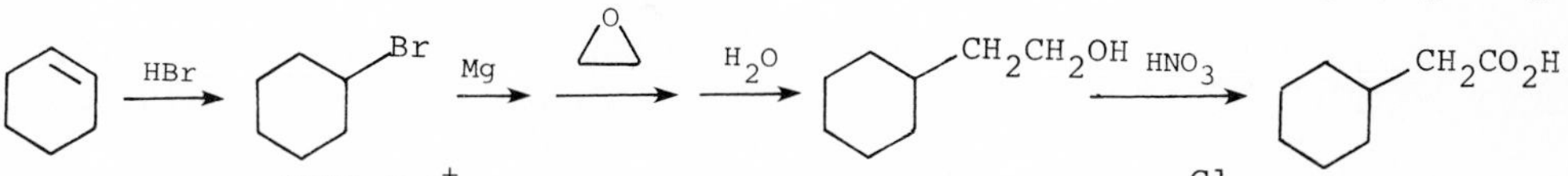

(f) $CH_3CH_2CH_2CH_2Cl \xrightarrow{NaCN} \xrightarrow{H_3O^+} CH_3CH_2CH_2CH_2CO_2H \xrightarrow{P + Cl_2} \xrightarrow{H_2O} CH_3CH_2CH_2CHClCO_2H \xrightarrow[CH_3OH]{H^+} CH_3CH_2CH_2CHClCO_2CH_3$

(g) $CH_3CH{=}CHCO_2H$ (from (b) above) $\xrightarrow{Br_2} CH_3CHBrCHBrCO_2H$

(h) 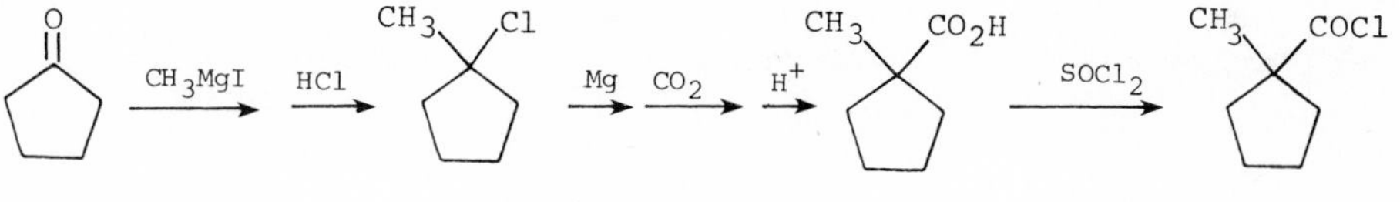

(i) $(CH_3)_3CCH_2COCH_3 \xrightarrow{CH_3CO_3H} (CH_3)_3CCH_2OCOCH_3$

(j) $(CH_3)_3CCH_2COCH_3 \xrightarrow[Br_2]{NaOH} (CH_3)_3CCH_2COOH \xrightarrow{CH_2N_2} (CH_3)_3CCH_2COOCH_3$

S5.

(a) $(CH_3)_3CCH_2CO_2CH(CH_3)_2 \xleftarrow[H^+]{(CH_3)_2CHOH} (CH_3)_3CCH_2CO_2H \xleftarrow{H^+} \xleftarrow{CO_2} \xleftarrow{Mg} (CH_3)_3CCH_2Br$

(Note: $(CH_3)_3CCH_2Br \xrightarrow{NaCN}$ (crossed out) very slow, because of hindrance from β-branching.)

(b) $(CH_3)_2CHCH_2CH_2CH_2CH_2CO_2H \xleftarrow[\Delta]{H_3O^+} \xleftarrow{NaCN} \xleftarrow{PBr_3} (CH_3)_2CHCH_2CH_2CH_2CH_2OH \xleftarrow{H_2O}$  $\xleftarrow{Mg} (CH_3)_2CHCH_2CH_2Br$

(c) $CH_3CH_2CH_2\overset{\overset{\displaystyle CO_2CH_3}{|}}{CH}(CH_2)_4CH_3$ $\xleftarrow[H^+]{CH_3OH}$ $CH_3CH_2CH_2\overset{\overset{\displaystyle CO_2H}{|}}{CH}(CH_2)_4CH_3$ $\xleftarrow[Pt]{H_2}$ $\underset{Ag_2O}{}$

$2\ CH_3CH_2CH_2CH_2\overset{O}{\overset{\|}{C}}H$ $\xrightarrow{NaOH}$ $CH_3CH_2CH_2\overset{}{C}=CHCH_2CH_2CH_2CH_3$

(d) [structure: cyclopentene-COCl] $\xleftarrow{SOCl_2}$ [cyclopentene-CO_2H] $\xleftarrow{H_3O^+}$ $\xleftarrow[\Delta]{H_2SO_4}$ [cyclopentane with OH and CN] $\xleftarrow{HCN}$ [cyclopentanone =O]

(e) $CH_3CH_2CH_2\underset{H}{\overset{}{C}}=\underset{H}{\overset{}{C}}\overset{CO_2H}{}$ $\xleftarrow[\substack{Lindlar \\ catalyst}]{H_2}$ $CH_3CH_2CH_2C\equiv CCO_2H$ $\xleftarrow{H_2O}$ $\xleftarrow{CO_2}$ $\xleftarrow[\Delta]{CH_3MgBr}$ $CH_3CH_2CH_2C\equiv CH$

(f) [lactone structure with O and CH3] $\xleftarrow{CH_3CO_3H}$ [2-methylcyclopentanone] $\xleftarrow{CH_3I}$ [cyclopentene-OLi] $\xleftarrow[\substack{THF \\ -78°C}]{LDA}$ [cyclopentanone]

(g) $CH_3CH_2\overset{\overset{\displaystyle CH_3}{|}}{CH}CHC\equiv CCO_2CH_3$ $\xleftarrow{CH_2N_2}$ $CH_3CH_2\overset{\overset{\displaystyle CH_3}{|}}{CH}CHC\equiv CCO_2H$ $\xleftarrow{H^+}$ $\xleftarrow{CO_2}$ $\xleftarrow{n\text{-BuLi}}$ $CH_3CH_2\overset{\overset{\displaystyle CH_3}{|}}{CH}CHC\equiv CH$
(with OCH3 groups as shown)

$\uparrow$ NaOH $(CH_3)_2SO_4$

$CH_3CH_2\overset{\overset{\displaystyle CH_3}{|}}{CH}CH$ (with =O) $\xrightarrow{NaC\equiv CH}$ $\xrightarrow{H_2O}$ $CH_3CH_2\overset{\overset{\displaystyle CH_3}{|}}{CH}CHC\equiv CH$ (with OH)

(h) [cyclopropyl structure] $\xleftarrow{NaCN}$ $\xleftarrow{PBr_3}$ [cyclopropyl structure] $\xleftarrow{H_2O}$ $\xleftarrow{CH_2=O}$ $\xleftarrow{Mg}$ [cyclopropyl-CHCH3 with Br]

$CH_3-\overset{}{C}H-CH_2C\equiv N$ $CH_3-\overset{}{C}H-CH_2OH$

(i) [cyclopentane with H, CH3, O2CCH3, H] $\xleftarrow[H^+]{CH_3CO_2H}$ [cyclopentane with H, CH3, OH, H] $\xleftarrow{H_2O}$ $\xleftarrow{CH_3Li}$ [bicyclic epoxide]

j) $CH_3CH_2\overset{O}{\overset{\|}{C}}OCH_2CH_2CH_2CH_2O\overset{O}{\overset{\|}{C}}CH_2CH_3$ $\xleftarrow[H^+]{CH_3CH_2CO_2H}$ $HOCH_2CH_2CH_2CH_2OH$ $\xleftarrow{NaBH_4}$ $\xleftarrow{O_3}$ [cyclobutane]

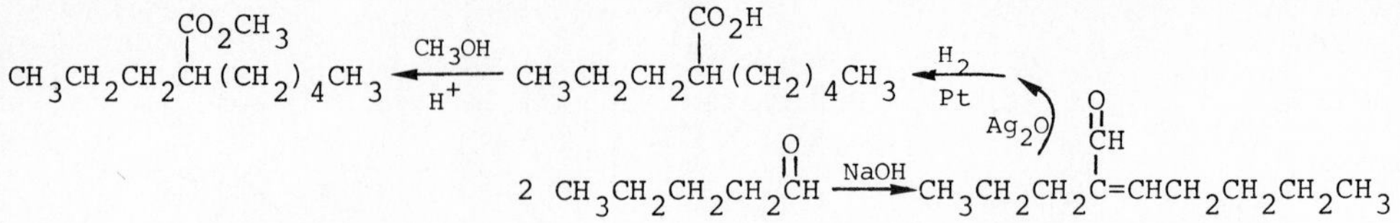

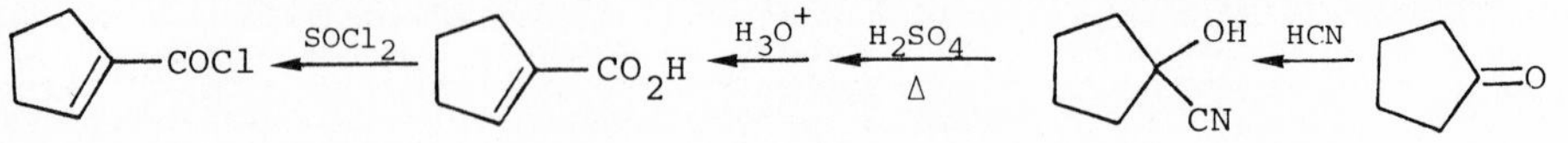

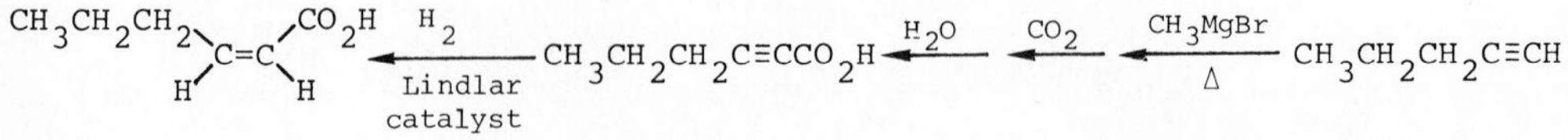

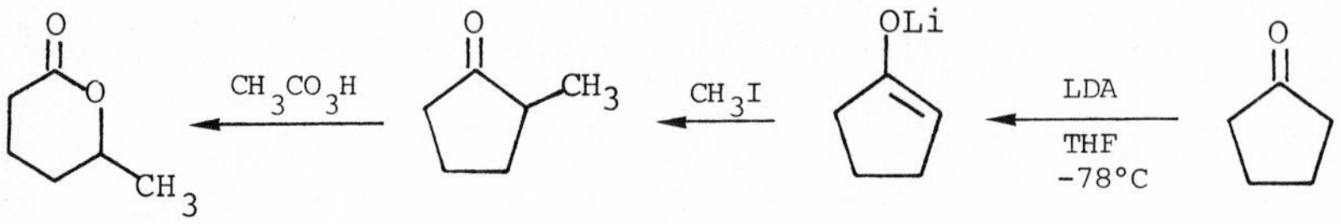

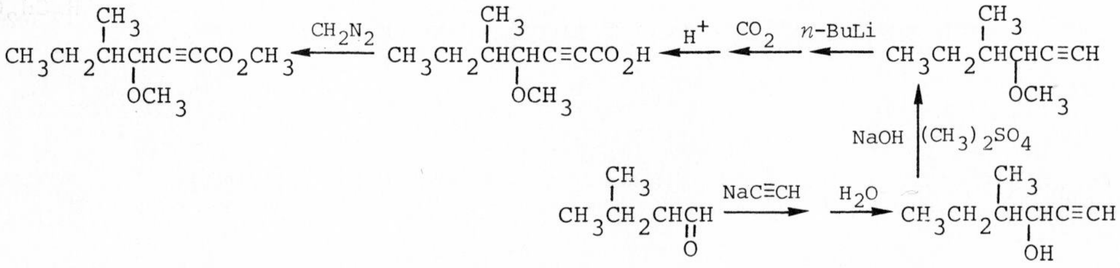

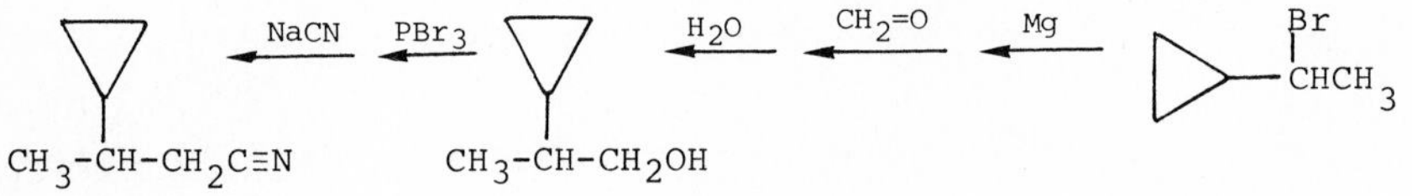

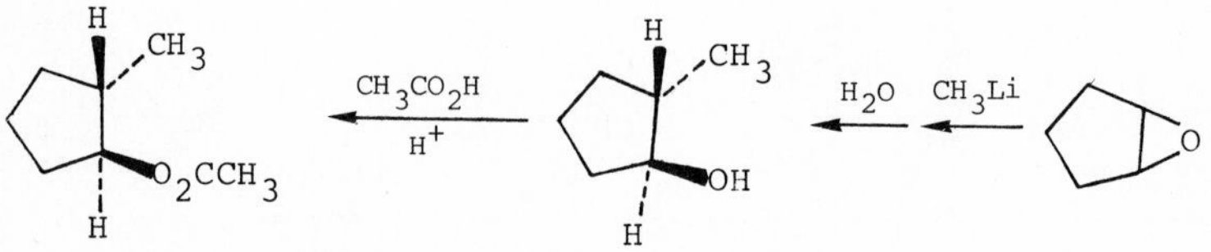

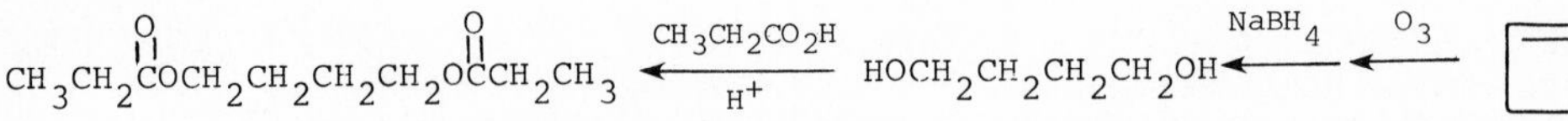

S6. K_a of chloroacetic acid is 1.4×10^{-3} $\underline{M}$ (Table 18.3).

at 1.0 $\underline{M}$ $\dfrac{[H^+][ClCH_2CO_2^-]}{[ClCH_2CO_2H]} = 1.4 \times 10^{-3}$ $[H^+] = [ClCH_2CO_2^-] = X$

$[ClCH_2CO_2H] = 1.0 - X$

$\dfrac{X^2}{1-X} = 1.4 \times 10^{-3}$

$X^2 + 1.4 \times 10^{-3}X - 1.4 \times 10^{-3} = 0;\quad X = \dfrac{-1.4 \times 10^{-3} \pm \sqrt{(1.4\times 10^{-3})^2 + 5.6 \times 10^{-3}}}{2} =$

$3.67 \times 10^{-2};\ pH = 1.44$

at 0.1 $\underline{M}$: $\dfrac{X^2}{0.1-X} = 1.4 \times 10^{-3};\ X^2 + 1.4 \times 10^{-3}X - 1.4 \times 10^{-4} = 0;\ X = 2.23 \times 10^{-2};$

$pH = 1.65$

at 0.001 M: $$\frac{X^2}{1\times10^{-3}-X} = 1.4\times10^{-3};\quad X^2 + 1.4\times10^{-3}X - 1.4\times10^{-6} = 0;$$

$$X = 1.35\times10^{-3};\quad pH = 2.87.$$

S7. Because dichloroacetic is a much stronger acid than acetic acid, only a very small amount of the acetic acid will be ionized; that is, essentially all of the protons will come from the dichloroacetic acid:

$$\frac{[H^+][Cl_2CHCO_2^-]}{[Cl_2CHCO_2H]} = 5.5\times10^{-2}\qquad [H^+]\simeq[Cl_2CHCO_2^-]=X,\quad [Cl_2CHCO_2H] = 1-X$$

$$\frac{X^2}{1-X} = 5.5\times10^{-2};\quad X=\frac{-5.5\times10^{-2}\pm\sqrt{(5.5\times10^{-2})^2+0.22}}{2} = 0.21\ \underline{M} =$$

$$[Cl_2CHCO_2^-] = [H^+];\quad pH = 0.68.$$

$$\frac{[H^+][CH_3CO_2^-]}{[CH_3CO_2H]} = 1.8\times10^{-5} = \frac{(0.21)y}{1-y};\quad y = 8.6\times10^{-5} = [CH_3CO_2^-]$$

$$\frac{[Cl_2CH_2CO_2^-]}{[CH_3CO_2^-]} = \frac{0.21}{8.6\times10^{-5}} = 2.44\times10^{3}$$

S8.

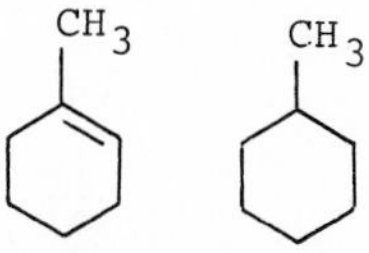

A B

$CH_3\overset{O}{\overset{\|}{C}}CH_2CH_2CH_2CH_2CO_2H$

C

$CH_3\overset{NNHC_6H_5}{\overset{\|}{C}}CH_2CH_2CH_2CH_2CO_2H$

D

$HO_2CCH_2CH_2CH_2CH_2CO_2H$

E

S9.

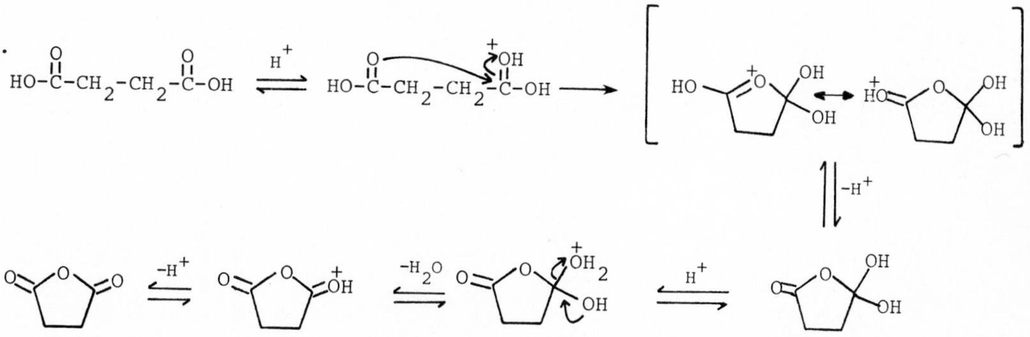

S10.

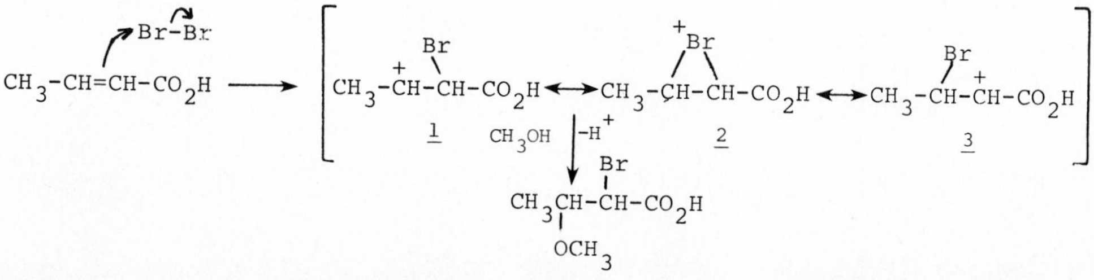

Resonance structure 3 is the least favored, because it puts a positive charge next to an already partially positive center: the carbonyl group. Therefore, "Markovnikov orientation" in this case favors nucleophilic attack at the β- and not the α-position.

S11. *least* acidic $CF_3CO_2CH_3$ (no ionizable hydrogens)

$CH_3CH_2CH_2CH_2OH$

CF_3CH_2OH

$CH_3CH_2C(CH_3)_2-CO_2H$

$CH_3CH_2CO_2H$

$HOCH_2CH_2CO_2H$

$ClCH_2CO_2H$

most acidic F_2CHCO_2H

19. DERIVATIVES OF CARBOXYLIC ACIDS

19.A Chapter Outline, Important Terms Introduced, and Reactions Discussed

19.1 Structure

esters
amides
acyl halides
acid anhydrides
nitriles

19.2 Nomenclature

alkyl alkanoate
alkanamide
alkanoyl halide
alkanoic anhydride
alkanenitrile
-carboxamide
-carbonyl halide
-carbonitrile

19.3 Physical Properties

19.4 Spectroscopy

A. Nuclear Magnetic Resonance

B. Infrared

$-CO_2R$	1735
$-COCl$	1800
$-\overset{O}{\overset{\|\|}{C}}O\overset{O}{\overset{\|\|}{C}}-$	1820 and 1760
$-CONR_2$	1650-1690

lactones
lactams

19.5 Basicity of the Carbonyl Oxygen

contribution of resonance structures

O-protonation of amides

19.6 Hydrolysis, Nucleophilic Addition-Elimination

the general reaction:

$$R-\overset{O}{\overset{\|}{C}}-X + H_2O \longrightarrow R-\overset{O}{\overset{\|}{C}}-OH + HX$$

X = halogen, OR', NR'_2, O_2CR'

(acid- vs. base-catalyzed mechanisms; tetrahedral intermediate; position of equilibrium)

unusual mechanism for t-butyl esters (via t-butyl cation)

nitrile hydrolysis:

$$R-C\equiv N \longrightarrow R\overset{O}{\overset{\|}{C}}NH_2 \longrightarrow RCO_2H + NH_3$$

(using either acid or base; can stop at amide)

sequence of bond-making vs. bond-breaking steps

nucleophilic addition-elimination mechanism

19.7 Other Nucleophilic Substitution Reactions

A. Reaction with Alcohols

ester formation:

$$\mathrm{R{-}\overset{O}{\overset{\|}{C}}{-}X + HOR' \longrightarrow R{-}\overset{O}{\overset{\|}{C}}{-}OR' + HX}$$

(usually catalyzed by tertiary amine or pyridine)

X = halogen or $\mathrm{O_2CR}$

transesterification:

$$\mathrm{R\overset{O}{\overset{\|}{C}}OR' + HOR'' \rightleftharpoons R\overset{O}{\overset{\|}{C}}OR'' + HOR'}$$

(acid- or base-catalyzed)

amide→ester:

$$\mathrm{R\overset{O}{\overset{\|}{C}}NH_2 + HOR' \xrightarrow{H^+} R\overset{O}{\overset{\|}{C}}OR' + \overset{+}{N}H_4}$$

(acid-catalyzed)

B. Reaction with Amines and Ammonia

amide formation:

$$\mathrm{R\overset{O}{\overset{\|}{C}}X + 2\ HNR'_2 \longrightarrow R\overset{O}{\overset{\|}{C}}NR'_2 + H_2\overset{+}{N}R'_2X}$$

(Schotten-Bauman conditions use 1 mole of NaOH)

X = halogen, $\mathrm{O_2CR}$

$$\mathrm{R\overset{O}{\overset{\|}{C}}OR' + HNR'_2 \longrightarrow R\overset{O}{\overset{\|}{C}}NR'_2 + HOR'}$$

(useful when $\mathrm{R\overset{O}{\overset{\|}{C}}X}$ is unavailable)

C. Reaction of Acyl Halides and Anhydrides with Carboxylic Acids and Carboxylate Salts. Synthesis of Anhydrides.

$$\mathrm{R\overset{O}{\overset{\|}{C}}O\overset{O}{\overset{\|}{C}}R + 2\ R'CO_2H \underset{\Delta}{\rightleftharpoons} R'\overset{O}{\overset{\|}{C}}O\overset{O}{\overset{\|}{C}}R' + 2\ RCO_2H}$$

(equilibrium process)

$$\mathrm{R\overset{O}{\overset{\|}{C}}Cl + \overset{+}{Na}\overset{-}{O}\overset{O}{\overset{\|}{C}}R' \longrightarrow R\overset{O}{\overset{\|}{C}}O\overset{O}{\overset{\|}{C}}R' + NaCl}$$

D. Reaction with Organometallic Compounds

ketones from acyl halides:

$$\mathrm{R\overset{O}{\overset{\|}{C}}Cl \longrightarrow R\overset{O}{\overset{\|}{C}}R'}$$

(with $\mathrm{R'MgX}$ (low temperature; excess gives tertiary alcohol), or $\mathrm{R'_2CuLi}$ (cuprate))

tertiary alcohols from esters:

$$\mathrm{R\overset{O}{\overset{\|}{C}}OCH_3 + 2\ R'M \longrightarrow \xrightarrow{H_2O} R{-}\underset{R'}{\overset{OH}{C}}{-}R'}$$

$$\mathrm{CH_3O\overset{O}{\overset{\|}{C}}OCH_3 + 3\ R'M \longrightarrow \xrightarrow{H_2O} R'{-}\underset{R'}{\overset{OH}{C}}{-}R'}$$

(M = Li or MgX)

ketones from nitriles:

$$\mathrm{R{-}C{\equiv}N + R'M \longrightarrow \xrightarrow{H_2O} R\overset{O}{\overset{\|}{C}}R'}$$

(M = Li, MgX; via imine salt)

19.8 Reduction

acid chloride to aldehyde:

$$\mathrm{R\overset{O}{\overset{\|}{C}}Cl \longrightarrow R\overset{O}{\overset{\|}{C}}H}$$

(with $\mathrm{H_2/Pd\text{-}BaSO_4}$-quinoline (Rosenmund reduction) or $\mathrm{LiAlH(O\text{-}\underline{t}\text{-}Bu)_3}$)

ester to alcohol:

$$\mathrm{R\overset{O}{\overset{\|}{C}}OR' \longrightarrow RCH_2OH}$$

(with $\mathrm{LiAlH_4}$ or $\mathrm{LiBH_4}$, or $\mathrm{Na/C_2H_5OH}$ (Bouveault-Blanc reaction)) amide to amine:

$$RC(=O)NR'_2 \xrightarrow{LiAlH_4} \xrightarrow{H_2O} RCH_2NR'_2 \qquad (R' = H \text{ or alkyl})$$

secondary amide to aldehyde:

$$RC(=O)NR'_2 \xrightarrow{LiAlH(OC_2H_5)_3} \xrightarrow{H_2O} RC(=O)H \qquad (R' = \text{alkyl, } \underline{\text{not}} \text{ H})$$

nitrile to amine:

$$R\text{-}C{\equiv}N \longrightarrow RCH_2NH_2 \qquad (\text{with } LiAlH_4 \text{ or } H_2/Ni/2000 \text{ psi})$$

nitrile to aldehyde:

$$R\text{-}C{\equiv}N \longrightarrow RC(=O)H \qquad (\text{with } (\underline{i}\text{-Bu})_2AlH \text{ (DIBAL) or } SnCl_2, HCl \text{ (R = aryl, Stephen reaction)})$$

19.9 Acidity of the α-Protons

$CH_3C(=O)CH_3$ pK_a 20 $CH_3C(=O)OCH_3$, CH_3CN pK_a 25 $CH_3C(=O)N(CH_3)_2$ pK_a 30

enolate formation:

$$CH_3CO_2R + LiNR'_2 \xrightarrow[-78°\ C]{THF} CH_2{=}C(OLi)(OR) + HNR'_2$$

$$BrCH_2CO_2R + Zn \xrightarrow[\Delta]{} CH_2{=}C(OZnBr)(OR) \qquad (\text{Reformatsky reagent})$$

addition to carbonyl compounds:

$$CH_2{=}C(OM)(OR) + {>}C{=}O \longrightarrow \xrightarrow{H_2O} {-}C(OH){-}CH_2CO_2R \qquad (M = ZnBr\text{: Reformatsky reaction})$$

$$2\ CH(R)HCO_2R' \xrightarrow{R'O^-} CH_2(R)C(=O)CH(R)CO_2R' \qquad (\text{Claisen condensation, acetoacetic ester condensation})$$

19.10 Reactions of Amides Which Occur on Nitrogen

amide ionization:

$$RC(=O)NH_2 \rightleftharpoons RC(=O)\underset{-}{N}H + H^+ \qquad pK_a\ 15$$

nitrile formation:

$$RC(=O)NH_2 \longrightarrow RC{\equiv}N \qquad (\text{using } SOCl_2, POCl_3, P_2O_5)$$

Hoffman degradation:

$$RC(=O)NH_2 \xrightarrow[Br_2]{NaOH} RNH_2 \qquad (\text{see Section 24.6.H.})$$

19.11 Pyrolytic Eliminations

ester pyrolysis:

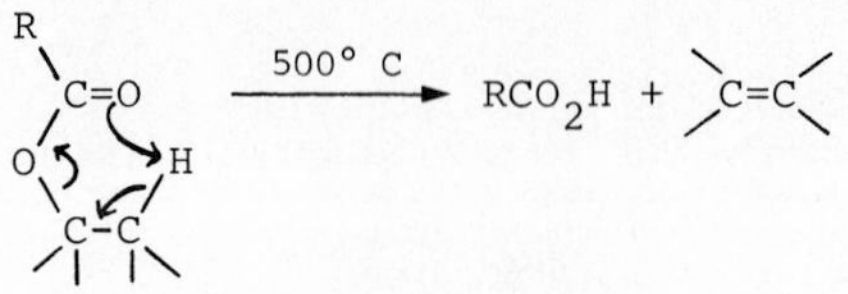

xanthate pyrolysis (Chugaev reaction):

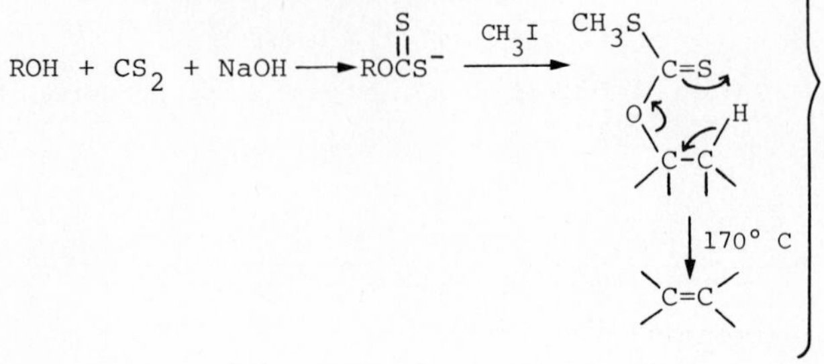

syn elimination stereochemistry

19.12 Waxes and Fats

A. Waxes (spermaceti, bee's wax, carnuba wax)

B. Fats

fatty acids	phosphatidic acids
triglycerides	lipid bilayer
saponification	

19.B Important Concepts and Hints

Addition-Elimination Mechanism

The addition-elimination mechanism for substitution reactions of carboxylic acid derivatives was introduced in Chapter 18, and its importance is emphasized by the number of reactions discussed in this chapter which proceed by it. It holds the same importance in the chemistry of carboxylic acid derivatives as the S_N2 mechanism holds for substitution reactions of alkyl systems.

The addition-elimination mechanism applies to both acidic and basic reaction conditions, and examples can be found in which the same overall transformation is catalyzed by either one. Ester and amide hydrolysis are two examples; for both of them you should compare closely the acid- and base-catalyzed mechanisms (see Section 19.6., and the answer to the Exercise at the end of that section), and be familiar with their similarities and differences (see Section 18.B. of this Study Guide for a similar comparison in another reaction sequence).

You should also be completely familiar with the general aspects of the mechanisms which hold under acidic and basic conditions, as depicted below.

Acidic Conditions:

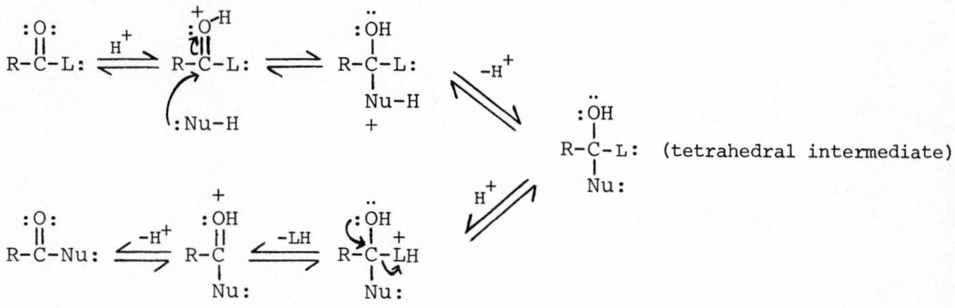

Basic Conditions:

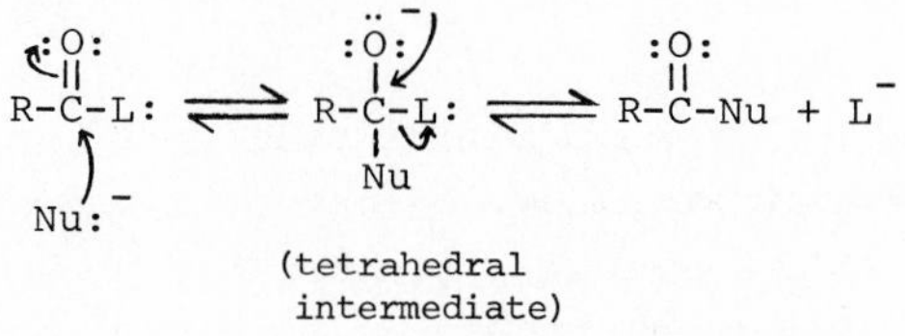

Notice the symmetry of these sequences: the steps leading to the formation of the tetrahedral intermediate are simply reversed in going from there to the product.

There are some reactions in which the distinction is somewhat blurred; for example, in the addition of ammonia to an acid chloride. In this case the nucleophile attacks before it loses its proton (as in the "acidic mechanism") and the leaving group departs before it gains one (as in the "basic mechanism"):

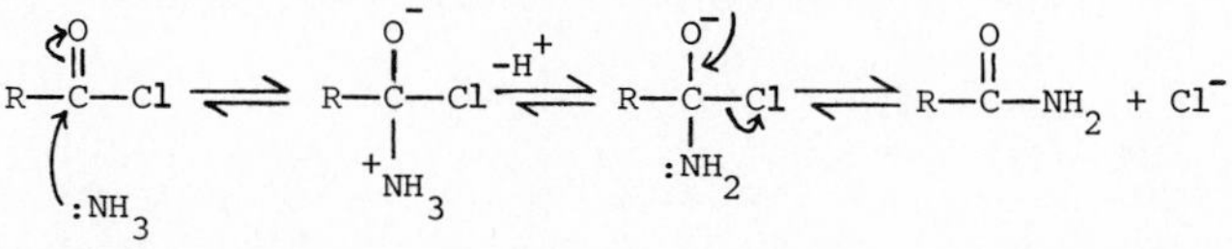

Because of the symmetrical nature of these sequences, the reactions could conceivably go in either direction. In fact, the reaction: (ester + $H_2O \overset{H^+}{\rightleftharpoons}$ acid + alcohol) is such an equilibrium process (see Section 18.7.C.2.). There is a convenient way to predict which way the equilibrium lies for reactions which proceed by the *basic* mechanism: compare the pK_a's of "Nu:$^-$" and "L$^-$"; the reaction will go in the direction which generates the least basic (most stable) leaving group:

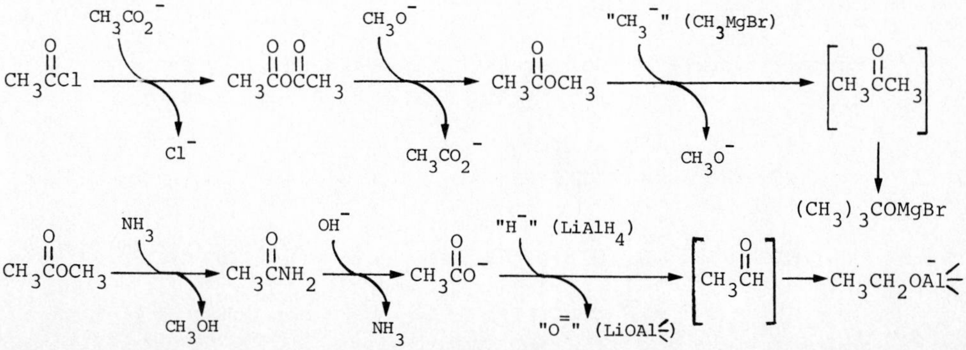

For acid-catalyzed amide hydrolysis, and for reactions with hydroxide ion, additional reactions of the products serve to remove one of the components of the equilibrium, and thereby drive it completely to one side:

reaction	equilibrium	additional reaction
acid-catalyzed hydrolysis of amides	$RC(=O)NH_2 + H_2O \overset{H^+}{\rightleftharpoons} RC(=O)OH + NH_3$	$\overset{H^+}{\rightleftharpoons} RC(=O)OH + NH_4^+$
base-catalyzed hydrolysis of esters, amides	$RC(=O)L + H_2O \overset{OH^-}{\rightleftharpoons} RC(=O)OH + HL$	$\overset{OH^-}{\rightleftharpoons} RC(=O)O^- + HL$
	$(L = OR' \text{ or } NR'_2)$	

19.C Answers to Exercises

(19.2) (a) methyl hexanoate, hexanamide, hexanoyl chloride, hexanoic anhydride, hexanenitrile.

(b) methyl 4-methylpentanoate, 4-methylpentanamide, 4-methylpentanoyl chloride, 4-methylpentanoic anhydride, 4-methylpentanenitrile.

(c) methyl 4-pentenoate, 4-pentenamide, 4-pentenoyl chloride, 4-pentenoic anhydride, 4-pentenenitrile

(d) methyl 3-bromopropanoate, 3-bromopropanamide, 3-bromopropanoyl chloride, 3-bromopropanoic anhydride, 3-bromopropanenitrile

(e) 2-methylpropyl 4-methylpentanoate.

(f) N,N-diethyl-4,4-dimethylpentanamide.

(19.4.B.)

ethyl propionate

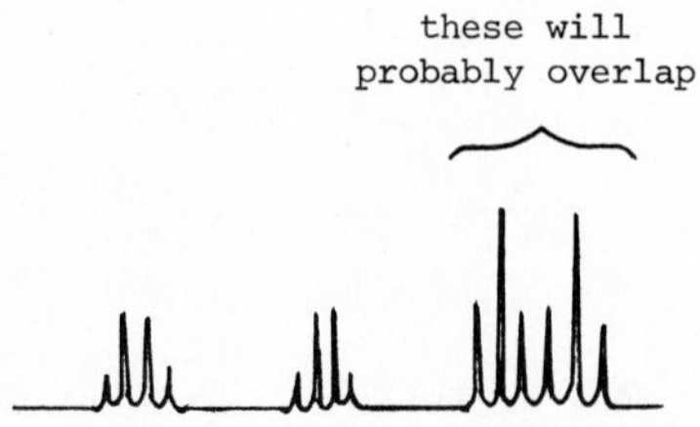

δ	4.2	2.3	1.3	1.2
area	2	2	3	3

isopropyl acetate

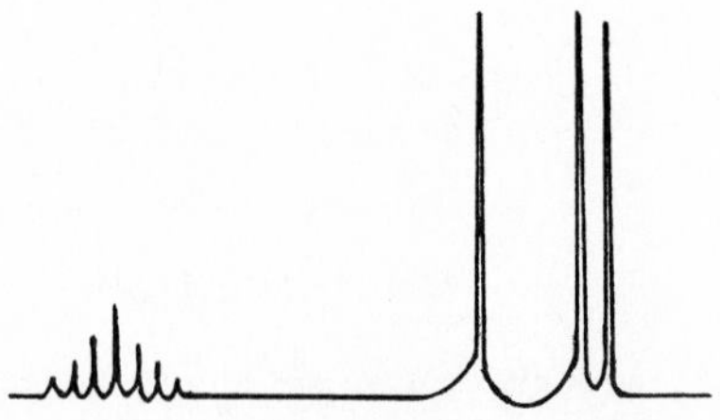

δ	5	2.1	1.3
area	1	3	6

(19.5) The dipolar resonance structure has a double bond between the carbonyl and the Y group: 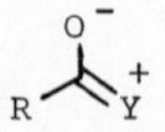 . Therefore, the more important this resonance structure is, the greater the degree of shortening of the carbonyl-to-Y bond relative to the single bond model, CH_3-Y.

CH_3-NH_2	vs.	$CH_3C(=O)-NH_2$	$C(=O)-N$ bond shorter by 0.11 Å
CH_3-OCH_3	vs.	$CH_3C(=O)-OCH_3$	$C(=O)-O$ bond shorter by 0.06 Å
CH_3-F	vs.	$CH_3C(=O)-F$	$C(=O)-F$ bond shorter by 0.01 Å

(19.6) Base-catalyzed:

$$CH_3-\overset{O}{\overset{||}{C}}-NH_2 \;\rightleftharpoons\; CH_3-\overset{O^-}{\underset{OH}{\overset{|}{C}}}-NH_2 \;\xrightarrow{slow}\; CH_3-C\overset{O}{\underset{OH}{\diagdown}} \;\;^-NH_2 \longrightarrow CH_3C\overset{O}{\underset{O^-}{\diagdown}} \;\; NH_3$$

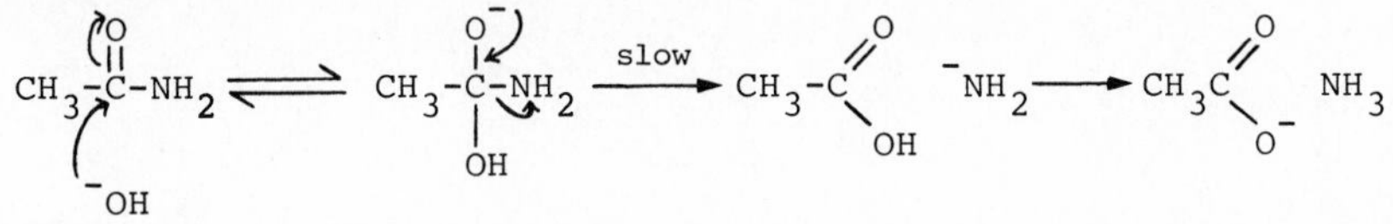

Formation of the amide ion (NH_2^-) is difficult because it is of such high energy. It is probably never expelled as the free amide ion, but is undergoing protonation (to give NH_3) at the same time as the carbonyl-to-N bond is breaking.

Acid-catalyzed:

$$CH_3-\overset{O}{\overset{||}{C}}-NH_2 \;\xrightarrow{H^+}\; CH_3-\overset{+OH}{\overset{||}{C}}-NH_2 \;\rightleftharpoons\; CH_3-\overset{OH}{\underset{+OH_2}{\overset{|}{C}}}-NH_2 \;\rightleftharpoons\;$$

$$H_2O:$$

$$\Big\Downarrow -H^+$$

$$CH_3-\overset{OH}{\underset{OH}{\overset{|}{C}}}-NH_2$$

$$\Big\Uparrow H^+$$

$$CH_3C\overset{O}{\underset{OH}{\diagdown}} \;\;^+NH_4 \;\rightleftharpoons\; CH_3-C\overset{+OH}{\underset{OH}{\diagdown}} \;\; NH_3 \;\rightleftharpoons\; CH_3-\overset{OH}{\underset{OH}{\overset{|}{C}}}-\overset{+}{N}H_3$$

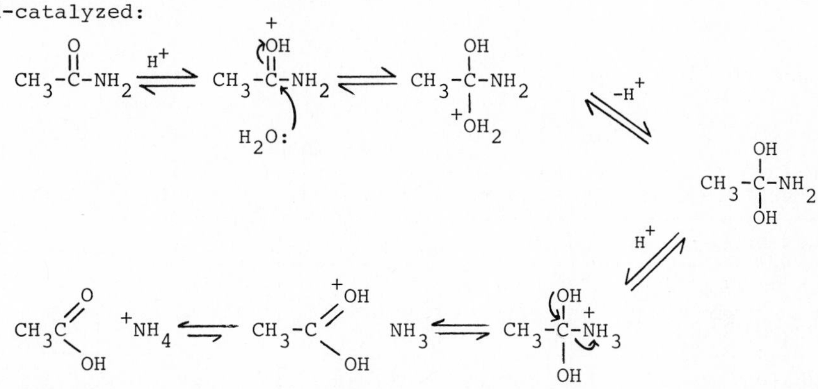

Note the similarity to the base-catalyzed reaction *from the electronic* point of view. The only difference is the number of protons involved at each stage.

(19.7)

	C_2H_5OH	NH_3	H_2O	CH_3MgBr	$\overset{+-}{Na}O_2CCH_3$
$(CH_3CO)_2O$	19.7.A.	19.7.B.	19.6.		(19.7.C.)
CH_3COCl	19.7.A.	19.7.B.	19.6.	19.7.D.	19.7.C.
$CH_3CO_2C_2H_5$	19.7.A.	19.7.B.	19.6.	19.7.D.	
CH_3CONH_2	19.7.A.		19.6.		
$CH_3C{\equiv}N$			19.6.	19.7.D.	

(19.8) (a) H_2/Pd-$BaSO_4$-quinoline (Rosenmund reduction) or $LiAlH(t\text{-BuO})_3$

(b) $(i\text{-Bu})_2AlH$ (DIBAL) (best) or $SnCl_2$/HCl (Stephen reduction)

(c) $LiAlH_4$ or $LiBH_4$ (d) $LiAlH_4$ (e) $LiAlH(OC_2H_5)_3$

(f) $LiAlH_4$ or H_2/Ni/2000 psi

(19.9)

(a) $CH_3CO_2C_2H_5 \xrightarrow{LiAlH_4} \xrightarrow{H_2O} C_2H_5OH \xrightarrow{PBr_3} CH_3CH_2Br$

$$CH_3CO_2C_2H_5 \xrightarrow{LiNiPr_2} \overset{OLi}{\underset{CH_2 \quad OC_2H_5}{\diagup\diagdown}} C \xrightarrow{CH_3CH_2Br} CH_3CH_2CH_2CO_2C_2H_5 \xrightarrow{LiAlH_4}$$

$$\Big\downarrow H_2O$$

$$CH_3CH_2CH_2CH_2C\overset{O}{\underset{OC_2H_5}{\diagdown}} \xleftarrow{CH_2=C\overset{OLi}{\underset{OC_2H_5}{\diagdown}}} CH_3CH_2CH_2CH_2Br \xleftarrow{PBr_3} CH_3CH_2CH_2CH_2OH$$

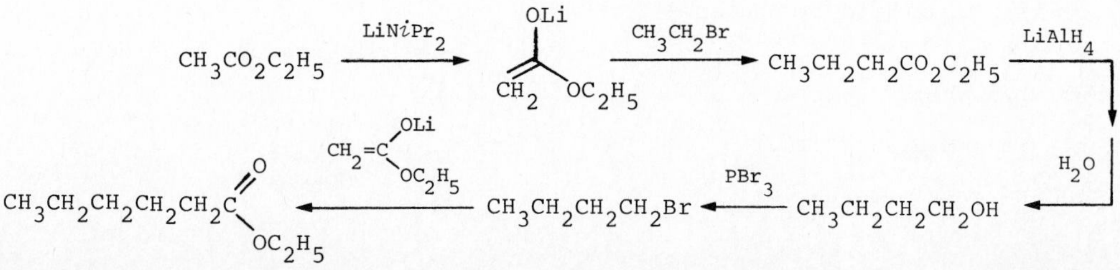

(b) $CH_3CH_2CH_2CH_2OH \xrightarrow[\text{pyridine}]{CrO_3} CH_3CH_2CH_2\overset{\overset{O}{\|}}{C}H \xrightarrow[]{CH_2=C\begin{smallmatrix}OLi\\OC_2H_5\end{smallmatrix}} \xrightarrow[]{H_2O}$

$$CH_3CH_2CH_2\overset{\overset{OH}{|}}{C}HCH_2CO_2C_2H_5$$

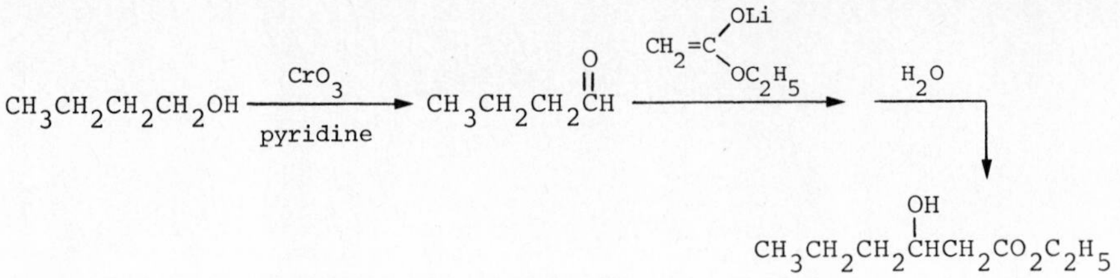

(c) $HCO_2C_2H_5 + CH_3CO_2C_2H_5 \xrightarrow[CH_3CH_2OH]{NaOC_2H_5} H\overset{\overset{O}{\|}}{C}CH_2CO_2C_2H_5$

(d) $\xrightarrow[DOC_2H_5]{NaOC_2H_5}$

(19.10)

| | Product | | | |
Starting Material	C_2H_5Br	$CH_3CH_2CO_2H$	$CH_3CH_2CONH_2$	$CH_3CH_2C{\equiv}N$
C_2H_5Br		18.6.		8.7.
$CH_3CH_2CO_2H$	18.7.D.		18.7.C.1.	
$CH_3CH_2CONH_2$		19.6		19.10.
$CH_3CH_2C{\equiv}N$		19.6	19.6.	

(19.11)

$t\text{-}C_4H_9O^-$ (E$_2$, *anti* elimination)

(S$_N$2 with inversion) | CH$_3$CO$_2^-$

500° $-CH_3CO_2H$ (*syn* elimination)

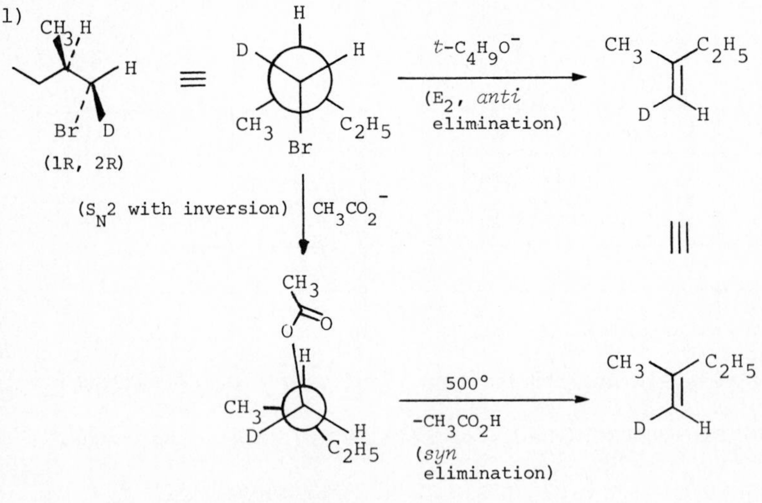

19.D Answers and Explanations to Problems

1. (a) propyl 3-ethylpentanoate

    (b) ethyl cyclopentanecarboxylate

    (c) N,N-dimethylcyclopentane-
              carboxamide

    (d) propanoyl chloride

    (e) butanoic anhydride

    (f) N-methylpropanamide

    (g) N-cyclohexyl-2-methyl-
              propanamide

    (h) pentanenitrile

    (i) 3-methylpentanoyl bromide

    (j) cyclohexyl ethanoate
              (cyclohexyl acetate is used
              more commonly)

2. (a) $(CH_3CH_2CH_2)(CH_3CH_2)CHCH_2\overset{O}{\overset{\|}{C}}NHCH_2CH_3$ (b) $H\overset{O}{\overset{\|}{C}}N(CH_3)_2$ (c) $CH_3CH_2CH_2\overset{O}{\overset{\|}{C}}OC_2H_5$

(d) $ClCH_2CH_2CO_2CH_3$ (e) $(CH_3)_2CHCH_2CH_2CH_2C{\equiv}N$ (f) $CH_3\overset{O}{\overset{\|}{C}}O\overset{O}{\overset{\|}{C}}H$

(g) 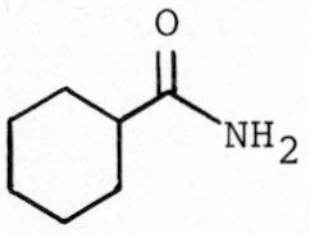

(h) 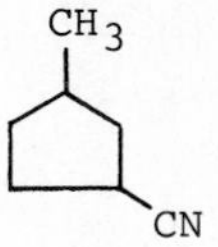

(i) 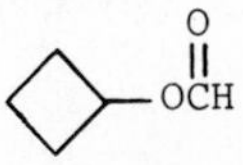

(j) $CH_3CH_2CH_2\overset{CHO}{\overset{|}{C}}HCH_2CO_2H$ (k) $CH_3\overset{O}{\overset{\|}{C}}CH_2CO_2C_2H_5$ (l) $CH_3\overset{O}{\overset{\|}{C}}NHBr$

(m) $CH_3CH_2CH_2\overset{O}{\overset{\|}{C}}Br$ (n) $CH_3CH_2\overset{O}{\overset{\|}{C}}O\overset{O}{\overset{\|}{C}}CH_2CH_3$

3. Acetic acid is a weak acid and a weak base as shown by:

$$CH_3CO_2H \rightleftharpoons H^+ + CH_3CO_2^-$$

$$K_2 = \frac{[CH_3CO_2^-][H^+]}{[CH_3CO_2H]} = 1.8 \times 10^{-5}$$

$$CH_3C(OH)_2^+ \rightleftharpoons H^+ + CH_3CO_2H$$

$$K_1 = \frac{[H^+][CH_3CO_2H]}{[CH_3C(OH)_2^+]} = 10^6$$

Make the assumption that in 0.1M HCl, <u>most</u> of the acetic acid is un-ionized; that is, present as CH_3CO_2H.

(a)

$$\frac{[CH_3CO_2^-]\cdot 10^{-1}}{[CH_3CO_2H]} = 1.8 \times 10^{-5}$$

$$\frac{[CH_3CO_2^-]}{[CH_3CO_2H]} = 1.8 \times 10^{-4}$$

% as $CH_3CO_2^-$ = 0.018%

(b)

$$\frac{[CH_3CO_2H]\cdot 10^{-1}}{[CH_3C(OH)_2^+]} = 10^6$$

$$\frac{[CH_3CO_2H]}{[CH_3C(OH)_2^+]} = 10^7$$

$$\frac{[CH_3C(OH)_2^+]}{[CH_3CO_2H]} = 10^{-7}$$

% as $CH_3C(OH)_2^+$ = 0.00001%

(% as $CH_3CO_2H \simeq$ 99.98%)

4.

$$CH_3\overset{O}{\overset{\|}{C}}OH_2^+ \rightleftharpoons H^+ + CH_3CO_2H$$

$$K' = \frac{[H^+][CH_3CO_2H]}{\left[CH_3\overset{O}{\overset{\|}{C}}OH_2^+\right]} = 10^{12} \quad \text{(eq. 1)}$$

$$CH_3\overset{^+OH}{\overset{\|}{C}}OH \rightleftharpoons H^+ + CH_3CO_2H$$

$$K_1 = \frac{[H^+][CH_3CO_2H]}{\left[CH_3\overset{^+OH}{\overset{\|}{C}}OH\right]} = 10^6 \quad \text{(eq. 2)}$$

Rearranging:
$$\frac{\left[CH_3\overset{+OH}{\underset{\|}{C}OH}\right]}{[H^+][CH_3CO_2H]} = 10^{-6} \qquad \text{(eq. 3)}$$

Multiply eq. 1 × eq. 3:

$$\frac{[H^+][CH_3CO_2H]}{\left[CH_3\overset{O}{\underset{\|}{C}}OH_2^+\right]} \cdot \frac{\left[CH_3\overset{+OH}{\underset{\|}{C}OH}\right]}{[H^+][CH_3CO_2H]} = 10^{12}\cdot 10^{-6}$$

$$\frac{\left[CH_3\overset{+OH}{\underset{\|}{C}OH}\right]}{\left[CH_3\overset{O}{\underset{\|}{C}}OH_2^+\right]} = 10^6$$

5. The formula $C_7H_{13}O_2Br$ shows that the molecule has one ring or double
bond. The infrared band at 1740 cm^{-1} strongly suggests an ester,
which would account for both oxygens: $-\overset{O}{\underset{\|}{C}}-O-$

We make the hypothesis that the compound is a bromine-containing ester.

Some of the nmr bands may be:

        1.0 (3H) triplet: $\underline{CH_3}-CH_2-$
        1.3 (6H) doublet: $(\underline{CH_3})_2CH-$

A possible structure which fits all this is:

$$\underset{\delta 2.1(2H)\text{ multiplet}}{} CH_3CH_2\overset{Br}{\underset{}{C}}H-\overset{O}{\underset{\|}{C}}-O-\overset{CH_3}{\underset{CH_3}{C}}H \quad \delta 4.6\ (1H)\ \text{multiplet}$$

δ 2.1(2H) multiplet   δ 4.2(1H) triplet

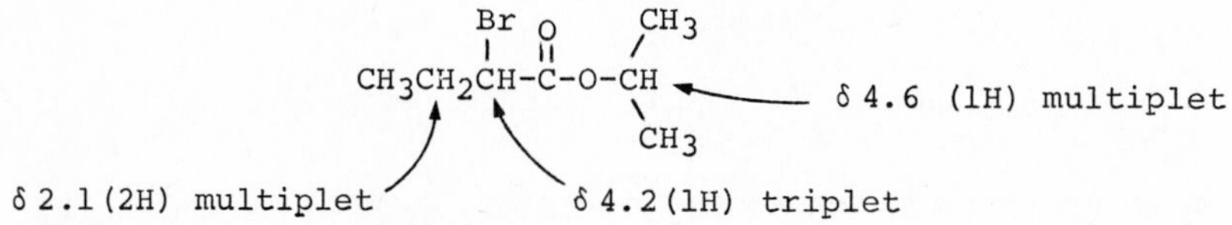

6. The ir band at about 1730 cm^{-1} suggests an ester carbonyl. The
double ions at 194 and 196 suggest the presence of Br. Our working
hypothesis that these are molecular ions gives a molecular weight
which, together with the nmr, leads to the formula $C_6H_{11}BrO_2$.

For example, 194 - 79(bromine) - 32(2 oxygens) = 83(C_6H_{11}).
Similarly, 196 - 81(bromine) -32(2 oxygens) = 83(C_6H_{11}). The nmr
spectrum shows three absorptions, a doublet (δ=4.3), a singlet
(δ=1.9), and a triplet (δ=1.3). The doublet and triplet comprise
a typical ethyl pattern. From the low field position of the
CH_2 quartet, we may conclude that we have CH_3CH_2O-. The singlet
must be the other six hydrogens, which are present therefore as
two equivalent CH_3 groups. Hence, the fragments we have to
piece together are:

$$CH_3-\ CH_3-\ CH_3CH_2O-\ -\overset{O}{\underset{\|}{C}}-\ -\overset{|}{\underset{|}{C}}-\ Br-$$

The only possible combination which gives two equivalent methyls is:

$$CH_3CH_2O\overset{O}{\underset{}{C}}\overset{CH_3}{\underset{CH_3}{C}}-Br$$

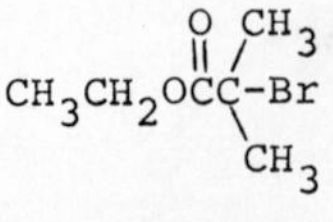

The important mass spectral fragment ions are consistent with this structural assignment:

$149,\ 151 = BrC(CH_3)_2C{\equiv}O^+$; $121,\ 123 = BrC(CH_3)_2^+$; $115 = {}^+C(CH_3)_2CO_2C_2H_5$

7. (a) $CH_3CH{=}CH_2 + CH_3CO_2H$ (b)  $+ CH_3SH$ (c) 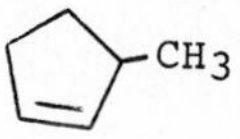 $+ CH_3SH$

(d) $CH_3CH_2CH_2CO_2^-\ Na^+$ (e) 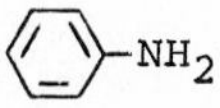 (f) 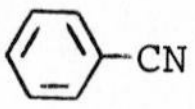 (g) $(CH_3)_3C{-}\overset{O}{\overset{\|}{C}}{-}ND_2$

(h) 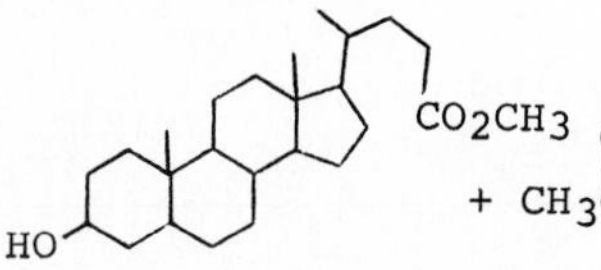 $+ CH_3\overset{O}{\overset{\|}{C}}{-}OCH_3$ (i) 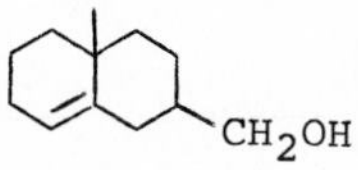

(j) $CH_3CH_2\overset{O}{\overset{\|}{C}}CH(CH_3)CO_2C_2H_5$ (k) $CH_3CH_2CH(OH)CH_2CO_2CH_3$ (l) $CH_3CH_2CH(OH)CH(CH_3)CO_2C_2H_5$

8. (a) $CH_3CH_2CH_2CO_2H \xrightarrow[\text{or } PCl_5]{SOCl_2} CH_3CH_2CH_2\overset{O}{\overset{\|}{C}}Cl$

(b) $CH_3CH_2CH_2\overset{O}{\overset{\|}{C}}OH \xrightarrow{NaOH} CH_3CH_2CH_2\overset{O}{\overset{\|}{C}}O^-\ Na^+ \xrightarrow{CH_3CH_2CH_2\overset{O}{\overset{\|}{C}}{-}Cl}$

(c) $CH_3CH_2CH_2CO_2H \xrightarrow{SOCl_2} \xrightarrow{CH_3CH_2CH_2CH_2OH}$

or $\xrightarrow[H^+]{CH_3CH_2CH_2CH_2OH}$

$\left(CH_3CH_2CH_2CO_2H \xrightarrow{LiAlH_4} CH_3CH_2CH_2CH_2OH \right)$

(d) $CH_3CH_2CH_2CO_2H + (CH_3)_2NH \xrightarrow{170°}$

or

$\xrightarrow{SOCl_2} \xrightarrow[OH^-]{(CH_3)_2NH}$

(e) $CH_3CH_2CH_2\overset{O}{\overset{\|}{C}}OH \xrightarrow{NH_3} \xrightarrow{170°} CH_3CH_2CH_2\overset{O}{\overset{\|}{C}}NH_2 \xrightarrow[\Delta]{P_2O_5} CH_3CH_2CH_2C{\equiv}N$

(f) $CH_3CH_2CH_2CN$ (from e) $\xrightarrow[NH_3]{H_2/\text{cat.}} CH_3CH_2CH_2CH_2NH_2$

or

$CH_3CH_2CH_2\overset{O}{\overset{\|}{C}}NH_2 \xrightarrow[\text{ether}]{LiAlH_4} CH_3CH_2CH_2CH_2NH_2$

(g) $CH_3CH_2CH_2CO_2H \xrightarrow[Br_2]{HgO} CH_3CH_2CH_2Br$

(h) $CH_3CH_2CH_2\overset{O}{\overset{\|}{C}}OH \xrightarrow{SOCl_2} \xrightarrow[OH^-]{HN(C_2H_5)_2} CH_3CH_2CH_2\overset{O}{\overset{\|}{C}}N(C_2H_5)_2 \xrightarrow[\text{ether}]{LiAlH_4}$

(i) $CH_3CH_2CH_2\overset{O}{\overset{\|}{C}}OH \xrightarrow{SOCl_2} CH_3CH_2CH_2\overset{O}{\overset{\|}{C}}Cl \xrightarrow{(CH_3CH(CH_3)CH_2)_2CuLi}$

(j) $CH_3CH_2CH_2\overset{\overset{O}{\|}}{C}OH \xrightarrow[H^+]{CH_3OH} CH_3CH_2CH_2CO_2CH_3 \xrightarrow{CH_3MgBr} \xrightarrow{H_3O^+} CH_3CH_2CH_2\underset{\underset{CH_3}{|}}{\overset{\overset{CH_3}{|}}{C}}OH$

(k) $CH_3CH_2CH_2CO_2H \xrightarrow{SOCl_2} \xrightarrow{LiAlH(O\underline{t}Bu)_3} CH_3CH_2CH_2CHO$

(l) $CH_3CH_2CH_2CO_2H \xrightarrow{LiAlH_4} CH_3CH_2CH_2CH_2OH$

(m) $CH_3CH_2CH_2CO_2H \xrightarrow[C_2H_5OH]{H^+} CH_3CH_2CH_2CO_2C_2H_5 \xrightarrow{NaOC_2H_5} CH_3CH_2CH_2\overset{\overset{O}{\|}}{C}\underset{\underset{CH_2CH_3}{|}}{C}HCO_2C_2H_5 \xrightarrow{LiAlH_4}$

(n) $CH_3CH_2CH_2CO_2C_2H_5 \xrightarrow[THF,\ -78°C]{LiN(\underline{i}-C_3H_7)_2} \xrightarrow{CH_3CH_2CH_2\overset{\overset{O}{\|}}{C}H} \xrightarrow{H_2O} CH_3CH_2CH_2\overset{\overset{OH}{|}}{C}H\underset{\underset{CH_2CH_3}{|}}{C}HCO_2C_2H_5$

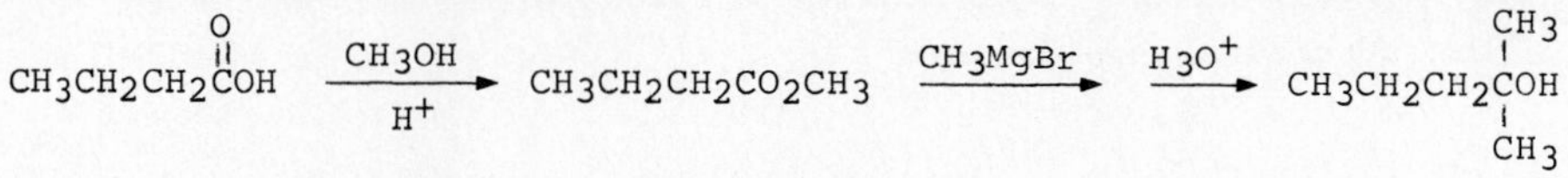

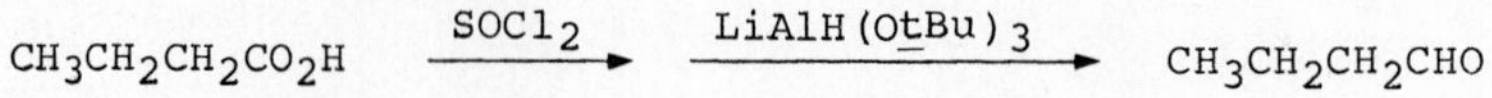

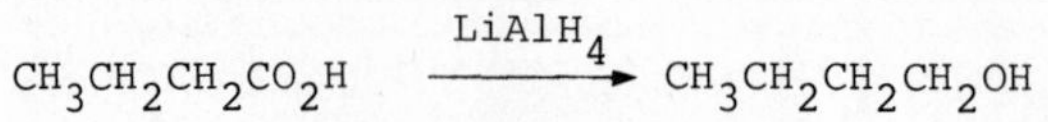

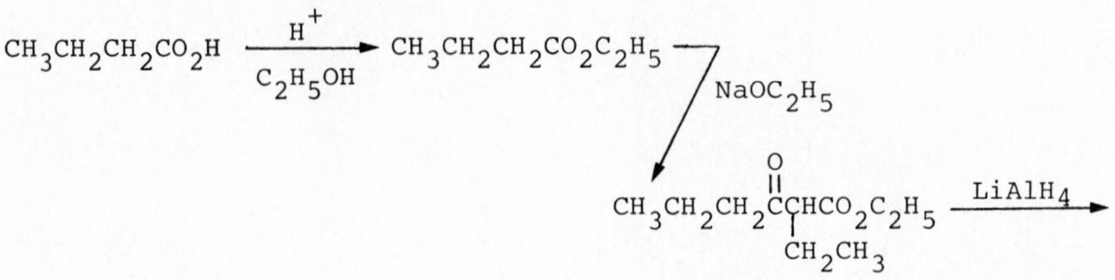

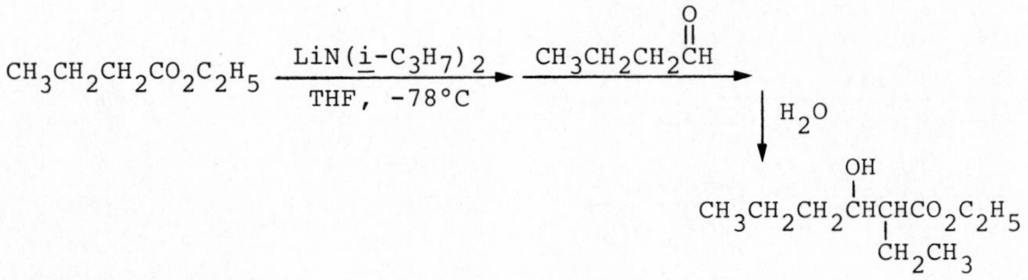

(o) See answer to (m).

9. (Note that the starting material and the product both have the S configuration, but one is dextrorotatory, and the other is levorotatory.)

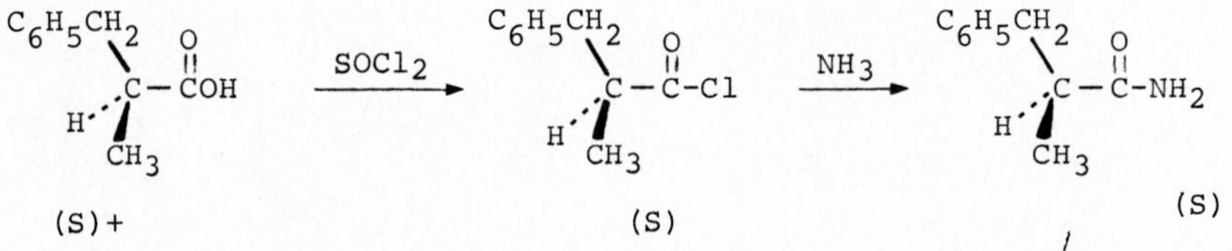

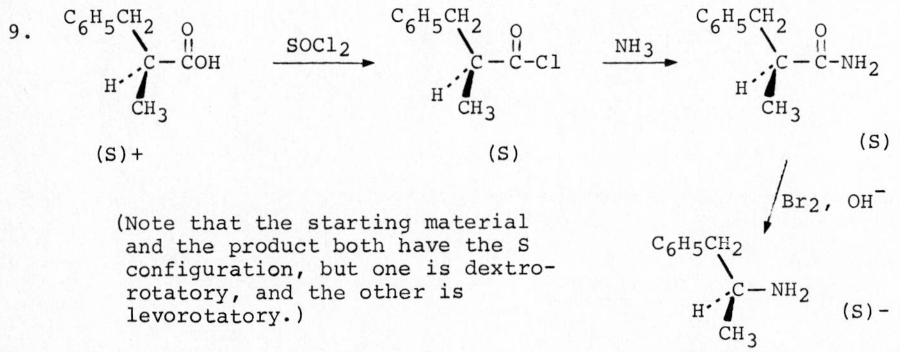

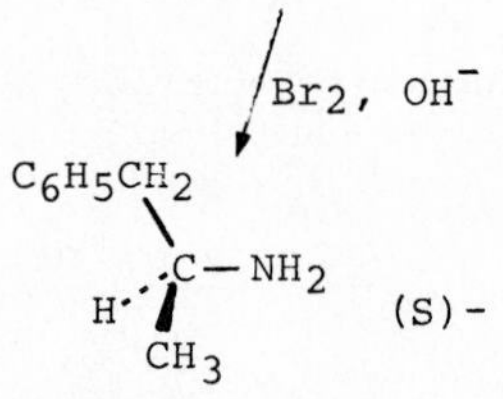

10. (a)

$$H_2N\text{-}\overset{\overset{O}{\|}}{C}\text{-}Cl + CH_3O^- \longrightarrow H_2N\text{-}\overset{\overset{O}{\|}}{C}\text{-}OCH_3 + Cl^-$$

After addition of CH_3O^-, the intermediate is: $H_2N\text{-}\underset{\underset{OCH_3}{|}}{\overset{\overset{O^-}{|}}{C}}\text{-}Cl$

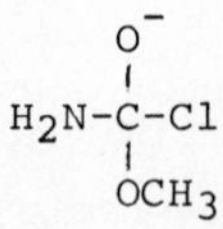

This may decompose in three ways. The three leaving groups are NH_2^-, CH_3O^-, Cl^-. The best leaving group is Cl^-, so the product is:

$$H_2N\text{-}\underset{\underset{OCH_3}{|}}{\overset{\overset{O^-}{|}}{C}}\text{-}Cl \longrightarrow H_2N\text{-}\overset{\overset{O}{\|}}{C}\text{-}OCH_3 + Cl^-$$

(b)

$$CH_3O\text{-}\overset{\overset{O}{\|}}{C}\text{-}Cl + NH_2^- \longrightarrow CH_3O\text{-}\overset{\overset{O}{\|}}{C}\text{-}NH_2 + Cl^-$$

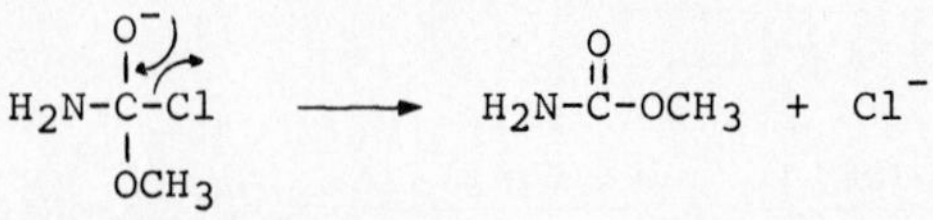

same reasoning as above.

11. (a)

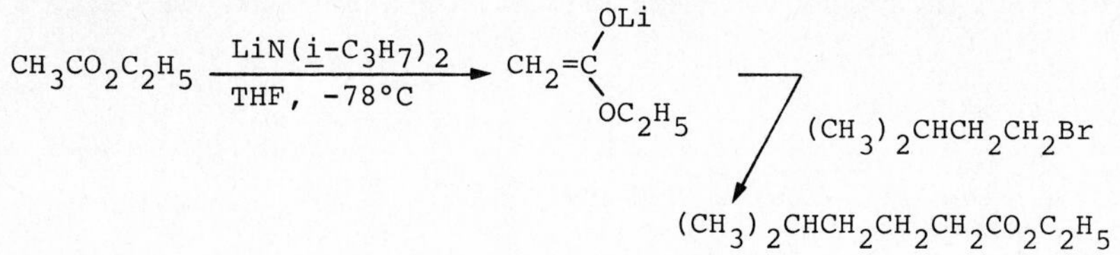

(b)

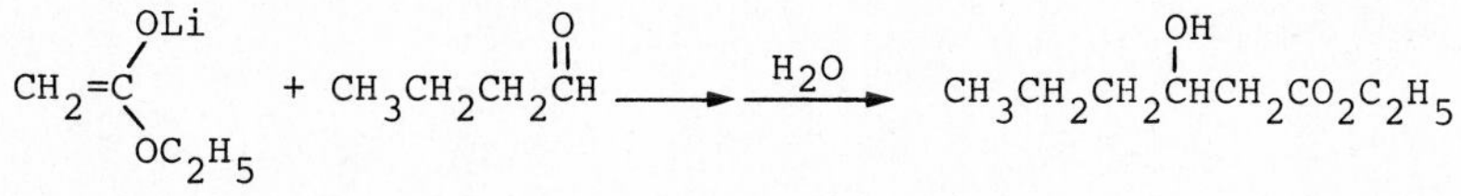

(c)

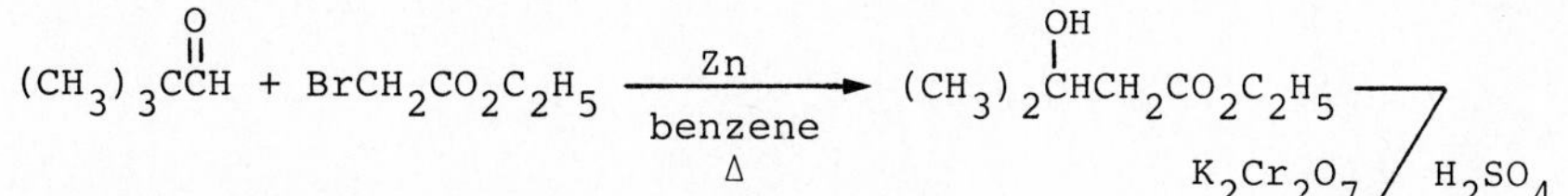

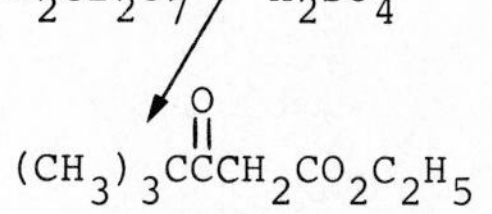

(Note: although only one ester would be enolizable, a mixed Claisen condensation would not be a good choice, because the ethyl acetate enolate would condense faster with ethyl acetate itself than with the sterically more congested ethyl 2,2-dimethylpropanoate.)

(d)

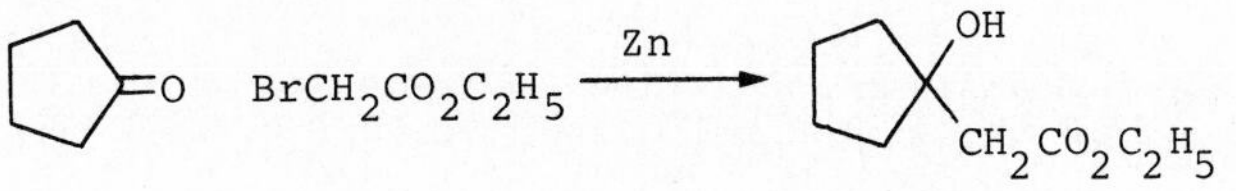

(or via lithium enolate)

12.

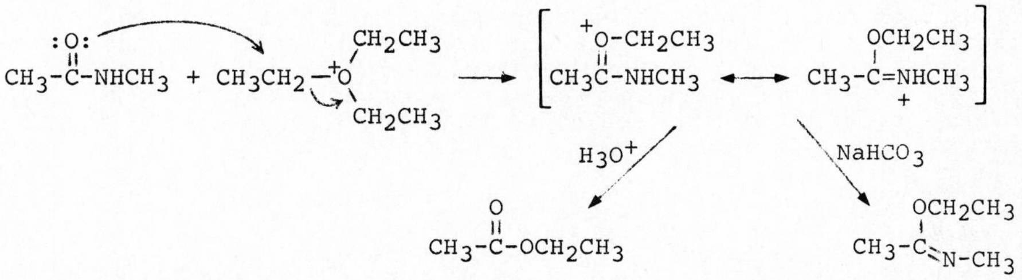

13.

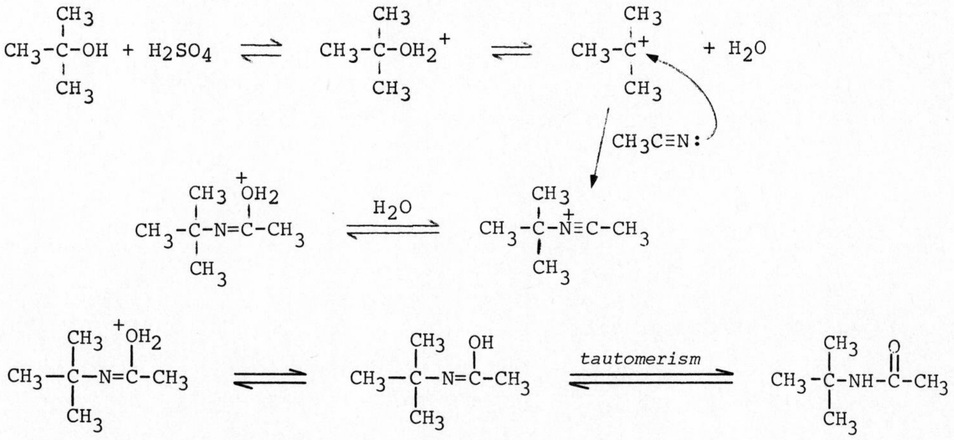

With 2-methyl-2,4-pentanediol, the tertiary carbocation forms more easily than the secondary carbocation. After addition of acetonitrile, the resulting cation reacts *intramolecularly* with the other hydroxyl group. The product is a heterocyclic compound, called a *tetrahydrooxazine*.

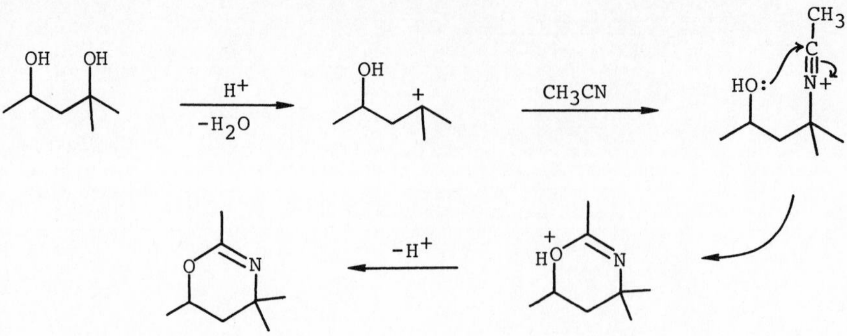

14. 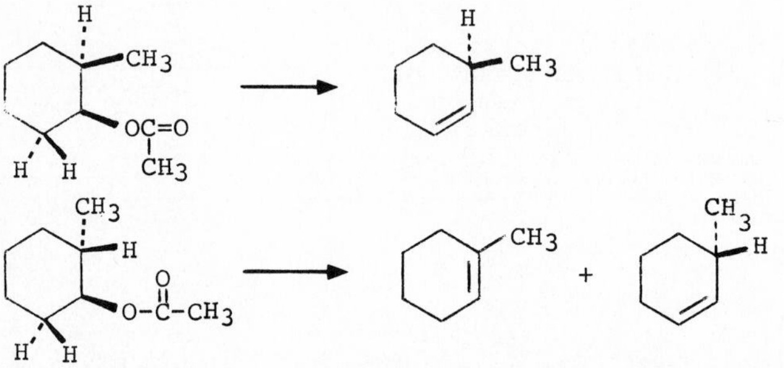

<u>syn</u> elimination; only H's on the same side of the ring can be removed.

15. The reaction has a favorable entropy (positive $\Delta S°$) because one molecule gives rise to two. The reaction has a relatively high energy of activation; hence, a high temperature is required. Furthermore, from $\Delta G° = \Delta H° - T\Delta S°$, the favorable entropy can give rise to a favorable equilibrium constant at sufficiently high temperature despite being endothermic.

16. 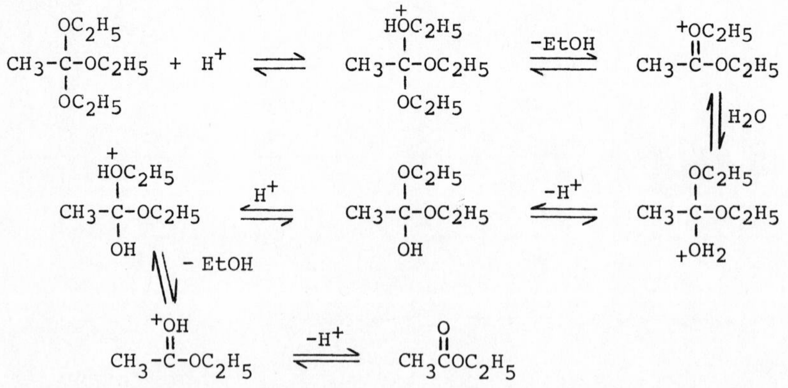

17. McLafferty rearrangement:

$$[CH_3(CH_2)_3COOCH_3]^{+\cdot} \longrightarrow CH_3CH{=}CH_2 + [CH_2{=}C(OH)OCH_3]^{+\cdot}$$

m/e 74

$$[CH_3CH_2CH(CH_3)COOCH_3]^{+\cdot} \longrightarrow CH_2{=}CH_2 + [CH_3CH{=}C(OH)OCH_3]^{+\cdot}$$

m/e 88

18.

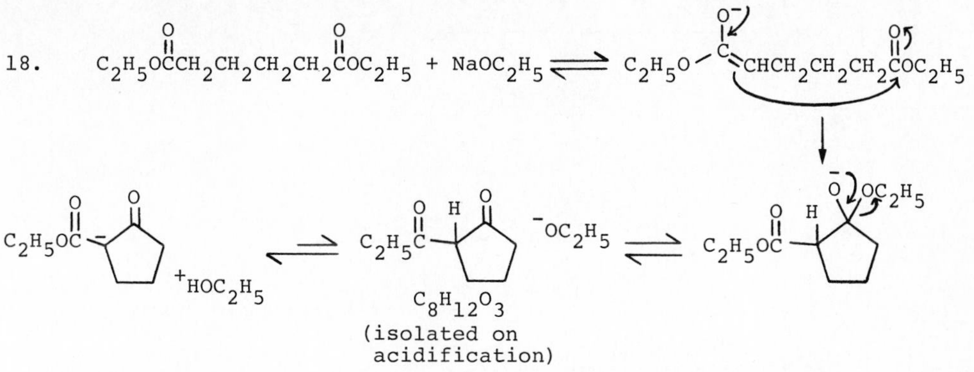

19.

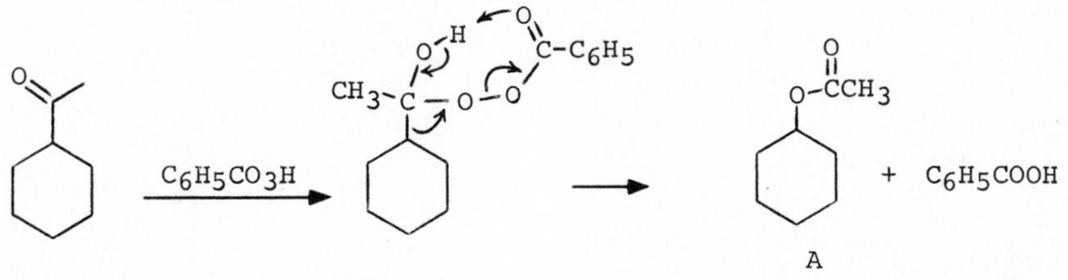

Note that the Baeyer-Villiger oxidation occurs exclusively with migration of the cyclohexyl group.

20. (a) S_N2 reaction with inversion of configuration.

(b) Convert to a sulfonate ester and displace with sodium methoxide.

19.E Supplementary Problems

S1. Give the IUPAC name of each of the following compounds.

(a) $CH_3CH_2CH(CH_3)C(=O)NHCH_3$

(b) 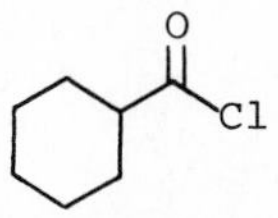

(c) $(CH_3)_2CHC(=O)OC(=O)CH(CH_3)_2$

(d) $CH_2{=}CH{-}CH{=}CH{-}OC(=O)CH_3$

(e) $CH_3CH_2CH_2C(=O)C{\equiv}N$

(f) $CH_3CH_2C(=O)NHBr$

(g) 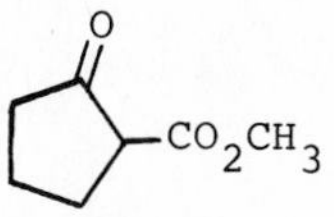

(h) 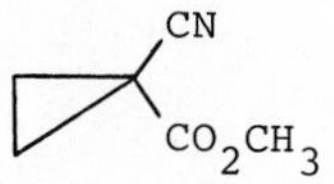

S2. Provide IUPAC names for juvenile hormone and juvabione (Section 18.8.) *(Don't forget stereochemistry.)*

S3. What is the major product of each of the following reaction sequences?

(a) $CH_3CH_2C(=O)OCH_3 \xrightarrow{NH_3} \xrightarrow{LiAlH_4} \xrightarrow{CH_3CH_2C(=O)Cl}$

(b) $(CH_3)_2CHC(=O)H \xrightarrow{Ag_2O} \xrightarrow[\Delta]{P\ +\ Br_2} \xrightarrow{HOCH(CH_3)_2}$

(c)

$$\text{[cyclohexane with } CO_2CH_3\text{]} \xrightarrow[25°\ C]{CH_3MgBr} \xrightarrow{HBr} \xrightarrow[C_2H_5OH]{NaCN}$$

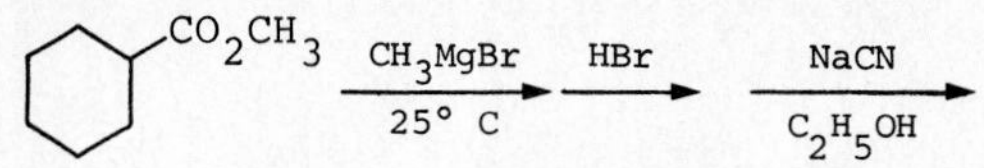

(d) $CH_3CH_2CH_2CH_2CO_2H \xrightarrow[Br_2]{AgO} \xrightarrow{Na^+ {}^-O_2CCH_3} \xrightarrow{500°\ C}$

(e) $CH_3CH_2CO_2C(CH_3)_3 \xrightarrow[\substack{THF \\ -78°\ C}]{LDA} \xrightarrow{CH_3CH_2\overset{\overset{\displaystyle O}{\|}}{C}H} \xrightarrow{CH_3CH_2\overset{\overset{\displaystyle O}{\|}}{C}Cl}$

(f)

$$\text{[cyclopentane with } CO_2H\text{]} \xrightarrow{SOCl_2} \xrightarrow{NH_3} \xrightarrow{P_2O_5}$$

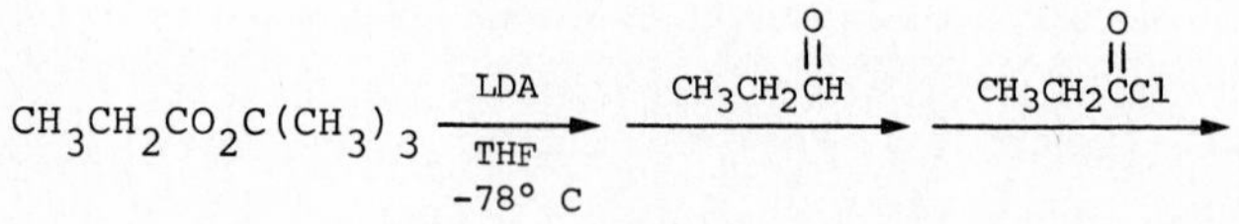

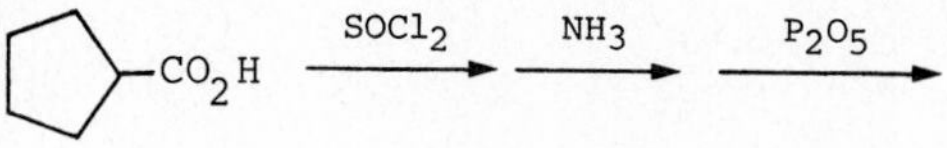

(g) $(CH_3)_2CHCH_2CO_2H \xrightarrow{SOCl_2} \xrightarrow{(CH_3)_2NH} \xrightarrow{LiAlH(O\text{-}t\text{-}Bu)_3} \xrightarrow{H_2O}$

(h)

$$\text{[cyclohexane with } \overset{\overset{\displaystyle O}{\|}}{C}H\text{]} \xrightarrow[Zn,\ benzene,\ \Delta]{BrCH_2CO_2C_2H_5} \xrightarrow[\Delta]{H_2SO_4}$$

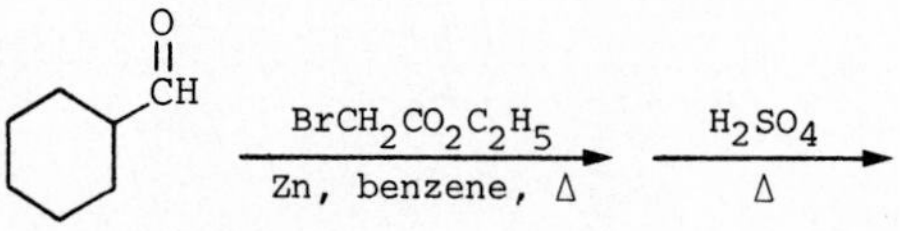

S4. Show how to carry out the following transformations.

(a)

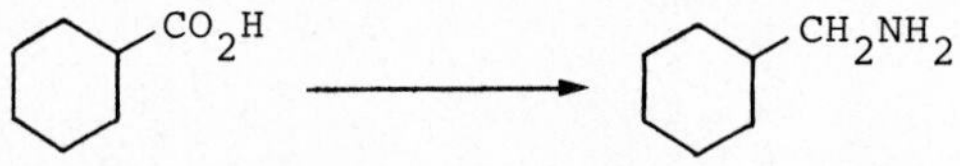

$$\text{[cyclohexane-} CO_2H\text{]} \longrightarrow \text{[cyclohexane-} CH_2NH_2\text{]}$$

(b) $(CH_3)_2CHCH_2CH_2OH \longrightarrow (CH_3)_2CHCH_2CH_2\underset{\underset{\displaystyle CH_3}{|}}{C}HCO_2CH(CH_3)_2$

(c)

$$\text{[cyclopentane-} CO_2H\text{]} \longrightarrow \text{[cyclopentane-}\underset{\underset{\displaystyle CHCH_2CO_2C_2H_5}{}}{\overset{\overset{\displaystyle OH}{|}}{}}\text{]}$$

(d)

$$\text{[}\gamma\text{-butyrolactone]} \longrightarrow CH_3\overset{\overset{\displaystyle O}{\|}}{C}OCH_2CH_2CH_2CH_2O\overset{\overset{\displaystyle O}{\|}}{C}CH_3$$

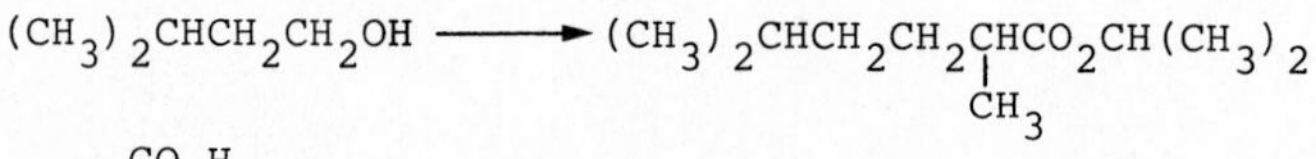

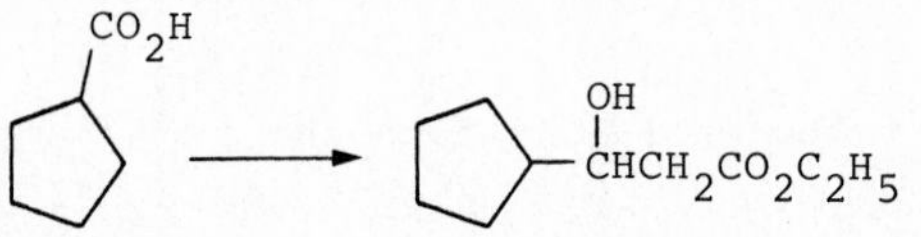

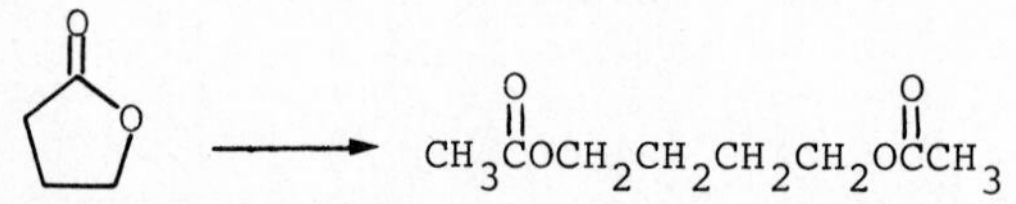

(e) $(CH_3)_2CHCH_2OH \longrightarrow ((CH_3)_2CHCH_2)_3CBr$

(f)

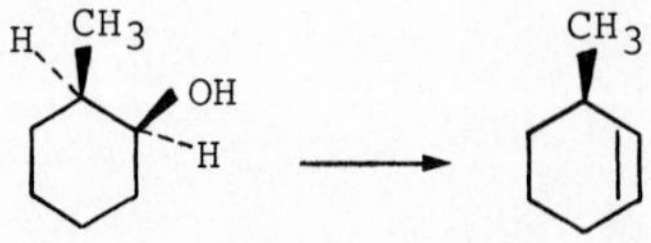

$$\text{[cyclohexane with } H, CH_3, OH, H\text{]} \longrightarrow \text{[cyclohexene with } CH_3\text{]}$$

(g) $BrCH_2CH_2CH_2Br \longrightarrow \text{[glutaric anhydride]}$

(h) $CH_3CH_2CH_2CO_2C_2H_5 \longrightarrow CH_3CH_2CH_2\underset{\underset{\displaystyle CH_2OH}{|}}{\overset{\overset{\displaystyle OH}{|}}{C}}HCHCH_2CH_3$

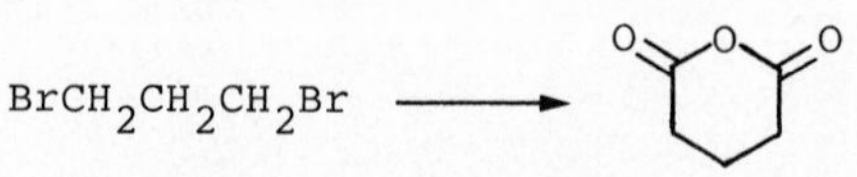

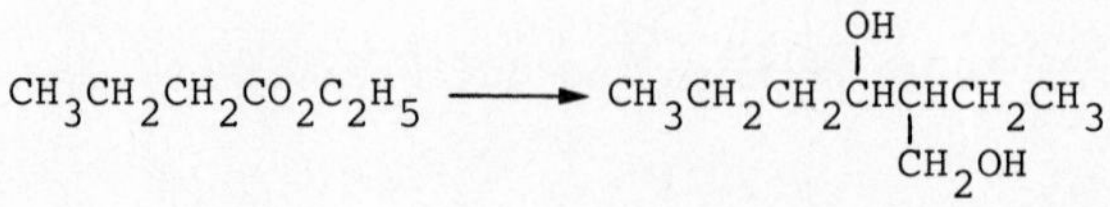

S5. (a) Provide structures for A-D below, and assign all the spectral
         information.

$$A \xrightarrow[2.\ H^+]{1.\ NaOCH_3} B \xrightarrow{SOCl_2} C \xrightarrow[quinoline\ poison]{H_2\ Pd\text{-}BaSO_4} D$$

IR: 1725,1820 cm^{-1} NMR: δ 2.0(quintet, 2H), 2.8(t,4H)	IR: 1740,1710, 2500-3000(br) cm^{-1} NMR: δ 3.8(3H), 13(s,1H) (and resonances for 6 other H's)	IR: 1735,1785 cm^{-1}	IR: 1725,1740 cm^{-1}

(b) When compound D is treated with HCN and a trace of base, compound E is formed. The nmr spectrum of E reveals the presence of only 7 hydrogens; the IR spectrum shows bands at 1735 and 2130 cm^{-1}. Propose a structure for E and write a reasonable mechanism for its formation.

S6. Reaction of acetonitrile in methanol with dry HCl gives initially methyl acetimidate hydrochloride (1) and then methyl orthoacetate (2). Write a mechanism for this reaction.

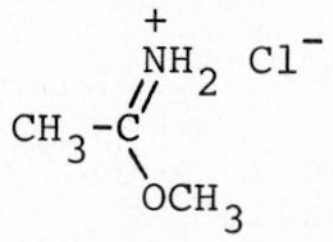

1

$CH_3C(OCH_3)_3$

2

S7. Methyl acetimidate (3) is hydrolyzed in aqueous sodium hydroxide to give mainly acetamide and methanol. In aqueous acid it hydrolyzes to give primarily methyl acetate and ammonia. Write mechanisms for these reactions and explain why different products are seen in acid and base.

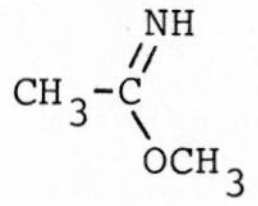

3

S8. Write a mechanism for the following transformation. *(Hint: see Problem 18 in this chapter.)*

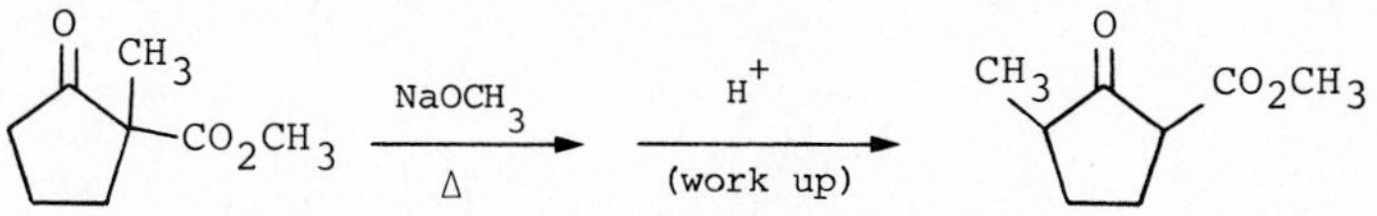

S9. Starting with benzoic acid 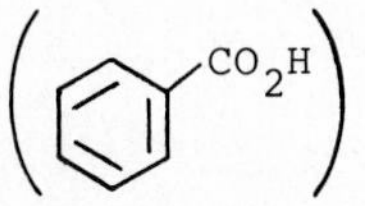 , propanoic acid, dimethylamine ($(CH_3)_2NH$) and any needed reagents, outline a synthesis of propoxyphene (the active ingredient in Darvon®).

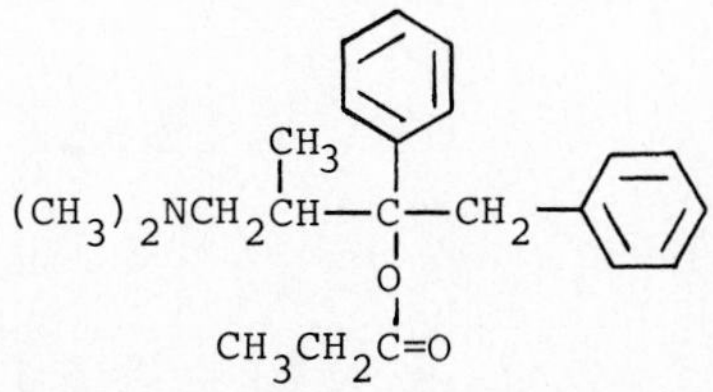

S10. A useful preparation of deuterioethanol, C_2H_5OD, involves refluxing diethyl carbonate with D_2O and a small amount of strong acid. Why is this preparation so convenient?

S11. Trimyristin is a white crystalline fat, mp 54-55°, obtainable from nutmeg, and is the principal constituent of nutmeg butter. Hydrolysis of trimyristin with hot aqueous sodium hydroxide gives an excellent yield of myristic acid, mp 52-53°, as the only fatty acid. What is the structure of trimyristin?

19.F Answers to Supplementary Problems

S1. (a) N,2-dimethylbutanamide
(b) Cyclohexanecarbonyl chloride
(c) 2-Methylpropanoic anhydride
(d) 1,3-Butadienyl acetate
(e) 2-Oxopentanenitrile
(f) N-bromopropanamide
(g) Methyl 2-oxocyclopentanecarboxylate
(h) Methyl 1-cyanocyclopropanecarboxylate

S2. Juvenile hormone: methyl (Z)-10,11-epoxy-7-ethyl-3,11-dimethyl-(E,E)-2,6-tridecadienoate

Juvabione: methyl (R)-4-((R)-1,5-dimethyl-3-oxohexyl)-1-cyclohexenecarboxylate

S3. (a) $CH_3CH_2CH_2NH\overset{O}{\overset{\|}{C}}CH_2CH_3$

(b) 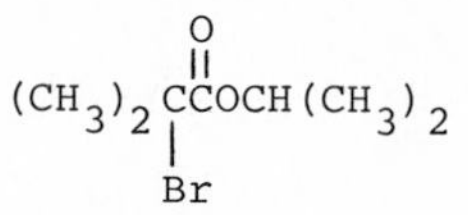

(c) 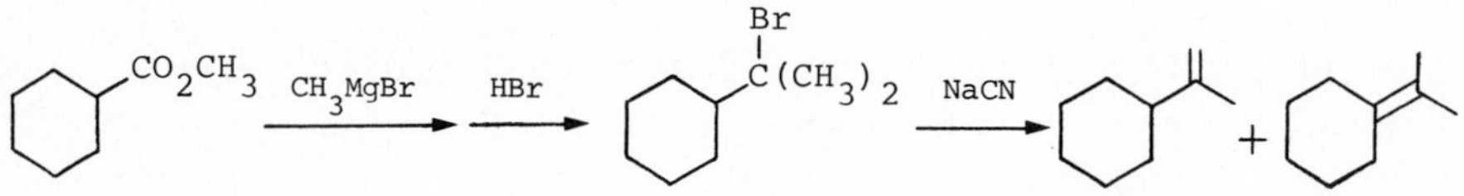

(E2, not S_N2. . .)

(d) $CH_3CH_2CH{=}CH_2$

(e) 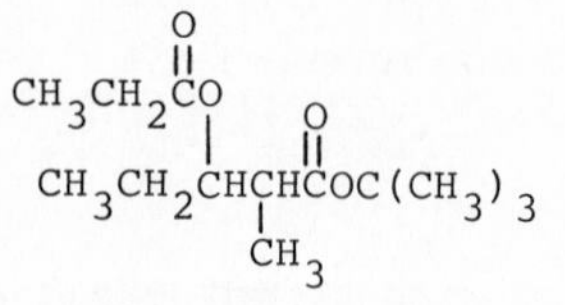

(f) 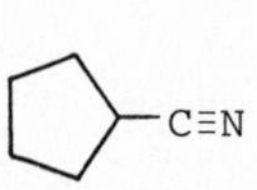

(g) $(CH_3)_2CHCH_2\overset{O}{\overset{\|}{C}}H$

(h) 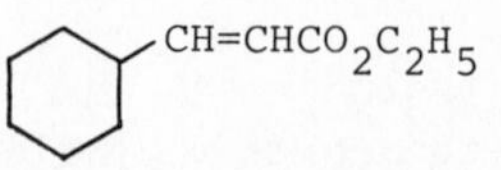

S4. (a) 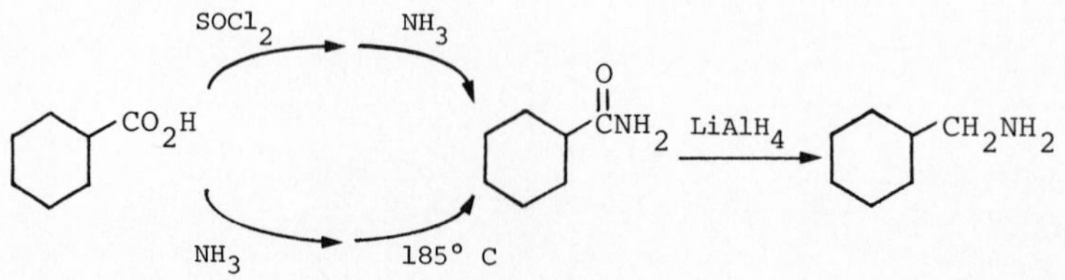

(b)

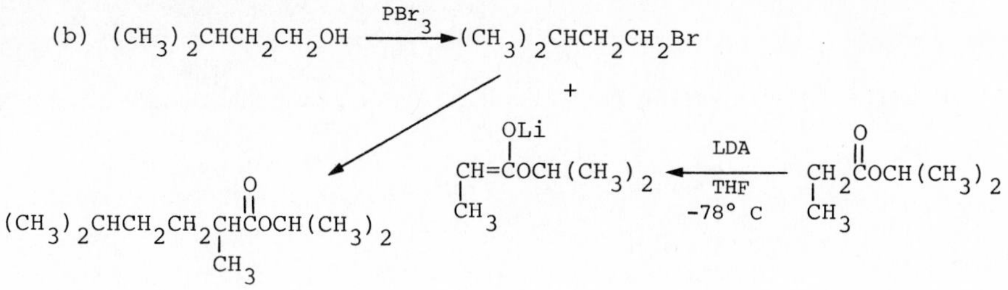

(c)

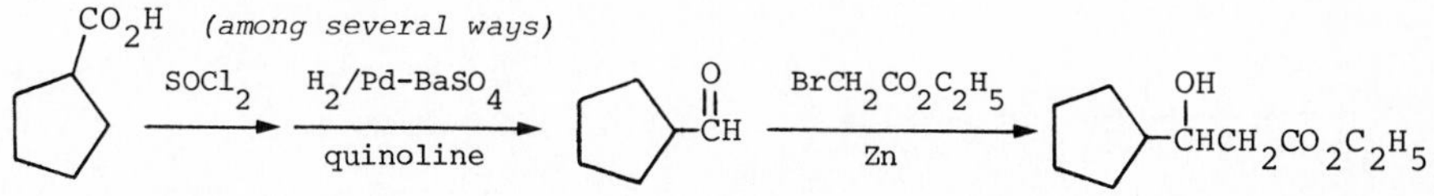

(d)

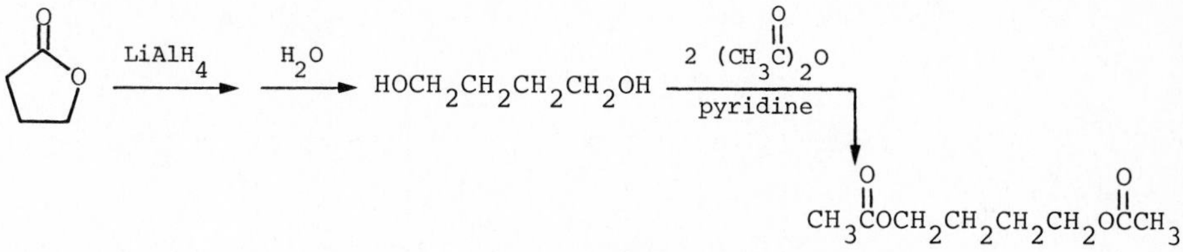

(e)

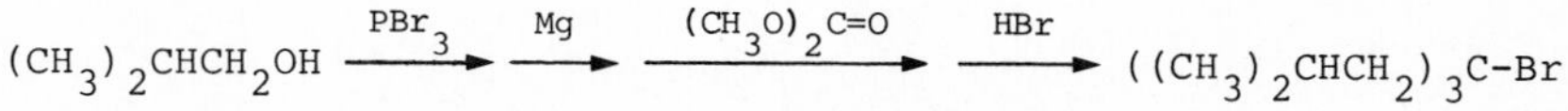

(f)

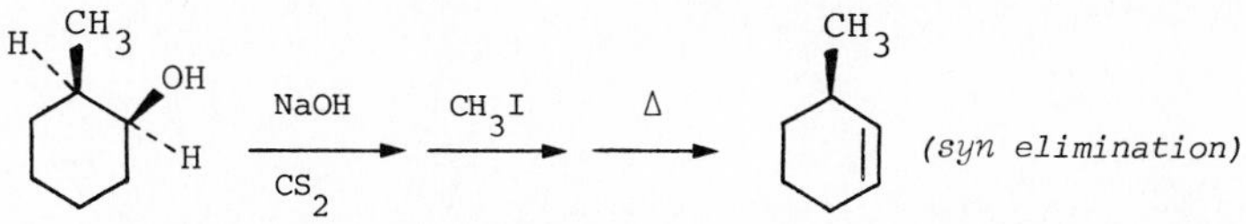

(g)

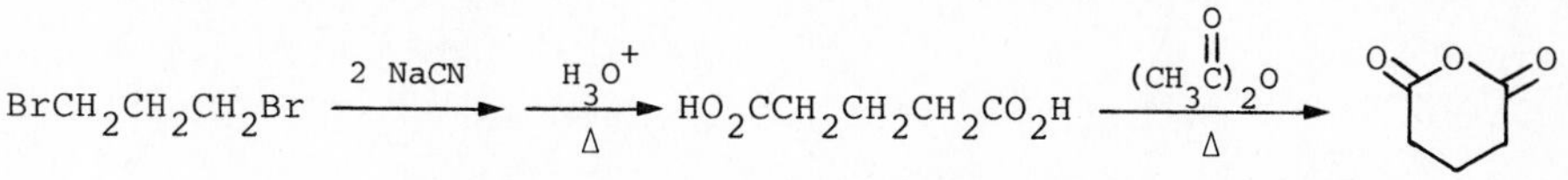

(h)

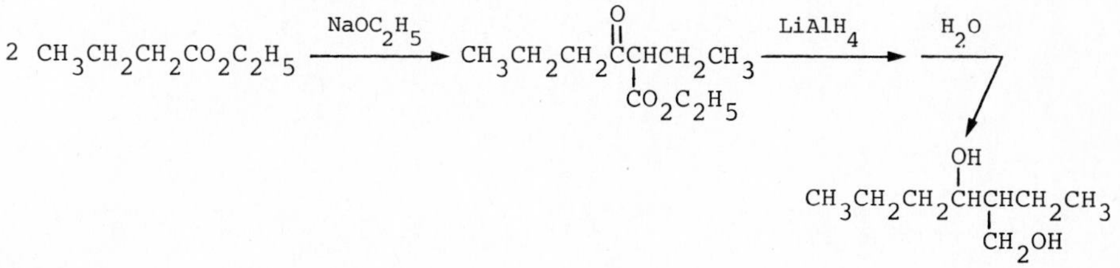

S5. (a)

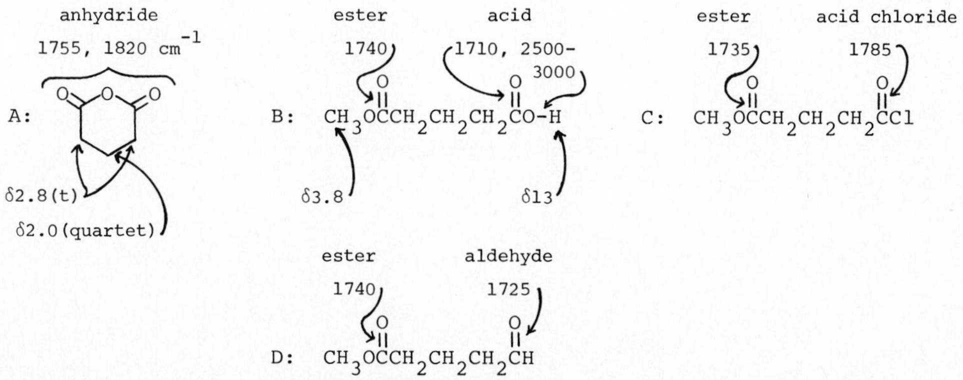
anhydride
1755, 1820 cm-1
A:
δ2.8(t)
δ2.0(quartet)
ester
1740
acid
1710, 2500-3000
B:
δ3.8
δ13
ester
1735
acid chloride
1785
C:
ester
1740
aldehyde
1725
D:

(b) $CH_3OCCH_2CH_2CH_2CH$ + HCN $\longrightarrow$ $CH_3OCCH_2CH_2CH_2CHCN$ $\rightleftharpoons$ $\overset{base}{}$ $CH_3OCCH_2CH_2CH_2CHCN$

6-ring lactone   nitrile
1735   2130

S6. $CH_3C\equiv N$ $\overset{H^+}{\rightleftharpoons}$ $CH_3C\equiv\overset{+}{N}H$ $\rightleftharpoons$ $CH_3\overset{NH}{\underset{\overset{+}{HOCH_3}}{C}}$ $\overset{-H^+}{\rightleftharpoons}$ $CH_3\overset{NH}{\underset{OCH_3}{C}}$ $\overset{H^+}{\rightleftharpoons}$ $CH_3\overset{\overset{+}{NH_2}}{\underset{OCH_3}{C}}$ $\rightleftharpoons$ $CH_3\overset{NH_2}{\underset{\overset{+}{HOCH_3}}{COCH_3}}$

$CH_3\ddot{O}H$ $CH_3\ddot{O}H$

$CH_3\overset{OCH_3}{\underset{\overset{+}{HOCH_3}}{COCH_3}}$ $\overset{CH_3OH}{\rightleftharpoons}$ $\left[CH_3\overset{\overset{+}{OCH_3}}{\underset{OCH_3}{C}} \longleftrightarrow CH_3\overset{OCH_3}{\underset{\overset{+}{OCH_3}}{C}} \right]$ $\overset{-NH_3}{\rightleftharpoons}$ $CH_3\overset{\overset{+}{NH_3}}{\underset{\ddot{:}OCH_3}{COCH_3}}$ $\overset{H^+}{\rightleftharpoons}$ $CH_3\overset{NH_2}{\underset{OCH_3}{COCH_3}}$

$\overset{-H^+}{\rightleftharpoons}$

$CH_3C(OCH_3)_3$

S7. With OH⁻: $CH_3-\overset{NH}{\underset{OCH_3}{C}}$ $\rightleftharpoons$ $CH_3-\overset{NH_2}{\underset{OH}{C}-OCH_3}$ $\rightleftharpoons$ $CH_3-\overset{NH_2}{\underset{O}{C}-OCH_3}$ + H_2O

H₂O ⟋ ⁻OH

HO⁻

(best leaving group lost:
pK_a *(CH₃OH) = 16,*
pK_a *(NH₃) = 35)*

$CH_3-\overset{NH_2}{\underset{O}{C}}$ + CH_3O^-

With H₃O⁺: $CH_3\overset{NH}{\underset{OCH_3}{C}}$ $\overset{H^+}{\rightleftharpoons}$ $CH_3\overset{\overset{+}{NH_2}}{\underset{OCH_3}{C}}$ $\rightleftharpoons$ $CH_3-\overset{NH_2}{\underset{\overset{+}{OH_2}}{C}-OCH_3}$ $\overset{-H^+}{\rightleftharpoons}$

$H_2\ddot{O}:$

$CH_3-\overset{NH_2}{\underset{OH}{C}-OCH_3}$

$\overset{-H^+}{\rightleftharpoons}$ $\left[CH_3-\overset{\overset{+}{OCH_3}}{\underset{OH}{C}} \longleftrightarrow CH_3-\overset{OCH_3}{\underset{\overset{+}{OH}}{C}} \right]$ $\overset{-NH_3}{\rightleftharpoons}$ $CH_3-\overset{\overset{+}{NH_3}}{\underset{\ddot{:}OH}{C}-OCH_3}$

$\overset{O}{\underset{}{CH_3COCH_3}}$

(In acid, the amine is
protonated (pK_a R$\overset{+}{N}$H₃ ≃
9), and NH₃ is now a
better leaving group
than ⁻OCH₃.)

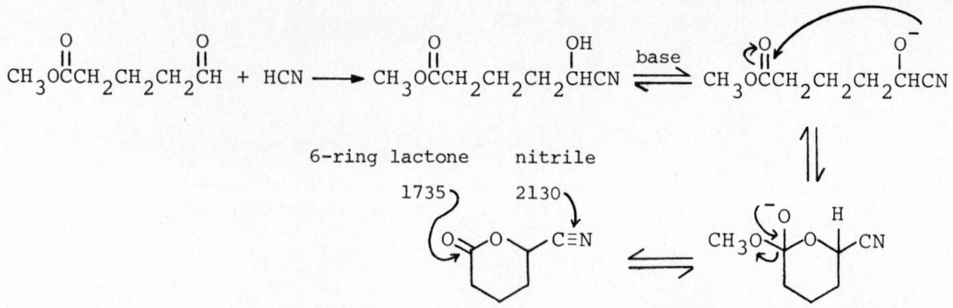

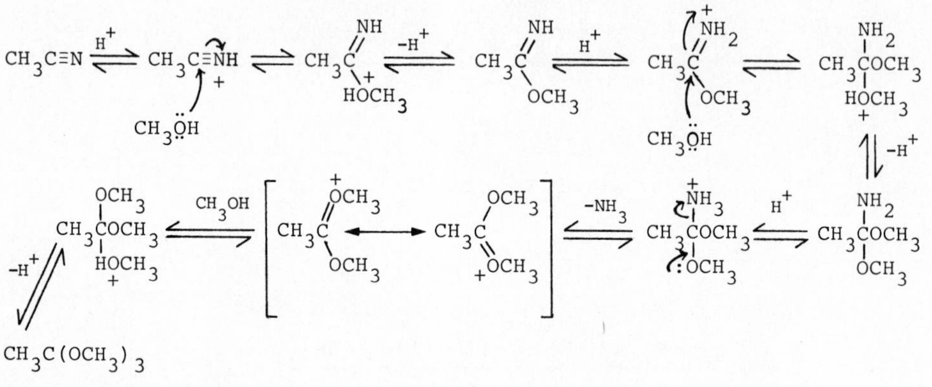

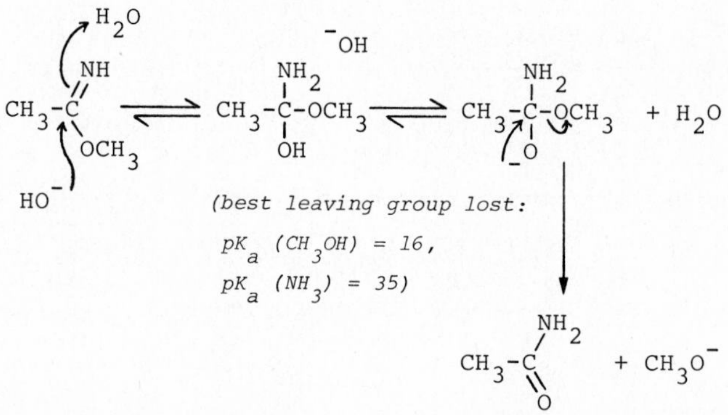

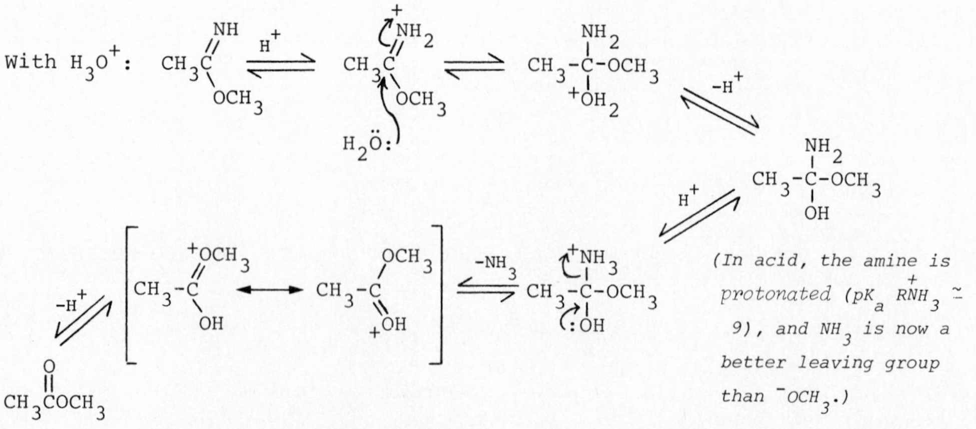

S8.

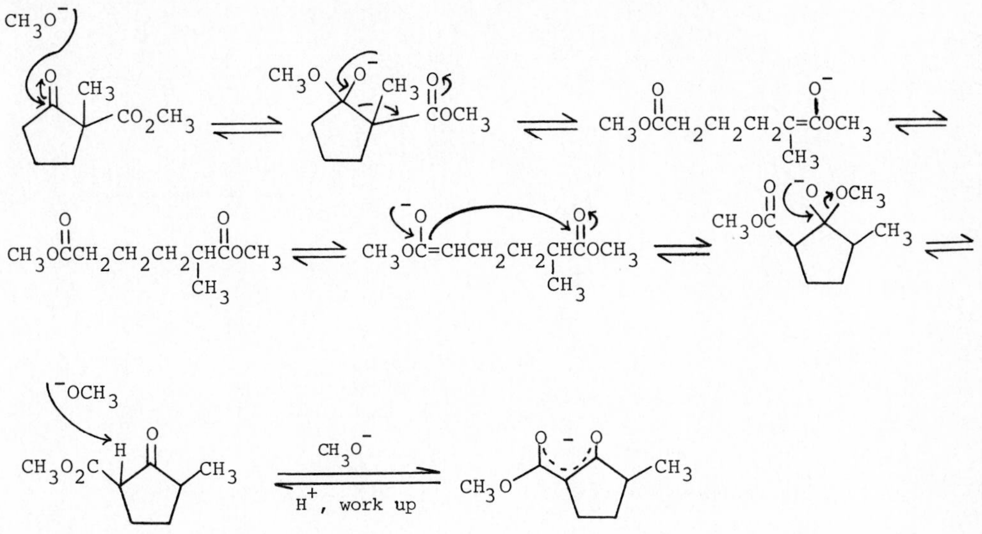

S9. This is clearly a case where you have to work backward!

1) What part of the target came from which starting material?

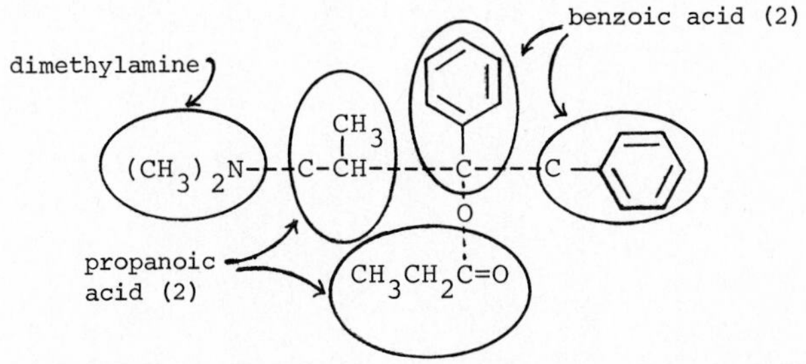

2) The CH_2 next to the nitrogen could come from $LiAlH_4$ reduction of the amide, and the propanoate ester from the 3° alcohol:

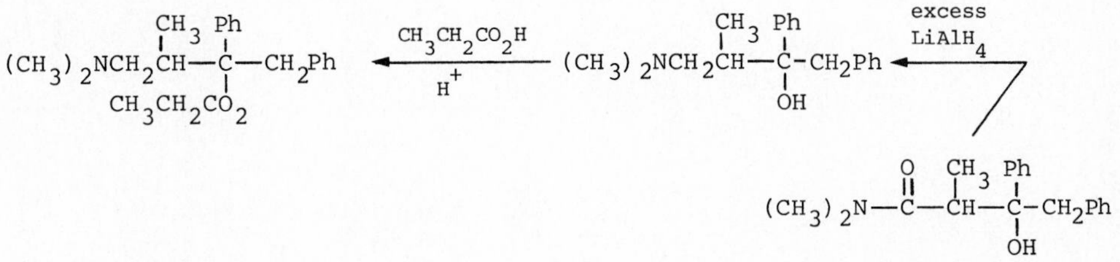

3) The hydroxyl β to the carbonyl suggests that an ester-enolate or Reformatsky reaction could be used to form the carbon-carbon bond.

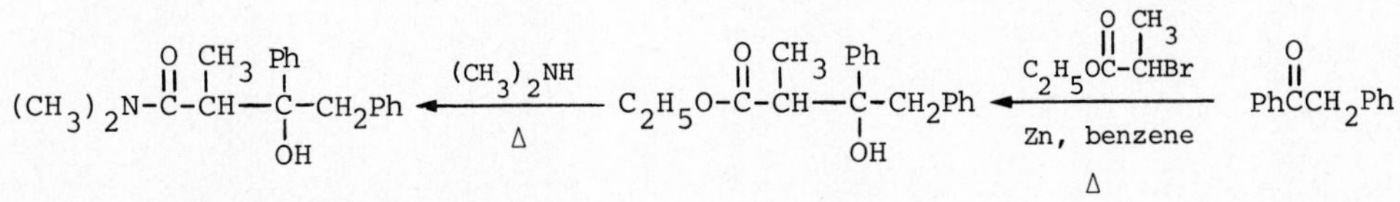

4) And so on:

$$CH_3CH_2CO_2H \xrightarrow{P,\ Br_2} \xrightarrow{C_2H_5OH} CH_3\overset{\displaystyle Br}{\overset{|}{C}}HCO_2C_2H_5$$

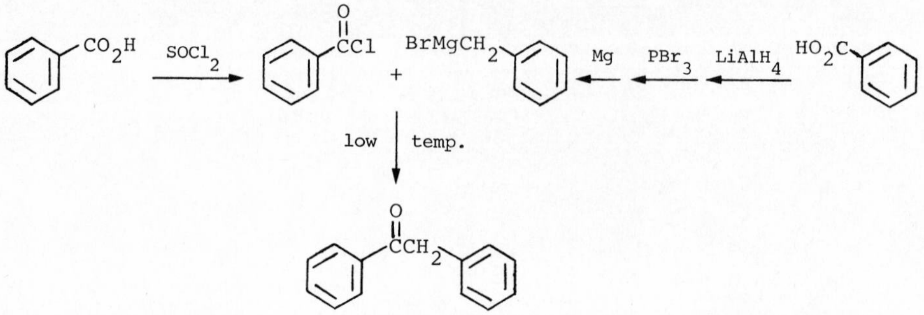

S10. $C_2H_5O\overset{O}{\overset{\|}{C}}OC_2H_5 + D_2O \longrightarrow C_2H_5OD + \left[C_2H_5O\overset{O}{\overset{\|}{C}}OD \right] \longrightarrow C_2H_5OD + CO_2$

The only other product is gaseous CO_2.

S11.
$$\begin{array}{l} CH_3(CH_2)_{12}COOCH_2 \\ \qquad\qquad\qquad\quad\ | \\ CH_3(CH_2)_{12}COOCH \\ \qquad\qquad\qquad\quad\ | \\ CH_3(CH_2)_{12}COOCH_2 \end{array}$$

20. CONJUGATION

20.A Chapter Outline, Important Terms Introduced, and Reactions Discussed

(**** *NOTE: Except for the Diels-Alder reaction, most of the reactions discussed in this chapter are not "new". What is different about them is how they proceed in conjugated systems.*)

20.1 Allylic Systems

A. Allylic Cations (resonance structures)

$$CH_3CH=CHCH_2OH$$

and

$$CH_3CHCH=CH_2$$
$$\quad\;\; |$$
$$\quad\;\; OH$$

$\xrightarrow{\;HX\;}$

$\xleftarrow{\;Ag^+,\;H_2O\;}$

$$CH_3CH=CHCH_2X$$

and

$$CH_3CHCH=CH_2$$
$$\quad\;\; |$$
$$\quad\;\; X$$

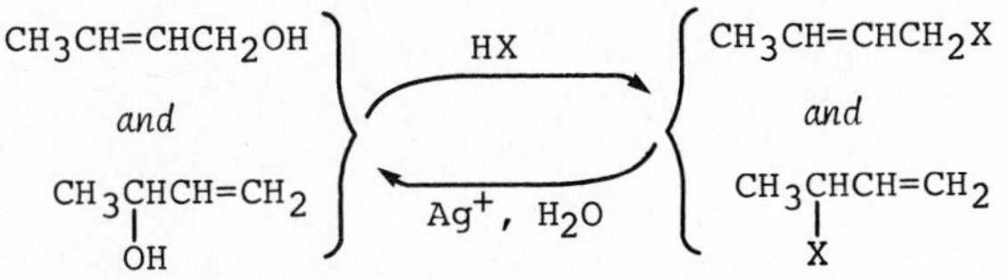

(allylic rearrangement)

B. S_N2 Reactions

with or without rearrangement

S_N2' reactions

C. Allylic Anions

"E^+" = *electrophile*

$$CH_3CH=CHCH_2Br$$

and

$$CH_3CHCH=CH_2$$
$$\quad\;\; |$$
$$\quad\;\; Br$$

$\xrightarrow{\;Mg\;}$ $\xrightarrow{\;"E^+"\;}$

$$CH_3CH=CHCH_2-E$$

and

$$CH_3CHCH=CH_2$$
$$\quad\;\; |$$
$$\quad\;\; E$$

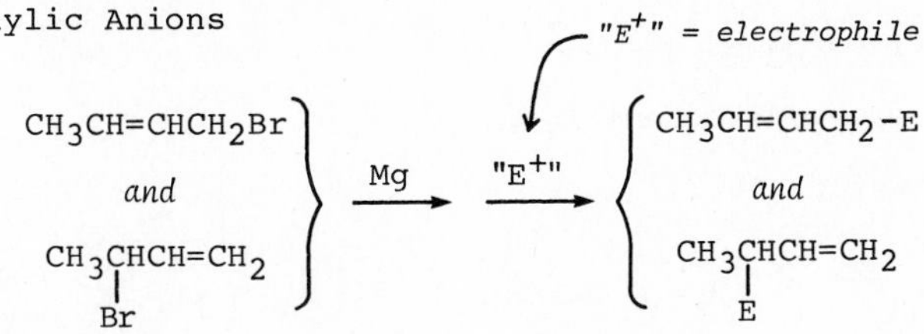

major product

dilution principle

conjugated carbons

D. Allylic Radicals

allylic bromination

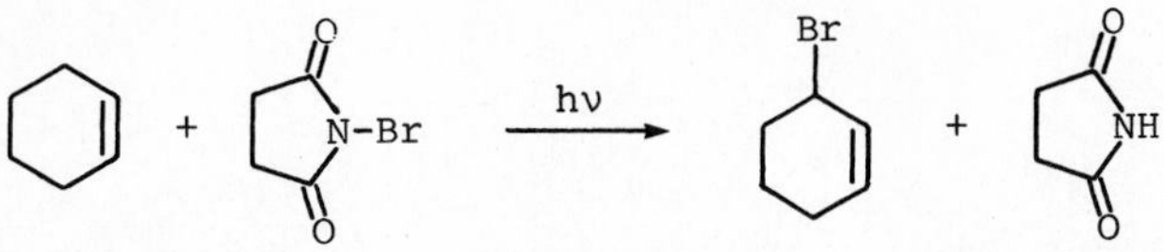

E. Molecular Orbital Description of Allylic Systems

bonding, non-bonding, and antibonding π molecular orbitals

20.2 Dienes

A. Structure and Stability

conjugated vs. unconjugated dienes

isolated double bonds

B. Addition Reactions

$$CH_2=CHCH=CH_2 + Br_2 \xrightarrow{-15°C}$$

BrCH$_2$CH=CHCH$_2$Br
46%

$$CH_2=CHCHCH_2Br$$
Br   54%

$$\xrightarrow{60°C}$$

90%

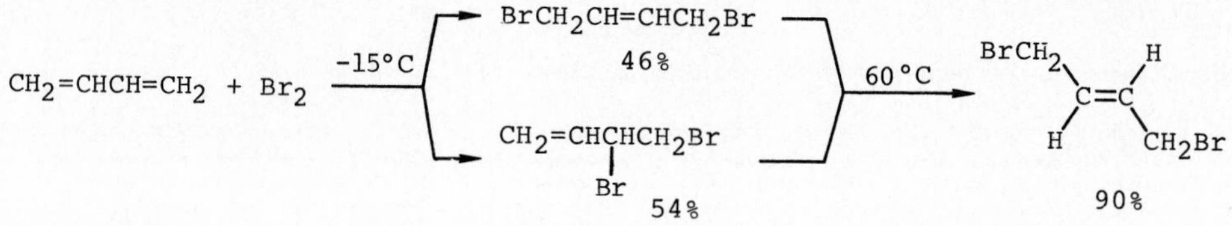

kinetic vs. thermodynamic control

C. 1,2-Dienes: Allenes
sp-hybridization of central carbon
capable of being chiral
cumulated double bonds

D. Preparation of Dienes
(dehydration, Grignard coupling, Wittig reaction, etc.)

E. Diene Polymers
elastomer
vulcanization, crosslinks

20.3 Unsaturated Carbonyl Compounds

A. Unsaturated Aldehydes and Ketones

α,β- vs. β,γ-unsaturation:

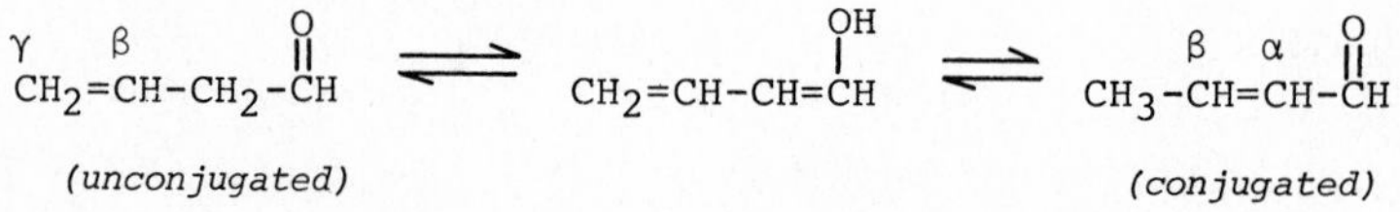

$$\underset{\text{(unconjugated)}}{\overset{\gamma\quad\beta\qquad\quad O}{CH_2=CH-CH_2-\overset{\|}{C}H}} \rightleftharpoons \overset{OH}{CH_2=CH-CH=\overset{|}{C}H} \rightleftharpoons \underset{\text{(conjugated)}}{\overset{\beta\quad\alpha\quad O}{CH_3-CH=CH-\overset{\|}{C}H}}$$ ("move into conjugation")

formation via aldol condensation:

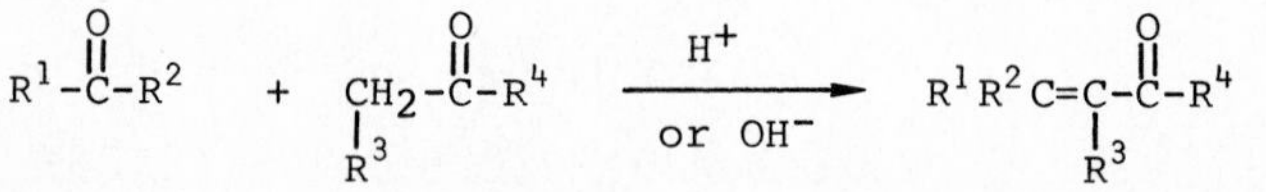

$$R^1-\overset{O}{\overset{\|}{C}}-R^2 + \underset{R^3}{CH_2-\overset{O}{\overset{\|}{C}}-R^4} \xrightarrow[\text{or OH}^-]{H^+} R^1R^2C=\underset{R^3}{\overset{O}{C-\overset{\|}{C}}-R^4}$$ (acid- or base-catalyzed)

ease of oxidation of allylic alcohols:

$$\overset{OH}{C=C-\overset{|}{C}H-} \xrightarrow{MnO_2} \overset{O}{C=C-\overset{\|}{C}-}$$

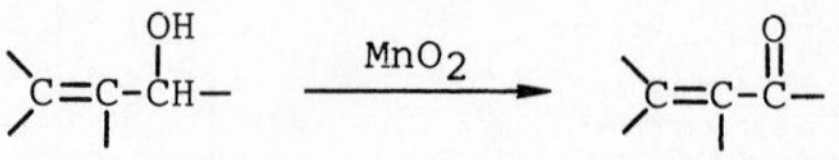

1,2-additions (normal additions) vs. 1,4-additions (conjugate additions):

$$\overset{O}{C=C-\overset{\|}{C}-} + HCN \longrightarrow NC-\overset{|}{C}-CH-\overset{O}{\overset{\|}{C}}-$$

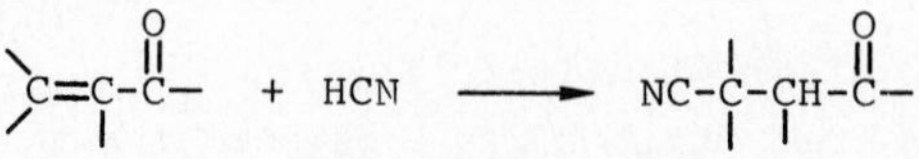

cuprates vs. organolithium (and Grignard) reagents:

$$R-\overset{|}{\underset{|}{C}}-CH-\overset{O}{\overset{\|}{C}}- \xleftarrow{H_2O} \xleftarrow{R_2CuLi} \overset{O}{C=C-\overset{\|}{C}-} \xrightarrow[\substack{\text{or} \\ \text{(RMgBr)}}]{RLi} \xrightarrow{H_2O} \overset{OH}{C=C-\overset{|}{C}-R}$$

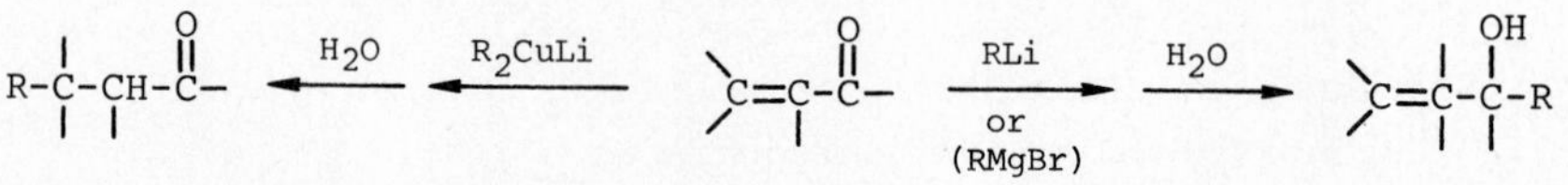

reduction methods

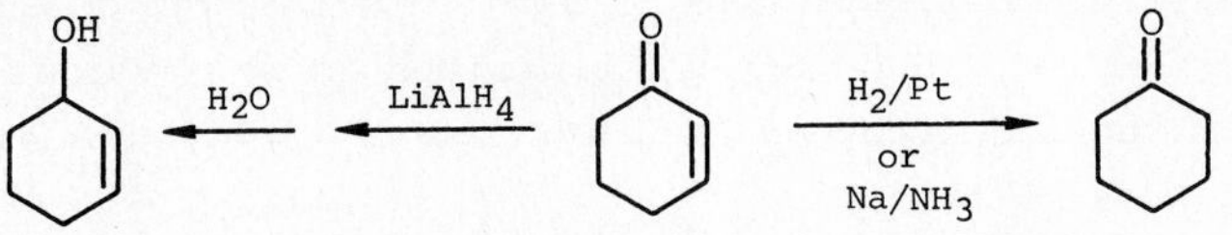

B. Unsaturated Carboxylic Acids and Derivatives

cross-conjugated

Horner-Emmons reaction:

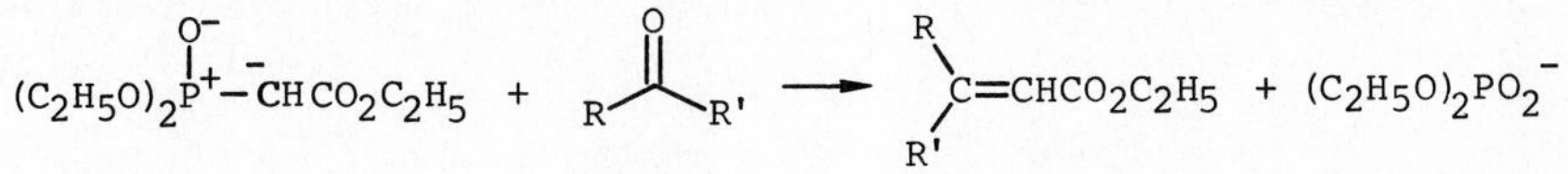

Perkin reaction:

$$RCH{=}O + (CH_3\overset{O}{\overset{\|}{C}})_2O \xrightarrow{CH_3CO_2Na} R{-}CH{=}CHCO_2H$$

C. Ketenes

$$CH_2{=}C{=}O + HNu: \longrightarrow CH_3{-}\overset{O}{\overset{\|}{C}}{-}Nu$$

20.4 Higher Conjugated Systems (trienes)

20.5 The Diels-Alder Reaction

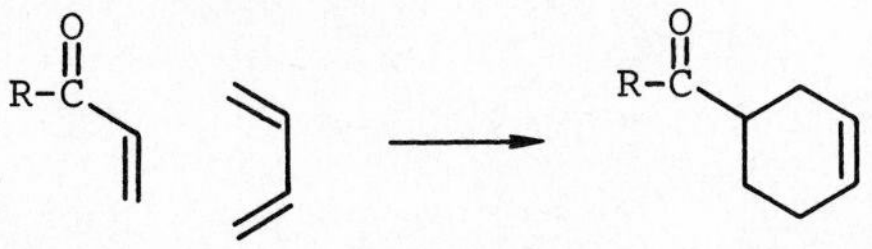

cycloaddition reaction

dienophile

head-to-head vs.
head-to-tail orientation

4+2 vs. 2+2 or 4+4 cycloadditions

exo vs. endo stereochemistry

bicyclic products available

20.B Important Concepts and Hints

Conjugation, or the Double Bond Relay

You are familiar by now with many of the reactions of carbon-carbon double bonds and of carbonyl groups. When both functional groups are present in the same molecule, the same reactions can usually be observed. However, when the p-orbitals of a double bond overlap with those of an adjacent double bond or carbonyl group, special chemical behavior is often seen. Unusual reactivity is also observed if an intermediate or transition state involves the formation of an sp^2-hybridized carbon adjacent to a double bond (allylic system). All of this comes under the heading of conjugation. One of the ways to understand conjugation intuitively is to think of it in the following way: any chemical behavior which involves an sp^2-hybridized carbon can be relayed two carbons away by an adjacent double bond. The following summary illustrates this point:

S_N1 Substitution (carbocation intermediate)

alkyl system:

$$-\overset{|}{\underset{|}{C}}-X \xrightarrow{-X^-} \left[\overset{|}{\underset{|}{\overset{+}{C}}} \right] \xrightarrow{Nu^-} -\overset{|}{\underset{|}{C}}-Nu$$

allylic system:

$$\underset{}{>}C=C-\overset{X}{\underset{|}{C}}- \xrightarrow{-X^-} \left[>C=C-\overset{+}{C}< \longleftrightarrow >\overset{+}{C}-C=C< \right] \xrightarrow{Nu^-}$$

$$>C=C-\overset{Nu}{\underset{|}{C}}- \quad (S_N1)$$
$$and/or$$
$$-\overset{Nu}{\underset{|}{C}}-C=C< \quad (S_N1')$$

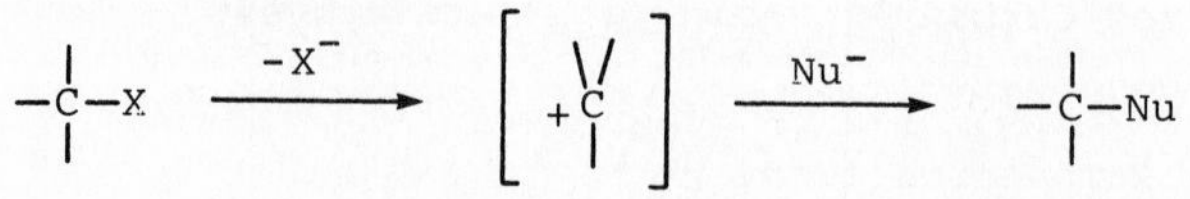

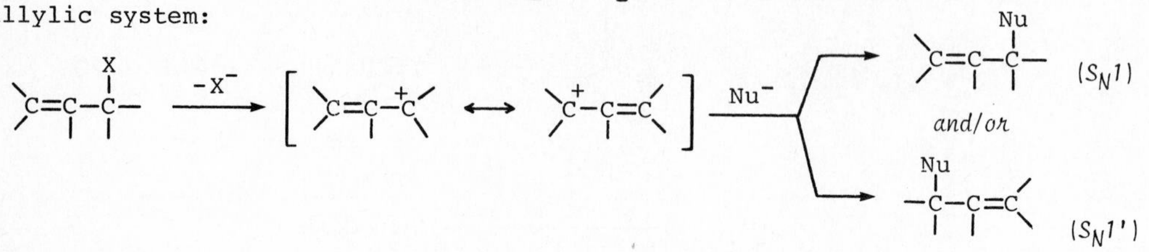

S_N2 Substitution

alkyl system:

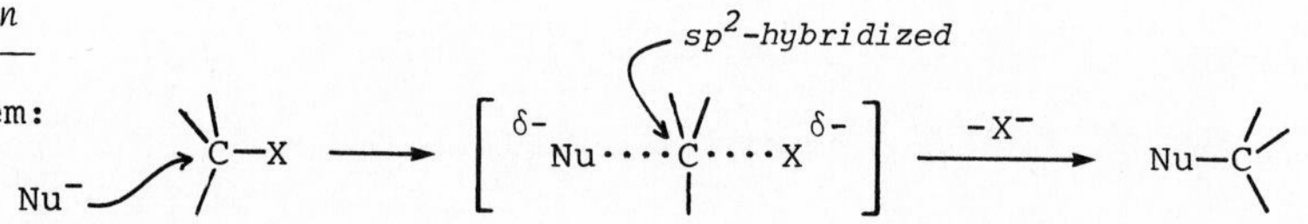

$$\underset{Nu^-}{\overset{}{\nearrow}} \overset{\backslash}{\underset{/}{C}}-X \longrightarrow \left[\delta^- Nu \cdots \overset{sp^2\text{-}hybridized}{\overset{\backslash/}{C}} \cdots X \delta^- \right] \xrightarrow{-X^-} Nu-\overset{/}{\underset{\backslash}{C}}$$

allylic system:

$$\underset{Nu^-}{} >C=C-\overset{X}{\underset{|}{C}}- \longrightarrow -\overset{|}{\underset{|}{C}}-C=C< \ + \ X^- \quad S_N2'$$

(also S_N2 without allylic rearrangement)

Grignard Reaction (carbanion intermediate)

alkyl system:

$$-\overset{|}{\underset{|}{C}}-X \xrightarrow{Mg} -\overset{|}{\underset{|}{C}}-MgX \xrightarrow{E^+} -\overset{|}{\underset{|}{C}}-E$$

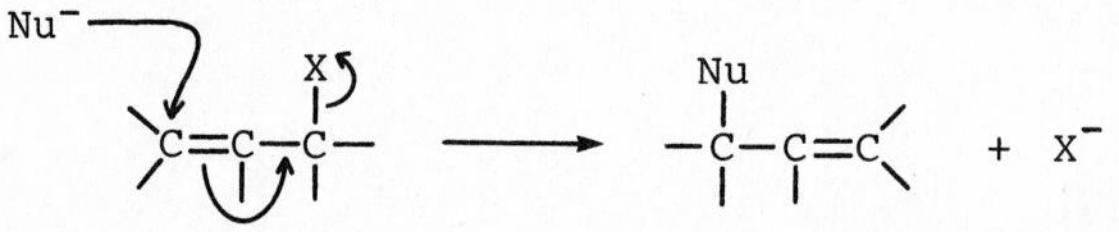

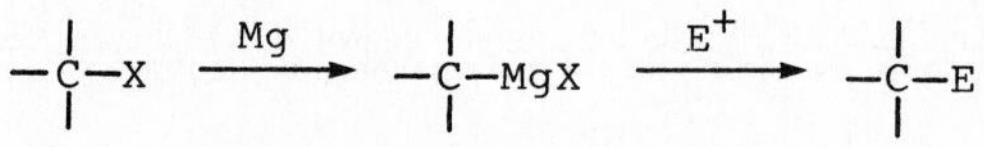

$"E^+"$ = electrophile

allylic system:

$$>C=C-\overset{X}{\underset{|}{C}}- \xrightarrow{Mg} \left[>C=C-\overset{MgX}{\underset{|}{C}}- \rightleftharpoons -\overset{MgX}{\underset{|}{C}}-C=C< \right] \xrightarrow{E^+}$$

$$>C=C-\overset{E}{\underset{|}{C}}-$$
$$and/or$$
$$-\overset{E}{\underset{|}{C}}-C=C<$$

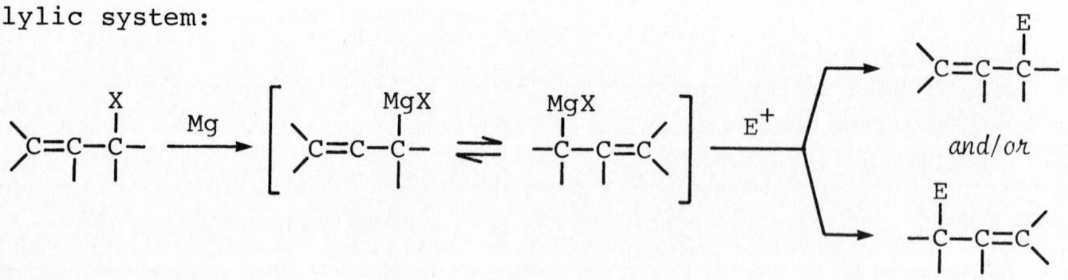

Free-Radical Halogenation (radical intermediate)

alkyl system:

$$-\overset{|}{\underset{|}{C}}-H \xrightarrow{X\cdot} -\overset{|}{\underset{|}{C}}\cdot \xrightarrow{X_2} -\overset{|}{\underset{|}{C}}-X$$

allylic system:

$$>C=C-\overset{H}{\underset{|}{C}}- \xrightarrow{X\cdot} \left[>C=C-\overset{\cdot}{\underset{|}{C}}- \longleftrightarrow -\overset{\cdot}{\underset{|}{C}}-C=C< \right] \xrightarrow{X_2}$$

$$>C=C-\overset{X}{\underset{|}{C}}-$$
$$and/or$$
$$-\overset{X}{\underset{|}{C}}-C=C<$$

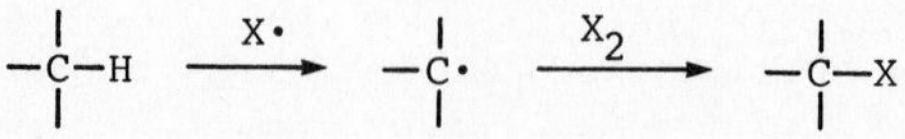

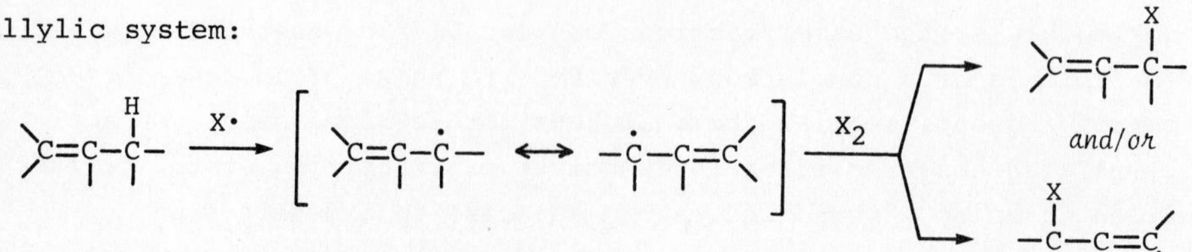

Electrophilic Addition

isolated double bond:

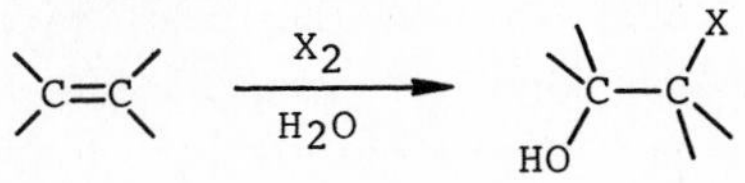

conjugated diene:

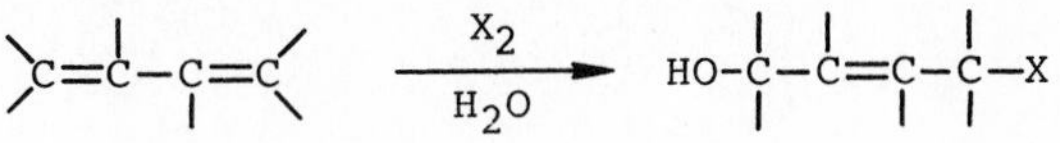

(normal addition can also occur)

Ketone Enolization

isolated carbonyl group:

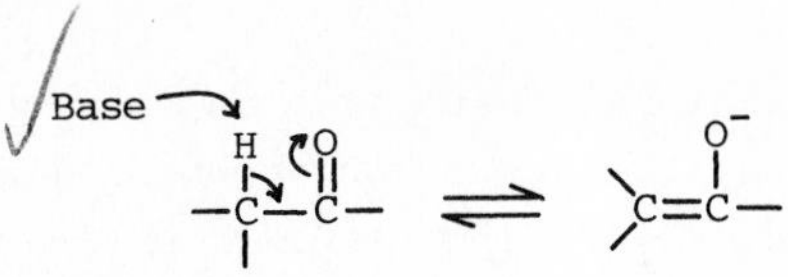

α,β-unsaturated carbonyl group:

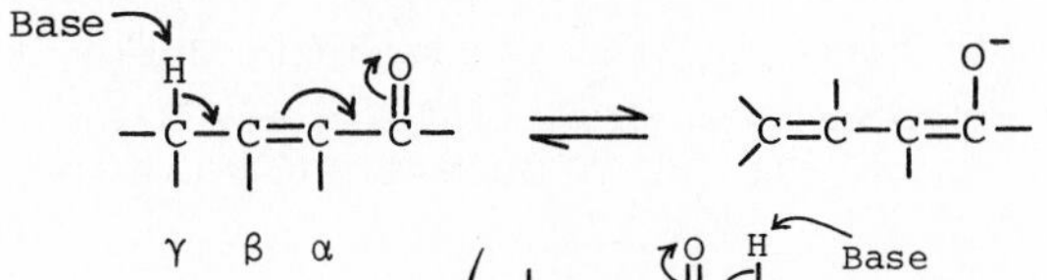

can still occur

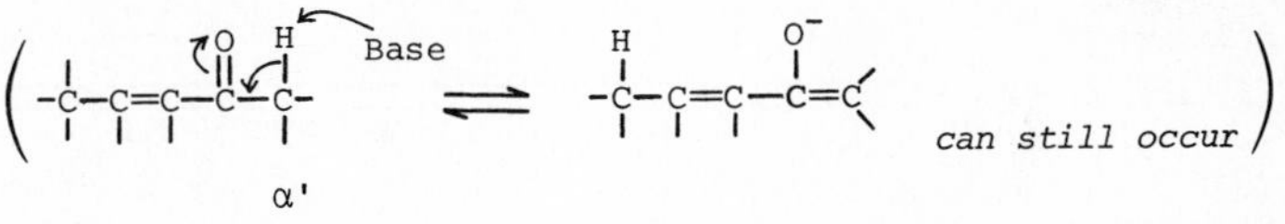

Addition to a Carbonyl

isolated carbonyl group:

(1,2-addition)

α,β-unsaturated carbonyl group:

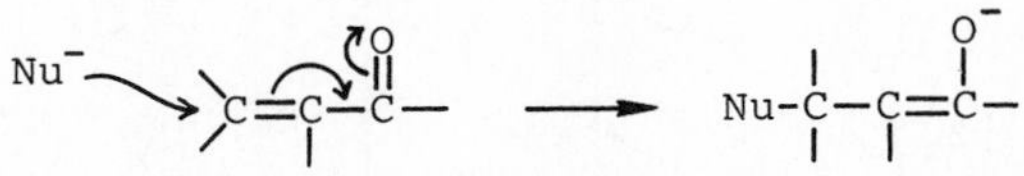

(1,4-addition; 1,2-addition can still occur)

Notice how in each case a reaction which can occur at one carbon atom can take place at the other end of a double bond which is conjugated to it. This does not mean that <u>all</u> reactions of allylic systems involve rearrangement, or that all additions to α,β-unsaturated carbonyl compounds are 1,4-; the "normal" modes of reaction are observed as well. Often by choosing specific reaction conditions or reagents, you can favor one over the other.

The Diels-Alder Reaction, or Electrons-going-around-in-a-circle

In contrast to the Diels-Alder reaction, reactions which look similar but involve four or eight electrons moving around in a circle (rather than six) occur only in exceptional circumstances:

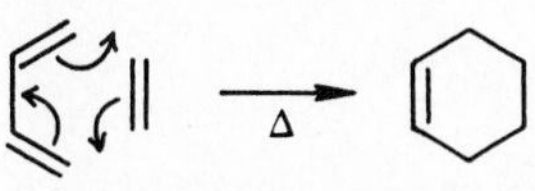

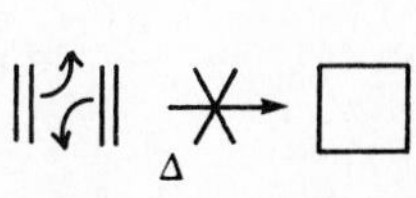

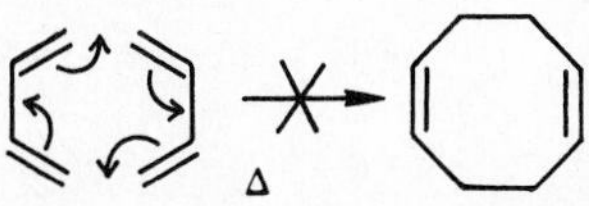

You have seen one other reaction which involves "six-electrons-in-a-cyclic-system": the pyrolysis of acetate (and xanthate) esters. The analogous four-electron transformation does not occur:

You will encounter additional cases such as this one, in which systems involving six electrons in a cyclic arrangement (in π-bonds (for example, benzene) or in transition states (the pyrolysis illustrated above is an example)) are favored relative to the analogous four- or eight-electron systems. The discovery of a unifying explanation for this phenomenon by R.B. Woodward and R. Hoffman was one of the most exciting events to occur in organic chemistry during this century. *(Section 34.2 of the text describes and outlines this discovery, which is now known as the Woodward-Hoffman rules.)*

20.C Answers to Exercises

(20.1.A)

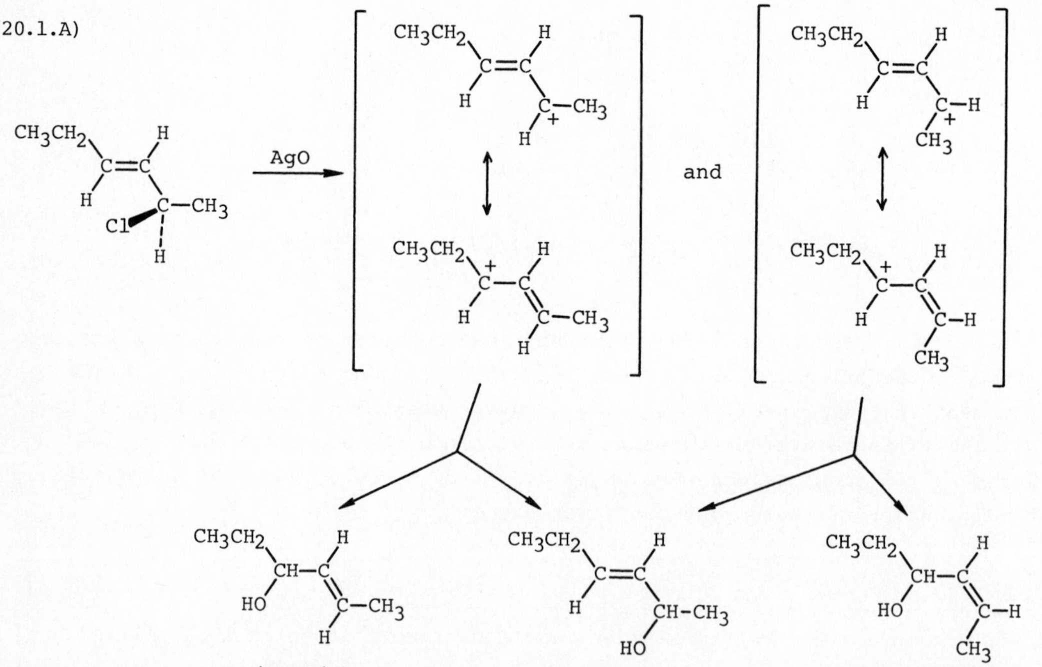

(racemic) *(racemic)* *(racemic)*

(20.1.B) $(CH_3)_3CMgBr$ + $BrCH_2CH{=}CH_2$ ⟶ $(CH_3)_3CCH_2CH{=}CH_2$

$(CH_3)_2CHCH_2MgBr$ + $BrCH_2CH{=}CH_2$ ⟶ $(CH_3)_2CHCH_2CH_2CH{=}CH_2$

(20.1.C)

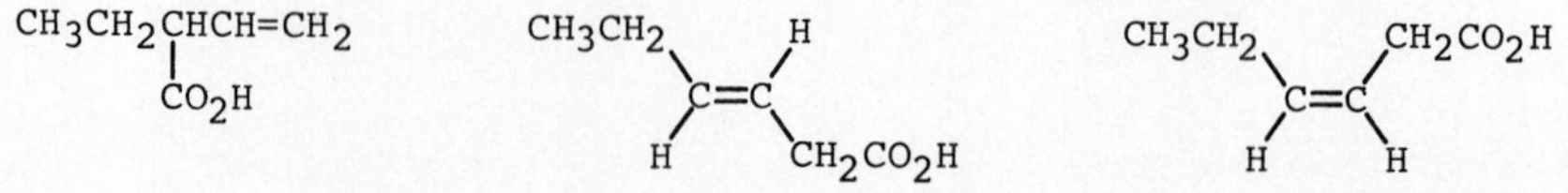

(20.1.D)

(a)

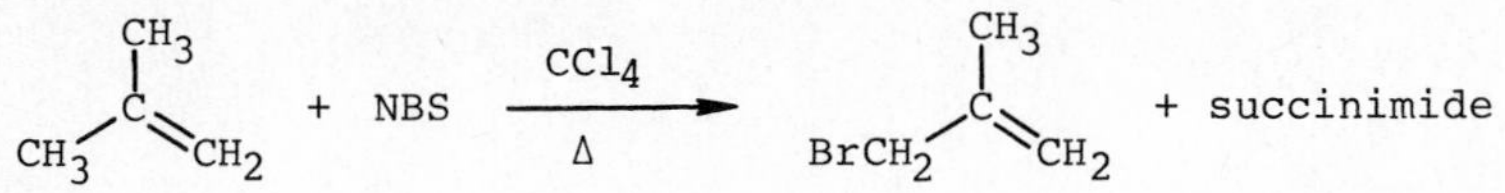

$$CH_3-C(CH_3)=CH_2 + NBS \xrightarrow[\Delta]{CCl_4} BrCH_2-C(CH_3)=CH_2 + succinimide$$

(b)

+ NBS $\xrightarrow[\Delta]{CCl_4}$ —Br + succinimide

(c)

$$CH_3CH_2CH=CHCH_3 + NBS \xrightarrow[\Delta]{CCl_4}$$

These two are the same compound

$$CH_3\underset{Br}{CH}CH=CHCH_3 \quad + \quad CH_3CH=CH\underset{Br}{CH}CH_3$$

— major product —

$$CH_3CH_2\underset{Br}{CH}CH=CH_2 \quad + \quad CH_3CH_2CH=CHCH_2Br$$

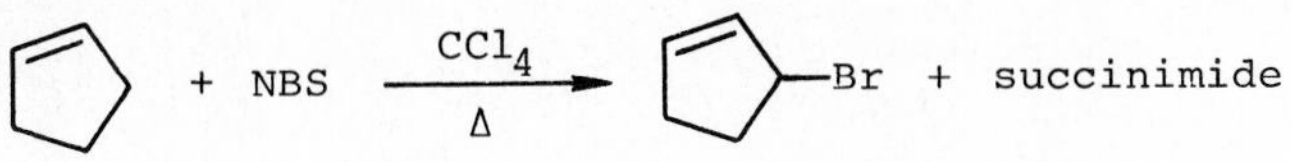

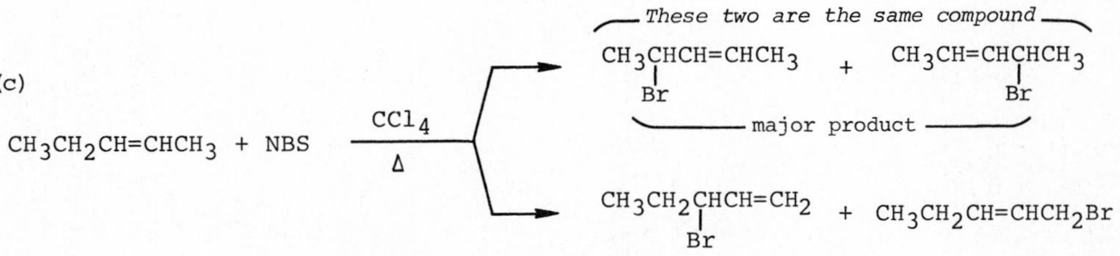

(20.1.E)

$$\left[CH_2=CH-CH=CH-\overset{-}{CH}_2 \longleftrightarrow CH_2=CH-\overset{-}{CH}-CH=CH_2 \longleftrightarrow {}^{-}CH_2-CH=CH-CH=CH_2 \right]$$

HIGHEST ENERGY
(least stable)

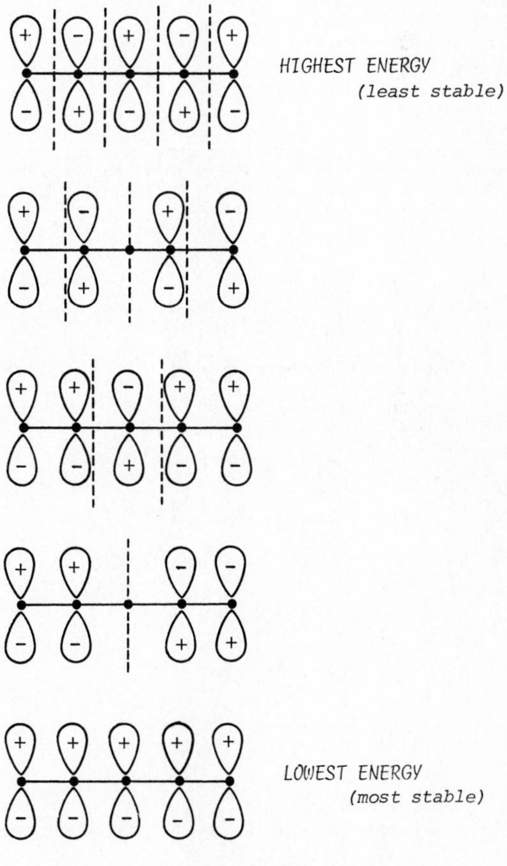

There are six π electrons
in the pentadienyl anion,
and they fill the lowest
three molecular orbitals.

LOWEST ENERGY
(most stable)

(20.2.B)

$$BrCH_2\underset{OH}{CH}CH=CH_2 \xrightarrow{NaOH} \underset{O}{CH_2-CH}CH=CH_2 \quad ;$$

$$BrCH_2\overset{}{C}=\overset{}{C}CH_2OH \text{ (H, H)} \xrightarrow{NaOH} \text{ }$$

$$BrCH_2\overset{H}{\underset{H}{C}}=\overset{H}{\underset{CH_2OH}{C}} \xrightarrow{NaOH} HOCH_2\overset{H}{\underset{H}{C}}=\overset{H}{\underset{CH_2OH}{C}}$$

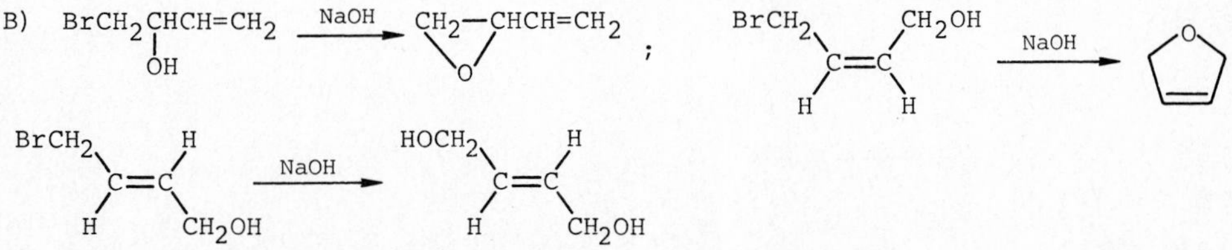

(20.3.A)   (a)

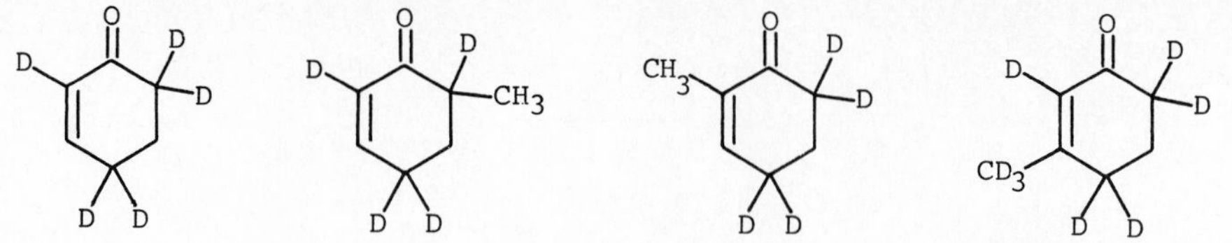

(b)

(2) $CH_3CH_2CH_2CH_2-\overset{\overset{\displaystyle CH_3}{|}}{\underset{\underset{\displaystyle CH_3}{|}}{C}}-CH_2\overset{\overset{\displaystyle O}{||}}{C}CH_3$

(1)

$(CH_3)_2C=CH-\overset{\overset{\displaystyle OH}{|}}{\underset{\underset{\displaystyle CH_3}{|}}{C}}-CH_2CH_2CH_2CH_3$

$\xleftarrow{H_2O}\ \underset{\underline{n}-BuLi}{\longleftarrow}$

$\underset{CuBr}{\underline{n}-C_4H_9MgBr}$

$(CH_3)_2C=CHCCH_3$

$\overset{H_2/Pd}{\longrightarrow}$ $(CH_3)_2CHCH_2\overset{\overset{\displaystyle O}{||}}{C}CH_3$ (3)

$\overset{Li/NH_3}{\longrightarrow}\ \overset{H_2O}{\longrightarrow}$

$(CH_3)_2CHCH_2\overset{\overset{\displaystyle O}{||}}{C}CH_3$

(4)

(6)

$(CH_3)_2-\overset{\overset{\displaystyle Br}{|}}{\underset{\underset{\displaystyle Br}{|}}{C}}-\overset{\overset{\displaystyle O}{||}}{CHCCH_3}$

$\underset{Br_2,\ CCl_4}{\longleftarrow}$

$\overset{HCN,}{\underset{Et_3Al}{\longrightarrow}}$

$(CH_3)_2\overset{}{\underset{\underset{\displaystyle CN}{|}}{C}}-CH_2\overset{\overset{\displaystyle O}{||}}{C}CH_3$ (5)

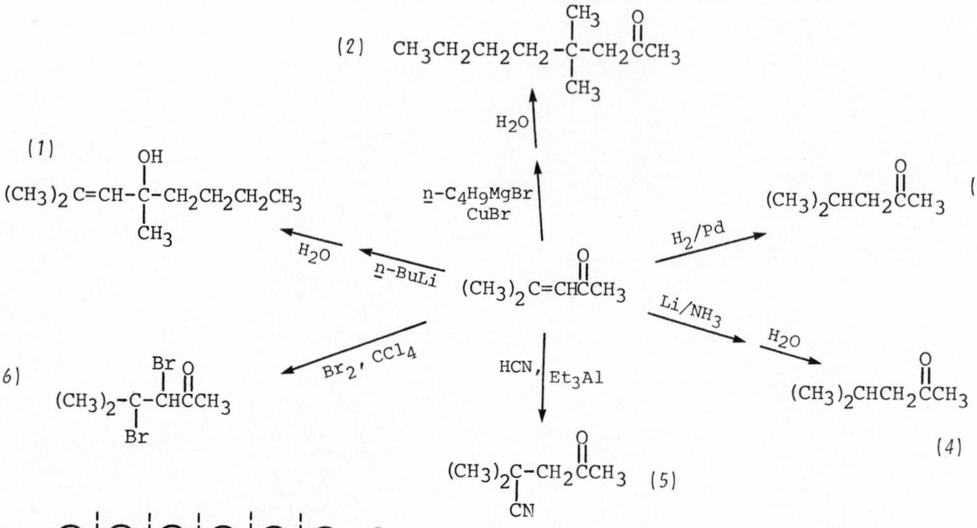

(20.4)

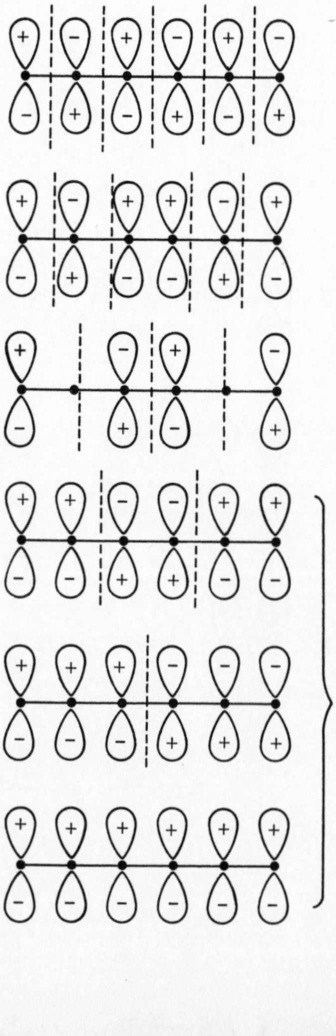

From occupied molecular orbitals:

$CH_2-CH-CH-CH-CH-CH_2$

three π bonds

two π bonds, one antibond

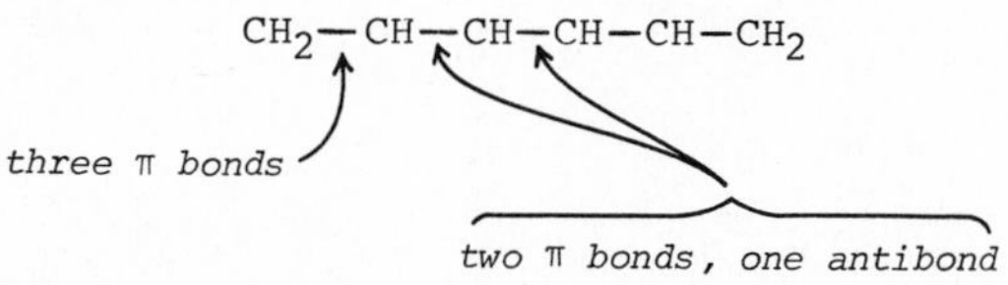

} occupied with six π electrons

$$Br^+ \curvearrowleft CH_2=CH-CH=CH-CH=CH_2 \longrightarrow BrCH_2-\overset{\delta+}{CH}\text{===}CH\text{===}\overset{\delta+}{CH}\text{===}CH\text{===}\overset{\delta+}{CH}\text{===}CH_2 \longrightarrow$$

$$BrCH_2CH=CHCH=CHCH_2Br \quad + \quad \underset{\underset{Br}{|}}{BrCH_2CH=CHCHCH=CH_2} \quad + \quad \underset{\underset{Br}{|}}{BrCH_2CHCH=CHCH=CH_2}$$

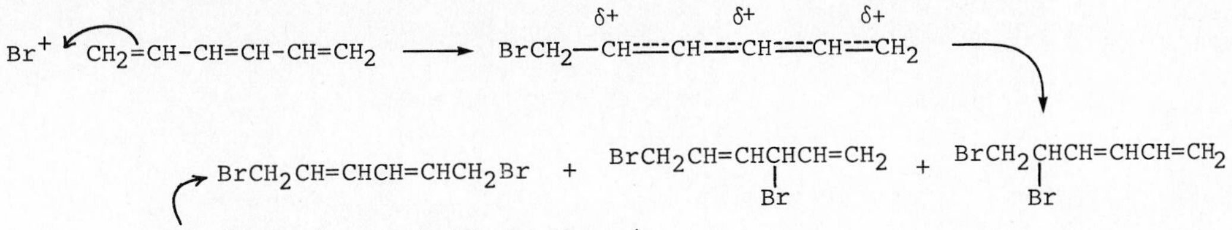

This isomer is expected to be the major
product at equilibrium, because it is conjugated and has the more highly
substituted double bonds.

(20.5)

$$CH_2=CHO_2CCH_3 \longrightarrow$$

endo-5-acetoxybicyclo[2.2.1]hept-2-ene

$$CH_2=CHCO_2H \longrightarrow$$

endo-bicyclo[2.2.1]hept-5-ene-2-
carboxylic acid

$$CH_3O_2CC\equiv CCO_2CH_3 \longrightarrow$$

dimethyl bicyclo[2.2.1]hepta-2,5-
diene-2,3-dicarboxylate

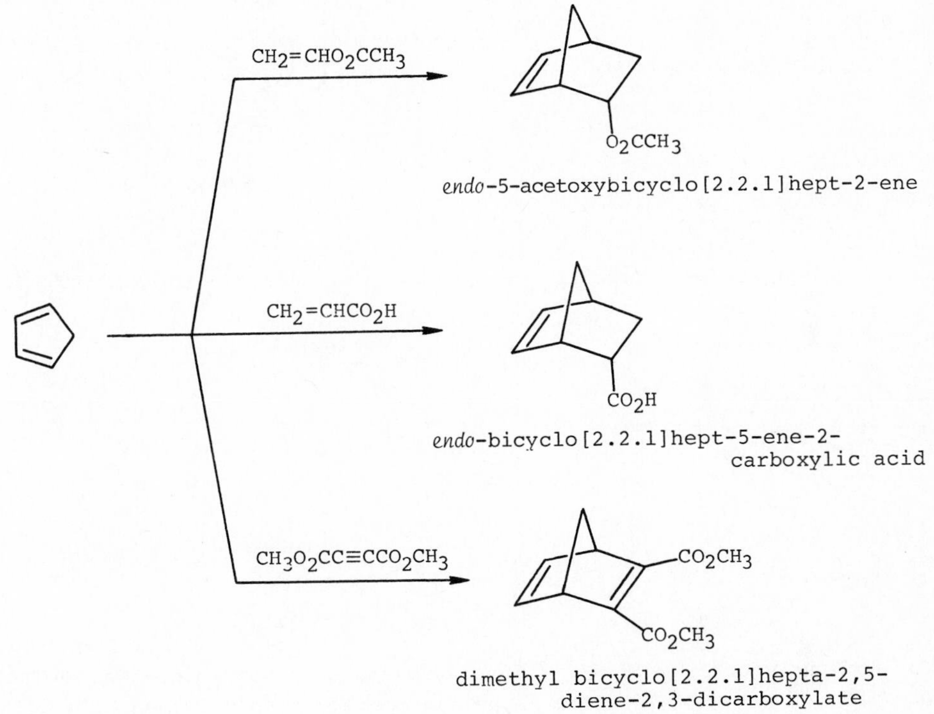

20.D Answers and Explanations for Problems

1. (a)

   (b)

   (c)

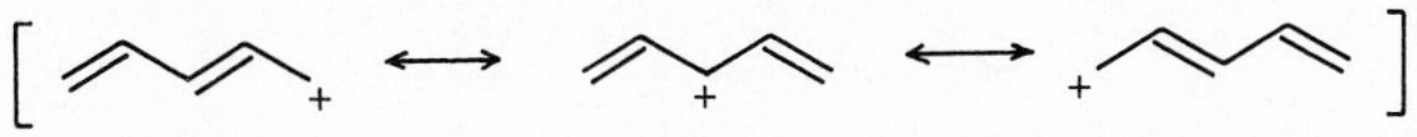

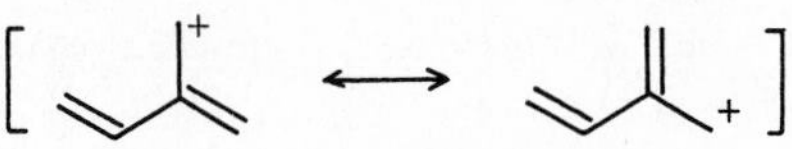

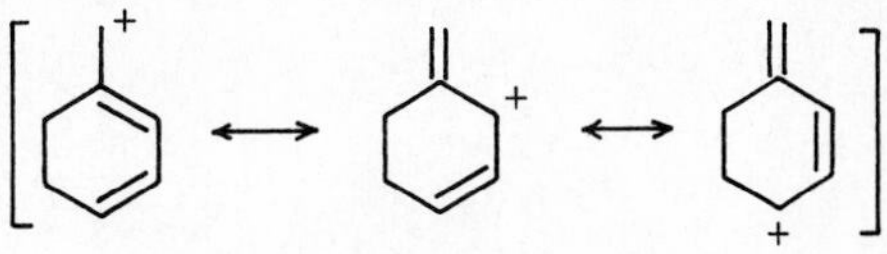

2. (a) $$\left[^+CH_2-CH=CH-CH_3 \longleftrightarrow CH_2=CH-\overset{+}{C}H-CH_3 \right]$$

      more important; secondary carbocation

   (b) $$\left[CF_3\overset{-}{C}H-CH=CH-CH_3 \longleftrightarrow CF_3CH=CH-\overset{-}{C}HCH_3 \right]$$

      more important; inductive effect of CF_3

(c) $\left[CF_3\overset{+}{C}H-CH=CHCH_3 \longleftrightarrow CF_3CH=CH-\overset{+}{C}HCH_3 \right]$

(CF_3 destabilizes cation) — more important (second structure)

(d) $\left[CH_2=CH-O^- \longleftrightarrow {}^-CH_2-CH=O \right]$

more important (first structure); negative charge on electronegative oxygen

(e) $\left[CH_3-\overset{O}{\overset{\|}{C}}-NH_2 \longleftrightarrow CH_3-\overset{O^-}{\overset{|}{C}}=NH_2^+ \right]$

more important (first structure); no charge separation

(f) $\left[CH_2=CH-OCH_3 \longleftrightarrow {}^-CH_2-CH=\overset{+}{O}CH_3 \right]$

more important (first structure); no charge separation

(g) 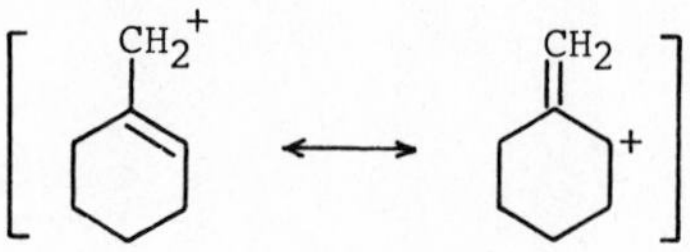

more important; secondary carbocation

3. (a) $Ba^{14}CO_3 + H_2SO_4 \longrightarrow BaSO_4 + H_2O + {}^{14}CO_2$

$CH_2=CHBr + Mg \xrightarrow{\text{ether}} CH_2=CHMgBr \xrightarrow{^{14}CO_2} \xrightarrow{H_3O^+} CH_2=CH^{14}CO_2H$

$\xrightarrow{LiAlH_4} \xrightarrow{H_2O} CH_2=CH-{}^{14}CH_2OH$

~~~~~~~~~~~~~~~~~~~~~~~~

$CH_2=CH-{}^{14}CH_2OH \xrightarrow{SOCl_2} CH_2=CH-{}^{14}CH_2Cl$ *or* ${}^{14}CH_2=CH-CH_2Cl$

1. Ozone; 2. Zn, HOAc

$CH_2=O + O=CHCH_2Cl$

Collect and count the $CH_2=O$. If rearrangement is complete, the $CH_2=O$ will have 100% of the ^{14}C. If there is no rearrangement, it will have 0% of the ^{14}C.

(b) $CH_2=CHCO_2CH_3 + LiAlD_4 \longrightarrow \xrightarrow{H_2O} CH_2=CH-CD_2OH$

Nmr will show only the vinyl H's and the OH. Recall that δ (OH) is concentration-dependent.

$CH_2=CHCD_2OH \xrightarrow{SOCl_2} CH_2=CHCD_2Cl + CD_2=CHCH_2Cl$

not rearranged — *rearranged*

The nmr spectrum will show a mixture of two chlorides. One component shows only vinyl H's as a complex multiplet, while the other shows a vinyl H (triplet and $-CH_2Cl$ (doublet) with an area ratio of 1:2. The $-CH_2Cl$ group will appear at about δ=4 (3 for $-CH_2Cl$ plus 1 for allylic)

4. (a) $CH_3CH_2MgBr + Br-CH_2CH=CH_2 \longrightarrow$ 

(b) 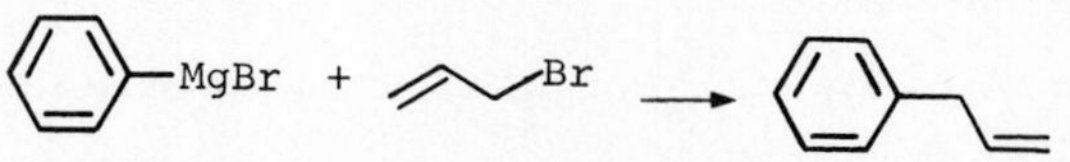

(c) 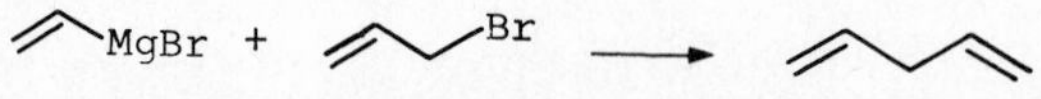
~~~~~~~~~~~~~~~~~~~~~~~~

(d) $(CH_3)_3CMgBr$ + ⟶ $(CH_3)_3CCH_2CH=CH_2$

5. (a) $CH_3CH=CHCH_2OH \xrightarrow{MnO_2} CH_3CH=CHCHO$

(b) $CH_3CH_2CH_2CH_2OH \xrightarrow{MnO_2}$ no reaction

(c) $HOCH_2CH_2CH=CHCH_2OH \xrightarrow{MnO_2} HOCH_2CH_2CH=CHCHO$

(d)

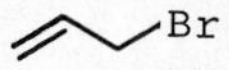

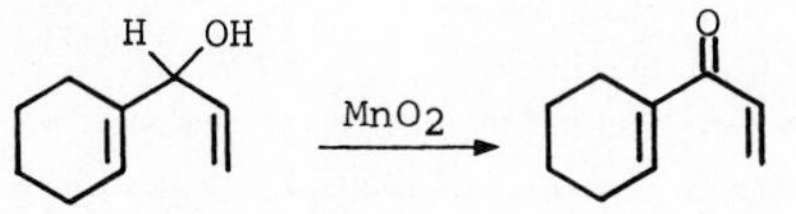

(e)

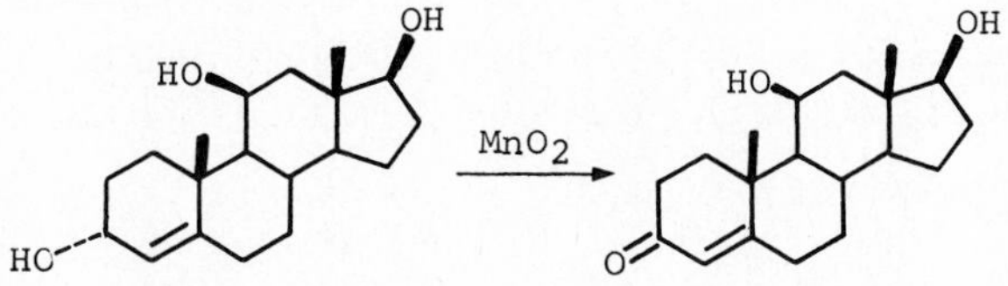

(f) $CH_3C{\equiv}C\overset{OH}{\overset{|}{C}}HCH_3 \xrightarrow{MnO_2} CH_3C{\equiv}C\overset{O}{\overset{\|}{C}}CH_3$

6.

δ 5.6 triplet (1H) (vinyl H)

δ 0.6 doublet (2H) (upfield because of negative charge in carbanion)

The two peaks of the doublet at δ 1.60 are due to these two methyls, which are not equivalent. One is *cis* to CH_2MgBr and one is *trans* to it.

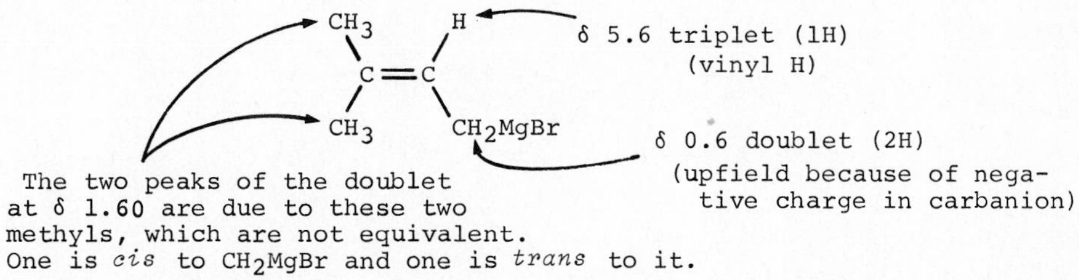

At room temperature, the two methyls become equivalent on the nmr time scale by the following mechanism.

present only in small amount

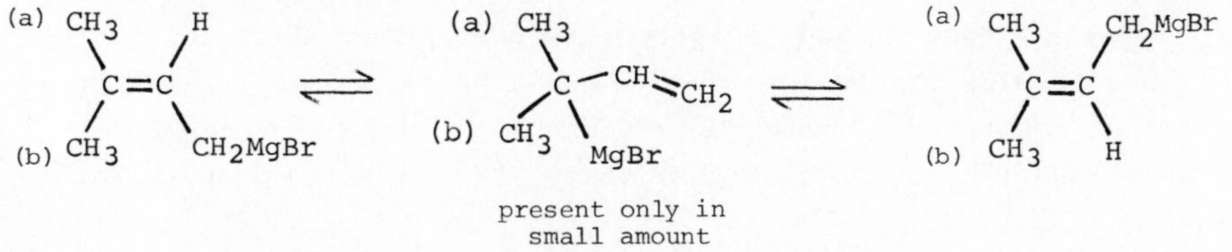

The equilibria are rapid at room temperature.

7. Abstraction of the allylic H gives the allyl free radical.

$$C_5H_{11}CH_2CH=CH_2 + Br^{\cdot} \longrightarrow C_5H_{11}\dot{C}HCH=CH_2$$

There are two isomeric allylic radicals:

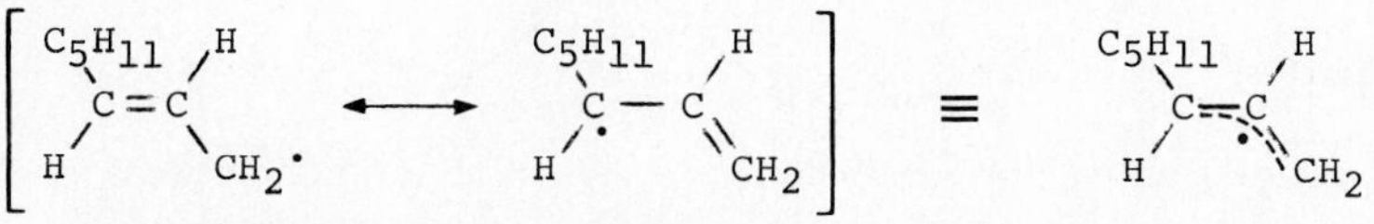

allylic radical A, *transoid*

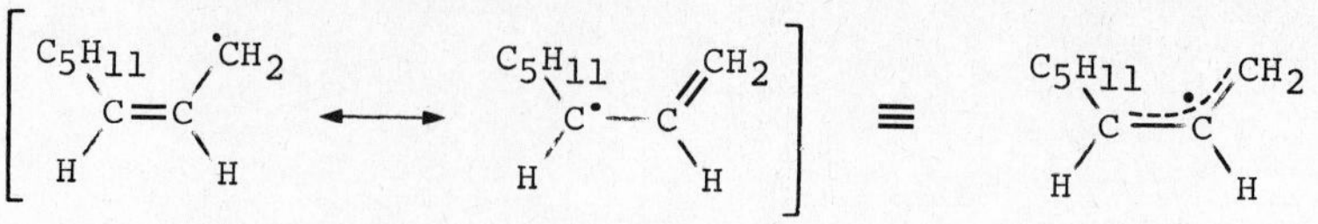

allylic radical B, *cisoid*

If either A or B reacts with Br_2 at C-3, the same product is produced:

$$\text{A or B} + Br_2 \longrightarrow C_5H_{11}\overset{\text{Br}}{\overset{|}{C}}HCH{=}CH_2 + Br\cdot$$

However, if reaction with Br_2 occurs at C-1, then the *transoid* allylic radical A gives *trans*-1-bromo-2-octene whereas the *cisoid* allylic radical B gives *cis*-1-bromo-2-octene. The barrier to interconversion of these two isomeric radicals is about 10 kcal $mole^{-1}$, much higher than the activation energy for reaction of either with Br_2 (see Chap. 6).

8. (a) $(CH_3)_3COH + OH^- \rightleftharpoons (CH_3)_3CO^- + H_2O$

$$(CH_3)_3CO^- + Cl{-}Cl \rightleftharpoons (CH_3)_3COCl + Cl^-$$

(b)

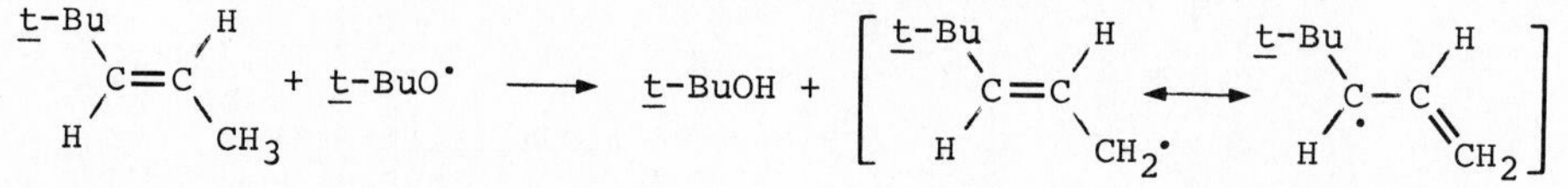

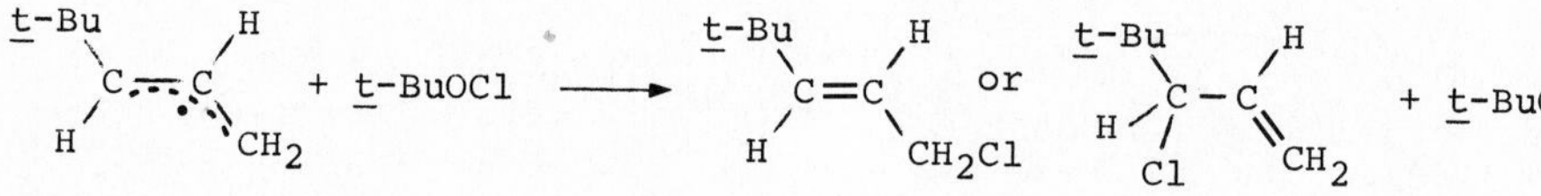

(c) This experiment shows that the *transoid* $\rightleftharpoons$ *cisoid* equilibration at -78° must be slower than reaction of the radical with t-BuOCl.

(d) The first experiment shows either that the two types of allylic radical (see problem #7) do not significantly interconvert at -78°, or that the *transoid* one is more stable than the *cisoid*. The second experiment establishes that the rates of interconversion of the radicals are slow.

(e) The resonance structures show that there is "double-bond character" between C-2 and C-3:

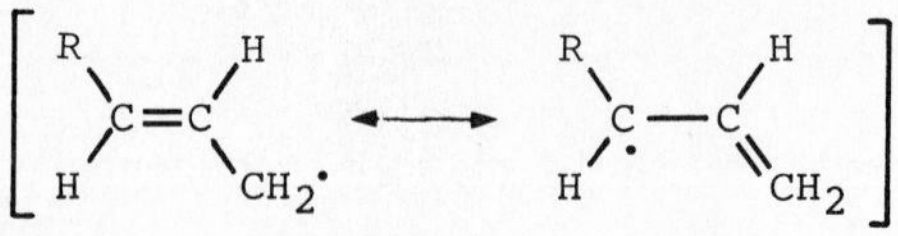

9.  $\xrightarrow{H^+}$ $\left[CH_2{=}CH\overset{+}{C}HCH_3 \longleftrightarrow {}^{+}CH_2CH{=}CHCH_3\right]$

Cl^- (reaction at C-3) 78% | Cl^- (reaction at C-1) 22%

Kinetic control

reaction occurs faster at site of greater (+) charge (secondary)

$CH_2{=}CH\overset{\text{Cl}}{\overset{|}{C}}HCH_3$ — *less stable* (monosubstituted double bond)

$ClCH_2CH{=}CHCH_3$ — *more stable* (disubstituted double bond)

If equilibration is allowed to occur, the *thermodynamic mixture* is produced. Equilibration occurs by ionization <u>back</u> to the carbocation.

10. This is a radical chain <u>addition</u> occurring <u>via</u> the allylic radical.

$$CCl_4 + RO\cdot \longrightarrow ROCl + CCl_3\cdot \quad \textit{(Initiation)}$$

$$\cdot CCl_3 + CH_2=CHCH=CH_2 \longrightarrow \left[\begin{array}{c} CH_2=CH\overset{\cdot}{C}HCH_2CCl_3 \\ \updownarrow \\ \cdot CH_2CH=CHCH_2CCl_3 \end{array} \right] \left.\begin{array}{c} \\ \\ \\ \end{array}\right\} \textit{Propagation}$$

$$CH_2\overset{\cdot}{=}\overset{\cdot}{CH}\overset{\cdot}{=}CHCH_2CCl_3 + CCl_4 \longrightarrow ClCH_2CH=CHCH_2CCl_3 + CCl_3\cdot \left.\right)$$

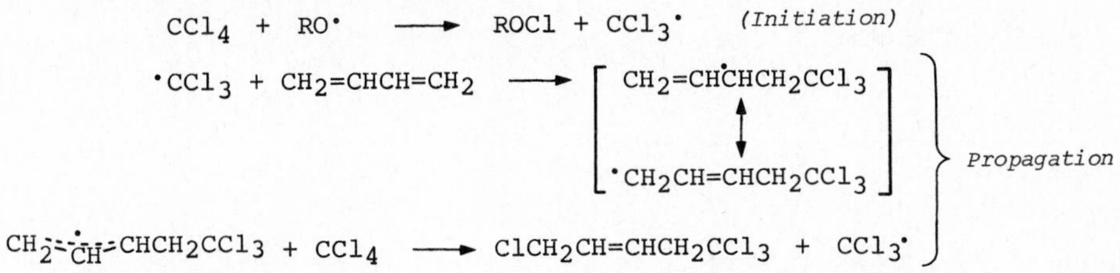

Reaction occurs mainly at the terminal (less hindered) position to afford the more stable product. The yield is low, since telomerization is a competing side reaction.

11.
$$\begin{array}{cc} CH_3 & CH_3 \\ | & | \\ CH_2=C-C=CH_2 \end{array} + Cl_2 \xrightarrow{CCl_4} \begin{array}{c} ClCH_2 \quad CH_3 \\ \diagdown \quad \diagup \\ C=C \\ \diagup \quad \diagdown \\ CH_3 \quad CH_2Cl \\ 45\% \end{array} + A + B \\ \quad\quad 54\% \quad 1\%$$

$$A = \begin{array}{c} CH_2 \\ \| \\ ClCH_2\overset{}{C}-C=CH_2 \\ | \\ CH_3 \end{array}$$

nmr: δ 4.20 (2H)

δ 1.90(3H)

The rest are vinyl H's.

mass spec: 116 + 118 molecular ions with ^{35}Cl or ^{37}Cl.

Fragment at $^m/e$ 81 due to:

$$\begin{array}{c} CH_2 \\ \| \\ ^+CH_2\overset{}{C}-C=CH_2 + Cl\cdot \\ | \\ CH_3 \end{array}$$

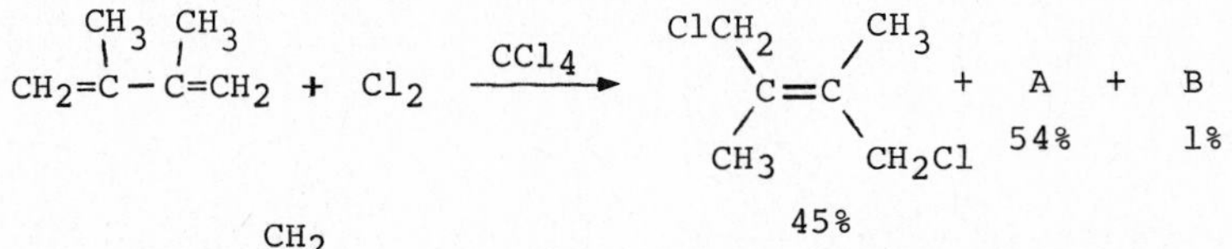

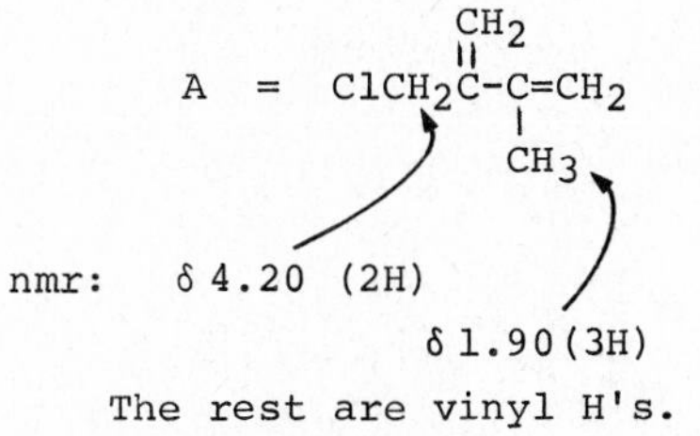

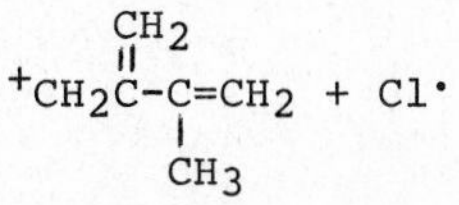

Facile fragmentation to an allylic cation and chlorine atom

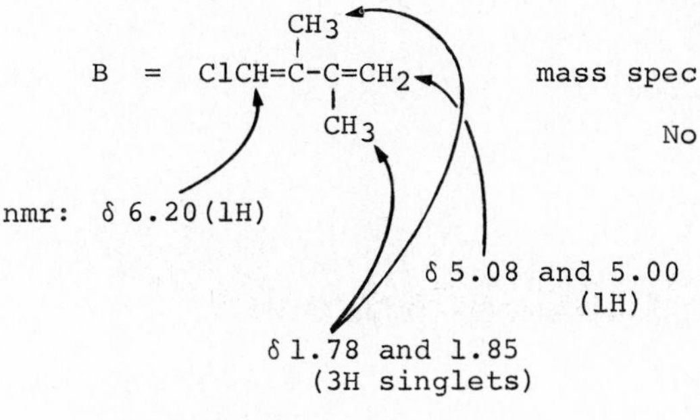

$$B = \begin{array}{c} CH_3 \\ | \\ ClCH=C-C=CH_2 \\ | \\ CH_3 \end{array}$$

mass spec: 116 + 118 molecular ion doublet.

No really good way to fragment.

nmr: δ 6.20(1H)

δ 5.08 and 5.00 (1H)

δ 1.78 and 1.85 (3H singlets)

<u>Mechanism</u> (A + B):

$$\begin{array}{c} CH_3 \\ | \\ CH_2=C-C=CH_2 \\ | \\ CH_3 \end{array} + Cl_2 \longrightarrow \begin{array}{c} CH_3 \\ | \\ ClCH_2-C-C=CH_2 \\ + | \\ CH_3 \end{array}$$

$-H^+$ ↙   ↘ $-H^+$

$$\begin{array}{c} CH_2 \\ \| \\ ClCH_2\overset{}{C}-C=CH_2 \\ | \\ CH_3 \end{array} \qquad\qquad \begin{array}{c} CH_3 \\ | \\ ClCH=C-C=CH_2 \\ | \\ CH_3 \end{array}$$

A   B

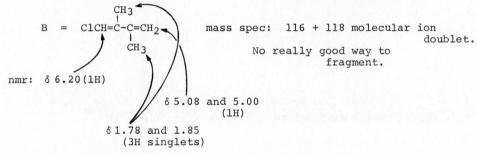

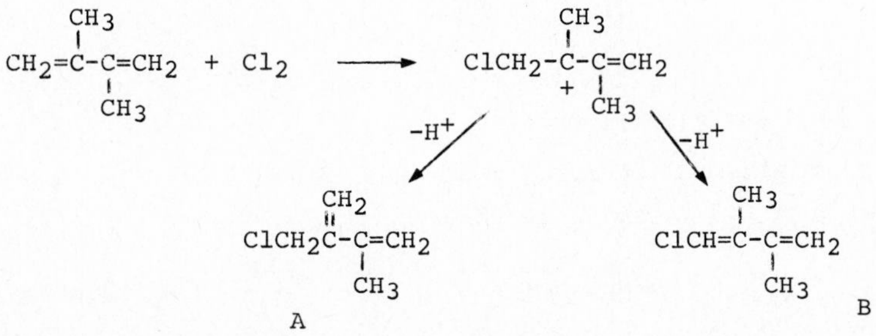

12. 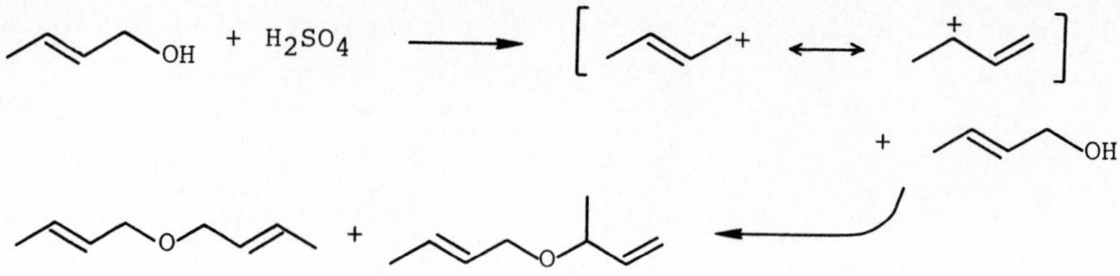

Of course the allylic cation can also react with water to give the isomeric alcohol:

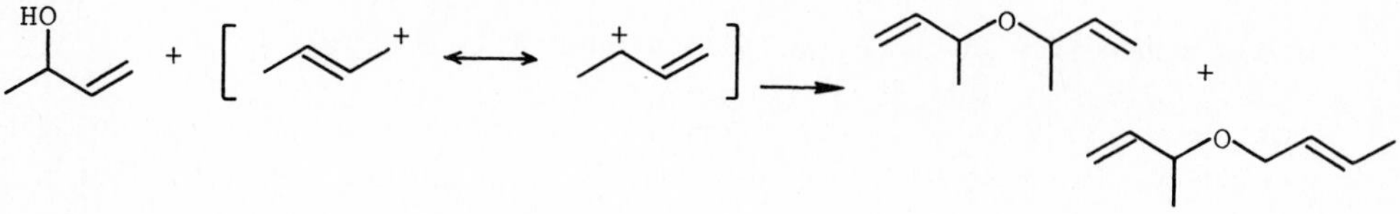

13. 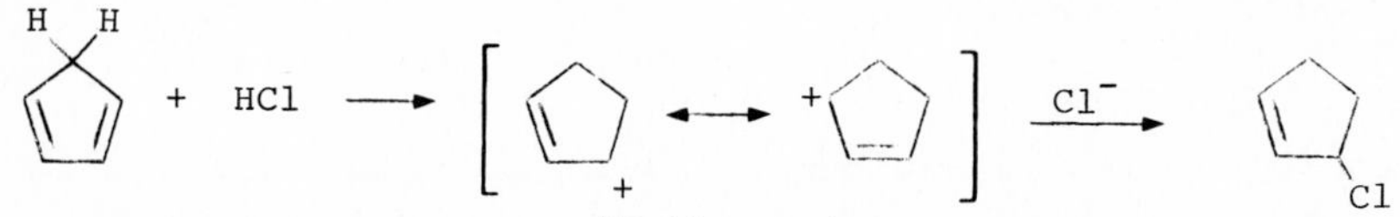

allylic cation

The 4-chloro isomer can only occur by way of the normal secondary carbocation, which is not as stable as the allylic cation which gives the 3-chloro isomer.

i.e. 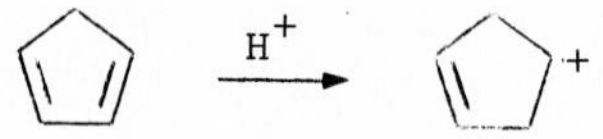 higher energy; not formed

14. (a)

$$CH_3CH_2CH_2C{\equiv}CH \rightleftharpoons CH_3CH_2C{\equiv}CCH_3 \rightleftharpoons CH_3CH_2CH{=}C{=}CH_2 \quad (175^\circ\ C;\ 448^\circ\ K)$$

1.3% A   95.2% B   3.5% C

Take B, the most stable, as the point of reference. The equilibrium constant for the B $\rightleftharpoons$ A equilibrium is:

$$K = \frac{[A]}{[B]} = \frac{1.3}{95.2} = 0.0136$$

$$\Delta G^\circ = -2.3\ RT \log K = -2.3(1.987)(448)(-1.865)\ \text{cal mole}^{-1}$$

$$\Delta G^\circ = +3.82\ \text{kcal mole}^{-1}$$

Similarly, the B $\rightleftharpoons$ C constant is:

$$K = \frac{[C]}{[B]} = \frac{3.5}{95.2} = 0.0367$$

$$\Delta G^\circ = -2.3(1.987)(448)(-1.434)\ \text{cal mole}^{-1}$$

$$\Delta G^\circ = +2.94\ \text{kcal mole}^{-1}$$

Note that allene C is more stable than the 1-alkyne and less stable than the internal alkyne.

Mechanism:

$$RCH_2C\equiv CH + OH^- \rightleftharpoons \left[R-\overset{..}{\underset{-}{C}}H-C\equiv CH \longleftrightarrow R-CH=C=\overset{..}{\underset{-}{C}}H \right]$$

$$\left[R-CH{=\!=\!=}C{\equiv\!\equiv}CH\right]^- + H_2O \rightleftharpoons R-CH=C=CH_2 + OH^-$$

$$R-CH=C=CH_2 + OH^- \rightleftharpoons \left[R-\overset{..}{\underset{-}{C}}=C=CH_2 \longleftrightarrow R-C\equiv C-\overset{..}{\underset{-}{C}}H_2 \right]$$

$$\left[R-C{\equiv\!\equiv}C{=\!=\!=}CH_2\right]^- + H_2O \rightleftharpoons R-C\equiv CCH_3 + OH^-$$

(b) In the case of $Na^+NH_2^-$, the amide ion is so basic that it converts
     the terminal alkyne completely to the carbanion, thus shifting the
     equilibrium quantitatively in this direction.

$$R-CH_2C\equiv CH + NH_2^- \rightleftharpoons RCH_2C\equiv C^- + NH_3 \qquad K \approx 10^{10}$$
$$\text{p}K_a\ 25 \qquad\qquad\qquad\qquad \text{p}K_a\ 35$$

15. Radical chain mechanism. The propagation steps are:

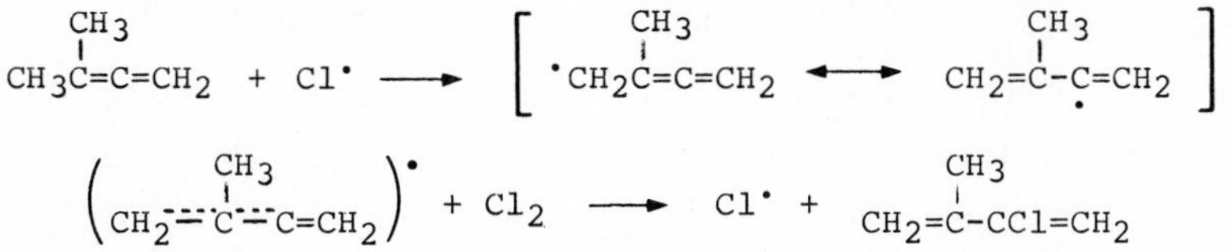

$$CH_3\overset{\overset{\textstyle CH_3}{|}}{C}=C=CH_2 + Cl^{\textstyle\cdot} \longrightarrow \left[{}^{\textstyle\cdot}CH_2\overset{\overset{\textstyle CH_3}{|}}{C}=C=CH_2 \longleftrightarrow CH_2=\overset{\overset{\textstyle CH_3}{|}}{C}-\overset{\textstyle\cdot}{C}=CH_2 \right]$$

$$\left(CH_2{=\!=}\overset{\overset{\textstyle CH_3}{|}}{C}{=\!=}C=CH_2 \right)^{\textstyle\cdot} + Cl_2 \longrightarrow Cl^{\textstyle\cdot} + CH_2=\overset{\overset{\textstyle CH_3}{|}}{C}-CCl=CH_2$$

16. $$HOCH_2\overset{\overset{\textstyle{}}{\underset{\textstyle OH}{|}}}{C}HCH_2OH + H^+ \rightleftharpoons HO-CH_2\overset{\overset{\textstyle{}}{\underset{\textstyle +OH_2}{|}}}{C}HCH_2OH \rightleftharpoons HO-CH_2\overset{+}{C}HCH_2-OH$$

$$\Big\Updownarrow \text{-}H^+$$

$$HO-CH=CH-CH_2OH_2{}^+ \underset{\text{}}{\overset{H^+}{\rightleftharpoons}} HO-CH=CH-CH_2OH \rightleftharpoons O=\overset{\overset{\textstyle H}{|}}{C}-CH_2CH_2OH$$

$$\Big\Updownarrow \text{- }H_2O$$

$$H\overset{+}{O}=CH-CH=CH_2 \overset{-H^+}{\rightleftharpoons} O=CH-CH=CH_2$$

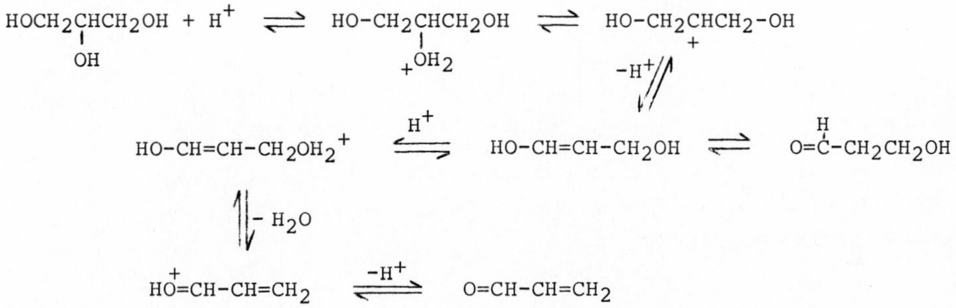

17.

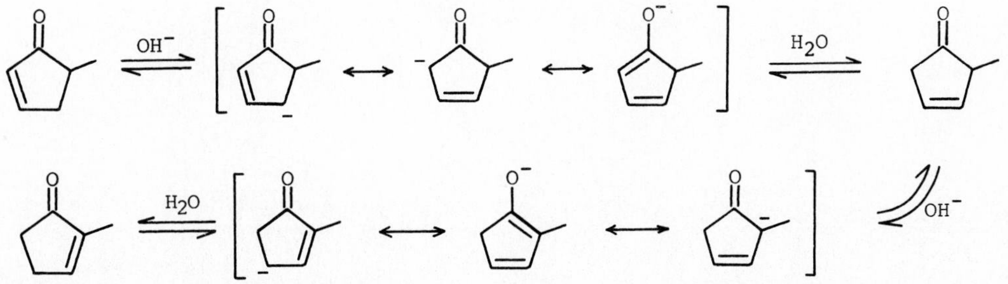

The β,γ- form is common to two α,β- forms in this case. Note that the
analogous transformation is not
possible for a cyclohexenone:

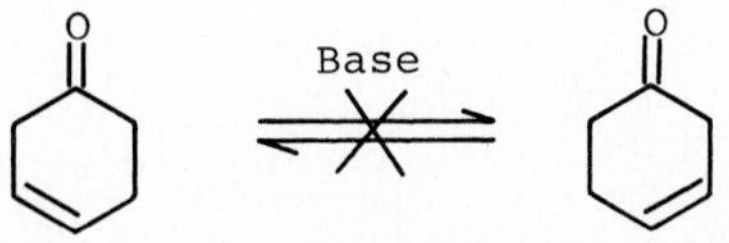

18.

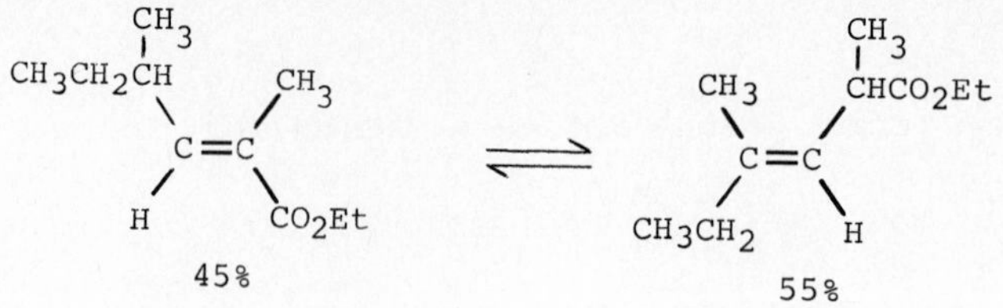

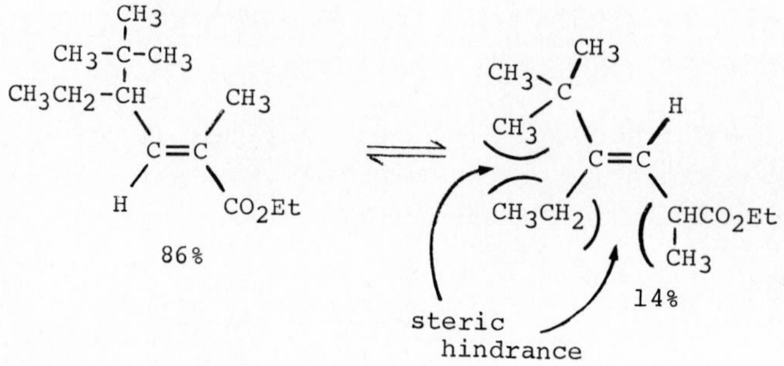

Steric hindrance is greater in the R = t-Bu case and is clearly evident using molecular models.

19.

$\Delta H° = +161$ kcal mole^{-1}

$\Delta H° = +169$ kcal mole^{-1}

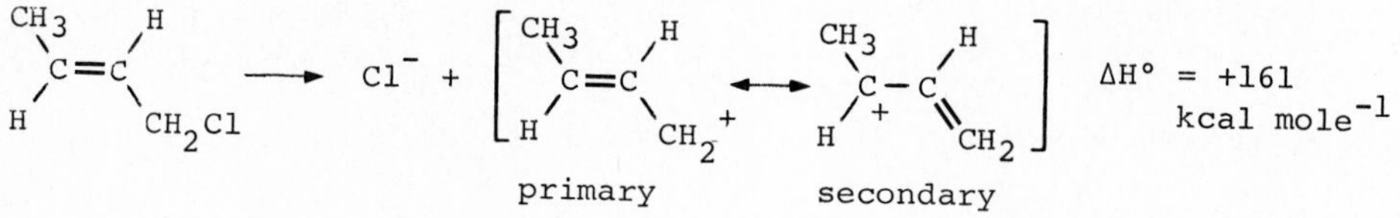

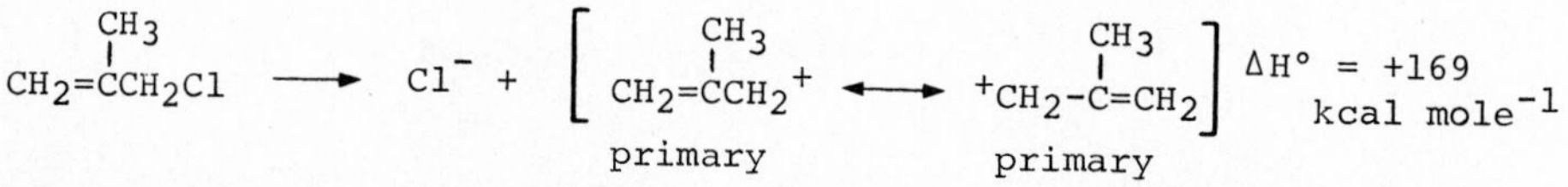

The allylic cation from 1-chloro-2-butene has the dual character of primary and secondary carbocations. That from 3-chloro-2-methyl-1-propene is primary-primary.

20.

(a)

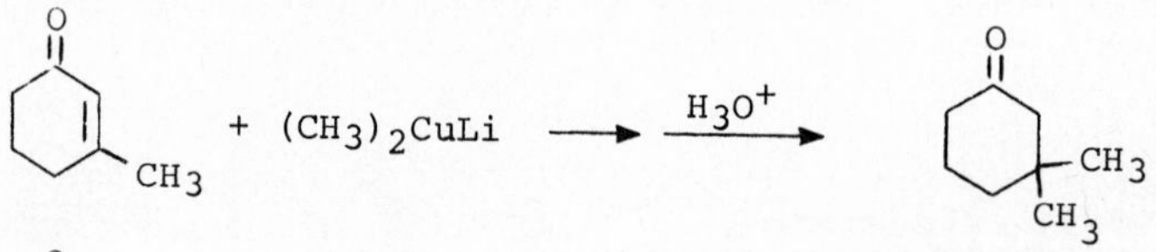

(b)

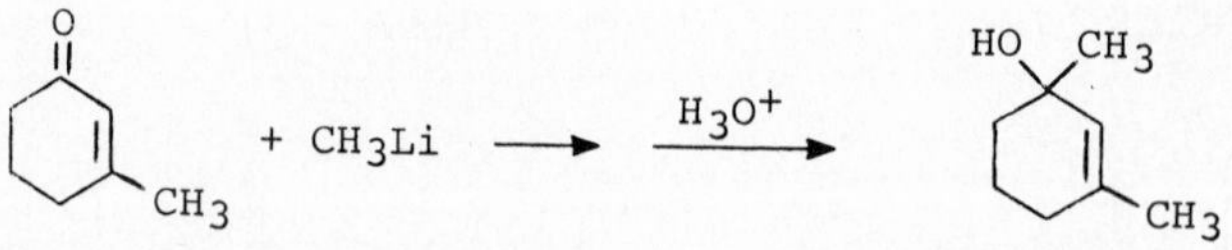

(c)

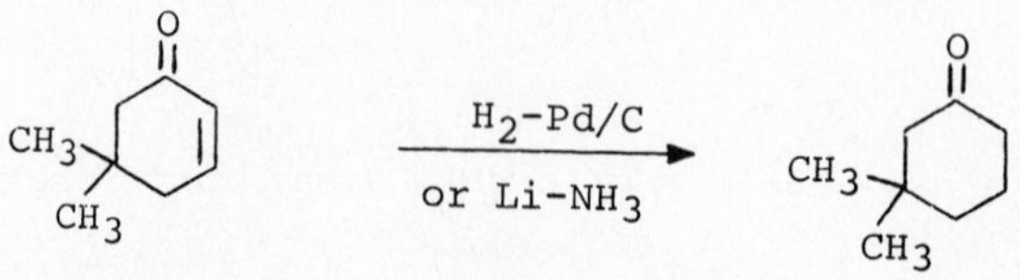

(d)

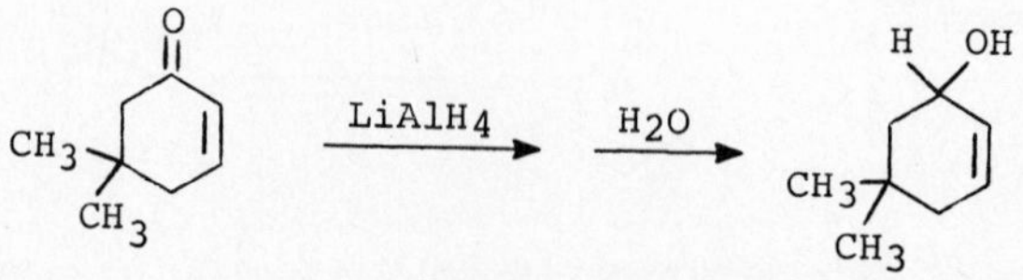

(e)

3-methylcyclohex-2-enone $+$ HCN $\xrightarrow{\text{Et}_3\text{Al}}$ (cyclohexanone with CH$_3$ and CN) $\xrightarrow[\Delta]{\text{H}_3\text{O}^+}$ (cyclohexanone with CH$_3$ and COOH)

21.

$$\text{H}_2\text{O} + \text{CH}_2=\text{C}=\text{O} \longrightarrow \left[\begin{array}{c} \text{CH}_2=\overset{\ }{\text{C}}-\text{O}^- \\ \ \ \ \ {}^+\text{OH}_2 \end{array} \longleftrightarrow \begin{array}{c} {}^-\text{CH}_2-\text{C}=\text{O} \\ \ \ {}^+\text{OH}_2 \end{array} \right]$$

$$\begin{array}{c} {}^-\text{CH}_2-\text{C}=\text{O} \\ \ \ \ {}^+\text{OH}_2 \end{array} + \text{H}_2\text{O} \rightleftharpoons \begin{array}{c} \text{CH}_3-\text{C}=\text{O} \\ \ \ \ {}^+\text{OH}_2 \end{array} + \text{OH}^- \rightleftharpoons \begin{array}{c} \text{CH}_3-\text{C}=\text{O} \\ \ \ \ \text{OH} \end{array} + \text{H}_2\text{O}$$

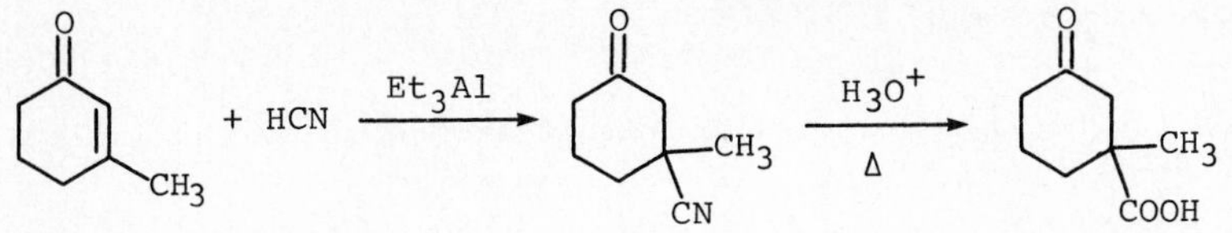

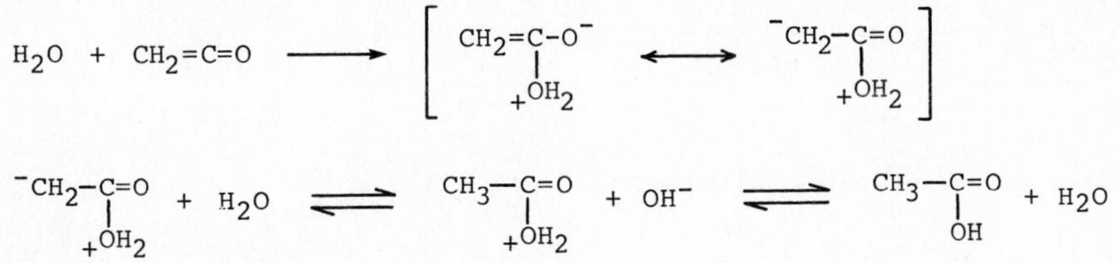

22.

(a)

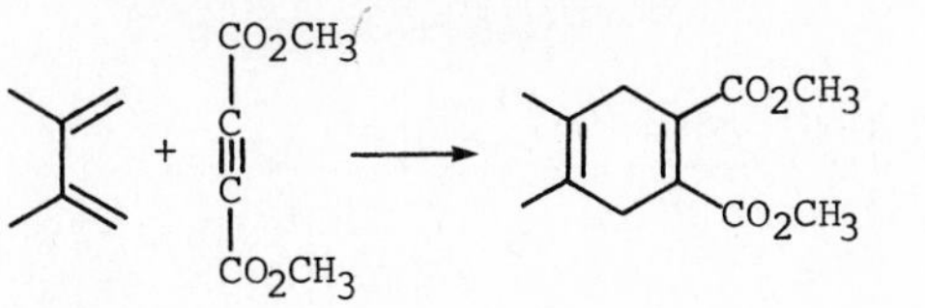

(b)

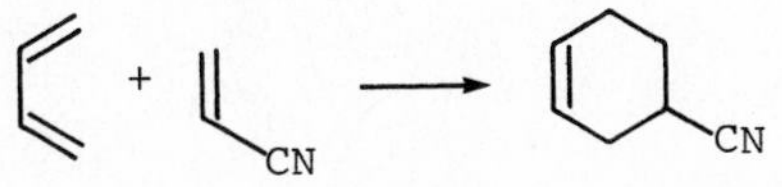

(c)

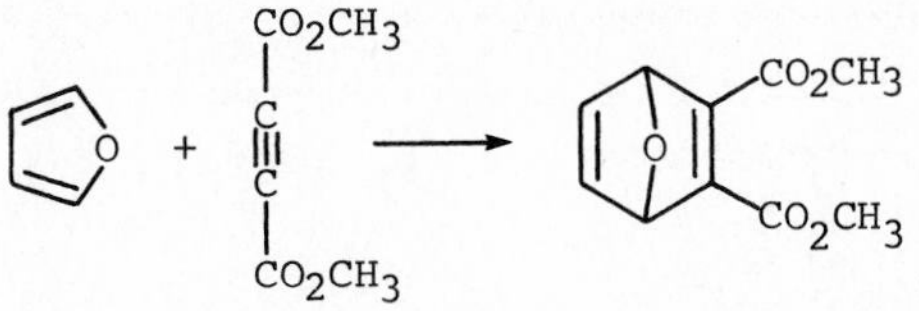

(d)

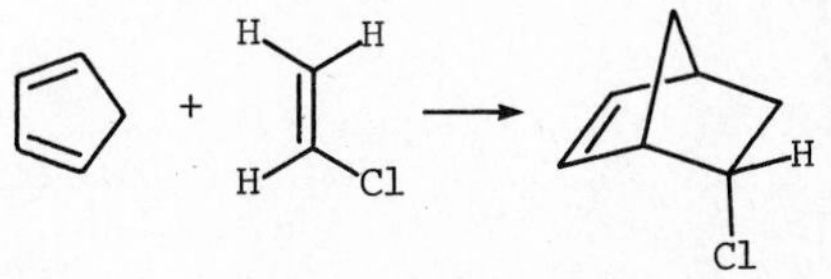

(e)

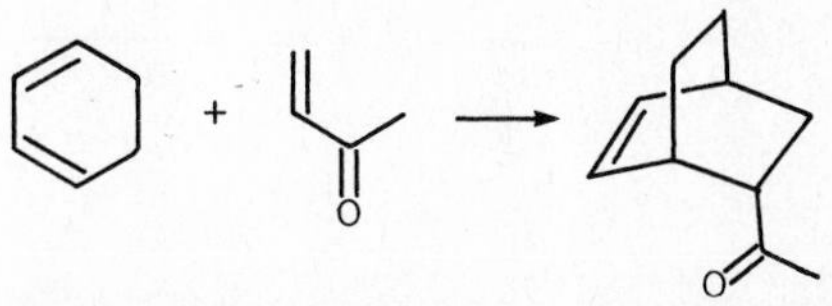

(f)

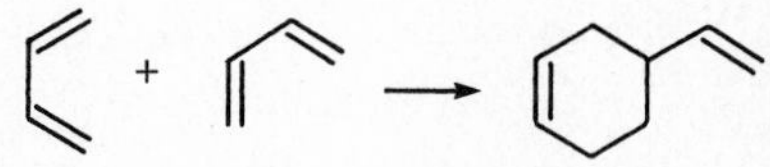

(g)

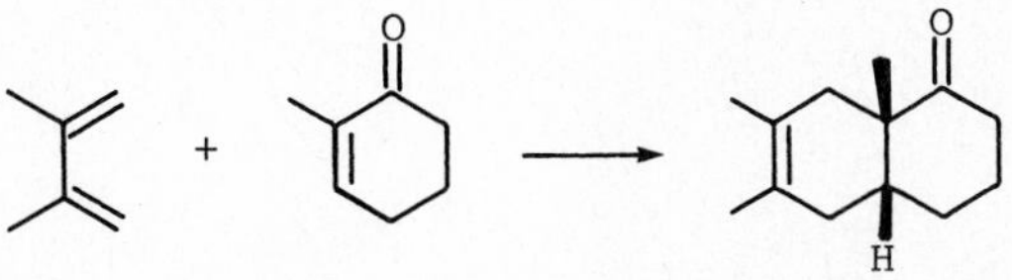

(h)

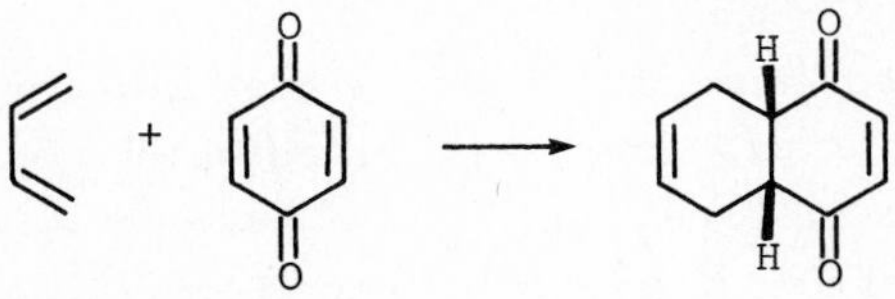

23.

(a)

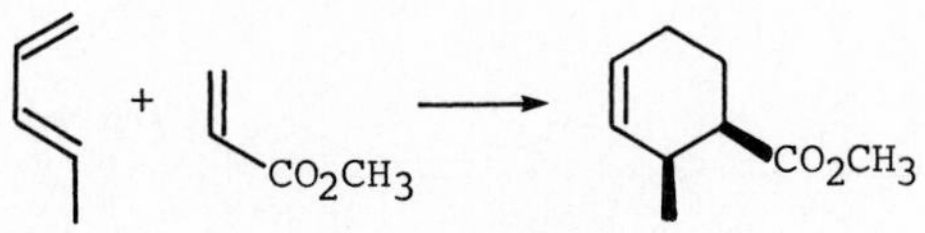

(b)

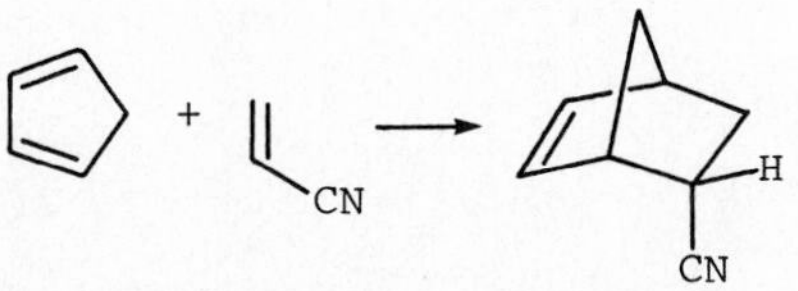

(c)

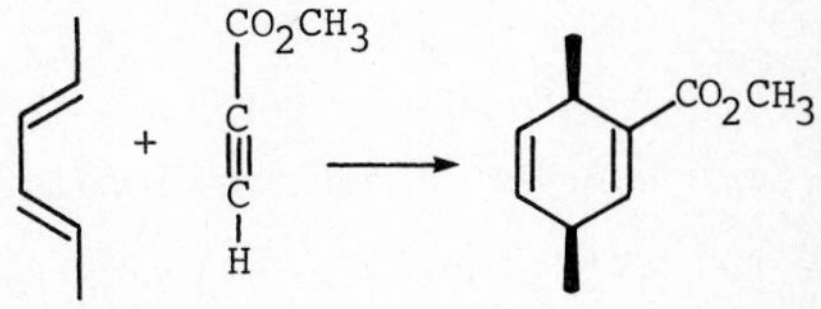

20.E Supplementary Problems

S1. What is the major product to result from each of the following reaction
sequences?

(a)

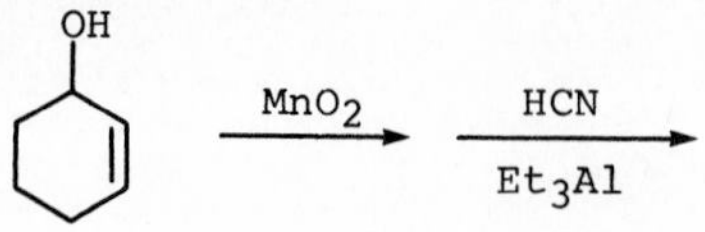

$\xrightarrow{MnO_2}$ $\xrightarrow[Et_3Al]{HCN}$

(b)

$+$ $CH_2{=}C\underset{CH_3}{\overset{CO_2CH_3}{}}$ $\longrightarrow$

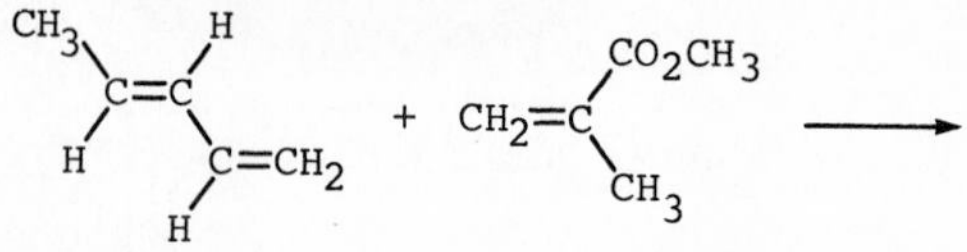

(c) $(CH_3)_2CHCHO$ $+$ $(C_2H_5)_2O_3\overset{-}{P}CHCO_2CH_3$ $\longrightarrow$ $\xrightarrow{Br_2}$

(d) $(CH_3)_3C\overset{O}{\overset{\|}{C}}H$ $+$ $\xrightarrow{NaOH}$ $\xrightarrow[0\,°C]{LiAlH_4}$ $\xrightarrow{H_2O}$

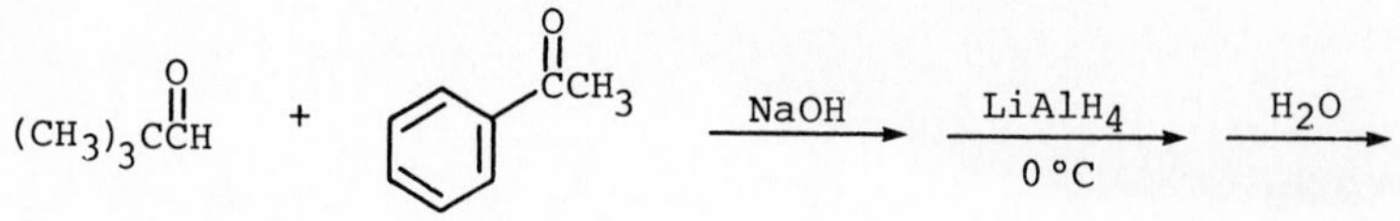

(e) $CH_3CH{=}CHCH_2Br$ $\xrightarrow[ether]{Mg}$ $\xrightarrow{CO_2}$ $\xrightarrow[H^+]{CH_3OH}$ $\xrightarrow{CH_3O^-}$

(f) $(CH_3)_2C{=}CHCH{=}CH_2$ $\xrightarrow[0\,°C]{HCl}$ $\xrightarrow{50\,°C}$

(g) $+$ $\underset{H}{\overset{HOOC}{}}C{=}C\underset{COOH}{\overset{H}{}}$ $\xrightarrow{\Delta}$

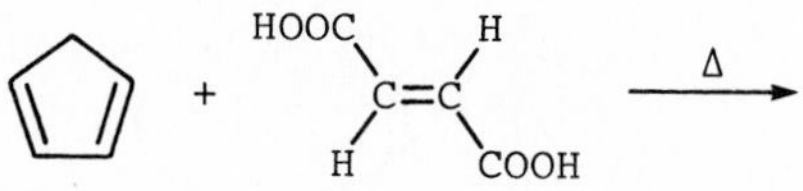

S2. Of the following three Diels-Alder reactions, one gives only a single product
and the other two each give a mixture of two isomeric products. Write the
products of each reaction, and indicate the major product for the reactions
which give mixtures.

(a) $+$ $\longrightarrow$

(c) $+$ $\longrightarrow$

(b) $+$ $\overset{CO_2Me}{}$ $\longrightarrow$

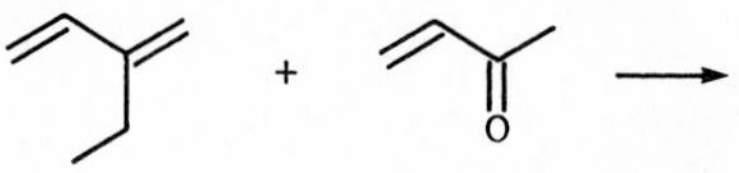

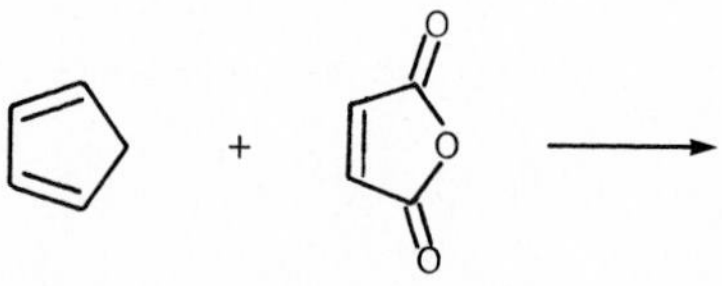

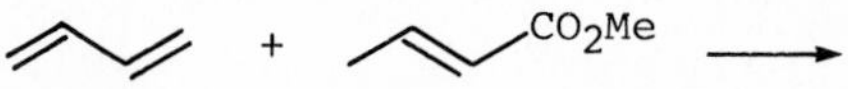

S3. The reaction sequence illustrated below was carried out to prepare methyl
1-cyclohexenecarboxylate. A product was obtained which shows an α,β-unsatur-
ated ester function in the infrared (ν_{max} = 1710, 1660 cm^{-1}). However, the
elemental analysis of the product was incorrect (both %C and %H were too
high). What is wrong with the sample? How may the synthesis be modified
to eliminate the problem?

$\xrightarrow{Br_2,\ P}$ $\xrightarrow{CH_3OH}$ $\xrightarrow[\substack{C_2H_5OH \\ \Delta}]{NaOC_2H_5}$

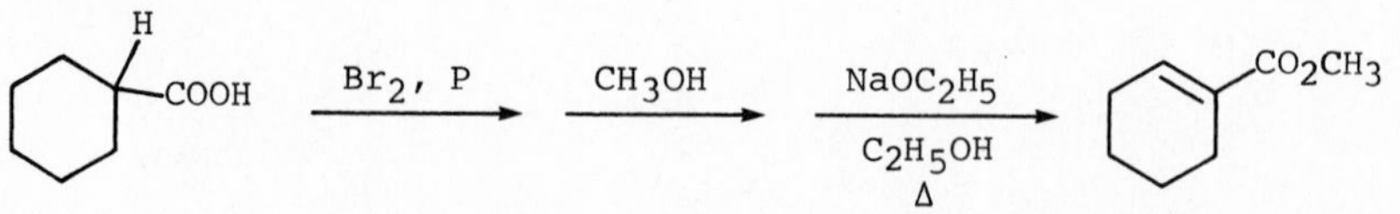

S4. Provide an explanation for the following differences in chemical behavior.

(a) $ClCH{=}CHCH_3$ $+$ $NaCN$ $\longrightarrow$ no reaction

$ClCH{=}CH\overset{O}{\overset{\|}{C}}CH_3$ $+$ $NaCN$ $\longrightarrow$ $NC{-}CH{=}CH\overset{O}{\overset{\|}{C}}CH_3$

(b) $CH_3OCH_2CH_2CH_3 \xrightarrow{KO\underline{t}\text{-Bu}}$ no reaction

$CH_3OCH_2CH_2CO_2CH_3 \xrightarrow{KO\underline{t}\text{-Bu}} CH_2{=}CHCO_2CH_3 + CH_3OH$

(c)

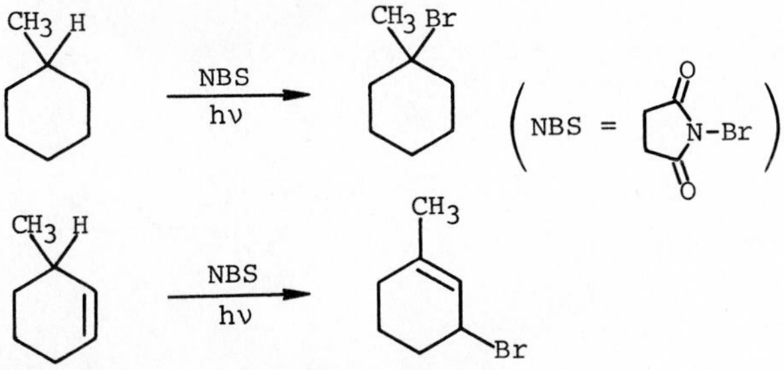

(d)

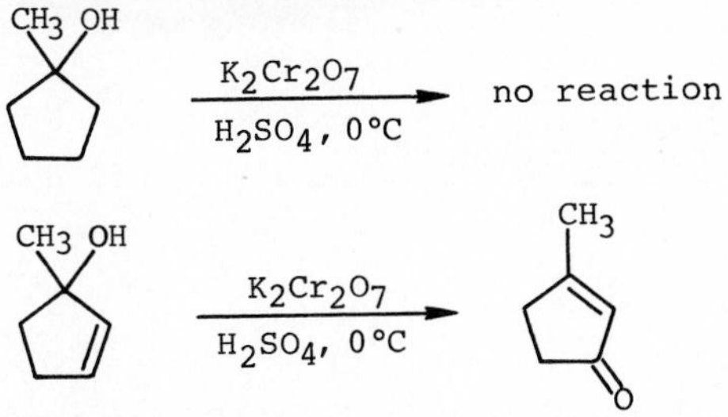

S5. The C=O stretch in the infrared spectrum of conjugated ketones comes at lower frequency than that for the analogous saturated systems. Explain why.

S6. Using Appendices I and II, calculate the change in enthalpy expected for the propagation steps of the free-radical chlorination of ethane, ethylene, and propene to give ethyl chloride, vinyl chloride, and allyl chloride, respectively.

S7. The prostaglandins are a class of compounds whose occurrence, structures, and potent biological effects have been studied and elucidated only within recent years. They are found throughout the body, and are implicated in many diverse biological processes, often at nanomolar (10^{-9} $\underline{M}$) concentrations. PGE_2, the most potent of the prostaglandins, is unstable in the presence of base: it loses water to give PGA_2, which then isomerizes to the physiologically inactive PGB_2 isomer.

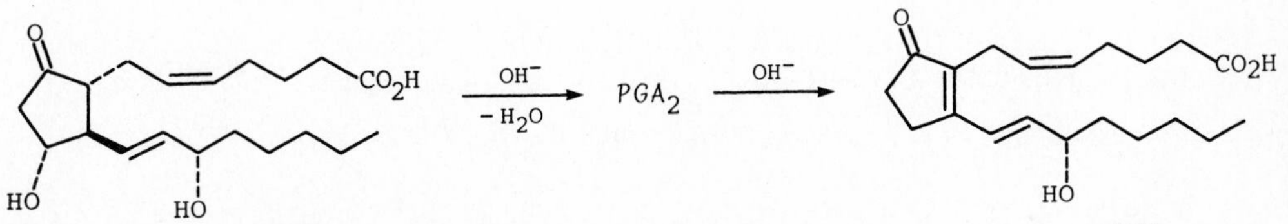

PGE_2 PGB_2

Write a reasonable structure for PGA_2, as well as step-by-step mechanisms for these two transformations.

S8. When 2-methyl-3-cyclohexenone is treated with base, it readily isomerizes to 2-methyl-2-cyclohexenone. Similar treatment of bicyclo[2.2.2]oct-5-en-2-one does not produce any reaction, however. Why do these two compounds differ so much in reactivity?

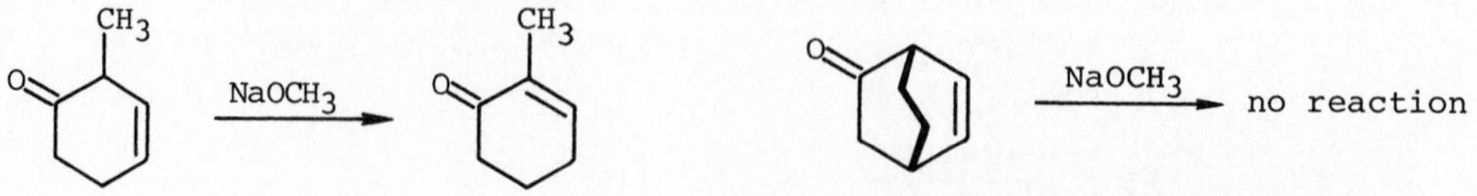

S9. Write a mechanism for the following transformation.

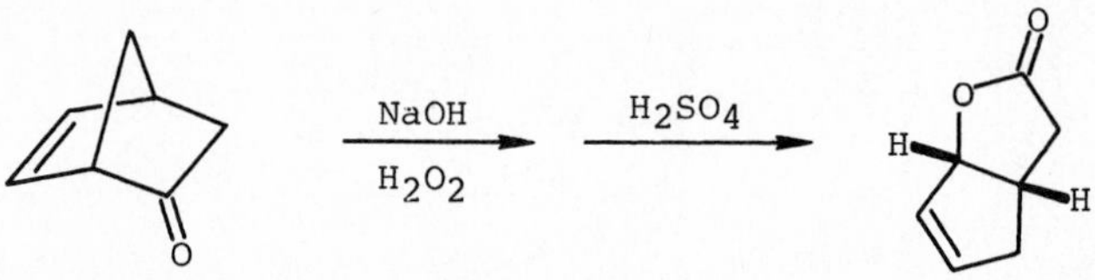

S10. Compound A ($C_7H_{14}O$) has a strong absorption in its infrared spectrum at 3400 cm^{-1}. It reacts with acetic anhydride to give a new compound (B, $C_9H_{16}O_2$), which shows an infrared absorption at 1735 cm^{-1}. Compound A reacts with Na_2CrO_4 in acetic acid to give C, which has an infrared band at 1710 cm^{-1}. Compound C reacts with bromine in acetic acid to give D ($C_7H_{11}BrO$). With excess bromine in aqueous NaOH, C gives a tetrabromo compound (E, $C_7H_8Br_4O$). Compound D reacts with potassium t-butoxide in refluxing t-butyl alcohol to give F, which has infrared absorptions at 1685 and 1670 cm^{-1}. When either C or F is treated with $NaOCH_3$ in CH_3OD, it is found to exchange four of its protons for deuterium. What are compounds A-F?

20.F Answers to Supplementary Problems

Hi, Stevie!

S1. (a)

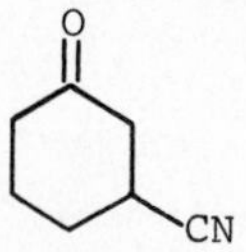

(c) $(CH_3)_2CHCH(Br)CH(Br)CO_2CH_3$

(b)

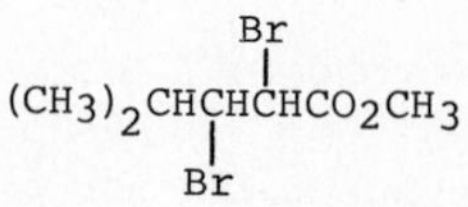

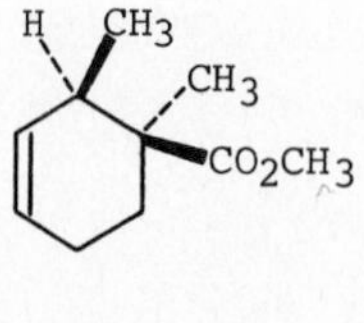

(d)

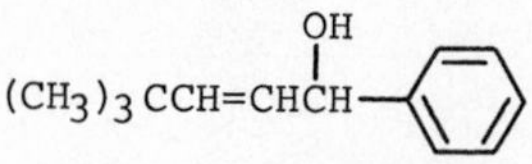

(e) $CH_3CH{=}CHCH_2Br \xrightarrow{Mg} [CH_3CH{=}CHCH_2MgBr \rightleftharpoons CH_3CH(MgBr)CH{=}CH_2] \xrightarrow{CO_2} CH_3CH(CO_2MgBr)CH{=}CH_2 \xrightarrow[H^+]{CH_3OH} \xrightarrow{CH_3O^-} CH_3C(CO_2CH_3){=}CHCH_3$

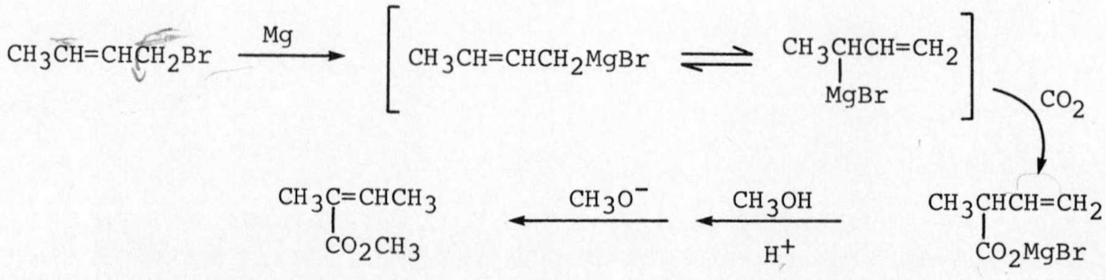

(f) $(CH_3)_2C=CHCH=CH_2$ $\xrightarrow[0°C]{HCl}$ $(CH_3)_2\underset{\underset{Cl}{|}}{C}CH=CHCH_3$ $\xrightarrow{50°C}$ $(CH_3)_2C=CHCH\underset{\underset{Cl}{|}}{C}H_3$

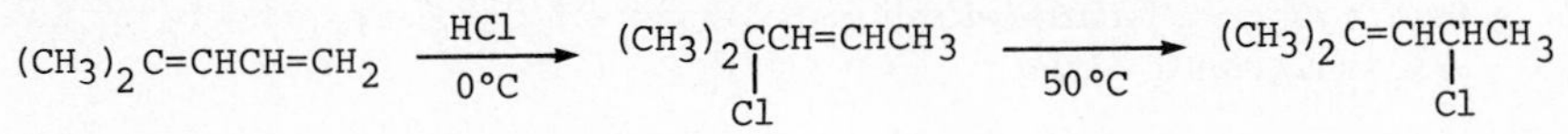

(g)

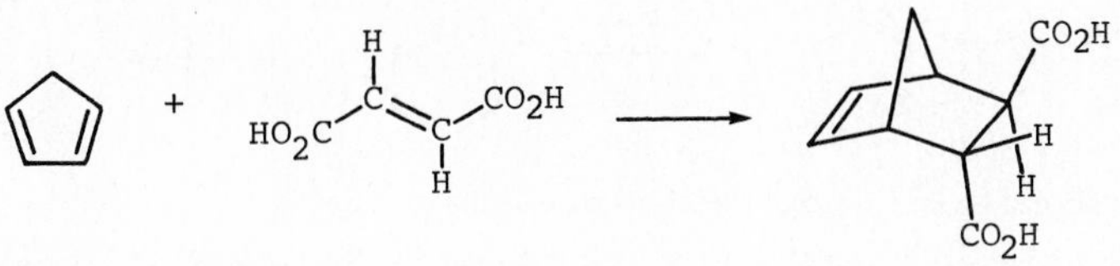

S2. (a)

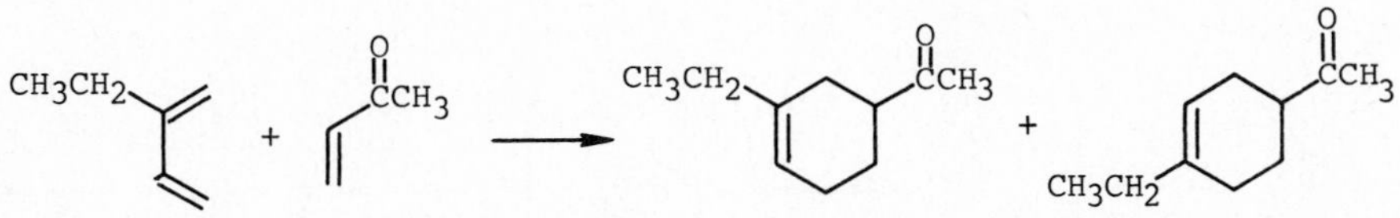

major   *minor*

(b)

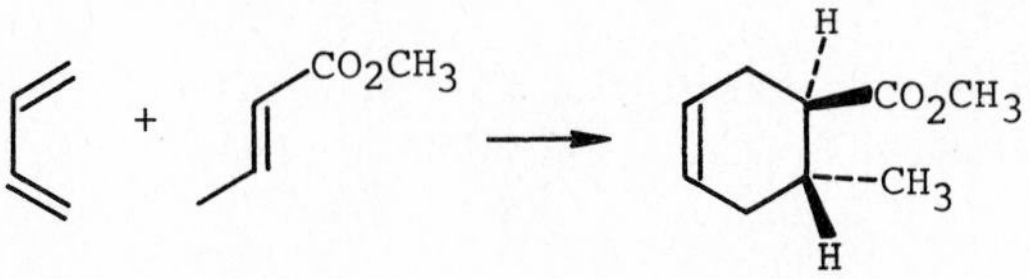

only product

(c)

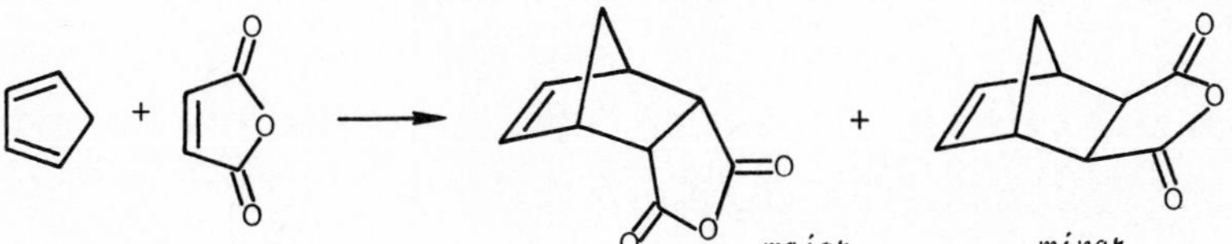

major   *minor*

S3.   The methyl group has been lost through ester exchange with the $NaOC_2H_5/C_2H_5OH$
       used in the last step (see next page). This can be avoided by using
       $NaOCH_3/CH_3OH$ to accomplish the elimination.

$+ \; C_2H_5OH$ $\xrightleftharpoons{NaOC_2H_5}$

excess

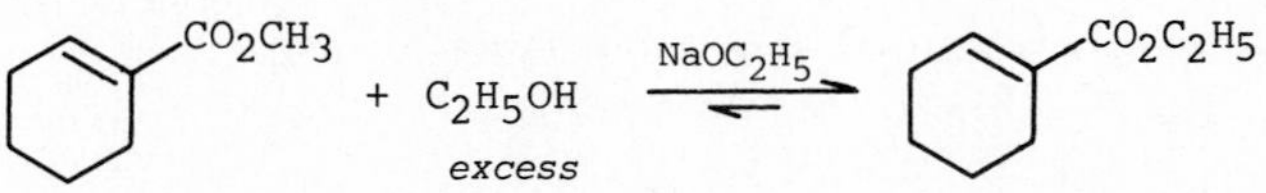

S4. (a) The chloroketone can undergo substitution by a conjugate addition–elimi-
       nation sequence. Such a mechanism is not possible for 1-chloropropene.

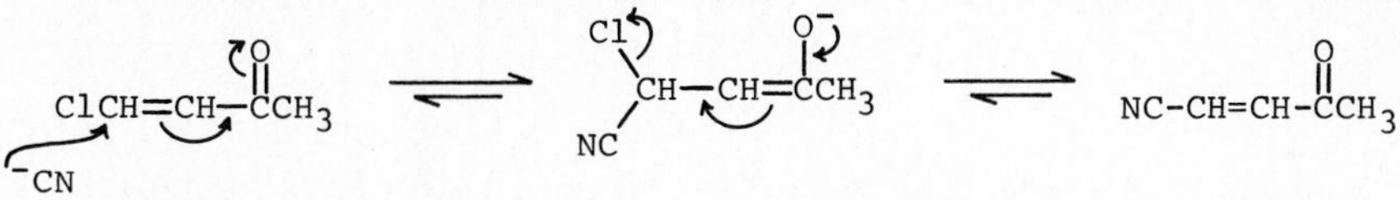

(b) The β-methoxyester can undergo elimination via an enolate, as in the
       second step of an aldol condensation. Again, the simple alkyl system
       cannot react in this manner.

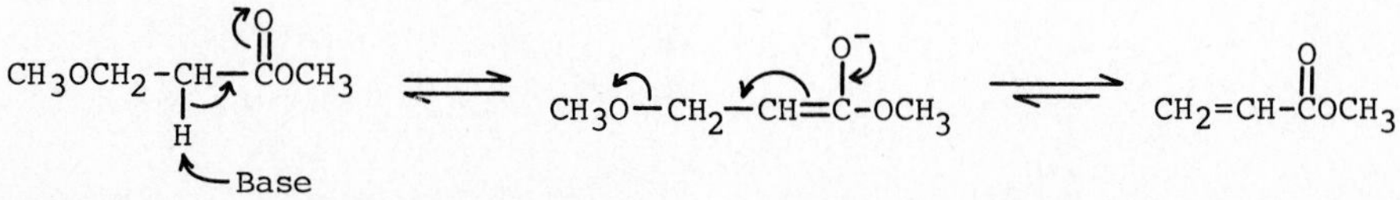

(c) In both free radical reactions, the hydrogen atom-abstraction is selective
       for the tertiary hydrogen. In the case of 3-methylhexene, this produces

an allylic radical, which can subsequently react at the other end of the original double bond.

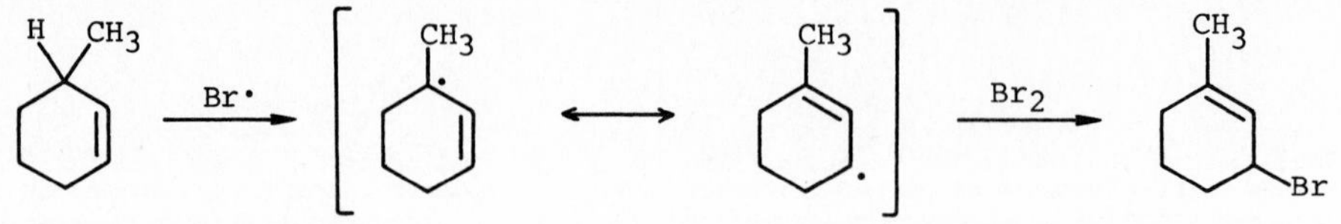

(d) Tertiary alcohols are not oxidized under mild conditions, but the tertiary allylic alcohol can undergo ionization, allylic isomerization, and then oxidation:

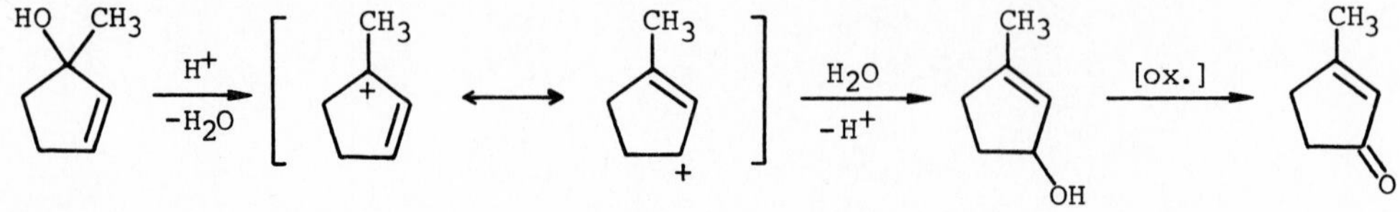

S5. Because of the contribution of resonance structures which have a C-O single bond, there is slightly less double-bond character in the carbonyl bond of an enone:

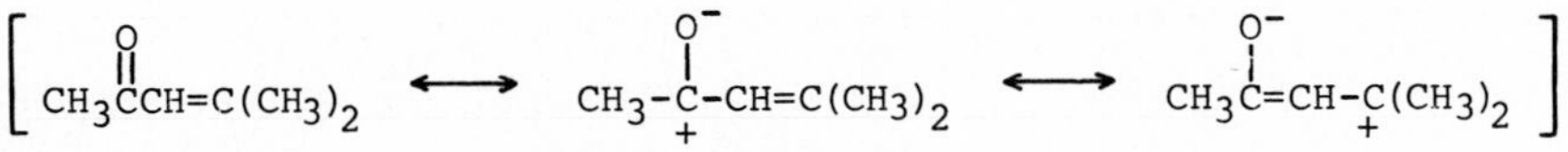

S6.

	CH_3-CH_3	+ Cl·	⟶	CH_3CH_2·	+ HCl	
ΔH°_f =	-20.2	28.9		26	-22.1	ΔH° = 5 kcal mole^{-1}
	CH_3-CH_2·	+ Cl_2	⟶	CH_3CH_2Cl	+ Cl·	
ΔH°_f =	26	0		-26.1	28.9	ΔH° = -23 kcal mole^{-1}
	CH_2=CH_2	+ Cl·	⟶	CH_2=CH·	+ HCl	ΔH° (kcal mole^{-1})
ΔH°_f =	12.5	28.9		68	-22.1	+4.5
	CH_2=CH·	+ Cl_2	⟶	CH_2=CHCl	+ Cl·	
ΔH°_f =	68	0		8.6	28.9	-30.5
	CH_2=CH-CH_3	+ Cl·	⟶	CH_2=CH-CH_2·	+ HCl	
ΔH°_f =	4.9	28.9		40	-22.1	-16
	CH_2=CH-CH_2·	+ Cl_2	⟶	CH_2=CH-CH_2Cl	+ Cl·	
ΔH°_f =	40	0		0*	28.9	-11

*Calculate ΔH°_f (CH_2=$CHCH_2Cl$) = ΔH°_f (CH_2=CH-CH_2·) + ΔH°_f (Cl·) - DH° (allyl-Cl)

= 40 + 28.9 - 69 = 0

S7.

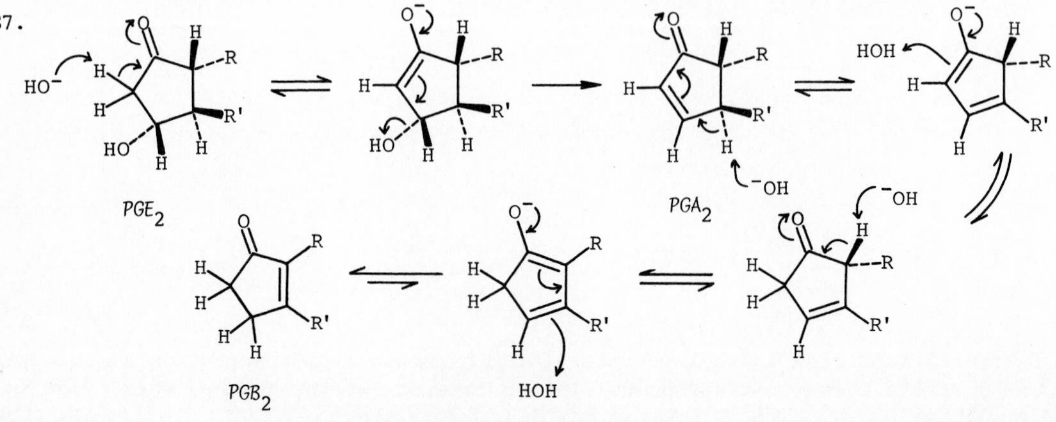

S8. Enolization and subsequent conjugation of the enone system both require that all the atoms involved are able to line up their p-orbitals:

This sort of configuration is not possible for the bicyclic β,γ-enone:

This orbital is perpendicular to the other p-orbitals and cannot overlap with them

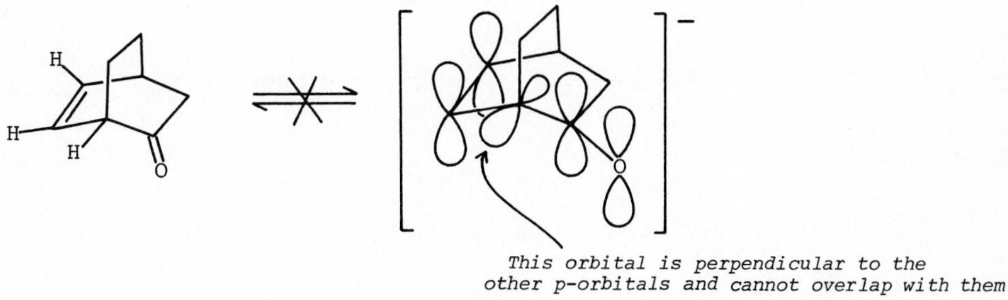

(The generalization that bicyclic systems cannot have a double bond at the bridgehead carbon is known as Bredt's rule. 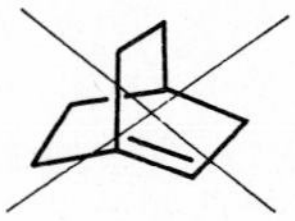)

S9.

This alkaline Baeyer-Villiger reaction is possible because of the strain in the bicyclo[2.2.1]heptane skeleton

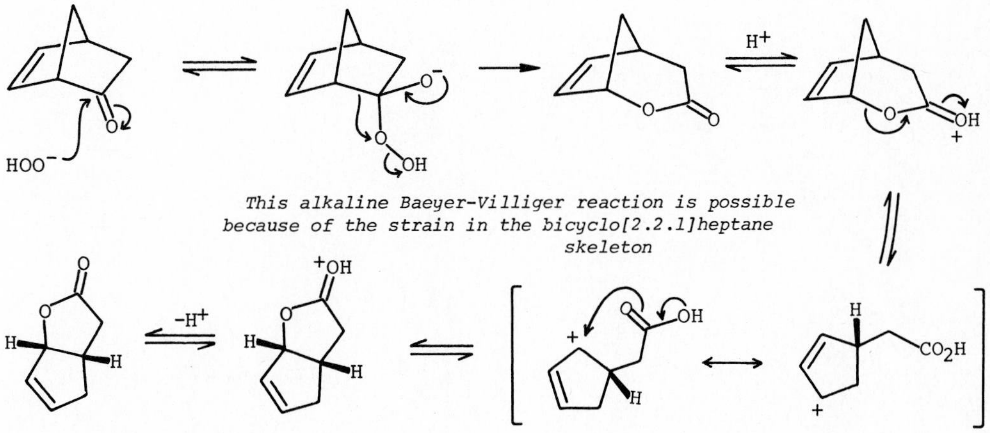

S10.

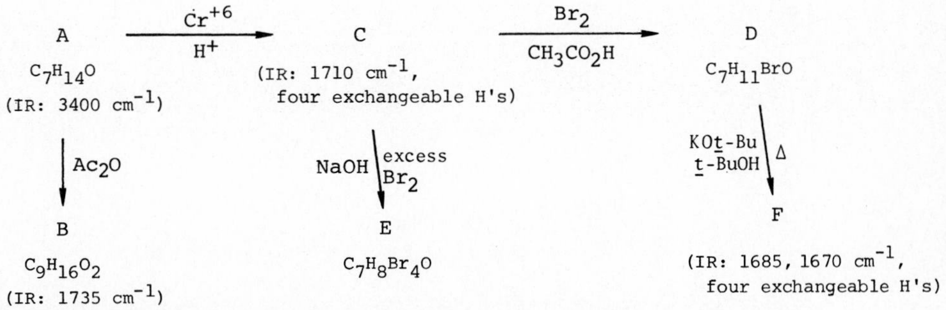

1) Formula for A ($C_7H_{14}O$): indicates one degree of unsaturation.

2) IR of A (3400 cm^{-1}), as well as the fact that reaction of A with Ac$_2$O gives B (C$_7$H$_{14}$O + CH$_3$CO$_2$H - H$_2$O = C$_9$H$_{16}$O$_2$ with IR 1735 cm^{-1} (indicates ester)):
   indicates A is an alcohol.

3) A reacts with Cr^{+6}, H$^+$ to give C (IR 1710 cm^{-1}): indicates B is an acyclic or six-membered ring ketone.

4) C has four exchangeable H's: indicates C is $-CH_2-\overset{\overset{\textstyle O}{\|}}{C}-CH_2-$ or $\overset{\diagdown}{\underset{\diagup}{}}CH-\overset{\overset{\textstyle O}{\|}}{C}-CH_3$

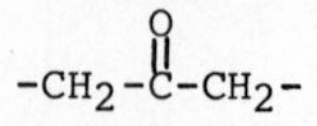

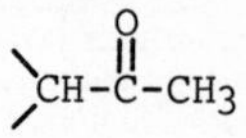

5) C reacts with excess Br$_2$ and NaOH to give E (C$_7$H$_8$Br$_4$O):
   indicates that C is <u>not</u> a methyl ketone (which would give $-CO_2H + HCBr_3$).

6) C $\xrightarrow[\text{H}^+]{\text{Br}_2}$ $\xrightarrow[\text{t-BuOH}]{\text{KO}\underline{t}\text{-Bu}}$ F (IR 1685, 1670 cm^{-1}): indicates that F is a conjugated
                                                                                              ketone.

7) F has four exchangeable H's: indicates F must be $\overset{\diagdown}{\underset{\diagup}{}}CH-C\overset{(H)}{=}CH-\overset{\overset{\textstyle O}{\|}}{C}-CH_2-$

   exchangeable

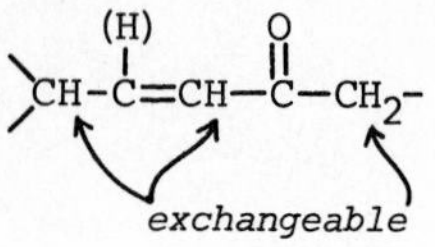

8) There is only one way to put all these facts together with seven carbons:

A

B

C

D

E

F

Ⓗ = exchangeable H's

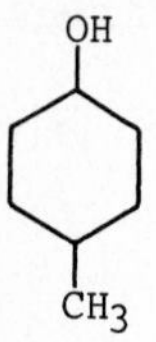

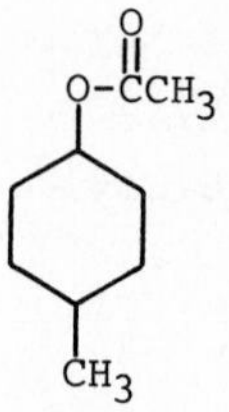

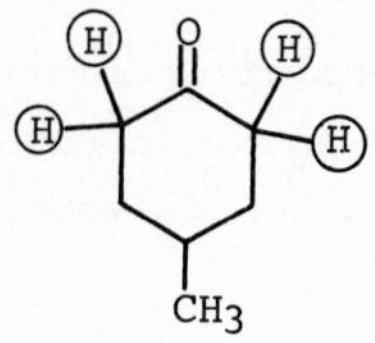

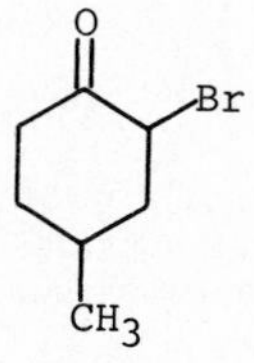

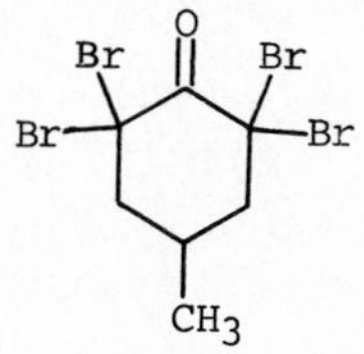

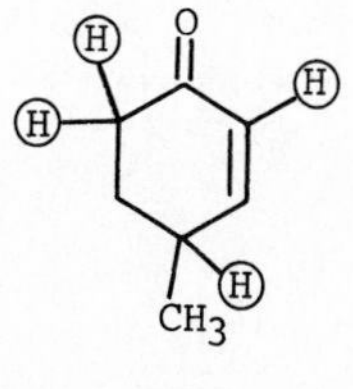

21. ULTRAVIOLET SPECTROSCOPY

21.A Chapter Outline and Important Terms Introduced

21.1 Electronic Transitions

excited state ground state

21.2 $\pi \rightarrow \pi^*$ Transitions

longer conjugation $\Rightarrow$ longer wavelength

21.3 $n \rightarrow \pi^*$ Transitions

extinction coefficient

21.4 Alkyl Substituents

hyperconjugation Woodward's rules

21.5 Other Functional Groups

21.6 Photochemical Reactions

21.B Important Concepts

As a useful spectroscopic method, UV spectroscopy is limited to conjugated systems. Therefore, it is not as generally applicable as nmr or IR. Nevertheless it can provide important information on structure and conformation in such systems.

The process of electronic excitation, on which UV spectroscopy is based, is also important in photochemistry. Many reactions which do not occur thermally with a molecule in its ground state can be made to go when the molecule is in an excited state. For example, the cyclization of two olefins to give a cyclobutane can be made to occur under irradiation, but not thermally, as pointed out in Chapter 20.

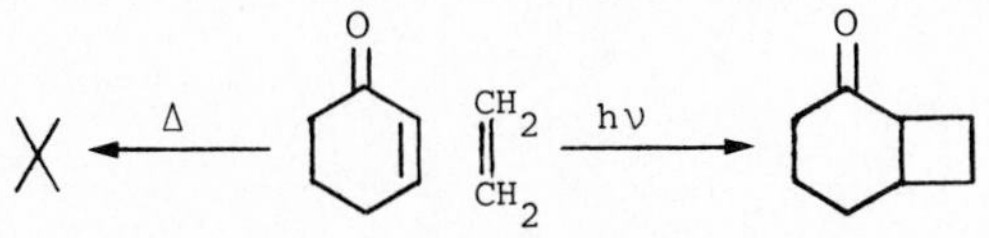

21.C Answers to Exercises

(21.3) $n \rightarrow \pi^*$: 0.00731 g crotonic acid/10 mL = 8.5×10^{-3} M

$$\varepsilon_{250} = \frac{0.77}{(8.5 \times 10^{-3})(1)} = 91$$

$\pi \rightarrow \pi^*$: 8.5×10^{-3} M diluted 100-fold = 8.5×10^{-5} M

$$\varepsilon_{200} = \frac{0.86}{(8.5 \times 10^{-5})(1)} = 10{,}120$$

(21.4) (a) $CH_2=CH-\overset{\displaystyle CH_3}{\overset{|}{C}}=CHCH_3$: 214 (pentadiene) + 10 (two alkyl groups) = 224

(b) : 253 (parent) + 20 (four alkyl groups) = 273

(c) $CH_3CH_2\overset{\displaystyle O}{\overset{\|}{C}}CH=CHCH_3$: 215 (enone) + 10 (one alkyl group) = 225

(d) : 215 (enone) + 20 (two alkyl groups) = 235

both of these count as substituents

21.D Answers and Explanations to Problems

1. Predicted λ_{max} values:

(a) 215 + 3 × 10 (three alkyl substituents) = 245
∴ 249 bottle

(b) 215 + 10 (one alkyl substituent) = 225
∴ 221 bottle

(c) 215 + 3 × 10 (three alkyl substituents) + 2 × 5 (two exocyclic double bond structures) = 255
∴ 258 bottle

(d) 215 + 2 × 10 (two alkyl substituents) = 235
∴ 233 bottle

2.

$$RX \xrightarrow{h\nu} R\cdot + X\cdot$$

$$RX + X\cdot \longrightarrow R\cdot + X_2$$

$$2\,R\cdot \longrightarrow R\text{-}R$$

The brown is due to Br_2, the violet to I_2.

3.

CH_3OH
$(CH_3CH_2)_2O$
$\underline{n}$-C_4H_9Cl
C_3F_8
CH_3CN

} Suitable; no absorption in accessible UV region

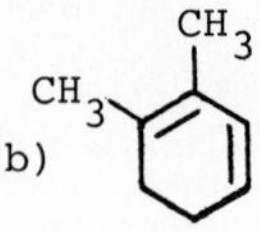

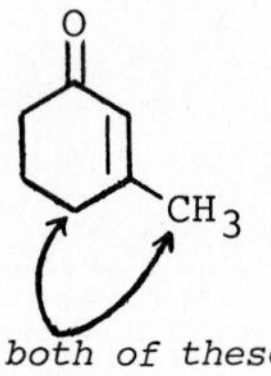

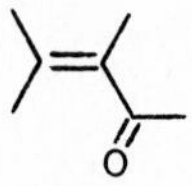

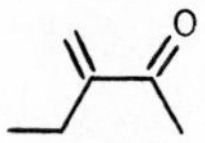

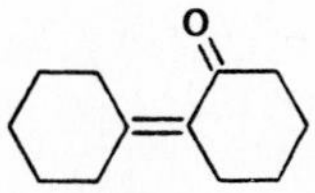

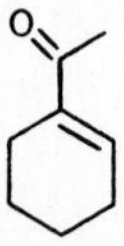

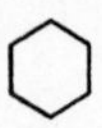

$CH_3SC_4H_9$
C_2H_5I

CH_2Br_2

} Unsuitable; all absorb at wavelength > 200 nm

4. The calculations follow. Experimental values are given in parentheses.

(a) 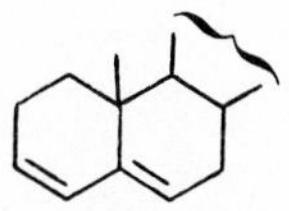 214 + 15 + 5 = 234 nm (234)

(b) 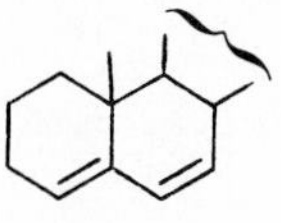 214 + 15 + 5 = 234 nm (235)

(c) 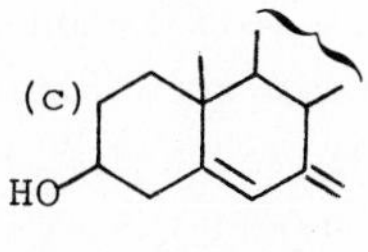 214 + 15 + 10 = 239 nm (236)

(d) 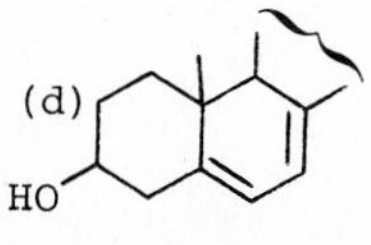 253 + 20 + 10 = 283 nm (282)

(e) 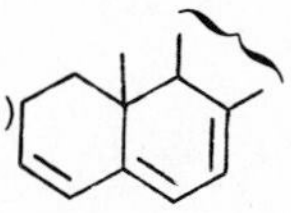 253 + 20 + 10 + 10 = 313 nm (315)

(f) 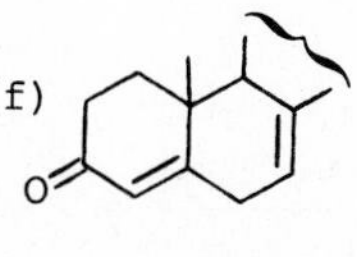 215 + 20 + 5 = 240 nm (235)

(g) 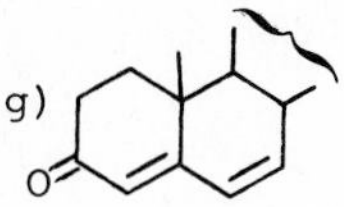 215 + 30 + 20 + 5 = 270 nm (274)

(h) 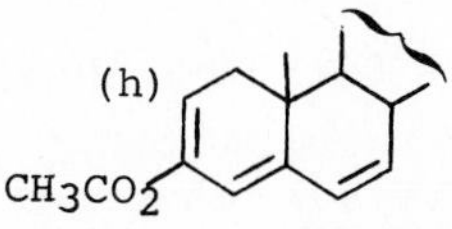 253 + 30 + 15 + 5 = 303 nm (306)

(i) 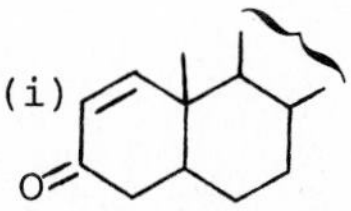 215 + 10 = 225 nm (230)

(j) 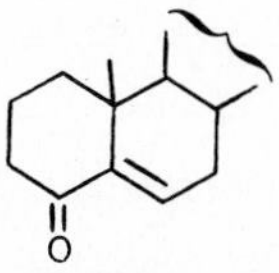 215 + 20 + 5 = 240 nm (241)

5. The linear correlation is especially good for $n \geq 3$. The interpolated value for $n = 9$ ($1/n = 0.111$) is $1/\lambda = 0.00236$ or $\lambda = 424$ nm.

6. 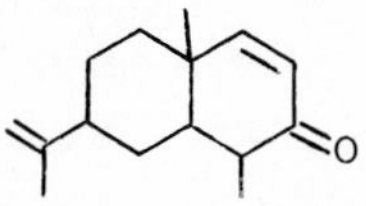

λ_{max}^{CALC} 215 + 10 = 225

The structure is not reasonable

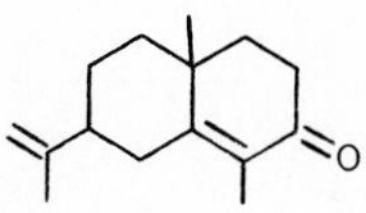

λ_{max}^{CALC} 215 + 30 + 5 = 250

fits much better

7. 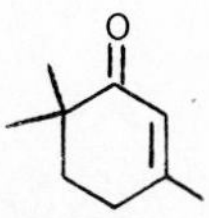

λ_{max}^{CALC} 215 + 20 = 235 nm

$$\varepsilon = \frac{1.0}{1.08 \times 10^{-4}} = 9{,}260$$

8. 

λ_{max}^{CALC} 212 + 10 = 222

The experimental value is 225 nm.

21.E Supplementary Problems

S1. What are plausible structures for A and B?

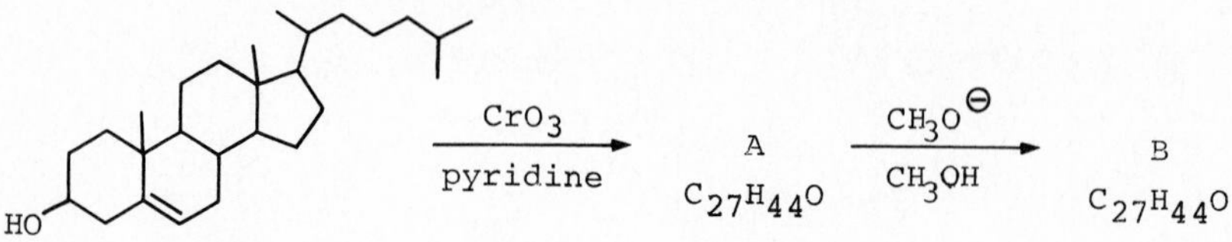

cholesterol $C_{27}H_{46}O$ λ_{max} = < 200 nm λ_{max} = 243 nm

S2. The Diels-Alder reaction between 1,3-dimethoxy-1,3-butadiene and 3-buten-2-one gives C as the major product. In aqueous acid under mild conditions, compound C reacts to give D, $C_8H_{10}O_2$. On treatment of D with $NaOCH_3/CH_3OH$ it isomerizes to E. The UV spectrum of D shows λ_{max} = 225 nm and that of E, λ_{max} = 232 nm. What are C, D, and E? Write a mechanism for the conversion of C → D.

S3. (a) The λ_{max} for (Z)-2,5-dimethyl-1,3,5-hexatriene is 237 nm. What is the value predicted from Woodward's rules? How do you explain this discrepancy?

(b) Heating (Z)-2,5-dimethyl-1,3,5-hexatriene leads to 1,4-dimethyl-1,3-hexadiene. What changes do you expect to see in the UV spectrum as this reaction progresses?

(c) What reaction do you expect to see on ultraviolet irradiation of the triene?

S4. Reaction of 6,6-dimethylcyclohex-2-enone with lithium diisopropylamide in THF at -78° C, followed by CH_3I and immediate work up, affords compound F, which shows no strong UV absorbance at wavelengths greater than 200 nm. If compound F is treated with acid or base, it is isomerized to G, which has a strong UV absorption in the region 210-250 nm.

(a) What are F and G?

(b) Predict more precisely the λ_{max} of G.

(c) How else could you tell the difference between F and G?

21.F Answers to Supplementary Problems

S1.

A B

S2.

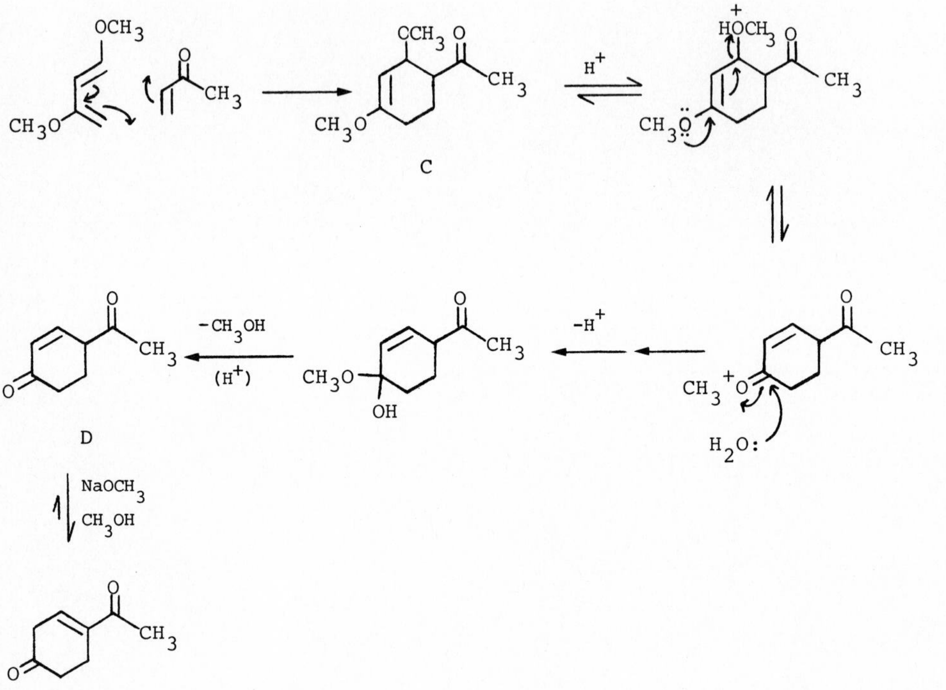

S3. (a) The calculated λ_{max} for 2,5-dimethyl-1,3,5-hexatriene is 254 nm. Because of the *cis* geometry of the central double bond, both of the terminal double bonds cannot simultaneously be coplanar with it for steric reasons, and the molecule behaves more like a diene than a triene.

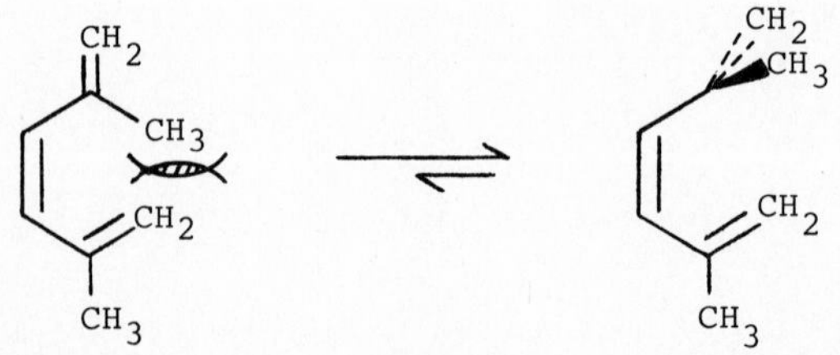

(b) The expected λ_{max} for 1,4-dimethyl-1,3-cyclohexadiene is 263 nm (actual: 264 nm), so an increase in λ_{max} will be observed as the reaction progresses.

(c)

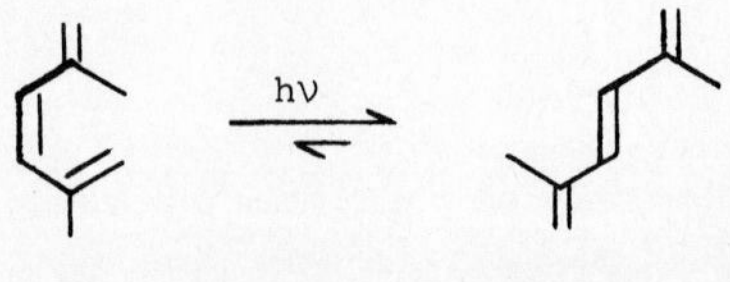

(among others)

S4.

(a) 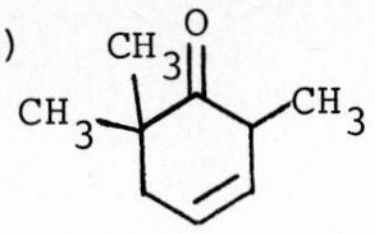

F

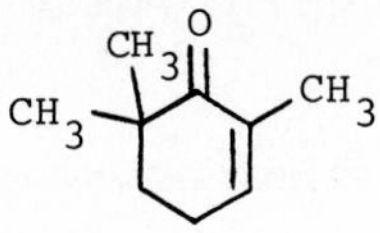

G

(b) 215 + 2 x 10 = 235 nm

(c) nmr of F: two methyl singlets and one methyl doublet; two vinyl H's
nmr of G: two singlet methyls, one with twice the area of the other; one vinyl H

IR of F: C=O ∿1710 cm^{-1}, weak C=C
IR of G: C=O ∿1675 cm^{-1}, strong C=C ∿1640 cm^{-1}

22. BENZENE AND THE AROMATIC RING

22.A Chapter Outline, Important Terms Introduced, and Reactions Discussed

22.1 <u>Benzene</u>

A. The Benzene Enigma
   phenyl
   aromatic

B. Resonance Energy of Benzene
   delocalization energy
   empirical resonance energy

C. Molecular Orbital Theory of Benzene
   degenerate molecular orbitals
   cyclic system of six π-electrons
   aromatic stability

D. Symbols for the Benzene Ring

E. Formation of Benzene
   dehydrogenation of cyclohexane:

(using Pd or Pt/Δ; or S, Δ)

   hydroforming process

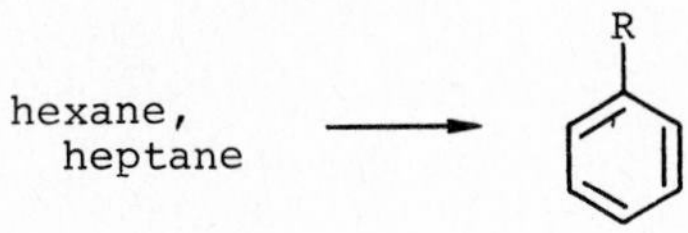

(using Cr_2O_3 or Pt as catalyst; industrial process)

22.2 <u>Substituted Benzenes</u>

A. Nomenclature
   ortho-, *meta*-, and *para*- (o-, m-, p-)
   arene
   tolyl-, xylyl-, mesityl-
   benzyl

B. Körner's Absolute Method
   (for telling o-, m-, and p- apart)

22.3 <u>Spectra</u>

A. Nmr Spectra
   ring current

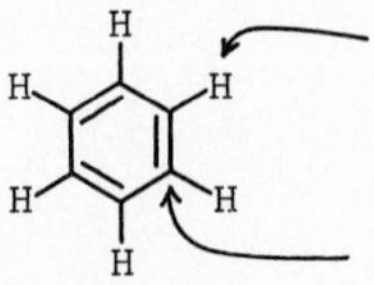

B. Infrared Spectra

C. Ultraviolet Spectra
   symmetry-forbidden transitions

22.4 <u>Dipole Moments in Benzene Derivatives</u>

22.5 Side-Chain Reactions

A. Free Radical Halogenation

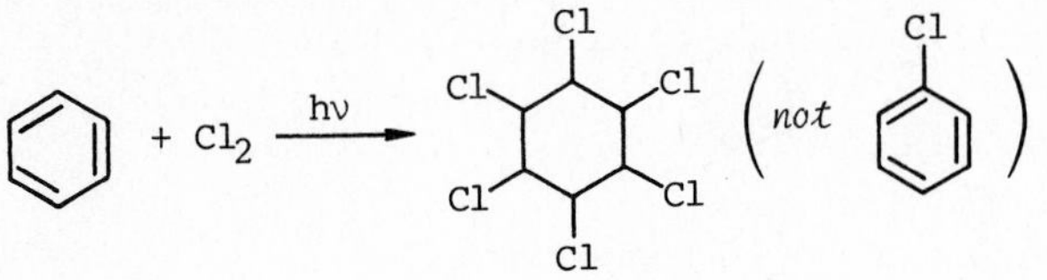

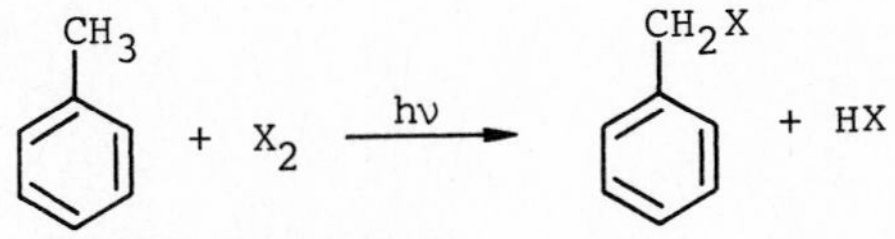

(very easily at benzylic positions)

benzyl radical

B. Benzylic Displacement and Carbocation Reactions

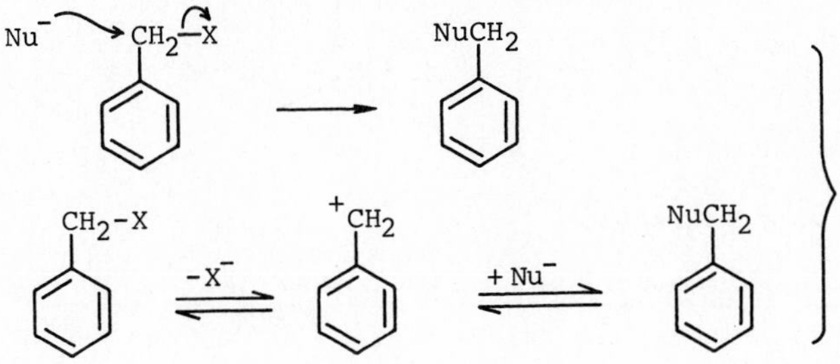

both S_N1 and S_N2 are fast

C. Oxidation

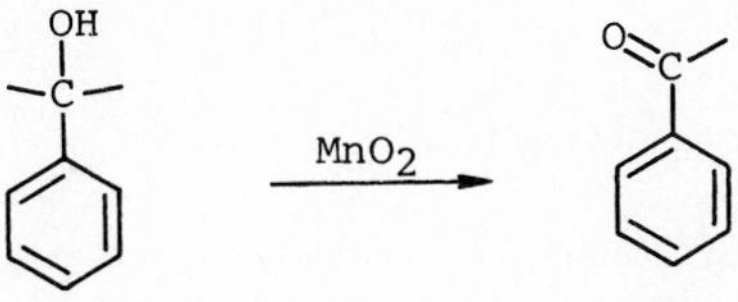

(as easy as allylic alcohol)

(using $Na_2Cr_2O_7/H_2SO_4$ or $KMnO_4$; at least one benzylic H is needed)

D. Acidity of Alkylbenzenes

benzylic carbanion

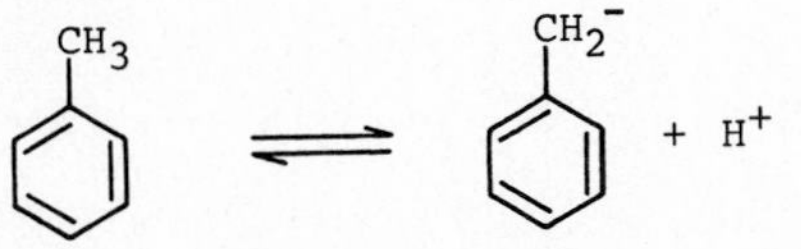

$pK_a = 41$

22.6 Reduction

A. Catalytic Reduction

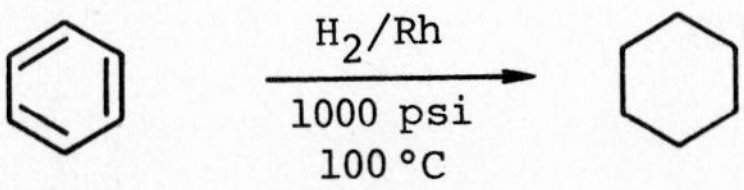

(can't stop short of complete hydrogenation)

B. Hydrogenolysis of Benzylic Groups

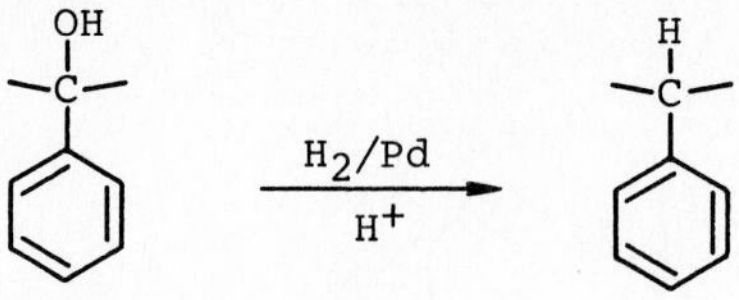

C. Birch Reduction (dissolving metal reduction)

alkyl-substituted benzenes:

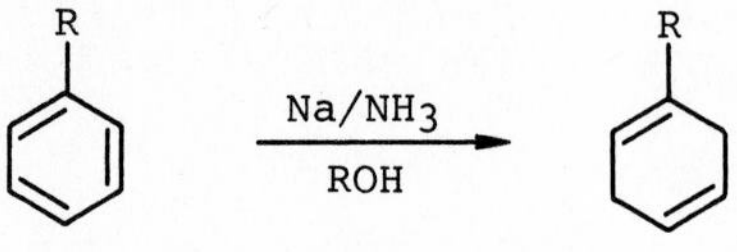

(via cyclohexadienyl anion)

alkoxy-substituted benzenes:

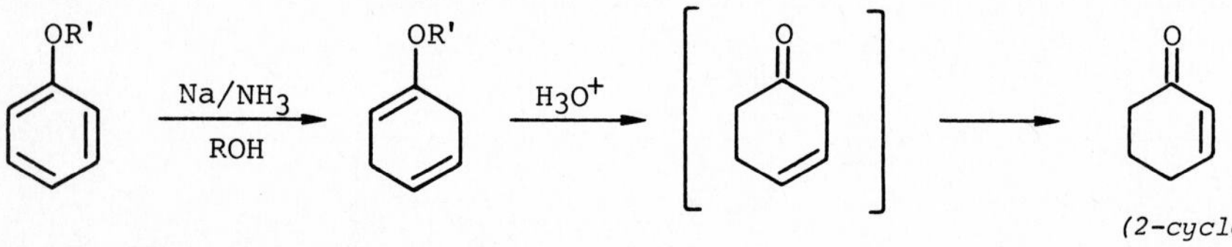

(2-cyclohexenone synthesis)

carbonyl-substituted benzenes:

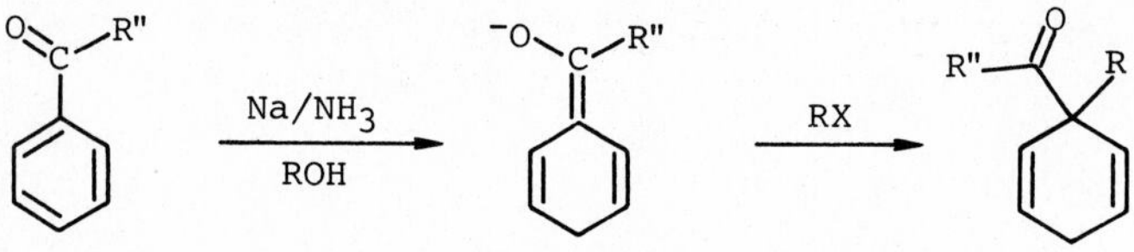

(R" = O^-, alkyl)

alkenyl-substituted benzenes (conjugated):

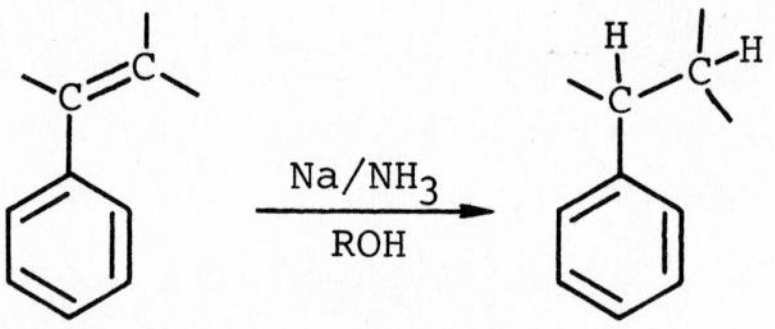

22.7 Aromaticity

A. Cyclooctatetraene: The Hückel 4n + 2 Rule

aromatic vs. non-aromatic or antiaromatic

B. Two-Electron Systems

cyclopropenyl cation

cyclopropenone

C. Six-Electron Systems

not cyclobutadiene (antiaromatic)

cyclopentadienyl anion

cycloheptatrienyl cation

D. Ten-Electron Systems

cyclononatetraenyl anion

E. Larger Cyclic π-Systems
annulenes

F. Metallocenes
ferrocene
sandwich compounds

22.B Important Concepts and Hints

This chapter introduces you to the benzene ring and some of its chemistry. You will notice in this and subsequent chapters that derivatives of benzene (aryl compounds) react quite differently from the alkyl compounds discussed in earlier chapters. To give you an overview of these differences, and a brief justification as to why they occur, we can divide the reactions into three groups: reactions of aryl σ-bonds, reactions on the ring which involve the π-electrons, and reactions on carbons directly attached to the ring (benzylic positions).

I. *Reactions of Aryl σ-Bonds:* Because such reactions involve an sp^2- instead of an sp^3-hybrid orbital, they are usually more difficult; for example, free radical halogenation is not successful with benzene.

II. *Reactions on the Ring which Involve the π-Electrons:* Hydrogenation of the π-system of the benzene ring is substantially more difficult than it is for the π-bond of an alkene. Furthermore, electrophilic addition, which is so important for alkenes, does not occur in aryl compounds except under unusual conditions. This behavior reflects the extra stability, known as aromatic stabilization, of the benzene ring's π-system. In the next chapter, you will see that *substitution* is the most important reaction of electrophiles with aromatic systems, rather than addition.

The Birch reduction is an important reaction of aryl compounds (and acetylenes) which simple alkenes do not undergo. At first glance, the fact that aryl compounds undergo the Birch reduction while alkenes are stable to such conditions would seem to conflict with the idea that the benzene π-system is more stable than that of an alkene. However, the first, and hardest, step in these reductions is the addition of an electron to the lowest-energy antibonding molecular orbital. For benzene, this orbital is lower in energy than that of an alkene, and benzene can accept the electron more readily. It is not until the next step that aromatic stabilization of the π-system is lost by protonation.

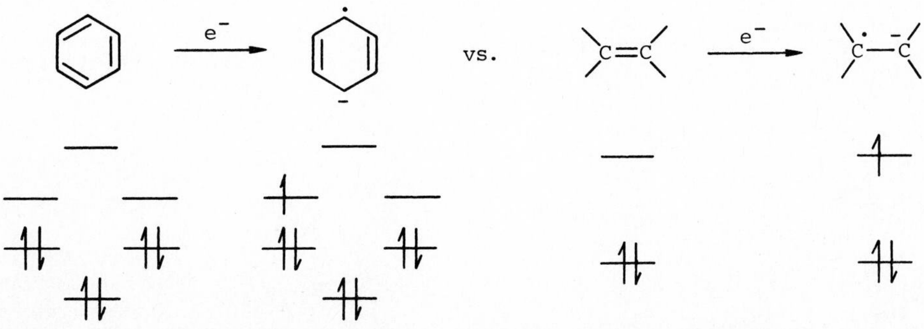

III. *Reactions Occurring at Benzylic Positions:* This class of reactions receives the most attention in Chapter 22, and it serves to tie in chemistry that you have learned from previous chapters with aromatic compounds. Because the π-system of the benzene ring can stabilize a p-orbital at the benzylic position via conjugation, it makes virtually every reaction at such a position easier (i.e., faster) that the alkyl counterpart. Reactions which involve cations (S_N1 reactions; oxidation of alkyl side chains), radicals (free radical halogenations; MnO_2 oxidation of benzylic alcohols; hydrogenolysis of benzylic groups), anions (acidity), or even sp^2-hybridized transition states (S_N2 reactions) are all accelerated by the overlap of the π-system of the ring with the p-orbital on the benzylic carbon. This overlap is depicted in Figure 22.12 for the benzyl radical.

The subject of aromaticity extends beyond derivatives of benzene alone, as pointed out in the last part of this chapter. Hückel's 4n+2 rule, and our more sophisticated understanding of the molecular orbital interactions which underly it, help to explain the special stability or instability of many cyclic conjugated systems. These range from the cyclopentadienyl anion and cycloheptatrienyl cation to cyclobutadiene and the aromatic heterocyclic molecules which you will encounter at the end of the course.

22.C Answers to Exercises

(22.1.D)

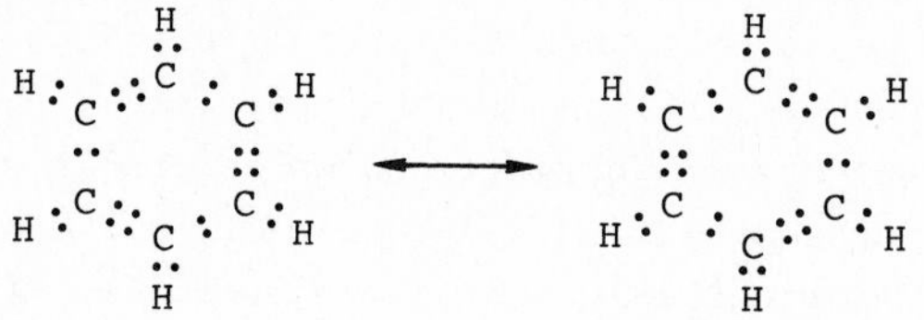

(22.2.A)

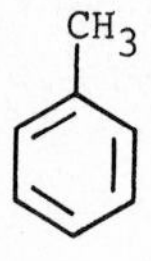

toluene
(methylbenzene)

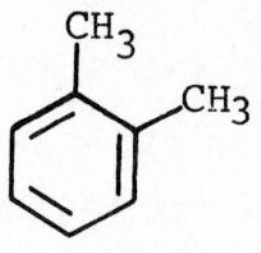

o-xylene
(1,2-dimethylbenzene)

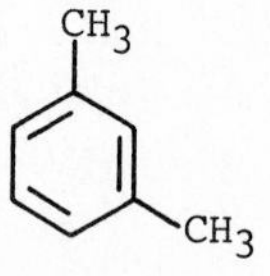

m-xylene
(1,3-dimethylbenzene)

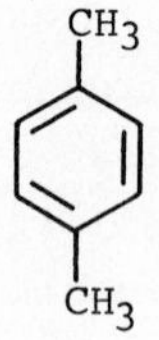

p-xylene
(1,4-dimethylbenzene)

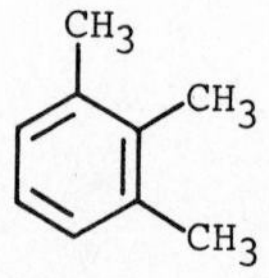

1,2,3-trimethylbenzene
(hemimellitene)

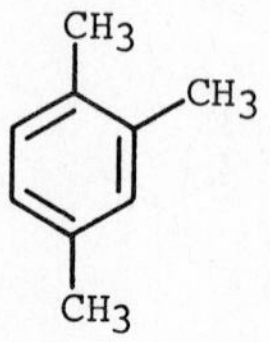

1,2,4-trimethylbenzene

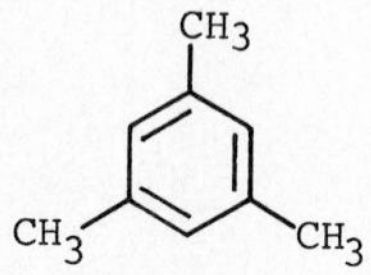

1,3,5-trimethylbenzene
(mesitylene)

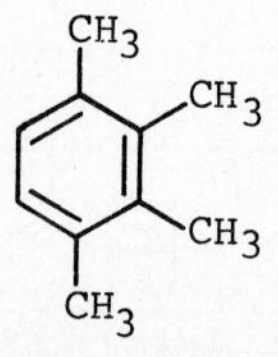

1,2,3,4-tetramethylbenzene
(prehitnene)

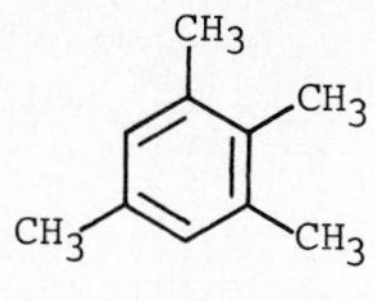

1,2,3,5-tetramethyl-benzene
(isodurene)

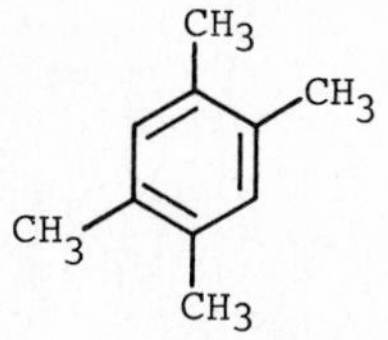

1,2,4,5-tetramethylbenzene
(durene)

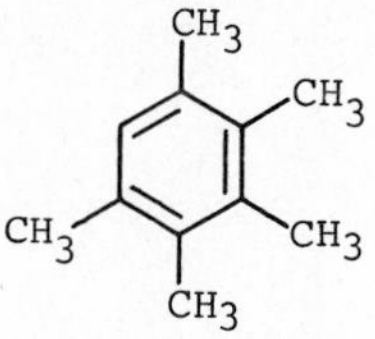

pentamethylbenzene

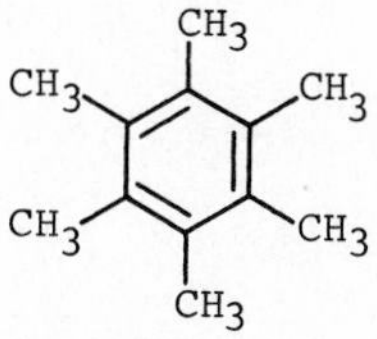

hexamethylbenzene

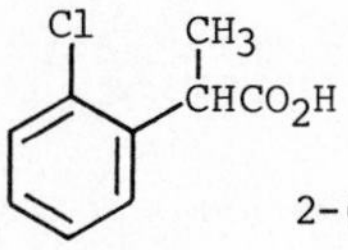

2-(2-chlorophenyl)-propanoic acid

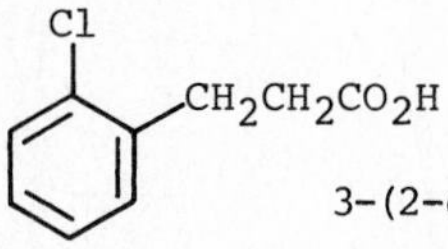

3-(2-chlorophenyl)propanoic acid

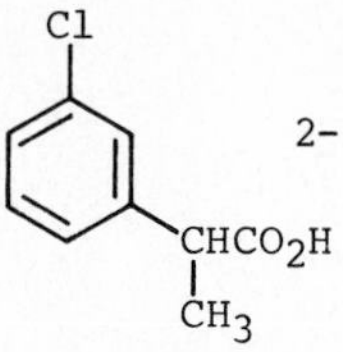

2-(3-chlorophenyl)-propanoic acid

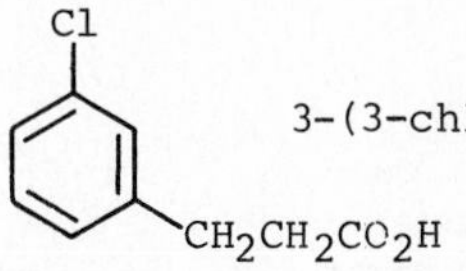

3-(3-chlorophenyl)propanoic acid

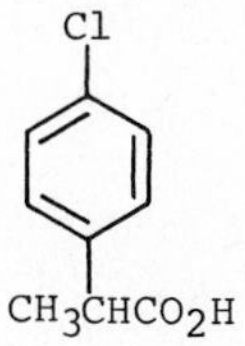

2-(4-chlorophenyl)-propanoic acid

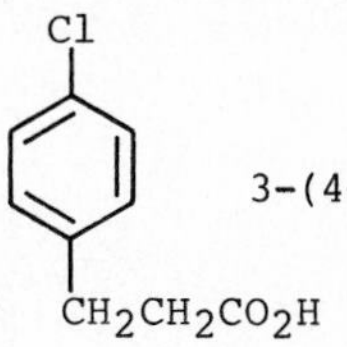

3-(4-chlorophenyl)propanoic acid

(common names: α-(o-chlorophenyl)propionic acid, etc.)

(22.2.B)

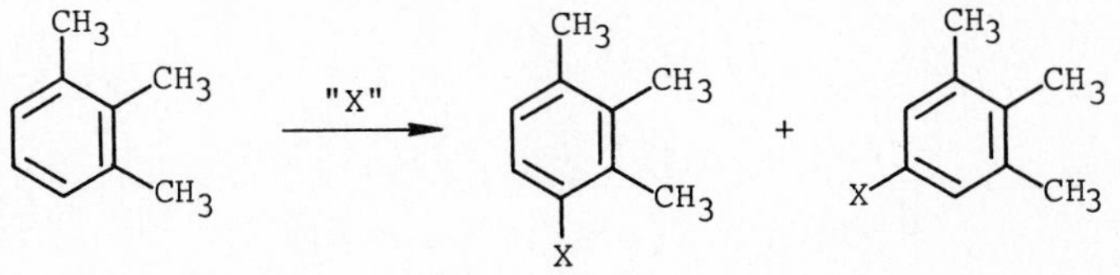

two possible products

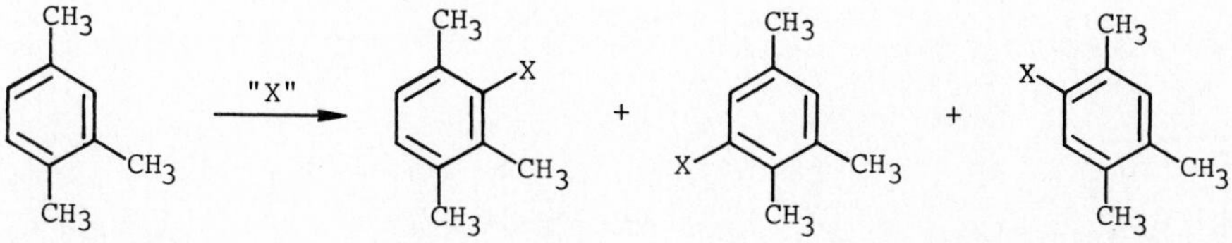

three possible products

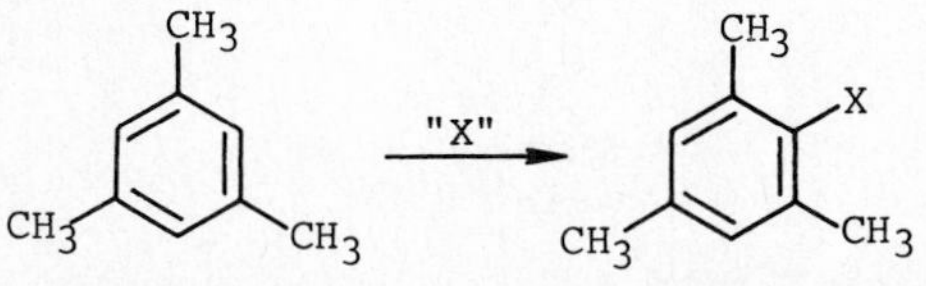

only one product is possible

(22.3.A)

(a) 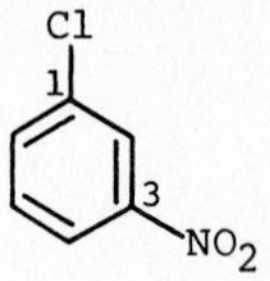

nmr: 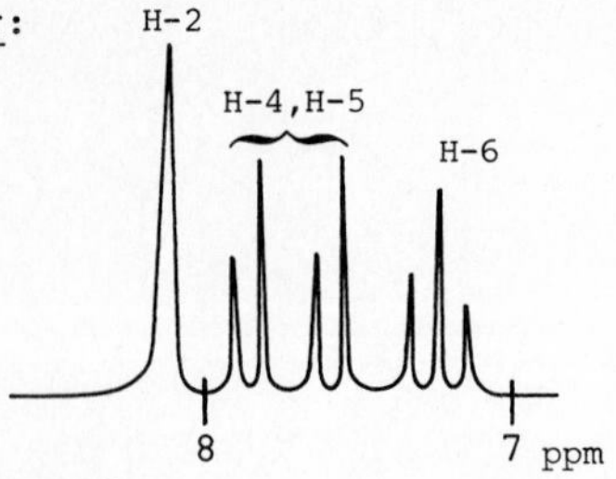

cmr:
six lines between 125 and 150 ppm.

(On a 60-MHz instrument,
there would be overlap, and a significantly more complex spectrum.)

(b) 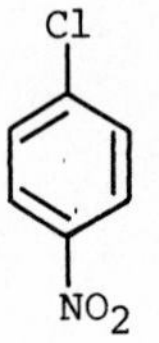

nmr: 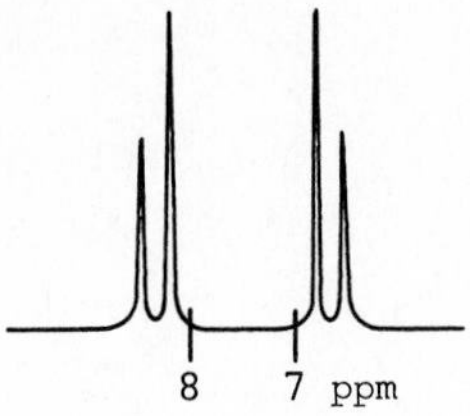

cmr:
four lines between 125 and 150 ppm

(22.4) 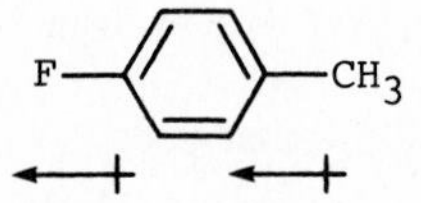

1.63 D + 0.37 D = 2.00 D (actual value = 2.01 D)

(22.5.A) 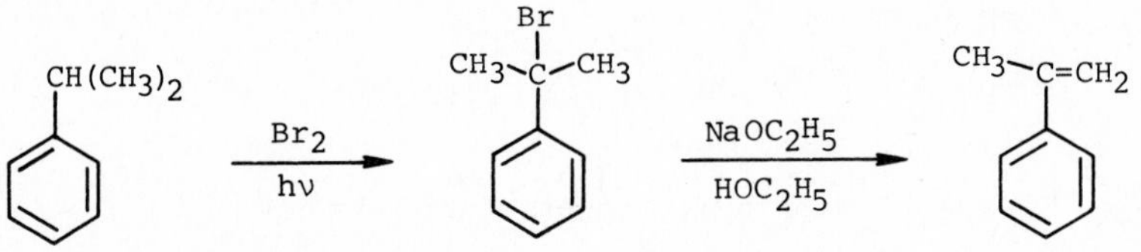

(22.5.B) 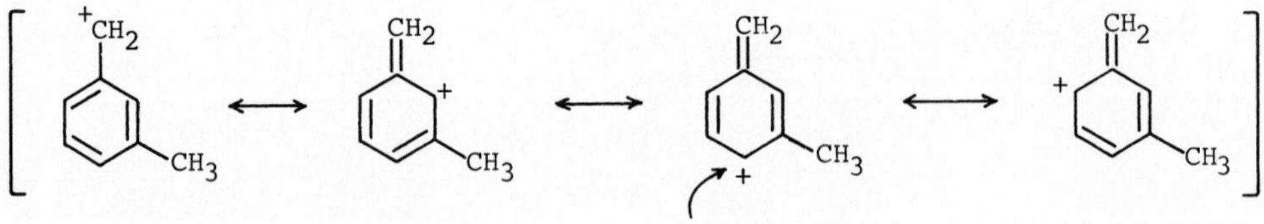

"secondary" cation (less stable)

Relative stability:	nearly equal	nearly equal	"secondary" cation (less stable)	nearly equal

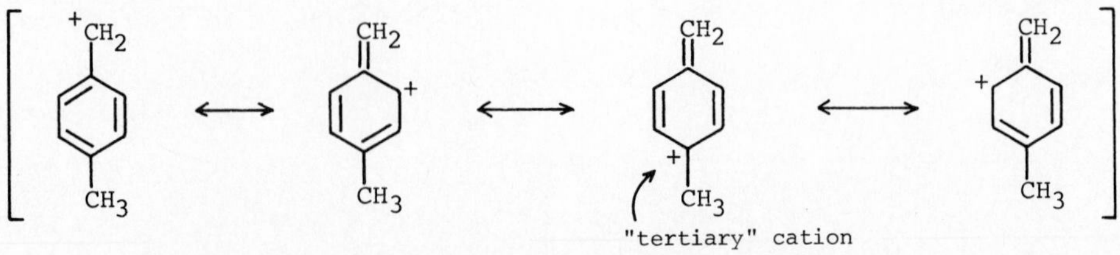

"tertiary" cation (more stable)

The p-methylbenzyl cation is more stable.

(22.5.C)

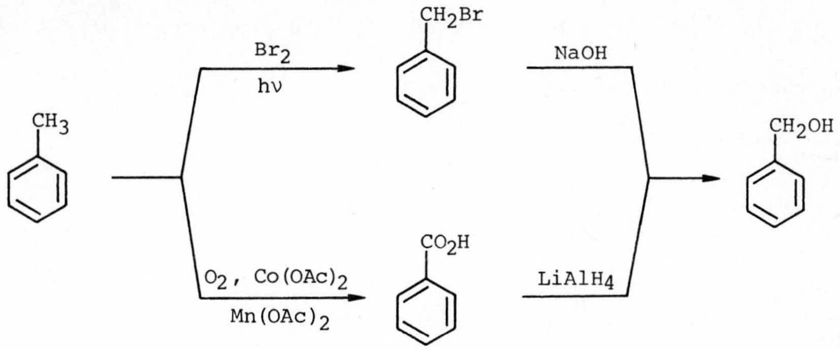

(22.6.C)

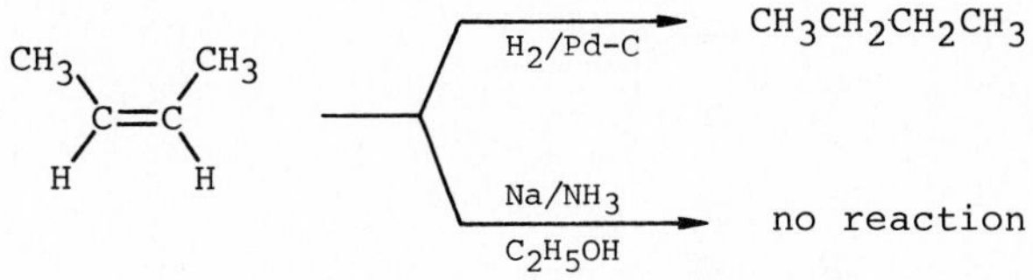

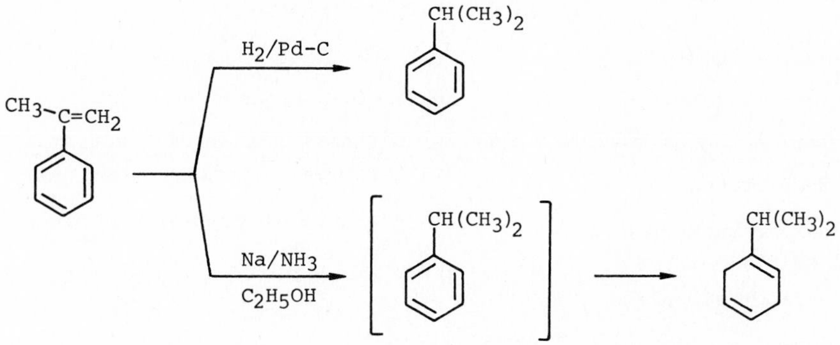

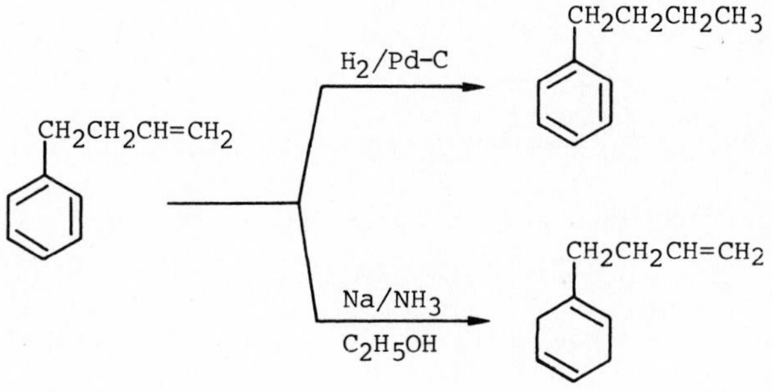

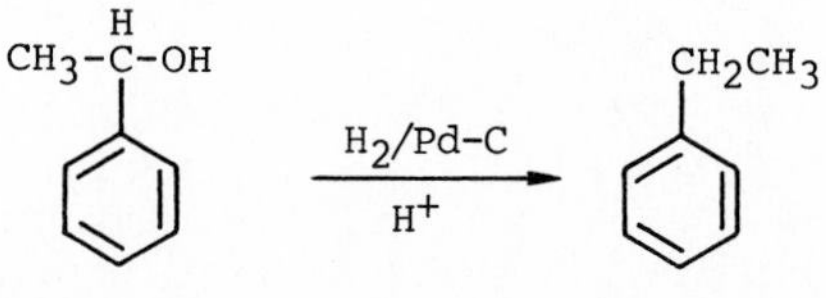

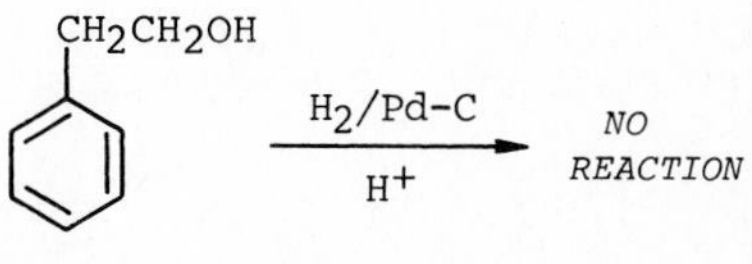

(22.7.C)

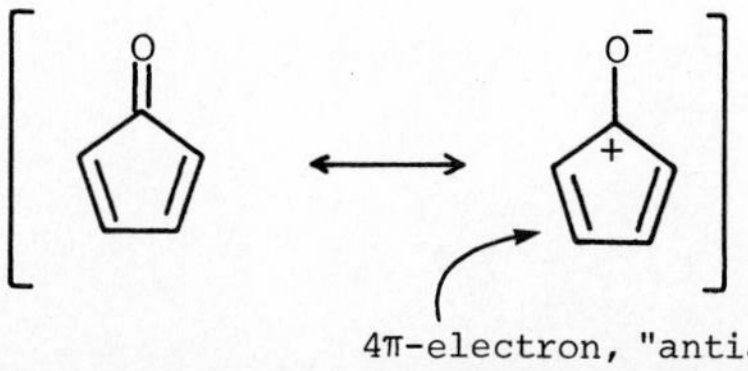

4π-electron, "antiaromatic"
system; <u>very</u> reactive

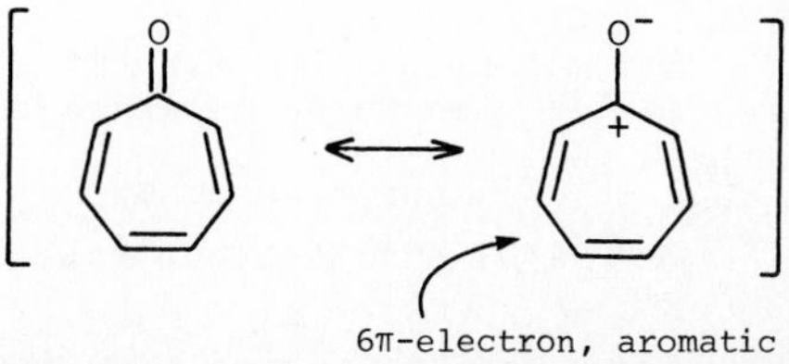

6π-electron, aromatic
system; stabilized

(22.7.E) (b) (42 electrons); (d) (6 electrons); (e) (26 electrons); (f) (10 electrons)

(22.7.F) All adhere to the 18-electron rule.

22.D Answers and Explanations for Problems

1. (a) 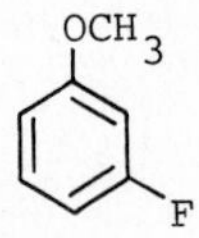

(b) 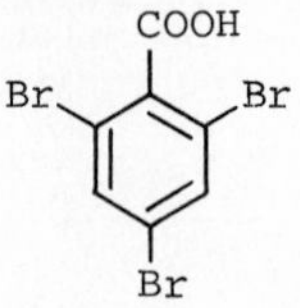

(c) 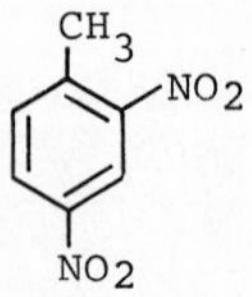

(d) 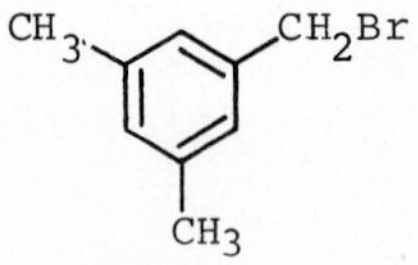

(e) 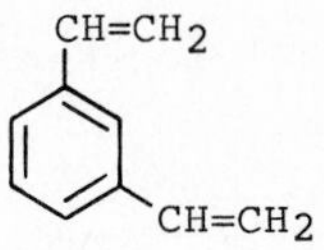

(f) 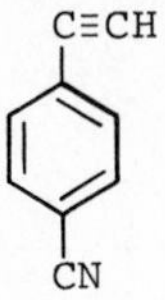

(g) 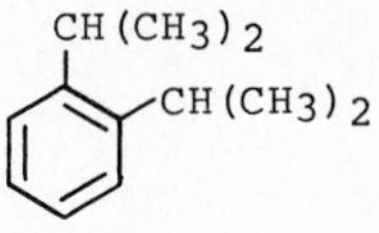

(h) 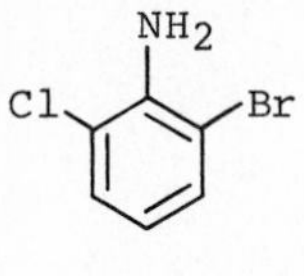

2. (a) 1,1,1-trichloro-2,2-di(4-chlorophenyl)ethane
(the abbreviation "DDT" comes from the old,non-systematic name: dichlorodiphenyltrichloroethane)

(b) p-bromopropylbenzene

(c) 3-bromo-4-iodocumene

(d) 2-bromo-3-ethyltoluene

(e) 2-(p-nitrophenyl)butane

(f) 2-(m-chlorophenyl)-3-methylbutane

(g) diphenyl-p-tolylcarbinol or diphenyl-(4-methylphenyl)methanol

(h) p-bromochlorobenzene

(i) 4-bromo-3-fluoro-2-iodotoluene

(j) m-methoxybenzaldehyde or m-anisaldehyde

(k) 4-bromo-2,6-dimethylbenzoic acid

(ℓ) 1,2,4-trimethylbenzene

3. (a) No. There would be two isomers of the form

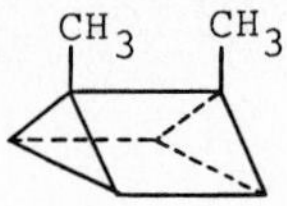

and

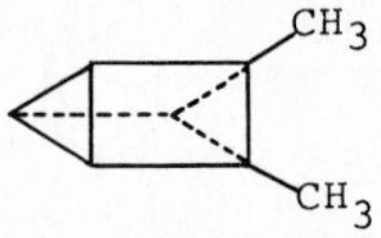

(b) The two structures are resonance structures as symbolized by

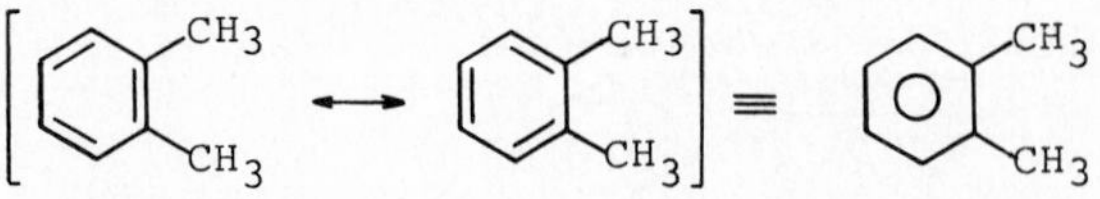

4. ΔH_f°

CH≡CH 54.3 kcal $mole^{-1}$

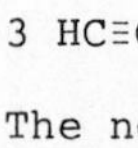 19.8 kcal $mole^{-1}$

3 HC≡CH ⇌ 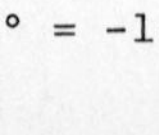 $\Delta H^\circ = -3(54.3)+19.8 = -143.1$ kcal $mole^{-1}$

The negative value for ΔS° reflects the loss in freedom of motion when three separate compounds form one.

$$\Delta G^\circ = \Delta H^\circ - T\Delta S^\circ$$

at 298°K $\Delta G^\circ = -143.1$ kcal $mole^{-1}$ $-(298$ deg. $\times -79.7$ cal $deg^{-1}mole^{-1})$

$\Delta G^\circ = -143.1$ kcal $mole^{-1} + 23.8$ kcal $mole^{-1} = -119.3$ kcal $mole^{-1}$

The equilibrium lies far to the right, but the probability is very small that three acetylenes can collide at the same time with the proper orientation for reaction.

5. (a) [cyclohexene structure] + H_2 $\longrightarrow$ [cyclohexane structure] $\Delta H° = -28.4$ kcal mole^{-1}

∴ [benzene structure] + $3H_2$ $\longrightarrow$ [cyclohexane structure] $\Delta H°_{calc.} = 3 \times (-28.4) = -85.2$ kcal mole^{-1}

$-85.2 - (-49.3) = -35.9$ kcal mole^{-1}

Empirical resonance energy = 35.9 kcal mole^{-1}

(b) [cyclooctene structure] + H_2 $\longrightarrow$ [cyclooctane structure] $\Delta H° = -23.3$ kcal mole^{-1}

[cyclooctatetraene structure] + $4H_2$ $\longrightarrow$ [cyclooctane structure] $\Delta H° = -100.9$ kcal mole^{-1}

$-93.2 - (-100.9) = +7.7$ kcal mole^{-1}

Empirical resonance energy = -7.7 kcal mole^{-1}

The positive value implies that the four double bonds in the tetraene are less stable than 4×1 double bonds; i.e., this value represents a <u>destabilization</u> energy or negative resonance energy. This means that not only is there no stabilization energy (resonance), but that cyclo-octatetraene is probably more strained because of the four double bonds.

(c) [benzene structure with H's labeled]

$6 \times E(C-H) = 6 \times 99 = 594$ kcal mole^{-1}
$3 \times E(C-C) = 3 \times 83 = 249$ kcal mole^{-1}
$3 \times E(C=C) = 3 \times 146 = \underline{438}$ kcal mole^{-1}
calc. $\Delta H°_{atom}$ $= 1281$ kcal mole^{-1}

Empirical resonance energy = 1318-1281 = 37 kcal mole^{-1}

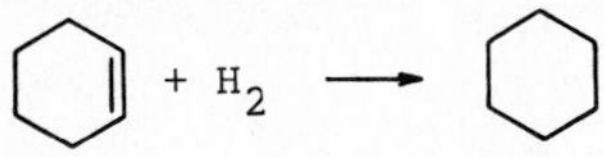

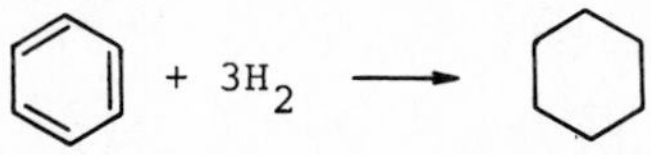

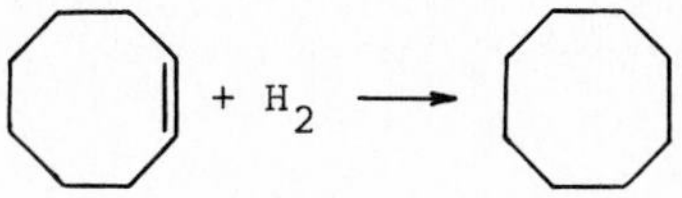

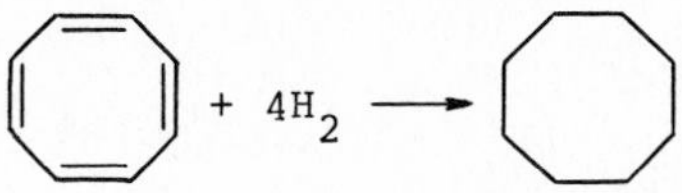

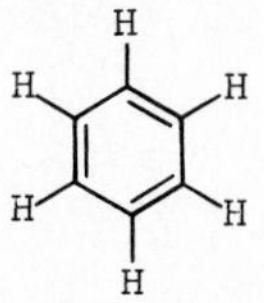

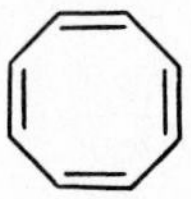

[cyclooctatetraene structure]

$8 \times 99 = 792$
$4 \times 83 = 332$
$4 \times 146 = \underline{584}$
        $= 1708$

Empirical resonance energy = 1713-1708 = 5 kcal mole^{-1}

(d) [cyclooctatetraene structure] + $10\, O_2$ $\longrightarrow$ $8CO_2 + 4H_2O$ $\Delta H°_C = -1054.7$ kcal mole^{-1}

$$\frac{\Delta H°_C}{n(CH)} = -\frac{1054.7}{8} = -131.8 \text{ kcal mole}^{-1}$$

est. $\Delta H°_C$ for $C_6H_6 = 6 \times (-131.8) = -790.8$ kcal mole^{-1}

Resonance Energy = 790.8-757.5 = 33.3 kcal mole^{-1}

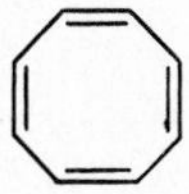

(e)

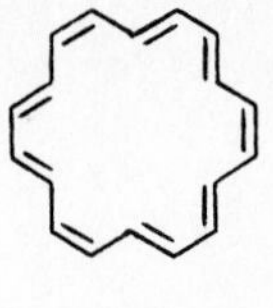

[18]annulene

Calculated empirical resonance energy =

$18 \times 99 = 1782$
$9 \times 83 = 747$
$9 \times 146 = 1314$
$\quad 3843$

$3890 - 3843 = 47 \text{ kcal mole}^{-1}$

The empirical resonance energy per CH:

$C_6H_6 \quad \frac{37}{6} = 6.17 \text{ kcal mole}^{-1}$ per CH

$C_{18}H_{18} \quad \frac{47}{18} = 2.61 \text{ kcal mole}^{-1}$ per CH

Although the *total* resonance energy of [18]annulene is greater than that of benzene itself, the resonance energy per CH is not quite one-half that of benzene.

6.
$$C_6H_6 + Cl\cdot \longrightarrow C_6H_5\cdot + HCl$$
$\Delta H^\circ_f =$ 19.8   28.9   80   −22.1 $\Delta H^\circ = +9 \text{ kcal mole}^{-1}$

$$C_6H_5\cdot + Cl_2 \longrightarrow C_6H_5Cl + Cl\cdot$$
$\Delta H^\circ_f =$ 80   0   12.2   28.9 $\Delta H^\circ = -39 \text{ kcal mole}^{-1}$

Although the overall reaction is highly exothermic, the first step is endothermic and will not support a chain reaction. Instead of reacting with benzene, the chlorine atoms will eventually react with each other.

7. The nmr spectrum clearly shows an ethyl group (triplet methyl and quartet methylene), and only aromatic protons. From the relative areas of the peaks ($CH_3 = 3$, $CH_2 = 2$, Ar-H = 2), it must be a diethylbenzene. The important absorption in the ir spectrum is the single strong band at 760 cm^{-1}, indicative of *ortho* substitution (Table 22.2); hence, o-diethylbenzene.

(*NOTE:* the alkyl groups have little effect on the chemical shift of the ring protons, so the formally non-equivalent Ar-H's resonate at the same frequency and appear as a singlet.)

8. One xylene (m-) is given by three different Br isomers:

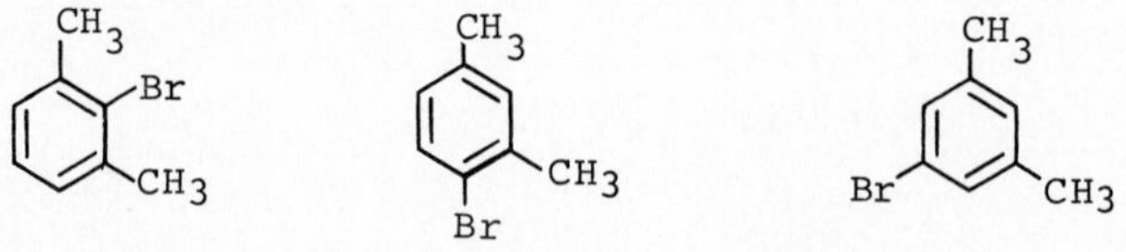

One (o-) is given by two isomers:

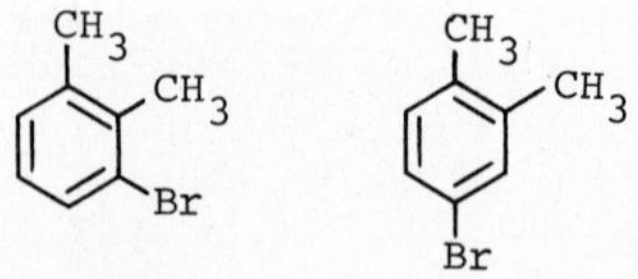

One (p-) is given by only a single isomer:

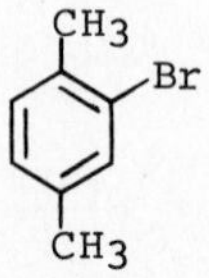

This method also serves to establish the structure of that bromoxylene which gave the unique p-xylene. Note how this method is simply a variation of the Körner absolute method.

9. (a) 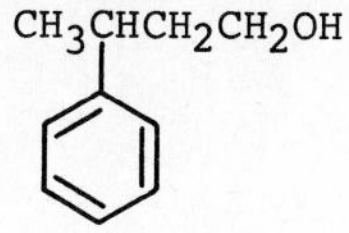

(b) 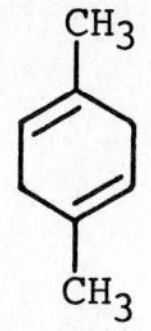

(c) 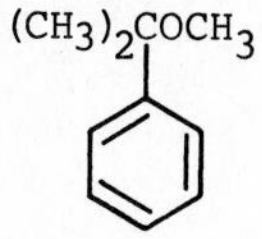

(d) 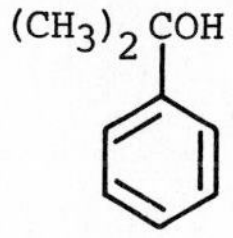

(e) 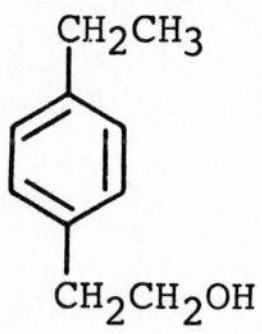

(f) 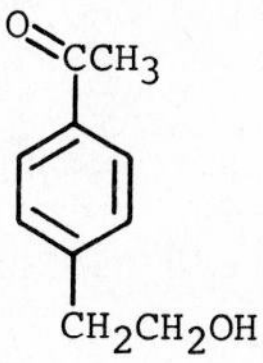

(g) 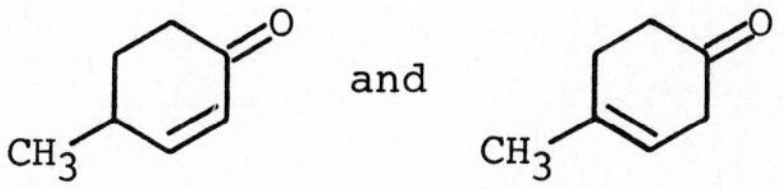

10.

(a) 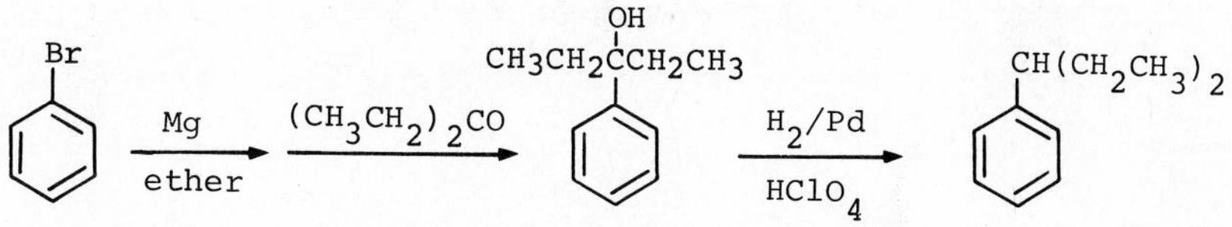

(b) 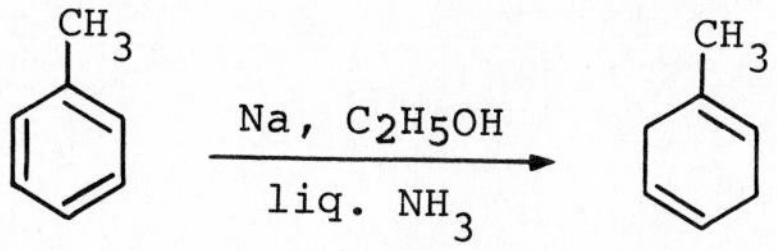

(c) 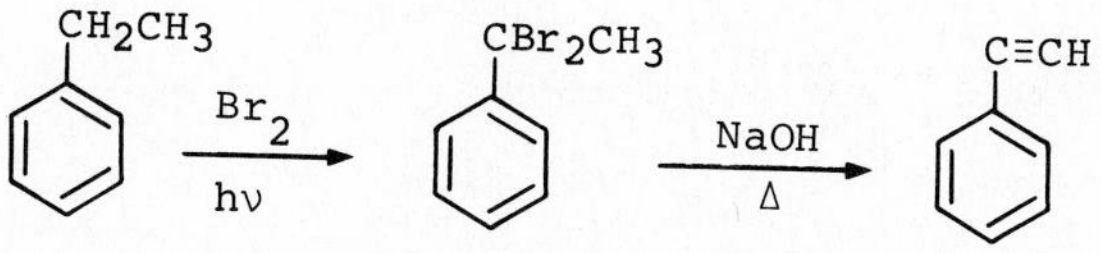

(d) 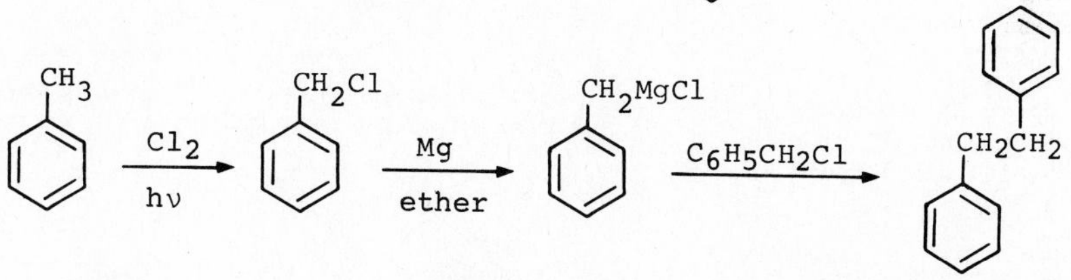

(e) 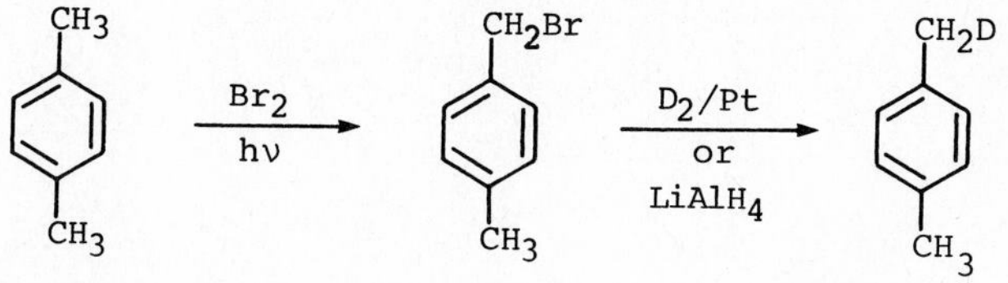

(f) 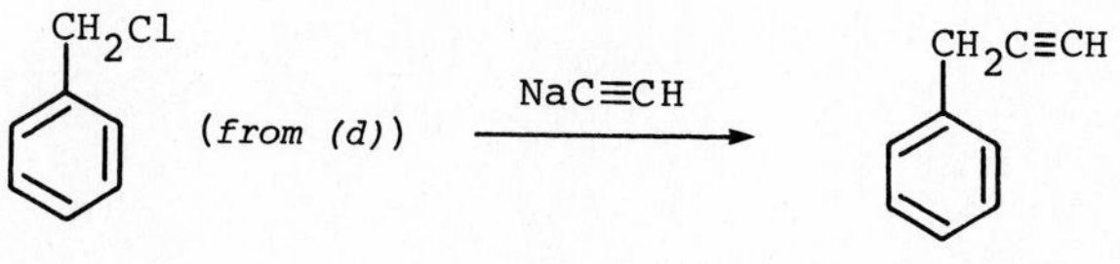

11. $Br_2 \longrightarrow 2\ Br^{\cdot}$

$\Delta H°$

-0.5 kcal mole^{-1}

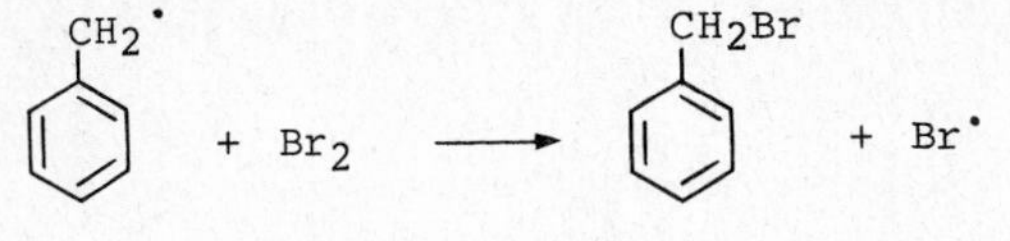

-11 kcal mole^{-1}

both steps are exothermic

From the Table of Bond Dissociation Energies:

DH° (benzyl-H) = 87 DH° (H-Br) = 87.5

DH° (benzyl-Br) = 57

From Tables of Heats of Formation, we calculate:

$Br_2 \longrightarrow 2\,Br\cdot$ DH° = 46

7.4 2 × 26.7

For ethane: $CH_3CH_2{-}H + Br\cdot \longrightarrow CH_3CH_2\cdot + HBr$ $\Delta H° = +10.5$ kcal mole^{-1}

This step is quite endothermic and so is relatively slow.

For isobutane: $(CH_3)_3CH + Br\cdot \longrightarrow (CH_3)_3C\cdot + HBr$ $\Delta H° = +3.5$ kcal mole^{-1}

This step is slightly endothermic, but still feasible.

However, bromination of toluene should be faster than for either of the alkanes.

12.

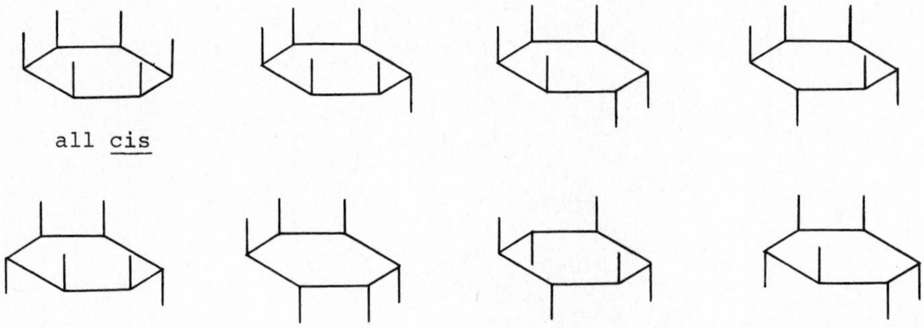

In the above drawings, only Cl bonds are shown. Each position also has a hydrogen.

slowest in E_2 eliminations, since there are no H's anti to a Cl

chiral; all others have a plane of symmetry

13.

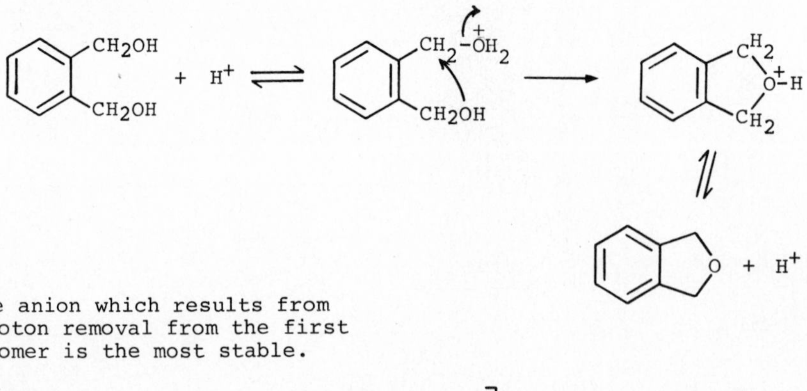

14. The anion which results from proton removal from the first isomer is the most stable.

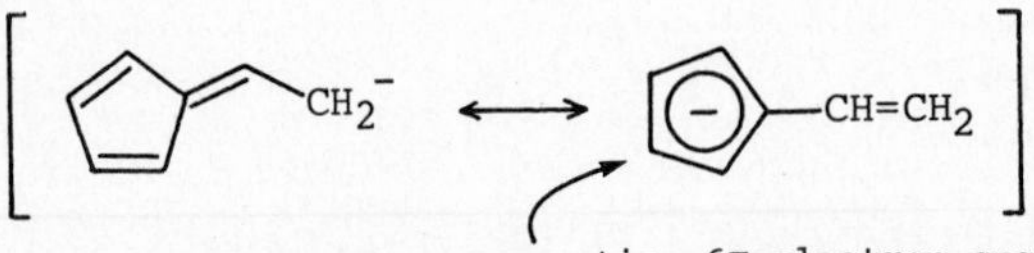

aromatic, 6π-electron cyclopentadienyl anion

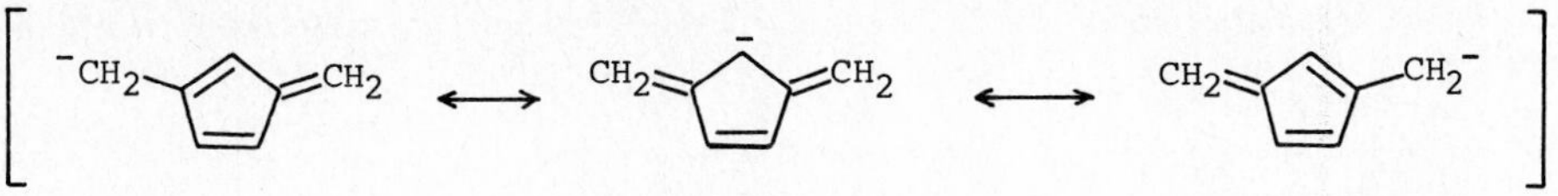

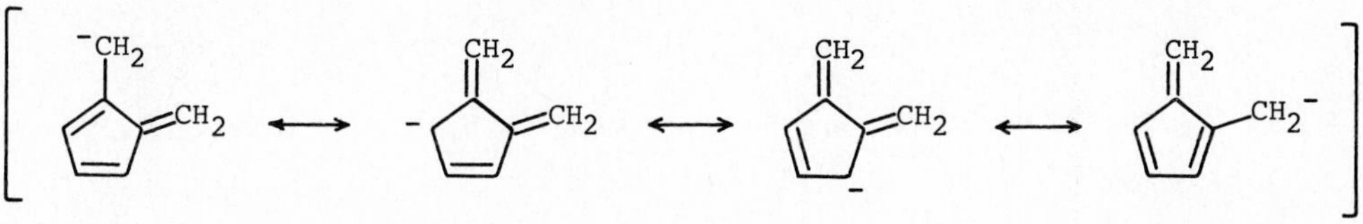

15. To satisfy the 18-π electron rule:

(a) x = 2 ; (b) x = 3 ; (c) x = 2 ($C_6H_5C{\equiv}CC_6H_5$ is coordinated via a π-complex with the acetylene; 2 electrons)

(d) x = 3 ; (e) x = 4 ; (f) x = 3 ; (g) x = 2

16. (a)

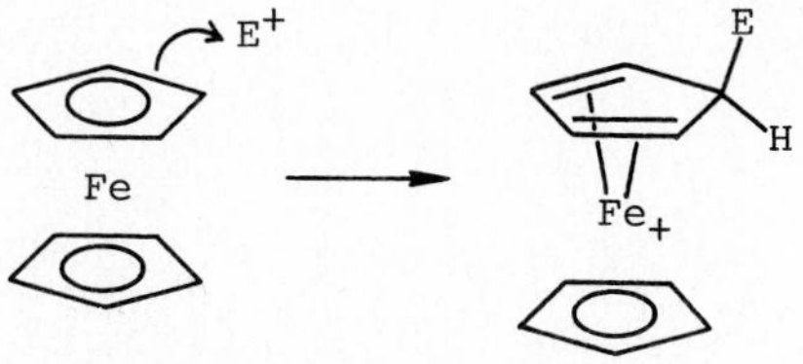

d-electrons	
Fe:	8
positive charge:	-1
diene ligand:	4
cyclopentadienyl:	5
	16 electrons

(b) Same ring:

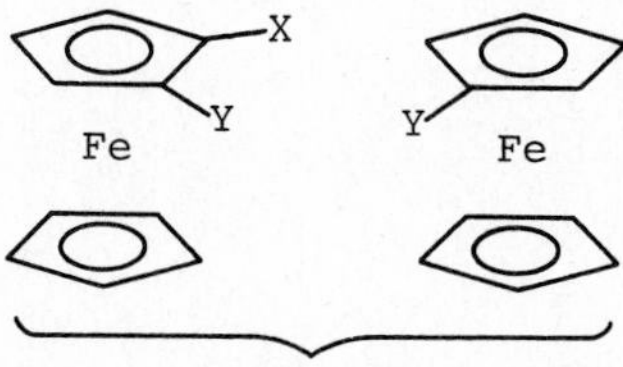

(chiral)

Different rings:

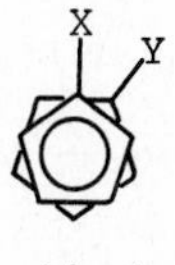

chiral

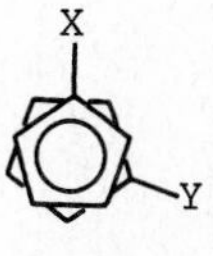

chiral

achiral

These will all interconvert with free rotation

22.E Supplementary Problems

S1. Provide an acceptable name for each of the following compounds.

(a)

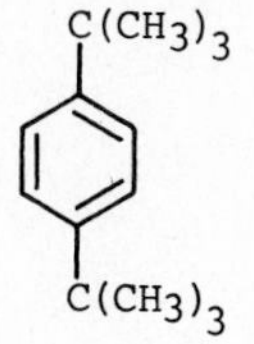

(b)

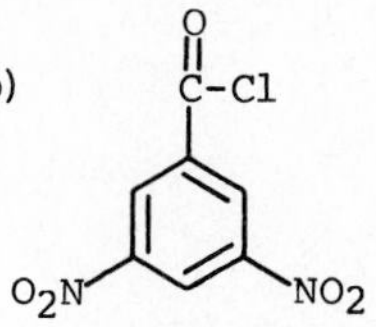

(c)

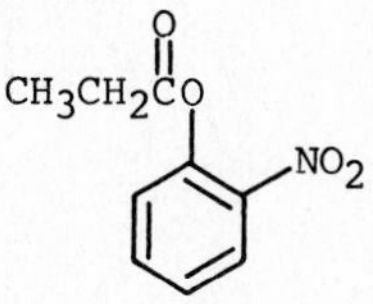

(d)

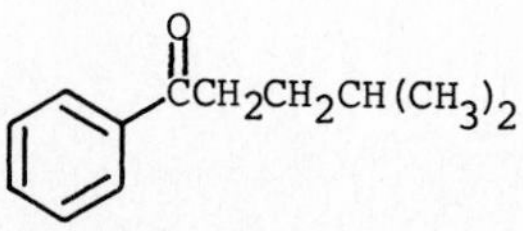

(e)

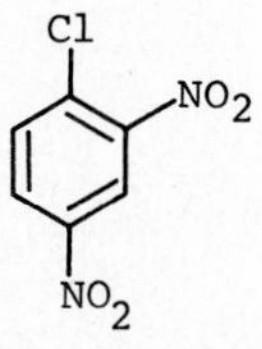

(f)

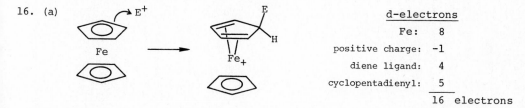

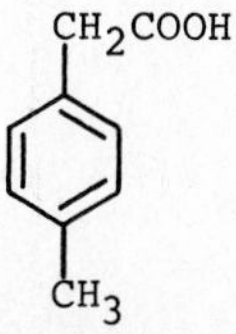

S2.   Predict the major product from each of the following reaction sequences:

(a)

    $\dfrac{Br_2}{h\nu}$ $\dfrac{H_2O}{\Delta}$ $\xrightarrow{MnO_2}$

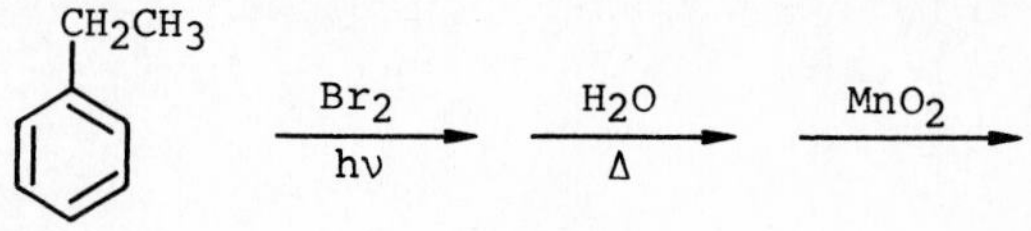

(b)

    $\xrightarrow{Mg}$ $\xrightarrow{H_2O}$ $\dfrac{H_2/Pd}{H^+}$

(c)

    $\dfrac{Na/NH_3}{C_2H_5OH}$ $\xrightarrow{H_3O^+}$

(d)

    $\xrightarrow{Li}$ $\xrightarrow{CH_3C\equiv N}$ $\xrightarrow{H_3O^+}$

(e)

    $\dfrac{KMnO_4}{\Delta}$ $\dfrac{CH_3OH}{H^+}$

(f)

    $\dfrac{H_2NNH_2}{KOH,\Delta}$ $\dfrac{Br_2}{h\nu}$ $\xrightarrow{Mg}$ $\xrightarrow{CO_2}$

(g)

    $\dfrac{Pt}{\Delta}$

(h)

    $\dfrac{D_2/Pd}{D^+}$

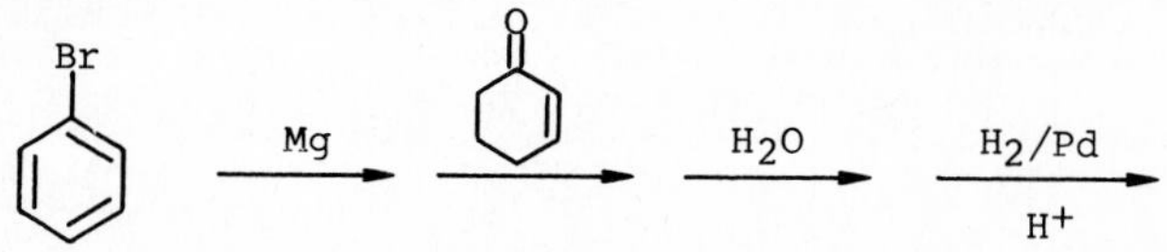

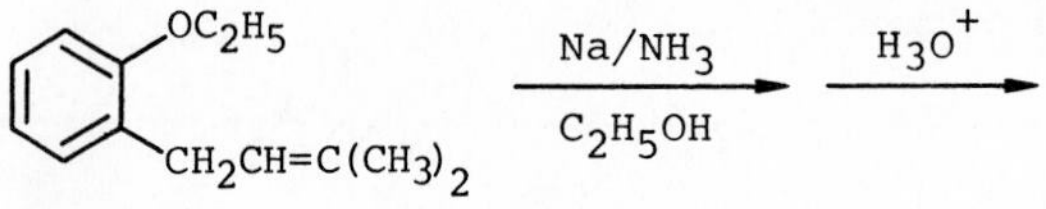

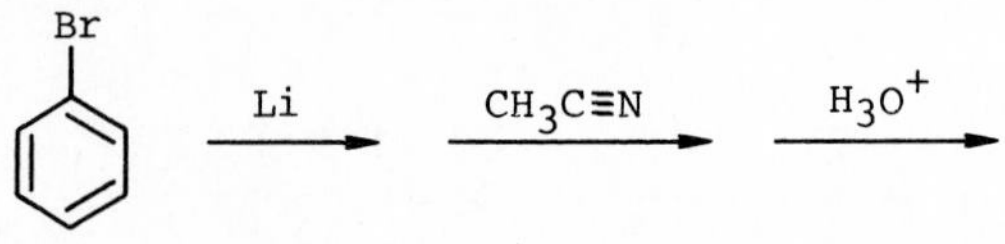

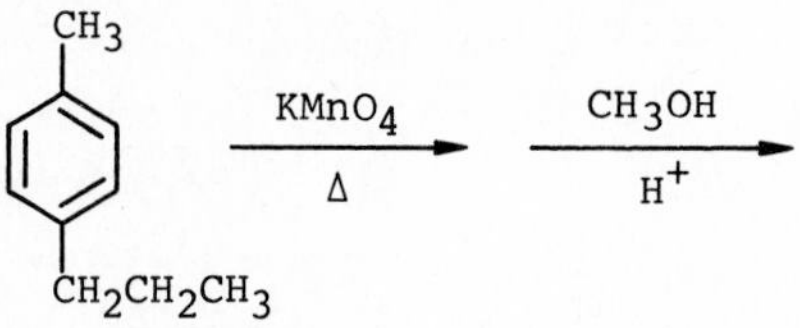

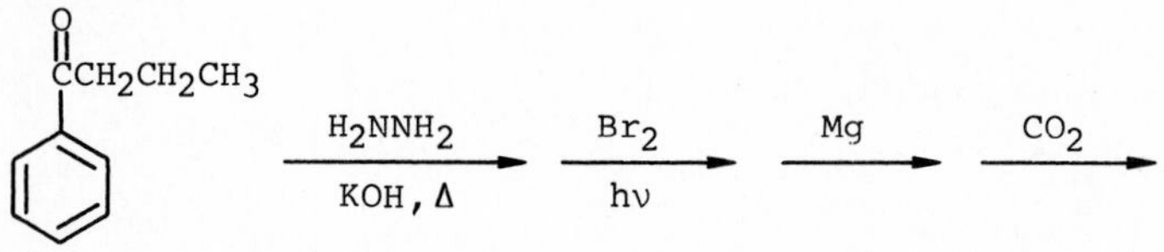

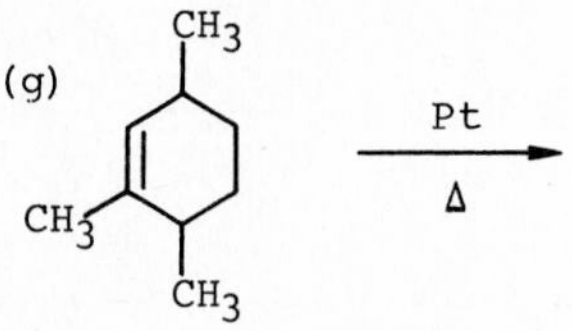

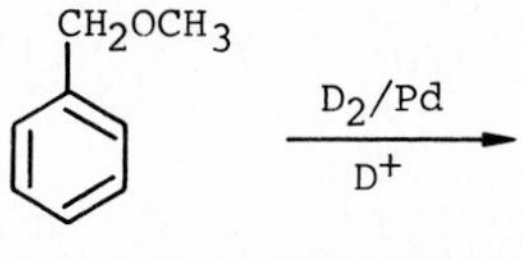

S3.   Predict the major product from reaction of *cis*-1-phenylpropene with each of the following reagents, and explain your choice.

    (a)   HCl   (c)   HBr, peroxides

    (b) Br_2/CH_3OH (d) $Hg(OAc)_2/H_2O$; then $NaBH_4$

S4.   Which of the two compounds illustrated below is more acidic?   Why?

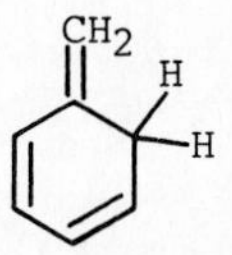

S5. Which of the following retro-Diels-Alder reactions will take place most easily? Which one will be the most difficult? Why?

(a)

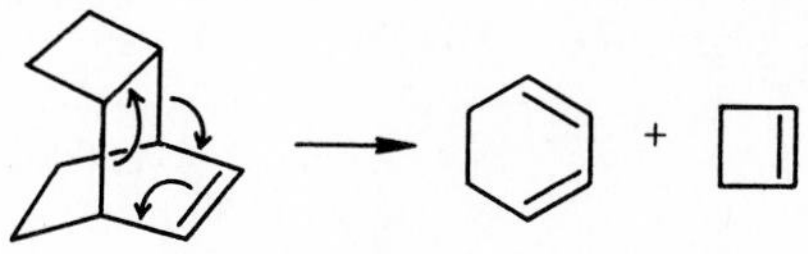

(b)

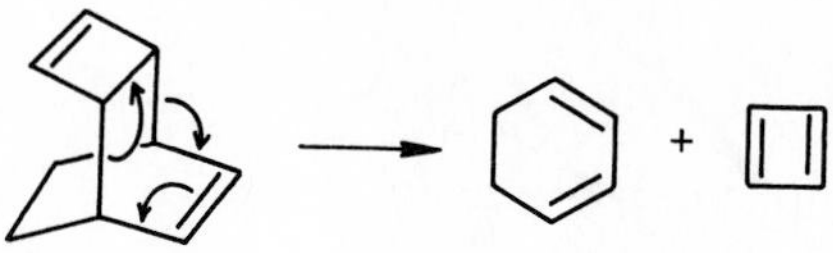

(c)

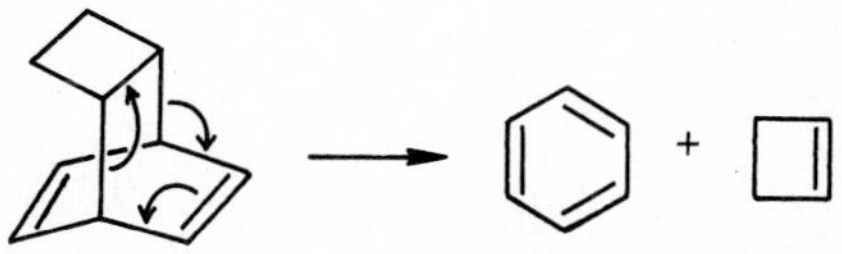

S6. The stabilization that a phenyl group provides to a radical center can be determined by comparing the bond dissociation energies of $C_6H_5CH_2$-H and CH_3-H bonds.

(a) How does this stabilization compare with that provided by a vinyl group in the allyl radical?

(b) Perform the same comparison for the cations *(see Sections 20.1.A and 22.5.B in the text.)*

(c) Explain any significant difference you see in radical vs. cation stabilization.

S7. Benzylic alcohols and ethers are cleaved under the conditions of the Birch reduction:

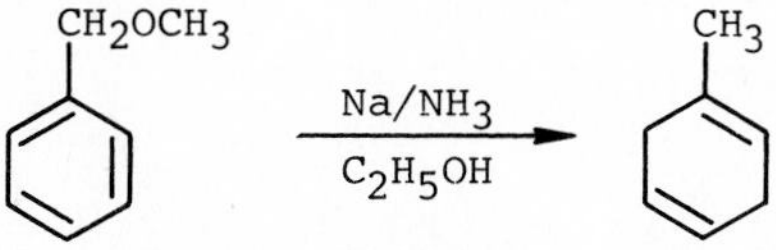

(a) Write a mechanism for this reaction.

(b) Explain why the following ether undergoes Birch reduction without cleavage:

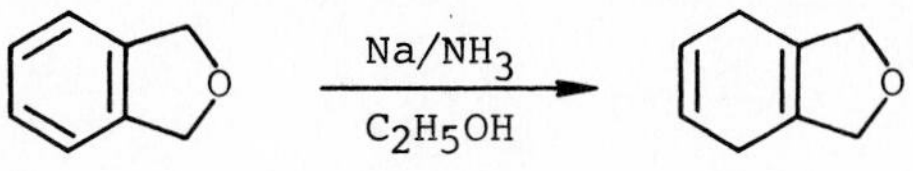

S8. Rank the following compounds in order of acidity.

(a)

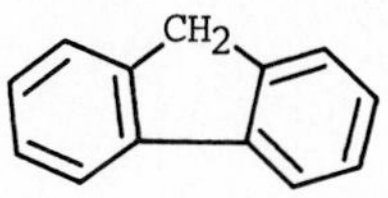

(b)

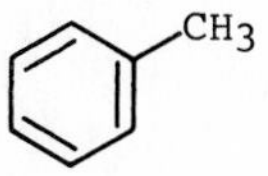

(c)

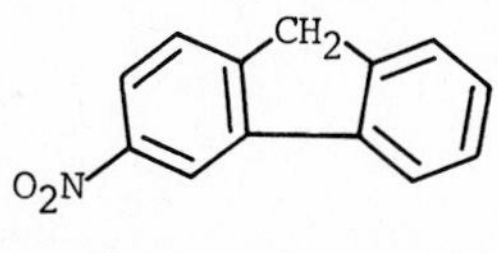

(d)

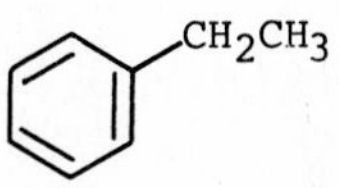

(e)

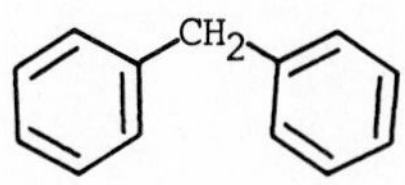

S9. Explain why 7-bromo-1,3,5-cycloheptatriene is extremely rapidly hydrolyzed.

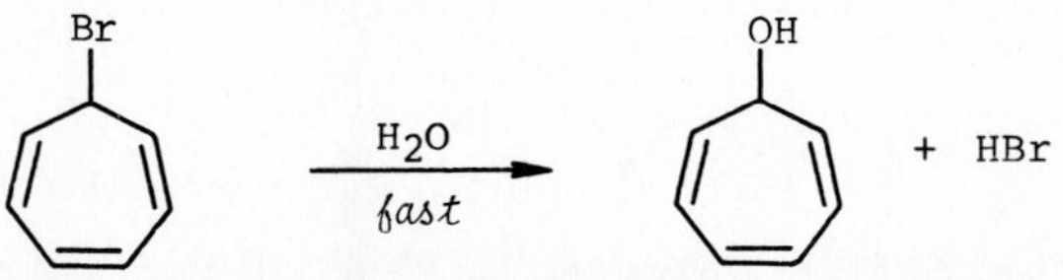

S10. *Para*-methoxybenzyl bromide, illustrated below, hydrolyzes in water many times faster than the *meta* isomer. Explain why this is so.

$$CH_3O\!-\!\!\langle\!\!\langle\,\rangle\!\!\rangle\!-\!CH_2Br + H_2O \longrightarrow CH_3O\!-\!\!\langle\!\!\langle\,\rangle\!\!\rangle\!-\!CH_2OH + HBr$$

22.F Answers to Supplementary Problems

S1. (a) p-di-t-butylbenzene

(b) 3,5-dinitrobenzoyl chloride

(c) o-chlorophenyl propionate, or 2-chlorophenyl propanoate

(d) 4-methyl-1-phenyl-1-pentanone

(e) 1-chloro-2,4-dinitrobenzene

(f) p-methylphenylacetic acid

S2.

(a)

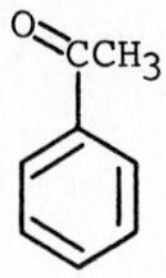

(b)

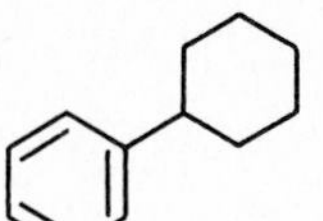

(c)

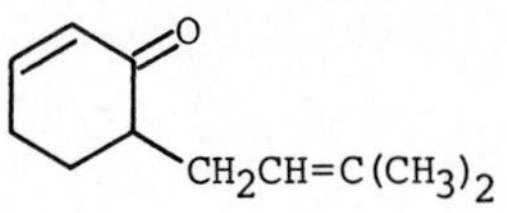

(d)

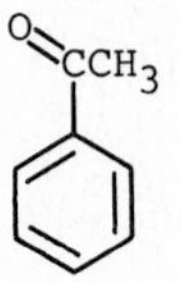

(e)

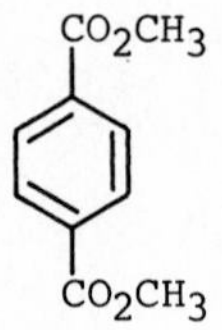

(f) $HO_2CCHCH_2CH_2CH_3$

(g)

(h) CH_2D

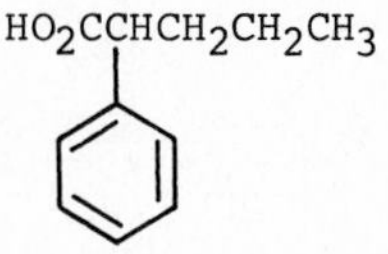

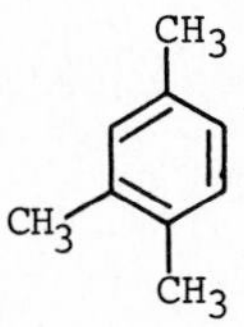

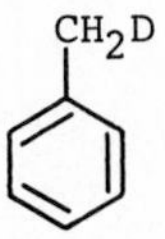

S3. (a)

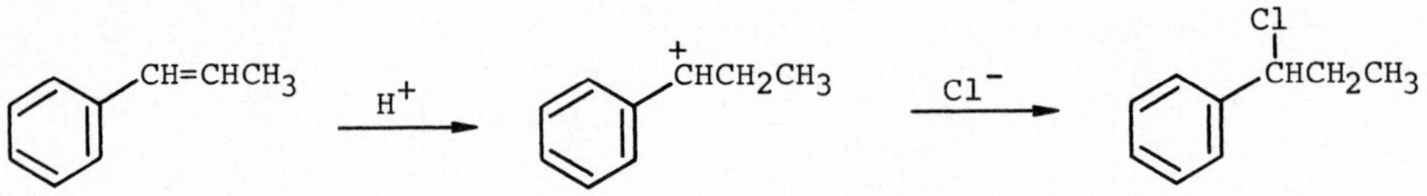

(b)

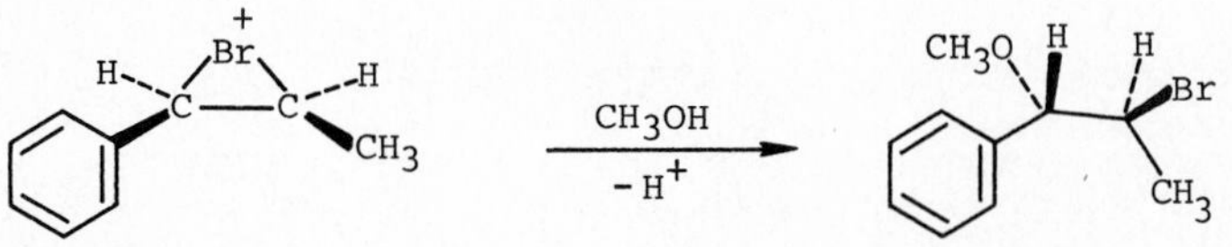

(c)

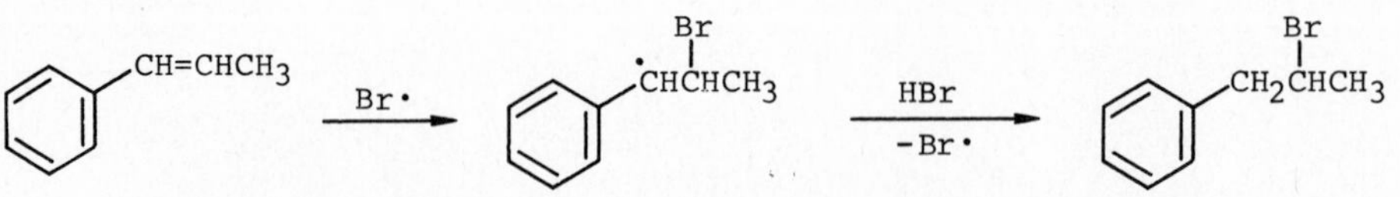

(d)

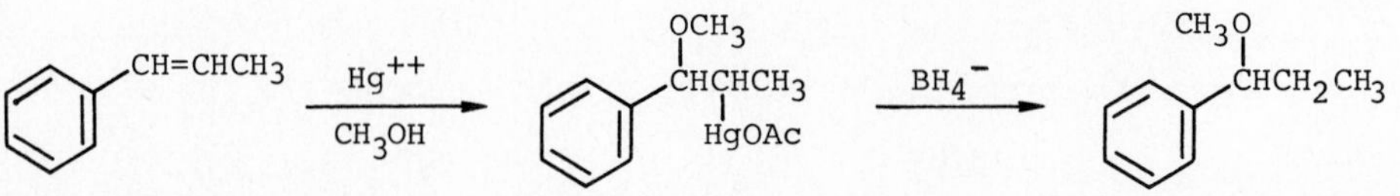

In (a), (b), and (d), the additions proceed in the Markovnikov sense (the most stable carbocation is formed); in (c), Br· adds so as to afford the most stable radical. Phenyl stabilizes radicals and carbocations better than a simple alkyl group does.

S4.

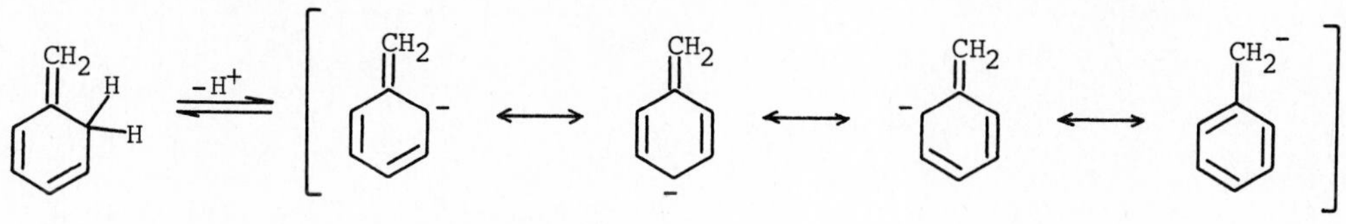

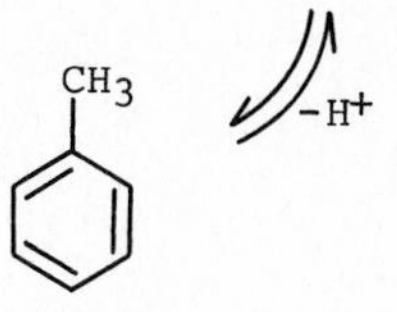

Both isomers ionize to give the same carbanion, so the difference in equilibrium will depend only on the difference in stability of the starting materials. Toluene is more stable than the methylenecyclohexadiene isomer, and therefore it will be less acidic.

S5. Reaction (c) will occur most rapidly because the aromatic stabilization of benzene is gained. Reaction (b) will be the most difficult because the very unstable (high-energy) cyclobutadiene is being formed. In reaction (a), no aromatic or antiaromatic compounds are being produced.

S6. (a)

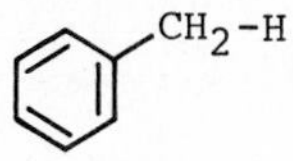

	$C_6H_5CH_2$–H	CH_3–H	CH_2=CH–CH_2–H	
DH° =	87	104	87	kcal $mole^{-1}$
Stabilization:	17		17	kcal $mole^{-1}$

Stabilization of a radical is about the same for phenyl and vinyl.

(b)

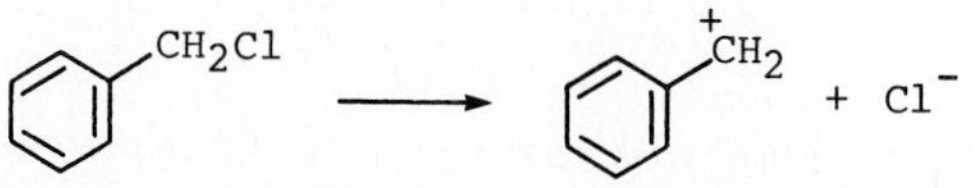

Reaction	ΔH° (kcal $mole^{-1}$)	*Stabilization:*
$C_6H_5CH_2Cl \longrightarrow C_6H_5\overset{+}{C}H_2 + Cl^-$	154	37 kcal $mole^{-1}$
$CH_3CH_2Cl \longrightarrow CH_3CH_2^+ + Cl^-$	191	
$CH_2{=}CHCH_2Cl \longrightarrow CH_2{=}CHCH_2^+ + Cl^-$	172	19 kcal $mole^{-1}$

(c) The phenyl and vinyl groups provide essentially the same stabilization to a radical center. (Although more carbons can share the radical in the benzylic case, those resonance structures lack the cyclic valence bond structure of an aromatic ring:

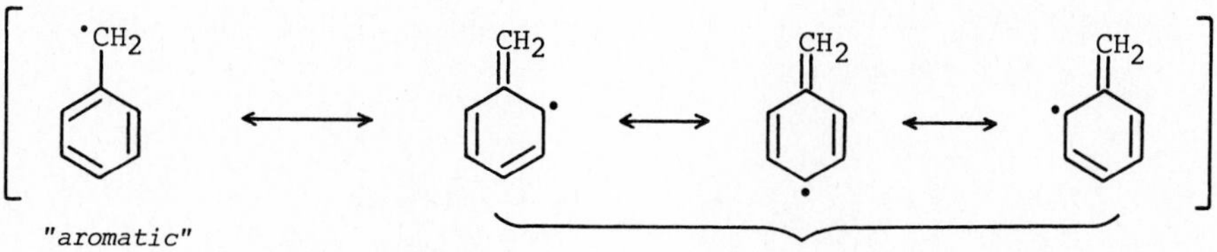

"aromatic" *"non-aromatic"*

In the case of a cation, the ability to distribute the positive charge over several atoms becomes more important, and the benzylic cation therefore becomes more stabilized than the allyl cation.

S7. (a)

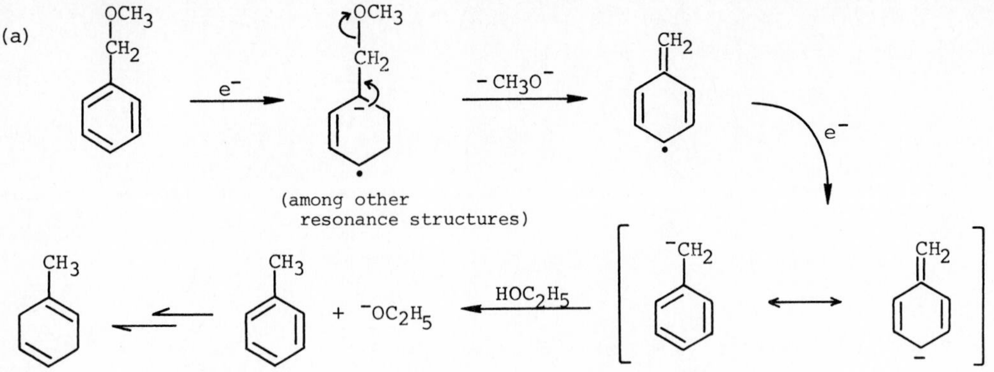

(b)

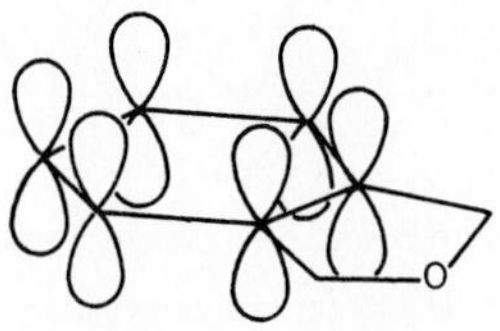

The C-O bonds are perpendicular to the p-orbitals of the π-system. Therefore the orbitals of the C-O bond cannot overlap with the aromatic p-orbitals, and bond cleavage cannot take place.

S8. *Most acidic:* (c), because of stabilization of the carbanion by both coplanar rings and electron-withdrawing nitro group.

(a) is more acidic than (e) because the tricyclic structure holds both rings coplanar for maximum overlap with the p-orbital of the benzylic carbanion.

(e) is more acidic than (b) and (d) because it has two phenyl rings which stabilize the carbanion.

(b) is more acidic than (d) because it does not have the electron-releasing methyl substituent on the carbanionic carbon.

Least acidic: (d)

S9. S_N1 substitution is very easy because the carbocation intermediate is an aromatic, 6π-electron system:

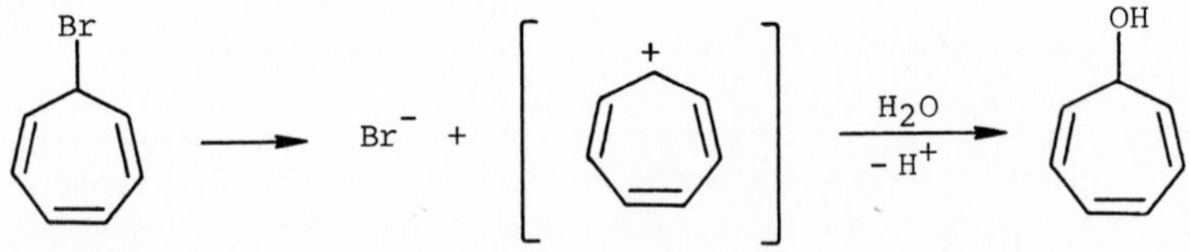

S10. The benzylic cation from S_N1-type cleavage of the *para*-isomer can be stabilized by the oxygen lone pair electrons by resonance:

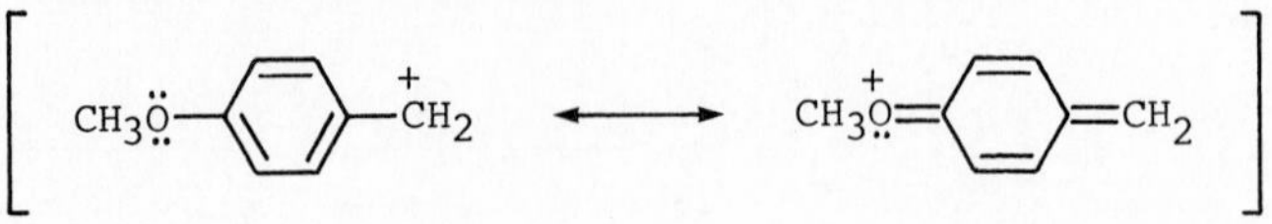

No such resonance structure is possible for the *meta*-isomer.

23. ELECTROPHILIC AROMATIC SUBSTITUTION

23.A Chapter Outline, Important Terms Introduced, and Reactions Discussed

23.1 Halogenation

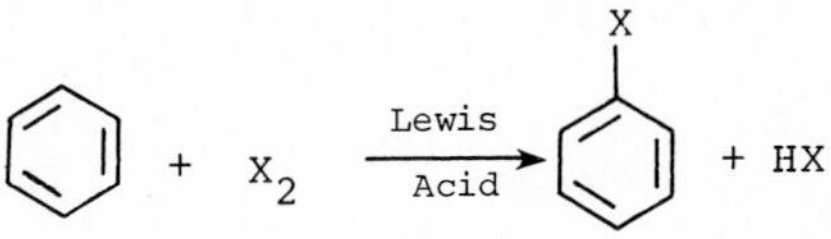

(Lewis acid: AlX_3, FeX_3; pentadienyl cation intermediate)

23.2 Protonation

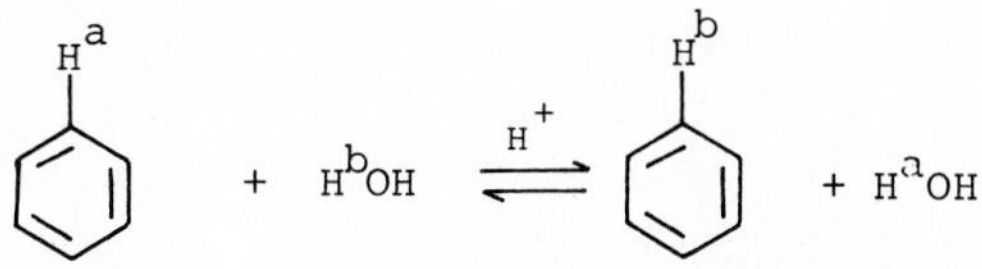

(proton exchange: H^a, H^b = H,D, or T)

tracer isotope vs. macroscopic isotope   liquid scintillation counter

23.3 Nitration

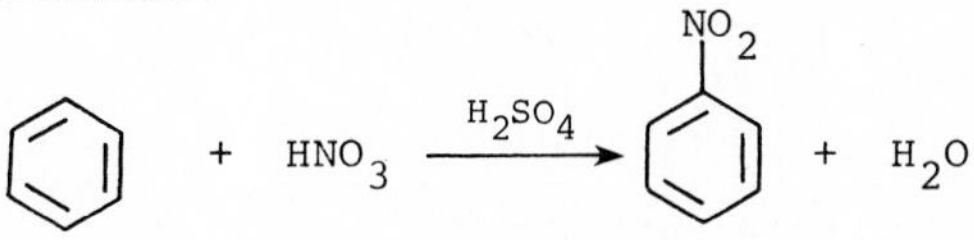

(attack of nitronium ion)

23.4 Friedel-Crafts Reactions (will not work on strongly deactivated rings)

A. Acylations

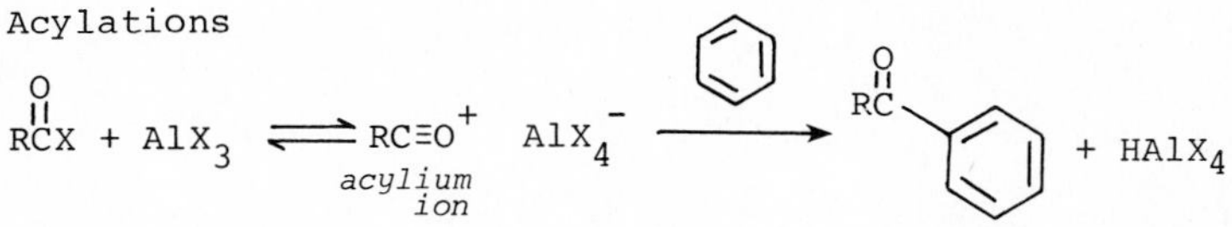

(X=Cl, Br; stops at monosubstitution)

Gattermann-Koch

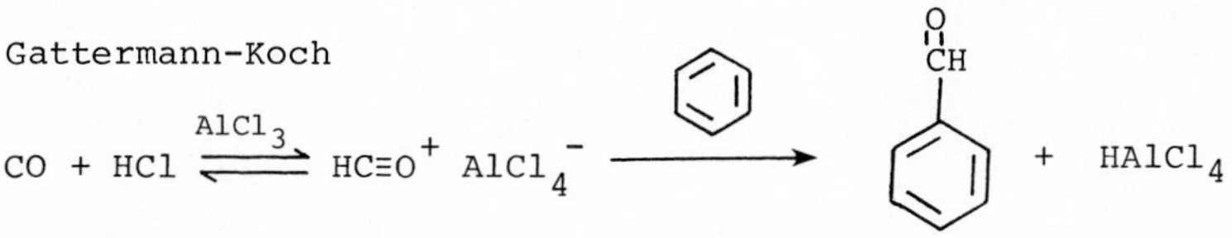

B. Alkylations

with alkyl halide:

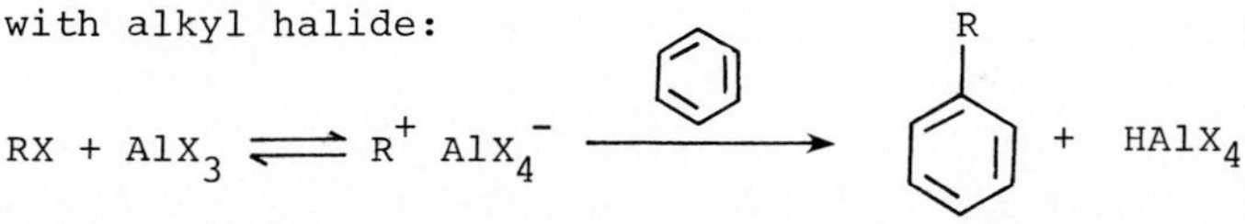

(poor for primary RX; carbocation rearrangement possible; polyalkylation a problem)

with olefin:

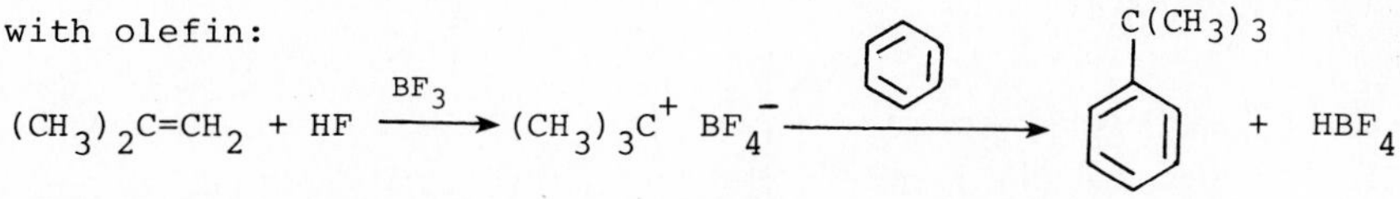

chloromethylation:

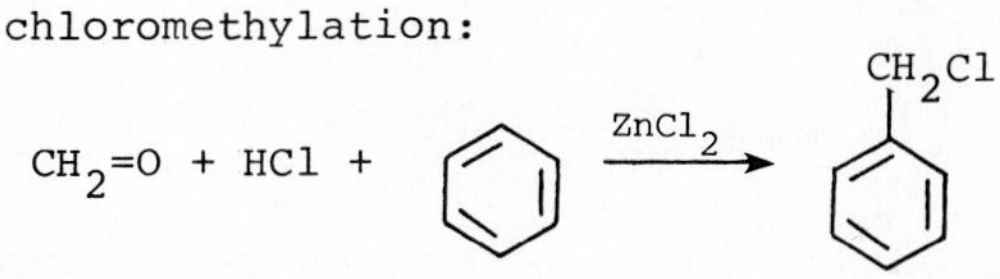

(carginogenic by-products; over-reaction to give diarylmethanes)

23.5 Orientation in Electrophilic Aromatic Substitution

ortho, para- vs. *meta*-directors   activating vs. deactivating

1. *o,p* with activation
2. *o,p* with deactivation
3. *m* with deactivation

23.6 Theory of Orientation in Electrophilic Aromatic Substitution

stabilization of resonance structures

23.7 Quantitative Reactivities: Partial Rate Factors

23.8 Effects of Multiple Substituents

23.9 Synthetic Utility of Electrophilic Aromatic Substitution

acylation + deoxygenation = "alkylation"
limitations of orientation and reactivity
avoidance of inseparable isomers

23.B Important Concepts and Hints

This entire chapter is devoted to the reaction depicted below: electrophilic aromatic substitution. We have drawn it in such a way as to show how the p-orbitals and electron distribution of the ring change throughout the process:

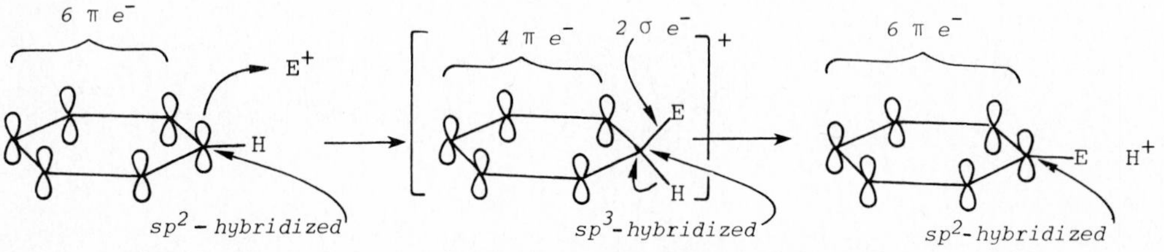

In the course of forming the intermediate, the π-system goes from neutral (6 electrons/6 orbitals) to positively charged (4 electrons/5 orbitals). Two electrons and one orbital are taken up in the new bond to the electrophile E^+. As pointed out in the text, instead of a nucleophile *adding* to the cationic intermediate, a proton is lost. This is simply the reverse of the initial attack, with the two electrons and the carbon orbital of the C-H bond going to reform the aromatic, 6π electron system.

A substituent attached to the ring can affect this process in two ways: by adding or withdrawing electron density through the σ bond:

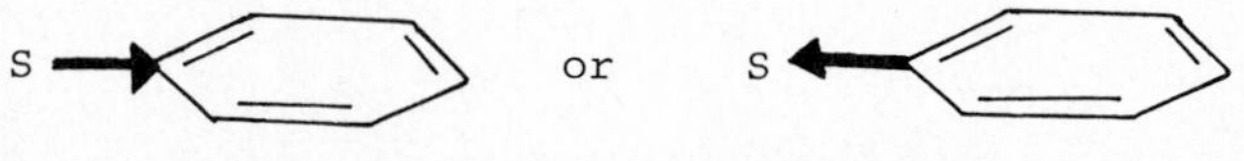

(This is usually called the "inductive" effect)

or by adding or withdrawing electron density via a π-type interaction:

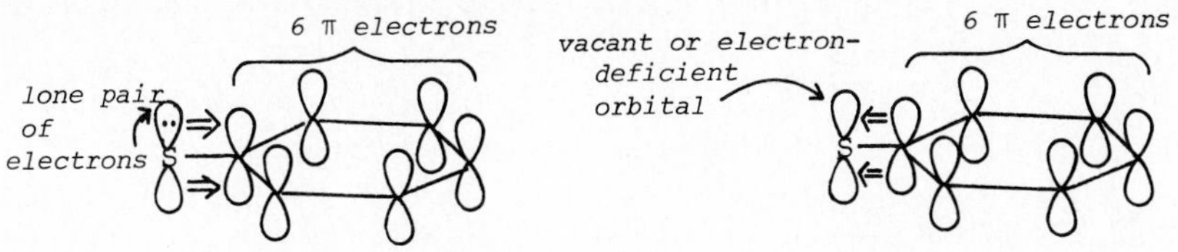

(This is usually called the "resonance" or "mesomeric" effect.)

The effect through the σ framework decreases in the order *o* >*m*>*p*, but it does not influence greatly the position of electrophilic attack. On the other hand, the effect through the π-system is only felt in the *ortho* and *para* positions, so it is the determining factor in orientation. The table below illustrates the influence of σ and π-effects for a variety of substituent types:

Type of Substituent	Group	π-effect	σ-effect	Substitution rate, relative to benzene: o-,p-positions	m-positions
RÖ, R_2N̈, RC(=O)N̈H	I	strongly electron-releasing	electron-withdrawing	+++	-
Alkyl	II	electron-releasing (via hyperconjugation)	electron-releasing	++	+
Halogen	II	electron-releasing	electron-withdrawing	-	--
NO_2, RC(=O), RS(=O)(=O)	III	strongly electron-withdrawing	electron-withdrawing	---	--

Notice how the *meta* reactivity (relative to benzene) parallels the σ effect, because the resonance (π) effect can only influence the *ortho* and *para* positions. The π effect combines with the σ effect in influencing the *ortho* and *para* rates. For strongly activating substituents the π-effect dominates the σ effect and the *ortho* and *para* positions are strongly activated. For alkyl substituents, the two effects act together, and *ortho*, *para* substitution gets an extra boost relative to *meta*. For halogens, the π- and σ-effects work in opposite directions again, but this time the π effect is weaker: substitution at all positions is deactivated, but the *ortho-*, *para* ones are less so. Finally, for strongly deactivating groups, the π- and σ-effects combine to deactivate *ortho* and *para* the most, resulting in *meta*-direction.

In predicting the results of competition between the directive effects of two substituents on the same ring, it is best to remember *three* groups of substituents (instead of the four above). Group I includes the strongly activating, *ortho*, *para*-directors like RÖ- and R_2N̈-; their influence dominates that of the other groups. Group II includes the "moderate" *ortho*, *para*-directors like alkyl and halogen; these substituents will yield to group I effects, and will win out only over those of group III. Group III includes all the *meta*-directing substituents; they will control orientation only in the absence of group I or II substituents. If the competition is between members of the same group, no "winner" is predictable, and you can expect to see mixtures of isomeric products.

23.C Answers to Exercises

(23.1)

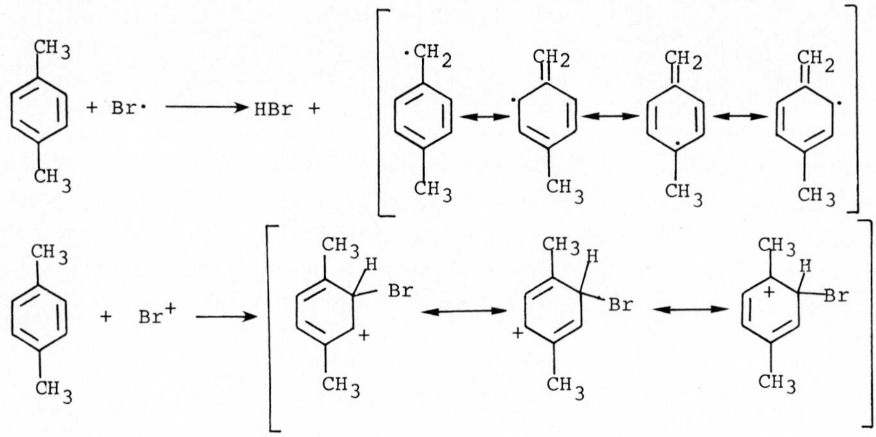

(23.2)

and

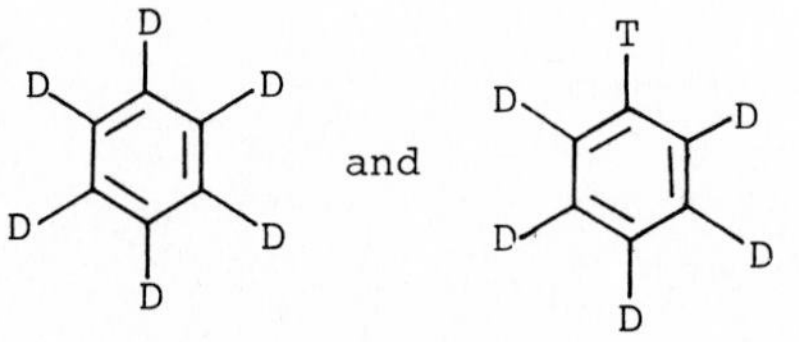

(23.3)

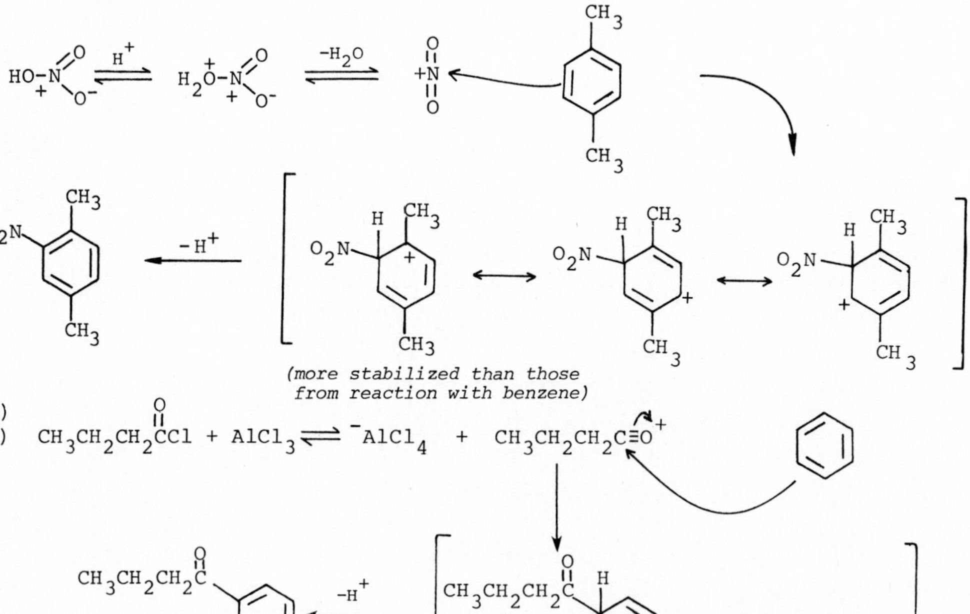

(more stabilized than those
from reaction with benzene)

(23.4.B)

(a) $CH_3CH_2CH_2CCl + AlCl_3 \rightleftharpoons {}^-AlCl_4 + CH_3CH_2CH_2C{\equiv}O^+$

+ other resonance
structures

(b) $HF + BF_3 \rightleftharpoons BF_4{}^- H^+ \xrightarrow{CH_2=C(CH_3)_2} (CH_3)_3C_+$

etc.

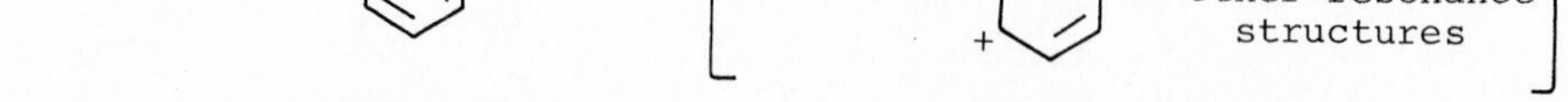

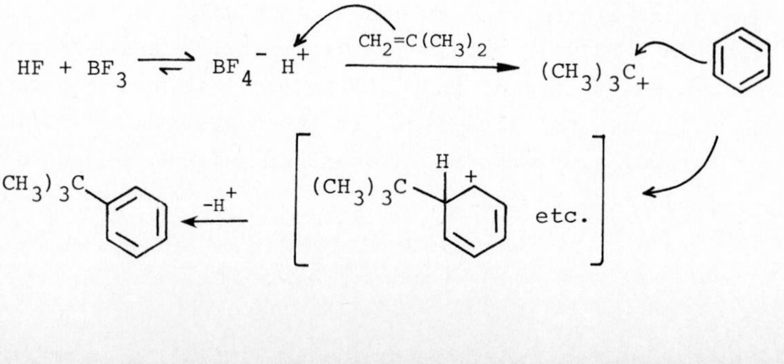

(c) $CH_3CH_2\overset{OH}{\underset{}{CHCH_3}} \xrightarrow{H_2SO_4} CH_3CH_2\overset{+OH_2}{\underset{}{CHCH_3}} \underset{-H_2O}{\rightleftharpoons} CH_3CH_2\overset{+}{CHCH_3}$

$CH_3CH_2\overset{CH_3}{\underset{}{CH}}\text{---}\bigcirc \xleftarrow{-H^+} \left[CH_3CH_2\overset{CH_3}{\underset{}{CH}}\overset{H}{\underset{}{\overset{+}{\bigcirc}}} \right]$ etc.

(d) $H_2C{=}O + ZnCl_2 \rightleftharpoons Cl_2\overset{-}{Zn}\text{-}\overset{+}{O}{=}CH_2 \quad \bigcirc \longrightarrow \left[Cl_2ZnOCH_2\overset{H}{\underset{}{\overset{+}{\bigcirc}}} \right]$ etc.

$ClCH_2\text{-}\bigcirc \underset{Cl^-}{\rightleftharpoons} {}^+CH_2\text{-}\bigcirc \underset{-H_2O}{\rightleftharpoons} H_2\overset{+}{O}CH_2\text{-}\bigcirc \underset{H^+}{\rightleftharpoons} HOCH_2\text{-}\bigcirc$

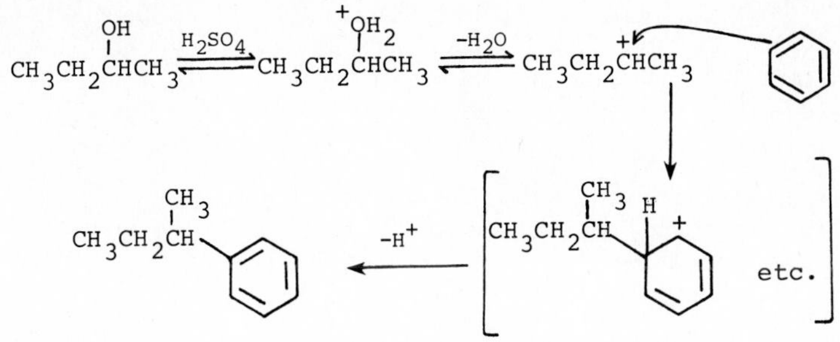

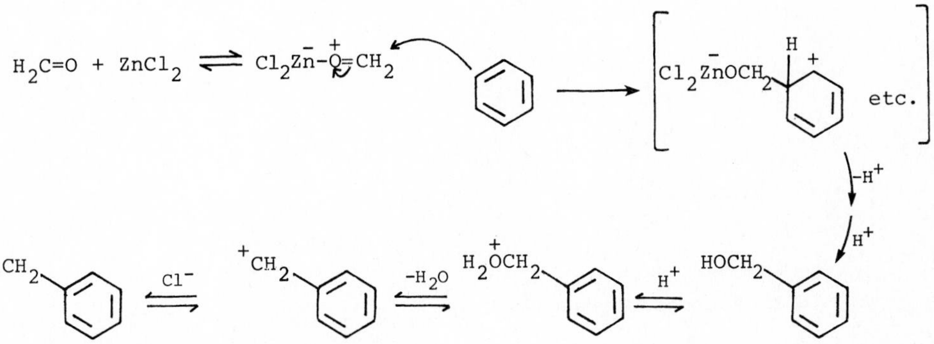

(23.5)

	$Br_2 + FeBr_3$	HNO_3, H_2SO_4	$(CH_3)_3CCl + AlCl_3$	$CH_3\overset{O}{\overset{\|}{C}}Cl + AlCl_3$
$C_6H_5CH_3$	$o\text{-}BrC_6H_4CH_3$ and p-isomer; *faster*	$o\text{-}O_2NC_6H_4CH_3$ and p-isomer; *faster*	$o\text{-}(CH_3)_3CC_6H_4CH_3$ and p-isomer (plus rearr.); *faster*	$o\text{-}CH_3\overset{O}{\overset{\|}{C}}C_6H_4CH_3$ and p-isomer; *faster*
$C_6H_5\overset{O}{\overset{\|}{C}}CH_3$	$(\ C_6H_5\overset{O}{\overset{\|}{C}}CH_2Br\ )$	$m\text{-}O_2NC_6H_4\overset{O}{\overset{\|}{C}}CH_3$; *slower*	*no reaction*	*no reaction*
C_6H_5Br	$o\text{-}C_6H_4Br_2$ and p-isomer; *slower*	$o\text{-}O_2NC_6H_4Br$ and p-isomer; *slower*	$o\text{-}(CH_3)_3CC_6H_4Br$ and p-isomer; *slower*	$o\text{-}CH_3\overset{O}{\overset{\|}{C}}C_6H_4Br$ and p-isomer; *slower*
$C_6H_5OCH_3$	$o\text{-}BrC_6H_4OCH_3$ and p-isomer; *faster*	$o\text{-}O_2NC_6H_4OCH_3$ and p-isomer; *faster*	$o\text{-}(CH_3)_3CC_6H_4OCH_3$ and p-isomer; *faster*	$o\text{-}CH_3\overset{O}{\overset{\|}{C}}C_6H_4OCH_3$ and p-isomer; *faster*
$C_6H_5NO_2$	$m\text{-}BrC_6H_4NO_2$; *slower*	$m\text{-}C_6H_4(NO_2)_2$; *slower*	*no reaction*	*no reaction*
$C_6H_5\overset{O}{\overset{\|}{N}H C}CH_3$	$o\text{-}BrC_6H_4\overset{O}{\overset{\|}{NHC}}CH_3$ and p-isomer; *faster*	$o\text{-}O_2NC_6H_4\overset{O}{\overset{\|}{NHC}}CH_3$ and p-isomer; *faster*	$o\text{-}(CH_3)_3CC_6H_4\overset{O}{\overset{\|}{NHC}}CH_3$ and p-isomer; *faster*	$o\text{-}CH_3CC_6H_4\overset{O}{\overset{\|}{NHC}}CH_3$ and p-isomer; *faster*

"faster" = reaction occurs faster than with benzene;   *"slower"* = reaction occurs slower than with benzene

(23.6)

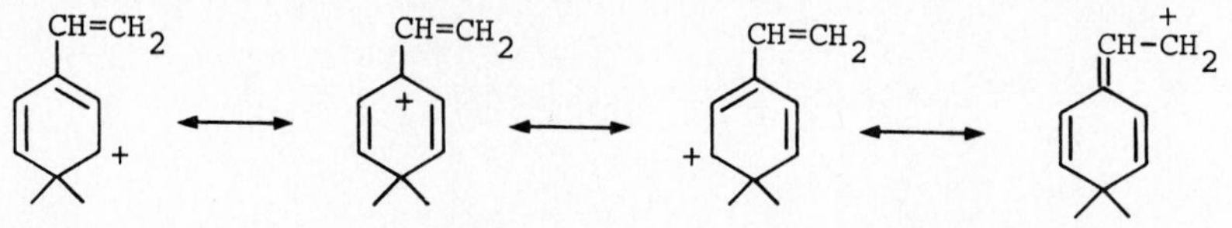

In the o- and p- positions, the vinyl group helps to distribute positive charge; thus, it is o,p-directing and activating. In the m-position, the vinyl group does not conjugate with the charge:

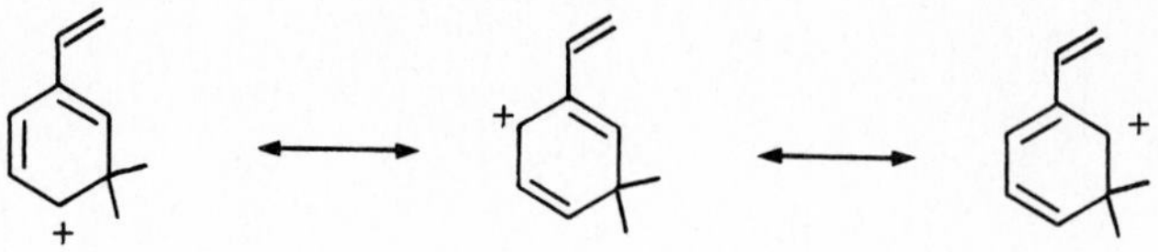

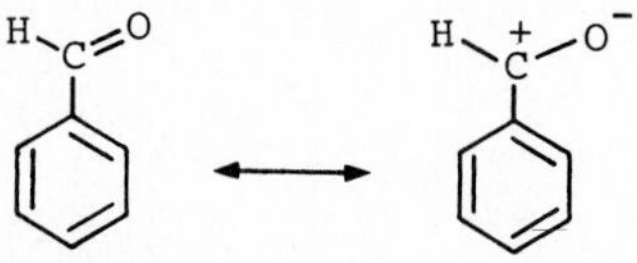

∴ large dipole directed away from ring:

m-Substitution keeps charge away from dipole in all three structures:

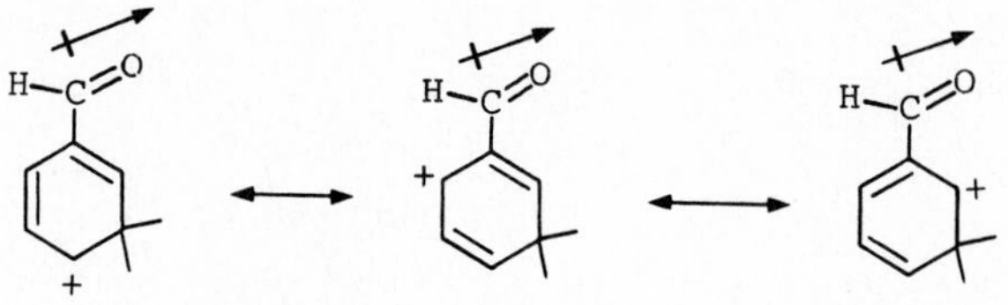

In o,p-substitution, one structure has (+) next to dipole:

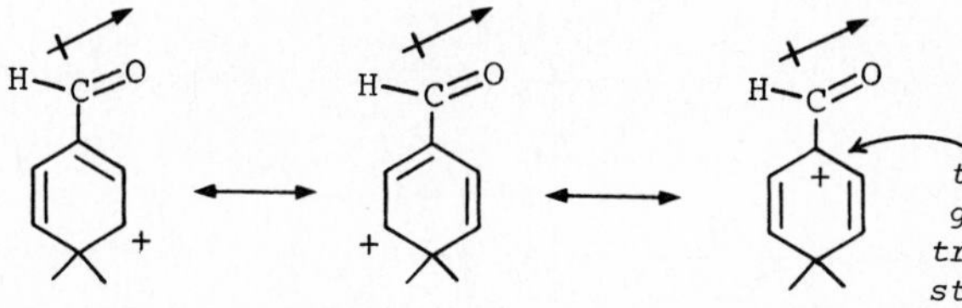

the positive charge is next to the positive carbon of the CO group, so that this structure contributes less; the resonance stabilization of intermediate and the transition state leading to it are reduced.

(23.7)

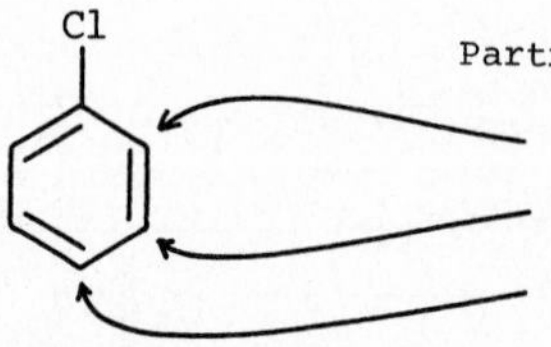

Partial rate factor	×	number of positions	=	relative amounts		
47	×	2	=	94	=	58%
3	×	2	=	6	=	4%
62	×	1	=	62	=	38%

(23.8)

(a)

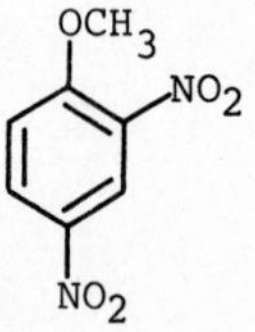

(b)

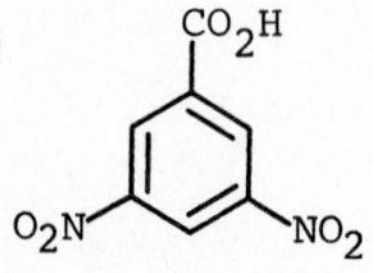

(c)

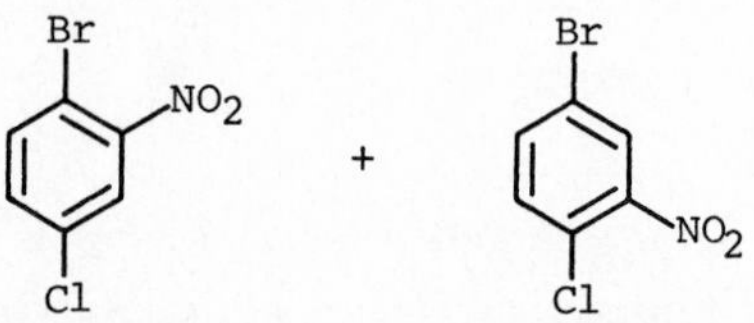

(d)

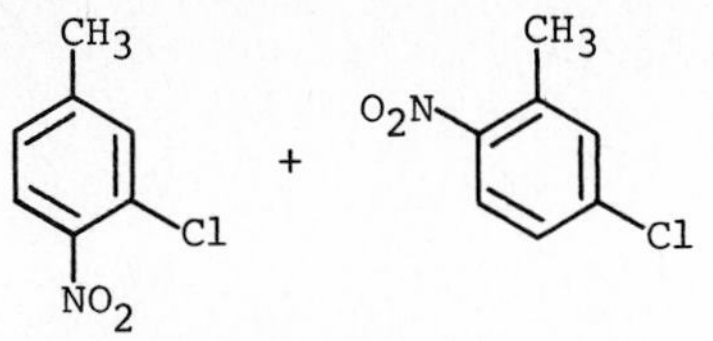

(23.9) All of these transformations of Ar-Br require formation of a Grignard reagent, which cannot be generated in the presence of the -CH=O, $-CO_2H$, or $-NO_2$ functional groups. The mono-Grignard reagents can be prepared from *m*- and *p*-dibromobenzene, however.

23.D Answers and Explanations for Problems

1. $I_2 + H_2O_2 \longrightarrow 2\ HOI \rightleftharpoons H_2\overset{+}{O}I + {}^{-}OI$

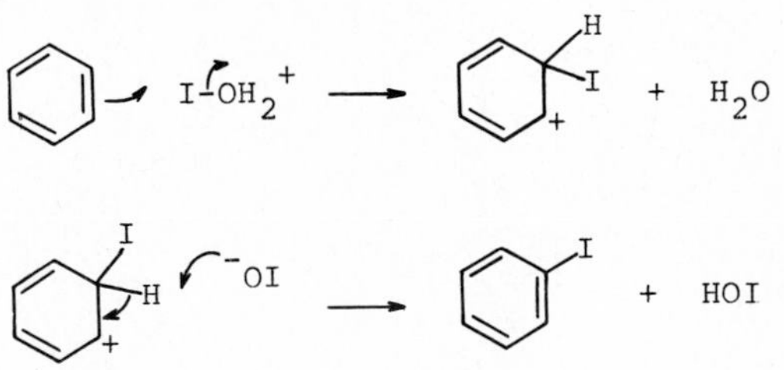

2. (a) Mechanism A:

1) $CH_2{=}O + ZnCl_2 \rightleftharpoons CH_2{=}\overset{+}{O}{-}\overset{-}{Z}nCl_2$

2) 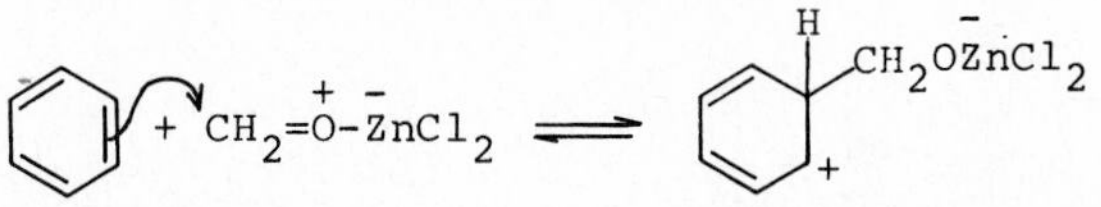

3) 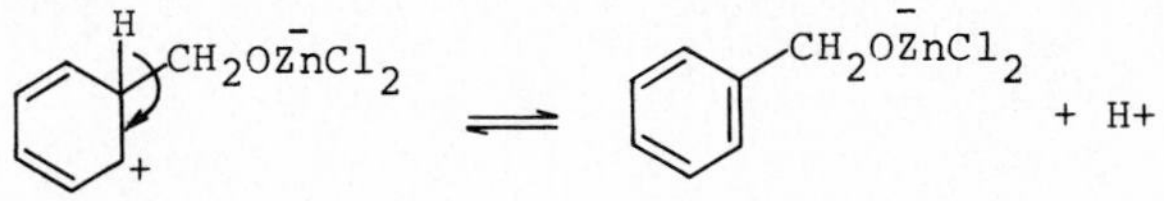

4) 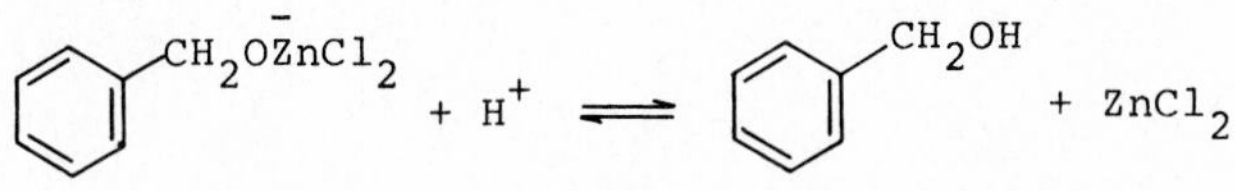

5) 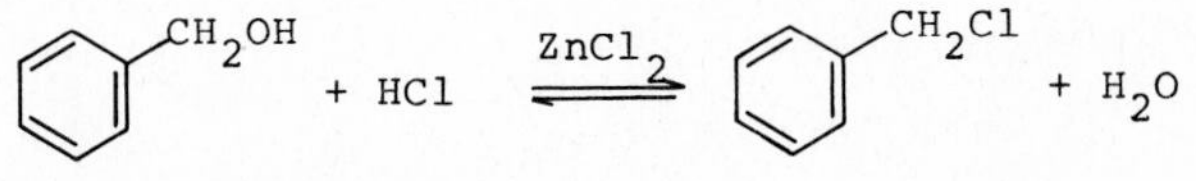

Mechanism B:

1) $CH_2{=}O + ZnCl_2 \rightleftharpoons CH_2{=}\overset{+}{O}{-}\overset{-}{Z}nCl_2$

2) $CH_2{=}\overset{+}{O}{-}\overset{-}{Z}nCl_2 + Cl^- \rightleftharpoons ClCH_2O\overset{-}{Z}nCl_2$

3) $ClCH_2O\overset{-}{Z}nCl_2 + H+ \rightleftharpoons ClCH_2\overset{H}{\underset{+}{O}}\overset{-}{Z}nCl_2$

4) $ClCH_2\overset{H}{\underset{+}{O}}\overset{-}{Z}nCl_2 \rightleftharpoons ClCH_2{+} + HO\overset{-}{Z}nCl_2$

5)
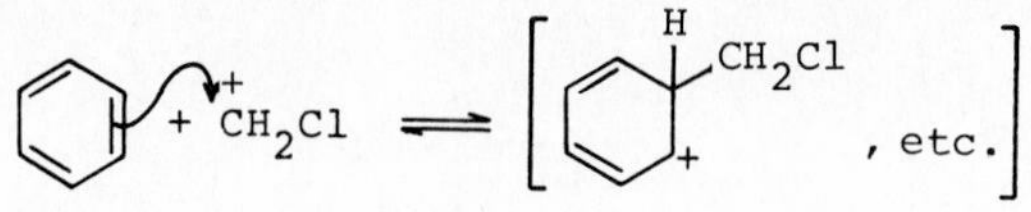

6)
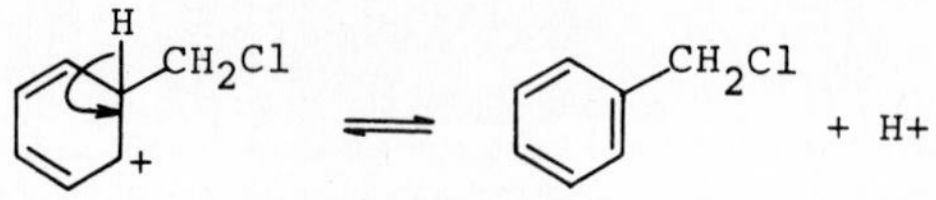

(b) 1) $CH_2{=}O + {}^{+}CH_2Cl \rightleftharpoons CH_2{=}\overset{+}{O}{-}CH_2Cl$

2) $Cl^- + CH_2{=}\overset{+}{O}CH_2Cl \longrightarrow ClCH_2OCH_2Cl$

or

1) $CH_2{=}O + CH_2{=}\overset{+}{O}{-}\overset{-}{Zn}Cl_2 \rightleftharpoons CH_2{=}\overset{+}{O}{-}CH_2{-}O\overset{-}{Zn}Cl_2$

2) $Cl^- + CH_2{=}\overset{+}{O}CH_2O\overset{-}{Zn}Cl_2 \rightleftharpoons ClCH_2OCH_2O\overset{-}{Zn}Cl_2$

3) $ClCH_2OCH_2O\overset{-}{Zn}Cl_2 \overset{H^+}{\rightleftharpoons} ClCH_2OCH_2\overset{H}{\underset{+}{O}}\overset{-}{Zn}Cl_2$

4) $ClCH_2\ddot{O}{-}CH_2{-}\overset{H}{\underset{+}{O}}\overset{-}{Zn}Cl_2 \rightleftharpoons ClCH_2\overset{+}{O}{=}CH_2 \quad HO\overset{-}{Zn}Cl_2$

5) $ClCH_2\overset{+}{O}{=}CH_2 \quad Cl^- \longrightarrow ClCH_2OCH_2Cl$

3. $CH_3\overset{O}{\overset{\|}{C}}OHgO\overset{O}{\overset{\|}{C}}CH_3 + H^+ \rightleftharpoons CH_3\overset{\overset{+}{O}H}{\overset{\|}{C}}OHgO\overset{O}{\overset{\|}{C}}CH_3$

$CH_3\overset{+OH}{\overset{\|}{C}}OHgO\overset{O}{\overset{\|}{C}}CH_3 \rightleftharpoons CH_3\overset{O}{\overset{\|}{C}}OH + \overset{+}{Hg}OAc$

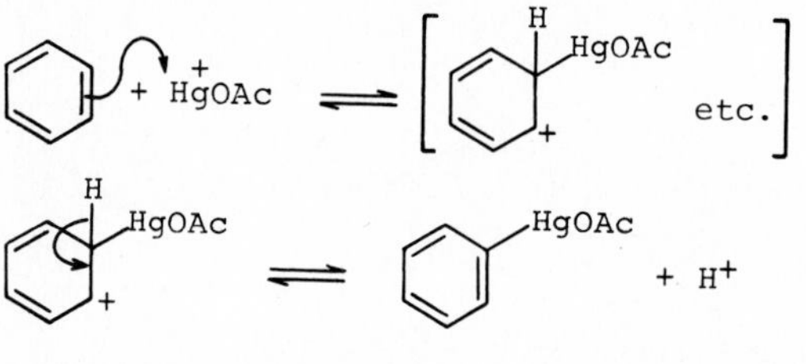

4.
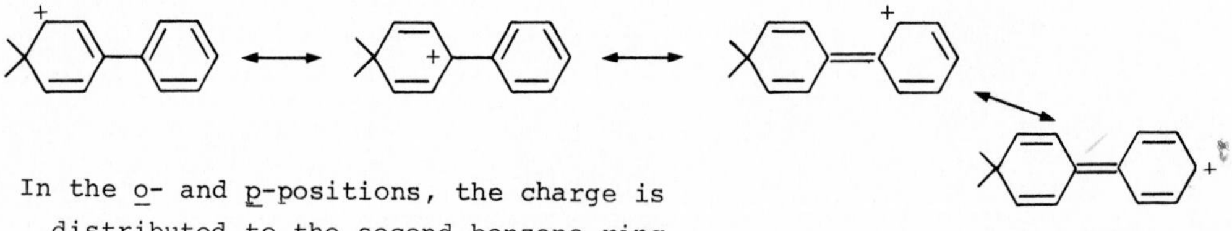

In the o- and p-positions, the charge is distributed to the second benzene ring.

etc.

5. The carboxy group is *meta*-directing and deactivating for the same reason that the formyl group is (see answer to Exercise at the end of Section 23.6).

6. (a)
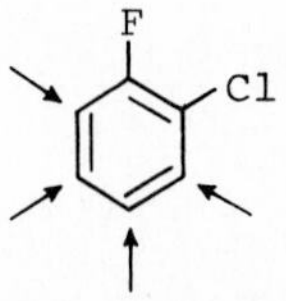

serious mixture!
o,p to F and Cl

(b)
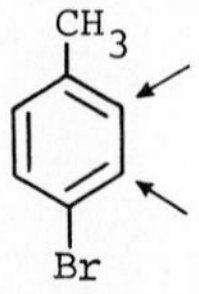

o- to both Br and CH_3

(c)
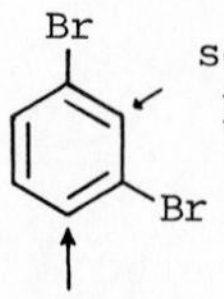

some o- to both Br

mostly
o- to one Br,
p- to the other

(d) 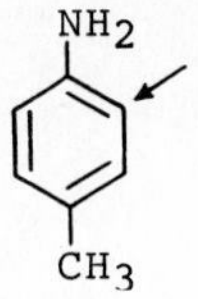

NH_2 dominates

(e) 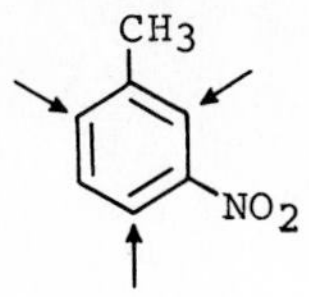

o,p- to CH_3

(f) 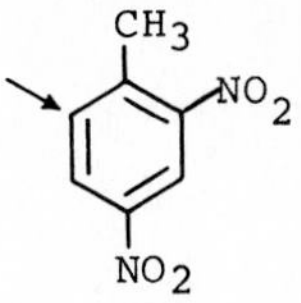

only available position o,p to methyl

(g) 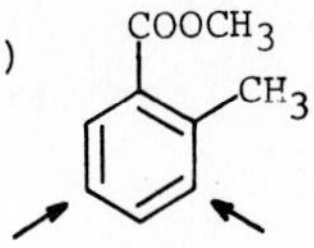

o,p- to CH_3

(h) 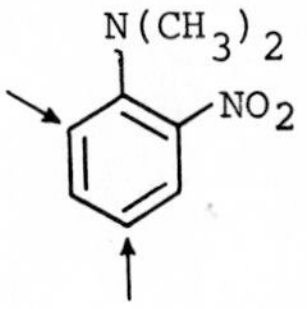

(i) 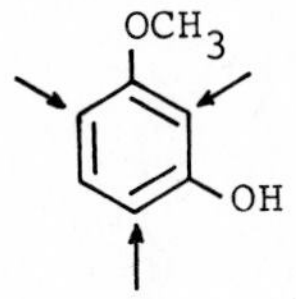

o,p to OCH_3 and to OH

(j) 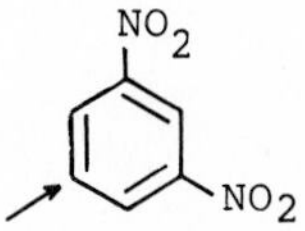

7. (a) $f_o = 6 \times \dfrac{(0.329)(605)}{2} = 597 \qquad f_m = 6 \times \dfrac{(0.003)(605)}{2} = 5.4$

$$f_p = 6(0.668)(605) = 2425$$

(b) 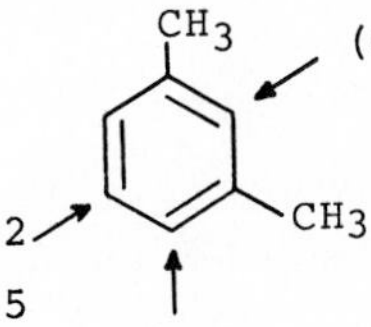

$(620)^2 = 3.84 \times 10^5$

$(5.0)^2 = 25$

$2 \times 820 \times 620 = 10.17 \times 10^5$ [2 because 4- and 6-positions are equivalent]

2-position: $(3.84 \times 10^5)/[3.84 \times 10^5 + 10.17 \times 10^5 + 25] = 27.4\%$

4-position: $(10.17 \times 10^5)/[$ " + " + "] = 72.6%

5-position: 25 /["] = 0.0018%

(c) 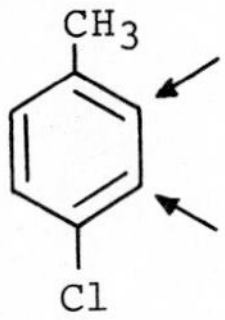

620(0.002) = 1.24 or 71%

5.0(0.1) = 0.5 or 29%

8. (a) No; both Cl and Br are o,p-directors.

(b) No; Friedel-Crafts acylation cannot be applied to a ketone.

The sequence 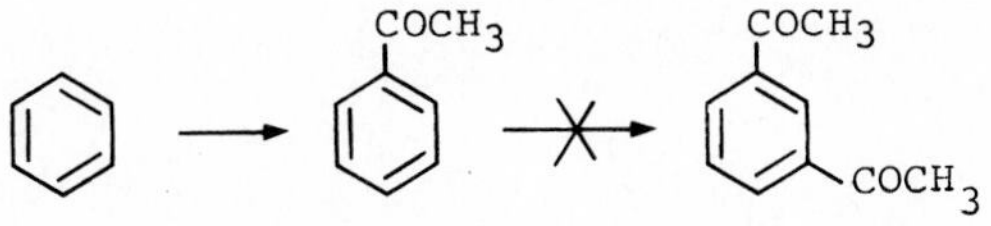 will not work.

(c) OK; 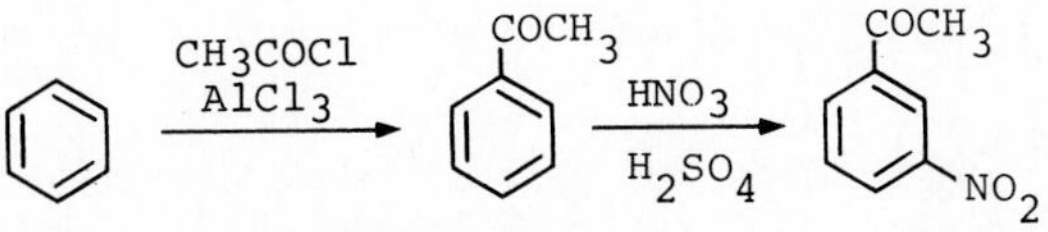

Note that nitration first, followed by Friedel-Crafts acylation, will not work. Acylation cannot be applied to $ArNO_2$ compounds.

(d) No;

the sequence

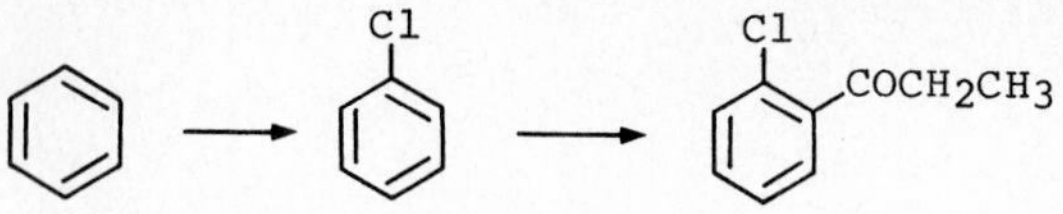

gives mostly *para*. The *ortho*-isomer is difficult to isolate pure. This method is OK for the *para*-isomer, since it is higher-melting and less soluble and can be separated pure from the *ortho*-isomer by crystallization.

(e) OK;

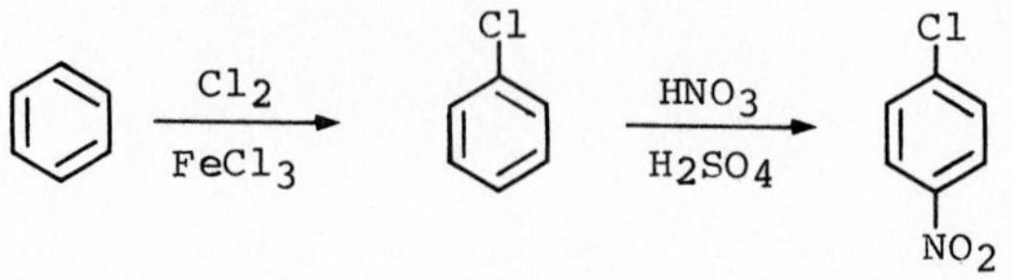

can be separated from *ortho*-isomer by crystallization (see Table 23.4)

(f) OK; nitration of *t*-butylbenzene strongly favors *para*-substitution.

9. (a) OK:

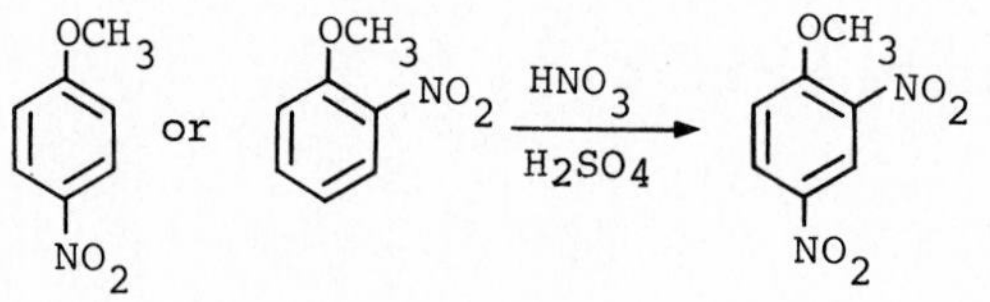

(b) OK;

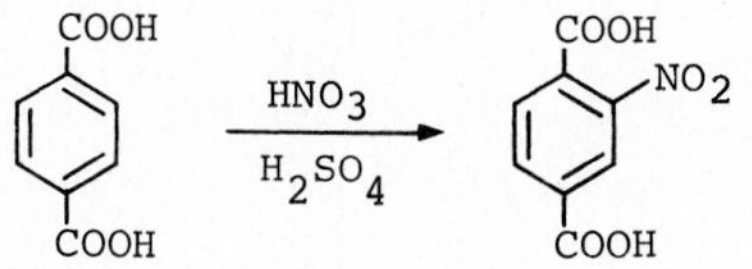

(no isomers possible)

(c) OK;

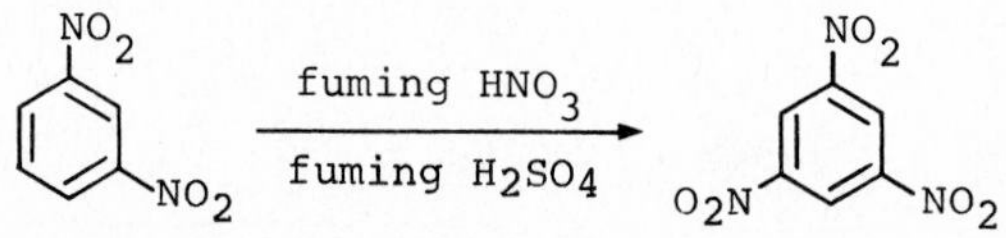

(both NO_2 groups direct *meta*-)

(d) OK;

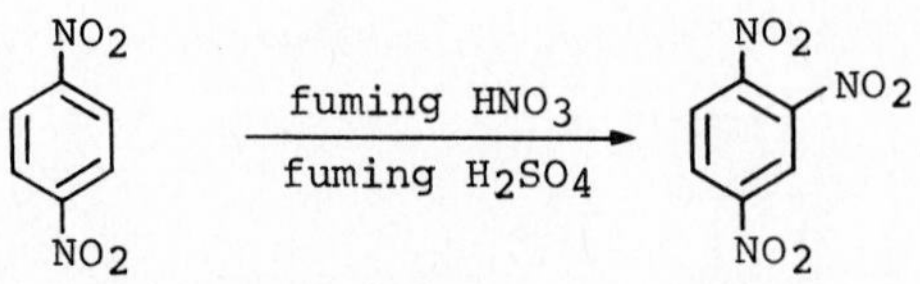

(no isomers possible)

(e) No; nitration of *o*-dichlorobenzene gives mostly 4-nitro

(f) OK;

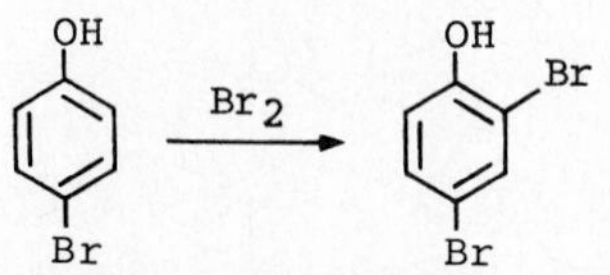

(OH dominates)

(g) No. Chlorination of

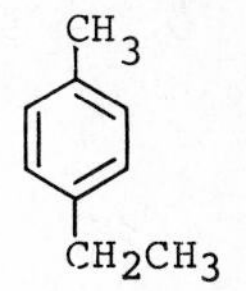

goes <u>ortho</u> to both CH_3 and
CH_2CH_3 and gives a mixture.

(h) No.

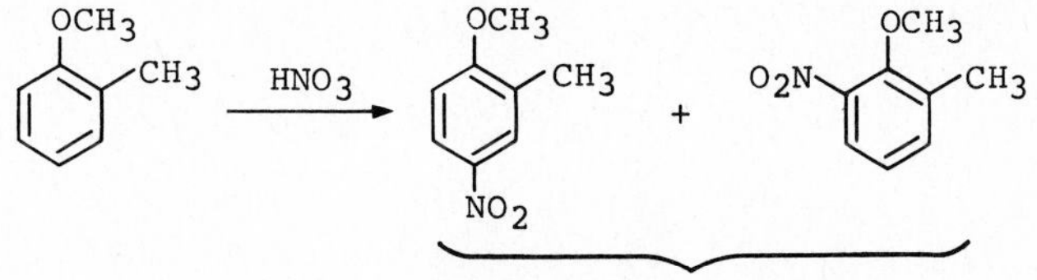

difficult to separate
(compare $O_2NC_6H_5OCH_3$ isomers in Table 23.6)

(i) OK;

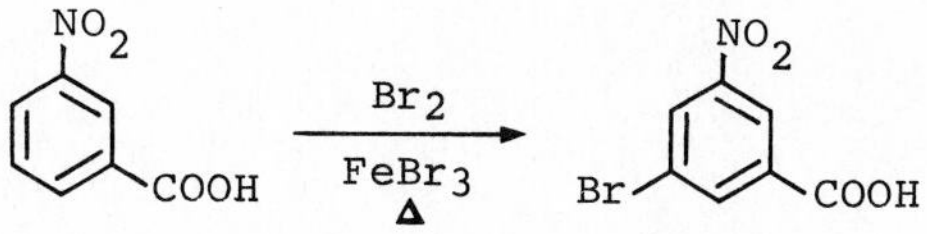

both groups orient *meta-*.

(j) No. Nitration of

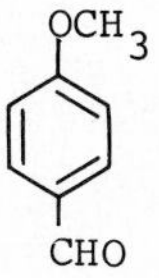

gives

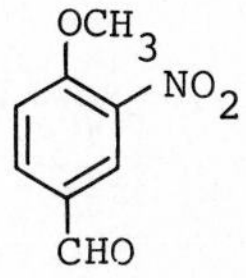

Friedel-Crafts acylation
reactions cannot be
applied to nitro com-
pounds, so *m*-nitroanisole
cannot be formylated.

10. The trifluoromethyl group is electron-withdrawing
because of the C-F bond dipole $\left(\overset{+\;\longrightarrow}{-CF_3} \right)$. The explanation is analogous to
that given for the COOH group in the answer to problem #5.

11. (a)

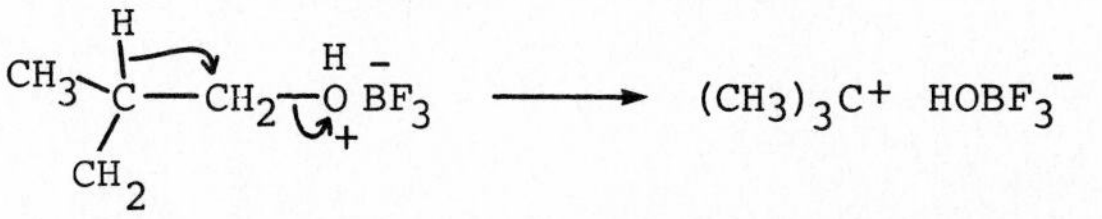

$(CH_3)_3C^+$ $HOBF_3^-$ *rearrangement of primary*
carbocation to tertiary

(b) $CH_3CDOHCH_2CH_3 + BF_3 \rightleftharpoons$

$- BF_3OH^-$

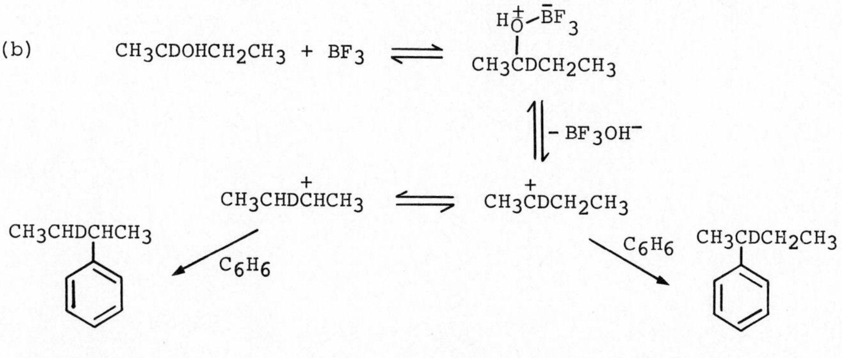

(c) Scrambling would require a primary carbocation:

$$CD_3\overset{+}{C}HCH_3 \rightleftharpoons {}^+CD_2CHDCH_3 \rightleftharpoons CD_2H\overset{+}{C}DCH_3$$

Primary carbocations are so unstable relative to secondary ones that
the reaction does not occur.

(d) Carbocations have a planar central carbon which is achiral. Consequently, a racemic reaction product is anticipated. In practice, the
reaction product is 99% racemized.

12. Aluminum chloride, unless specially purified and handled on a vacuum line, always has traces of H_2O and HCl.

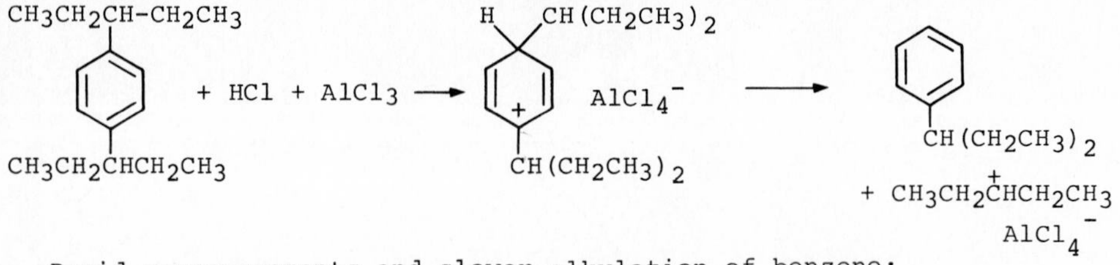

Rapid rearrangements and slower alkylation of benzene:

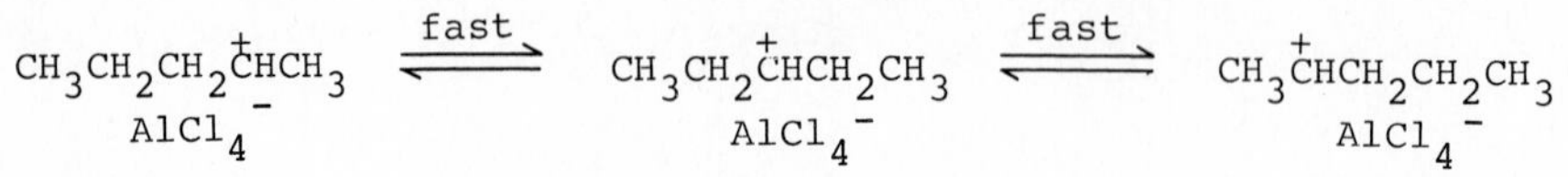

$$CH_3CH_2CH_2\overset{+}{C}HCH_3\ AlCl_4^- \underset{}{\overset{fast}{\rightleftharpoons}} CH_3CH_2\overset{+}{C}HCH_2CH_3\ AlCl_4^- \overset{fast}{\rightleftharpoons} CH_3\overset{+}{C}HCH_2CH_2CH_3\ AlCl_4^-$$

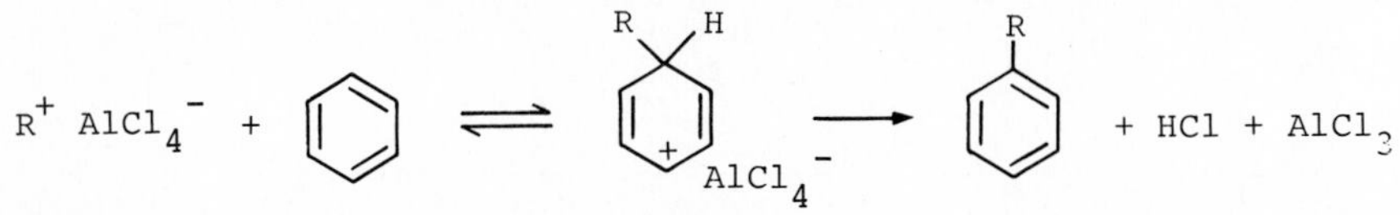

$CH_3\overset{+}{C}HCH_2CH_2CH_3$ and $CH_3CH_2\overset{+}{C}HCH_2CH_3$ are so similar in structure, we may confidently expect that their rates of alkylating benzene will be closely similar. Thus, the relative rates of formation of 2-phenylpentane and 3-phenylpentane will be the same as the relative populations of the two carbocations, or 2:1.

Consequently, one mole of 1,4-di(3-pentyl)benzene gives one mole of 3-phenylpentane (after cleavage of one pentyl group), and the cleaved pentyl group gives: 0.33 moles 3-phenylpentane
0.67 moles 2-phenylpentane

Total: 1.33 moles of 3-phenylpentane
and 0.67 moles of 2-phenylpentane,
or a ratio of 2:1 for 3-isomer/2-isomer.

13.

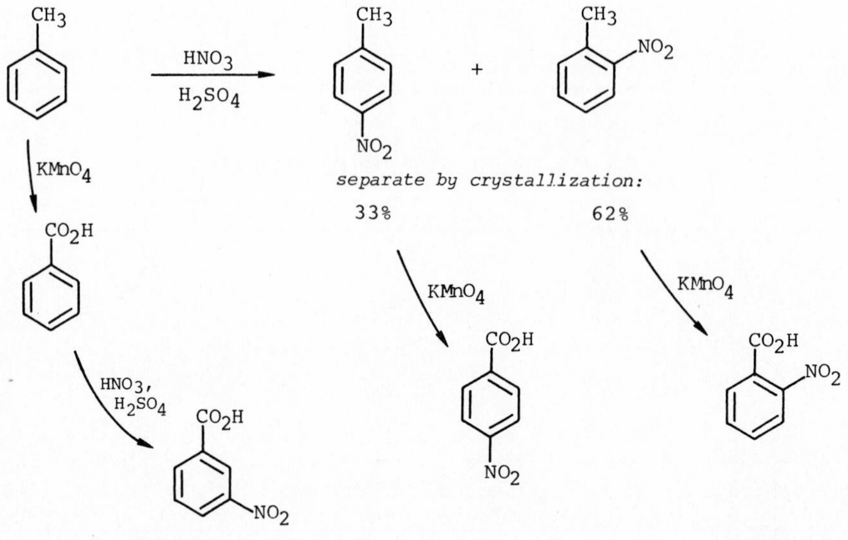

14.

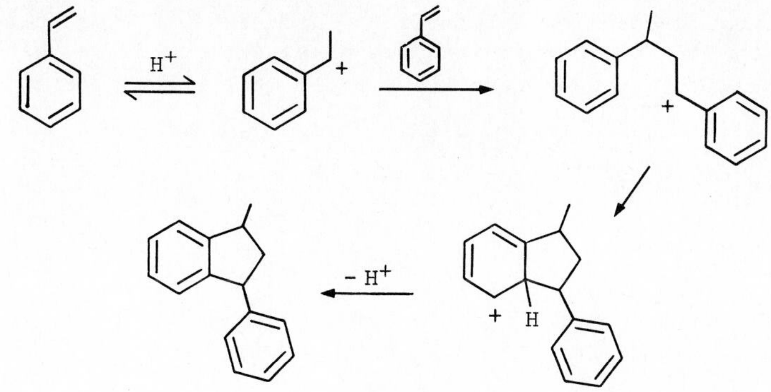

15. (a)

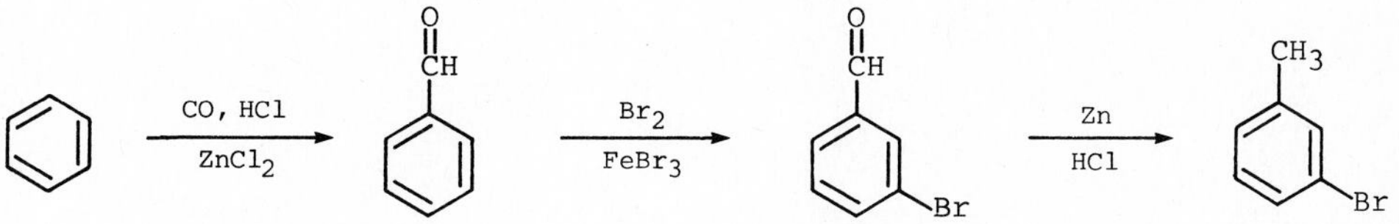

(b)

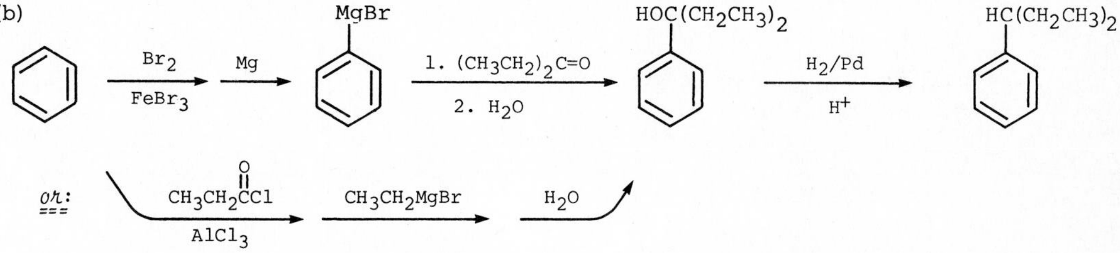

$\underline{or}$: $CH_3CH_2\overset{O}{\overset{\|}{C}}Cl$ $\xrightarrow{AlCl_3}$ CH_3CH_2MgBr $\xrightarrow{H_2O}$

(c)

(plus *ortho*-isomer)

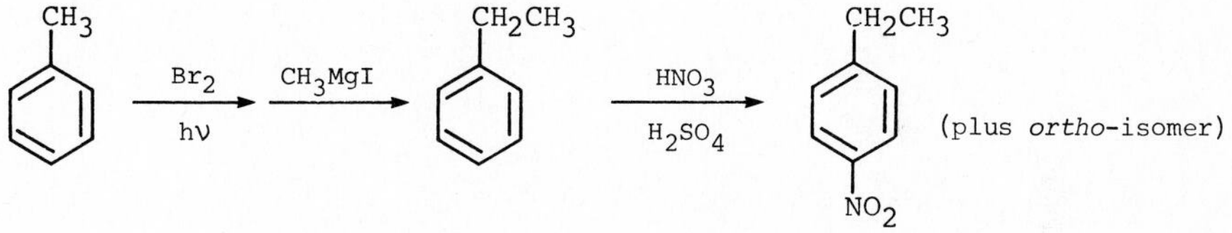

(d)

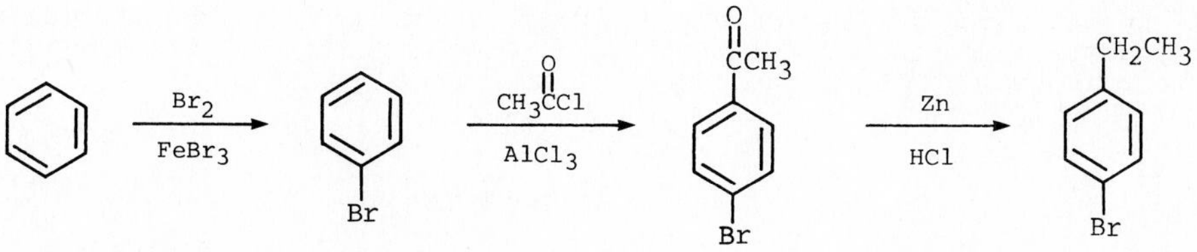

(e)

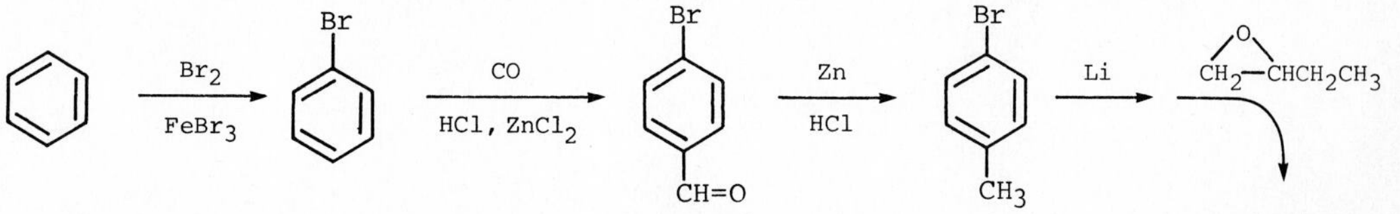

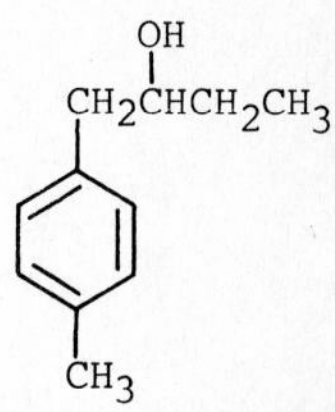

16. (a)

$$C_6H_5C(CH_3)_2Cl \xrightarrow{\text{slow}} C_6H_5\overset{+}{C}(CH_3)_2 + Cl^-$$

$$C_6H_5\overset{+}{C}(CH_3)_2 + H_2O \xrightarrow{\text{fast}} C_6H_5\overset{CH_3}{\underset{CH_3}{\overset{|}{\underset{|}{C}}}}-\overset{+}{O}H_2 \rightleftharpoons C_6H_5C(CH_3)_2OH + H^+$$

(b) *p*-CH_3 stabilizes carbocation intermediate and the transition state leading to it because one of the contributing resonance structures is that of a tertiary carbocation. *p*-CH_3O stabilizes still more because of an oxonium ion structure:

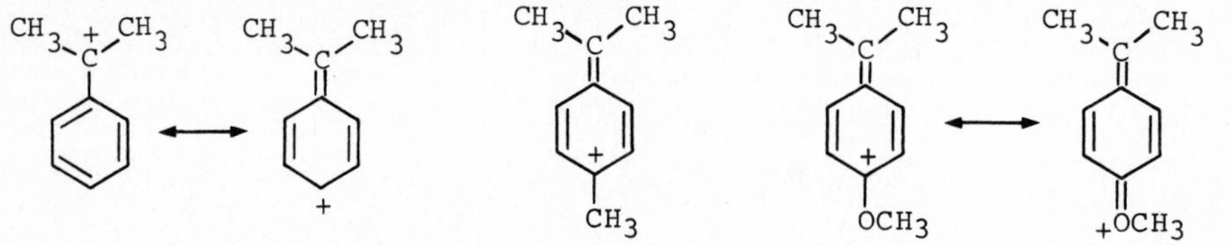

unsubstituted *tertiary* *oxonium ion*

The *p*-NO_2 group destabilizes the carbocation by electrostatic repulsion:

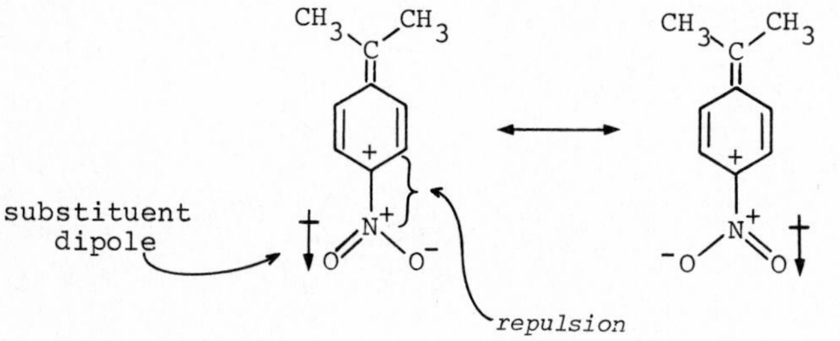

17. *Lowest dissociation constant:* (*p*-$O_2NC_6H_4)_3CCl$
(*m*-$ClC_6H_4)_3CCl$
$(C_6H_5)_3CCl$
(*m*-$CH_3C_6H_4)_3CCl$
(*p*-$CH_3C_6H_4)_3CCl$
Highest dissociation constant: (*p*-$CH_3OC_6H_4)_3CCl$

Substituents which activate the ring toward electrophilic substitution will also stabilize the carbocation. The influence of these substituents is greatest in the *para* position because of resonance.

18. (a)

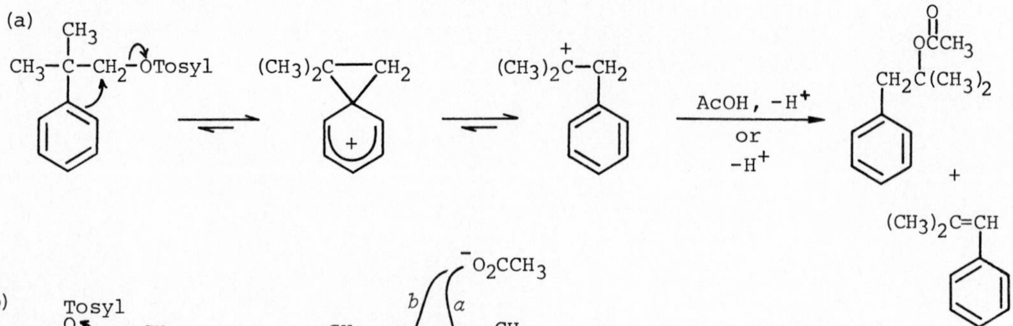

(b)

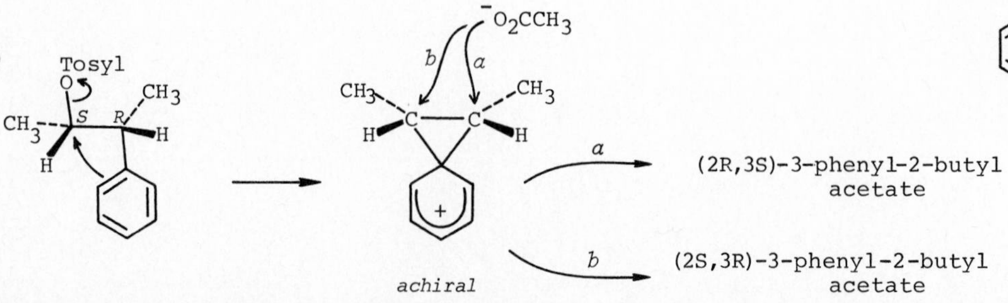

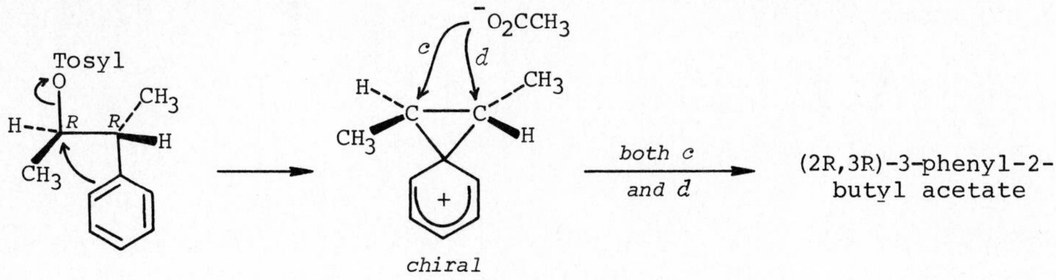

23.E Supplementary Problems

S1. Predict the products from the following reactions. If more than one product is anticipated, indicate which (if any) will predominate.

(a)

excess

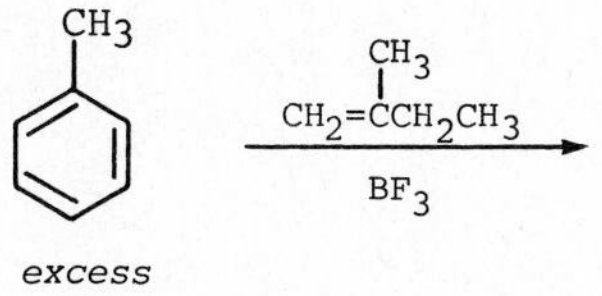

(b)

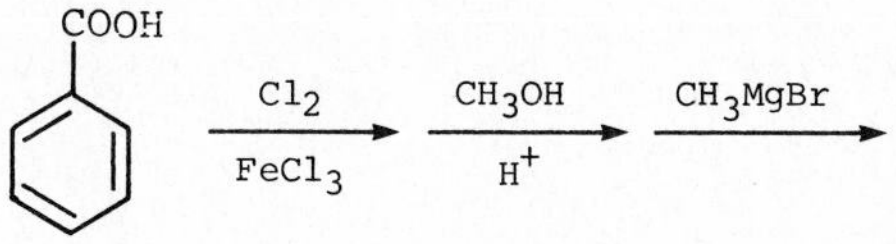

(c)

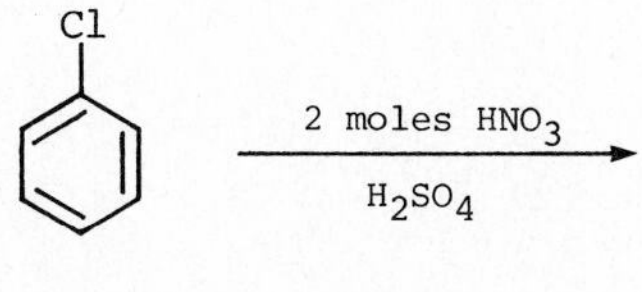

(d)

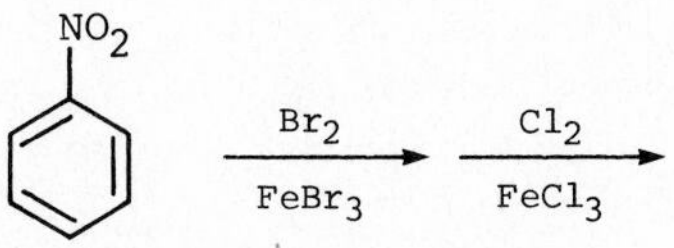

(e)

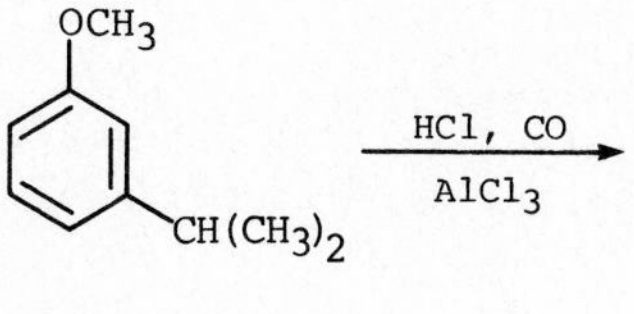

(f)

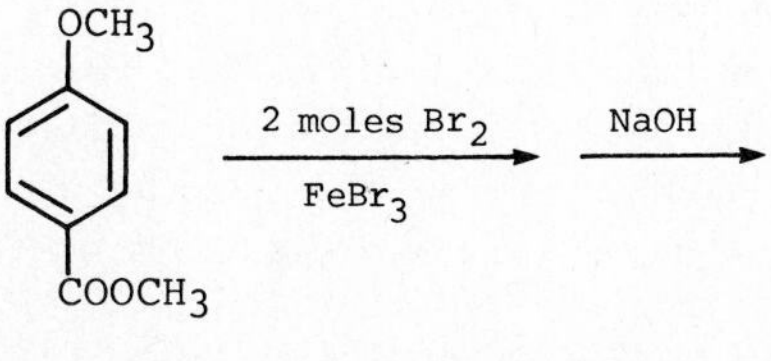

(g)

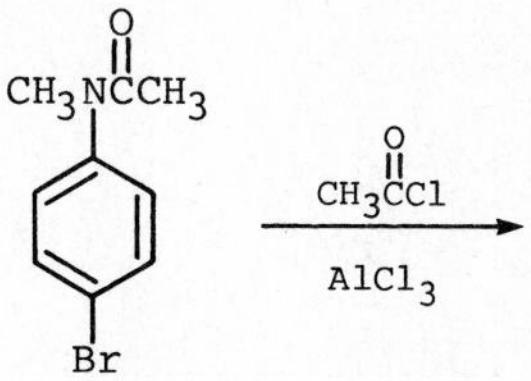

(h)

excess

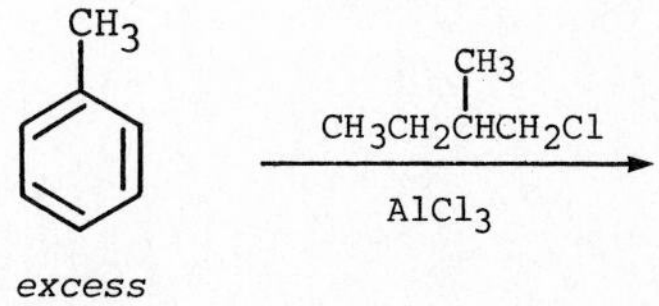

(i)

excess

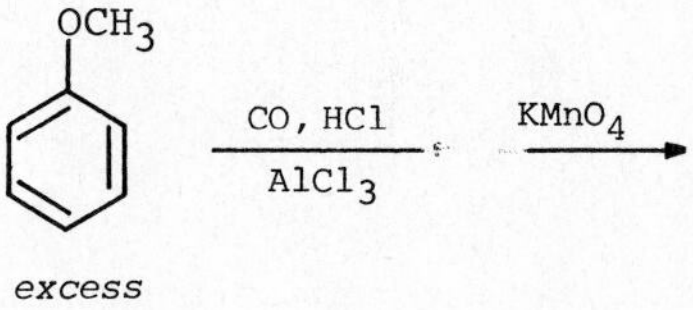

(j)

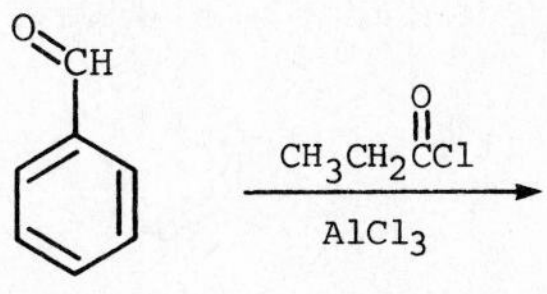

(k)

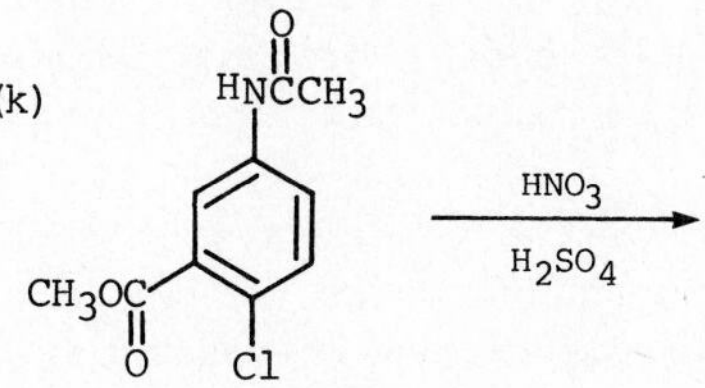

(ℓ)

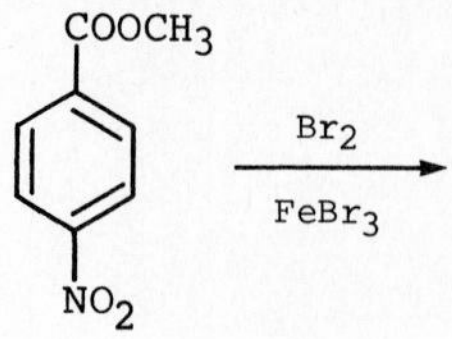

(m)

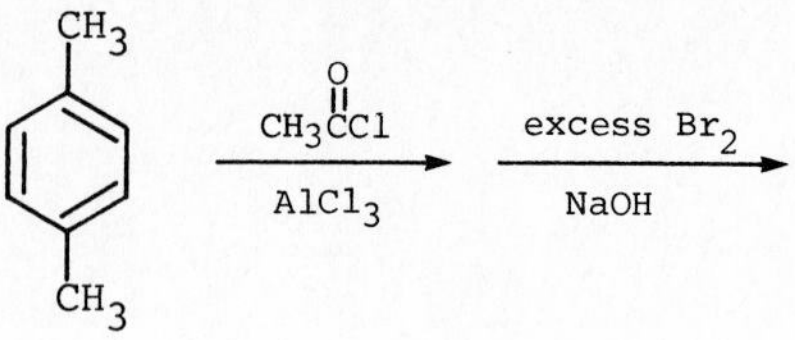

S2. Show how to make each of the following compounds (as free of isomers as possible), starting with benzene or toluene and any other reagents.

(a)

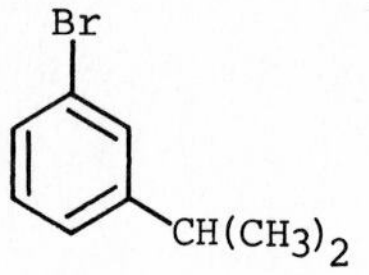

(b)

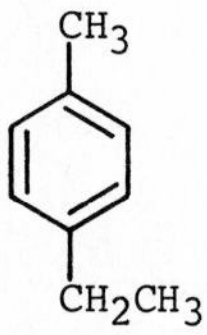

(c)

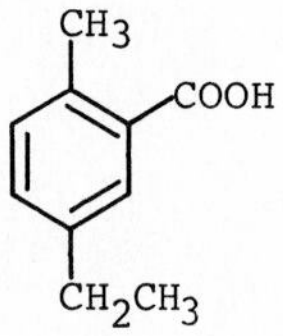

(d)

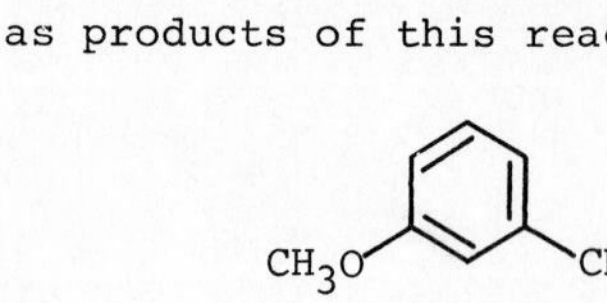

(e)

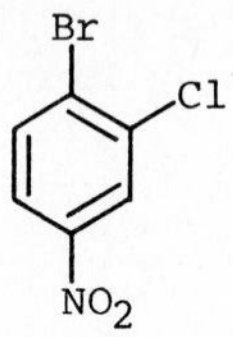

(f)

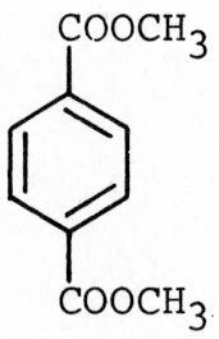

(g)

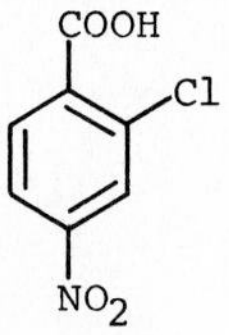

(h)

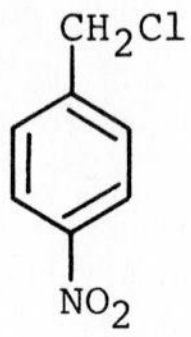

(i)

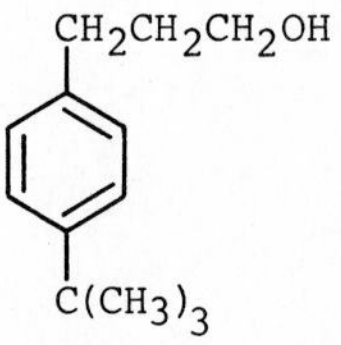

(j)

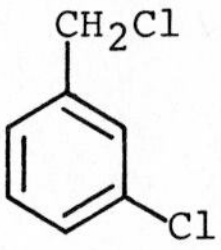

S3. Write a reasonable mechanism for the cyclization illustrated below, showing all of the intermediates involved. Do you expect to see any other compound(s) as products of this reaction?

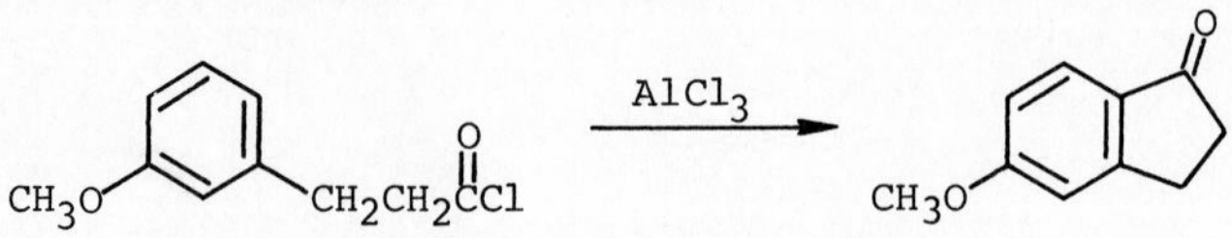

S4. Treatment of phenol with sodium nitrite and HCl results in the formation of *p*-nitrosophenol. Write a reasonable mechanism for the formation of the active electrophilic species and its reaction with phenol.

S5. None of the reaction sequences outlined on the next page will lead to the compounds shown as the major products. For each case, show what the main product(s) actually would be, and provide an alternative sequence that will produce the desired compound as the major isomer, *using the same starting materials.*

(a)

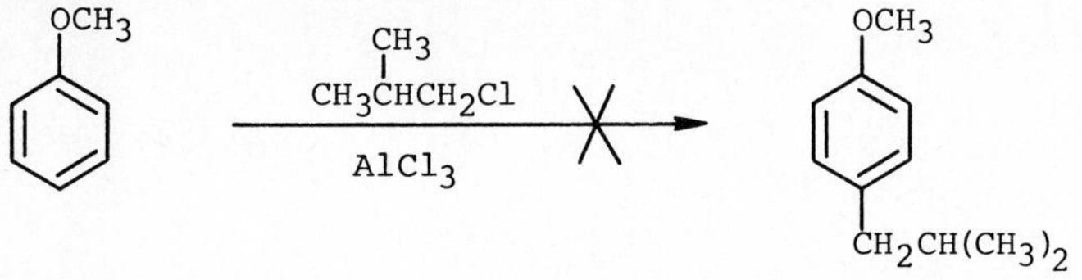

(b)

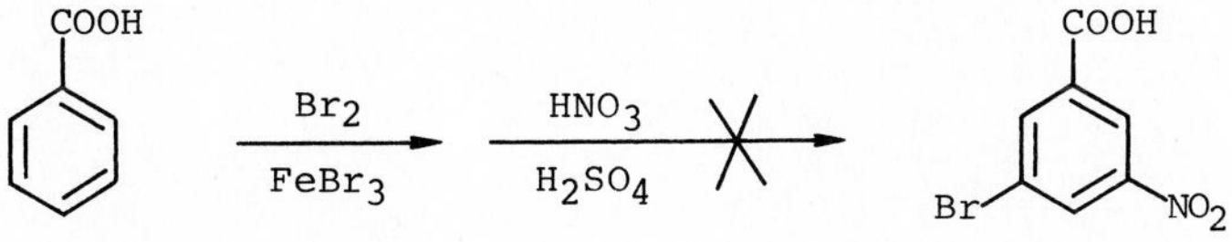

(c)

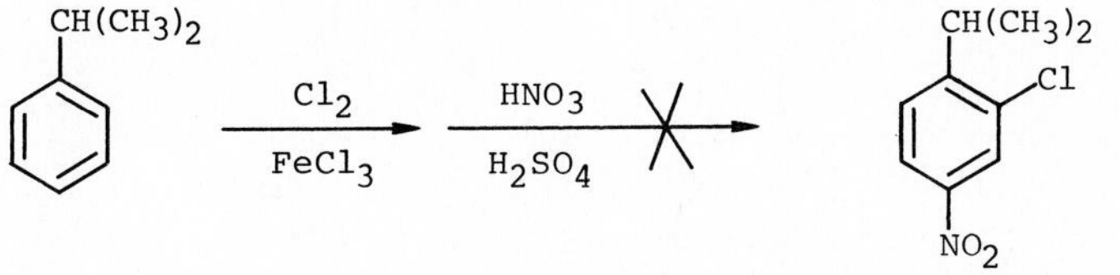

(d)

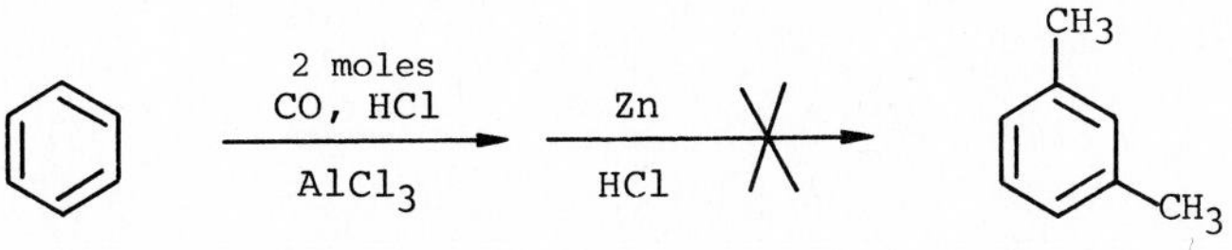

S6. Predict the favored position of electrophilic aromatic substitution of the following compounds, and justify your answer.

(a) 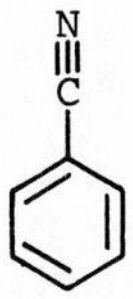

(b) 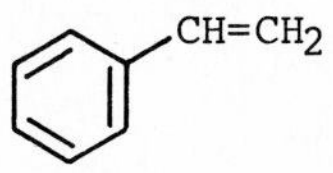

(c) 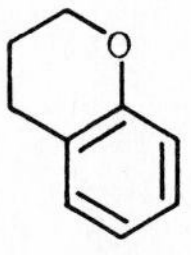

(d) 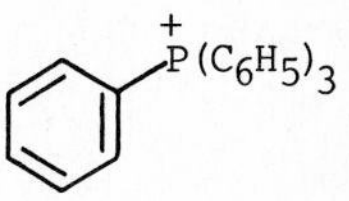

S7. If sodium triethylphenylborate is treated with D_2SO_4/D_2O, it undergoes cleavage to give deuteriobenzene and a triethylboron compound:

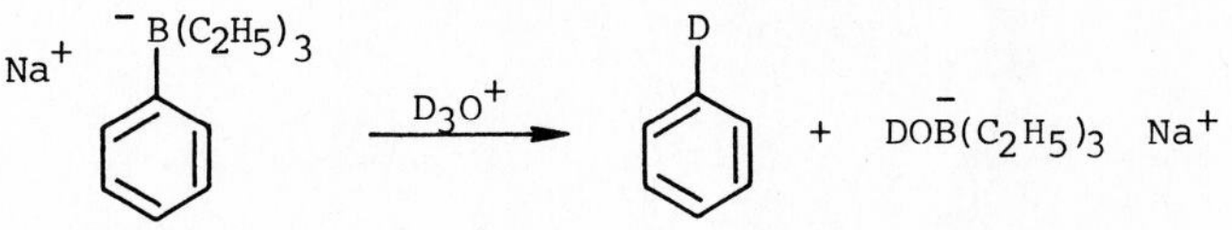

(a) Write a mechanism for this transformation.

(b) What product would you expect if you treated the same compound with Br_2? Write a mechanism for the reaction you predict.

S8. Write a mechanism for the following transformation:

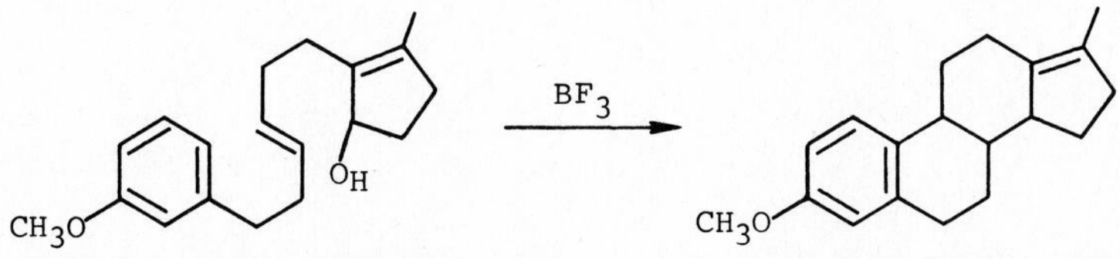

23.F Answers to Supplementary Problems

S1. (a) 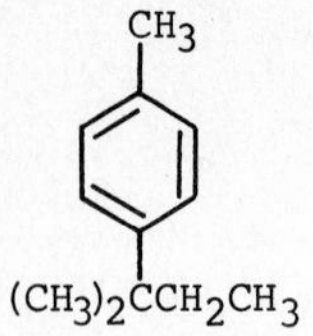 ; 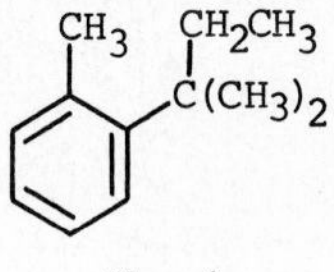
(less)

(b) 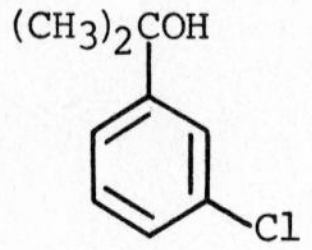

(c) 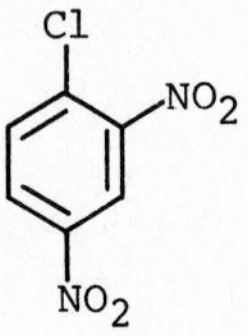 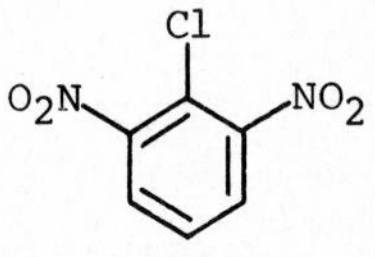
(less)

(d) 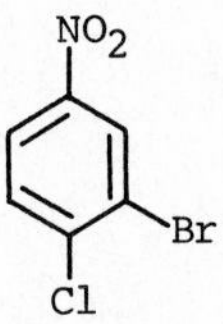 and 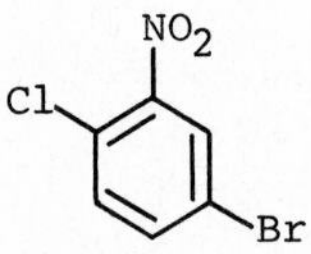

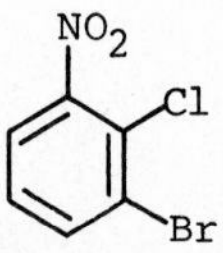
(less)

(e) 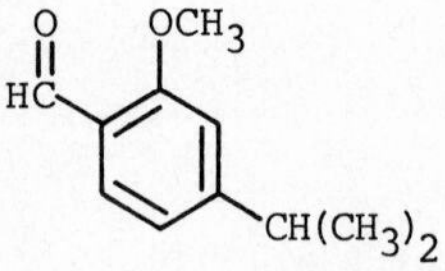 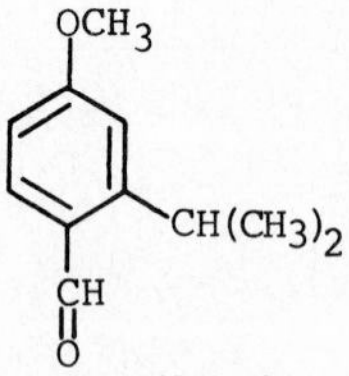
(less)

(f) 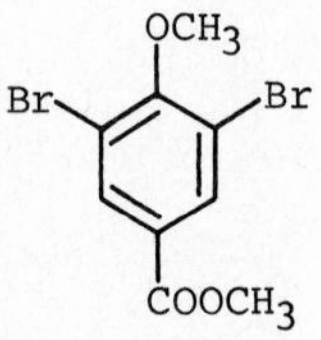

(g) 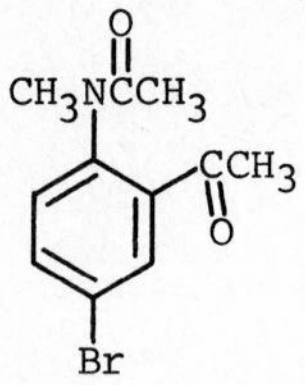

(h) 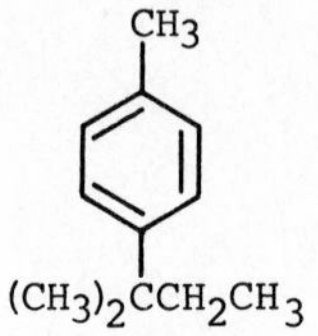 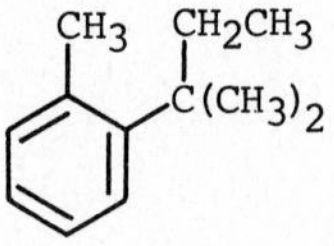
(less)

(i) 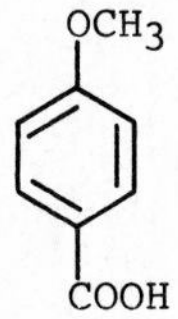 and 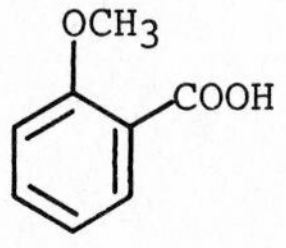

(j) 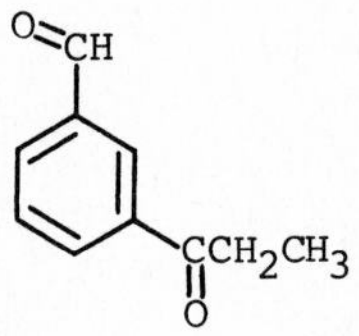

(k) 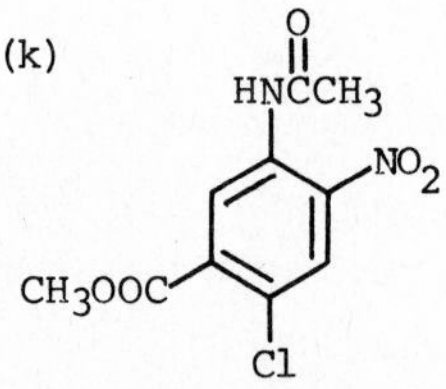 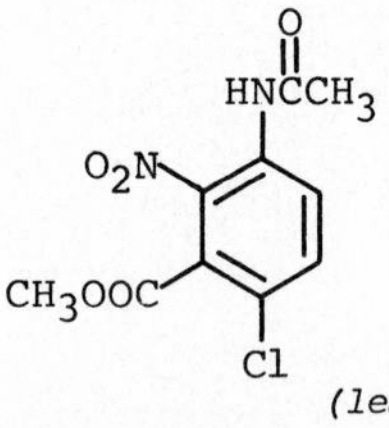
(less)

(ℓ) 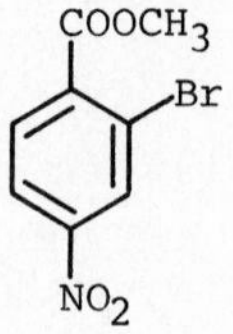 and 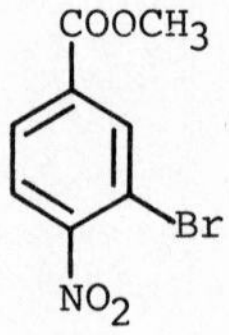

(m) 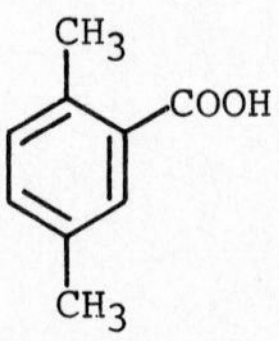

(Haloform reaction of a methyl ketone; note alkaline conditions)

S2. (a)

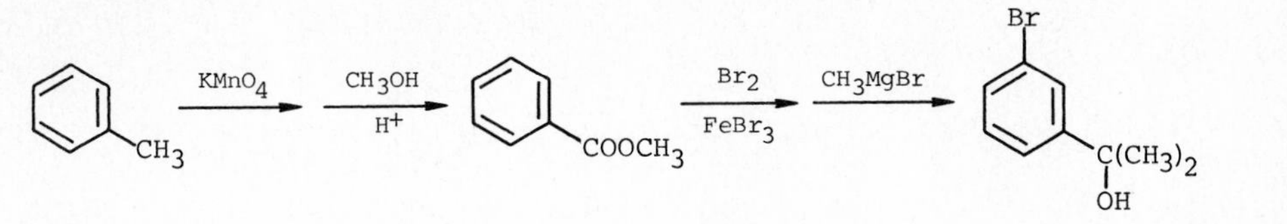
KMnO4
CH3OH
H+
COOCH3
Br2
FeBr3
CH3MgBr
Br
C(CH3)2
OH
H2/Pd
Br
CH(CH3)2
CH3

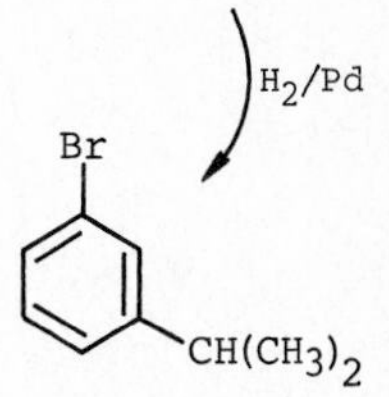

(b)

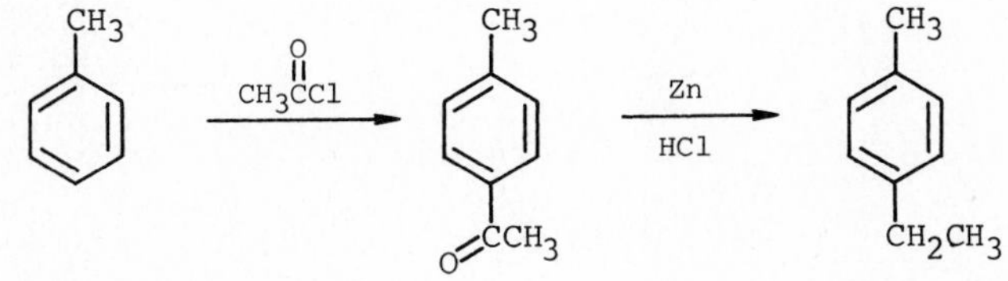
CH3
O
CH3CCl
CH3
CCH3
O
Zn
HCl
CH3
CH2CH3

(c)

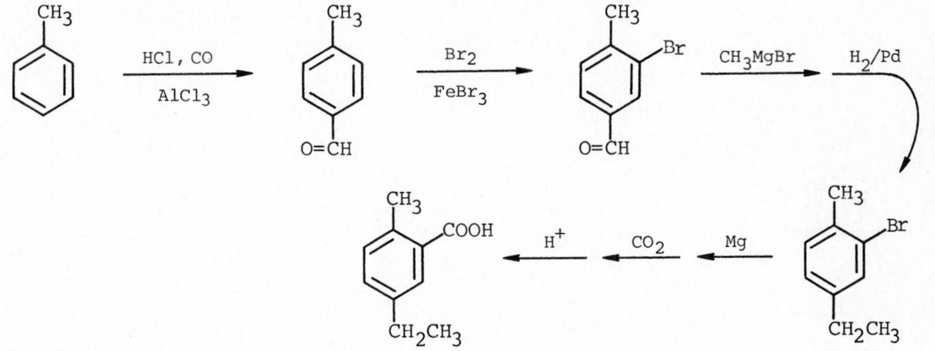
CH3
HCl, CO
AlCl3
CH3
O=CH
Br2
FeBr3
CH3
Br
O=CH
CH3MgBr
H2/Pd
CH3
Br
CH2CH3
Mg
CO2
H+
CH3
COOH
CH2CH3

(d)

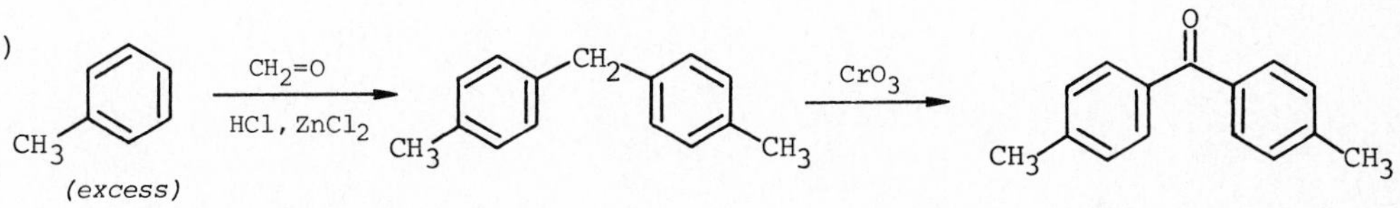
CH3
(excess)
CH2=O
HCl, ZnCl2
CH2
CH3
CH3
CrO3
O
CH3
CH3

(e)

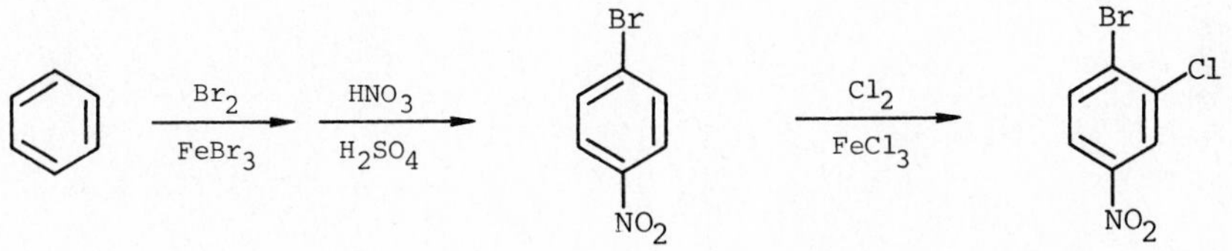
Br2
FeBr3
HNO3
H2SO4
Br
NO2
Cl2
FeCl3
Br
Cl
NO2

(f)

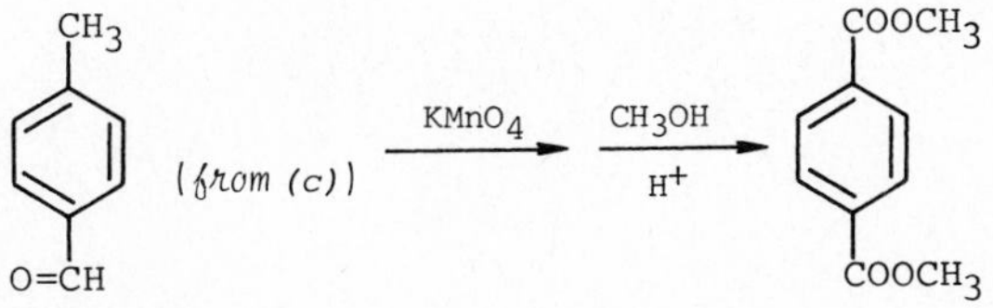
CH3
O=CH
(from (c))
KMnO4
CH3OH
H+
COOCH3
COOCH3

(g)

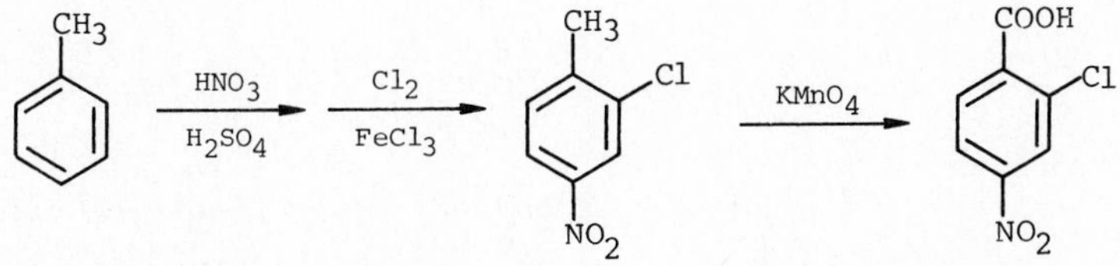
CH3
HNO3
H2SO4
Cl2
FeCl3
CH3
Cl
NO2
KMnO4
COOH
Cl
NO2

(h)

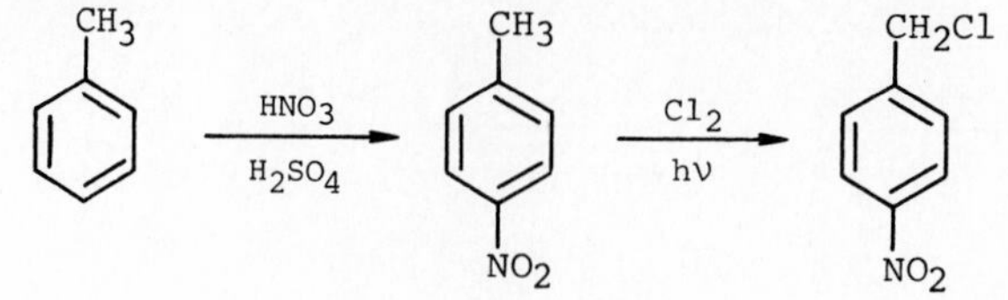

(separate from ortho)

(i)

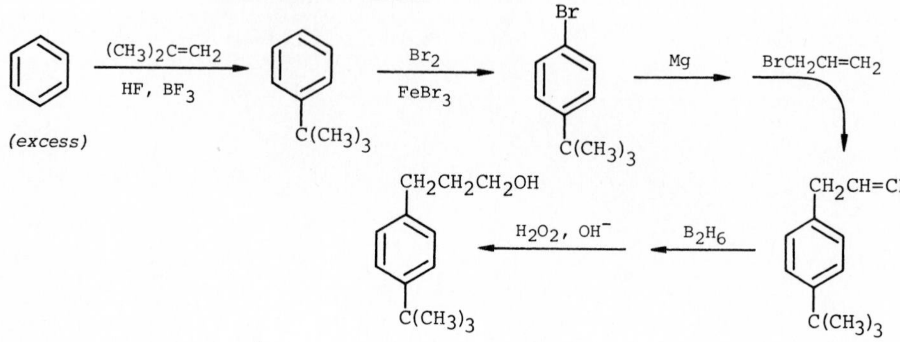

(j)

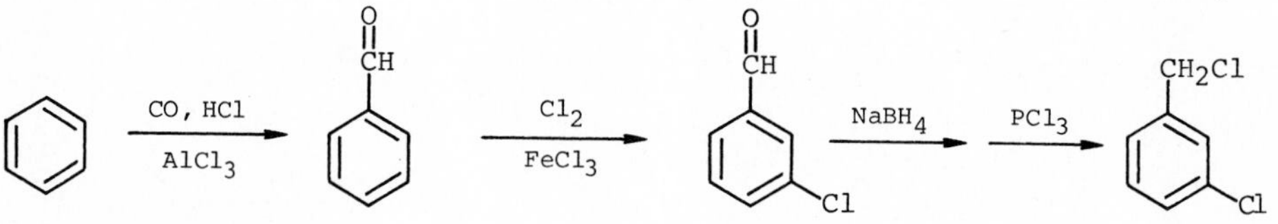

S3.

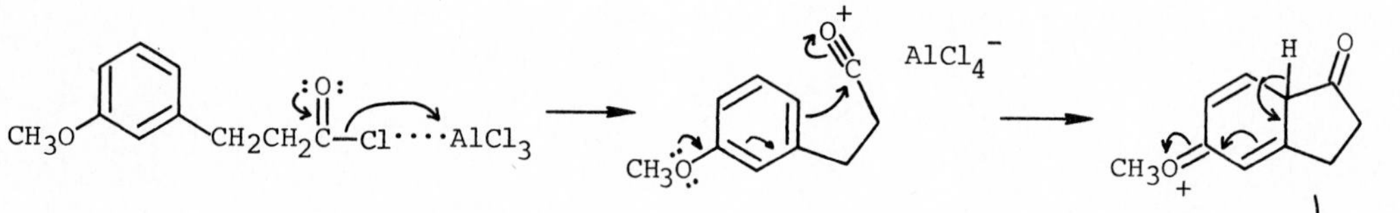

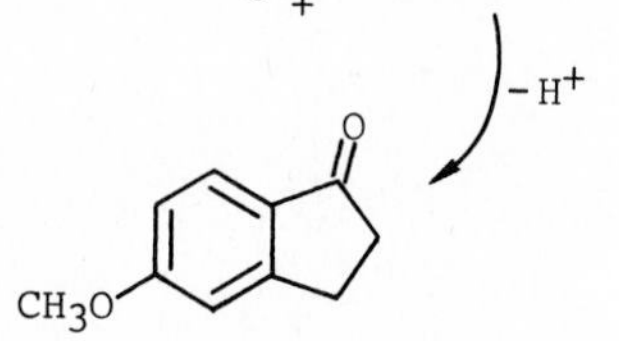

Some product from *ortho*
attack will also be seen:

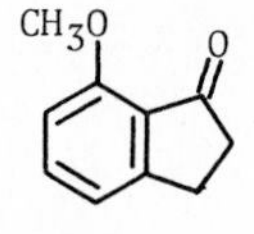

S4.

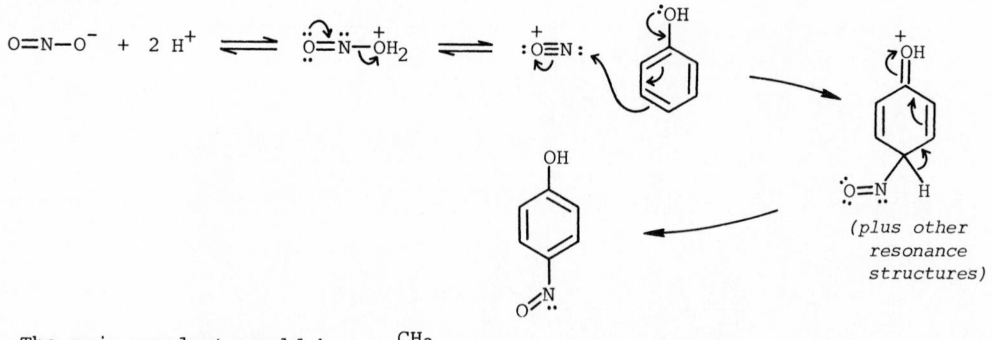

S5. (a) The main product would be

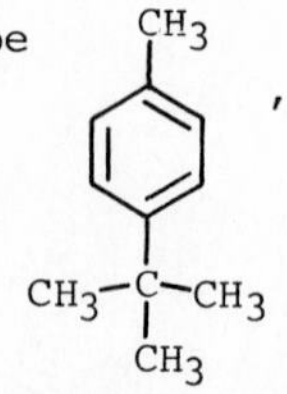

, from rearrangement of the primary alkyl halide before substitution.

To make the desired compound, employ an acylation/deoxygenation sequence:

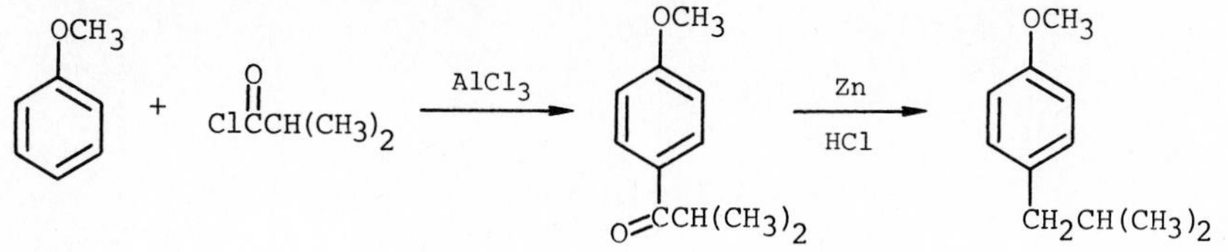

(b) Main products would be

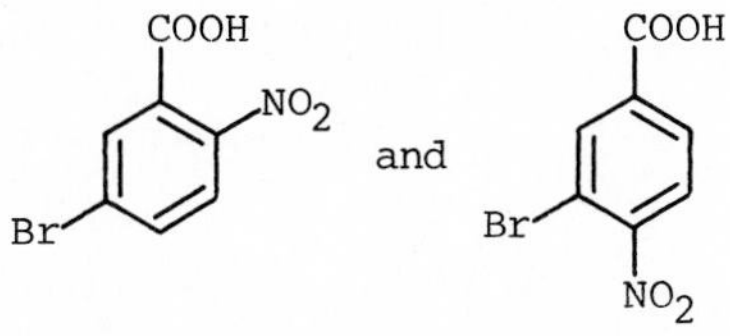

because Br (Group II) dominates COOH (Group III) in directing power.

To obtain the desired product, reverse the sequence of steps:

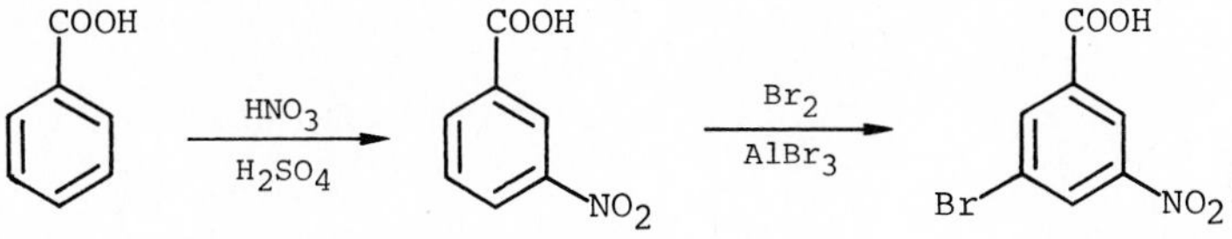

(c) The chlorine will go preferentially *para* (with some *ortho*), and then will compete with the isopropyl group to give a mixture of isomers:

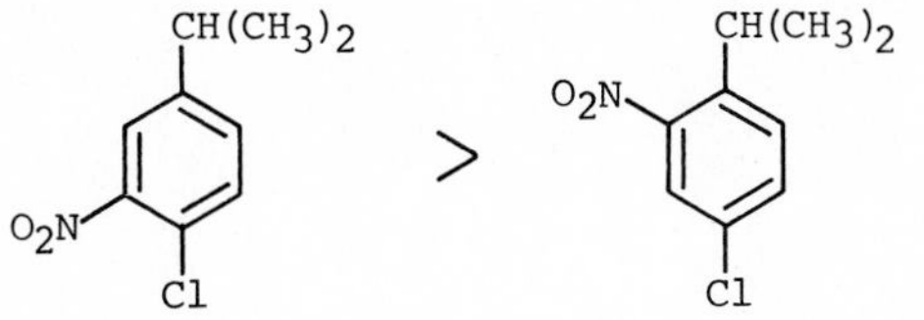

(for steric reasons)

Again, reversing the order of the reactions will furnish the correct product

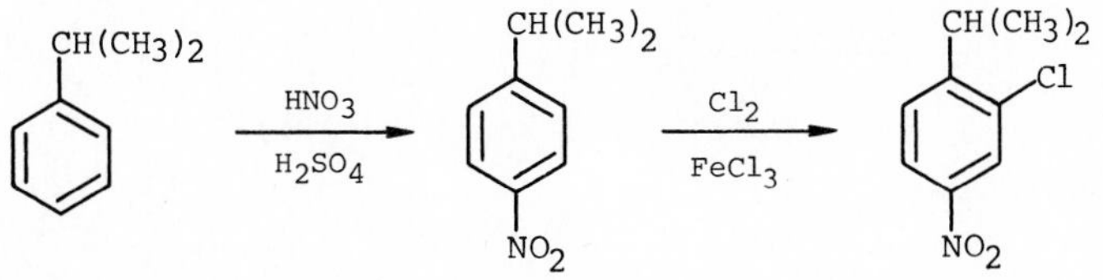

(d) Only one formyl group can be attached by a Gattermann-Koch reaction (as in Friedel-Crafts acylation), so toluene will be the overall product. To get around this requires a somewhat more involved sequence:

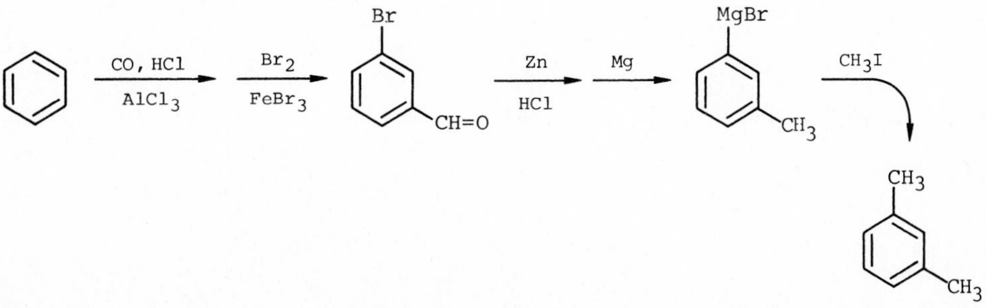

S6. (a)

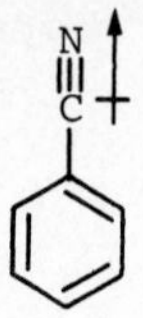

Dipole moment and resonance

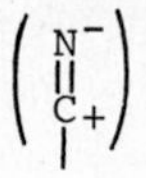

both disfavor adjacent positive charge, so this group will be deactivating, *meta*-directing.

(b) A vinyl group is *o,p*-directing because conjugation with the double bond helps to stabilize the positive charge:

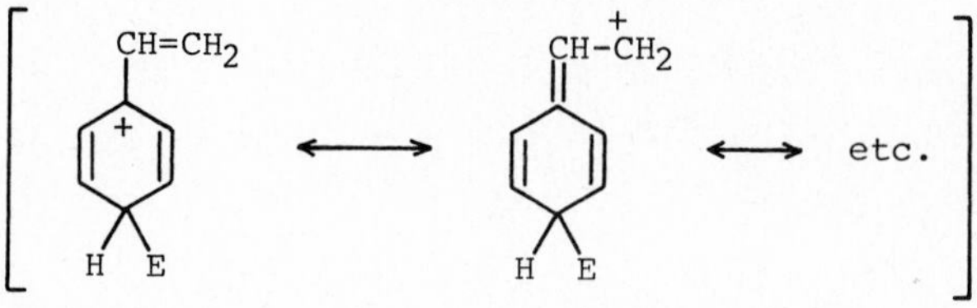

(*NOTE:* electrophilic attack on the vinyl group itself is preferred:

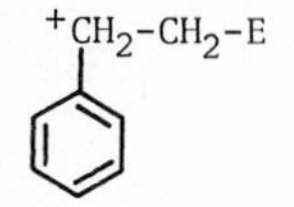

)

(c)

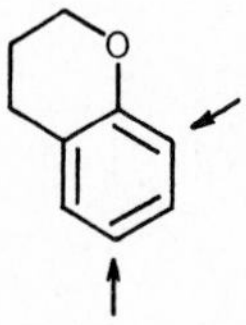

This is simply an aryl ether, and substitution will occur *ortho* and *para* to the oxygen.

(d)

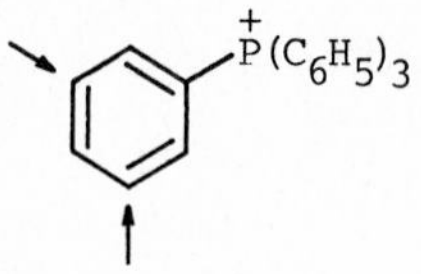

Because of the cationic phosphorus, the ring is deactivated, with the greatest effect at the *ortho* and *para* positions; this substituent is deactivating, *meta*-directing.

S7. (a)

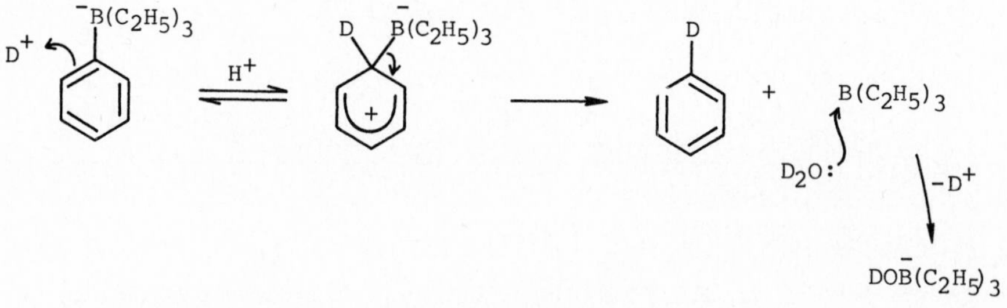

(This is simply the reverse of the electrophilic aromatic substitution reactions which we have been focussing on.)

(b)

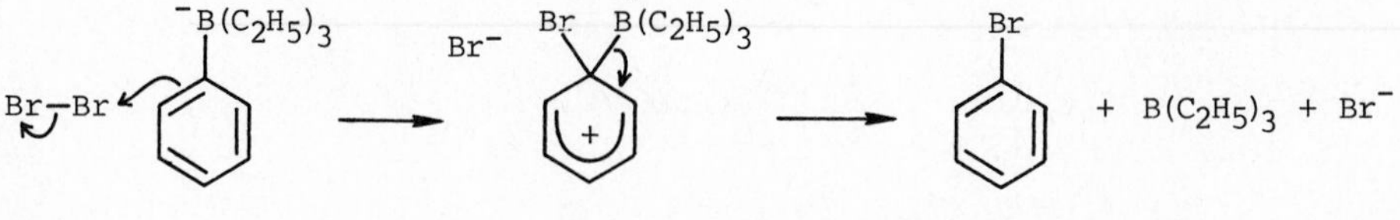

S8.

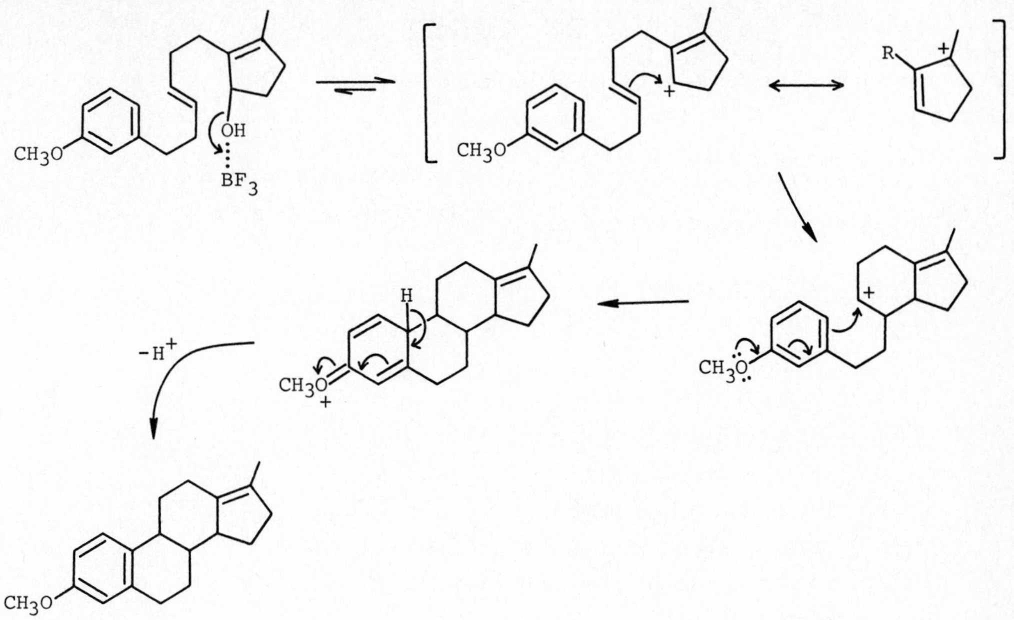

24. AMINES

24.A Chapter Outline, Important Terms Introduced, and Reactions Discussed

24.1 Structure

primary, secondary, and tertiary amines

quaternary ammonium compounds

nonbonding electron pair

nitrogen inversion

24.2 Nomenclature of Amines

-amine

amino-, alkylamino-

aniline

N-

24.3 Physical Properties of Amines

A. Colligative Properties

B. Spectroscopic Properties

1. Infrared spectra (weak N-H stretch)
2. Nuclear magnetic resonance spectra

   $C\underline{H}_3NR_2$: 2.2 ppm $\qquad$ $N\underline{H}$: 0.6-3.0 ppm

3. Mass spectra

   M^+ = odd m/e (for odd number of nitrogens)

   immonium ions

24.4 Basicity

$$R_3N: + H^+ \rightleftharpoons R_3\overset{+}{N}H$$

pK_a of alkylammonium compounds: 9-11

pK_a of anilinium ion: 4.6

resolution of racemic carboxylic acids via formation of diastereomeric salts

24.5 Quaternary Ammonium Compounds

A. Tertiary Amines as Nucleophiles

$$R_3N\colon \curvearrowright R'\!-\!X \longrightarrow R_3\overset{+}{N}R' \; X^-$$

B. Phase-Transfer Catalysis

salts soluble in organic solvents

greater reactivity because of reduced solvation

24.6 Synthesis of Amines

A. Direct Alkylation of Ammonia or other Amines

$$H_3N: + RX \longrightarrow H_3\overset{+}{N}R\; X^- \underset{}{\overset{NH_3}{\rightleftharpoons}} H_2NR \xrightarrow{RX} H_2\overset{+}{N}R_2$$

(polyalkylation problems)

B. Indirect Alkylation: The Gabriel Synthesis

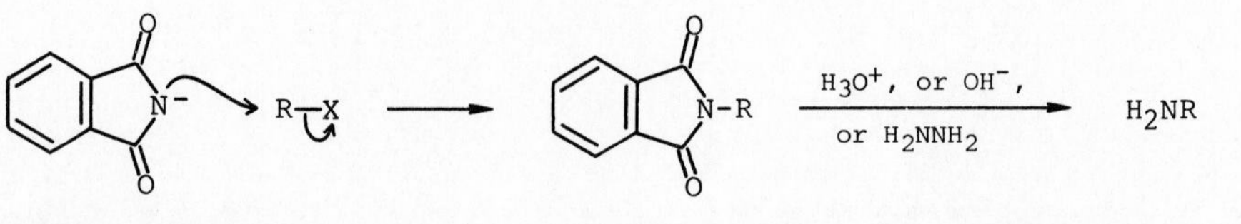

phthalimide anion

C. Reduction of Nitro Compounds

$$Ar-NO_2 \longrightarrow Ar-NH_2$$ *(using H_2/Ni, Fe/HCl, $SnCl_2/HCl$, $NaSH$, Zn or Sn/HCl, etc.; this method is best for aromatic compounds)*

D. Reduction of Nitriles

$$RC{\equiv}N \longrightarrow RCH_2NH_2$$ *(using H_2/Ni or $LiAlH_4$; possible secondary amine formation)*

E. Reduction of Oximes

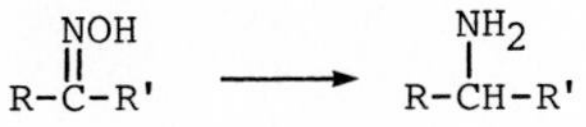

$$\overset{\overset{\displaystyle NOH}{\|}}{R-C-R'} \longrightarrow \overset{\overset{\displaystyle NH_2}{|}}{R-CH-R'}$$ *(using H_2/Ni or $LiAlH_4$)*

F. Reduction of Imines: Reductive Amination

$$\overset{\overset{\displaystyle O}{\|}}{R-C-R'} + H_2NR'' \rightleftharpoons \overset{\overset{\displaystyle H\overset{+}{N}-R''}{}}{R-C-R'} \longrightarrow \overset{\overset{\displaystyle HN-R''}{|}}{R-CH-R'}$$

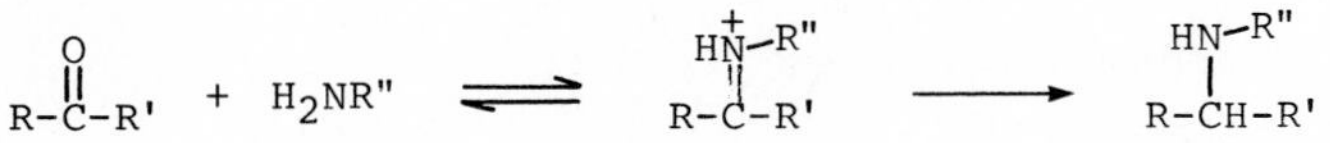

immonium ion

(using H_2/Ni; $H\overset{\overset{\displaystyle O}{\|}}{C}NR_2$, Δ (Leuckart); $CH_2{=}O/HCOOH$ (Eschweiler-Clarke))

G. Reduction of Amides

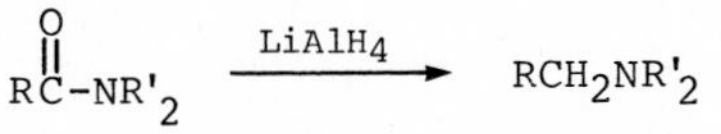

$$\overset{\overset{\displaystyle O}{\|}}{RC-NR'_2} \xrightarrow{LiAlH_4} RCH_2NR'_2$$ *(for primary, secondary, or tertiary amines)*

H. Preparation of Amines from Carboxylic Acids: The Hofmann, Curtius, and Schmidt Rearrangements

Hofmann:

$$\overset{\overset{\displaystyle O}{\|}}{R-C-NH_2} + X_2 + OH^- \longrightarrow \left[\overset{\overset{\displaystyle O}{\|}}{RCNHX}\right] \xrightarrow[-HX]{OH^-}$$

$$X = Br, Cl$$

Curtius:

$$\overset{\overset{\displaystyle O}{\|}}{R-C-Cl} + NaN_3 \longrightarrow \overset{\overset{\displaystyle O}{\|}}{R-C-N_3} \xrightarrow{\Delta} \overset{\overset{\displaystyle O}{\|}}{R-C-N:} \longrightarrow RN{=}C{=}O$$

acyl nitrene   *isocyanate*

Schmidt:

$$\overset{\overset{\displaystyle O}{\|}}{R-C-OH} + NaN_3 \xrightarrow{H_2SO_4}$$

$$RNH-\overset{\overset{\displaystyle O}{\|}}{C}-OR'$$ *carbamate*

$$\big\uparrow R'OH$$

$$\big\downarrow H_2O$$

$$\left[RNH-\overset{\overset{\displaystyle O}{\|}}{C}-OH\right]$$ *carbamic acid*

$$\swarrow -CO_2$$

$$RNH_2$$

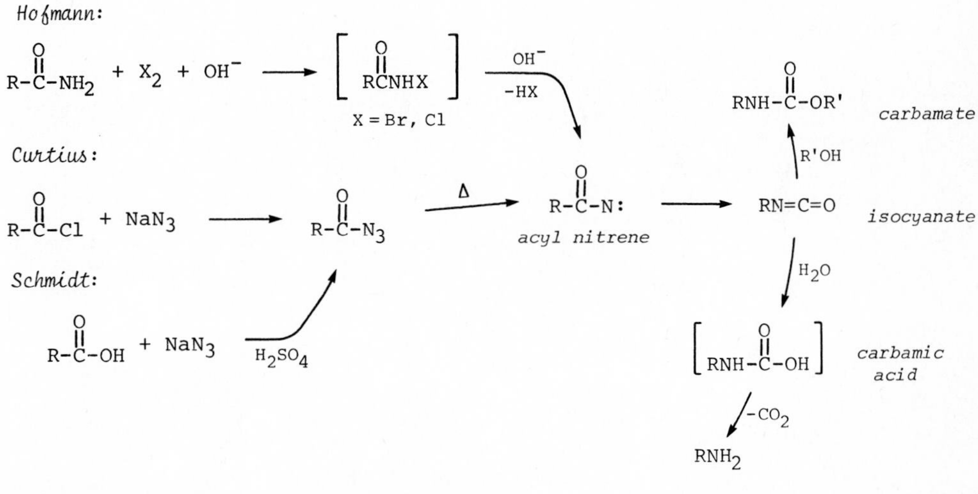

24.7 Reactions of Amines   *(see Sections 24.4, 24.5, and 24.6, too)*

A. Formation of Amides   *(see Section 19.4.B)*

$$\overset{\overset{\displaystyle O}{\|}}{R-C-Cl} + 2\ HNR'_2 \longrightarrow R\overset{\overset{\displaystyle O}{\|}}{C}-NR'_2 + H_2\overset{+}{N}R'_2\ Cl^-$$

B. Reaction with Nitrous Acid

nitrosation:

$$R_2NH + HONO \longrightarrow R_2N-N{=}O$$ *(via nitrosonium ion)*

diazotization:

$RNH_2 + HONO \rightleftharpoons R\overset{+}{N}{\equiv}N \xrightarrow{R = alkyl}$ "R+" $\longrightarrow$ *reactions of carbocations*

$R\overset{+}{N}{\equiv}N \xrightarrow{R = aryl}$ *arenediazonium ion*

C. Oxidation

hydroxylamines:

$$R_2NH + H_2O_2 \longrightarrow R_2NOH + H_2O$$

amine oxides:

$$R_3N: + H_2O_2 \longrightarrow R_3\overset{+}{N}{-}O^- + H_2O$$

nitro compounds:

$$Ar{-}NH_2 \xrightarrow{CF_3CO_3H} Ar{-}NO_2$$

D. Electrophilic Aromatic Substitution
(strong *ortho,para*-directing)

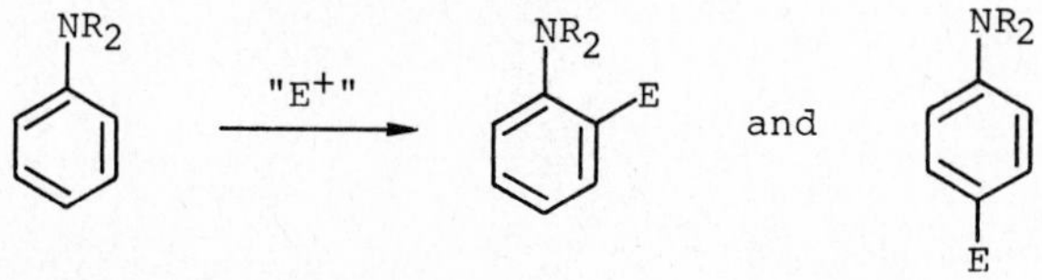

Vilsmeier reaction:

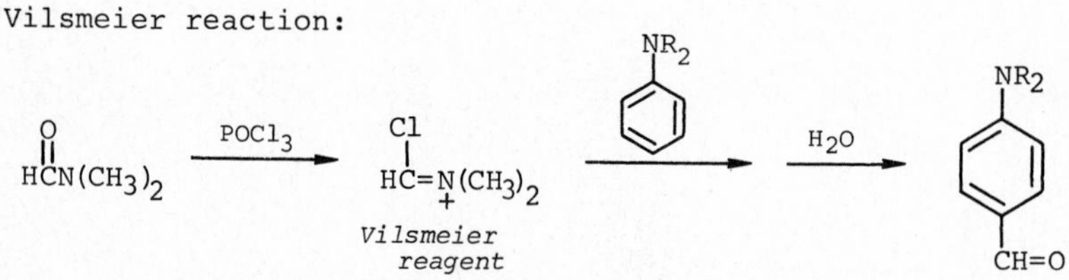

E. Elimination of the Amino Group: the Cope and Hofmann Elimination Reactions

Hofmann degradation:

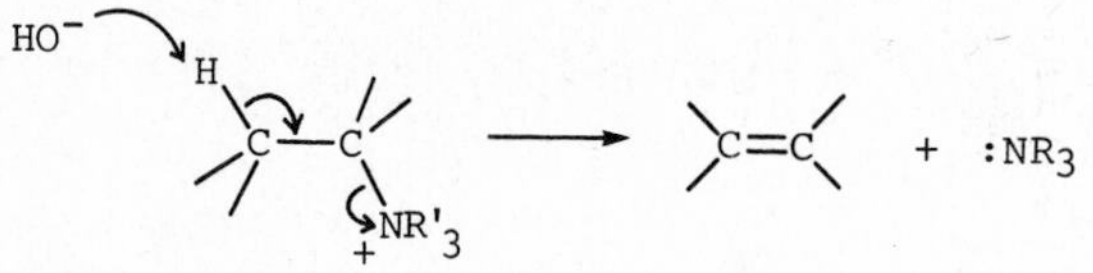

(Hofmann rule: favors formation of least-substituted olefin; anti elimination)

R' ≠ H; usually CH_3

Cope elimination:

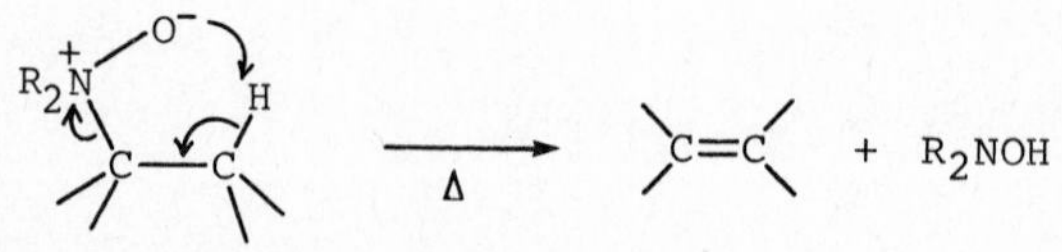

(syn elimination)

24.8 Enamines

formation, alkylation:

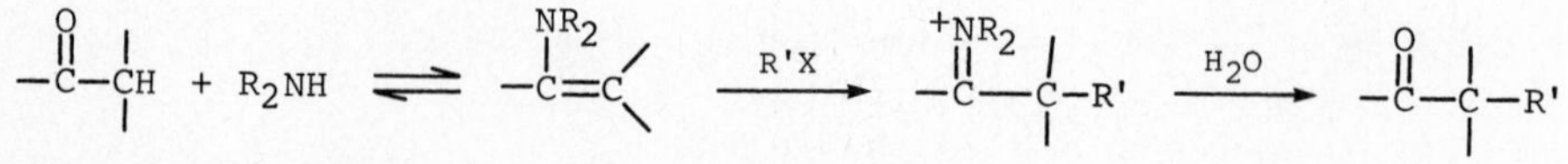

Mannich reaction:

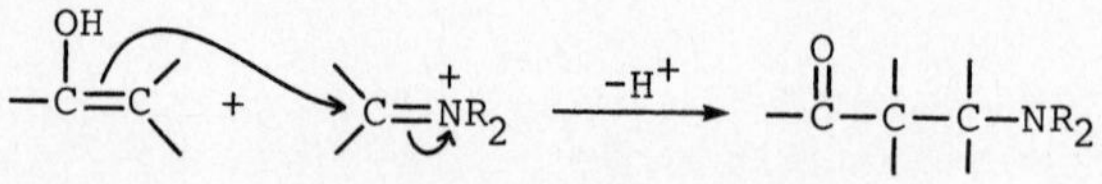

24.B Important Concepts and Hints

The amino group is different from the other functional groups you have encountered because, in its neutral form, it is basic and appreciably nucleophilic. Other functional groups require strong acid to be completely protonated, and usually it is only when they are in a deprotonated, anionic form that they are good nucleophiles (for instance: acetylide anions from acetylenes, enolates from ketones, etc.). As you know, acidity and basicity are commonly indicated by referring to the "pK_a" of a compound. The pK_a value indicates the position of the following equilibrium:

$$\text{H-Y} \underset{}{\overset{K_a}{\rightleftharpoons}} \text{Y}^- + \text{H}^+ \;; \qquad K_a = \frac{[\text{H}^+][\text{Y}^-]}{[\text{HY}]} \;; \qquad \text{p}K_a = -\log K_a$$

It is clearly convenient to use the same term to indicate both how acidic H-Y is and how basic Y^- is, since they are related. But confusion can arise if you forget that pK_a literally refers to a compound functioning as an acid; i.e., losing a proton. This has not been a problem when discussing the functional groups presented in the text previously, but it can arise in the chemistry of amines. In Chapter 4 of this Study Guide, we pointed out what was wrong with the common statement: "The pK_a of ammonia is 9." Two correct statements are: "The pK_a of ammonium ion is 9", or "The pK_a of ammonia is 35." Because acid-base chemistry plays such an important role in the reactions of amines, it would be a good idea for you to review Chapter 4 of the text and of this Study Guide.

Much of the chemistry of amines involves the formation and reactions of imines and immonium ions. These reactions are analogous to those of carbonyl compounds to a great extent, as you can see from the list below.

Starting Material	*Reaction*	*Product*
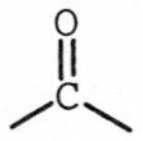	reduction (H_2/catalyst; $LiAlH_4$; $NaBH_4$; ($\underline{i}$-PrO)$_3$Al; etc.)	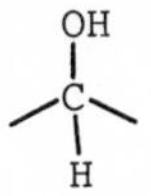
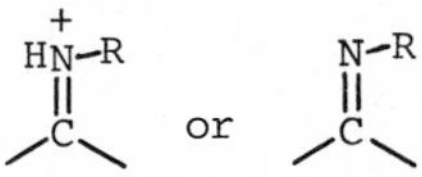	reduction (H_2/catalyst; $LiAlH_4$; $NaBH_4$ or $NaBH_3CN$; Eschweiler-Clarke and Leuckart)	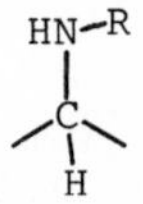
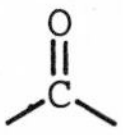	aldol condensation 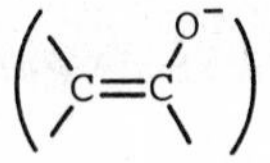	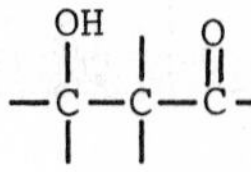
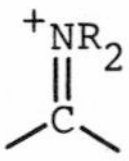	Mannich reaction 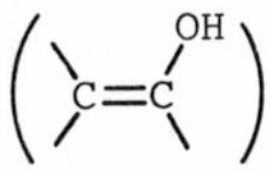	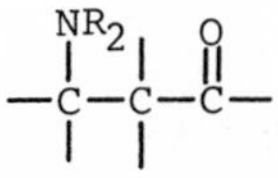
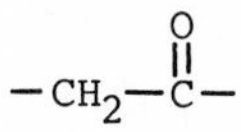	enol formation (acid- or base-catalyzed)	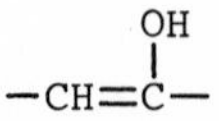
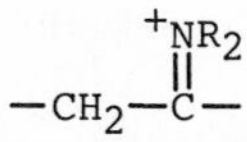	enamine formation	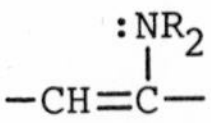

In general, imines (the neutral forms) are less reactive toward nucleophilic attack than are their carbonyl counterparts, and immonium ions (the cationic forms) are more reactive than their carbonyl analogs.

24.C Answers to Exercises

(24.2)   methanamine   N-ethylethanamine   N,N-dipropylpropanamine

N-methylethanamine   N-ethyl-N-methylethanamine   N-ethyl-N-methylcyclopropanamine

1-methylpropanamine   1-ethyl-3-methylbutanamine

N,N-dimethylethanamine   N,N-diethyl-1,3-dimethylbutanamine

N-ethyl-N,1-dimethylpropanamine   N,3-dimethylpentanamine

benzenamine   3-bromobenzenamine   4-nitrobenzenamine   N,N-dimethylbenzenamine

4-methylaminobutanoic acid   4-aminobenzoic acid

4-methylbenzenamine   3-methoxybenzenamine   2-ethoxybenzenamine

(24.3.B)

	IR (cm^{-1}) (N-H wag is 666-909 cm^{-1})	NMR (δ) (NH will come as a singlet, ~0.6-3.0)	Mass spec (m/e) (M$^+$ = 73 if visible)
CH$_3$CH$_2$CH$_2$CH$_2$NH$_2$	3400, 3500 (doublet, weak)	2.2 (t, 2H)	30 (CH$_2$=$\overset{+}{\text{N}}$H$_2$)
(CH$_3$)$_2$CHCH$_2$NH$_2$	3400, 3500 (doublet, weak)	2.4 (d, 2H)	30 (CH$_2$=$\overset{+}{\text{N}}$H$_2$)
CH$_3$CH$_2$$\overset{\text{CH}_3}{\underset{\mid}{\text{CH}}}$-NH$_2$	3400, 3500 (doublet, weak)	2.8 (m, 1H)	58 (CH$_3$CH$_2$CH=$\overset{+}{\text{N}}$H$_2$) 44 (CH$_3$CH=$\overset{+}{\text{N}}$H$_2$)
(CH$_3$)$_3$CNH$_2$	3400, 3500 (doublet, weak)	1.2 (s, 9H)	58 ((CH$_3$)$_2$C=$\overset{+}{\text{N}}$H$_2$)
CH$_3$CH$_2$CH$_2$NHCH$_3$	3310-3350 (very weak)	2.2 (s, 3H) 2.4 (t, 2H)	44 (CH$_2$=$\overset{+}{\text{N}}$HCH$_3$)
(CH$_3$)$_2$CHNHCH$_3$	3310-3350 (very weak)	2.2 (s, 3H) 2.8 (m, 1H)	58 (CH$_3$CH=$\overset{+}{\text{N}}$HCH$_3$)
(CH$_3$CH$_2$)$_2$NH	3310-3350 (very weak)	ethyl group	58 (CH$_2$=$\overset{+}{\text{N}}$HCH$_2$CH$_3$)
CH$_3$CH$_2$N(CH$_3$)$_2$	no N-H bands	2.2 (s, 6H) ethyl group	58 (CH$_2$=$\overset{+}{\text{N}}$(CH$_3$)$_2$)

(24.4)

CH$_3$CH$_2$$\overset{+}{\text{N}}H_3$ +

pK$_a$ = 10.64

$\ce{->[K] <-}$

CH$_3$CH$_2$NH$_2$ +

pK$_a$ = 4.60

$$K = \frac{10^{4.60}}{10^{10.64}} = 10^{-6.04} = 9.1 \times 10^{-7}$$

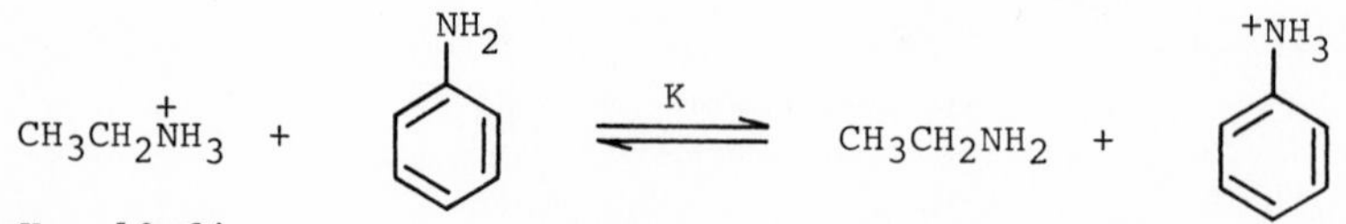

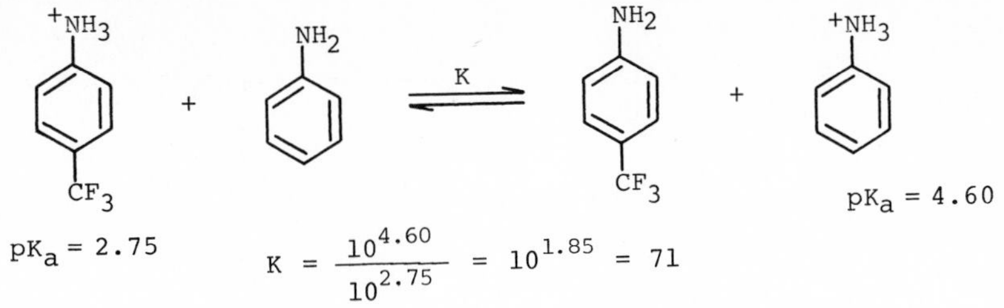

(24.5.B)

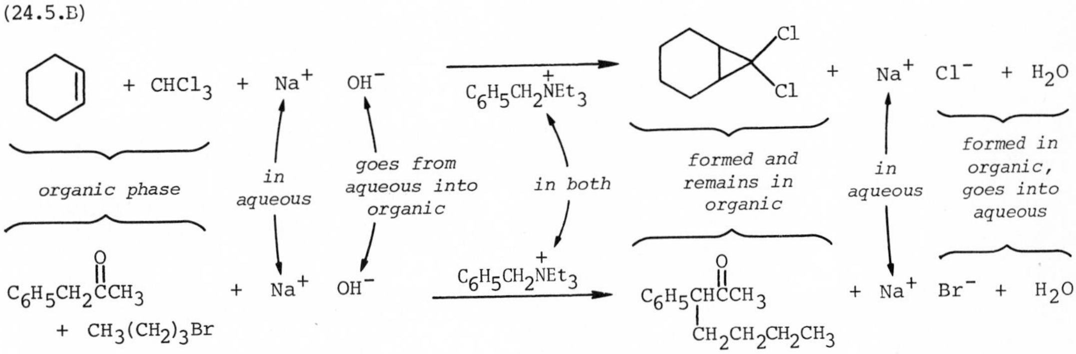

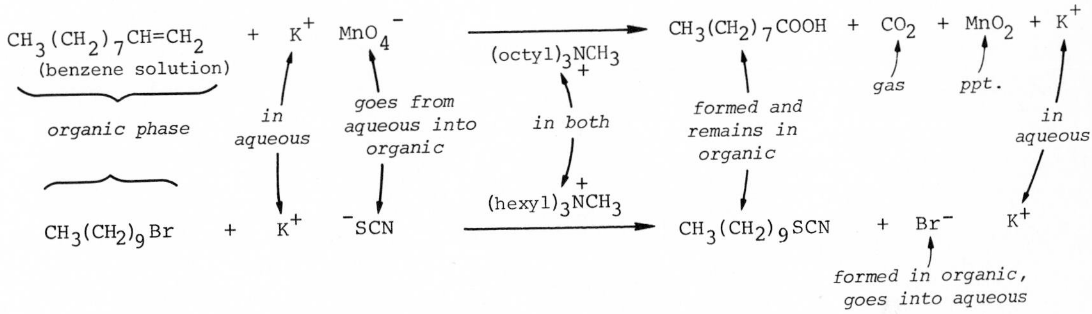

(24.6.B) Alkylation of the phthalimide ion involves S_N2 displacement, which will not occur with a neopentyl halide (too sterically hindered) or a t-butyl halide (E2 elimination instead). In addition, the sequence can only give primary amines, and cannot provide di-n-propylamine.

(24.6.C)

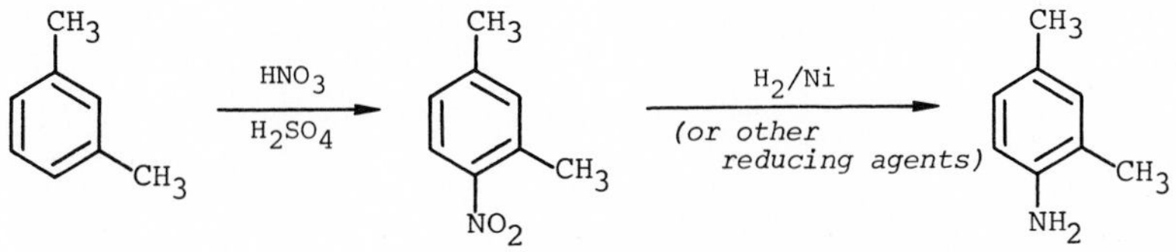

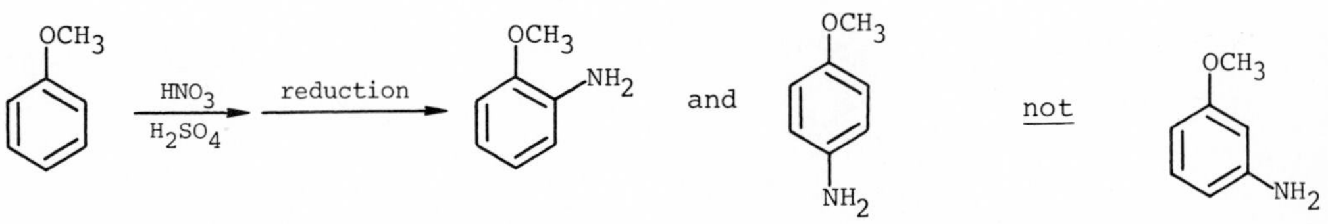

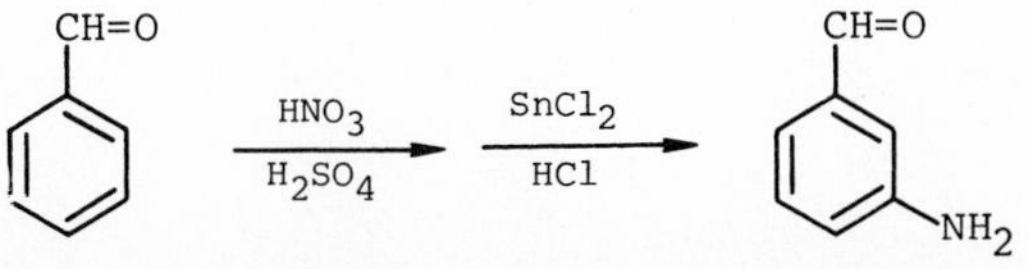

(not *ortho*)

(24.6.D)

(a) $(CH_3)_2CHCH_2CH_2Br$ $\xrightarrow{NaCN}$ $(CH_3)_2CHCH_2CH_2C\equiv N$ $\xrightarrow{LiAlH_4}$ $(CH_3)_2CHCH_2CH_2CH_2NH_2$

(b) $(CH_3)_2CHCH=O$ $\xrightarrow{HCN}$ $(CH_3)_2CH\overset{OH}{\underset{}{CH}}C\equiv N$ $\xrightarrow{H_2/Ni}$ $(CH_3)_2CH\overset{OH}{\underset{}{CH}}CH_2NH_2$

(24.6.F)

(a)

(b)

(c)

(d)

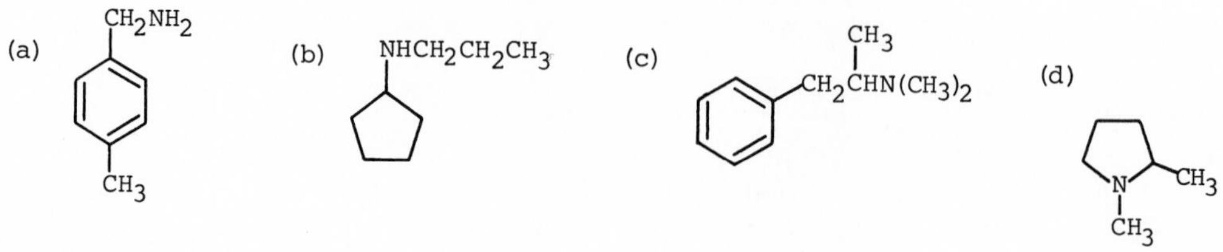

(24.6.G)

(a) $CH_3CH_2CH_2CH_2NH_2$ $\xrightarrow[Et_3N]{Cl\overset{O}{C}CH(CH_3)_2}$ $CH_3CH_2CH_2CH_2NH\overset{O}{C}CH(CH_3)_2$ $\xrightarrow{LiAlH_4}$ $CH_3CH_2CH_2CH_2NHCH_2CH(CH_3)_2$

(b) $(CH_3)_2CHNH_2$ $\xrightarrow{CH_3\overset{O}{C}Cl}$ $\xrightarrow{LiAlH_4}$ $(CH_3)_2CHN\underset{CH_2CH_3}{\overset{H}{\mid}}$ $\xrightarrow[2.\ LiAlH_4]{1.\ Cl\overset{O}{C}CH_2CH_2CH_3}$ $(CH_3)_2CHN\underset{CH_2CH_3}{\overset{}{\mid}}CH_2CH_2CH_2CH_3$

(24.6.H)

$CH_3CH_2\underset{}{\overset{CH_3}{\mid}}CH-CH_2CH_2COOH$

or

$\xrightarrow{SOCl_2}$ $R-\overset{O}{C}Cl$ $\xrightarrow{NH_3}$ $R-\overset{O}{C}NH_2$ $\xrightarrow{NaOH,\ Cl_2}$

$\xrightarrow{NaN_3}$ $R-\overset{O}{C}-N_3$ $\xrightarrow{\Delta}$ $R-NH_2$

$\xrightarrow[H_2SO_4]{NaN_3,}$ $R-\overset{O}{C}-N_3$

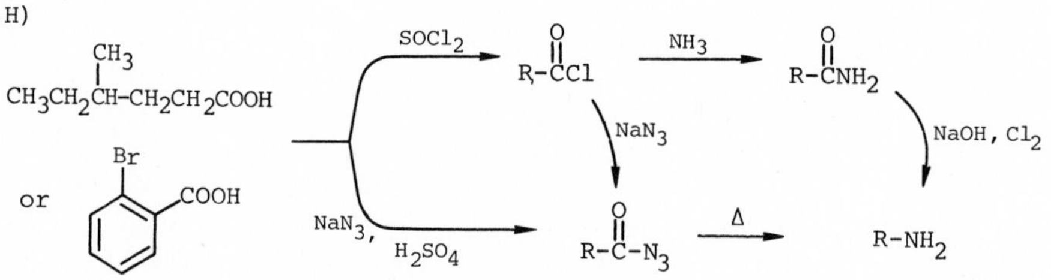

(24.7.B)

(a) $CH_3CH_2CH_2NH_2$ $\xrightarrow[HCl]{NaNO_2}$ $CH_3CH_2CH_2Cl$ + $CH_3CH_2CH_2OH$ + $CH_3\underset{OH}{\overset{}{CH}}CH_3$

(b) $(CH_3CH_2CH_2)_2NH$ $\xrightarrow[HCl]{NaNO_2}$ $(CH_3CH_2CH_2)_2NN=O$

(c) $(CH_3CH_2CH_2)_3N$ $\xrightarrow[HCl]{NaNO_2}$ $\left[(CH_3CH_2CH_2)_3\overset{+}{N}H \quad Cl^- \right]$ *only product*

(d)

(e)

(f)

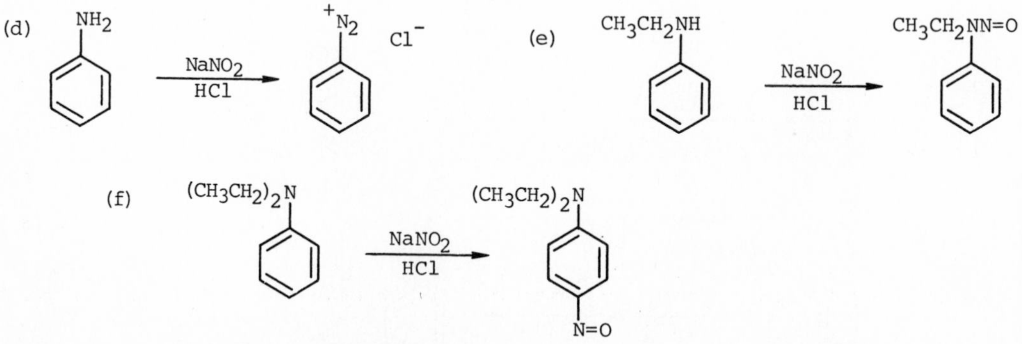

(24.7.D)

(a)

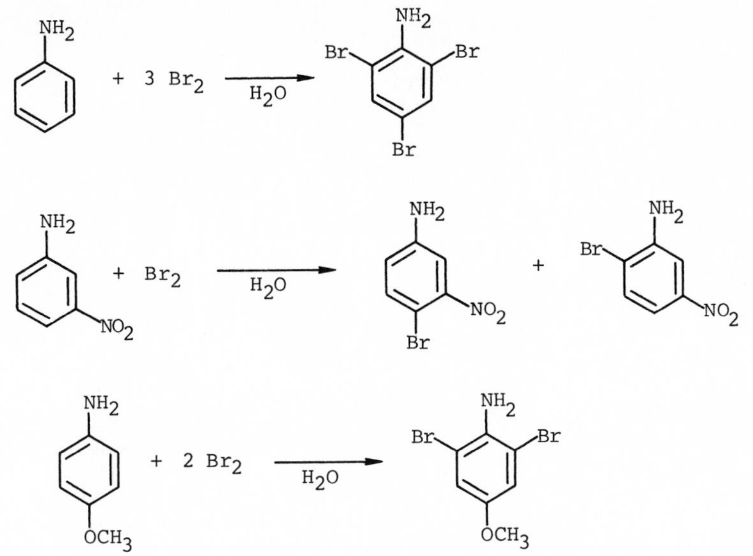

(b)

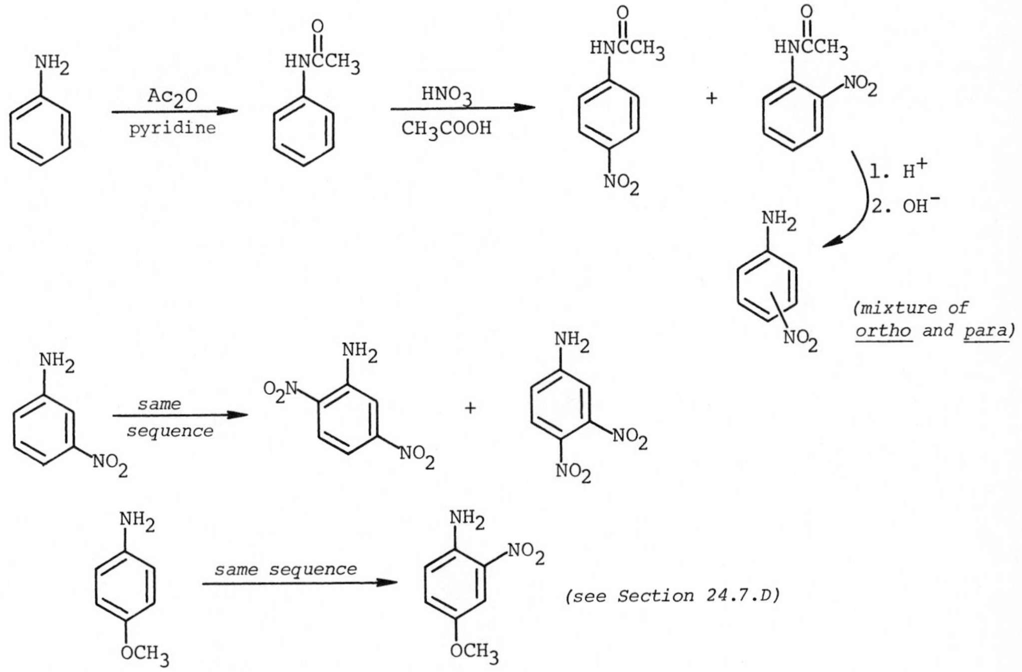

(mixture of ortho and para)

(see Section 24.7.D)

(24.7.E)

(a) $CH_3(CH_2)_4N(CH_3)_2 \xrightarrow{CH_3I} CH_3(CH_2)_4\overset{+}{N}(CH_3)_3\ I^- \xrightarrow{AgOH} CH_3(CH_2)_4\overset{+}{N}(CH_3)_3\ {}^-OH \xrightarrow{\Delta} CH_3CH_2CH_2CH{=}CH_2 + (CH_3)_3N$

(b) $CH_3(CH_2)_7N(CH_2CH_3)_2 \xrightarrow[2.\ AgOH]{1.\ CH_3I} \xrightarrow{\Delta} CH_3(CH_2)_7N(CH_3)CH_2CH_3 + CH_2{=}CH_2$

(c) $(CH_3CH_2)_3N \xrightarrow{H_2O_2} (CH_3CH_2)_3\overset{+}{N}{-}O^- \xrightarrow{\Delta} (CH_3CH_2)_2NOH + CH_2{=}CH_2$

(d)

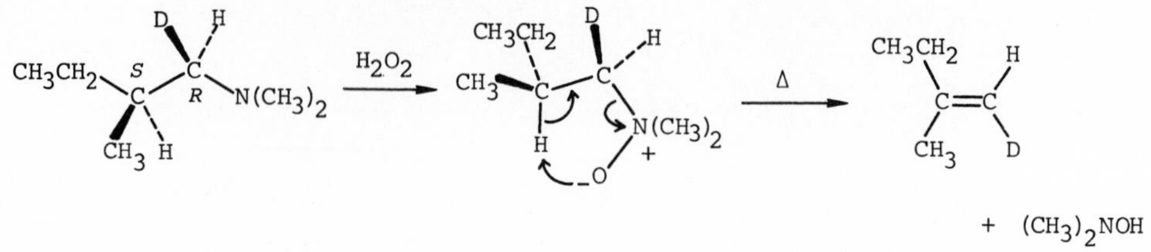

(24.8) $CH_3CH_2\overset{O}{\overset{\|}{C}}CH_2CH_3 + CH_2{=}O + (CH_3)_2\overset{+}{N}H_2\ Cl^- \longrightarrow CH_3CH_2\overset{O}{\overset{\|}{C}}\underset{CH_3}{\underset{|}{C}H}CH_2N(CH_3)_2$

(*NOTE:* this product should actually be named 2-methyl-1-dimethylamino-3-pentanone)

24.D Answers and Explanations for Problems

1. (a) 2-methylbutanamine
   (b) trimethylamine (N,N-dimethylmethanamine)
   (c) N-ethyl-N-methyl-2-propenamine
   (d) *p*-bromophenyltrimethylammonium chloride
   (e) N-ethyl-N-nitrosobenzenamine
   (f) N-ethyl-4-nitrosobenzenamine
   (g) ethyltrimethylammonium iodide
   (h) N,N-dimethylpropanamine oxide
   (i) 1-ethyl-N,N-dimethylpropanamine
   (j) N-ethyl-3-methylbenzenamine
   (k) N,4-diisopropyl-N-methyl-benzenamine
   (ℓ) 2,4,6-trichloroaniline

2. $R_1R_2R_3R_4N^+ + X^- \rightleftharpoons R_1X + R_2R_3R_4N$ (inverts)

   In this specific case, the most reactive group for S_N2 reaction by X^- is allyl:

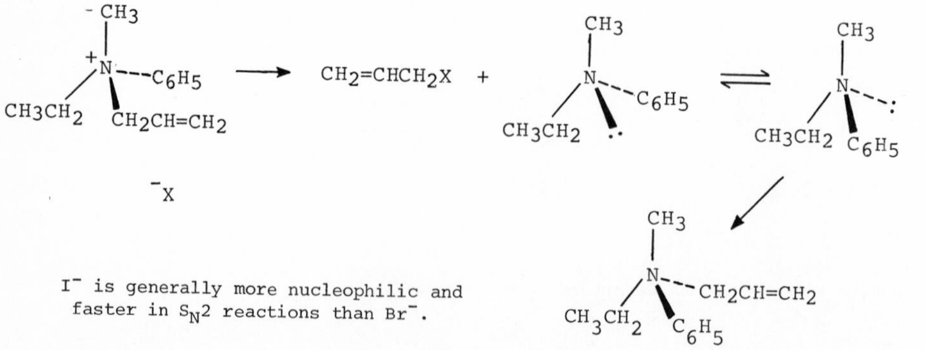

I^- is generally more nucleophilic and faster in S_N2 reactions than Br^-.

3. Looking first at the ir spectrum and Table 24.2, we find a doublet at 3290 and 3370 cm^{-1} which suggests NH_2, although the frequencies are somewhat lower than the Table indicates. The presence of primary amine ($-NH_2$) is further suggested by the bands at 1600 cm^{-1} (NH_2 bend), 1160 cm^{-1} (C-N stretch), and 850 cm^{-1} (N-H wag).

The sextuplet at δ 2.8 suggests $\overset{H}{C}-N$ adjacent to five hydrogens, perhaps $RCH_2-\overset{H}{\underset{CH_3}{C}}-N$. The 10:1 ratio suggests $CH_3CH_2\overset{H}{\underset{CH_3}{C}}NH_2$. This structure fits the mass spectrum: parent = 73, and two principal fragments: $CH_3CH_2-CH(CH_3)NH_2$ [fragments: 58 (loss of CH_3CH_2), 44 (loss of CH_3)]

4. Looking first at the ir to identify the functional groups present, we find a weak band at 3300 cm^{-1}, suggesting R_2NH. This assignment is consistent with the strong band at 700 cm^{-1} (N-H wag).

In the nmr spectrum, the sharp doublet at δ 1.0 looks suspiciously like methyl next to a single hydrogen: $CH_3-\overset{H}{C}$. The area of 12 would then indicate four equivalent methyls. The septuplet at δ 2.9 corresponds to $CH(CH_3)_2$. An area of 2 suggests two such groups, with the resonance at δ 0.6 corresponding to NH. Hence, the probable structure becomes $[(CH_3)_2CH]_2NH$. This structure is confirmed by the mass spectrum: parent = 101. The most important fragment is $101 - CH_3 \equiv (CH_3)_2CH\overset{+}{N}H{=}CHCH_3$.

5. In a planar amine, the bonding orbitals from N are sp^2 hybrids which have a natural angle of 120°. The ring strain compared to a three-membered ring is greater than the difference between the angle in the aziridine ring and the 107° angle in the open amine.

An alternative but equivalent explanation focuses on the lone pair. In an ordinary amine, this lone pair is in an orbital having some s-character; in the planar amine the lone pair is in a pure p-orbital. Removing s-character from the lone pair electrons takes some energy. In aziridine, the narrow bond angle of the three-membered ring requires more p-character and leaves more s-character for the lone pair. This greater degree of s-character requires more energy to form the planar system with a p-lone pair.

6. (a) $K_a = \dfrac{[RNH_2][H^+]}{[RNH_3^+]}$ $\qquad -\log K_a = -\log [H^+] - \log \dfrac{[RNH_2]}{[RNH_3^+]}$

At equal $[RNH_2]$ and $[RNH_3^+]$, the last term = 0 $\therefore pK_a = pH$

for CH_3NH_2, $pK_a = pH = 10.62$

(b) $\log [RNH_2]/[RNH_3^+] = -pK_a + pH$

for CH_3NH_2:

pH	log $[RNH_2]/[RNH_3^+]$	$[RNH_2]/[RNH_3^+]$
6	-4.62	2.4×10^{-5}
8	-2.62	2.4×10^{-3}
10	-0.62	0.24
12	1.38	24

7. (a)

$$CH_3NH_2 + H^+ \rightleftharpoons CH_3NH_3^+$$

$$CH_3COOH \rightleftharpoons CH_3CO_2^- + H^+$$

$$CH_3NH_2 + CH_3COOH \rightleftharpoons CH_3NH_3^+ + CH_3CO_2^-$$

$$K = \frac{[CH_3NH_3^+][CH_3CO_2^-]}{[CH_3NH_2][CH_3COOH]} = \frac{[CH_3NH_3^+]}{[CH_3NH_2][H^+]} \cdot \frac{[H^+][CH_3CO_2^-]}{[CH_3COOH]}$$

$$= \frac{K_a(CH_3COOH)}{K_a(CH_3NH_3^+)} = \frac{1.75 \times 10^{-5}}{2.40 \times 10^{-11}} = 7.29 \times 10^5$$

(b) Assume equal concentrations of amine and acid to start,

$$\text{then}\quad 7.29 \times 10^5 = \frac{x^2}{(a-x)^2} \qquad \frac{x}{a-x} = 854$$

From 6(b): $-\log 854 = -pK_a + pH$

$$pH = 10.62 - \log 854 = 7.69$$

8. (a) Extract with dilute HCl to remove amine; extract with dilute NaOH or Na_2CO_3 to remove the carboxylic acid. Hydrocarbon remains.

(b) A mixture of both enantiomers of the hydrogen phthalate results. Reaction with an optically pure amine such as naturally-occurring brucine or strychnine gives two diastereomeric salts, e.g.,

(-)brucine-H$^+$ (+) 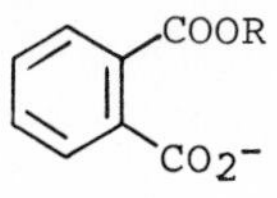 and (-)brucine-H$^+$ (-) 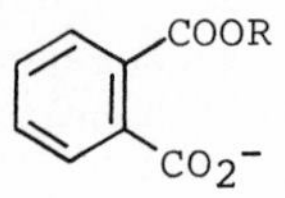

These salts are separated by crystallization. Actually, the brucine (-)(+) salt is usually less soluble in acetone. The individual salts are treated with dilute HCl (which removes the alkaloid as the soluble hydrochloride). Heating each hydrogen phthalate with aqueous NaOH hydrolyzes the ester and gives each enantiomer of the alcohol.

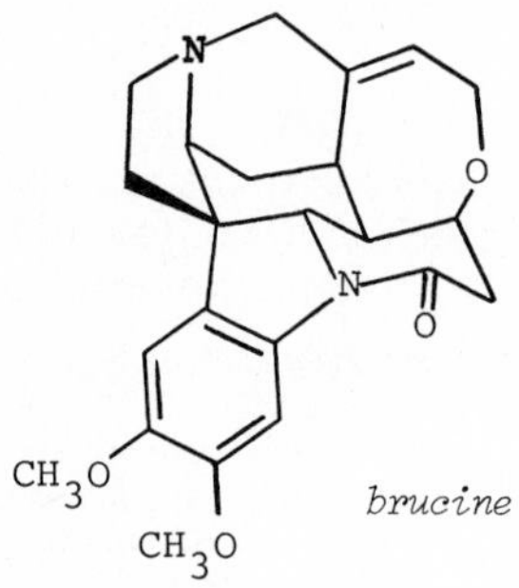

brucine

9. (a) 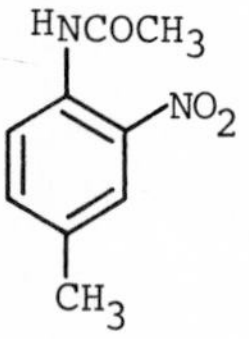 (b) 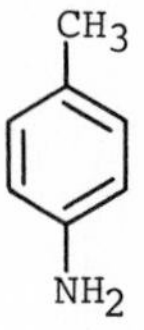 (c) 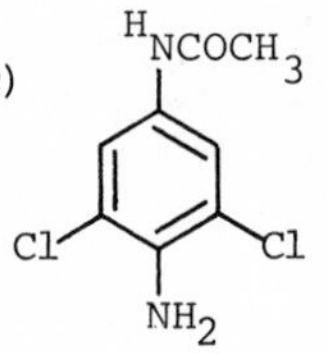

(d) 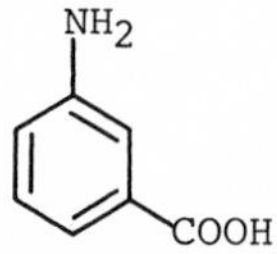 (e) 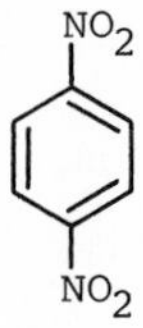 (f) 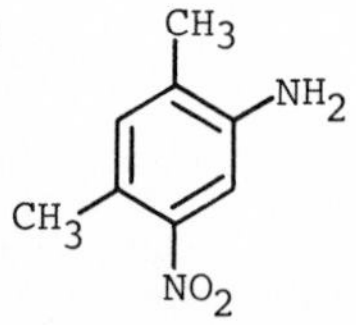

10. $CH_3\overset{O}{\overset{\|}{C}}CH_3 + CH_2{=}O + (CH_3)_2\overset{+}{N}H_2\ Cl^- \longrightarrow CH_3\overset{O}{\overset{\|}{C}}CH_2CH_2N(CH_3)_2 \xrightarrow{CH_3I}$

$CH_3\overset{O}{\overset{\|}{C}}CH{=}CH_2 + (CH_3)_3N \xleftarrow[\Delta]{Ag_2O} CH_3\overset{O}{\overset{\|}{C}}CH_2CH_2\overset{+}{N}(CH_3)_3\ I^-$

11. (a) primary arylamine ⟶ diazonium ion: $C_6H_5N_2^+$

(b) secondary amine ⟶ N-nitrosamine: $C_6H_5N(NO)CH_3$

(c) substituted amide ⟶ N-nitroso amide: $C_6H_5N(NO)COCH_3$

N-nitroso amides are best prepared by reaction of the amide with NOCl. They rearrange readily to provide diazoesters, $RN{=}NO\overset{O}{\overset{\|}{C}}R'$, which then undergo further radical or carbocation reactions involving R• or R$^+$.

(d) tertiary arylamine ⟶ electrophilic substitution: 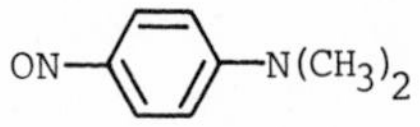

(e) primary alkylamine ⟶ diazonium ion which decomposes: $C_6H_5CH_2OH$

(f) 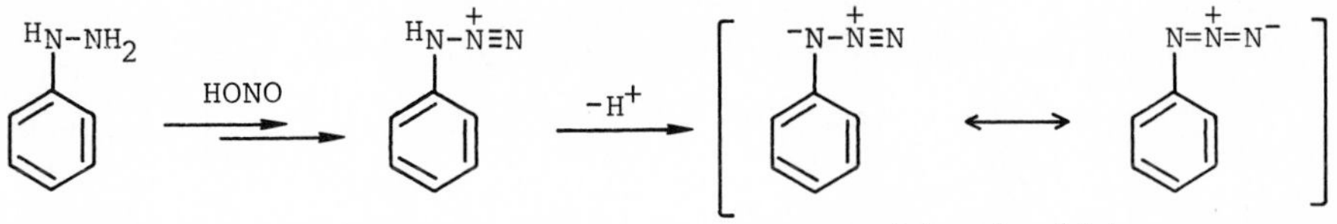

(phenyl azide)

(g) secondary amine ⟶ N-nitrosamine: $C_6H_5CH_2N(N{=}O)CH_3$.

(h) tertiary alkylamine ⟶ no reaction (except salt formation: $C_6H_5CH_2\overset{+}{N}H(CH_3)_2\ Cl^-$)

(i) same answer as for part (c): $C_6H_5CH_2N(N{=}O)COCH_3$

12. Dipole moments are oriented because of the conjugation effects indicated:

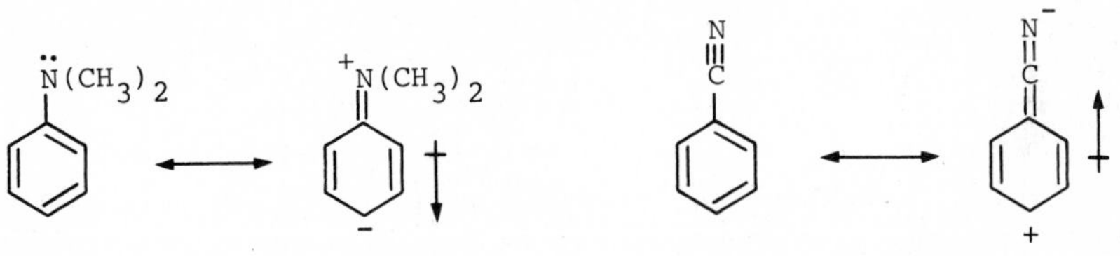

When the functional groups are conjugated in the same molecule, the dipolar effects are enhanced by the structure drawn at the right. Since the charges are now far apart, even a small contribution by this structure has an important effect on the dipole moment.

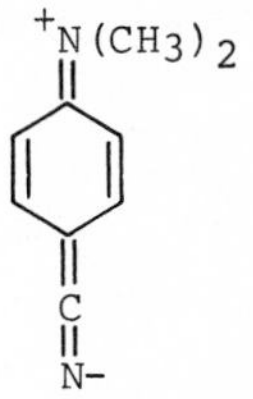

13. 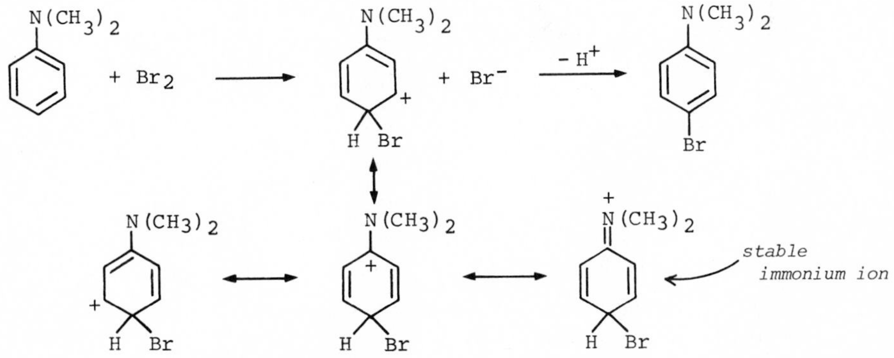

The transition state for reaction with benzene has no such immonium cation structure, and hence has much more energy and is much less stable.

14. Aniline is a weaker base than aliphatic amines, in part because of conjugation of the nitrogen lone pair electrons with the aromatic π-system:

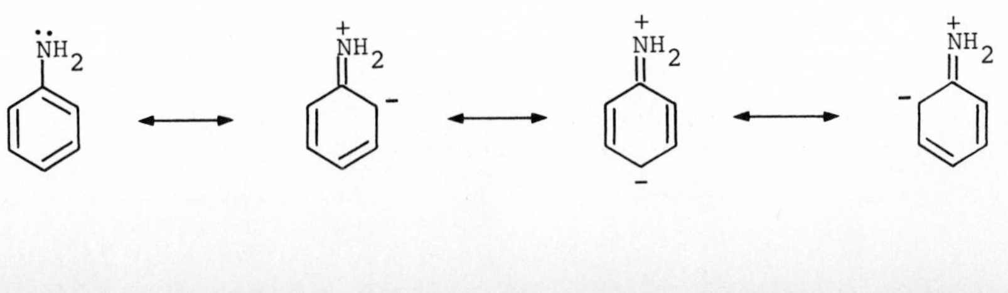

This conjugation is not present in the ammonium ion, 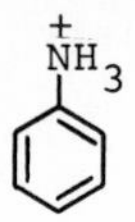 ;

hence, the amine has additional stabilization and the equilibrium, $RNH_3^+ \rightleftharpoons H^+ + RNH_2$, is displaced more to the right for R = aromatic ring compared to R = alkyl.

However, in *o*-methyl-N,N-dimethylaniline, steric hindrance prevents effective conjugation (*see (I) below*). The $(CH_3)_2N$ group must twist at right angles to the ring (*see (II) below*).

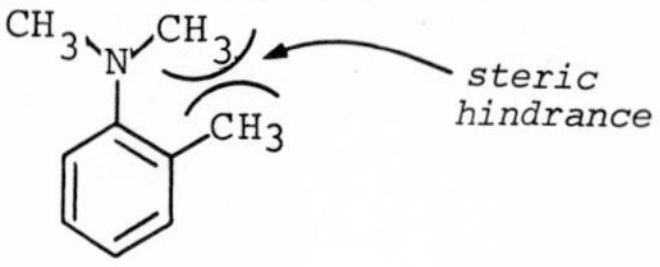

(*I*)

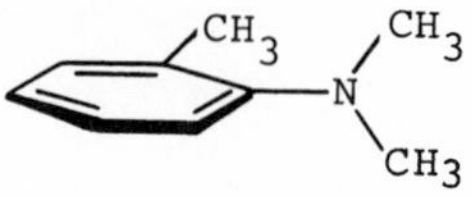

(*II*)

The nitrogen lone pair now cannot overlap with the π-system. This amine does not have additional conjugation stabilization, thus it is more basic than the primary amine, in which the smaller hydrogens are not as involved in steric hindrance.

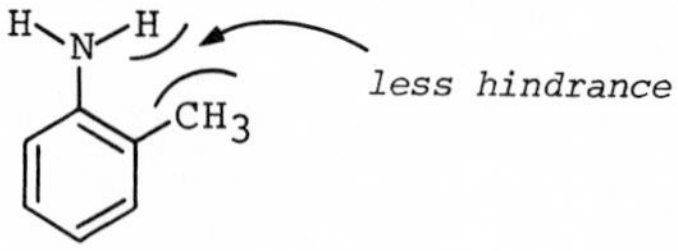

15. (a) 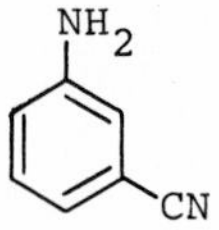 The basicity of *p*-cyanoaniline is reduced by 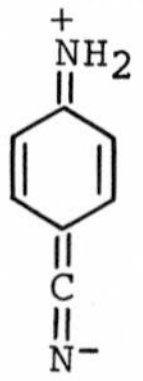

(b) 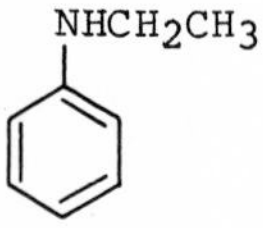 The basicity of the amide is reduced by 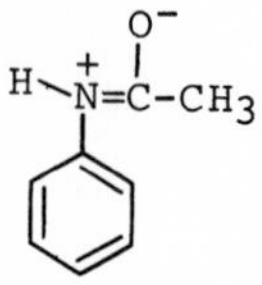

(c) 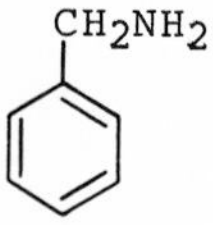 The basicity of the aromatic amine is reduced by 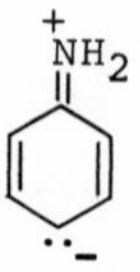 , etc.

(d) 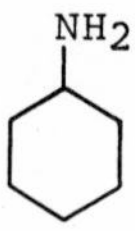 Same reason as for part (c) above.

(e) 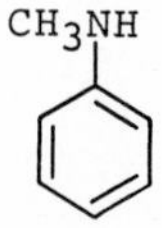 The lone pair of the imine is more sp^2 in character; more s-character = more stable electrons = less basic.

(f) 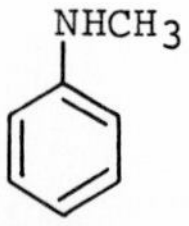 The Cl group is more electron-withdrawing than CH_3, because Cl is more electronegative than carbon. Therefore, the structure illustrated at the right is less stable. 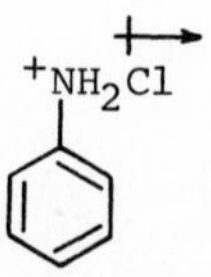

(g)

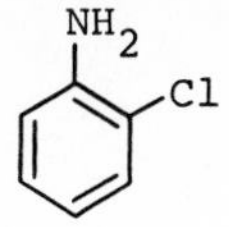

The effect of the electronegative Cl substituent is
much greater when it is attached directly to the nitrogen.

16. (a) $CH_3(CH_2)_3CH_2OH \xrightarrow[H_2SO_4]{HBr} CH_3(CH_2)_3CH_2Br \xrightarrow[\substack{\text{large} \\ \text{excess}}]{NH_3} CH_3(CH_2)_3CH_2NH_2$

or $\underbrace{\qquad\qquad \text{Gabriel synthesis} \qquad\qquad}$

(b) $CH_3(CH_2)_3CH_2NH_2$ [from (a)] $\xrightarrow[\substack{HCOOH \\ 100°}]{CH_2=O} CH_3(CH_2)_3CH_2N(CH_3)_2$

(c) $CH_3CH_2CH_2OH \xrightarrow[\text{pyridine}]{CrO_3} CH_3CH_2CHO \xrightarrow[H_2/Pt]{\substack{NH_3 \\ C_2H_5OH}} \left[CH_3CH_2CH_2NH_2 \right]$

$(CH_3CH_2CH_2)_2NCH_3 \xleftarrow[\substack{HCOOH \\ \Delta}]{CH_2=O} (CH_3CH_2CH_2)_2NH$

(d) $(CH_3)_2CHCH_2CH_2OH \xrightarrow[H_2SO_4]{HBr} (CH_3)_2CHCH_2CH_2Br \xrightarrow[\substack{\text{Gabriel} \\ \text{synthesis}}]{NH_3 \text{ or}} (CH_3)_2CHCH_2CH_2NH_2$

$\xrightarrow[\substack{\text{(CH}_3\text{CO)}_2\text{O}}]{\substack{CH_3COOH, \Delta \\ \text{or}}}$

$(CH_3)_2CHCH_2CH_2NHCH_2CH_3 \xleftarrow{LiAlH_4} (CH_3)_2CHCH_2CH_2NHCOCH_3$

17. (a) Requires a single inversion of configuration:

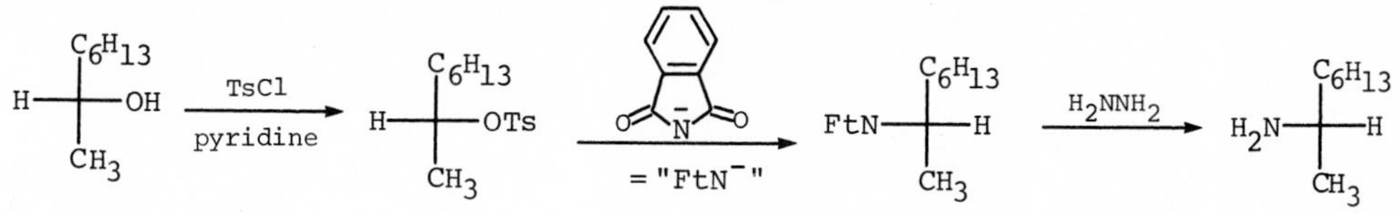

(R)   (S)

(b) Requires two inversions, to give overall retention:

(R)

18. (a) $CH_3CH_2NO_2 + CH_3CH_2CHO \xrightarrow{OH^-} O_2NCHCHCH_2CH_3 \xrightarrow[\substack{Fe, FeSO_4 \\ aq. H_2SO_4}]{Pt/H_2 \text{ or}} H_2NCHCHCH_2CH_3$

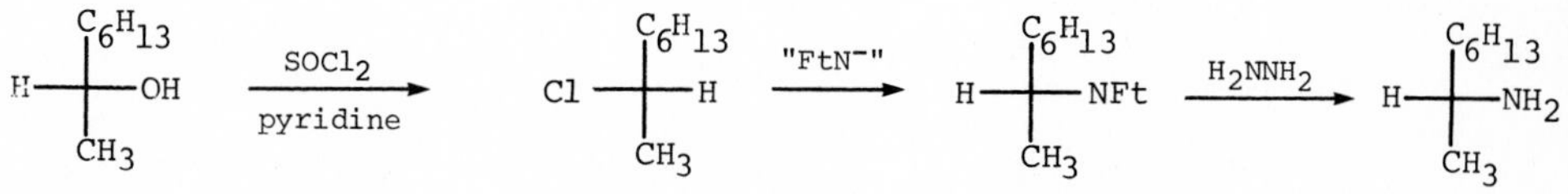

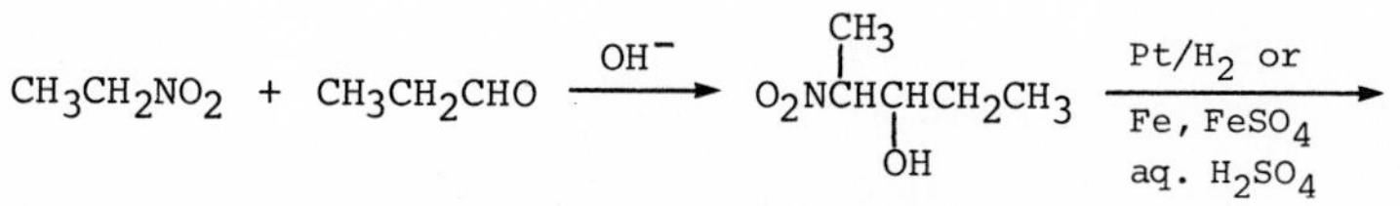

*mixture of
diastereomers*

(b) Best method is by epoxide opening:

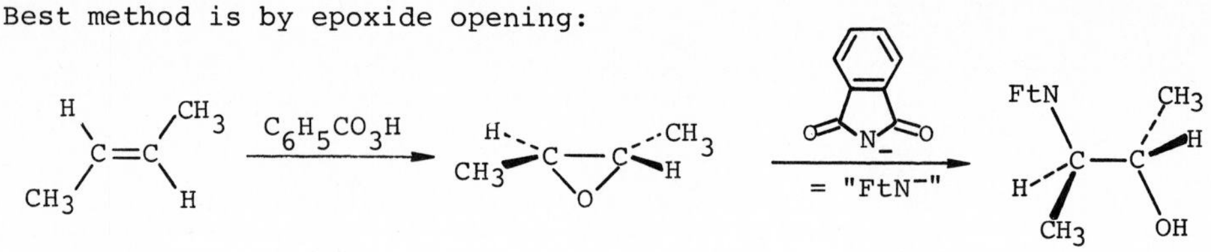

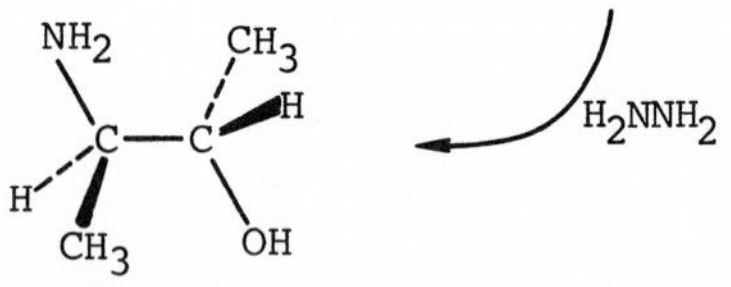

NOTE: what happens if phthalimide attacks the other epoxide carbon?

(c)

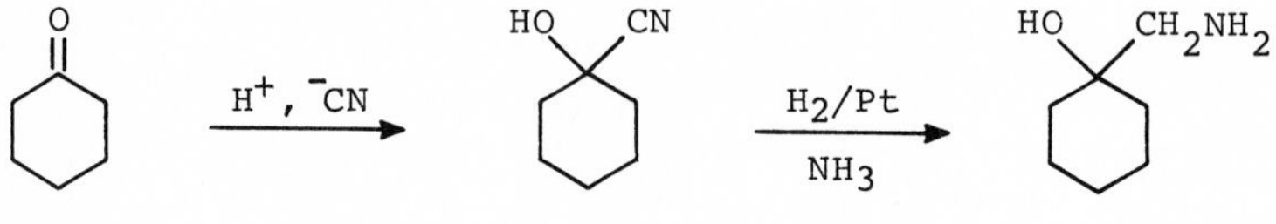

(d) $CH_3CH_2\overset{O}{\overset{\|}{C}}CH_2CH_3 + CH_2{=}O + (CH_2CH_3)_2\overset{+}{N}H_2\ Cl^- \longrightarrow CH_3CH_2\overset{O}{\overset{\|}{C}}\underset{CH_3}{\underset{|}{C}}HCH_2N(CH_2CH_3)_2$

(e)

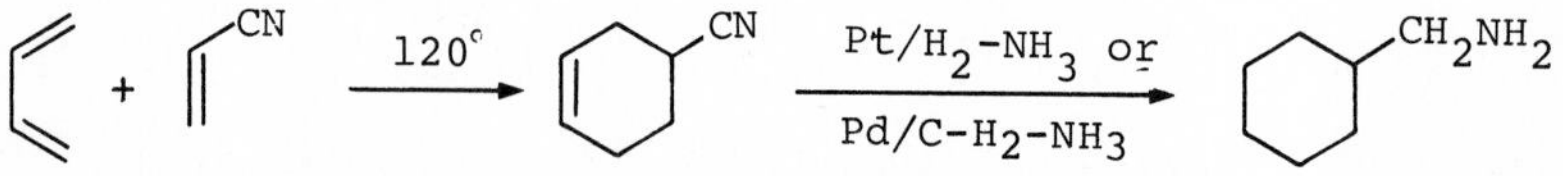

Catalytic hydrogenation reduces both the double bond and C≡N. (Alternatively, use $LiAlH_4$ first *(to reduce CN)*, then Pt/H_2 *(to reduce double bond)*).

(f)

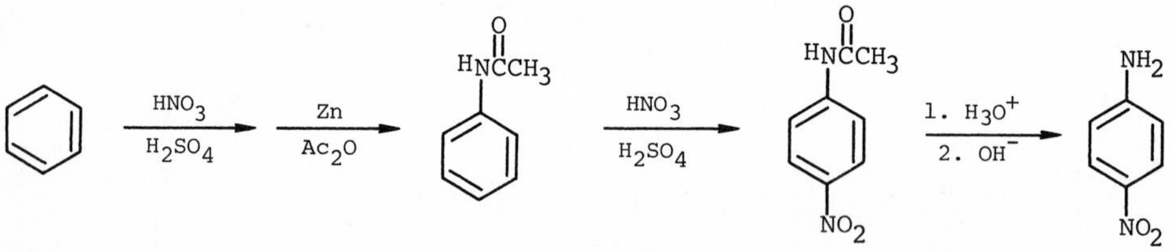

(g)

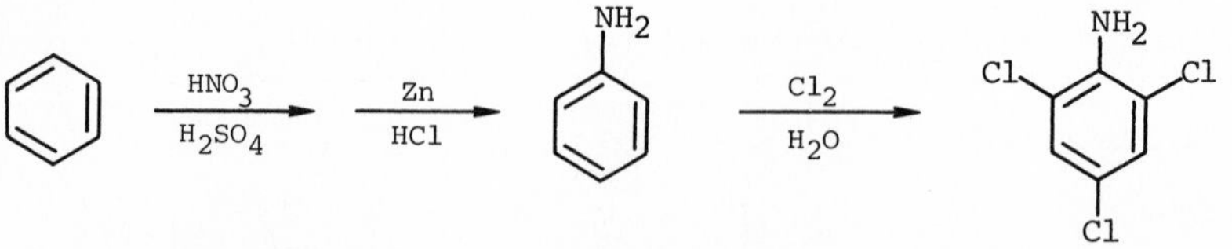

19. The basicity of aniline is lower than that of aliphatic amines because of conjugation of the nitrogen lone pair with the benzene ring. In diphenylamine, the lone pair can conjugate with two rings; hence, the basicity is still lower.

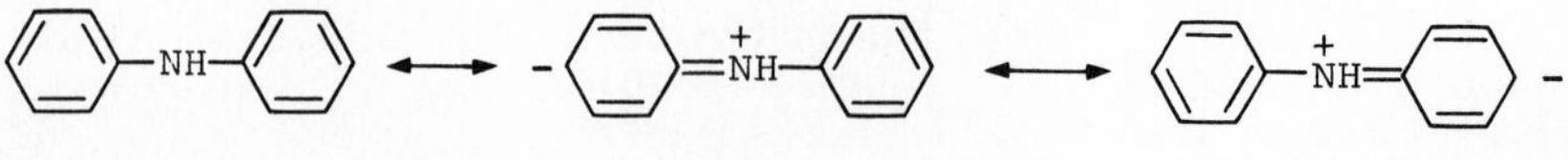

, etc.

20. *ortho-*

Adjacent positive charges; thus high electrostatic repulsion. These structures are of high energy and contribute little to the resonance hybrid.

para-

meta-

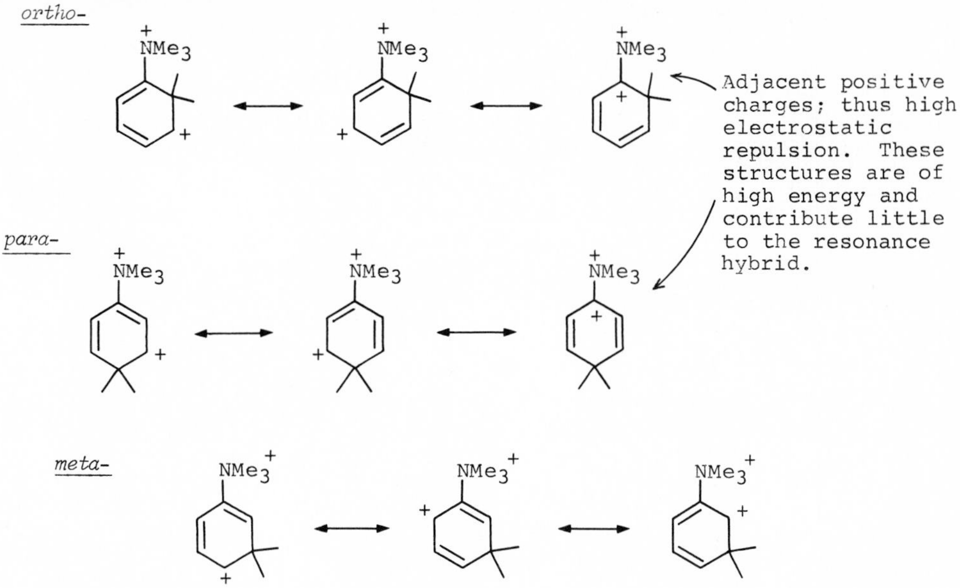

The positive charges in the *meta*-isomer are separated by at least one atom; thus, the *meta* transition state is of lowest energy. However, two positive charges in the same molecule represent substantial electrostatic repulsion. For this reason the trimethylammonium group is highly deactivating, but *m*-directing.

21. Addition of N-C is best accomplished by a Michael reaction:

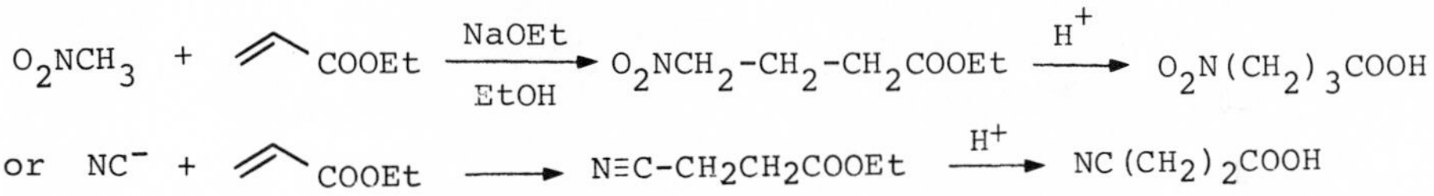

$$O_2NCH_3 \; + \; \text{\Large/}\!\!=\!\!\text{\Large\backslash}_{COOEt} \xrightarrow[\text{EtOH}]{\text{NaOEt}} O_2NCH_2-CH_2-CH_2COOEt \xrightarrow{H^+} O_2N(CH_2)_3COOH$$

or $\quad NC^- \; + \; \text{\Large/}\!\!=\!\!\text{\Large\backslash}_{COOEt} \longrightarrow N\equiv C-CH_2CH_2COOEt \xrightarrow{H^+} NC(CH_2)_2COOH$

$$RCOOH \xrightarrow{SOCl_2} RCOCl \xrightarrow{(CH_3)_2NH} RCON(CH_3)_2$$

$$O_2N(CH_2)_3CON(CH_3)_2 \xrightarrow[\text{aq. } H_2SO_4]{Fe, \; FeSO_4} H_2N(CH_2)_3CON(CH_3)_2 \xrightarrow{LiAlH_4} H_2N(CH_2)_4N(CH_3)_2$$

$\quad\quad\quad\quad$ ⤷ *(does not reduce amide)*

or $\quad NC(CH_2)_3CON(CH_3)_2 \xrightarrow{LiAlH_4}$

(reduces both −CN and amide)

(b)

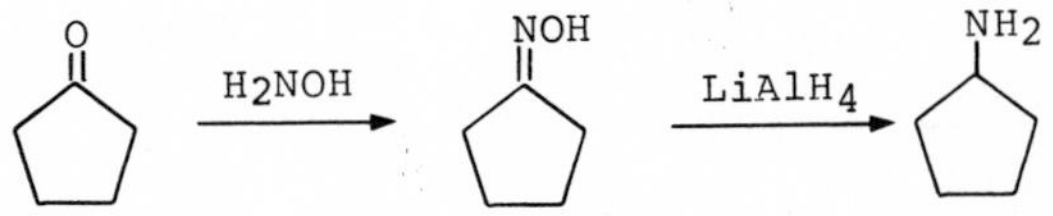

an oxime

(c)

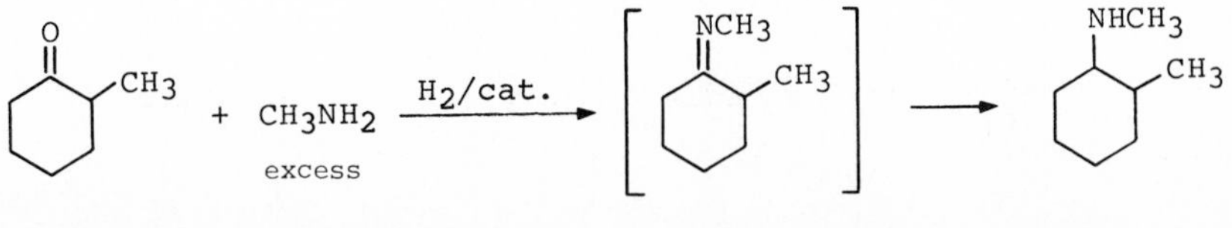

(d) 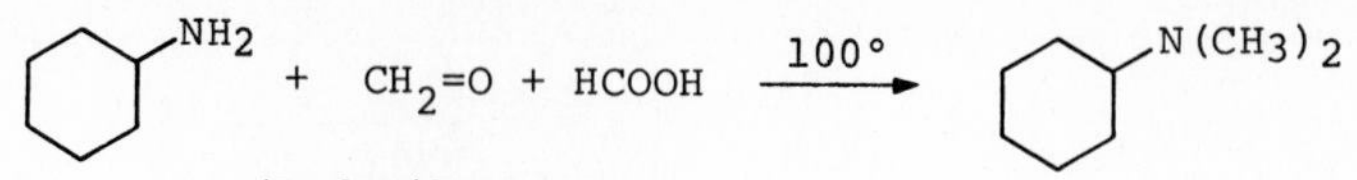

(*Eschweiler-Clarke*)

(e) 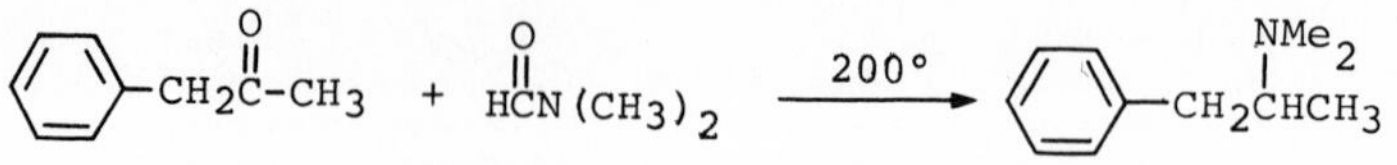

(*Leuckart*)

22. (a) 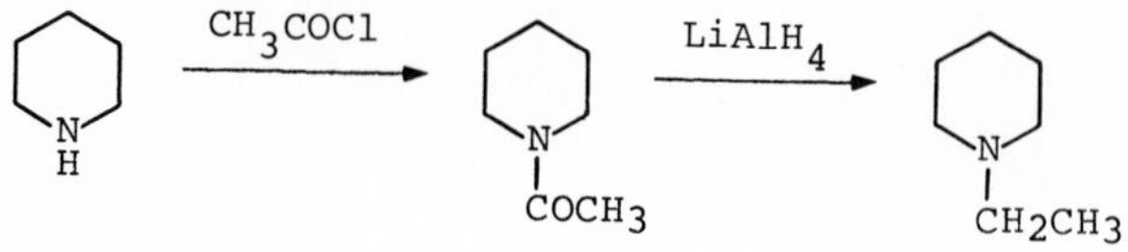

(b) 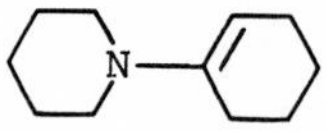

an enamine

(c) 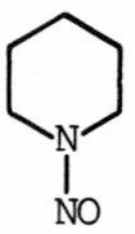

(d) 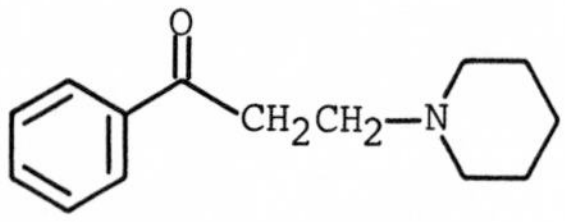

(e) 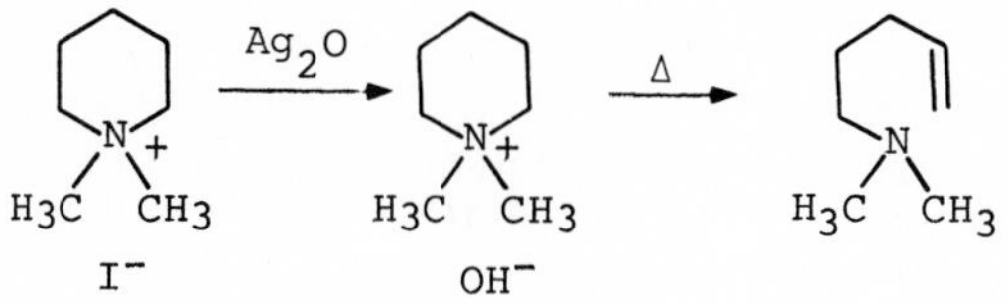

(f) 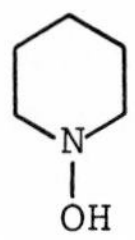

23. 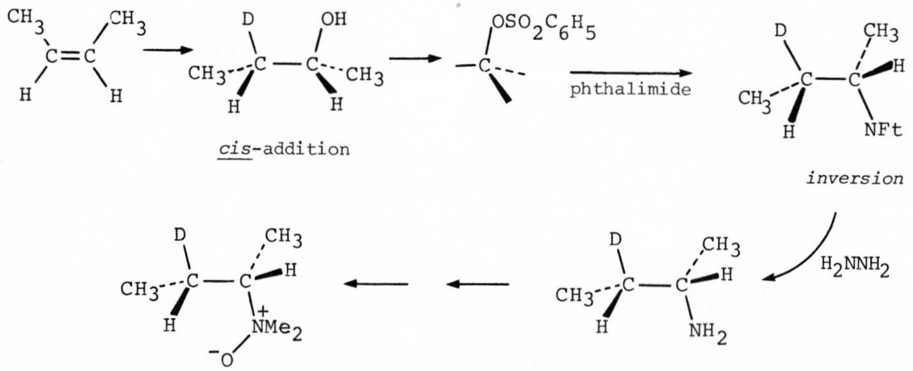

There are three possible pathways for elimination. Loss of H from CH_3 gives $CH_3CHDCH{=}CH_2$ (one deuterium).

From methylene group, the molecule may lose H or D by *syn*-elimination:

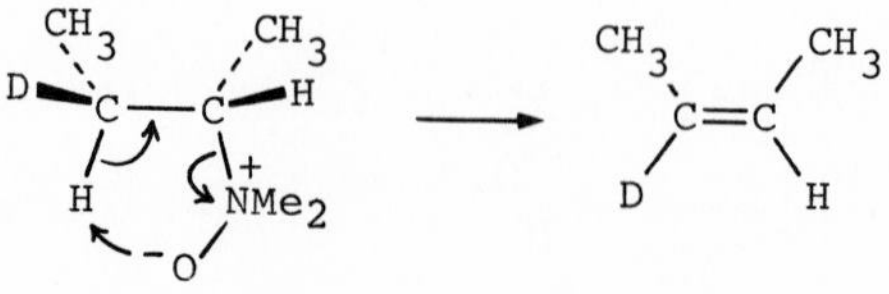

cis-2-butene has one deuterium

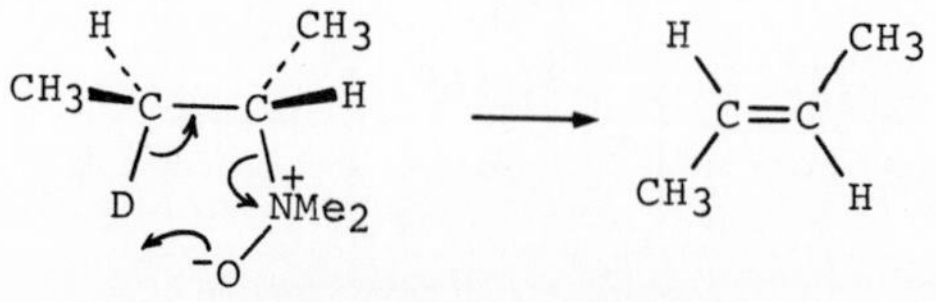

trans-2-butene has no deuterium

24. This elimination occurs with least hindered or most acidic hydrogen:

(a) $CH_2{=}CH_2$ + $Me_2NCH_2CH(CH_3)_2$ (least hindered)

(b) $(CH_3)_3N$ + $(CH_3)_2CHCH{=}CH_2$ (least hindered)

(c) The hydrogen α to the CO group is most acidic; it eliminates rather than the less hindered hydrogen on CH_3 of the ethyl group:

$$(CH_3CH_2)_2N + CH_2{=}CHCOCH_3$$

(d) This problem is best solved by the use of models. Recall that the Hofmann elimination proceeds by the E2 mechanism, involving the hydrogen which is *anti* to the leaving nitrogen. The preferred conformation is the one having the two methyl groups equatorial to the six-membered rings. Only the bridgehead proton has the necessary *anti*-coplanar relationship to the C-N bond which is being broken:

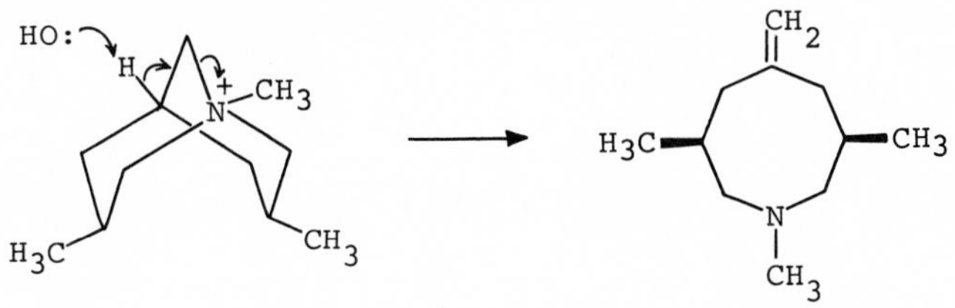

(e) *syn*-elimination:

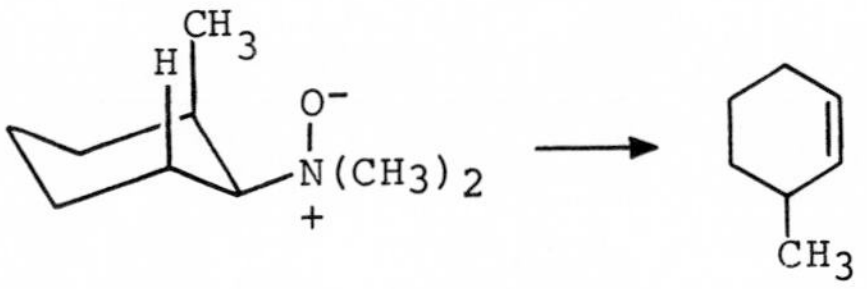

25. (a) requires *syn*-elimination; ∴ use amine oxide

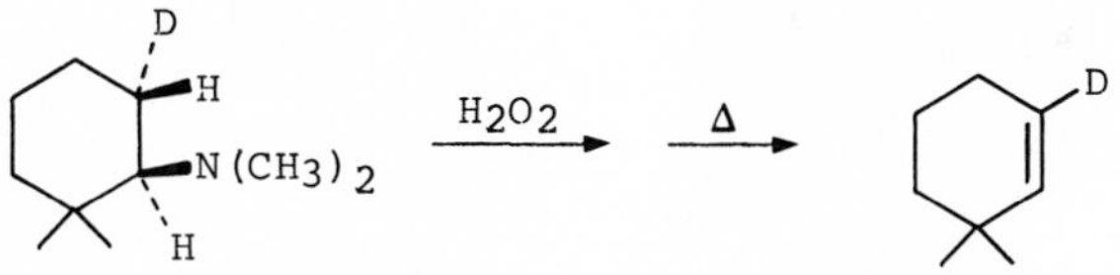

(b) requires *anti*-elimination; use Hofmann reaction

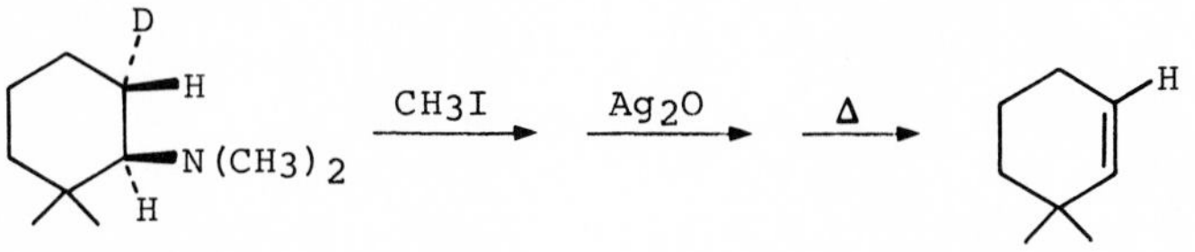

26.

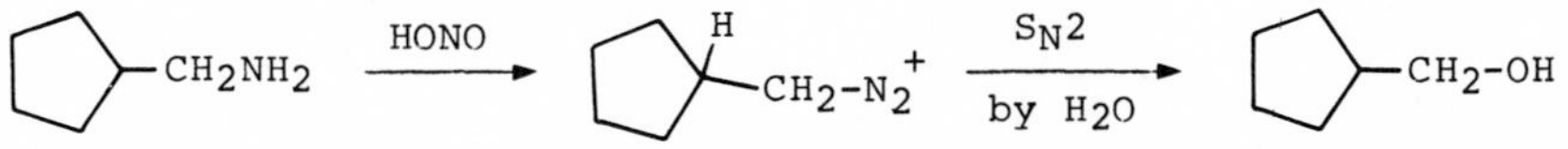

elimination of β-H →

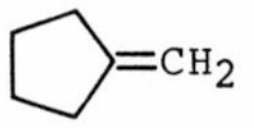

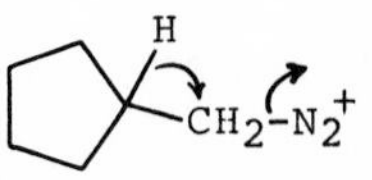

rearrangement of β-H gives:

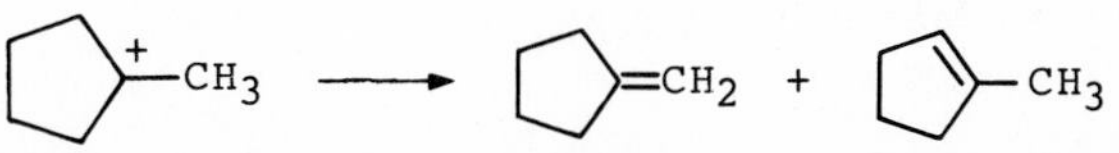

Rearrangement of ring:

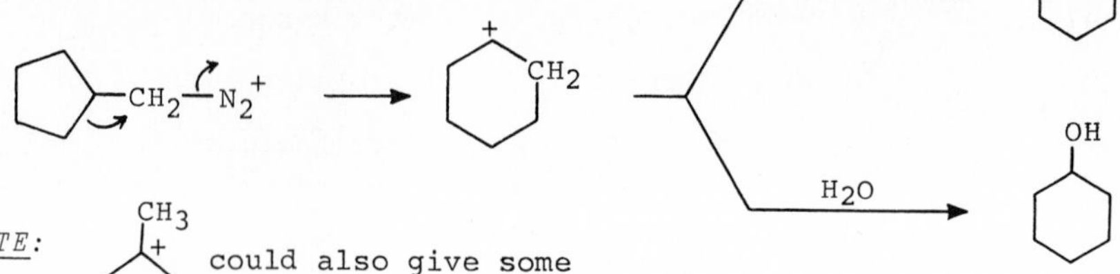

NOTE: 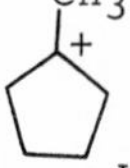 could also give some

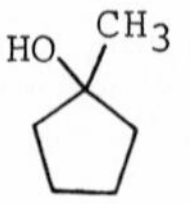 , and a small amount is probably present.

27. (a) $\underline{n}$-$C_{10}H_{21}Br \xrightarrow[\text{aq. } NaOOCCH_3]{C_6H_5CH_2\overset{+}{N}Et_3\ Cl^-} \underline{n}$-$C_{10}H_{21}O\overset{O}{\overset{\|}{C}}CH_3$

($CH_3CO_2^-$ solubilized in organic phase)

(b) $3,4\text{-}(CH_3)_2C_6H_3\overset{O}{\overset{\|}{C}}CH_3 \xrightarrow[NaOD,\ D_2O]{C_6H_5CH_2\overset{+}{N}Et_3\ Cl^-} 3,4\text{-}(CH_3)_2C_6H_4\overset{O}{\overset{\|}{C}}CD_3$

($R\overset{O}{\overset{\|}{C}}CH_3 + OD^- \rightleftharpoons R\overset{O^-}{\overset{|}{C}}{=}CH_2 + HOD$ in organic phase)

(c) $(CH_3)_2CH(CH_2)_4OH + (CH_3)_2SO_4 \xrightarrow[50\%\ NaOH]{C_6H_5CH_2\overset{+}{N}Et_3\ Cl^-} (CH_3)_2CH(CH_2)_4OCH_3 + Na^+\ {}^-O_3SOCH_3$

($ROH + OH^- \rightleftharpoons RO^- + H_2O$ in organic phase)

(d) $(CH_3)_2CH\overset{O}{\overset{\|}{C}}H + C_6H_5CH_2Cl \xrightarrow[50\%\ NaOH]{C_6H_5CH_2\overset{+}{N}Et_3\ Cl^-} (CH_3)_2\underset{CH_2C_6H_5}{\underset{|}{C}}\overset{O}{\overset{\|}{C}}H + NaCl$

($(CH_3)_2CH\overset{O}{\overset{\|}{C}}H + OH^- \rightleftharpoons (CH_3)_2C{=}\overset{O^-}{\overset{|}{C}}H + H_2O$ in organic phase)

(e) $C_6H_5CH{=}CH_2 + CHCl_3 \xrightarrow[50\%\ NaOH]{C_6H_5CH_2\overset{+}{N}Et_3\ Cl^-}$ 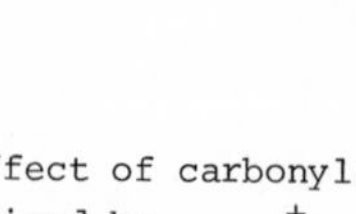 $+ NaCl$

($CHCl_3 + OH^- \longrightarrow :CCl_2 + H_2O + Cl^-$ in organic phase)

28. (a) $ClCH_2CH_2NH_3^+$; inductive effect of Cl

(b) $CH_3ONH_3^+$; inductive effect of oxygen

(c) $CH_3\overset{O}{\overset{\|}{C}}NH_3^+$; electron-withdrawing inductive effect of carbonyl; in addition, the product amide is resonance-stabilized by $R\text{-}\underset{O_-}{\underset{|}{C}}{=}\overset{+}{N}H_2$

(d) $CH_2{=}\overset{+}{N}H_2$; lone pair in $CH_2{=}\ddot{N}H$ has more s-character than that in $CH_2CH_2\ddot{N}H_2$

(e) $CH_3OOCCH_2NH_3^+$; inductive effect of COOR group

(f) $CH_2{=}CHNH_3^+$; amine is resonance-stabilized: $\left[CH_2{=}CH\text{-}\ddot{N}H_2 \longleftrightarrow {}^-CH_2\text{-}CH{=}\overset{+}{N}H_2\right]$

29. Since the mixture of cyclooctadienes contains no 1,3-isomer, it must consist of a mixture of the 1,4- and 1,5-isomers. Hence, the starting $C_9H_{17}N$ amine

must have the symmetrical bicyclic structure:

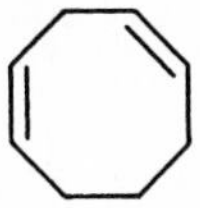

1,4-cyclooctadiene

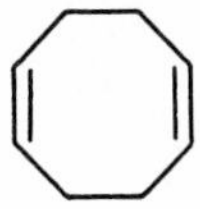

1,5-cyclooctadiene

granatine

30.

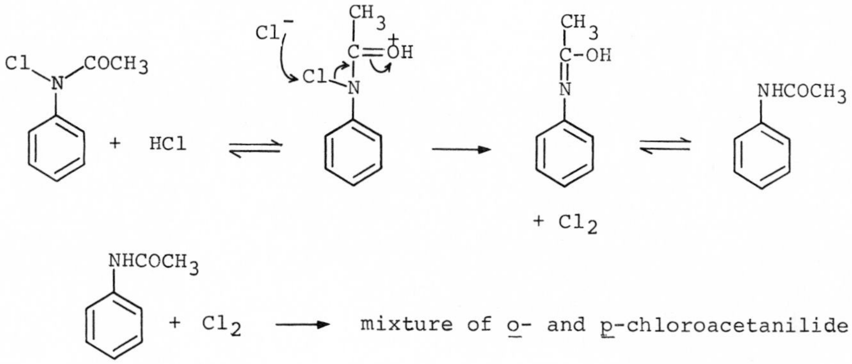

24.E Supplementary Problems

S1. Draw the structure of each of the following compounds.

(a) triisobutylamine
(b) *cis*-1,4-diaminocyclohexane
(c) 3-(aminomethyl)aniline
(d) N,3-dimethylbutanamine
(e) ethyltripropylammonium chloride
(f) ethyldiisopropylamine oxide
(g) N,O-diethylhydroxylamine
(h) anilinium bromide

S2. For each pair of compounds below, explain how you would distinguish between the two without using NMR.

(a) $CH_3CH_2NHCH_2CH(CH_3)_2$ and $(CH_3)_2CHNHCH(CH_3)_2$

(b) $(CH_3CH_2CH_2)_2NH$ and $CH_3CH_2CH(NH_2)CH_2CH_2CH_3$

(c) $(CH_3)_3CCH_2NH_2$ and $CH_3CH_2CH(CH_3)CH_2NH_2$

(d)

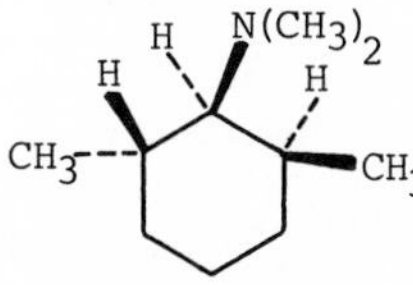

and

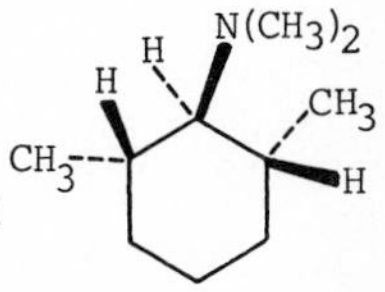

S3. Rank the following compounds in order of increasing basicity.

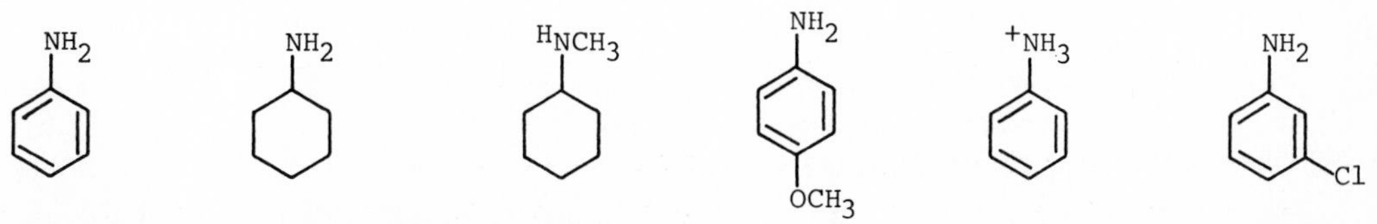

I II III IV V VI

S4. Do you expect p-aminoacetophenone ($H_2NC_6H_4\overset{\overset{\displaystyle O}{\|}}{C}CH_3$) to be a stronger or weaker base than aniline itself? Justify your answer with resonance structures.

S5. What is the principal organic product to result from each of the following reaction sequences?

(a)

$\xrightarrow{\text{NaCN}}$ $\xrightarrow{\text{LiAlH}_4}$ $\xrightarrow{(CH_3\overset{\overset{\displaystyle O}{\|}}{C})_2O}$

(b)

$+$ $H_2NCH(CH_3)_2$ $\xrightarrow{H_2/Pt}$ $\xrightarrow{H_2O_2}$

(c)

$\xrightarrow{D_2/Pt}$ $\xrightarrow[\substack{\text{HCOOH} \\ \text{HCl, } \Delta}]{CH_2=O}$ $\xrightarrow[\text{2. } \Delta]{\text{1. } H_2O_2}$

(d)

$\xrightarrow{CF_3CO_3H}$ $\xrightarrow[H_2SO_4]{HNO_3}$ $\xrightarrow{H_2/Ni}$

(e) $CH_3CH_2CH_2CH=CH_2$ $\xrightarrow[\text{(peroxides)}]{HBr}$

$\xrightarrow[\Delta]{H_3O^+}$

(f) $CH_3CH_2CH_2CH=O$ $\xrightarrow{H_2NOH}$ $\xrightarrow{H_2/Ni}$ $\xrightarrow{HCl}$

(g)

$\xrightarrow{SOCl_2}$ $\xrightarrow{NaN_3}$ $\xrightarrow[\text{2. } H_2O]{\text{1. } \Delta}$

(h) $(CH_3)_3CCH_2CH_2\overset{\overset{\displaystyle O}{\|}}{C}CH_3$ $\xrightarrow[\Delta]{H\overset{\overset{\displaystyle O}{\|}}{C}N(CH_3)_2}$ $\xrightarrow{CH_3I}$ $\xrightarrow{Ag_2O}$ $\xrightarrow{\Delta}$

(i)

$\xrightarrow[HCl]{Sn}$ $\xrightarrow[\substack{CH_2=O \\ HCl, \Delta}]{HCOOH}$ $\xrightarrow[POCl_3]{H\overset{\overset{\displaystyle O}{\|}}{C}N(CH_3)_2}$ $\xrightarrow{H_3O^+}$

(j) $(CH_3)_2CHCH_2NH_2$ $\xrightarrow{(CH_3)_2CHCCl}$ $\xrightarrow{LiAlH_4}$ $\xrightarrow[HCl]{NaNO_2}$

(k)

$+$ $CH_3CH_2\overset{\overset{\displaystyle O}{\|}}{C}CH_2CH_3$ $\xrightarrow[-H_2O]{\Delta}$ $\xrightarrow[\text{2. } H_3O^+]{\text{1. } CH_3I}$

(ℓ)

$\xrightarrow[\substack{CH_2=O, HCl \\ \Delta}]{C_6H_5\overset{\overset{\displaystyle O}{\|}}{C}CH_3}$

(m)

$\xrightarrow{SOCl_2}$ $\xrightarrow{NH_3}$ $\xrightarrow[\text{NaOH}]{Br_2}$

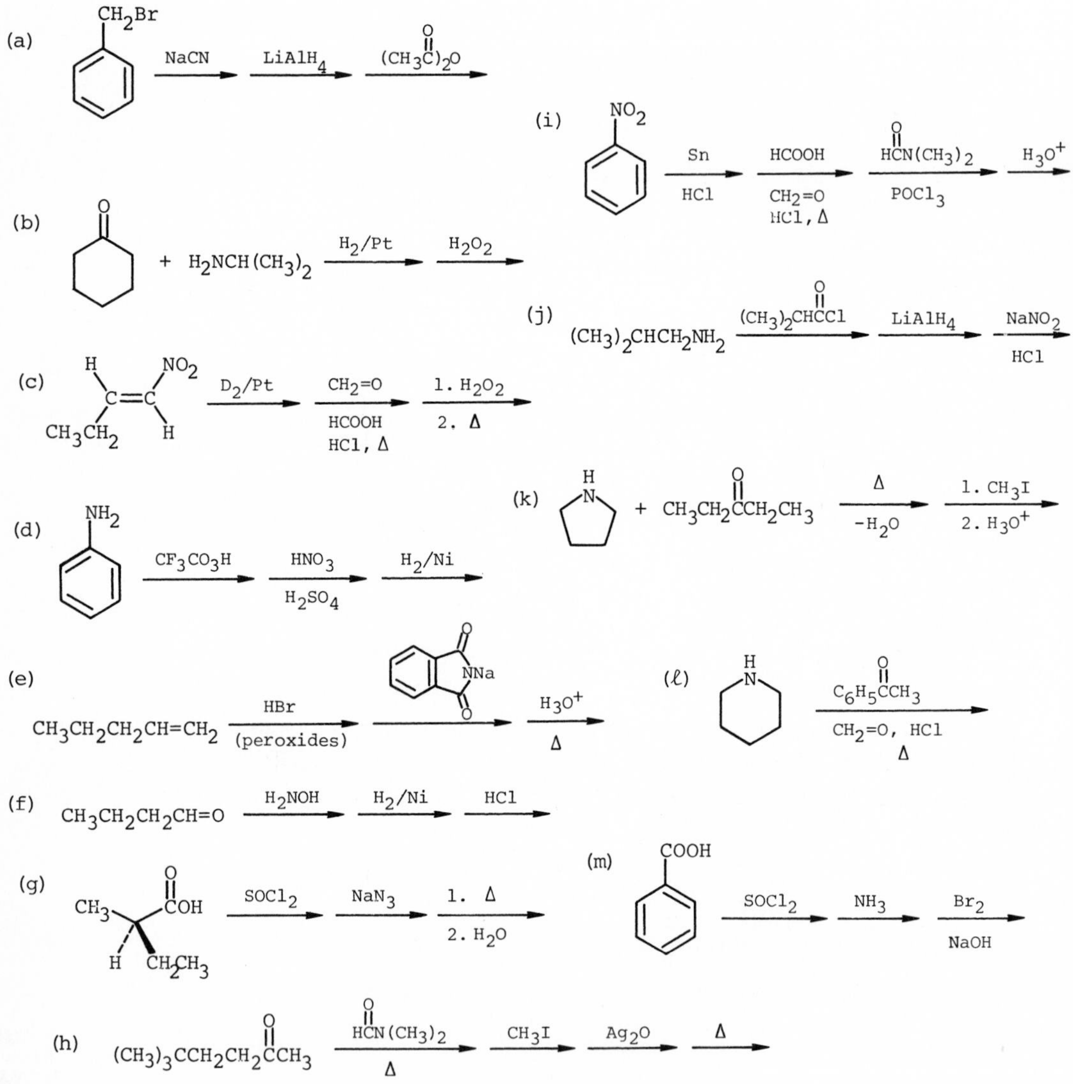

S6. Outline syntheses of the following compounds, using 3-methylbutanoic acid as starting material.

(a) $(CH_3)_2CHCH_2NH_2$ (b) $(CH_3)_2CHCH_2CH_2NH_2$ (c) $(CH_3)_2CHCH_2CH_2CH_2NH_2$

S7. Write the intermediates formed during the following reaction sequence, as well as a mechanism for the last reaction.

$CH_3CH_2CH_2\overset{\overset{\displaystyle O}{\|}}{C}H$ $\xrightarrow{HCN}$ $\xrightarrow{H_2/Ni}$ $\xrightarrow[K_2CO_3, \Delta]{\text{excess } CH_3I}$ $CH_3CH_2CH_2CH\overset{\displaystyle O}{\overset{\displaystyle \diagdown\diagup}{-}}CH_2$

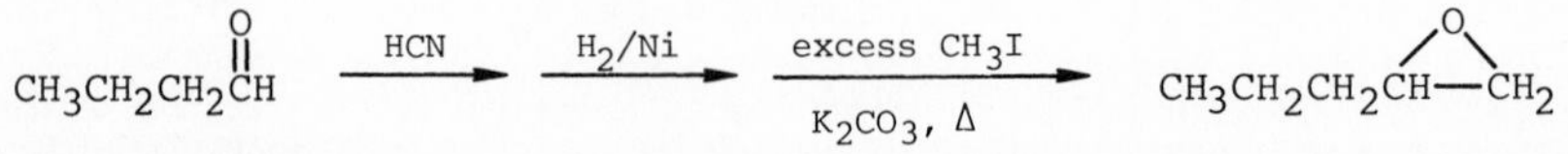

S8. Provide an explanation for the following behavior, and write a detailed mechanism for the conversion of B to A.

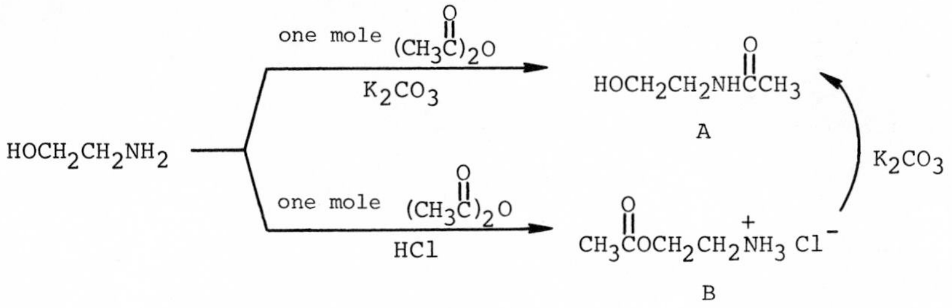

S9. Show how to synthesize the following compounds, using only inorganic compounds (e.g., $NaNO_2$, NH_3, NaCN, etc.) as a source of nitrogen.

(a) 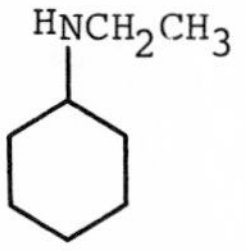

(b) 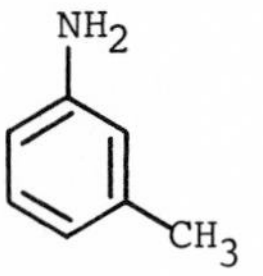

(c) 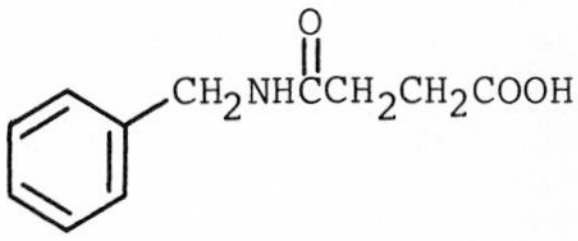

(d) 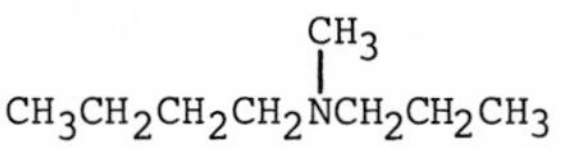

(e) 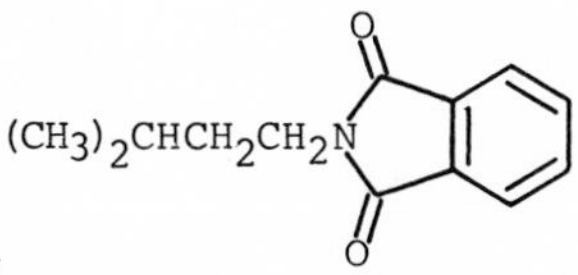

(f) 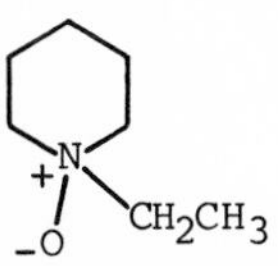

(g) 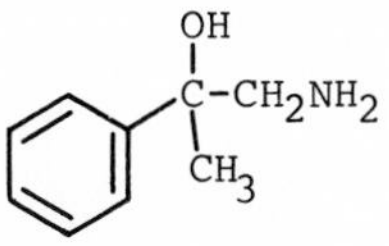

(h) $(CH_3CH_2)_2N-N=O$

S10. Write a step-by-step mechanism for the following transformations:

(a) 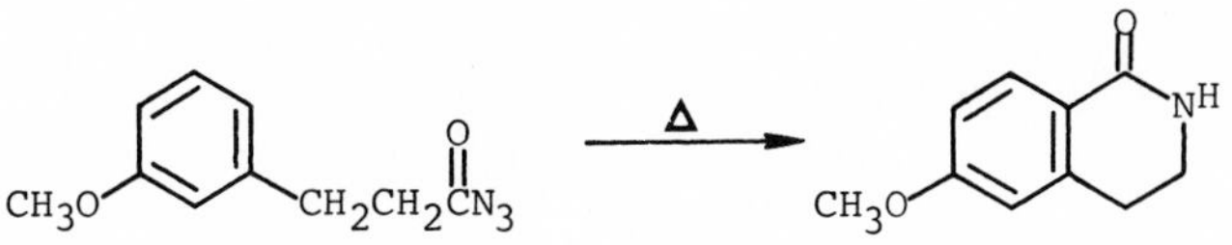

(b) 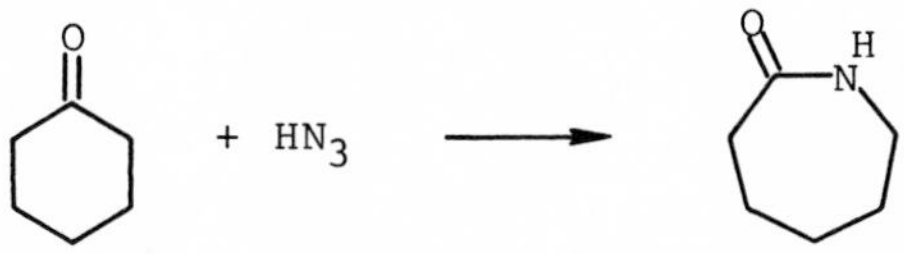

S11. Offer an explanation for the degree of reactivity of immonium ions, carbonyl compounds, and imines toward nucleophilic attack:

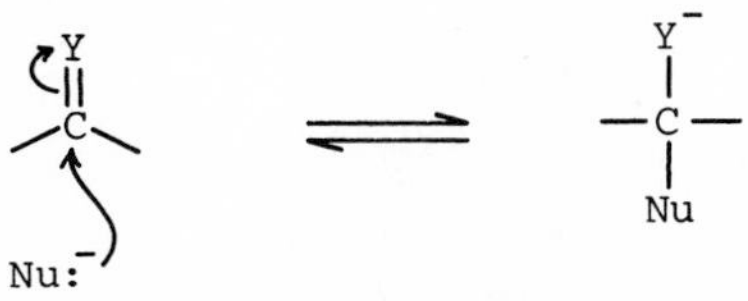

$Y = \overset{+}{N}R_2 > O > NR$

S12. Write a mechanism for the following transformation:

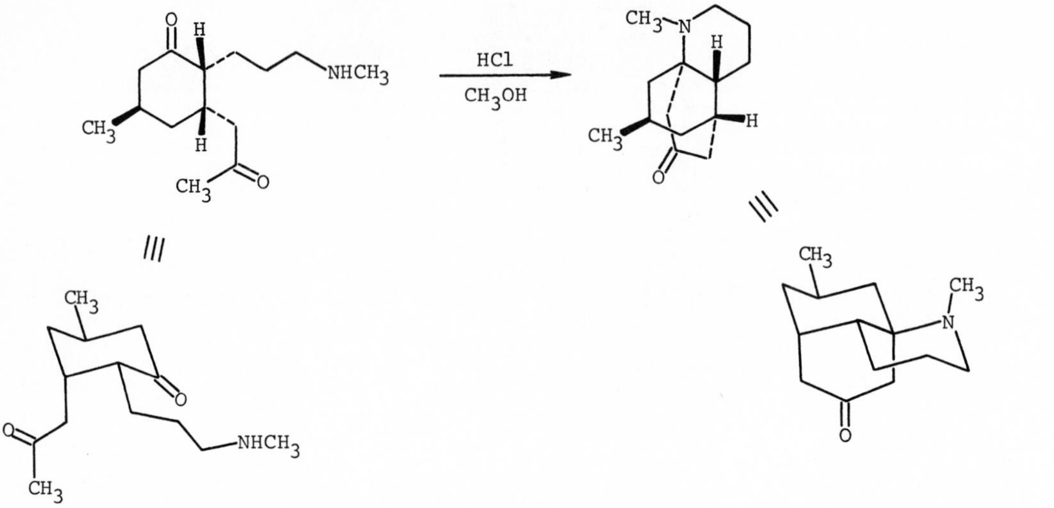

24.F Answers to Supplementary Problems

S1.

(a) $((CH_3)_2CHCH_2)_3N$

(b)

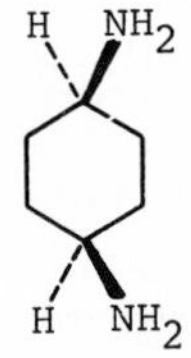

(c)

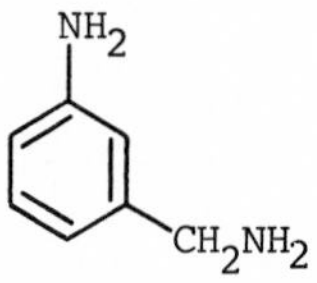

(d) $(CH_3)_2CHCH_2CH_2NHCH_3$

(e) $(CH_3CH_2CH_2)_3\overset{+}{N}CH_2CH_3$

(f) $CH_3CH_2\overset{+}{\underset{O_-}{N}}(CH(CH_3)_2)_2$

(g) $CH_3CH_2NHOCH_2CH_3$

(h)

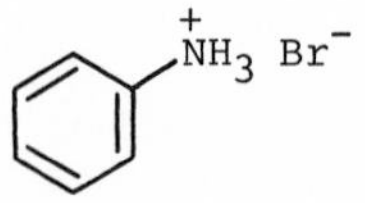

S2.   (a) Mass spectra:

$CH_3\overset{86}{\overset{\frown}{-}}CH_2NHCH_2\overset{\frown}{-}CH(CH_3)_2$
$\underset{58}{}$

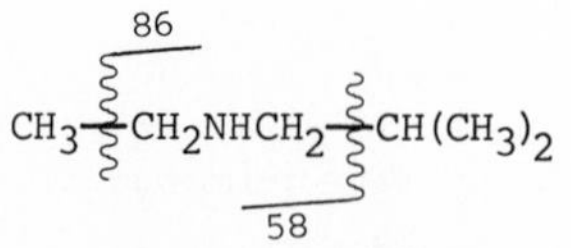

(*strong* m/e 58 expected)

$CH_3\overset{CH_3}{\overset{\frown}{-}}CHNHCH\overset{CH_3}{\overset{\frown}{-}}CH_3$
$\underset{86}{} \quad \underset{86}{}$

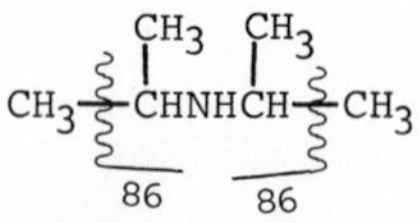

(*weak* m/e 58 expected)

(b) $(CH_3CH_2CH_2)_2NH$

very weak N-H stretch:
      3310-3350 cm^{-1}

(or mass spectrometry)

$\underset{NH_2}{CH_3CH_2CHCH_2CH_2CH_3}$

weak N-H stretch: doublet 3400, 3500 cm^{-1}

N-H band: 1580-1650 cm^{-1}

(c) $RNH_2 \xrightarrow[CH_2{=}O,\ HCl]{HCOOH} RN(CH_3)_2 \xrightarrow{H_2O_2} R\text{-}\overset{+}{N}(O^-)(CH_3)_2 \xrightarrow{\Delta}$

$R = (CH_3)_3CCH_2-$: no elimination takes place

$R = CH_3CH_2CH(CH_3)CH_2-$: products are $CH_3CH_2C(CH_3){=}CH_2 + (CH_3)_2NOH$

(d) $RN(CH_3)_2 \xrightarrow[Ag_2O]{CH_3I} R\overset{+}{N}(CH_3)_2\ OH^- \xrightarrow{\Delta}$

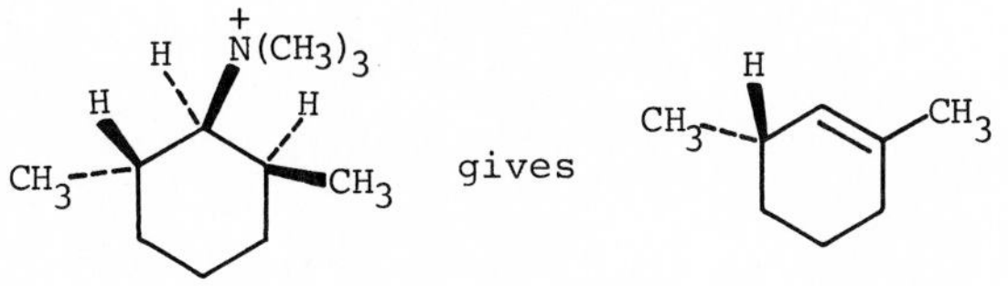

(resulting from *anti* elimination)

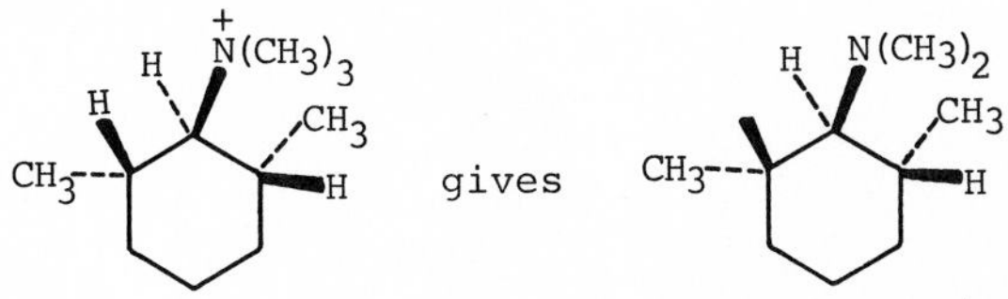

starting material

(Demethylation occurs because there is no hydrogen *anti* to the nitrogen)

S3. *Weakest base:* V (to act as a base, it must gain a proton, and an *additional* positive charge)

VI (Cl is electron-withdrawing)

I

IV (CH_3O is electron-releasing, through conjugation of the lone pair electrons)

II (alkylamines are stronger bases than arylamines)

Strongest base: III (dialkylamines are even better than alkylamines)

S4.

Because of some contribution from this resonance structure, the nitrogen already has positive character and will be a weaker base.

S5. (a)

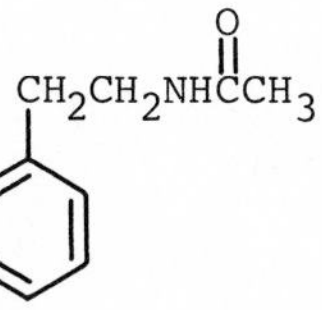

(b)

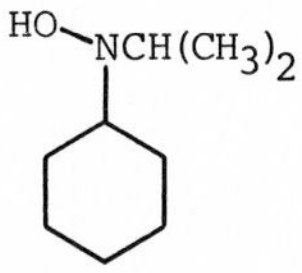

(c)

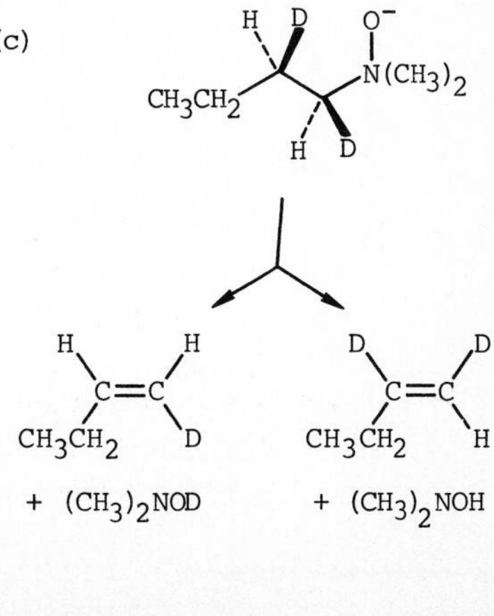

(d)

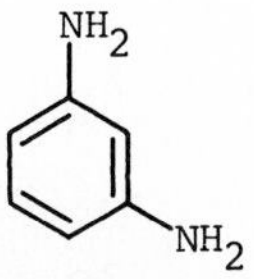

(e) $CH_3CH_2CH_2CH_2CH_2NH_2$

(f) $CH_3CH_2CH_2CH_2\overset{+}{N}H_3$
           Cl^-

(g)

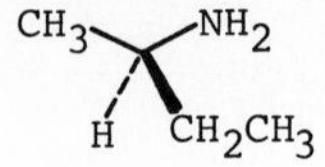

(h) $(CH_3)_3CCH_2CH_2CH=CH_2$
      $+ (CH_3)_3N:$

(i)

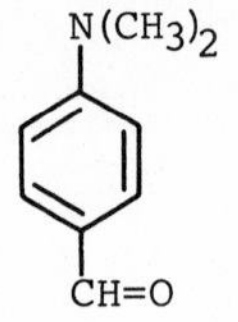

(j) $(CH_3)_2CHCH_2NCH_2CH(CH_3)_2$
                    $N=O$

(k)
$+ (CH_3)_2CH\overset{O}{\overset{\|}{C}}CH_2CH_3$

(ℓ) $C_6H_5\overset{O}{\overset{\|}{C}}CH_2CH_2N$

(m)

S6.   (a) $(CH_3)_2CHCH_2COOH \xrightarrow[H_2SO_4]{NaN_3} \left[(CH_3)_2CHCH_2\overset{O}{\overset{\|}{C}}N_3 \right] \longrightarrow (CH_3)_2CHCH_2\overset{O}{\overset{\|}{C}}\ddot{N}:$

$$(CH_3)_2CHCH_2NH_2 \longleftarrow \left[(CH_3)_2CHCH_2N=C=O \right]$$

(b) $(CH_3)_2CHCH_2COOH \xrightarrow[\Delta]{NH_3} (CH_3)_2CHCH_2CONH_2 \xrightarrow{LiAlH_4} (CH_3)_2CHCH_2CH_2NH_2$
     or $\xrightarrow{SOCl_2}$ $\xrightarrow{NH_3}$

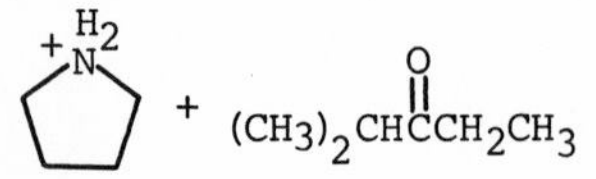

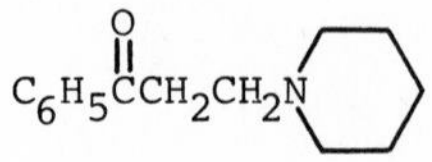

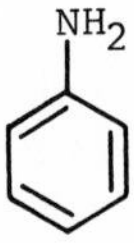

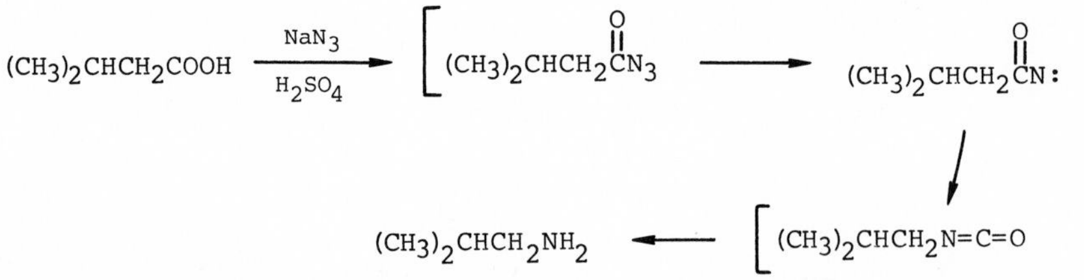

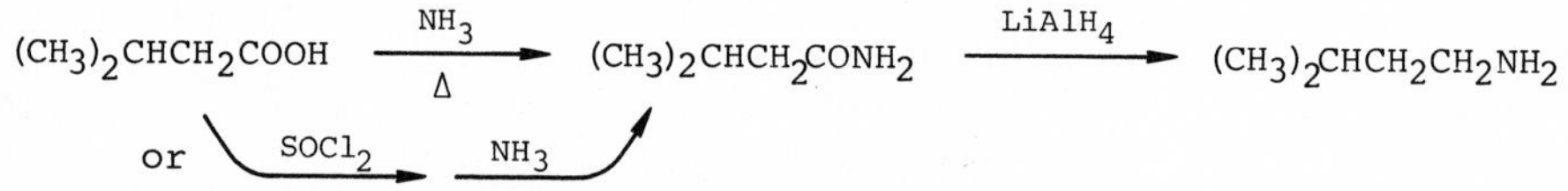

(c) $(CH_3)_2CHCH_2COOH \xrightarrow[2.\ PBr_3]{1.\ LiAlH_4} (CH_3)_2CHCH_2CH_2Br \xrightarrow[2.\ LiAlH_4]{1.\ NaCN} (CH_3)_2CHCH_2CH_2CH_2NH_2$

S7.

$CH_3CH_2CH_2\overset{O}{\overset{\|}{C}}H + HCN \longrightarrow CH_3CH_2CH_2\overset{OH}{\overset{|}{C}}H-C\equiv N \xrightarrow{H_2/Ni} CH_3CH_2CH_2\overset{OH}{\overset{|}{C}}HCH_2NH_2 \xrightarrow{CH_3I}$

$\xleftarrow[K_2CO_3]{} \overset{I^-}{RNH(CH_3)_2} \xleftarrow{CH_3I} RNHCH_3 \xleftarrow[K_2CO_3]{} \overset{+}{R}NH_2CH_3 \ I^-$
KI, KHCO$_3$ KI, KHCO$_3$

$RN(CH_3)_2 \xrightarrow{CH_3I} \left[CH_3CH_2CH_2\overset{OH}{\overset{|}{C}}H-CH_2-\overset{+}{N}(CH_3)_3 \right] \xrightarrow{K_2CO_3} CH_3CH_2CH_2\overset{O^-}{\overset{|}{C}}H-CH_2$
                                                               $\overset{+}{N}(CH_3)_3$

$\overset{I^-}{(CH_3)_4N^+} \xleftarrow{CH_3I} N(CH_3)_3 + CH_3CH_2CH_2\overset{O}{\overset{\diagdown}{CH-CH_2}} \longleftarrow$

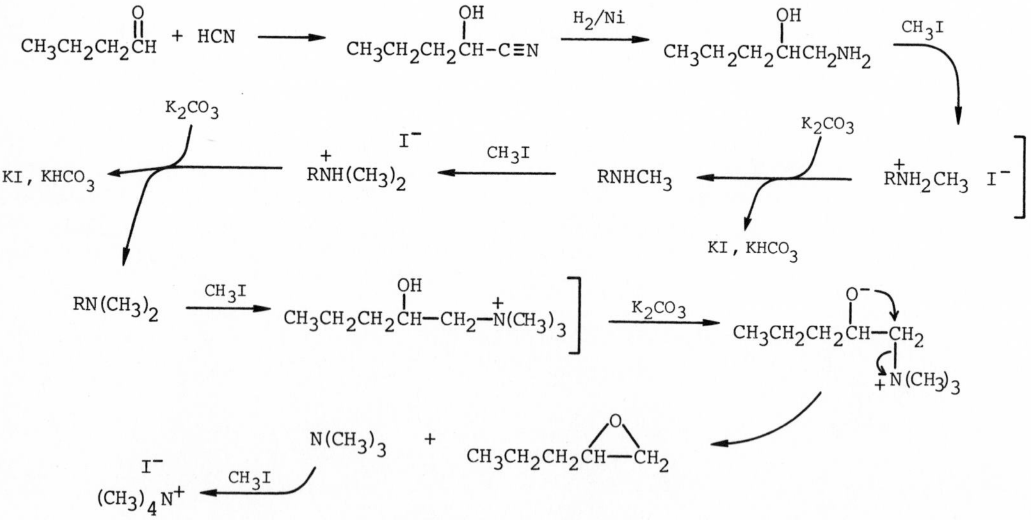

S8.   Under basic conditions, the amine is more nucleophilic than the hydroxy group, and acylation takes place preferentially on nitrogen to give the amide. In acid, the amine is unreactive because it is protonated, and acylation occurs on oxygen to furnish the ester. However, on deprotonation of the amine again, a transacylation reaction takes place to provide the more stable amide:

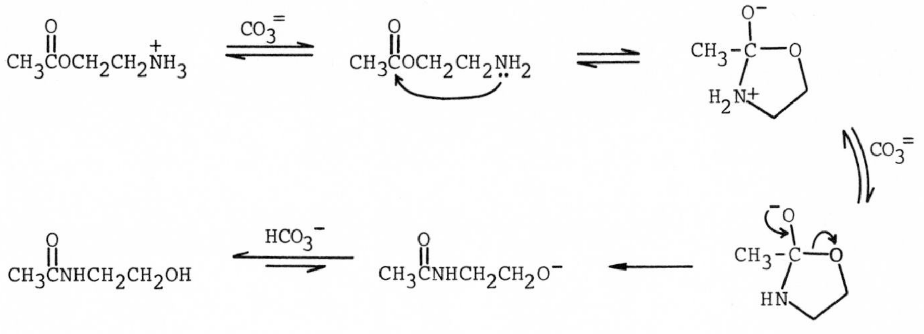

S9. (a)

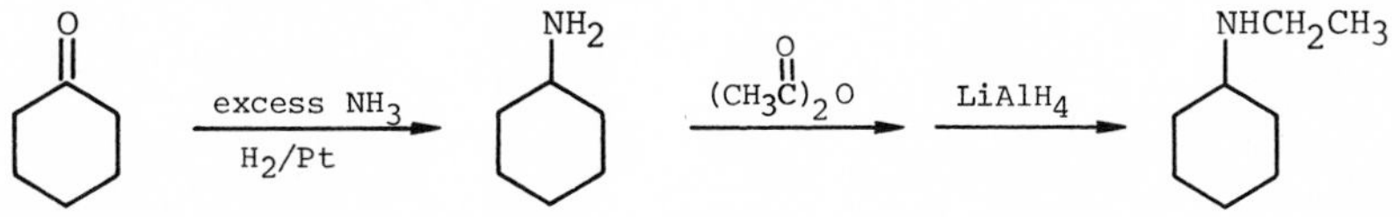

Alternatively,

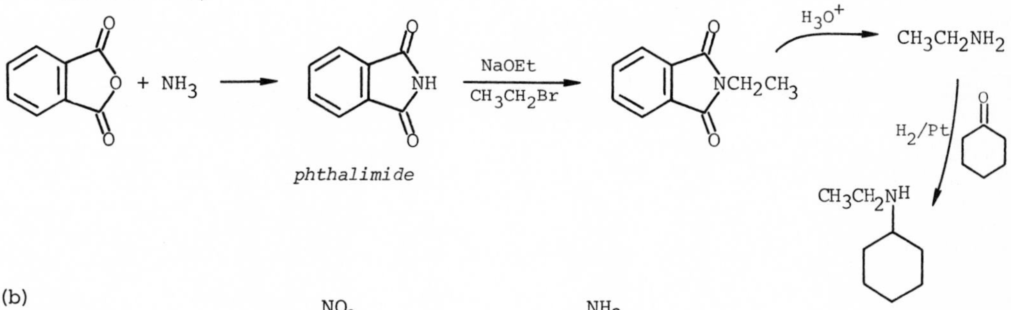

(b)

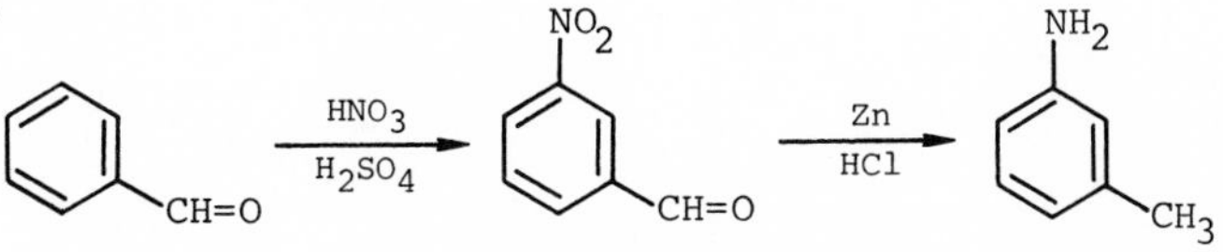

(Note that both reductions can be accomplished in one step.)

(c) phthalimide *(see part (a))*

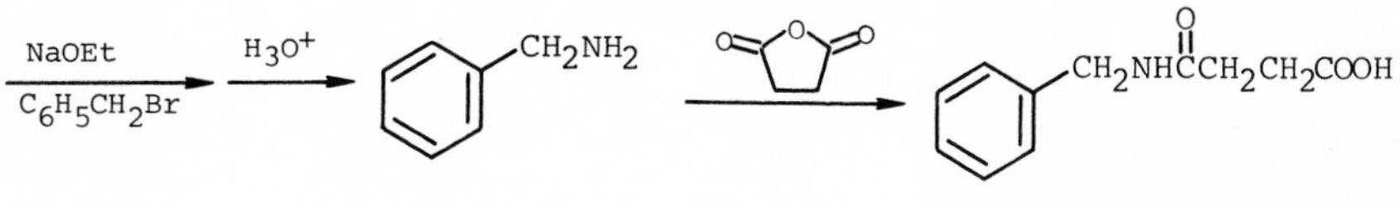

(d) phthalimide

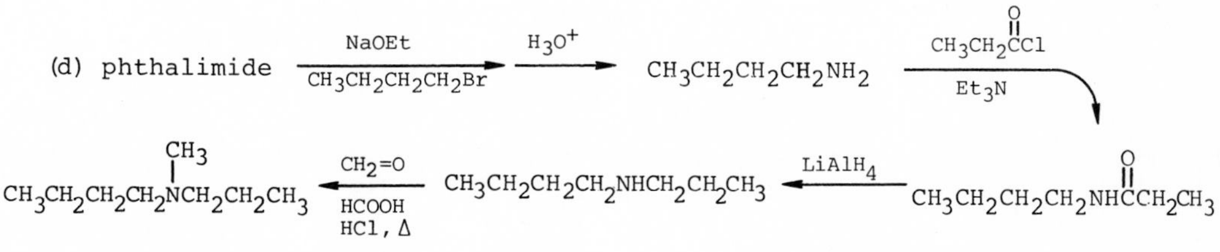

(e) phthalimide

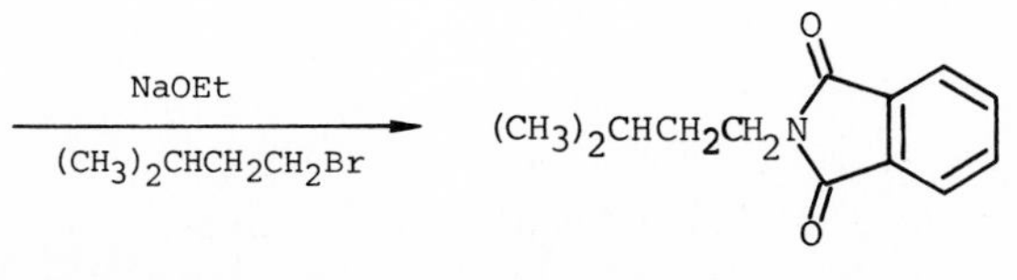

(f) $CH_3CH_2NH_2$ $\xrightarrow[\text{K}_2\text{CO}_3]{\text{BrCH}_2\text{CH}_2\text{CH}_2\text{CH}_2\text{CH}_2\text{Br}}$ $\xrightarrow{\text{H}_2\text{O}_2}$

(see part (a))

(g) $C_6H_5\overset{\overset{\text{O}}{\|}}{C}CH_3$ $\xrightarrow{\text{HCN}}$ $C_6H_5\overset{\overset{\text{OH}}{|}}{\underset{\overset{|}{CH_3}}{C}}C{\equiv}N$ $\xrightarrow{\text{LiAlH}_4}$ $C_6H_5\overset{\overset{\text{OH}}{|}}{\underset{\overset{|}{CH_3}}{C}}CH_2NH_2$

(h) $CH_3CH_2NH_2$ $\xrightarrow[\text{LiAlH}_4]{\overset{\overset{\text{O}}{\|}}{(CH_3C)_2O}}$ $(CH_3CH_2)_2NH$ $\xrightarrow[\text{HCl}]{\text{NaNO}_2}$ $(CH_3CH_2)_2N{-}N{=}O$

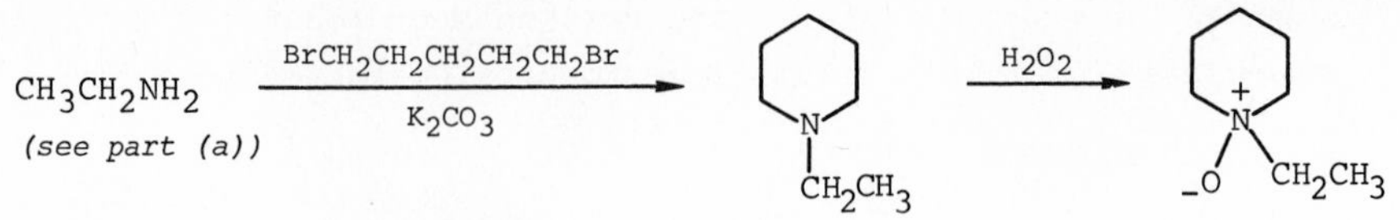

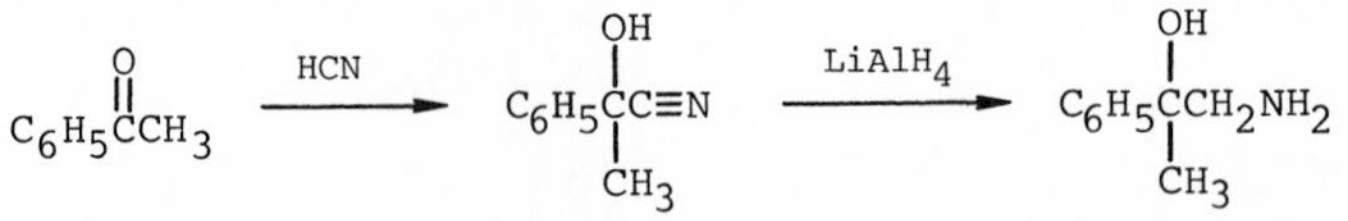

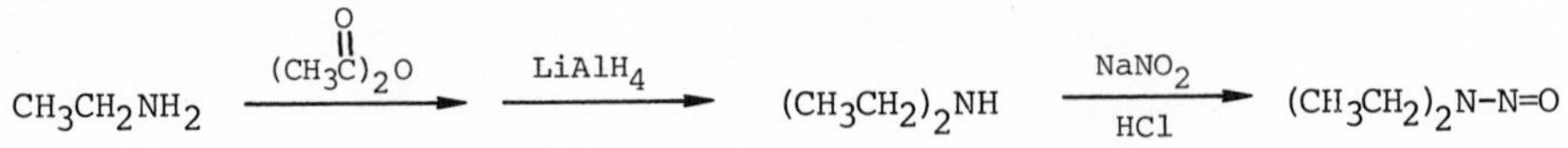

S10. (a)

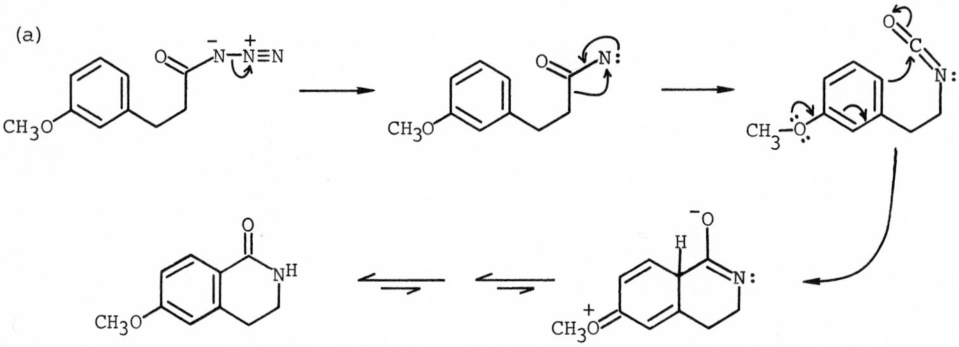

(b)

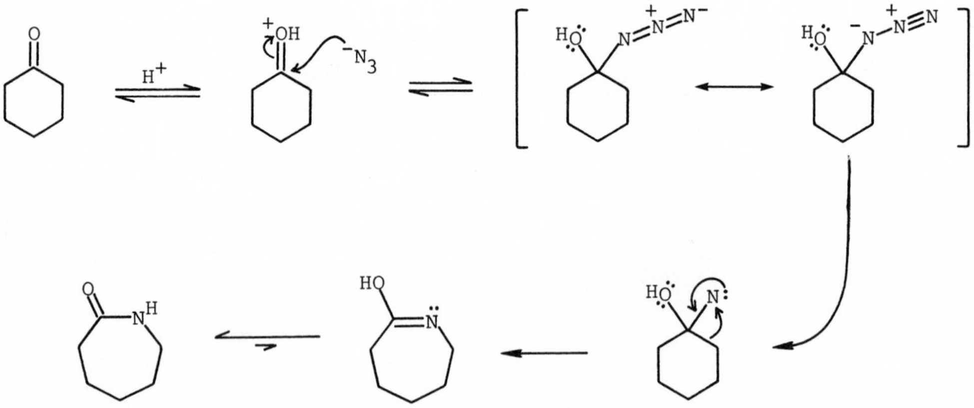

S11. The ability of Y to stabilize the electron pair it gains during the reaction is indicated by the pK_a of the conjugate acid:

p$K_a \cong$ 9–10   p$K_a \cong$ 16–17   p$K_a \cong$ 35–40

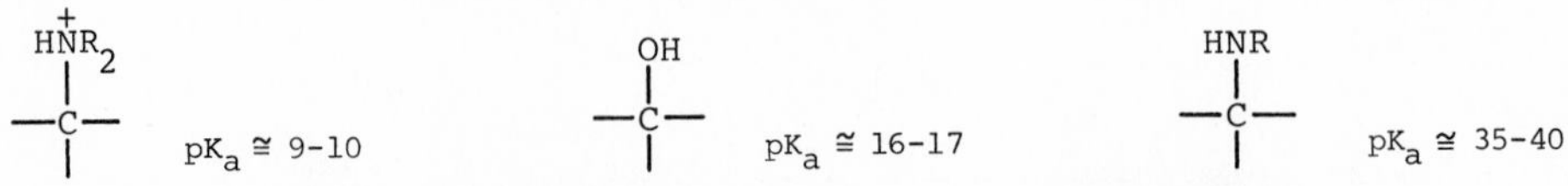

S12.

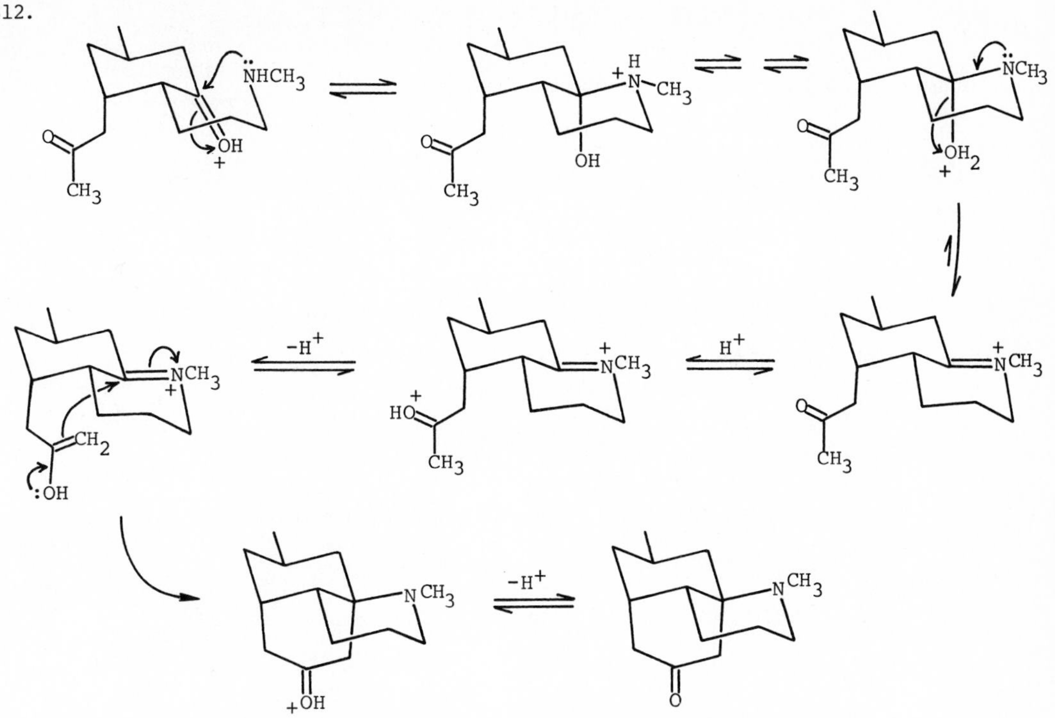

25. OTHER NITROGEN FUNCTIONS

25.A Outline of Chapter 25, Important Terms Introduced, and Reactions Discussed

25.1   <u>Nitro Compounds</u>

   A. Nitroalkanes

      from free radical nitration:

$$RH + HNO_3 \xrightarrow{400°} RNO_2 + H_2O \quad \text{(Industrial reaction)}$$

      from displacement by nitrite:

$$RX + NO_2^- \longrightarrow RNO_2 + X^- \text{ (Na}^+ \text{ or Ag}^+ \text{ salt; nitrite ester a byproduct)}$$

      carbanion reactions of nitroalkanes:

$$CH_3NO_2 \rightleftharpoons \overset{-}{}CH_2NO_2 \xrightarrow{\overset{\overset{O}{\parallel}}{RCH}} RCH\overset{OH}{\overset{|}{C}}H_2NO_2 \longrightarrow RCH=CHNO_2 \quad \text{(like aldol)}$$

      reduction of $-NO_2$ to $-NH_2$: see section 24.6.C

   B. Nitroarenes (preparation covered in Chapters 23 and 24)

   C. Reactions of Nitroarenes

      reduction of nitro group:

         to amine

$$ArNO_2 \longrightarrow ArNH_2 \quad \text{(using Zn, Sn, or SnCl}_2 \text{ in HCl; } H_2/\text{catalyst)}$$

         to hydroxylamine

$$ArNO_2 \longrightarrow ArNHOH \quad \text{(Zn, } NH_4Cl\text{)}$$

         to azoxy compound

$$2 \; ArNO_2 \longrightarrow Ar\overset{O^-}{\underset{+}{N}}=NAr \quad (As_2O_3, \text{ aq NaOH: via ArN=O +ArNH}_2)$$

         to azo compound

$$2 \; ArNO_2 \longrightarrow ArN=NAr \quad \text{(Zn, alcoholic NaOH)}$$

         to hydrazo compound

$$2 \; ArNO_2 \longrightarrow ArNHNHAr \quad (H_2NNH_2, \text{ Ru/C, alcoholic KOH)}$$

    other interconversions:

         hydroxylamine to nitroso compound

$$ArNHOH \longrightarrow ArN=O \quad (Na_2Cr_2O_7, \; H_2SO_4)$$

         azoxy compound ⇄ azo compound

$$Ar\overset{O^-}{\underset{+}{N}}=NAr \underset{\underset{H_2O_2, \; HOAc}{\longleftarrow}}{\overset{\overset{(EtO)_3P:}{\longrightarrow}}{}} ArN=NAr$$

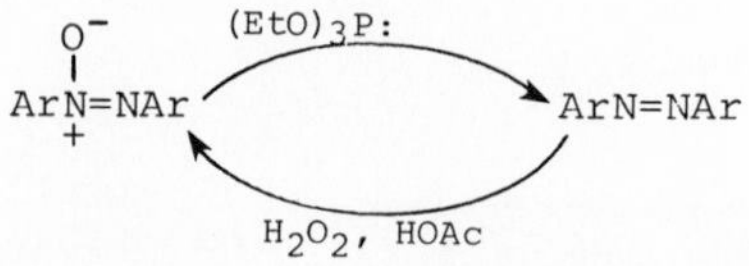

azo compound ⇌ hydrazo compound

$$ArN{=}NAr \underset{O_2 \text{ or } NaOBr}{\overset{H_2NNH_2,\ Pd/C}{\rightleftharpoons}} ArNHNHAr$$

azo compound to amine

$$ArN{=}NAr \longrightarrow 2\ ArNH_2$$ *($Na_2S_2O_4/H_2O$, or any reagents which take $ArNO_2$ to $ArNH_2$)*

benzidine rearrangement

$$ArNHNHAr \xrightarrow[\Delta]{H^+} H_2N\text{-}Ar\text{-}Ar\text{-}NH_2$$

25.2 Isocyanates, Carbamates, and Ureas

isocyanates from Hofmann rearrangement: see section 24.6.H

isocyanates from displacement:

$$RX + NaN{=}C{=}O \longrightarrow RN{=}C{=}O + NaX$$

carbamates (urethanes) and ureas from isocyanates:

$$RN{=}C{=}O \xrightarrow{HOR'} RNH\overset{O}{\overset{\|}{C}}OR'$$ *($\longrightarrow RNH_2 + CO_2$ if $R'{=}H$; polyurethane from diisocyanates and diols)*

$$RN{=}C{=}O \xrightarrow{H_2NR'} RNH\overset{O}{\overset{\|}{C}}NHR'$$

25.3 Azides

alkyl and acyl azides by displacement:

$$RX + N_3^- \longrightarrow RN_3 + X^-$$

$$R\overset{O}{\overset{\|}{C}}Cl + N_3^- \longrightarrow R\overset{O}{\overset{\|}{C}}N_3 + Cl^-$$ *(see Curtius rearrangement, Section 25.6.H)*

reduction of alkyl azides:

$$RN_3 \longrightarrow RNH_2$$ *(with $LiAlH_4$ or H_2/catalyst; alternative to Gabriel synthesis)*

25.4 Diazo Compounds

preparation of esters with diazomethane: see Section 18.7.A

preparation of diazomethane from N-nitroso amides:

$$R\overset{O}{\overset{\|}{C}}\underset{N=O}{\underset{|}{N}}CH_3 \xrightarrow{OH^-} RCO_2^- + CH_2{=}N{=}N$$

reaction of acyl halides with diazomethane:

$$R\overset{O}{\overset{\|}{C}}Cl + 2\ CH_2N_2 \longrightarrow R\overset{O}{\overset{\|}{C}}CHN_2 + CH_3Cl + N_2$$

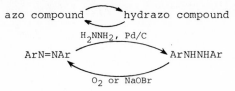

reactions of α-diazoketones:

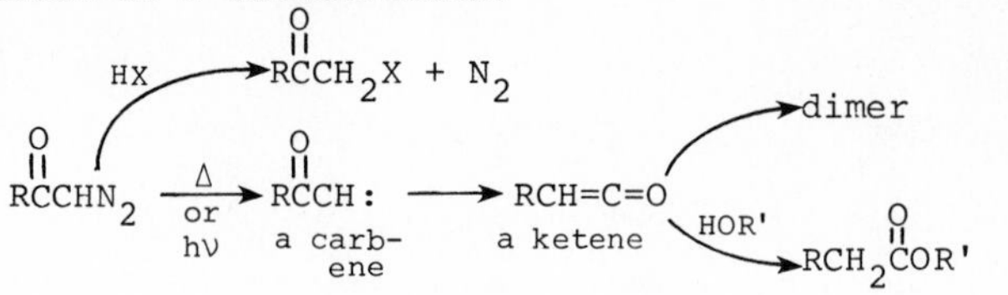

carbene addition from α-diazoesters:

$$N_2=CHCO_2Et \xrightarrow{\Delta} \left[:CHCO_2Et\right] \longrightarrow \text{———} CO_2Et$$

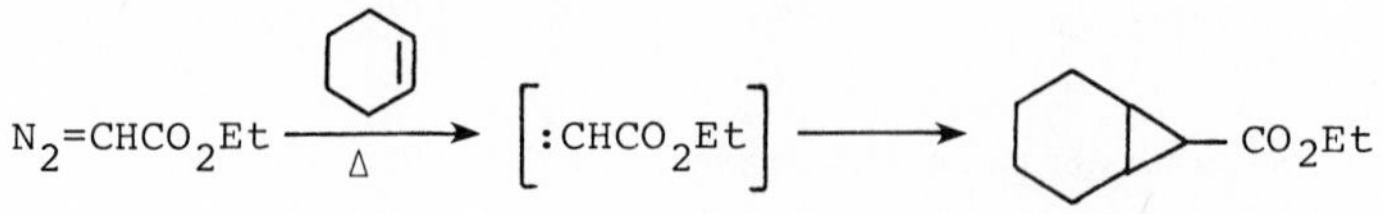

25.5 Diazonium Salts

preparation from aryl amines: see Section 24.7.B

A. Acid-Base Equilibria of Arenediazonium Ions

$$ArN \equiv N^+ + H_2O \underset{K_1}{\overset{-H^+}{\rightleftharpoons}} \underset{\substack{arenediazo-\\ hydroxide}}{ArN=NOH} \underset{K_2}{\overset{OH^-}{\rightleftharpoons}} \underset{\substack{arene-\\ diazotate}}{ArN=NO^-} \quad (K_2 > K_1)$$

synthesis of substituted benzene derivatives via diazonium
        displacement

B. Thermal Decomposition of Diazonium Salts; Formation cf AroH, ArI,
    and ArSH

$$ArN_2^+ \begin{cases} \xrightarrow{HOR} ArOR \\ \xrightarrow{NaI} ArI \\ \underline{or} \xrightarrow{NaSH} \\ \xrightarrow[\substack{S\\ \parallel \\ NaSCOCH_3}]{} ArSR' \end{cases}$$

C. The Sandmeyer Reaction: Preparation of ArCl, ArBr, and ArCN

$$ArN_2^+ + CuX \xrightarrow{\Delta} ArX + Cu^+ \quad (X = Cl, Br, or CN)$$

D. Preparation of Fluoro- and Nitroarenes

$$ArN_2^+BF_4^- \begin{cases} \xrightarrow{\Delta} ArF + N_2 + BF_3 \quad \textit{(Schiemann reaction; via } PF_6^- \textit{ salt too)} \\ \xrightarrow[NaNO_2]{\Delta} ArNO_2 + NaBF_4 + N_2 \quad \textit{(Gatterman reaction)} \end{cases}$$

E. Replacement of the Diazonium Group by Hydrogen

$$ArN_2^+ \xrightarrow{H_3PO_2} ArH$$

F. Arylation Reactions

$$ArN_2^+ + NaOH + Ar'H \xrightarrow{\Delta} Ar-Ar' \quad \textit{(biphenyl synthesis; radical inter-mediate; } \underline{o}, \underline{p} \textit{ with } a\underline{ll} \textit{ substituents)}$$

$$\underset{NO}{\overset{O}{\underset{|}{ArNCCH_3}}} + Ar'H \xrightarrow{\Delta} Ar-Ar'$$

G. Diazonium Ions as Electrophiles: Azo Compounds

$$ArN_2^+ + Ar'H \longrightarrow Ar{-}N{=}N{-}Ar' \quad (Ar'H = \text{electron-rich})$$

triazene intermediate with $ArNH_2$

H. Synthetic Utility of Arenediazonium Salts

25.B Important Concepts and Hints

This chapter covers a true potpourri of reactions, with the common factor that they all have to do with nitrogen compounds in which the nitrogen is not an amine.

Much of this chemistry involves arene derivatives, including the various ways you can reduce nitroarenes and all the intermediates that can be obtained on the way to aryl amines, and the chemistry of arenediazonium salts. The former topic is full of detail, and is hard to learn in a way that doesn't ultimately rely on memorization. On the other hand, the reactions of arenediazonium salts have the common goal of replacing the diazonium group with something else. This is an important group of reactions because it enables the introduction of a wide range of substituents onto an aromatic ring by a method other than electrophilic aromatic substitution. This can be useful when normal directive effects in electrophilic substitution make it impossible to obtain a specific isomer. Keep in mind the fact that RS, HS, RO, HO, F, Cl, Br, I, CN, CH=O (from CN), CO_2H (from CN), and even H can be represented by amino or nitro substituents at various stages of the synthesis.

25.C Answers to Exercises

(25.1.C)

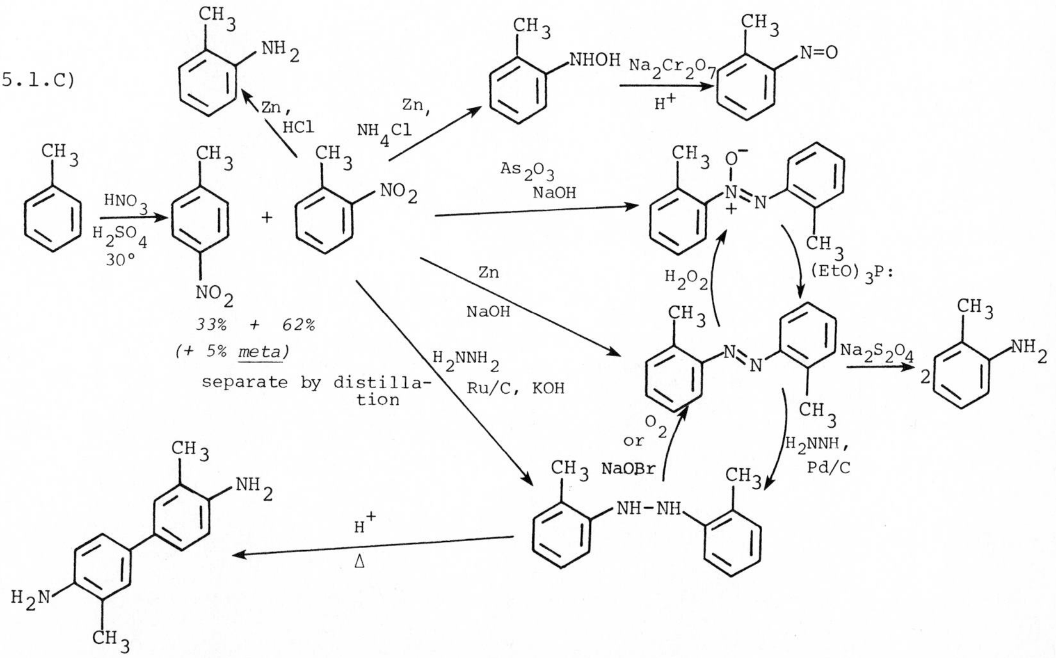

(25.3) $R:\ddot{N}::C::\ddot{O}:$ $\quad$ $R:\ddot{N}::\overset{-}{N}::\overset{+}{N}:$

$$C_6H_5CH_2Cl + NaN{=}C{=}O \xrightarrow{CH_3OH} \left[C_6H_5CH_2N{=}C{=}O\right] \longrightarrow C_6H_5CH_2NH\overset{O}{\overset{\|}{C}}OCH_3 + NaCl$$

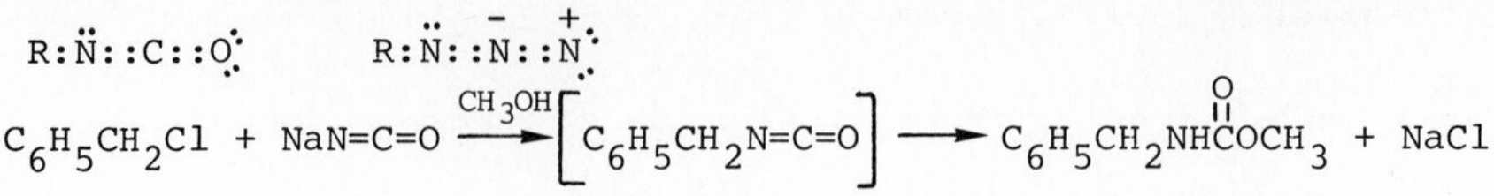

(25.4)

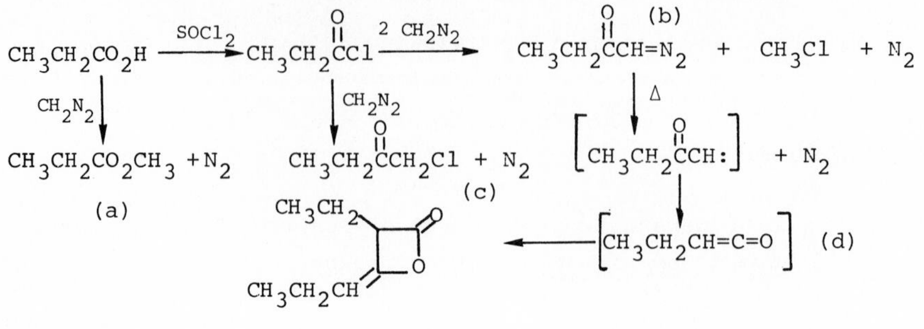

(25.5.B)

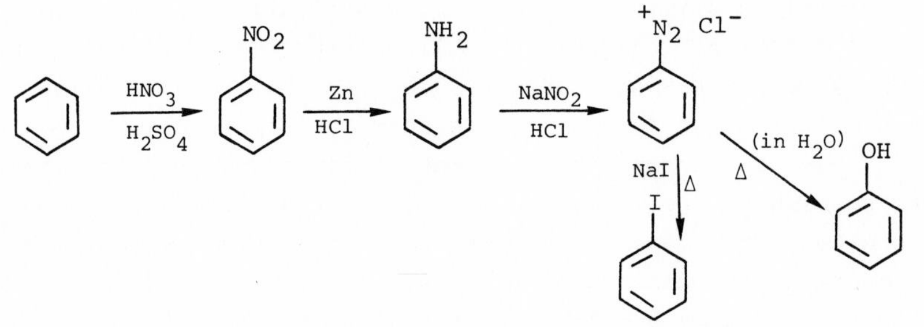

(25.5.C)

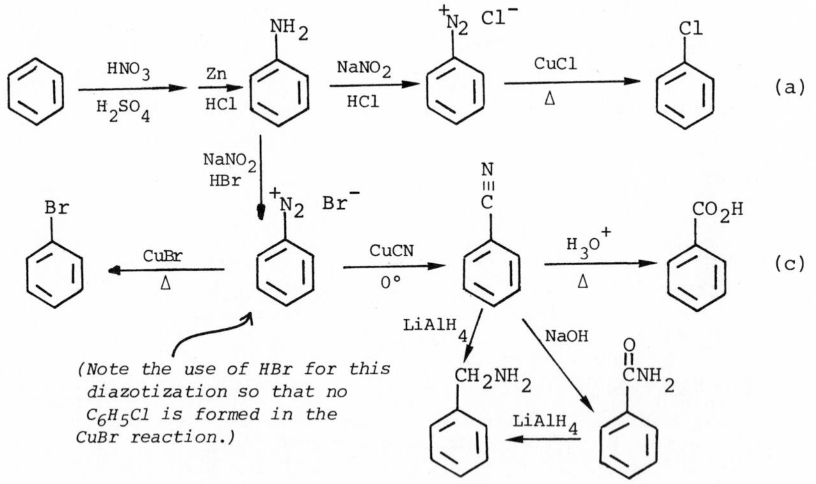

(25.5.E)

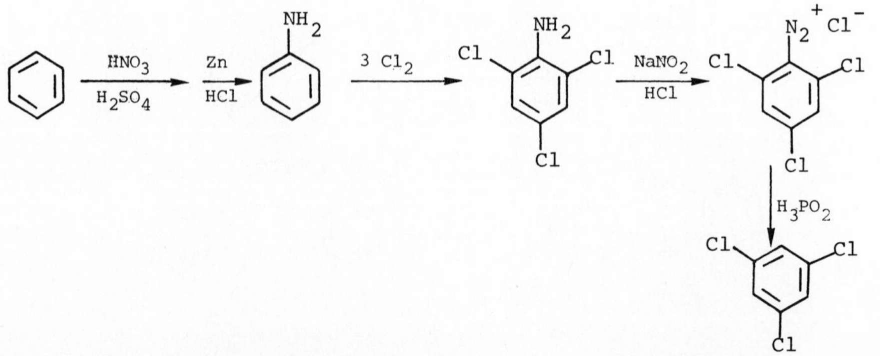

(25.5.F)

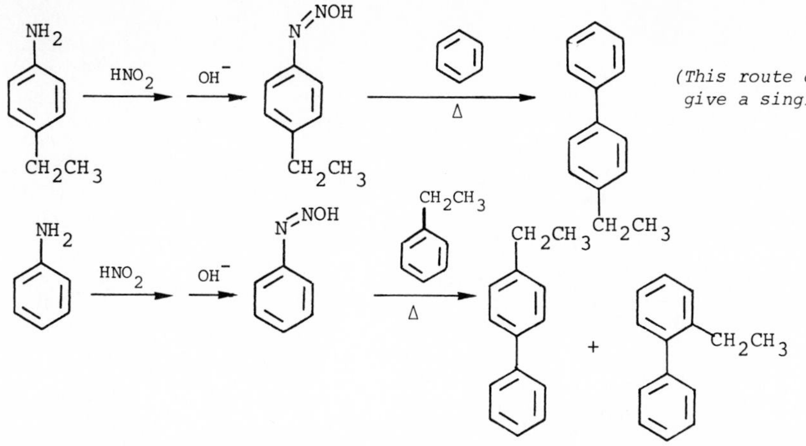

(This route can only give a single isomer)

25.D Answers and Explanations to Problems

1. (a)

p-iodotoluene

(b)

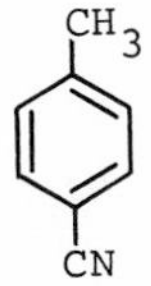

p-tolunitrile

(c)

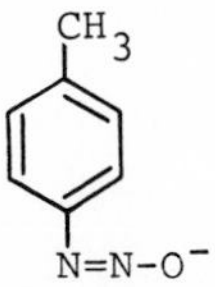

p-toluenediazotate ion

(d)

p-methylphenol
p-cresol

(e)

p-bromotoluene

(f)

p-nitrotoluene

(g)

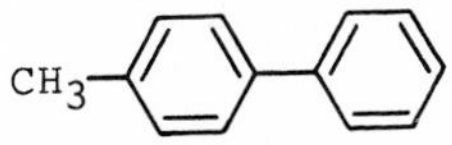

p-methylbiphenyl

(h)

p-fluorotoluene

(i)

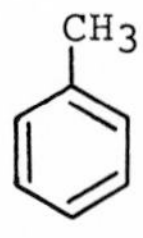

toluene

(j)

p-chlorotoluene

(k)

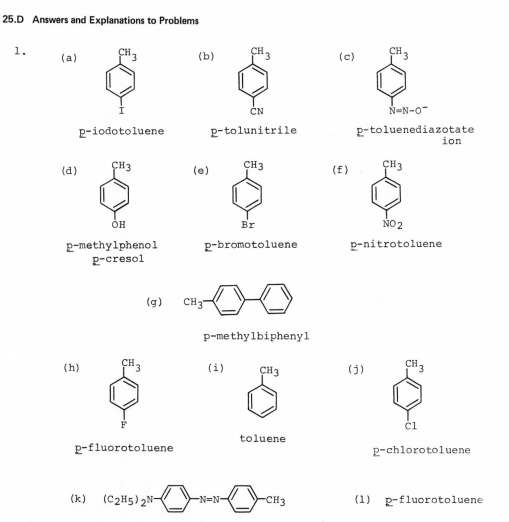

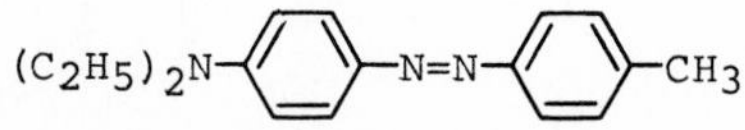

p-N,N-diethylaminophenylazo-p-toluene
or p-tolylazo-p-N,N-diethylaniline

(l) p-fluorotoluene

2. (a)

CH_3 | HNO_3 / H_2SO_4 → CH_3, NO_2 | H_2/cat. or Fe, HCl → CH_3, NH_2

separate from
<u>para</u> (distillation)

(b)

CH_3 | HNO_3 / H_2SO_4 → CH_3, NO_2 | H_2/cat. or Fe, HCl → CH_3, NH_2 | $(CH_3CO)_2O$ → CH_3, $NHCOCH_3$

separate
from ortho

CH_3, NO_2 ← H_3PO_2 | HCl NaNO$_2$ cold ← CH_3, NO_2, NH_2 ← HCl Δ ← CH_3, NO_2, $NHCOCH_3$ ← HNO_3 CH_3COOH

(c)   SEE (b)

(d)

CH_3 | $KMnO_4$ → CO_2H | fuming HNO_3 / H_2SO_4 → CO_2H, NO_2 | H_2/cat. or Fe, HCl → CO_2^-, NH_3^+

(e)

CH_3, NO_2 (from (a) above) $\xrightarrow{Cl_2, FeCl_3}$ CH_3, NO_2, Cl

(f)

CH_3, NH_2 (from (a)) $\xrightarrow{(CH_3CO)_2O}$ CH_3, $NHCOCH_3$ $\xrightarrow{HNO_3, CH_3COOH}$ O_2N, CH_3, $NHCOCH_3$ $\xrightarrow{H_2/\text{catalyst}}$

Cl_2 / CH_3COOH (alternative paths)

CH_3, NH_2, Cl ← HCl Δ ← CH_3, $NHCOCH_3$, Cl ← Cu_2Cl_2 ← $NaNO_2$ HCl ← CH_3, $NHCOCH_3$, H_2N

(g)

⬡ $\xrightarrow{\text{fuming } HNO_3,\ H_2SO_4}$ NO_2, NO_2 $\xrightarrow{H_2/\text{cat. or Fe, HCl}}$ NH_2, NH_2

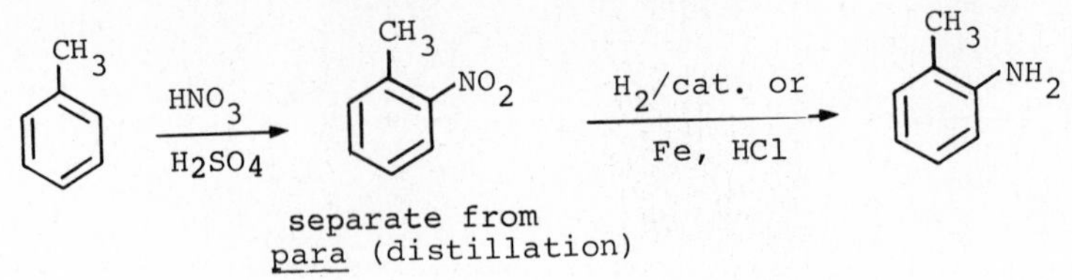

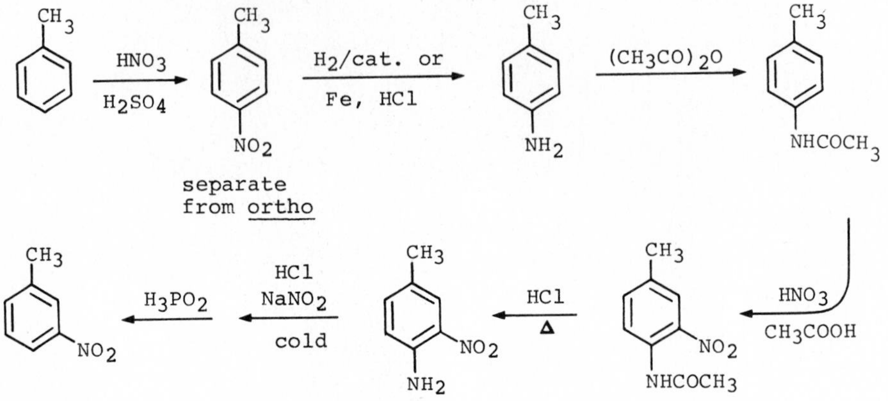

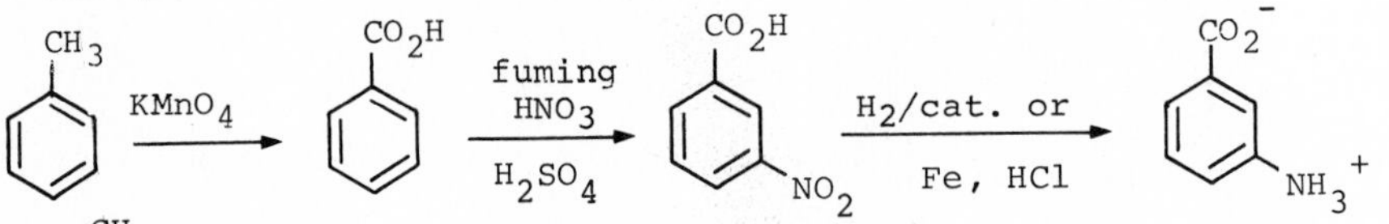

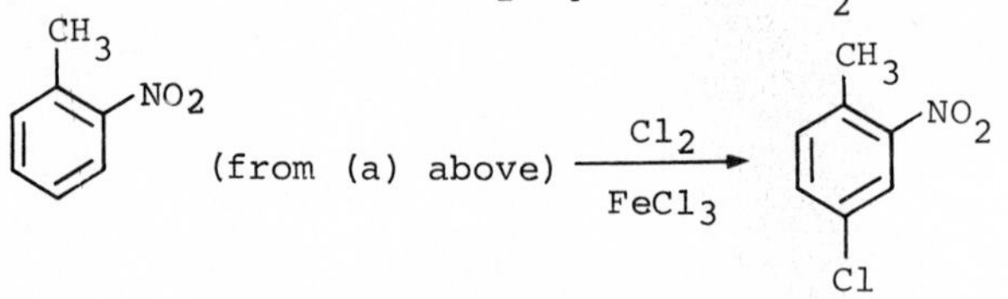

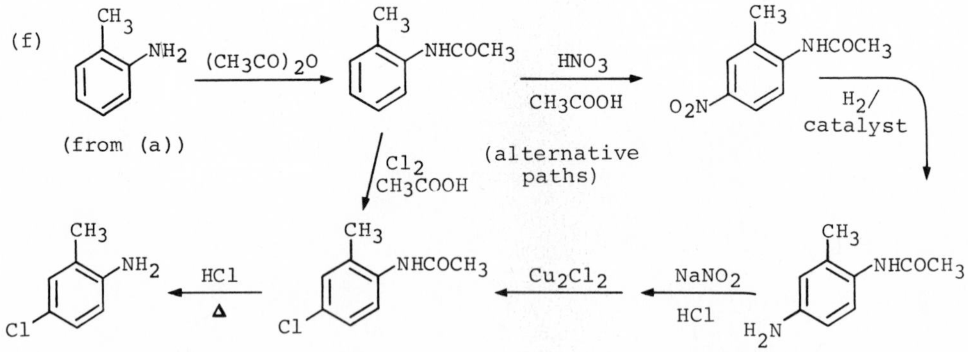

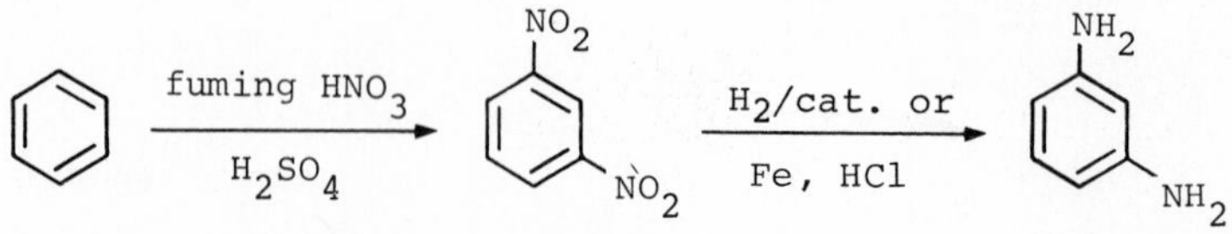

(h)

NO_2
Fe, HCl
NH_2
$(CH_3)_2SO_4$
or CH_3Br
$N(CH_3)_2$
$NaNO_2$
HCl
$N(CH_3)_2$
NO
$SnCl_2$, HCl
or H_2/cat.
$N(CH_3)_2$
NH_2

(i)

NH_2
(from (h))
$(CH_3CO)_2O$
$NHCOCH_3$
HNO_3
CH_3COOH
$NHCOCH_3$
NO_2
HCl
Δ
NH_2
NO_2

3. (a)

CH_3
HNO_3
H_2SO_4
CH_3
NO_2
separate
from ortho
H_2/cat. or
Fe, HCl
CH_3
NH_2
$(CH_3CO)_2O$
CH_3
$NHCOCH_3$
Br_2
CH_3COOH
CH_3
Br
$NHCOCH_3$
HCl
Δ
CH_3
Br
NH_2

(b)

CH_3
Br
NH_2
[from (a)]
$NaNO_2$
HCl, cold
CH_3
Br
N_2^+
KI
H_2O
CH_3
Br
I

(c)

CH_3
NH_2
(from (a))
Br_2
H_2O
CH_3
Br
Br
NH_2
$NaNO_2$,
HCl
cold
H_3PO_2
CH_3
Br
Br

(d)

fuming
HNO_3,
H_2SO_4
NO_2
NO_2
NaSH
alc.
NH_2
NO_2
$NaNO_2$,
HCl
cold
NaOH
cold
NO_2

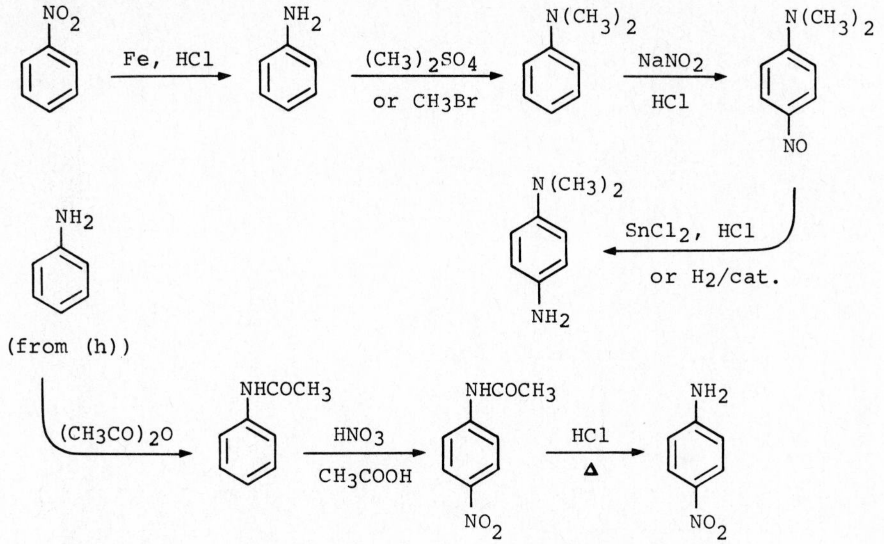

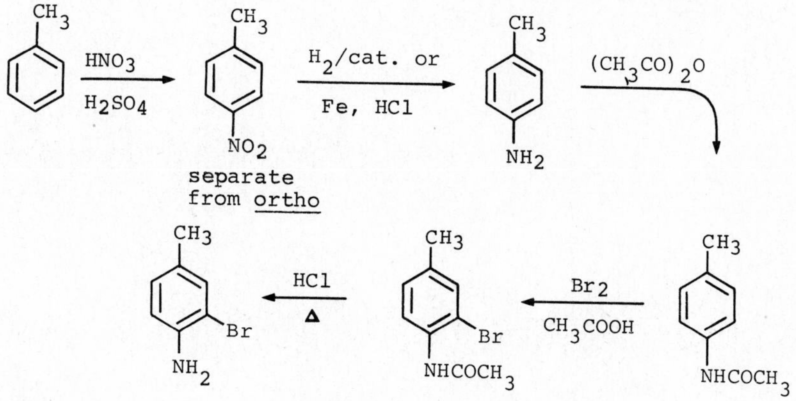

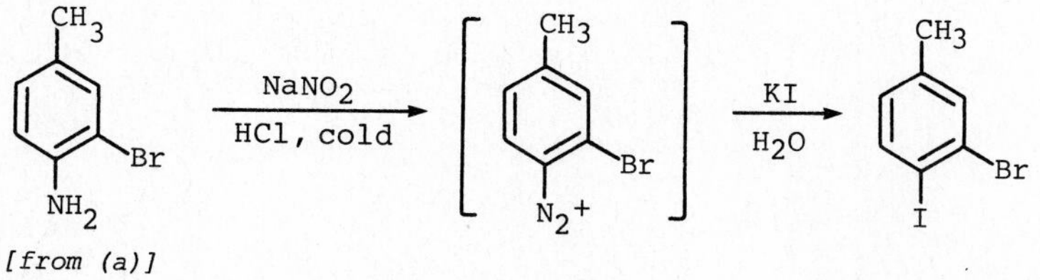

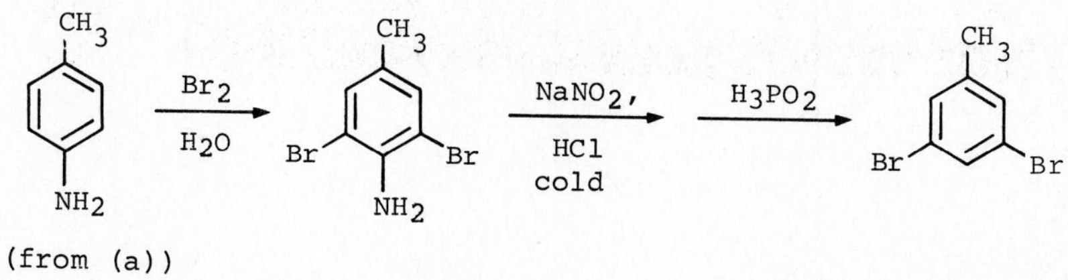

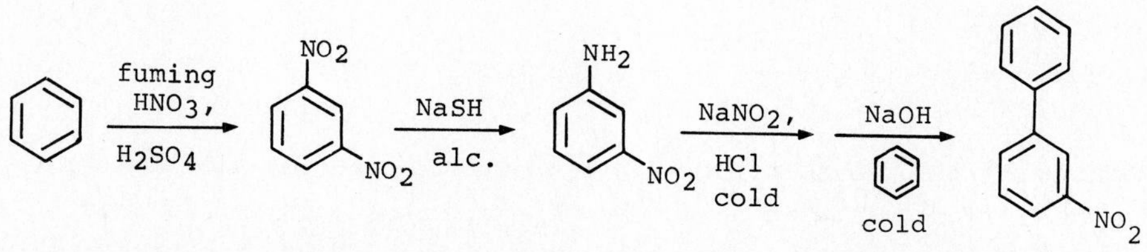

(e) 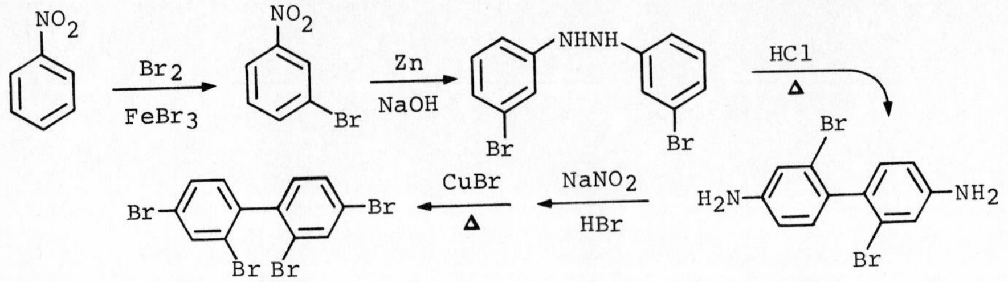

(f) 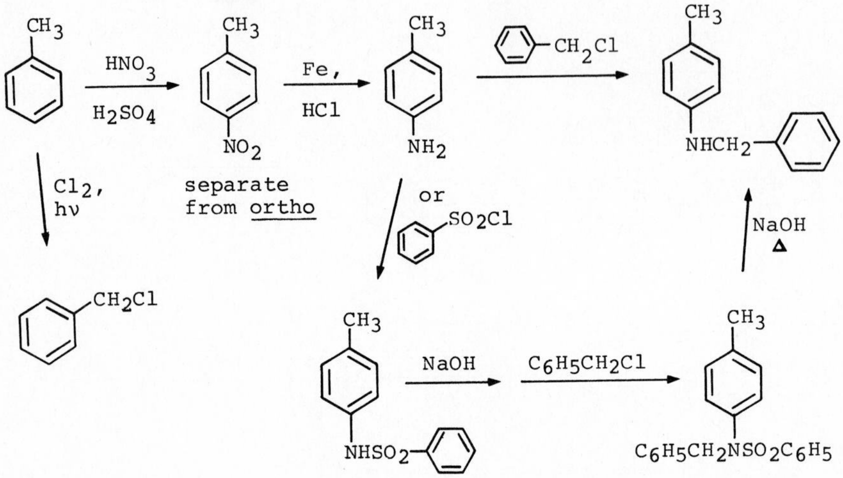

(g) 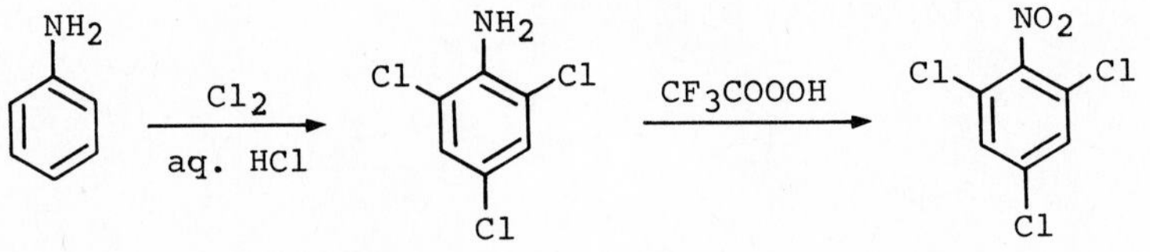

(h) 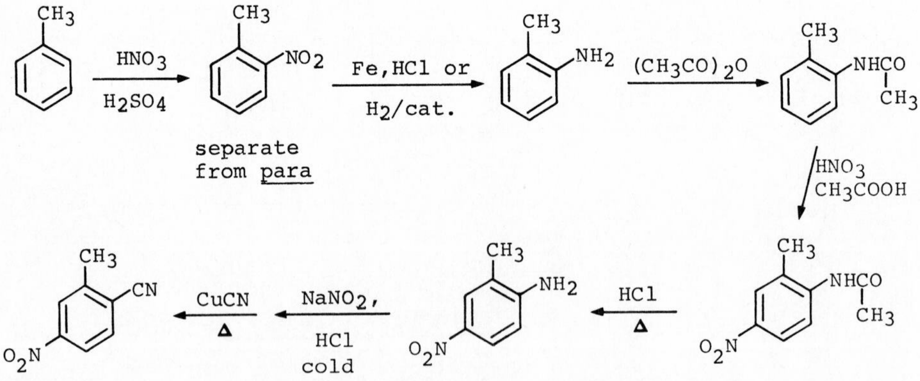

4. From $pK_a = -\log \dfrac{[H^+][AcO^-]}{[AcOH]}$, for $[AcO^-] = [AcOH]$,

$$pK_a = pH$$

Hence, the pH of a solution with equimolar amounts of AcO^- and AcOH is 4.74. This value is above the pK_a of methyl orange, 3.5 (see Sect. 25.5.H); hence, methyl orange is in the yellow, unprotonated, form.

5. On reaction at the p-position, the odd electron can conjugate with the nitro group:

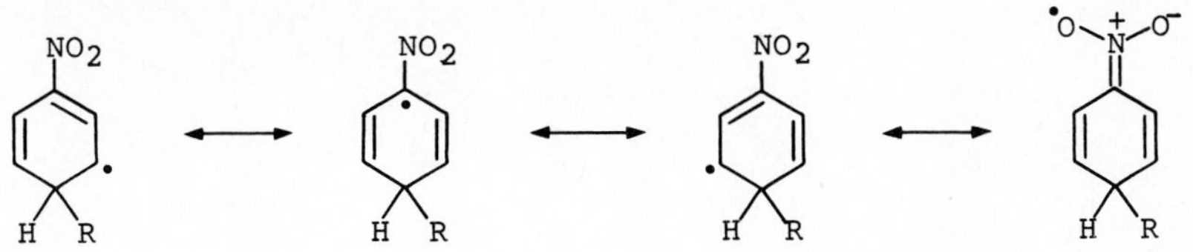

The nitro group helps to stabilize the odd electron and directs o,p-, just as in nucleophilic aromatic substitution.

6. $CuBr + Cl^- \rightleftharpoons CuCl + Br^-$

The use of CuBr + HCl gives a mixture of bromocumene and chlorocumene.

7.

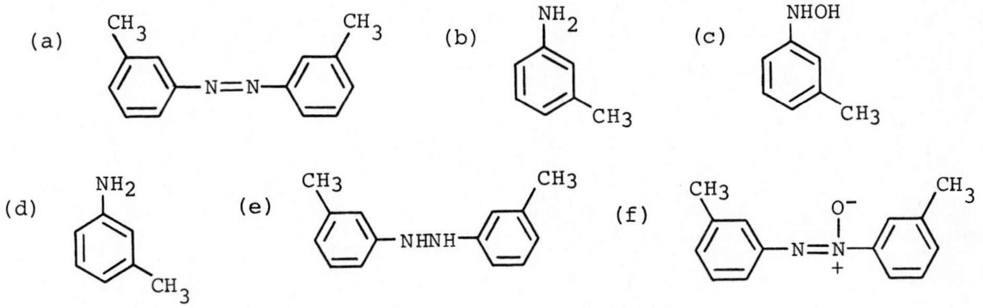

8. (a) Loss of N_2 gives a carbene which rearranges to a ketene:

$$R\text{-}\overset{O}{\overset{\|}{C}}\text{-}CHN_2 \xrightarrow{-N_2} R\text{-}\overset{O}{\overset{\|}{C}}\text{-}\ddot{C}\text{-}H \longrightarrow R\text{-}CH{=}C{=}O$$

This rearrangement is similar to that which leads to a nitrene in the Curtius reaction. The ketene rapidly adds water to give a carboxylic acid.

$$R\text{-}CH{=}C{=}O + H_2O \longrightarrow R\text{-}CH{=}\underset{{}^{+}OH_2}{\underset{|}{C}}\text{-}O^- \longrightarrow RCH_2COOH$$

(b) $$CH_3COCHN_2 \xrightarrow[-N_2]{h\nu} CH_3CO\ddot{C}H \longrightarrow CH_3CH{=}C{=}O \xrightarrow{CH_3OH} CH_3CH_2COOCH_3$$

9.

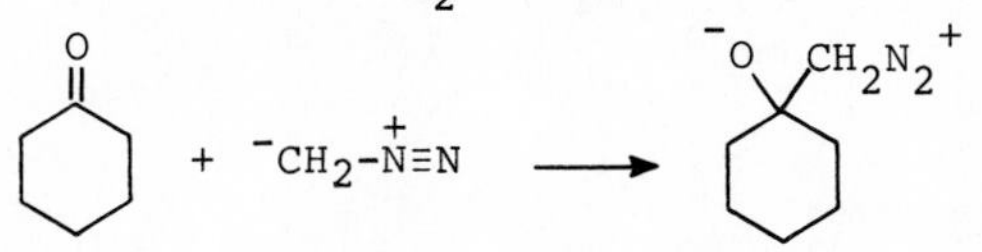

Two modes of further reaction:

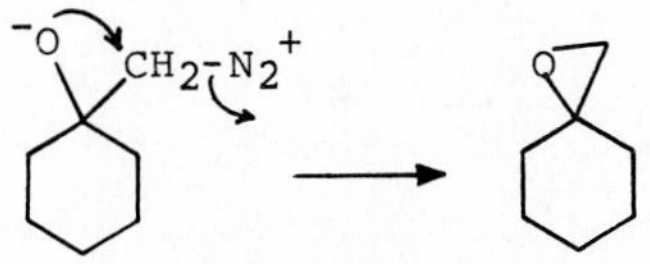

displacement of N_2 by $-O^-$

rearrangement:

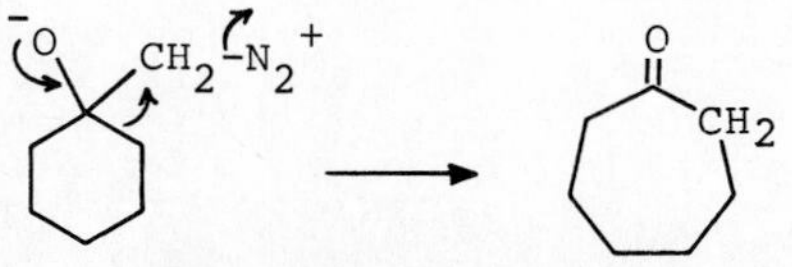

10. This is an example of the Hofmann degradation. When one gets to the isocyanate stage, the nucleophilic species now present is methanol.

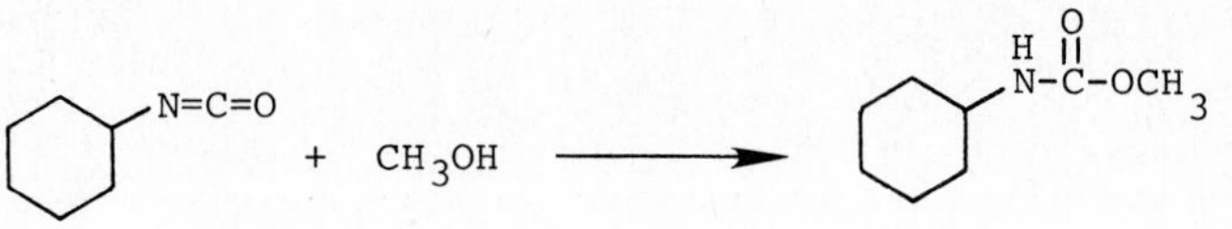

The product, a carbamic acid ester, is stable.

11. The predominant base in methanolic potassium hydroxide is actually methoxide ion because of the equilibrium:

$$HO^- + CH_3OH \rightleftharpoons CH_3O^- + H_2O$$

(large excess)

The reaction is slow, since the ethyl carbamate is also an amide. The main reaction observed is transesterification:

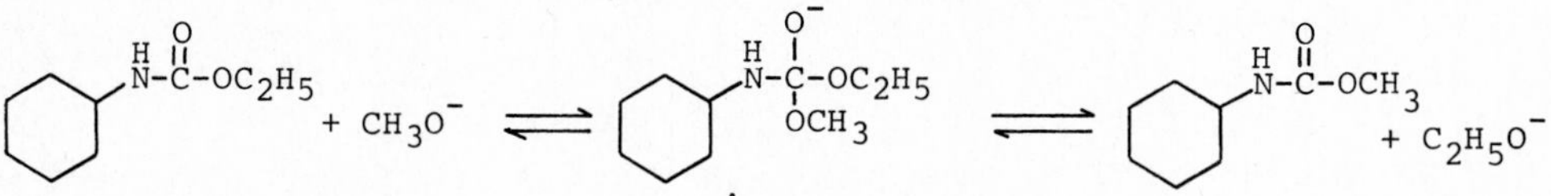

A

The alternative path, elimination of amide from A, is not observed, since 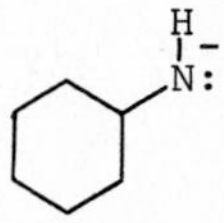

is such a poor leaving group. After a long time, of course, cyclohexylamine will be produced, since occasionally a hydroxide ion will attack to give an irreversible hydrolysis:

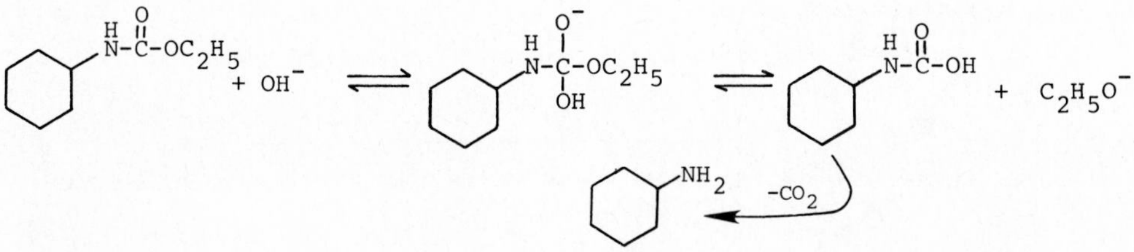

12.

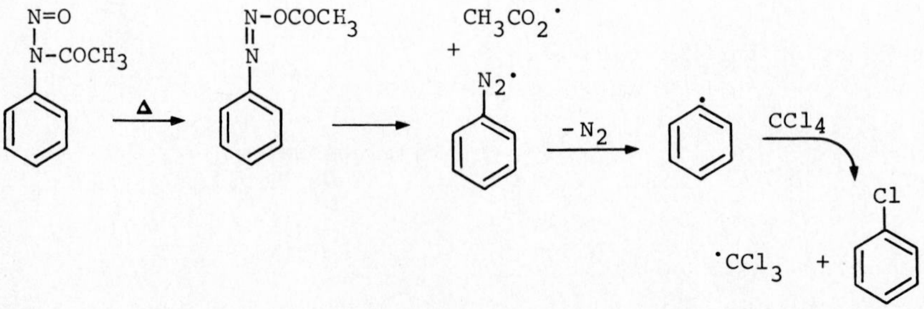

25.E Supplementary Problems

S1. (a) Nitroethylene can be prepared from 2-nitroethanol by distillation from phthalic anhydride. Write a mechanism for this reaction.

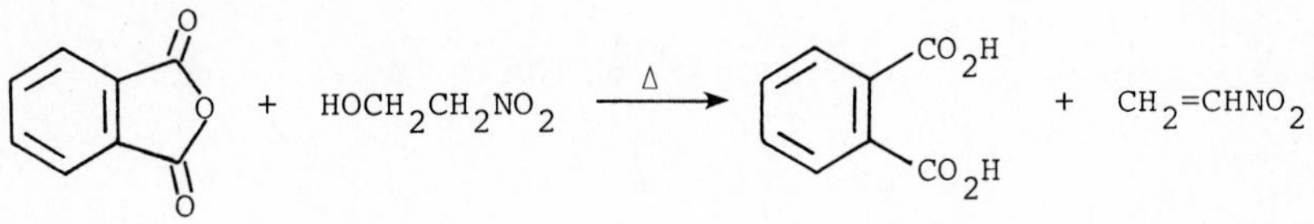

(b) Nitroethylene is a powerful electrophile and reacts rapidly with dimethylamine to give N,N-dimethyl-2-nitroethanamine, for example. Write a reasonable mechanism for this reaction.

S2. (a) If nitrocyclohexane is treated with a strong base to form the anion, and then protonated with acid and isolated rapidly, the so-called <u>aci</u> tautomer is isolated. Propose a structure for this compound.

(b) If the <u>aci</u> form of nitrocyclohexane is kept in the presence of aqueous acid, it undergoes hydrolysis to give cyclohexanone (the Nef reaction). Write a reasonable mechanism for this transformation.

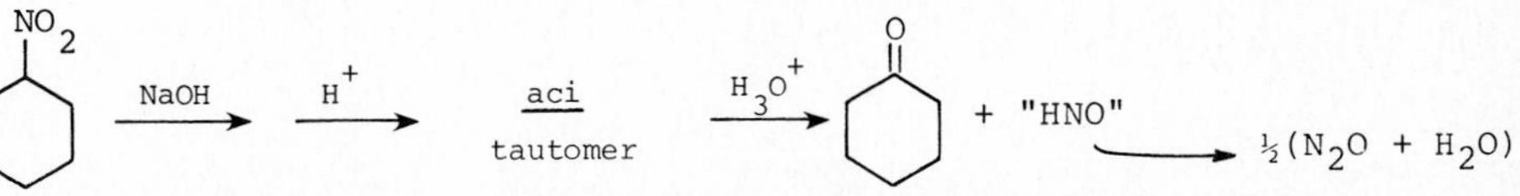

S3. What side product would you expect if the reaction depicted in problem S10 (a), chapter 24 of this Study Guide were carried out in ethanol?

S4. Phenyl azide can be prepared by treatment of benzenediazonium ion with sodium azide or by reaction of phenylhydrazine with nitrous acid. What intermediates are involved in this latter method?

S5. Show how to synthesize the following compounds from benzene or any monosubstituted derivative.

(a)

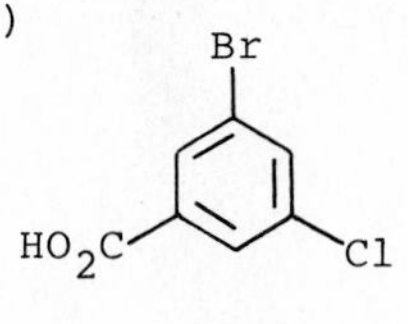

(b)

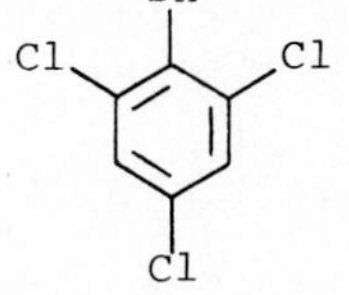

(c)

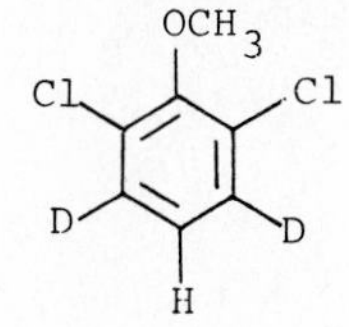

(d)

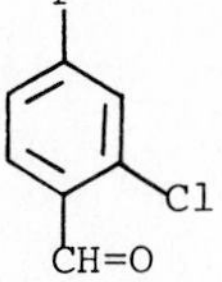

(e)

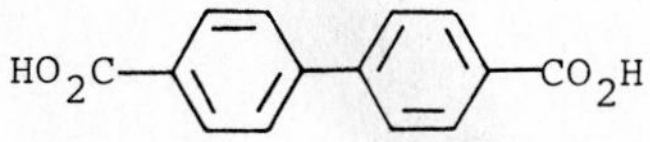

(f)

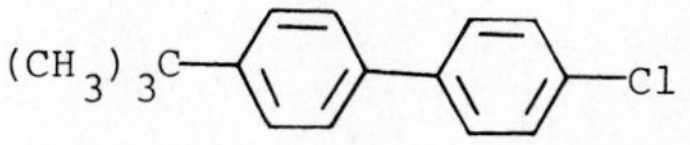

S6. Write a reasonable mechanism for the formation of diazomethane on treatment of N-methyl-N-nitrosoacetamide with sodium hydroxide.

S7. In contrast to ethyl diazoacetate, the diazoacetate anion is very unstable, decomposing in aqueous solution to give diazomethane and CO_2. Propose a mechanism for this transformation. (Note, the diazomethane anion N_2CH^- is not involved.)

$$N_2CHCO_2^- \xrightarrow{H_2O} N_2CH_2 + CO_2 + OH^-$$

S8. What is the major product from each of the following reactions.

(a)

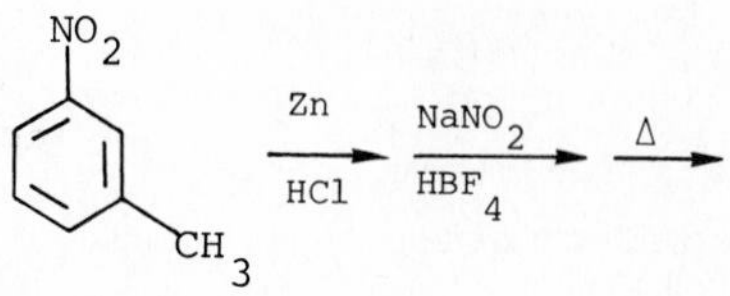

(d)

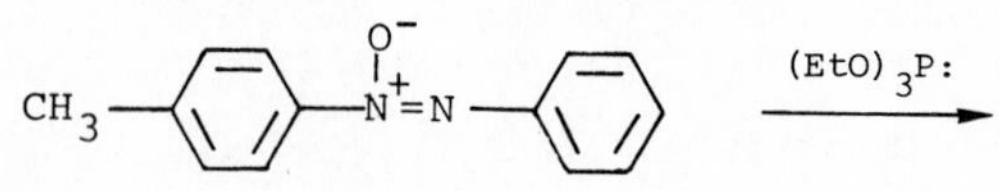

(b)

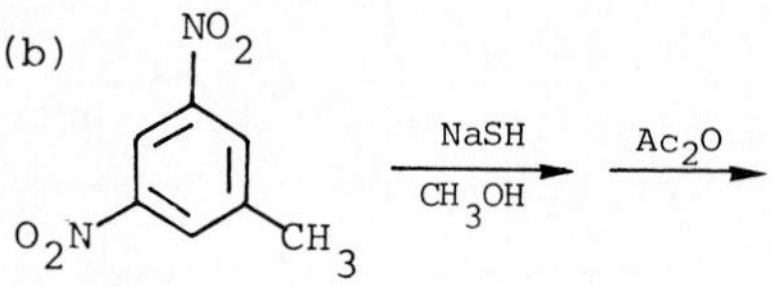

(e)

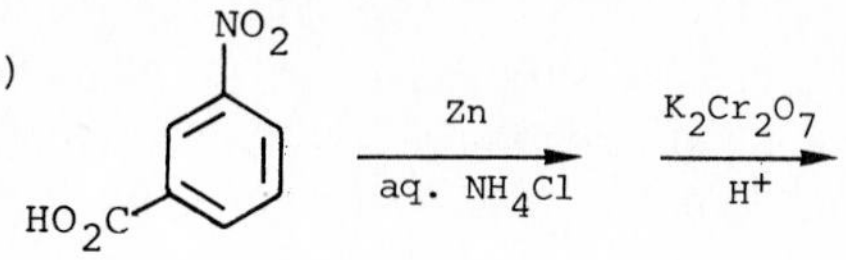

(c)

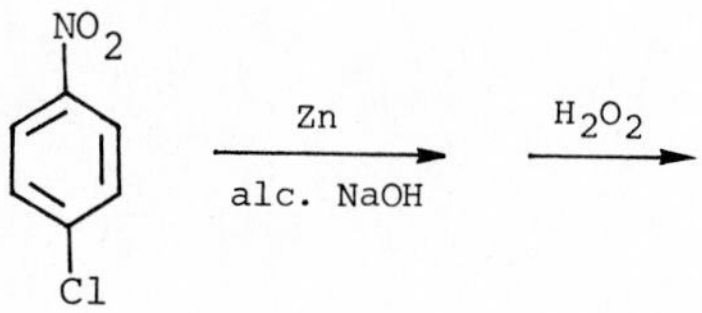

(f)

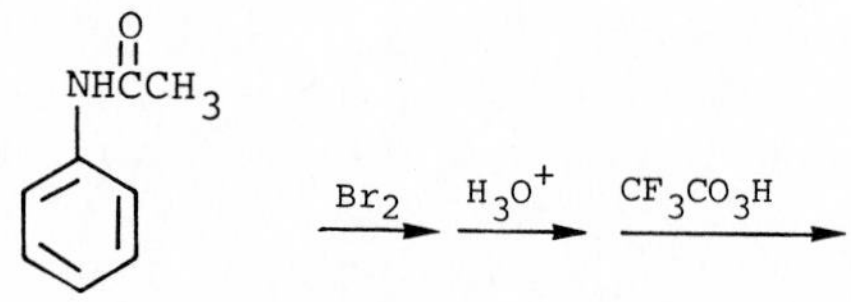

25.F Answers to Supplementary Problems

S1. (a)

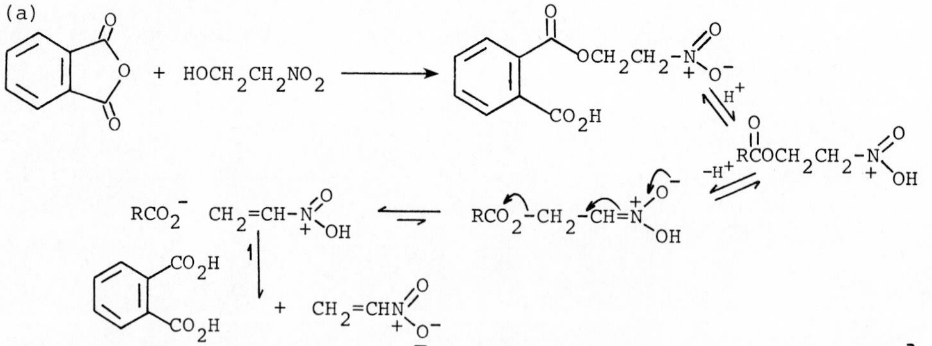

(b)

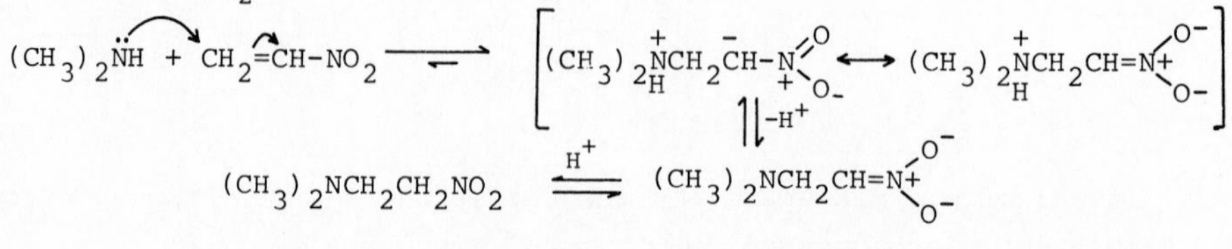

S2.

aci form

S3.

$$\text{CH}_3\text{O} \overset{\Delta}{\underset{\text{EtOH}}{\longrightarrow}} \left[\text{ArCH}_2\text{CH}_2\text{N=C=O} \right] \overset{\text{EtOH}}{\longrightarrow} \text{ArCH}_2\text{CH}_2\text{NHCOEt}$$

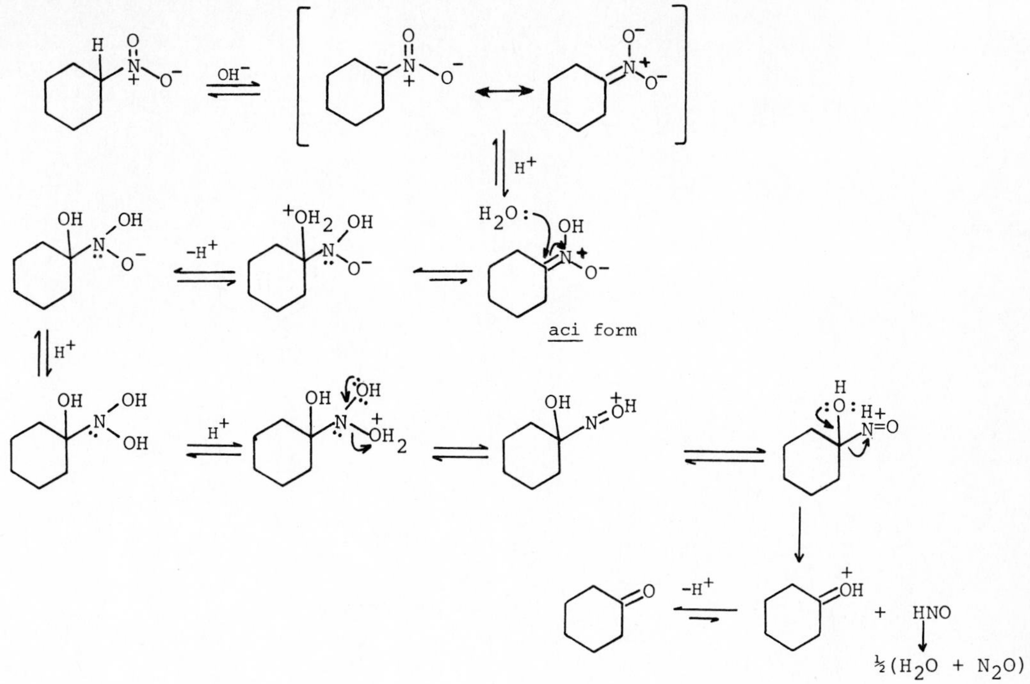

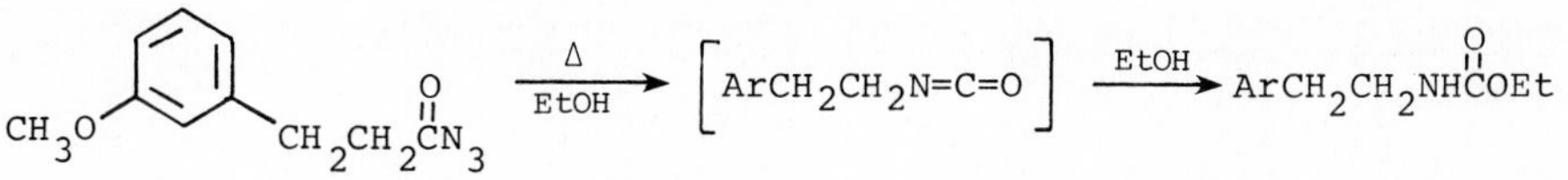

S4.

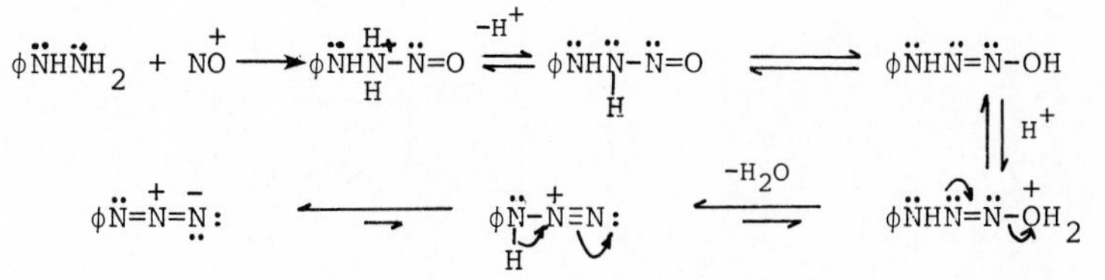

S5. (a)

(b)

(c)

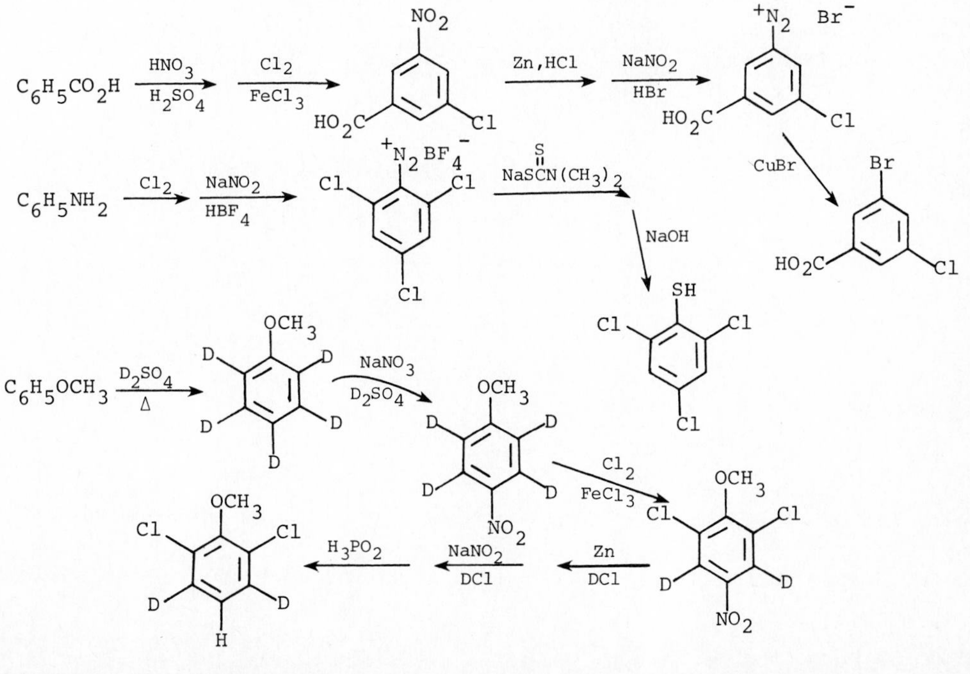

(d) $C_6H_5NO_2$ $\xrightarrow[\text{FeCl}_3]{\text{Cl}_2}$ $\xrightarrow[\text{Ac}_2\text{O}]{\text{Zn}}$ $\xrightarrow[\text{CO, HCl}]{\text{AlCl}_3}$

$\xleftarrow{\Delta}$ $\xrightarrow[\text{HBF}_4]{\text{NaNO}_2}$ $\xleftarrow{\text{H}_3\text{O}^+}$

(e) $2\ C_6H_5NO_2$ $\xrightarrow[\substack{\text{Ru-C}\\\text{KOH}}]{\text{H}_2\text{NNH}_2}$ $C_6H_5NHNHC_6H_5$ $\xrightarrow[\Delta]{\text{H}^+}$

$\xrightarrow[\text{2 CuCN}]{\text{NaNO}_2\ \text{HBF}_4}$

$\xleftarrow{\text{H}_3\text{O}^+}$

(f) $C_6H_5C(CH_3)_3$ $\xrightarrow{\text{HNO}_3}$ $\xrightarrow[\text{Ac}_2\text{O}]{\text{Zn}}$ $\xrightarrow{\text{NOCl}}$

Δ C_6H_5Cl

(and some ortho)

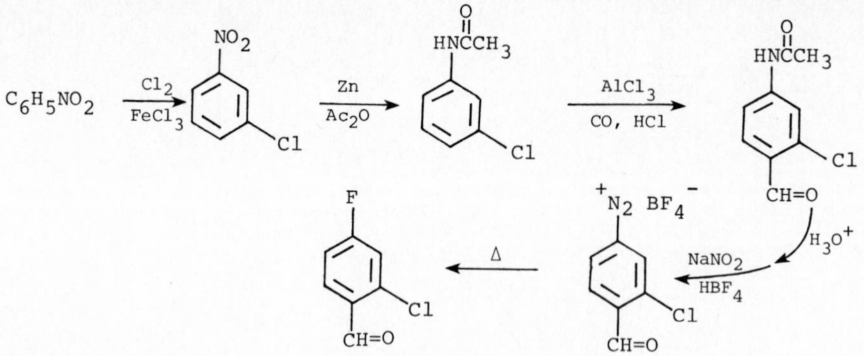

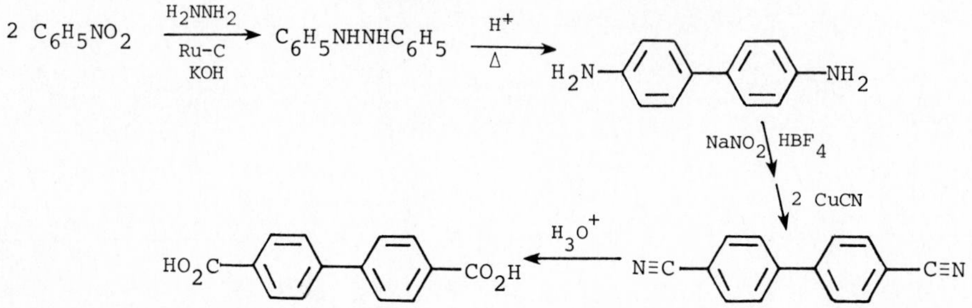

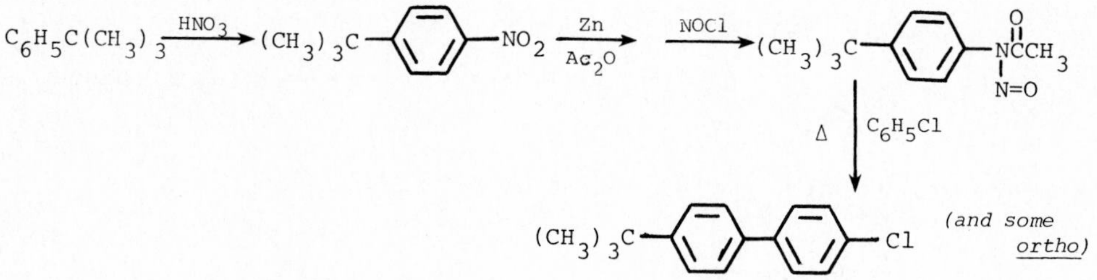

S6.

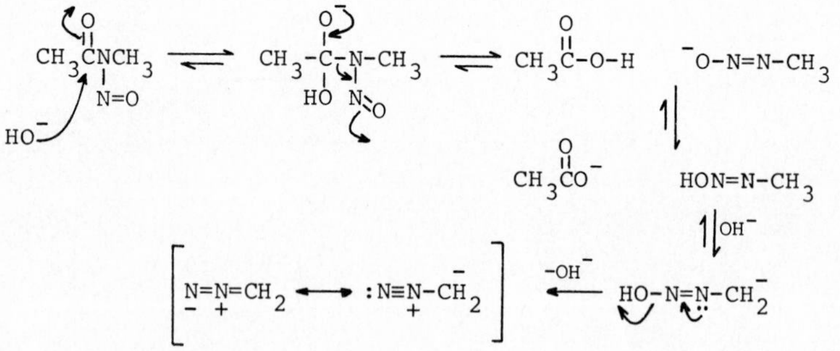

S7.

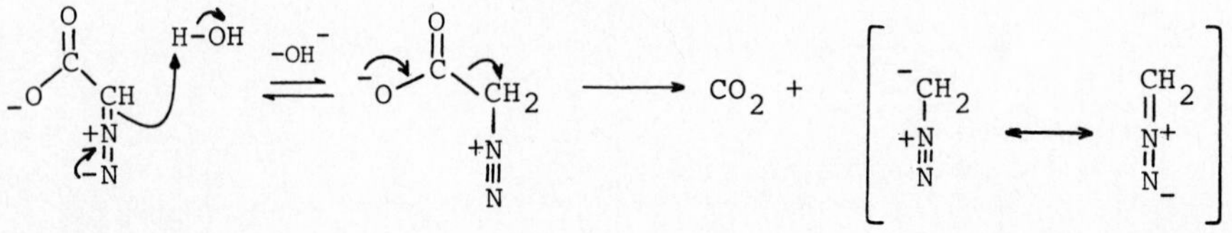

S8. a)

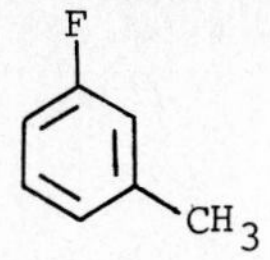

b)

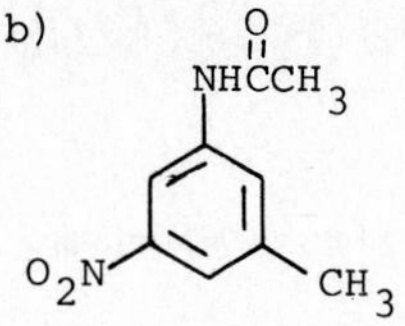

c)

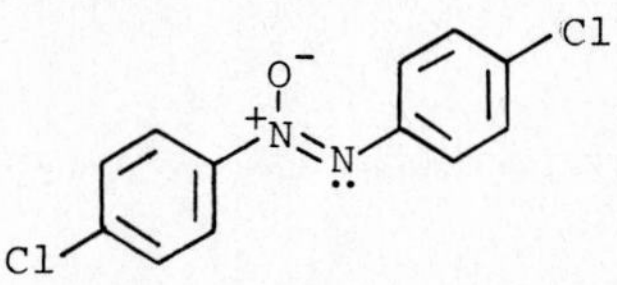

d)

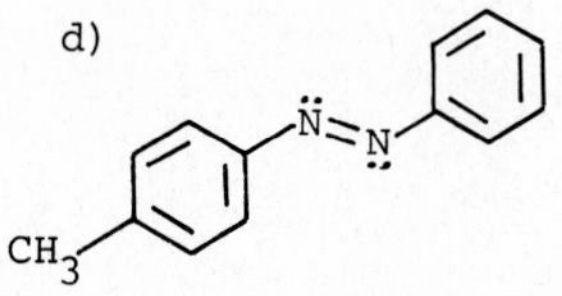

e)

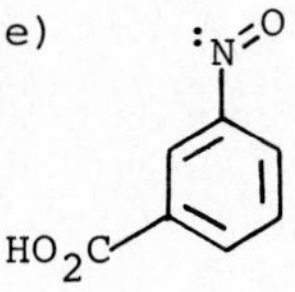

f)

26. SULFUR AND PHOSPHORUS COMPOUNDS

26.A Chapter Outline, Important Terms Introduced, and Reactions Discussed

26.1 <u>Thiols and Sulfides</u> (stink!)

    alkanethiol   dialkyl sulfide   alkylthio-

26.2 <u>Preparation of Thiols and Sulfides</u>

    by displacement:

$$RX + R'S^- \longrightarrow RSR' + X^- \quad \textit{(if R' = H, problem with sulfide formation)}$$

$$RX + NH_2\overset{\overset{\displaystyle S}{\|}}{C}NH_2 \longrightarrow R\text{-}S\text{-}\overset{+}{C}(NH_2)_2 \underset{X^-}{} \xrightarrow{OH^-} RSH + urea$$

    via Grignard reaction:

$$RMgX + S \longrightarrow \xrightarrow{H^+} RSH$$

26.3 <u>Reactions of Thiols and Sulfides</u>

    acidity:

$$RSH + OH^- \rightleftharpoons RS^- + H_2O \quad (pK_a \simeq 10.5)$$

    thiols ⇌ disulfides:

$$2\ RSH \underset{Li/NH_3}{\overset{I_2}{\rightleftarrows}} RSSR$$

    thiolation of enolates with disulfides:

$$\overset{\overset{\displaystyle O^-}{|}}{-C}=C\diagdown \quad RS\text{-}SR \longrightarrow -\overset{\overset{\displaystyle O}{\|}}{C}-\overset{|}{\underset{|}{C}}\text{-}SR + RS^-$$

    oxidation of thiols and disulfides to sulfonic acids:

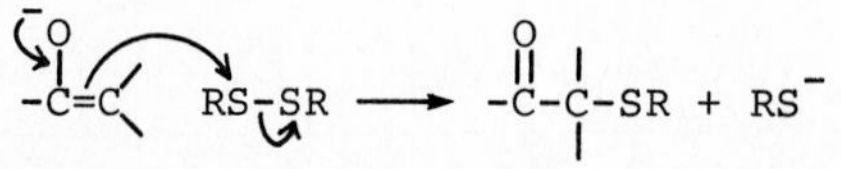

$$RSH \text{ or } RSSR \longrightarrow RSO_3H \quad \textit{(with KMnO}_4\textit{, HNO}_3\textit{, or Cl}_2\textit{/HNO}_3\textit{)}$$

    oxidation of sulfides to sulfoxides and sulfones:

$$RSR \xrightarrow{H_2O_2} R\overset{\overset{\displaystyle O^-}{|}}{\underset{}{S}}{}^+R \xrightarrow[\Delta]{RCO_3H} R\text{-}\overset{\overset{\displaystyle O^-}{|}}{\underset{\underset{\displaystyle O}{|}}{S}}{}^{++}\text{-}R$$

        dimethylsulfoxide as important dipolar aprotic solvent

    pyrolysis of sulfoxides:

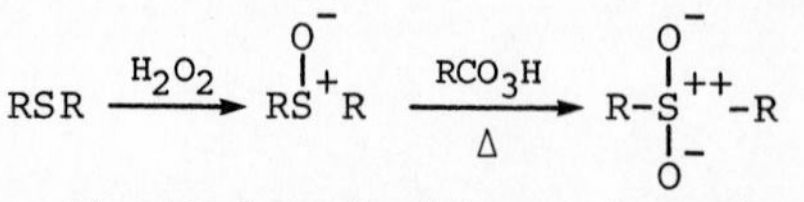

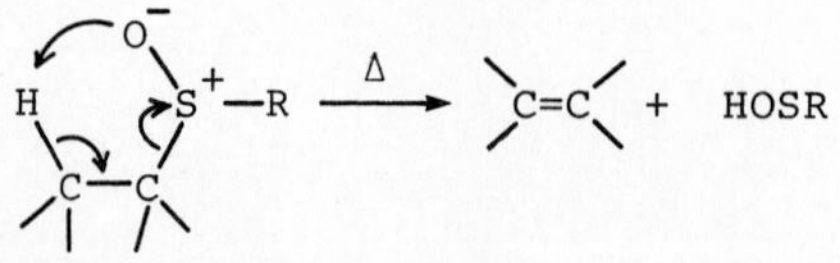

$$H\underset{\underset{\displaystyle \overset{|}{C}-\overset{|}{C}}{}}{\overset{\overset{\displaystyle O^-}{|}}{S^+}}\text{-}R \xrightarrow{\Delta} \diagup C=C\diagdown + HOSR \qquad \textit{(syn elimination stereochemistry, instability of sulfenic acids)}$$

alkylation of sulfides:

$$R_2S + R'X \longrightarrow \underset{X^-}{R_2\overset{+}{S}R'}$$ *(sulfonium salts)*

formation and desulfurization of dithioacetals:

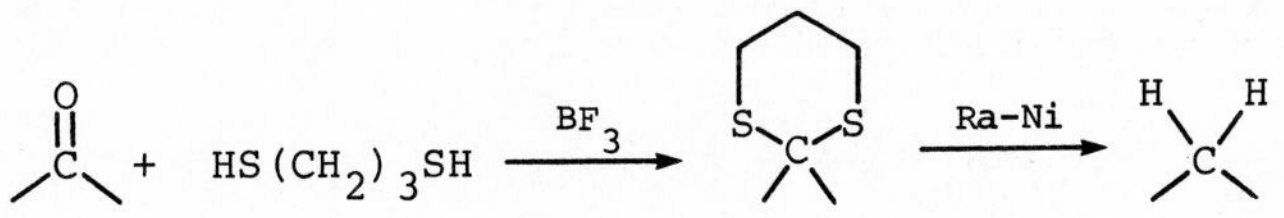

(a mild deoxygenation process)

26.4 Sulfate Esters

alkylsulfuric acid, $ROSO_3H$

(strong acid = good leaving group)

dialkyl sulfate, $RO\overset{\overset{O}{\|}}{\underset{\underset{O}{\|}}{S}}OR$

(good alkylating agent)

26.5 Sulfonic Acids

A. Alkanesulfonic acids

from displacement with sulfite ion:

$$RX + SO_3^{=} \longrightarrow RSO_3^- + X^-$$

formation of bisulfite adducts:

$$R\overset{\overset{O}{\|}}{C}H + HSO_3^- \longrightarrow R\text{-}\overset{\overset{OH}{|}}{C}HSO_3^-$$ *(seldom works for ketones)*

preparation of sulfonate esters via the sulfonyl chlorides:

$$RSO_3^- + PCl_5 \longrightarrow RSO_2Cl \xrightarrow{R'OH} RSO_3R' + HCl$$ *(R = CH_3: "mesylates")*

B. Arenesulfonic acids

1. Electrophilic Aromatic Sulfonation

$$ArH + SO_3 \xrightarrow{H_2SO_4} ArSO_3H$$ *(reversible at higher temperatures; strongly deactivating substituent)*

$$ArH + ClSO_3H \longrightarrow ArSO_2Cl$$

2. Reactions of Arenesulfonic Acids

nucleophilic aromatic substitution:

$$ArSO_3^- + NaOH\text{-}KOH \xrightarrow[300^\circ\ C]{} ArOH$$ *(possible with $\bar{C}N$ too)*

formation of sulfonate esters and sulfonamides via sulfonyl chloride:

$$ArSO_2Cl \xrightarrow{ROH} ArSO_3R + HCl$$ *(excellent alkylating agents)*

$$ArSO_2Cl \xrightarrow{RNH_2} ArSO_2NHR + HCl$$

formation of diaryl sulfones by Friedel-Crafts reaction:

$$ArSO_2Cl + Ar'H \xrightarrow{AlCl_3} ArS(=O)_2Ar' + HAlCl_4$$

reduction of sulfonyl halides to sulfinic acids:

$$ArSO_2Cl \xrightarrow{Zn} ArSO_2H$$

alkylation of sulfinate salts to give alkyl aryl sulfones:

$$ArSO_2^- + RX \longrightarrow ArS(=O)_2R + X^-$$

26.6 Phosphines and Phosphonium Salts

alkyl-, dialkyl-, trialkylphosphines

formation via Grignard reagent:

$$3\ RMgX + PCl_3 \longrightarrow R_3P + 3\ MgXCl$$

alkylation of phosphines:

$$R_3P: + R'X \longrightarrow R_3\overset{+}{P}R'\ X^-$$ *(R = alkyl or aryl; phosphonium salts)*

phosphines as organometallic ligands

26.7 Phosphate and Phosphonate Esters

phosphoric acids and esters pK_a's of H_3PO_4: 2.15, 7.2, and 12.4

formation of esters from acid chlorides:

$$3\ ROH + POCl_3 \longrightarrow (RO)_3P{=}O + 3\ HCl$$ *(monoester dichlorides, diester monochlorides)*

hydrolysis via C-O bond cleavage in acid

phosphorous acid and esters

formation of esters from acid chlorides:

$$3\ ROH + PCl_3 \xrightarrow{pyridine} (RO)_3P: + 3\ pyridine{\cdot}HCl$$ *(trialkyl phosphite)*

alkylphosphonic acids and esters

Arbuzov-Michaelis Reaction:

$$(RO)_3P: + R'X \xrightarrow[\Delta]{} (RO)_3\overset{+}{P}R'\ X^- \longrightarrow RO\text{-}P(=O)(OR)\text{-}R' + RX$$

hydrolysis of phosphonate esters:

$$R'P(=O)(OR)_2 \xrightarrow{OH^-} R'P(=O)(O^-)\text{-}OR \xrightarrow[\Delta]{H_3O^+} RPO_3H_2$$

26.8   Sulfur- and Phosphorus-Stabilized Carbanions

phosphorus ylids, the Wittig reagent:

$$(C_6H_5)_3\overset{+}{P}CH_2R \xrightarrow{\textit{n}-BuLi} (C_6H_5)_3\overset{+}{P}\overset{-}{C}HR \xrightarrow{\underset{C}{\overset{O}{\parallel}}} \underset{/}{\overset{\backslash}{C}}=CHR + (C_6H_5)_3P=O$$

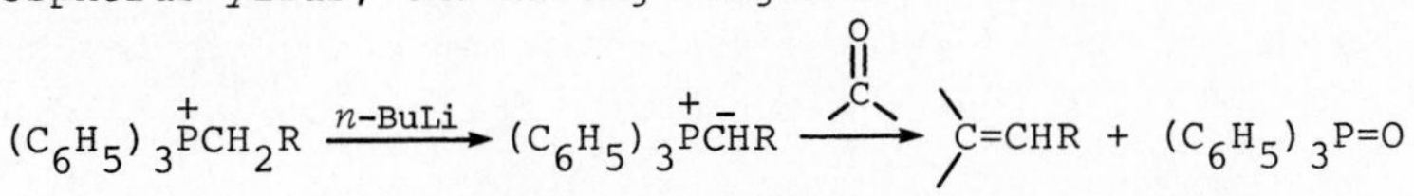

(see Section 13.7.G.)

phosphonate carbanions, Horner-Emmons reaction:

$$(EtO)_2\overset{O}{\overset{\parallel}{P}}CH_2CO_2Et \rightleftharpoons (EtO)_2\overset{O}{\overset{\parallel}{P}}\overset{-}{C}HCO_2Et \xrightarrow{\underset{C}{\overset{O}{\parallel}}} \underset{/}{\overset{\backslash}{C}}=CHCO_2Et + (EtO)_2PO_2^-$$

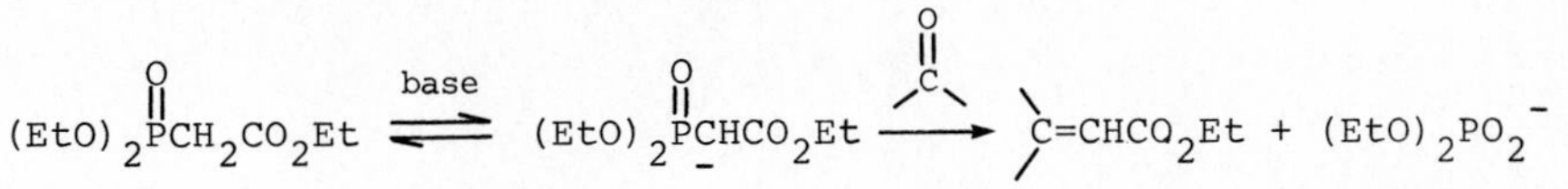

sulfur-stabilized carbanions:

$$C_6H_5SCH_3 \xrightarrow{\textit{n}-BuLi} C_6H_5SCH_2Li$$

reactions with electrophiles

$(RX, \overset{O}{\overset{\parallel}{C}})$

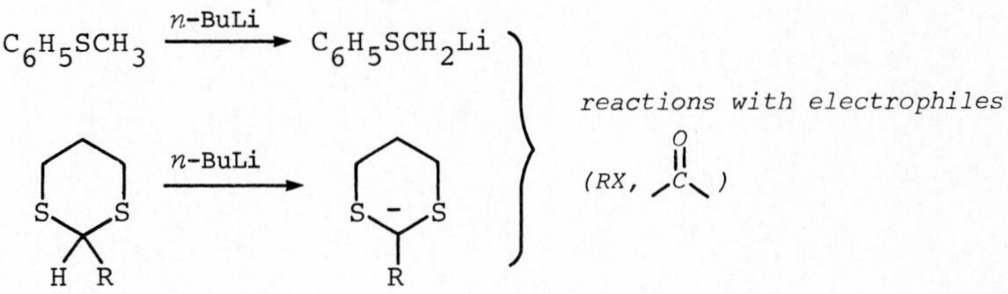

sulfonium ylids, epoxide formation:

$$(CH_3)_3S^+ \xrightarrow{\textit{n}-BuLi} (CH_3)_2\overset{+}{S}\overset{-}{C}H_2 \xrightarrow{\underset{C}{\overset{O}{\parallel}}} \overset{CH_2\!-\!O}{\underset{C}{\diagdown}} + (CH_3)_2S$$

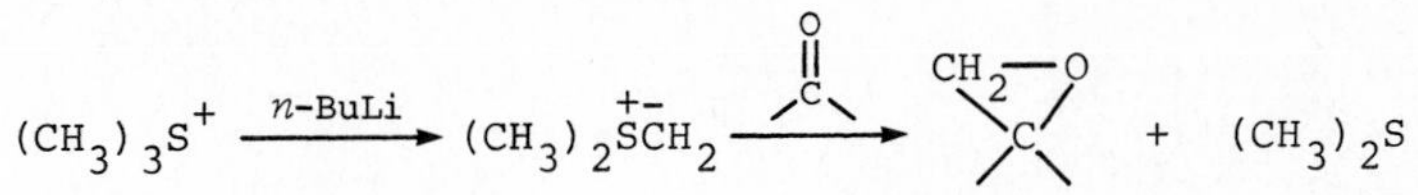

anions of sulfones (pK$_a$ = 31) and sulfoxides (pK$_a$ = 35):

$$CH_3\overset{O}{\underset{(O)}{\overset{\parallel}{\underset{\parallel}{S}}}}CH_3 \xrightarrow{NaH} CH_3\overset{O}{\underset{(O)}{\overset{\parallel}{\underset{\parallel}{S}}}}\overset{-}{C}H_2 \ Na^+ \xrightarrow{RCO_2CH_3} RC\overset{O}{\overset{\parallel}{}}CH_2\overset{O}{\underset{(O)}{\overset{\parallel}{\underset{\parallel}{S}}}}CH_3$$

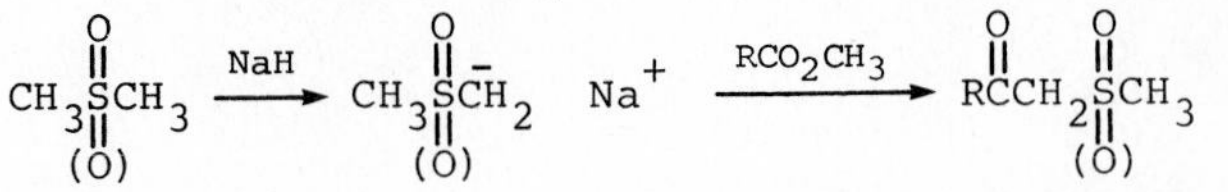

26.B Important Concepts and Hints

The chemistry of sulfur and phosphorus compounds is becoming
increasingly important because of the invention of many functional group
interconversions and carbon-carbon bond-forming reactions which involve
sulfur- and phosphorus-containing intermediates. In many respects, analogies can be drawn between thiols/sulfides/disulfides and alcohols/ethers/
peroxides. However, there are no oxygen counterparts to the more highly
oxidized sulfur compounds: sulfoxides, sulfones, and sulfinic and sulfonic
acids. In a similar manner, the organic chemistry of phosphorus ranges
from the reduced derivatives (phosphines) to the higher oxidation states
(phosphonic and phosphoric acids).

The major utility of sulfur- and phosphorus-containing compounds as
carbon-carbon bond-forming reagents lies in the diverse ways that these
functional groups can stabilize adjacent carbanions. With strong base
(n-butyllithium or LDA) a proton can be removed from carbons adjacent to

sulfide, sulfoxide, sulfone, sulfonium ion, phosphonium ion, and phosphonate groups (among others!). These carbanionic reagents react with
alkyl halides or carbonyl derivatives to form new carbon-carbon bonds.
Particularly important examples are the Wittig reaction, and the use of

dithioacetal anions as synthetic equivalents to acyl anions ("R$\overset{\overset{\displaystyle O}{\|}}{C}$-").

More "synthetic methods" which use sulfur and phosphorus functional
groups are being invented all the time--perhaps you can think of some new
ones yourself!

26.C Answers to Exercises

(26.1) (a) $CH_3CH_2SCH(CH_3)_2$ (b) $CH_3CH_2CH_2CH_2SH$ (c) $CH_3(CH_2)_4\overset{\overset{\displaystyle SCH_3}{|}}{CH}CH_2CH_3$

(d) $C_6H_5SC_6H_5$ (e) $CH_3CH_2\overset{\overset{\displaystyle SH}{|}}{CH}CH_2CH_3$ (f) $CH_3\overset{\overset{\displaystyle S}{\|}}{CH}$ (g) $CH_3\overset{\overset{\displaystyle O}{\|}}{C}SC_2H_5$

(26.2) (a) $(CH_3)_3CCH_2Br$ $\xrightarrow{Mg}$ $\xrightarrow{S_8}$ $\xrightarrow{H^+}$ $(CH_3)_3CCH_2SH$ *(neopentyl bromide is too hindered to undergo displacement reaction)*

(b) 2 $CH_3CH_2CH_2OH$ $\xrightarrow{PBr_3}$ 2 $CH_3CH_2CH_2Br$ $\xrightarrow{Na_2S}$ $(CH_3CH_2CH_2)_2S$

(26.3) (a)

(i) $CH_3CH_2CH=CH_2$ + (CH_3SOH)

$\Big\uparrow \Delta$

$CH_3CH_2CH_2CH_2\overset{\overset{\displaystyle O}{\|}}{S}CH_3$

$\Big\uparrow H_2O_2$

$CH_3CH_2CH_2CH_2SCH_3$ $CH_3CH_2CH_2CH_2SSCH_2CH_2CH_2CH_3$ (ii)

$\overset{NaOH}{\underset{CH_3Br}{\nwarrow}}$ $\overset{I_2, KI}{\nearrow}$ $\underset{Li/NH_3}{\searrow}$

$CH_3CH_2CH_2CH_2SH$

$\Big\downarrow \overset{NaOH}{CH_3CH_2I}$

$CH_3CH_2CH_2CH_2SCH_2CH_3$

$\Big\downarrow CH_3I$

$\downarrow I^-$

$CH_3CH_2CH_2CH_2\underset{\underset{\displaystyle CH_3}{|}}{\overset{+}{S}}CH_2CH_3$

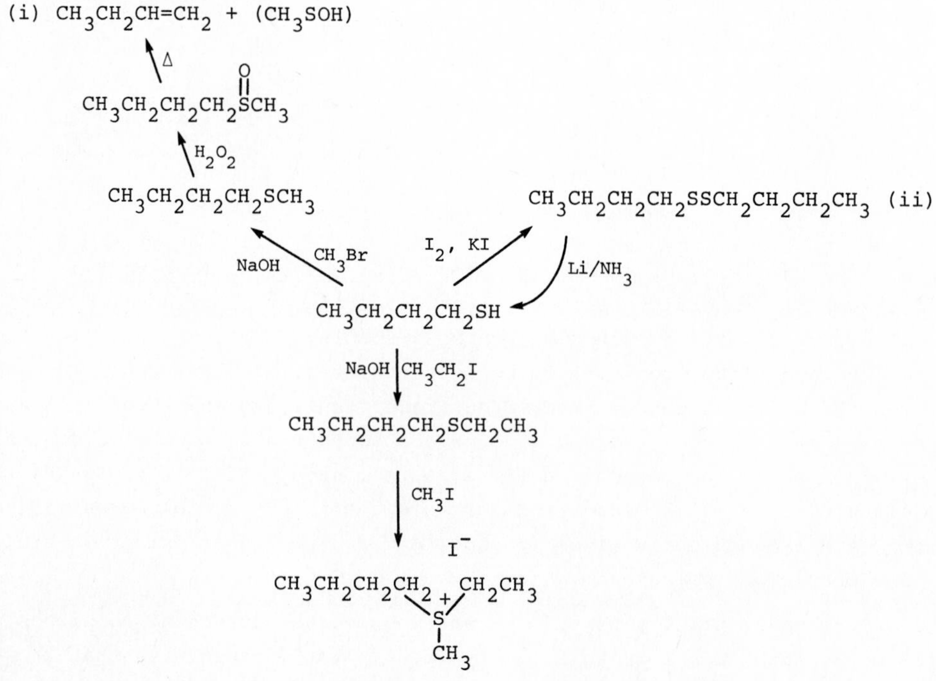

(iii)

(b)

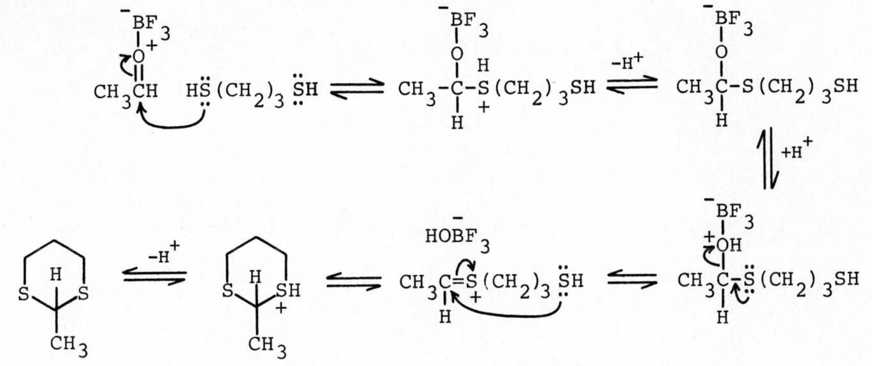

(26.5.A.)

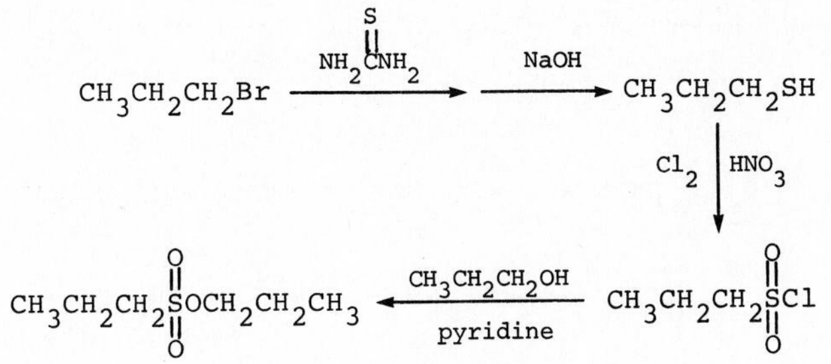

(26.5.B.1.)

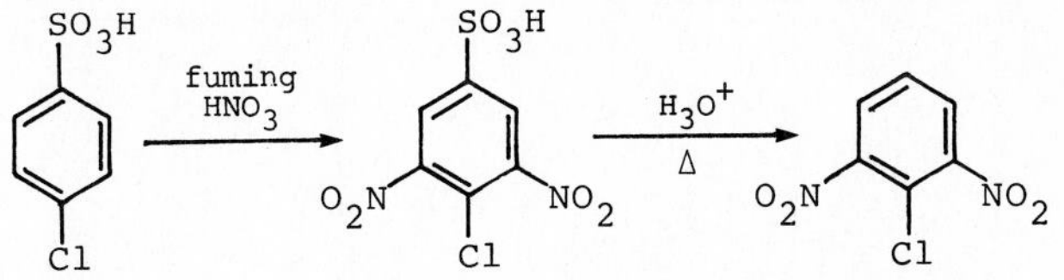

(26.5.B.2.)

(a)

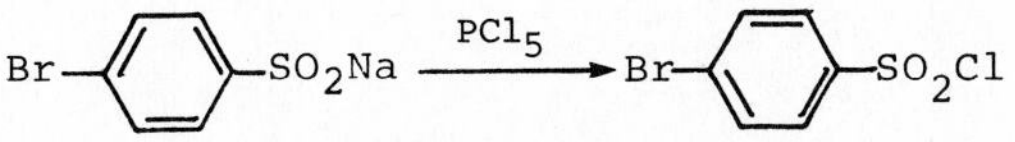

(i)

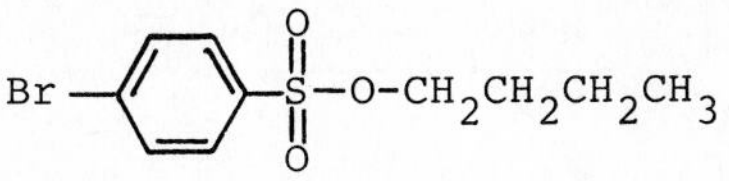

(ii)

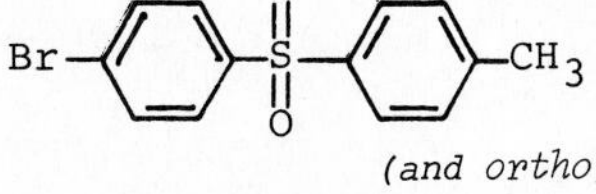

(and ortho)

(iii)

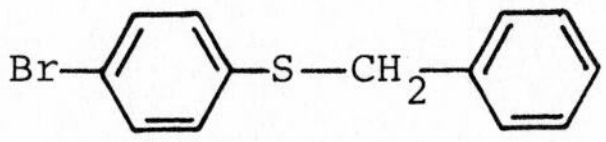

(iv)

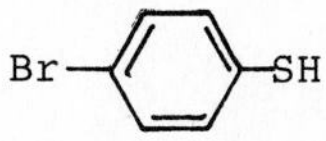

(b)

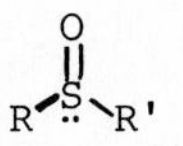

(26.6)

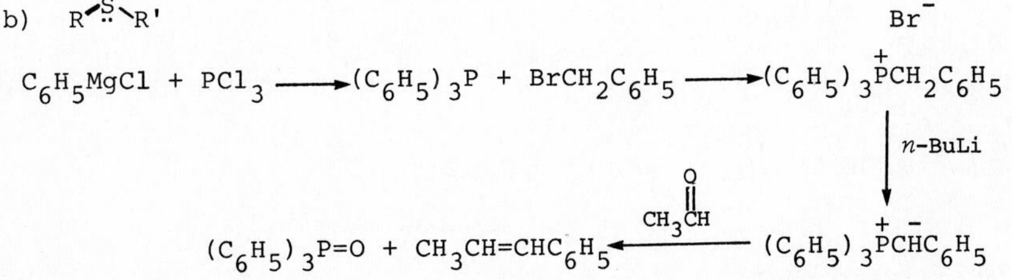

(26.7) (a) $2\ C_2H_5OH + POCl_3 \xrightarrow{\text{pyridine}} (C_2H_5O)_2\overset{O}{\overset{\|}{P}}Cl \xrightarrow{CH_3ONa} (C_2H_5O)_2\overset{O}{\overset{\|}{P}}OCH_3$

(b) $3\ i\text{-}C_4H_9OH + POCl_3 \xrightarrow{\text{pyridine}} (i\text{-}C_4H_9O)_3P{=}O$

(c) $3\ C_2H_5OH + PCl_3 \xrightarrow{\text{pyridine}} (C_2H_5O)_3P{:} \xrightarrow[\Delta]{CH_3CH_2CH_2I} CH_3CH_2CH_2\overset{O}{\overset{\|}{P}}(OC_2H_5)_2 + C_2H_5I$

(d) $(C_2H_5O)_3P \xrightarrow[\Delta]{C_2H_5I\ \text{(catalytic)}} C_2H_5\overset{O}{\overset{\|}{P}}(OC_2H_5)_2 \xrightarrow[2)\ H^+]{1)\ NaOH} C_2H_5\overset{O}{\overset{\|}{P}}(OH)OC_2H_5$

[from (c)]

(26.8) (a) $CH_3(CH_2)_6\overset{O}{\overset{\|}{C}}H + (C_6H_5)_3\overset{+}{P}\overset{-}{C}H_2 \longrightarrow CH_3(CH_2)_6CH{=}CH_2 + (C_6H_5)_3P{=}O$

(b) $CH_3(CH_2)_3\overset{O}{\overset{\|}{C}}H + (CH_3)_2\overset{+}{S}\overset{-}{C}H_2 \longrightarrow CH_3(CH_2)_3\overset{O}{\widehat{CH-CH_2}} + CH_3SCH_3$ 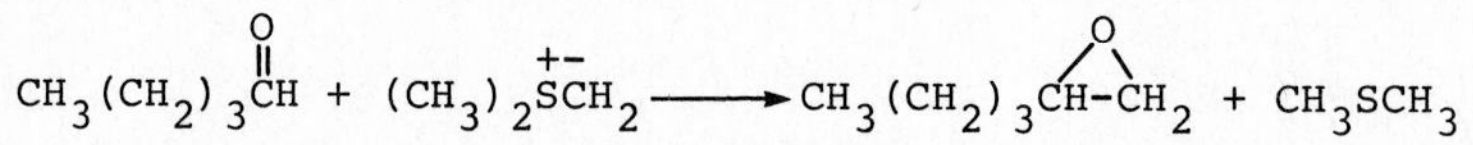

(c) 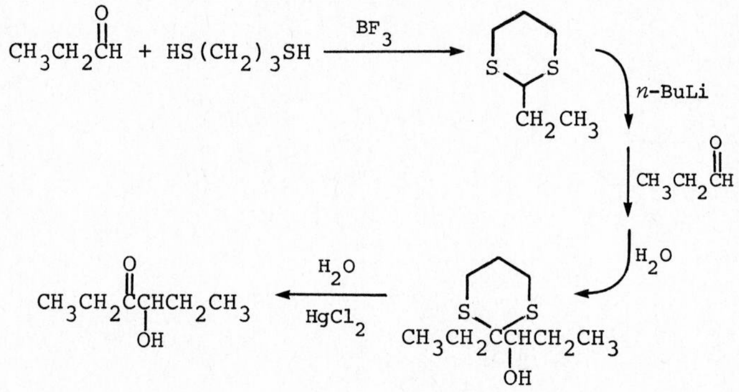

(d) 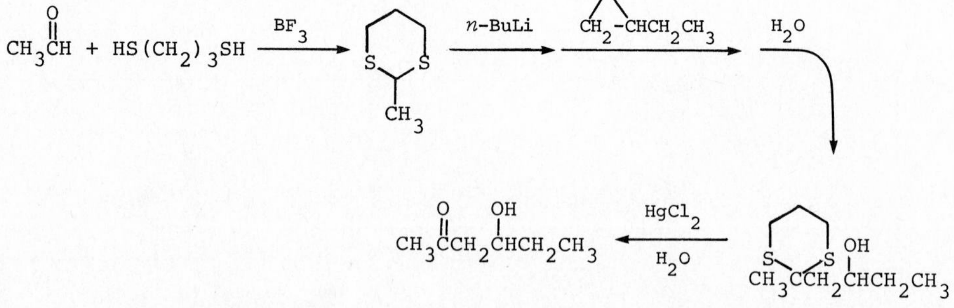

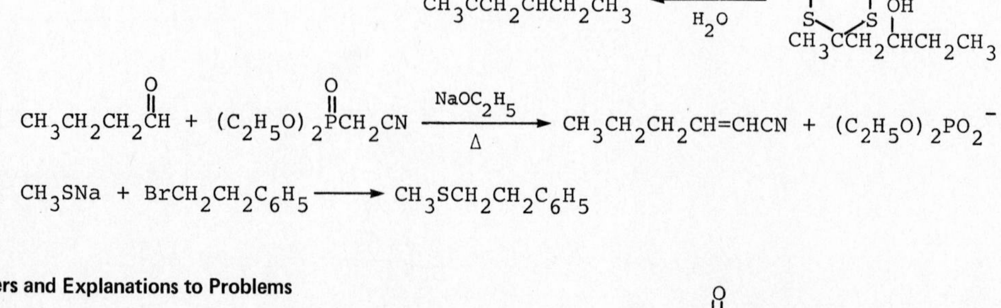

(e) $CH_3CH_2CH_2\overset{O}{\overset{\|}{C}}H + (C_2H_5O)_2\overset{O}{\overset{\|}{P}}CH_2CN \xrightarrow[\Delta]{NaOC_2H_5} CH_3CH_2CH_2CH{=}CHCN + (C_2H_5O)_2PO_2^-$

(f) $CH_3SNa + BrCH_2CH_2C_6H_5 \longrightarrow CH_3SCH_2CH_2C_6H_5$

26.D Answers and Explanations to Problems

1. (a) $C_2H_5SCH_2C(CH_3)_3$ (b) $(CH_3)_2CHCH_2SH$ (c) 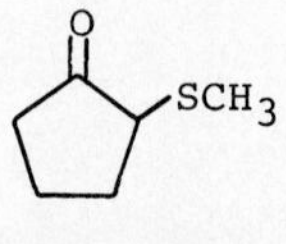

(d) $(CH_3CH_2CH_2CH_2)_2S$ (e) $\overset{\displaystyle SH}{\bigcirc}$ (f) $(CH_3)_2CHCH_2OSO_3H$

(g) $C_2H_5O\overset{O}{\underset{O}{\overset{\|}{\underset{\|}{S}}}}OC_2H_5$ (h) $O_2N-\!\!\bigcirc\!\!-SO_3CH_3$ (i) $(CH_3CH_2CH_2CH_2O)_3P=O$

(j) $C_2H_5\overset{O}{\overset{\|}{P}}(OC_2H_5)_2$ (k) $\bigcirc\!-O\overset{O}{\underset{O}{\overset{\|}{\underset{\|}{S}}}}CH_3$ (l) $(C_6H_5O)_3P:$

(m) $CH_2=CH-CH_2O-\overset{O}{\underset{\underset{O^-}{\|}}{\overset{\|}{P}}}-O-\overset{O}{\underset{\underset{O^-}{}}{\overset{\|}{P}}}-O^-$

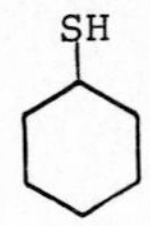

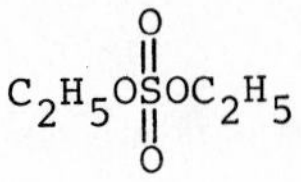

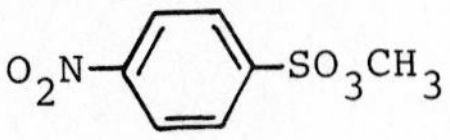

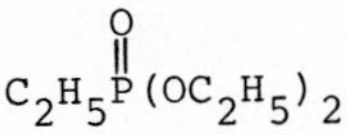

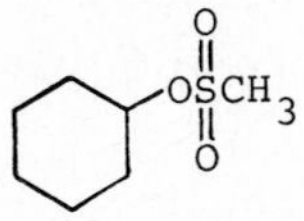

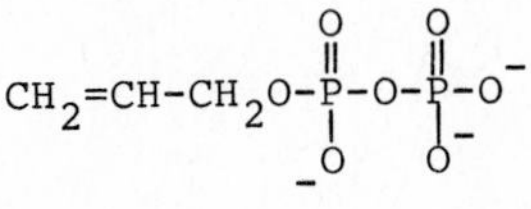

2. (a) 4-methyl-2-pentanethiol   (b) 2-methyl-3-methylthiopentane

(c) tetracyclopentylphosphonium   (d) N-methyl-4-isopropylbenzene-
     bromide   sulfonamide

(e) 3-(2,2-dimethylpropyl)oxycarbonyl-4-bromobenzenesulfonic acid

   *In a complex case such as this the sulfonic acid group can be expressed
   as a prefix, sulfo; hence: neopentyl 2-bromo-5-sulfobenzoate.*

(f) *p*-nitrobenzenesulfonyl chloride   (g) methyl *p*-bromophenyl sulfone

(h) cyclopropyldisulfide   (i) diethyl ethylphosphonate

3. (a) $CH_3CH_2CH_2CH_2\overset{O}{\overset{\|}{P}}(OCH_2CH_2CH_2CH_3)_2$ (b) $t\text{-}BuOCH_3 + CH_3OSO_3^- K^+$

(c) $CH_3CH_2CH_2\overset{O}{\underset{\underset{OH}{|}}{\overset{\|}{P}}}(OCH_3)$ (d) $(C_2H_5O)_2\overset{O}{\overset{\|}{P}}OCH_2CH_2CH(CH_3)_2$ (e) $(CH_3O)_3P=O$

(f) $CH_3CH_2SCH_3 + CH_3I$ (g) $(CH_3)_3CCH_2SCH_2CH_3 + CH_3CH_2Br$

(h) $\overset{CH_3}{\underset{SCH_2CH_3}{\bigcirc}} + CH_3CH_2Cl$ (i) $\bigcirc\!\!\bigcirc\!-OH$ (j) $C_2H_5\overset{O}{\underset{O}{\overset{\|}{\underset{\|}{S}}}}CH_2C_6H_5$

(k) $\overset{S}{\underset{S}{\bigcirc}}$ (l) $C_6H_5CH\overset{S}{\underset{S}{\diagup\!\diagdown}}$ (m) $\overset{S}{\underset{S}{\bigcirc}}CHC(CH_3)_2\!\!-\!\!OH$

(n) $\overset{OLi}{\underset{CH_2\overset{O}{\overset{\|}{P}}(OCH_3)_2}{\bigcirc}}$ (o) $C_6H_5\overset{O}{\overset{\diagup\!\diagdown}{CH-CH_2}} + CH_3SCH_3$

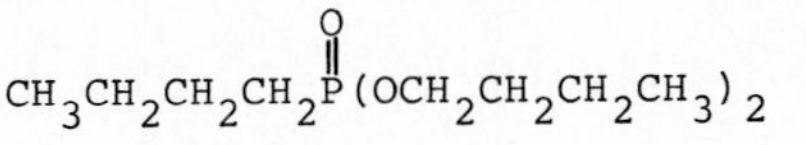

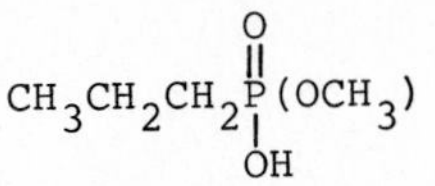

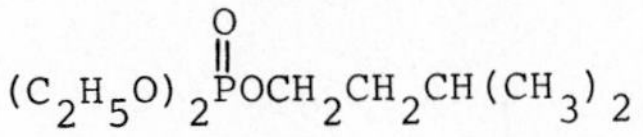

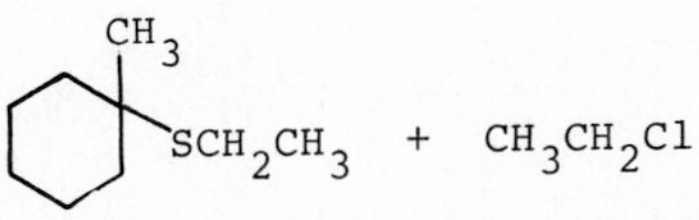

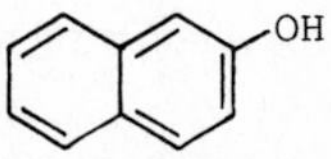

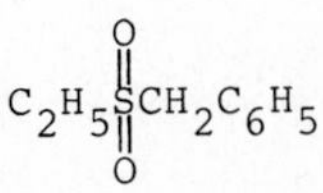

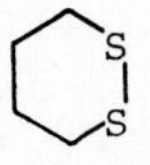

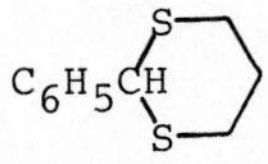

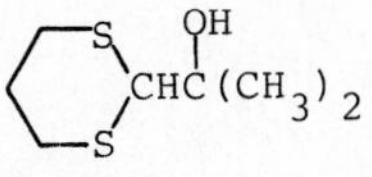

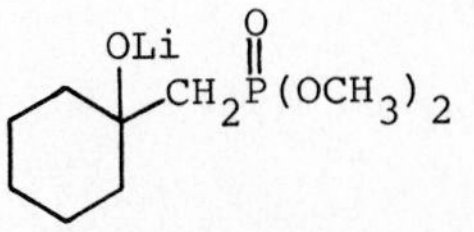

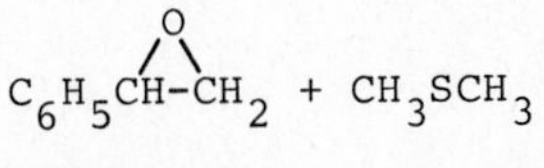

4. (a) $CH_3SH + 2\ KMnO_4 = CH_3SO_3^- + OH^- + 2\ K^+ + 2\ MnO_2$

(b) $3\ (CH_3)_2S + 4\ KMnO_4 + 2\ H_2O = 3\ (CH_3)_2SO_2 + 4\ MnO_2 + 4\ KOH$

(c) $(CH_3)_2S + 4\ HNO_3 \longrightarrow (CH_3)_2SO_2 + 4\ NO_2 + 2\ H_2O$

5. (a) E2 elimination would occur in either step to give isobutylene rather than the desired phosphite triester or alkylphosphonate.

(b) Friedel-Crafts reactions do not apply with *meta*-directing groups such as $-SO_3H$.

(c) The ketone group is sensitive to strong base. Hydroxide promotes aldol condensation reactions; hence the final step with fused KOH will give a mess.

6. (a) $(CH_3)_3CBr \xrightarrow{Mg} \xrightarrow{S_8} \xrightarrow{CH_3I} (CH_3)_3CSCH_3$

(*NOTE:* $(CH_3)_3CBr + NaSCH_3 \longrightarrow (CH_3)_2C{=}CH_2 + HSCH_3 + NaBr$)

(b)

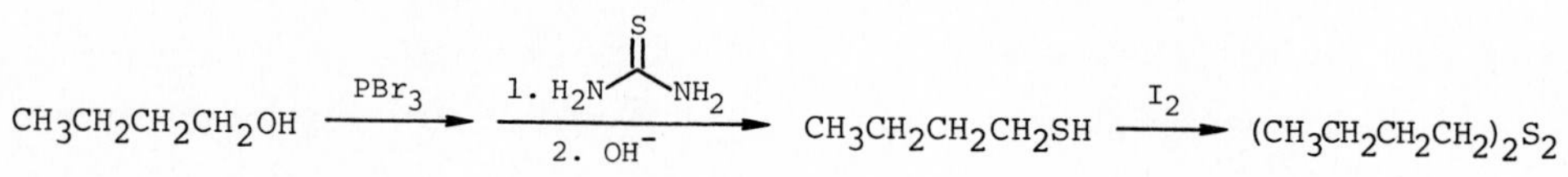

(c)

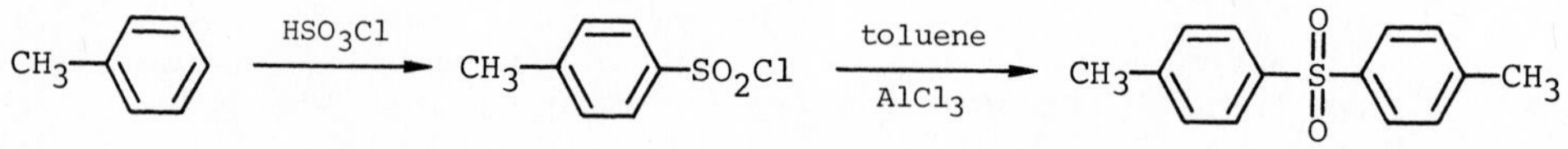

(d)

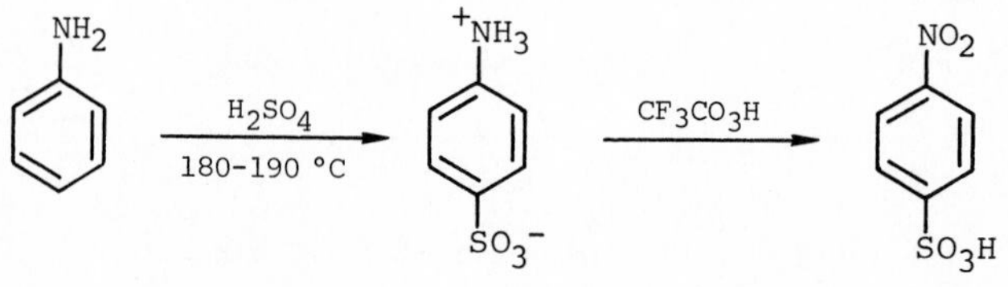

(e)

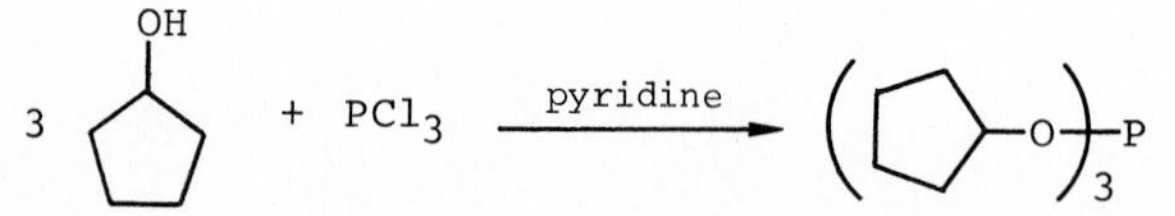

(f) $C_6H_5CH_3 \xrightarrow[h\nu]{Br_2} C_6H_5CH_2Br \xrightarrow{Na_2SO_3} C_6H_5CH_2SO_3^-\ Na^+ \xrightarrow{H^+} C_6H_5CH_2SO_3H$

(g)

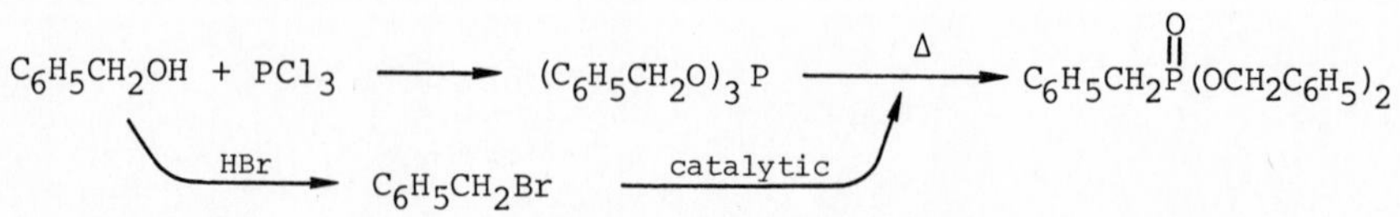

(h)

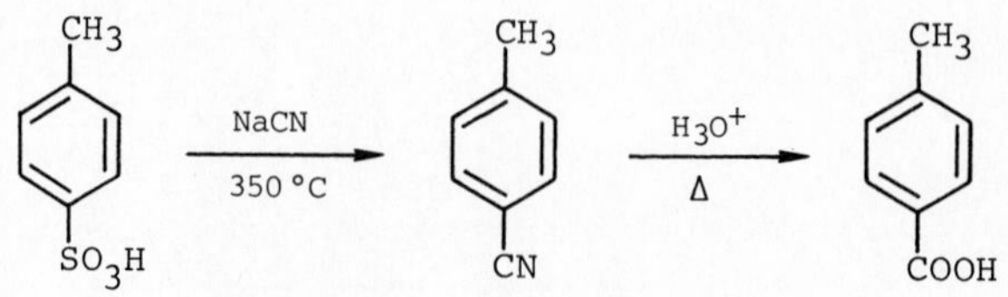

(i) 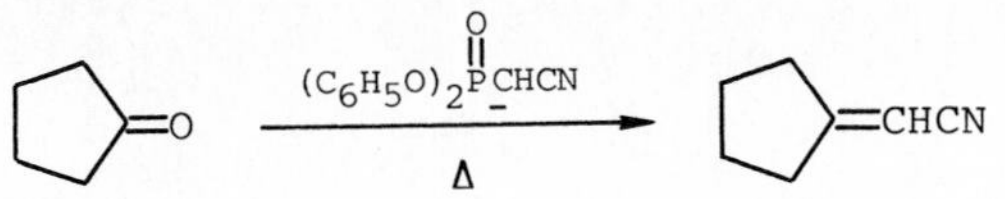

(j) 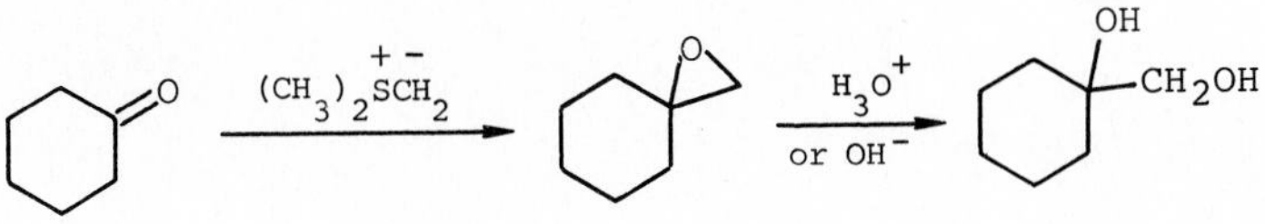

(k) 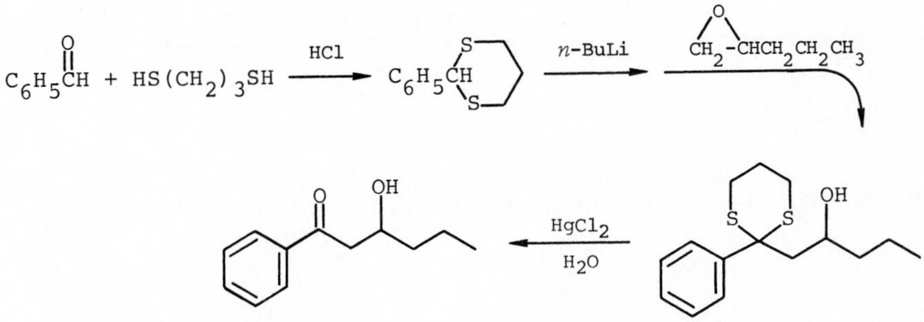

(ℓ) 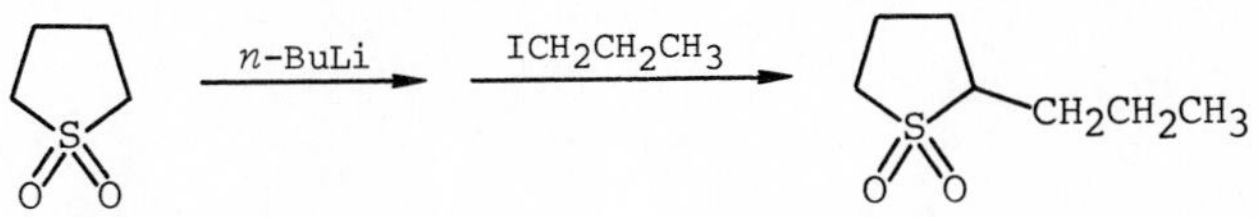

7. 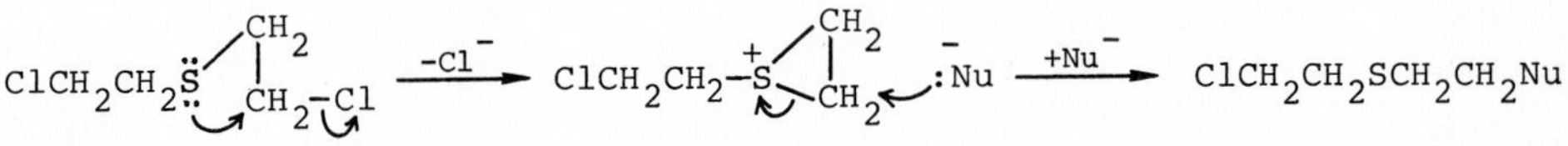

8.

$$\begin{array}{ll} CH_3S\cdot + H\cdot = CH_3SH & \Delta H = -75 \\ CH_4 = CH_3\cdot + H\cdot & \Delta H^\circ = DH^\circ = 104 \\ \hline CH_4 + CH_3S\cdot = CH_3SH + CH_3\cdot & \Delta H^\circ = +29 \text{ kcal mole}^{-1} \end{array}$$

$\Delta S°$ is probably about zero, therefore $\Delta G° \simeq \Delta H°$, and the equilibrium lies far to the left. Hence, CH_3SH works as an inhibitor by reacting with alkyl radicals to stop propagation of the radical chains. The $CH_3S\cdot$ formed cannot abstract hydrogen atoms from carbon; hence nothing happens until two $CH_3S\cdot$ radicals come together to form the disulfide.

9. RSO_2OH are much more acidic than RCOOH; the ΔpK_a is about 10. RSO_2NH_2 are more acidic than $RCONH_2$; the ΔpK_a is about 5. With these analogies, we would expect RSO_2CH_3 to be more acidic than $RCOCH_3$; actually, RSO_2CH_3 are less acidic than $RCOCH_3$ in aqueous solution, but the difference is only a few pK_a units.

10. The shorter bond results from the increased coulombic attraction of the dipolar dative bond: $\geq P \rightarrow O = \geq \overset{+}{P}-\overset{-}{O}$. This attraction also results in a stronger bond (*bond strengths:* dative P→O bond ≈ 130 kcal $mole^{-1}$, single P-O bond ≈ 90 kcal $mole^{-1}$). The conversion of a P-O single bond into a P→O dative bond provides most of the 47 kcal $mole^{-1}$ which is the driving force of the rearrangement.

11.

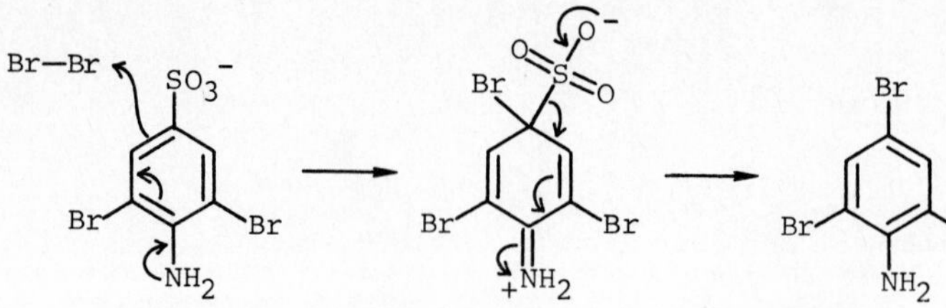

Ipso is the Latin
word for "the same"

12.

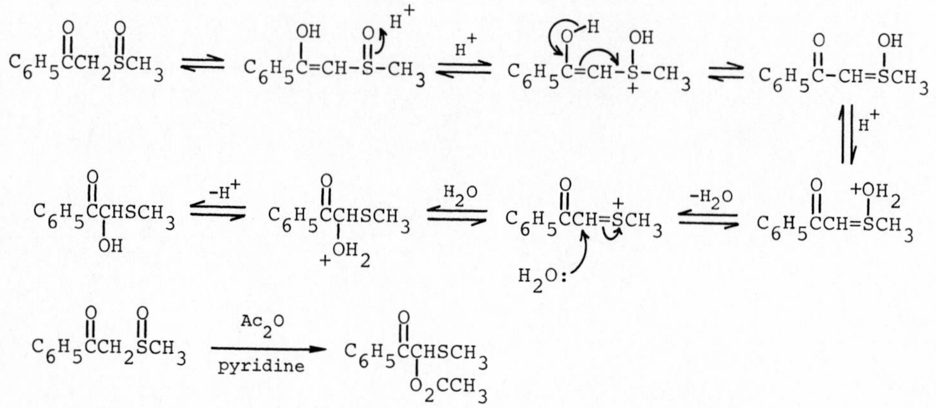

$$C_6H_5\overset{O}{\underset{}{C}}CH_2\overset{O}{\underset{}{S}}CH_3 \xrightarrow[\text{pyridine}]{Ac_2O} C_6H_5\overset{O}{\underset{O_2CCH_3}{C}}CHSCH_3$$

13.

$$+ (C_6H_5SOH)$$

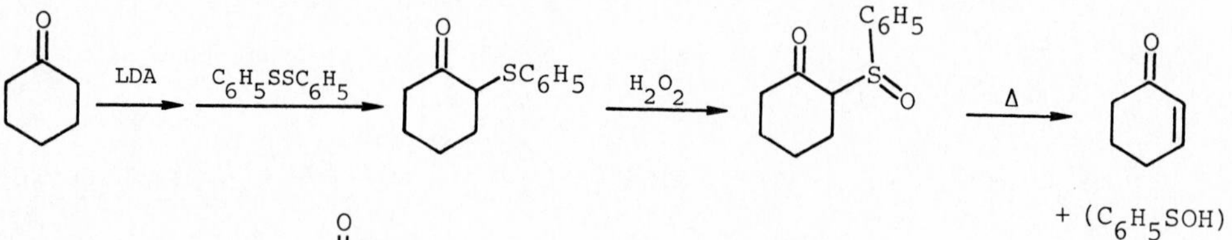

14. $CH_3SCH_3 + \frac{1}{2} O_2 \longrightarrow CH_3\overset{O}{\underset{}{S}}CH_3 \qquad \Delta H° = -27.2 \text{ kcal mole}^{-1}$

   $CH_3\overset{O}{\underset{}{S}}CH_3 + \frac{1}{2} O_2 \longrightarrow CH_3\overset{O}{\underset{O}{S}}CH_3 \qquad \Delta H° = -53.0 \text{ kcal mole}^{-1}$

The second oxidation step is much more exothermic than the first. Assuming that the analogy
holds true for the disulfide systems, one would expect the thiosulfonic ester to be thermodynamically favored over the disulfoxide.

15.

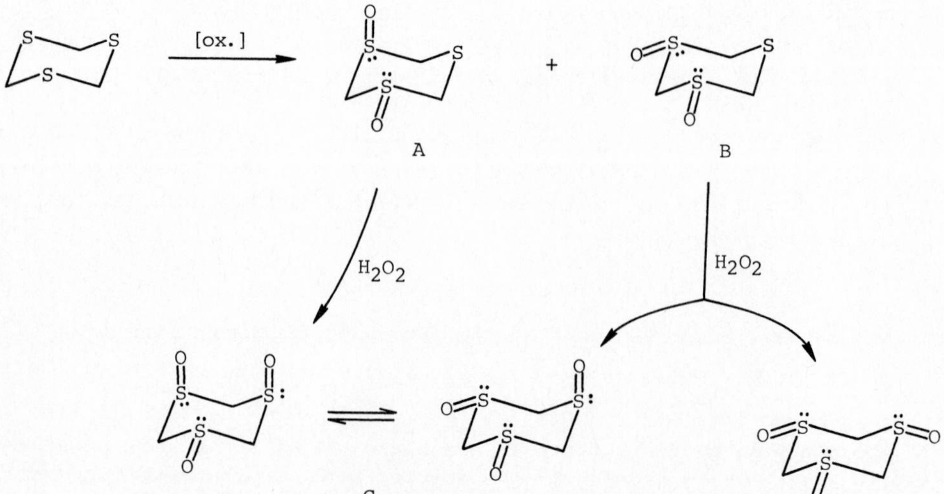

equivalent by chair ⇌ chair interconversion

26.E Supplementary Problems

S1. Name the following compounds:

(a) $(CH_3O)_3P$

(b) $Br-\langle\bigcirc\rangle-\overset{O}{\underset{..}{\overset{||}{S}}}-OCH_3$

(c) $(CH_3CH_2)_2\overset{+}{S}-\langle\bigcirc\rangle$
  I^-

(d) $CH_3(CH_2)_{10}CH_2O\overset{O}{\underset{O}{\overset{||}{S}}}O^-$ Na^+

(e) $(CH_3CH_2O)_2\overset{O}{\overset{||}{P}}-O-\overset{O}{\overset{||}{P}}(OCH_2CH_3)_2$

(f) $(CH_3CH_2CH_2CH_2)_3\overset{+}{P}CH_3$ Cl^-

(g) $\langle\bigcirc\rangle-\overset{O}{\underset{OCH_3}{\overset{||}{P}}}-OCH_2CH_3$

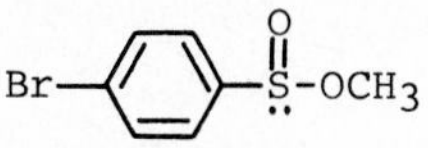

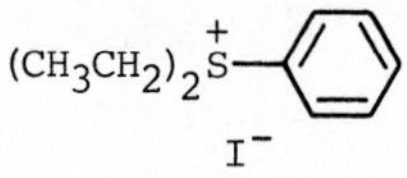

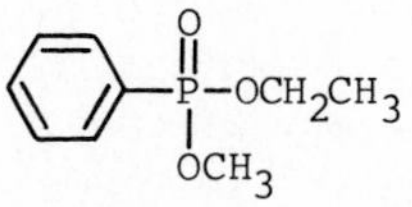

(h) $H_2NCH_2PO_3H_2$

(i) $CCl_3-\overset{O}{\underset{O}{\overset{||}{S}}}-C_6H_5$

(j) $(CH_3CH_2CH_2)_2S=O$

S2. What is the major product to result from each of the following reaction sequences?

(a) $(C_6H_5)_3P$ $\xrightarrow{CH_3CH_2I}$ $\xrightarrow{C_6H_5Li}$ $\xrightarrow{\text{(cyclohexanone)}}$

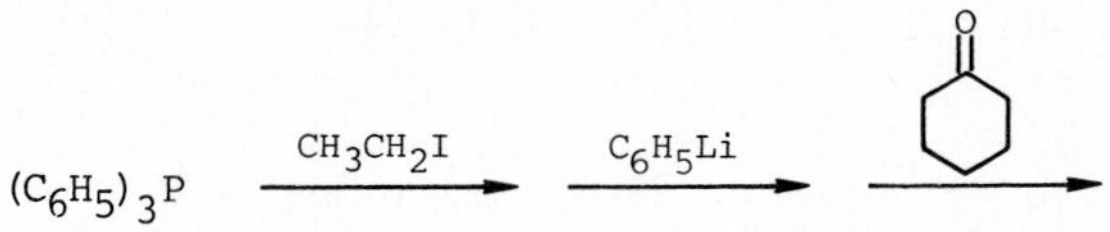

(b) $(CH_3)_2CHCH=O$ $\xrightarrow[BF_3]{HS(CH_2)_3SH}$ $\xrightarrow{n-BuLi}$ $\xrightarrow{CH_3CH_2O-\overset{O}{\underset{O}{\overset{||}{S}}}-C_6H_5}$ $\xrightarrow[HgCl_2]{H_2O}$

(c) $(CH_3O)_3P:$ $\xrightarrow[\Delta]{BrCH_2COOCH_3}$ $\xrightarrow{NaH}$ $\xrightarrow[\Delta]{\text{(cyclohexanone)}}$

(d) $\langle\bigcirc\rangle-CH_3$ $\xrightarrow{ClSO_3H}$ $\xrightarrow{Zn}$ $\xrightarrow{NaOH}$ $\xrightarrow{BrCH_2CH=CH_2}$

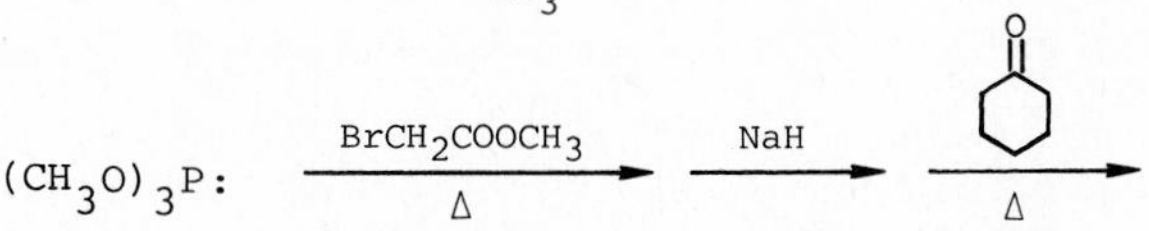

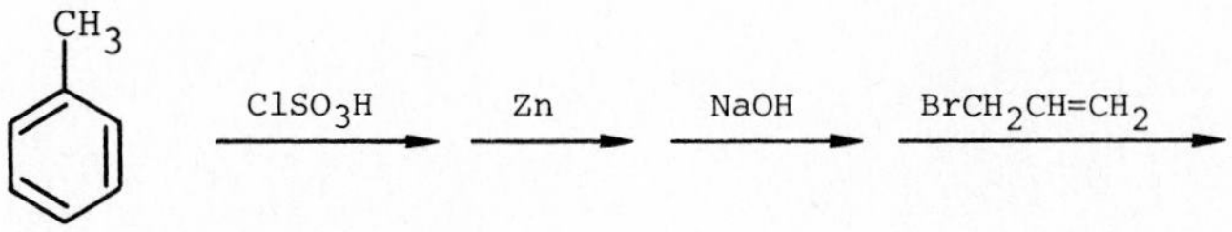

(e) $BrCH_2CH_2CH_2CH_2Br$ $\xrightarrow[C_2H_5OH]{NaSH}$

(f) $BrCH_2CH_2CH_2CH_2Br$ $\xrightarrow{2\ (NH_2)_2C=S}$ $\xrightarrow{NaOH}$

(g) $(CH_3O)_2\overset{O}{\overset{||}{P}}-O-\overset{O}{\overset{||}{P}}(OCH_3)_2$ $\xrightarrow[CH_3CH_2OH]{CH_3CH_2ONa}$

(h) $C_6H_5SCH_3$ $\xrightarrow{n-BuLi}$ $\xrightarrow[2.\ 2\ CH_3CO_3H]{1.\ (CH_3)_2C=O}$

S3. Show how to accomplish the following transformations in a practical manner:

(a) $CH_3CH_2CH_2\overset{O}{\overset{||}{C}}CH_2CH_3$ $\longrightarrow$ $CH_3CH_2CH_2-\overset{O-CH_2}{\underset{}{\overset{\diagup\diagdown}{C}}}-CH_2CH_3$

(b) $CH_3CH_2CH_2\overset{O}{\overset{||}{C}}H$ $\longrightarrow$ $CH_3CH_2CH_2\overset{O}{\overset{||}{C}}D$

(c) $(CH_3)_2CHCH_2\overset{O}{\overset{||}{C}}CH_3$ $\longrightarrow$ $(CH_3)_2CHCH_2\overset{CH_3}{\underset{}{\overset{|}{C}}}=CHCOOCH_2CH_3$

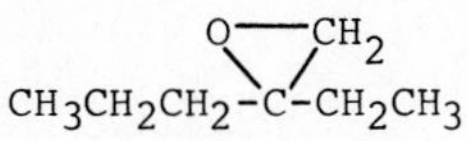

(d) $C_6H_5CH_3$ $\longrightarrow$ $C_6H_5CH_2-\overset{\overset{O}{\parallel}}{\underset{\underset{O}{\parallel}}{S}}-\!\!\!\!\bigcirc\!\!\!\!-CH_3$

(e) $\bigcirc$ $\longrightarrow$ $(C_6H_5O)_3P=O$

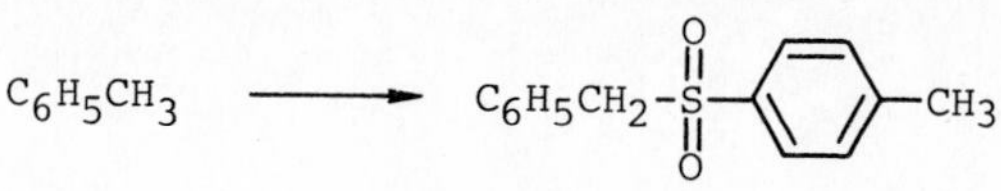

S4.   The Arbuzov-Michaelis reaction is a very important one for the preparation of dialkyl alkylphosphonates from trialkylphosphites, but it fails if applied to the synthesis of diphenyl alkylphosphonates from triphenylphosphite. Why? What products do you expect to see instead?

S5.   A chemist tried to synthesize 1-methyl-2-(phenylthio)ethanamine from propylene oxide by sequential treatment with sodium benzothiolate, p-toluenesulfonyl chloride, and ammonia. However, 2-(phenylthio)propanamine was the actual product. Draw the intermediates involved in this transformation and explain why the unanticipated isomer was produced.

$$CH_3\overset{\overset{O}{\diagup\diagdown}}{CH-\!CH_2} \quad \xrightarrow{C_6H_5SNa} \quad \xrightarrow{CH_3C_6H_4SO_2Cl} \quad \xrightarrow{NH_3} \quad \underset{CH_3\overset{\overset{SC_6H_5}{\mid}}{CHCH_2NH_2}}{}$$

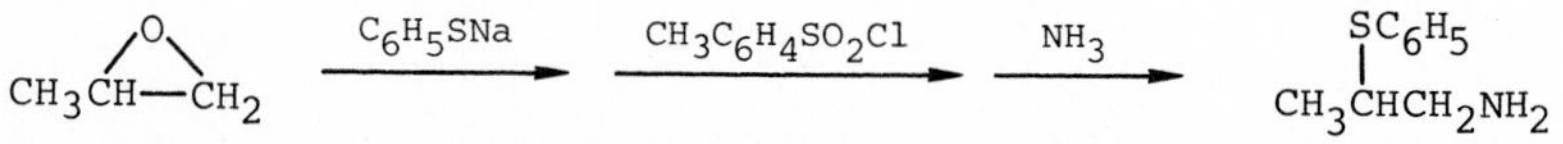

S6.   If a carboxylic ester is to be converted to a β-ketosulfone by reaction with the carbanion derived from dimethyl sulfone, two equivalents of the carbanion are required; when one equivalent is used, only 50% of the ester reacts.

(a)   Why are two equivalents necessary for complete conversion?

(b)   β-Ketosulfones are "desulfonylated", to give the ketone itself, using aluminum amalgam in wet THF:

$$-\overset{\overset{O}{\parallel}}{C}-\overset{\mid}{\underset{\mid}{C}}-SO_2R \quad \xrightarrow[\substack{H_2O \\ THF}]{Al(Hg)} \quad -\overset{\overset{O}{\parallel}}{C}-\overset{\diagup}{\underset{\diagdown}{CH}} \quad + \quad {}^-O_2SR$$

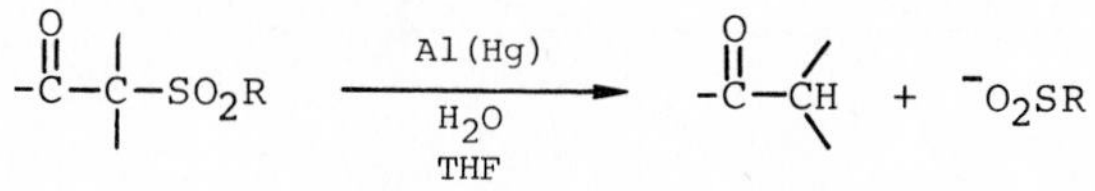

Using this reaction, devise a method for the conversion of a carboxylic ester into a ketone using dimethyl sulfone and an alkyl halide:

$$RCOOCH_3 \; + \; CH_3\overset{\overset{O}{\parallel}}{\underset{\underset{O}{\parallel}}{S}}CH_3 \; + \; BrR' \quad \xrightarrow{?} \quad R\overset{\overset{O}{\parallel}}{C}CH_2R'$$

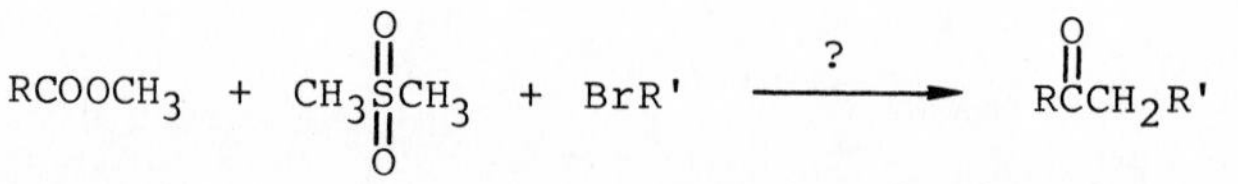

S7.   Dithioketals are often difficult to hydrolyze, even using mercuric chloride as catalyst, particularly if there is an acid-sensitive functional group in the molecule. A method for accomplishing this hydrolysis under neutral conditions using methyl iodide in aqueous acetone with sodium bicarbonate has been devised. Write a step-by-step mechanism to illustrate how this reaction occurs.

$$\underset{\underset{R}{}\overset{}{S}\diagdown\underset{C}{}\diagup S\underset{R}{}}{\bigcirc} \quad \xrightarrow[\substack{NaHCO_3,\ H_2O \\ acetone}]{excess\ CH_3I} \quad R\overset{\overset{O}{\parallel}}{C}R$$

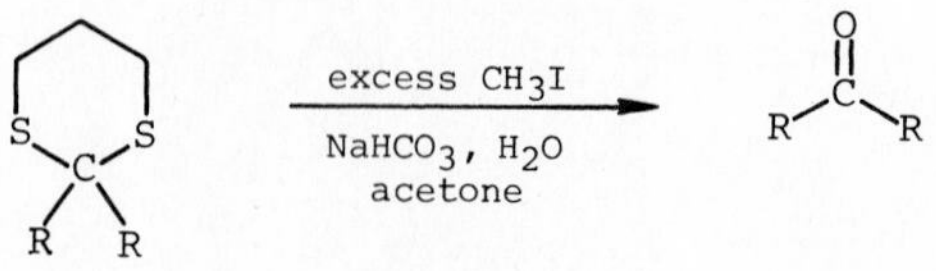

S8.   Depending on the order in which substituents are introduced onto the bicyclic lactone illustrated below, the final sulfoxide elimination reaction produces one or the other double bond isomer. How do you account for this difference in behavior?

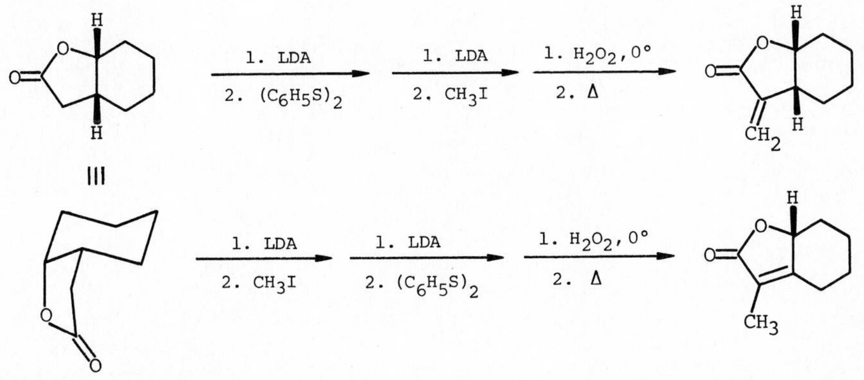

26.F Answers to Supplementary Problems

S1. (a) trimethyl phosphite
(b) methyl *p*-bromobenzenesulfinate
(c) diethylphenylsulfonium iodide
(d) sodium dodecyl sulfate (an important detergent)
(e) tetraethyl pyrophosphate
(f) tributylmethylphosphonium chloride
(g) ethyl methyl phenylphosphonate
(h) aminomethylphosphonic acid
(i) trichloromethyl phenyl sulfone
(j) dipropyl sulfoxide

S2. (a) 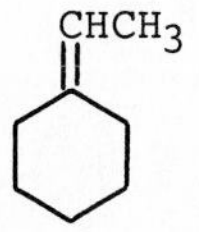 + $(C_6H_5)_3P{=}O$

(b) $(CH_3)_2CH\overset{O}{\overset{\|}{C}}CH_2CH_3$

(c) 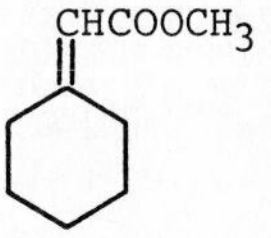

(d) 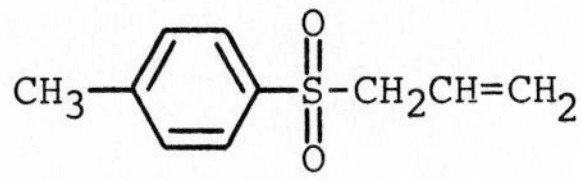

(e) 

$(HSCH_2CH_2CH_2CH_2Br + HS^- \rightleftharpoons H_2S + {}^-SCH_2CH_2CH_2CH_2Br$ 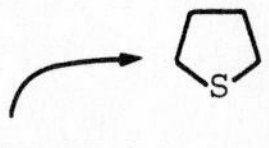

is faster than intermolecular displacement by a second ^-SH group

(f) $HSCH_2CH_2CH_2CH_2SH$

(g) $(CH_3O)_2PO_2^-$ + $CH_3CH_2OP(O)(OCH_3)_2$

(h) $C_6H_5-S(O_2)-CH_2C(OH)(CH_3)_2$

S3. (a)  $R-\overset{O}{\overset{\|}{C}}-R'$ + $(CH_3)_2S^+\,\bar{C}H_2$ ⟶ 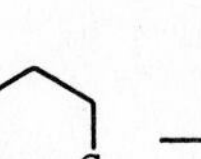 + $(CH_3)_2S$

(b) 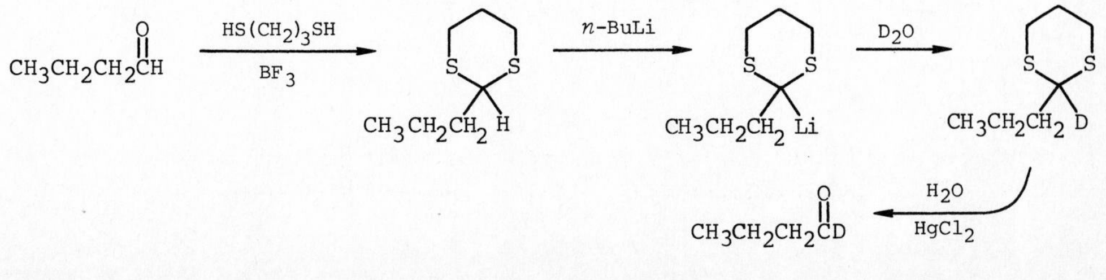

(c) 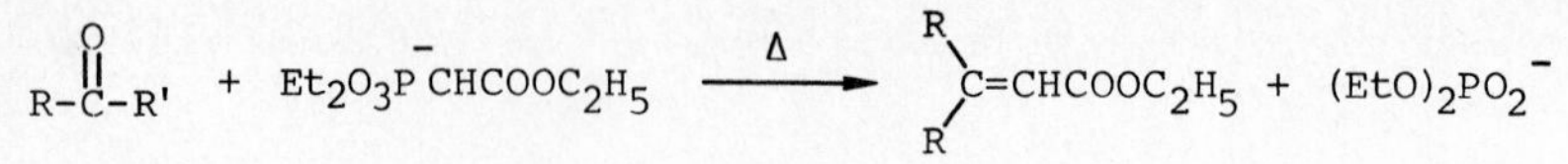

(d) 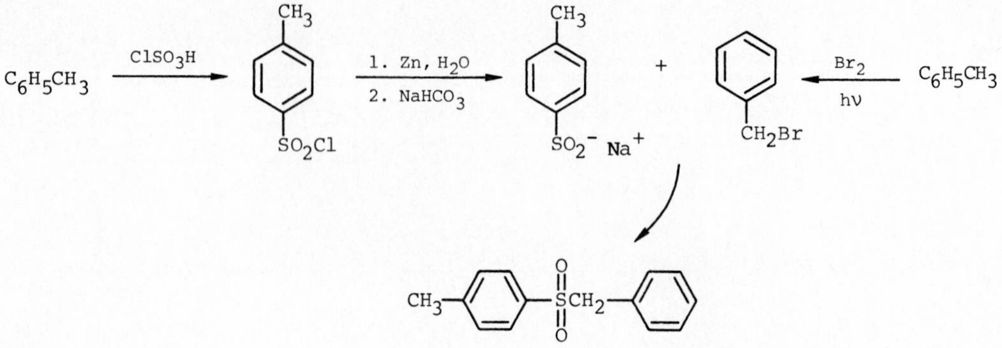

(e) 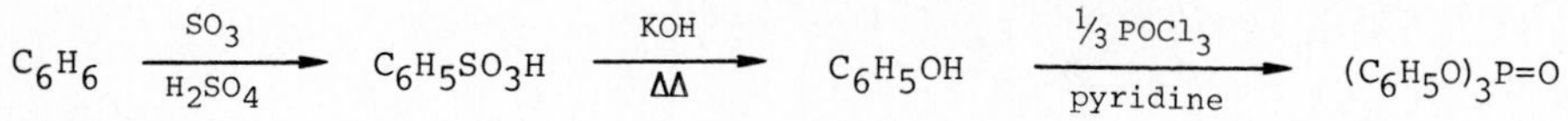

S4. 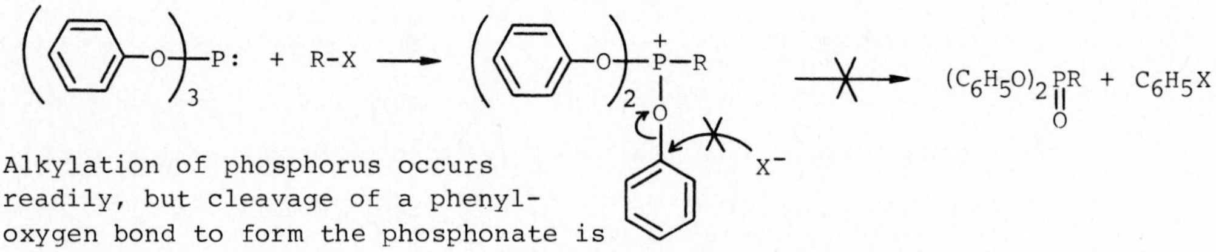

Alkylation of phosphorus occurs readily, but cleavage of a phenyl-oxygen bond to form the phosphonate is not possible, due to the difficulty of nucleophilic substitution on the benzene ring.

S5. 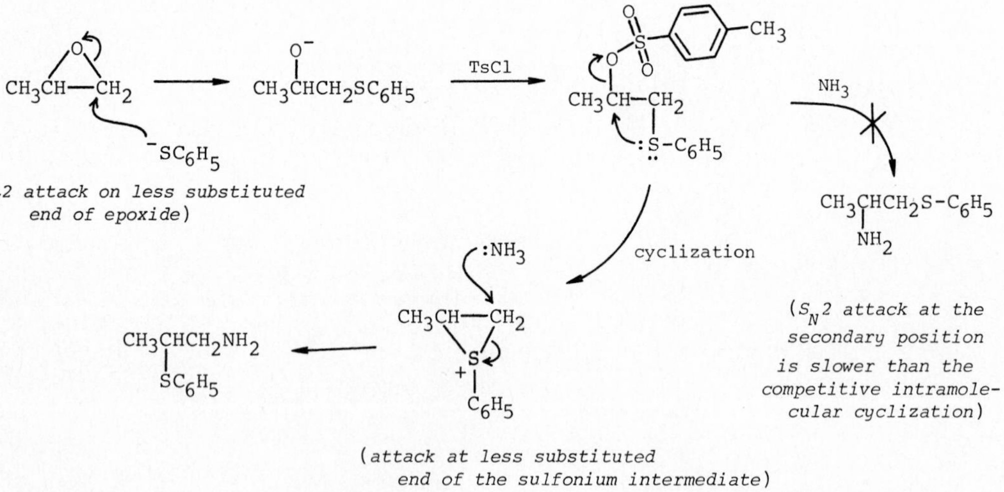
TsCl
NH3
cyclization
(SN2 attack on less substituted end of epoxide)
(SN2 attack at the secondary position is slower than the competitive intramolecular cyclization)
(attack at less substituted end of the sulfonium intermediate)

S6. (a) 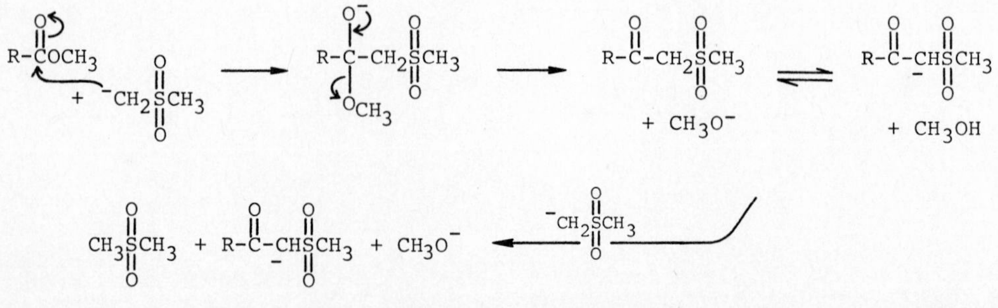

The dimethyl sulfone carbanion is more basic than both methanol and the β-keto sulfone product, and one mole will be protonated for each mole of β-keto sulfone which is formed. The necessity of using two equivalents of a carbanion whenever the product is more acidic than the conjugate acid of the carbanion is a common one. Formation of the enolate of the product also prevents addition of another equivalent of carbanion to the ketone carbonyl.

(b)

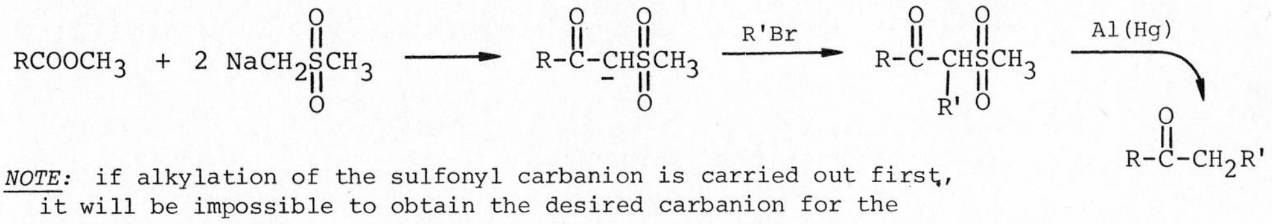

(*NOTE:* if alkylation of the sulfonyl carbanion is carried out first, it will be impossible to obtain the desired carbanion for the acylation reaction:

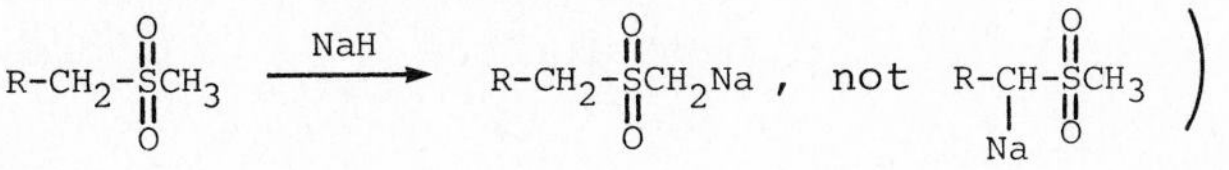

)

S7.

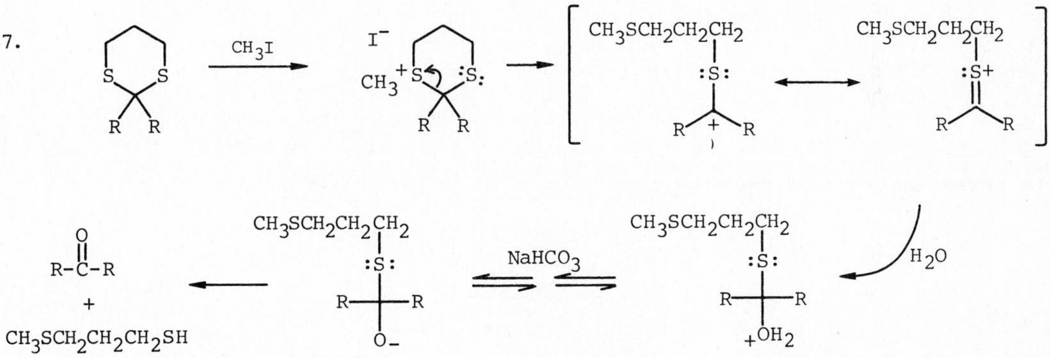

S8. Because of the folded structure of the bicyclic lactone, all reactions of the enolates (alkylations and sulfenylations) occur from the top side of the molecule (*exo* face), as illustrated below, for steric reasons.

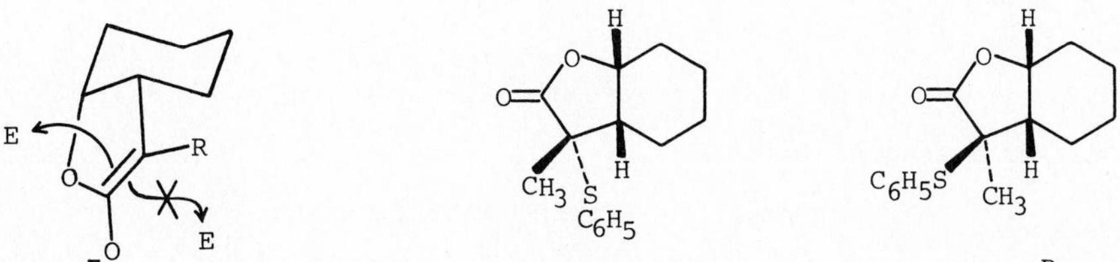

A B

If alkylation follows sulfenylation, compound A will result. The subsequent sulfoxide elimination can occur only toward the methyl group, because the bridgehead hydrogen is *trans* to it. If sulfenylation follows alkylation, diastereomer B will be produced. In this case, the sulfoxide can eliminate in either direction because the bridgehead hydrogen is *cis* to the sulfur group; as it turns out, elimination to give the more highly substituted double bond is preferred.

27. DIFUNCTIONAL COMPOUNDS

27.A Chapter Outline, Important Terms Introduced, and Reactions Discussed

27.1 Introduction

Scope of chapter: diols, dicarbonyl compounds, hydroxy carbonyl compounds

27.2 Nomenclature of Difunctional Compounds

diene, diyne, diol, dione, dicarboxylic acid

glycols hydroxyalkanone hydroxyalkanoic acid

27.3 Diols

A. Preparation of Diols

1. 1,2-diols

hydroxylation (see sections 11.6.E. and 10.12)

syn ($KMnO_4$ or OsO_4) or *anti* addition (via epoxide and hydrolysis)

erythro and *threo*

reductive dimerization of ketones (pinacol reaction):

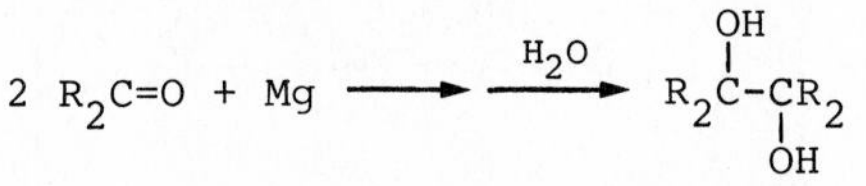

(via ketyl radical anion)

2. 1,3-diols

via aldol reaction and reduction:

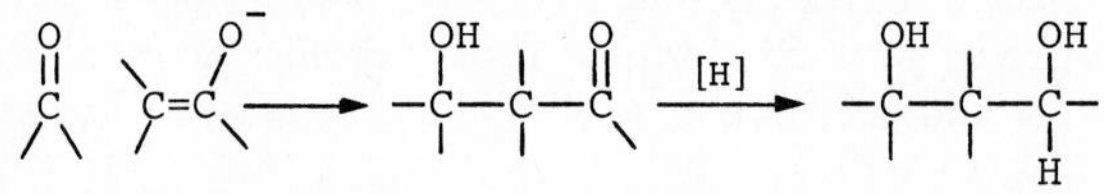

3. 1,4- and higher diols

(from same reactions as in monofunctional systems)

B. Reactions of Diols

1. Dehydration: The Pinacol Rearrangement

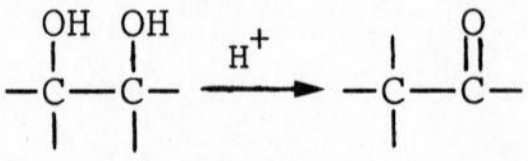

cyclic ethers from 1,4- and 1,5-diols

2. Oxidation

cleavage of 1,2-diols:

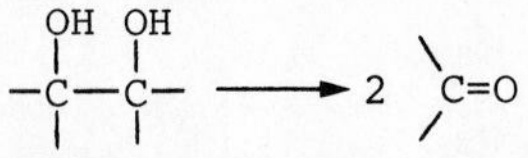

(with $Pb(OAc)_4$ or HIO_4)

3. Ketal and acetal formation

with 1,2- and 1,3-diols (see section 13.7.B.)

27.4 Hydroxy Aldehydes and Ketones

A. Synthesis of Hydroxy Aldehydes and Ketones

acyloin condensation:

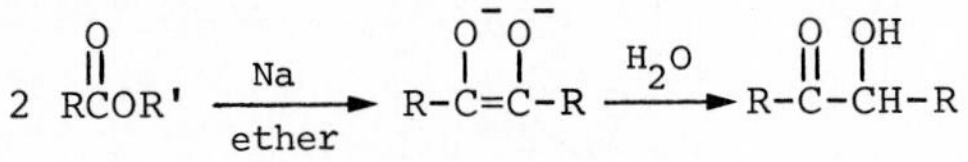

(good for making large ring compounds)

benzoin condensation:

$$2 \ Ar\overset{\overset{\displaystyle O}{\|}}{C}H \ + \ NaCN \ \longrightarrow \ Ar\overset{\overset{\displaystyle O}{\|}}{C}-\overset{\overset{\displaystyle OH}{|}}{C}H-Ar$$

aldol condensation (see section 13.7.F.)

B. Reactions of Hydroxy Aldehydes and Ketones

1. Dehydration (of β-hydroxy carbonyl compounds)

(with acid (via enol) or base (via enolate))

2. Cyclic hemiacetals and hemiketals

(when 5- or 6-membered rings are formed)

3. Oxidation

periodate cleavage:

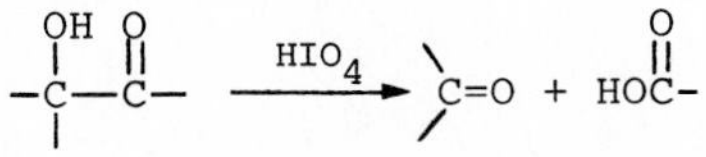

27.5 Hydroxy Acids

(lactic, citric, malic, tartaric acids, etc.)

A. Synthesis of Hydroxy Acids

hydrolysis of α-halo acids:

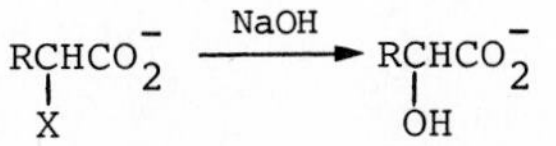

via hydrolysis of cyanohydrins:

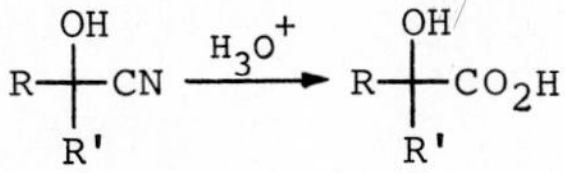

aldol-like condensation (to give β-hydroxy acids):

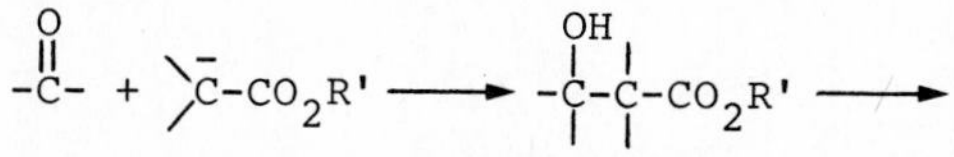

hydrolysis of lactones from Baeyer-Villiger reaction:

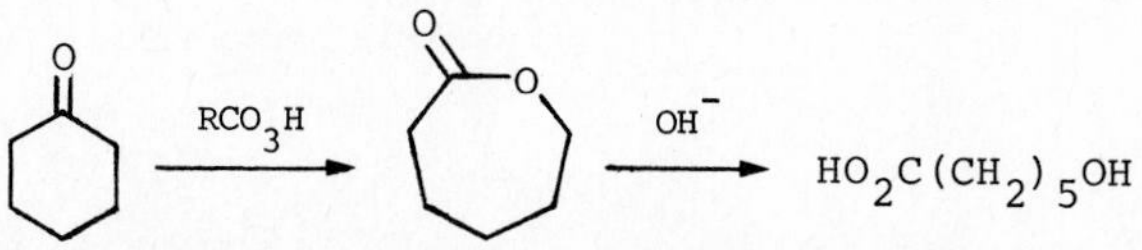

B. Reactions of Hydroxy Acids

    1. Formation of lactones

            equilibrium with hydroxy acids:

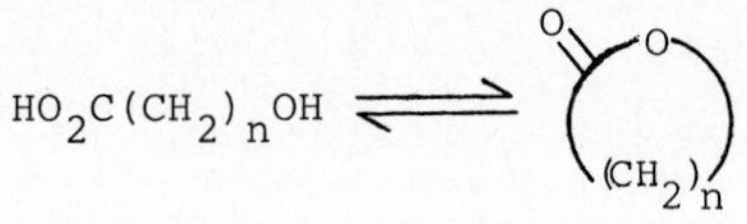

$$HO_2C(CH_2)_nOH \rightleftharpoons \qquad \text{(best for } n = 3 \text{ or } 4)$$

            synthesis by intramolecular alkylation:

$$^-O_2C(CH_2)_nX \longrightarrow \qquad + \; X^-$$

    2. Polymerization; Lactides

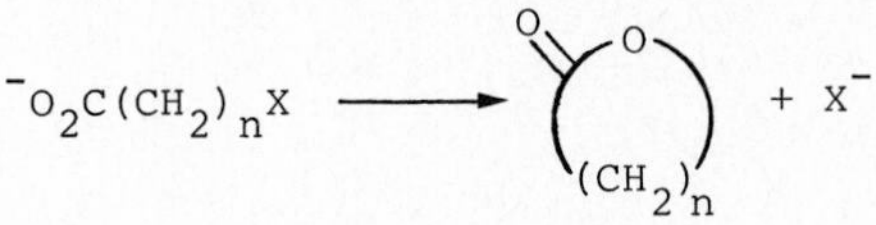

$$HO_2C(CH_2)_nOH \xrightarrow[H^+]{-H_2O} \text{polymer } (n > 4)$$

$$2 \; RCHCO_2H \xrightarrow[H^+]{-2 \; H_2O}$$

(with OH on the $RCHCO_2H$)

    3. Dehydration

            β-hydroxy acids similar to β-hydroxy ketones

                (conjugation of double bond with carbonyl less important for
                acids than ketones)

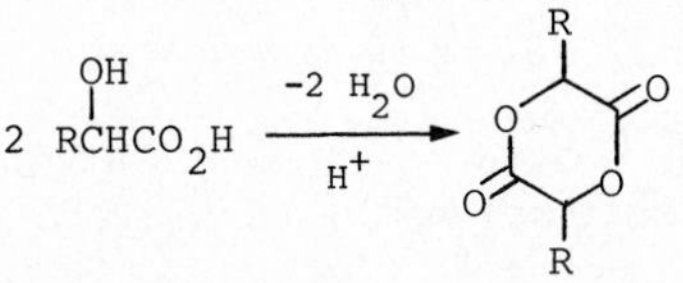

27.6 Dicarboxylic Acids

    A. Synthesis of Dicarboxylic Acids

    (from same reactions as monofunctional systems)

            cleavage of olefins:

$$\begin{array}{c} HC \\ \| \\ HC \end{array} \xrightarrow{KMnO_4} \begin{array}{c} HO_2C \\ HO_2C \end{array}$$

            cleavage of ketones:

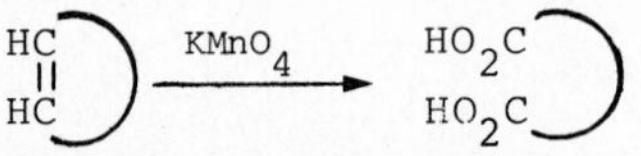

$$\xrightarrow{HNO_3} HO_2C \qquad CO_2H$$

            chemistry of carbonic acid

                (phosgene, carbonates, ureas)

            oxidation of aromatic systems:

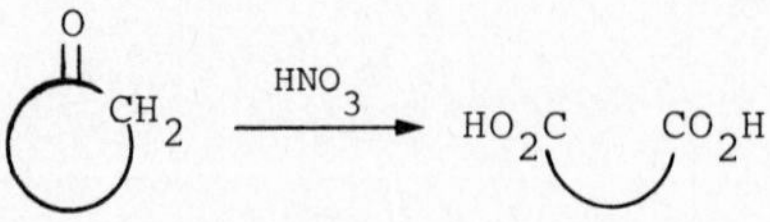

$$\xrightarrow[Co^{II}]{O_2}$$

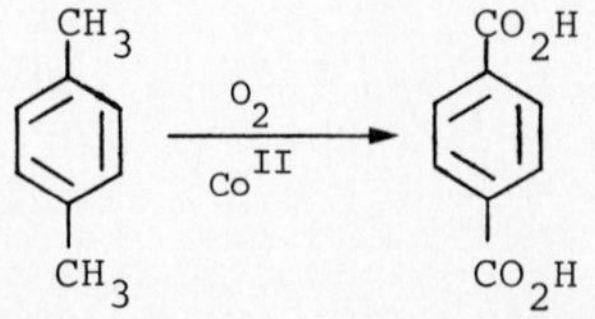

(industrial synthesis of terephthalic
acid)

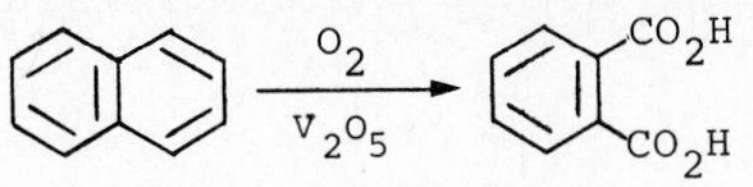

(industrial synthesis of phthalic acid)

B. Reactions of Dicarboxylic Acids and Their Derivatives

1. Acidity of dicarboxylic acids

   electrostatic effects

2. Formation of polyesters and polyamides

   condensation polymers vs. addition polymers

   Dacron and nylon copolymer

3. Behavior on heating

   decarboxylation of malonic acids:

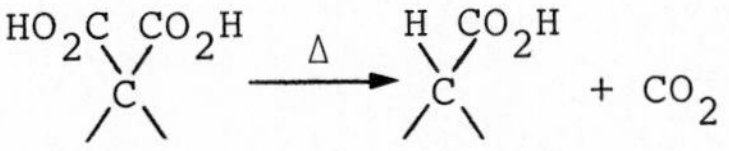

   anhydride formation from succinic and glutaric acids:

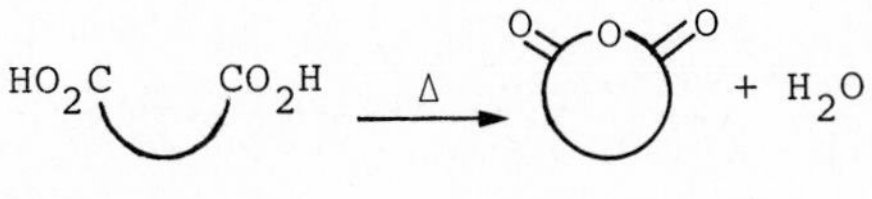

(5- or 6-membered ring)

   cyclic imides:

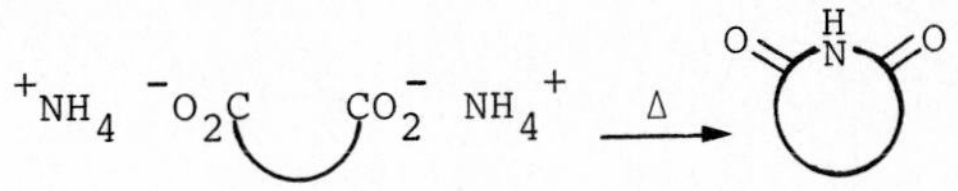

(5- or 6-membered ring)

   use of cyclic anhydrides in Friedel-Crafts reactions:

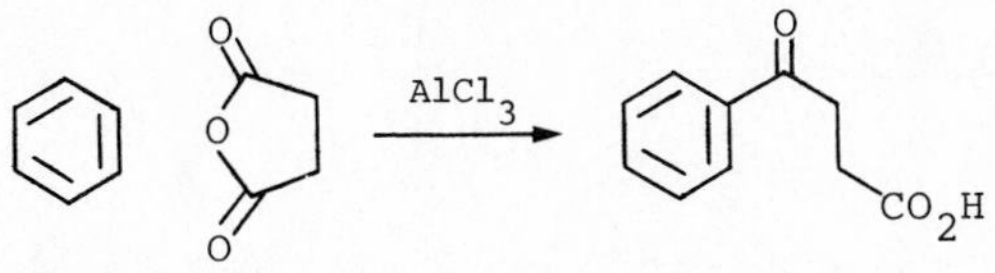

   intramolecular aromatic acylation:

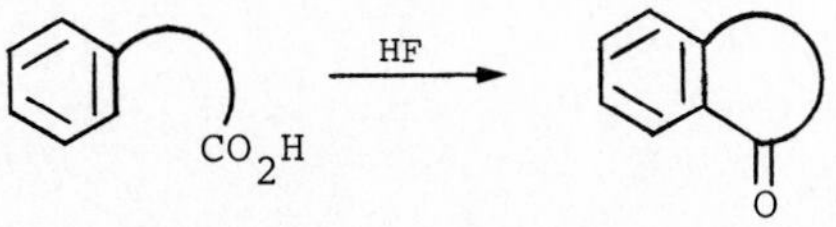

(5- or 6-membered ring)

4. Dieckman condensation

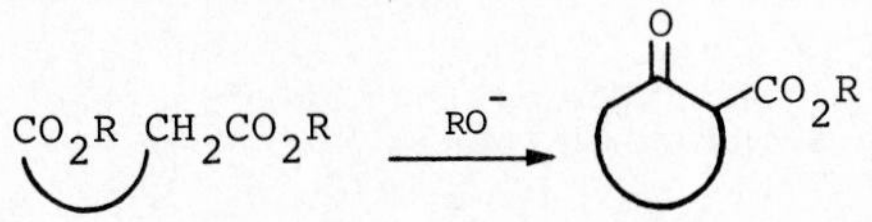

(5- or 6-membered rings; see Claisen condensation, section 19.9)

27.7 Diketones, Keto Aldehydes, Keto Acids, and Keto Esters

A. Synthesis

1. 1,2-diketones (α-diketones)

oxidation:

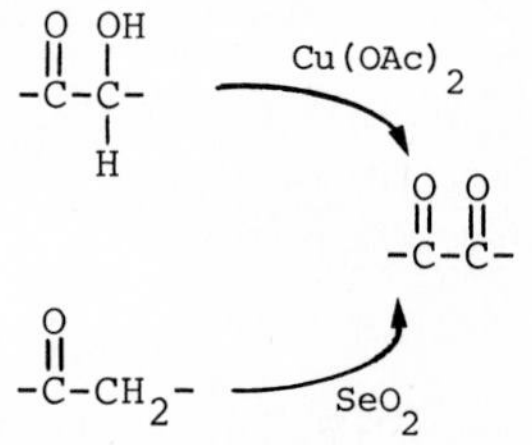

2. 1,3-diketones (β-diketones)

via mixed Claisen condensation:

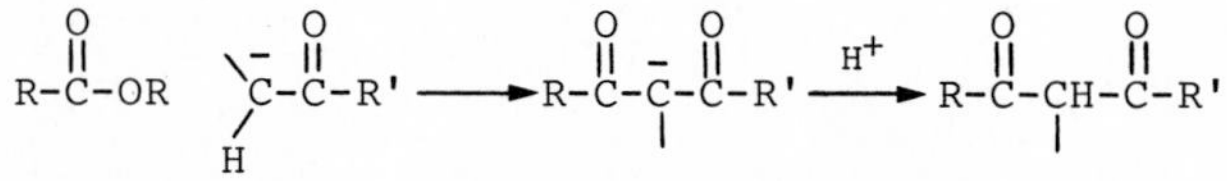

B. Properties

1. Keto-enol equilibria

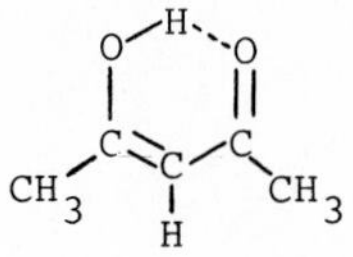

2. 1,3-Dicarbonyl compounds as carbon acids

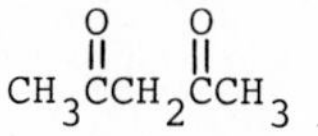

$$CH_3\overset{O}{\overset{\|}{C}}CH_2\overset{O}{\overset{\|}{C}}CH_3 \quad pK_a = 9$$

C. Reactions

1. The benzilic acid rearrangement

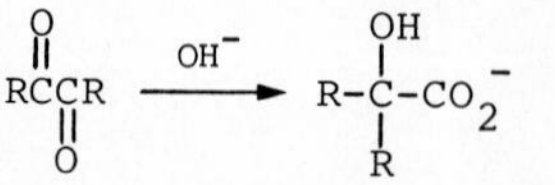

(ring contraction in cyclic systems)

2. Decarboxylation of β-keto acids

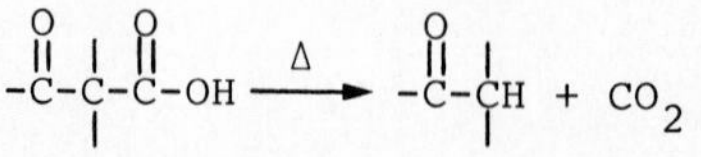

(must be possible to form enol)

3. Alkylation of 1,3-dicarbonyl compounds: the malonic ester and acetoacetic ester syntheses

malonic ester synthesis:

$$EtO_2CCHCO_2Et \xrightarrow{RX} EtO_2CCHCO_2Et \xrightarrow[R'X]{NaOEt} EtO_2C\underset{R}{\overset{R'}{\underset{|}{\overset{|}{C}}}}-CO_2Et$$

$$\Delta \downarrow H_3O^+ \qquad\qquad \Delta \downarrow H_3O^+$$

$$RCH_2CO_2H \qquad\qquad \underset{R'}{\overset{R}{\underset{/}{\overset{\backslash}{C}}}}HCO_2H$$

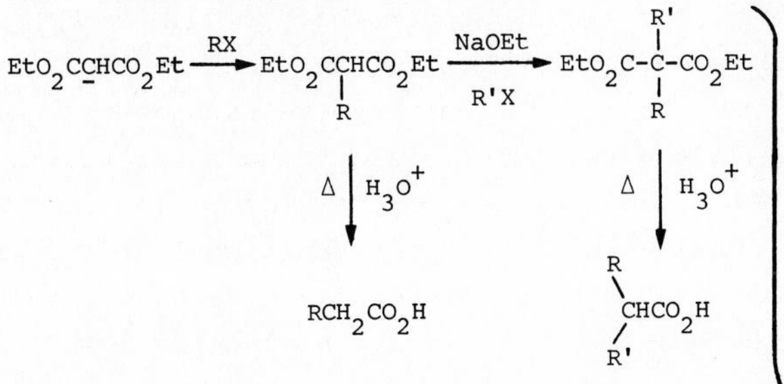

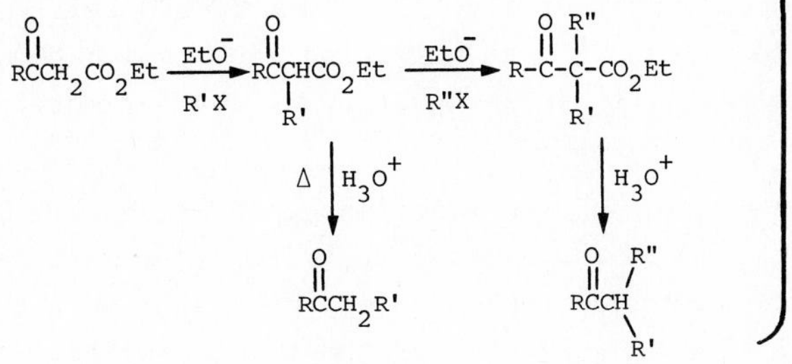

(subject to limitations of S_N2 reactions; works in cyclic systems, too)

acetoacetic ester synthesis:

$$\overset{O}{\overset{||}{R}CCH_2CO_2Et} \xrightarrow[R'X]{EtO^-} \overset{O}{\overset{||}{R}C\underset{R'}{C}HCO_2Et} \xrightarrow[R''X]{EtO^-} R-\overset{O}{\overset{||}{C}}-\underset{R'}{\overset{R''}{\underset{|}{\overset{|}{C}}}}-CO_2Et$$

$$\Delta \downarrow H_3O^+ \qquad\qquad H_3O^+$$

$$\overset{O}{\overset{||}{R}CCH_2R'} \qquad\qquad \overset{O}{\overset{||}{R}C}\underset{R'}{\overset{R''}{\underset{/}{\overset{\backslash}{C}}}}H$$

4. The Knoevenagel condensation

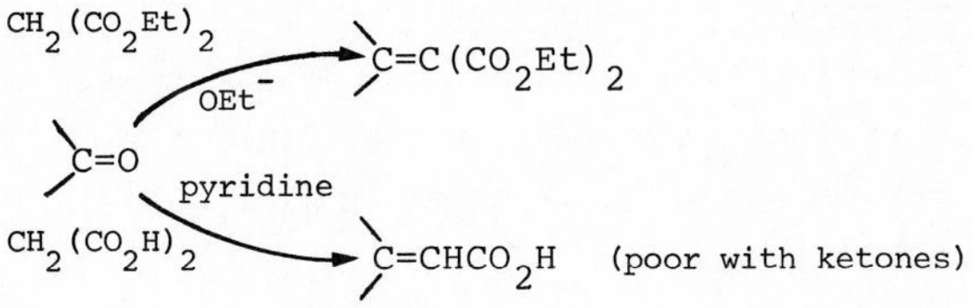

5. The Michael addition

$$\underset{CO_2Et}{\overset{\backslash}{\underset{/}{C}}=C\overset{/}{\underset{\backslash}{}}} + CH_2(CO_2Et)_2 \xrightarrow{EtO^-} (EtO_2C)_2CH-\underset{}{\overset{|}{C}}-\underset{H}{\overset{|}{C}}-CO_2Et$$

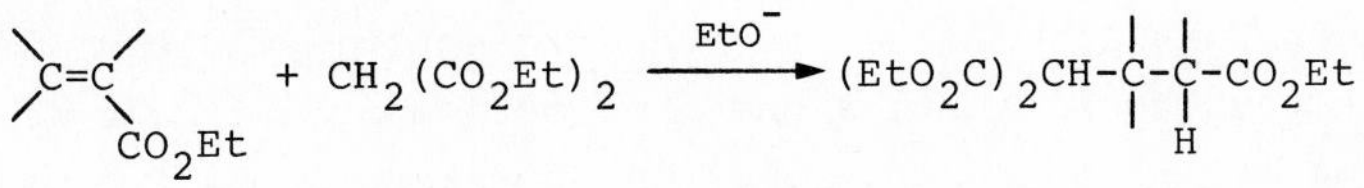

Robinson annelation:

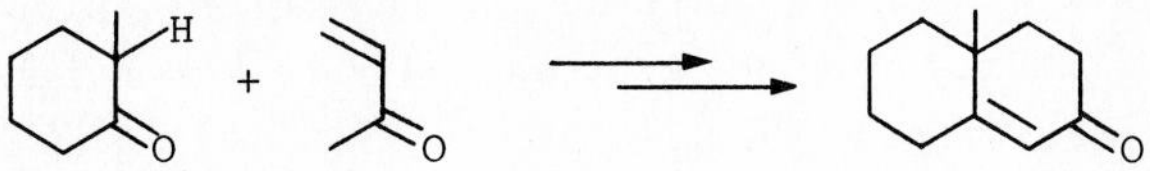

6. Reverse Claisen condensation

of non-enolizable β-keto esters:

$$R-\overset{O}{\overset{||}{C}}-\underset{R'}{\overset{R'}{\underset{|}{\overset{|}{C}}}}-CO_2Et \xrightarrow{EtO^-} RCO_2Et + \underset{R'}{\overset{R'}{\underset{/}{\overset{\backslash}{C}}}}HCO_2Et$$

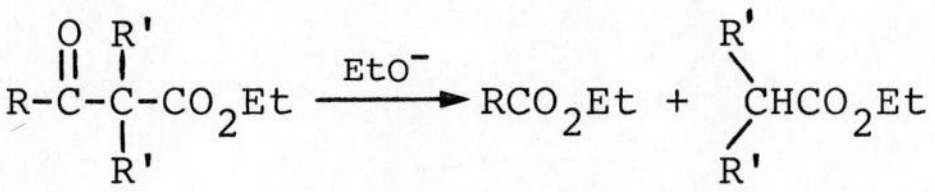

27.B Important Concepts and Hints

For the most part, previous chapters have focused on the chemistry of one functional group at a time. When two are present in a molecule, the reactions the compound undergoes often involve each group independently. However, in many cases the two functional groups interact, and together undergo reactions not typical of either one alone. In a sense, such a combination can be considered to be a new functional group, with its own set of reactions and ways to be synthesized. Some examples of this sort were presented in the chapter on conjugation (Chapter 20).

This chapter is organized in a logical fashion, presenting the various pairwise combinations of alcohol, aldehyde and ketone, and carboxylic acid functional groups, and outlining the special behavior which results when they are in 1,2- (adjacent), 1,3-, 1,4- or more remote relationships.

You may have seen many of the reactions in this chapter before; for instance, hydroxylation of olefins to give 1,2-diols and the cleavage of olefins to give dicarboxylic acids were both discussed in the chapter on alkenes (Chapter 11). Similarly, the synthesis of β-hydroxy ketones and their dehydration were discussed in Chapter 13 in connection with the aldol condensation. The reactions of difunctional compounds frequently involve the formation or cleavage of cyclic compounds, by reactions which you have previously learned for the intermolecular case. For example, the hydrolysis or formation of a lactone is simply the hydrolysis or formation of an ester in which the hydroxy and carboxylic acid groups are part of the same molecule.

From the synthetic standpoint, the synthesis of β-keto esters by the Claisen condensation, its intramolecular counterpart the Dieckman condensation, the alkylation of β-keto esters and malonic esters and their subsequent decarboxylation (the "acetoacetic ester synthesis" and the "malonic ester synthesis") are among the most important reactions in organic synthesis. (The decarboxylation of β-keto acids and malonic acids is another example of 6-electrons going around a circle--see section 20.B of this Study Guide.) The Robinson annelation is also an important sequence, involving a Michael addition and subsequent intramolecular aldol condensation, and it has been used extensively in the synthesis of natural products (see Special Topics Chapter 34).

27.C   Answers to Exercises

(27.2)   For example:

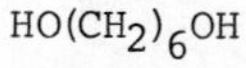

$HO(CH_2)_6OH$

1,6-hexanediol

$CH_3\overset{\displaystyle OH}{\underset{|}{C}}HCH=C(CH_3)_2$

4-methyl-3-penten-2-ol

$(CH_3CH_2)_2\overset{\displaystyle OH}{\underset{|}{C}}CH=O$

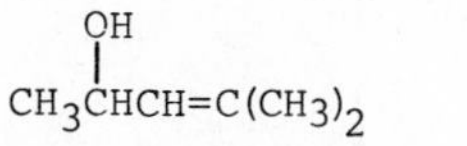

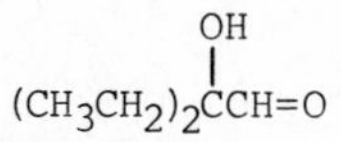

2-ethyl-2-hydroxybutanal

$CH_3CH_2\overset{\displaystyle OH}{\underset{|}{C}}H\overset{\displaystyle}{\underset{\underset{\displaystyle O}{||}}{C}}CH_2CH_3$

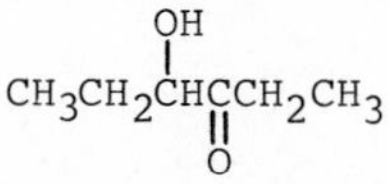

4-hydroxy-3-hexanone

$(CH_3)_2CH\overset{\displaystyle OH}{\underset{|}{C}}HCH_2COOH$

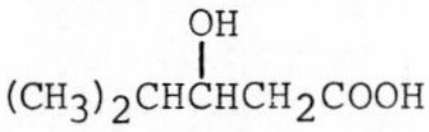

3-hydroxy-4-methyl-
pentanoic acid

$(CH_3)_2C=CHCH=CH_2$

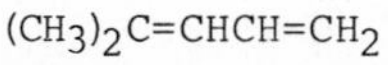

4-methyl-1,3-pentadiene

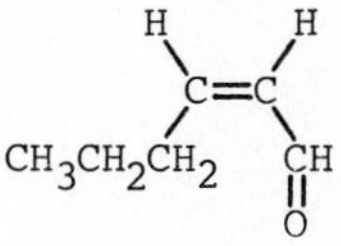

(Z)-2-hexanal

$CH_2=CHCH_2CH_2CH_2COOH$

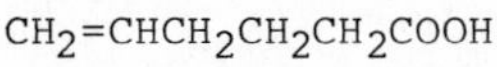

5-hexenoic acid

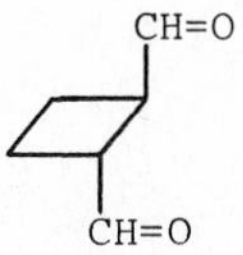

trans-1,2-cyclobutanedicarbaldehyde
(or -dicarboxaldehyde)

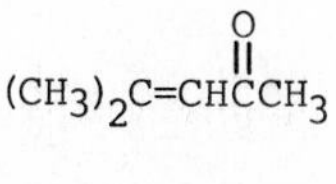

$(CH_3)_2C=CH\overset{\displaystyle O}{\overset{\displaystyle ||}{C}}CH_3$

4-methyl-3-penten-2-one

$H\overset{\displaystyle O}{\overset{||}{C}}CH_2\overset{\displaystyle CH_3}{\underset{|}{C}}H\overset{\displaystyle}{\underset{\underset{\displaystyle O}{||}}{C}}CH_3$

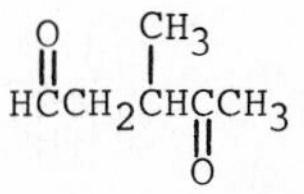

3-methyl-4-oxopentanal

$H\overset{\displaystyle O}{\overset{||}{C}}CH_2\overset{\displaystyle CH_3}{\underset{\underset{\displaystyle CH_3}{|}}{C}}COOH$

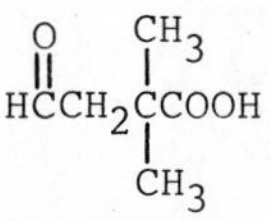

2,2-dimethyl-4-oxobutanoic
acid

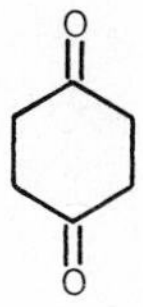

1,4-cyclohexanedione

$CH_3CH_2CH_2\overset{\displaystyle O}{\overset{||}{C}}CH_2COOH$

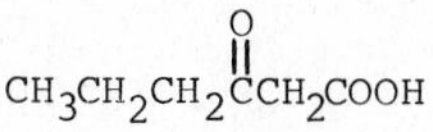

3-oxohexanoic acid

$HOOC\overset{\displaystyle}{\underset{\underset{\displaystyle CH_2CH_3}{|}}{C}}HCH_2COOH$

2-ethylbutanedioic acid

(27.3.A)

racemic
mixture

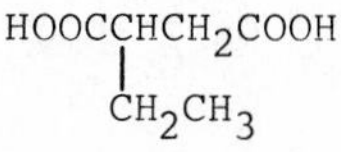

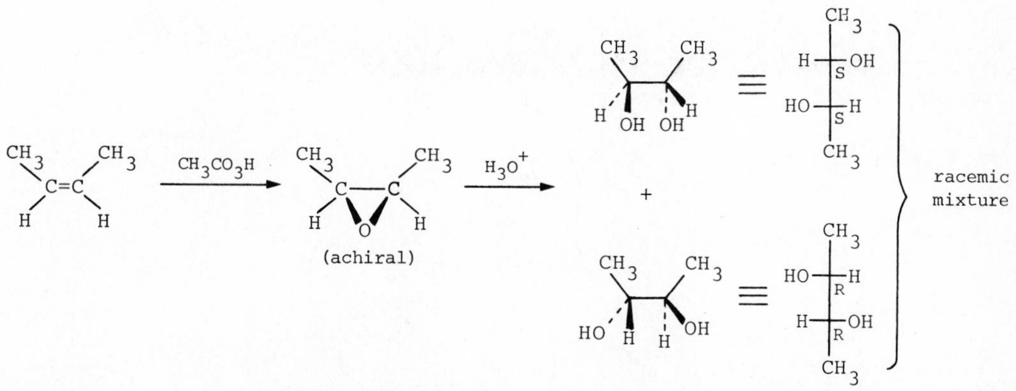

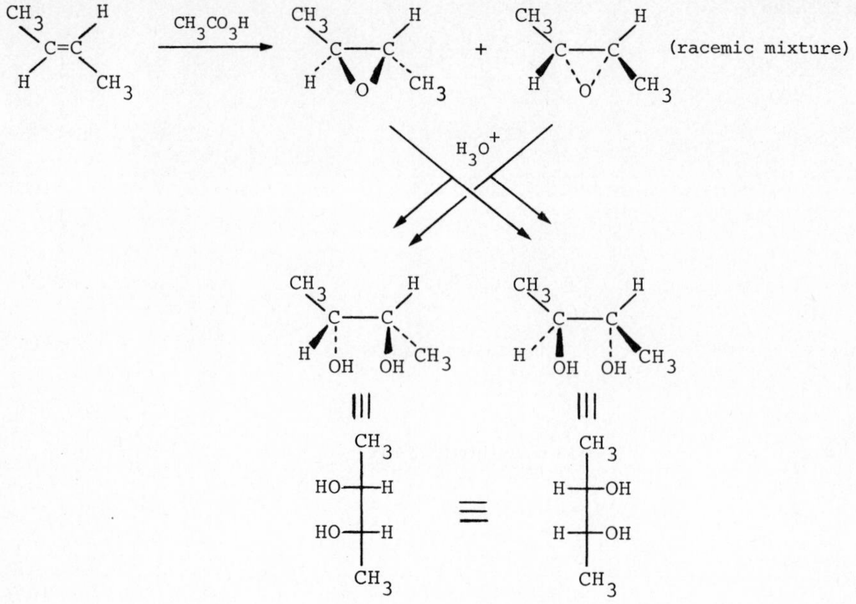

(meso compound)

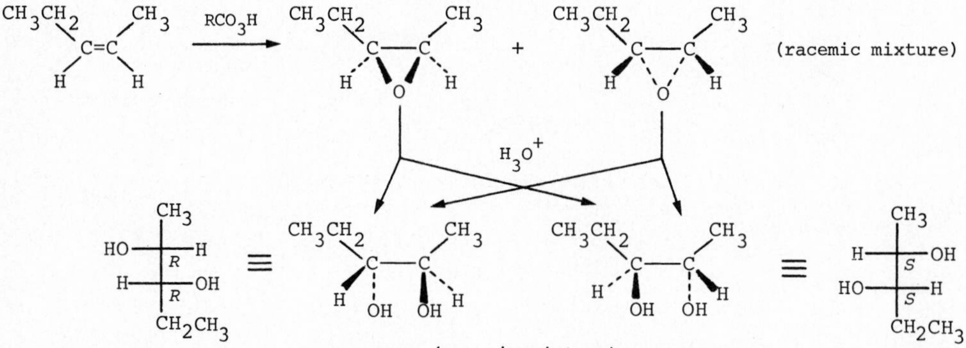

(racemic mixture)

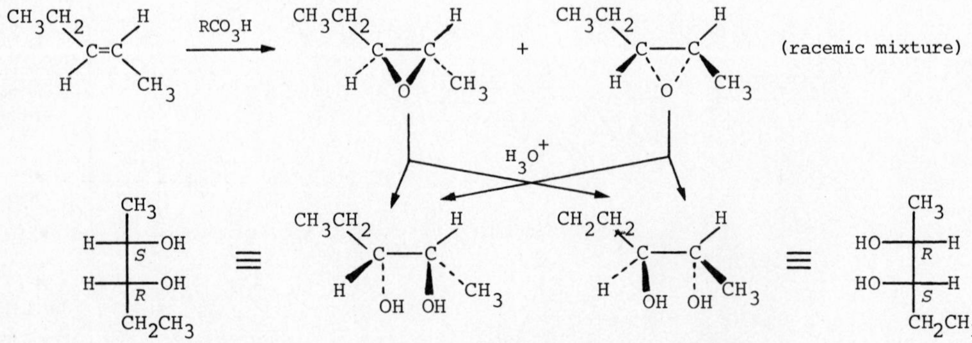

(racemic mixture)

(27.3.B.1.)

Favored relative to phenyl migration when Y = CH₃, disfavored for Y = NO₂

$C_6H_5CC-\left(\left(\text{(C}_6H_4\text{)}\right)-CH_3\right)_2$ and $O_2N-\text{(C}_6H_4\text{)}-\overset{O}{\overset{\|}{C}}-C(C_6H_5)_2$ are the major products.

(27.3.B.3.)

(a)

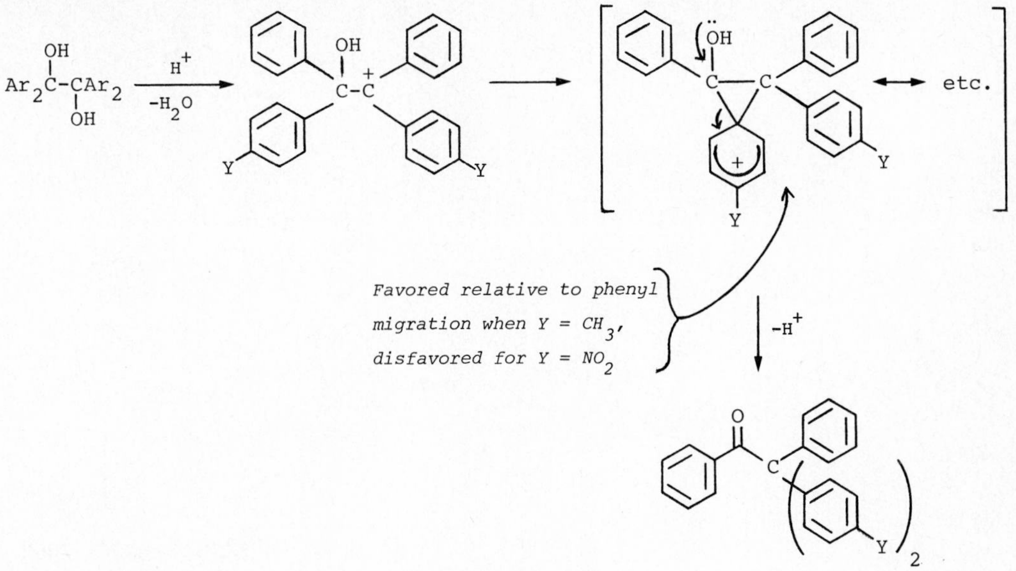

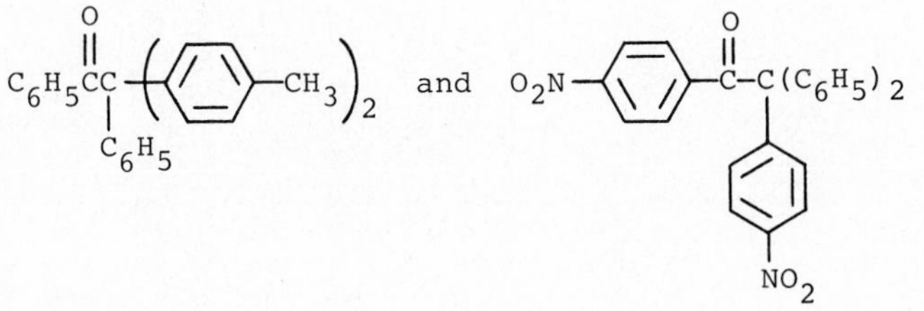

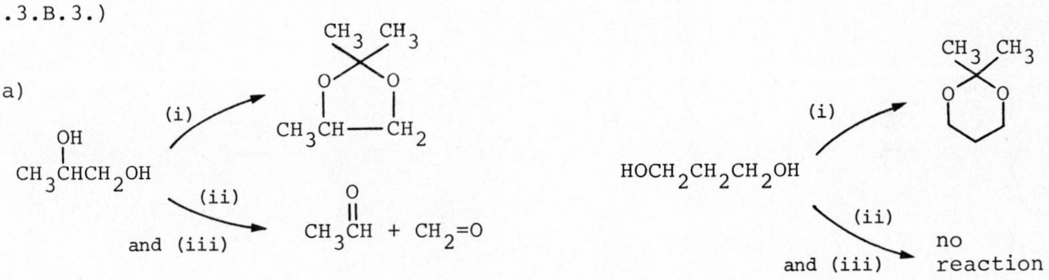

(b)

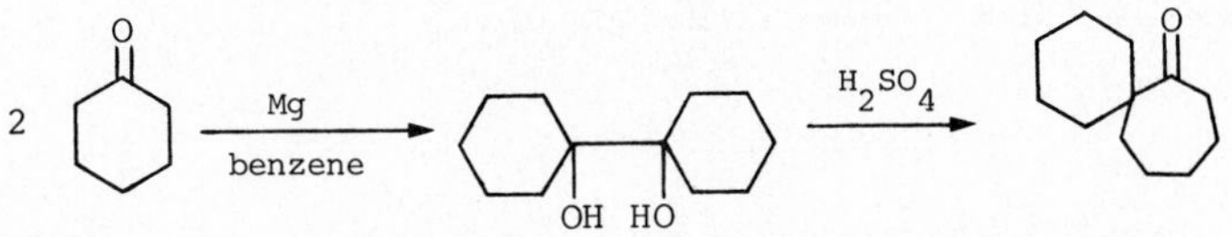

(27.4.B.3.)

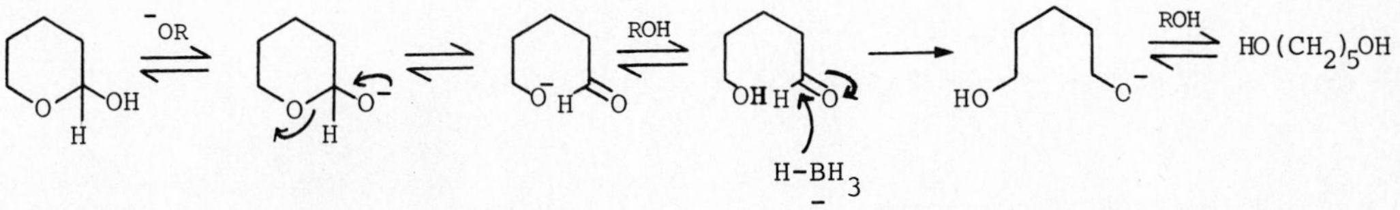

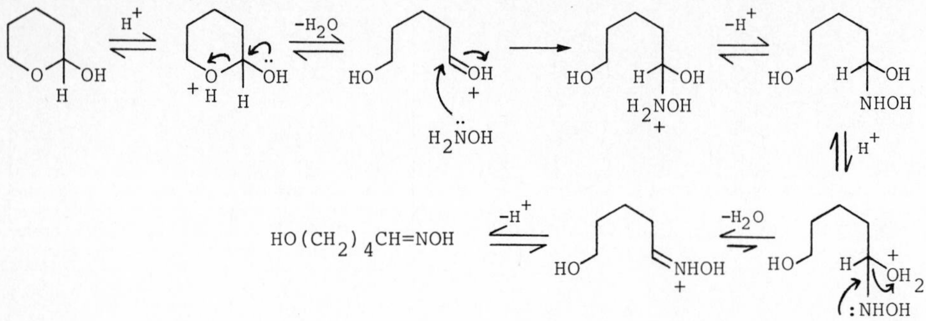

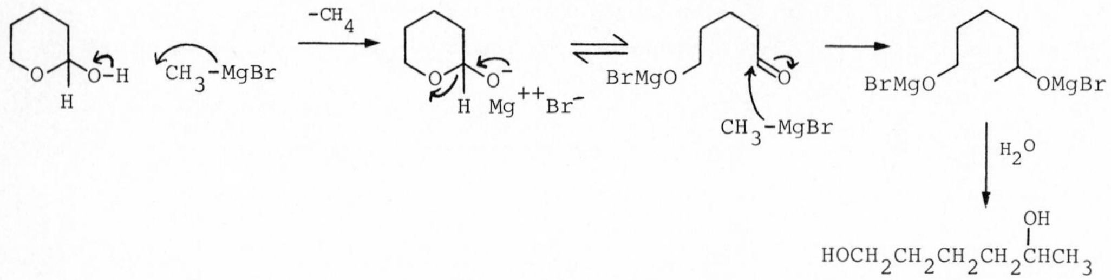

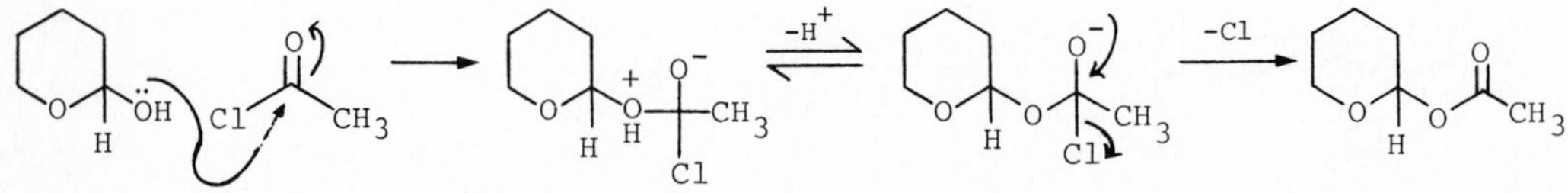

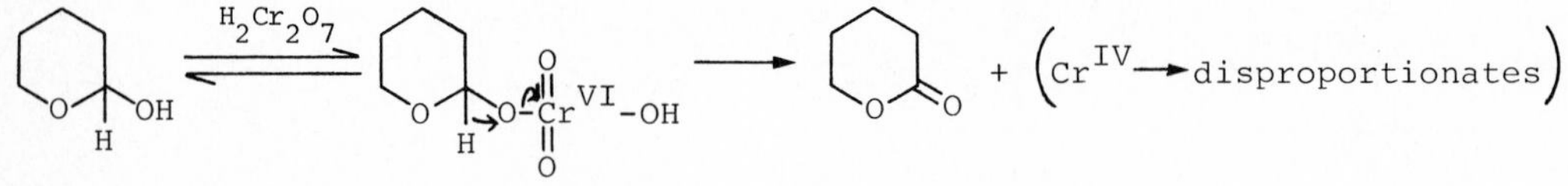

(27.5.A.)

(a) $CH_3CH_2CH_2CH_2CO_2H \xrightarrow{P,\ Br_2} \xrightarrow{H_2O} CH_3CH_2CH_2\underset{}{\overset{Br}{C}}HCO_2H \xrightarrow{NaOH} CH_3CH_2CH_2\overset{OH}{C}HCO_2H$

(b) $CH_3CH_2CH_2\overset{O}{\overset{\|}{C}}H \xrightarrow{HCN} CH_3CH_2CH_2\overset{OH}{C}HC{\equiv}N \xrightarrow{H_2SO_4}$ (product of (a))

$CH_3\overset{O}{\overset{\|}{C}}OC_2H_5 \xrightarrow[THF,\ -70°]{LDA} \xrightarrow{CH_3CH_2\overset{O}{\overset{\|}{C}}H} \xrightarrow{OH^-} CH_3CH_2\overset{OH}{C}HCH_2CO_2H$

(27.5.B.3.)

(a) $2\ CH_3CH_2\overset{OH}{C}HCO_2H \xrightarrow{H^+}$ 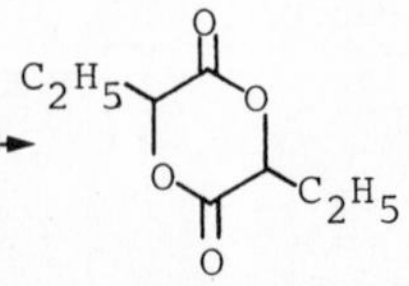

(b) $CH_3\overset{OH}{C}HCH_2CO_2H \xrightarrow{H^+} CH_3CH{=}CHCO_2H$

(c) $HOCH_2CH_2CH_2CO_2H \xrightarrow{H^+}$ 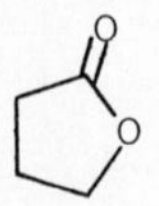

(27.6.B.3.)

(a) 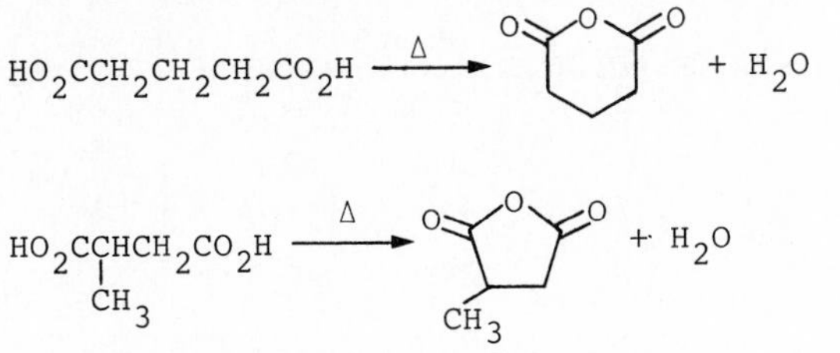

$$CH_3CH_2CH(CO_2H)_2 \xrightarrow{\Delta} CH_3CH_2CH_2CO_2H + CO_2$$

$$(CH_3)_2C(CO_2H) \xrightarrow{\Delta} (CH_3)_2CHCO_2H + CO_2$$

(b) 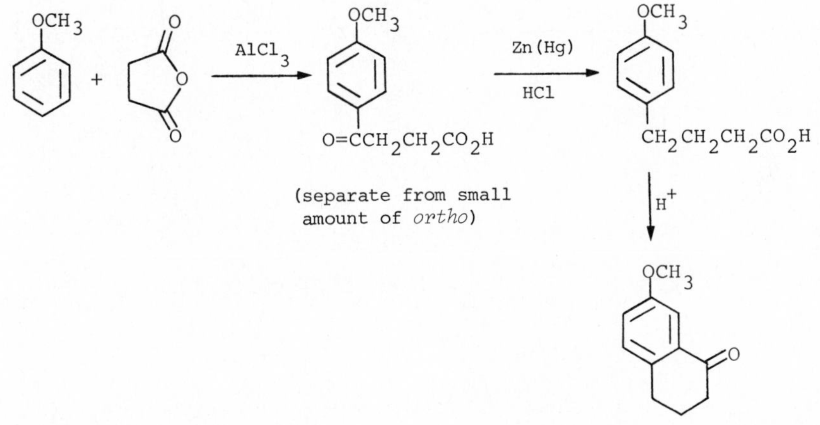

(27.6.B.4.)

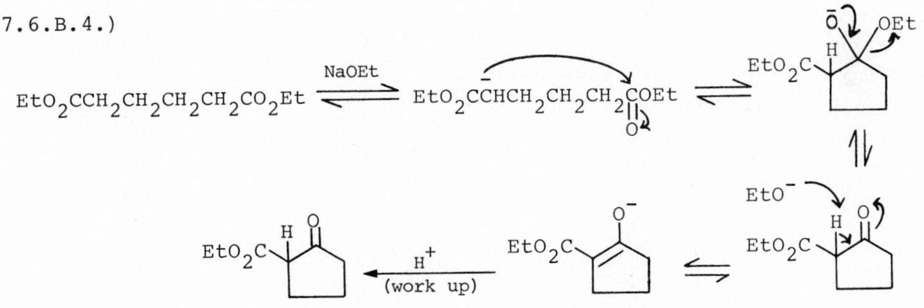

(among other resonance structures)

(27.7.A.)

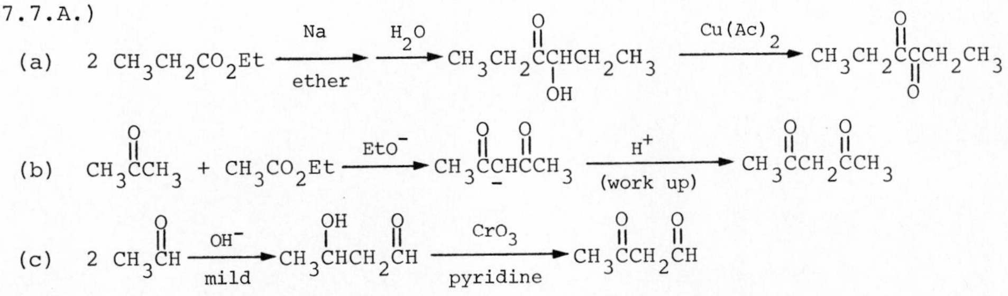

(a) $$2\ CH_3CH_2CO_2Et \xrightarrow[\text{ether}]{\text{Na}} \xrightarrow{H_2O} CH_3CH_2\overset{O}{\overset{\|}{C}}\underset{OH}{\underset{|}{C}}HCH_2CH_3 \xrightarrow{Cu(Ac)_2} CH_3CH_2\overset{O}{\overset{\|}{C}}\underset{O}{\underset{\|}{C}}CH_2CH_3$$

(b) $$CH_3\overset{O}{\overset{\|}{C}}CH_3 + CH_3CO_2Et \xrightarrow{EtO^-} CH_3\overset{O}{\overset{\|}{C}}\underset{-}{C}H\overset{O}{\overset{\|}{C}}CH_3 \xrightarrow[\text{(work up)}]{H^+} CH_3\overset{O}{\overset{\|}{C}}CH_2\overset{O}{\overset{\|}{C}}CH_3$$

(c) $$2\ CH_3\overset{O}{\overset{\|}{C}}H \xrightarrow[\text{mild}]{OH^-} CH_3\overset{OH}{\overset{|}{C}}HCH_2\overset{O}{\overset{\|}{C}}H \xrightarrow[\text{pyridine}]{CrO_3} CH_3\overset{O}{\overset{\|}{C}}CH_2\overset{O}{\overset{\|}{C}}H$$

(27.7.C.2.)

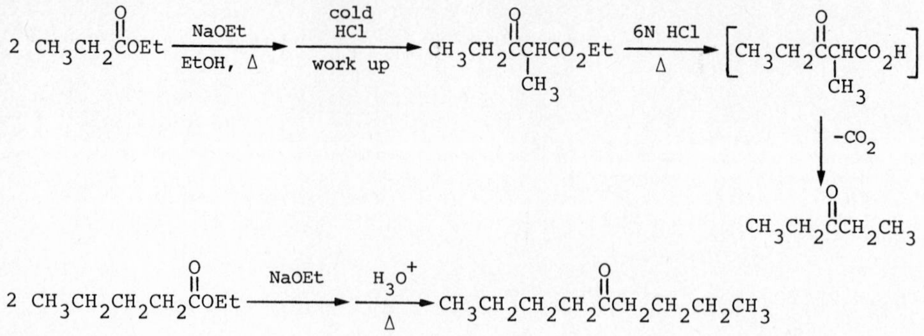

(27.7.C.3.)

(a)

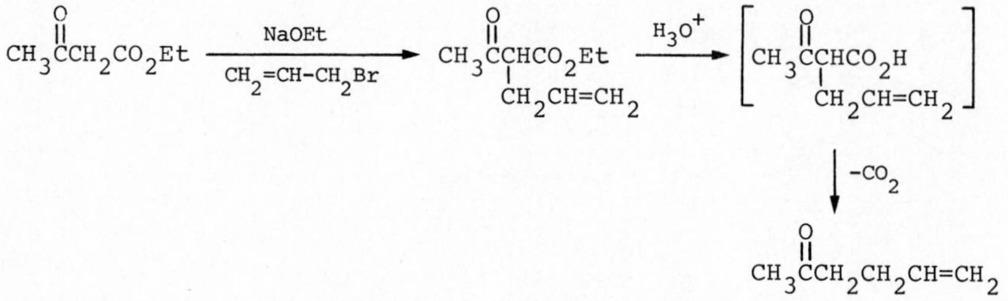

(b)

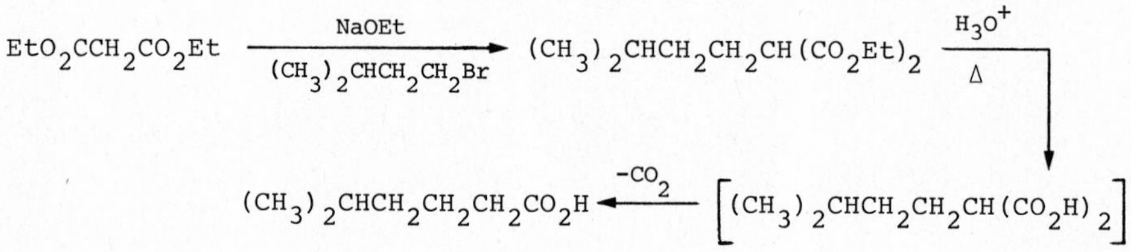

(27.7.C.4.)

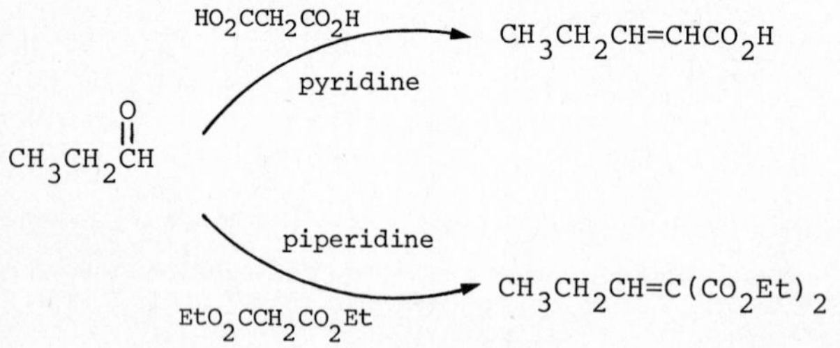

(27.7.C.5.)

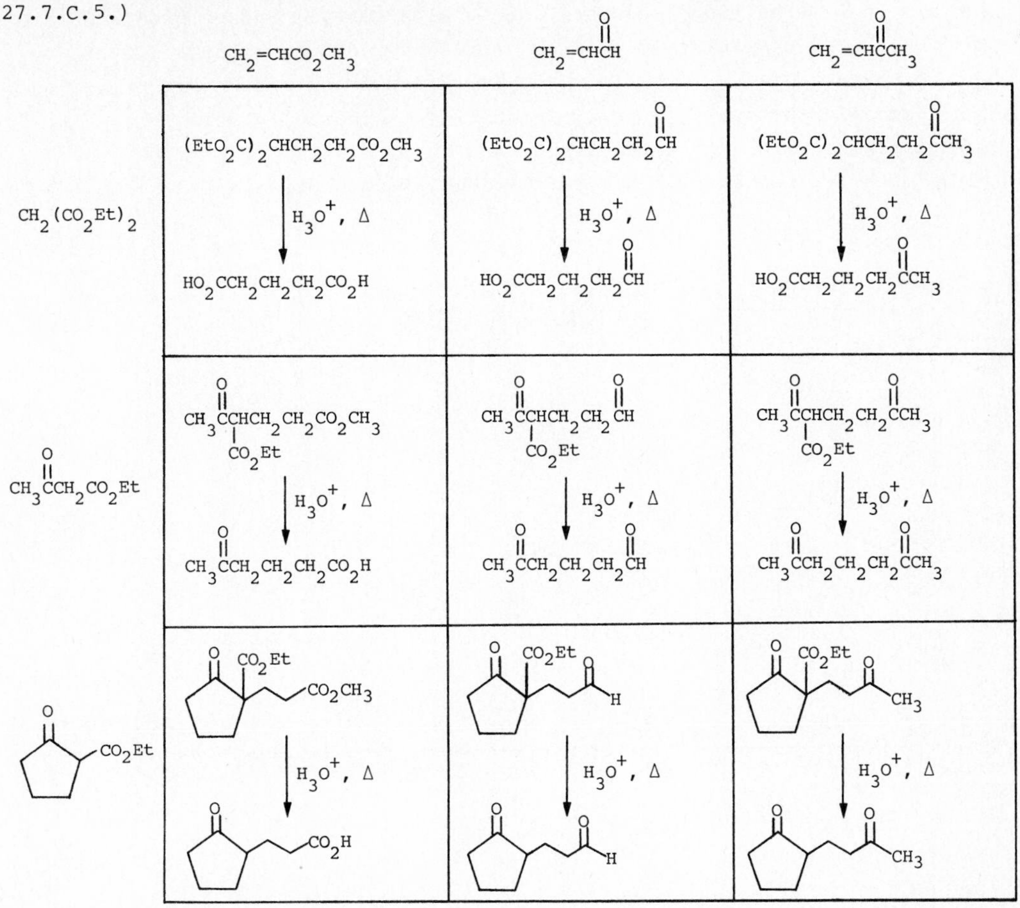

	$CH_2=CHCO_2CH_3$	$CH_2=CHCH$ (with C=O)	$CH_2=CHCCH_3$ (with C=O)
$CH_2(CO_2Et)_2$	$(EtO_2C)_2CHCH_2CH_2CO_2CH_3$ $\downarrow H_3O^+, \Delta$ $HO_2CCH_2CH_2CH_2CO_2H$	$(EtO_2C)_2CHCH_2CH_2CH$ (terminal C=O) $\downarrow H_3O^+, \Delta$ $HO_2CCH_2CH_2CH_2CH$ (terminal C=O)	$(EtO_2C)_2CHCH_2CH_2CCH_3$ (C=O) $\downarrow H_3O^+, \Delta$ $HO_2CCH_2CH_2CH_2CCH_3$ (C=O)
$CH_3CCH_2CO_2Et$ (C=O)	$CH_3CCHCH_2CH_2CO_2CH_3$ CO_2Et $\downarrow H_3O^+, \Delta$ $CH_3CCH_2CH_2CH_2CO_2H$	$CH_3CCHCH_2CH_2CH$ CO_2Et $\downarrow H_3O^+, \Delta$ $CH_3CCH_2CH_2CH_2CH$	$CH_3CCHCH_2CH_2CCH_3$ CO_2Et $\downarrow H_3O^+, \Delta$ $CH_3CCH_2CH_2CH_2CCH_3$

(27.7.C.6.)

$$(CH_3)_2CHCCH_2CH_2CO_2H \xleftarrow{H^+} (CH_3)_2CHCCH_2CH_2CO_2^-$$

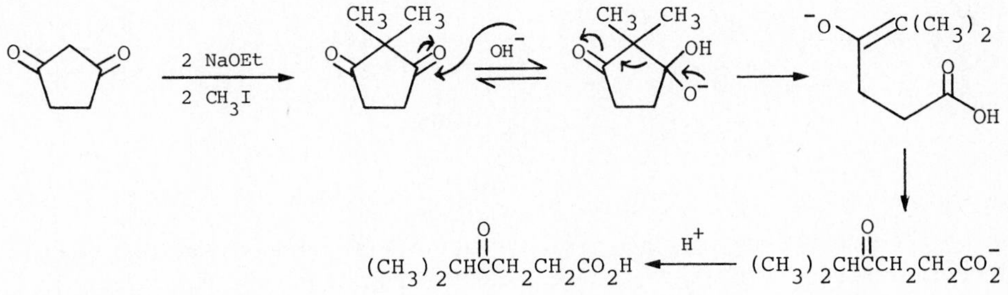

27.D Answers and Explanations to Problems

1.   (a) methyl 4-methyl-3-oxopentanoate
    (b) 2,2-dimethyl-1,3-propanediol
    (c) 4-hydroxybutanoic acid
    (d) 4-oxopentanenitrile

    (e) β-methylglutaric acid;
        3-methylpentanedioic acid
    (f) α-methyladipic acid;
        2-methylhexanedioic acid

(g) dimethyl methylpropylmalonate; dimethyl 2-methyl-2-propylpropanedioate

(h) 3-methyl-5-oxopentanoic acid   (i) 4-hydroxybutanamide

(j) hex-5-en-2-one   (k) 2,2-dimethylcyclohexane-1,4-dione

(l) α-methyl-γ-ketovaleric acid; 2-methyl-4-oxopentanoic acid

(m) glutaronitrile; pentanedinitrile

(n) ε-methyl-γ-ketoenanthaldehyde; 6-methyl-4-oxoheptanal

2.

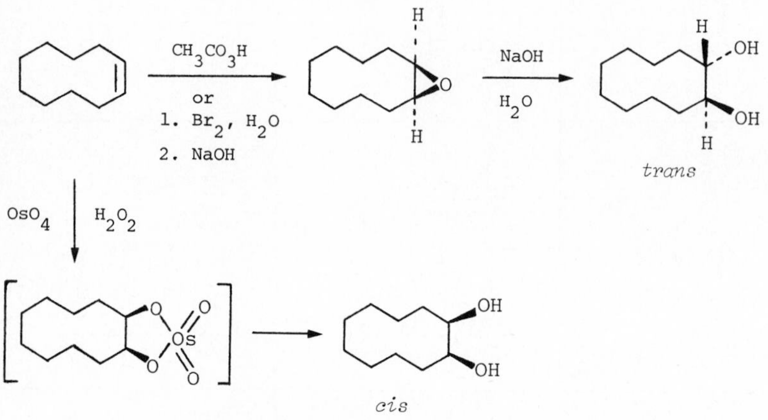

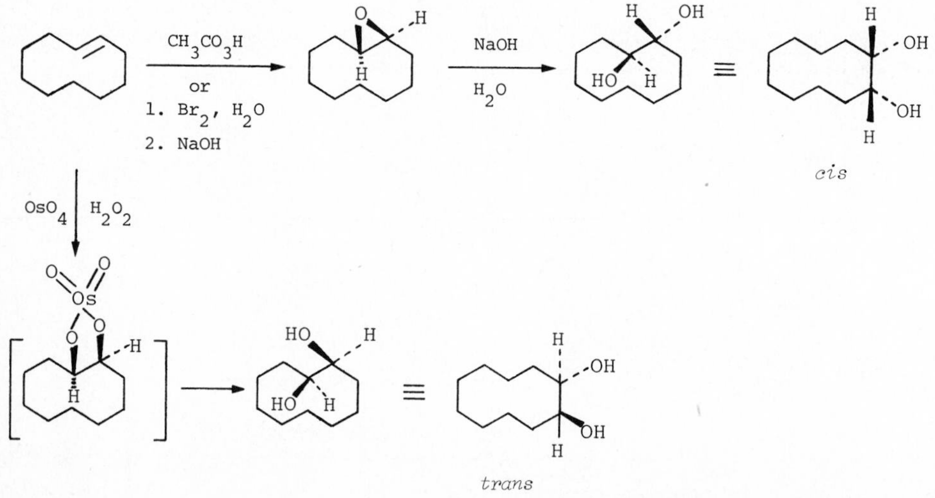

3. Loss of the two hydroxy groups can give a primary carbocation (unfavored) or
    a tertiary carbocation which can be further stabilized by delocalization
    into the two benzene rings.

$$(C_6H_5)_2\overset{OH}{\underset{|}{C}}CH_2OH \overset{H^+}{\underset{}{\rightleftarrows}} (C_6H_5)_2\overset{OH}{\underset{|}{C}}CH_2OH_2^+ \;\;\bcancel{\rightleftarrows}\;\; (C_6H_5)_2\overset{OH}{\underset{|}{C}}-CH_2^+$$

resonance-stabilized
tertiary carbocation

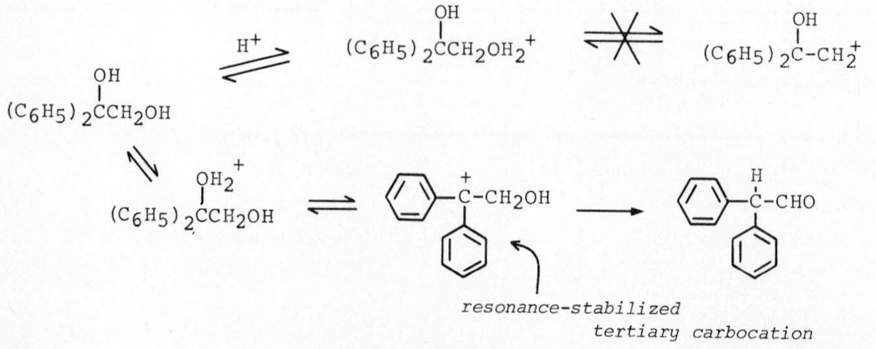

4.   (a)   (b) $HOCH_2-\underset{\underset{CH_3}{|}}{\overset{\overset{CH_3}{|}}{C}}-CHO$ + CH_2O (c)

(d) $\left[O-(CH_2)_6-\overset{\overset{O}{\|}}{C}\right]_n$ (e)   (f)
polymer

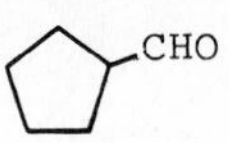

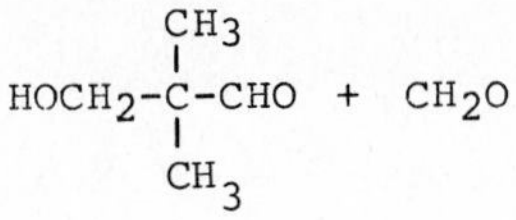

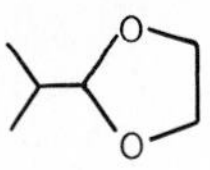

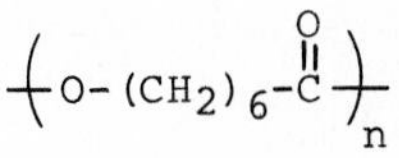

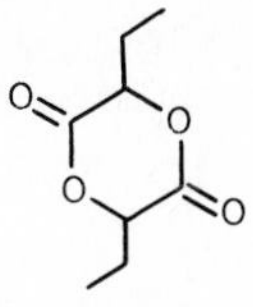

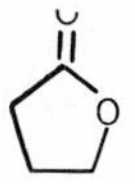

(g)

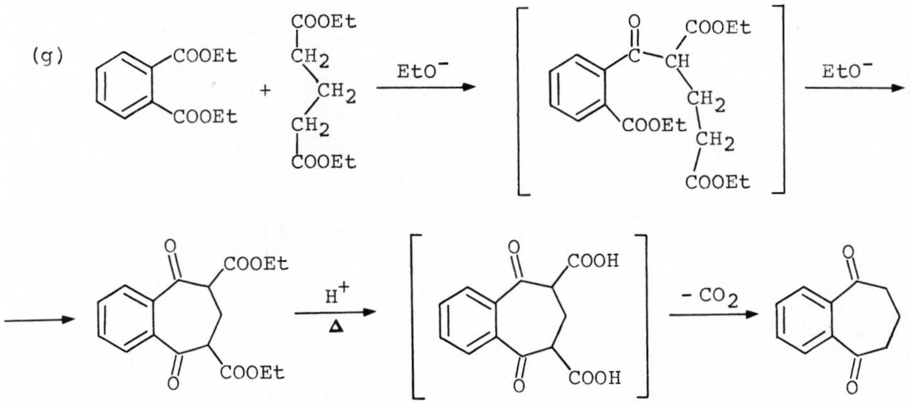

Note that both COOH groups are β-keto acids.

(h)

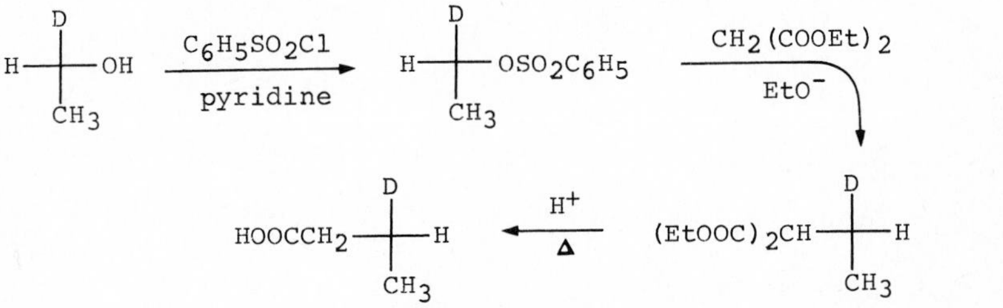

(R)   note inversion of configuration

(i) $(EtOOC)_2CH_2$ + EtO^- ⇌ $(EtOOC)_2\overset{-}{C}H$

    $(EtOOC)_2\overset{-}{C}H$ + C_6H_5CHO ⇌ $(EtOOC)_2CH\overset{\overset{O^-}{|}}{C}HC_6H_5$

    $(EtOOC)_2CH\overset{\overset{O^-}{|}}{C}HC_6H_5$ + $EtOH$ ⇌ $(EtOOC)_2CH\overset{\overset{OH}{|}}{C}HC_6H_5$ + EtO^-

            ⇅

$(EtOOC)_2C=CHC_6H_5$ $\overset{}{\underset{-OH^-}{\longleftarrow}}$ $(EtOOC)_2\overset{-}{C}\overset{\overset{OH}{|}}{C}HC_6H_5$

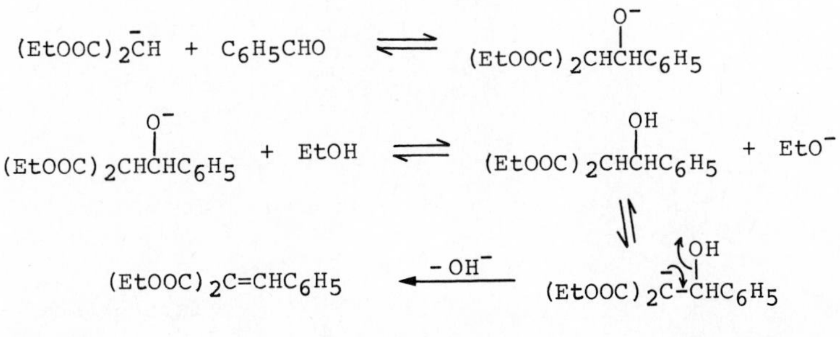

(i) continued next page

(i) (continued)

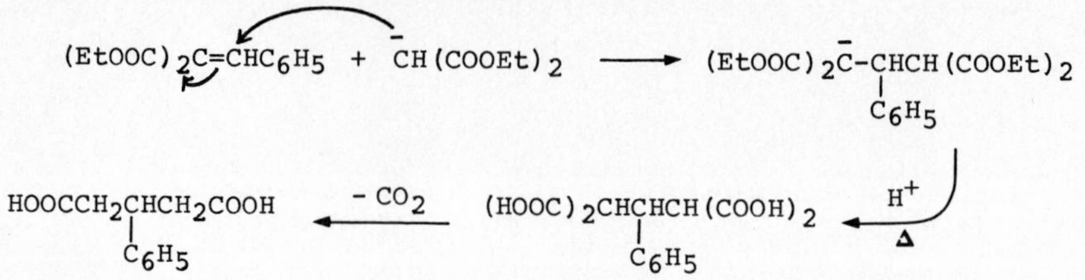

$(EtOOC)_2C=CHC_6H_5$ + $^-CH(COOEt)_2$ $\longrightarrow$ $(EtOOC)_2\overset{-}{C}$-$CHCH(COOEt)_2$
$\overset{|}{C_6H_5}$

$HOOCCH_2CHCH_2COOH$ $\overset{-CO_2}{\longleftarrow}$ $(HOOC)_2CHCHCH(COOH)_2$ $\overset{H^+}{\underset{\Delta}{\quad}}$
$\overset{|}{C_6H_5}$ $\qquad\qquad$ $\overset{|}{C_6H_5}$

(j) $C_6H_5COCH_2COOEt$ $\overset{EtO^-}{\underset{EtI}{\longrightarrow}}$ $C_6H_5COCHCOOEt$ $\overset{EtO^-}{\underset{CH_3I}{\longrightarrow}}$ $C_6H_5COCCOOEt$
$\qquad\qquad\qquad\qquad\qquad\quad \overset{|}{CH_2CH_3} \qquad\qquad\qquad \overset{CH_2CH_3}{\underset{CH_3}{|}}$

$C_6H_5COCHCH_2CH_3$ $\overset{H^+}{\underset{\Delta}{\quad}}$
$\overset{|}{CH_3}$

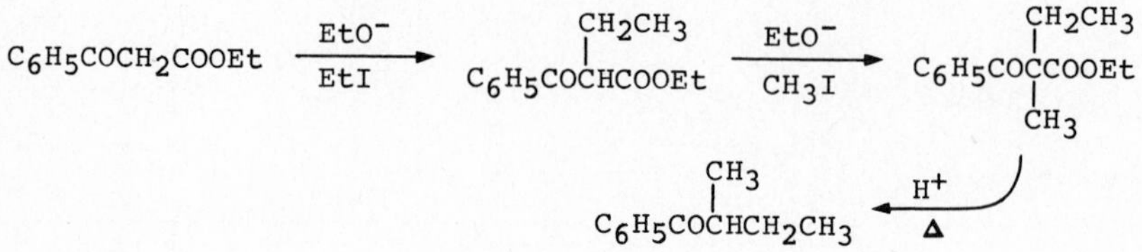

(k) CH_3COOEt + $EtOOCCOOEt$ $\overset{EtO^-}{\longrightarrow}$ $EtOOCCH_2COCOOEt$

5.

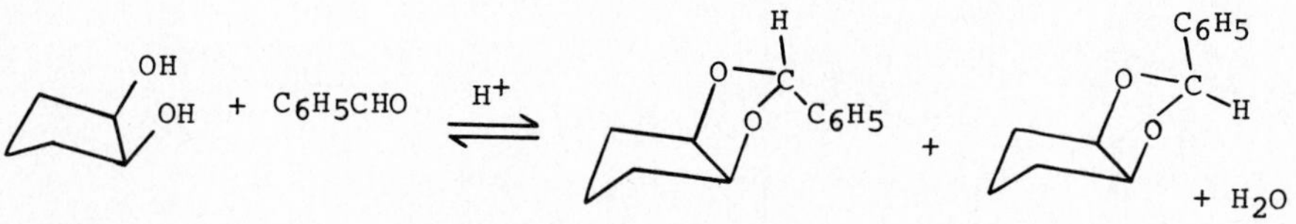

The $MgSO_4$ is a drying agent to remove H_2O and shift the equilibrium to the right.

The trans isomer does not react, since the product is too strained (examine molecular models).

6.   (a)

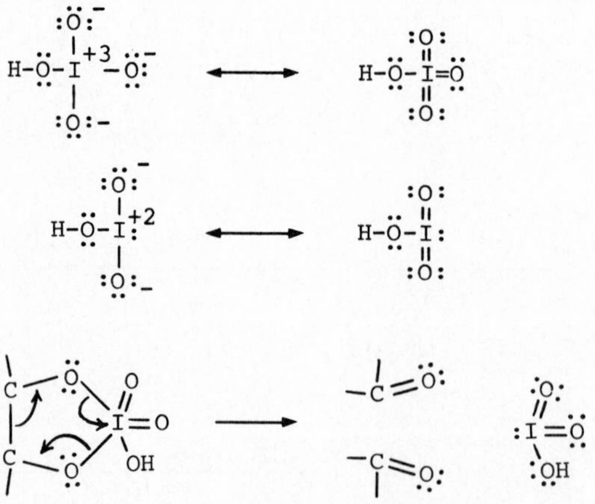

(b)   Ease of formation of the cyclic iodate intermediate with the cis isomer. This is easier to see with molecular models.

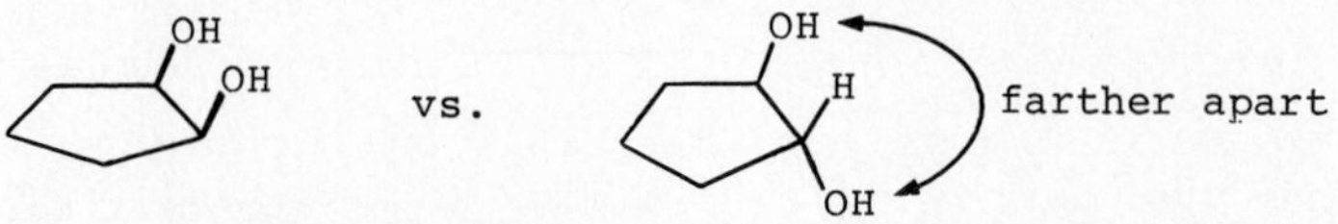

(c) 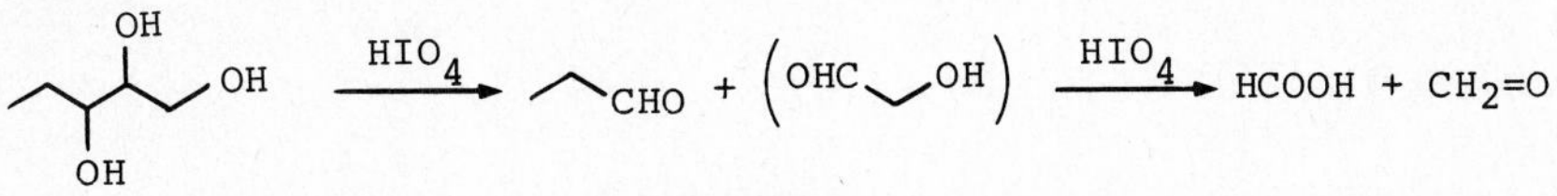

The 1,2,3-isomer will consume <u>two</u> equivalents of HIO_4.
The 1,2,4-isomer only reacts once.

7. (a) 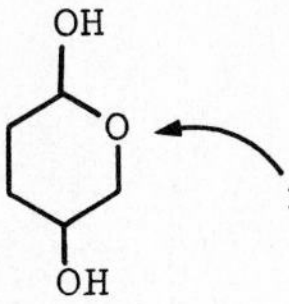

major isomer; six-membered saturated ring is preferred

(b) 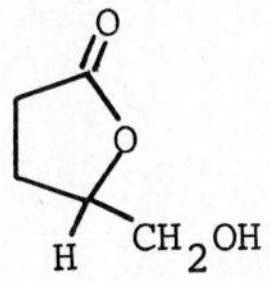

five-membered lactone ring is preferred

8. (a) 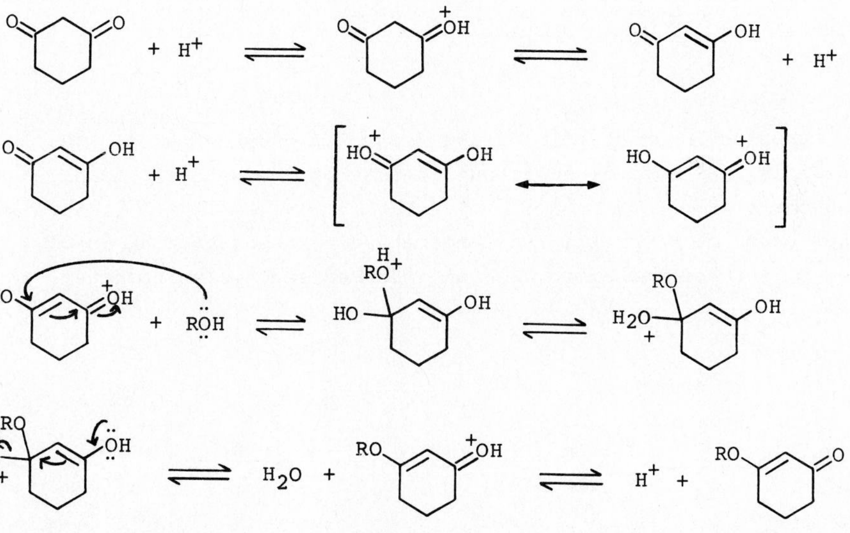

(b) 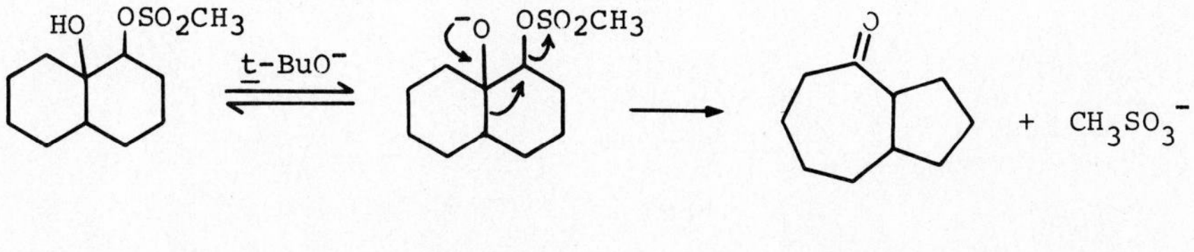

The reaction is analogous to the benzilic acid rearrangement. When the methanesulfonate group ionizes, rearrangement occurs so that the product is a neutral ketone, rather than a zwitterion:

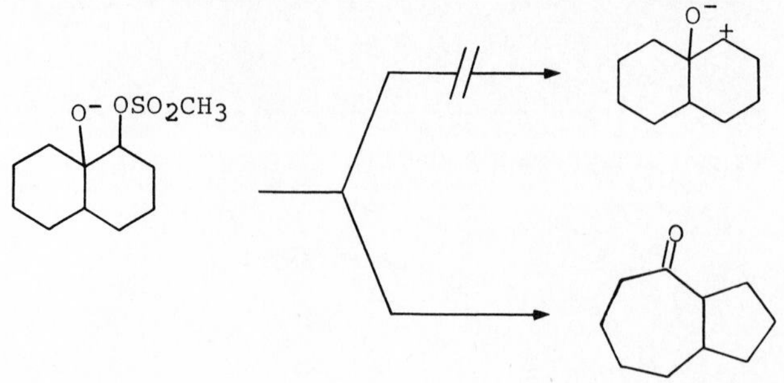

(c)

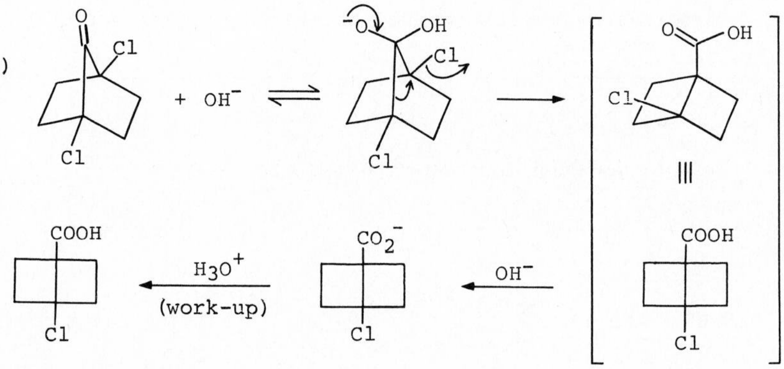

The reaction is an example of the *Favorskii rearrangement.* The mechanism in this case is analogous to that of the benzilic acid rearrangement.

Other cases of Favorskii rearrangement are known which apparently proceed by another mechanism, *via* an intermediate cyclopropanone.

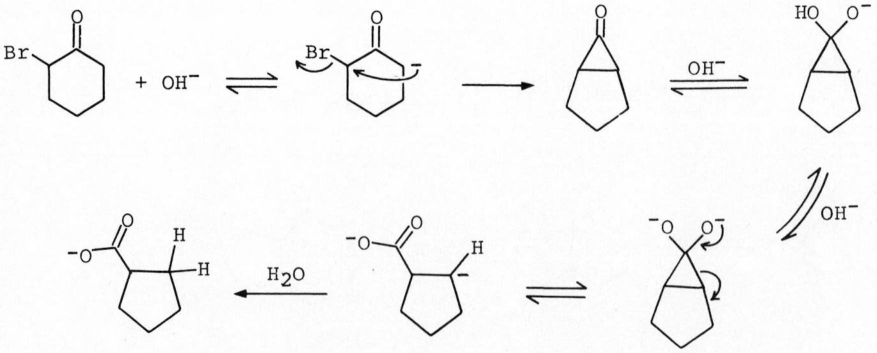

The cyclopropanone mechanism can only operate when the α-haloketone has α' hydrogens. In other cases, the "semi-benzilic" mechanism operates.

(d)

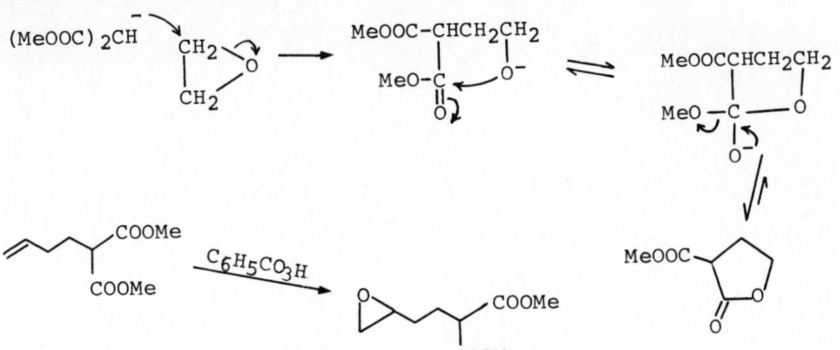

(e)

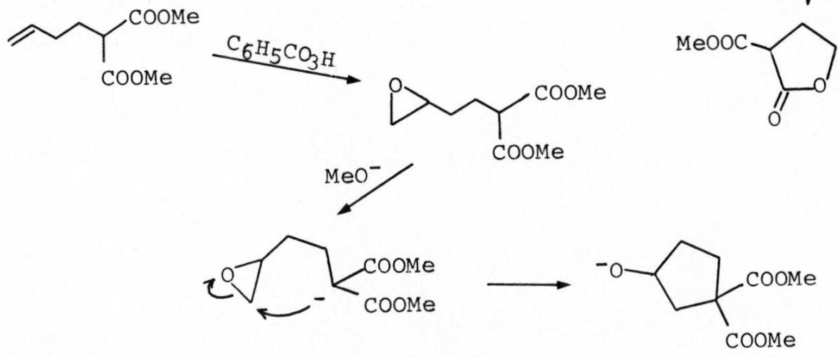

(f)

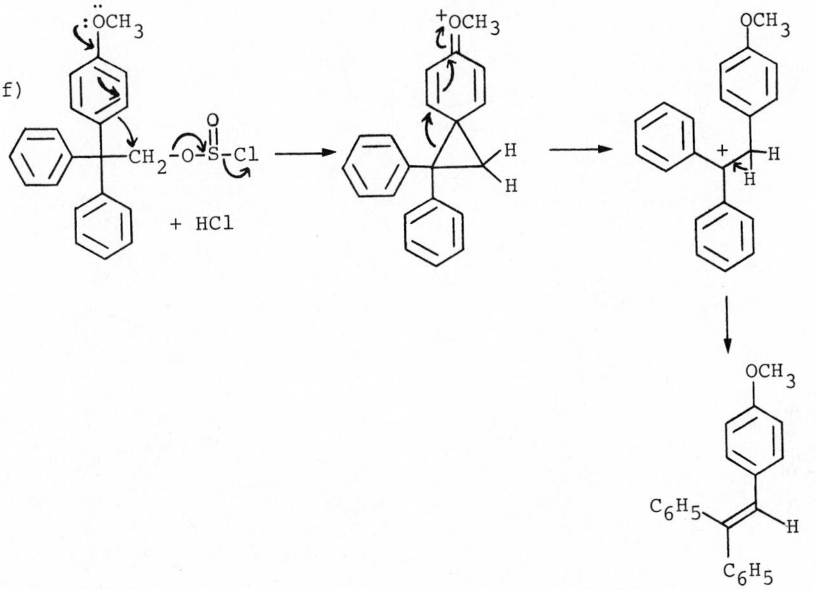

9. $CH_3CO\overset{-}{C}HCOOEt$ + $BrCH_2CH_2CH_2Br$ $\longrightarrow$ $CH_3COCHCOOEt$
                                                                                          |
                                                                                 $CH_2CH_2CH_2Br$

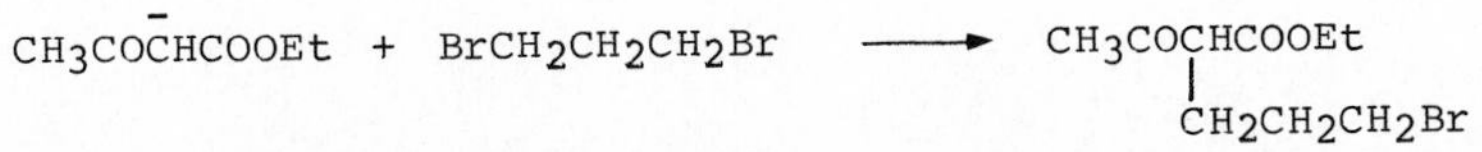

The first alkylation is normal. The second, however, would give a
strained four-membered ring:

however:

negative
charge also
on O

faster

gives six-
membered ring

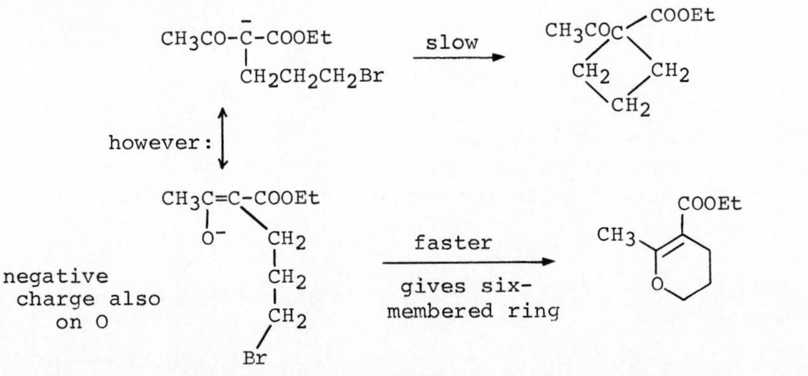

O-alkylation is a side reaction that is normally not important in alkylations of β-keto esters, but it can become dominant with some reagents.

10.

(a) 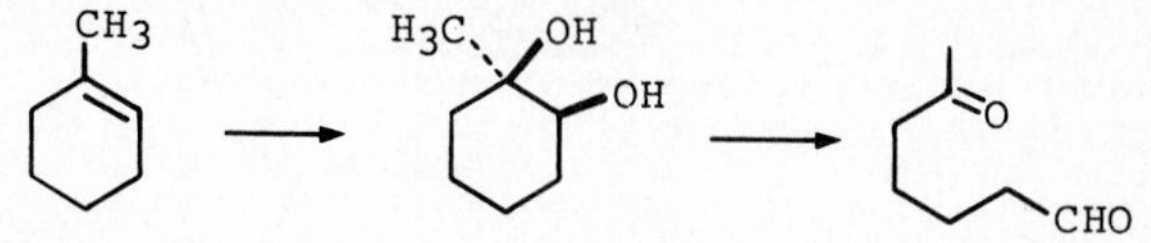

(b) 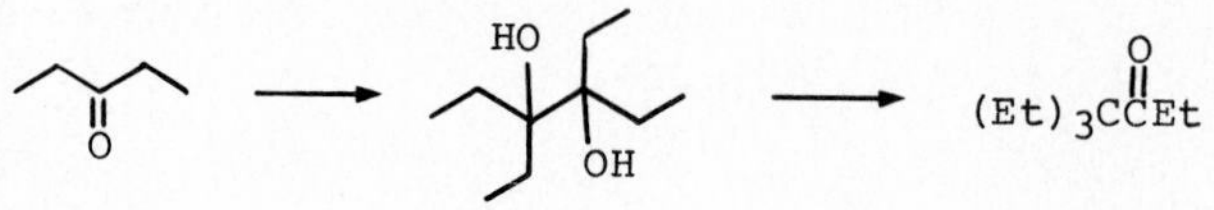

(c) 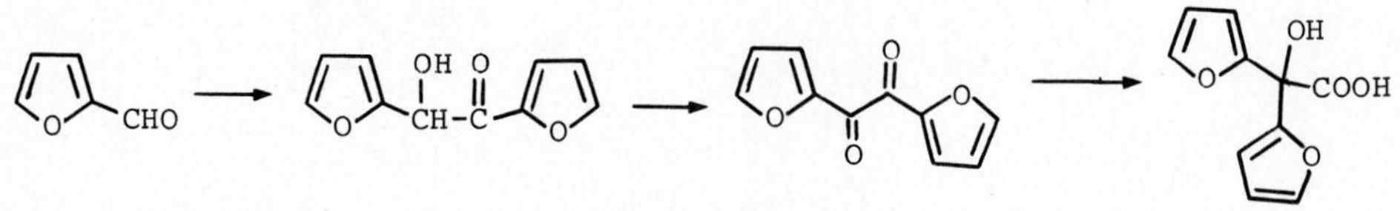

(d) 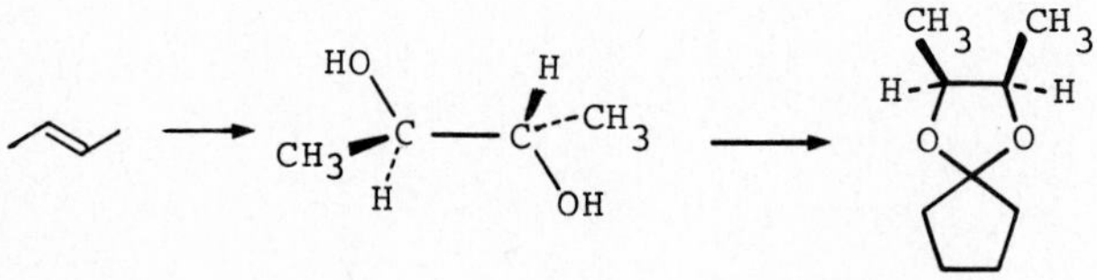

11. 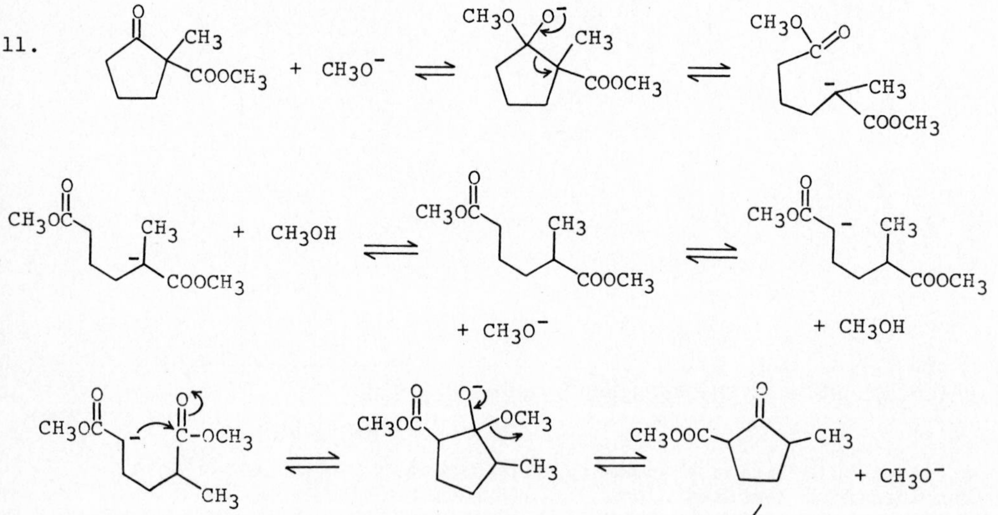

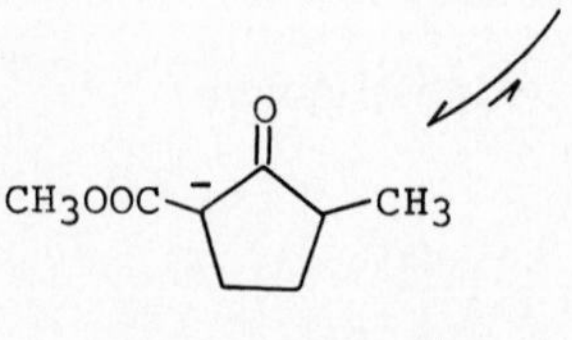

The product ion is a stabilized carbanion (anion of the β-keto ester), and pulls the entire set of equilibria.

12. (a) $CH_3COCH_3 \xrightarrow{Mg} (CH_3)_2\overset{HO}{C}-\overset{OH}{C}(CH_3)_2 \xrightarrow[\Delta]{H^+} (CH_3)_3C\overset{O}{\overset{\|}{C}}CH_3 \xrightarrow{NaOCl} \xrightarrow{H^+} (CH_3)_3CCOOH$

(b) $CH_3CH_2COOH \xrightarrow{SOCl_2} CH_3CH_2COCl \xrightarrow[Pd/BaSO_4]{H_2} CH_3CH_2CHO$

$\xrightarrow{\text{dil. } OH^-} CH_3CH_2CHOHCH(CH_3)CHO \xrightarrow{Ag_2O} CH_3CH_2CHOHCH(CH_3)COOH$

or: $CH_3CH_2COOH \xrightarrow[Br_2]{PBr_3} CH_3CHBrCOBr \xrightarrow{CH_3OH} CH_3CHBrCOOCH_3$

$\xrightarrow[CH_3CH_2CHO]{Zn} CH_3CH_2CHOHCH(CH_3)COOCH_3 \xrightarrow[H_2O]{OH^-} CH_3CH_2CHOHCH(CH_3)COOH$

(c) (d) (e) $CH_3CH_2COOH \xrightarrow[H^+]{MeOH} CH_3CH_2COOCH_3 \xrightarrow[ether]{Na} \xrightarrow{H^+} CH_3CH_2CHOHCOCH_2CH_3$

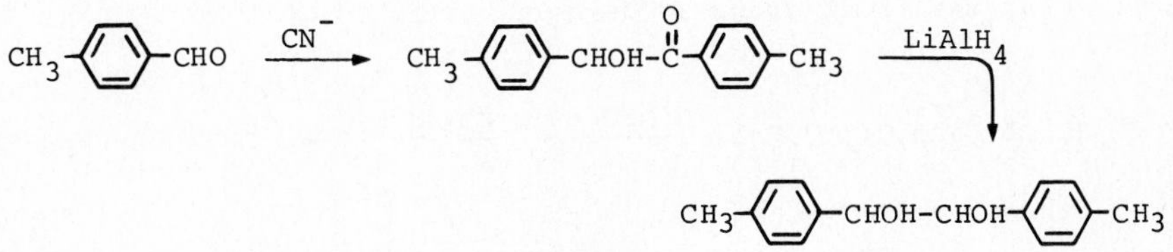

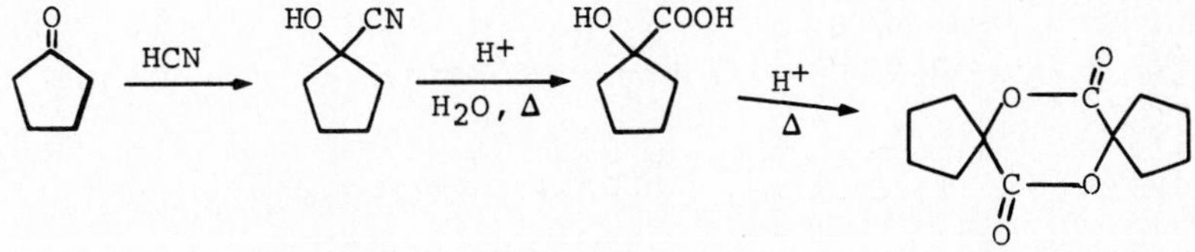

$\xrightarrow{LiAlH_4} CH_3CH_2CHOHCHOHCH_2CH_3$

(f) (g) $C_6H_5CH_2COOEt \xrightarrow[NaOEt]{CO(OEt)_2} C_6H_5CH(COOEt)_2 \xleftarrow{H^+} C_6H_5\bar{C}(COOEt)_2$

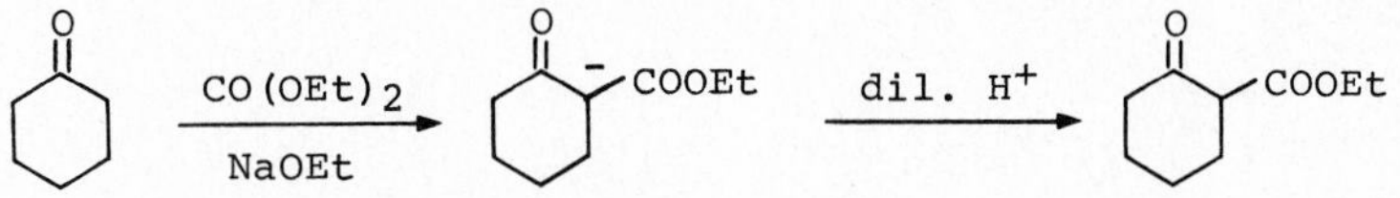

$C_6H_5CH_2COOEt \xrightleftharpoons{NaOEt} C_6H_5\bar{C}HCOOEt \xrightarrow{EtOCOOEt} C_6H_5CH(COOEt)\text{—}C(OEt)_2O^- \rightleftharpoons C_6H_5CH(COOEt)COOEt + EtO^- \rightleftharpoons C_6H_5\bar{C}(COOEt)_2$

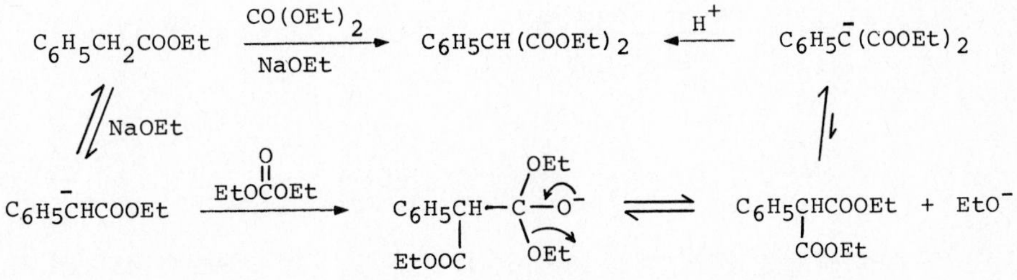

(h) $C_6H_5CH(COOEt)_2 + EtO^- \rightleftharpoons C_6H_5\bar{C}(COOEt)_2 \xrightarrow{ClCH_2COOEt} C_6H_5C(COOEt)_2(CH_2COOEt) + Cl^-$

$\xrightarrow[2.\ H^+]{1.\ OH^-} C_6H_5C(COOH)_2(CH_2COOH) \xrightarrow[-CO_2]{\Delta} C_6H_5CH(COOH)CH_2COOH$

13. (a) $(CH_3)_2CHCH_2Br + CH_2(COOEt)_2 \xrightarrow{NaOEt} \xrightarrow{OH^-} \xrightarrow[\Delta]{H^+} (CH_3)_2CHCH_2CH_2COOH$

(b) a methylallylacetic acid:

$CH_2{=}CHCH_2Br + CH_2(COOEt)_2 \xrightarrow{NaOEt} \xrightarrow[CH_3I]{NaOEt} CH_2{=}CHCH_2C(CH_3)(COOEt)_2$

$\xrightarrow{OH^-} \xrightarrow[\Delta]{H^+} CH_2{=}CHCH_2CH(COOH)CH_3$

Note that the allyl group is added first. Some dialkylation occurs as a side reaction. Separation of mono- from di-alkylated ester is easier when the first group is larger.

(c) $CH_2(COOEt)_2 + CH_2{=}C(CH_3)CH_2Cl \xrightarrow{NaOEt} \xrightarrow{OH^-} \xrightarrow[\Delta]{H^+} CH_2{=}C(CH_3)CH_2CH_2COOH$

(d) 1,5-keto acid; Michael addition:

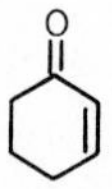 + $CH_2(COOEt)_2 \xrightarrow{NaOEt}$ etc.

(e) $EtOOCCH{=}CH_2 + CH_2(COOEt)_2 \xrightarrow{NaOEt} EtOOCCH_2CH_2CH(COOEt)_2$

$\xrightarrow[\Delta]{H^+} HOOCCH_2CH_2CH_2COOH$

or $CH_2(COOEt)_2 + CH_2O$:

$(EtOOC)_2CH^- + CH_2{=}O \longrightarrow (EtOOC)_2CHCH_2O^- \rightleftharpoons (EtOOC)_2C^-CH_2OH$

$\rightleftharpoons (EtOOC)_2C{=}CH_2 + OH^-$

$(EtOOC)_2C{=}CH_2 \xrightarrow{(EtOOC)_2CH^-} (EtOOC)_2C^-CH_2CH(COOEt)_2$

$\xrightarrow{H^+} (EtOOC)_2CHCH_2CH(COOEt)_2 \xrightarrow{OH^-} \xrightarrow[\Delta]{H^+} HOOCCH_2CH_2CH_2COOH$

(f) $CH_2(COOEt)_2 \xrightarrow[Br(CH_2)_5Br]{NaOEt\ (2\ moles)} [(EtOOC)_2CHCH_2CH_2CH_2CH_2CH_2Br]$

$\longrightarrow$ 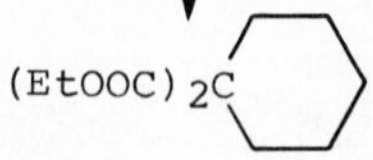 $\xrightarrow{OH^-} \xrightarrow[\Delta]{H^+}$ 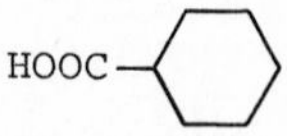

14. (a) "isopentylacetone", ∴ $CH_3COCH_2COOEt + BrCH_2CH_2CH(CH_3)_2$

$\xrightarrow{NaOEt} CH_3COCH(COOEt)CH_2CH_2CH(CH_3)_2 \xrightarrow{OH^-} \xrightarrow[\Delta]{H^+} CH_3CO(CH_2)_3CH(CH_3)_2$

(b) a 1,5-keto acid, ∴ Michael addition:

$CH_3COCH_2COOEt + CH_2{=}CHCOOEt \xrightarrow{NaOEt} CH_3COCH(COOEt)CH_2CH_2COOEt \longrightarrow$ etc.

(c) "propylethylacetone", ∴

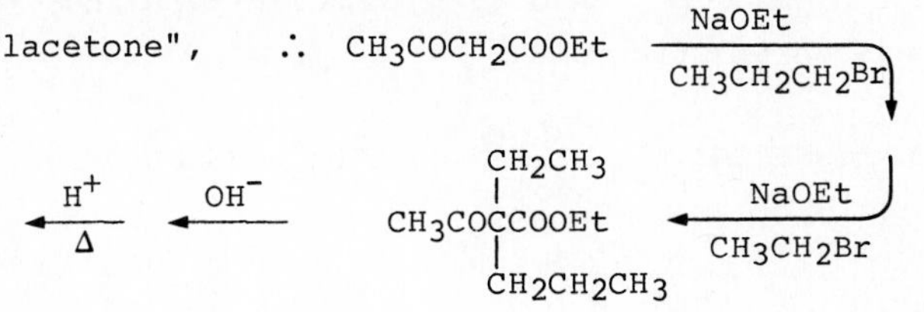

(d) Note symmetrical nature; both ends can be attached by an acetoacetic ester alkylation:

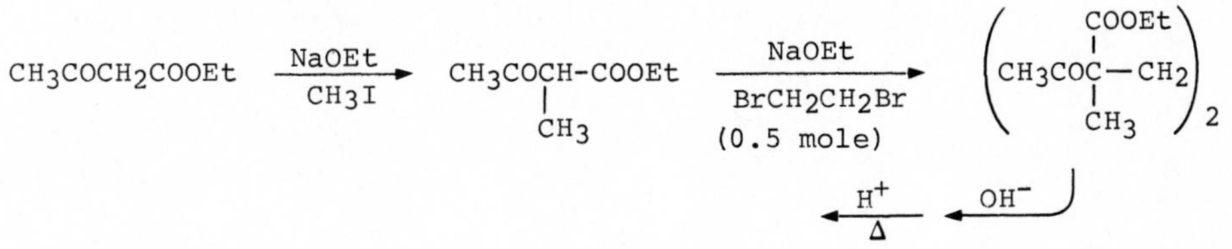

15. (a) This product is a substituted succinic acid. Two possible routes:

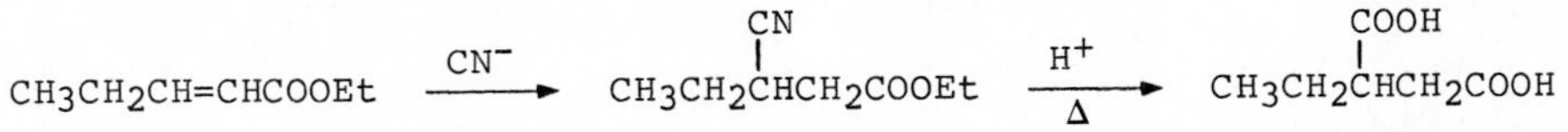

or

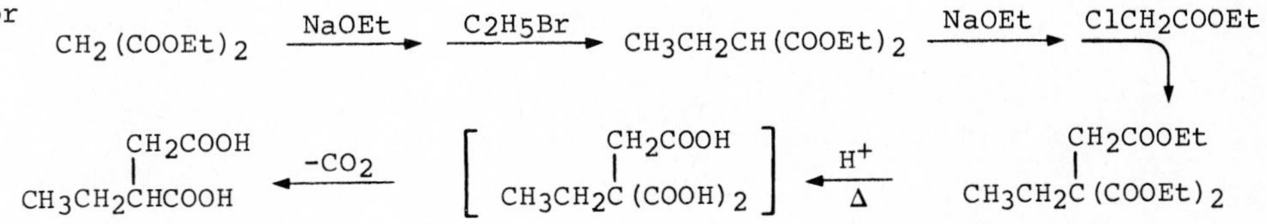

(b)

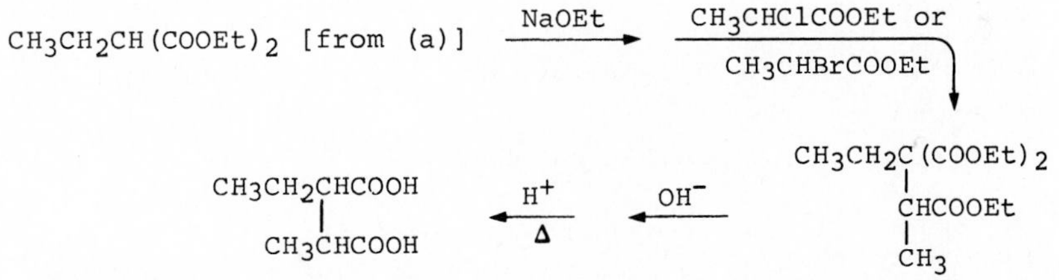

Note that a mixture of two diastereomers results.

(c)

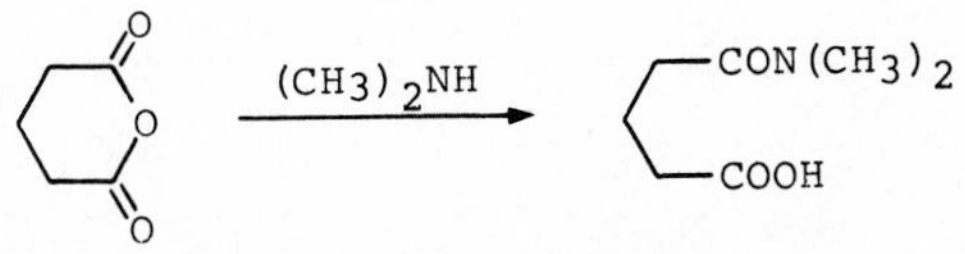

(d)

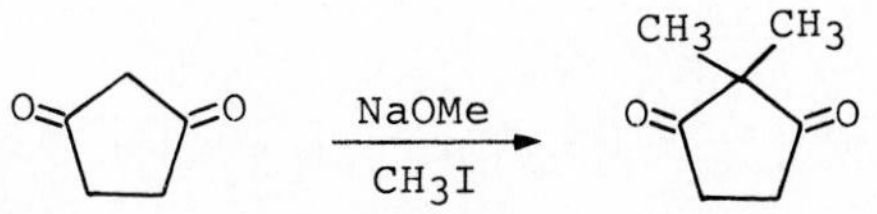

(e) This compound obviously can be made from $(CH_3)_2CHCOCH_2COOEt$ + $ClCH_2COOEt \xrightarrow{EtO^-}$ etc. The problem is to make the β-keto ester.

It cannot be made by the Claisen condensation, but it is available by a sequence starting with:

$$CH_3COCH(CH_3)_2 \xrightarrow[\substack{EtOH \\ HCOOEt}]{NaOEt} (CH_3)_2CHCOCH_2CHO$$

[why must the condensation go in the methyl group?]

$$ester \xleftarrow[\substack{or\ CH_2N_2}]{EtOH/H^+} R\text{-}COOH \xleftarrow{Ag_2O}$$

(f)

$$\text{(cyclopentanone)} + BrCH_2\overset{O}{\overset{\|}{C}}OCH_3 \xrightarrow{Zn} \xrightarrow{H_3O^+} \text{(cyclopentane-OH, CH_2CO_2CH_3)} \xrightarrow[\substack{H_2O}]{OH^-}$$

$$\xleftarrow{H_3O^+} \text{(cyclopentane-OH, CH_2COOH)}$$

(g)

$$\text{(}CO_2CH_3\text{)} \xrightarrow[toluene]{Na} \text{(H, OH, O structure)} \xrightarrow{Cu(OAc)_2} \text{(diketone)}$$

(h)

$$\text{(ketone)} \xrightarrow{Mg} \xrightarrow{H_2O} (CH_2CH_3)_2\overset{HO\ \ OH}{\overset{|\ \ \ |}{C\text{-}C}}(CH_2CH_3)_2 \xrightarrow{H^+} (CH_3CH_2)_3C\text{-}\overset{O}{\overset{\|}{C}}CH_2CH_3$$

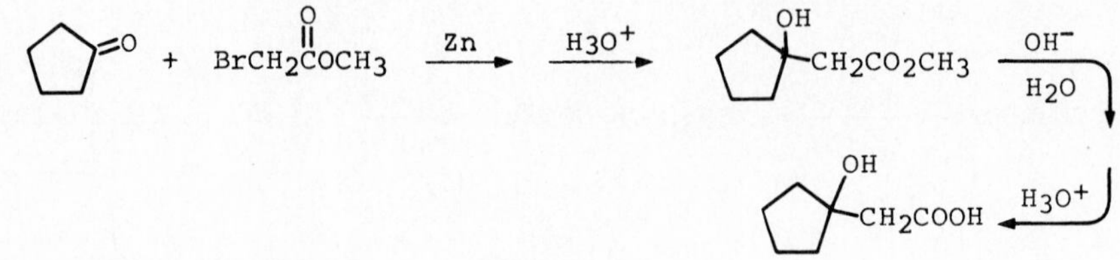

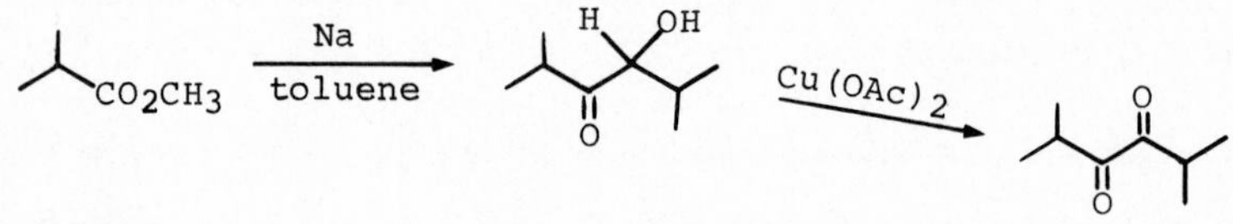

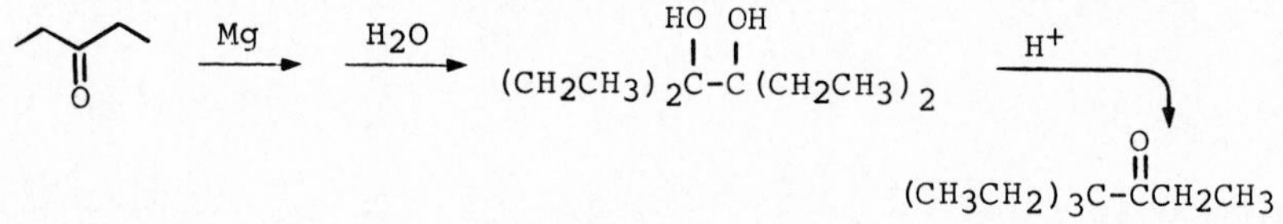

(i) This is a 1,6-dicarbonyl compound for which none of the special methods applies. Thus, treat it as a methyl ketone:

$$\boxed{CH_3\overset{O}{\overset{\|}{C}}\text{-}CH_2}\text{-}(CH_2)_3COOH$$

$$CH_3CO\overset{-}{C}HCOOEt + Br(CH_2)_3COOCH_3 \longrightarrow CH_3CO\overset{COOEt}{\overset{|}{C}}H(CH_2)_3COOMe$$

$$CH_3COCH_2(CH_2)_3COOH \xleftarrow{H^+,\Delta}$$

The bromoester can be made from commercially available γ-butyro-lactone or by:

$$\overset{O}{\underset{CH_2\text{-}CH_2}{\triangle}} + CH_2(COOEt)_2 \xrightarrow[\Delta]{EtO^-} \xrightarrow{H^+} \text{(lactone)} \xrightarrow[\substack{MeOH \\ \Delta}]{HBr} Br(CH_2)_3COOMe$$

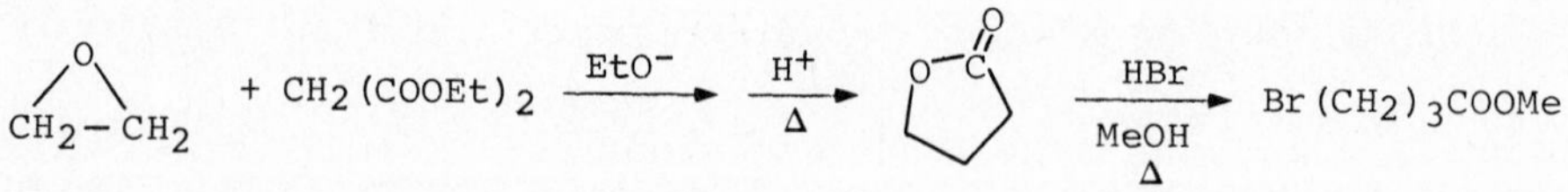

(j) General approach:

$$C\text{-}C\text{-}C\text{-}\overset{O}{\overset{\|}{C}}\text{-}\overset{COOR}{\overset{\vdots}{C}}\text{-}C\text{-}C \longrightarrow C\text{-}C\text{-}C\text{-}\overset{O}{\overset{\|}{C}}\text{-}\overset{\overset{|}{C}}{C}\text{-}C\text{-}C$$

Note: symmetrical ketone;
∴available from Claisen condensation

$$2\ CH_3CH_2CH_2COOEt \xrightarrow{EtO^-} CH_3CH_2CH_2\overset{O}{\overset{\|}{C}}\text{-}\overset{COOEt}{\overset{|}{C}}HCH_2CH_3 \xrightarrow{NaOEt} \xrightarrow{CH_3I}$$

$$CH_3CH_2CH_2CO\overset{}{\overset{}{C}}HCH_2CH_3 \xleftarrow[\Delta]{H^+} \xleftarrow{OH^-} CH_3CH_2CH_2\overset{O}{\overset{\|}{C}}\text{-}\overset{COOEt}{\overset{|}{C}}CH_2CH_3$$
$$\underset{CH_3}{|} \qquad\qquad\qquad\qquad\qquad\qquad\qquad \underset{CH_3}{|}$$

(k) α-keto esters are available by Claisen condensations with diethyl oxalate:

$$CH_3CH_2CH_2CH_2COOEt + EtOOCCOOEt \xrightarrow{NaOEt} CH_3CH_2CH_2\overset{\overset{\displaystyle COOEt}{|}}{CH}COCOOEt$$

$$CH_3CH_2CH_2CH_2COCOOH \xleftarrow[-CO_2]{\Delta} \left[CH_3CH_2CH_2\overset{\overset{\displaystyle COOH}{|}}{CH}COCOOH \right] \xleftarrow[\Delta]{H^+} \xleftarrow{OH^-}$$

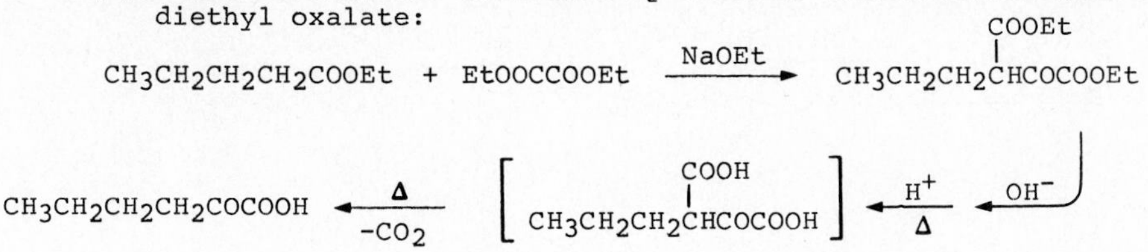

(an α-keto acid and a β-keto acid;
β-COOH decarboxylates)

(l) Substituted glutaric acid available by Michael addition:

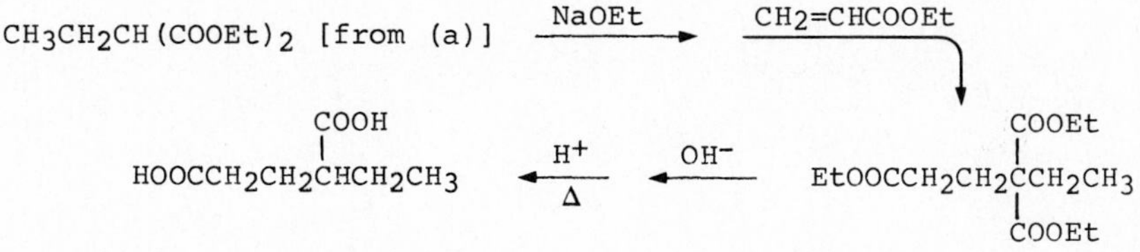

$$CH_3CH_2CH(COOEt)_2 \text{ [from (a)]} \xrightarrow{NaOEt} CH_2=CHCOOEt$$

$$HOOCCH_2CH_2\overset{\overset{\displaystyle COOH}{|}}{CH}CH_2CH_3 \xleftarrow[\Delta]{H^+} \xleftarrow{OH^-} EtOOCCH_2CH_2\overset{\overset{\displaystyle COOEt}{|}}{\underset{\underset{\displaystyle COOEt}{|}}{C}}CH_2CH_3$$

(m)

$$(EtOOC)_2CH_2 + \overset{O}{\overset{\diagup\diagdown}{CH_2-CHCH_2CH_3}} \xrightarrow{NaOEt} \left[(EtOOC)_2CHCH_2\overset{\overset{\displaystyle O^-}{|}}{CH}CH_2CH_3 \right]$$

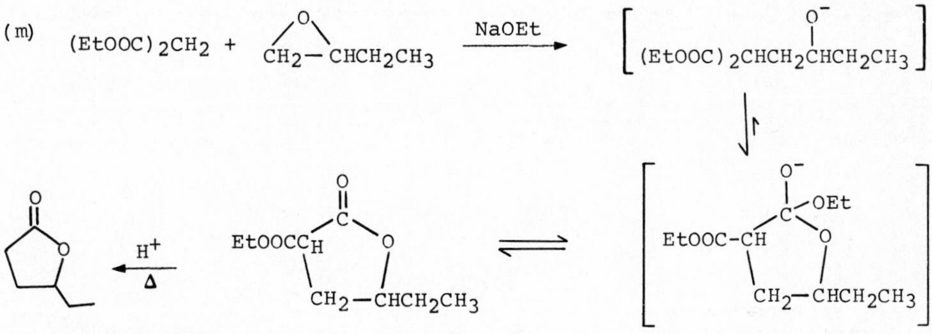

[see problem 8(d)]

(n)

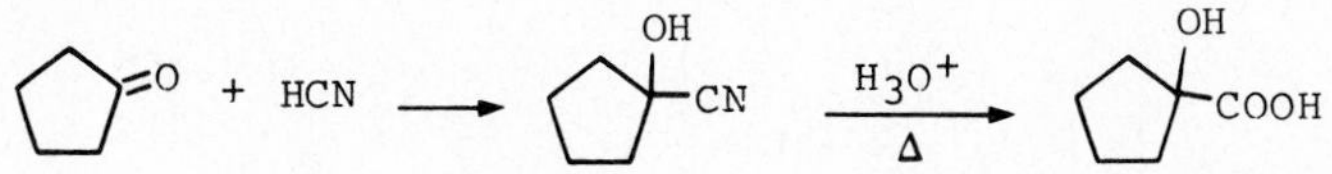

(o)

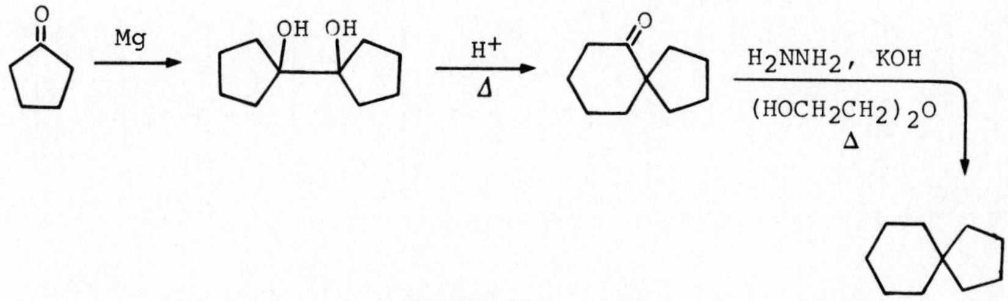

16. $EtO_2CCH_2CO_2Et + 2\ C_6H_5CH_2Cl \xrightarrow[50\% \text{ KOH}]{C_6H_5CH_2\overset{+}{N}Et_3Cl^-} (C_6H_5CH_2)_2C(CO_2Et)$

(In the organic phase, the following equilibrium occurs faster than ester hydrolysis:

$EtO_2CCH_2CO_2Et + OH^- \rightleftharpoons EtO_2C\overset{-}{C}HCO_2Et + H_2O)$

$$\downarrow NaOH\ |\ EtOH$$

$$\uparrow H^+\ |\ \Delta$$

$$(C_6H_5CH_2)_2CHCO_2H$$

17. (a)

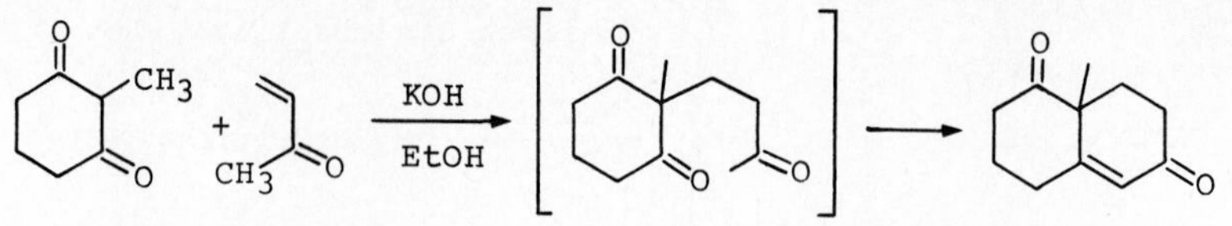

(b)

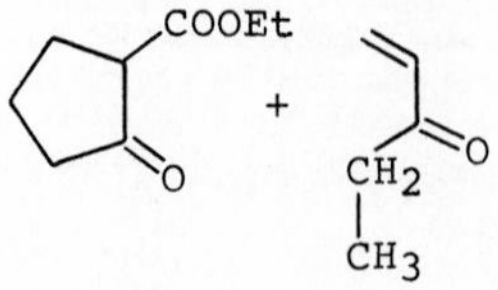

(c)

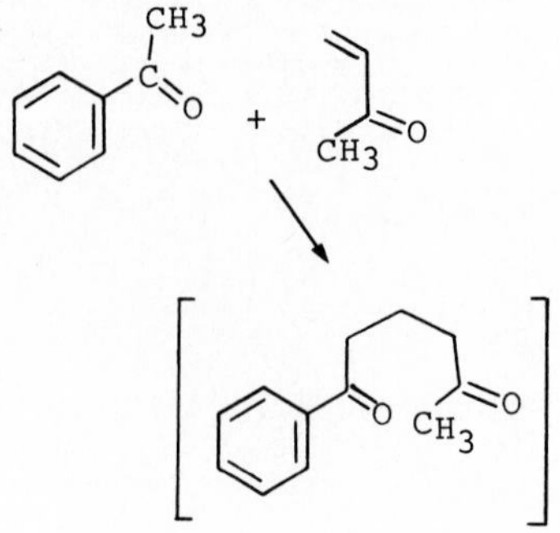

(d)

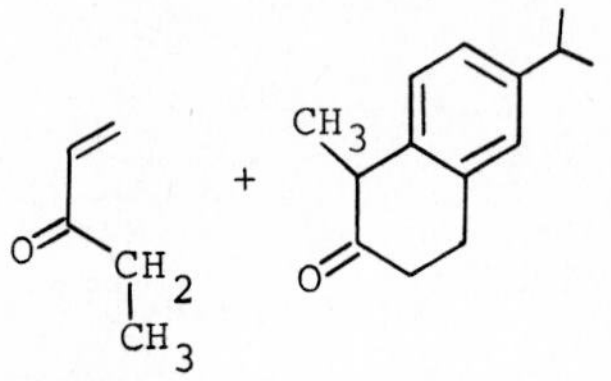

18.

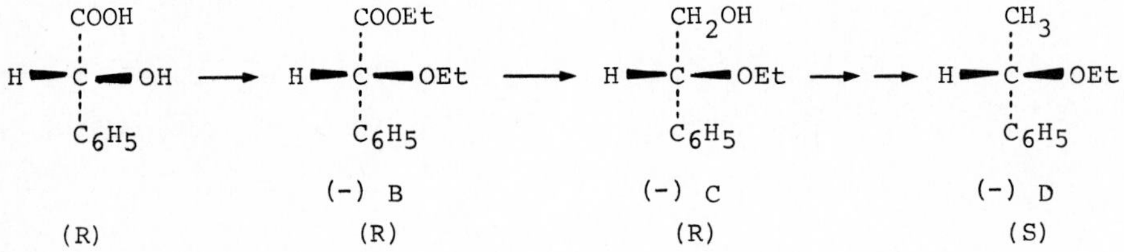

Note inversion of configuration in the conversion (−)E → (−)F

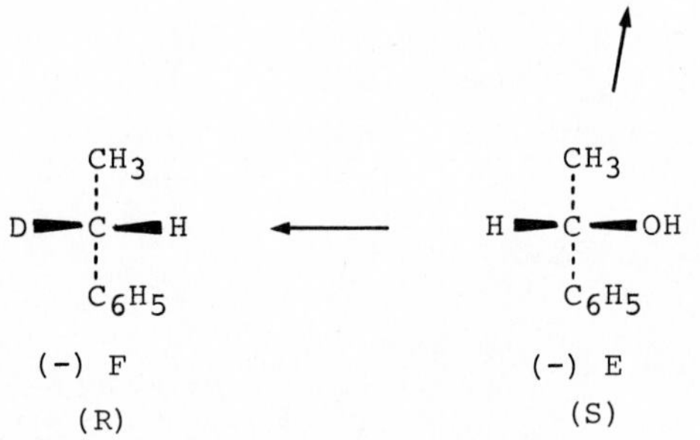

19.

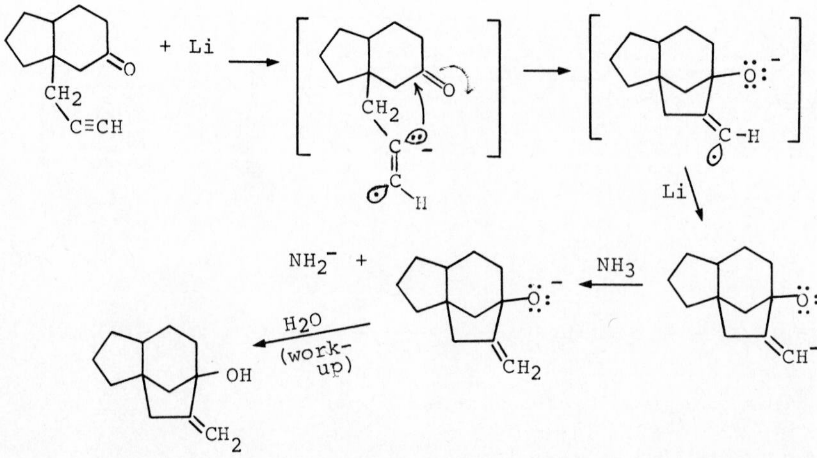

The first step is like the metal-ammonia reduction of an alkyne

(see Section 12.6 A). The initial radical anion undergoes cyclization, with the nucleophilic carbon attacking the carbonyl group.

27.E Supplementary Problems

S1. Write the structure of each of the following compounds:

(a) *trans*-4-cyclopenten-1,3-diol

(b) 4-hydroxy-3-hexanone

(c) 4-oxocyclohexanecarboxaldehyde

(d) (2R,4S)-2,4-dimethylpentanedioic acid

(e) (R)-2,3-dihydroxypropanal

(f) methyl 2-oxocyclopentanecarboxylate

(g) diethyl methylmalonate

(h) 1,1,2,2-tetraphenyl-1,2-ethanediol

S2. Give the major product of each of the following reaction sequences:

(a)

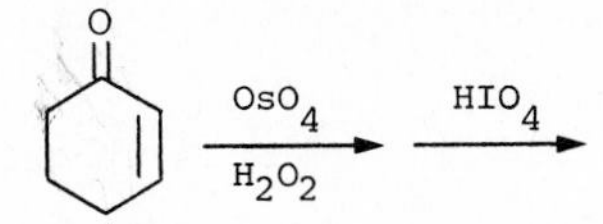

$$\xrightarrow[\text{H}_2\text{O}_2]{\text{OsO}_4} \xrightarrow{\text{HIO}_4}$$

(b) $CH_3CH_2\overset{\overset{\text{O}}{\|}}{C}OEt \xrightarrow[\text{EtOH}]{\text{NaOEt}} \xrightarrow{CH_3CH_2Br} \xrightarrow[\Delta]{\text{H}_3\text{O}^+}$

(c) $EtO_2CCH_2CO_2Et \xrightarrow[\text{NaOEt}]{CH_2=CHCO_2Et} \xrightarrow{\text{H}_3\text{O}^+} \xrightarrow{\Delta}$

(d) $CH_3O_2CCH_2CH_2CO_2CH_3 \xrightarrow{\text{Na}} \xrightarrow{\text{H}_2\text{O}}$

(e)

$$\xrightarrow[CH_3CO_2H]{Co^{II}, \ O_2} \xrightarrow{SOCl_2} \xrightarrow[Et_3N]{H_2N(CH_2)_4NH_2}$$

(f)

$$\xrightarrow{\text{HCN}} \xrightarrow[-\text{H}_2\text{O}]{\text{H}_3\text{O}^+} \xrightarrow{\Delta}$$

(g)

$$\xrightarrow[\substack{\text{cold,}\\ \text{dilute}}]{\text{KMnO}_4} \xrightarrow{\text{H}^+}$$

(h)

$$\xrightarrow{\text{H}^+}$$

(i)

$$\xrightarrow{\text{H}^+, \ \Delta}$$

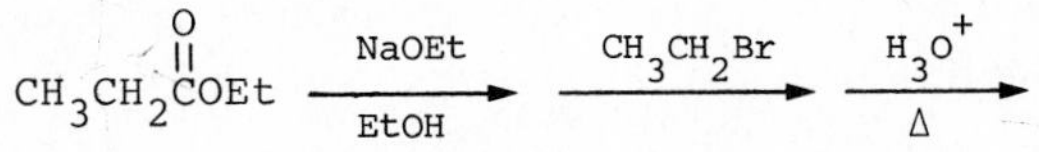

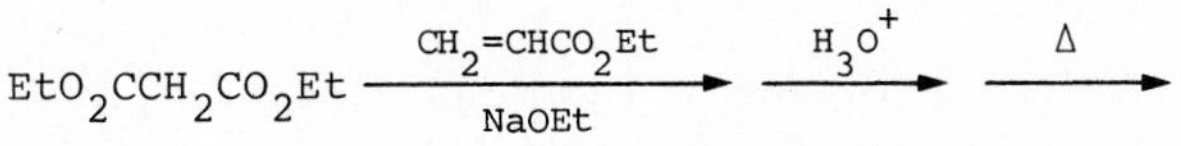

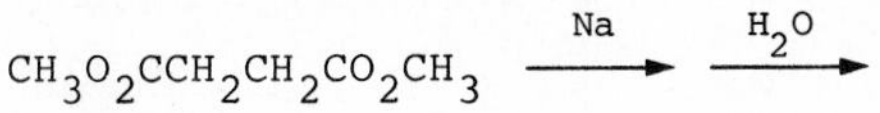

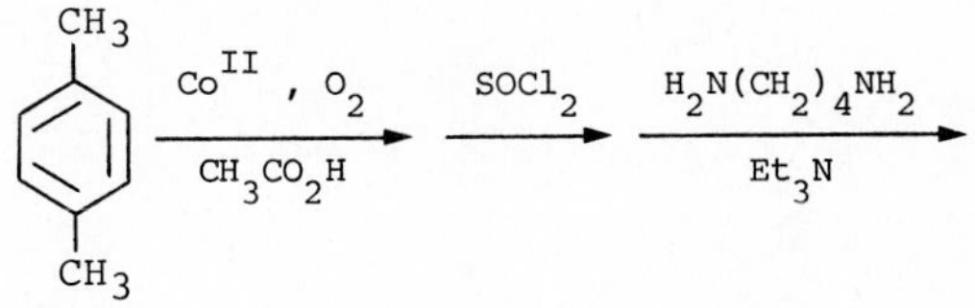

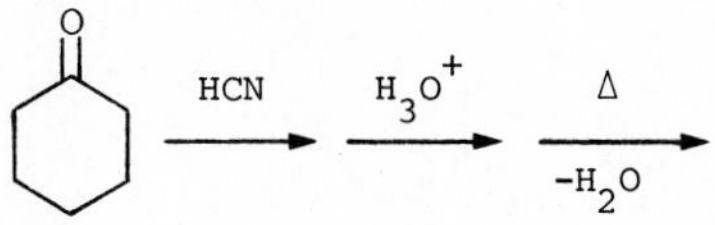

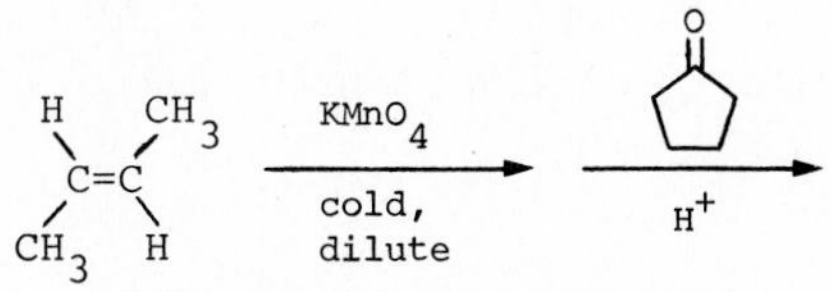

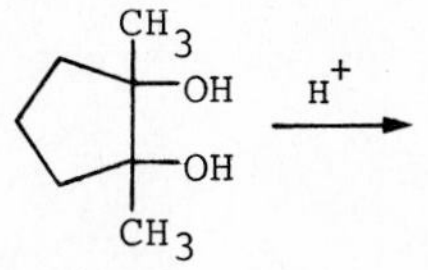

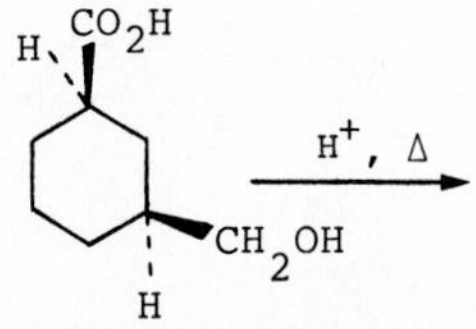

S3. Propose efficient syntheses of the following compounds, using any starting material containing five carbons or less:

(a) $CH_3CH_2\overset{}{\underset{}{C}}H\overset{}{C}H\overset{}{C}HCH_2CH_3$ with OH, OH and CH₃ substituents

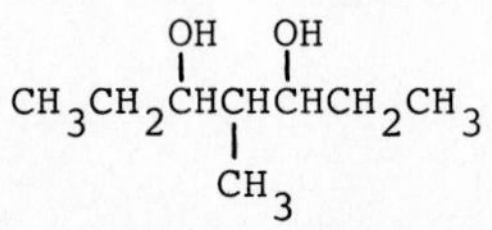

(b)

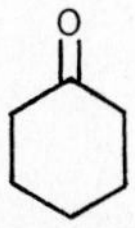

(c)

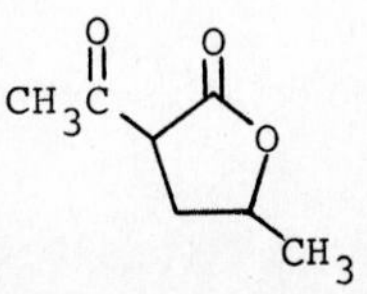

(d) $(CH_3CH_2)_3\overset{O}{\overset{||}{C}}CH_2CH_3$

(e) 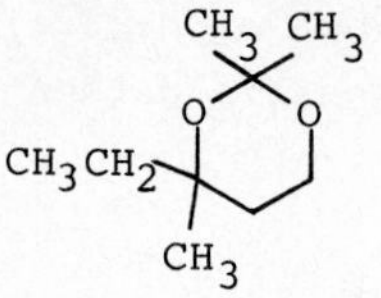

(f) 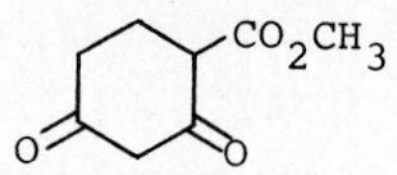

(g) $(CH_3)_2CHCH_2\overset{O}{\overset{||}{C}}\underset{O}{\underset{||}{C}}CH_2CH(CH_3)_2$

(h) 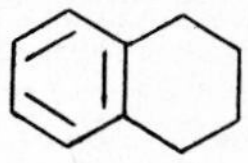

(You can use benzene as a starting material for this one.)

(i) 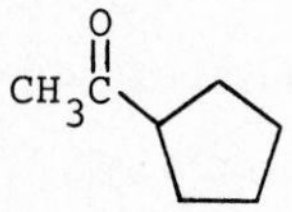

(j) 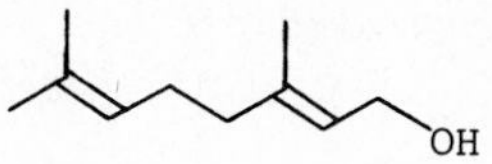

geraniol (odor of geraniums)

S4. Carbamates can be synthesized from an alcohol, NaN=C=O, and acid:

$ROH + NaNCO \xrightarrow{H^+} RO\overset{O}{\overset{||}{C}}NH_2$. Given this reaction, show how to synthesize the tranquilizer meprobamate:

$$NH_2\overset{O}{\overset{||}{C}}OCH_2\underset{CH_2CH_3}{\overset{CH_3}{C}}CH_2O\overset{O}{\overset{||}{C}}NH_2$$

S5. Provide structures which are consistent with the information given below.

$$\underset{C_4H_6O_4}{B} \xleftarrow[I_2]{NaOH} \underset{C_5H_8O_3}{A} \xrightarrow[CH_3OH]{H^+} \underset{C_8H_{16}O_4}{C} \xrightarrow{LiAlH_4} \underset{C_7H_{16}O_3}{D} \xrightarrow{H^+\text{ (catalytic)}} E$$

B: NMR: δ2.3(s, area 2), δ12(s, area 1)

A: IR (dilute solution): 1710, 1760, 2400-3400 cm^{-1}

D: IR: 1050, 1100, 3400cm^{-1}

E: IR: 1070, 1120 cm^{-1}; mass spectrum: $M^+ = 116$, major fragment at m/e 101

S6. The benzilic acid rearrangement is a useful way to contract the ring of a cyclic α-diketone. Show how it could be applied in a synthesis of bicyclo[2.1.1]-5-hexanone from bicyclo[2.2.1]-2-heptanone.

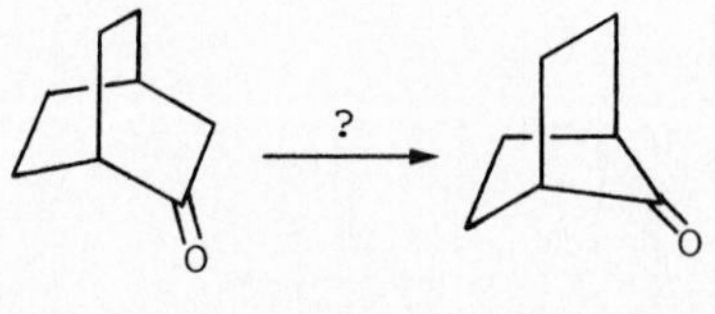

S7. α-Cyanocarboxylic acids also lose carbon dioxide on heating, although less readily than malonic or β-keto acids. Write a mechanism for this decarboxylation, showing all the intermediates involved.

$$N{\equiv}CCH_2CO_2H \xrightarrow{\Delta} N{\equiv}CCH_3 + CO_2$$

S8. Write mechanisms for each of the following transformations, showing all
     the steps involved.

(a) $2\ (EtO_2C)_2CH_2 + CH_2=O \xrightarrow[\text{EtOH}]{\text{NaOEt}} (EtO_2C)_2CHCH_2CH(CO_2Et)_2$

(b) $HOCH_2CH_2CH_2\overset{\overset{\displaystyle O}{\|}}{C}CH_2COCH_3 \xrightarrow{\ H^+\ }$

$+\ H_2O$

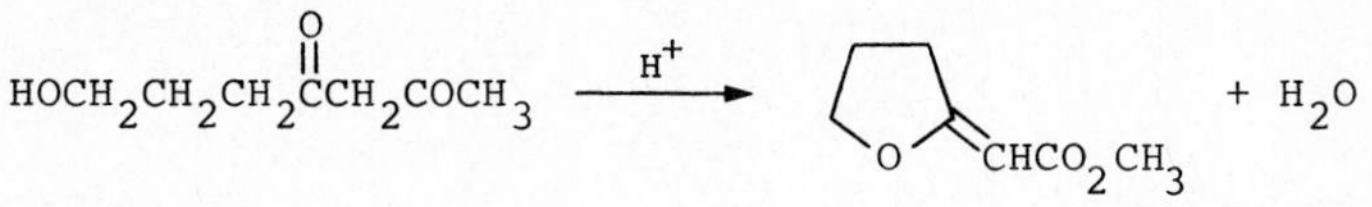

S9. The molecule shown below is very unstable, rapidly decomposing with loss
     of CO_2. What is the product and how is it formed?

$\longrightarrow\ ?\ +\ CO_2$

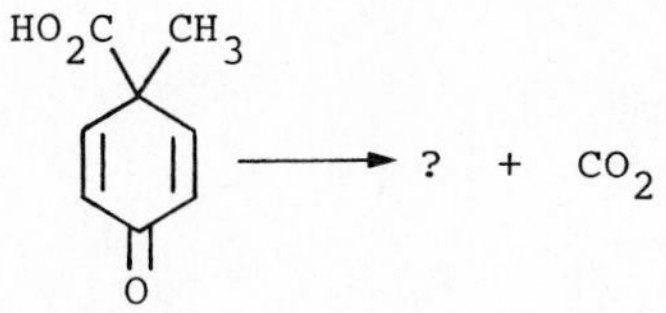

S10. A byproduct which can sometimes be isolated during the course of a Robin-
     son annelation is the β-hydroxy ketone shown below.

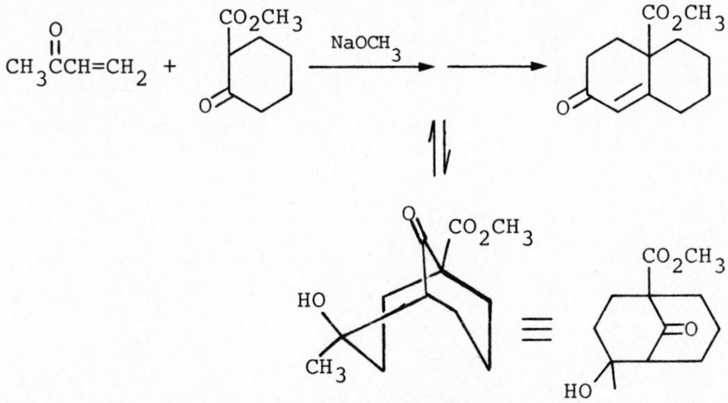

(a) Write a mechanism for the formation of this compound.
(b) The formation of this β-hydroxy ketone is reversible, and on con-
     tinued treatment with base it is converted to the normal annelated
     product. Why doesn't it undergo the normal dehydration reaction of
     β-hydroxy ketones and lead to an unsaturated ketone with the same
     carbon skeleton?

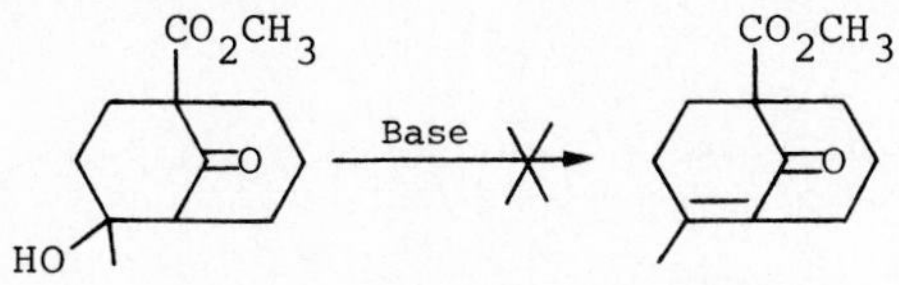

S11. Write a reasonable mechanism for the following transformation.

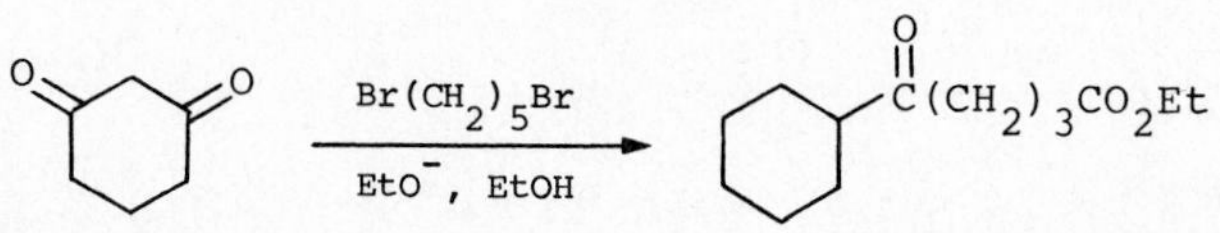

27.F Answers to Supplementary Problems

S1. (a)

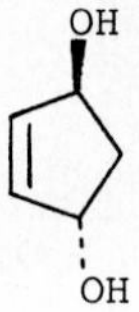

(b) $CH_3CH_2\overset{O}{\overset{\|}{C}}\underset{OH}{\underset{|}{C}H}CH_2CH_3$

(c)

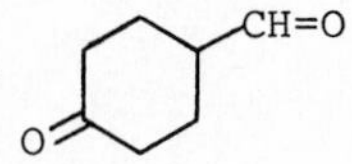

(d)

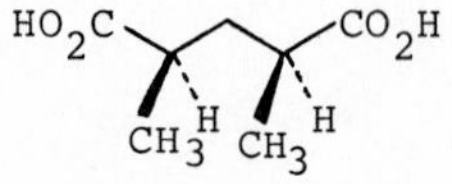

(e)

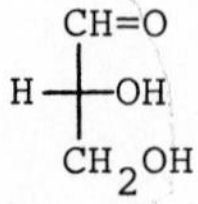

(f)

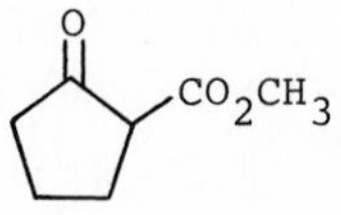

(g) $CH_3CH_2O_2C\underset{CH_3}{\underset{|}{C}H}CO_2CH_2CH_3$

(h) $(C_6H_5)_2\overset{OH}{\overset{|}{C}}-\overset{OH}{\overset{|}{C}}(C_6H_5)_2$

S2. (a) $HO_2CCH_2CH_2CH_2CH{=}O + HCO_2H$

(b) $CH_3CH_2\overset{O}{\overset{\|}{C}}\underset{CH_3}{\underset{|}{C}H}CH_2CH_3$

(c)

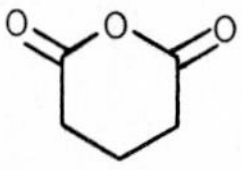

(d)

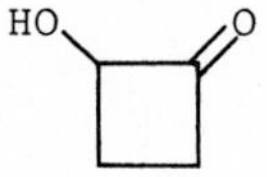

(e)

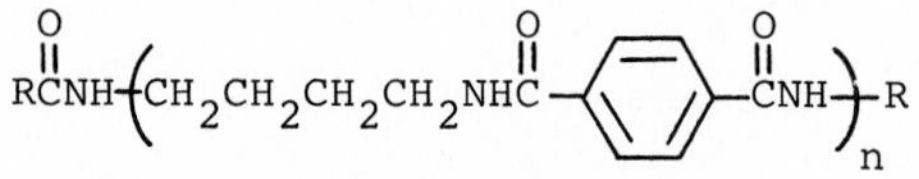

(f)

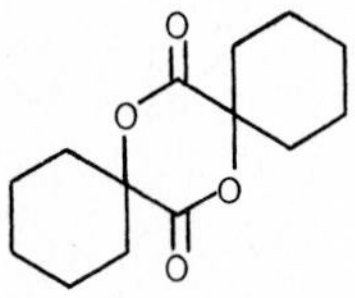

(g)

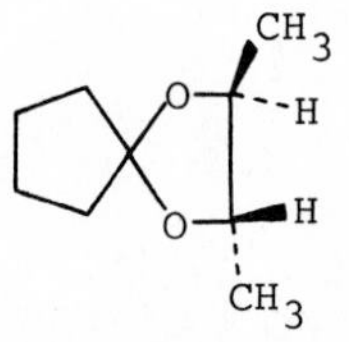

(h)

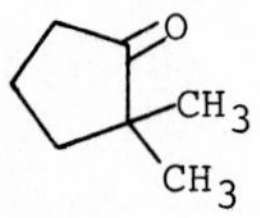

(i)

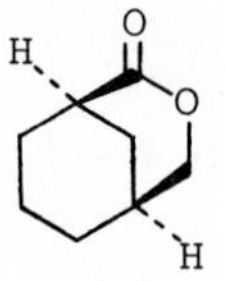

S3. (a) $CH_3CH_2CO_2CH_3 + CH_3CH_2\overset{O}{\overset{\|}{C}}CH_2CH_3 \xrightarrow{NaOCH_3} CH_3CH_2\overset{O}{\overset{\|}{C}}\underset{CH_3}{\underset{|}{C}H}\overset{O}{\overset{\|}{C}}CH_2CH_3 \xrightarrow{NaBH_4}$

(b) $BrCH_2CH_2CH_2CH_2CH_2Br \xrightarrow{2\ NaCN} \xrightarrow[\Delta]{H_3O^+} HO_2C(CH_2)_5CO_2H \xrightarrow{\Delta}$

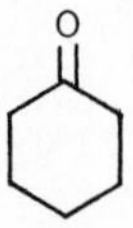

(c)

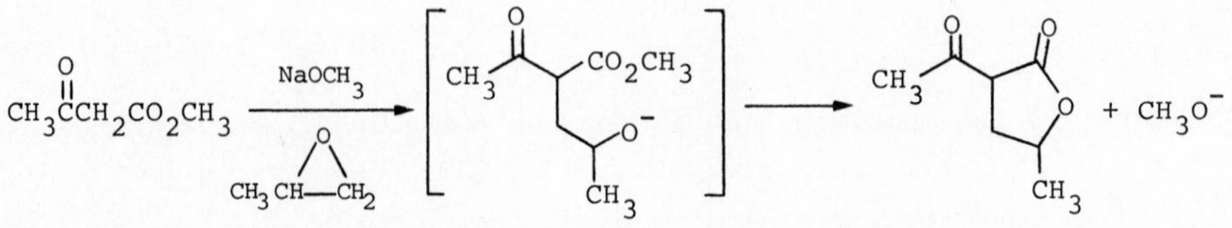

(d)   2 $(CH_3CH_2)_2C=O$ $\xrightarrow{Mg}$ $(CH_3CH_2)_2\underset{\underset{HO}{|}}{\overset{\overset{OH}{|}}{C}}-C(CH_2CH_3)_2$ $\xrightarrow{H^+}$ $(CH_3CH_2)_3C\overset{O}{\overset{||}{C}}CH_2CH_3$

(e) $CH_3CH_2\overset{O}{\overset{||}{C}}CH_3$ $+ \ BrCH_2COOEt$ $\xrightarrow[\underset{\Delta}{benzene}]{Zn}$ $CH_3CH_2\underset{\underset{CH_3}{|}}{\overset{\overset{OH}{|}}{C}}CH_2COOEt$ $\xrightarrow{LiAlH_4}$ $\xrightarrow[H^+, \ -H_2O]{(CH_3)_2C=O}$

(f)

$$\xrightarrow{NaOCH_3}$$

(g) $(CH_3)_2CHCH_2COOH$ $\xrightarrow[H^+]{CH_3OH}$ $\xrightarrow[\text{2. }H_2O]{\text{1. Na}}$ $(CH_3)_2CHCH_2\overset{O}{\overset{||}{C}}\underset{\underset{OH}{|}}{C}HCH_2CH(CH_3)_2$ $\xrightarrow{Cu(OAc)_2}$ $R\overset{O}{\overset{||}{C}}-\overset{O}{\underset{||}{C}}R$

(h)

$$\xrightarrow[HCl]{AlCl_3} \xrightarrow{Zn} \xrightarrow{HF} \xrightarrow[HCl]{Zn}$$

(i) $CH_3\overset{O}{\overset{||}{C}}CH_2CO_2CH_3$ $\xrightarrow[Br(CH_2)_4Br]{2 \ NaOCH_3}$ $\xrightarrow[\Delta]{H_3O^+}$

(j) $CH_3\overset{O}{\overset{||}{C}}CH_2CO_2CH_3$ $\xrightarrow[(CH_3)_2C=CHCH_2Br]{NaOCH_3}$ $\xrightarrow[\Delta]{H_3O}$

$\xrightarrow{\Delta}$ $(CH_3O)_2\overset{O}{\overset{||}{P}}CHCO_2CH_3$

$\xrightarrow{LiAlH_4}$ $CHCO_2CH_3$

(mixture of isomers)

S4. $(CH_3OOC)_2CH_2$ $\xrightarrow[(CH_3O)_2SO_2]{NaOCH_3}$ $\xrightarrow[CH_3CH_2Br]{NaOCH_3}$ $CH_3OOC\underset{\underset{CH_2CH_3}{|}}{\overset{\overset{CH_3}{|}}{C}}COOCH_3$ $\xrightarrow{LiAlH_4}$ $\xrightarrow[H^+]{2 \ NaNCO}$

$H_2N\overset{O}{\overset{||}{C}}OCH_2\underset{\underset{CH_2CH_3}{|}}{\overset{\overset{CH_3}{|}}{C}}CH_2O\overset{O}{\overset{||}{C}}NH_2$

S5. $CH_3\overset{O}{\overset{||}{C}}CH_2CH_2COOH$ $HOOCCH_2CH_2COOH$ $CH_3\underset{\underset{OCH_3}{|}}{\overset{\overset{OCH_3}{|}}{C}}CH_2CH_2COOCH_3$

      A   B   C

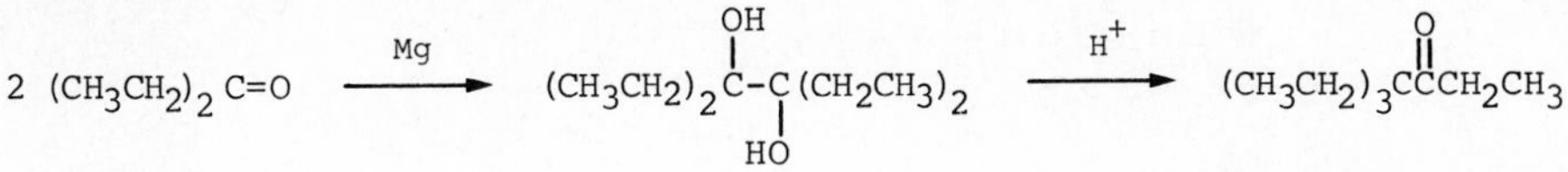

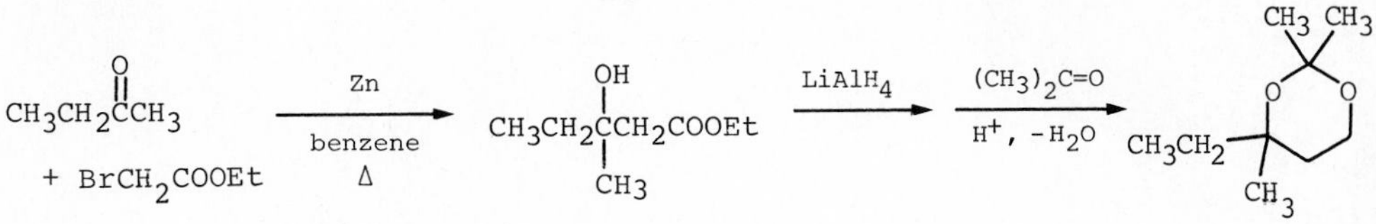

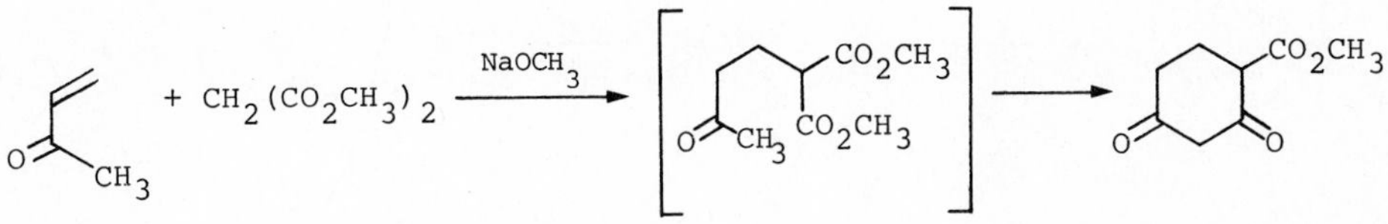

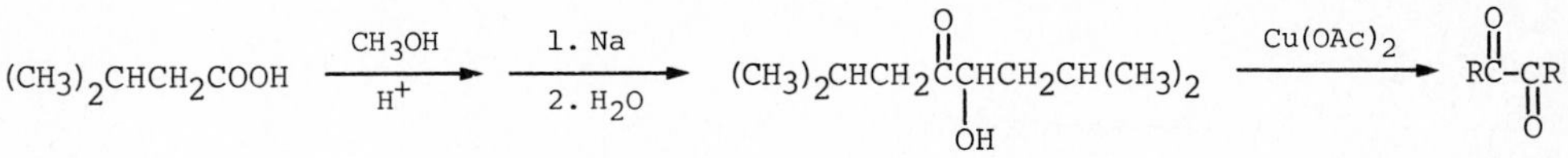

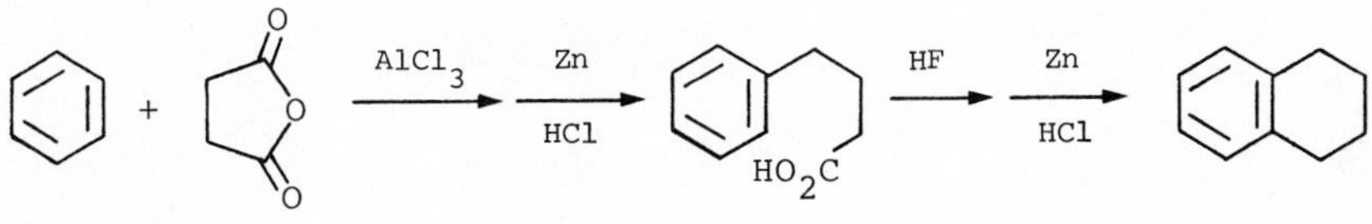

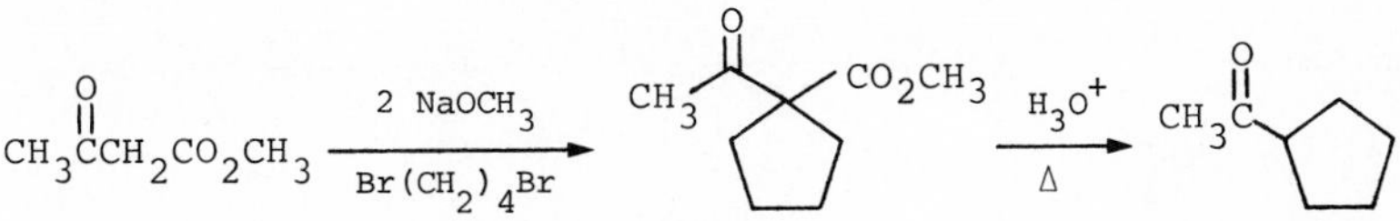

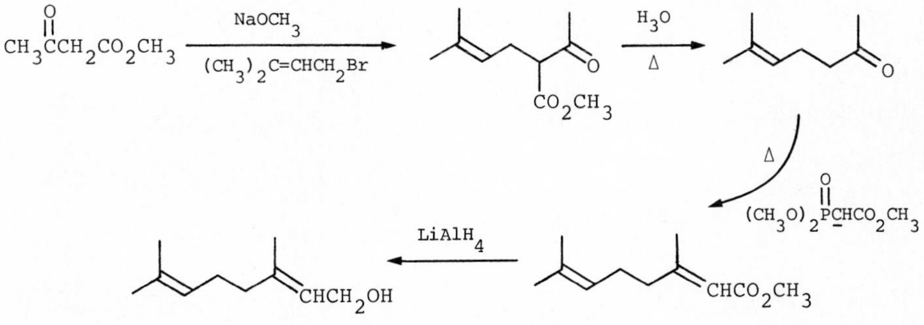

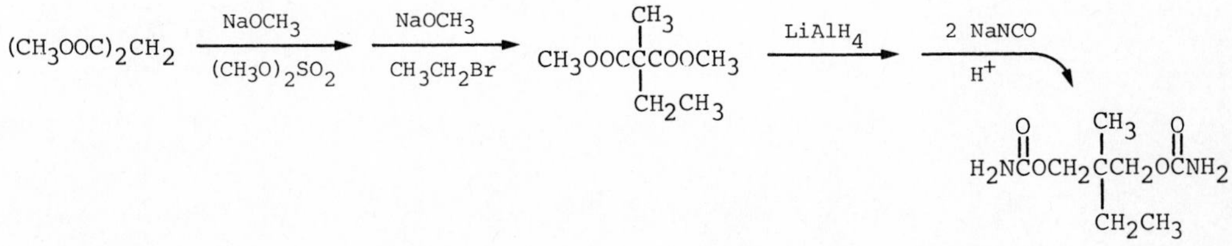

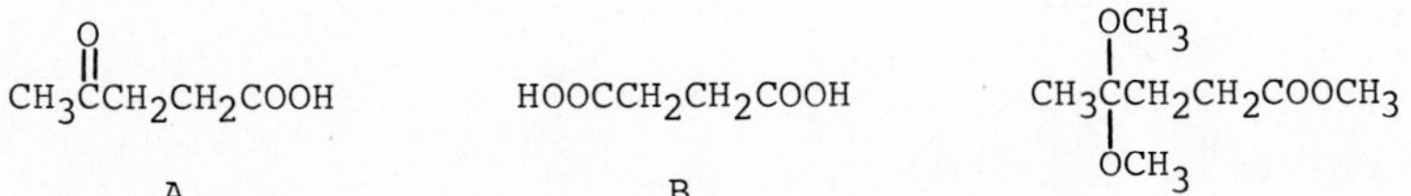

$$CH_3\overset{\displaystyle OCH_3}{\underset{\displaystyle OCH_3}{\overset{|}{\underset{|}{C}}}}CH_2CH_2CH_2OH$$

D

E

S6.

$$\xrightarrow[\Delta]{SeO_2}$$ $$\xrightarrow{NaOH}$$ $$\xrightarrow{HIO_4}$$

S7.

$$N\equiv C-CH_2 \longrightarrow HN=C=CH_2 \rightleftharpoons N\equiv CCH_3$$

$$CO_2$$

S8. (a) $(EtO_2C)_2CH_2 \rightleftharpoons (EtO_2C)_2\overset{-}{C}H \quad H_2C=\overset{..}{O} \rightleftharpoons (EtO_2C)_2CH-CH_2-O^-$

$(EtO_2C)_2\overset{-}{C}CH_2CH(CO_2Et)_2 \rightleftharpoons (EtO_2C)_2C=CH_2 \overset{-OH^-}{\rightleftharpoons} (EtO_2C)_2\overset{-}{C}-CH_2-OH$

$\big\Updownarrow$ ROH, $-RO^-$

$^-CH(CO_2Et)_2$

$(EtO_2C)_2CHCH_2CH(CO_2Et)_2$

(b) $\overset{..}{H\overset{..}{O}}CH_2CH_2CH_2\overset{\displaystyle O}{\overset{||}{C}}CH_2CO_2CH_3 \overset{H^+}{\rightleftharpoons} \overset{..}{H\overset{..}{O}}CH_2CH_2CH_2\overset{\overset{\displaystyle +OH}{||}}{C}CH_2CO_2CH_3 \rightleftharpoons$

S9.

$$\overset{-H^+}{\rightleftharpoons}$$ $$\xrightarrow{-CO_2}$$ $$\xrightarrow{H^+}$$

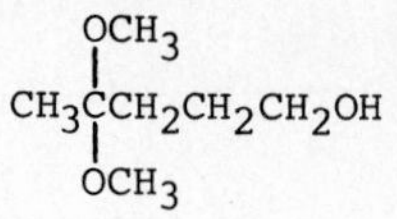

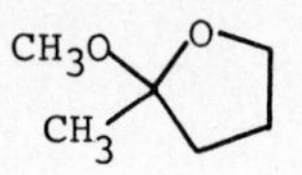

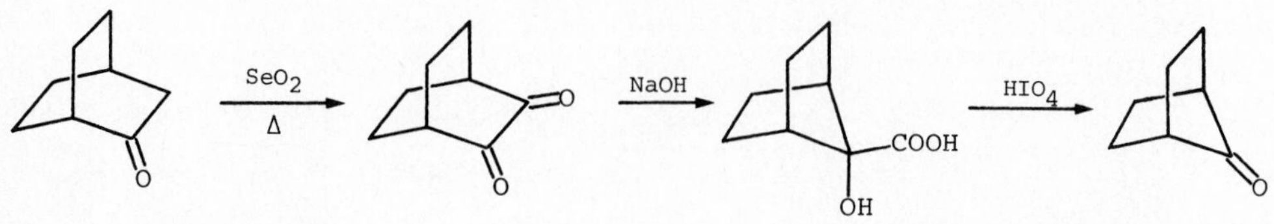

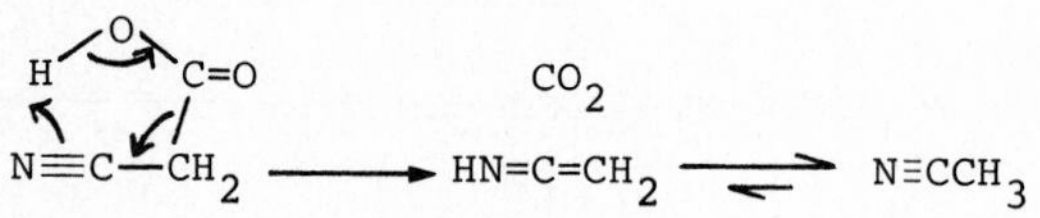

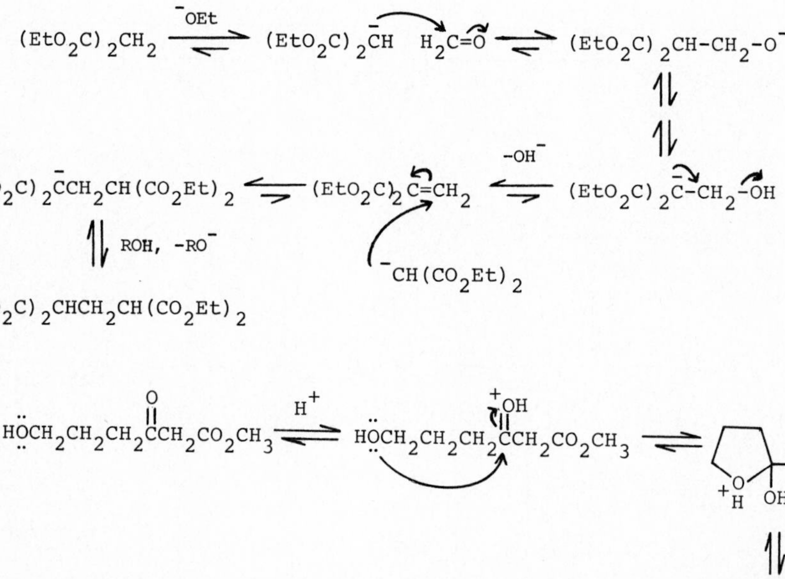

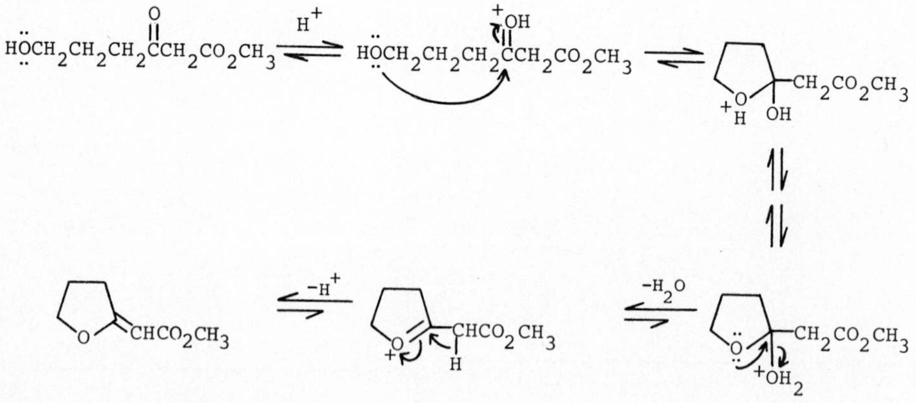

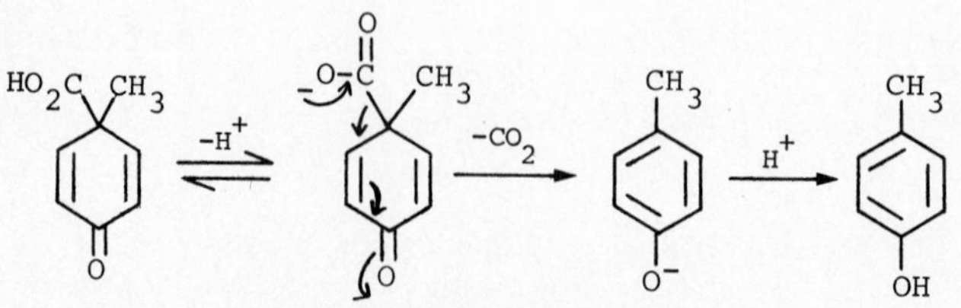

S10. (a)

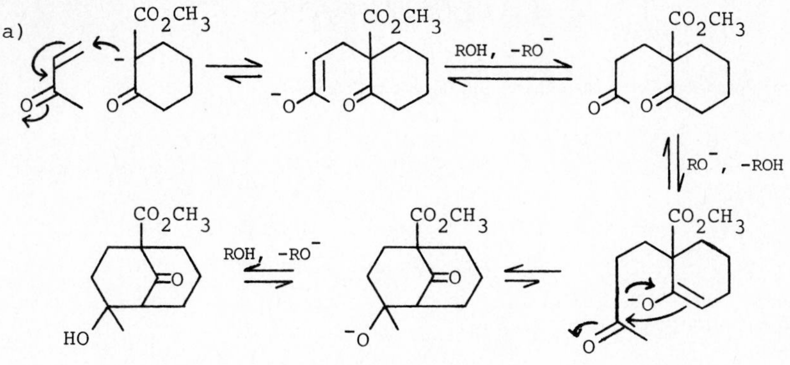

(b) Elimination of water from this bicyclic ketol would involve (1) formation of the enolate of the ketone, as well as (2) loss of hydroxide to generate the double bond. Neither of these occurrences is possible in this system, because either one would require the formation of a π-bond between two orbitals which are perpendicular to each other:

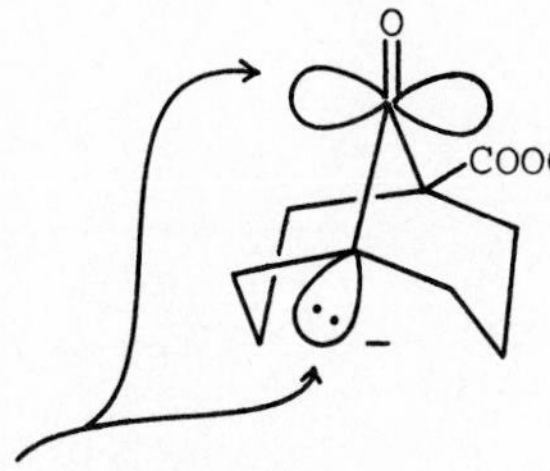

no overlap possible; the enolate is not formed

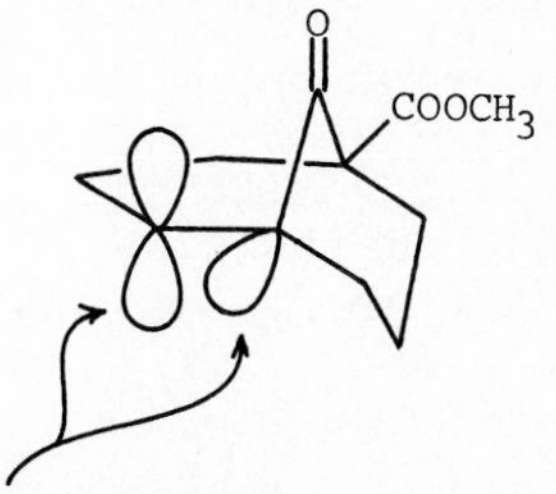

no overlap possible; no double bond is formed

S11.

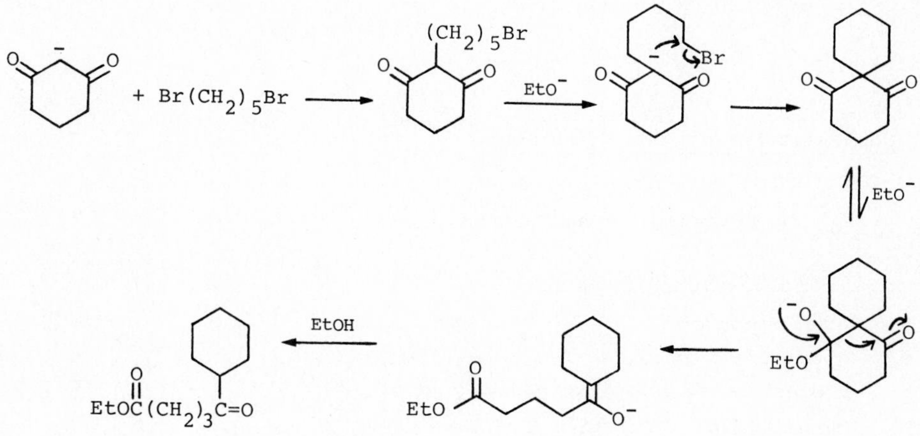

28. CARBOHYDRATES

28.A Chapter Outline, Important Terms Introduced, and Reactions Discussed

28.1 Introduction

carbohydrate: $(CH_2O)_n$
sugar
mono-, di-, tri-, tetra-, oligo-, and polysaccharides

aldose, aldopentose, etc.
ketose, ketohexose, etc.

28.2 Stereochemistry and Configurational Notation of Sugars

R,S *vs.* D,L
Fischer projections

meso compounds

28.3 Cyclic Hemiacetals; Anomerism; Glycosides

α- and β-anomers
Haworth projections

pyranose *vs.* furanose

mutarotation:

e.g.,

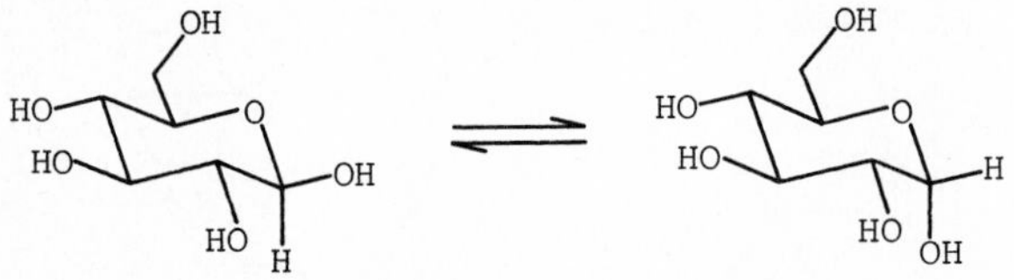

(acid- or base-catalyzed)

glycoside formation and hydrolysis:

e.g.,

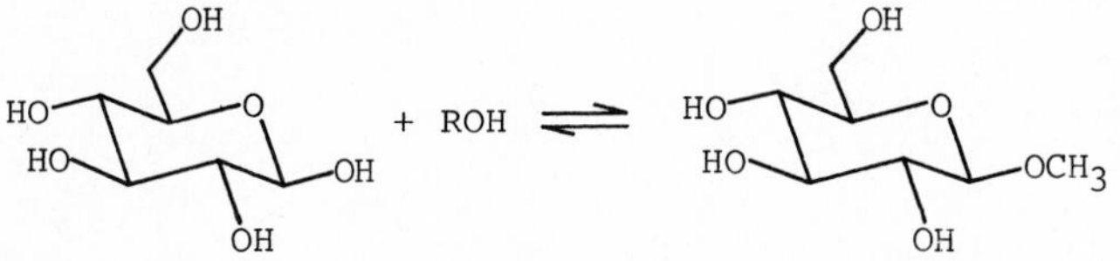

(α- and β-anomers)

(acid- or glycosidase enzyme-catalyzed)

28.4 Conformations of the Pyranoses

chair conformations

axial *vs.* equatorial substituents

28.5 Reactions of Monosaccharides

A. Ether Formation

glycoside formation *(see Section 28.3)*

protection of anomeric carbon

methylation:

e.g.,

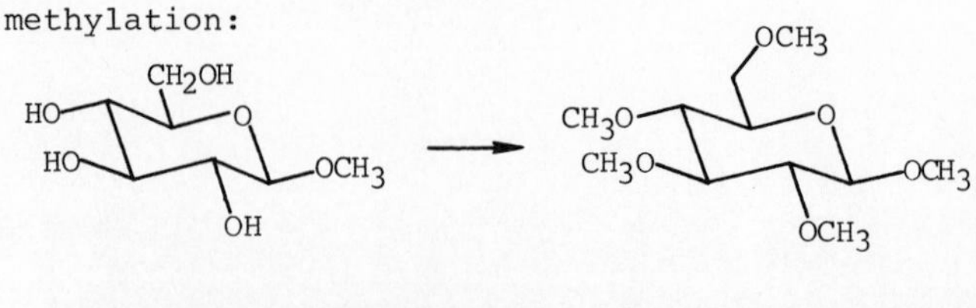

(using $NaOH/(CH_3)_2SO_4$ or AgO/CH_3I; not with hemiacetal)

B. Formation of Cyclic Acetals and Ketals

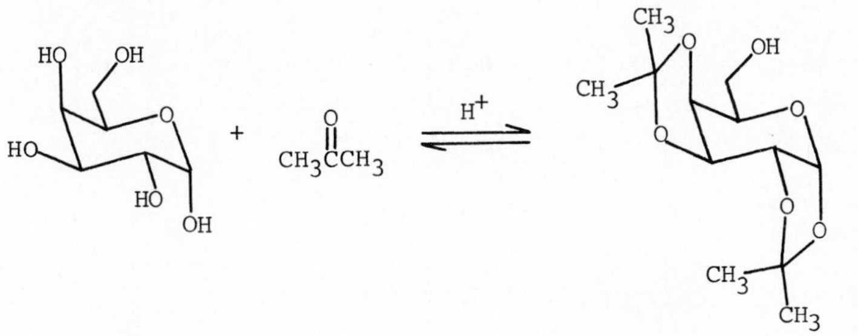

+ 2 H_2O

(exact product depends on specific saccharide; this method is useful for selective protection)

C. Esterification

equatorial anomeric hydroxyl faster than axial

D. Reduction: Alditols

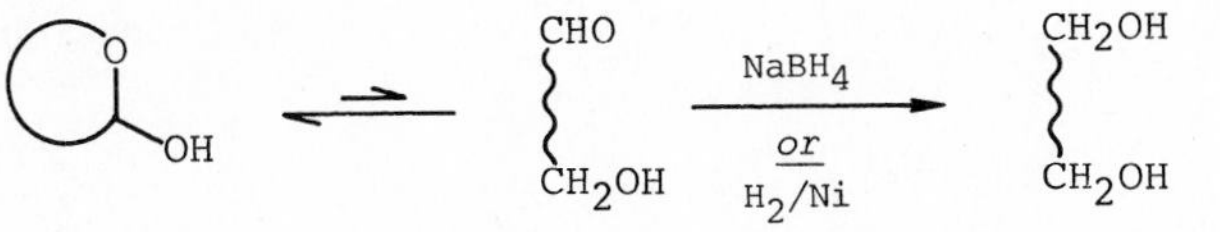

(via open-chain form)

E. Oxidation: Aldonic and Saccharic Acids

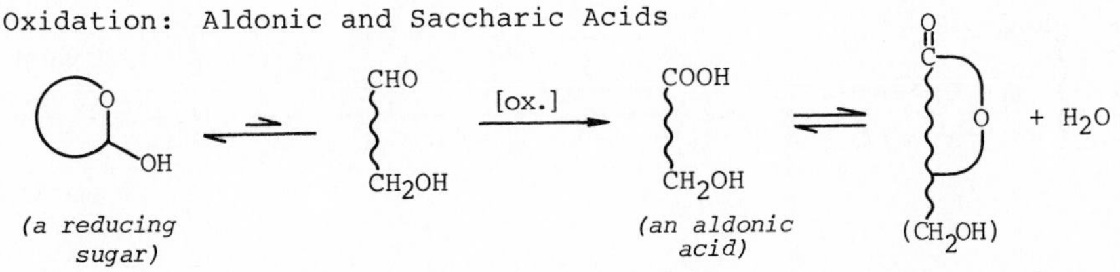

([ox.] = $^{+}Ag(NH_3)_2$ (*Tollens reagent*) or $Cu(OH)_2$ (*Fehling's reagent*) or Br_2; necessity of hemiacetal form; possible formation of five-membered lactones; reducing *vs*. non-reducing sugars)

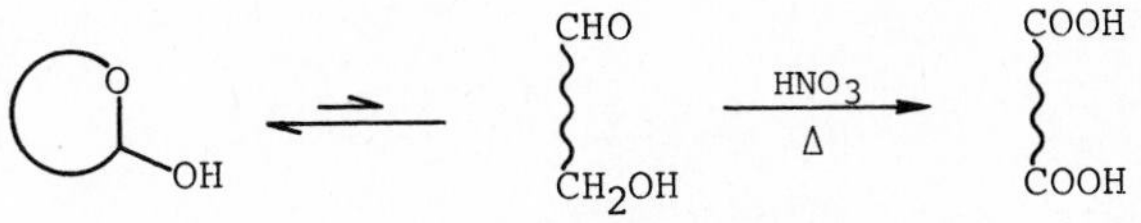

(a saccharic acid)

F. Oxidation by Periodic Acid

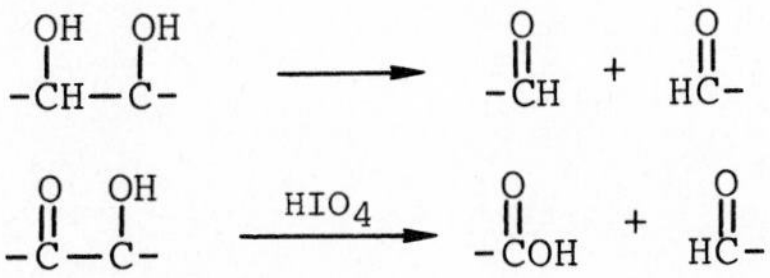

Formation of formic acid from:

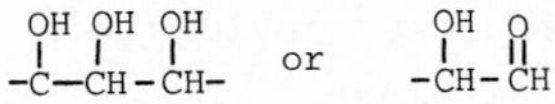

G. Phenylhydrazones and Osazones

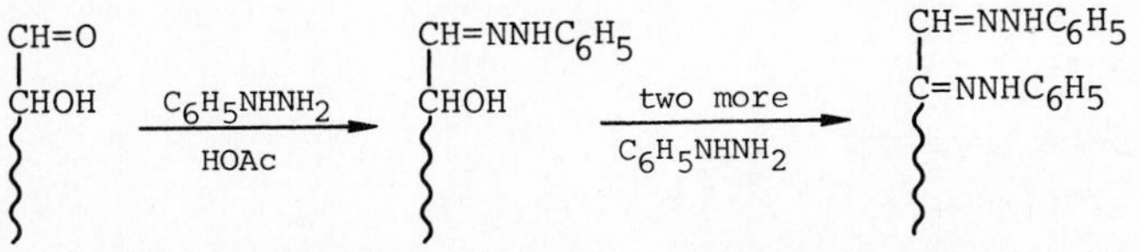

+ NH_3 and $C_6H_5NH_2$

(same osazone results from both C-2 epimers)

(stereochemical test; means of characterization)

H. Chain Extension: The Kiliani-Fischer Synthesis

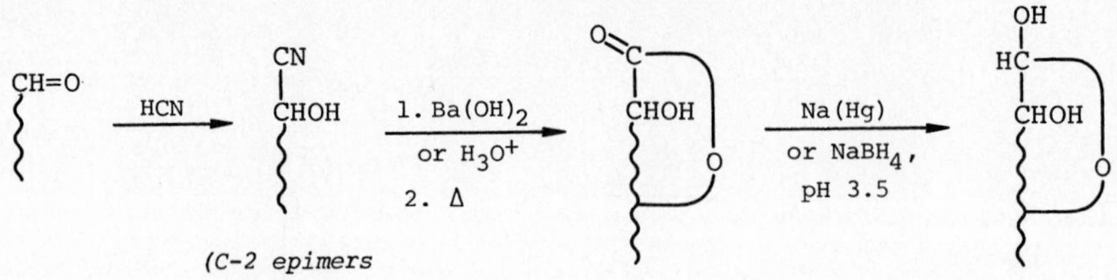

(C-2 epimers produced)

I. Chain Shortening: The Ruff and Wohl Degradations

Ruff degradation:

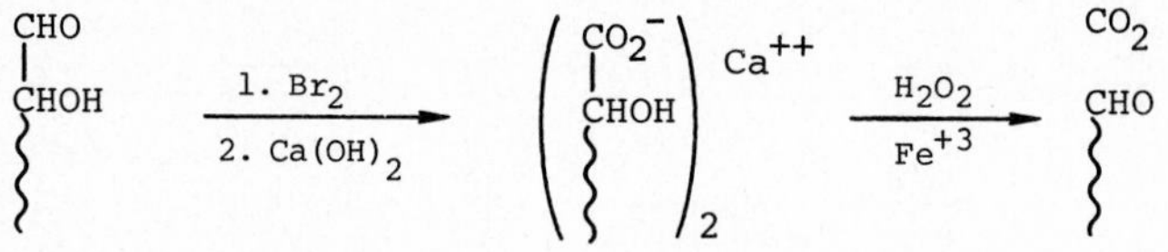

(same aldose is produced from both C-2 epimers)

Wohl degradation:

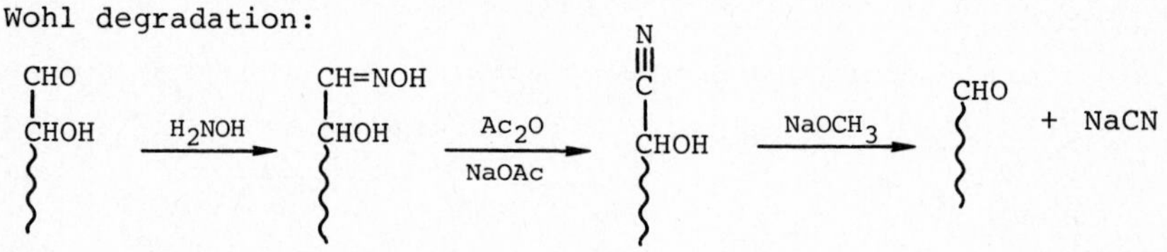

(same aldose from both C-2 epimers)

28.6 Relative Stereochemistry of the Monosaccharides: The Fischer Proof

use of symmetry

28.7 Oligosaccharides

structure proof via methylation/cleavage
invertase enzymes

28.8 Polysaccharides

starch
amylose, amylopectin
celluloses

28.9 Sugar Phosphates

28.10 Natural Glycosides

glycosyl residue *vs.* aglycon
laetrile *vs.* erythromycin

28.B Important Concepts and Hints

Appropriately, the chemistry of carbohydrates follows the discussion of difunctional compounds. With only a couple of exceptions, all of the reactions presented in this chapter are ones you have seen before. What makes them different is the fact that the molecules involved have a functional group on every carbon! Nonetheless, these molecules behave according to the principles outlined in the preceding chapter, and only a few transformations which are specific to carbohydrate chemistry have to be learned. In essence, this topic requires you to apply the knowledge you've already gained to more complicated systems.

The topic of sugar chemistry also requires you to review the subject of stereochemistry. You have to be able to manipulate Fischer projections, determine R and S configurations, and recognize the presence or absence of planes of symmetry and *meso* compounds. The presence or absence of optical activity has been the single most important observation in schemes for the determination of the relative stereochemistry of sugars. Many of the techniques devised for probing the stereochemistry of a sugar have involved bringing the two ends of the carbon chain to the same oxidation state (-CH_2OH or -COOH) and then looking for optical activity. An optically active compound cannot be *meso*, and an optically inactive compound is assumed to be *meso*.

Fischer projections are by convention drawn with the carbon chain extending vertically and the most oxidized end of the molecule at the top, if there is a difference. To determine whether a Fischer projection represents a *meso* compound, draw (or imagine) a line across the picture exactly half-way down. If there is an even number of carbons, this line will cross one of the carbon-carbon bonds; if there is an odd number of carbons, the line will cut across the middle carbon atom and its substituents. With this line in place (or in mind), compare the pattern above the line with that below: if they are mirror images, the compound is *meso*; if they differ in any way the compound is chiral, and thus capable of being optically active. Some examples are illustrated below:

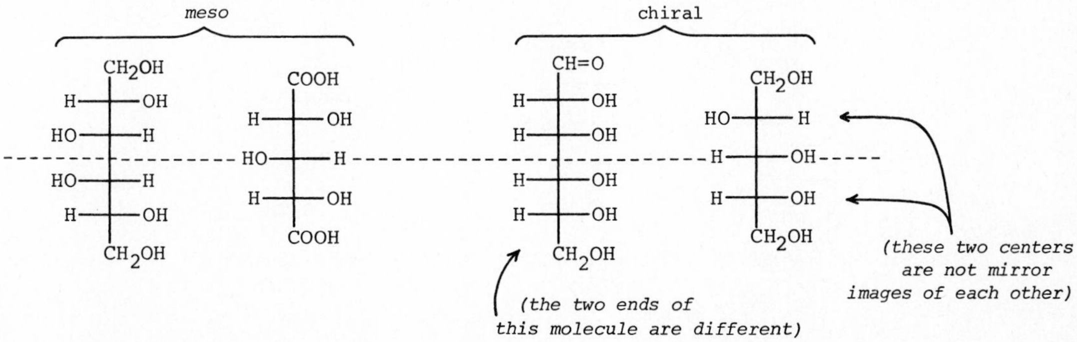

Many students have difficulty mentally interconverting the perspective diagrams of the chair forms of sugar hemiacetals with the linear Fischer projections of their open-chain isomers. Can you tell whether the two structures drawn on the next page represent the same sugar or not? (They do.)

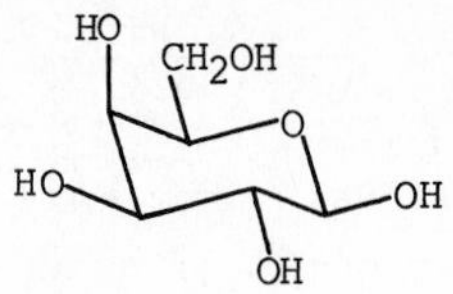

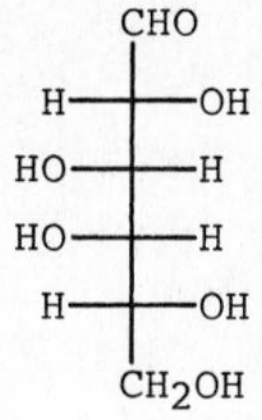

One hint that we can give you (in addition to *using models!!*) is to become familiar with Haworth projections. These are intermediate between the perspective formulas and Fischer projections, and serve to relate the two in a logical way. A Haworth projection is essentially a flattened form of the cyclic structure; the distinction between axial and equatorial is ignored and all that matters is whether substituents stick up or down. To go from the perspective formula to the Haworth projection, you simply look for up or down orientations. The Haworth projection of the sugar above is:

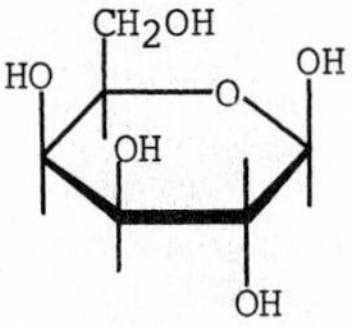

The substituents that stick up in a Haworth formula are on the left in a Fischer projection, with two exceptions: 1) the next-to-last carbon gets mixed up because the substituent in the Haworth formula is part of the carbon chain itself (if $-CH_2OH$ sticks up (D-sugar), the hydroxy group is on the right); and 2) the configuration of the anomeric carbon (α or β) is not represented in the open-chain Fischer projection. As usual, these correlations can be most easily visualized and understood with the help of models.

28.C Answers to Exercises

(28.2)

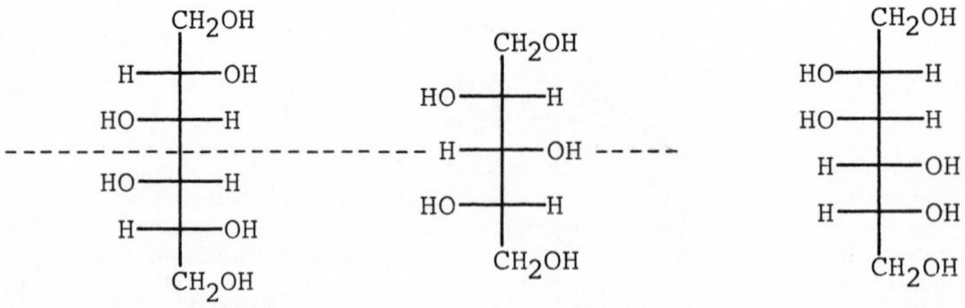

from D-galactose (*achiral*) — from L-xylose (*achiral*) — from D-mannose (*chiral*)

(28.3)

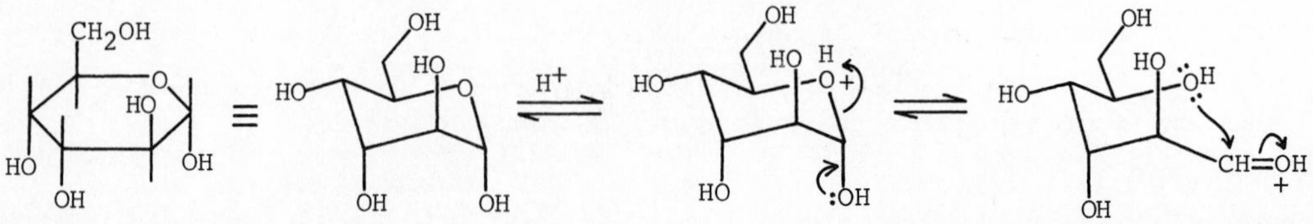

α-D-altrose

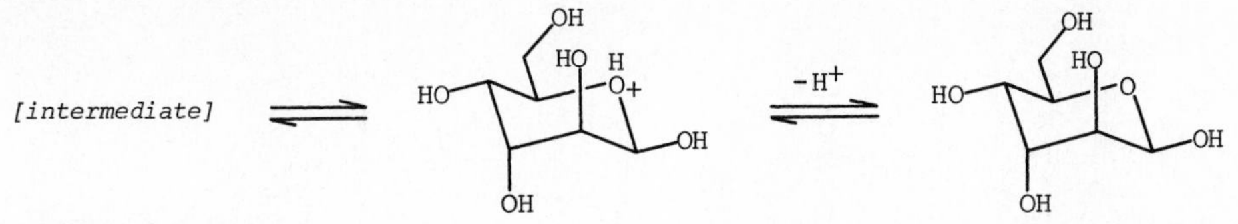

β-D-altrose

(28.4)

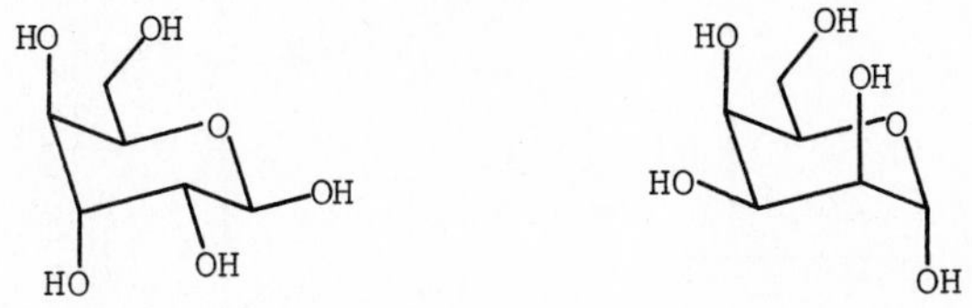

β-D-gulose   α-D-talose

(28.5.A)

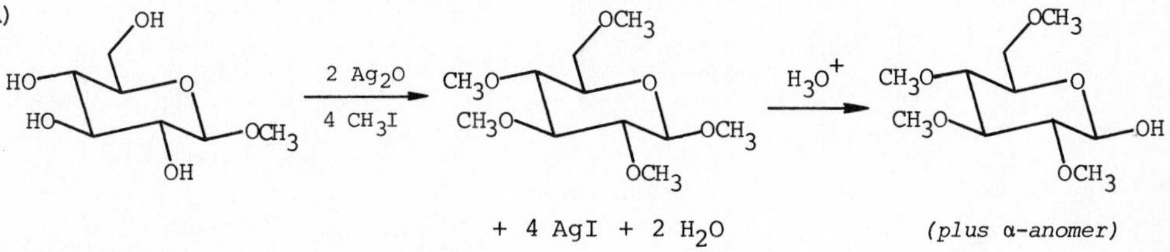

(28.5.B)

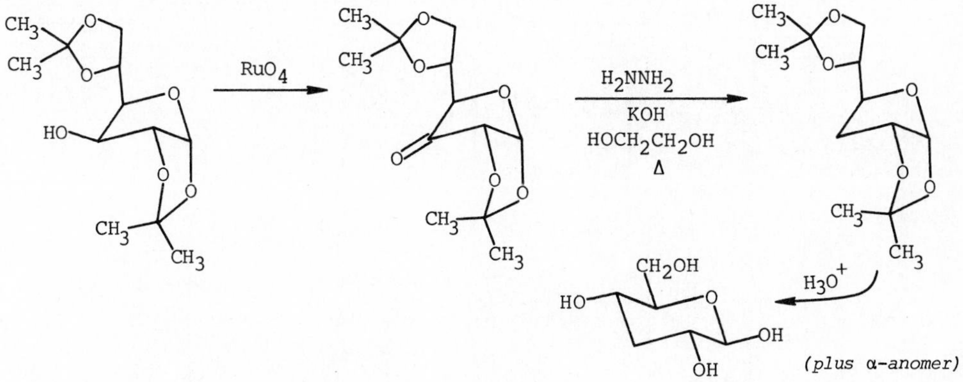

(28.5.E)

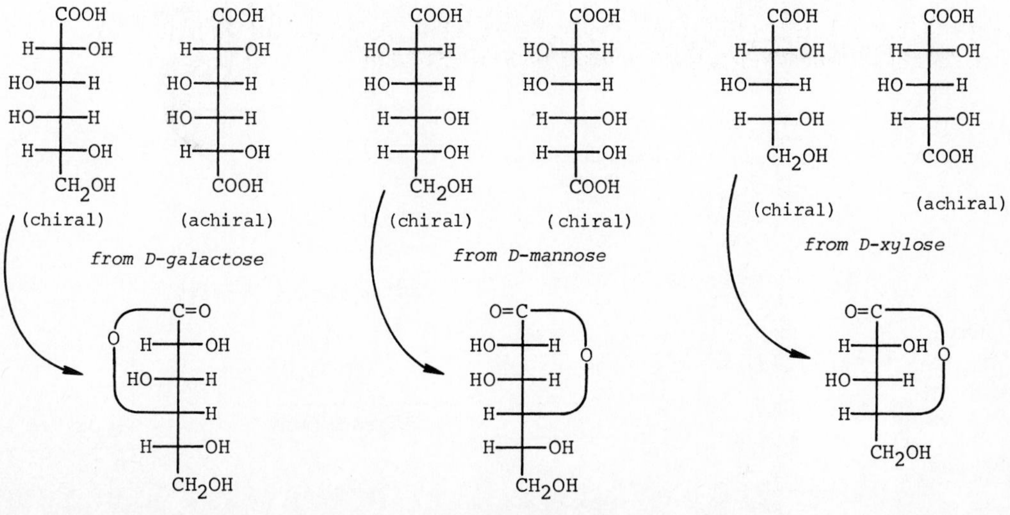

(28.5.F)

(a) $NaIO_4$ *(one equivalent of $NaIO_4$ consumed)*

(b) $NaIO_4$ +   *(two equivalents of $NaIO_4$ consumed)*

(c) $NaIO_4$ $2 \; HCOOH + 2 \; CH_2=O$ *(three equivalents of $NaIO_4$ consumed)*

(28.5.G)

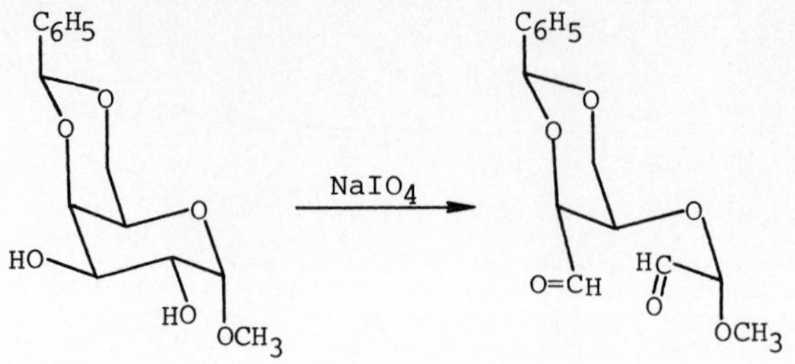

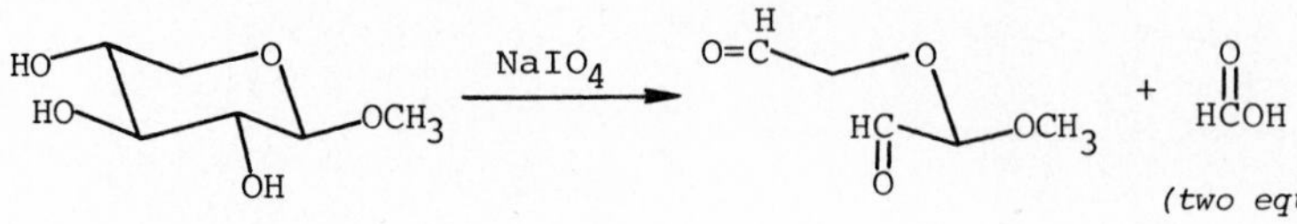

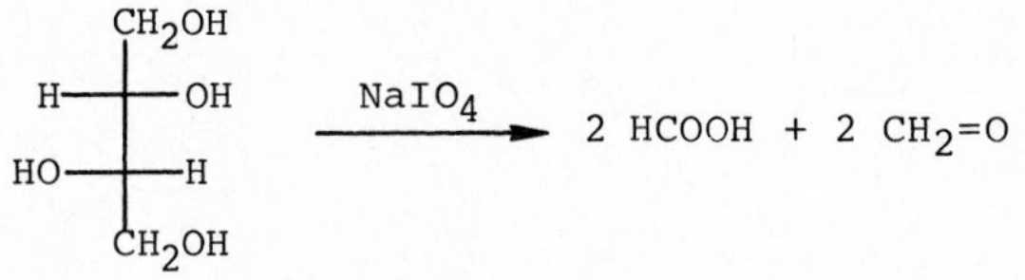

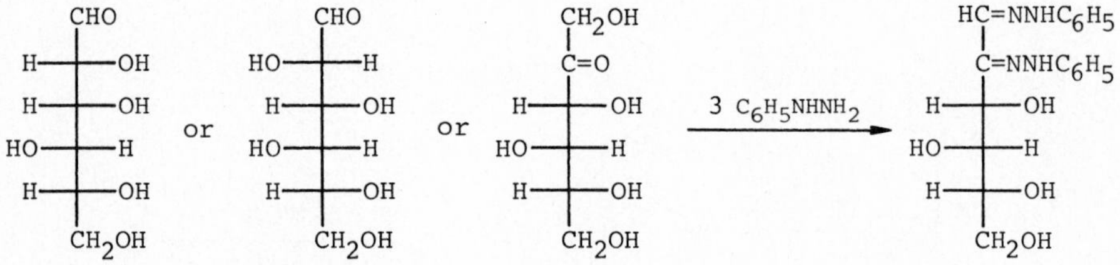

D-(−)-gulose   or   *D-(−)-idose*   or   *D-(+)-sorbose* $3 \; C_6H_5NHNH_2$

(28.5.H)

D-ribose   Kiliani–Fischer   *D-allose*   +   *D-altrose*

D-xylose   Kiliani–Fischer   *D-gulose*   +   *D-idose*

(28.5.I)

D-lyxose   and   *D-xylose*   Wohl degradation   *D-threose*

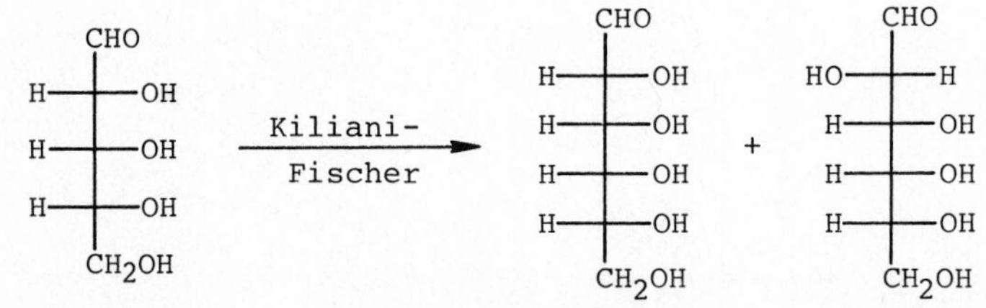

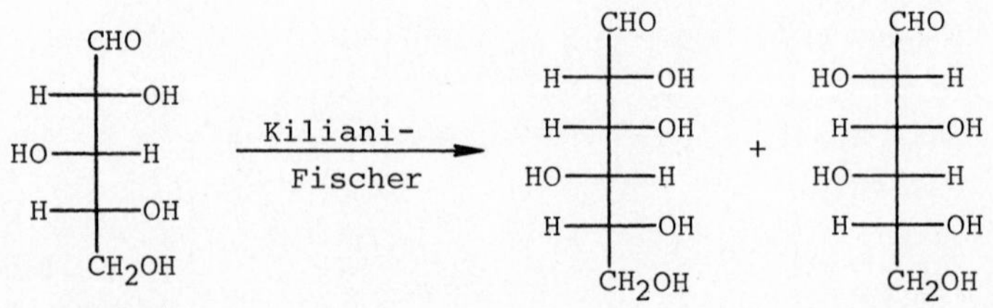

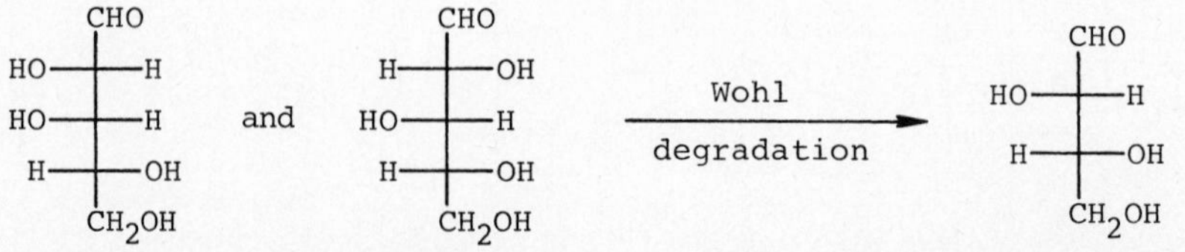

(28.7)

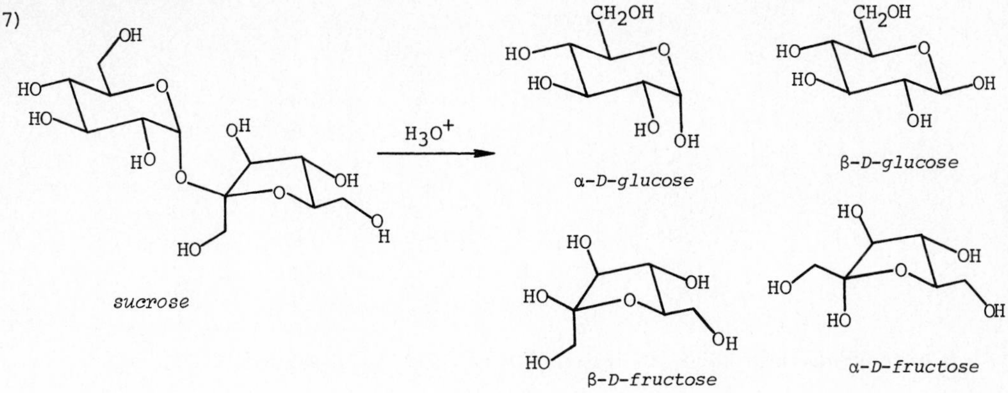

28.D Answers and Explanations for Problems

1.

D-glyceraldehyde	2R	D-allose	2R, 3R, 4R, 5R
D-erythrose	2R, 3R	D-altrose	2S, 3R, 4R, 5R
D-threose	2S, 3R	D-glucose	2R, 3S, 4R, 5R
D-ribose	2R, 3R, 4R	D-mannose	2S, 3S, 4R, 5R
D-arabinose	2S, 3R, 4R	D-gulose	2R, 3R, 4S, 5R
D-xylose	2R, 3S, 4R	D-idose	2S, 3R, 4S, 5R
D-lyxose	2S, 3S, 4R	D-galactose	2R, 3S, 4S, 5R
		D-talose	2S, 3S, 4S, 5R

2.

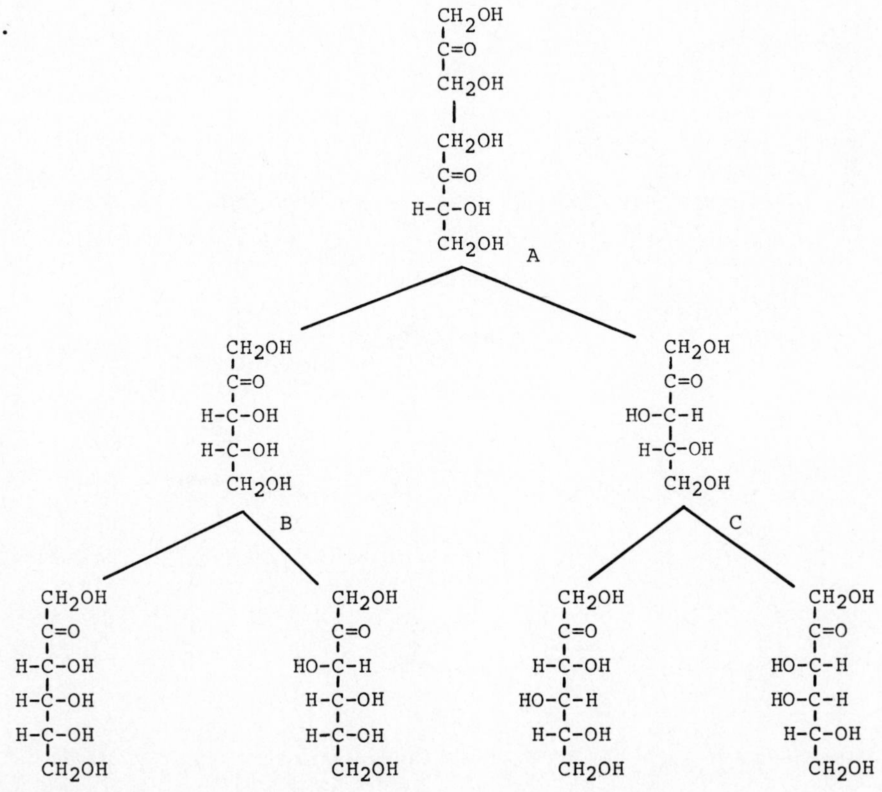

D E (D-fructose) F G

IDENTICAL OSAZONES

Ketose	Hexoses
A	D-erythrose, D-threose
B	D-ribose, D-arabinose
C	D-xylose, D-lyxose
D	D-allose, D-altrose
E	D-glucose, D-mannose
F	D-gulose, D-idose
G	D-galactose, D-talose

3. D-erythrose, D-ribose, D-xylose, D-allose, D-galactose

4.

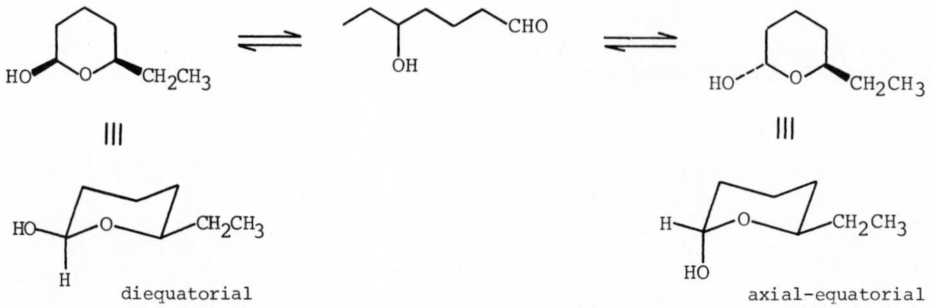

diequatorial *(more stable)* axial-equatorial *(less stable)*

Mechanism:

Acid:

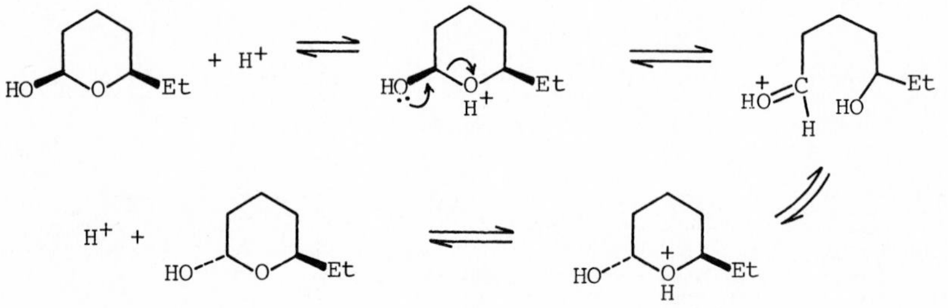

Base:

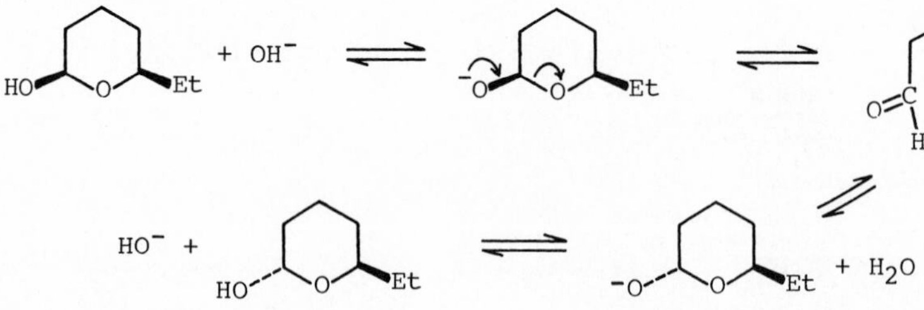

5. (a)

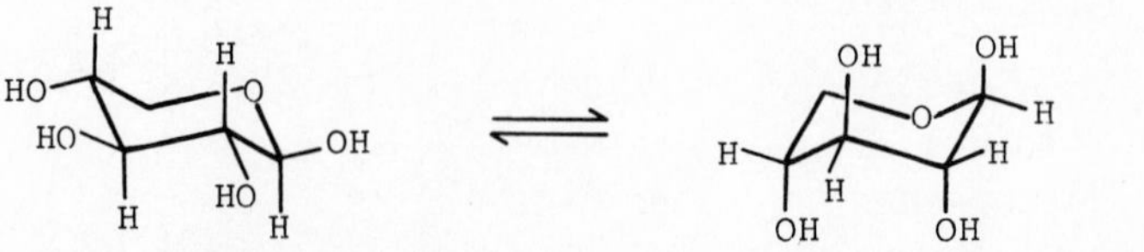

more stable *less stable* (four axial groups)

(b)

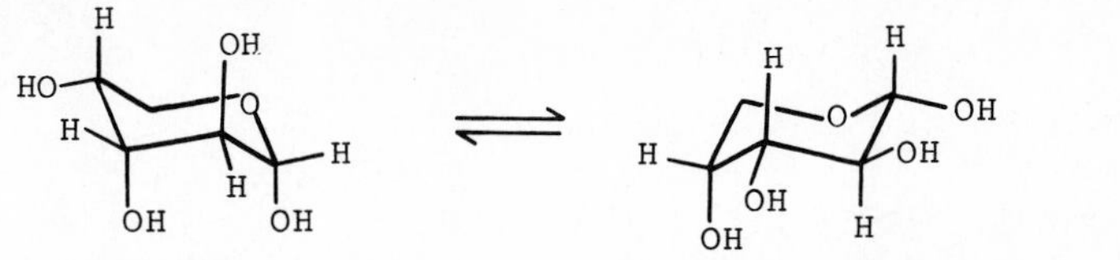

less stable (three axial groups)

more stable (one axial group)

6.

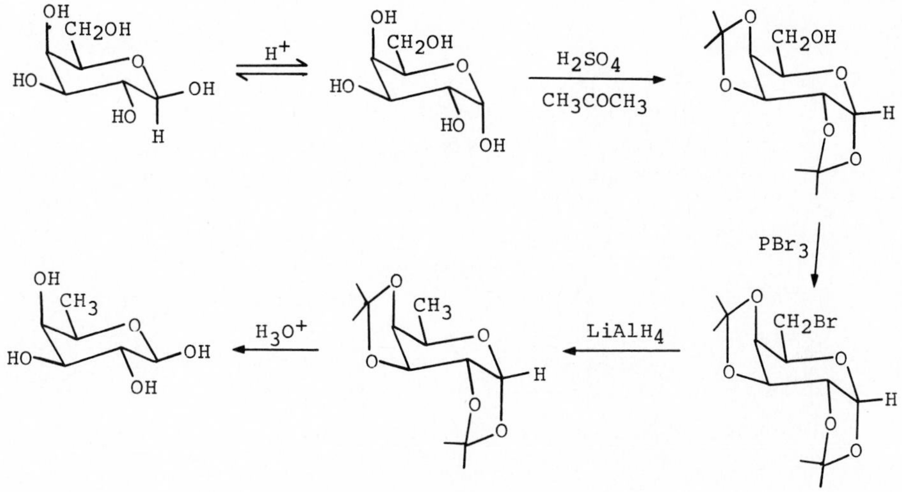

7.

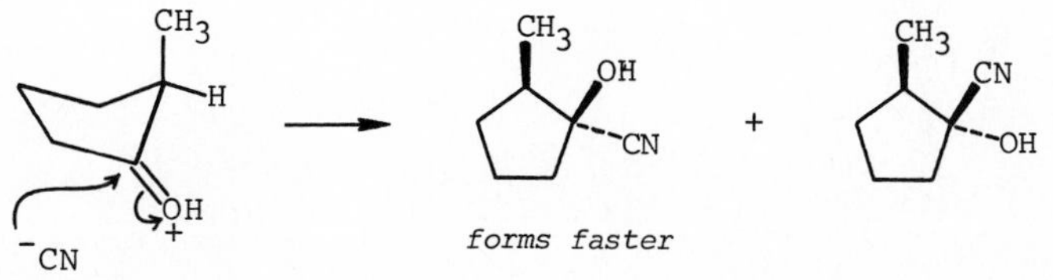

8. Kiliani-Fischer chain extension follows the aldose tree in Table 28.1.

9. Reverse of the aldose tree in Table 28.1.

10.

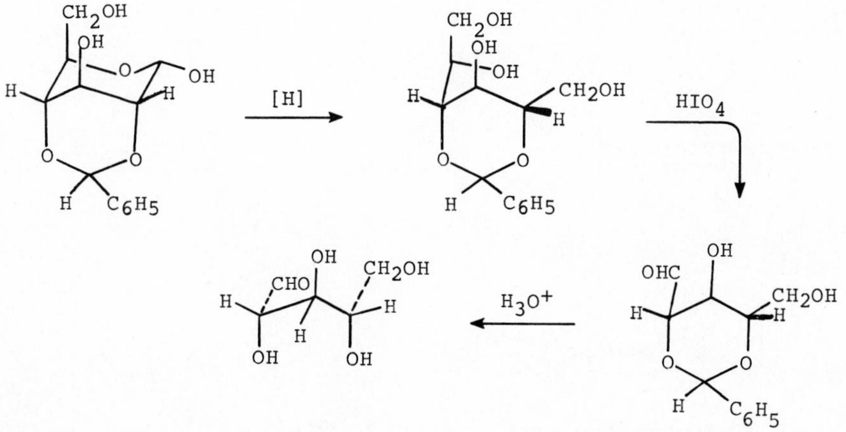

Overall process:

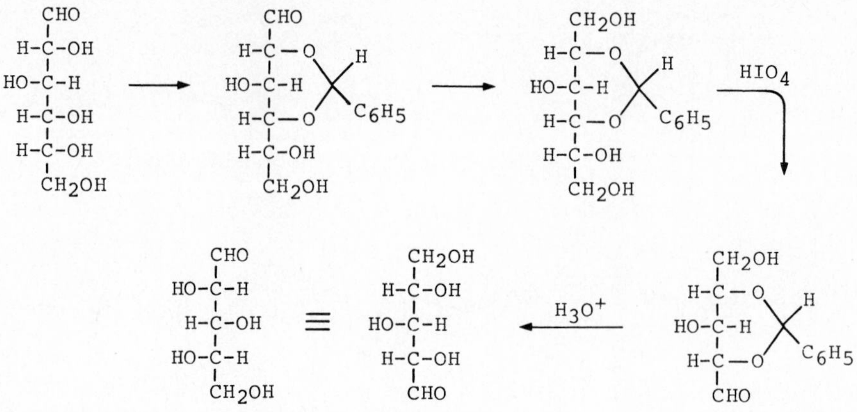

$$\begin{array}{c} CHO \\ H-C-OH \\ HO-C-H \\ H-C-OH \\ H-C-OH \\ CH_2OH \end{array} \longrightarrow \begin{array}{c} CHO \\ H-C-O\diagdown \\ HO-C-H \diagup^{H} \\ H-C-O\diagup C_6H_5 \\ H-C-OH \\ CH_2OH \end{array} \longrightarrow \begin{array}{c} CH_2OH \\ H-C-O\diagdown \\ HO-C-H \diagup^{H} \\ H-C-O\diagup C_6H_5 \\ H-C-OH \\ CH_2OH \end{array} \xrightarrow{HIO_4}$$

$$\begin{array}{c} CHO \\ HO-C-H \\ H-C-OH \\ HO-C-H \\ CH_2OH \end{array} \equiv \begin{array}{c} CH_2OH \\ H-C-OH \\ HO-C-H \\ H-C-OH \\ CHO \end{array} \xleftarrow{H_3O^+} \begin{array}{c} CH_2OH \\ H-C-O\diagdown \\ HO-C-H \diagup^{H} \\ H-C-O\diagup C_6H_5 \\ CHO \end{array}$$

L-xglose

11.

A:
$$\begin{array}{c} CHO \\ H-\!\!\!-OH \\ H-\!\!\!-OH \\ H-\!\!\!-OH \\ CH_2OH \end{array}$$
A

B:
$$\begin{array}{c} COOH \\ H-\!\!\!-OH \\ H-\!\!\!-OH \\ H-\!\!\!-OH \\ COOH \end{array}$$
B

C:
$$\begin{array}{c} CH=NOH \\ H-\!\!\!-OH \\ H-\!\!\!-OH \\ H-\!\!\!-OH \\ CH_2OH \end{array}$$
C

D:
$$\begin{array}{c} CN \\ H-\!\!\!-OAc \\ H-\!\!\!-OAc \\ H-\!\!\!-OAc \\ CH_2OAc \end{array}$$
D

E:
$$\begin{array}{c} CHO \\ H-\!\!\!-OH \\ H-\!\!\!-OH \\ CH_2OH \end{array}$$
E

F:
$$\begin{array}{c} COOH \\ H-\!\!\!-OH \\ H-\!\!\!-OH \\ COOH \end{array}$$
F

12.

$$\begin{array}{c} CHO \\ H-\!\!\!-OCH_3 \\ CH_3O-\!\!\!-H \\ H-\!\!\!-OCH_3 \\ H-\!\!\!-OH \\ CH_2OCH_3 \end{array}$$
*2,3,4,6-tetra-O-
methyl-D-glucose*

$$\begin{array}{c} COOH \\ H-\!\!\!-OCH_3 \\ H-\!\!\!-OCH_3 \\ H-\!\!\!-OCH_3 \\ CH_2OH \end{array} \xleftarrow[\;H_2O\;]{Br_2} \begin{array}{c} CHO \\ H-\!\!\!-OCH_3 \\ H-\!\!\!-OCH_3 \\ H-\!\!\!-OCH_3 \\ CH_2OH \end{array}$$
*2,3,4-tri-O-methyl-
D-ribonic acid*

Therefore, H must be:

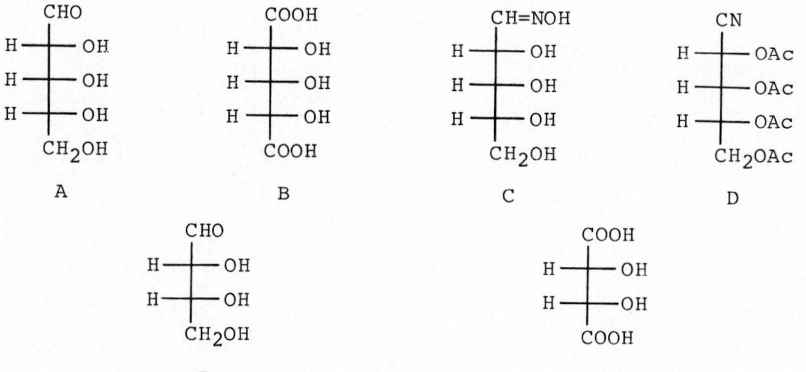

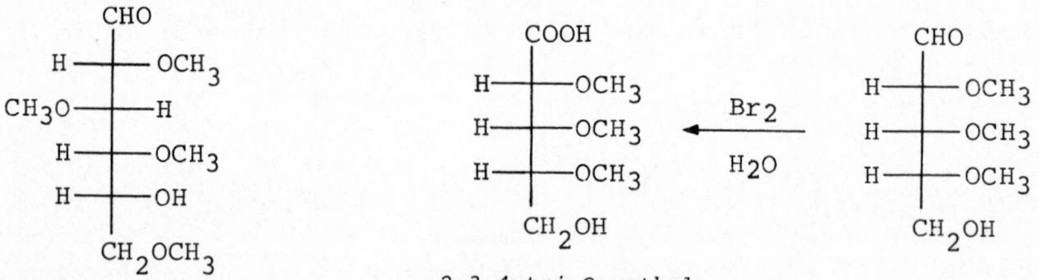

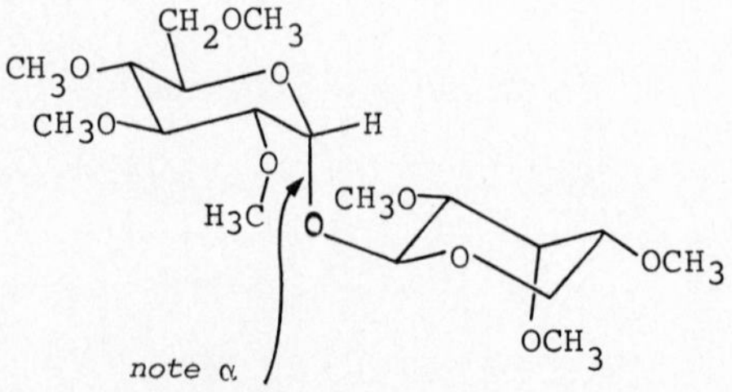

and G is:

*note α
linkage*

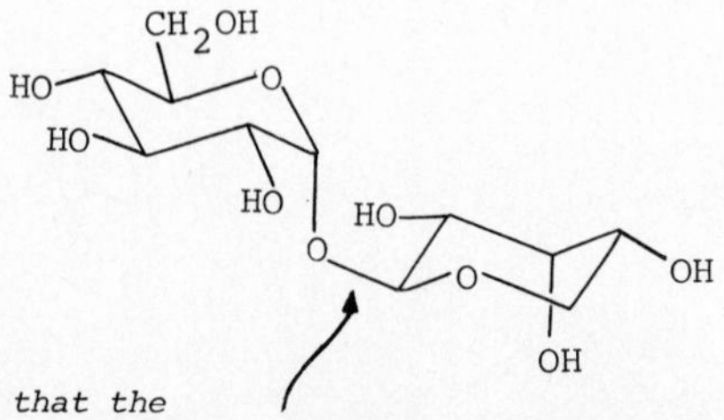

*Note that the
stereochemisty at the pentose anomeric
carbon is not established. Either α or β
is compatible with the data provided.*

13. (a) Optically active N must be:

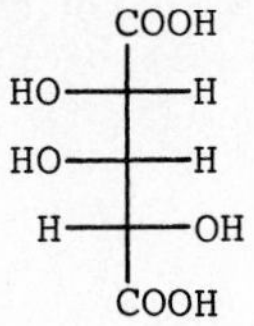

Hence, pentose M must be:

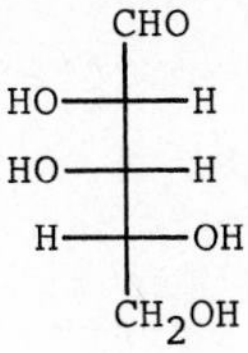

M'

or

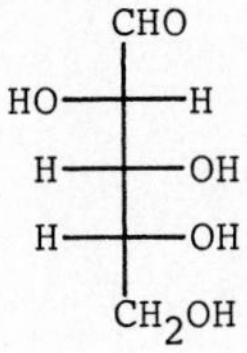

M"

Since M results from K by Ruff degradation, four possible structures are indicated for K, but only one can give an optically inactive diacid, L.

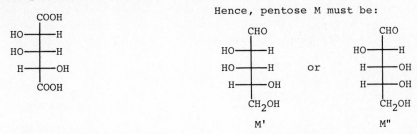

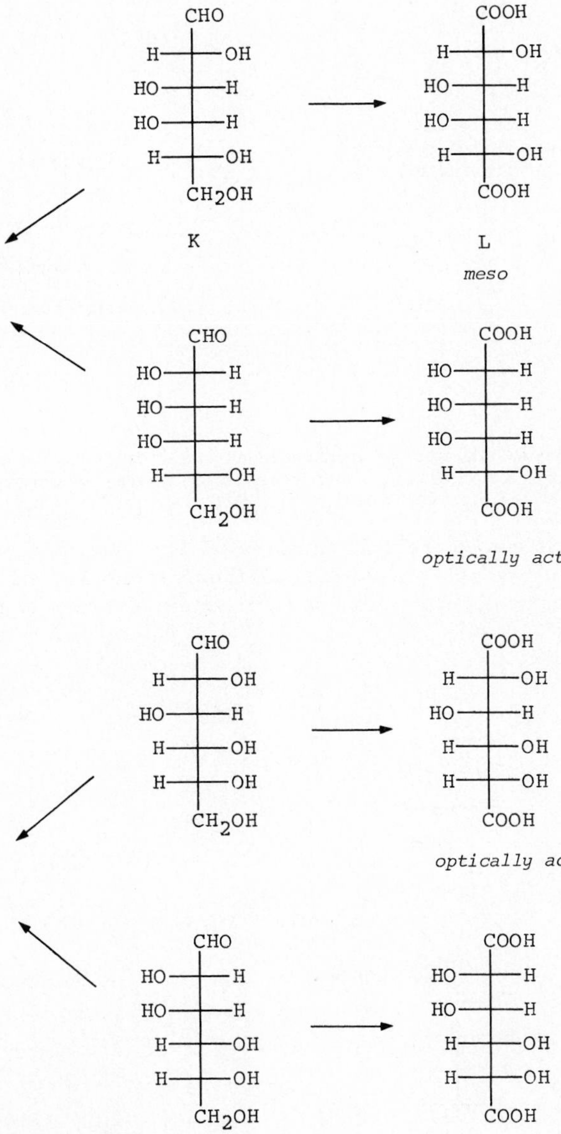

K

L

meso

M'

optically active

optically active

M"

optically active

Thus, M ≡ M'.

J is evidently a methyl glycoside of K. The reaction with $NaOH/(CH_3)_2SO_4$ converts all hydroxyl groups to methyl ether groups. The aqueous HCl hydrolyzes the glycoside acetal and nitric acid oxidizes at the resulting CHO and C-OH groups. From the products, we deduce the hydrolysis product as:

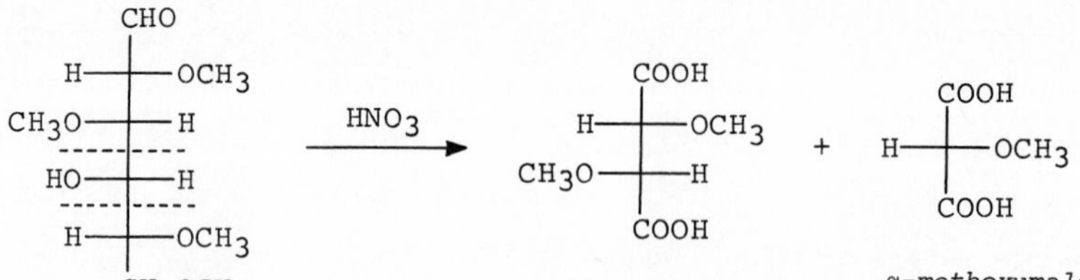

α,β-dimethoxy-succinic acid *α-methoxymalonic acid*

Thus, J must be:

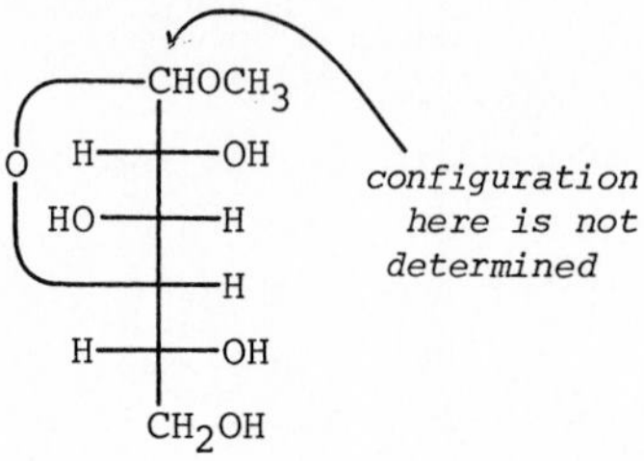

Note that

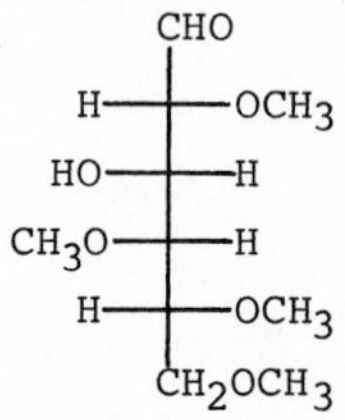

could give α,β-dimethoxysuccinic acid and α-methoxymalonic acid on oxidation, but the corresponding cyclic acetal has a four-membered ring and is not a reasonable structure.

NOTE also that the α-methoxymalonic acid could derive from further oxidation of the dimethoxysuccinic acid, rather than from oxidation of the 6-methoxy ether.

(b) Not only has the configuration (α or β) of the anomeric carbon not been established, but we have assumed D-configurations in the above structures. The available data do not allow a distinction between D and L.

14.

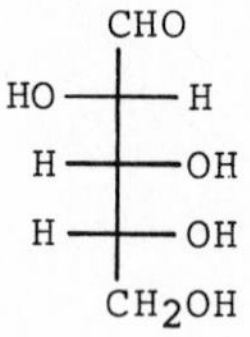

O

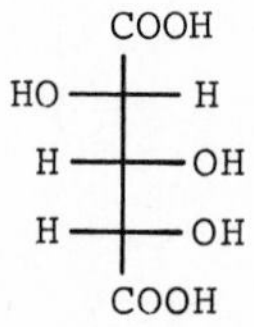

P

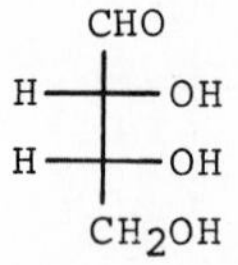

Q

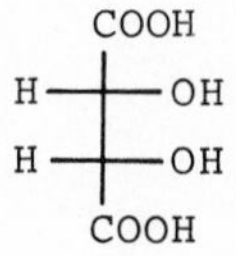

R

15.

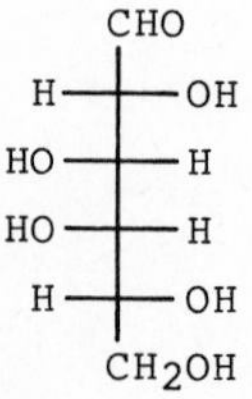

S

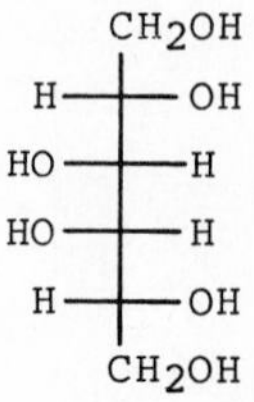

T

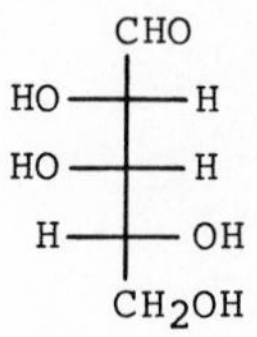

U

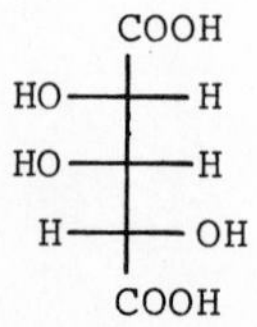

V

16.

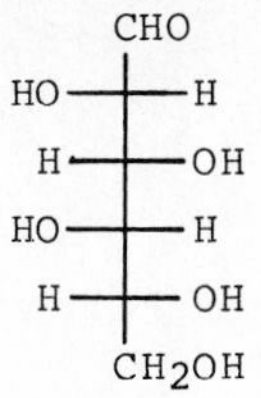

W

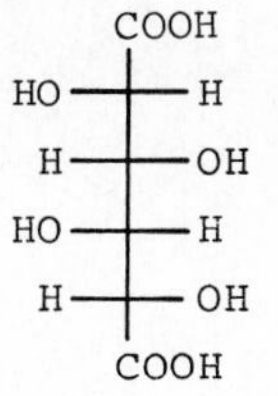

X

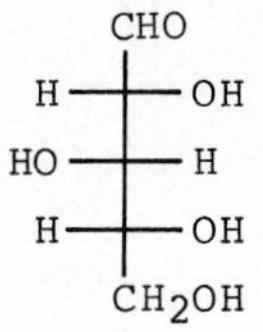

Y

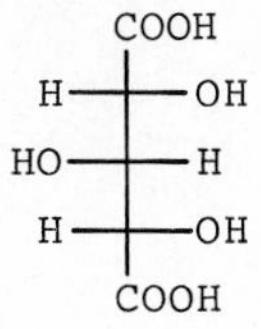

Z

17. Let n_α = fraction α, and $1 - n_\alpha$ = fraction β

$$n_\alpha(29.3) + (1 - n_\alpha)(-17.0) = 14.2$$

$n_\alpha = 0.674$, or 67.4% $n_\beta = 0.326$, or 32.6%

18.

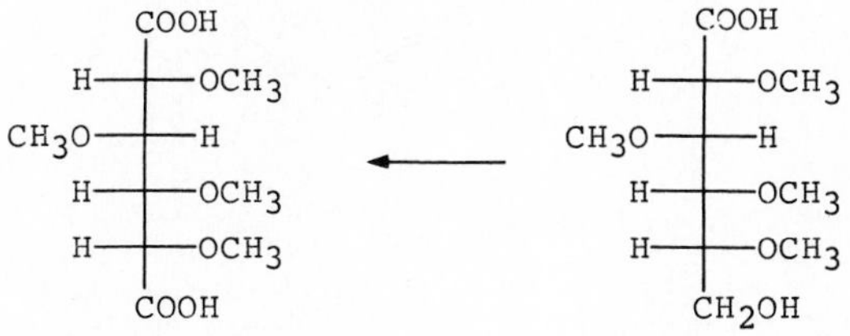

tetra-O-methylglucaric acid   AA, *tetra-O-methylgluconic acid*

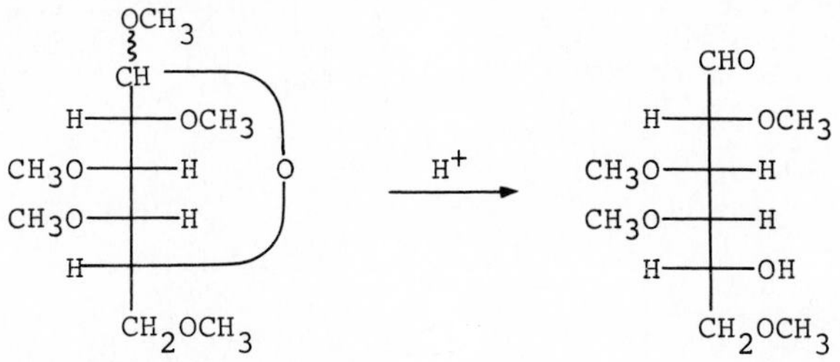

methyl-2,3,4,6-tetra-O-methylgalactopyranoside   BB, *tetra-O-methylgalactose*

Melibionic acid must join C-6 of gluconic acid to C-1 of galactose:

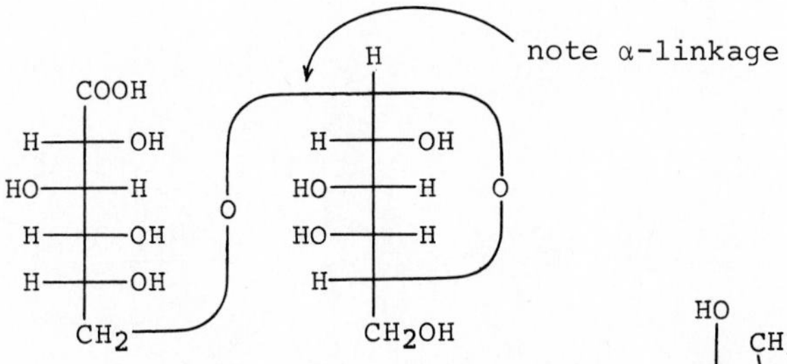

melibionic acid

or:

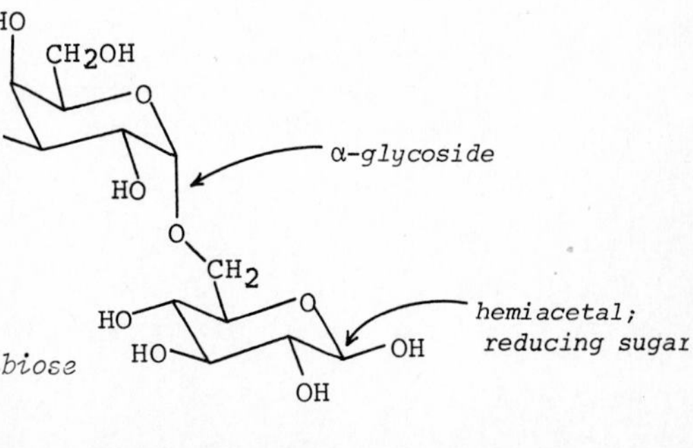

melibiose

19.

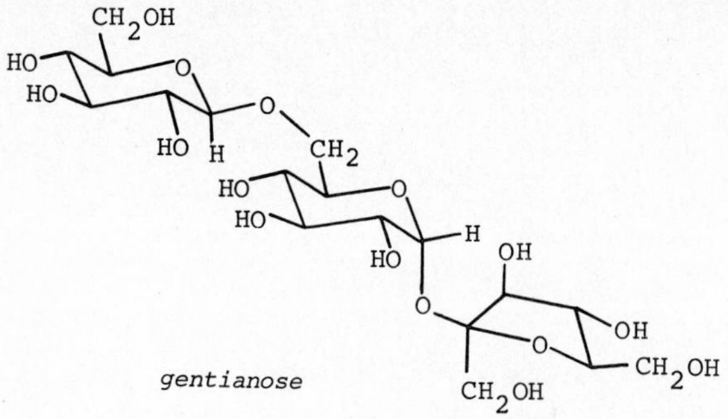

gentianose

20.

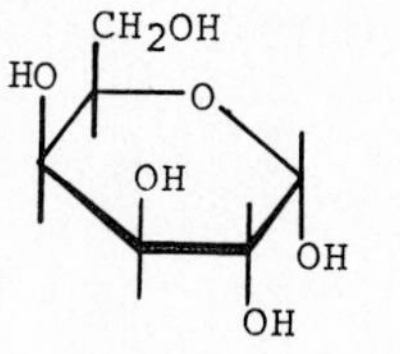

α-*D-galactopyranose*

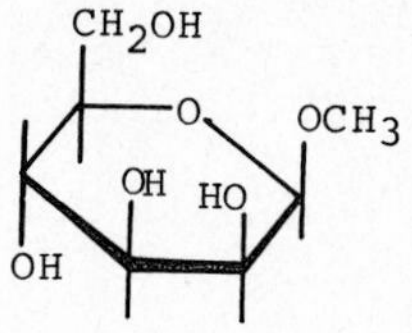

methyl β-D-mannoside

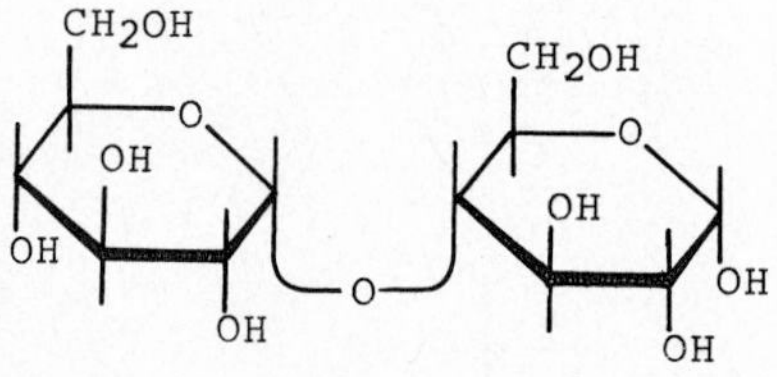

α-*maltose*

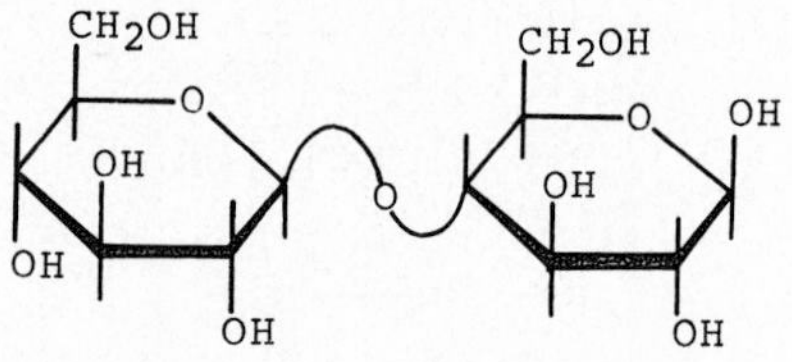

β-*cellobiose*

21.

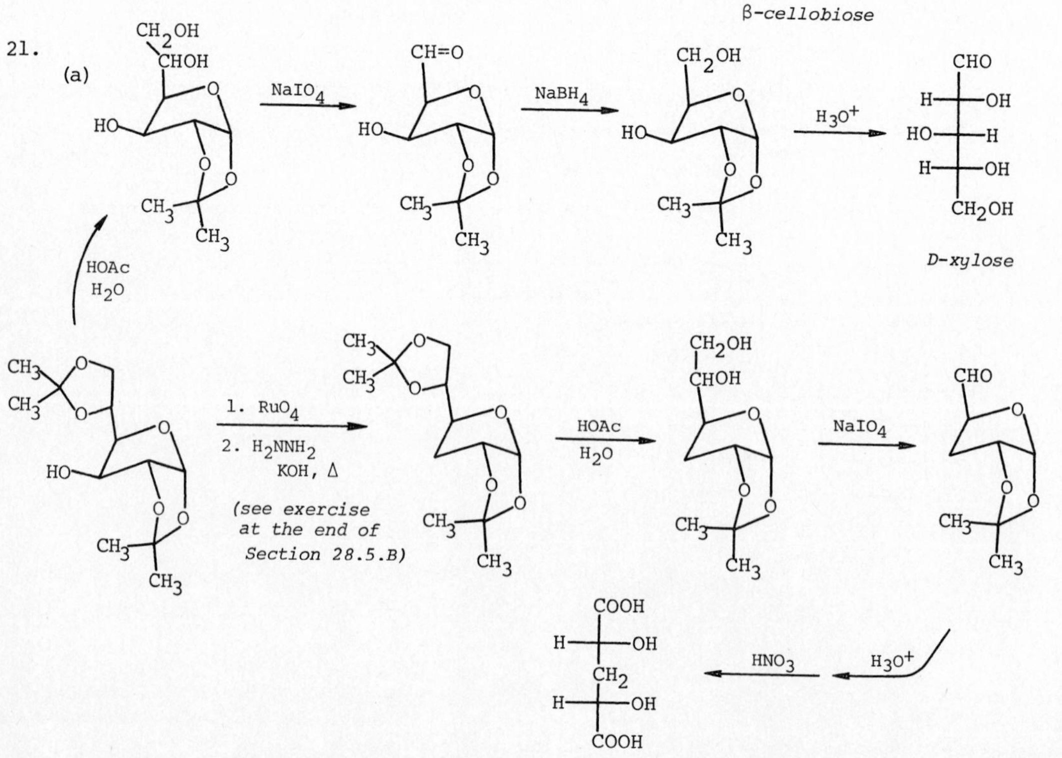

(c)

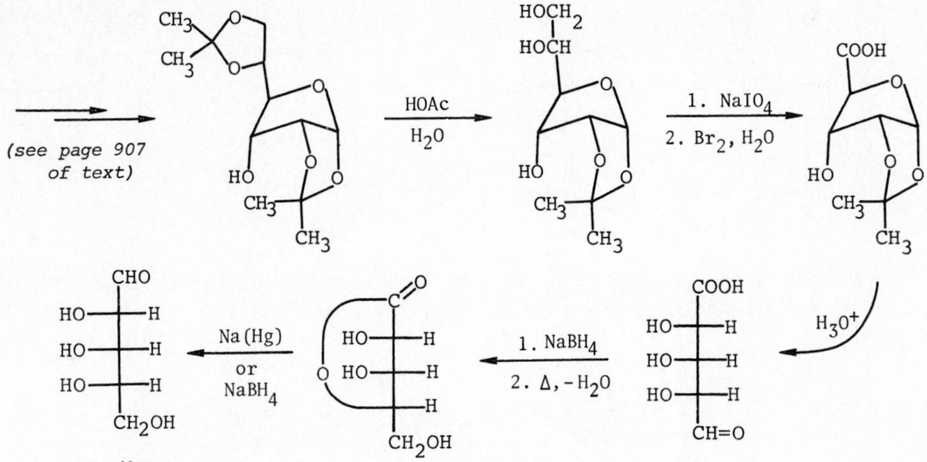

L-ribose

22.

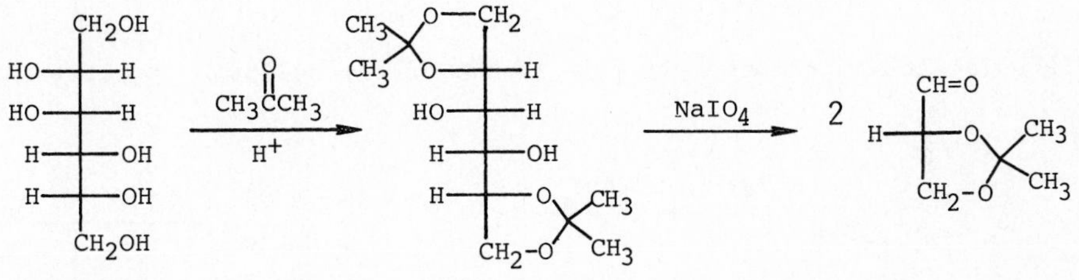

D-mannitol

28.E Supplementary Problems

S1. Write a Fischer projection of the open-chain form of each of the sugars illustrated below.

(a)

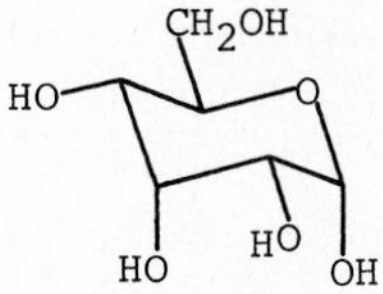

(b)

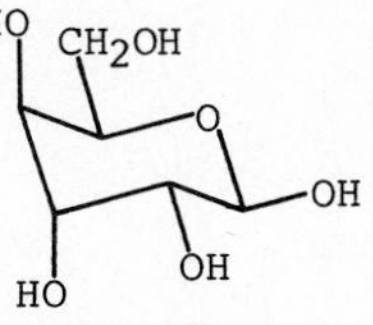

(c)

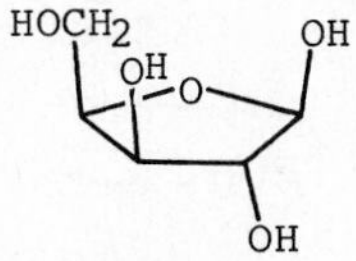

(d)

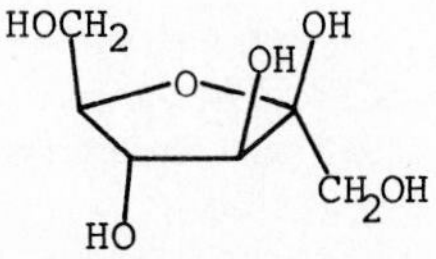

(e)

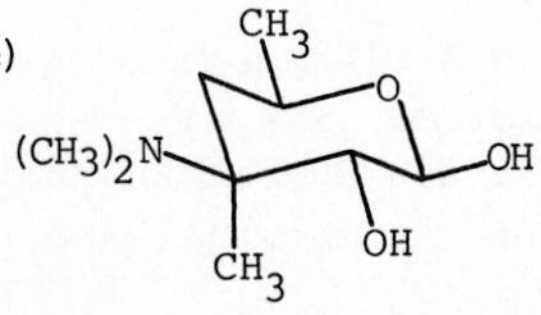

(f)

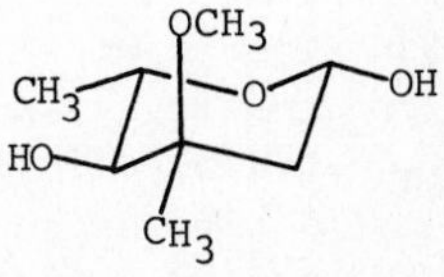

S2. Write perspective formulas of the pyranose β-anomers of each of the sugars illustrated on the following page, choosing the most stable conformation where appropriate.

(a)

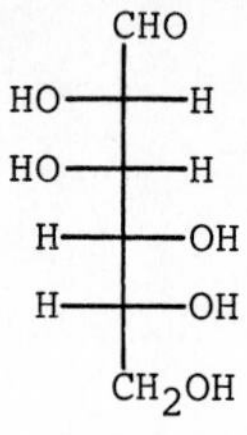

```
      CHO
HO ——— H
HO ——— H
H  ——— OH
H  ——— OH
     CH₂OH
```

(b)

```
      CHO
HO ——— H
H  ——— OH
HO ——— H
H  ——— OH
     CH₂OH
```

(c)

```
      CHO
H  ——— OH
      CH₂
HO ——— H
     CH₂OH
```

(d)

```
      CHO    O
H  ——— NHCCH₃
HO ——— H
H  ——— OH
H  ——— OH
     CH₂OH
```

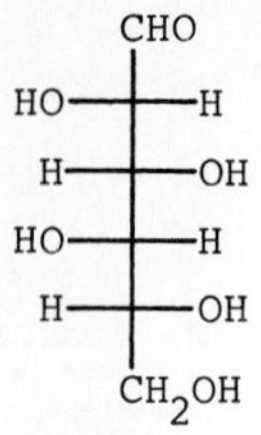

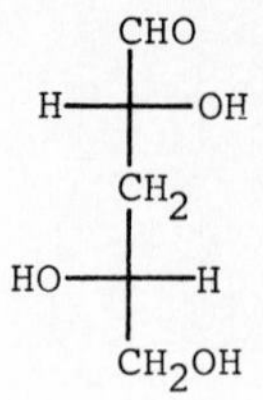

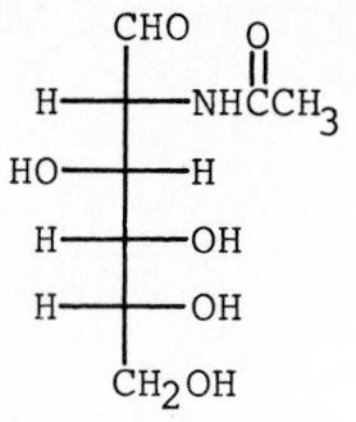

S3. Write a step-by-step mechanism for the base-catalyzed interconversion of the furanose and pyranose forms of fructose:

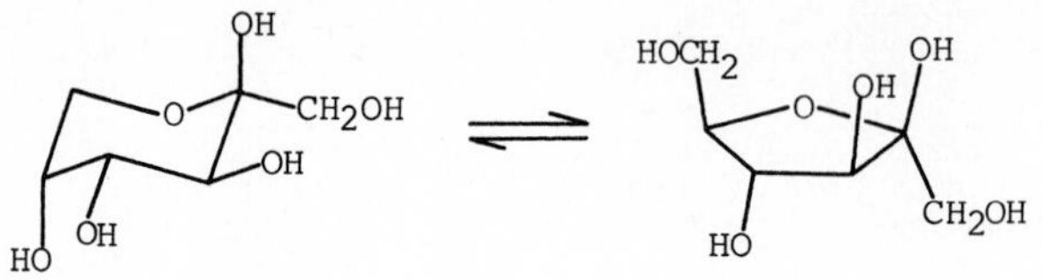

S4. Predict the major product(s) from each of the following reaction sequences:

(a)

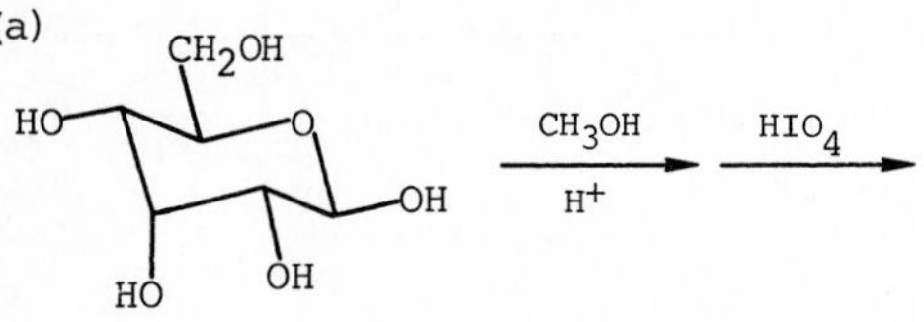

$$\xrightarrow[H^+]{CH_3OH} \quad \xrightarrow{HIO_4}$$

(d)

```
      CH=O
H  ——— OH
H  ——— OH
     CH₂OH
```

$$\xrightarrow{HCN} \quad \xrightarrow[2.\ \Delta]{1.\ H_3O^+} \quad \xrightarrow[pH\ 3.5]{NaBH_4}$$

(b)

$$\xrightarrow{Ag(NH_3)_2^+}$$

(e)

```
      CH=O
H  ——— OH
HO ——— H
      CH₃
```

$$\xrightarrow[H_2O]{Br_2} \quad \xrightarrow{CaCO_3} \quad \xrightarrow[Fe^{+3}]{H_2O_2}$$

(c)

```
      CH=O
HO ——— H
H  ——— OH
     CH₂OH
```

$$\xrightarrow[H_2O]{Br_2} \quad \xrightarrow[-H_2O]{\Delta}$$

(f)

$$\xrightarrow[(CH_3)_2SO_4]{NaOH} \quad \xrightarrow{H_3O^+}$$

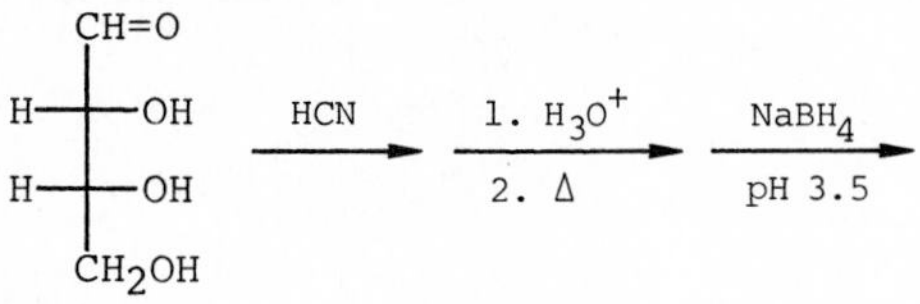

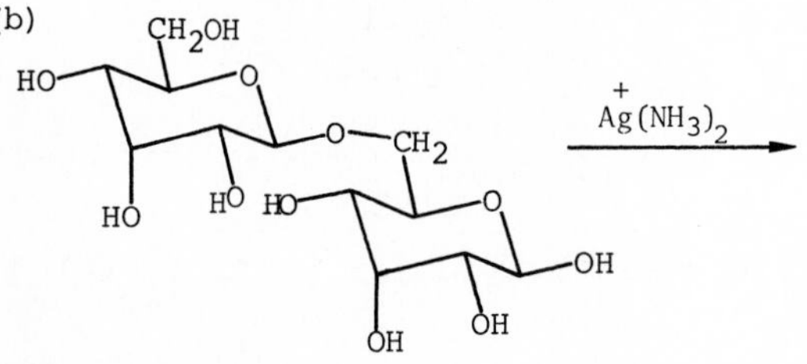

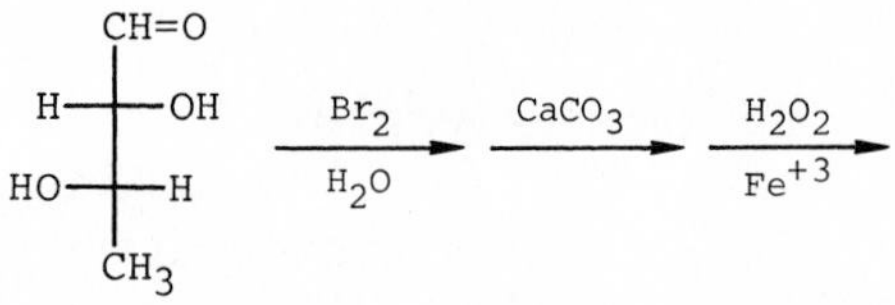

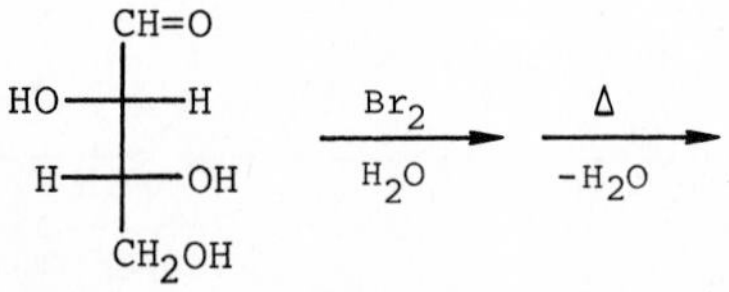

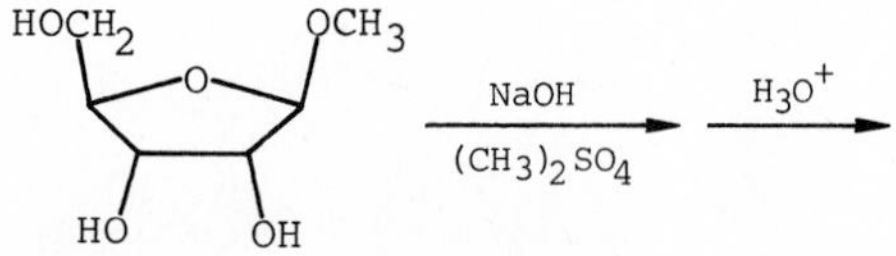

S5. Treatment of D-aldopentose A with sodium borohydride gives an optically inactive alditol, B. Reaction of A with HCN, followed by acid-catalyzed hydrolysis, produces two aldonic acid, C and D, both of which afford optically active saccharic acids after treatment with nitric acid. What are A, B, C, and D?

S6. Oxidation of L-aldohexose E with nitric acid leads to an optically active product. Ruff degradation of E provides an aldopentose F, which loses all optical activity on reaction with sodium borohydride. Kiliani-Fischer chain extension of F gives E back again, along with an isomeric aldohexose G. G reacts with nitric acid to afford an optically inactive diacid. What are E, F, and G?

S7.

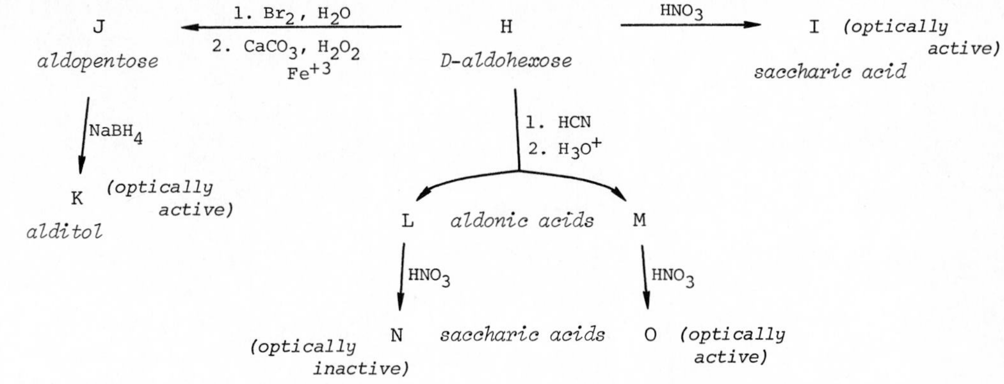

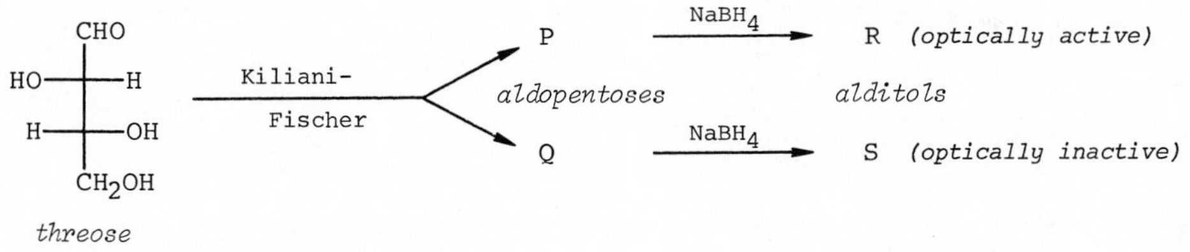

Draw Fischer projections for all the compounds (H through S) included in the scheme above. *(HINT: both P and Q are different from J.)*

S8. Hydrolysis of the disaccharide sophorose furnishes two moles of D-glucose; it is not hydrolyzed by α-glucosidase. Oxidation of sophorose with bromine water, followed by permethylation with sodium hydroxide and dimethyl sulfate, leads to an octamethylsophoronic acid derivative. Mild acid treatment of this compound produces a solution which reduces periodic acid. What is the structure of sophorose?

S9. Formation of a triphenylmethyl ("*trityl*") ether from an alcohol and triphenylmethyl chloride with pyridine is selective for the reaction of primary alcohols in carbohydrate derivatives, as illustrated below:

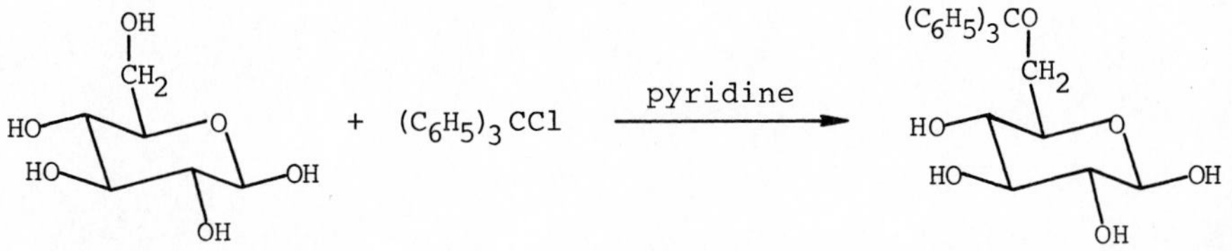

A trityl ether is stable to alkaline reaction conditions, but is hydrolyzed in dilute acid. Using this information, show how to synthesize L-fucose from D-galactose.

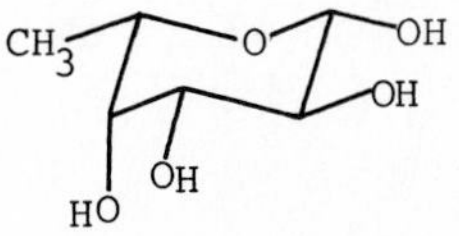

L-fucose

28.F Answers to Supplementary Problems

S1. (a)

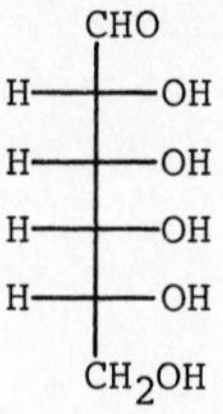

(b)

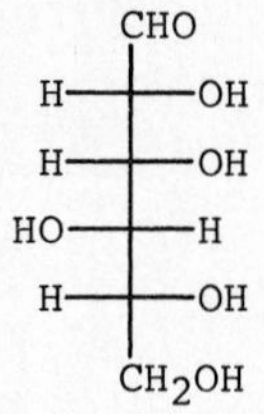

(c)

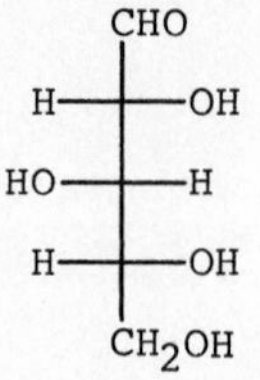

(d)

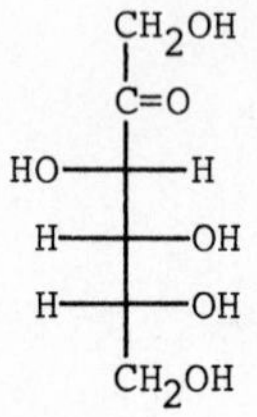

(e)

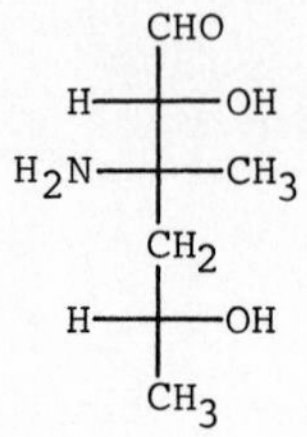

(f)

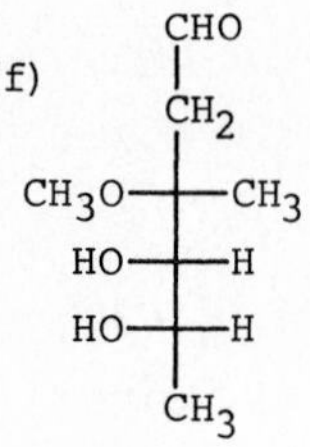

S2. (a)

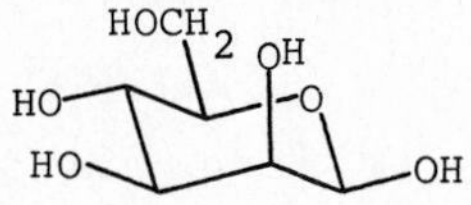

(b)

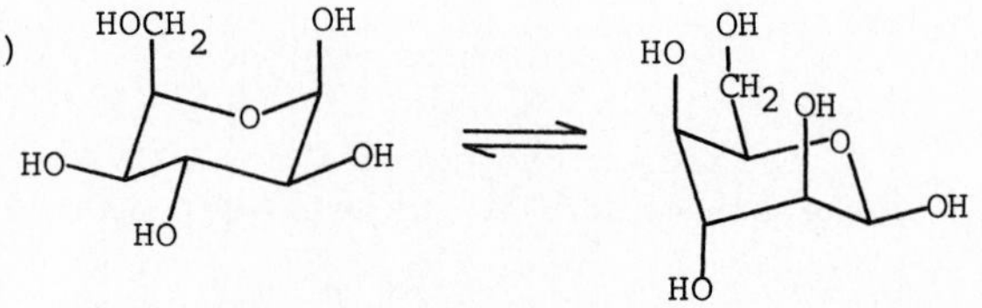

(c)

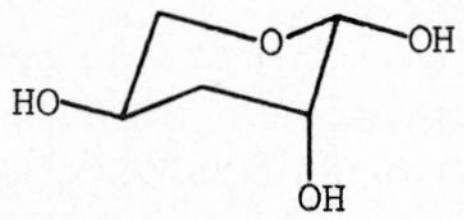

NOTE: for an L-sugar, this is the β-anomer

(d)

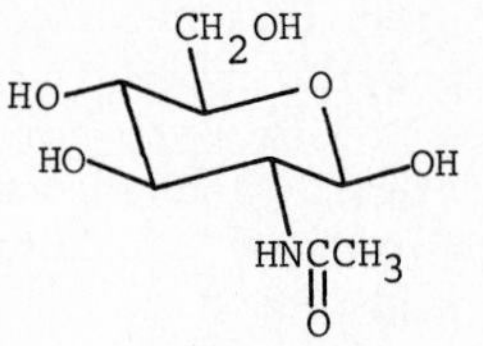

S3.

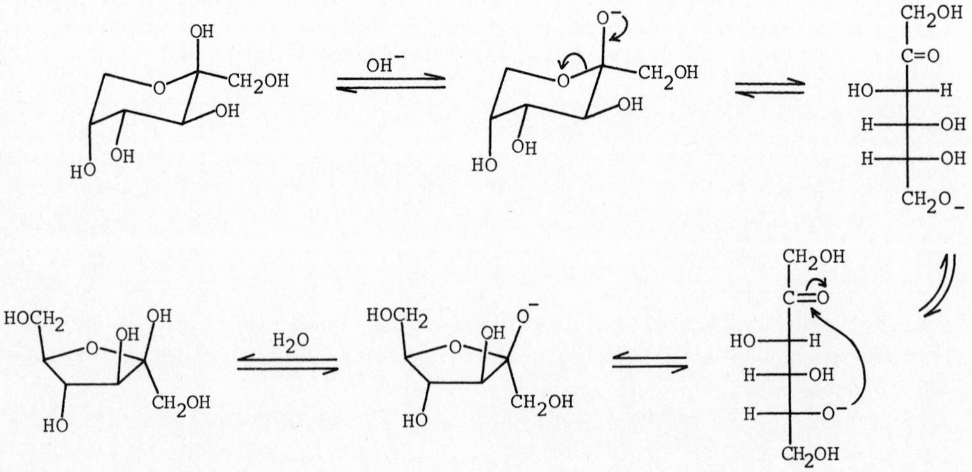

S4. (a)

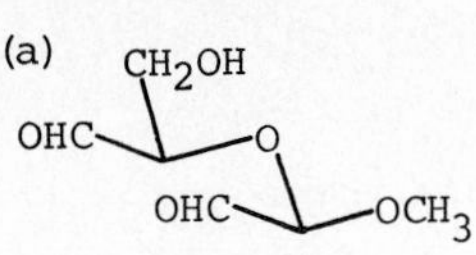

$+ \ HCOOH$

(b)

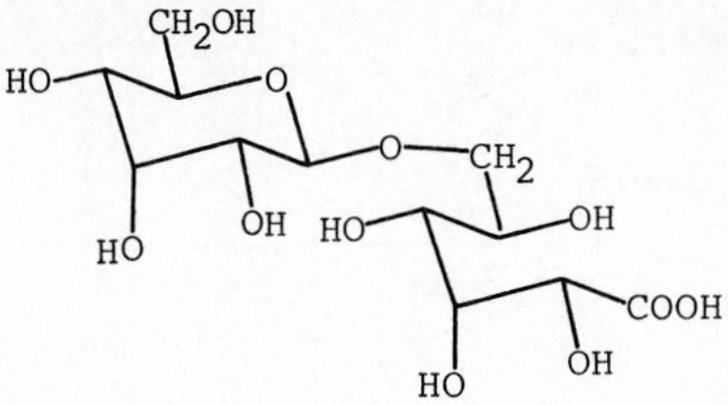

(c)

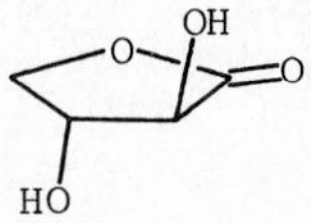

(d)

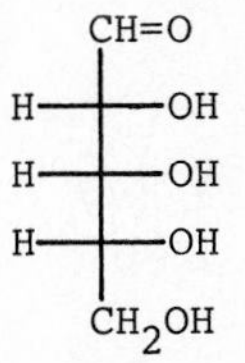

and

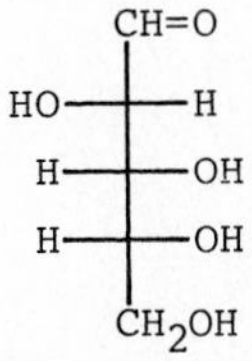

(e)

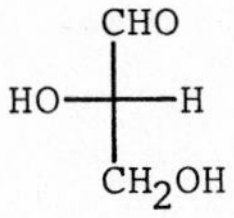

(f)

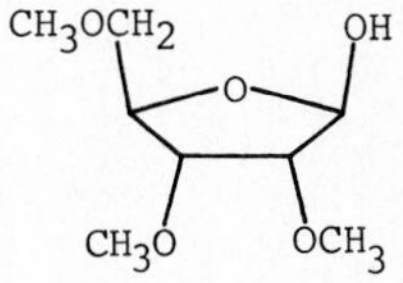

S5.

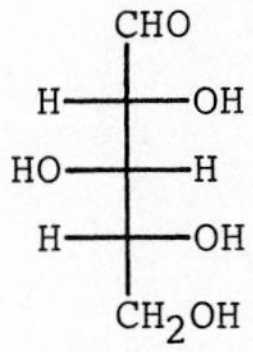

A

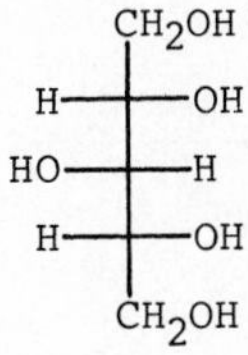

B

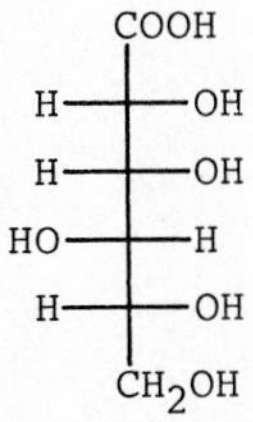

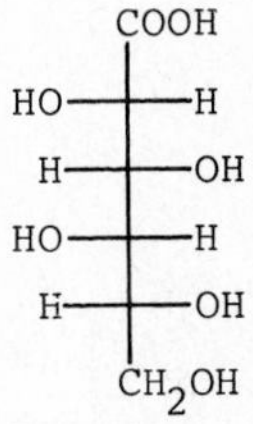

C and D

S6.

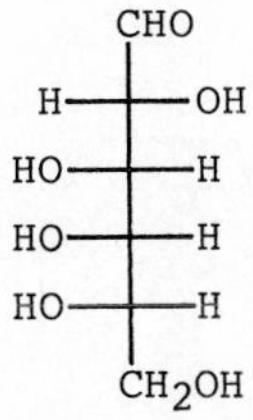

E

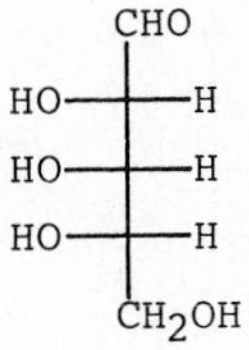

F

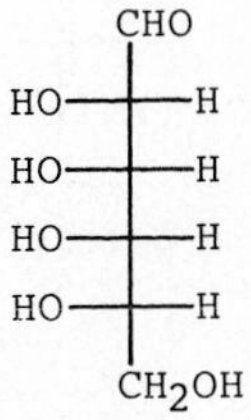

G

S7.

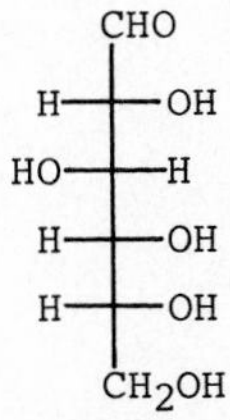

H *(Glucose)*

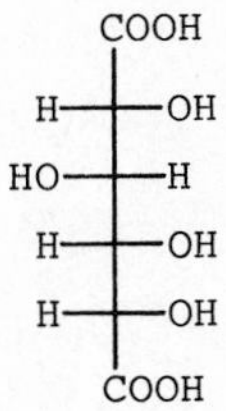

I

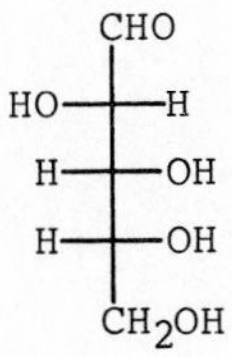

J

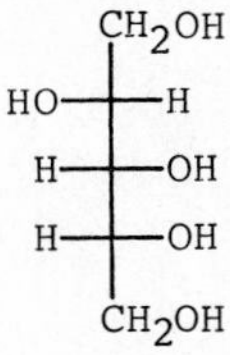

K

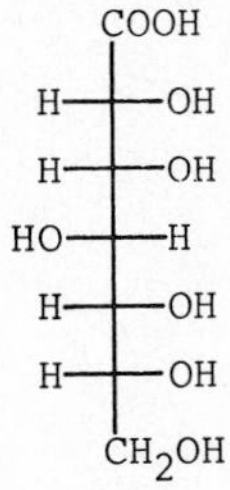

L

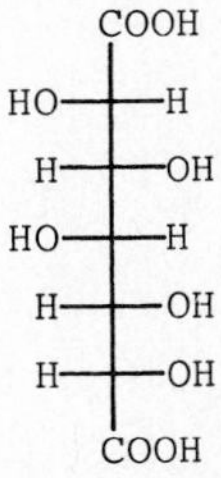

M

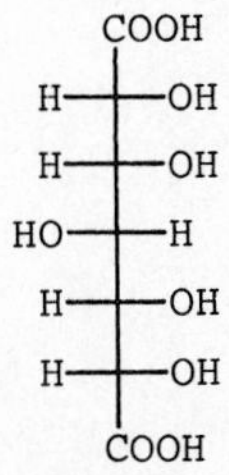

N

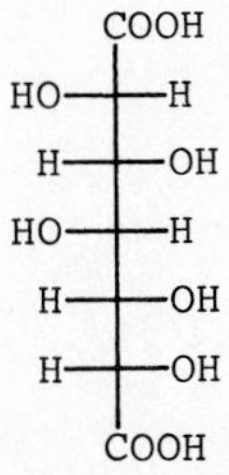

O

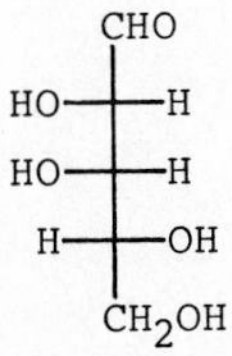

P

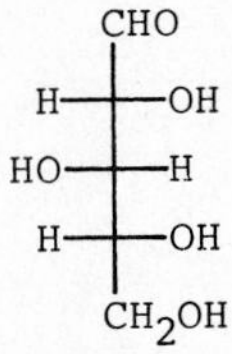

Q

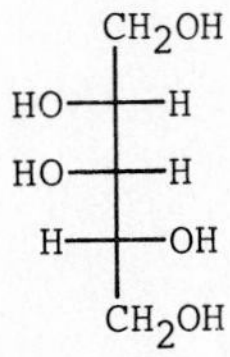

R

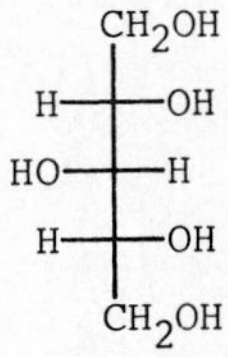

S

S8.

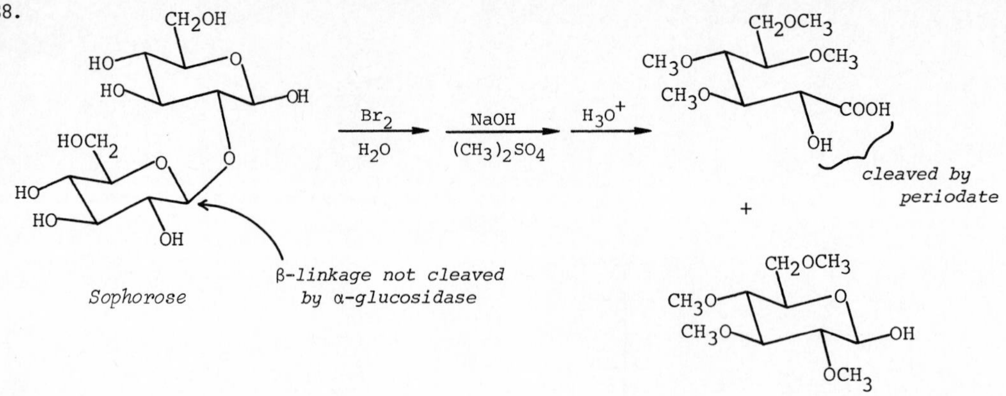
CH2OH
HO
HO
O
OH
HOCH2
O
O
HO
HO
OH
Sophorose
β-linkage not cleaved by α-glucosidase
Br2
H2O
NaOH
(CH3)2SO4
H3O+
CH2OCH3
CH3O
OCH3
CH3O
COOH
OH
cleaved by periodate
+
CH2OCH3
CH3O
O
CH3O
OH
OCH3

S9.

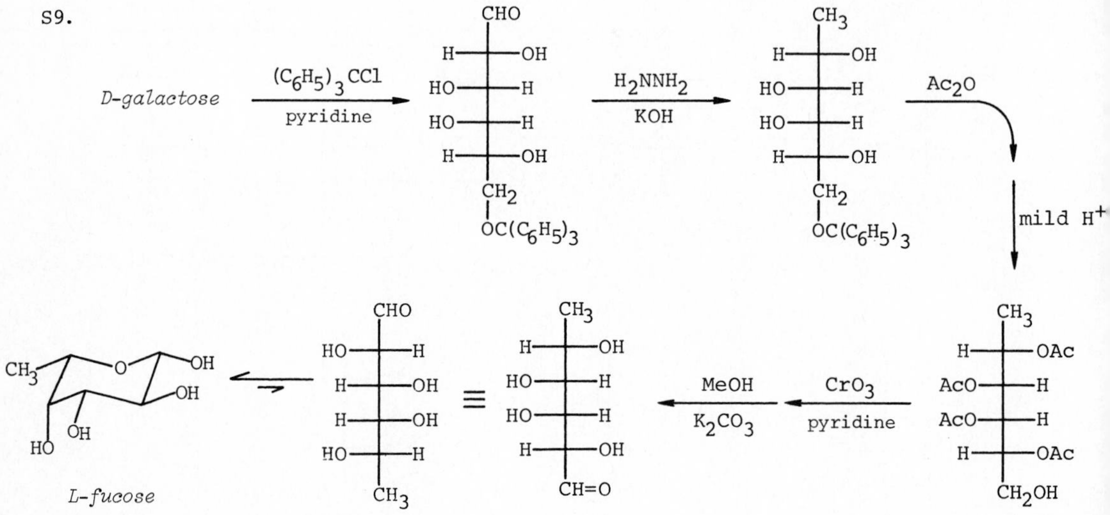
D-galactose
(C6H5)3CCl
pyridine
CHO
H
OH
HO
H
HO
H
H
OH
CH2
OC(C6H5)3
H2NNH2
KOH
CH3
CH2
OC(C6H5)3
Ac2O
mild H+
CH3
H
OAc
AcO
H
AcO
H
H
OAc
CH2OH
CrO3
pyridine
MeOH
K2CO3
CH3
CH=O
≡
CHO
CH3
CH3
O
OH
OH
OH
HO
L-fucose

29. AMINO ACIDS, PEPTIDES, AND PROTEINS

29.A Outline of Chapter 29, Important Terms Introduced, and Reactions Discussed

29.1 Introduction

amphoteric

zwitterion, inner salt

peptide bond

di-, tri-, tetra-, and polypeptides

29.2 Structure, Nomenclature, and Physical Properties of Amino Acids

D,L vs. R,S (D=R, L=S)

29.3 Acid-Base Properties of Amino Acids

amphoterism

isoelectric point

pk_a's $\simeq 2.4$, 9.8 for α-amino acids (plus side chain groups)

29.4 Synthesis of Amino Acids

A. Commercial Availability

L- cheaper than D-

B. Amination of α-Halo Acids

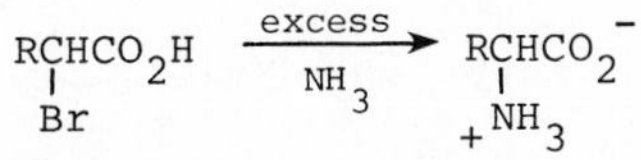

$$\underset{\text{Br}}{\text{RCHCO}_2\text{H}} \xrightarrow[\text{NH}_3]{\text{excess}} \underset{^{+}\text{NH}_3}{\text{RCHCO}_2^-}$$

C. Alkylation of N-substituted Aminomalonic Esters

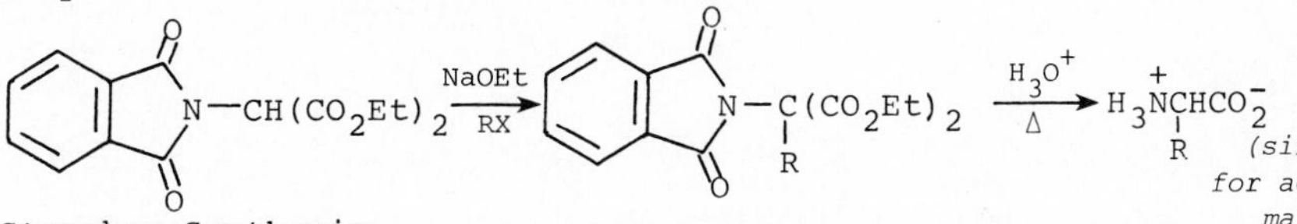

(similarly for acetamido-malonate)

D. Strecker Synthesis

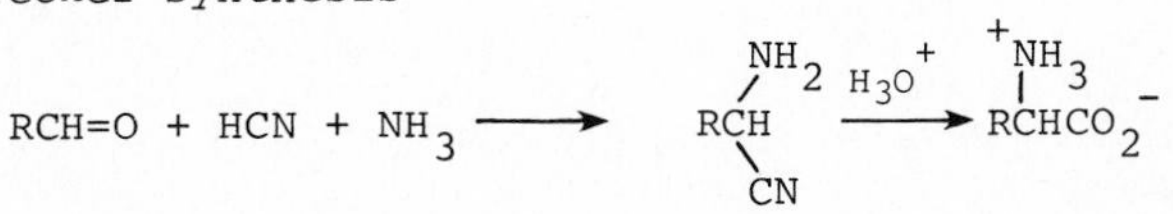

$$\text{RCH=O} + \text{HCN} + \text{NH}_3 \longrightarrow \text{RCH(NH}_2\text{)CN} \xrightarrow{\text{H}_3\text{O}^+} \text{RCH(}^{+}\text{NH}_3\text{)CO}_2^-$$

E. Miscellaneous Methods

for proline, lysine, etc.

F. Resolution

method of diastereomeric salts

29.5 Reactions of Amino Acids

A. Esterification

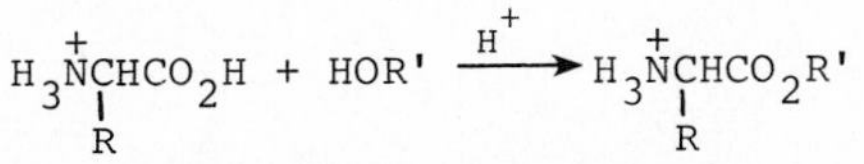

$$\text{H}_3\overset{+}{\text{N}}\underset{\text{R}}{\text{CH}}\text{CO}_2\text{H} + \text{HOR}' \xrightarrow{\text{H}^+} \text{H}_3\overset{+}{\text{N}}\underset{\text{R}}{\text{CH}}\text{CO}_2\text{R}'$$

B. Amide Formation

$$\text{H}_3\overset{+}{\text{N}}\underset{\text{R}}{\text{CH}}\text{CO}_2^- + \text{R}'\overset{\text{O}}{\overset{\|}{\text{C}}}\text{X} \xrightarrow{\text{base}} \text{R}'\overset{\text{O}}{\overset{\|}{\text{C}}}\text{NH}\underset{\text{R}}{\text{CH}}\text{CO}_2^-$$

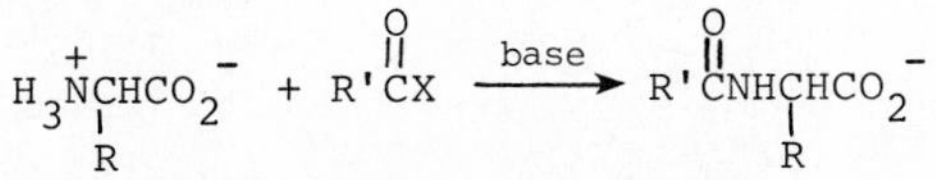

(X = Cl, O_2CR')

C. Ninhydrin Reaction

oxidative deamination   analytical method

29.6 Peptides

A. Structure and Nomenclature

amino acid sequence
N-terminal vs. C-terminal amino acids
disulfide bond

B. Synthesis of Peptides

homopolymer
2,5-diketopiperazine formation

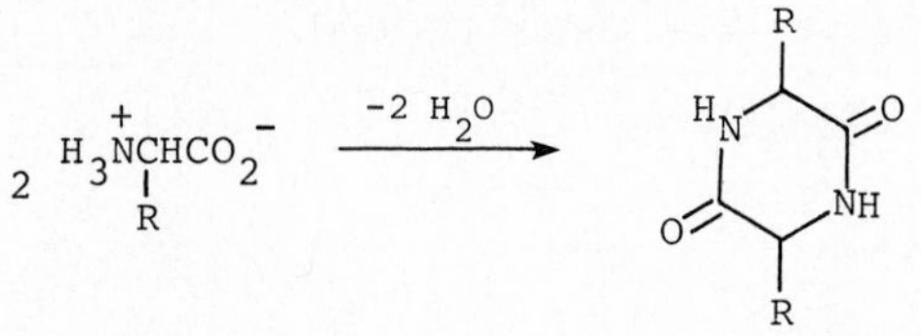

protecting groups: carbobenzoxy ("Cbz") and *t*-butoxycarbonyl ("Boc") groups

Cbz:

$$C_6H_5CH_2OCCl + H_3\overset{+}{N}CHCO_2^- \xrightarrow{\text{base}} C_6H_5CH_2OCNHCHCO_2H$$ *(removed by hydrogenolysis)*

(C=O above each OC; R below each CH)

Boc:

$$(CH_3)_3COCX + H_3\overset{+}{N}CHCO_2^- \xrightarrow{\text{base}} (CH_3)_3COCNHCHCO_2H$$ *(removed with HCl or CF_3COOH)*

coupling with dicyclohexylcarbodiimide ("DCC")

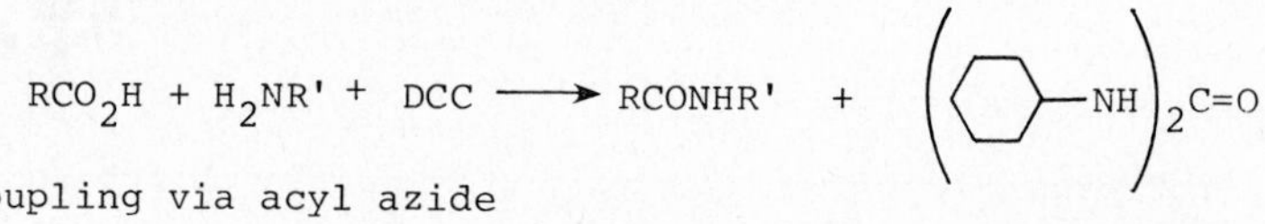

coupling via acyl azide

$$RCO_2CH_3 + H_2NNH_2 \longrightarrow RCNHNH_2 \xrightarrow{HNO_2} RCN_3 \xrightarrow{H_2NR'} RCNHR' + HN_3$$

(can avoid carboxy protection)

coupling with N-carboxyanhydrides (NCA's, Leuchs anhydride)

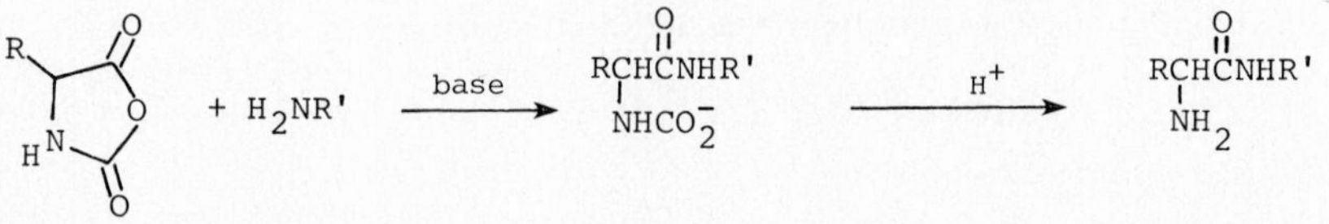

Merrifield solid-phase technique; polymer-bound peptides

C. Structure Determination

1. Amino Acid Analysis

amino acid analyzer

2. Identification of the N-Terminal Amino Acid

Sanger method

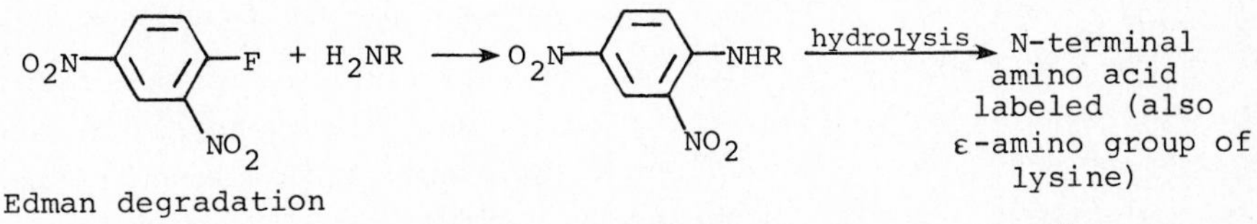

Edman degradation

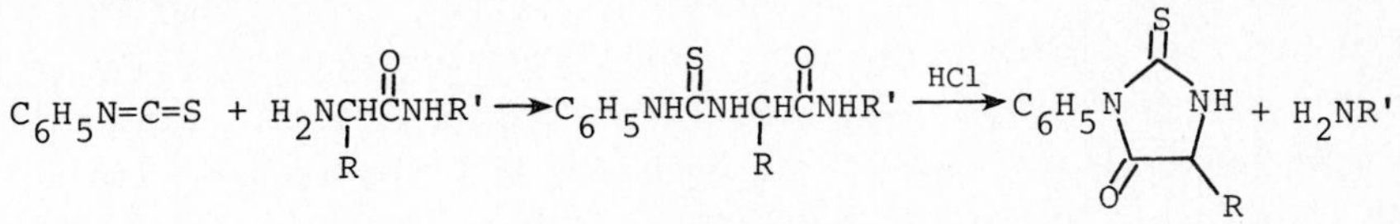

(stepwise, sequential removal and identification of N-terminal amino acids)

3. Identification of the C-Terminal Amino Acid

sequential removal of C-terminal amino acids with carboxypeptidase

4. Fragmentation of the Peptide Chain

proteases: trypsin, chymotrypsin, pepsin
cyanogen bromide (cleaves at methionine carbonyl)
partial degradation and peptide mapping

29.7 Proteins

A. Molecular Shape

fibrous vs. globular enzymes

B. Factors That Influence the Molecular Shape

1. Secondary Inter- or Intrachain Bonding

hydrogen bonds disulfide bridges

2. Electronic and Steric Properties of the Side Chain Groups

hydrophobic vs. hydrophilic

C. Structure of the Fibrous Proteins

α-helix super helix
random coil pleated sheet

D. Structure of the Globular Proteins

denaturation/renaturation

E. Biological Function of the Proteins--An Overview

29.B Important Concepts and Hints

Like the last chapter, this one discusses the special chemistry that arises when two familiar functional groups are present in the same molecule. Although most of the reactions of amino acids are ones you've seen before, complications can arise because of the juxtaposition of the acidic and basic groups. For example, a sequence of protection and deprotection steps is necessary for controlled formation of the amide linkage in a polypeptide, whereas it is a very straightforward process in monofunctional molecules.

The topic of this chapter is important not only for the chemistry it presents, but also because of the biological importance of amino acids, oligopeptides, and proteins. The frontiers of biology have reached the molecular level, and it is necessary to understand the chemistry of one of its most important groups of building blocks. It really is useful to know the structures, names, and abbreviations of the amino acids, and we urge you to learn them if you have any interest in the life sciences.

29.C Answers to Exercises

(29.2) (a) This is especially important for biology and biochemistry students.

(b) (D) = (R); (L) = (S) for the α-position of amino acids.

(29.3) $H_3\overset{+}{N}CH_2CO_2H$ (pH 2) $\qquad H_3\overset{+}{N}CH_2CO_2^-$ (pH 4 and 8) $\qquad H_2NCH_2CO_2^-$ (pH 11)

(29.4.B)

$$C_6H_5CH_2CH_2CO_2H \xrightarrow[\Delta]{P,Br_2} \xrightarrow{H_2O} C_6H_5CH_2CH(Br)CO_2H \xrightarrow{NH_3} C_6H_5CH_2CH(\overset{+}{N}H_3)CO_2^- + NH_4Br$$

$$(CH_3)_2CHCH_2CO_2H \xrightarrow[\Delta]{P,Br_2} \xrightarrow{H_2O} (CH_3)_2CHCH(Br)CO_2H \xrightarrow{NH_3} (CH_3)_2CHCH(\overset{+}{N}H_3)CO_2^- + NH_4Br$$

$$(CH_3)_2CHCH_2CH_2CO_2H \xrightarrow[\Delta]{P,Br_2} \xrightarrow{H_2O} (CH_3)_2CHCH_2CH(Br)CO_2H \xrightarrow{NH_3} (CH_3)_2CHCH_2CH(\overset{+}{N}H_3)CO_2^- + NH_4Br$$

for serine:

$$HOCH_2CH_2CO_2H \xrightarrow[\Delta]{P,Br_2} \xrightarrow{H_2O} BrCH_2CH(Br)CO_2H$$

for tyrosine:

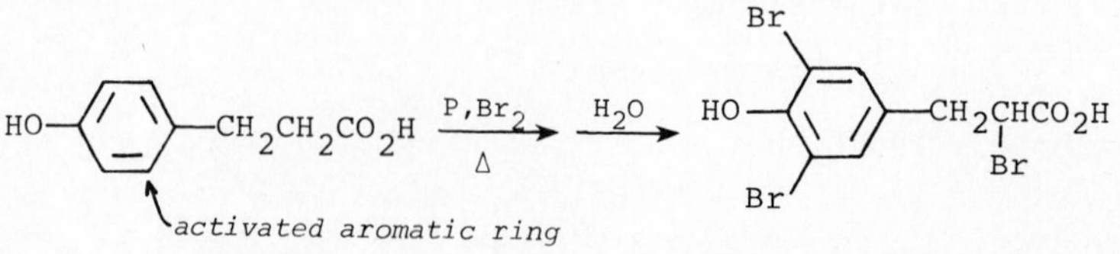

(29.4.C)

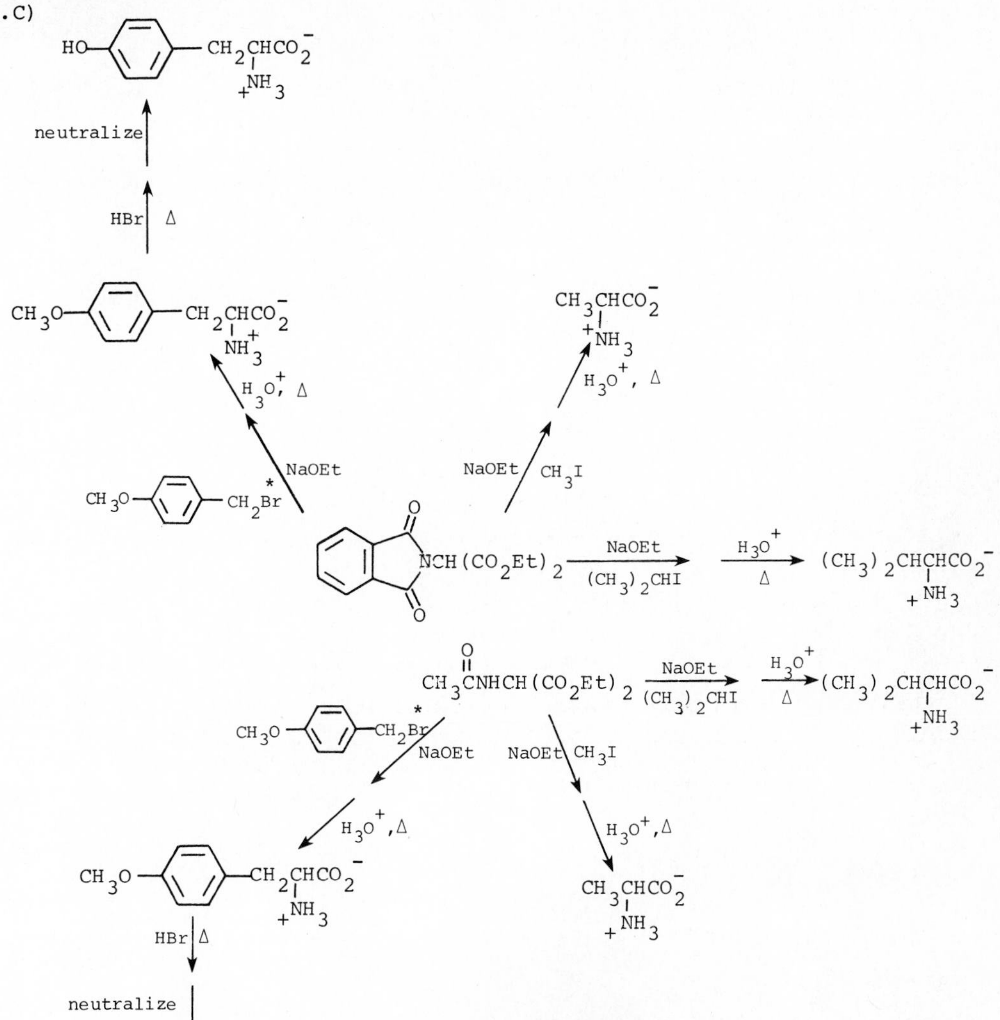

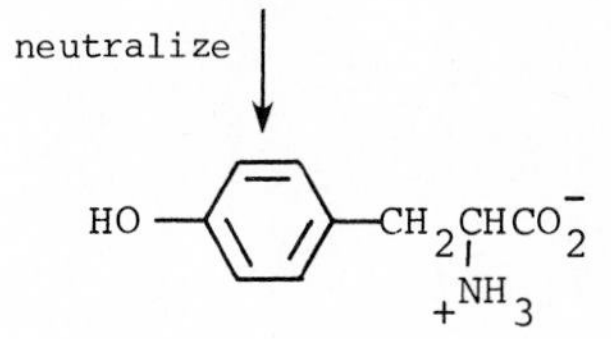

(* *The aromatic hydroxy group must be protected as the methyl ether during the alkylation step; otherwise, it too would ionize and be alkylated (polymerized).*)

(29.4.D)

for tyrosine:

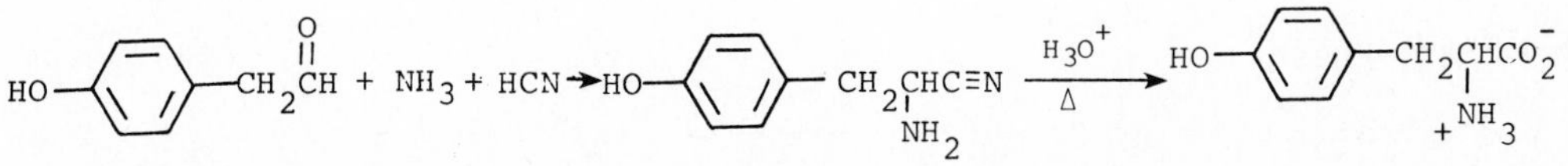

for lysine:

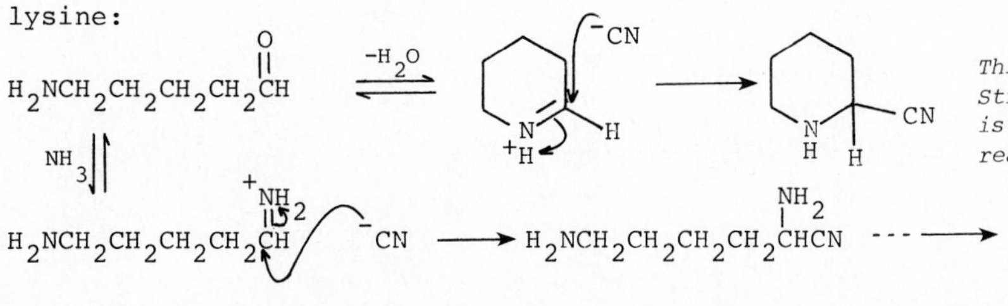

This intramolecular Strecker synthesis is the dominant reaction.

(29.4.E)

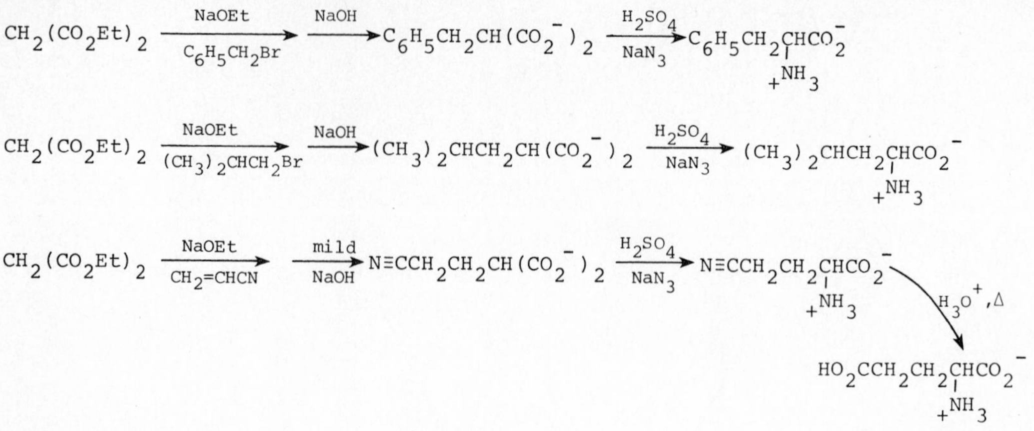

$$CH_2(CO_2Et)_2 \xrightarrow[C_6H_5CH_2Br]{NaOEt} \xrightarrow{NaOH} C_6H_5CH_2CH(CO_2^-)_2 \xrightarrow[NaN_3]{H_2SO_4} C_6H_5CH_2\underset{\overset{|}{+NH_3}}{CH}CO_2^-$$

$$CH_2(CO_2Et)_2 \xrightarrow[(CH_3)_2CHCH_2Br]{NaOEt} \xrightarrow{NaOH} (CH_3)_2CHCH_2CH(CO_2^-)_2 \xrightarrow[NaN_3]{H_2SO_4} (CH_3)_2CHCH_2\underset{\overset{|}{+NH_3}}{CH}CO_2^-$$

$$CH_2(CO_2Et)_2 \xrightarrow[CH_2=CHCN]{NaOEt} \xrightarrow[NaOH]{mild} N\equiv CCH_2CH_2CH(CO_2^-)_2 \xrightarrow[NaN_3]{H_2SO_4} N\equiv CCH_2CH_2\underset{\overset{|}{+NH_3}}{CH}CO_2^-$$

$$\xrightarrow{H_3O^+,\Delta} HO_2CCH_2CH_2\underset{\overset{|}{+NH_3}}{CH}CO_2^-$$

(29.5.B)

$$C_6H_5CH_2\underset{\overset{|}{+NH_3}}{CH}CO_2^- \xrightarrow[CH_3OH,\Delta]{HCl} \xrightarrow{neutralize} C_6H_5CH_2\underset{\overset{|}{NH_2}}{CH}CO_2CH_3 \xrightarrow[Et_3N]{Ac_2O} C_6H_5CH_2\underset{\overset{|}{CH_3\underset{\overset{\|}{O}}{C}NH}}{CH}CO_2CH_3$$

(29.6.B)

Cbz-Ala + H$_2$N$\underset{\overset{|}{CH_3}}{CH}$COOEt ($\equiv$ Ala-Et) $\xrightarrow{DCC}$ Cbz-Ala-Ala-Et $\xrightarrow{H_2-Pd/C}$ Ala-Ala-Et

$$\text{Phe-Ala-Ala-Et} \xleftarrow{H_2-Pd/C} \text{Cbz-Phe-Ala-Ala-Et} \xleftarrow[DCC]{Cbz-Phe} $$

(Cbz-Val, DCC) →

Cbz-Val-Phe-Ala-Ala-Et $\xrightarrow{H_2-Pd/C}$ Val-Phe-Ala-Ala-Et $\xrightarrow{Cbz-Ala}$

Ala-Val-Phe-Ala-Ala-Et $\xleftarrow{H_2-Pd/C}$ Cbz-Ala-Val-Phe-Ala-Ala-Et

(mild hydrolysis →)

Ala-Val-Phe-Ala-Ala

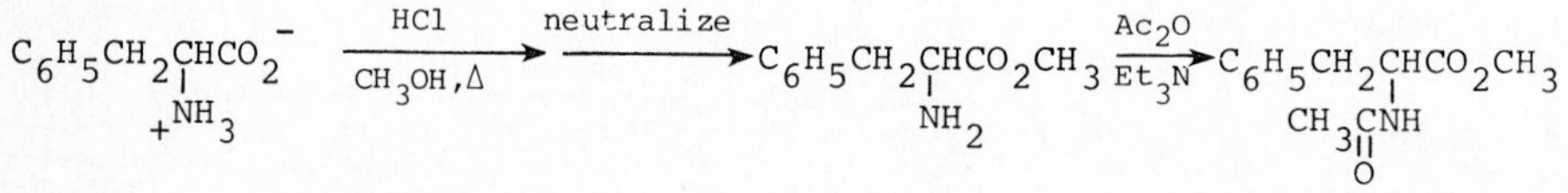

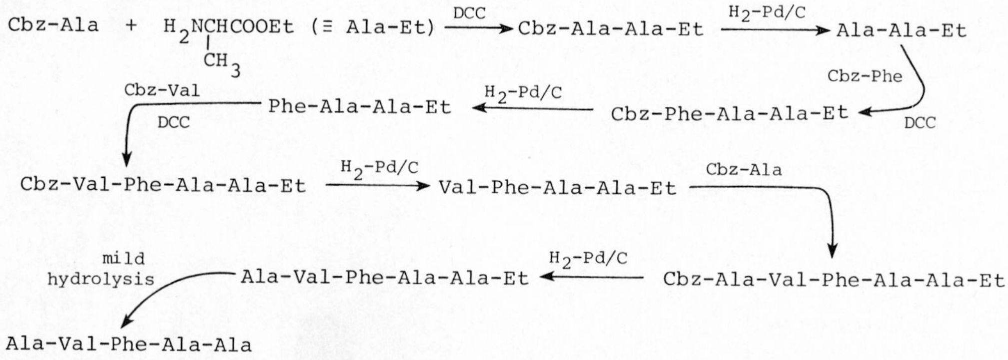

(29.6.C.2)

1) Edman degradation of tetrapeptide: N-terminal is Val: Val-( )-( )-( )

2) Edman degradation of tripeptide: next is Ser: Val-Ser-( )-( )

3) Sanger method on dipeptide: then Ala: Val-Ser-Ala-( )

4) Amino acid composition: all that is left is Gly: Val-Ser-Ala-Gly

1st Edman:

$$C_6H_5N=C=S + H_3\overset{+}{N}\underset{\overset{|}{CH(CH_3)_2}}{CH}\overset{\overset{\cdot}{\overset{O}{\|}}}{C}NH\text{-Ser-Ala-Gly} \xrightarrow{HCl} \text{(thiohydantoin)} + \text{Ser-Ala-Gly}$$

2nd Edman:

$$C_6H_5N=C=S + H_3\overset{+}{N}\underset{\overset{|}{CH_2OH}}{CH}\overset{\overset{O}{\|}}{C}NH\text{-Ala-Gly} \xrightarrow{HCl} \text{(thiohydantoin)} + \text{Ala-Gly}$$

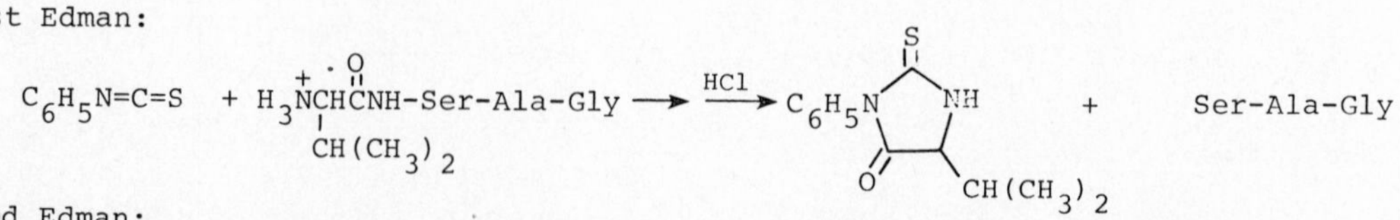

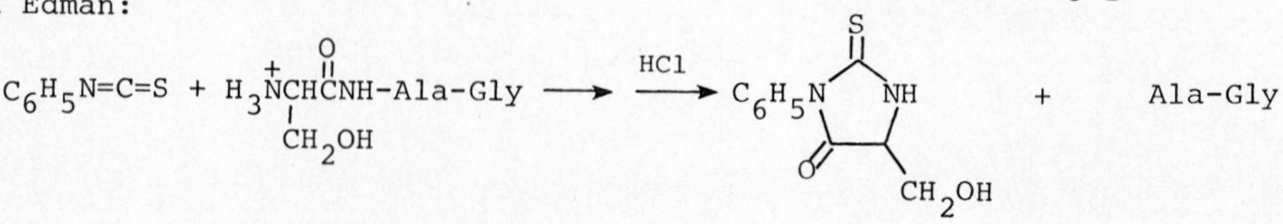

Sanger:

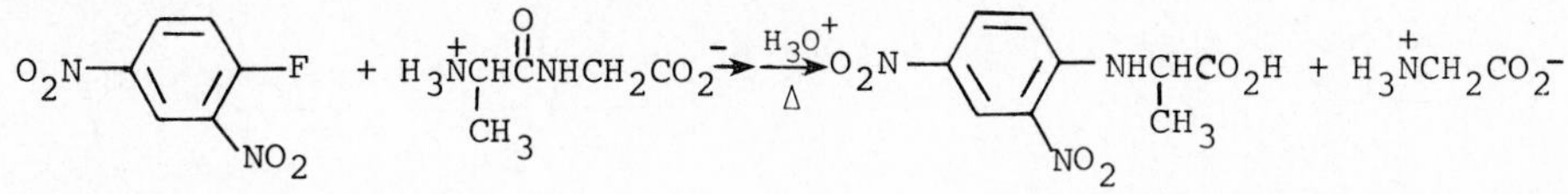

(29.6.C.3)

Gly-Ser-Phe

(29.6.C.4)

Gly-Ala-Leu-Leu-Phe

pepsin cleavages ↑ ↑ (between Ala-Leu and Leu-Leu)

29.D Answers and Explanations to Problems

1. This question requires Tables 29.2 and 29.4.

	pH=2	pH=7	pH=12
(a)	$CH_3CH_2CH(CH_3)\text{-}CH(NH_3^+)COOH$	$C_2H_5CH(CH_3)\text{-}CH(NH_3^+)CO_2^-$	$C_2H_5CH(CH_3)\text{-}CH(NH_2)CO_2^-$
(b)	$HOOCCH_2CH(NH_3^+)CO_2^-$	$^-O_2CCH_2CH(NH_3^+)CO_2^-$	$^-O_2CCH_2CH(NH_2)CO_2^-$
(c)	$H_3\overset{+}{N}(CH_2)_4CH(NH_3^+)CO_2^-$	$H_3\overset{+}{N}(CH_2)_4CH(NH_3^+)CO_2^-$	$H_2N(CH_2)_4CH(NH_2)CO_2^-$
(d)	$H_3\overset{+}{N}CH_2CONHCH_2COOH$	$H_3\overset{+}{N}CH_2CONHCH_2CO_2^-$	$H_2NCH_2CONHCH_2CO_2^-$

The remainder of the question requires two generalizations:

(1) The -COOH of the peptide is less acidic (higher pK) than that in the amino acid, because the $H_3\overset{+}{N}$- group with positive charge is farther away;

(2) the $-NH_3^+$ group of the peptide is more acidic (lower pK_a) than that in the amino acid, because the -CONH- group has greater electron-attracting inductive effect than $-CO_2^-$ group.

(e) terminal NH_2 of Lys in the peptide is approximately the same as in the amino acid; ∴

$H_3\overset{+}{N}(CH_2)_4CH(NH_3^+)CONHCH_2COOH$

ambiguous: α-NH_3^+ in Lys-Gly is close to 7 ∴ similar amounts of

$H_3\overset{+}{N}(CH_2)_4CH(NH_3^+)CONHCH_2CO_2^-$

and

$H_3\overset{+}{N}(CH_2)_4CH(NH_2)CONHCH_2CO_2^-$

$H_2N(CH_2)_4CH(NH_2)CONHCH_2CO_2^-$

(f)

$$
\overset{+}{\underset{\substack{| \\ CH_2COOH}}{CH_3CHCONHCH-CONH-CHCOOH}} \quad \overset{+}{\underset{\substack{| \\ CH_2CO_2^-}}{CH_3CHCONHCH-CONH-CHCO_2^-}} \quad \underset{\substack{| \\ CH_2CO_2^-}}{CH_3CHCONHCH-CONH-CHCO_2^-}
$$

 CH(CH_3)_2   CH(CH_3)_2   CH(CH_3)_2

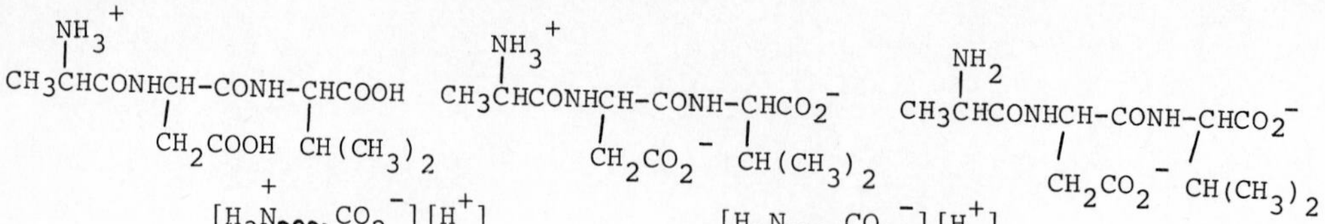

2.
$$K_1 = \frac{[\overset{+}{H_3N}\text{\raisebox{0pt}{\scriptsize$\sim\!\!\sim$}}CO_2^-][H^+]}{[\overset{+}{H_3N}\text{\raisebox{0pt}{\scriptsize$\sim\!\!\sim$}}COOH]} \qquad\qquad K_2 = \frac{[H_2N\text{\raisebox{0pt}{\scriptsize$\sim\!\!\sim$}}CO_2^-][H^+]}{[\overset{+}{H_3N}\text{\raisebox{0pt}{\scriptsize$\sim\!\!\sim$}}CO_2^-]}$$

$$K_1K_2 = \frac{[H_2N\text{\raisebox{0pt}{\scriptsize$\sim\!\!\sim$}}CO_2^-]}{[\overset{+}{H_3N}\text{\raisebox{0pt}{\scriptsize$\sim\!\!\sim$}}COOH]} \cdot [H^+]^2 \qquad$$ At the isoelectric point, $[H_2N\text{\raisebox{0pt}{\scriptsize$\sim\!\!\sim$}}CO_2^-] = [\overset{+}{H_3N}\text{\raisebox{0pt}{\scriptsize$\sim\!\!\sim$}}COOH]$

At this point, $[H^+] = (K_1K_2)^{1/2}$; or, $pH = \frac{1}{2}[pK_1 + pK_2]$

3. The $-COOH$ of Ala is more acidic than that of β-alanine; i.e., the distance to the $-NH_3^+$ group is greater in β-alanine; thus, lower electrostatic attraction in the conjugate base, $\overset{+}{H_3N}\text{\raisebox{0pt}{\scriptsize$\sim\!\!\sim$}}CO_2^-$.

The NH_3^+ group in $\underset{\substack{| \\ CH_3}}{H_3\overset{+}{N}CHCO_2^-}$ is more acidic than that in $H_3\overset{+}{N}CH_2CH_2CO_2^-$.

The difference is probably associated with solvation. When (+) and (−) are far apart they can be separately solvated, but when they are close together the solvation is less efficient.

solvent molecules
in this region
don't know which
way to point!

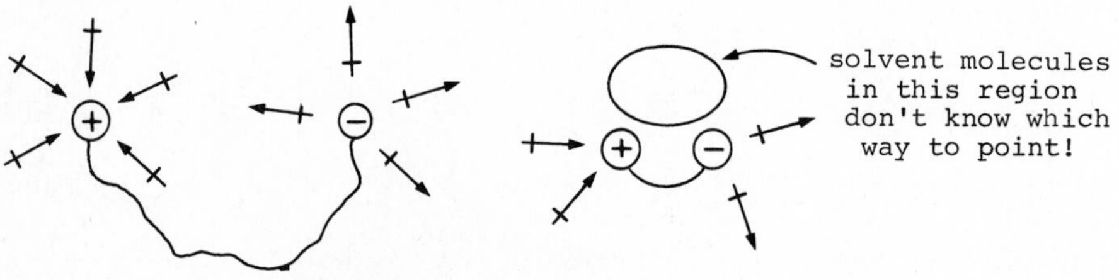

The same arguments apply to 4-aminobutanoic acid.

Isoelectric points: β-alanine   6.90

4-aminobutanoic acid   7.33

4.
$$\overset{\substack{8.60 \\ \downarrow \\ }}{} \qquad \overset{\substack{2.81 \\ \downarrow \\ }}{}$$
$$\underset{\substack{| \\ CH_2CH_2COOH \\ \uparrow \\ 4.45}}{H_3\overset{+}{N}CH_2CONHCHCOOH}$$

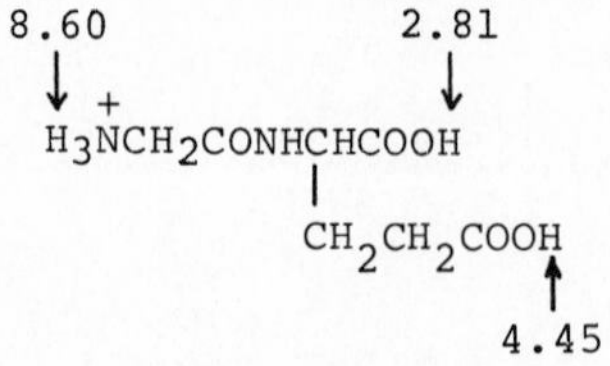

The 4.45 value is close to that for simple aliphatic carboxylic acids; 2.81 is typical of peptide $-COOH$.

5. (a) $CH_3(CH_2)_4COOH \xrightarrow{\text{P, Br}_2} CH_3(CH_2)_3CHBrCOOH \xrightarrow{\text{NH}_3} \underset{\substack{| \\ CH_3(CH_2)_3CHCO_2^-}}{CH_3(CH_2)_3\overset{+NH_3}{CHCO_2^-}}$

caproic acid

or $CH_3(CH_2)_3Br + {}^-\overset{\substack{NHCOCH_3 \\ |}}{C(COOEt)_2} \longrightarrow CH_3(CH_2)_3\overset{\substack{NHCOCH_3 \\ |}}{C(COOEt)_2} \xrightarrow{\text{OH}^-} \underset{\Delta}{\overset{\text{H}^+}{\longrightarrow}}$

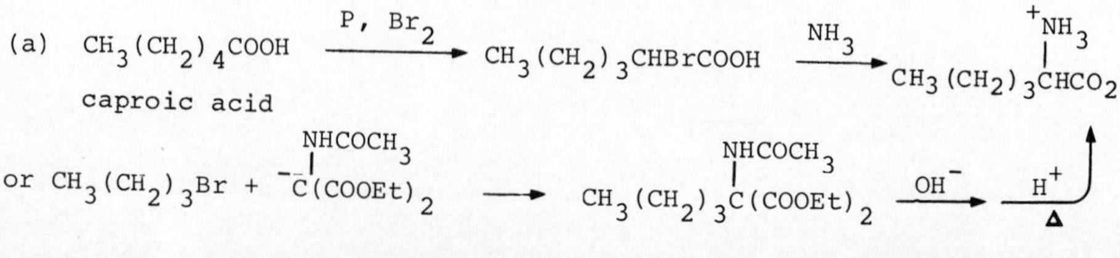

(b) (c) $(CH_3)_3CCHO + NH_3 + HCN \longrightarrow (CH_3)_3CCH(NH_2)CN \xrightarrow[\Delta]{\text{aq. NaOH}} \xrightarrow{H^+} (CH_3)_3CCH(\overset{+}{N}H_3)CO_2^-$

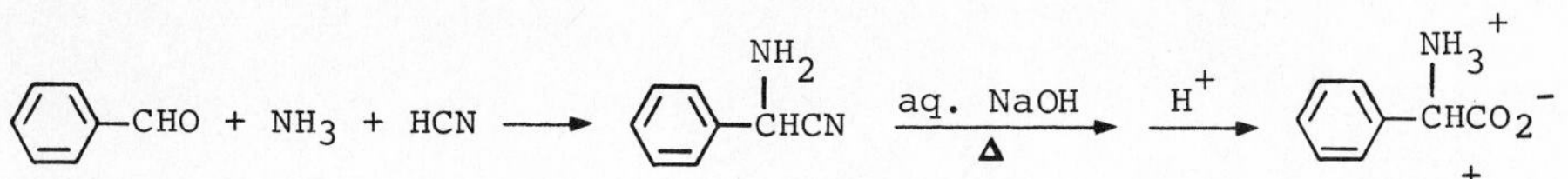

(d) $CH_2(COOEt)_2 \xrightarrow[\text{EtBr}]{\text{NaOEt}} EtCH(COOEt)_2 \xrightarrow[\text{MeI}]{\text{NaOEt}} CH_3CH_2C(CH_3)(COOEt)_2 \xrightarrow{OH^-} \xrightarrow{H^+} CH_3CH_2C(CH_3)(COOH)COOH \xrightarrow[H_2SO_4]{HN_3} CH_3CH_2C(CH_3)(\overset{+}{N}H_3)CO_2^-$

(Schmidt reaction)

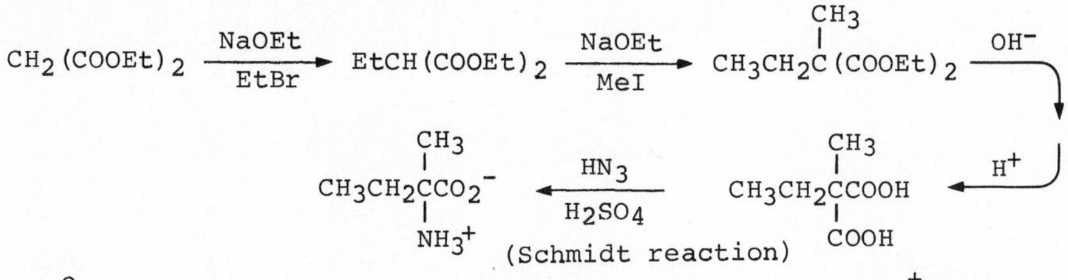

(e) (f) 6\. (a) $CH_3CHO + NH_3 + H^*CN \longrightarrow$ etc.

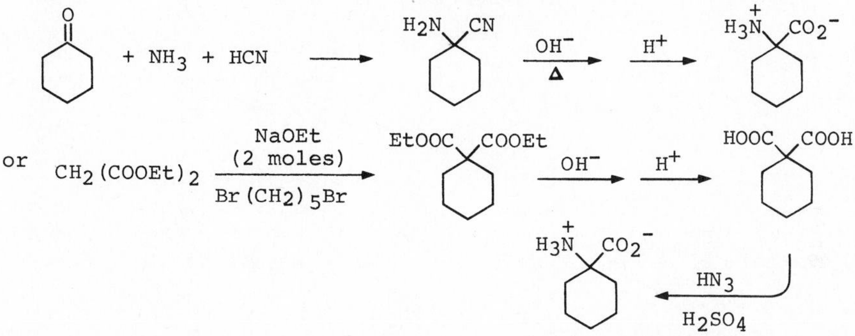

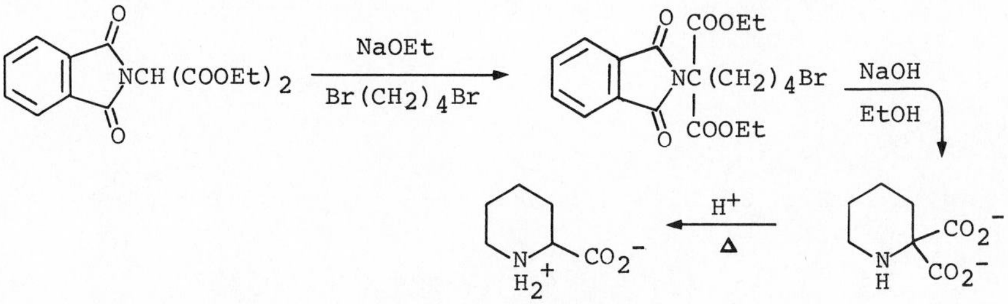

(b) $CH_3MgX + {}^*CO_2 \longrightarrow CH_3{}^*COOH \xrightarrow{LiAlH_4} CH_3{}^*CH_2OH \xrightarrow[H_2SO_4]{Na_2Cr_2O_7} CH_3{}^*CHO \xrightarrow[HCN]{NH_3}$ etc.

(c) $ROC(=O)OR \xrightarrow{LiAlD_4} CD_3OH \xrightarrow{HI} CD_3I \xrightarrow[\text{EtOH, NaOEt}]{CH_3CONHCH(COOEt)_2}$ etc.

a carbonate ester

or $CO + D_2 \xrightarrow[\text{catalyst}]{\Delta} CD_3OH$

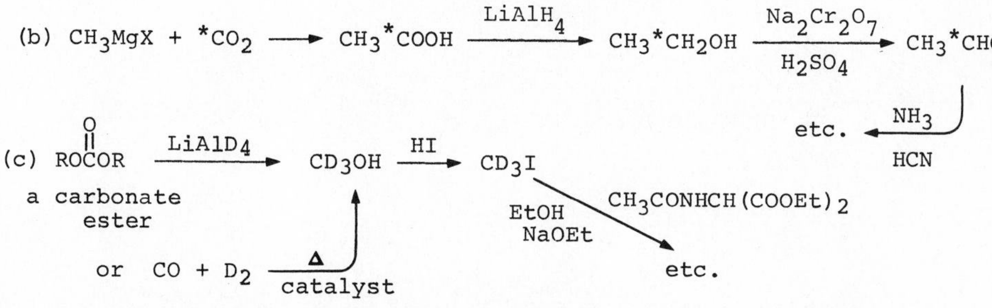

NOTE: would you use dimethyl carbonate for the first step?

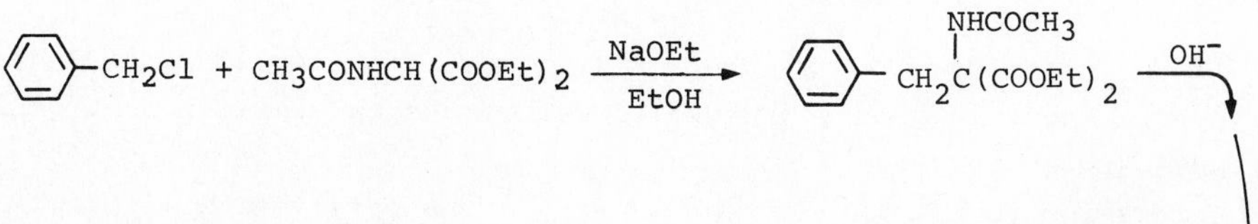

(d) *(continued on next page)*

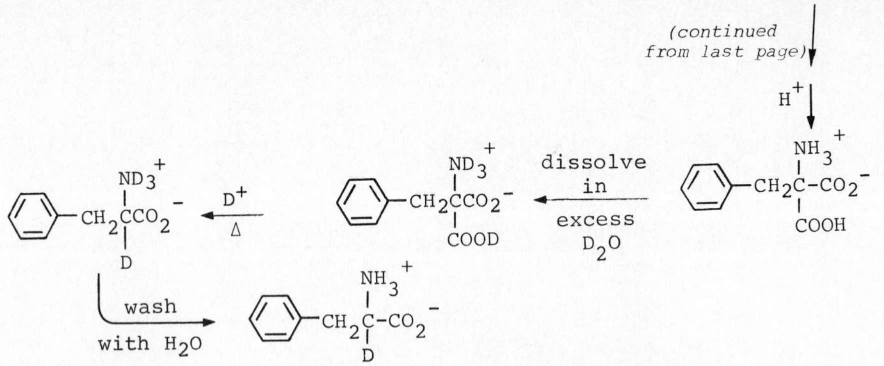

(e)

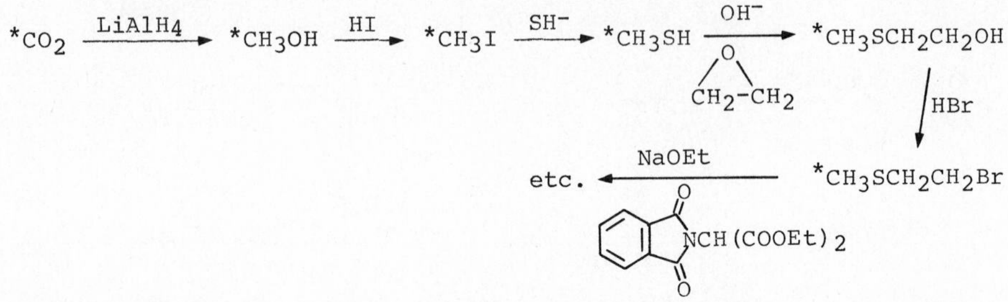

(f)

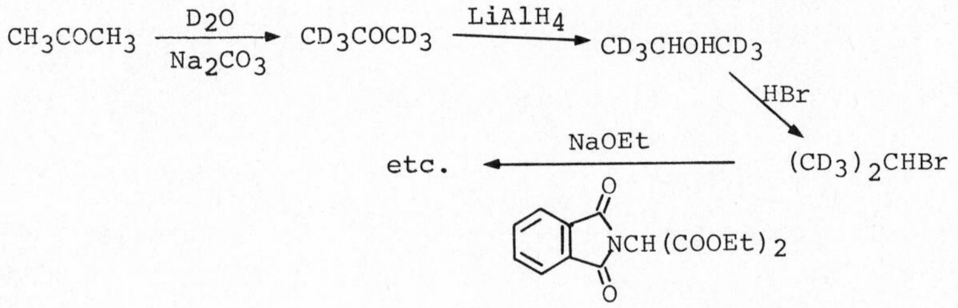

(g)

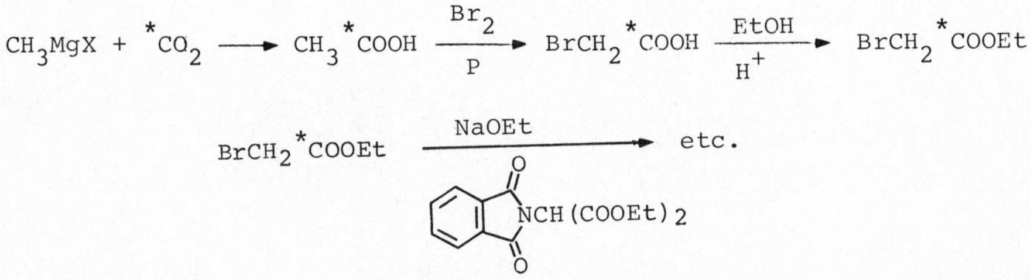

(h)

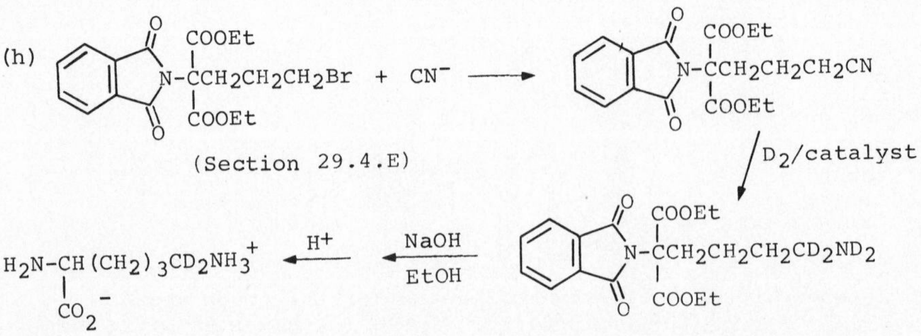

$LiAlD_4$ clearly cannot be used in the reduction step because of the other reducible groups present; however, the catalytic reduction is not straightforward either. If excess NH_3 is present to cut down secondary amine formation (Section 24.6.F), some catalysts will promote exchange:

$$NH_3 + D_2 \xrightarrow{\text{cat.}} ND_3 + H_2$$, etc., and the deuterium will become diluted. In practice, experiments would be required to find the best conditions for the hydrogenation with D_2.

7. Acid-catalyzed esterification starts with a protonated carbonyl,

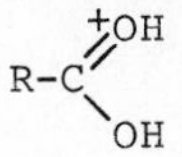

$$R-C\overset{\overset{+OH}{\|}}{\underset{OH}{}}$$. Protonation of an amino acid is more difficult because of electrostatic repulsion with the $-NH_3^+$ group:

repulsion

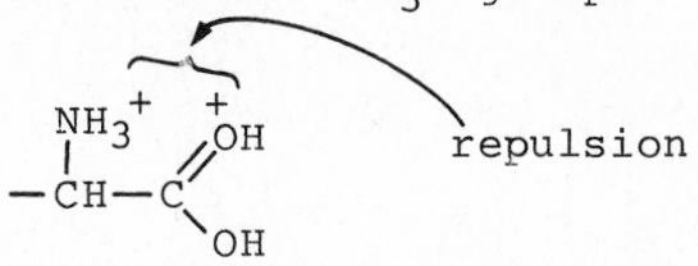

8. Hydrolysis of the benzoyl amide group requires conditions that also cause hydrolysis of the peptide bond; both are normal amide functions.

9. When the dimer, Gly-Gly, is formed, the next step of cyclization to form 2,5-diketopiperazine is much faster (six-membered ring formation) than intermolecular reaction with another glycine. Hence, the glycine is converted most rapidly to the diketopiperazine, and thereafter it is the diketopiperazine which polymerizes. Since this is then a polymerization of dimers, even-numbered peptides will predominate.

10. (a) (1) $BrCH_2COBr + CH_3\overset{\overset{NH_2}{|}}{CH}CO_2^- \longrightarrow BrCH_2CON\overset{\overset{CH_3}{|}}{H}CHCO_2^- \xrightarrow{NH_3} H_3\overset{+}{N}CH_2CONH\overset{\overset{}{|}}{\underset{CH_3CHCO_2^-}{}}$

(a) (2) $BrCH_2COBr + H_2NCH_2CON\overset{\overset{CH_3}{|}}{H}CHCO_2^- \longrightarrow BrCH_2CONHCH_2CON\overset{\overset{CH_3}{|}}{H}CHCO_2^- \xrightarrow{NH_3}$

Gly-Gly-Ala

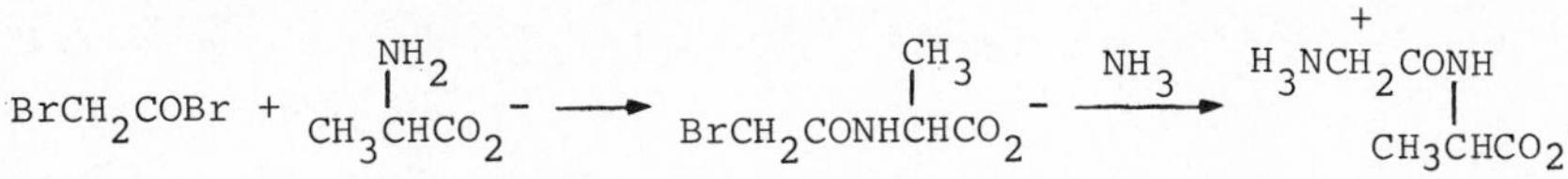

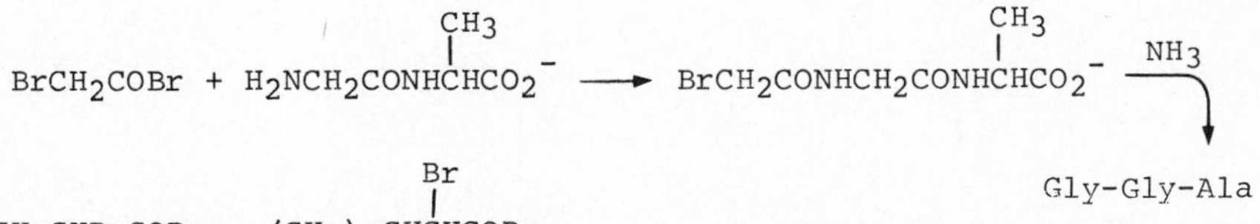

(b) $CH_3CHBrCOBr;$ $(CH_3)_2CH\overset{\overset{Br}{|}}{C}HCOBr$

(c) Val-Val or Ala-Ala   Note that the amino acid used forms the COOH end of the peptide

(d) Stereochemistry: would require optically active α-bromo acyl bromides, which are difficult to prepare and which racemize readily with mild base:

$$R-CHBrCOBr + base \rightleftharpoons R-\overset{-}{C}BrCOBr + base-H^+$$

configuration lost

Thus a mixture of diastereomers would result. As an example, in (c) above, the use of racemic acid bromides will produce L-Ala-L-Ala, D-Ala-L-Ala, L-Ala-D-Ala, D-Ala-D-Ala, etc.

11. $((CH_3)_3CO\overset{\overset{O}{\|}}{C})_2O + H_3\overset{+}{N}\overset{\overset{}{|}}{\underset{R}{C}}HCO_2^- \longrightarrow (CH_3)_2CO\overset{\overset{O}{\|}}{C}NH\overset{\overset{}{|}}{\underset{R}{C}}HCO_2H \equiv Boc\text{-amino acid}$

$$H_3\overset{+}{N}\overset{\overset{}{|}}{\underset{R}{C}}HCO_2^- \xrightarrow[H^+]{CH_3OH} H_3\overset{+}{N}\overset{\overset{}{|}}{\underset{R}{C}}HCO_2CH_3 \equiv \text{amino acid-}CH_3$$

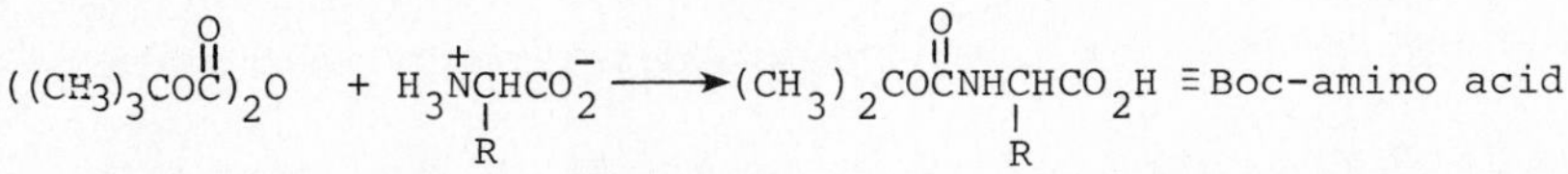

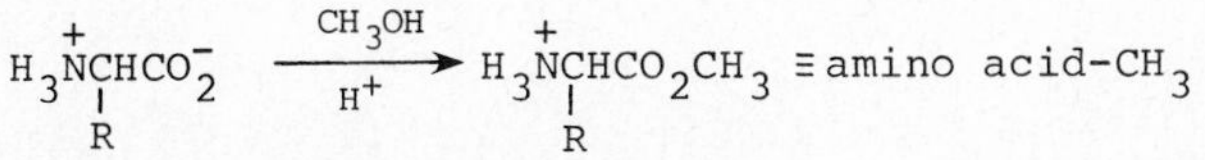

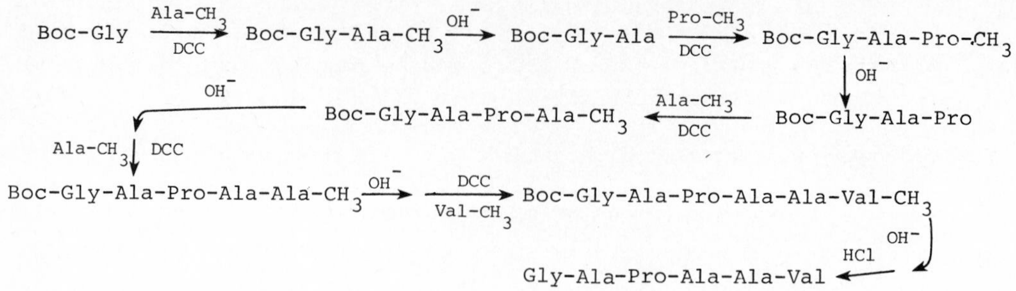

$$\text{Boc-Gly} \xrightarrow[\text{DCC}]{\text{Ala-CH}_3} \text{Boc-Gly-Ala-CH}_3 \xrightarrow{\text{OH}^-} \text{Boc-Gly-Ala} \xrightarrow[\text{DCC}]{\text{Pro-CH}_3} \text{Boc-Gly-Ala-Pro-CH}_3$$

$$\xleftarrow{\text{OH}^-} \qquad \text{Boc-Gly-Ala-Pro-Ala-CH}_3 \xleftarrow[\text{DCC}]{\text{Ala-CH}_3} \text{Boc-Gly-Ala-Pro} \xleftarrow{\text{OH}^-}$$

$$\text{Ala-CH}_3 \bigg\downarrow \text{DCC}$$

$$\text{Boc-Gly-Ala-Pro-Ala-Ala-CH}_3 \xrightarrow{\text{OH}^-} \xrightarrow[\text{Val-CH}_3]{\text{DCC}} \text{Boc-Gly-Ala-Pro-Ala-Ala-Val-CH}_3$$

$$\text{OH}^- \Big)$$
$$\text{HCl}$$
$$\text{Gly-Ala-Pro-Ala-Ala-Val} \xleftarrow{}$$

12.

$$\text{Val-CH}_3 + \cdots \xrightarrow{\text{H}^+} \text{Ala-Val-CH}_3 \cdots \xrightarrow{\text{H}^+} \text{Ala-Ala-Val-CH}_3$$

$$\text{Ala-Pro-Ala-Ala-Val-CH}_3 \xleftarrow{\text{H}^+} \xleftarrow{} \text{Pro-Ala-Ala-Val-CH}_3 \xleftarrow{\text{H}^+}$$

$$\xrightarrow{\text{H}^+} \text{Gly-Ala-Pro-Ala-Ala-Val-CH}_3 \xrightarrow{\text{OH}^-} \text{Gly-Ala-Pro-Ala-Ala-Val}$$

13. $\text{Boc-Gly} + \text{ClCH}_2-\langle\rangle\text{-polymer} \longrightarrow \text{Boc-Gly-CH}_2-\text{(P)} \xrightarrow{\text{H}^+} \text{Gly-CH}_2-\text{(P)}$

$$\text{DCC} \mid \text{Boc-Gly}$$

$$\text{Boc-Glu}(\delta\text{-Bz})\text{-Gly-Gly-CH}_2\text{(P)} \xleftarrow[\text{DCC}]{\text{BocNH-CHCO}_2\text{H}} \xleftarrow{\text{H}^+} \text{Boc-Gly-Gly-CH}_2\text{(P)}$$

$$\text{H}^+$$

$$\xrightarrow[\text{Boc-Lys}(\epsilon\text{-Cbz})]{\text{DCC}} \text{Boc-Lys}(\epsilon\text{-Cbz})\text{-Glu}(\delta\text{-Bz})\text{-Gly-Gly-CH}_2\text{(P)} \xrightarrow{\text{H}^+}$$

$$\bigg\downarrow \text{Boc-Ala}$$

$$\xrightarrow[\text{H}_2/\text{Pt}]{} \xleftarrow{\text{HF}} \xleftarrow{\text{H}^+} \text{Boc-Ala-Lys}(\epsilon\text{-Cbz})\text{-Glu}(\delta\text{-Bz})\text{-Gly-Gly-CH}_2-\text{(P)}$$

$$\text{Ala-Lys-Glu-Gly-Gly}$$

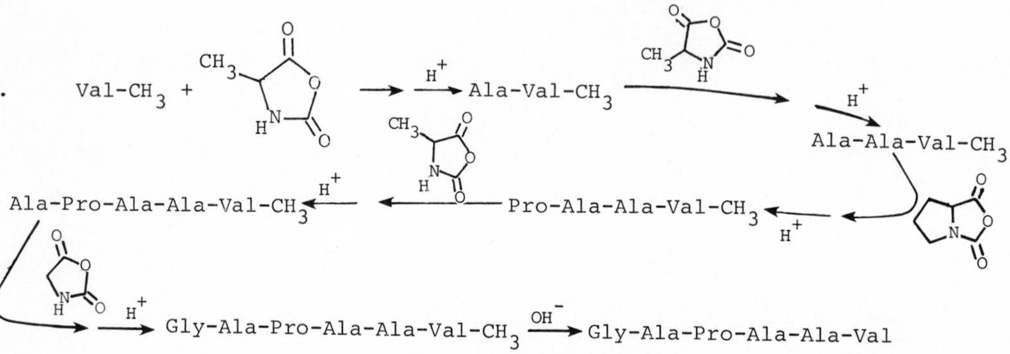

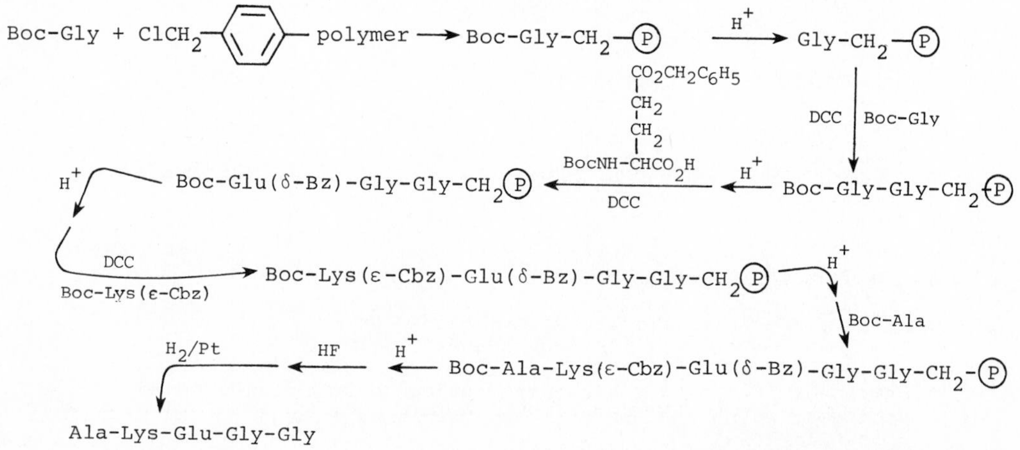

14.   Either of the methods in answers 11-13 can be used by continuing to
       add on groups in the sequence Ala, Ala, Phe, Val, Ala. However, the
       first and second halves of this decapeptide are the same. Hence, one
       could simply couple two of the pentapeptides:

$$\text{Cbz-Ala-Val-Phe-Ala-Ala-Et}$$
$$\xrightarrow{\text{H}_2\text{-Pd/C}} \text{Ala-Val-Phe-Ala-Ala-Et}$$
$$+$$
$$\xrightarrow[\text{hydrolysis}]{\text{mild}} \text{Cbz-Ala-Val-Phe-Ala-Ala}$$
$$\bigg\} \text{DCC}$$

$$\text{Cbz-Ala-Val-Phe-Ala-Ala-Ala-Val-Phe-Ala-Ala-Et} \xleftarrow{}$$

$$\bigg\downarrow \text{H}_2\text{-Pd/C}$$

$$\xrightarrow[\text{hydrolysis}]{\text{mild}} \text{Ala-Val-Phe-Ala-Ala-Ala-Val-Phe-Ala-Ala}$$

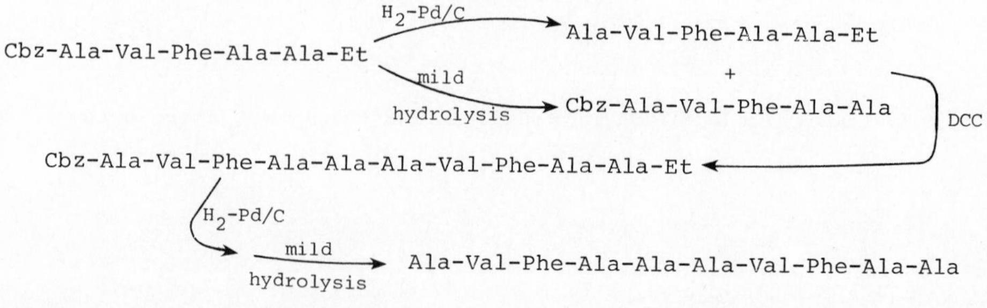

15.

TRH ≡ pyroGlu-His-Pro-NH$_2$ or

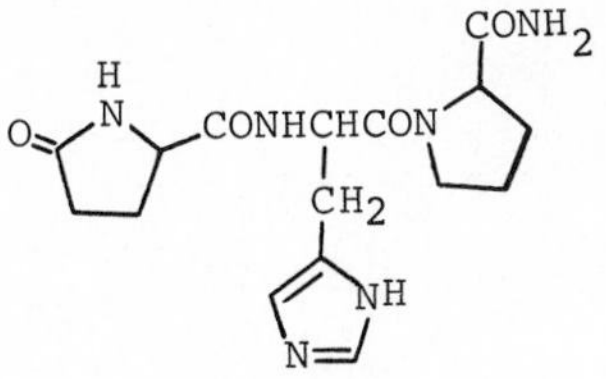

Synthesis:

Cbz-His + Pro-Et $\left(\equiv \right.$ $\left. \right)$

$\downarrow$ DCC

Cbz-His-Pro-Et $\xrightarrow{\text{H}_2\text{-Pd/C}}$ His-Pro-Et

pyroGlu-His-Pro-NH$_2$ $\xleftarrow[\substack{\text{1. mild} \\ \text{hydrolysis} \\ \text{2. NH}_3,\ \text{DCC}}]{\text{NH}_3\ \text{or}}$ pyroGlu-His-Pro-Et $\xleftarrow[\text{DCC}]{\text{pyroglu}}$

NOTE: conversion to the acid chloride with SOCl$_2$, followed by treatment with NH$_3$ to form the final amide is an alternative, but will probably cause some racemization at the proline α-carbon to give a mixture of diastereomers.

16.

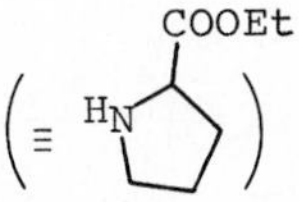

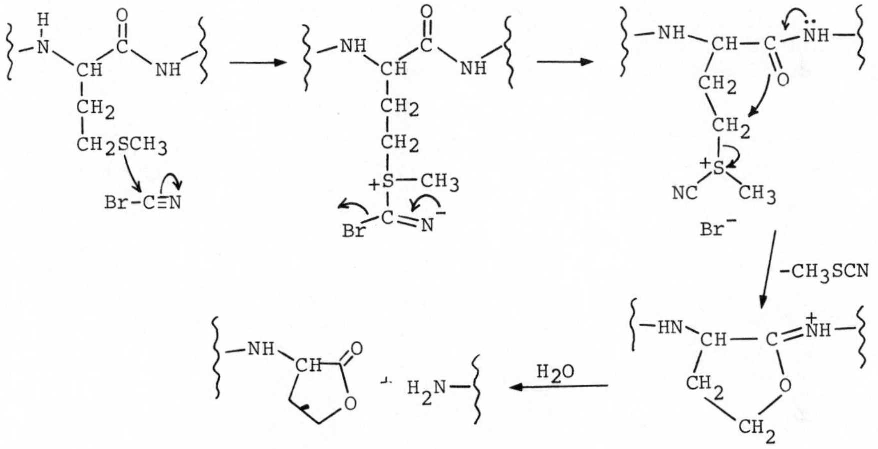

17. (a)

                                                                    15 20 25 30
Lys; Glu-Thr-Ala-Ala-Ala-Lys; Phe-Glu-Arg; Glu-His〜〜〜〜〜〜Met-Lys;
                                            45   50   55   60
Ser-Arg; Asn-Leu-Thr-Lys; Asp-Arg; Cys-Lys; Pro-Val〜〜〜〜〜〜Glu-Lys;
              70   75 80   85
Asn-Val-Ala-Cys-Lys; Asn-Gly〜〜〜〜〜〜Cys-Arg; Glu-Thr-Gly-Ser-Ser-Lys;
                                        105 110 115 120 124
Tyr-Pro-Asn-Cys-Ala-Tyr-Lys; Thr-Thr-Glu-Ala-Asn-Lys; His〜〜〜〜〜Val

   (b) Ten polypeptides: three ending with Phe, six with Tyr, and one
       with Val (terminal piece). No Trp is present.

   (c) BrCN cleaves only at Met of which four are present, two of which
        are joined. Five pieces will be formed, one of which
                          is cyclized Met itself:

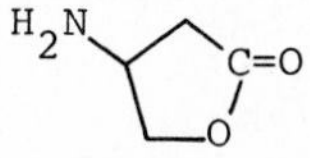

18.
   I-II-IV-III or I-IV-II-III. I must be the amino end because it is
   the only fragment starting with Glu; similarly, III must be the carboxy
   end. Positions of II and IV are not established.

19. Consider the following logic:

1) T-1 to T-8 contain all 56 amino acids. T-9 is clearly (T-7) - (T-6). T-7 must be a product of abnormal cleavage, since trypsin does not normally hydrolyze at Tyr. Thus, T-9 presumably is the primary hydrolysis product that partially hydrolyzed to T-6 and T-7.

2) Since the protein N-end starts with Thr and the C-end is Cys, the sequence starts with: (T-4)~~~~~~~~~~(T-3).

3) Furthermore, this also tells us that the chains from chymotrypsin digestion are in the sequence: (Ch-1)—(Ch-2)—(Ch-3).

4) Comparing the amino acid compositions, we find that Ch-2 is the same as T-7 + Tyr and Ile. Similarly, T-8 is part of Ch-1 (note that there is only one Ala in the protein). Since T-6 follows T-7, we can write the protein as:

$$(T\text{-}4)—(T\text{-}8)—\overbrace{(T\text{-}7)—(T\text{-}6)}^{(T\text{-}9)}—(T\text{-}5,1,2)—(T\text{-}3)$$

5) The only question left is where the single units, T-1 and T-2 fit with relation to T-5. For this we turn to the T* series. T*-2 contains the Cys and comes at the end; in fact, the composition of T*-2 is that of T-5 and T-3. Thus, T-1 and T-2 come between (T-6) and (T-5), but with CH_3NCS, Lys has been modified so that it will not cleave; hence T*-3 must terminate with Asp.

6) The complete sequence then is:

(T-4)—(T-8)—(T-7)—(T-6)—(T-1)—(T-2)—(T-5)—(T-3)

29.E Supplementary Problems

S1. What is the principal ionic form of the dipeptide histidyltyrosine (His-Tyr) at pH 2, 5, 8, and 11?

S2. What is the approximate isoelectric point of the following amino acids?

(a) Valine (b) Glutamic Acid (c) Lysine

S3. Show how to synthesize the following amino acids.

(a)
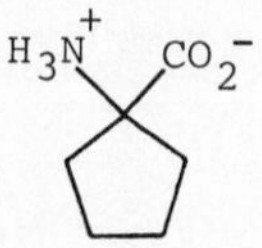

(c)
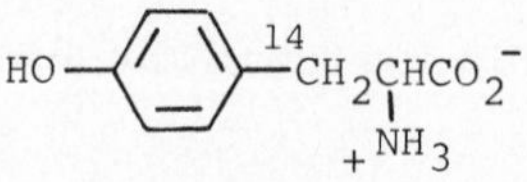

(*^{14}C available as $Ba^{14}CO_3$ or $Na^{14}CN$*)

(b) $HO_2CCH_2CH_2CHCO_2^-$ with $^+NH_3$ on the CH

(d)
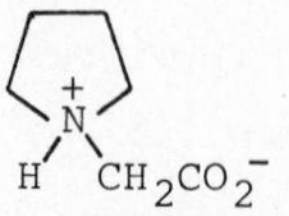

S4. Outline syntheses of the following tripeptides, using the coupling reagent shown and appropriate protecting groups as necessary.

(a) Gly-Val-Met (N-carboxy anhydrides) (c) Ser-Leu-Tyr (acyl azides)

(b) Phe-Ile-Asp (DCC)

S5. How many isomers are produced in the Strecker synthesis of isoleucine? What methods may be used to separate them from each other?

S6. Conversion of an N-acyl amino acid to the acid chloride often leads to formation of an azlactone and racemization. Write reasonable mechanisms for these reactions:

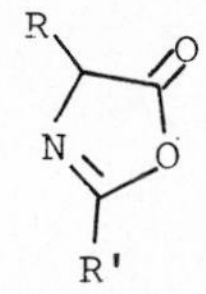

an azlactone

S7. Another method for peptide bond formation is the "mixed-anhydride" coupling method, as shown in the example below

$$CbzNHCH_2CO_2H \xrightarrow[2.\ H_3\overset{+}{N}CH(CH_3)CO_2^-,\ Et_3N]{1.\ ClC(=O)OCH_2CH_3,\ Et_3N} CbzNHCH_2C(=O)NHCH(CH_3)CO_2^-$$

(a) Write a balanced equation for this reaction sequence and indicate what intermediates are involved.

(b) What would happen if an unprotected amino acid were used as the starting material?

(c) What do you expect is the major side product in the sequence?

S8. The enkephalins are believed to be the natural compounds in the central nervous system whose activity is imitated by morphine. The amino acid composition of one of the enkephalins is (Gly_2MetPheTyr). Reaction of this pentapeptide with 8-dimethylaminonaphthalenesulfonyl chloride (dansyl chloride) and subsequent acid-catalyzed hydrolysis affords the dansyl derivative of tyrosine as the only modified amino acid. When the enkephalin is treated with cyanogen bromide, no cleavage of the chain takes place. Partial hydrolysis of the pentapeptide with chymotrypsin gives only tyrosine, methionine and a tripeptide.

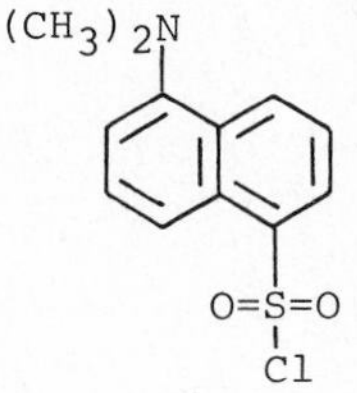

Dansyl Chloride

(a) What is the most likely structure of the dansyl derivative of tyrosine?

(b) What is the sequence of the enkephalin?

29.F Answers to Supplementary Problems

S1. pH 2:

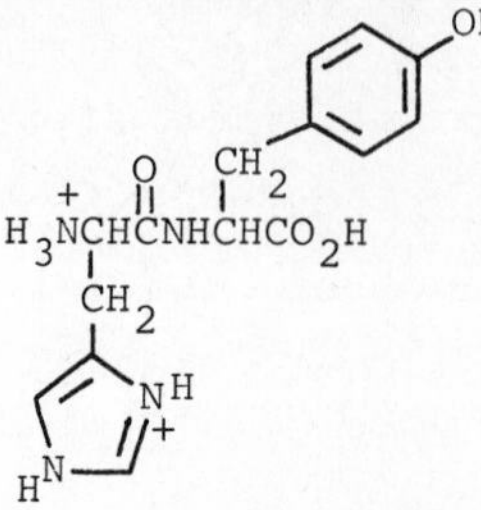

pH 5:

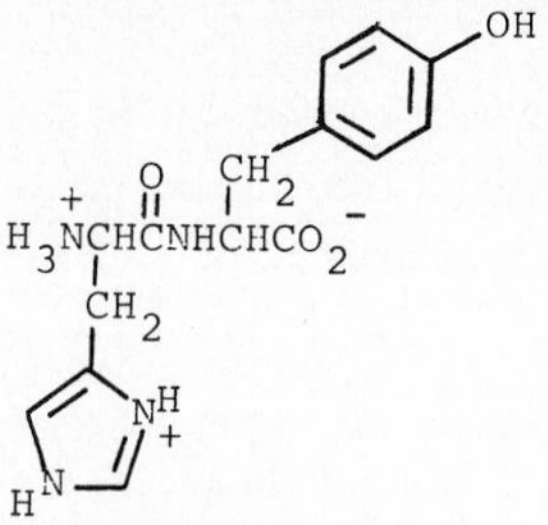

pH 8:

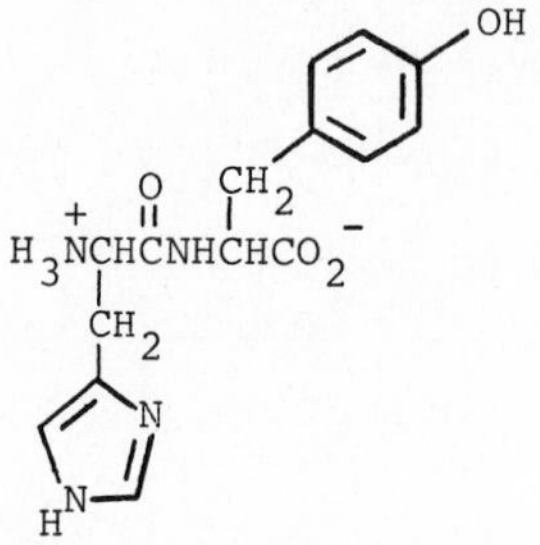

pH 11

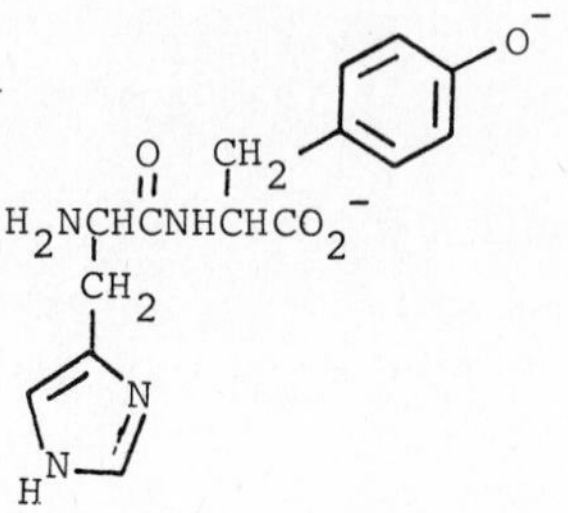

S2. (a) $\frac{2.29\ (CO_2H) + 9.72\ (\overset{+}{N}H_3)}{2} = 6.0$

(b) $\frac{2.13\ (\alpha\text{-}CO_2H) + 4.32\ (\gamma\text{-}CO_2H)}{2} = 3.23$

(c) $\frac{9.20\ (\alpha\text{-}NH_2) + 10.8\ (\varepsilon\text{-}NH_2)}{2} = 10.0$

(Write out the major ionic forms and their net charge at various pH's to see how the answers to (b) and (c) were obtained.)

S3. (a)

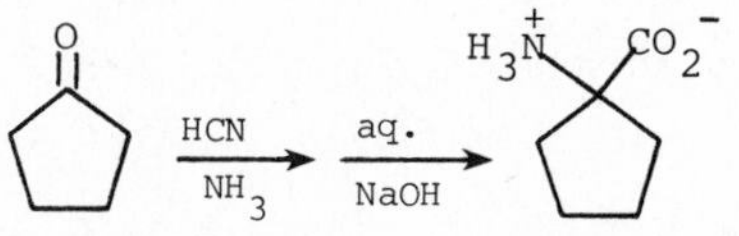

(b)

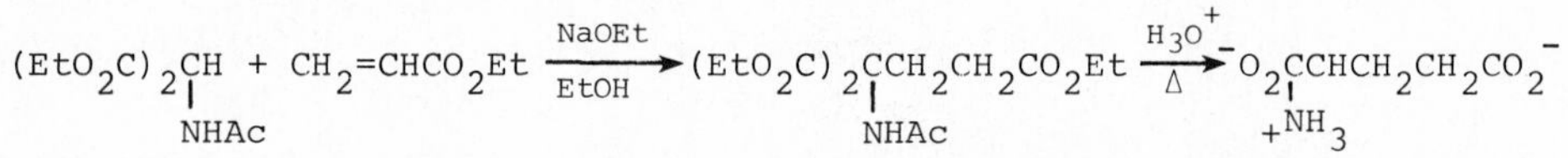

(c)

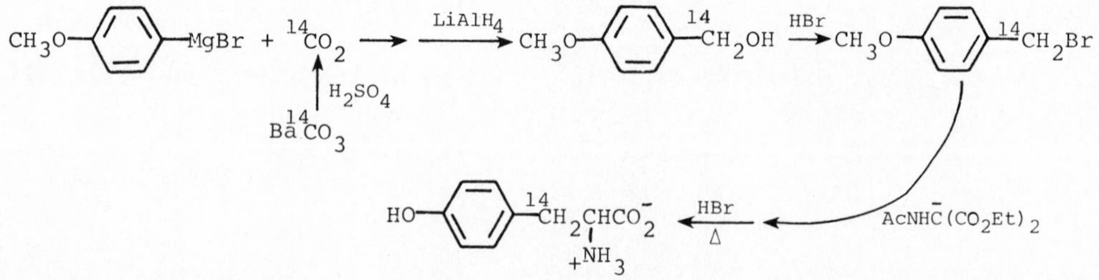

(d)

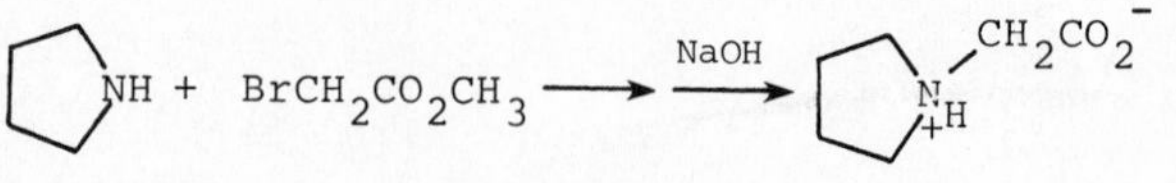

S4.

(a) $(CH_3)_2CH$— [oxazolidinedione structure] + Met $\xrightarrow{pH\ 10}$ $\xrightarrow[-CO_2]{H^+}$ Val-Met $\longrightarrow$ [oxazolidinedione structure]

$\xrightarrow{pH\ 10}$

$\xrightarrow{H^+,\ CO_2^-}$

Gly-Val-Met

(b) Cbz-Ile + $H_2NCHCO_2CH_3$ $\overset{CH_2CO_2CH_3}{|}$ (Asp-Me$_2$) $\xrightarrow{DCC}$ Cbz-Ile-Asp-Me$_2$ $\xrightarrow{H_2-Pd/C}$

Phe-Ile-Asp $\xleftarrow{OH^-}$ $\xleftarrow{H_2-Pd/C}$ $\xleftarrow[DCC]{Cbz-Phe}$ Ile-Asp-Me$_2$

(c) Boc-Leu-N$_3$ + Tyr $\xrightarrow{Et_3N}$ Boc-Leu-Tyr $\xrightarrow{HCl}$ Leu-Tyr $\xrightarrow{Boc-Ser-N_3,\ Et_3N}$

Ser-Leu-Tyr $\xleftarrow{HCl}$ Boc-Ser-Leu-Tyr

S5.

$CH_3CH_2CHCH=O$ $\overset{CH_3}{|}$ $\xrightarrow[NH_3]{HCN}$ $\xrightarrow{NaOH}$

chiral center

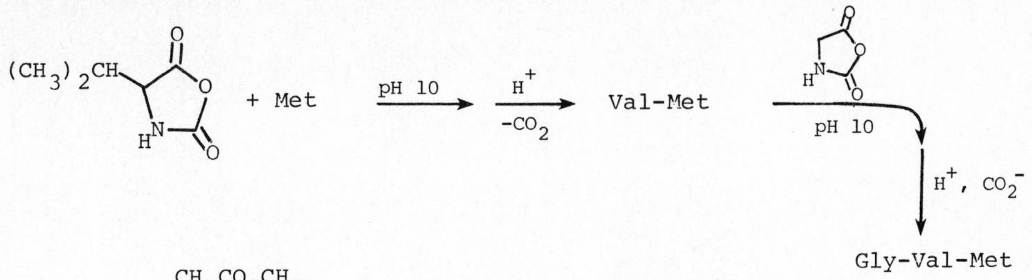

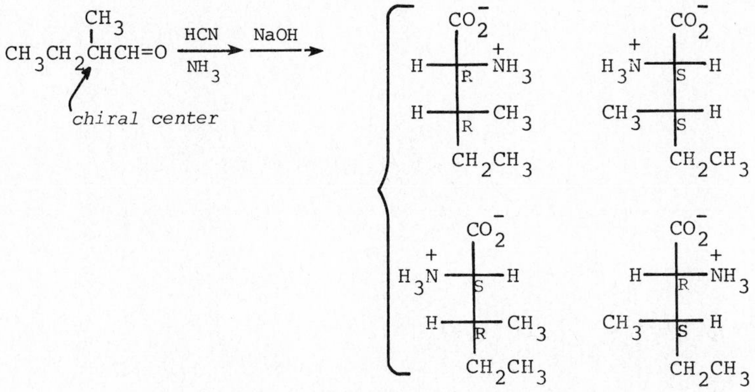

The racemic mixture of RS and SR isomers can be separated
from the RR,SS racemate by crystallization or chromatography
because they are diastereomeric. Separation of the RS enantiomer
from the SR enantiomer, and separation of the RR from the SS
enantiomer require resolution procedures such as those discussed
in Section 29.4.F.

S6.

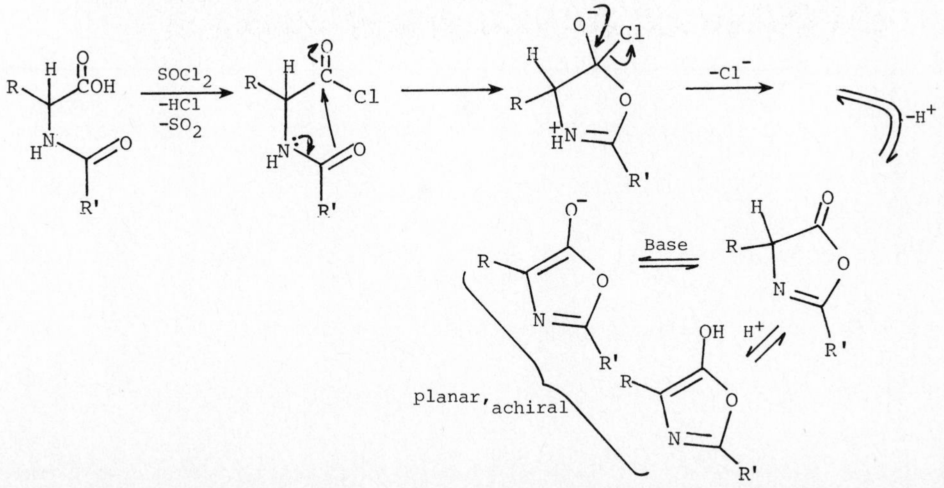

S7. (a) $CbzNHCH_2CO_2^-$ $Et_3\overset{+}{N}H$ $\xrightarrow{Cl\overset{O}{\overset{\|}{C}}OEt}$ $CbzNHCH_2\overset{O}{\overset{\|}{C}}O\overset{O}{\overset{\|}{C}}OEt$ + $Et_3\overset{+}{N}H$ Cl^-

major $\Big|$ minor

$H_2\overset{..}{N}CHCO_2^-$ $Et_3\overset{+}{N}H$
   |
   CH_3

$\downarrow$

$CbzNHCH_2\overset{O}{\overset{\|}{C}}NHCHCO_2^-$ $Et_3\overset{+}{N}H$ + CO_2 + $EtOH$
                   |
                   CH_3

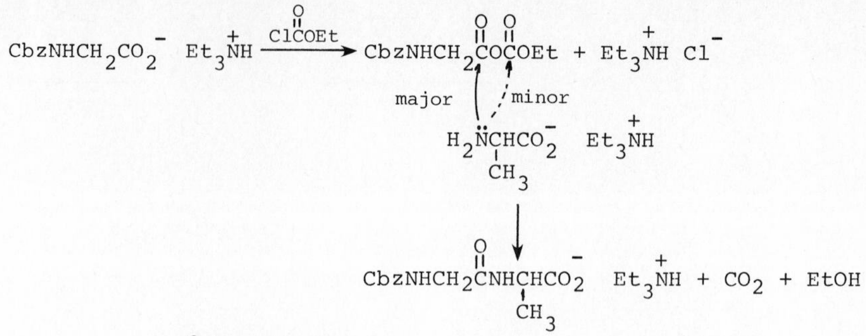

(b) $H_2NCH_2CO_2^-$ $Et_3\overset{+}{N}H$ $\xrightarrow{Cl\overset{O}{\overset{\|}{C}}OEt}$ $EtO\overset{O}{\overset{\|}{C}}NHCH_2CO_2H$ + $Et_3\overset{+}{N}H$ Cl^-

(c) $EtO\overset{O}{\overset{\|}{C}}NHCHCO_2^-$, from attack at the wrong carbonyl of the mixed anhydride.
         |
         CH_3

S8. (a) $(CH_3)_2\overset{+}{N}H$

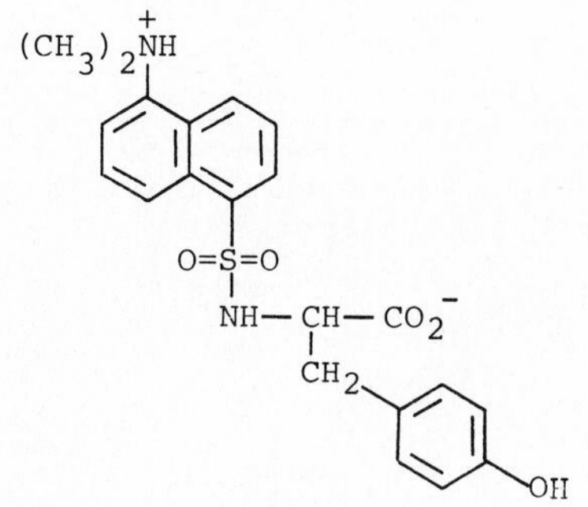

(b) 1) Enkephalin $\xrightarrow[chloride]{dansyl}$ $\xrightarrow{H_3O^+}$ modified Tyr: Tyr is N-terminal residue.

   2) Enkephalin $\xrightarrow{BrCN}$ no cleavage: Met must be C-terminal residue
                                        (otherwise cleavage would have occurred)

   3) Enkephalin $\xrightarrow[trypsin]{chymo-}$ Tyr, tripeptide, Met: C-terminal amino acid of
                                        tripeptide must be Phe.

   4) Therefore the sequence is : Tyr-Gly-Gly-Phe-Met

30. AROMATIC HALIDES, PHENOLS, PHENYL ETHERS, AND QUINONES

30.A Chapter Outline, Important Terms Introduced, and Reactions Discussed

30.1 Introduction

benzyl vs. aryl halides

30.2 Preparation of Halobenzenes

electrophilic aromatic substitution *(see Section 23.1)*

substitution of arenediazonium salts *(see Section 25.5)*

30.3 Reactions of Halobenzenes

A. Substitution Reactions

1. The Addition-Elimination mechanism

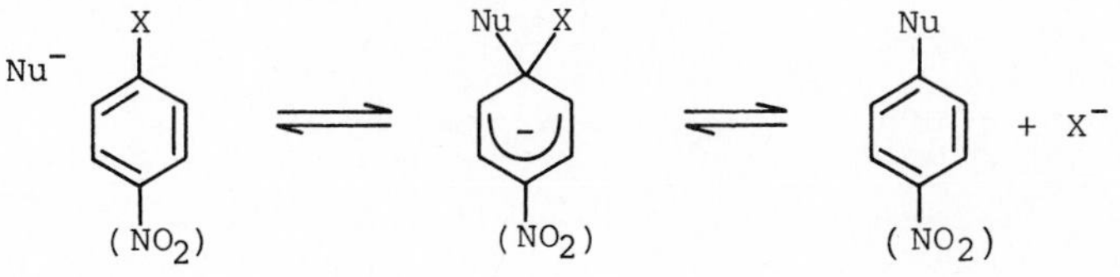

(requires electron-withdrawing groups such as nitro)

2. The Elimination-Addition mechanism; Benzyne

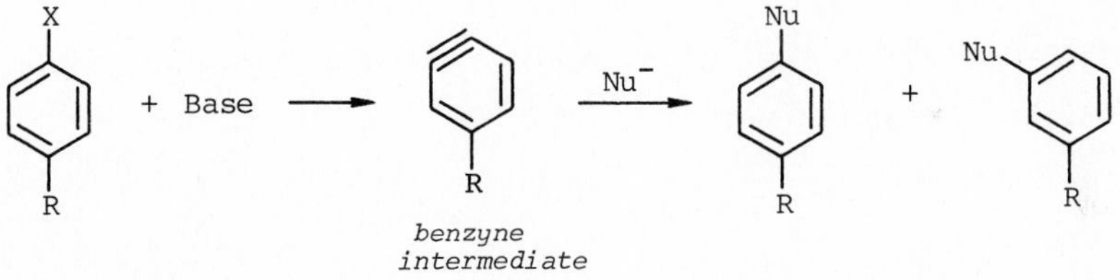

B. Metallation

$$ArX + M \longrightarrow ArMX$$

X = Br, I; M = Mg, 2 Li, 2 Na

$$ArX + RLi \longrightarrow ArLi + RX$$

X = Br, I; R = alkyl

$$ArX + RX \xrightarrow{2\ Na} ArR + 2\ NaX$$ *(Wurtz-Fittig reaction)*

X = Br, I

$$2\ ArX \xrightarrow[\Delta]{Cu} Ar\text{-}Ar$$ *(Ullmann reaction; is facilitated by electron-withdrawing groups (e.g., NO_2, CN); biaryl synthesis)*

X = Cl, Br, I

$$2\ ArLi \xrightarrow{CuBr} \xrightarrow{O_2} Ar\text{-}Ar$$ *(alternative to Ullmann coupling)*

30.4 Nomenclature of Phenols and Phenyl Ethers

phenol (benzenol) — phenyl ethers (alkoxyarenes)

30.5 Preparation and Properties of Phenols and Ethers

A. Preparation of Phenols

NaOH fusion of sulfonic acids *(see Section 26.2)*

hydrolysis of aryl halides and arenediazonium salts *(see Sections 30.3.A and 25.3)*

cumene oxidation:

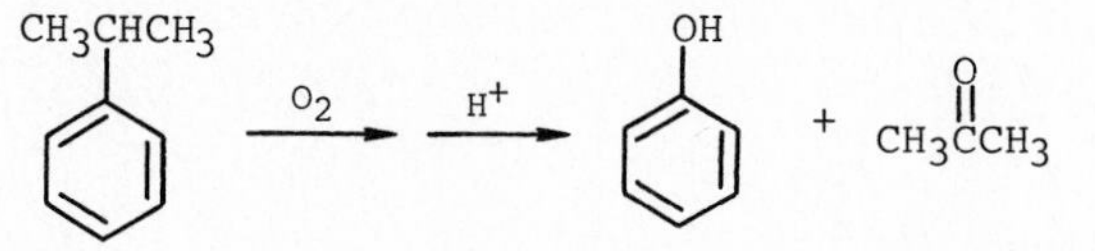

(industrial process)

B. Acidity of Phenols

$pK_a \cong 10$

effects of substitution

C. Preparation of Ethers

Williamson ether synthesis:

$$ArOH + OH^- \rightleftharpoons ArO^- + H_2O \xrightarrow{RX} ArOR$$

via nucleophilic aromatic substitution *(see Section 30.3.A)*

30.6 Reactions of Phenols and Ethers

A. Esterification

requires acid chloride or anhydride *(acid-catalyzed process)*

B. Reactions of Phenolate Ions

1. Halogenation

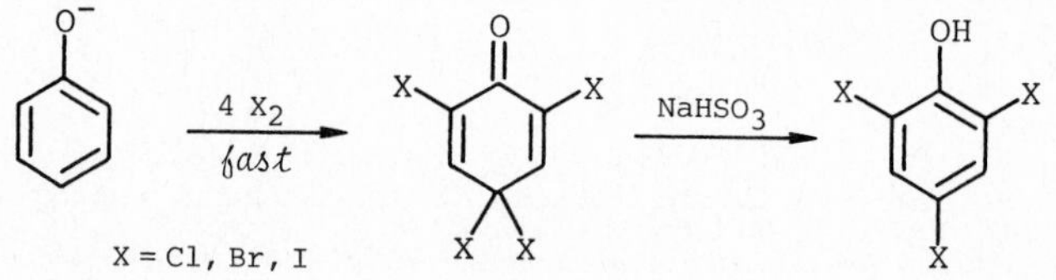

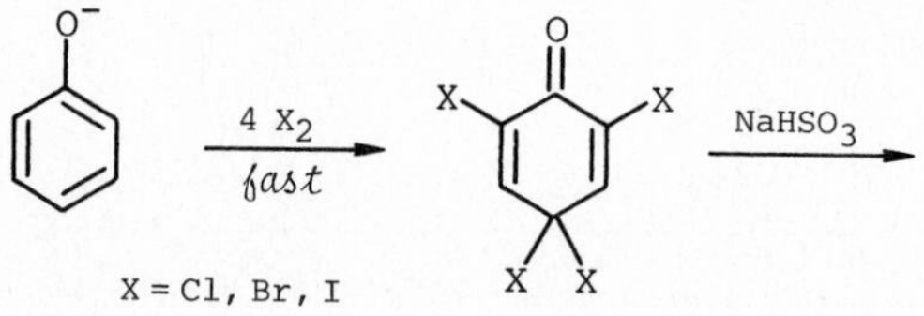

(controllable under acidic conditions)

2. Condensation with Aldehydes

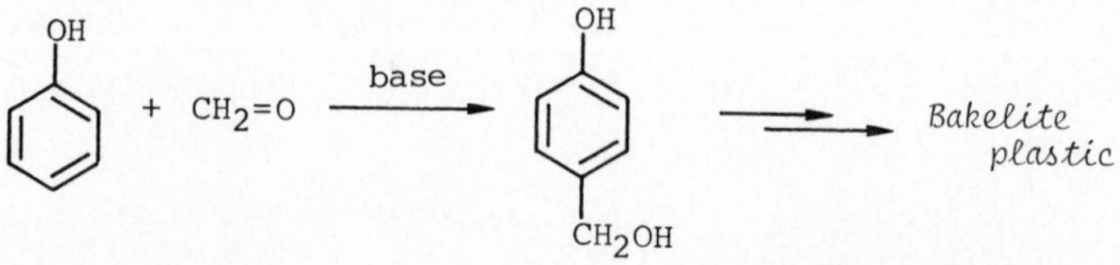

3. Kolbe synthesis

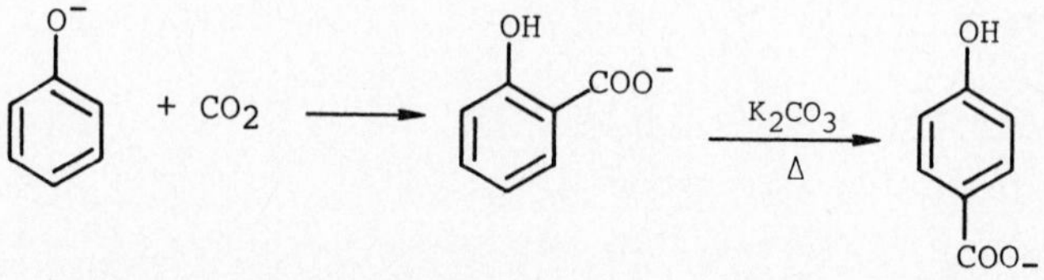

(kinetic vs. thermodynamic control)

4. Reimer-Tiemann reaction

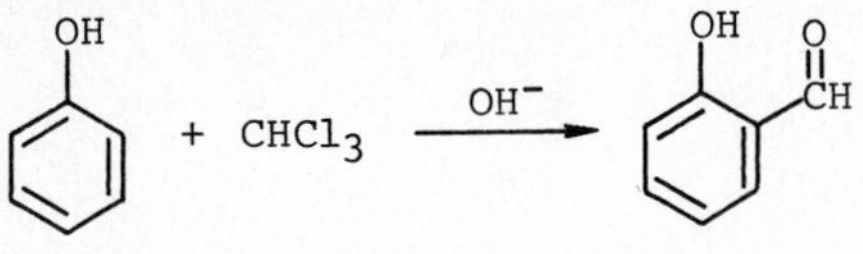

(via dichlorocarbene)

5. Diazonium coupling

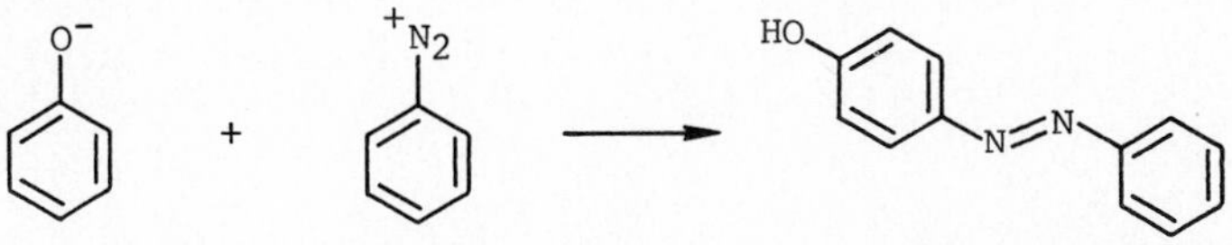

(azo dyes)

C. Electrophilic Substitution on Phenols and Phenyl Ethers

highly activated system

chelation effects in orientation

(many examples of electrophilic aromatic substitution discussed)

Houben-Hoesch synthesis:

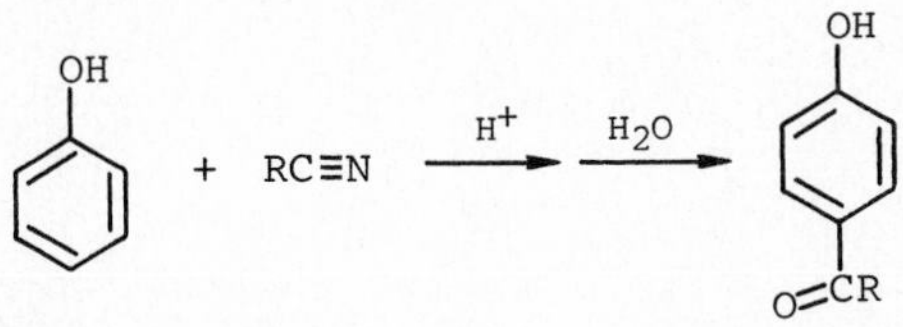

phthalein synthesis:

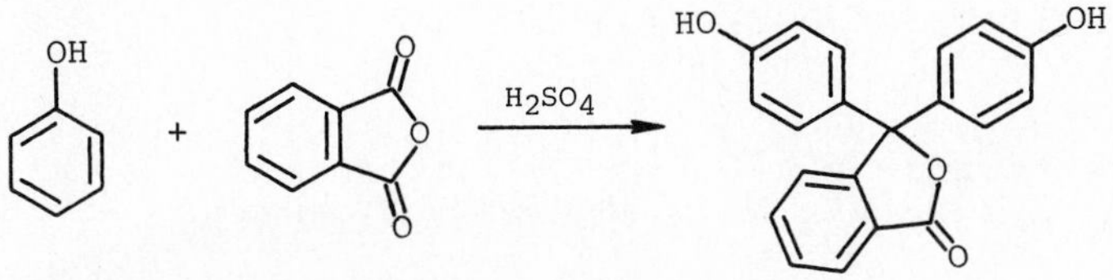

(pH indicators and laxatives...)

Fries rearrangement:

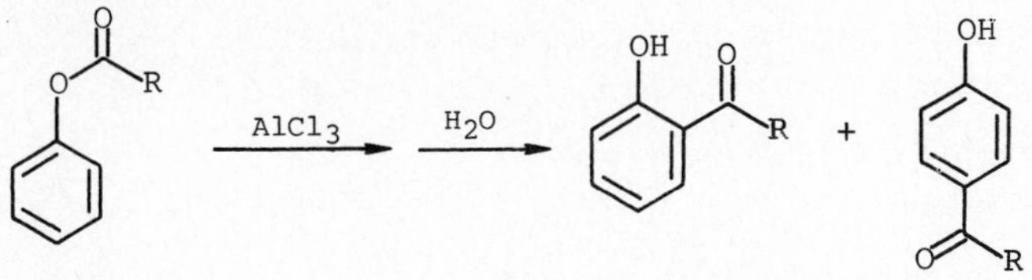

D. Reactions of Ethers

cleavage:

$$\text{ArOR} \xrightarrow[\Delta]{\text{HBr}} \text{ArOH} + \text{RBr}$$

Claisen rearrangement:

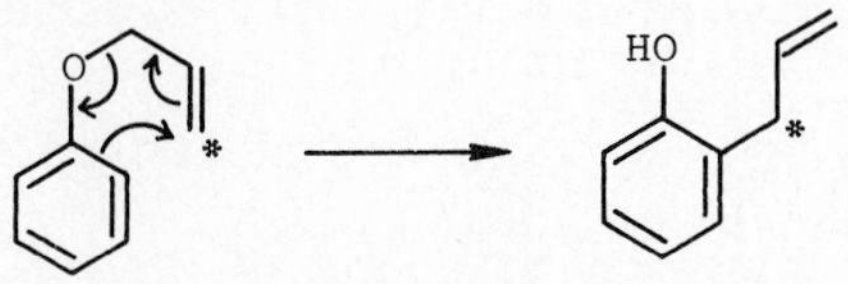

(with allylic rearrangement)

aliphatic Claisen rearrangement:

Cope rearrangement:

sigmatropic rearrangements

30.7 Quinones

A. Nomenclature

benzoquinone, 1,2-naphthoquinone, etc.

B. Preparation

oxidation of dihydroxyarenes:

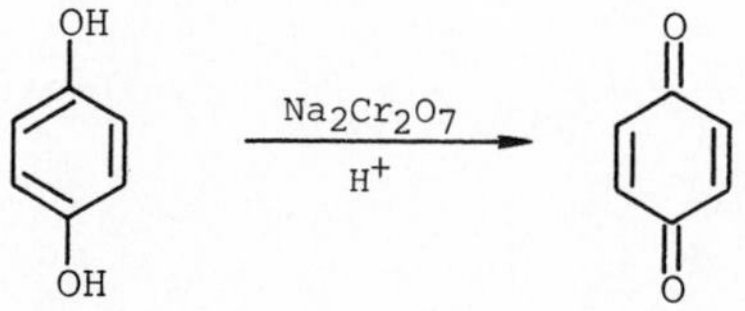

(hydroquinone)

oxidation of phenols and ethers

C. Reduction/Oxidation Equilibria

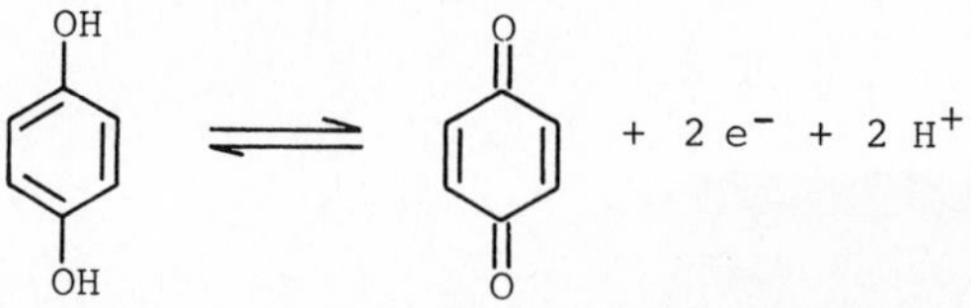

reversibility of electron transfer

dependence of reduction potentials on substituents

radical anions, antioxidants

D. Charge-Transfer Complexes

quinhydrone

E. Reactions of Quinones

addition:

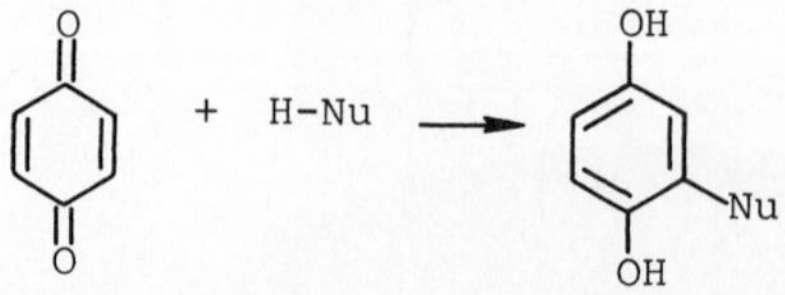

Diels-Alder reaction:

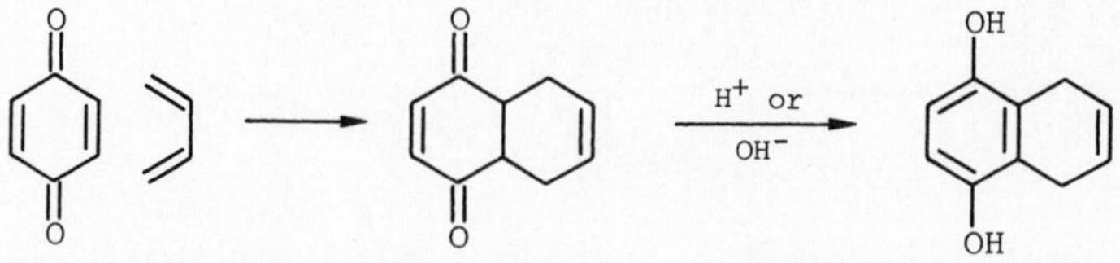

30.C   Answers to Exercises

(30.2)

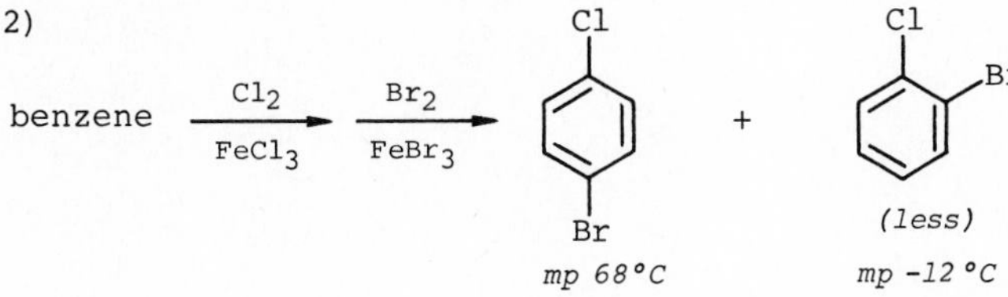

benzene $\xrightarrow[\text{FeCl}_3]{\text{Cl}_2} \xrightarrow[\text{FeBr}_3]{\text{Br}_2}$

separate *para* by crystallization

mp 68°C   *(less)* *mp –12 °C*

benzene $\xrightarrow[\text{FeBr}_3]{\text{Br}_2} \xrightarrow[\text{H}_2\text{SO}_4]{\text{HNO}_3}$

separate *para* by crystallization

mp 127°C   *(less)* *mp 43°C*

$\xrightarrow[\text{FeCl}_3]{\text{Cl}_2}$ $\xrightarrow{\text{Zn}}_{\text{HCl}}$ $\xrightarrow{\text{HNO}_2}$ $\xrightarrow{\text{H}_3\text{PO}_2}$

benzene $\xrightarrow[\text{H}_2\text{SO}_4]{\text{HNO}_3} \xrightarrow[\text{FeBr}_3]{\text{Br}_2}$ $\xrightarrow{\text{Zn}}_{\text{HCl}}$ $\xrightarrow{\text{HNO}_2}$ $\xrightarrow[\Delta]{\text{CuCl}}$

(30.3.A.1) $\xrightarrow[\text{H}_2\text{SO}_4]{\text{HNO}_3}$ $\xrightarrow[\Delta]{\text{CH}_3\text{O}^-}$ $\xrightarrow{\text{Zn}}_{\text{HCl}}$

(30.3.A.2) $\xrightarrow[\text{NH}_3]{\text{KNH}_2}$

$\xrightarrow[\text{NH}_3]{\text{KNH}_2}$

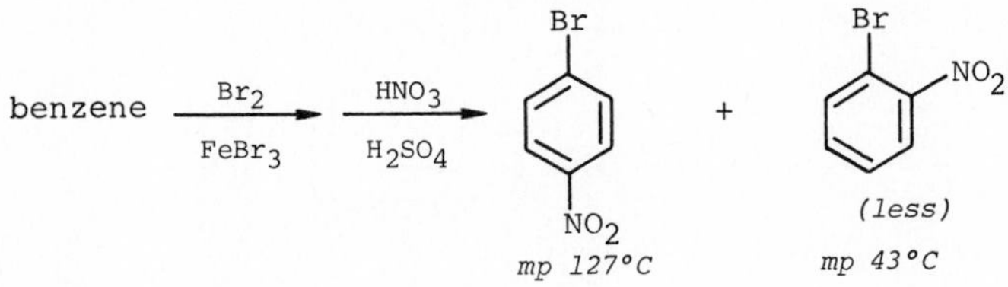

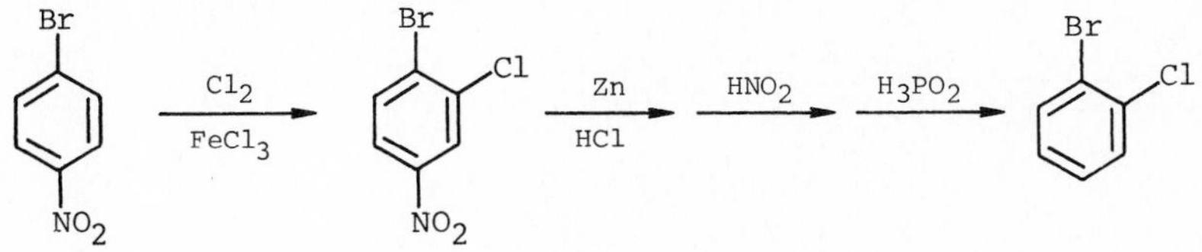

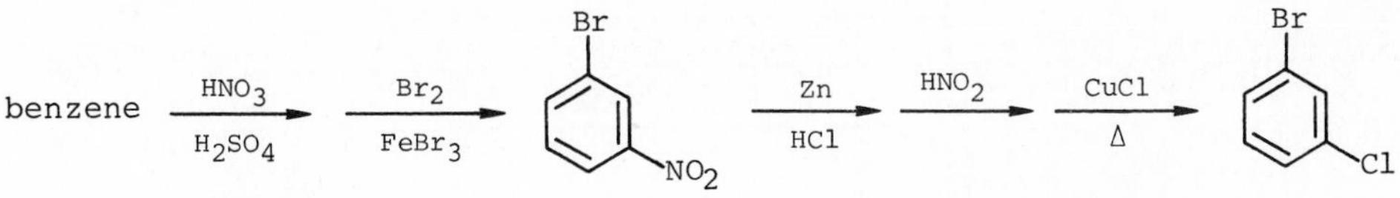

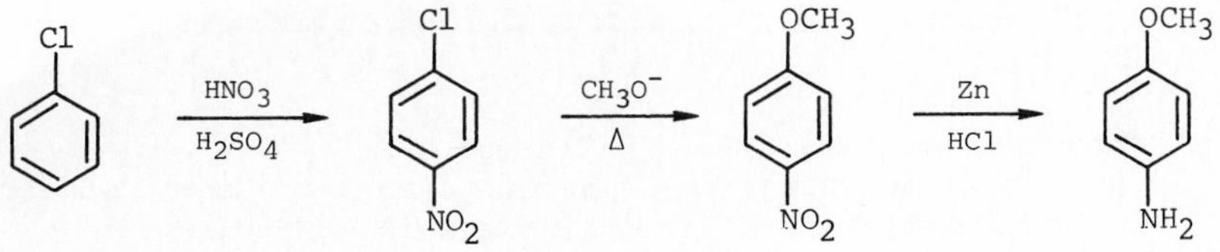

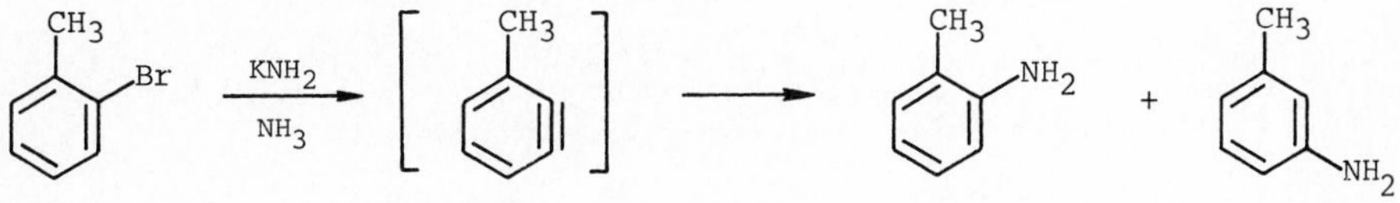

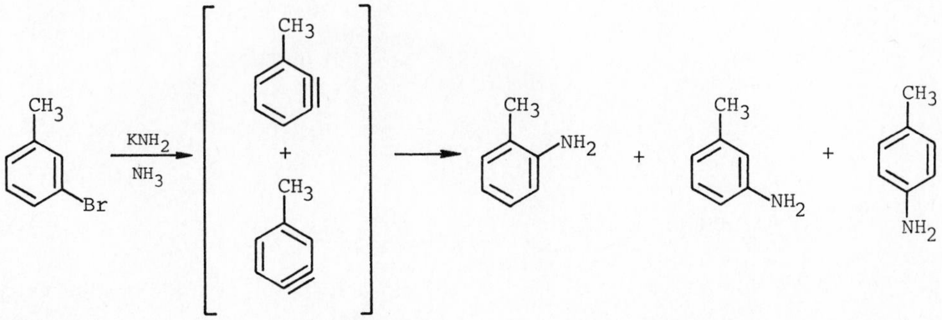

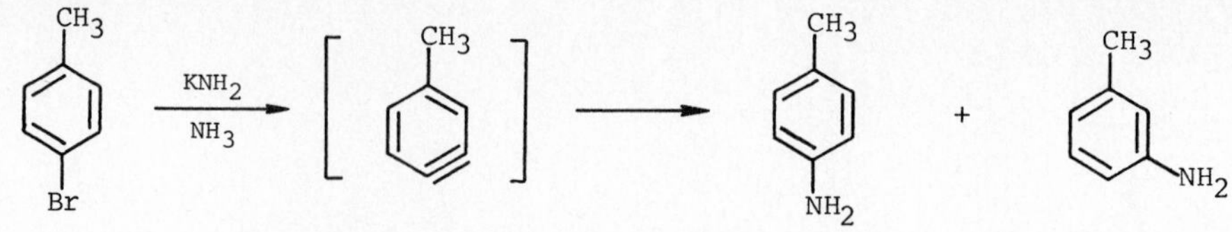

(30.3.B)

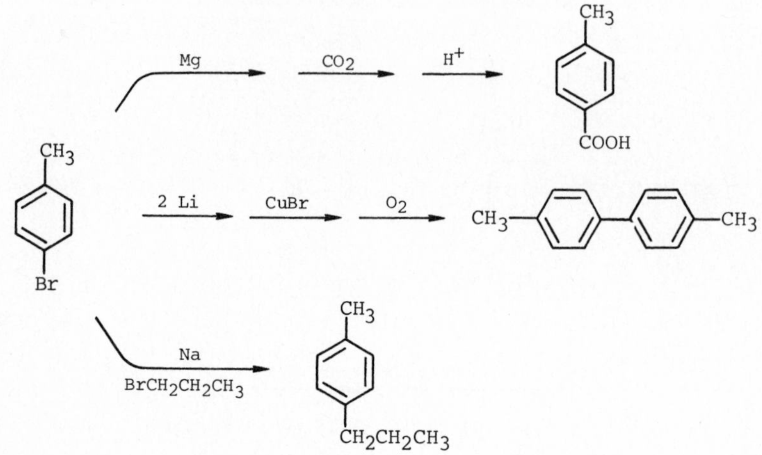

(30.5.A)

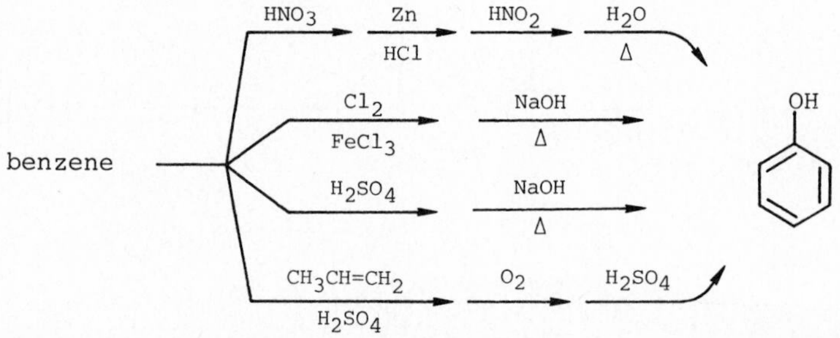

(30.5.B) (a) Because the acidity of phenols and anilinium ions is concerned with the loss of a proton from a hetero atom attached to an aromatic ring, similar substituent effects are seen in both cases.

(b)

pK_a of phenol = 10.0 ; so that $K = \frac{[C_6H_5O^-][H^+]}{[C_6H_5OH]} = 10^{-10}$

Assuming that a negligible amount of the phenol ionizes, and that all protons arise from ionization of phenol:

$[C_6H_5O^-] = [H^+]$, and $[C_6H_5OH] = 0.1$, so that $K = \frac{[H^+]^2}{0.1} = 10^{-10}$

therefore $[H^+] = 3.16 \times 10^{-6}\,\underline{M}$; $pK_a = 5.5$

When $[HA] = [A^-]$, $pH = pK_a$ of HA ; so that pH = 10.0

(30.5.C) $C_6H_5OH + BrCH_2CH{=}CH_2 + NaOH \longrightarrow C_6H_5OCH_2CH{=}CH_2 + NaBr + H_2O$

(30.6.B)

(a)

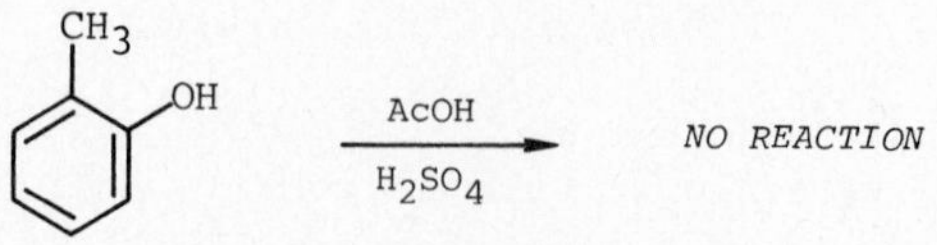

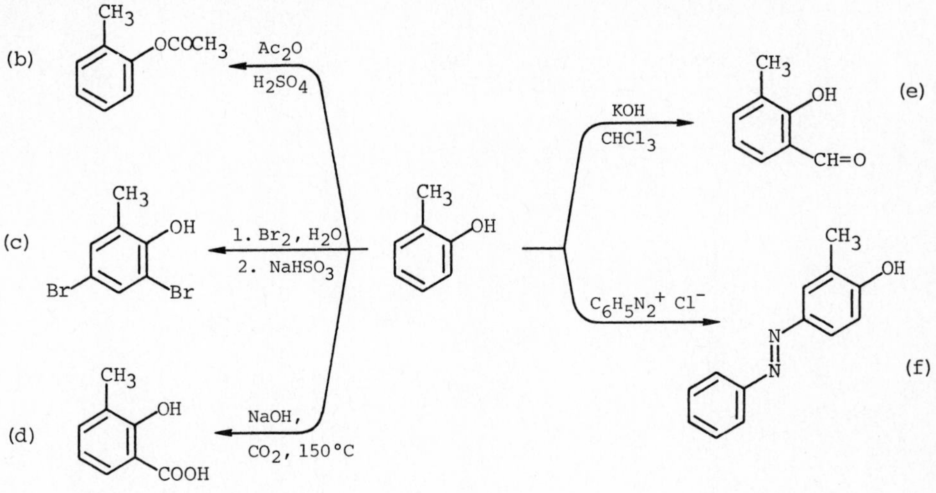
(b)
Ac2O
H2SO4
(c)
1. Br2, H2O
2. NaHSO3
(d)
NaOH,
CO2, 150°C
KOH
CHCl3
(e)
C6H5N2+ Cl−
(f)

(30.6.C)

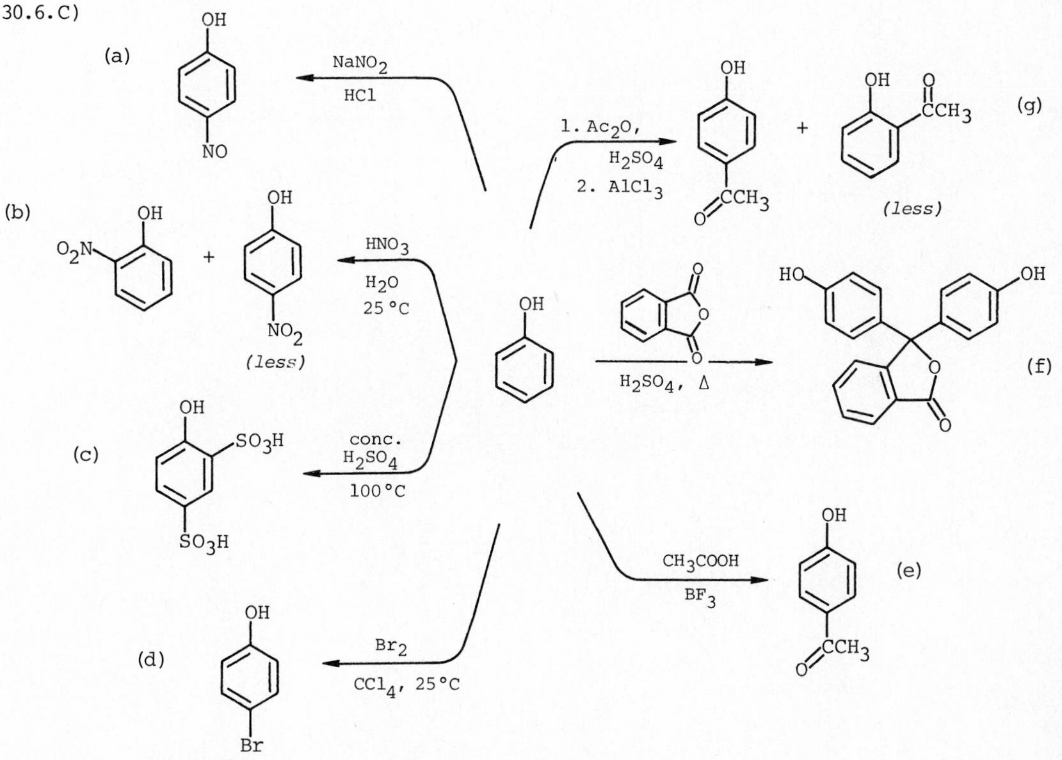
(a)
NaNO2
HCl
(b)
HNO3
H2O
25 °C
(less)
(c)
conc.
H2SO4
100°C
(d)
Br2
CCl4, 25°C
1. Ac2O,
H2SO4
2. AlCl3
(less)
(g)
H2SO4, Δ
(f)
CH3COOH
BF3
(e)

(30.6.D)

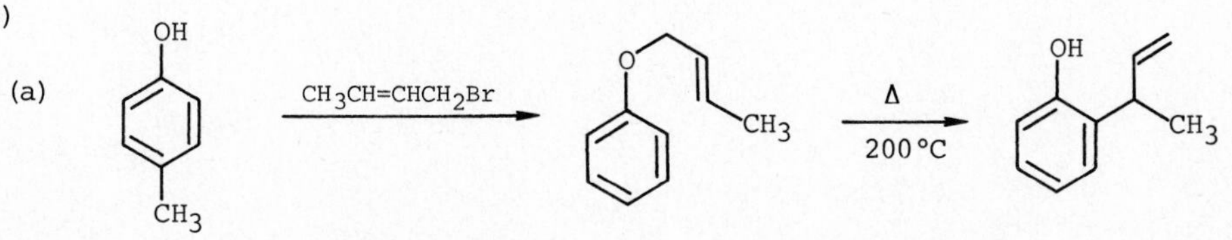
(a)
CH3CH=CHCH2Br
Δ
200°C

(b)

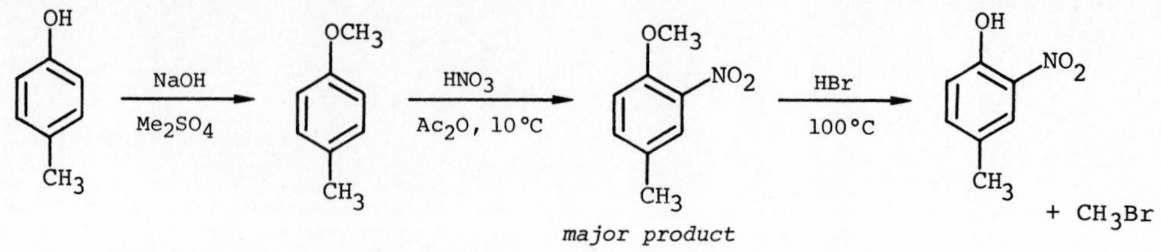

(30.7.C)

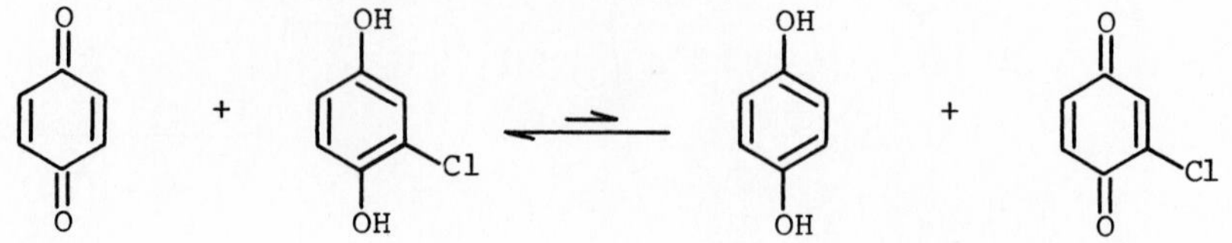

Reduction potentials: 0.699 0.713 *(stronger oxidizing agent)*

(30.7.E)

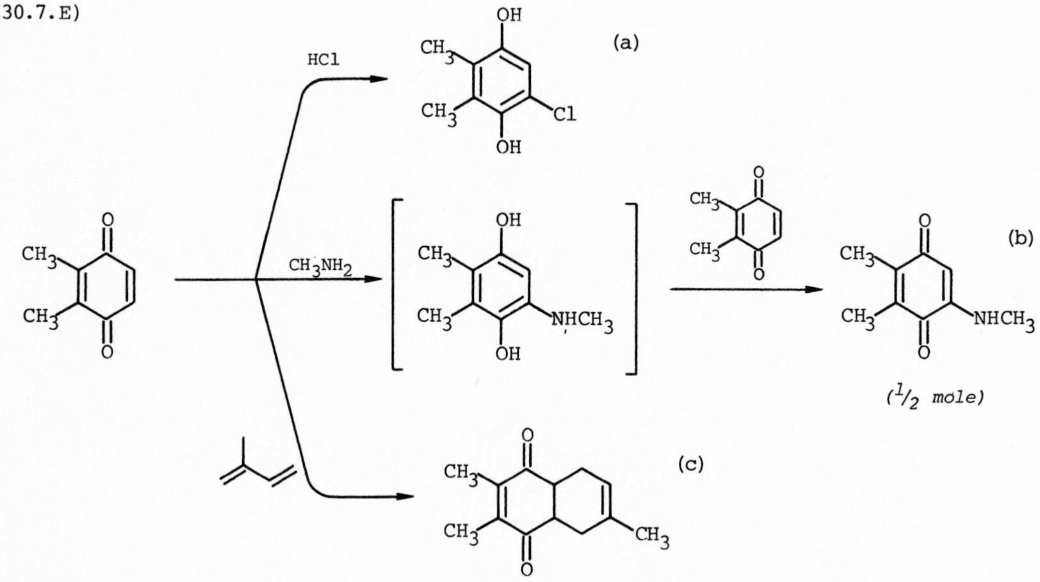

30.D Answers and Explanations for Problems

1. (a)

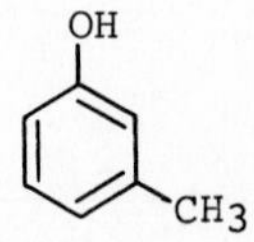

(b)

(c)

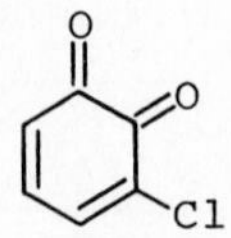

(d)

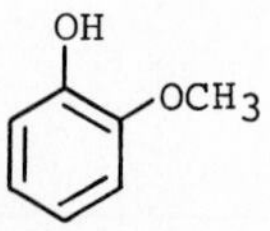

(e)

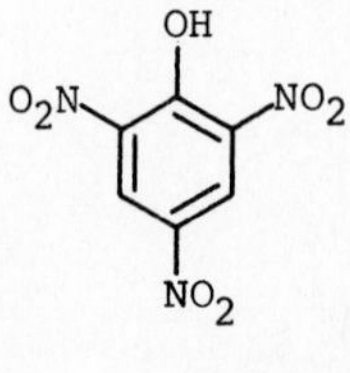

(f)

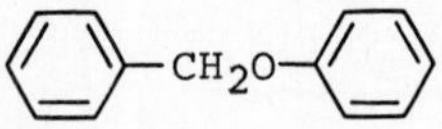

(g)

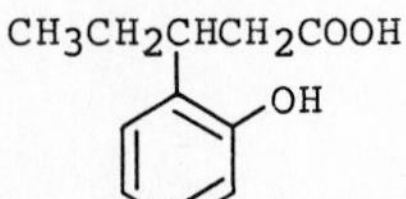

(h) 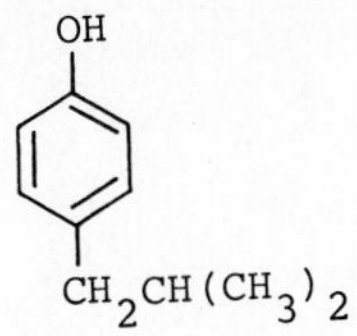

(i) 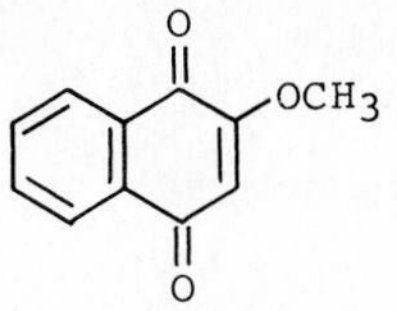

(j) 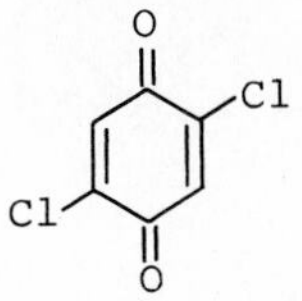

2. 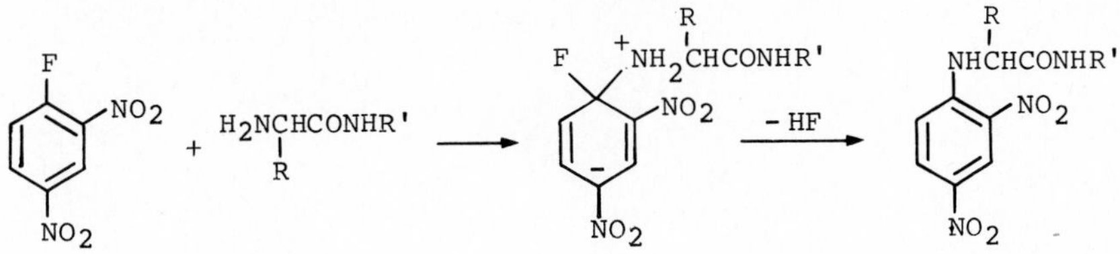

3. 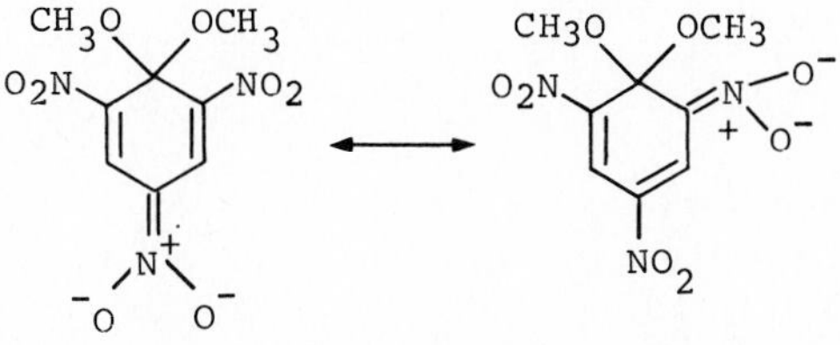

, etc. Many structures can be written

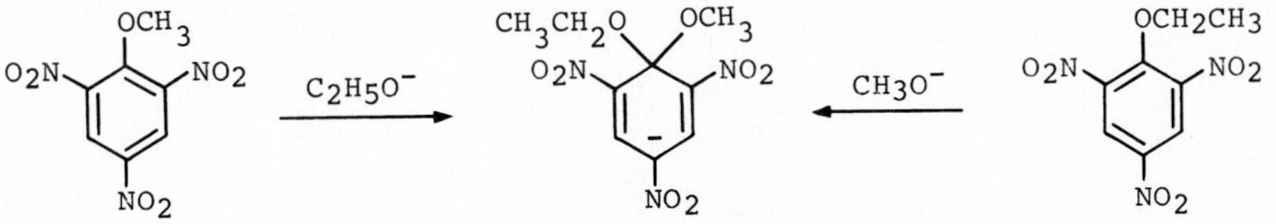

same compound from either route

4. 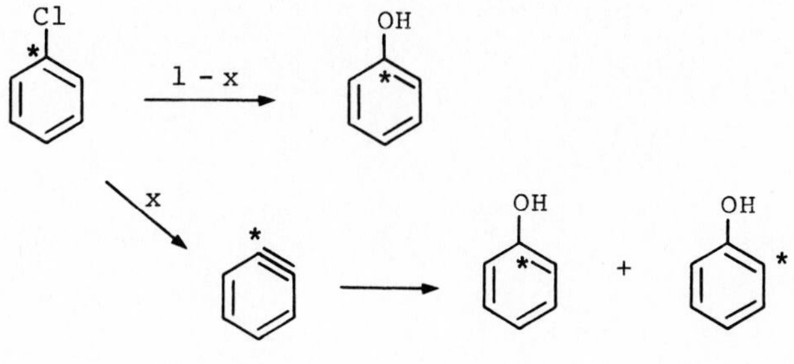

x ≡ fraction going by way of benzyne

% label in 2-position is $100(\frac{x}{2}) = 42\%$; thus, x = 0.84

only 16% goes <u>via</u> normal nucleophilic substitution

5.

(a) 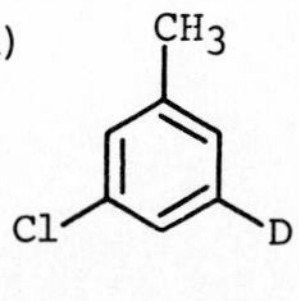

(b) 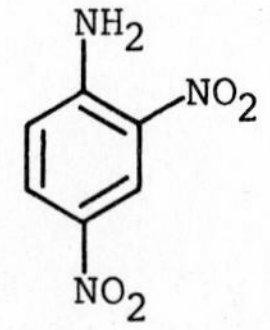

(c) 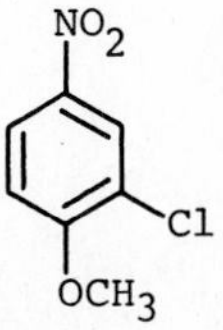

(d) 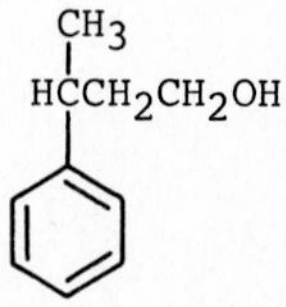

6. (a) 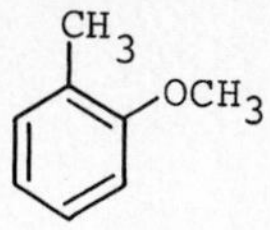

(b) no reaction

(c) no reaction

(d) 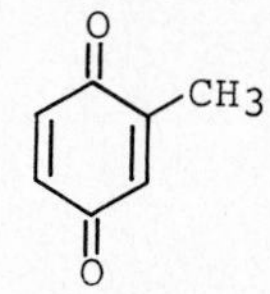

(e) 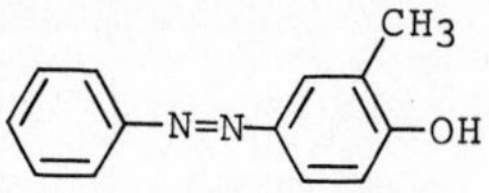

(f) no reaction

(g) no reaction

(h) 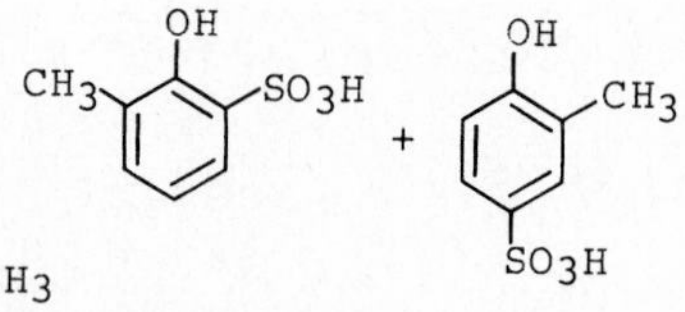

(i) CH_3COOH + CO_2 + H_2O

(j) 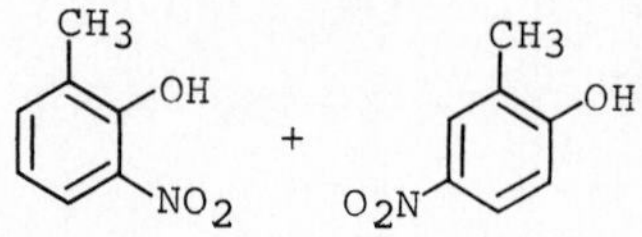

(k) 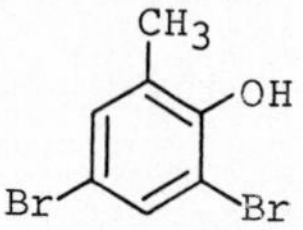

(l) 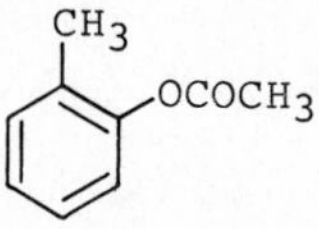

(m) 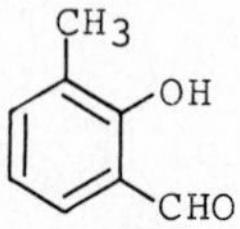

(n) 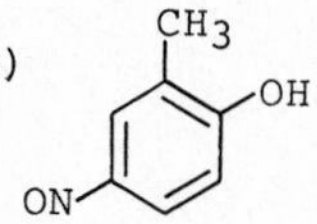

(o) 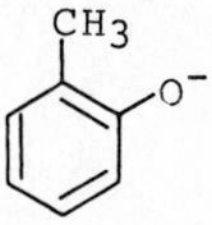

(p) 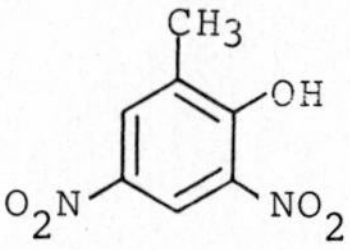

(q) no reaction

(r) 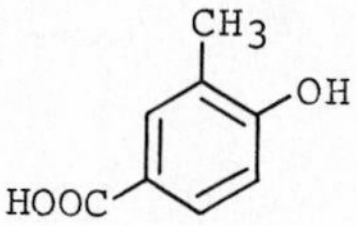

(s) 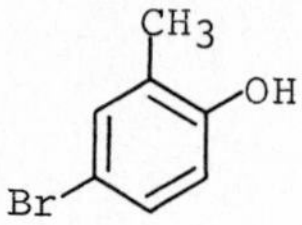

7. (a) no reaction

(b) 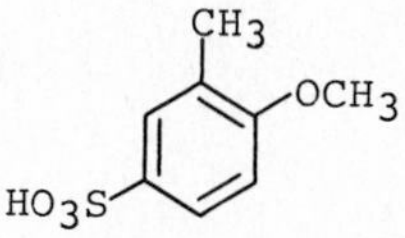

(+ ortho-)

(c) 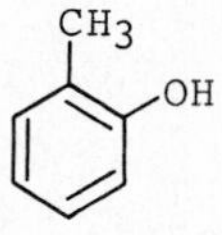

+ CH_3Br

(d) 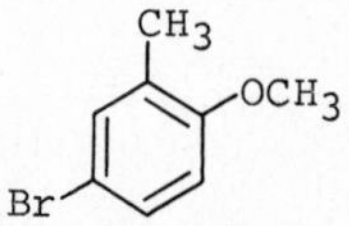

(e) 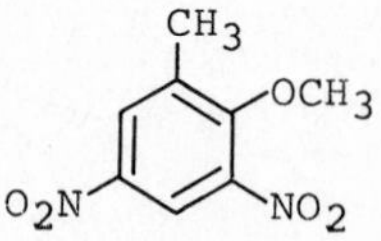

(f) no reaction

(g) no reaction

(h) 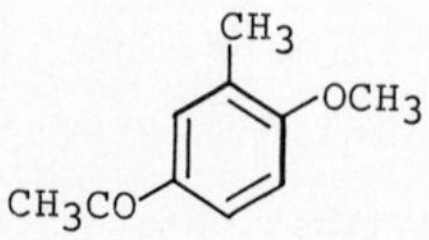

(i) 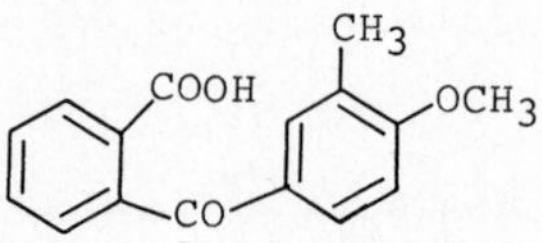

(j) From the following analogies:

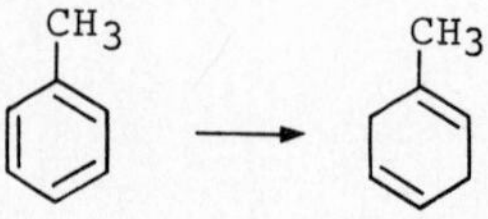

the expected reaction would be:

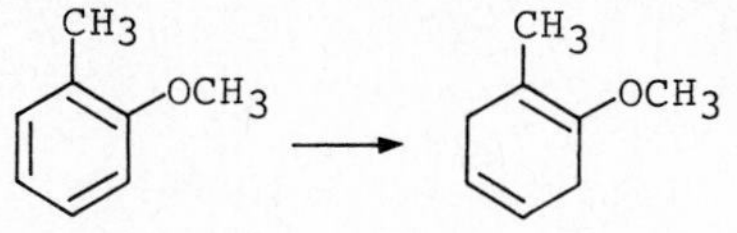

8. (a)

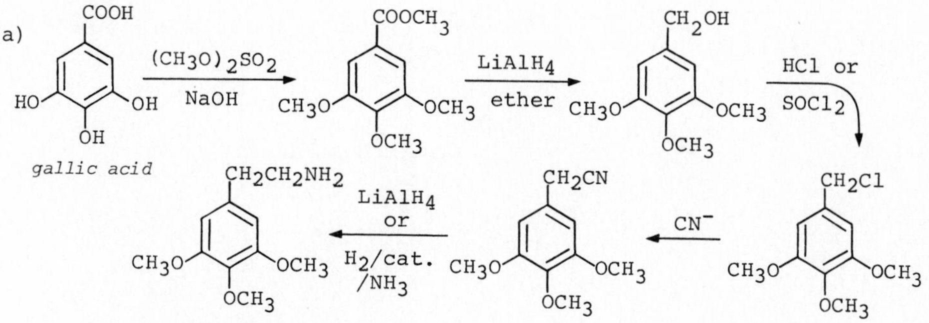

gallic acid

2-(3,4,5-trimethoxy-phenyl)ethanamine

(b)

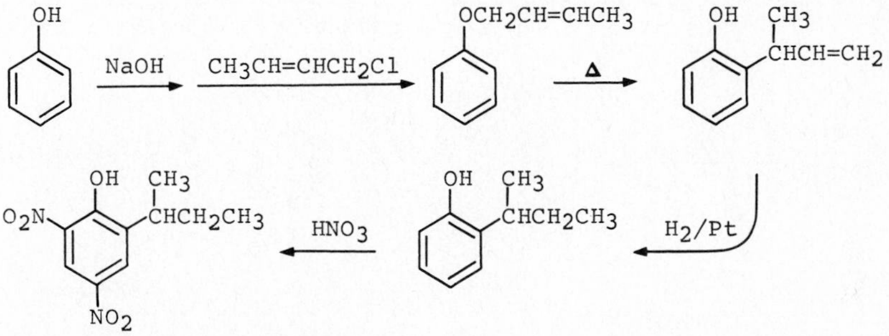

4,6-dinitro-2-*sec*-butylphenol

(c)

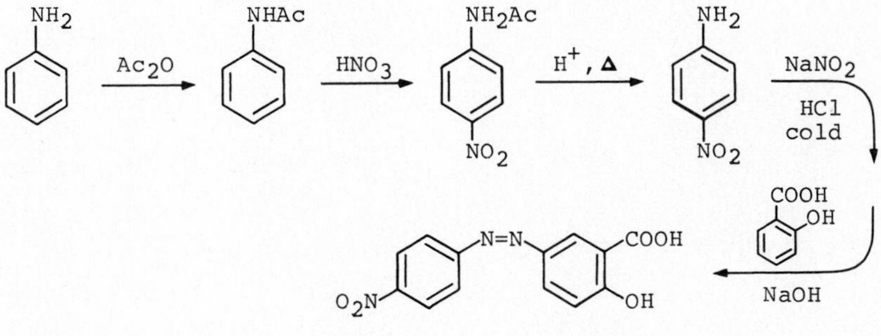

5-(*p*-nitrophenyl)azo-2-hydroxybenzoic acid

(d)

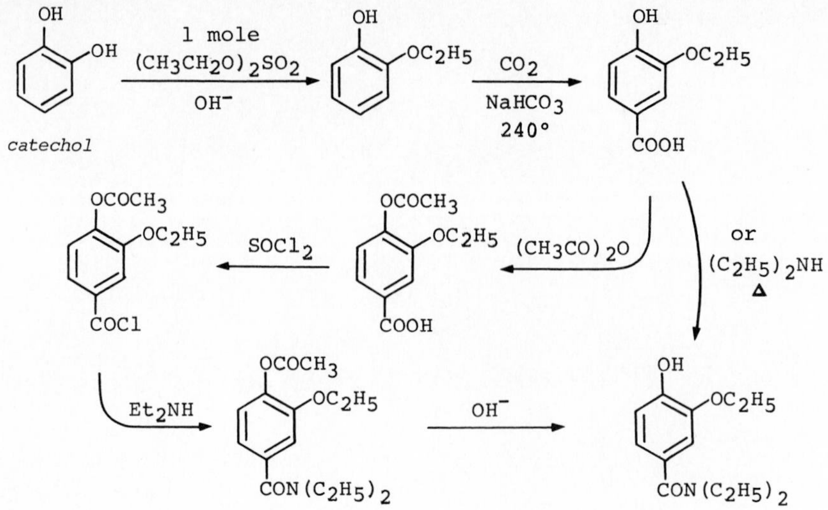

N,N-diethyl-3-ethoxy-4-hydroxybenzamide

(e)

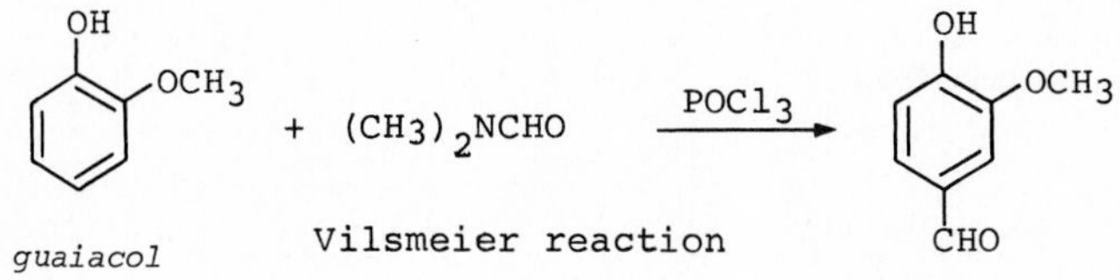

4-hydroxy-3-methoxybenzaldehyde

(f)

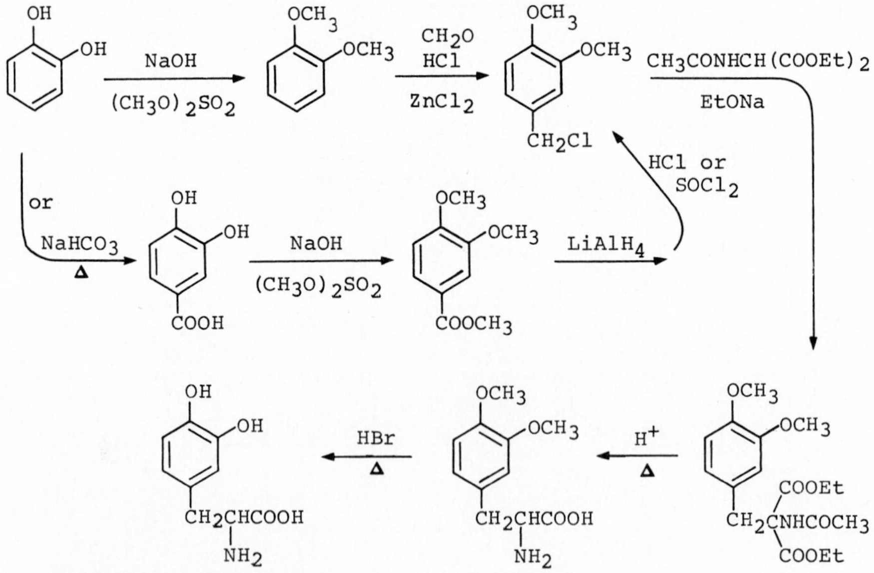

2-amino-3-(3,4-dihydroxyphenyl)propanoic acid

(g)

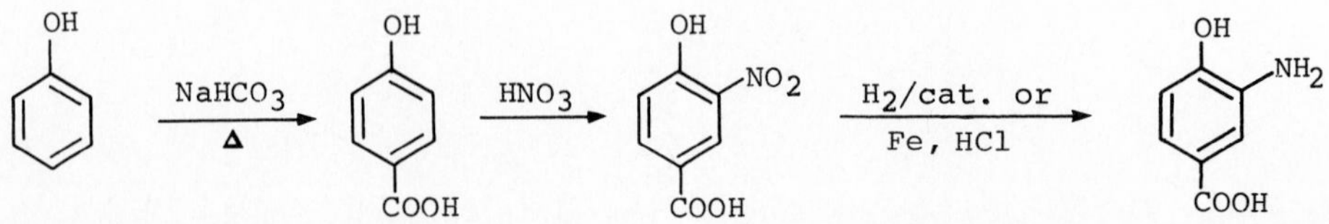

[more....]

methyl 3-amino-4-hydroxybenzoate

9. (a)

(b)

(c)

(d)

(e)

10. (a)

$$C_6H_5CH=CHCH_2Cl \longrightarrow C_6H_5CH=CHCH_2O-\text{⟨⟩} \longrightarrow$$

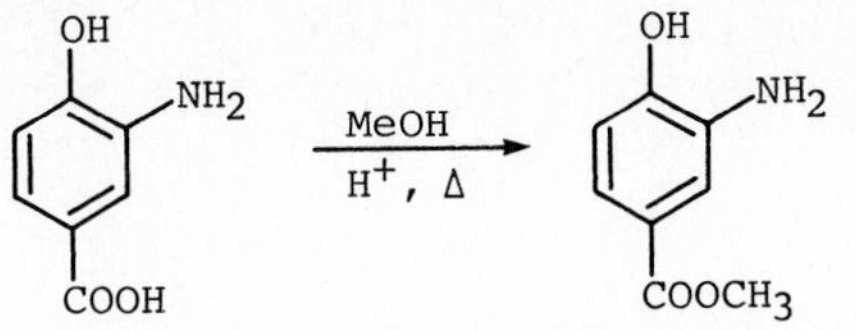

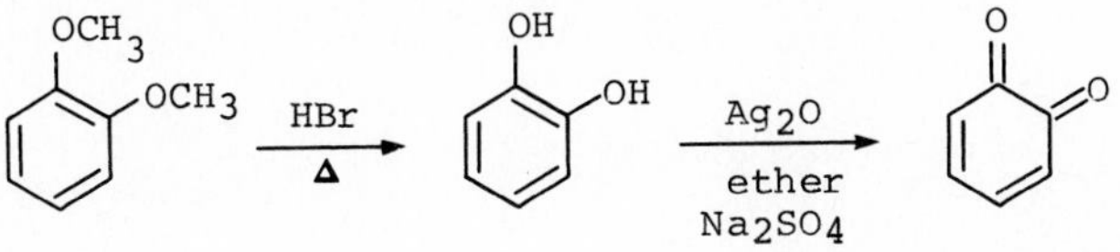

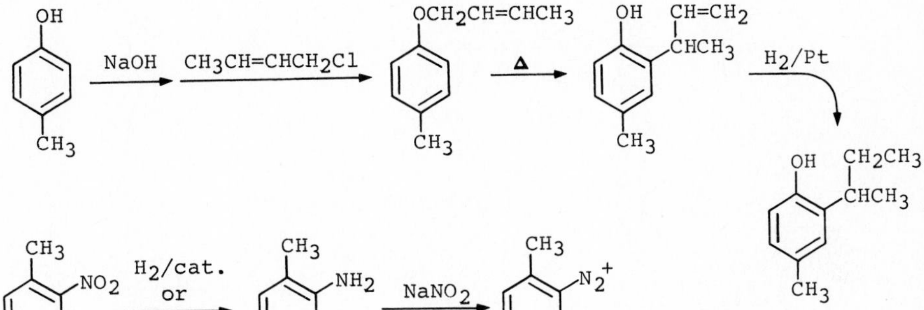

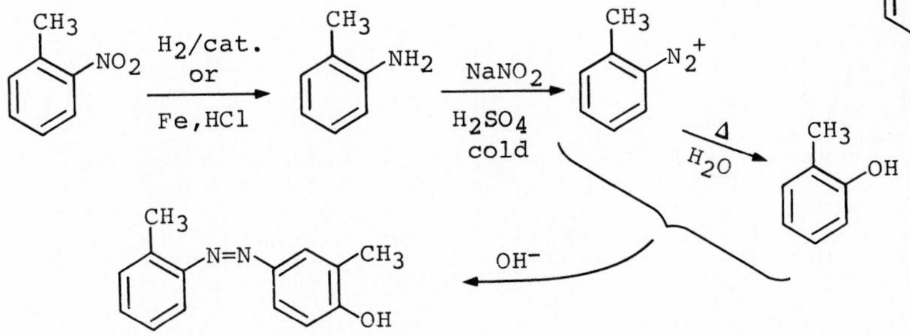

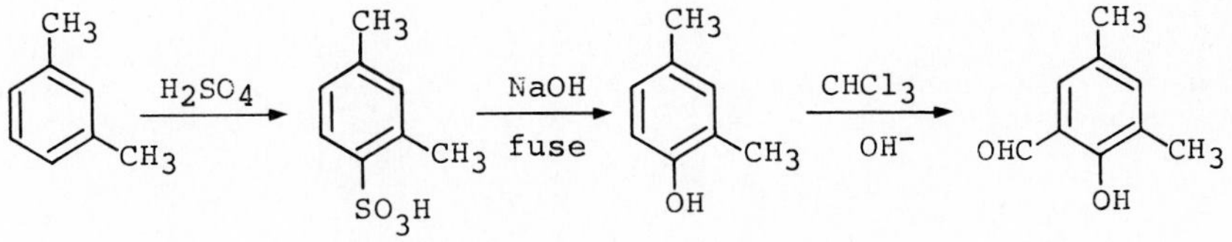

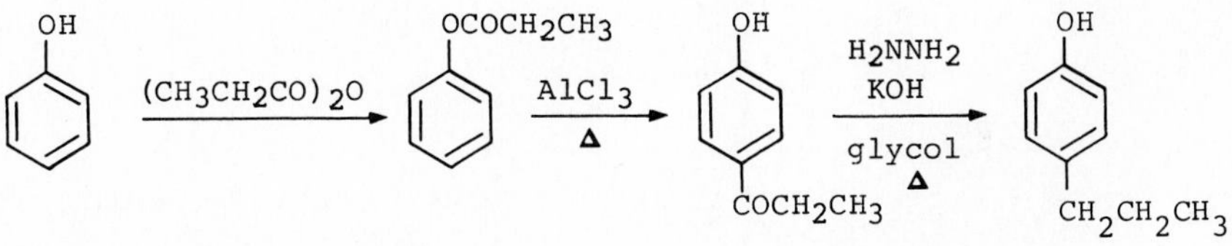

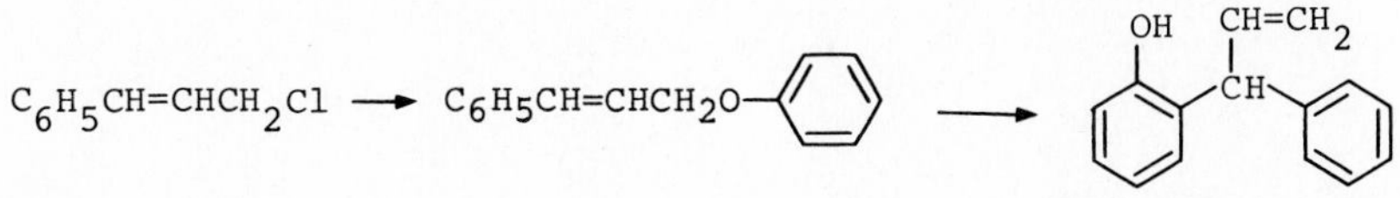

(b)

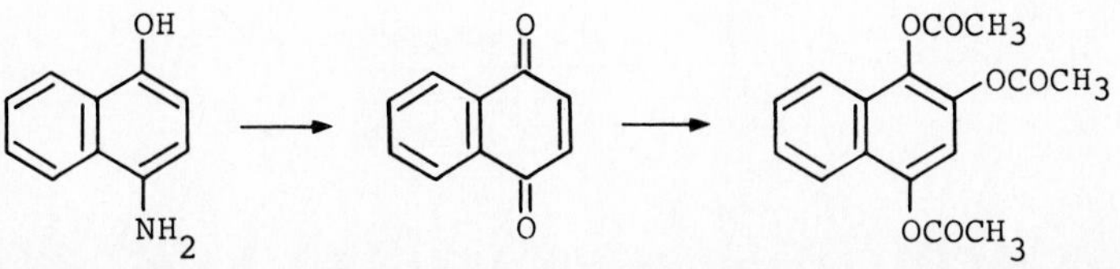

(c)

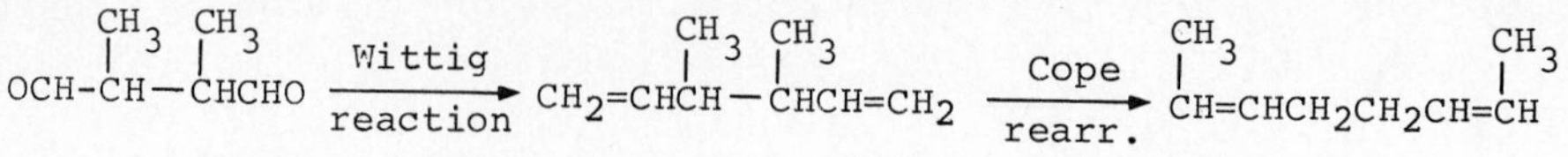
Wittig
reaction
Cope
rearr.

(d)

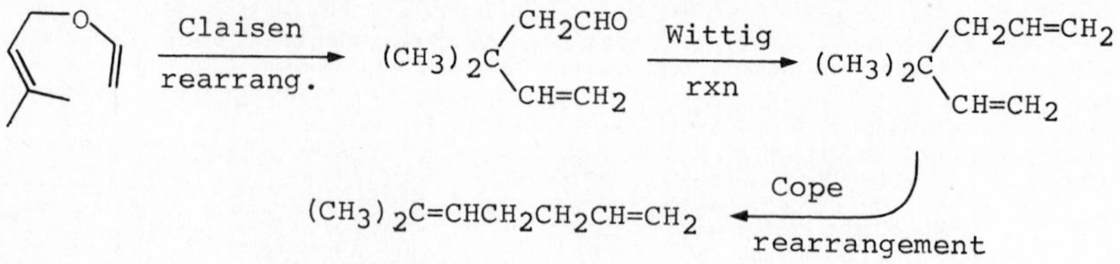
Claisen
rearrang.
Wittig
rxn
Cope
rearrangement

(e)

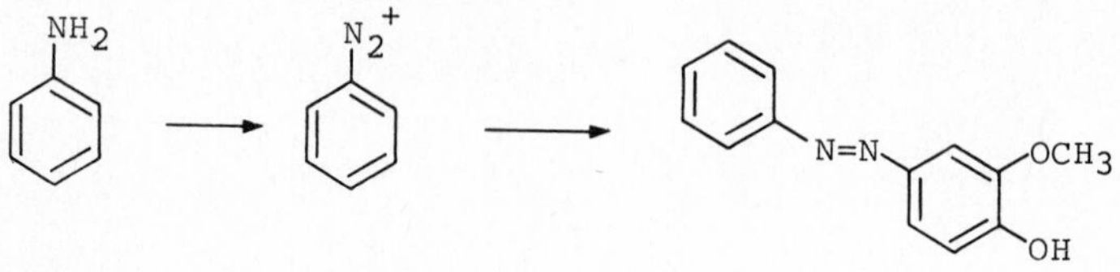

(f)

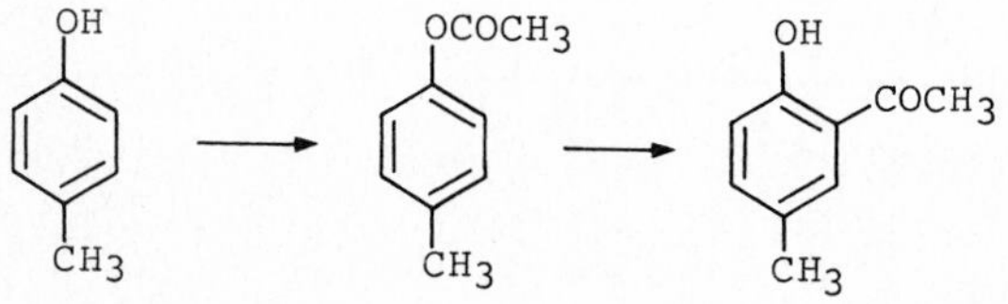

11.

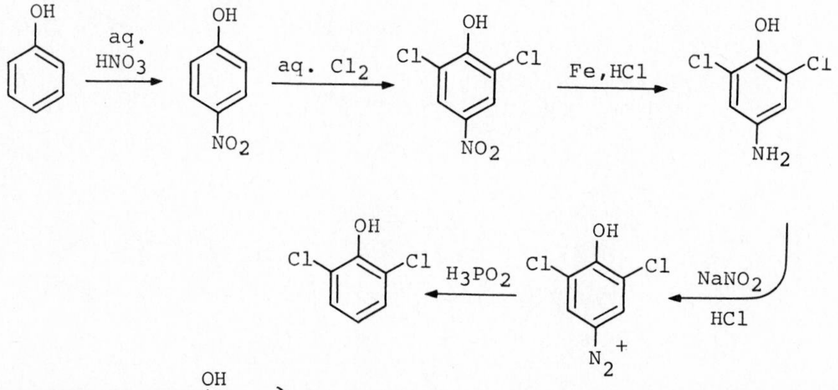
aq.
HNO_3
aq. Cl_2
Fe,HCl
$NaNO_2$
HCl
H_3PO_2

12.

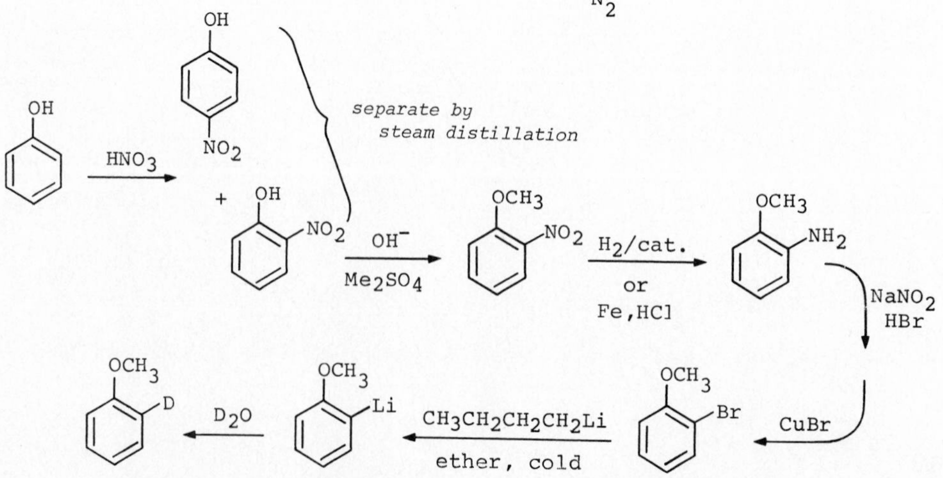
HNO_3
separate by
steam distillation
OH^-
Me_2SO_4
H_2/cat.
or
Fe,HCl
$NaNO_2$
HBr
CuBr
$CH_3CH_2CH_2CH_2Li$
ether, cold
D_2O

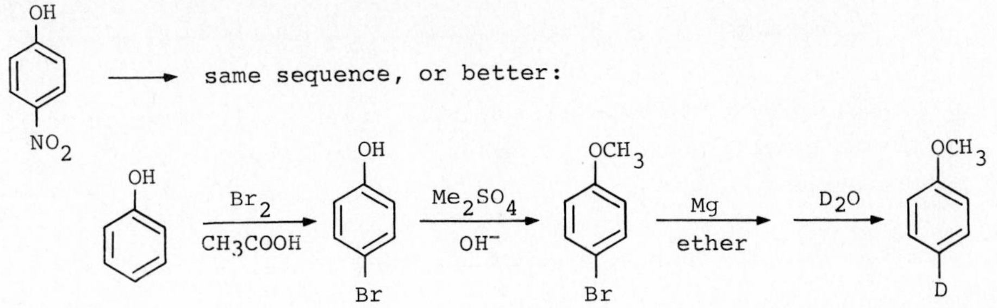

Actually, anisole itself can be metallated to give a far simpler sequence for obtaining the *ortho*-deuterated compound:

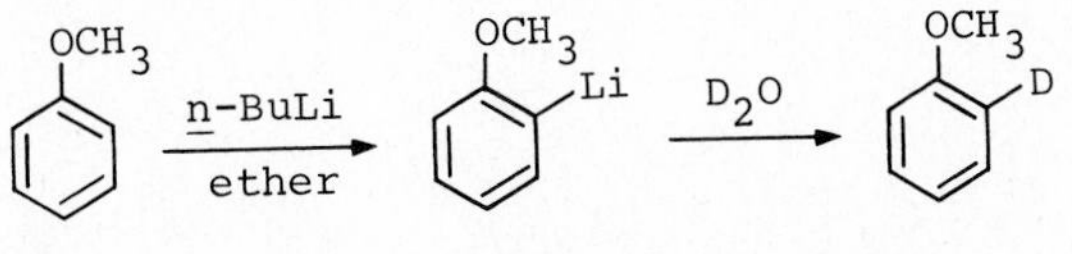

The only problem with this sequence is that it is difficult to accomplish complete metallation. Thus, the final deuterioanisole is accompanied by undeuterated anisole which cannot be separated. For many purposes, a partially deuterated compound will suffice. This procedure can be applied without problems for tracer-labeled anisole-2-t.

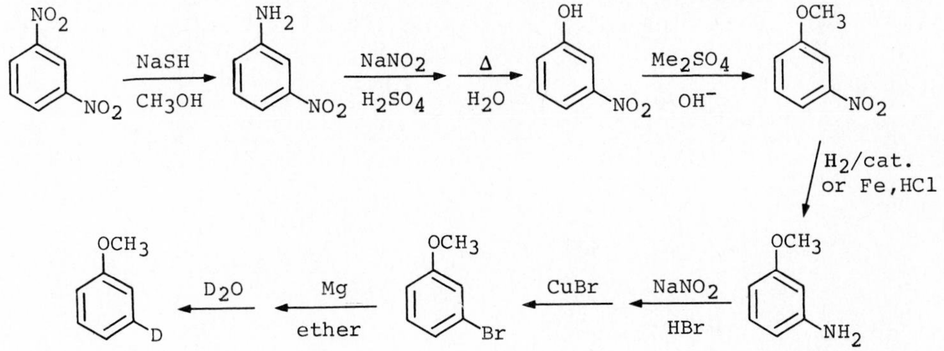

13.

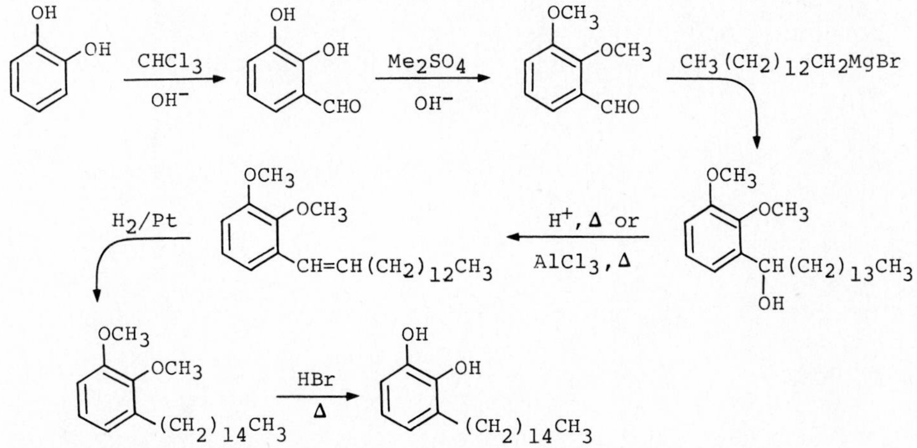

14. Benzyl alcohol is only slightly more acidic than ethanol; hence, a solution of potassium benzyloxide in ethanol contains substantial amounts of potassium ethoxide. Both can give S_N2 reactions to give a mixture of ethers:

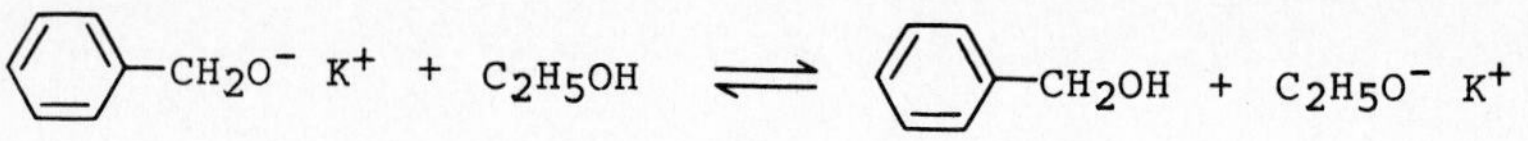

p-Cresol, however, is much more acidic than ethanol. A solution of potassium p-methylphenolate in ethanol contains very little potassium ethoxide.

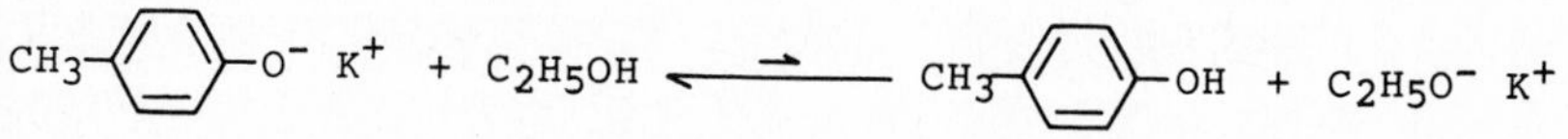

15.

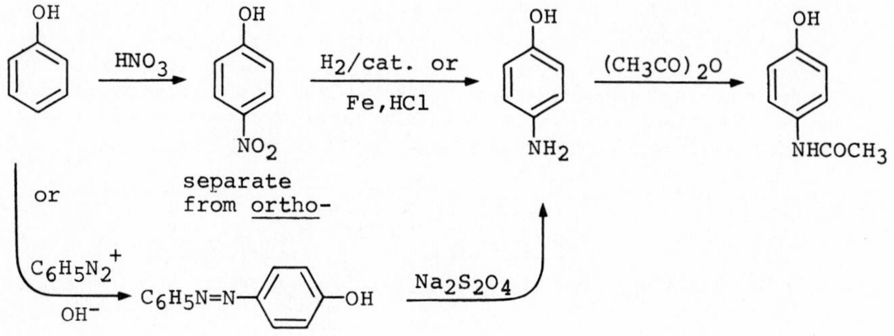

16.

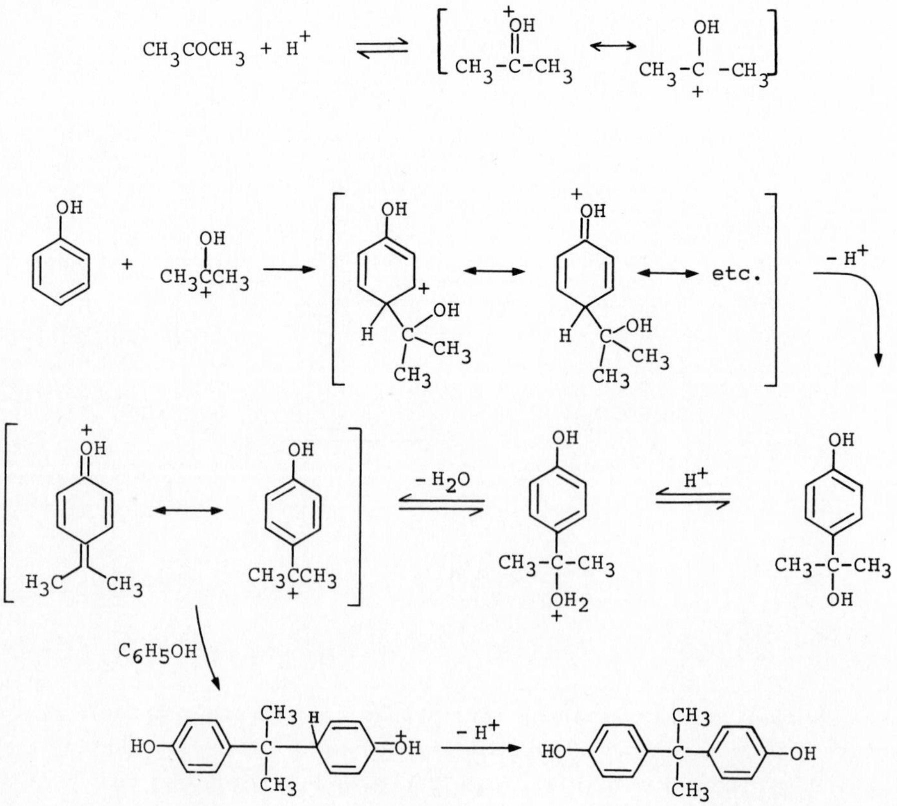

17. Yes. SO_3H is a bulky group and goes <u>ortho</u> to the smaller CH_3 group rather than to the larger $(CH_3)_2CH$ group.

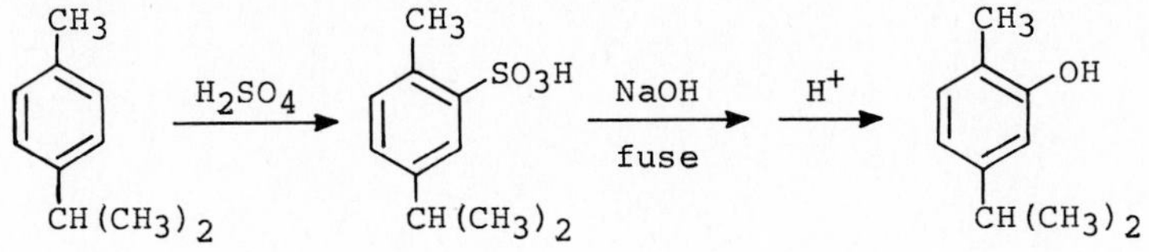

carvacrol

18.

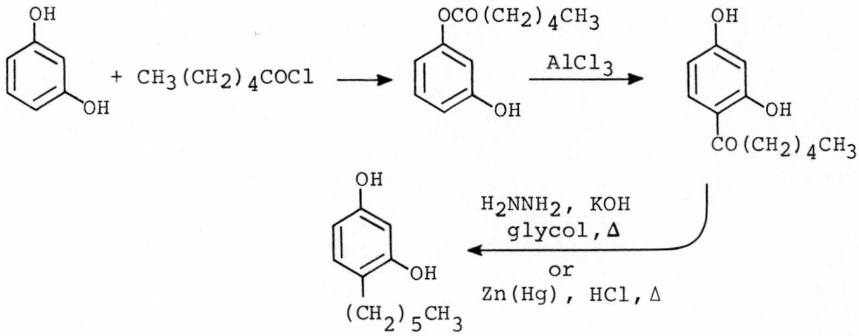

19. (a)

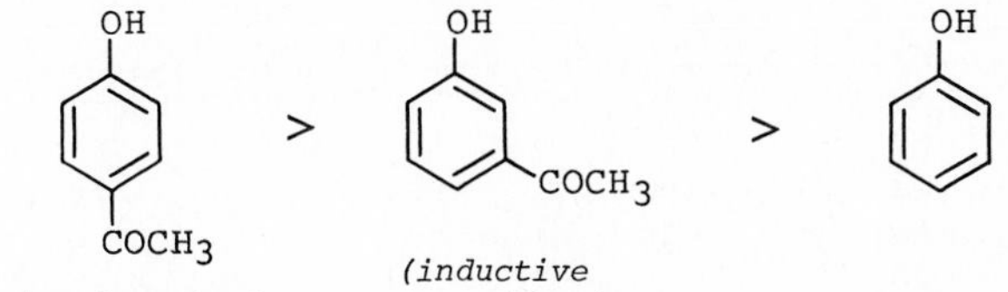

(conjugation) *(inductive effect)*

(b)

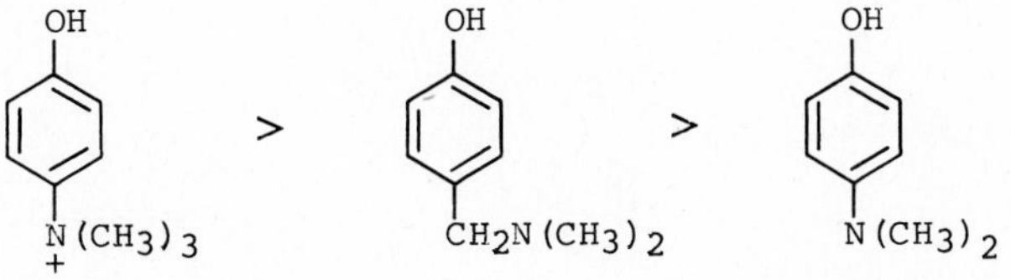

(inductive effect) *(electron-donating group)*

(c)

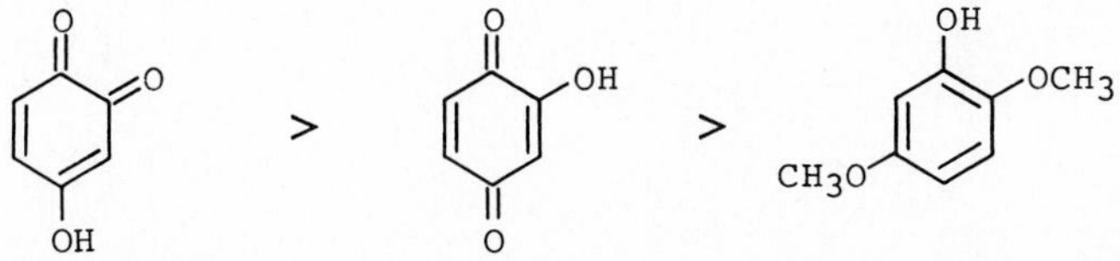

(<u>NOTE</u>: the two quinones give the same anion, but the *ortho*-quinone is less stable; thus, it is more acidic.)

20. Oxidation of a phenol gives the quinone.

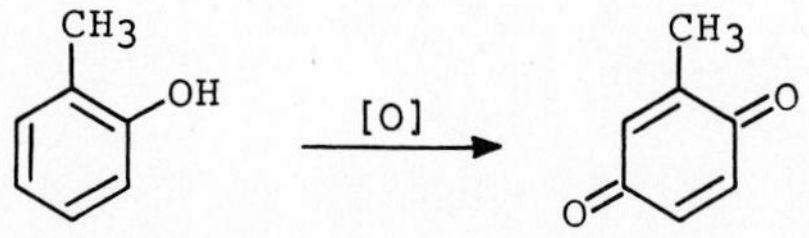

21.

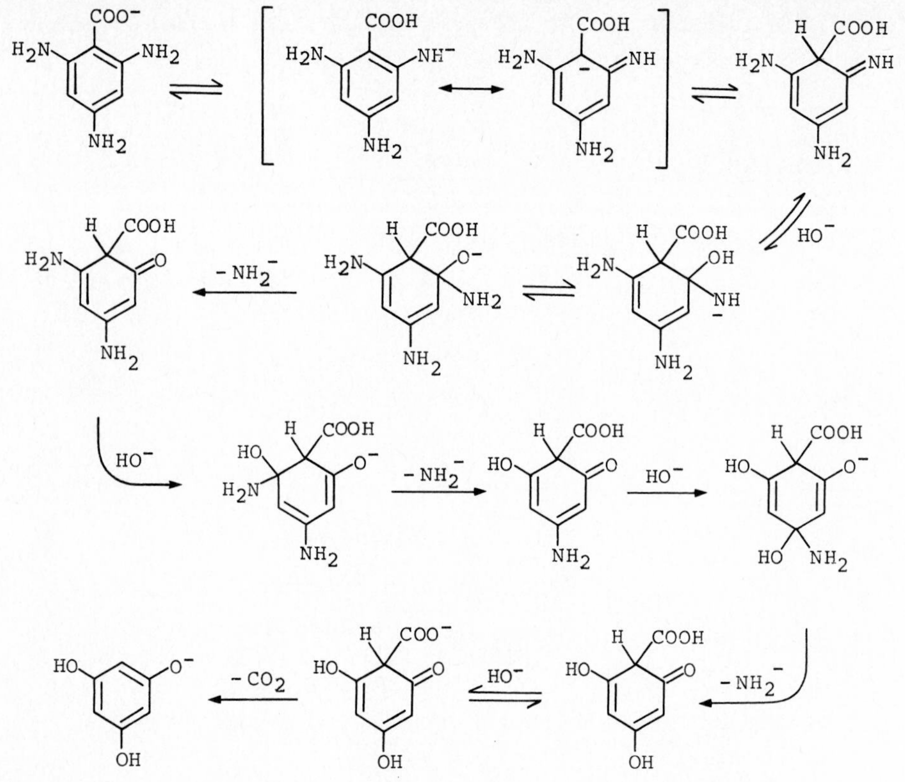

NOTE: this mechanism has many possibilities which depend on the timing of the steps for decarboxylation and hydrolysis of ammonia groups.

22.

Internal hydrogen-bonding:

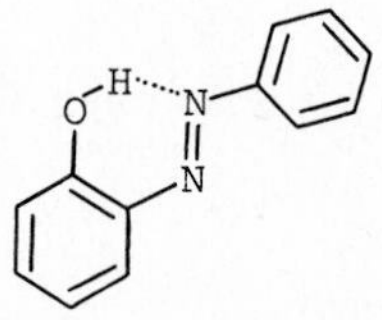

In the *ortho* case, the hydrogen-bonding is all internal. In the *para* case, such internal hydrogen bonding is not possible. Instead the -OH group hydrogen-bonds to water or to -N=N- groups in other molecules, and the volatility is decreased.

Yes, the carbonyl group of o-hydroxypropiophenone can also hydrogen-bond:

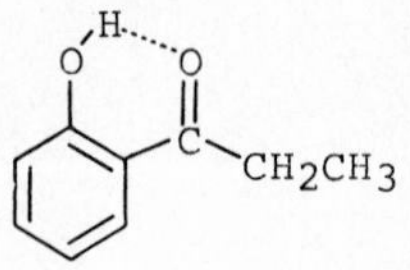

23.

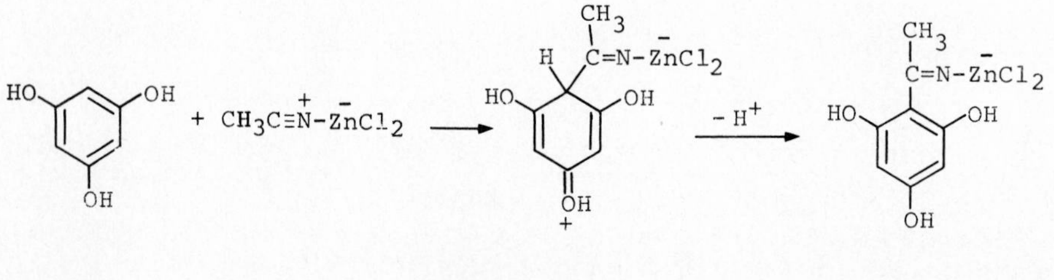

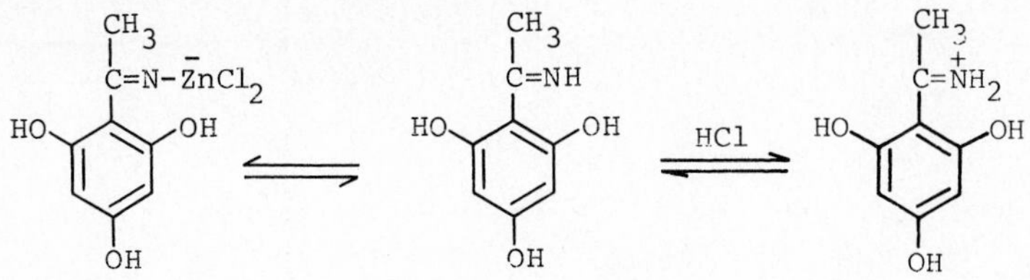

The ring of phenol is less highly activated, since it has one hydroxy group instead of three. Thus, reaction at the ring is slower, and reaction of the -OH group with the nitrile can compete:

$$C_6H_5OH + CH_3C\equiv\overset{+}{N}-\overset{-}{Z}nCl_2 \longrightarrow C_6H_5\overset{+}{O}(H)-\underset{CH_3}{C}=N-\overset{-}{Z}nCl_2 \rightleftharpoons C_6H_5O\underset{CH_3}{C}=NH \xrightarrow{HCl} C_6H_5O\underset{CH_3}{C}=\overset{+}{N}H_2 \quad Cl^-$$

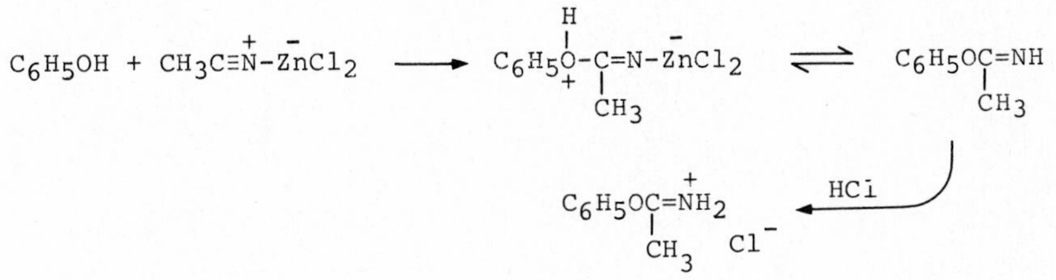

24. (a)

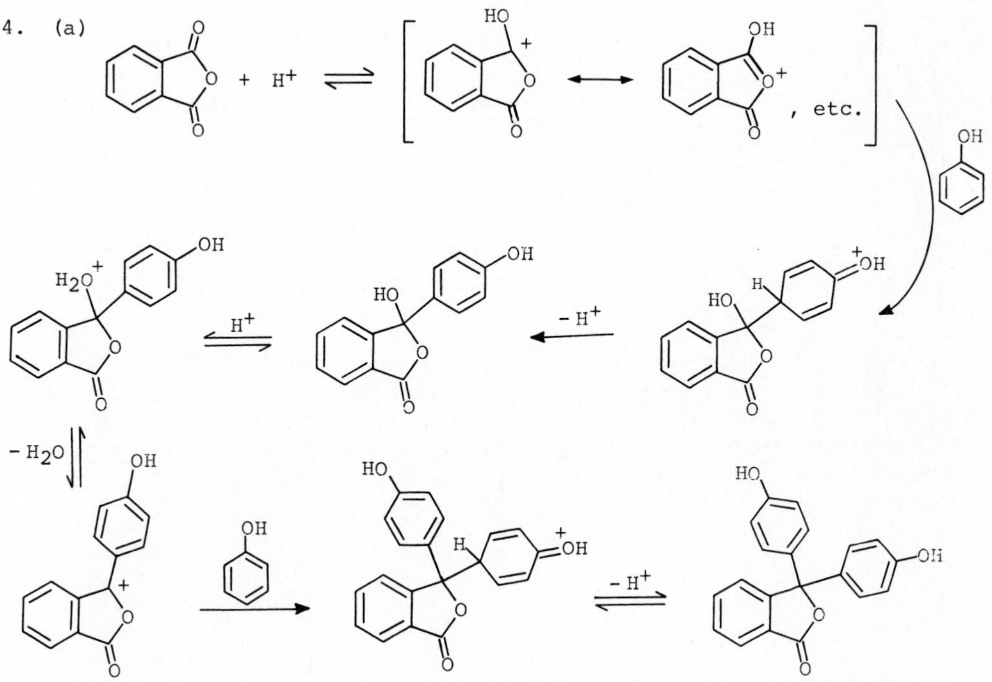

(b) The immediate product is:

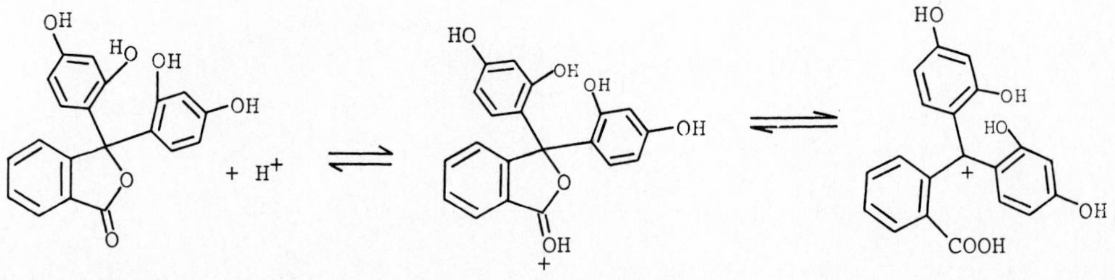

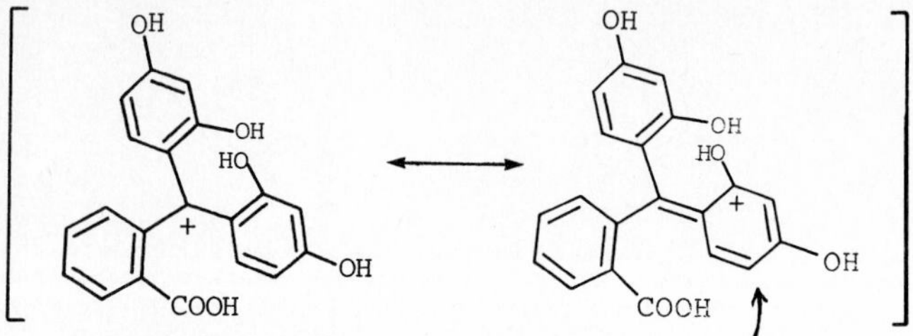

Note that the ring has a positive charge at each OH;

nucleophilic attack can occur:

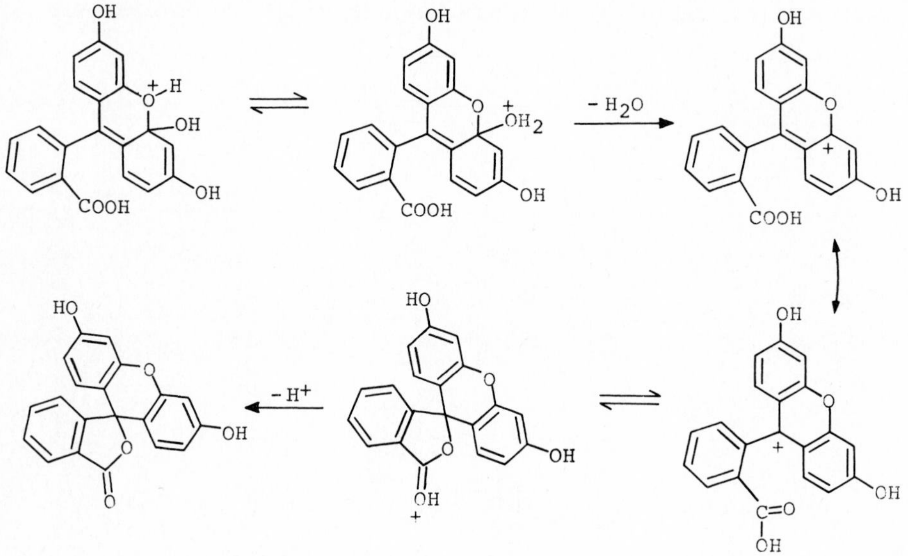

25. The potential difference is given directly by Table 30.2 as 0.713 - 0.699 = 0.014 volts, not a very powerful battery. When determining the signs, remember that the -Cl group destabilizes the quinone; hence, this side needs protons and electrons to reach equilibrium. Since it is drawing electrons, it must represent the anode, and the quinone-hydroquinone side is the cathode.

26.

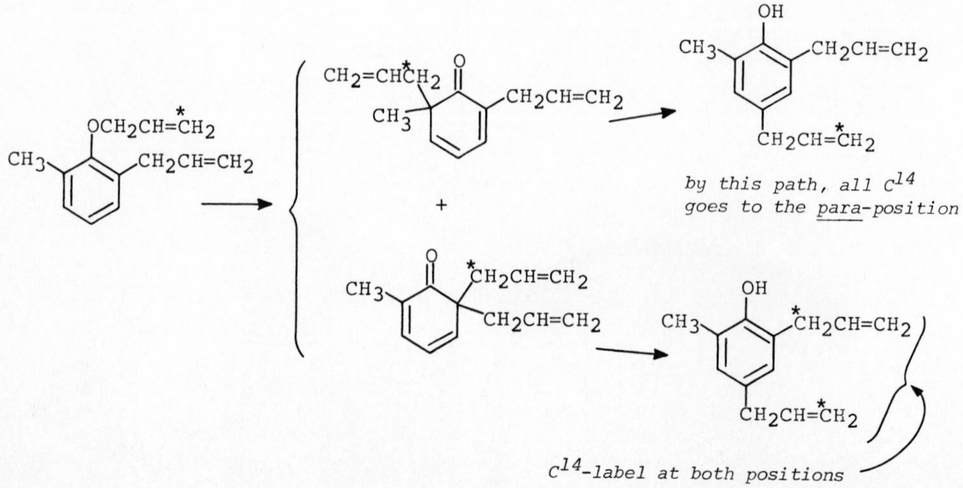

27. (a) $\quad C_2H_5OCH=CH_2 + Hg^{++} \;\rightleftharpoons\; C_2H_5O-\overset{+}{C}HCH_2Hg^+ \;\xrightarrow{ROH}\; C_2H_5O-\underset{\underset{+}{\overset{|}{ROH}}}{\overset{|}{C}}HCH Hg^+$

$$ROCH=CH_2 \atop + Hg^{++} \;\rightleftharpoons\; \underset{+}{ROCHCH_2Hg^+} \;\overset{C_2H_5OH}{\underset{}{+}}\;\rightleftharpoons\; C_2H_5\underset{\underset{OR}{|}}{\overset{\overset{H}{|}}{O}}-\overset{+}{C}HCH_2Hg^+ \;\rightleftharpoons\; C_2H_5O-\underset{\underset{RO}{|}}{C}HCH_2Hg^+ \;+\; H^+$$

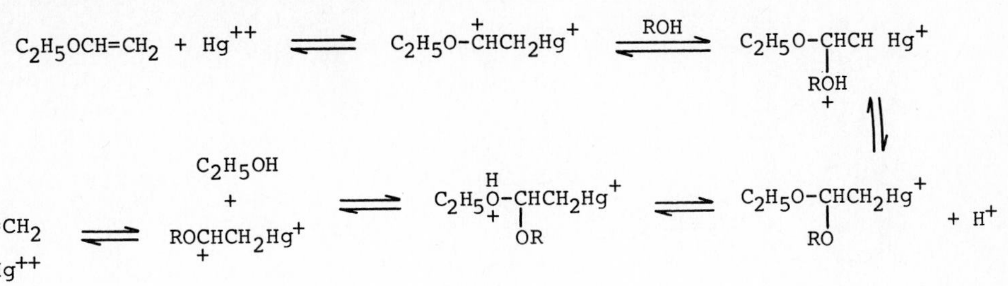

(b)

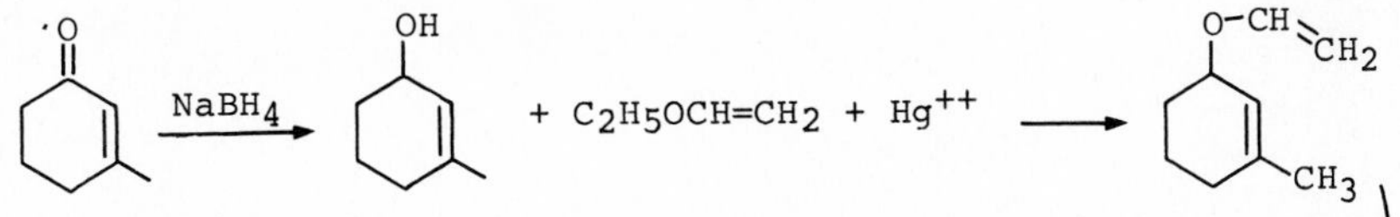

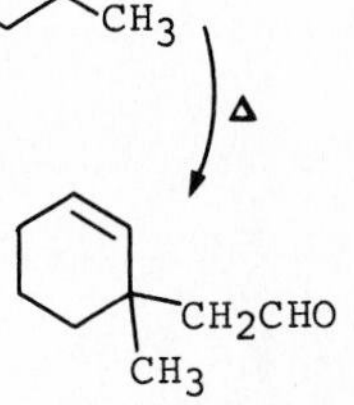

Preparation of the unsaturated ketone can be *via*:

$$\text{or}$$

$$CH_3COCH_2COOEt + CH_2=CHCOCH_3 \;\xrightarrow[\text{EtOH}]{\text{EtONa}}\; \left[\; EtOOC \cdots \;\right] \longrightarrow$$

$$EtOOC \cdots \;\xleftarrow{H^+,\,\Delta}\; \left[\; HOOC \cdots \;\right] \;\xrightarrow{-CO_2}\;$$

28.

$$\text{+ } POCl_3 \;\xrightarrow{\text{pyridine}}\; \left(\cdots -O- \right)_3 PO$$

29.

$$\text{+ } H_2O \longrightarrow \left[\cdots \longleftrightarrow \cdots \right]$$

$$\longrightarrow \cdots \text{+ } NO_2^- \;\rightleftharpoons\; \cdots \text{+ } HNO_2$$

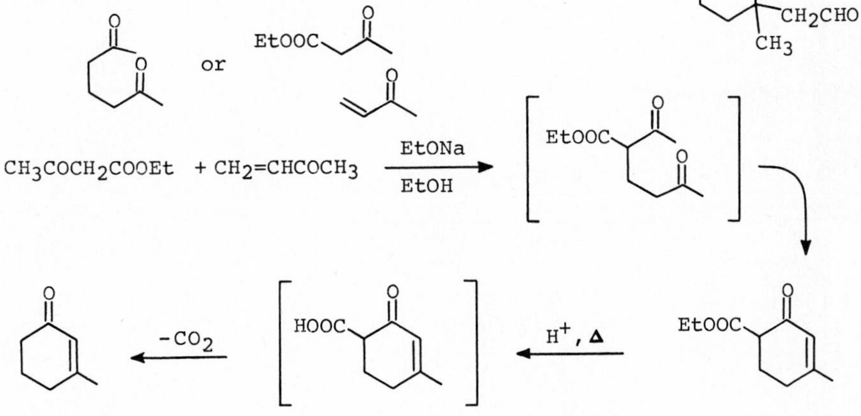

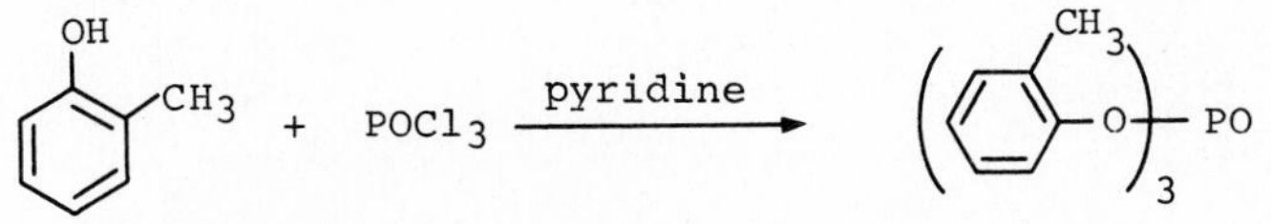

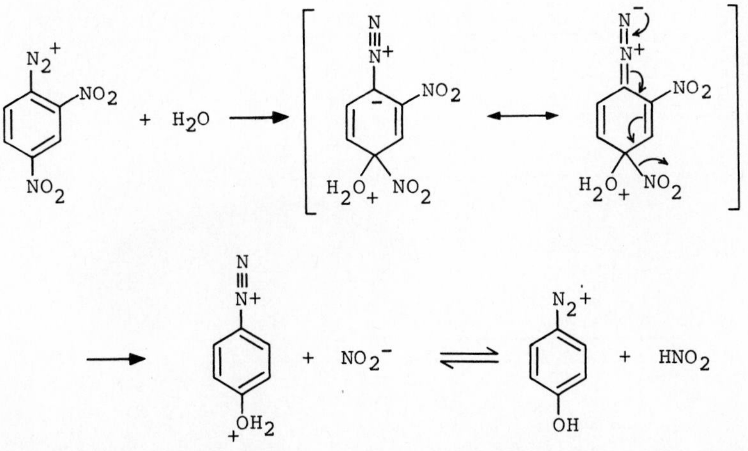

The $-N_2^+$ group deactivates nucleophilic aromatic substitution, much as the $-N=O$ and $-NO_2$ groups.

30.

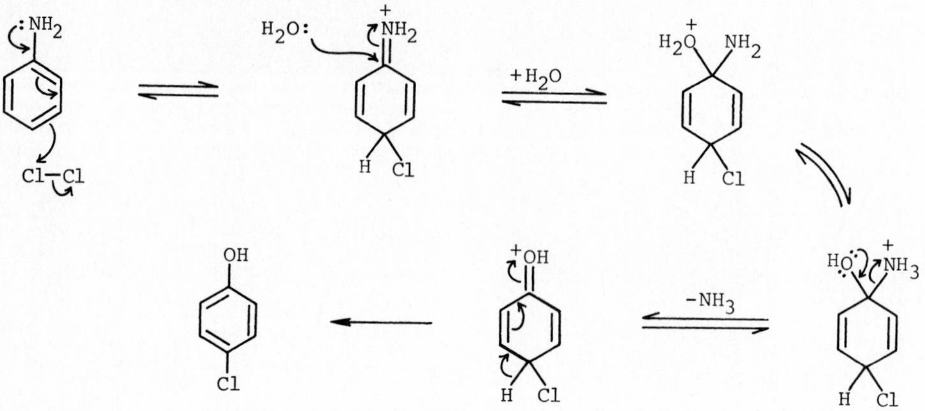

30.E Supplementary Problems

S1. Name the following compounds.

(a)

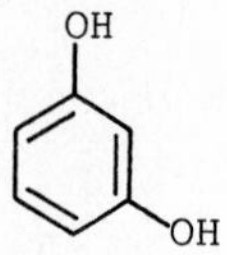

(c)

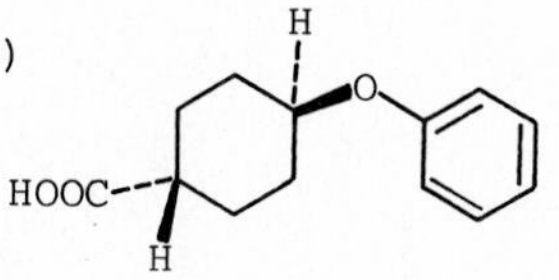

(e)

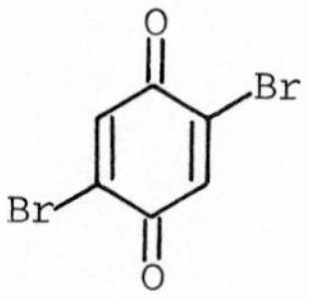

(b)

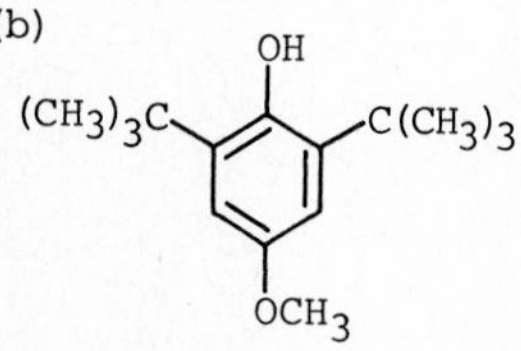

(d)

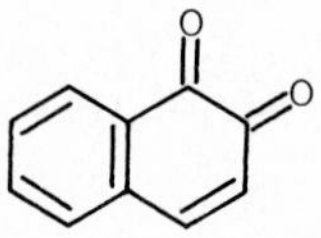

(f)

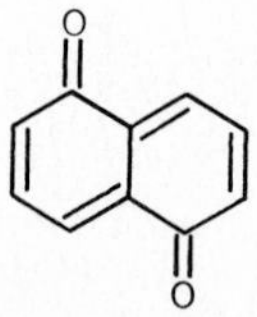

S2. Predict the major product to result from each of the following reaction sequences.

(a)

$$\xrightarrow[\text{HCl}]{\text{Sn}} \xrightarrow[\text{HCl}]{\text{NaNO}_2} \xrightarrow{\text{NaI}} \xrightarrow[\Delta]{\text{Cu}}$$

(b)

$$\xrightarrow[\text{350 °C}]{\text{KOH}} \xrightarrow[\substack{\text{CO}_2 \\ \text{150 °C}}]{\text{KHCO}_3}$$

(c)

$$\xrightarrow[\text{aq. NaOH, } \Delta]{\text{C}_6\text{H}_5\text{N}_2^+} \xrightarrow{\text{Na}_2\text{S}_2\text{O}_4} \xrightarrow[\text{H}_2\text{SO}_4]{\text{K}_2\text{Cr}_2\text{O}_7}$$

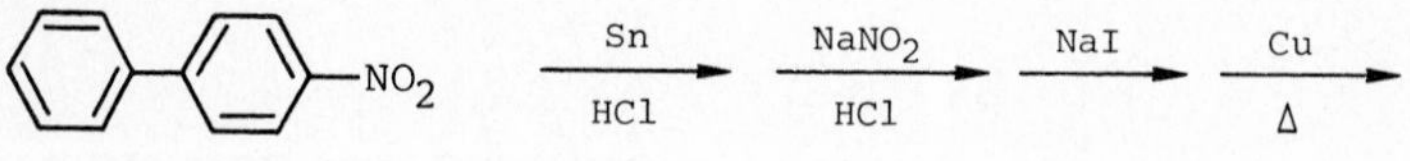

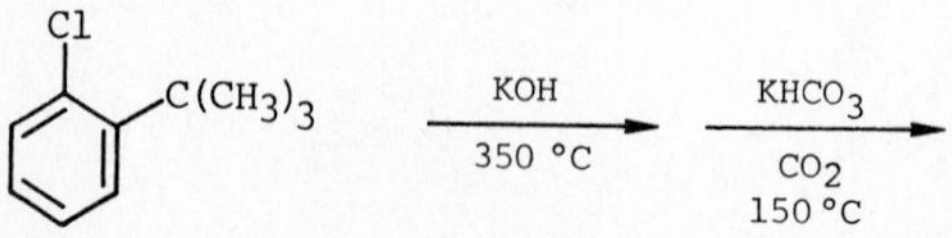

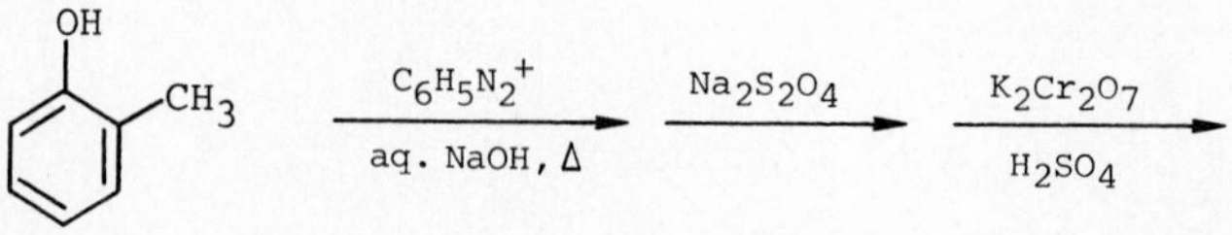

(d)   benzene $\xrightarrow[\text{BF}_3]{(\text{CH}_3)_3\text{COH}}$ $\xrightarrow[\text{FeBr}_3]{\text{Br}_2}$ $\xrightarrow[\text{ether}]{\underline{n}\text{-BuLi}}$ $\xrightarrow{(\text{CH}_3)_2\text{C=O}}$ $\xrightarrow{\text{H}_2\text{O}}$
      (excess)

(e) $\xrightarrow[\text{NaOH}]{(\text{CH}_3)_3\text{C=CHCH}_2\text{Br}}$ $\xrightarrow{\Delta}$ $\xrightarrow{\text{H}_2\text{SO}_4}$

(f)   2   + $\xrightarrow{\text{H}^+}$

(g) $\xrightarrow[\Delta]{\text{H}^+}$ $\xrightarrow[\text{H}_2\text{SO}_4]{\text{H}_2\text{Cr}_2\text{O}_7}$

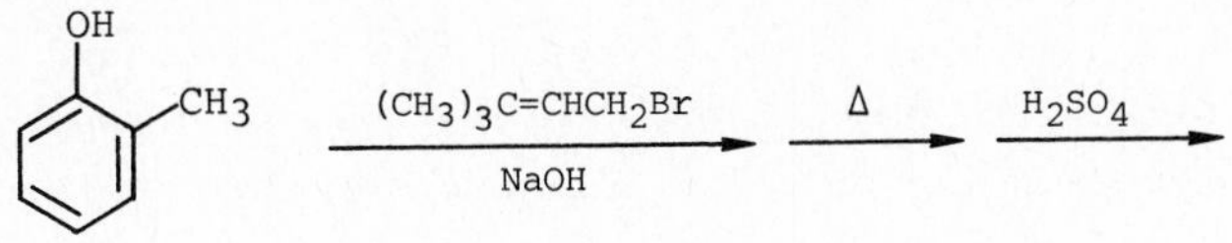

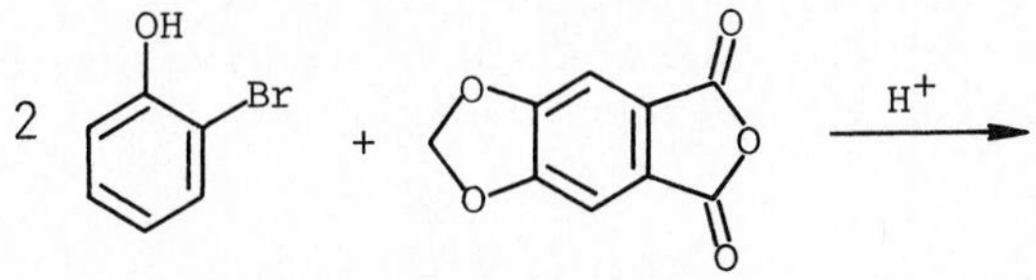

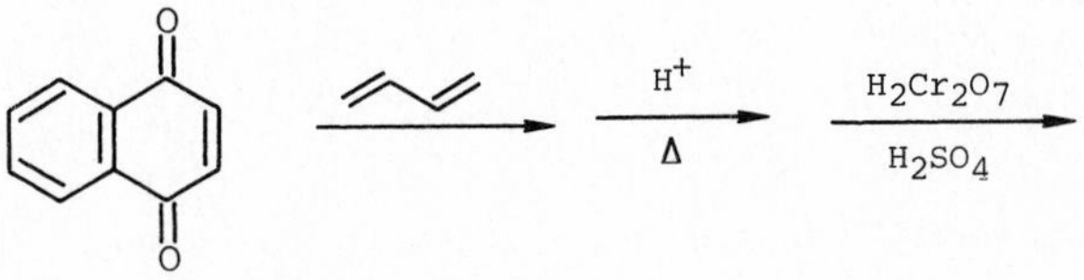

S3.   Show how to synthesize the following compounds, starting with any monosubstituted benzene derivatives and any other non-aromatic compounds.

(a)

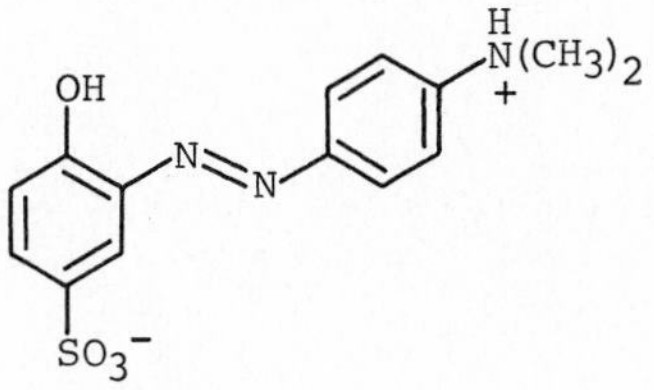

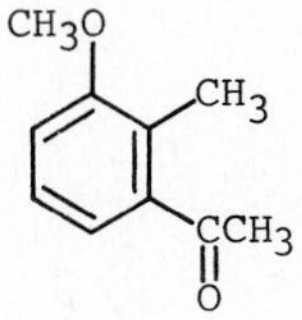

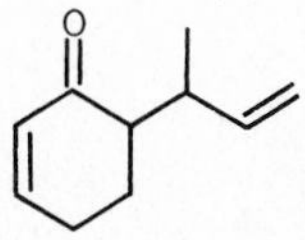

(b)

      Zingerone (a constituent of ginger)

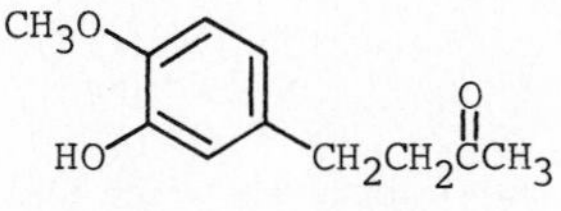

(c)

(d)

(e)

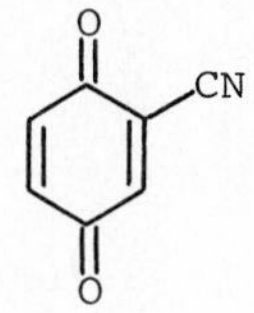

(f)

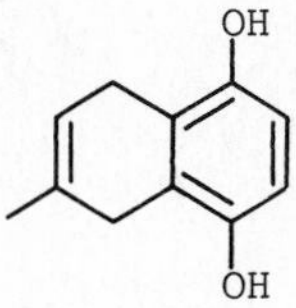

S4.   Write the expected products from reaction of 2-hydroxy-5-(hydroxymethyl)-benzoic acid with the following reagents:

(a)   NaOH, CO_2, Δ

(b)   excess CH_2N_2

(c)   1 mole of $CH_3\overset{\text{O}}{\underset{}{\text{C}}}O\overset{\text{O}}{\underset{}{\text{C}}}CH_3$

(d)   NaOH, $CHCl_3$

(e)   HBr

S5.   Write the expected products from reaction of 3-phenylpropanoic acid with the following reagents:

(a) Cl_2, catalytic amount of PCl_3 (c) Cl_2, hν

(b) Cl_2, $FeCl_3$ (d) $SOCl_2$

S6. Treatment of a protein with 2,4,6-trinitrobenzenesulfonic acid results in the attachment of 2,4,6-trinitrophenyl groups to the ε-amino groups of the lysine residues on the surface of the protein. Write a step-by-step mechanism for this reaction.

S7. Hydroquinone, and molecules such as BHA and BHT, are used as antioxidants because they interrupt propagation of the radical chain involved in the usual autoxidation mechanism. Suggest a mechanism for this interruption.

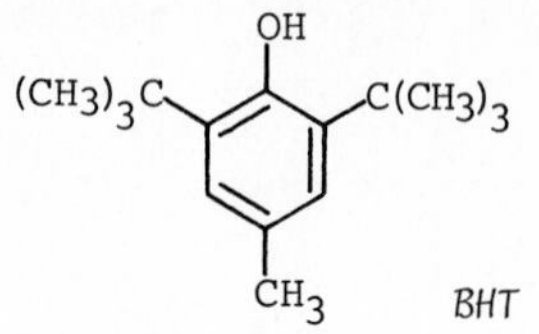

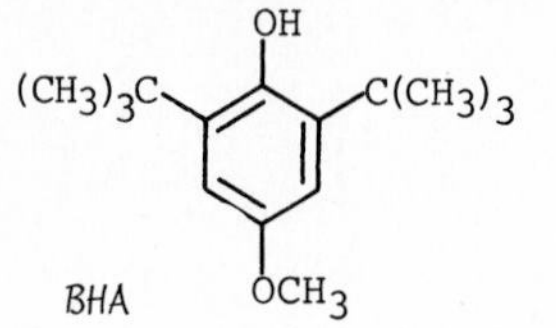

BHA *BHT*

S8. Write a step-by-step mechanism for the following reaction:

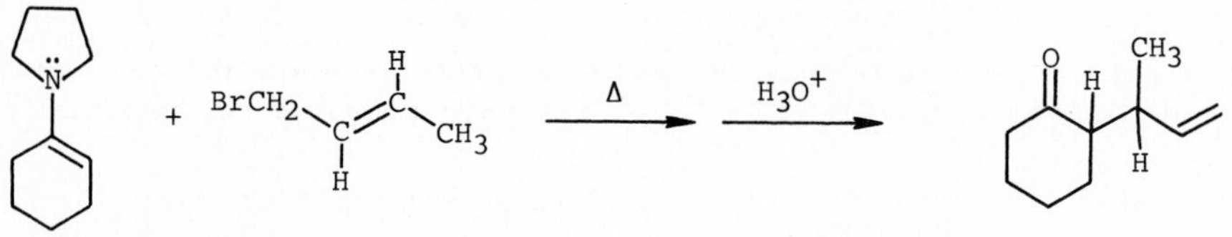

S9. The aliphatic Claisen rearrangement involves a "chair-like" transition state, as depicted below.

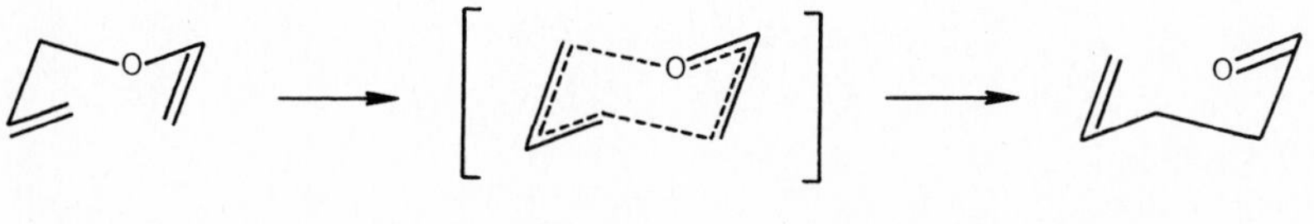

(a) If a secondary allylic vinyl ether is the substrate for the rearrangement, a *trans*-olefin is produced selectively. Show how this observation is consistent with the transition state illustrated above.

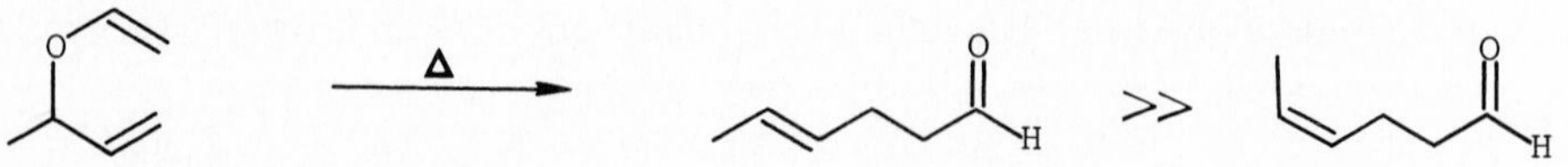

(b) What products would you expect from the Claisen rearrangement of the following enol ethers?

1)

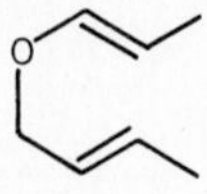

2)

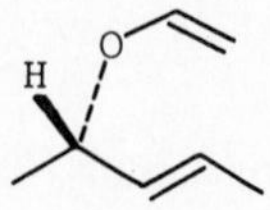

S10. (a) How do you account for the difference in the following reactions?

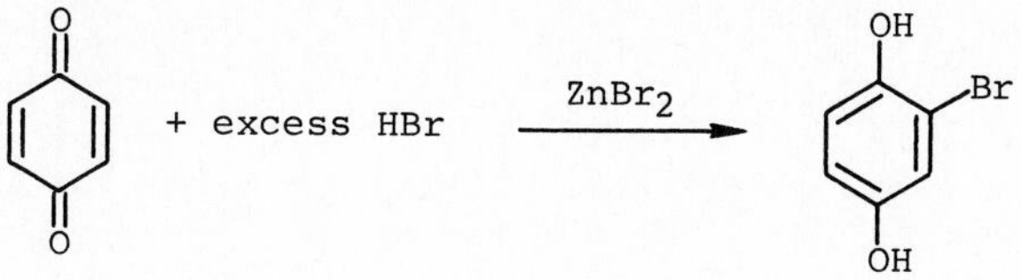

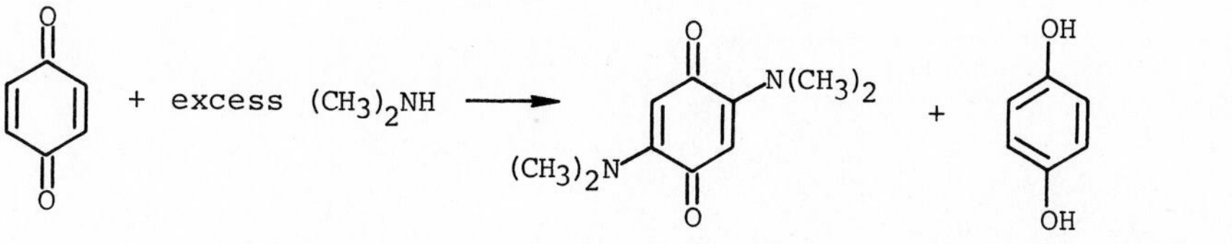

one-third mole two-thirds mole

(b) Why is the 2,5-di(dimethylamino) isomer produced in the second reaction above, rather than the 2,3- or the 2,6-isomers?

30.F Answers to Supplementary Problems

S1. (a) 1,3-dihydroxybenzene
(m-dihydroxybenzene, resorcinol)

(b) 2,6-di(t-butyl)-4-methoxyphenol
("butylated hydroxyanisole", BHA)

(c) *trans*-4-phenoxycyclohexanecarboxylic acid

(d) 1,2-naphthoquinone

(e) 2,5-dibromo-1,4-benzoquinone

(f) 1,5-naphthoquinone

S2.

(a)

(b)

no reaction occurs here because of steric hindrance

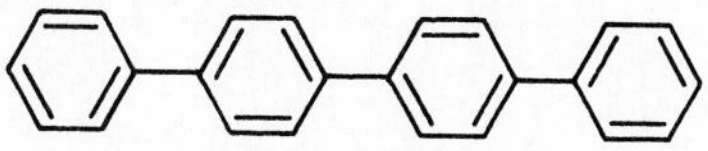

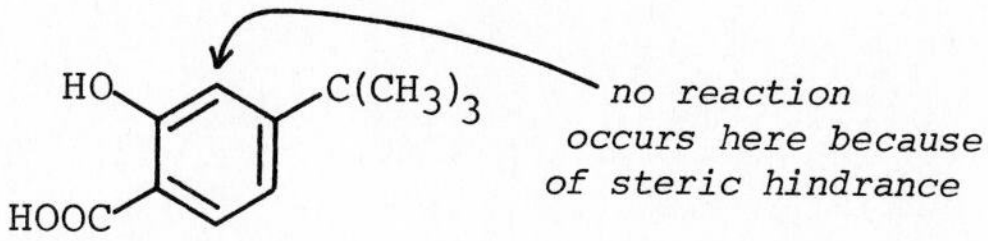

(c)

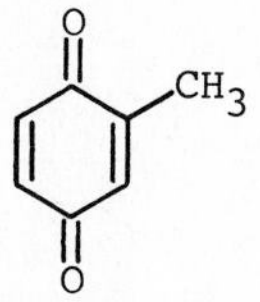

(d)

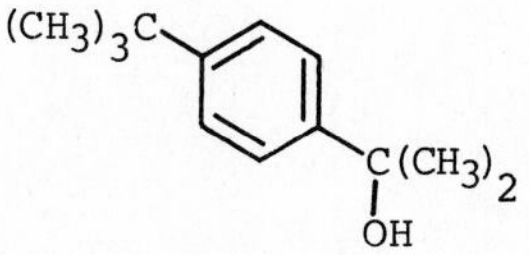

(e)

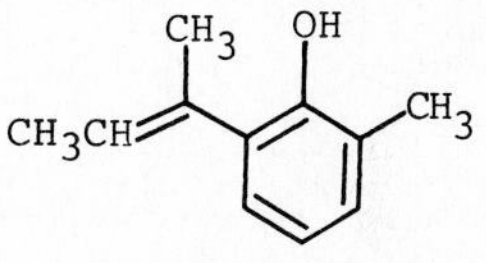

(from a carbocation rearrangement)

(f)

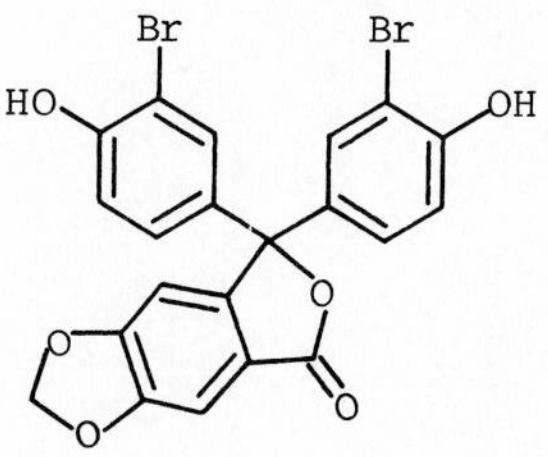

(g)

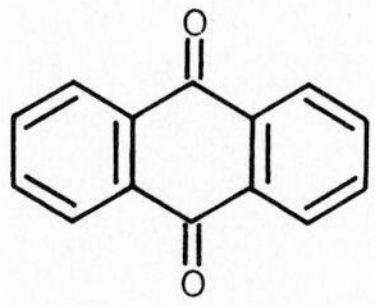

S3. (a)

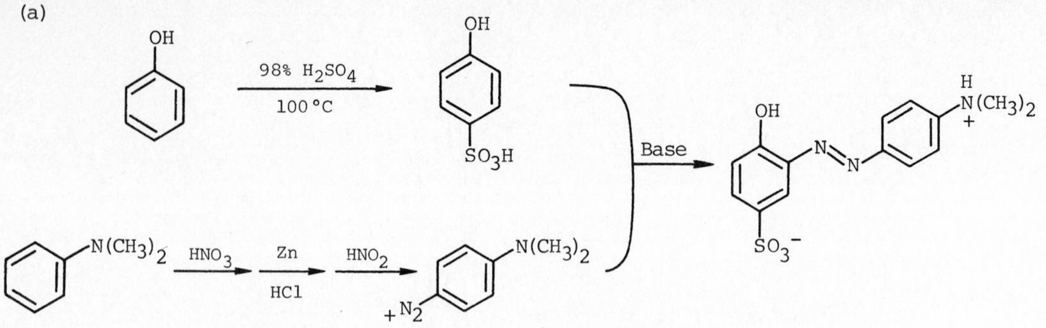
OH
98% H_2SO_4
100 °C
OH
SO_3H
Base
OH
N=N
H
$N(CH_3)_2$
+
SO_3^-
$N(CH_3)_2$
HNO_3
Zn
HCl
HNO_2
$N(CH_3)_2$
$^+N_2$

(b)

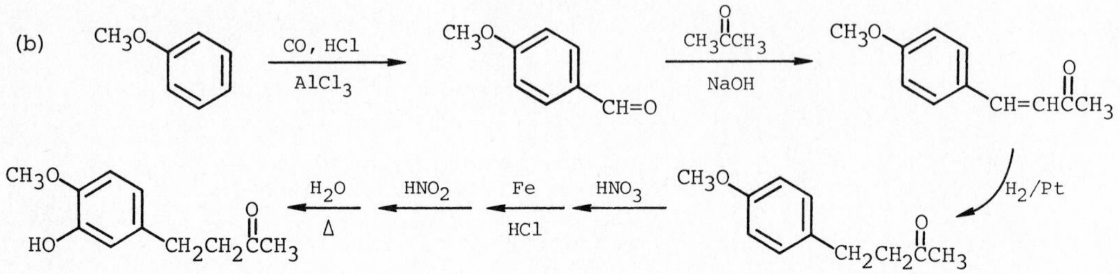
CH_3O
CO, HCl
$AlCl_3$
CH_3O
CH=O
CH_3CCH_3 (C=O)
NaOH
CH_3O
$CH=CHCCH_3$
H_2/Pt
CH_3O
$CH_2CH_2CCH_3$
HNO_3
Fe
HCl
HNO_2
H_2O
Δ
CH_3O
HO
$CH_2CH_2CCH_3$

(c)

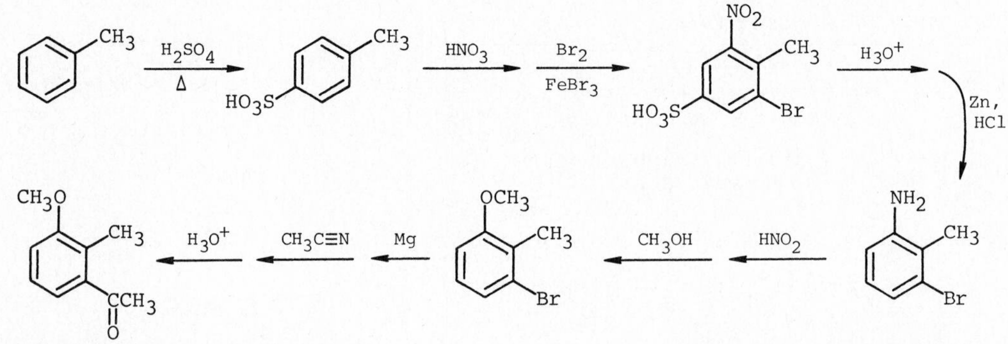
CH_3
H_2SO_4
Δ
CH_3
HO_3S
HNO_3
Br_2
$FeBr_3$
NO_2
CH_3
HO_3S
Br
H_3O^+
Zn, HCl
NH_2
CH_3
Br
HNO_2
CH_3OH
OCH_3
CH_3
Br
Mg
$CH_3C{\equiv}N$
H_3O^+
CH_3O
CH_3
CCH_3
O

(d)

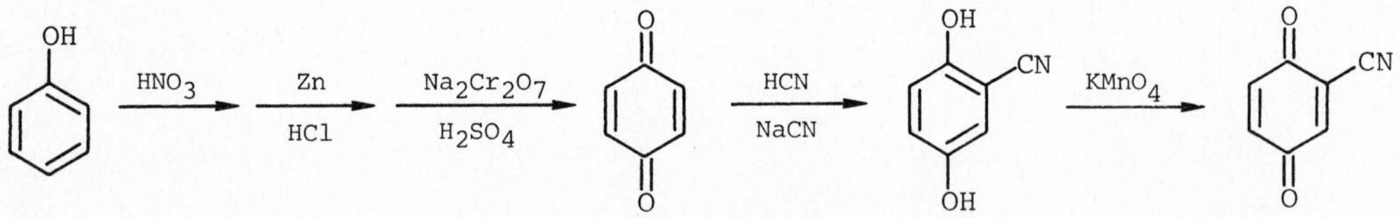
OH
HNO_3
Zn
HCl
$Na_2Cr_2O_7$
H_2SO_4
O
O
HCN
NaCN
OH
CN
OH
$KMnO_4$
O
CN
O

(e)

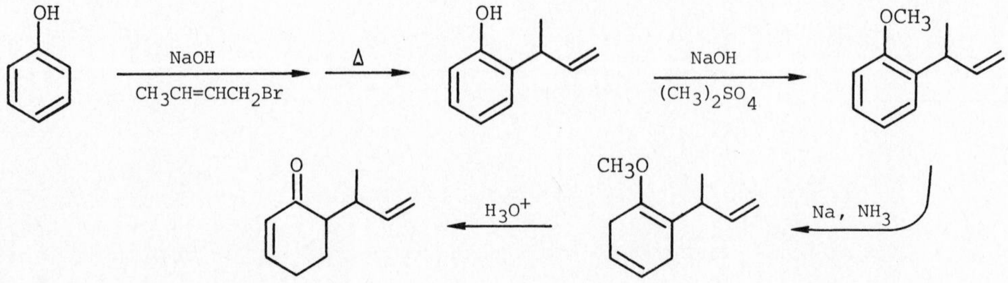
OH
NaOH
$CH_3CH{=}CHCH_2Br$
Δ
OH
NaOH
$(CH_3)_2SO_4$
OCH_3
Na, NH_3
CH_3O
H_3O^+
O

(f) 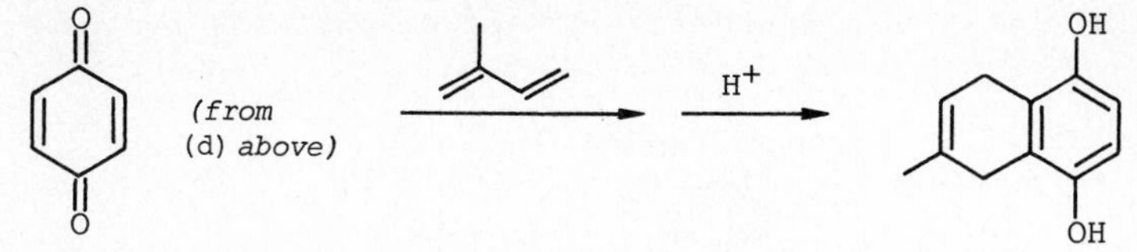

S4. (a) 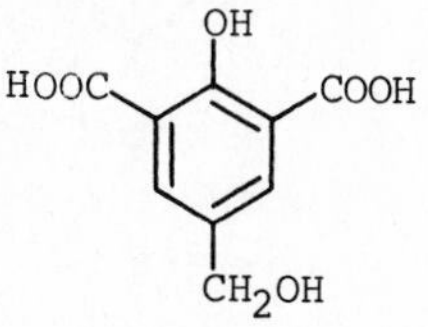

(b) 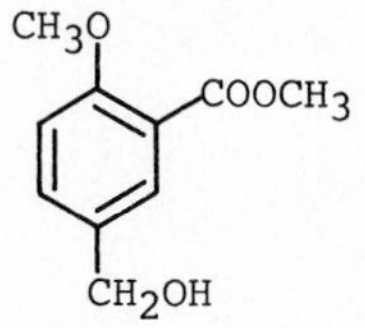

(c) 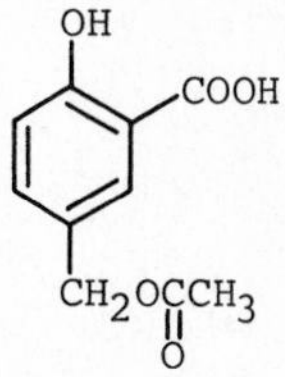

(d) 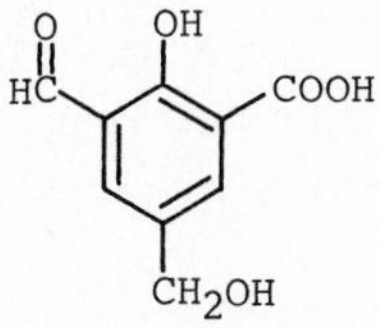

(e) 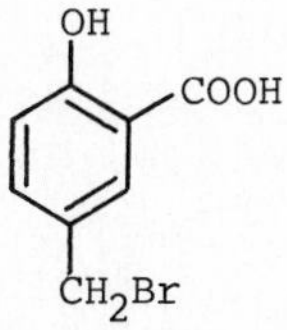

S5. (a) $C_6H_5CH_2\overset{Cl}{\overset{|}{C}}HCOOH$

(b) 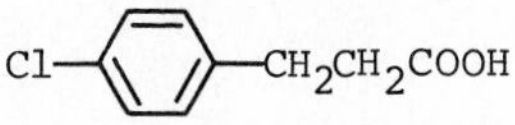

(c) $C_6H_5\overset{Cl}{\overset{|}{C}}HCH_2COOH$

(d) $C_6H_5CH_2CH_2\overset{O}{\overset{\|}{C}}-Cl$

S6. 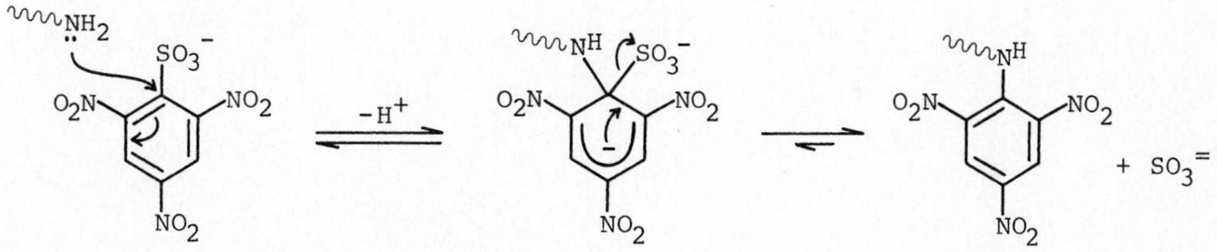

S7. The propagation steps of a free radical oxidation process are the following:

$$R\cdot + O_2 \longrightarrow ROO\cdot$$

$$ROO\cdot + RH \longrightarrow ROOH + R\cdot$$

Phenols such as hydroquinone, BHA, and BHT interrupt this chain because they lose a hydrogen atom more readily than RH. By doing so, they form a stable radical which is unable to continue the propagation chain.

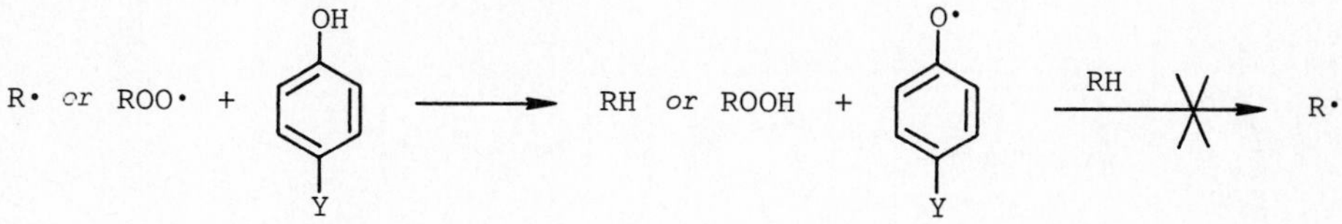

S8. The clue is the fact that the enamine alkylation proceeds with allylic rearrangement:

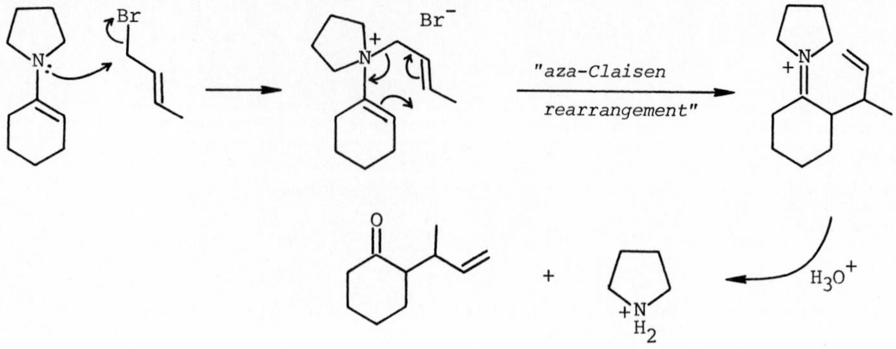

S9. (a) Since there is a chiral center in the molecule, there are two diastereomeric chair-like transition states possible:

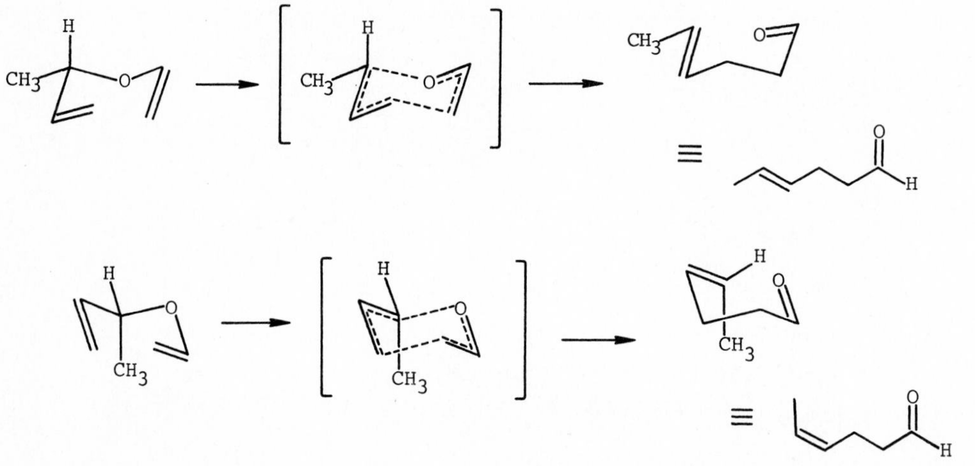

Rearrangement by the first transition state is preferred because the methyl group is in an equatorial-like position, rather than the axial-like orientation in the transition state which leads to the *cis*-product.

(b)

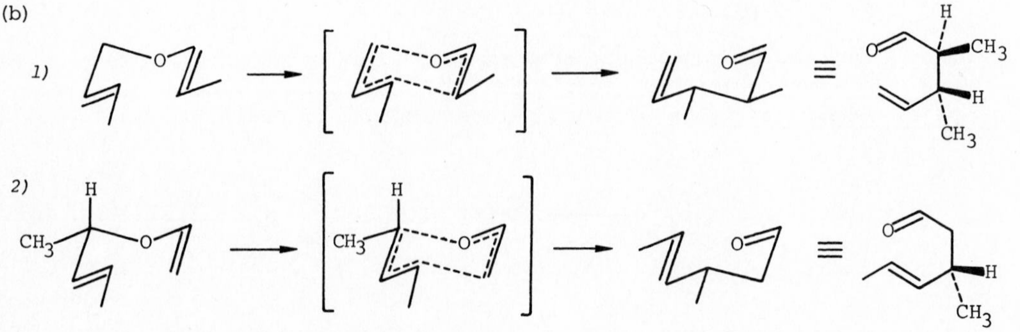

S10. (a) The product of HBr addition has a higher oxidation potential than hydroquinone itself, therefore only addition occurs:

The opposite is true for the dimethylamine adduct. As soon as it is formed, it is oxidized by the starting material and another amine adds. This second product in turn undergoes oxidation as well:

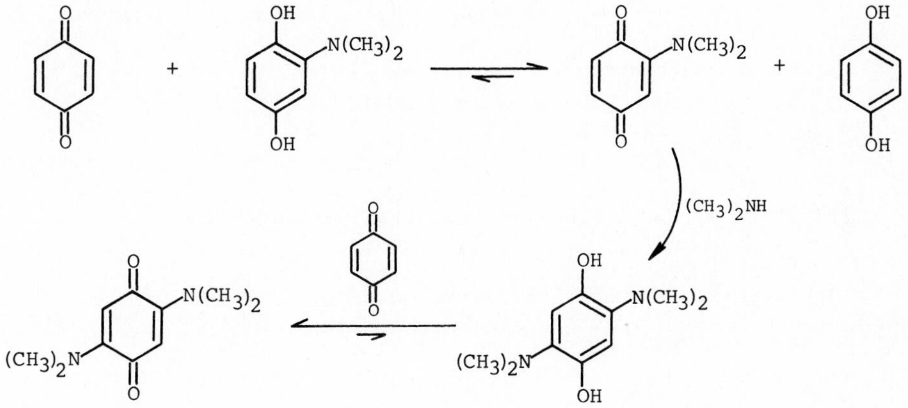

(b)

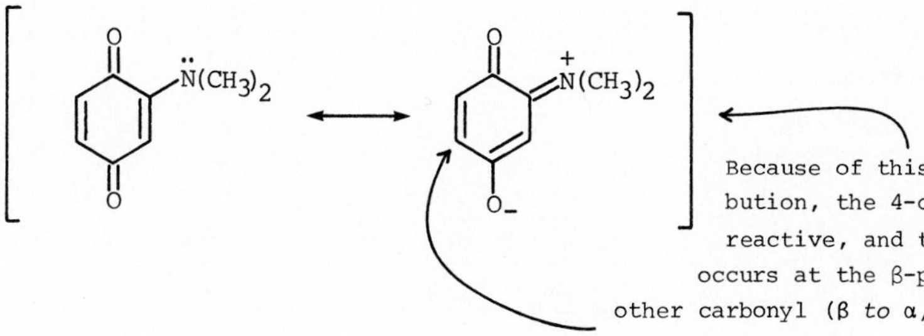

Because of this resonance contribution, the 4-carbonyl is much less reactive, and the second Michael occurs at the β-position of the other carbonyl (*β to α,β-unsaturated carbonyl*)

31. POLYCYCLIC AROMATIC HYDROCARBONS

31.A Outline of Chapter 31, Important Terms Introduced, and Reactions Discussed

31.1 Nomenclature

biphenyls

fused-ring systems: naphthalene, anthracene, phenanthrene

31.2 Biphenyl

A. Synthesis

pyrolysis of benzene:

$$2\ C_6H_6 \xrightarrow{\Delta} C_6H_5\text{-}C_6H_5 + H_2$$ *(industrial preparation)*

benzidine rearrangement *(see Section 25.1.C)*

Ullmann reaction *(see Section 30.3.B)*

Gomberg-Bachmann reaction *(see Section 25.6)*

B. Structure

chirality of 2,2',6,6'-tetrasubstituted derivatives

C. Reactions

electrophilic aromatic substitution *(usually favors para)*

D. Related Compounds

terphenyls

fluorene ($pK_a = 23$)

31.3 Naphthalene

A. Structure and Occurrence

resonance energy

cis and *trans* decalins

B. Synthesis

annelation routes

aromatization of hydroaromatics:

(using Pd/Δ, chloranil, or S or Se, Δ)

Diels-Alder reactions of *p*-benzoquinone

C. Reactions of Naphthalene

1. Electrophilic Substitution

   favors 1-position kinetically

   sulfonation can provide 2-naphthalenesulfonic acid thermodynamically

2. Oxidation

   naphthalene to 1,4-naphthoquinone or phthalic anhydride

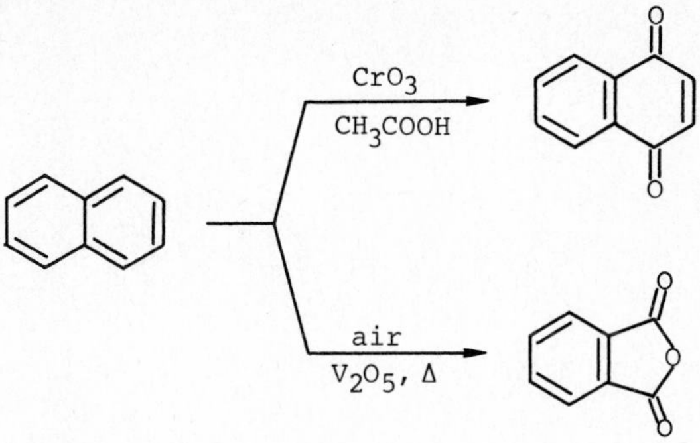

(for substituted derivatives, the most electron-rich ring is oxidized)

2. Reduction

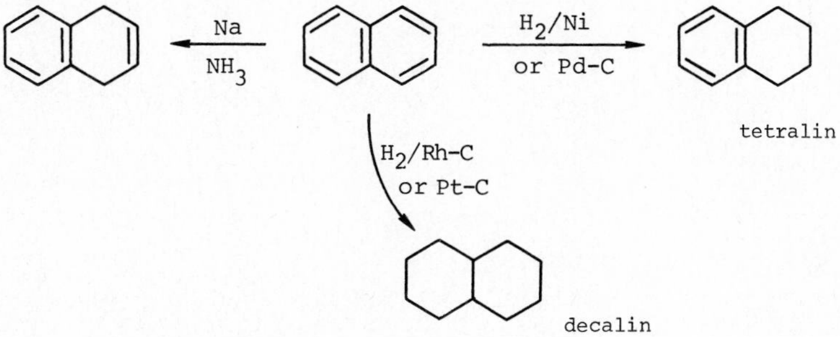

D. Substituted Naphthalenes
   transformations of substituent groups
   directive effects in electrophilic aromatic substitution reactions
   Bucherer reaction:

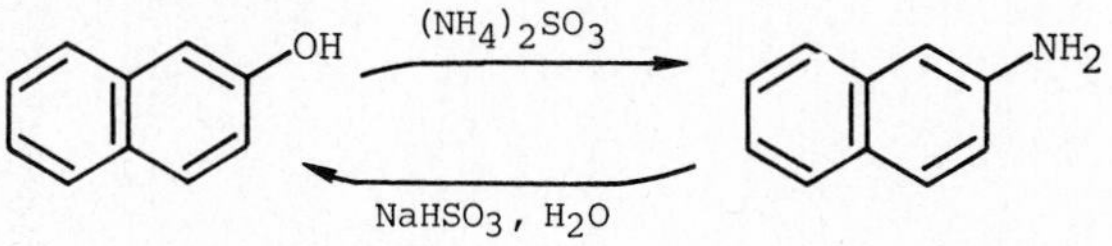

(also works for 1-substituted derivatives)

31.4 Anthracene and Phenanthrene

A. Structure and Stability

B. Preparation of Anthracenes and Phenanthrenes
   annelation methods

C. Reactions
   oxidation to quinones:

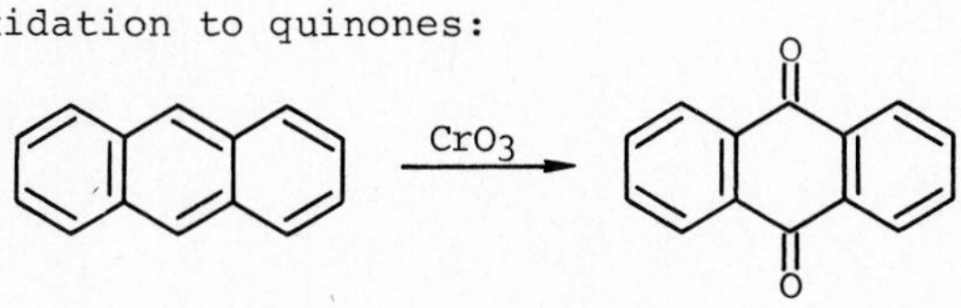

(occurs on central ring of phenanthrene, too)

reduction to dihydro compounds:

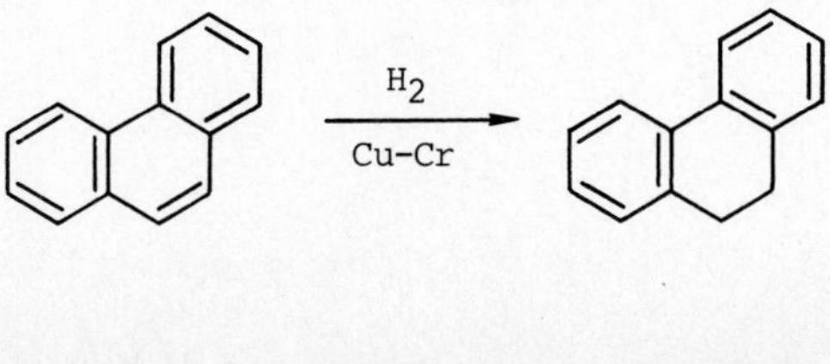

(occurs on central ring of anthracene, too)

Diels-Alder reactions of anthracene:

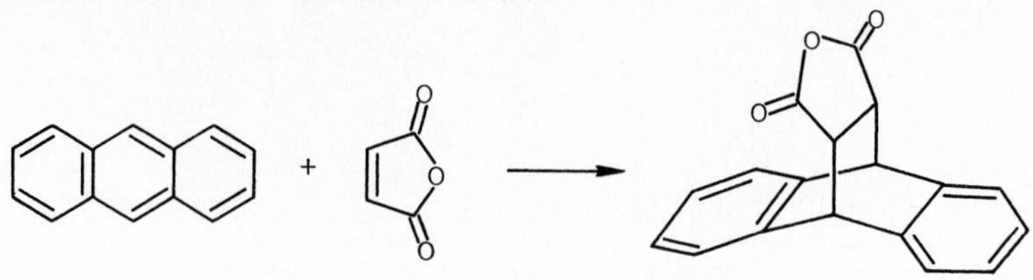

electrophilic aromatic substitution *(occurs on central ring)*

31.5 Higher Polybenzenoid Hydrocarbons

acene	benz- derivatives
graphite	carcinogens

31.C Answers to Exercises

(31.1) 1-bromo-2,5-dimethylanthracene

trans-7-isopropyl-4a-methyl-1,2,3,4,4a,9,10,10a-octahydrophenanthrene

(31.2.A)

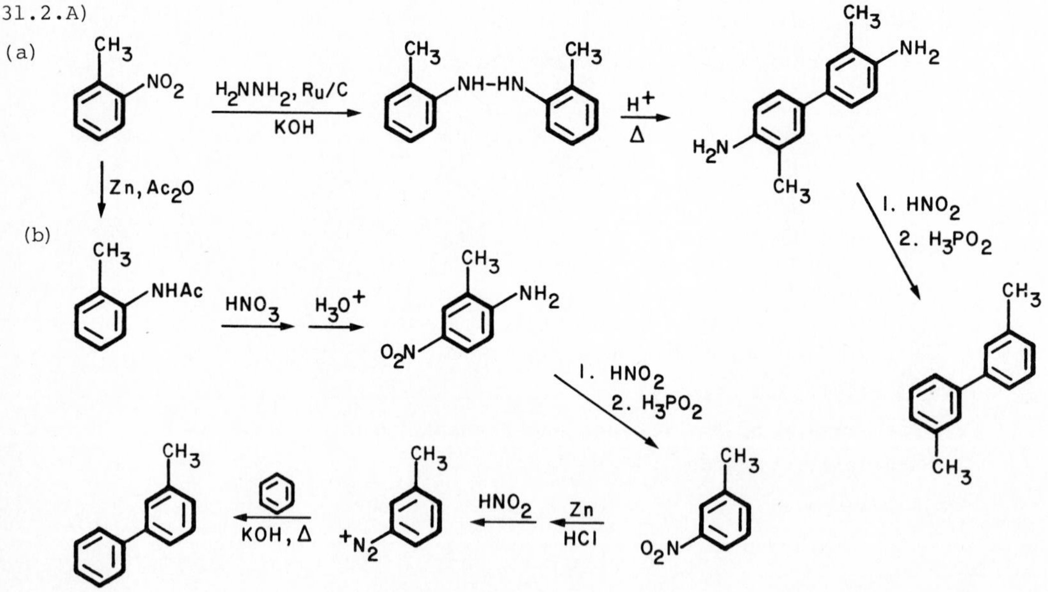

(31.2.C) *para* attack:

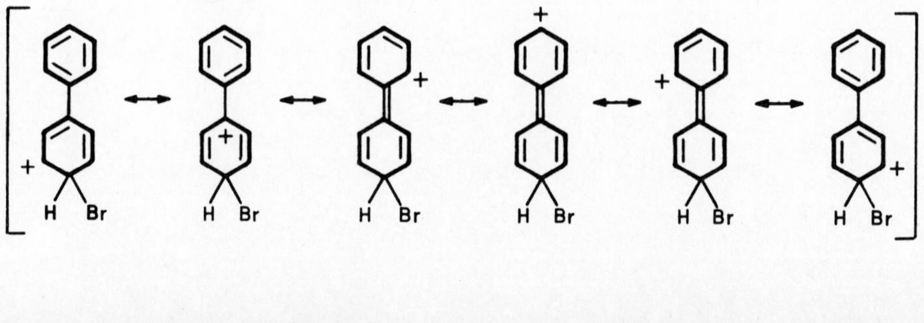

meta attack:

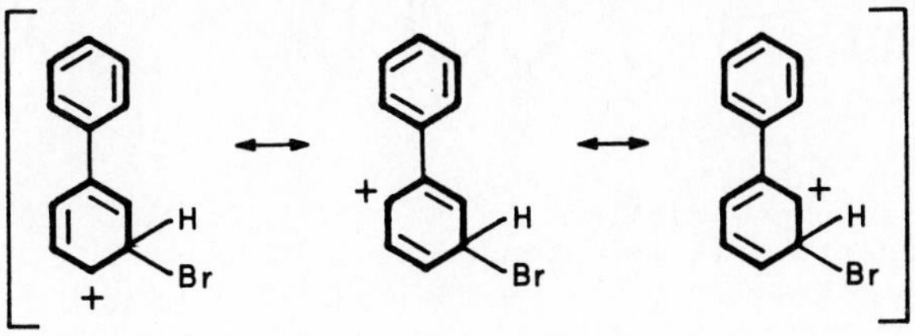

The phenyl substituent itself can stabilize the positive charge via resonance when it is in the *para* position.

(31.2.D)

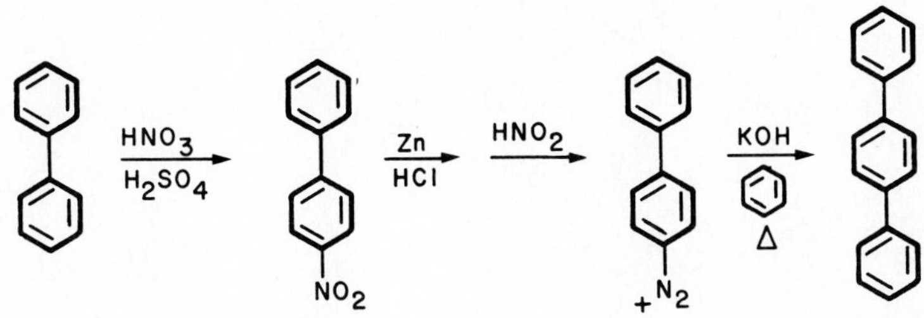

(31.3.A) Fractional double bond character = the number of resonance structures with a double bond divided by the total number of resonance structures:

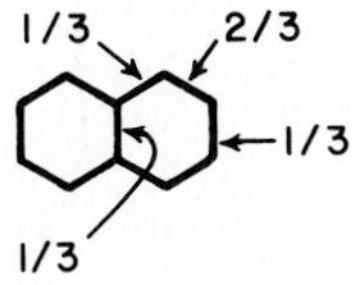

The prediction is roughly correct.

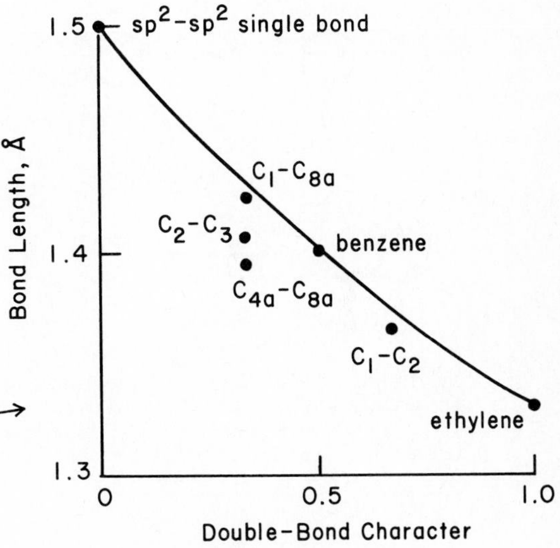

(31.3.B)

(a)

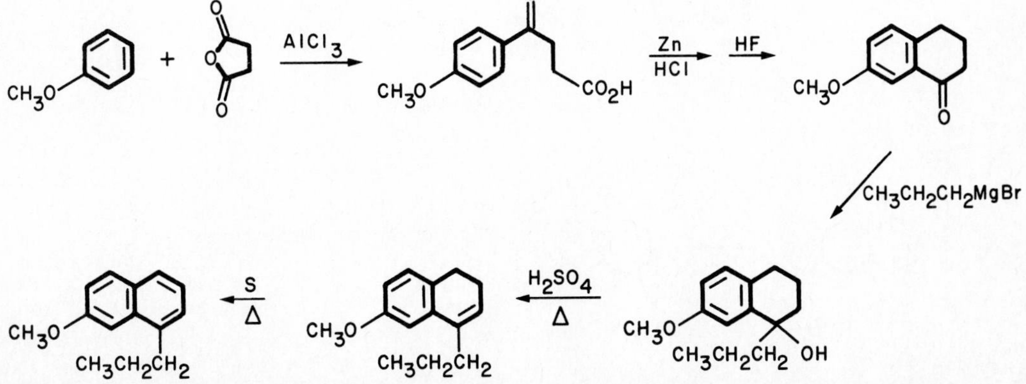

(b)

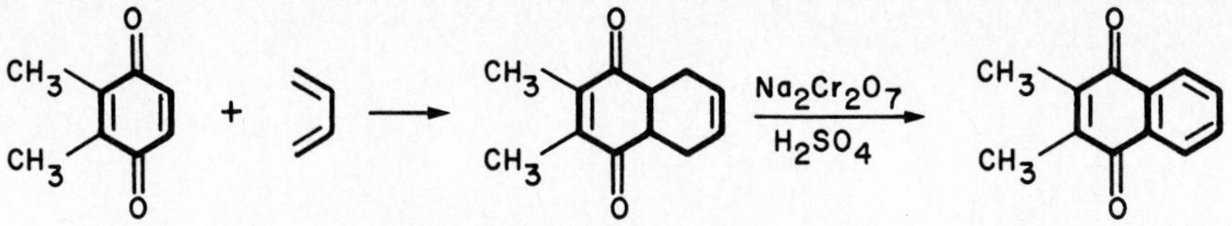

NOTE: the reaction sequence outlined below, which is depicted in the text, often confuses students. It is intended to illustrate that, with proper choice of reagents, it is possible to carry out the sequence in a stepwise fashion.

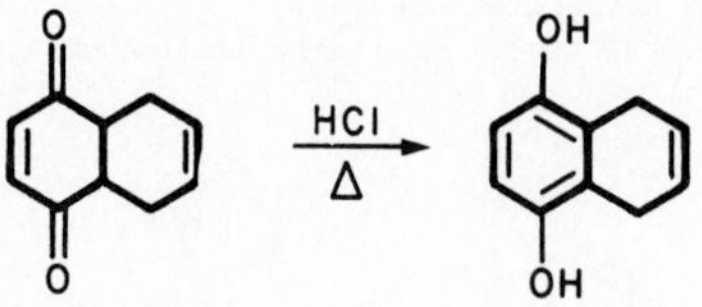

(This reaction is simply a double enolization.)

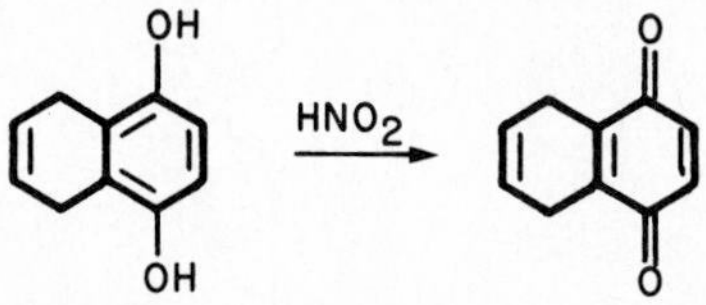

(This is an oxidation, which stops at this stage when HNO_2 is the oxidant. Note that HNO_2 is a mild oxidizing agent and a weak acid)

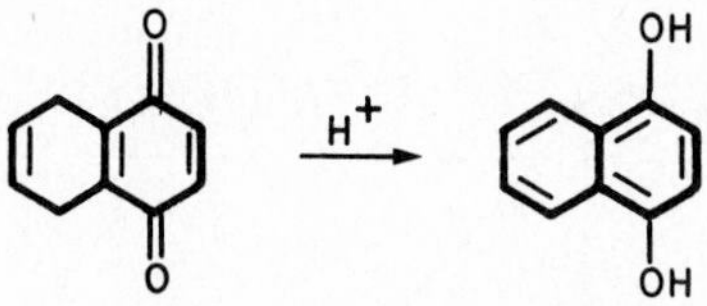

(This reaction is another double enolization. It occurs under the acidic conditions of $K_2Cr_2O_7/H_2SO_4$ oxidation.)

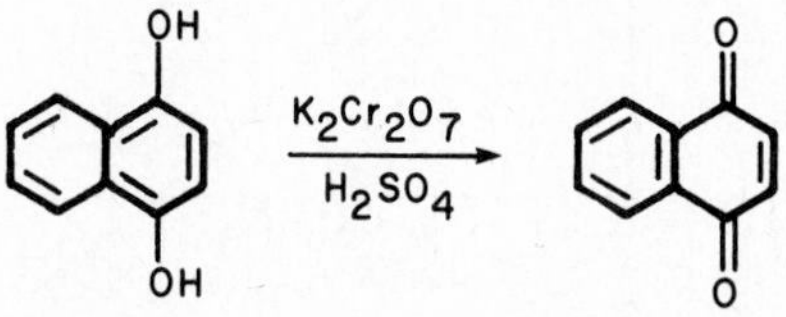

(This is the final oxidation step.)

It is not necessary to use each one of these reagents sequentially in order to achieve the overall transformation from the Diels-Alder adduct to the 1,4-naphthoquinone. The transformation can be accomplished all at once under the acidic conditions of the chromic acid oxidation (below). However, you should bear in mind that all of the steps depicted above are involved.

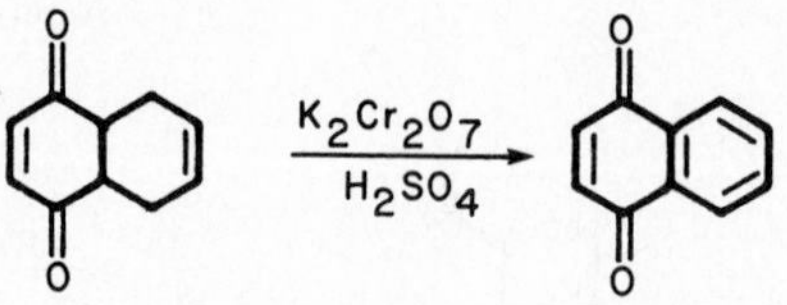

(31.3.C.1)

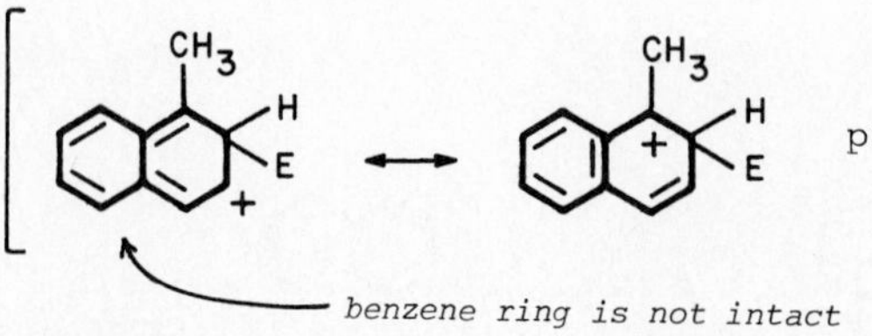

plus three structures involving the other ring

benzene ring is not intact

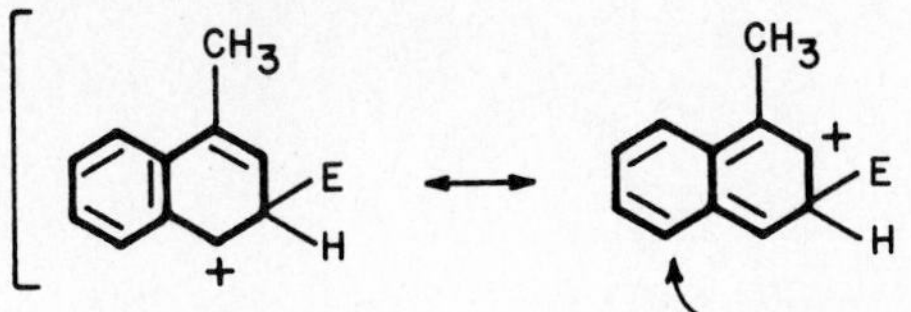

benzene ring is not intact

plus three structures involving the other ring

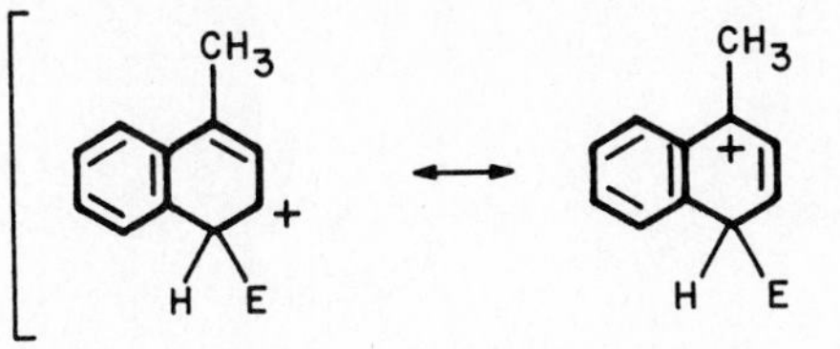

plus three structures involving the other ring

Only substitution at the 4-position will produce a tertiary carbocation resonance structure which maintains the aromaticity of the benzene ring. Therefore it is favored.

(31.3.C.3)

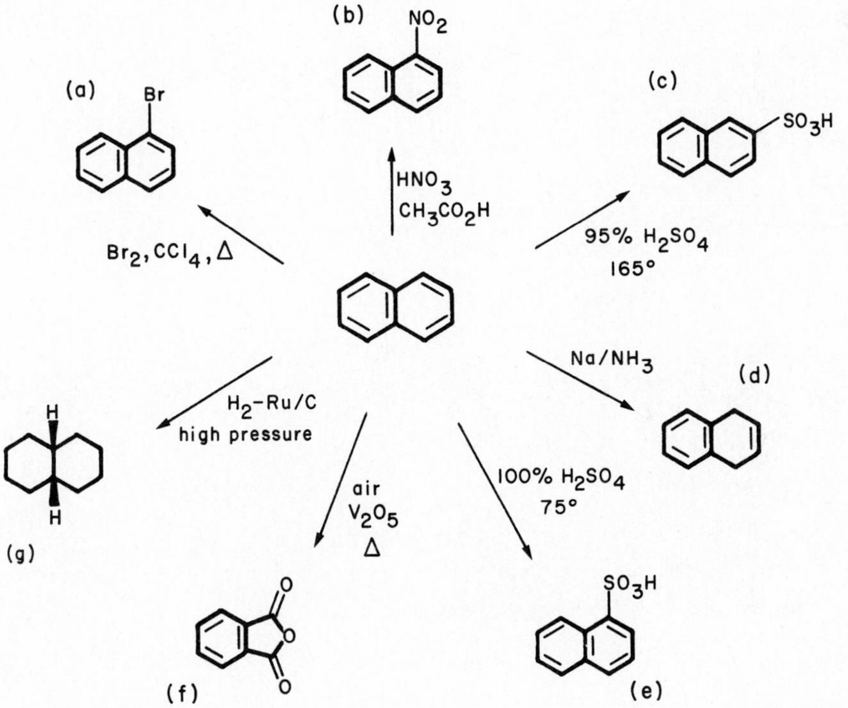

(31.3.D)

1: 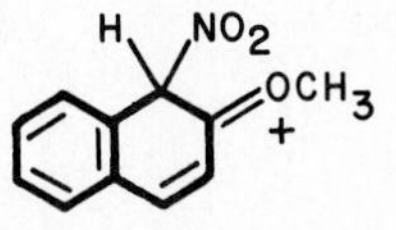

3: 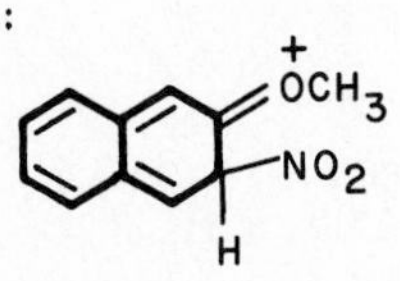

4: 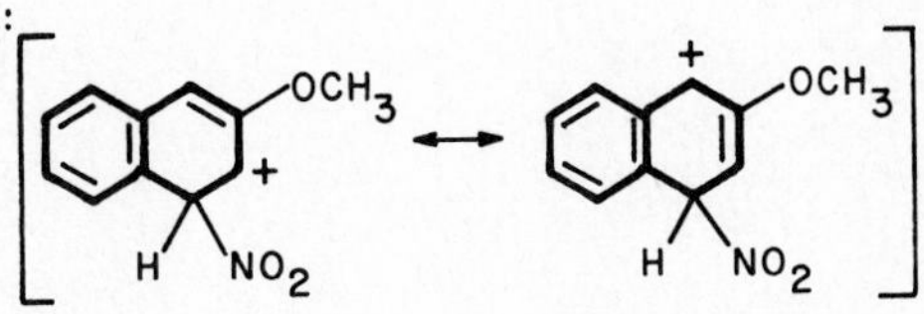

5: 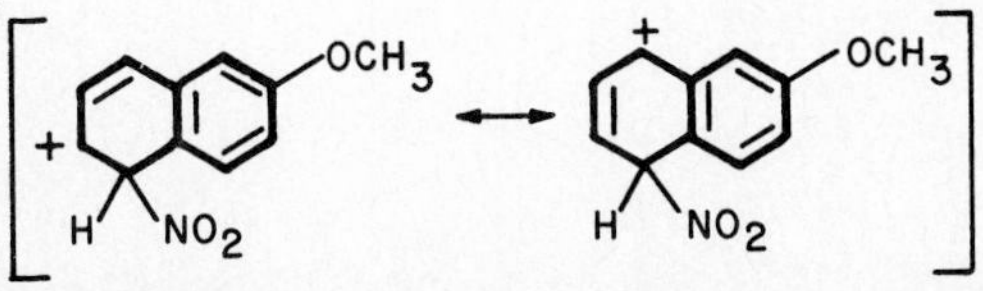

6: 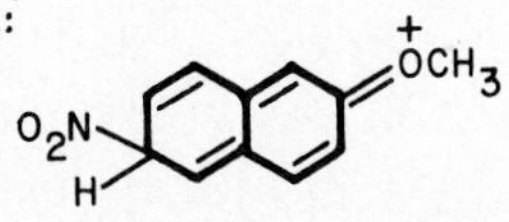

7: 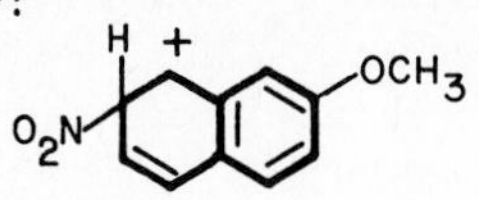

8: 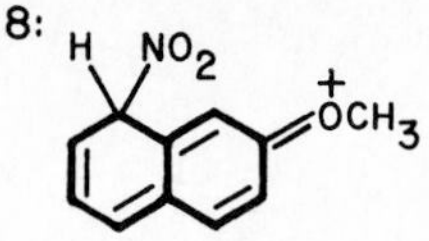

Only the 1-substitution intermediate can utilize the oxygen lone pair electrons to stabilize the positive charge and maintain an aromatic ring.

(31.4.B)

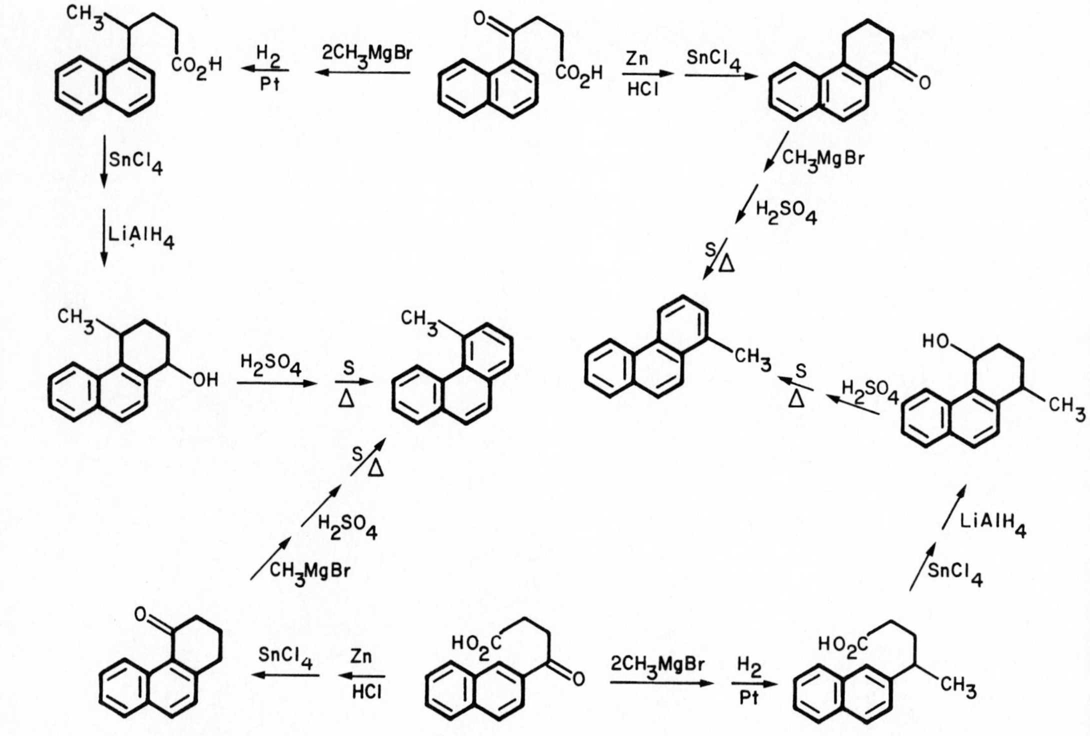

(31.4.C)

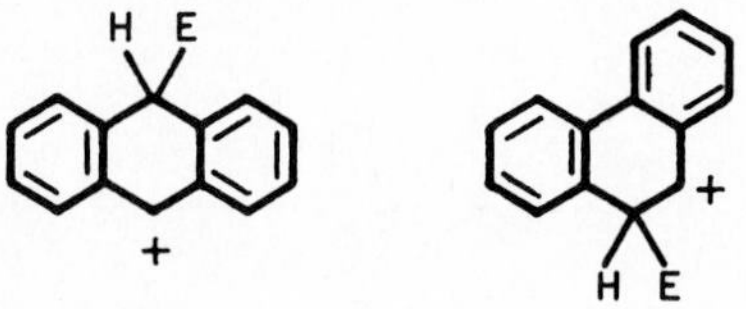

These are the two most favored resonance structures, because each allows two of the rings to remain fully aromatic as benzene rings.

(31.5)

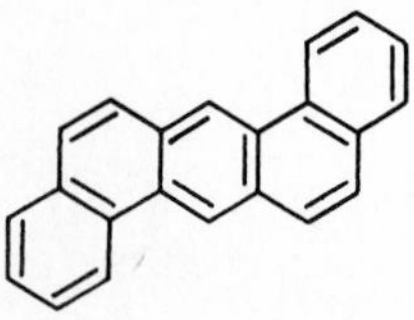

dibenz[a,h]-anthracene

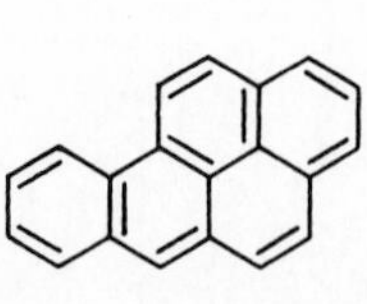

benzo[a]pyrene

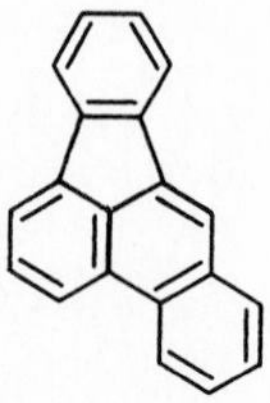

benzo[b]fluoranthene

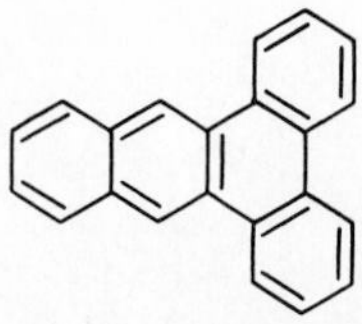

dibenz[a,c]anthracene

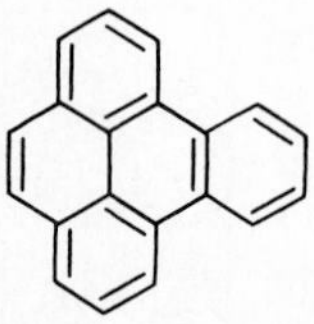

benzo[e]pyrene

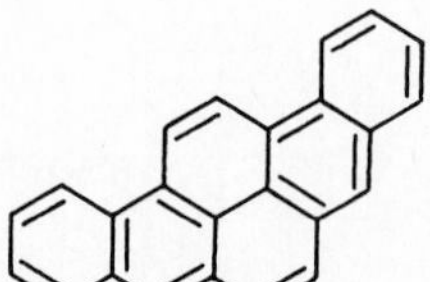

dibenzo[a,i]pyrene

31.D Answers and Explanations to Problems

1. (a)

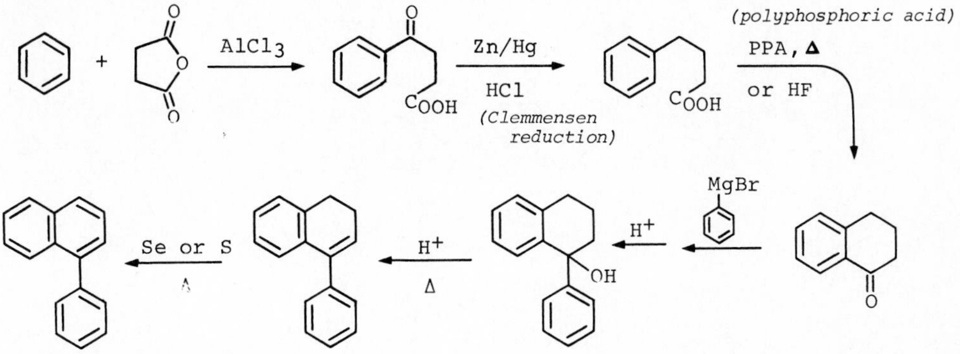

(b)

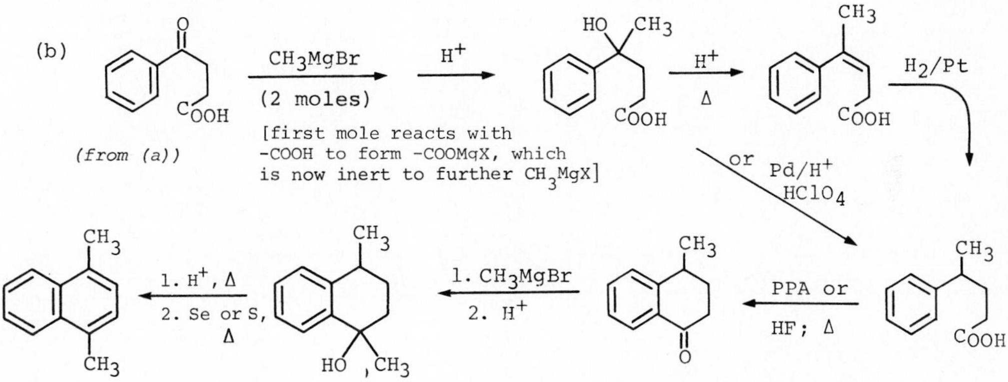

[*NOTE:* if not hydrogenated first, we would obtain a naphthol at this point]

Alternative:

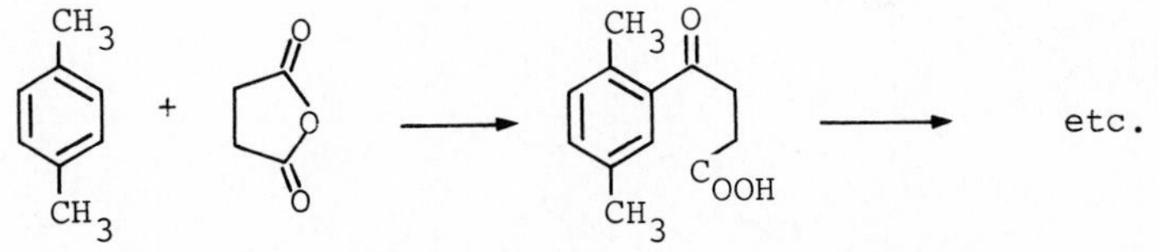

(c)

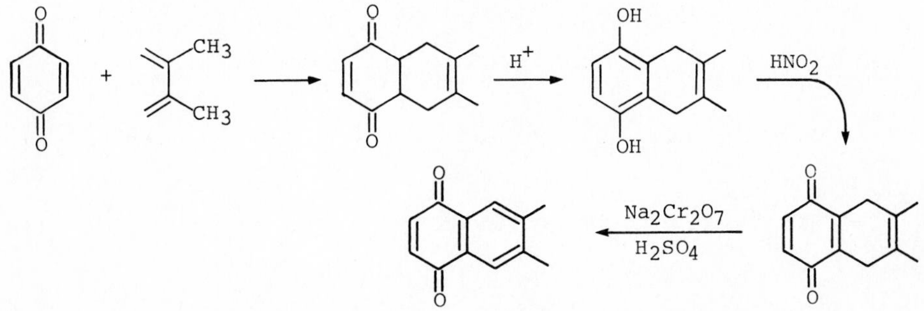

(d)

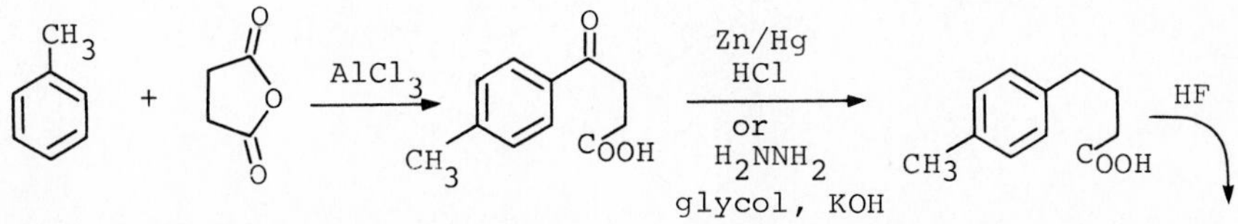

[cont'd...

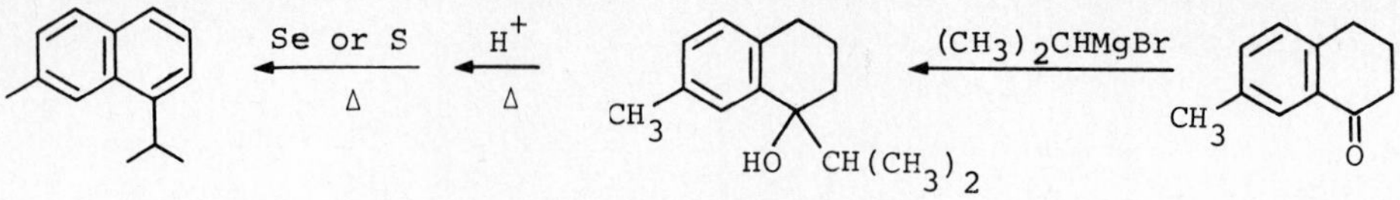
Se or S
Δ
H+
Δ
CH3
HO CH(CH3)2
(CH3)2CHMgBr
CH3
O

(e)

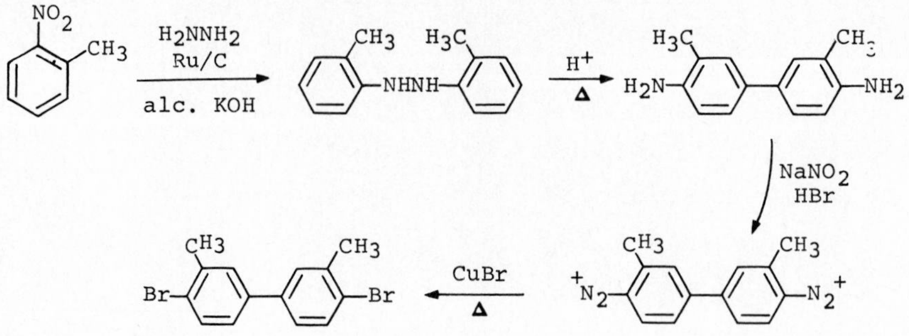
NO2
CH3
H2NNH2
Ru/C
alc. KOH
CH3 H3C
NHNH
H+
Δ
H3C CH3
H2N NH2
NaNO2
HBr
CH3 CH3
Br Br
CuBr
Δ
+N2 N2+

(f)

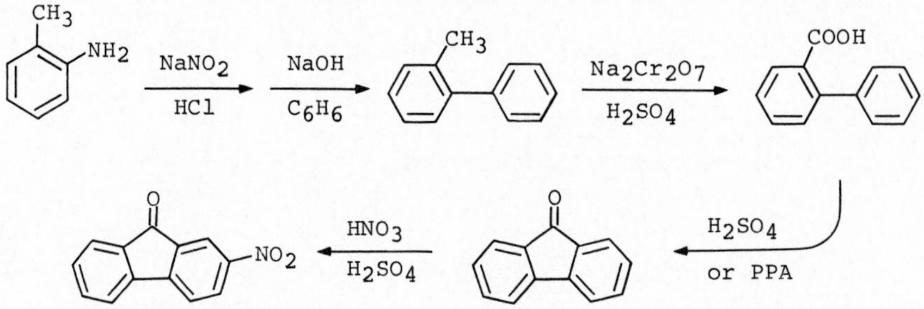
CH3
NH2
NaNO2
HCl
NaOH
C6H6
CH3
Na2Cr2O7
H2SO4
COOH
O
NO2
HNO3
H2SO4
O
H2SO4
or PPA

(g)

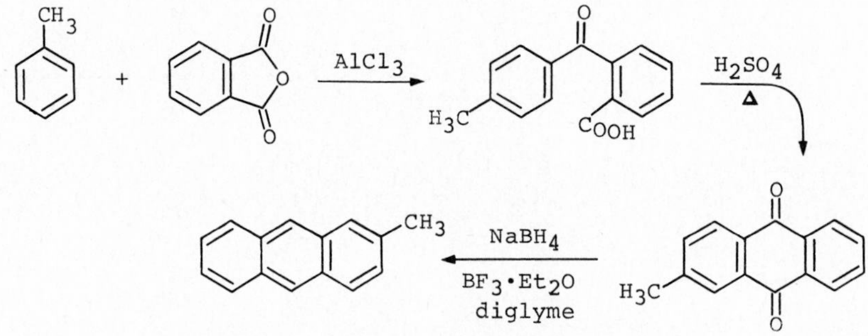
CH3
+
O
O
O
AlCl3
O
H3C
COOH
H2SO4
Δ
CH3
NaBH4
BF3•Et2O
diglyme
O
H3C
O

(h)

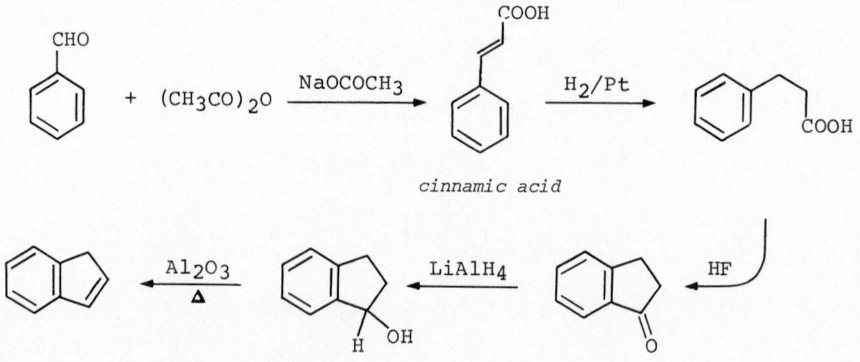
CHO
+ (CH3CO)2O
NaOCOCH3
COOH
H2/Pt
COOH
cinnamic acid
Al2O3
Δ
H OH
LiAlH4
O
HF

2.

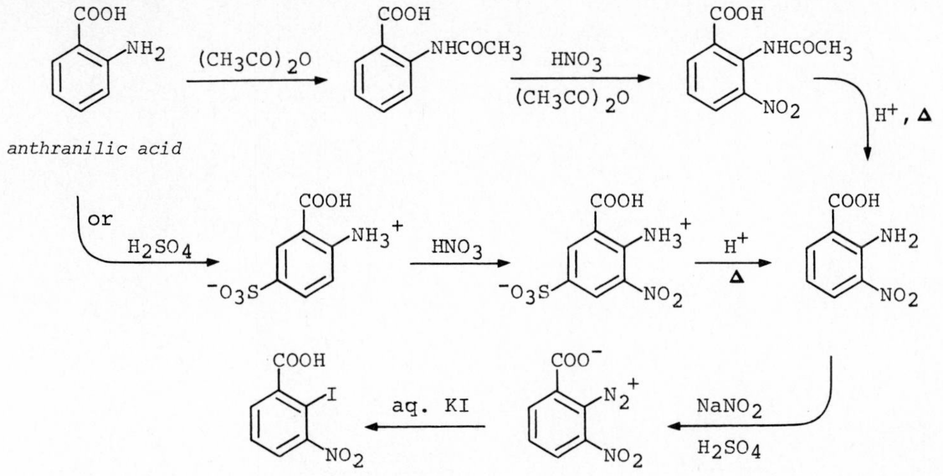

3. The 6-position conjugates with the 2-methyl. The 7-position does not.

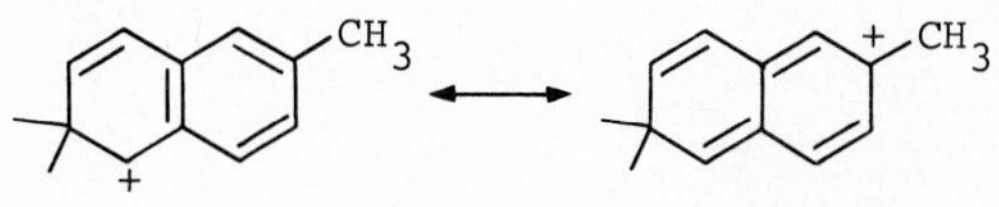

No such tertiary carbocation structure is possible for attack at the 7-position.

4.

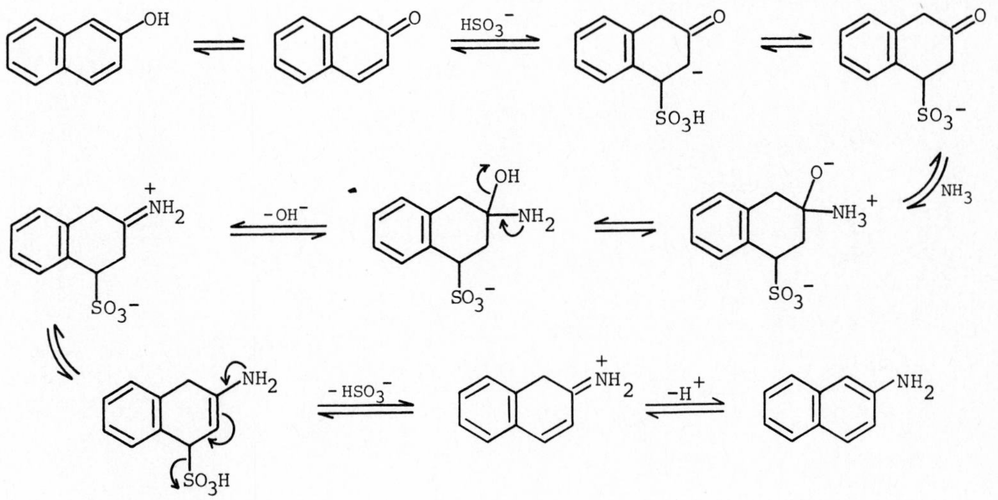

5. (a) $\Delta H^\circ_{hydrog.} = -43.5 - 36.1 = -79.6$ kcal mole^{-1}

(b) For cyclohexene, $\Delta H^\circ_{hydrog.} = -28.4$ kcal mole^{-1}

For five double bonds, the value for $\Delta H^\circ_{hydrog.}$ would be $5 \times (-28.4) = -142.0$ kcal mole^{-1}

(c) Empirical resonance energy = $142.0 - 79.6 = 62.4$ kcal mole^{-1}

6. (a) 1,6-dimethyl-4-isopropylnaphthalene

(b)

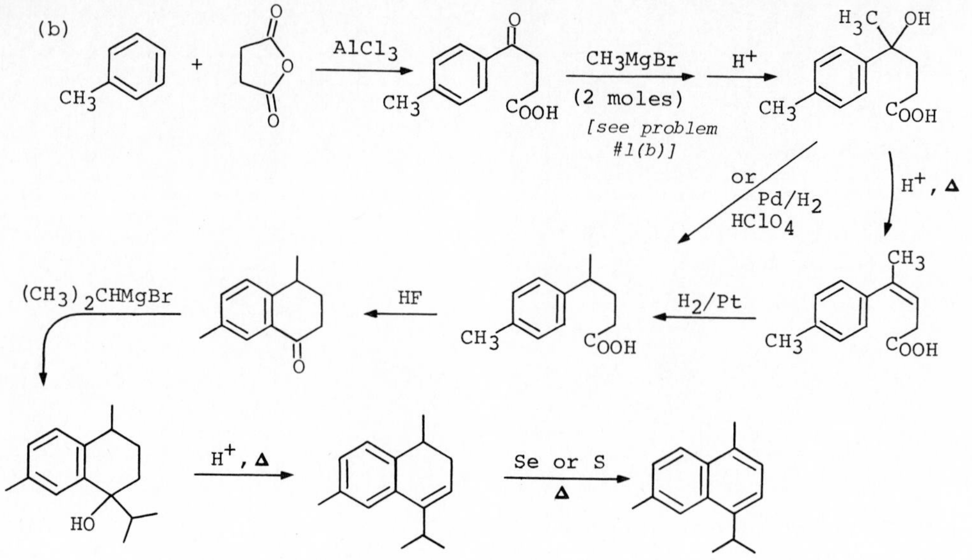

7\. (a)

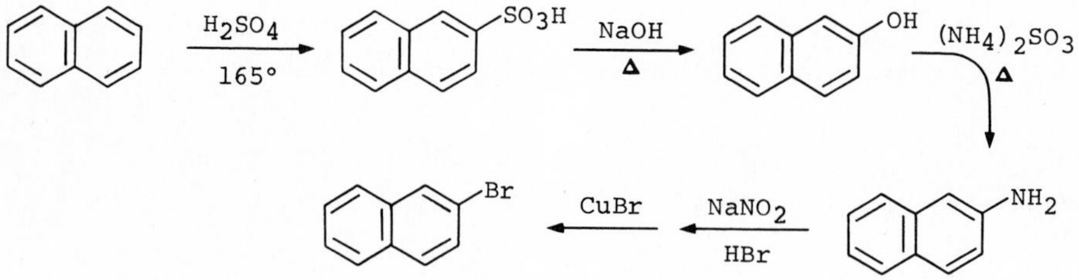

(b)

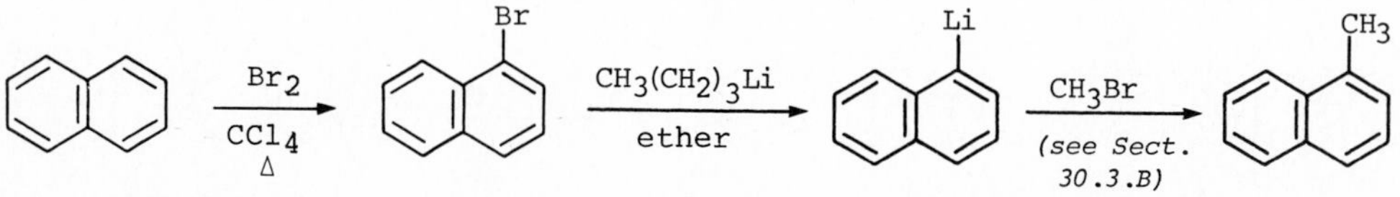

(c)

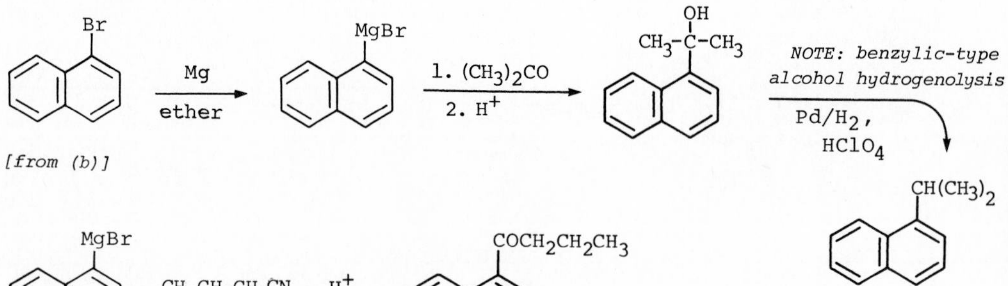

(d)

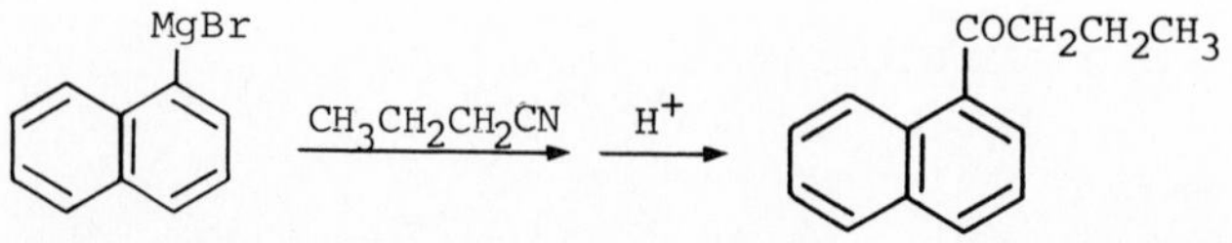

[*from (c)*]

(e)

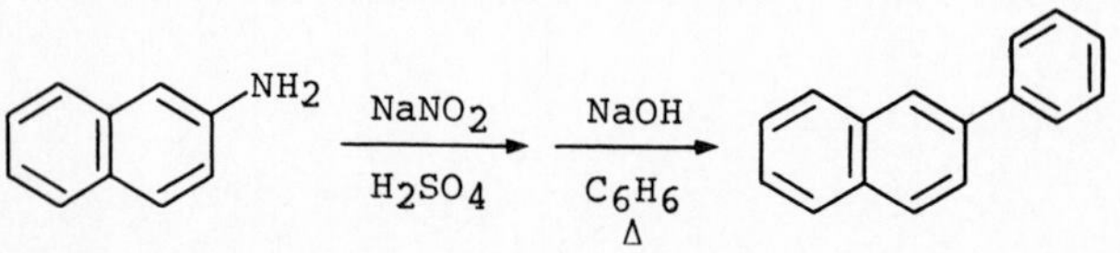

[from (a)]

(f)

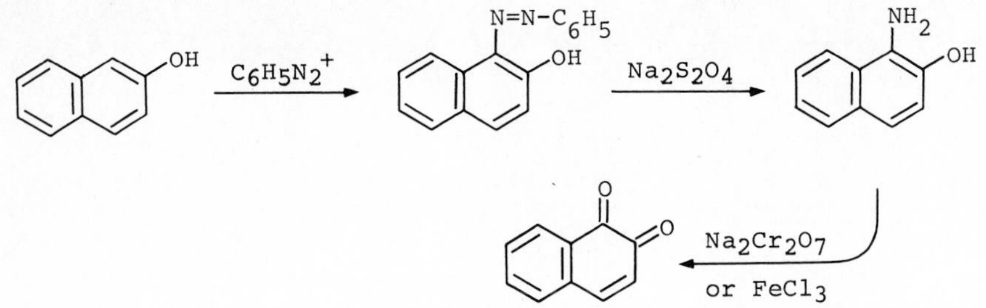

(g)

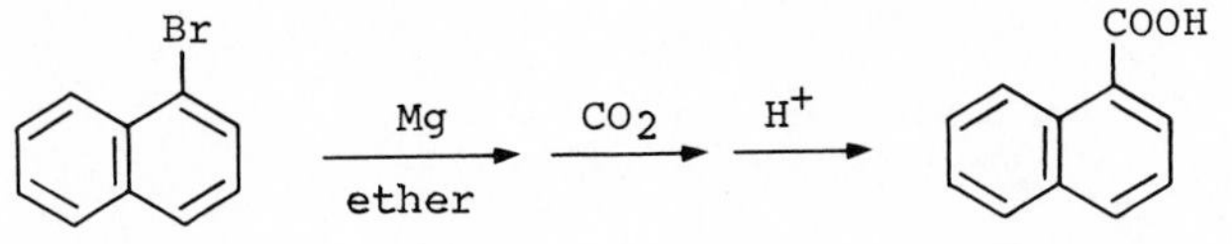

[from (b)]

(h)

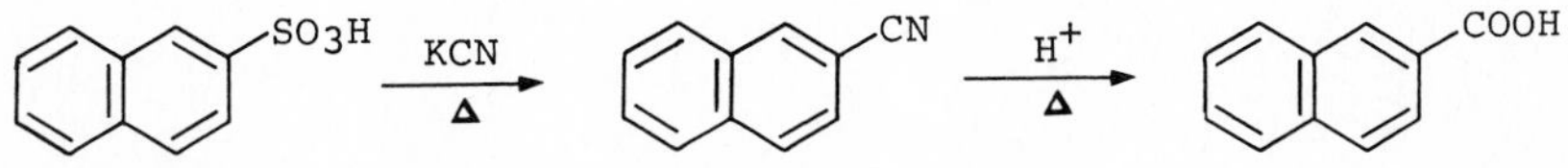

[from (a)]

(i)

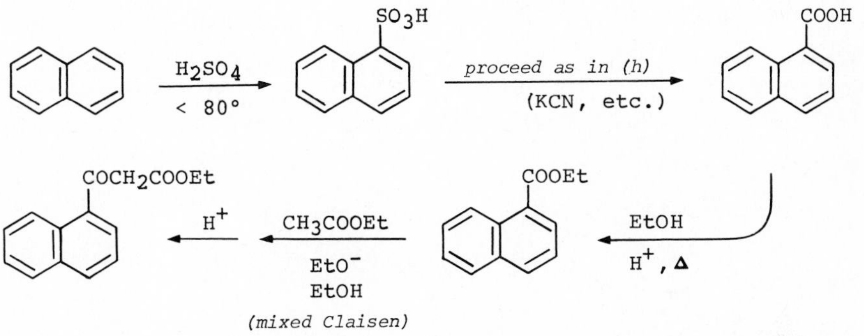

8.

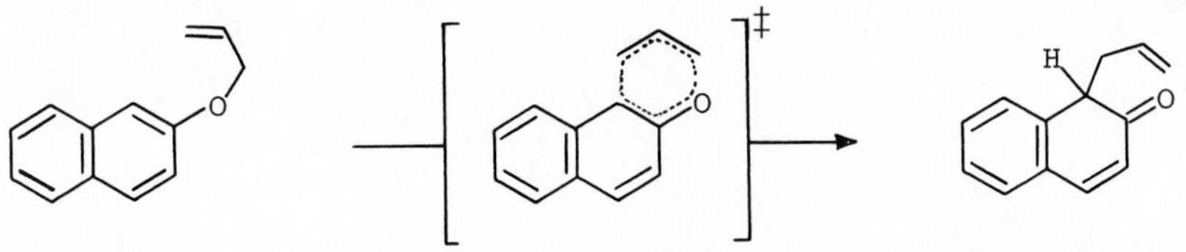

This transition state and intermediate both still have an intact benzene ring.

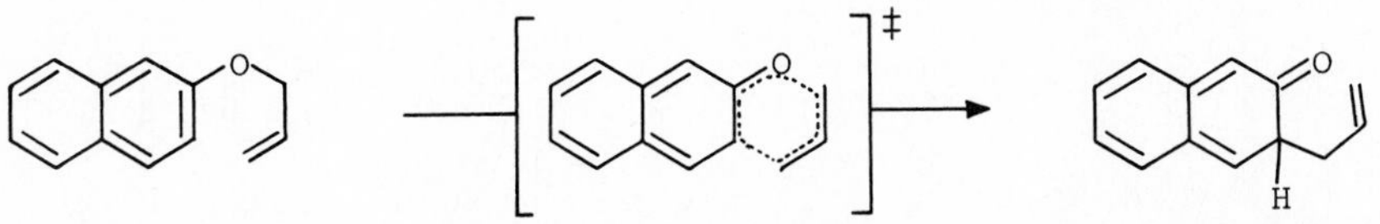

In this transition state and intermediate, the resonance stabilization of both benzene rings has been lost.

9. (a)

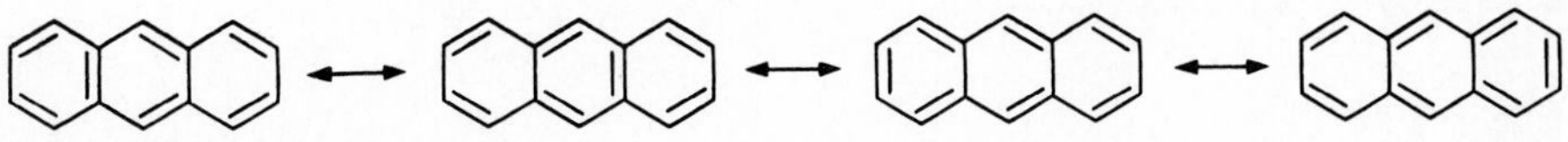

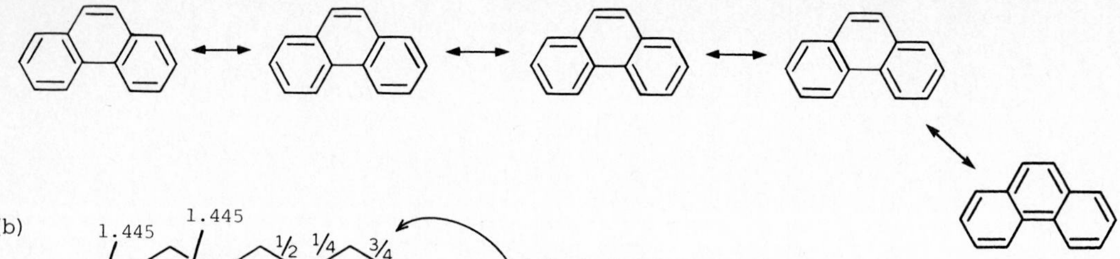

(b)

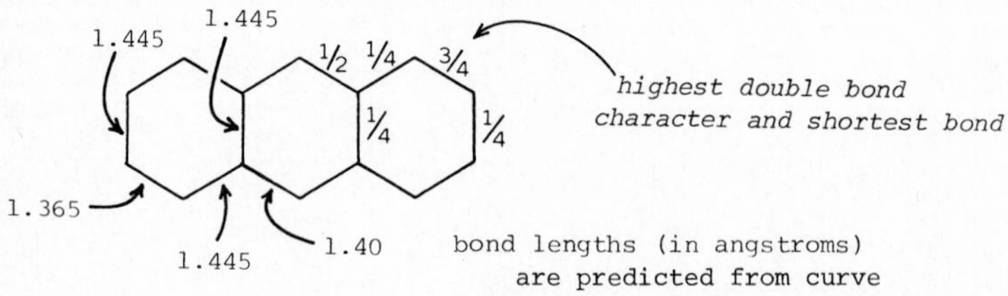

bond lengths (in angstroms) are predicted from curve

10. (a)

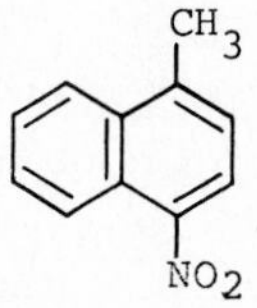

(b)

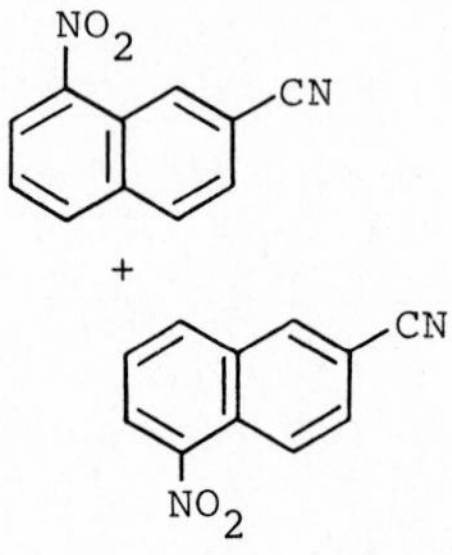

(c)

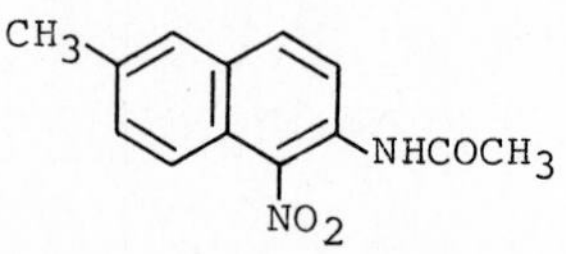

(d)

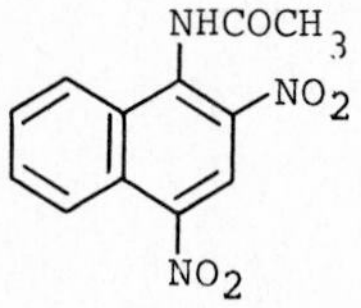

(e)

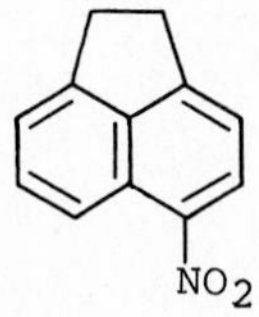

(α-alkylnaphthyl type)

(f)

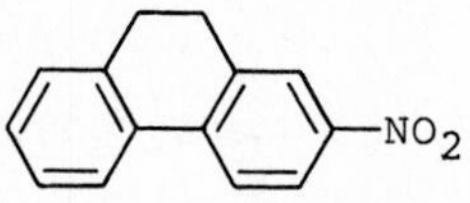

(biphenyl type)

(g)

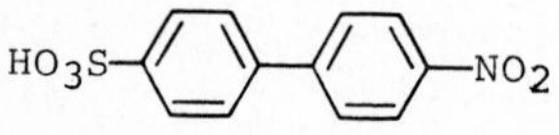

(h)

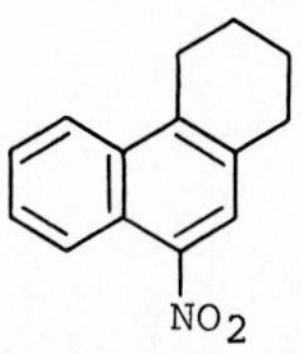

(i)

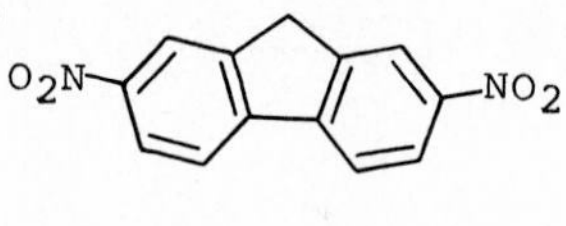

(j)

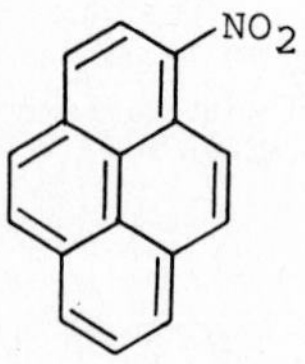

(k)

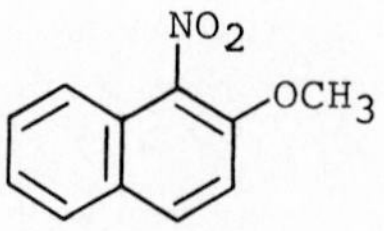

(l)

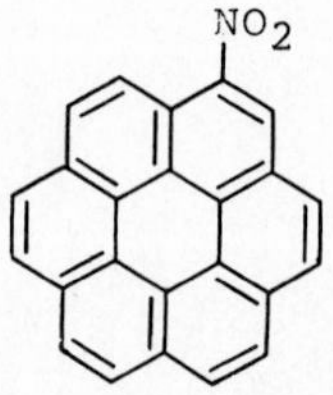

(*Note:* all positions are equivalent!)

11. (a)

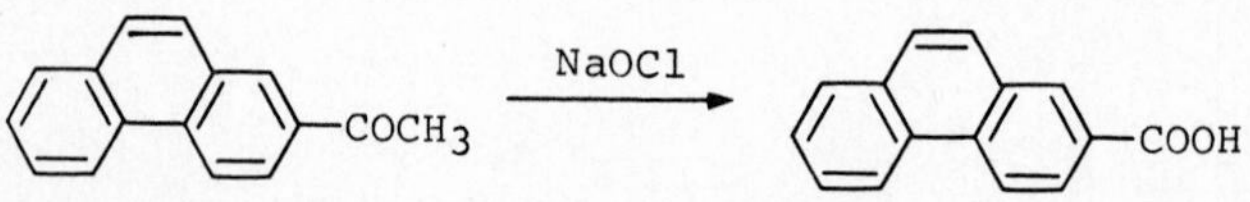

(b)

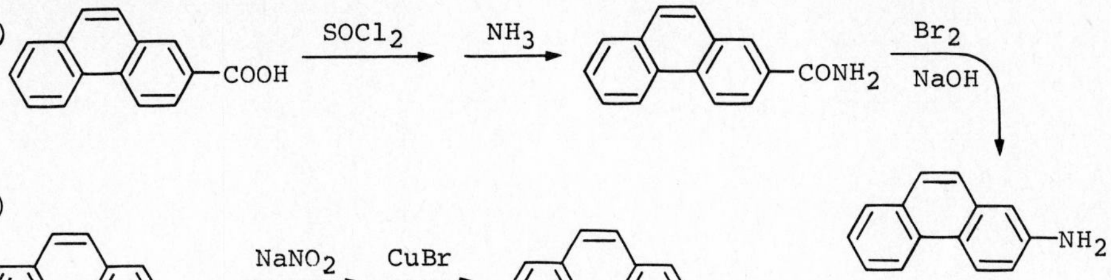

(c)

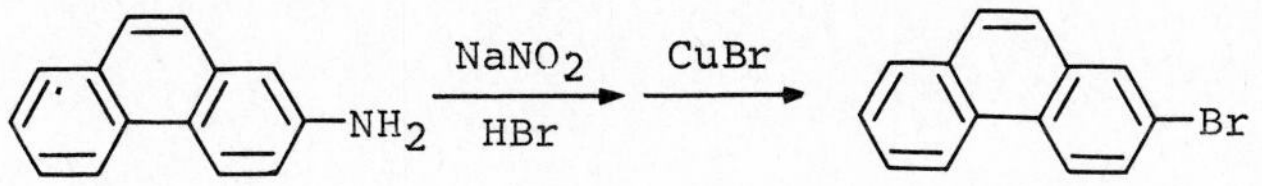

(d)

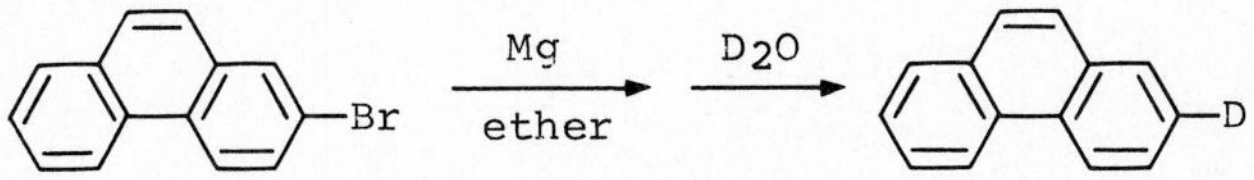

The same sequences apply, starting with the 3-acetyl compound.

12. (a)

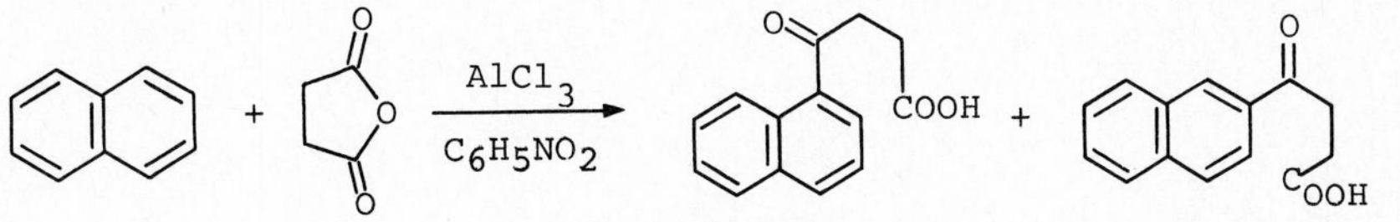

separate mixture; see Section 31.3.C.1

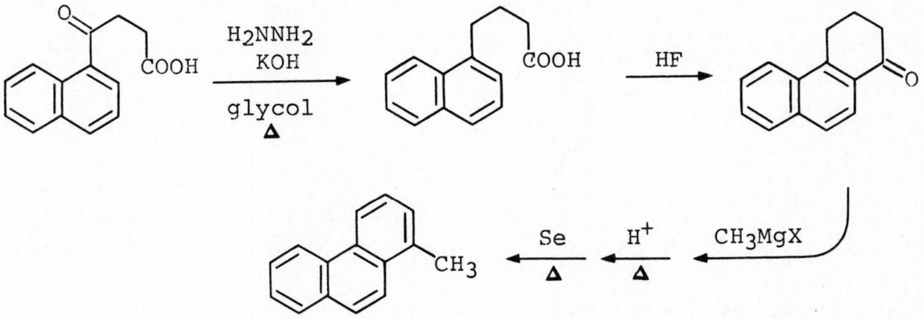

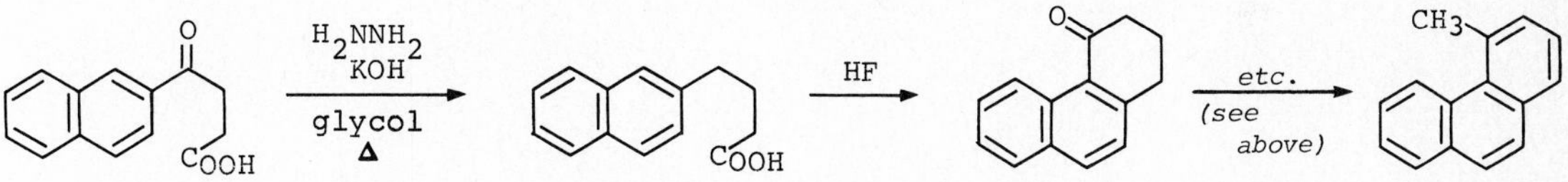

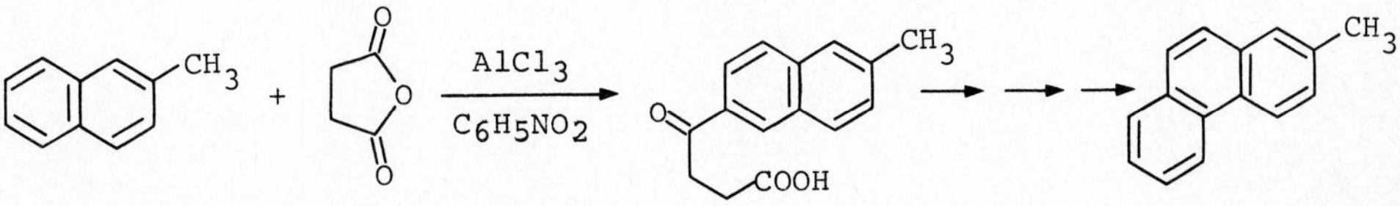

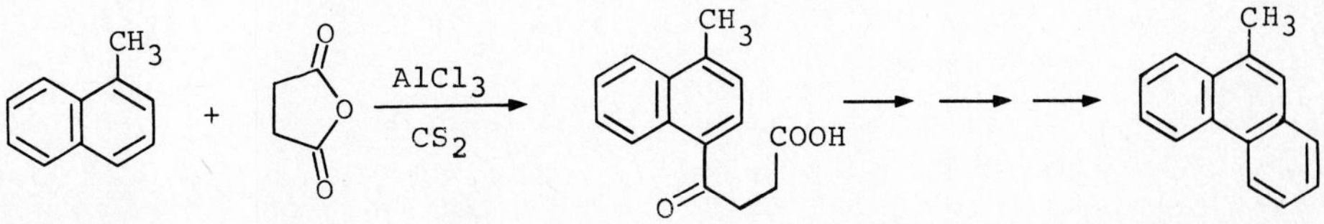

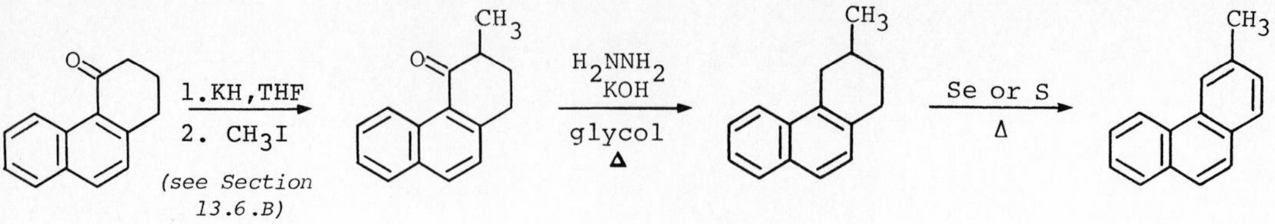

(b)

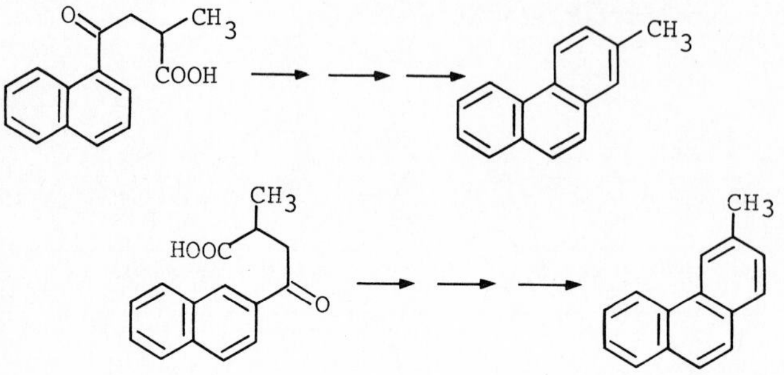

13.

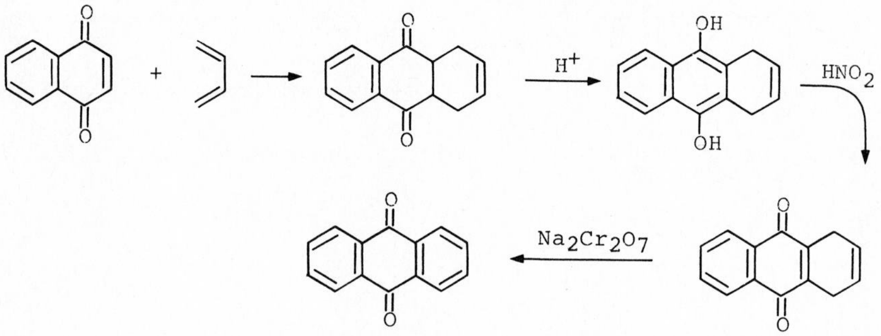

14. 1-methylpyrene; 1,2,3,4-tetramethylphenanthrene; 5,6-dimethylchrysene

Note that in benzo derivatives the numbering changes. For example, the numbering in benzo[a]anthracene is:

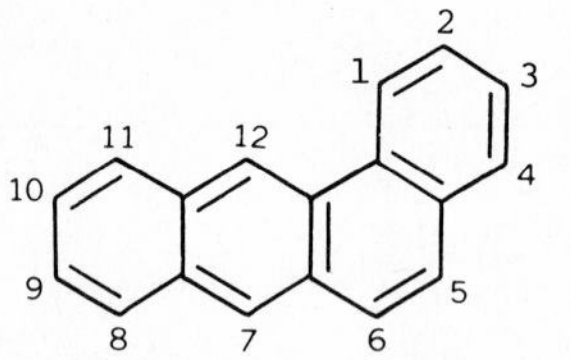

(see Section 31.5)

The name of a carcinogenic derivative is given:

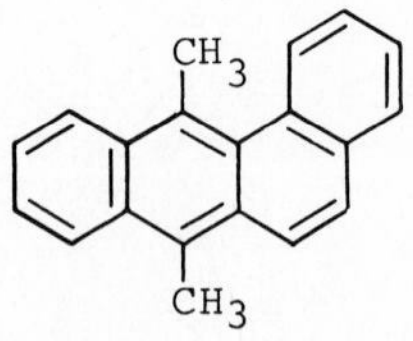

7,12-dimethylbenz[a]anthracene

15. (a)

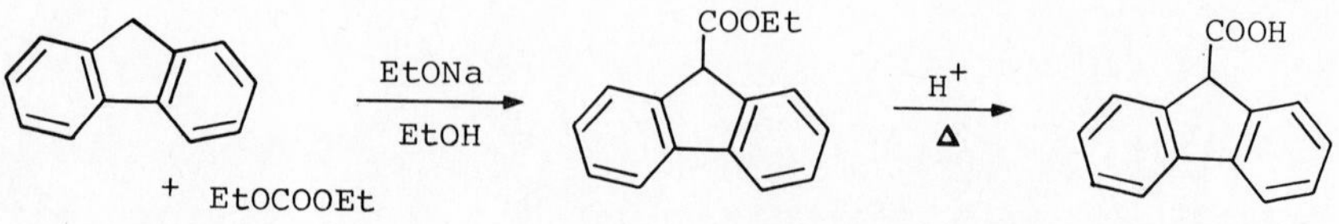

(b)

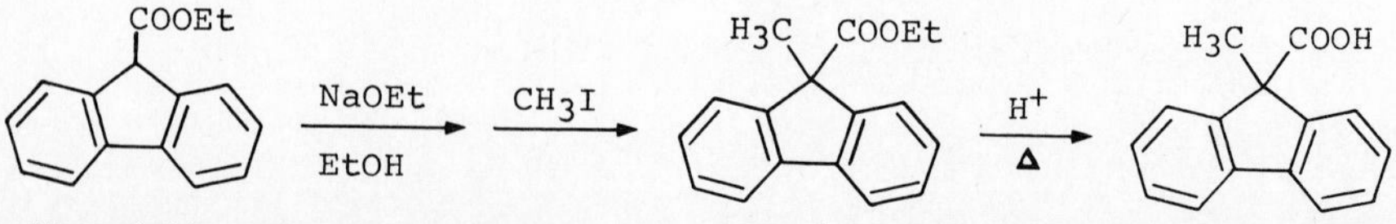

[from (a)]

(c) 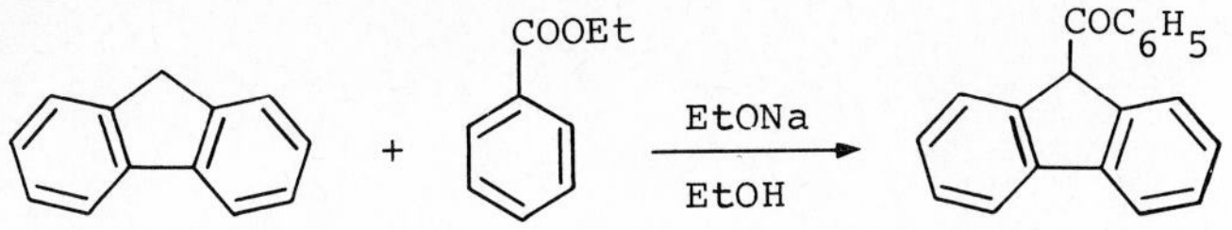

(d) 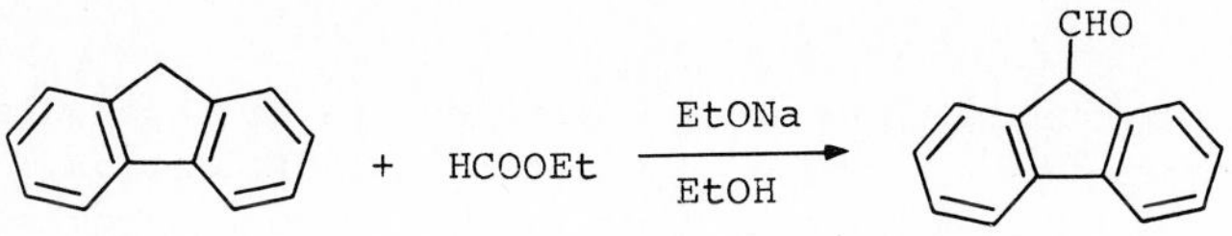

16. 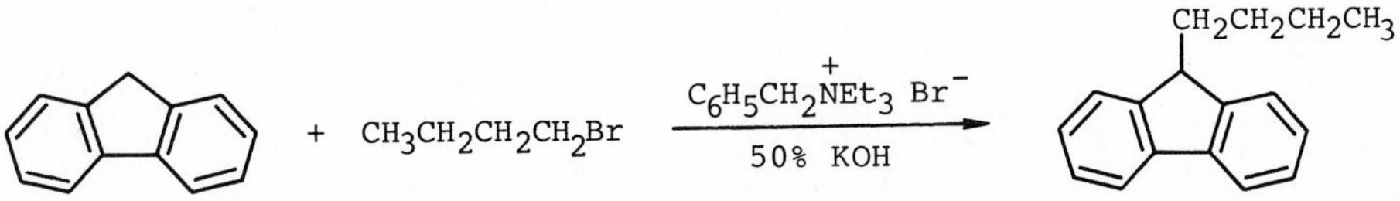

17. (a) 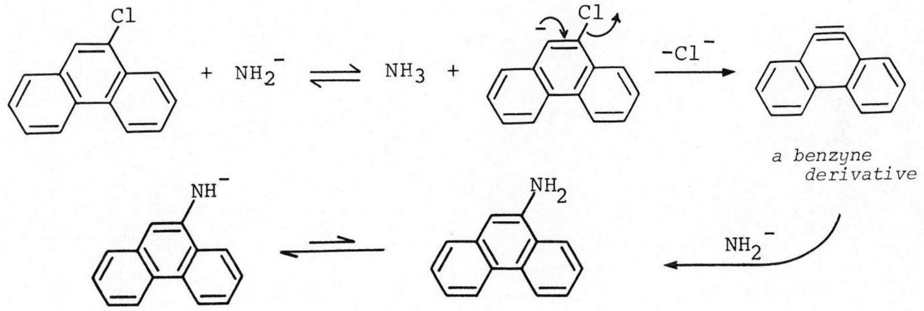

(b) 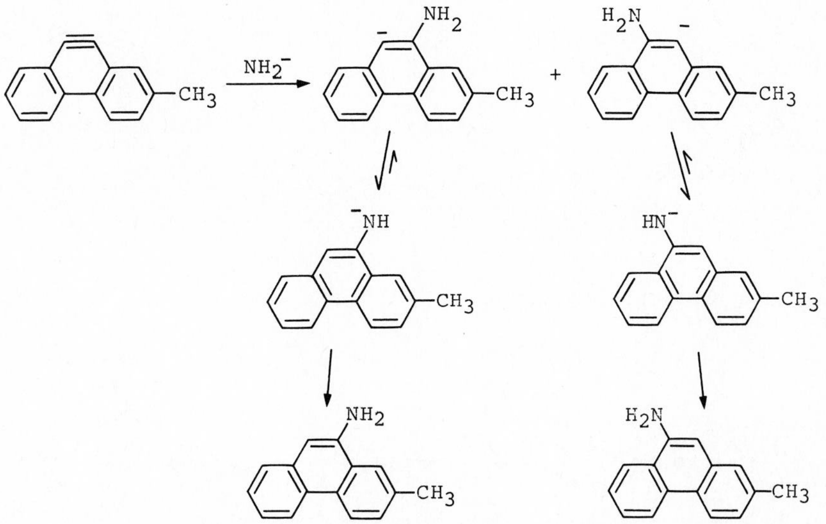

A mixture of two ismoers is obtained.

18. 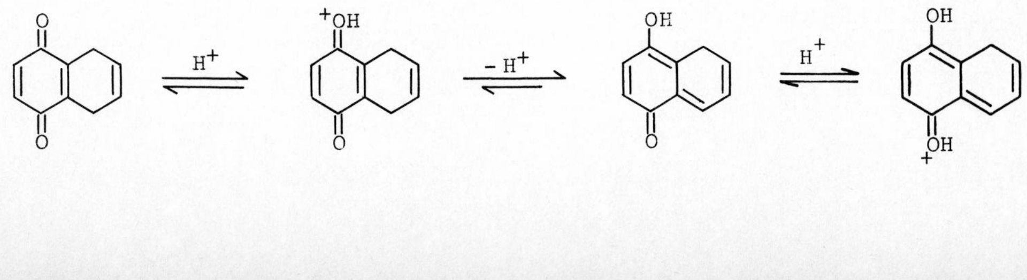

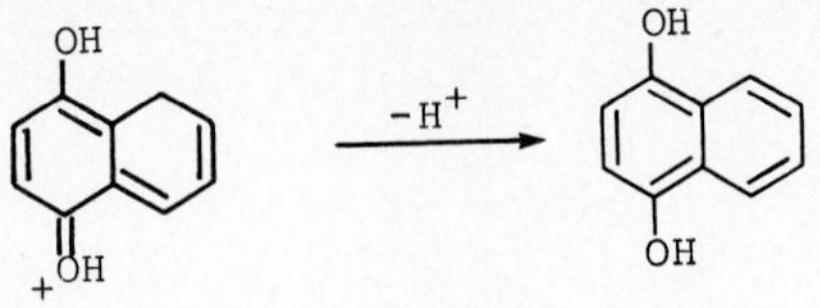

19. 1,2-Naphthoquinone has one benzene ring. On reduction, a naphthalene ring is generated, with a consequent increase in resonance stabilization. 2,6-Naphthoquinone has no benzene ring. On reduction to a naphthalene, the entire stabilization energy of the two aromatic rings is gained.

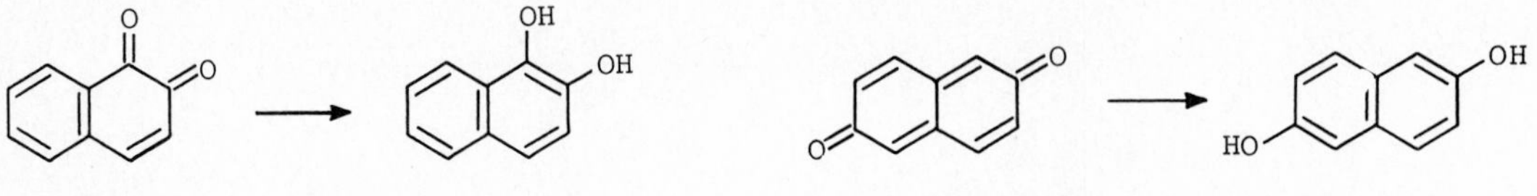

1,2-naphthoquinone

2,6-naphthoquinone
(no benzene conjugation)

These relationships may be summarized by the following energy diagram:

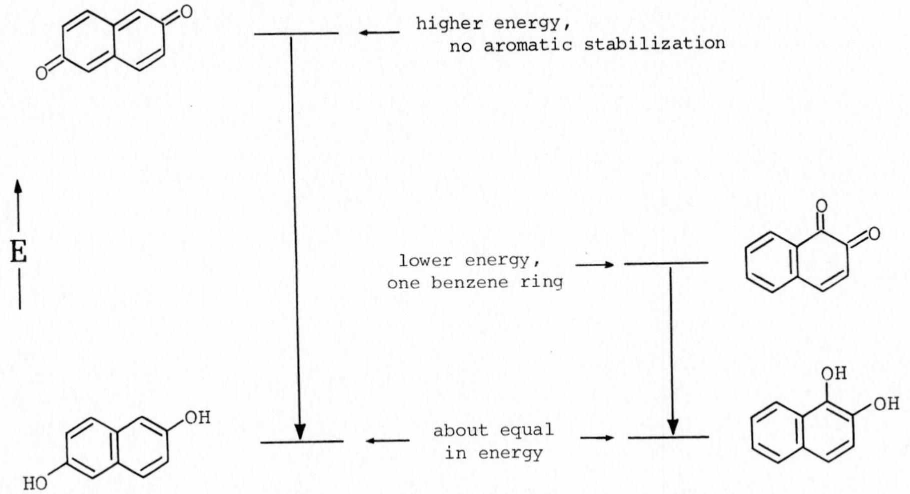

20. Steganone and isosteganone differ in their conformation about the biphenyl bond:

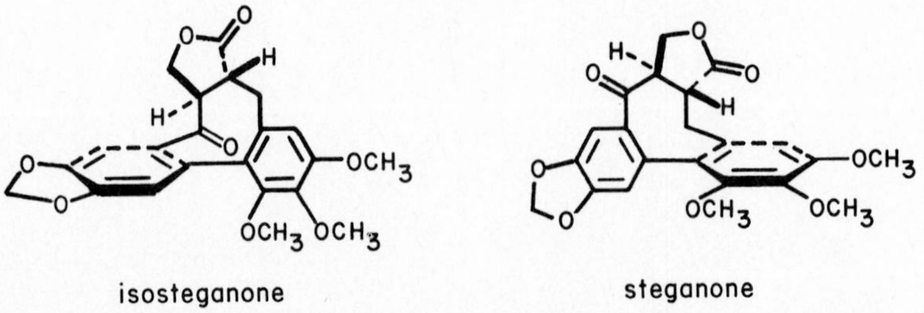

21. Removal of the bridgehead proton from triptycene places the negative charge in an orbital which cannot overlap with any of the p-orbitals of the aromatic rings. It lies in the nodal plane of all the π-systems, and so receives no stabilization by resonance:

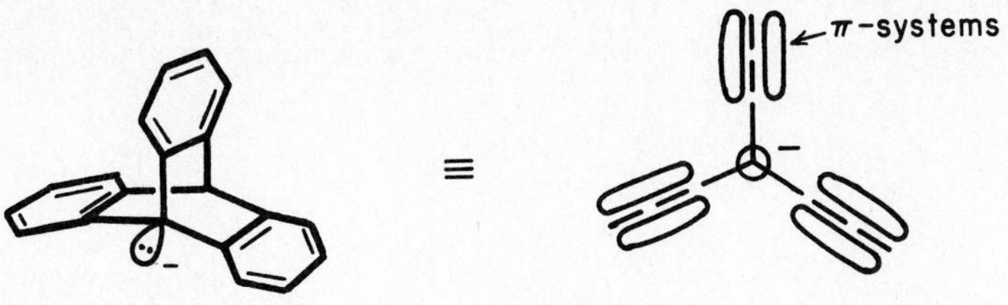

22.

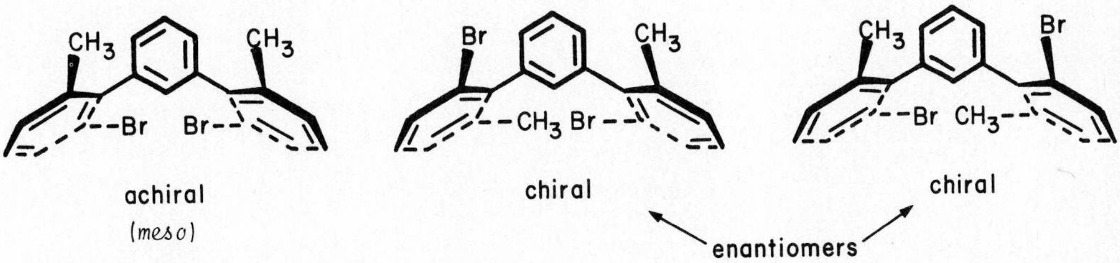

23.

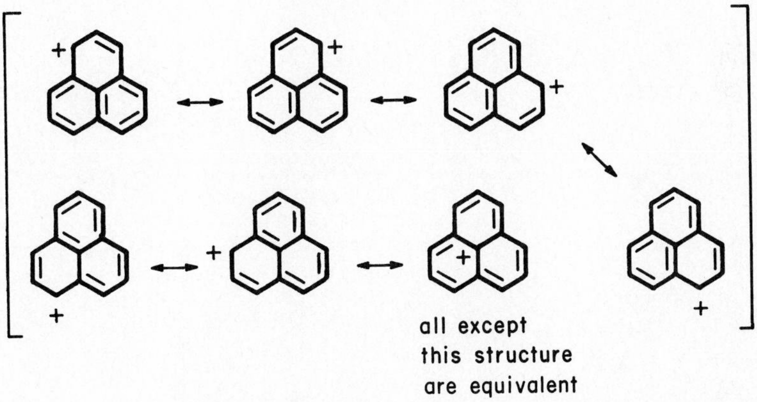

Substitution at the 1-position of pyrene results in the formation of a perinaphthenyl-like cation:

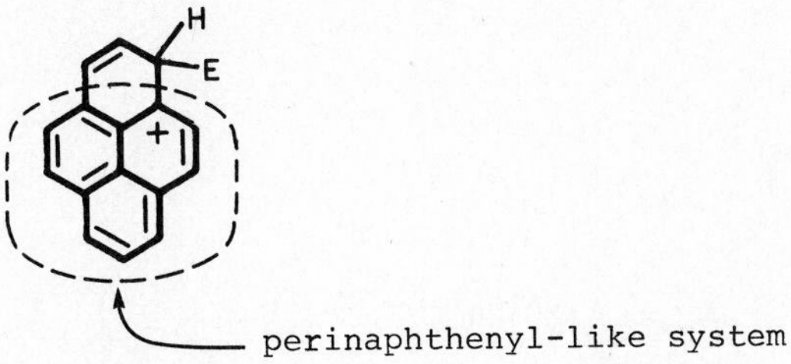

31.E Supplementary Problems

S1. Write the structure of each of the following compounds.

(a) 5-dimethylamino-1-naphthalenesulfonic acid

(b) 3-bromo-4,4'-dimethylbiphenyl

(c) 2,7-dinitrofluorene

(d) 1,4-phenanthraquinone

(e) benzo[a]chrysene

(f) dibenzo[b,e]fluoranthene

S2. Write the major product from each of the following reaction sequences:

(a)
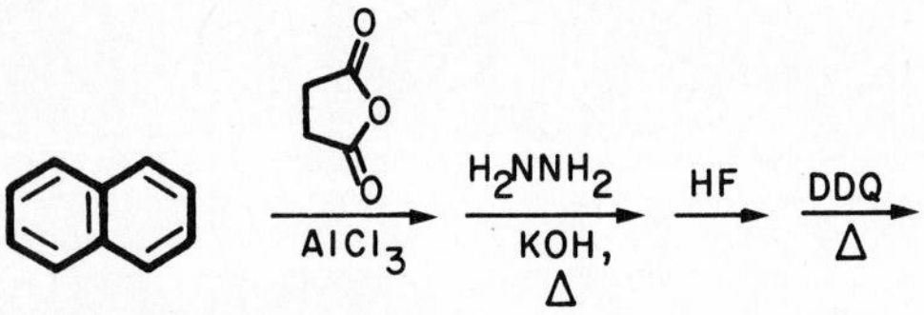

(b)
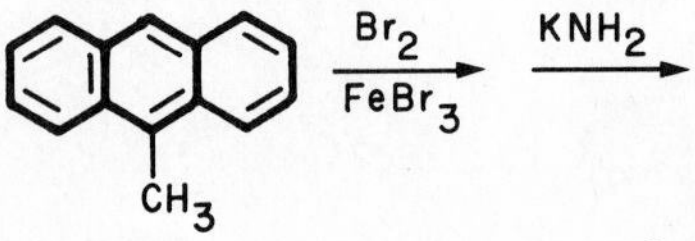

(c)
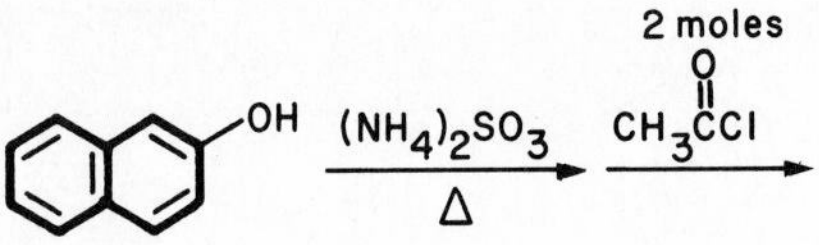

(d)
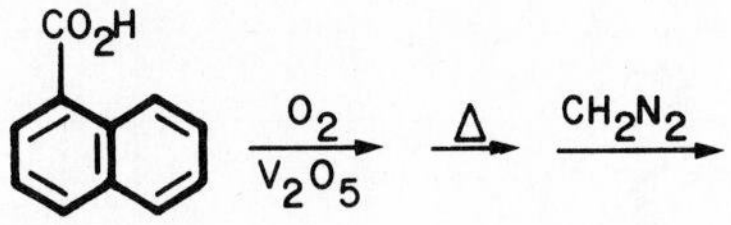

(e)
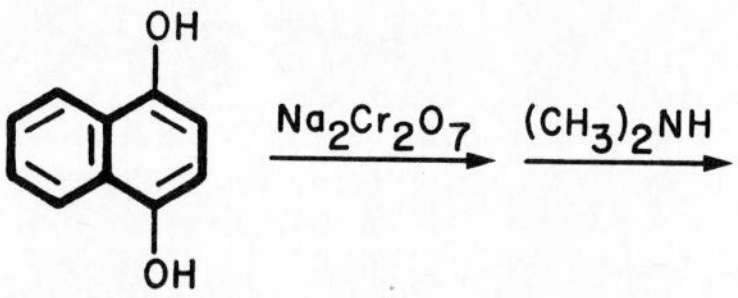

(f)
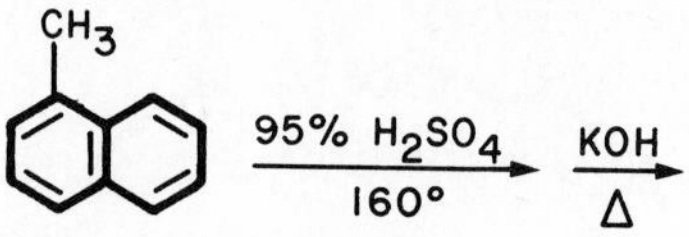

(g)
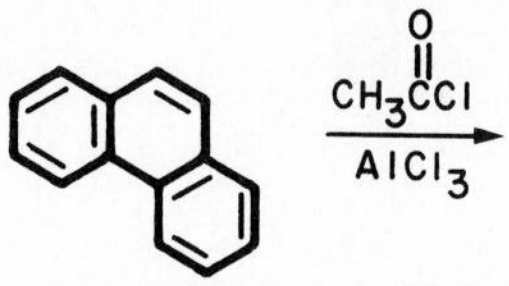

(h)
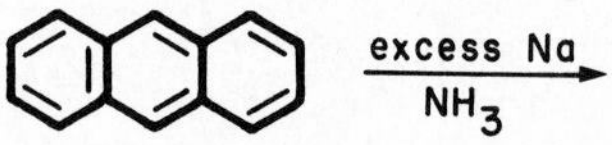

S3. Outline a synthesis for each of the compounds drawn below, starting from benzene, naphthalene, or any monosubstituted benzene, and any other non-aromatic compounds.

(a)
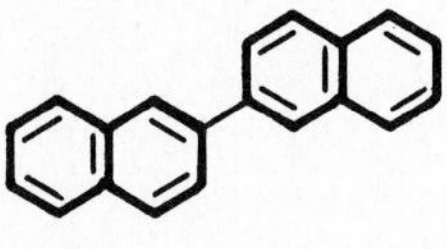

(b)
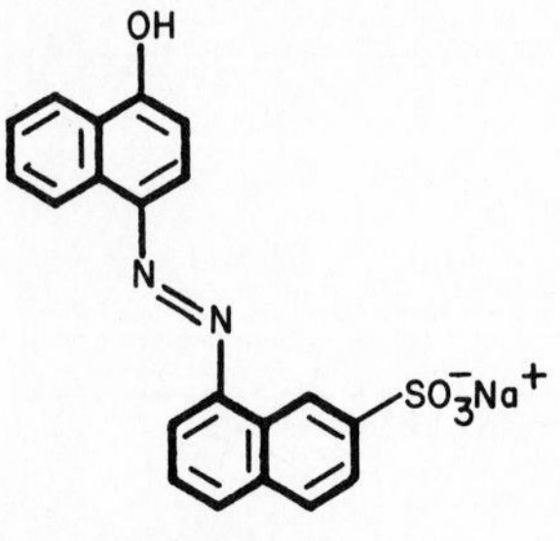

(c)
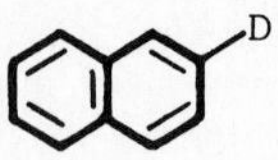

(d)
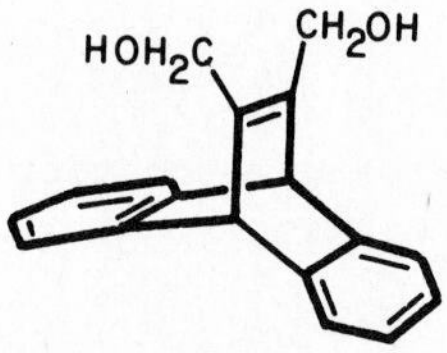

(e)
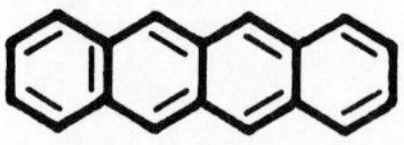

(f)
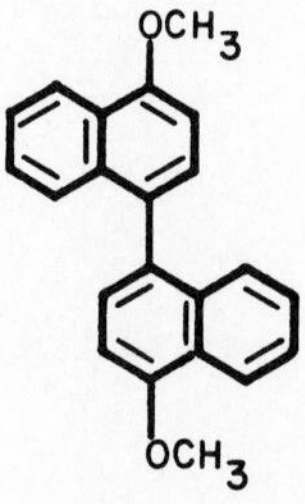

(g)
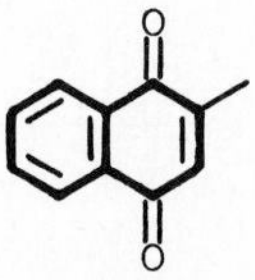

Menadione
(a vitamin K substitute)

(h)
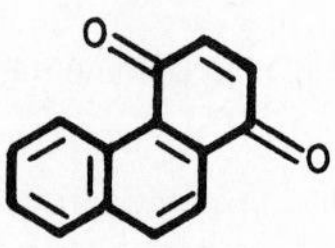

S4. Predict the positions of the following equilibria and justify your answers.

(a)

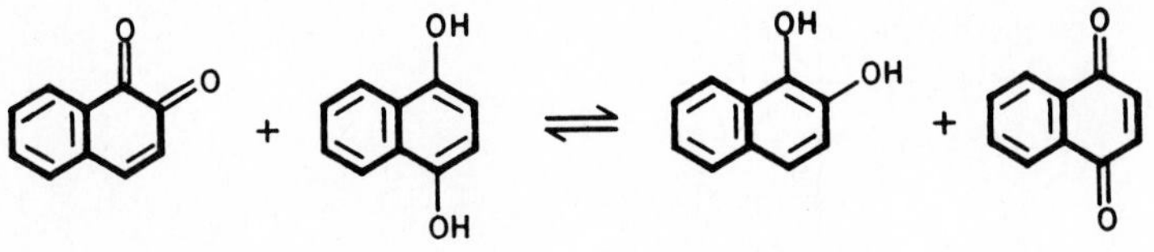

(b)

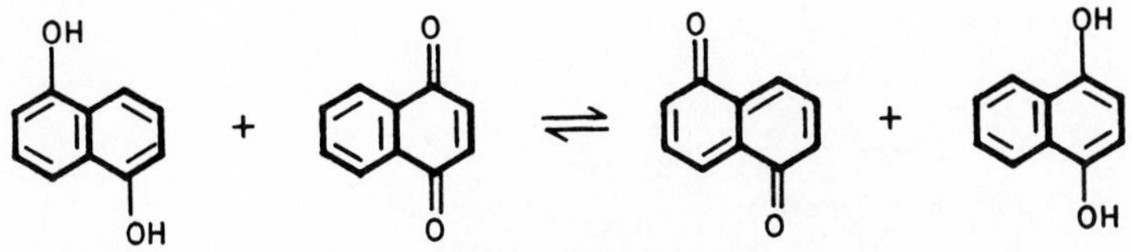

(c)

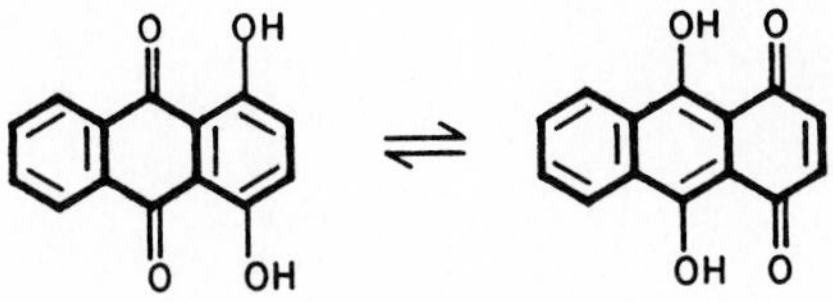

S5. Write the structures of the six intermediates which are produced during the following sequence of reactions.

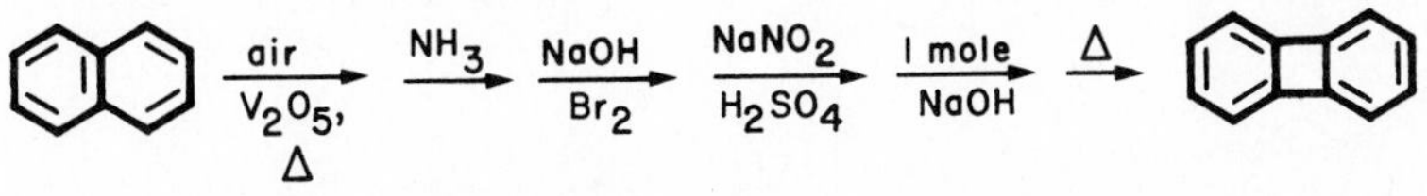

S6. 6-Methoxy-1-tetralone (1) is an important intermediate in an industrial synthesis of estrone (2), as well as a number of contraceptive drugs which are derived from it. Devise an efficient preparation of 1, using naphthalene as the starting material.

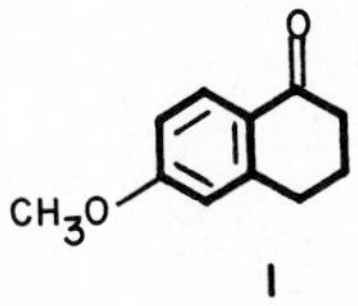

1

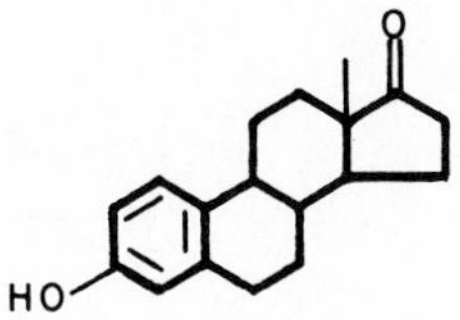

2

S7. Write a reasonable mechanism for the thermal decarboxylation of 2-hydroxy-1-naphthoic acid.

S8. The streptovaricins are a group of antibiotics which are produced by the micro-organism *Streptomyces spectabilis*. Heating streptovaricin C in refluxing toluene results in its partial isomerization to atropisostreptovaricin C. Reaction of either streptovaricin C or its isomer with $NaIO_4$ leads to the same compound. What is a likely structure of atropisostreptovaricin C (streptovaricin C is depicted on the following page).

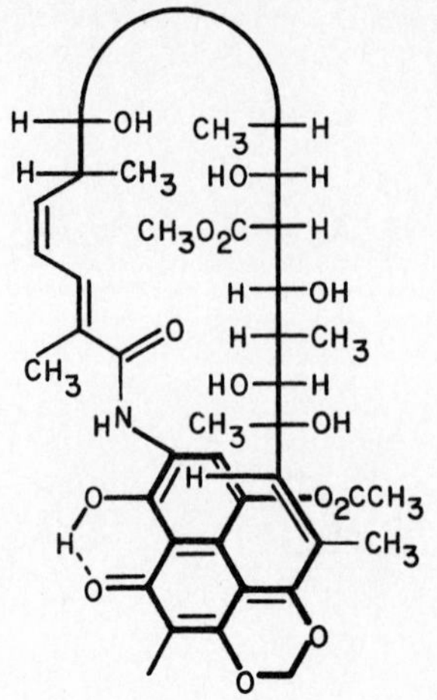

Streptovaricin C

31.F Answers to Supplementary Problems

S1.

(a)

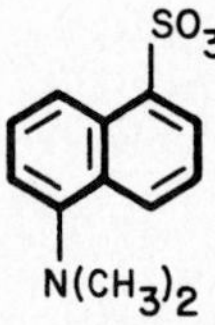

(b)

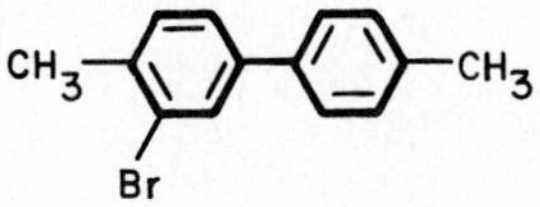

(c)

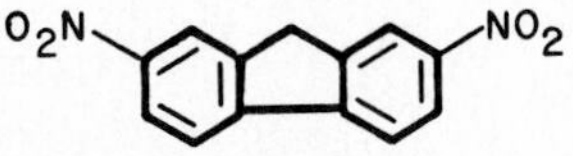

(d)

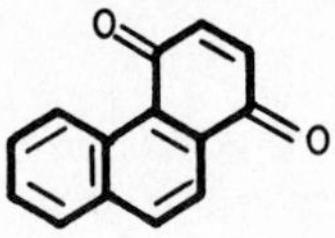

(e)

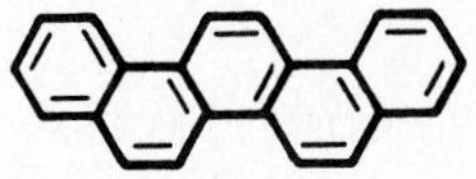

(f)

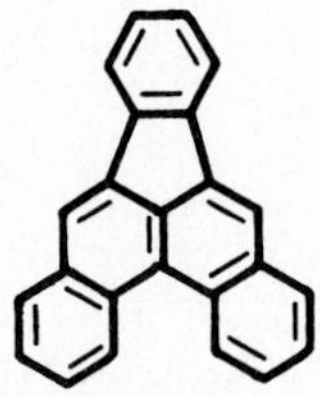

S2.

(a)

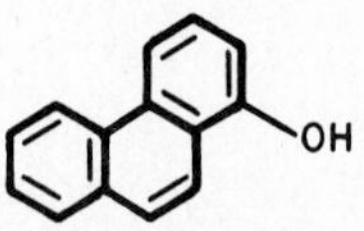

(b)

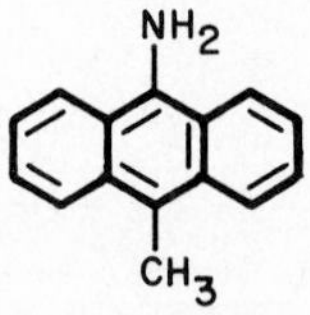

(c)

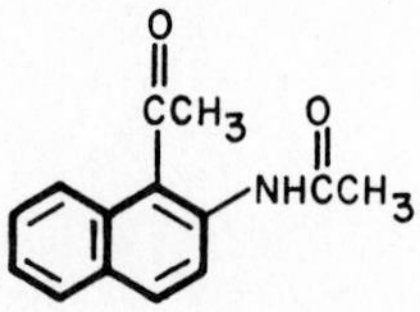

(d)

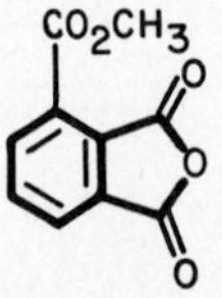

(e)

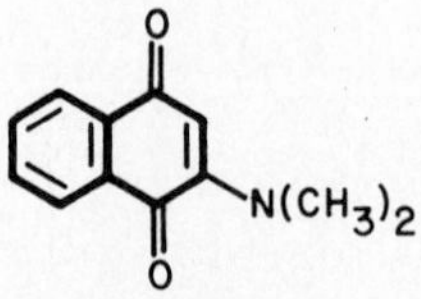

and

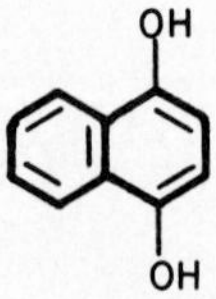

1/2 mole 1/2 mole

(see Problem #S10 in Chapter 30 of this Study Guide)

(f)

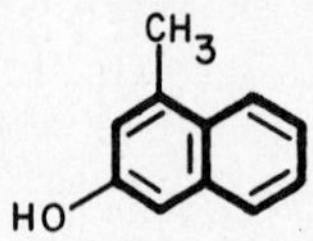

(g)

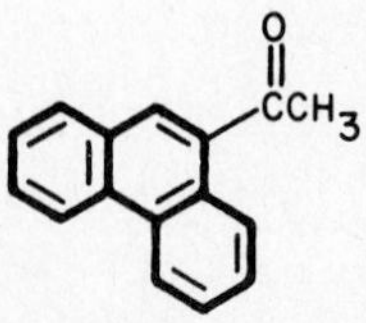

(h)

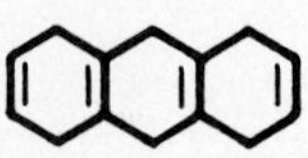

S3. (a)

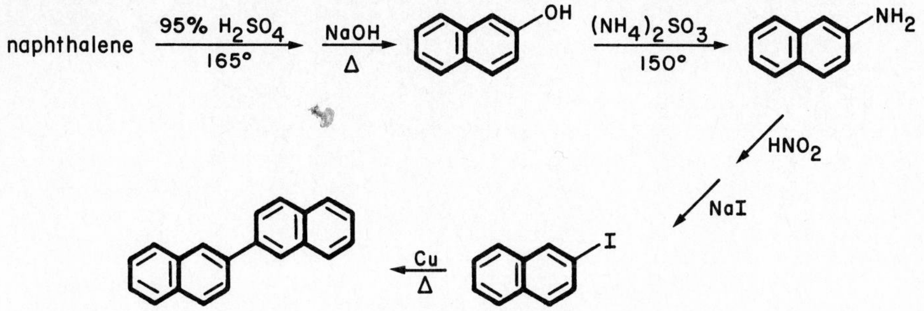
naphthalene
95% H2SO4
165°
NaOH
Δ
OH
(NH4)2SO3
150°
NH2
HNO2
NaI
I
Cu
Δ

(b)

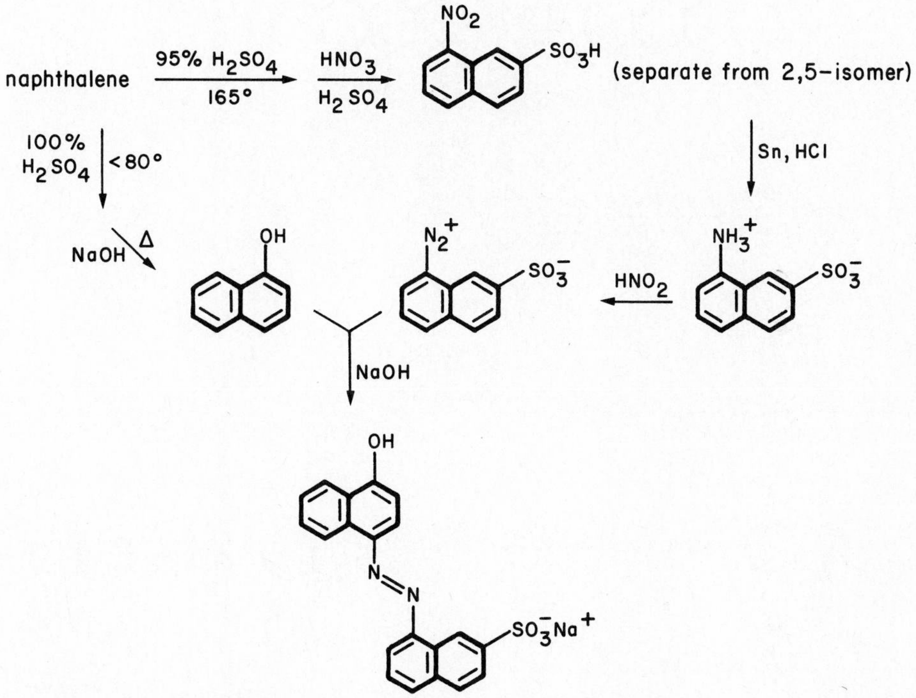
naphthalene
95% H2SO4
165°
HNO3
H2SO4
NO2
SO3H
(separate from 2,5-isomer)
100%
H2SO4
<80°
NaOH
Δ
OH
N2+
SO3−
HNO2
NH3+
SO3−
Sn, HCl
NaOH
OH
N
N
SO3−Na+

(c)

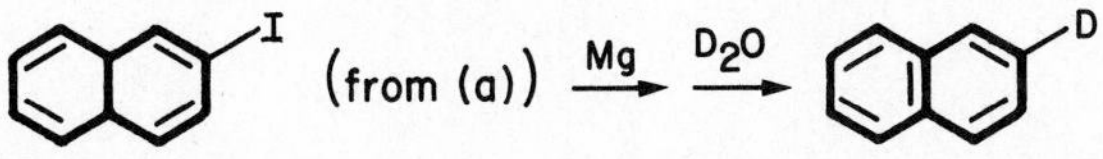
I
(from (a))
Mg
D2O
D

(d)

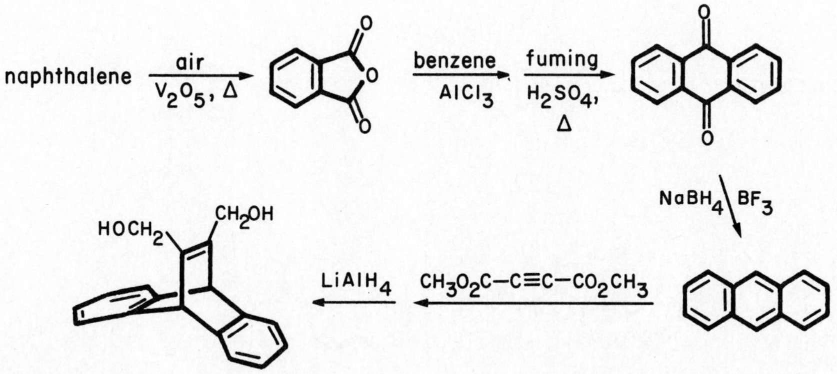
naphthalene
air
V2O5, Δ
O
O
O
benzene
AlCl3
fuming
H2SO4,
Δ
O
O
NaBH4
BF3
CH3O2C−C≡C−CO2CH3
LiAlH4
HOCH2
CH2OH

(e)

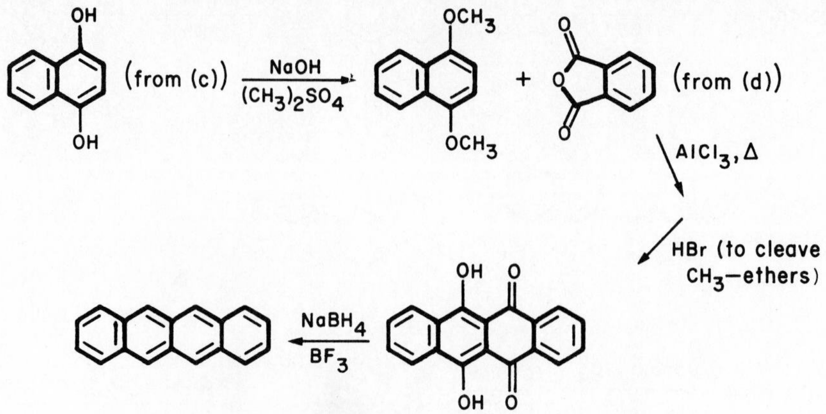

$$\text{(from (c))} \xrightarrow[\text{(CH}_3\text{)}_2\text{SO}_4]{\text{NaOH}} + \text{(from (d))}$$

$$\downarrow \text{AlCl}_3, \Delta$$

$$\downarrow \text{HBr (to cleave CH}_3\text{—ethers)}$$

$$\xleftarrow[\text{BF}_3]{\text{NaBH}_4}$$

(f)

naphthalene $\xrightarrow{\text{HNO}_3}{\text{H}_2\text{SO}_4}$ $\xrightarrow[\substack{\text{Ru/C} \\ \text{KOH}}]{\text{H}_2\text{NNH}_2}$ $\xrightarrow[\Delta]{\text{H}^+}$ $\xrightarrow[\substack{2.\ \text{CH}_3\text{OH} \\ \Delta}]{1.\ \text{HNO}_2}$

(g)

$$\text{(from (b))} \xrightarrow{\text{K}_2\text{Cr}_2\text{O}_7} \xrightarrow{\text{Na}_2\text{S}_2\text{O}_4} \xrightarrow[\text{CHCl}_3]{\text{NaOH}}$$

$$\downarrow \text{Zn, HCl}$$

$$\xleftarrow[\text{H}_2\text{SO}_4]{\text{K}_2\text{Cr}_2\text{O}_7}$$

(h)

naphthalene $\xrightarrow[\substack{\text{AlCl}_3 \\ \text{CS}_2}]{}$ $\xrightarrow{\text{Zn}}{\text{HCl}} \xrightarrow{\text{HF}}$

$$\downarrow \text{CrO}_3$$

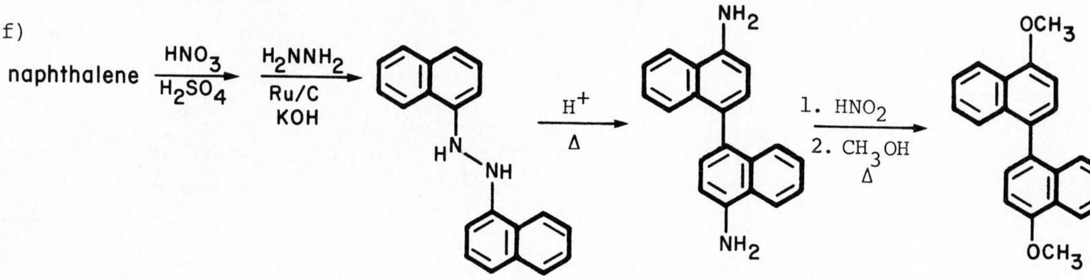

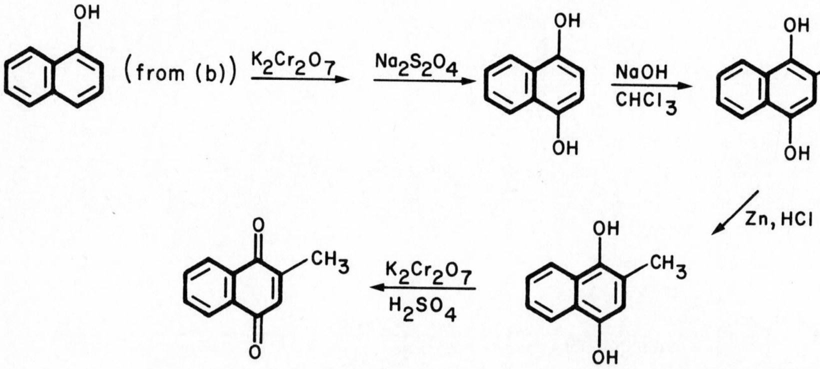

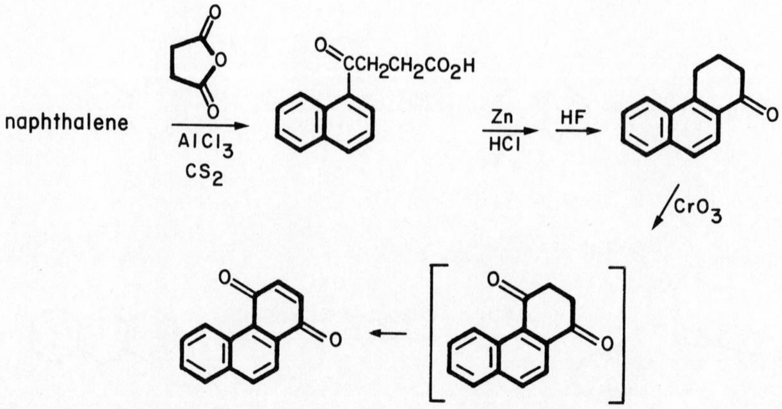

S4. (a) favors

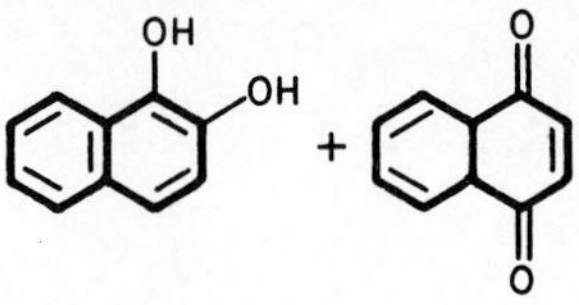

1,2-Naphthoquinone and 1,2-benzoquinone both have higher reduction potentials than their 1,4-isomers (see Table 30.2) because of unfavorable interaction between their aligned dipoles:

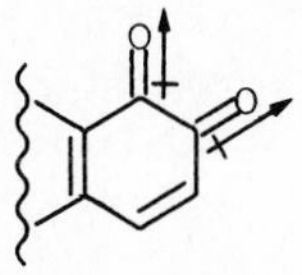

(b) favors

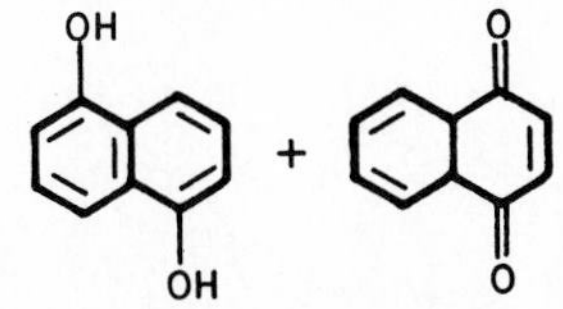

See answer to problem #19 of this chapter.

(c) favors

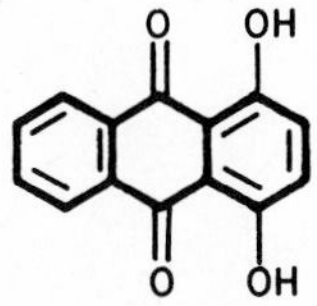

This structure has the aromatic stabilization of the two benzene rings, which is greater than the stabilization of the naphthalene system present in the 1,4-quinone tautomer.

S5.

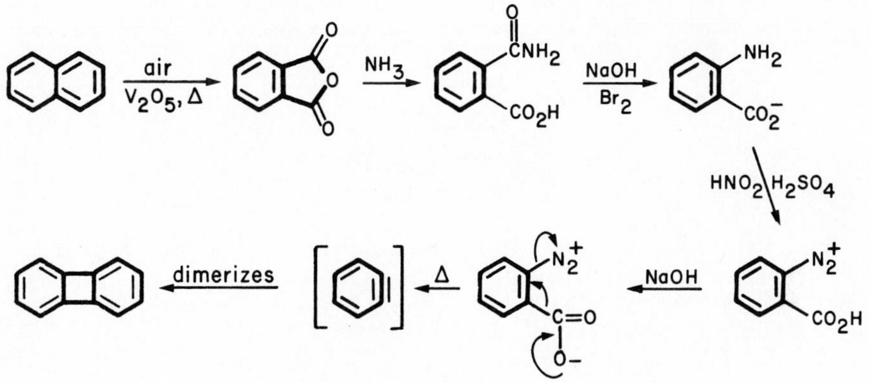

S6.

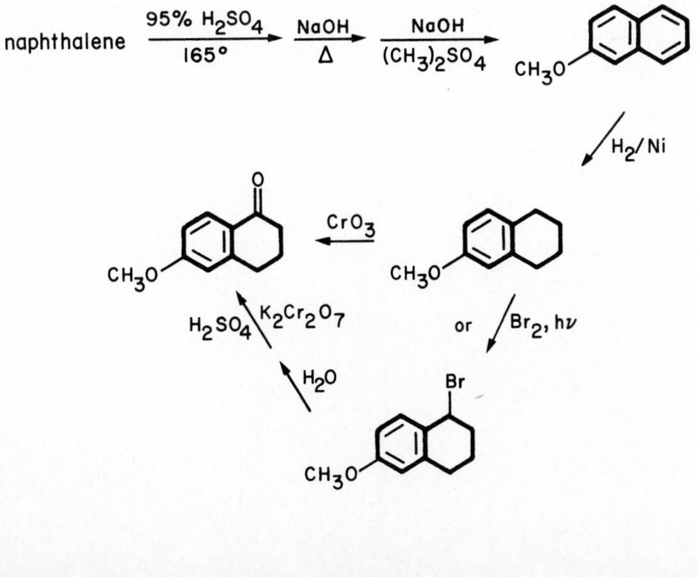

S7.

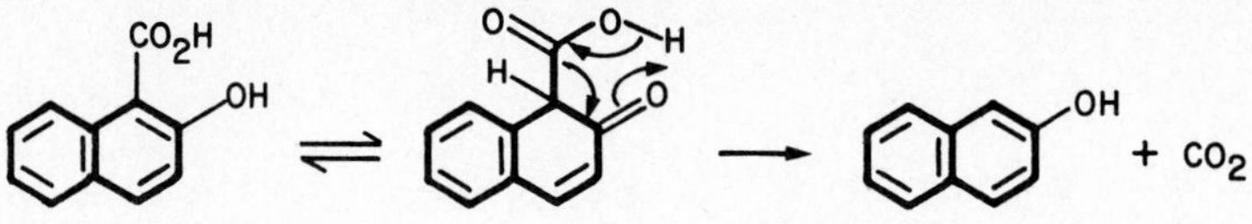

S8. The aromatic ring system and the highly-substituted bridging chain of streptovaricin C can be represented schematically, as illustrated below:

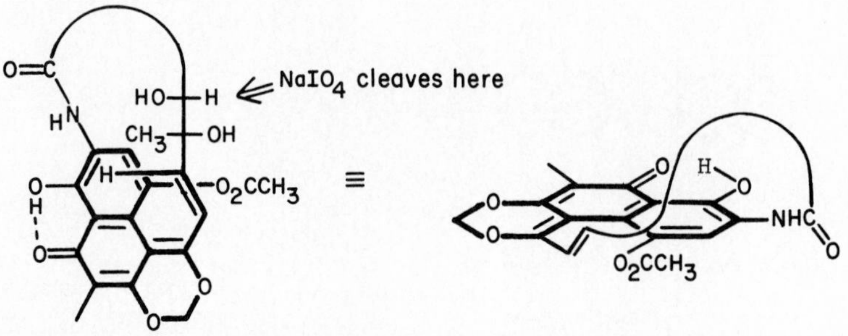

This compound reacts with $NaIO_4$ only at the vicinal diol position indicated at the left above, to cleave the bridging chain. Because atropisostreptovaricin gives the same product upon $NaIO_4$ cleavage, it can differ only in conformation, and most likely is the "ring-flipped" conformer drawn below:

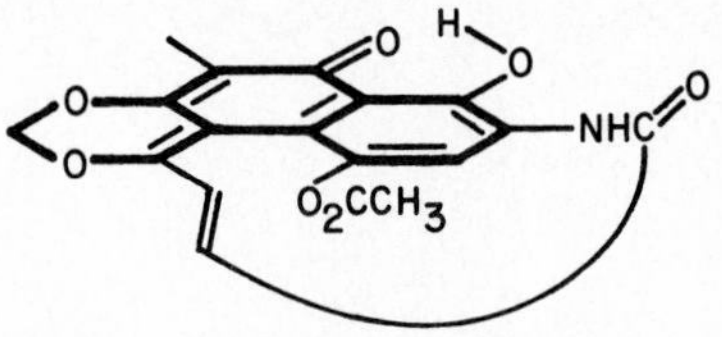

32. HETEROCYCLIC COMPOUNDS

32.A Chapter Outline, Important Concepts Introduced, and Reactions Discussed

32.1 Introduction

definition of heterocycles (-iran (3), -etan (4), -olan (5), -ane (6))

aza- (N), oxa- (O), thia- (S)

32.2 Non-aromatic Heterocycles

A. Nomenclature

B. Three-Membered Rings

epoxides = oxiranes *(see Sections 10.12.A and 11.4.E)*

aziridines via Wenker synthesis:

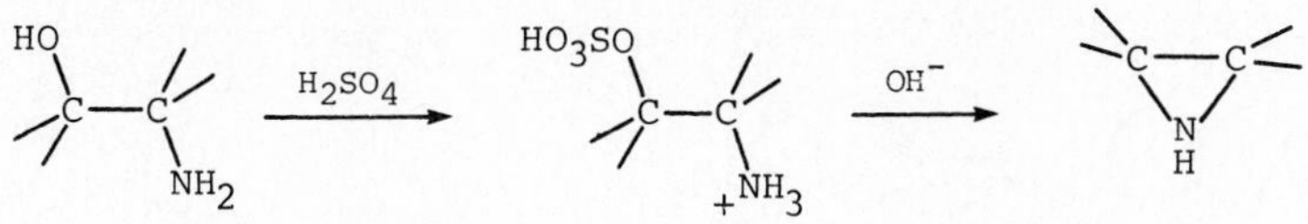

aziridines via iodoisocyanates:

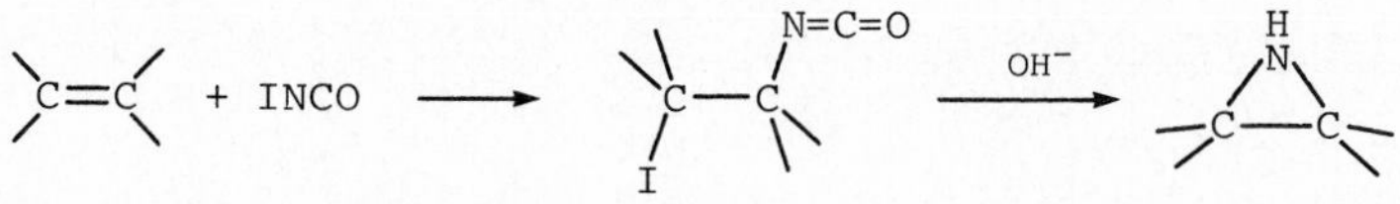

thiiranes from epoxides:

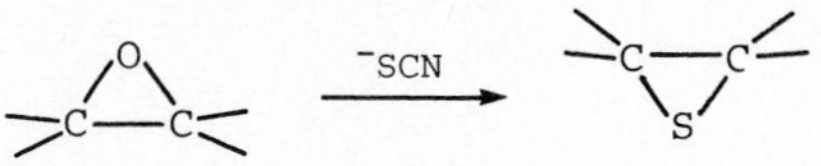

reactions: ring opening

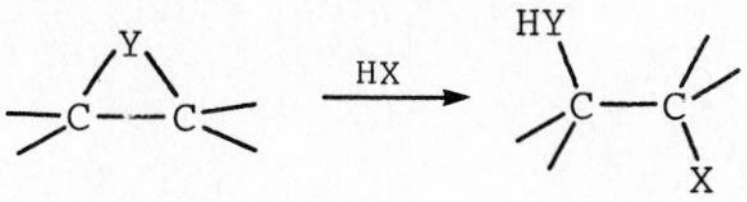

X = halogen, OR; Y = O, S, NR'

C. Four-Membered Rings

via ring closure:

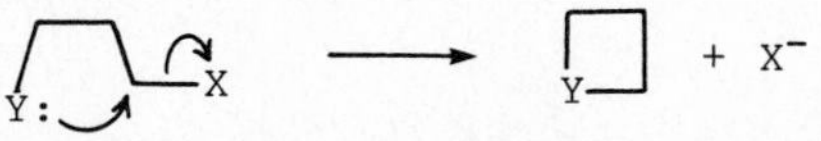

(poor yields)

X = good leaving group; Y: = O^-, S^-, or NHR

β-lactones and β-lactams via cycloaddition:

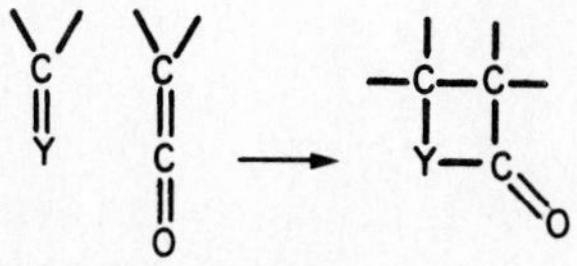

Y = O or NR

ring-opening reactions; similar to three-membered rings

D. Five- and Six-membered Rings

by hydrogenation of the aromatic heterocycles:

e.g.,

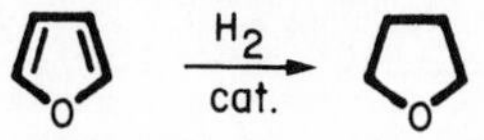

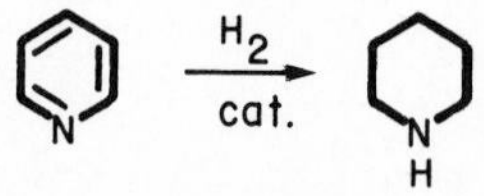

nucleophilic ring closure:

intramolecular reductive amination:

e.g.,

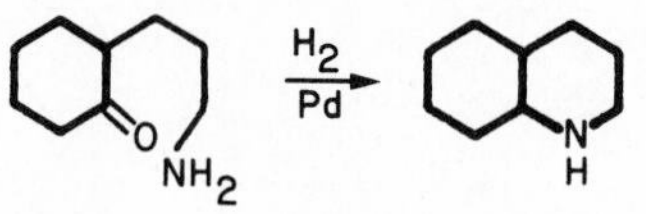

32.3 Furan, Pyrrole, and Thiophene

A. Structure and Properties

use of lone pair electrons of heteroatom to attain aromatic six-π-electron system

effect of aromaticity on pK_a

position of protonation of pyrrole

B. Synthesis

furan derivatives from dehydration of pentoses *(industrial synthesis)*

thiophene from pyrolysis of butenes and sulfur *(industrial synthesis)*

Paal-Knorr synthesis:

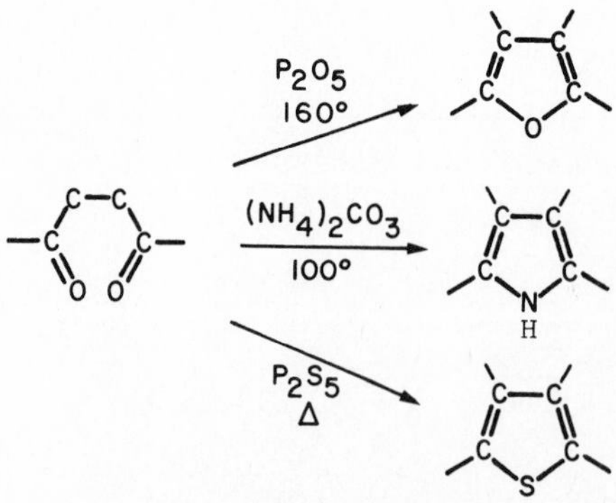

Knorr pyrrole synthesis:

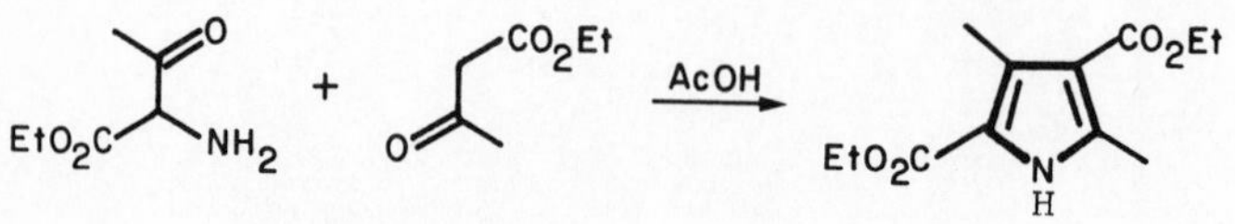

C. Reactions

reactivity toward nucleophiles:
pyrrole > furan > thiophene >> benzene

electrophilic aromatic substitution oriented toward 2-positions

hydrolysis (primarily of furans):

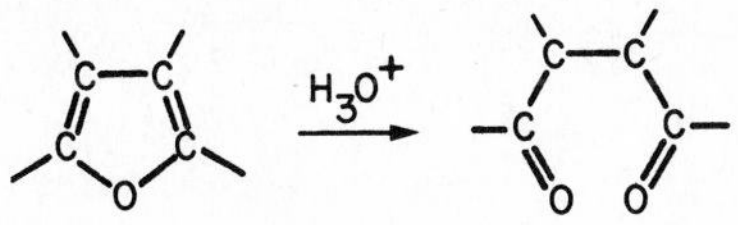

32.4 Condensed Furans, Pyrroles, and Thiophenes

A. Structure and Nomenclature

indole

B. Synthesis

Fischer indole synthesis:

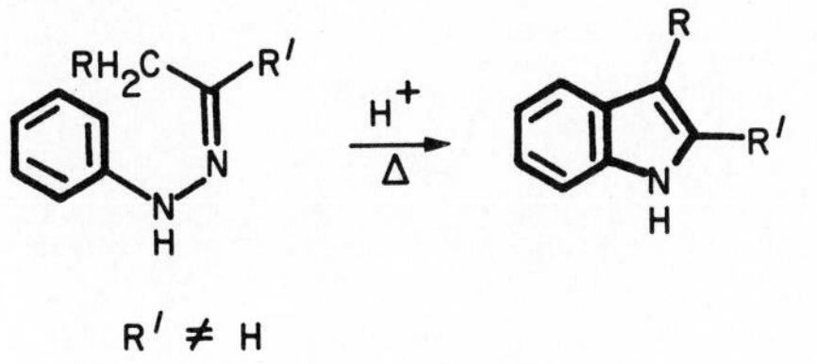

(use R = H, R′ = COOH, then decarboxylate to obtain indole itself)

benzofuran from coumarin:

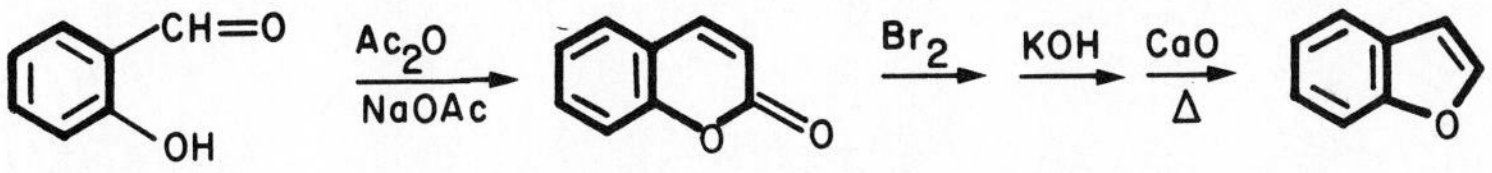

C. Reactions

electrophilic aromatic substitution:

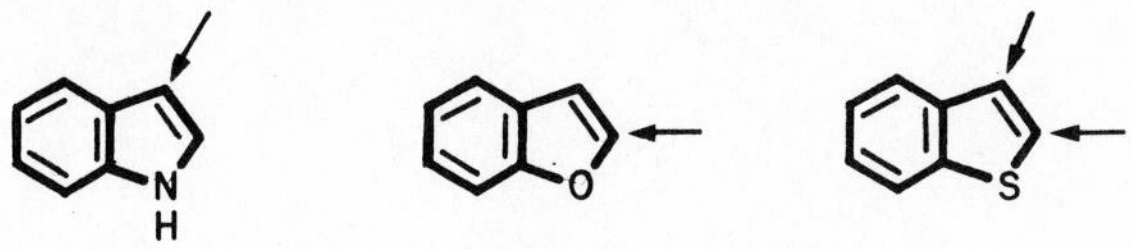

32.5 Azoles

A. Structure and Nomenclature

oxazole, imidazole, thiazole
isoxazole, pyrazole, isothiazole
basicity
occurrence in nature

B. Synthesis

isoxazoles and pyrazoles from 1,3-dicarbonyl compounds:

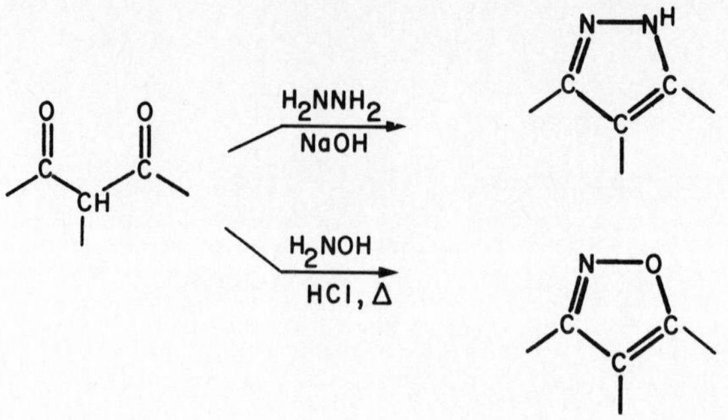

isoxazoles via nitrile oxide-1,3-dipolar cycloaddition:

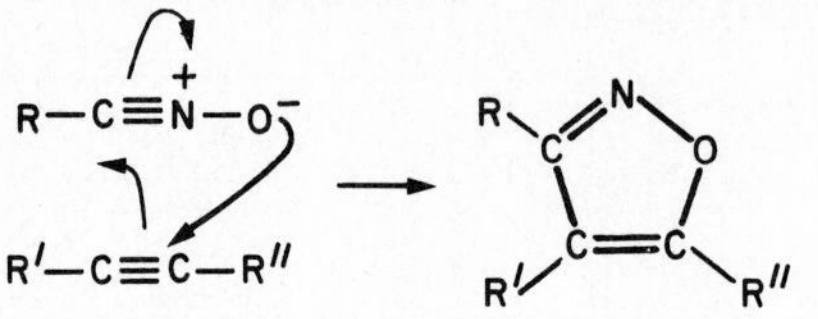

pyrazoles via diazomethane-1,3-dipolar cycloaddition:

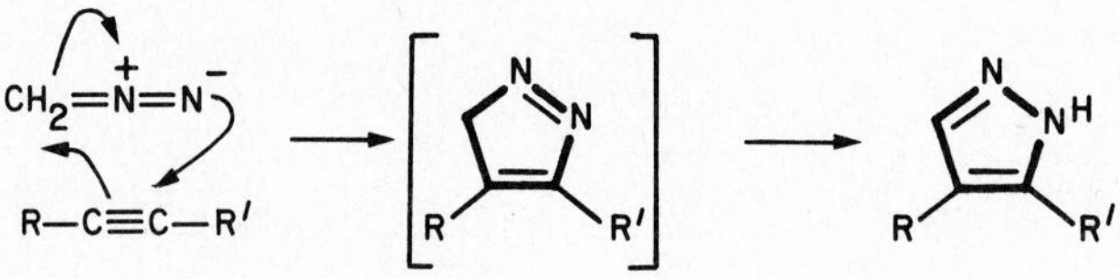

Paal-Knorr-like cyclization:

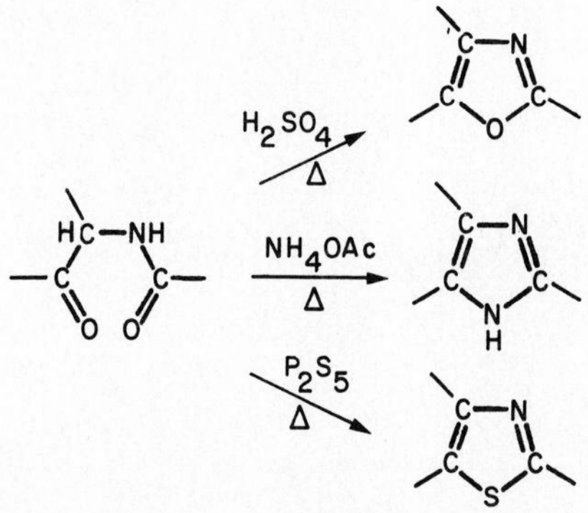

C. Reactions

less reactive toward electrophilic aromatic substitution than the monohetero analogs:

pyrazole > isothiazole > isoxazole

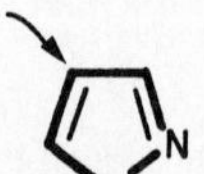

Y = O, S, NH

substitution occurs at C-4

imidazole > thiazole > oxazole

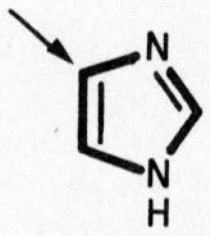

32.6 Pyridine

A. Structure and Physical Properties

basicity *(pK_a of pyridinium ion is 5.2)*

B. Synthesis

Hantzsch pyridine synthesis:

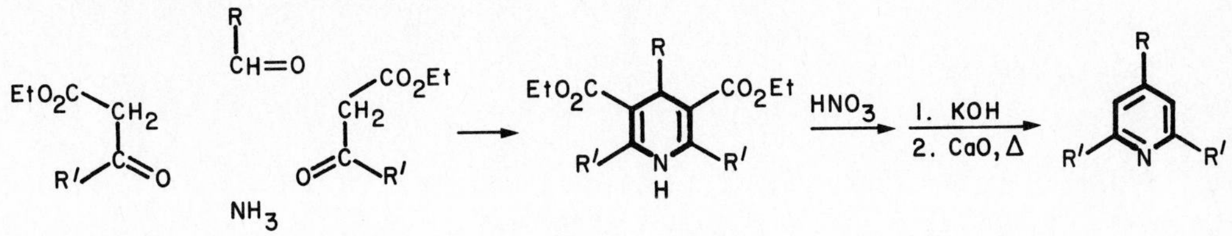

C. Reactions

as a base and nucleophile, and as solvent

resistant to oxidation and electrophilic aromatic substitution:

e.g.,

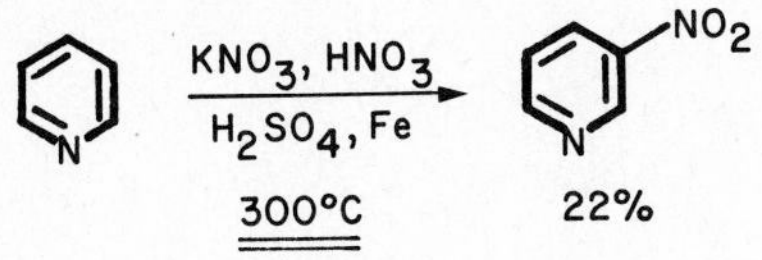

substitution facilitated by activating groups or by N-oxide:

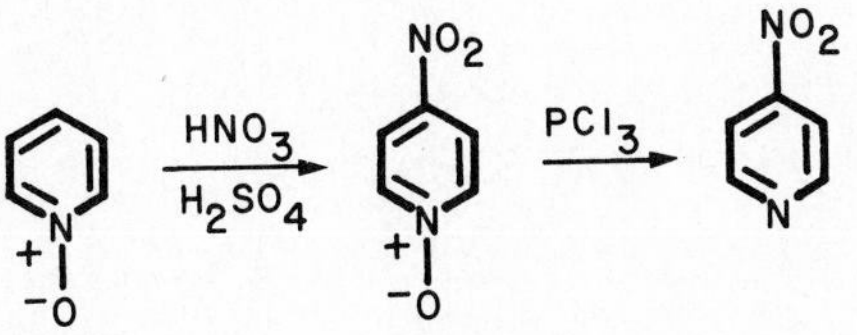

nucleophilic substitution; Chichibabin reaction:

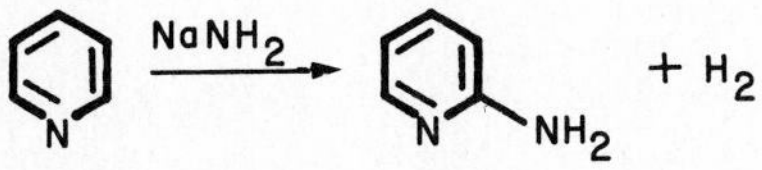

diazotization of aminopyridines to make pyridones:

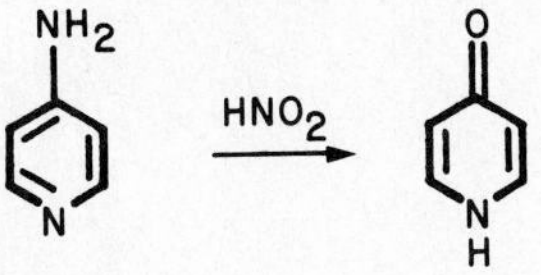

(major tautomer)

oxidation-substitution of N-alkylated pyridinium salts:

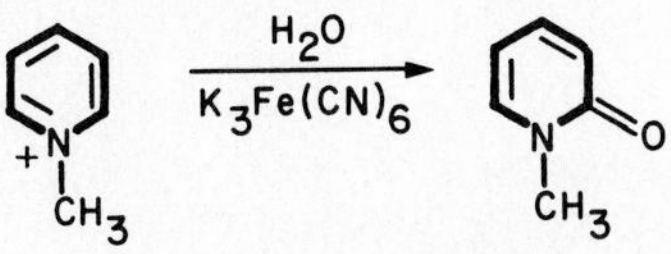

acidity of α- and γ-alkyl groups:

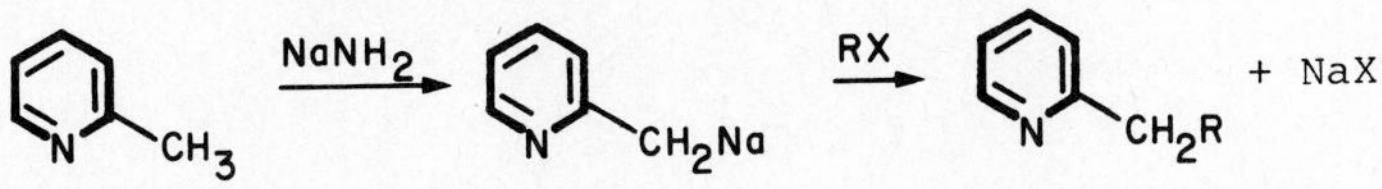

32.7 Quinoline and Isoquinoline

A. Structure and Nomenclature

B. Synthesis

Skraup reaction:

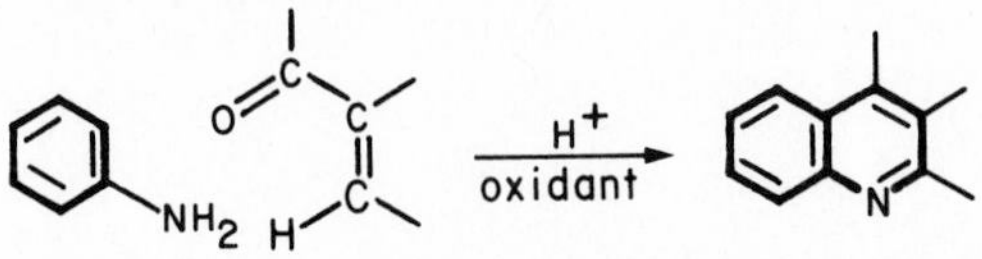

(in Döbner-Miller reaction, the unsaturated carbonyl component is synthesized in situ)

Friedländer synthesis:

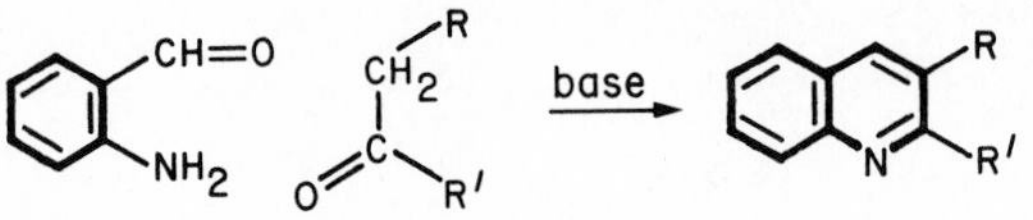

Bischler-Napieralski synthesis:

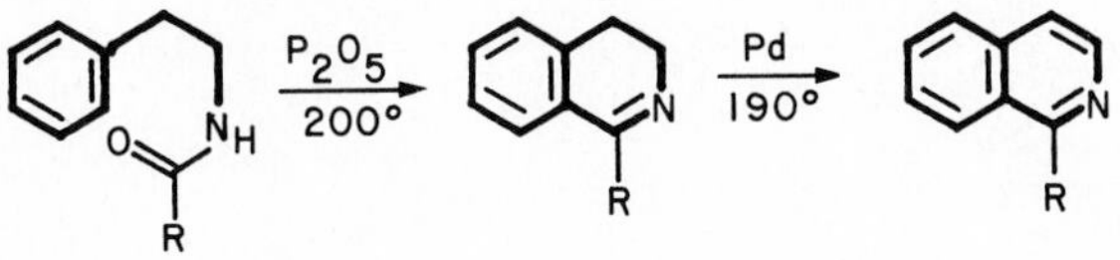

C. Reactions

electrophilic aromatic substitution (avoids pyridine ring):

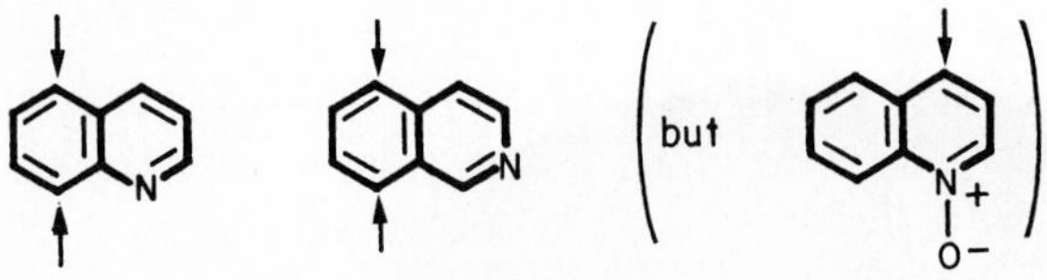

nucleophilic aromatic substitution:

acidity of alkyl derivatives:

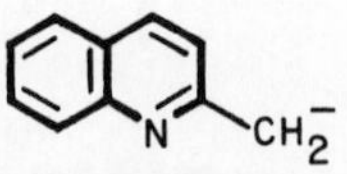

32.8 Diazines

A. Structure and Occurrence

pyridazine, pyrimidine, pyrazine

purine

nucleic acid components

B. Synthesis

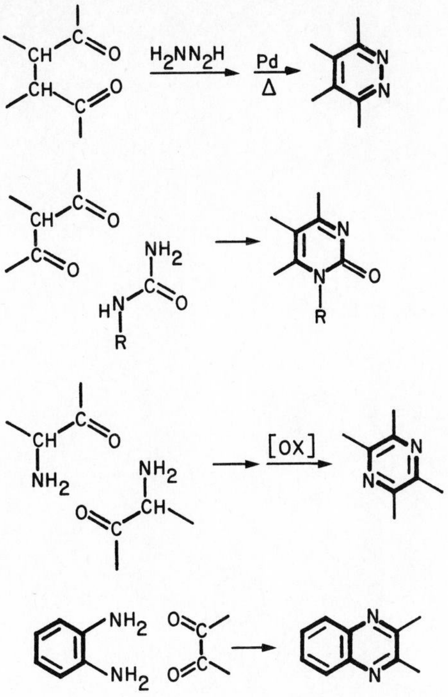

C. Reactions

electrophilic aromatic substitution requires activating groups

nucleophilic aromatic substitution is reasonably easy

32.9 Pyrones and Pyrylium Salts

A. Pyrones

α-pyrone from pyrolysis of malic acid

γ-pyrones from 1,3,5-triketones:

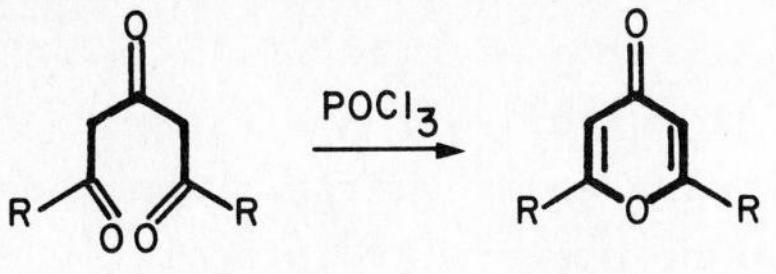

α-pyrones as Diels-Alder dienes:

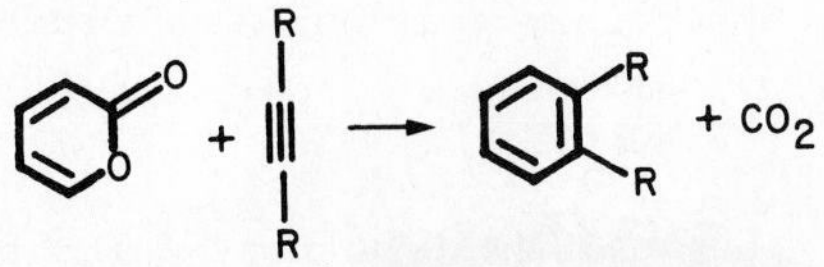

pyridones from γ-pyrones:

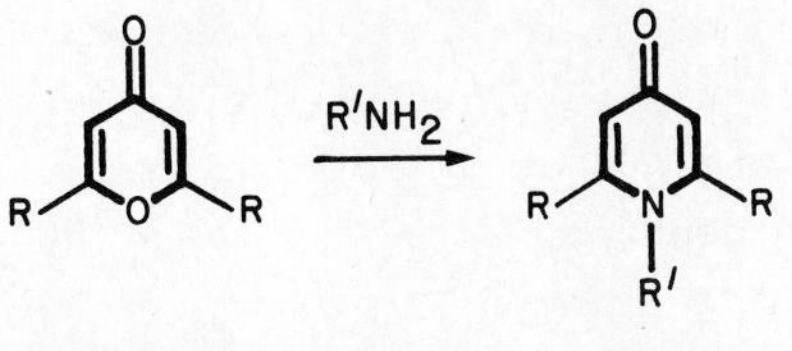

basicity of pyrones:

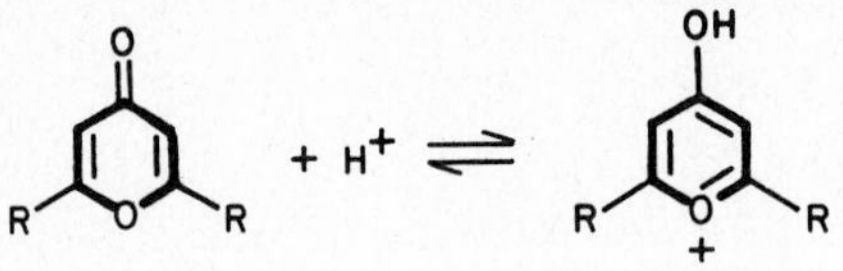

$pK_a = 0.4$

pyrylium salt

B. Pyrylium Salts

from pyrones and a Grignard reagent:

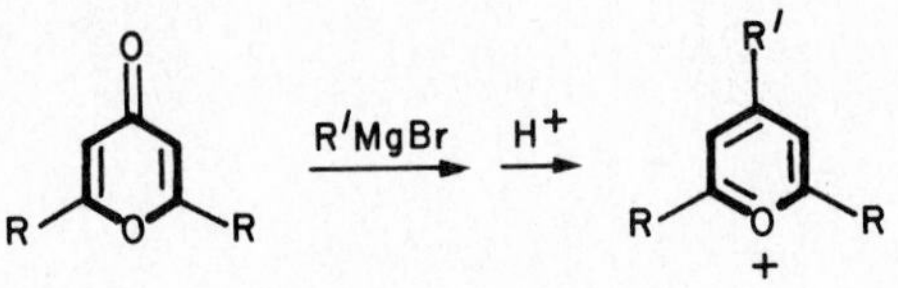

from enone condensation:

e.g.,

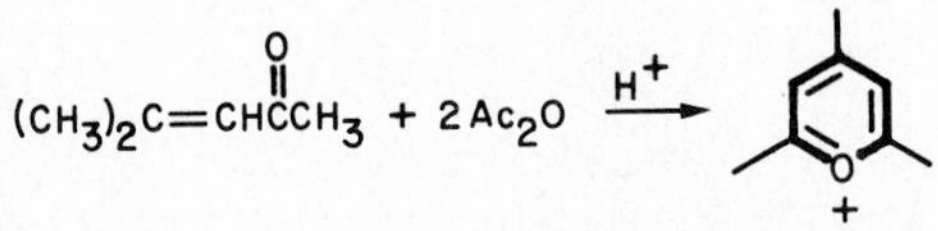

reactions with nucleophiles:

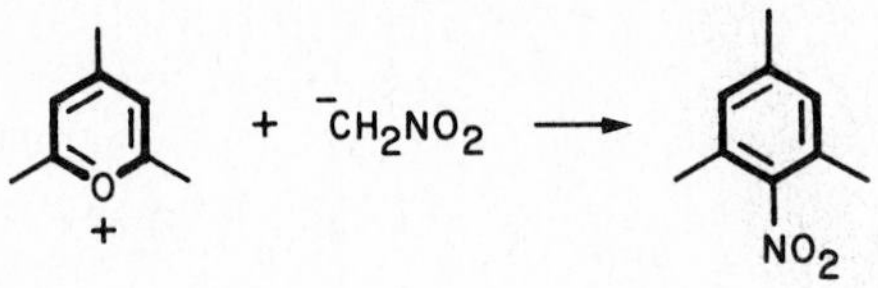

32.B Important Concepts and Hints

There is an astounding amount of material presented in this chapter on heterocyclic compounds. That the chapter is organized as it is reflects both your chemical sophistication as the end of your organic course approaches ("you know more so you can learn more") and the importance and breadth of the field of heterocyclic chemistry itself. Many biologically significant compounds, both naturally-occurring and man-made, are heterocycles, and most organic chemists encounter heterocyclic compounds either directly or indirectly during the course of their research.

The subject is divided between saturated and unsaturated (usually aromatic) heterocycles. The syntheses and reactions of the saturated heterocycles are almost the same as those of acyclic compounds which have the same functional groups. However, special syntheses and greater reactivity are seen for the three- and four-membered ring compounds. The saturated rings themselves are prepared either by hydrogenation of the aromatic analogs or by intramolecular alkylation reactions. (Peracid epoxidation is an exception.)

The classification of aromatic heterocycles encompasses a vast range of compounds, including five- and six-membered and polycyclic systems. and many combinations and orientations of one or more nitrogens, oxygens, and sulfurs (and others!). There is a correspondingly large number of methods for the synthesis of these compounds. A few can be made by cycloaddition reactions, but by far the greatest number arise from condensation reactions. The single most important characteristic of the condensation reactions that produce unsaturated heterocycles is the following: the ring carbons which are directly attached to the heteroatom were originally either carbonyl carbons or were adjacent to carbonyl carbons. To convince yourself of this fact, go through the chapter outline preceding this section of the Study Guide and look for *exceptions* to this generalization. In a sense, the synthesis of aromatic heterocycles is simply another aspect of the chemistry of carbonyl compounds.

32.C Answers to Exercises

(32.2.B)

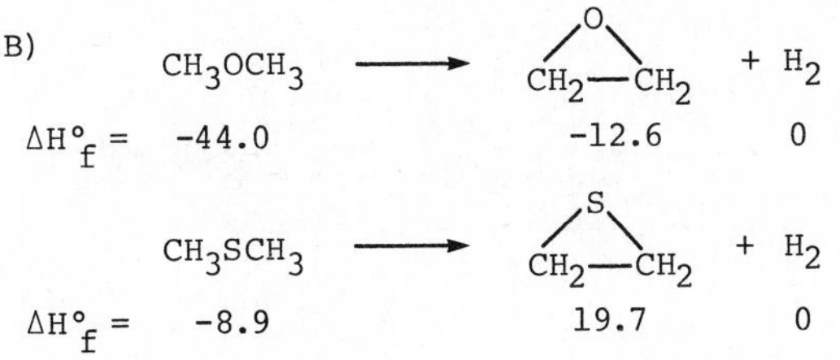

$$CH_3OCH_3 \longrightarrow \underset{CH_2-CH_2}{\overset{O}{\triangle}} + H_2$$

$$\Delta H^\circ_f = \quad -44.0 \qquad\qquad -12.6 \qquad 0 \qquad\qquad \Delta H^\circ = +31.4 \text{ kcal mole}^{-1}$$

$$CH_3SCH_3 \longrightarrow \underset{CH_2-CH_2}{\overset{S}{\triangle}} + H_2$$

$$\Delta H^\circ_f = \quad -8.9 \qquad\qquad 19.7 \qquad 0 \qquad\qquad \Delta H^\circ = +28.6 \text{ kcal mole}^{-1}$$

The thiirane ring is less strained because less distortion of the C–S–C bond angle is required to close the ring. In dimethyl sulfide the C–S–C bond angle is 98.9° (see Section 26.1) vs. the C–O–C bond angle of 111.7° of dimethyl ether (see Section 10.1).

(32.2.C)

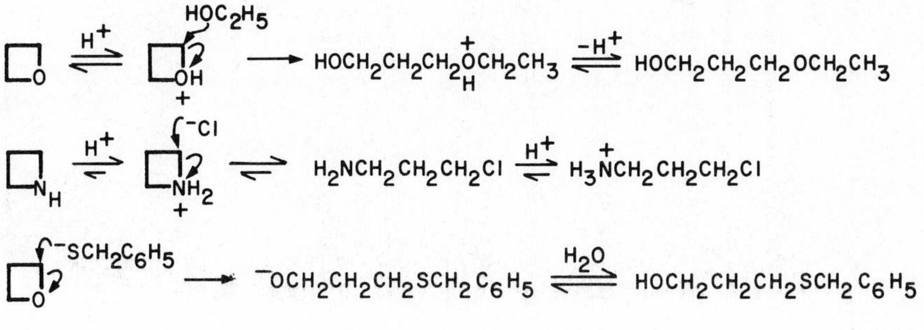

$$\square_O \overset{H^+}{\rightleftharpoons} \square_{\overset{|}{O}H}^+ \overset{HOC_2H_5}{\longrightarrow} HOCH_2CH_2CH_2\overset{+}{O}CH_2CH_3 \overset{-H^+}{\rightleftharpoons} HOCH_2CH_2CH_2OCH_2CH_3$$

$$\square_{N}{}_H \overset{H^+}{\rightleftharpoons} \square_{\overset{|}{N}H_2}^+ \overset{^-Cl}{\rightleftharpoons} H_2NCH_2CH_2CH_2Cl \overset{H^+}{\rightleftharpoons} H_3\overset{+}{N}CH_2CH_2CH_2Cl$$

$$\square_O \overset{^-SCH_2C_6H_5}{\longrightarrow} {}^-OCH_2CH_2CH_2SCH_2C_6H_5 \overset{H_2O}{\rightleftharpoons} HOCH_2CH_2CH_2SCH_2C_6H_5$$

(32.3.A)

	H° (kcal mole^{-1})	Aromatic Stabilization (kcal mole^{-1})

$$\underset{\substack{C \\ H\ H}}{\text{(cyclopentadiene)}} + 2\ H_2 \longrightarrow \text{(cyclopentane)}$$

$$\Delta H^\circ_f = \quad 31.9 \qquad\qquad 0 \qquad\qquad -18.4 \qquad\qquad -50.3 \qquad\qquad (0)$$

$$\underset{O}{\text{(furan)}} + 2\ H_2 \longrightarrow \underset{O}{\text{(tetrahydrofuran)}}$$

$$\Delta H^\circ_f = \quad -8.3 \qquad\qquad 0 \qquad\qquad -44.0 \qquad\qquad -35.7 \qquad\qquad 14.6$$

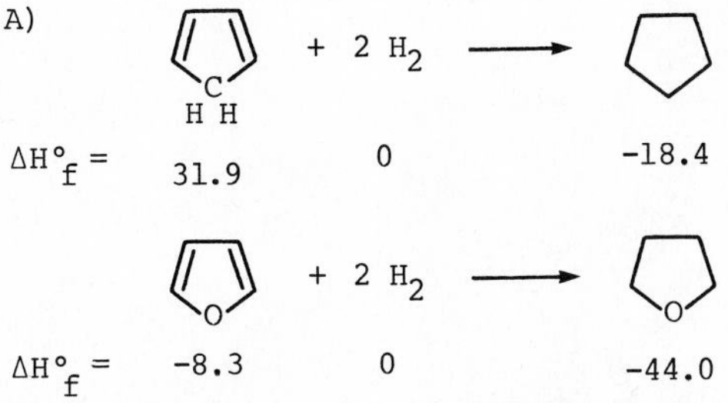

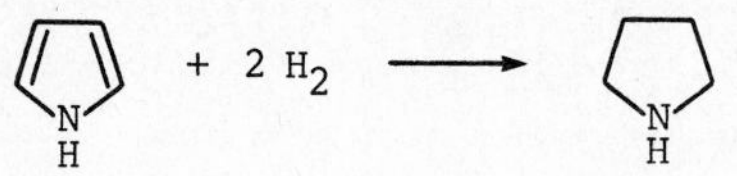

ΔH°_f = 25.9 0 -0.8

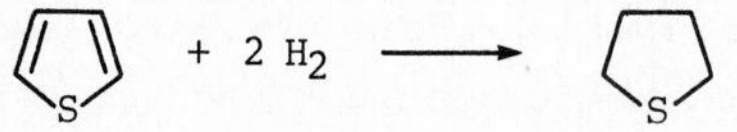

ΔH°_f = 27.6 0 -8.1

	ΔH° (kcal mole^{-1})	Aromatic Stabilization (kcal mole^{-1})
Pyrrole + 2 H_2	-26.7	23.6
Thiophene + 2 H_2	-35.7	14.6

The resonance structures which contribute to aromatic stabilization are those which involve a positive charge on the heteroatom:

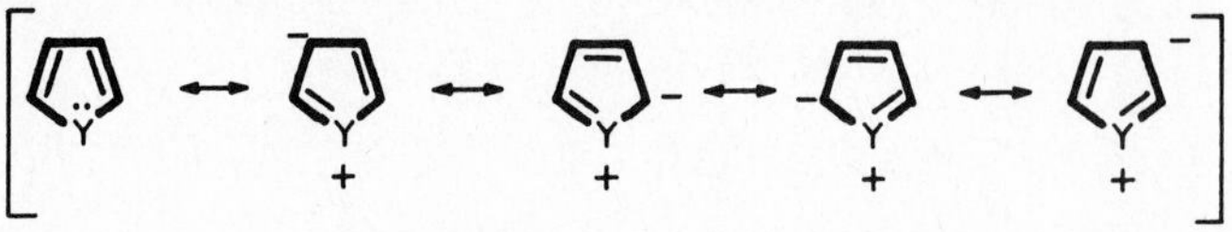

This is easier for the more basic nitrogen atom than for oxygen or sulfur.

(32.3.B) (b)

32-6:

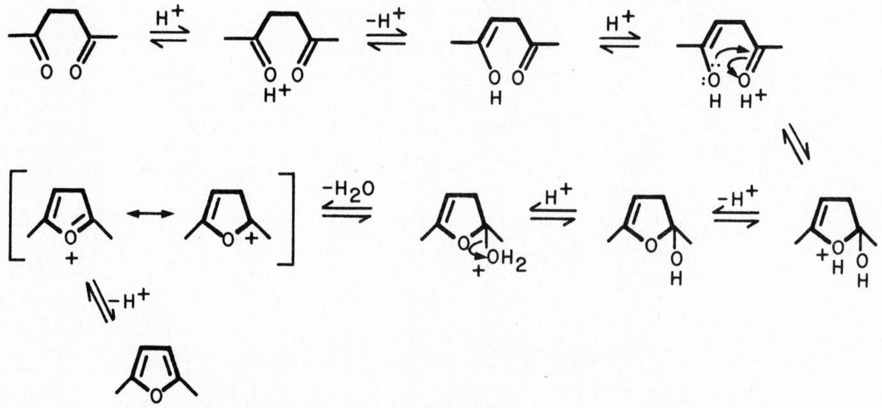

32-7:

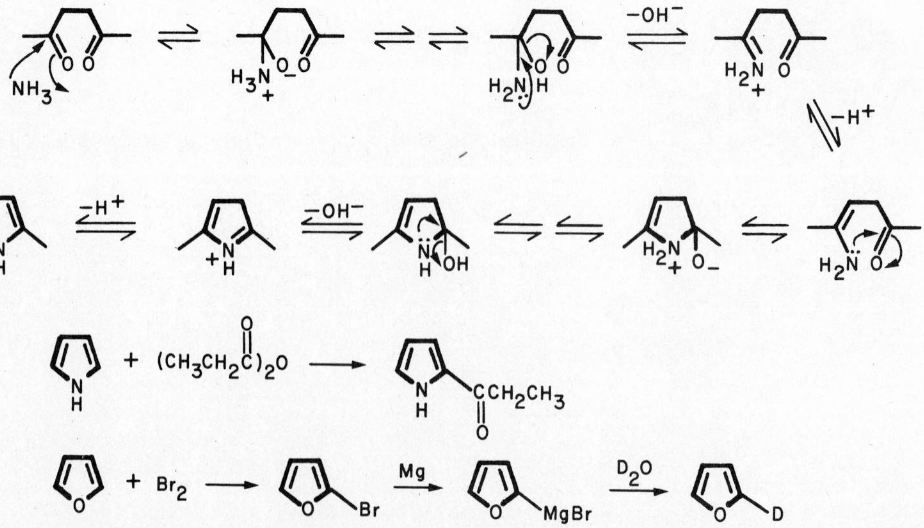

(32.3.C)

(32.4.C)

requires loss of aromaticity of benzene ring in order to stabilize positive charge with nitrogen

benzene ring remains intact

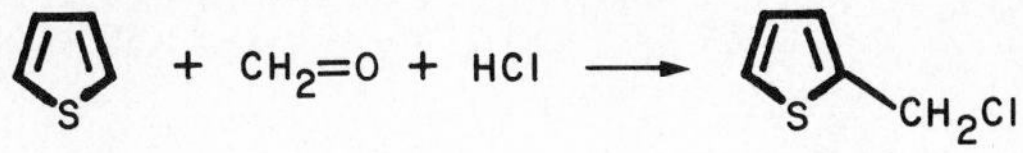

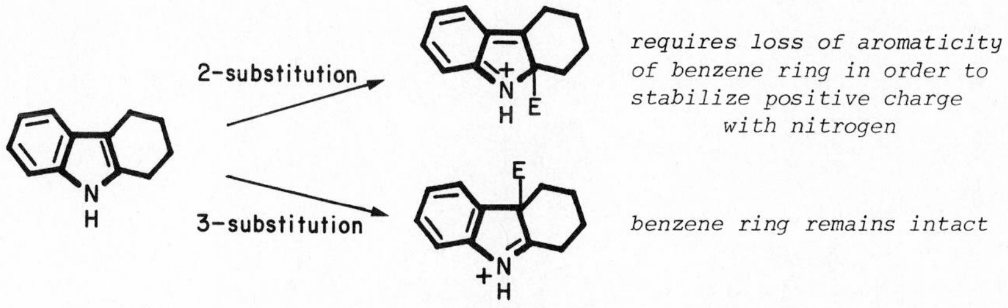

(32.5.B)

32-8:

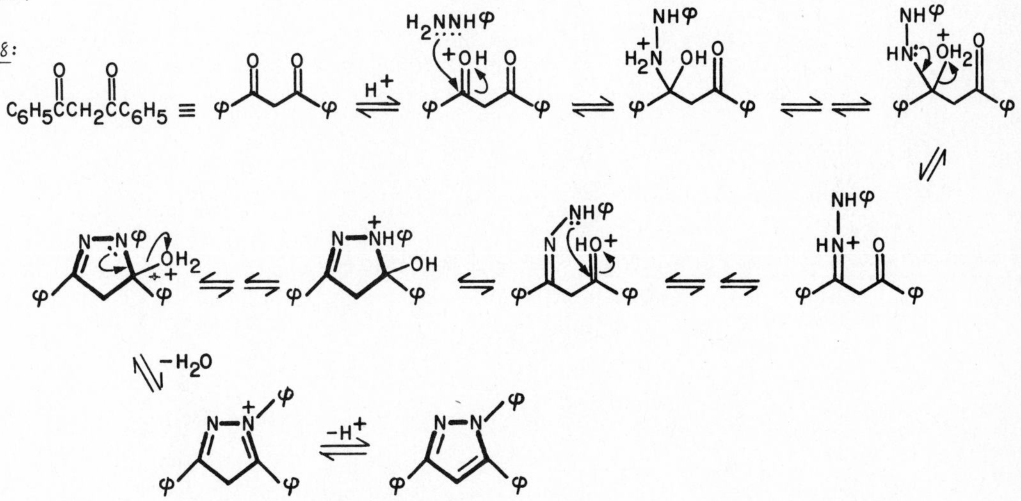

32-9:

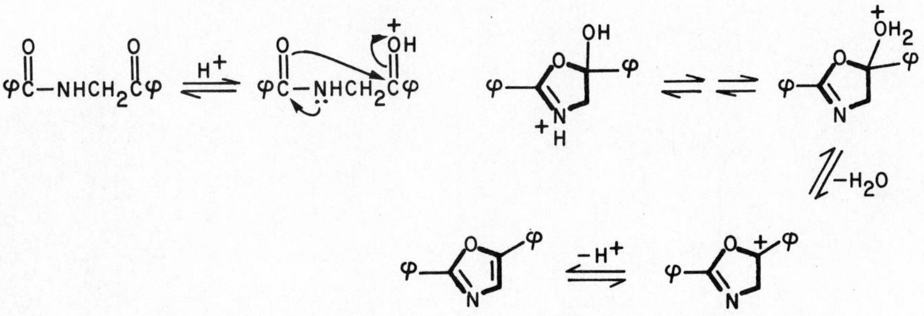

32-10:

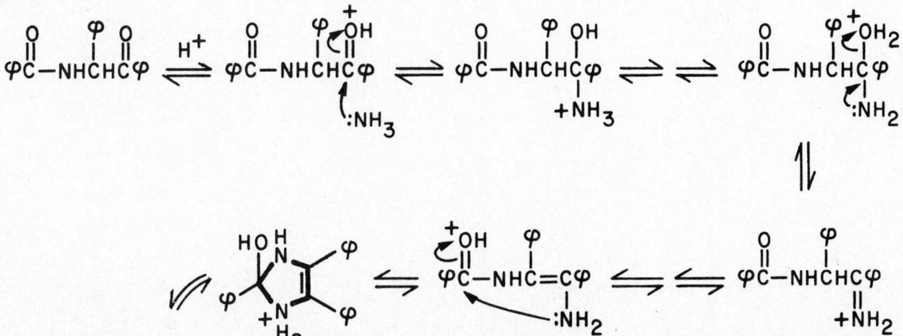

[cont'd...]

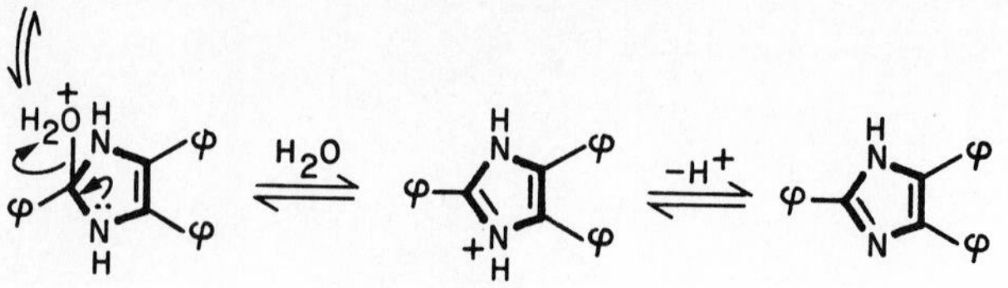

(32.5.C)

4-substitution:

5-substitution:

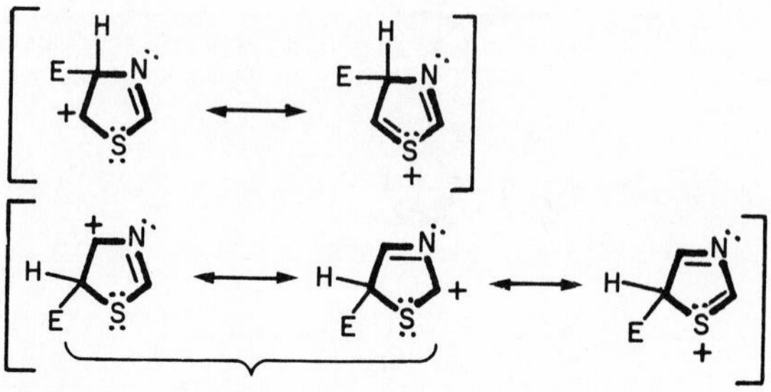

both of these are poor because the positive charge is next to an electron-withdrawing nitrogen

(32.6.B)

The proton source is NH_4^+; the base is $:NH_3$

"Enamine"

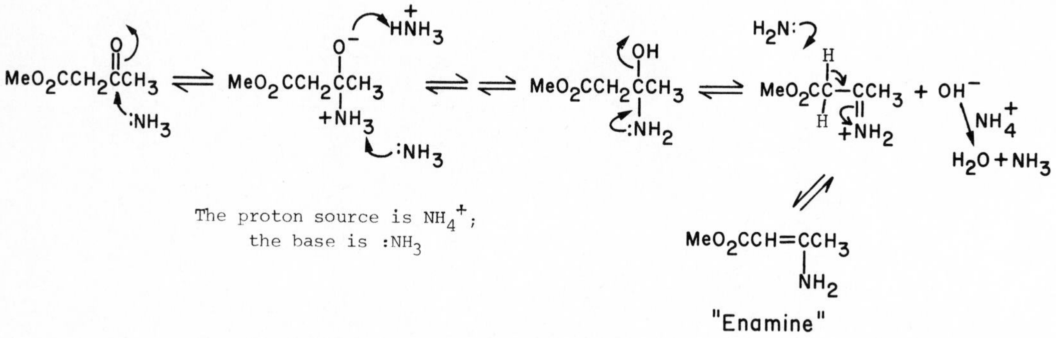

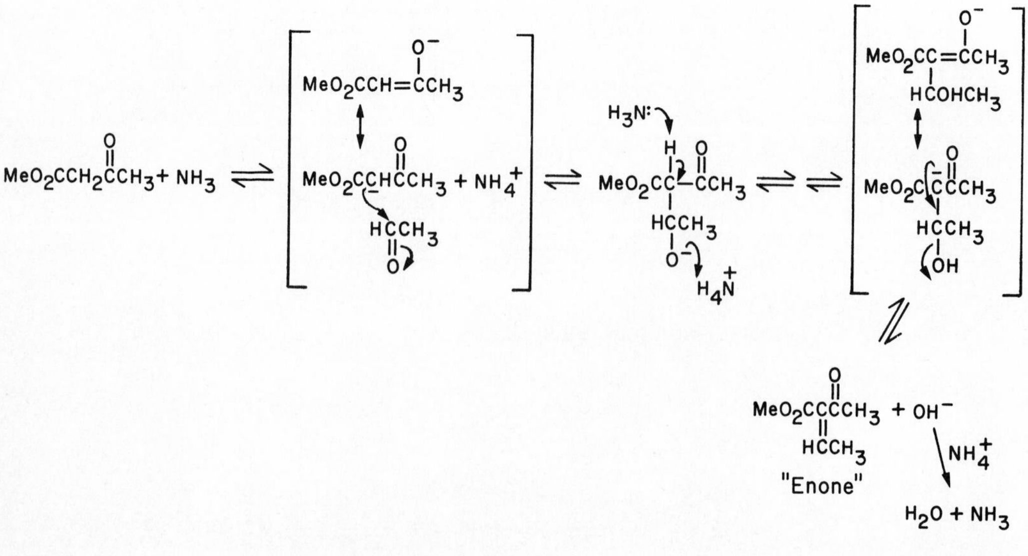

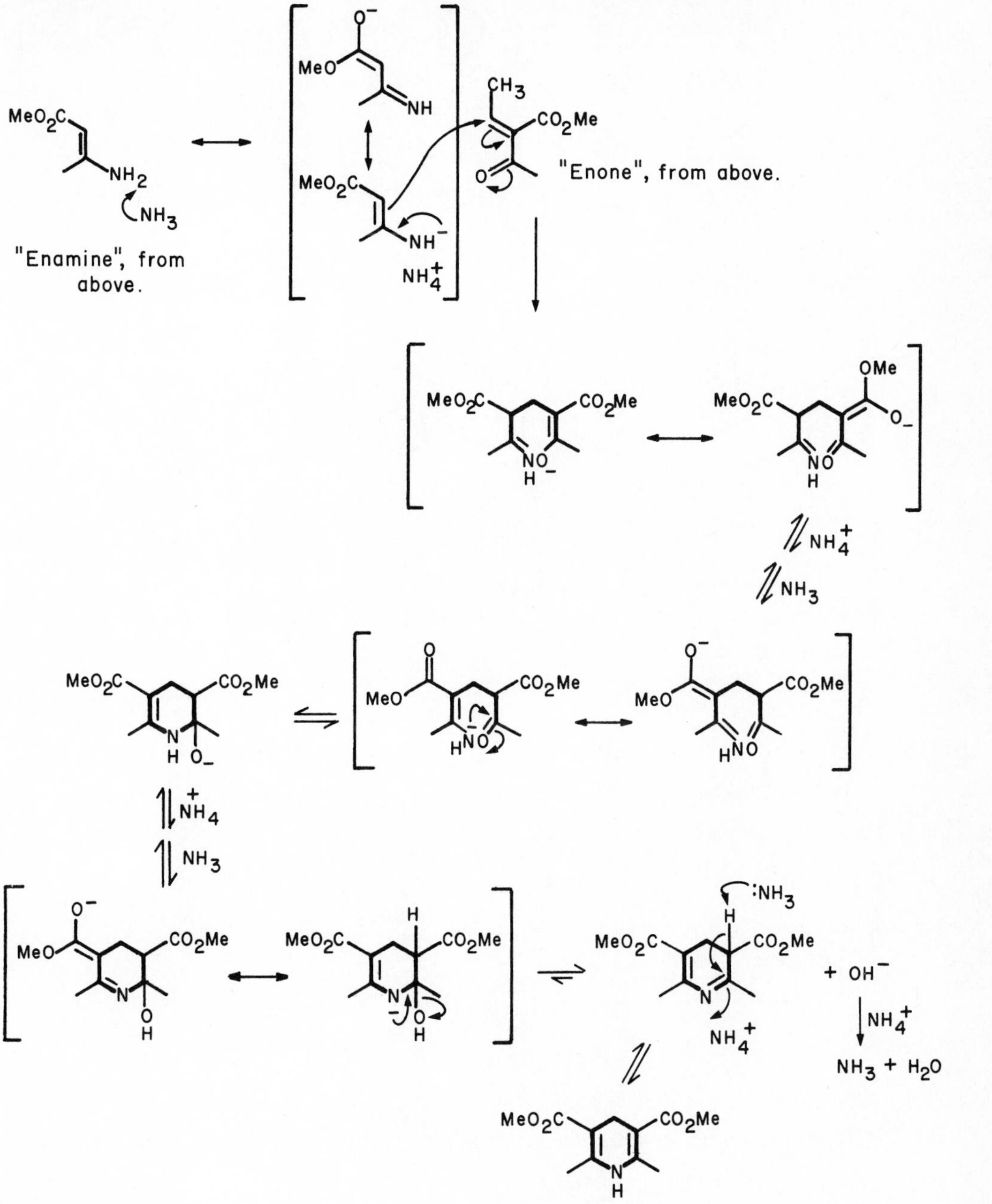
"Enamine", from above.
"Enone", from above.

(32.6.C)

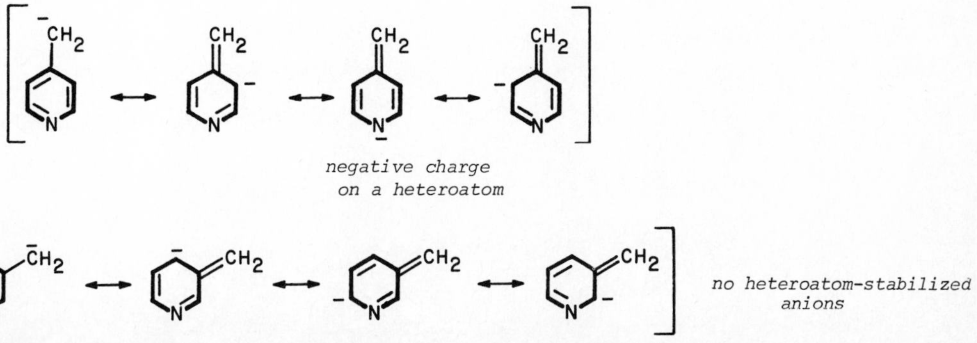
negative charge on a heteroatom
no heteroatom-stabilized anions

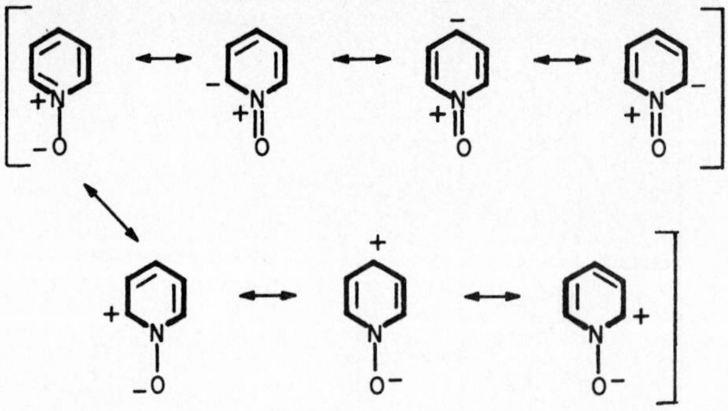

(32.7.B)

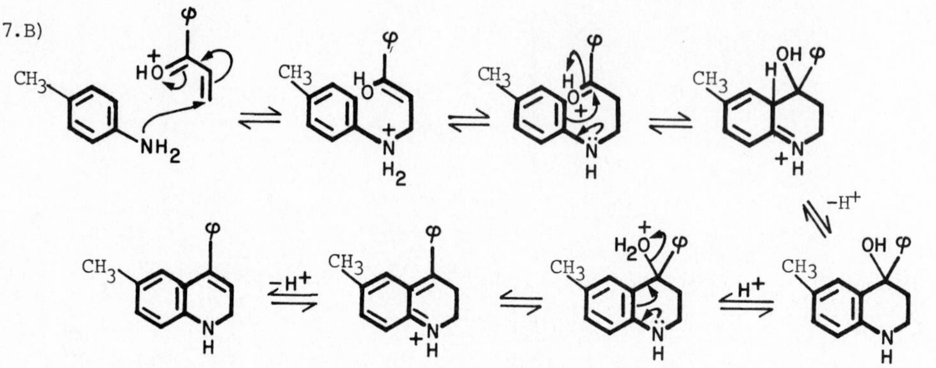

(32.7.C)

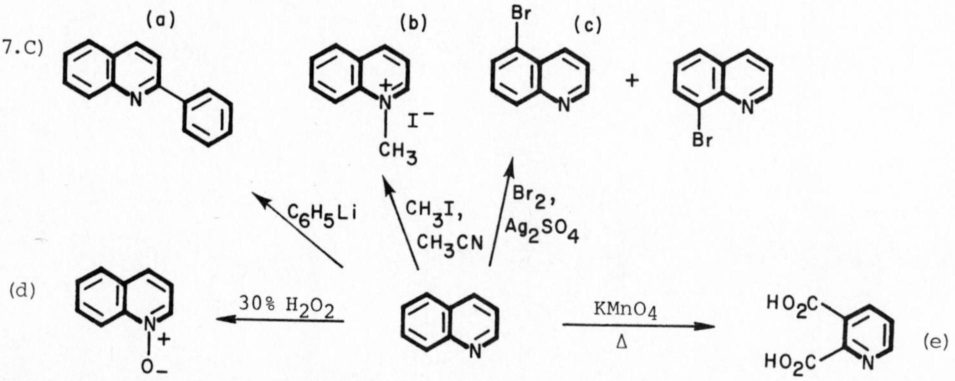

(32.9.A)

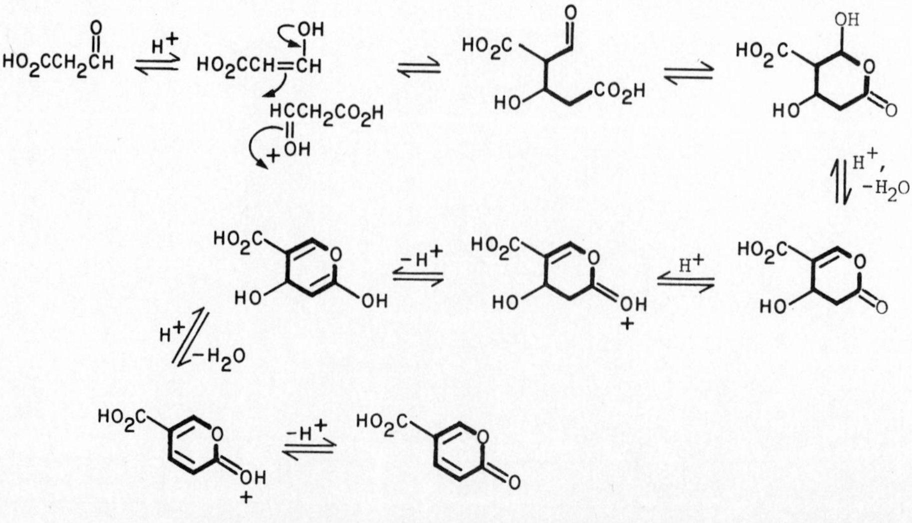

(32.9.B)

32-12:

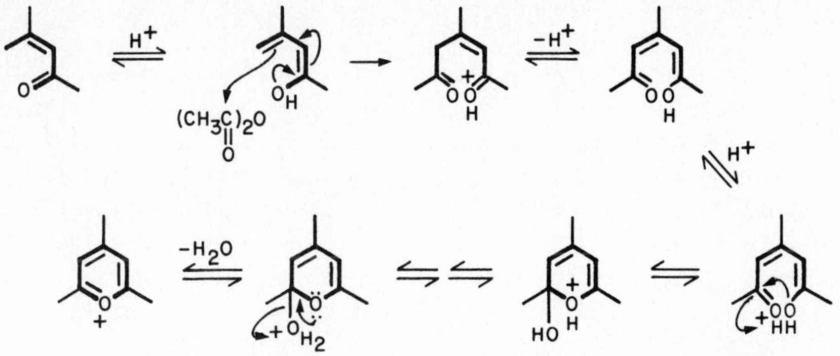

32-13:

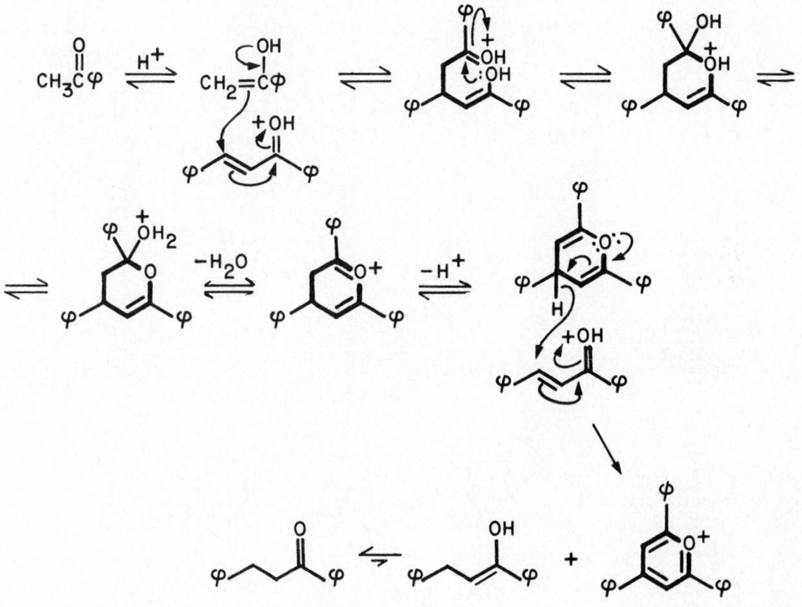

32-14:

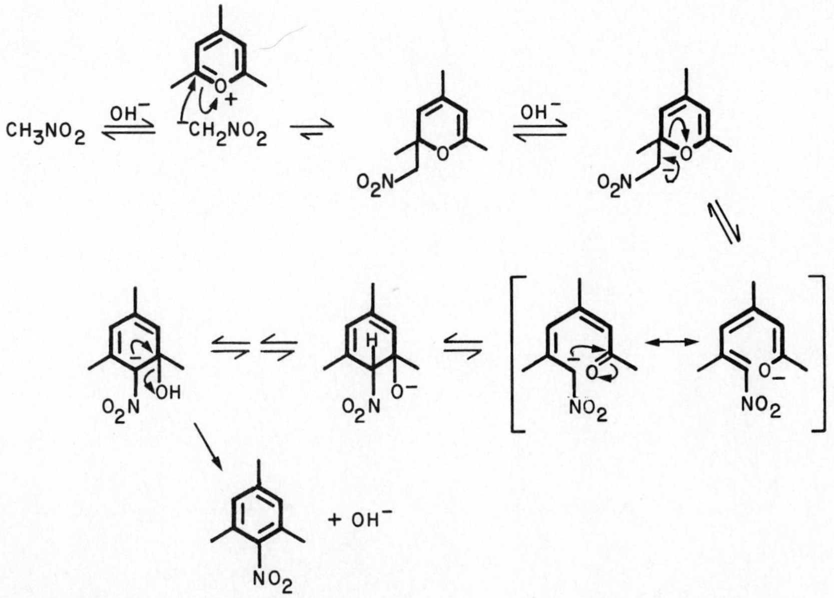

32.D Answers and Explanations for Problems

1. (a) 3-methyltetrahydropyran
   (b) 3-azetidinone
   (c) 2-methyl-2-ethyloxirane
   (d) 2-nitro-3-bromofuran
   (e) 5-chloro-2-furoic acid
   (f) 2-aminothiazole
   (g) 4-methyl-3-isoxazolecarboxylic acid
   (h) 5-nitroisothiazole
   (i) 4-nitro-1-phenylimidazole
   (j) 6-bromoindole-3-carboxylic acid
   (k) 3-pyridinecarboxylic acid
   (ℓ) 4-methylpyridine oxide
   (m) 7-chloro-1-methylisoquinoline
   (n) 2,3-dimethylquinoline
   (o) 2-amino-4-methylpyrimidine
   (p) 3,6-dimethylpyrazine
   (q) 3-chlorobenzofuran
   (r) 2-(2-hydroxyethyl)thiophene
   (s) 2-(4-methoxyphenyl)-6-phenyl-1,4-pyrone
   (t) 4-*t*-butyl-2,6-dimethylpyrylium tetrafluoroborate

2. (a) 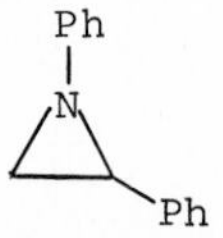

(b) 

(c) 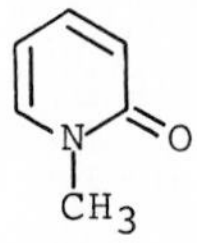

(d) 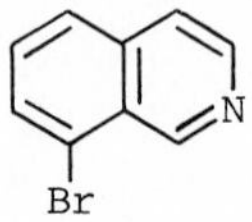

(e) 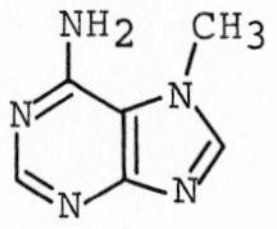

(f) 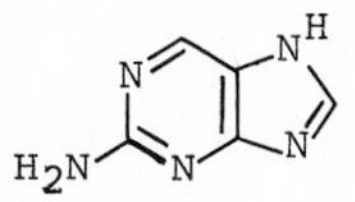

(g) 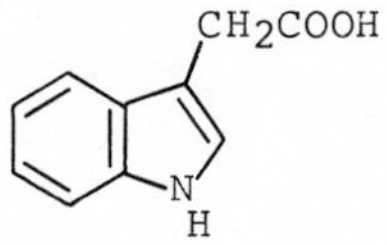

(h) 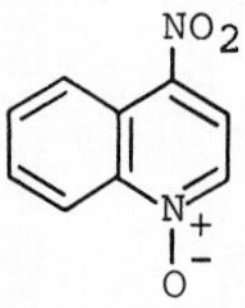

(i) 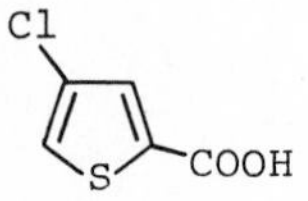

(j) 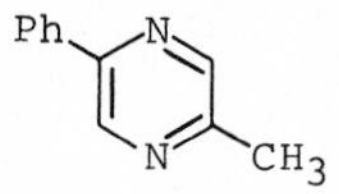

(k) 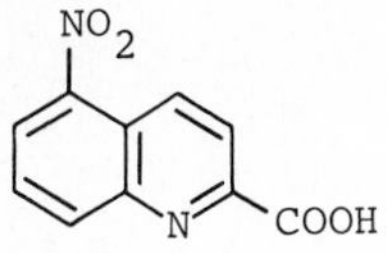

(l) 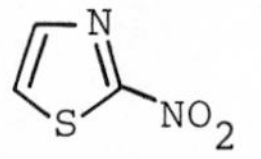

(m) 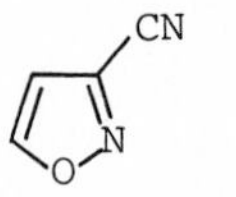

(n) 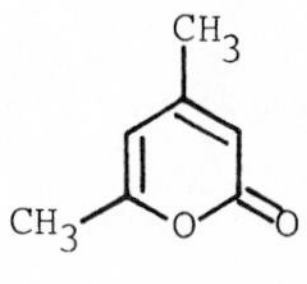

3. (a) 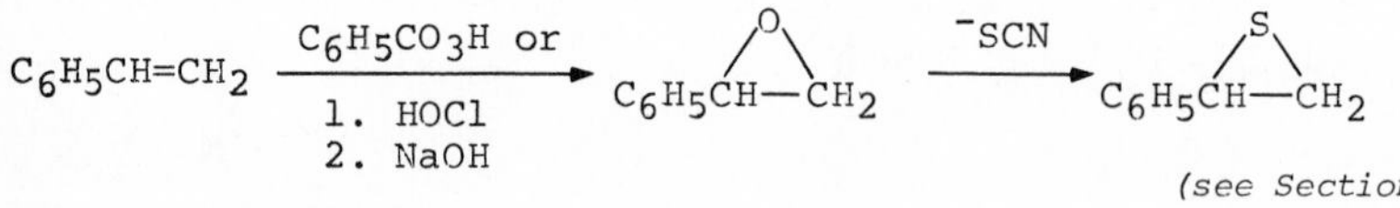

(see Section 32.2.B)

(b) 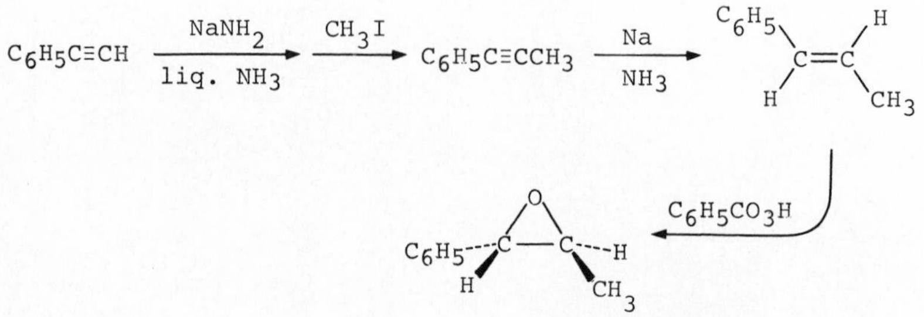

(c)

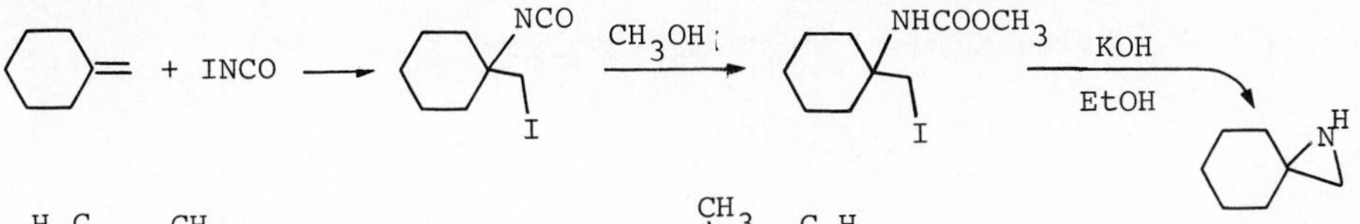

(d)

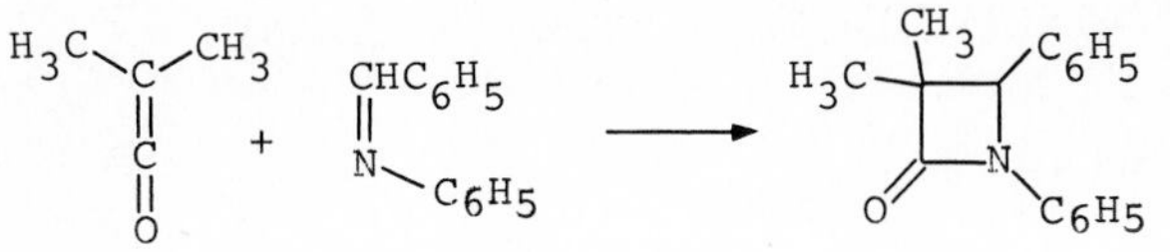

NOTE: $(CH_3)_2CBrCOBr \xrightarrow{Zn} (CH_3)_2C{=}C{=}O$ *and* $C_6H_5NH_2 + O{=}CHC_6H_5 \longrightarrow C_6H_5N{=}CHC_6H_5$

(Section 20.3.C)

(e)

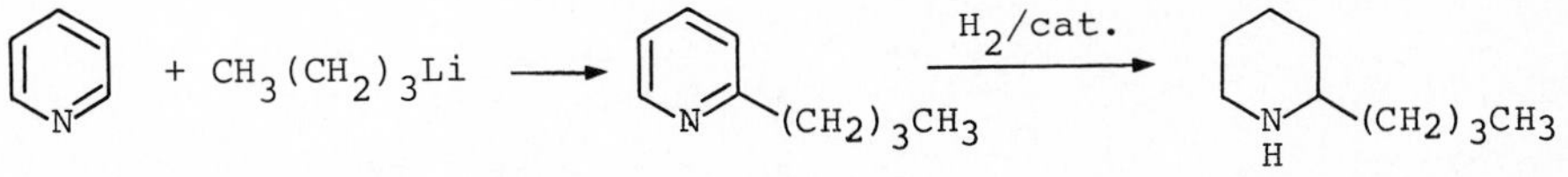

(f)

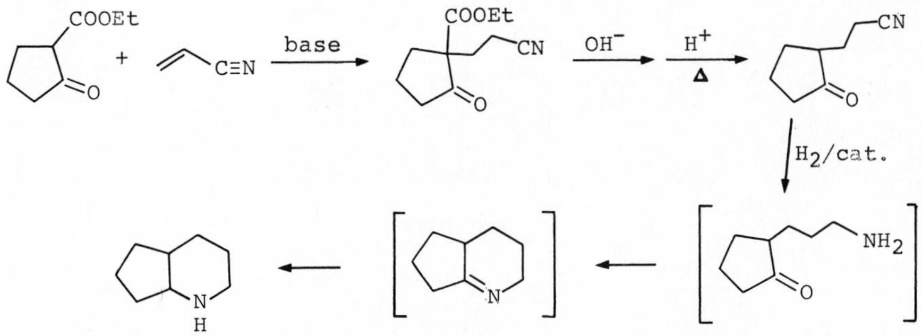

(compare Section 24.6.F)

4. In these syntheses, note the type of heterocyclic ring system present and use the appropriate synthetic route.

   (a) This problem requires a 1,4-diketone:

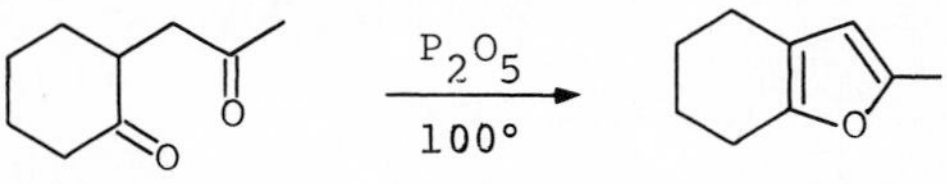

The 1,4-diketone may be prepared in several ways. One possibility is:

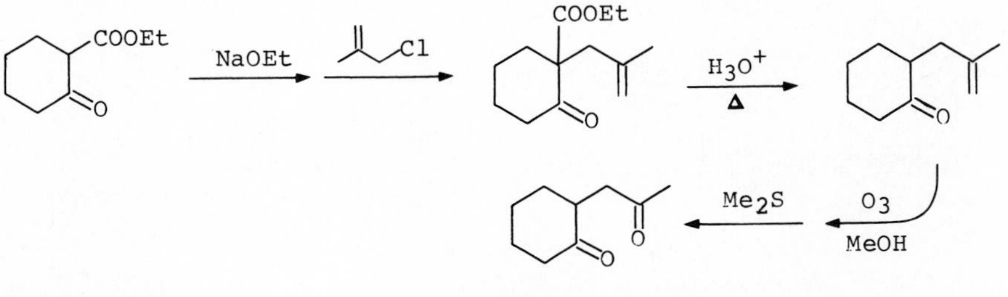

   (b) Knorr pyrrole synthesis *(Section 32.3.B)*:

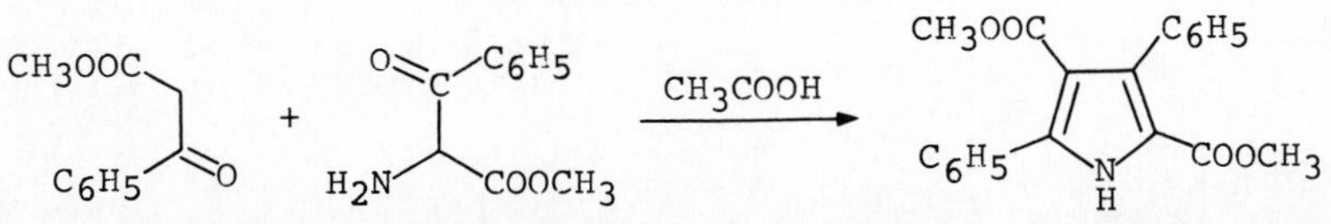

$$C_6H_5COOCH_3 + CH_3COOCH_3 \xrightarrow[NaOCH_3]{CH_3OH} C_6H_5COCH_2COOCH_3 \xrightarrow{HONO} C_6H_5CO\overset{\overset{\displaystyle NOH}{\|}}{C}COOCH_3$$

$$\underset{C_6H_5CO\overset{\overset{\displaystyle NH_2}{|}}{C}HCOOCH_3}{} \xleftarrow[H^+, \Delta]{MeOH} \underset{C_6H_5CO\overset{\overset{\displaystyle NHCOCH_3}{|}}{C}HCOOCH_3}{} \xleftarrow[(CH_3CO)_2O]{H_2/Pt}$$

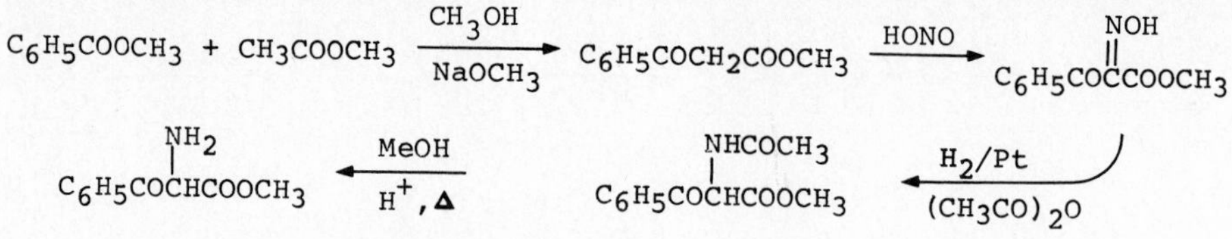

(c) Fischer indole synthesis *(Section 32.4.B)*:

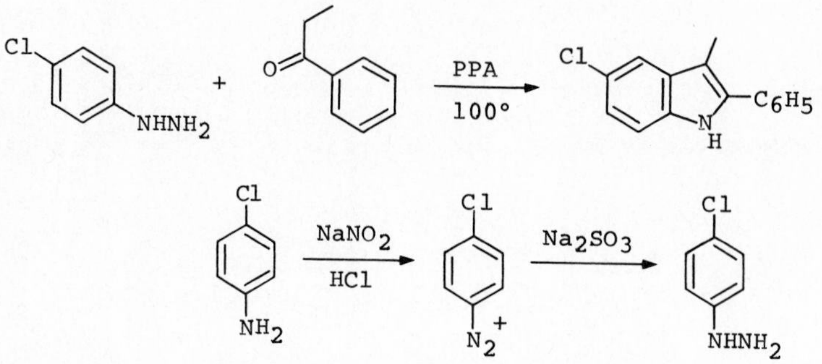

(d)

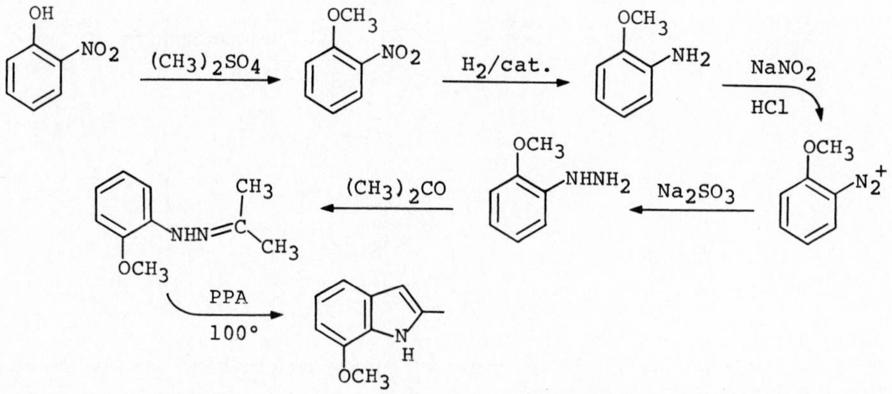

(e)

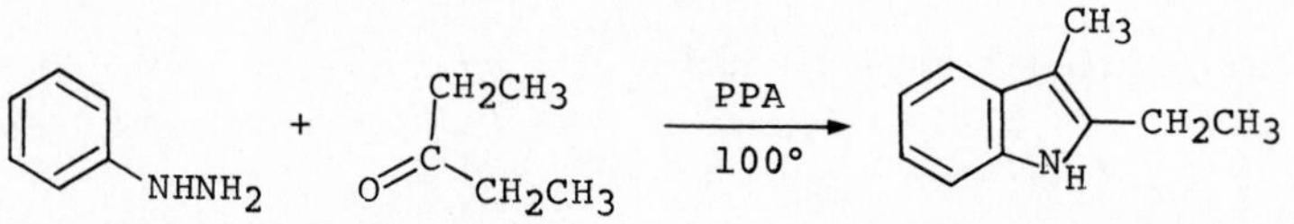

(f) hydroxylamine + 1,3-dicarbonyl compound *(Section 32.5.B)*:

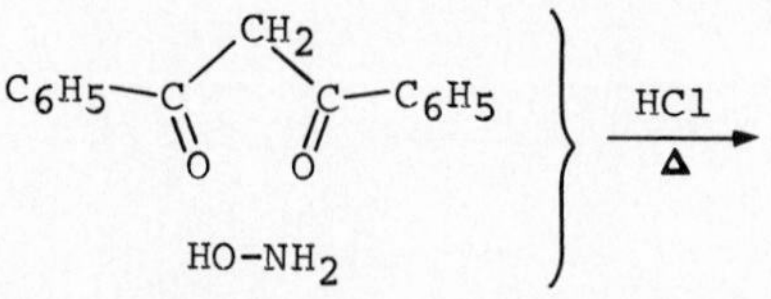

β-Diketones are prepared from esters + ketones:

$$C_6H_5COCH_3 + EtOOCC_6H_5$$
$$\xrightarrow[EtOH]{EtO^-} C_6H_5COCH_2COC_6H_5$$

(g) β-diketone + hydrazine:

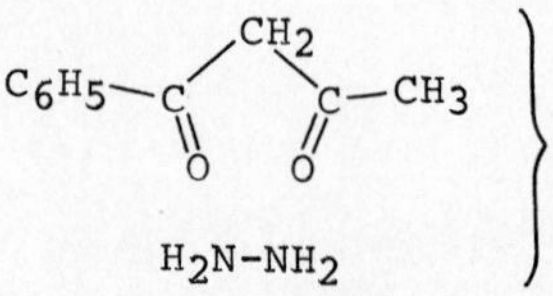

Note that unsymmetrical pyrazoles can be prepared because hydrazine is symmetrical. With hydroxylamine, this β-diketone would give a mixture of isoxazoles.

The β-diketone can be made by: $C_6H_5COCH_3$ + CH_3COOEt $\xrightarrow{EtO^-}$

or: C_6H_5COOEt + CH_3COCH_3 $\xrightarrow{EtO^-}$

(h) This isoxazole is unsymmetrical, and the required β-diketone is hard to make.
An alternative preparation is a cycloaddition with nitrile oxides
(see Section 32.5.B):

$C_6H_5CH=NOH$ $\xrightarrow{Cl_2}$ $C_6H_5CCl=NOH$ $\xrightarrow{NaOH}$ $C_6H_5C\equiv\overset{+}{N}-O^-$ $\xrightarrow{CH_3OOCC\equiv CCOOCH_3}$

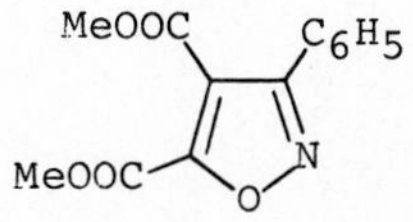

(i) Here also, the β-dicarbonyl approach
does not look promising. An alternative
preparation uses diazomethane *(Section 32.5.B):*

$CH_3C\equiv CCOOCH_3$
$CH_2=\overset{+}{N}=N^-$ $\left.\right\}$ $\xrightarrow[0°]{ether}$ $\left[\text{image}\right]$ $\longrightarrow$

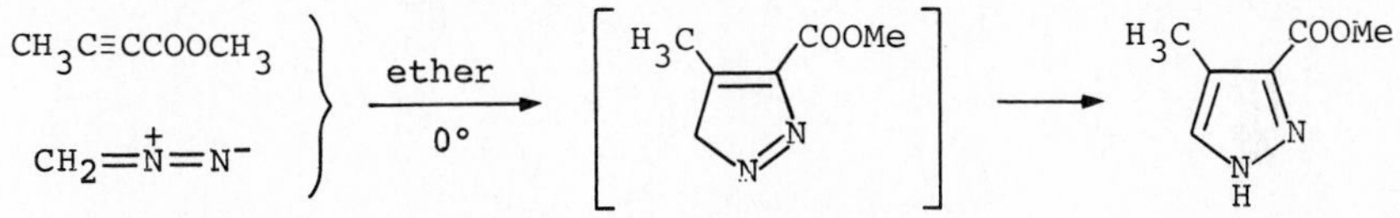

The acetylene compound may be prepared by:

$CH_3C\equiv CH$ $\xrightarrow{RMgX}$ $CH_3C\equiv CMgX$ $\xrightarrow{CO_2}$ $\xrightarrow{H^+}$ $CH_3C\equiv CCOOH$ $\xrightarrow[H^+]{CH_3OH}$

(j) Don't be fooled by the way this compound is written. Imidazoles
are in rapid tautomeric equilibrium (remember their basicity):

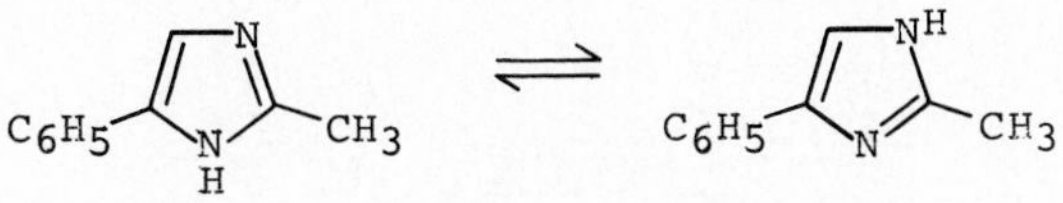

The Paal-Knorr cyclization can be designed in two ways *(Section 32.5.B):*

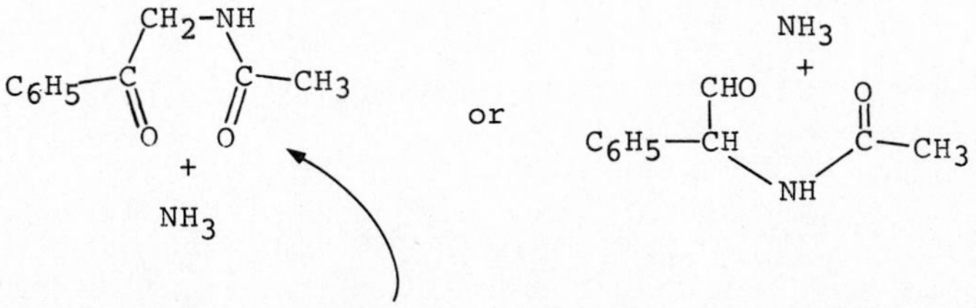

This is a better approach since it does not involve a sensitive aldehyde. Ketones are better than aldehydes in all of these cyclizations.

$C_6H_5COCH_2NH_2$ + $(CH_3CO)_2O$ $\longrightarrow$ $C_6H_5COCH_2NHCOCH_3$ $\xrightarrow[\substack{CH_3COOH \\ 120°}]{NH_4^+ \ OAc^-}$

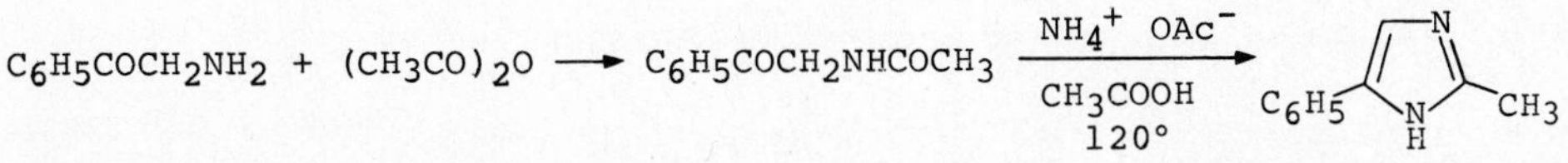

(k) Hantzsch pyridine synthesis *(Section 32.6.B)*:

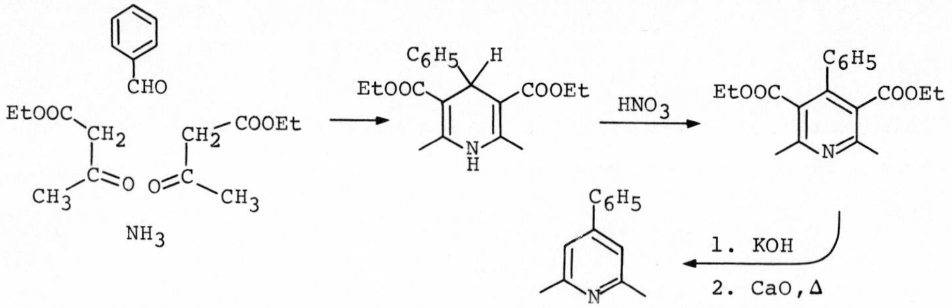

(l) Skraup reaction *(Section 32.7.B)*:

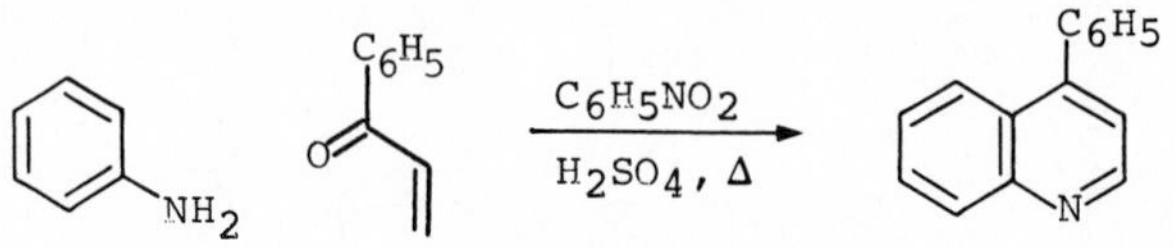

(m) Skraup reaction with:

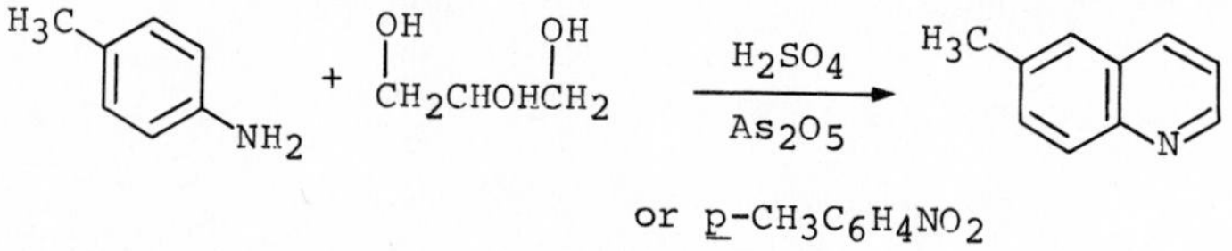

(n) This kind of quinoline is best prepared by the Friedländer method *(Section 32.7.B)*:

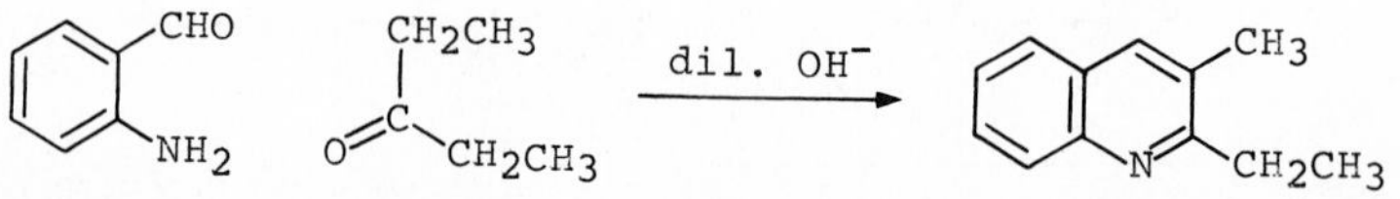

(o) Isoquinolines are prepared by Bischler-Napieralski synthesis *(Section 32.7.B)*:

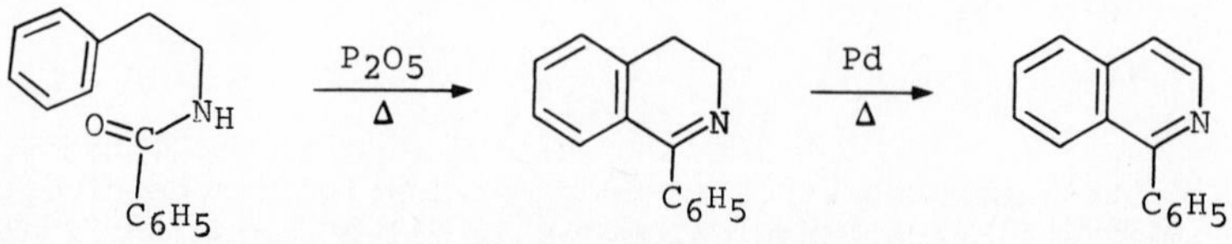

The amide is prepared from $C_6H_5CH_2CH_2NH_2$ + $ClC(=O)-C_6H_5$ (benzoyl chloride)

$C_6H_5CH_2Cl + CN^- \longrightarrow C_6H_5CH_2CN \xrightarrow[\text{or } H_2/\text{cat.}/NH_3]{LiAlH_4}$ (to $C_6H_5CH_2CH_2NH_2$)

(p) Section 32.8.B:

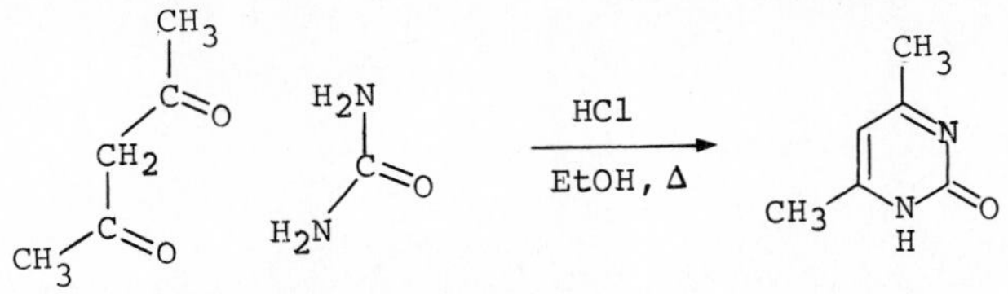

(q) Pyrazines can be prepared by dimerization of α-aminoketones *(Section 32.8.B)*, but this method is useful only for symmetrical pyrazines. This example is symmetrical:

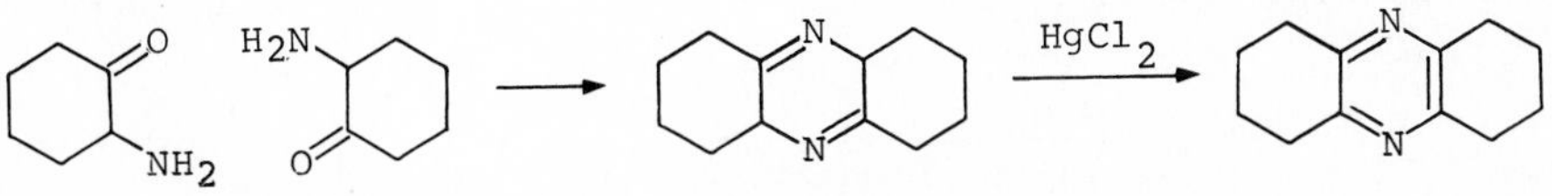

(r) This unsymmetrical pyrazine is of the quinoxaline type *(Section 32.8.B)*:

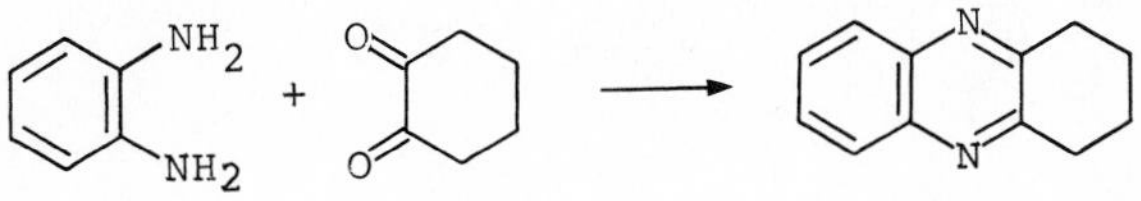

The diketone can be prepared in several ways; one method is given in Section 27.7.A.

(s) This compound is a barbituric acid derivative, prepared from a β-keto ester and urea.

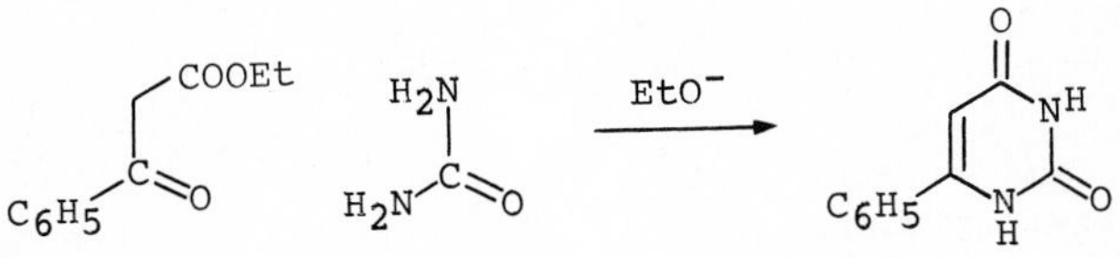

(t)

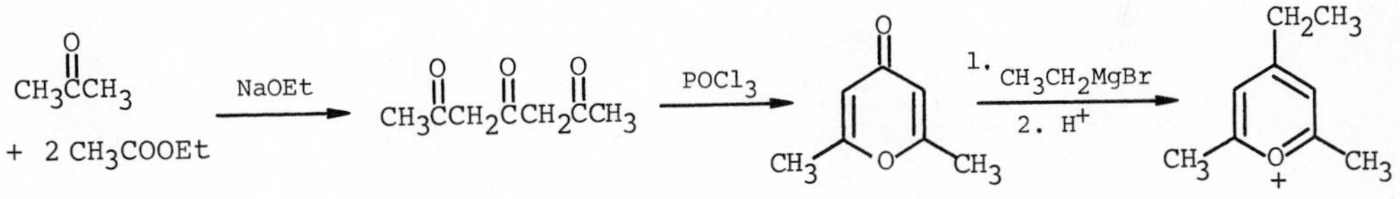

(u)

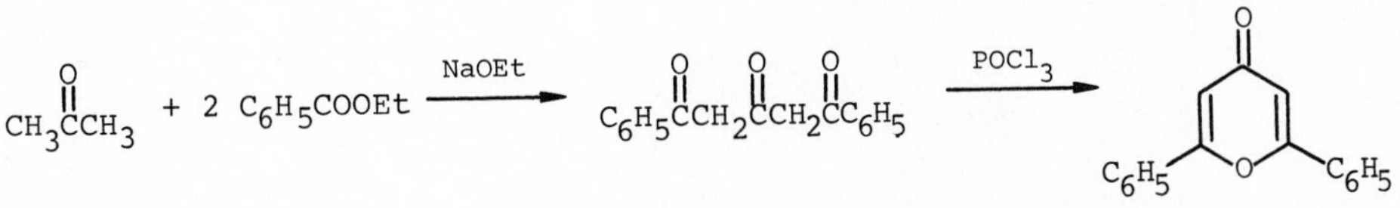

5. (a)

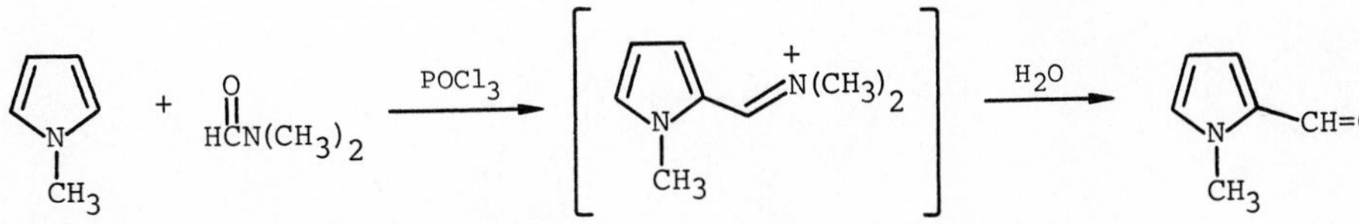

(b)

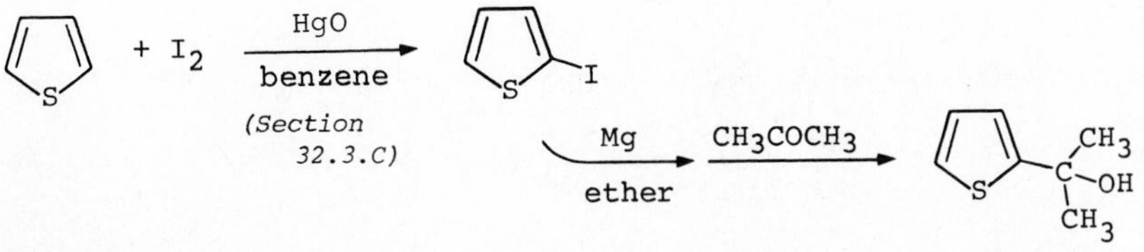

(c)

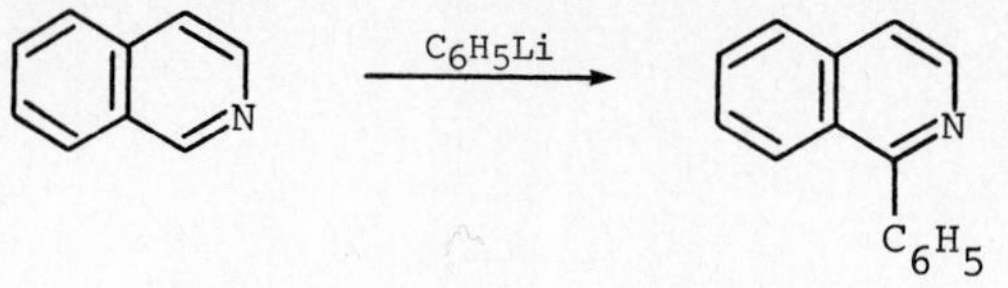

(d)

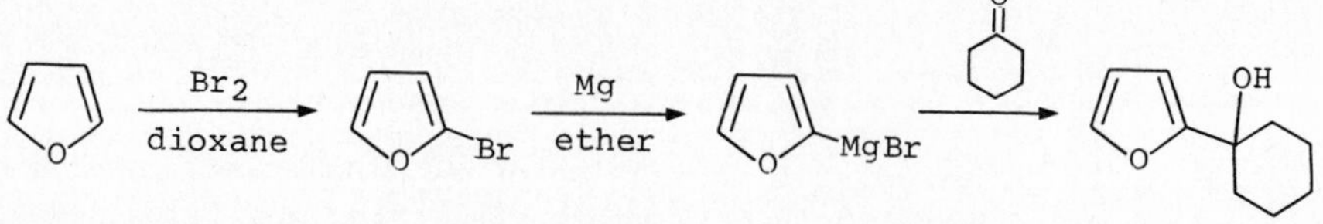

(e) Friedel-Crafts acylations occur readily on furan; only mild Lewis acids are required, if at all:

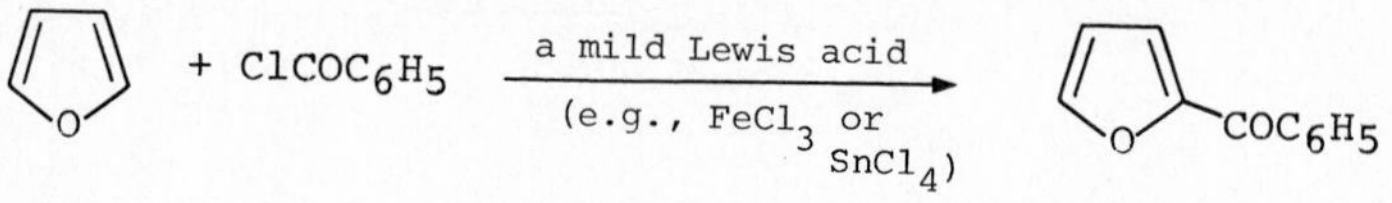

(f)

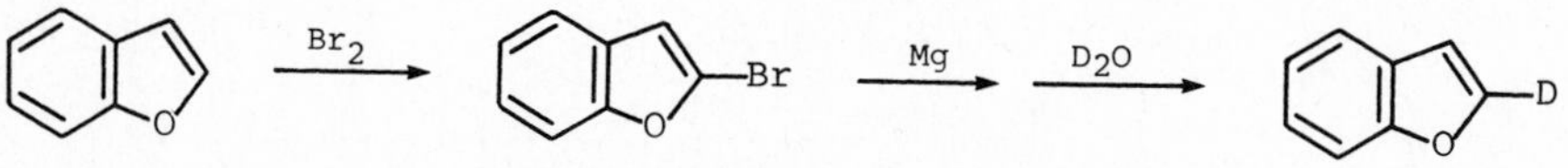

(g)

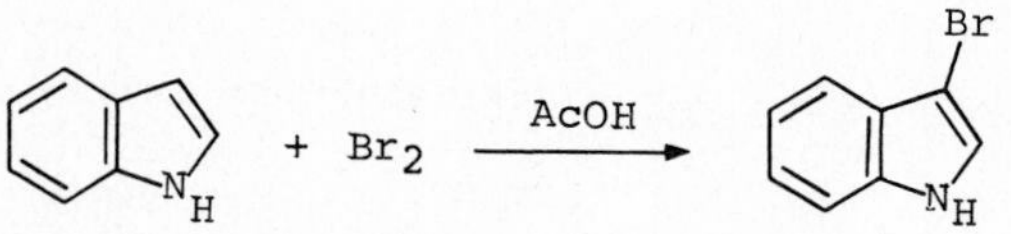

(h)

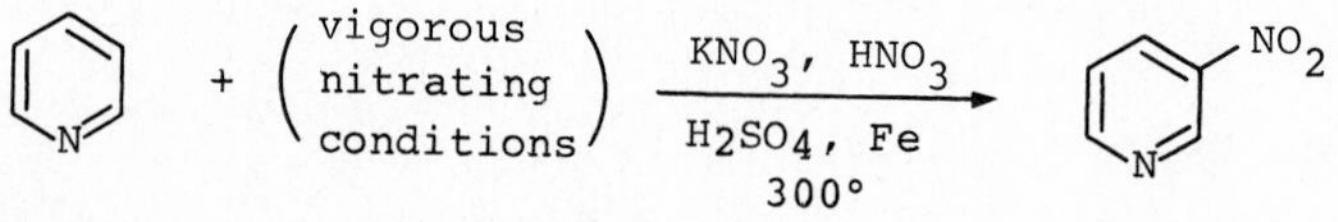

(i) α-Picoline must first be converted into the N-oxide so that nitration will occur at the γ-position:

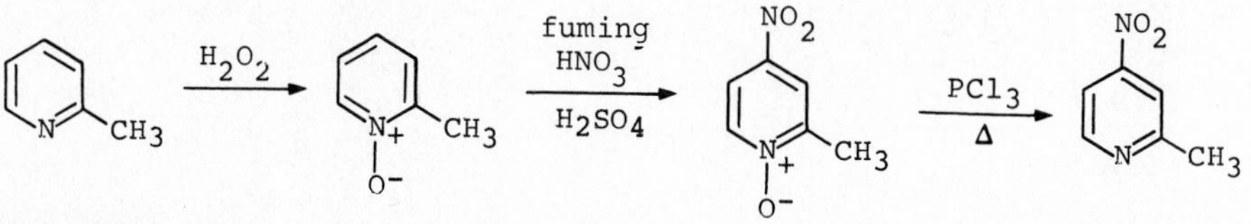

(j) Chichibabin reaction:

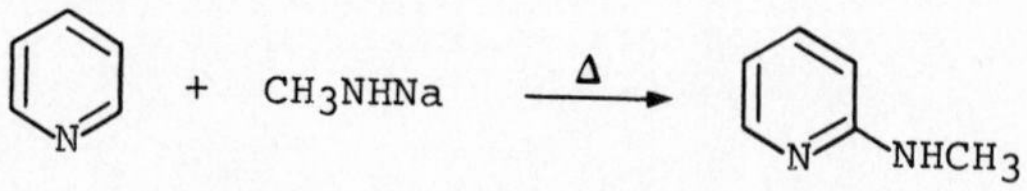

(Section 32.6.C)

(k)

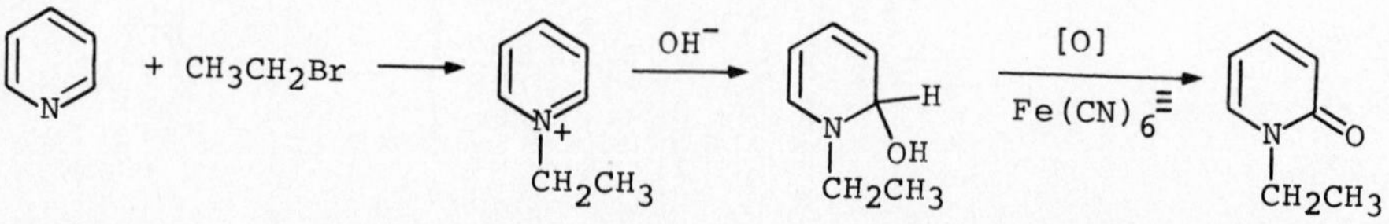

(ℓ)

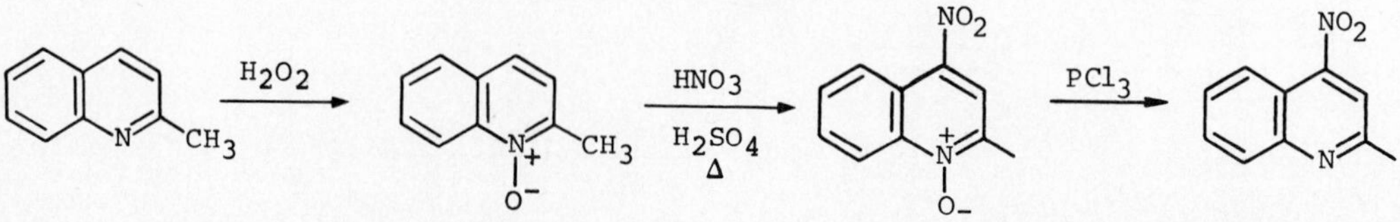

(m)

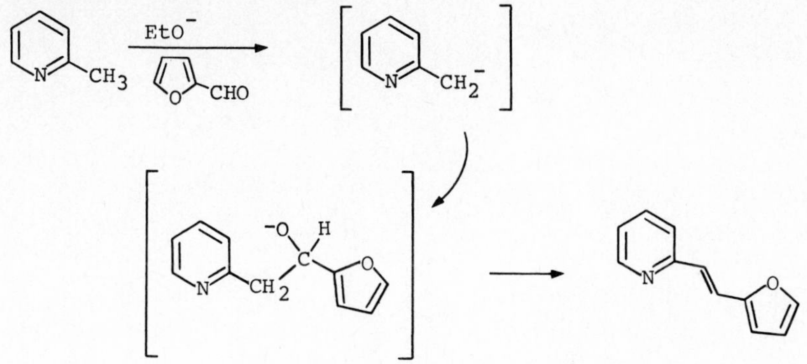

(n)

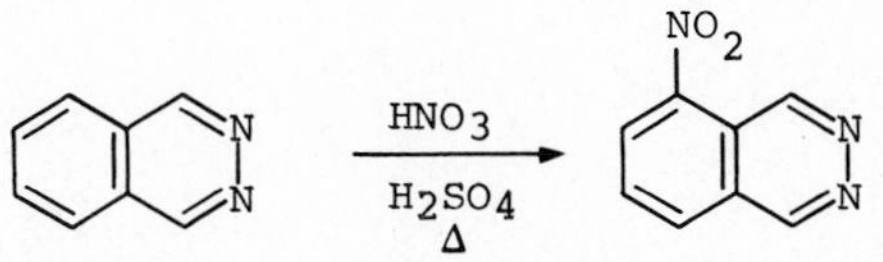

6. Michael addition reaction:

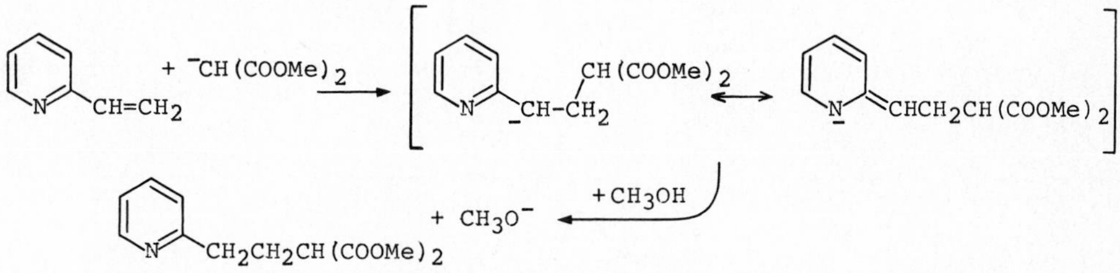

7. (a)

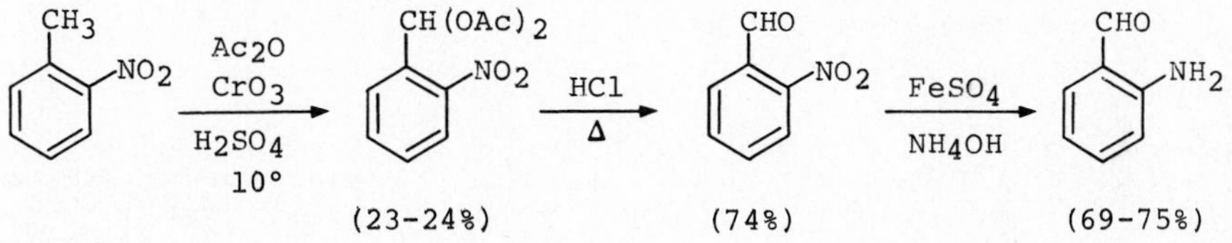

(b)

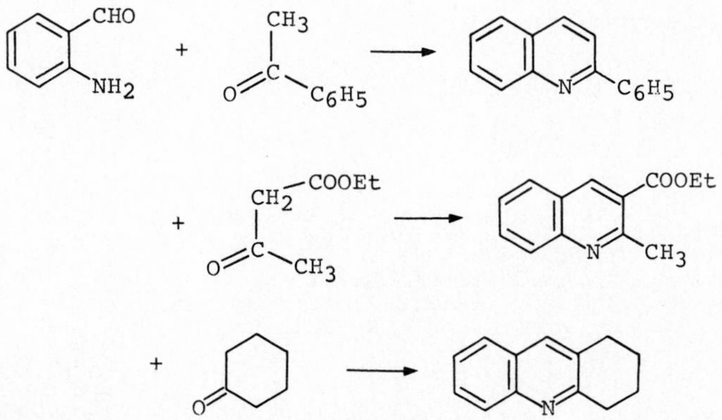

(c)

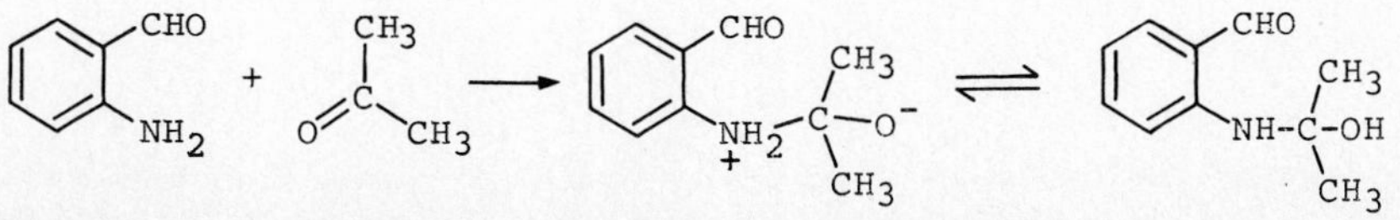

[cont'd...]

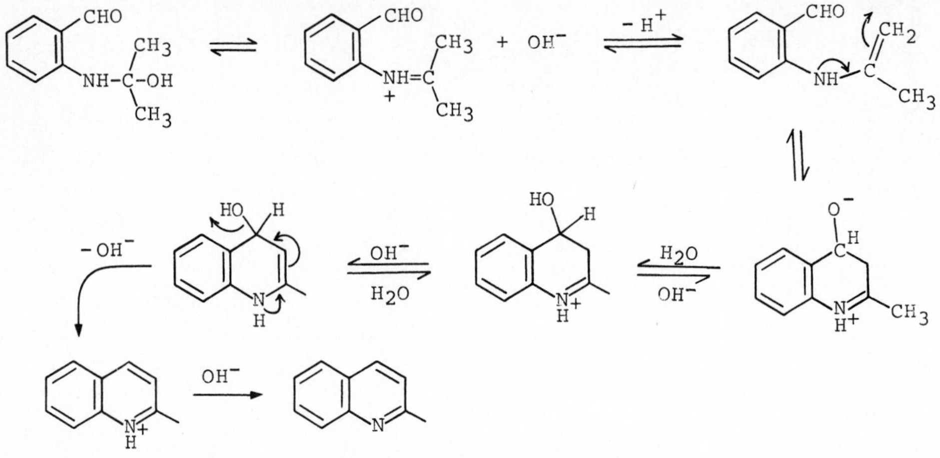

8. 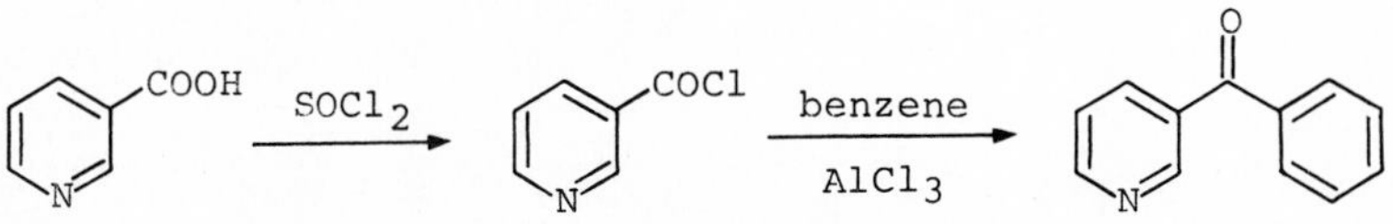

Although Friedel-Crafts acylations cannot be performed on pyridine, a β-pyridinecarboxylic halide can be utilized to acylate benzene. This procedure, however, does not work with the α- or γ-acids.

9. (a) 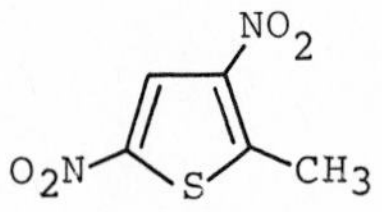 (b) 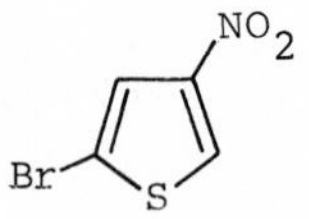 (c) 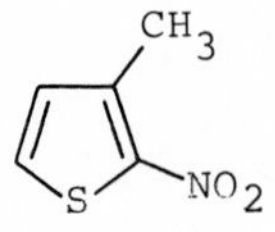

For (a)-(c), see Section 32.3.C.

(d) 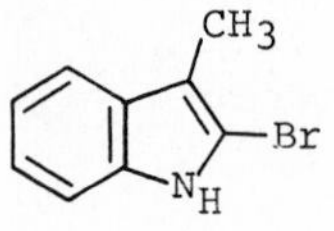

See Section 32.4.C

(e) 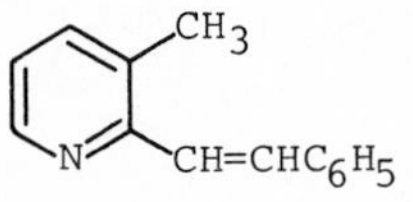

only the methyl which is conjugated to the nitrogen will react (α- or γ-methyl)

(see Exercise at the end of Section 32.6.C)

(f) a Fischer indole synthesis:

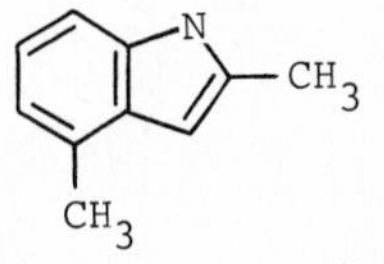 and 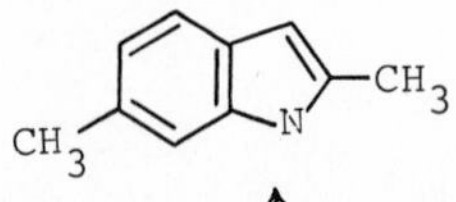

We expect a greater amount of this product, since there is less steric hindrance for the cyclization reaction.

(g) 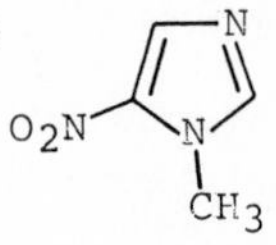

10. 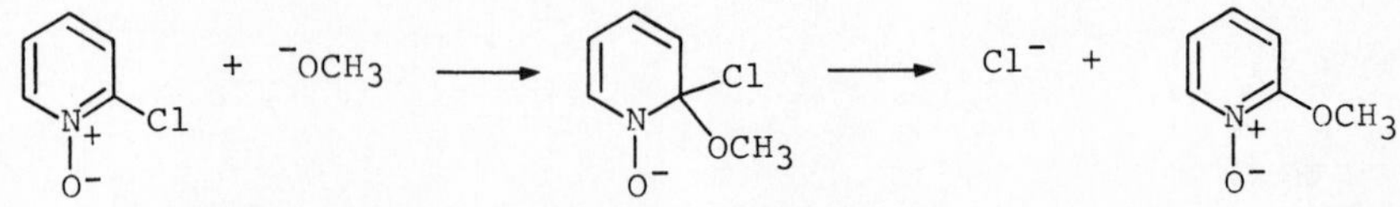

11. 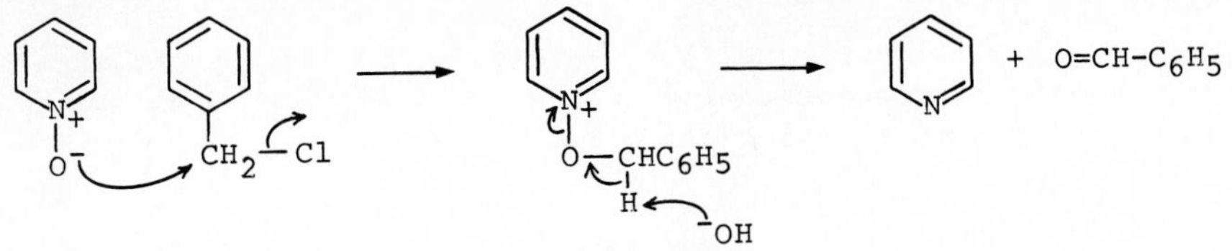

Pyridine serves as a leaving group in a reaction that is essentially an E2 reaction, and which constitutes a new aldehyde synthesis.

12. 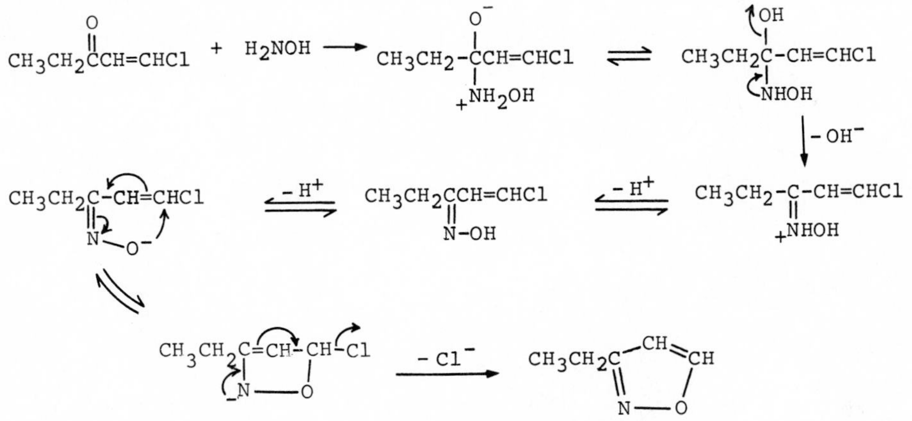

or conjugate addition to the double bond:

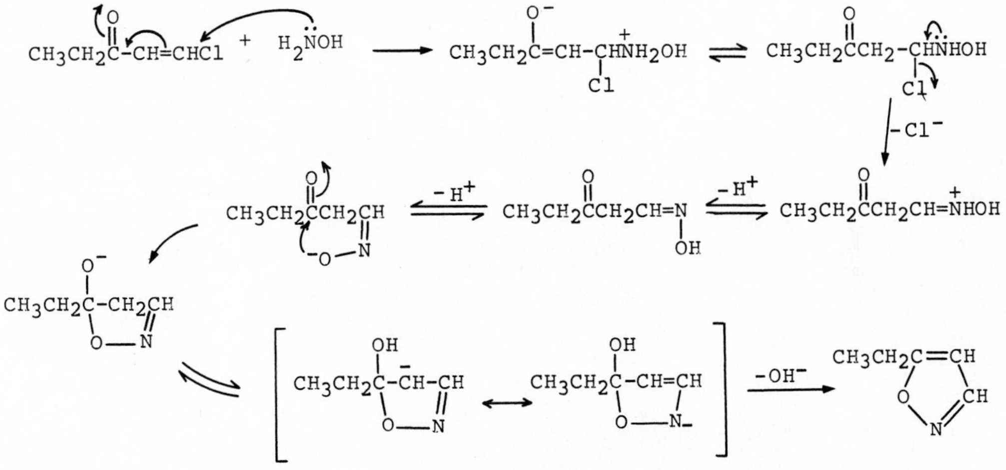

Note that in both competing steps. the more basic nitrogen in H_2NOH attacks first, rather than the oxygen.

13. 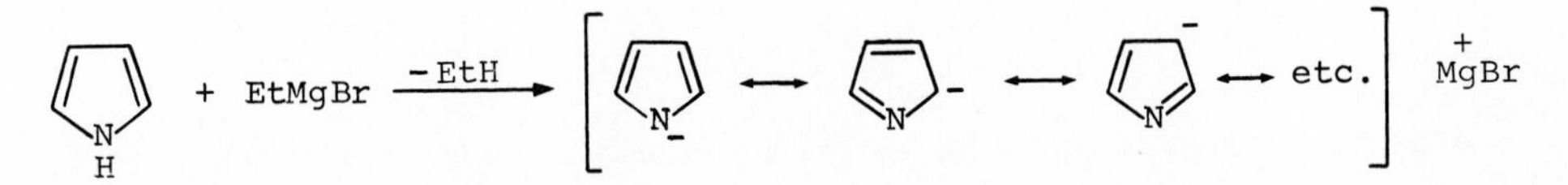

Pyrrole is rather acidic; recall cyclopentadiene (in Table 31.1). The pyrrole anion is an ambident anion with negative charges distributed among the nitrogen and all four ring carbons. The carbon is more nucleophilic and displaces on CH_3I.

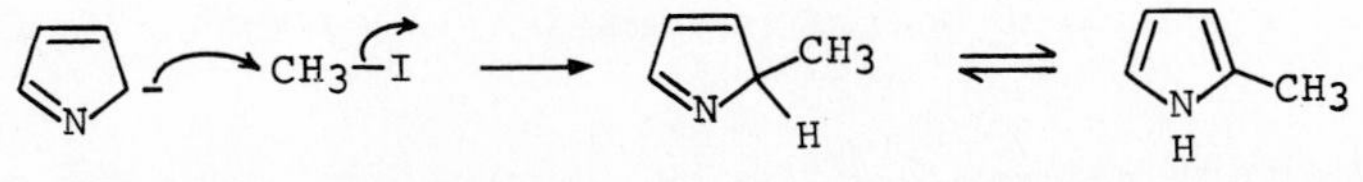

14. This one is rather subtle. The first reaction involves the carbonyl of the chloroketone, <u>not</u> displacement of chloride. Cyclization to an oxirane follows, then ring opening and cyclization to the furan:

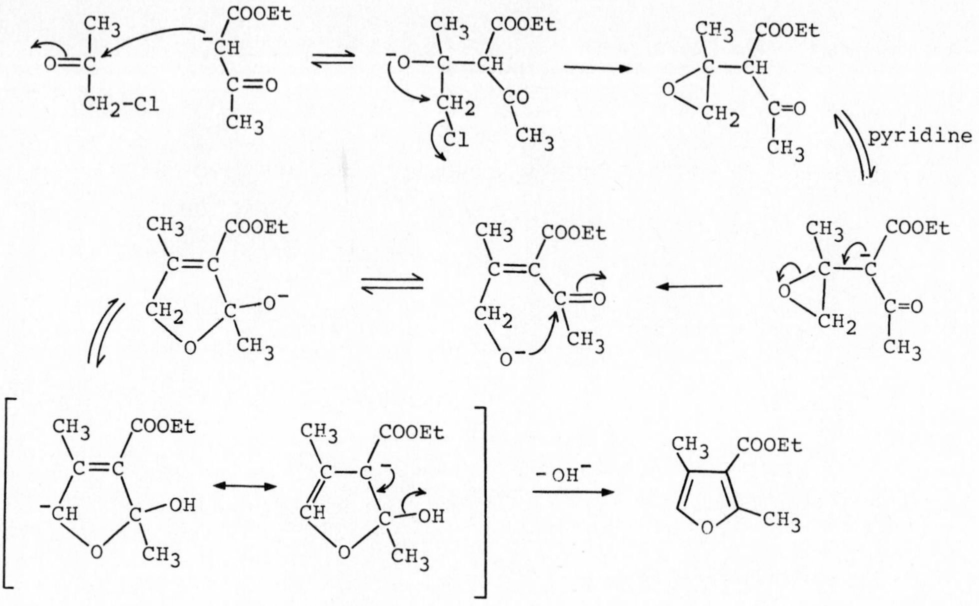

15.

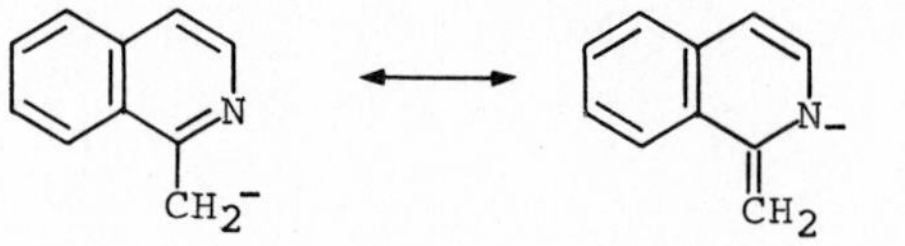

The negative charge is readily distributed to a more electronegative nitrogen; the benzene ring is still intact.

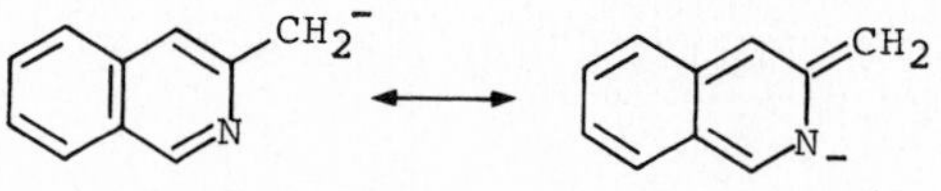

In order to delocalize the negative charge onto the nitrogen, the benzene ring must be disrupted, with an attendant loss in resonance stabilization.

16. (a)

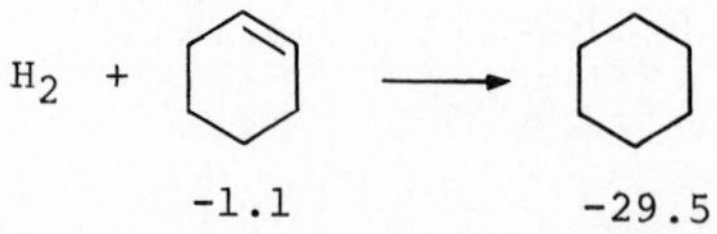

$\Delta H° = -28.4$ kcal mole^{-1}

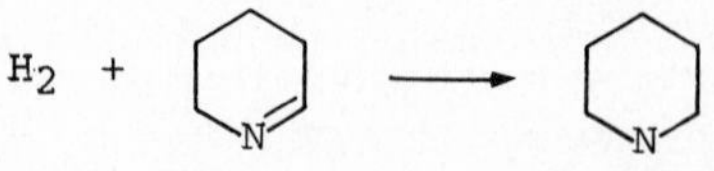

$\Delta H° = -21$ kcal mole^{-1}

Therefore, for pyridine (two C=C bonds and one C=N bond), we would expect:

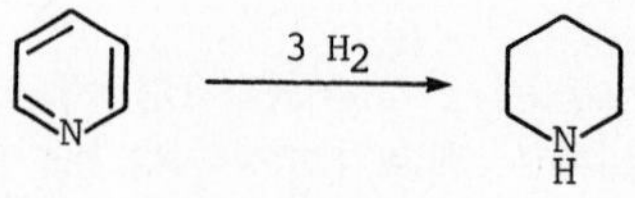

predicted $\Delta H° = 2 \times (-28.4) - 21 = -77.8$ kcal mole^{-1}

actual $\Delta H° = -11.8 - (34.6) = -46.4$ kcal mole^{-1}

The resonance energy is $-46.4 - (-77.8) = 31.4$ kcal mole^{-1}

This empirical resonance energy is similar to that of benzene.

(b) 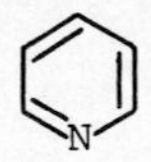 $\longrightarrow$ 5 C + 5 H + N $\Delta H^\circ_{atomiz.} = 1193.4$ kcal mole^{-1}

+34.6   5 × 170.9   5 × 52.1   113.0

Bond energies: 5 C-H + 2 C=C + 2 C-C + N-C + N=C

$= 5 \times 99 + 2 \times 146 + 2 \times 83 + 73 + 147 = 1173$ kcal mole^{-1}

Therefore, the empirical R.E. = 1193 - 1173 = 20 kcal mole^{-1}

Note that using bond energies,

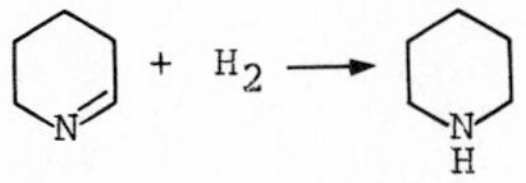

(C-H) + (N-H) + (C-N) - (C=N) - (H-H)

= 99 + 93 + 73 - 147 - 104

$\Delta H^\circ = -14$ kcal mole^{-1}

The experimental value used in (a) is -21 kcal mole^{-1}.

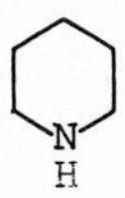 $\longrightarrow$ 5 C + 11 H + N   experimental $\Delta H^\circ = 1552.4$ kcal mole^{-1}

calculated from bond energy Table: 1561 kcal mole^{-1}

The Table of bond energies gives $\Delta H^\circ_{atomiz.}$ that are accurate to ±1% or less, but this still amounts to several kcal mole^{-1}; i.e., in practice, the ΔH° we calculate only amount to a few percent of the total atomization energies. In general, energy differences derived from average bond energies can have substantial errors.

17. Compare the bond dipole of $\overset{\longmapsto}{\text{C—Br}}$ in 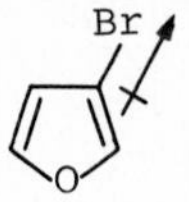 with that for 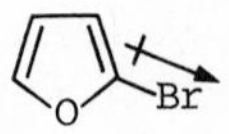

0.91 D < 1.46 D; therefore, 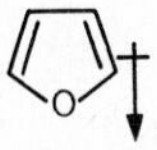

The similarity between 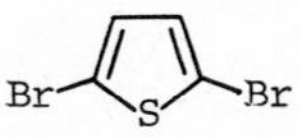 -  = 0.6 D

1.1 D   0.51 D

and 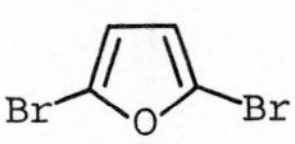 -  = 0.9 D

1.63 D   0.70 D

suggests 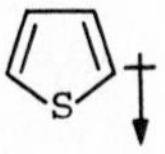 for thiophene. If the thiophene dipole were in the opposite direction, the effect of the bromines could not be rationalized.

In the case of pyrrole,

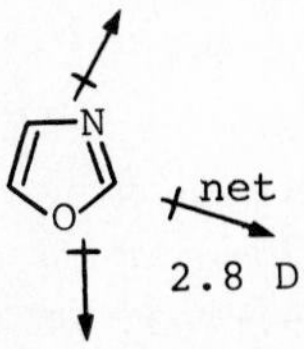

2.8 D, but 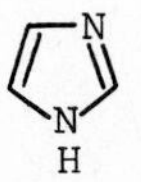 strongly suggests 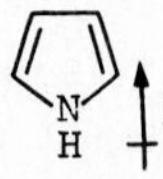 1.81 D

6.2 D

18. 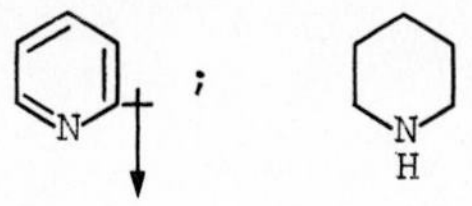 is also ↓ , and pyridine is expected to be enhanced in this direction by polarization, as suggested by resonance structures such as: 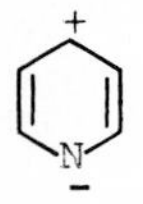

19. 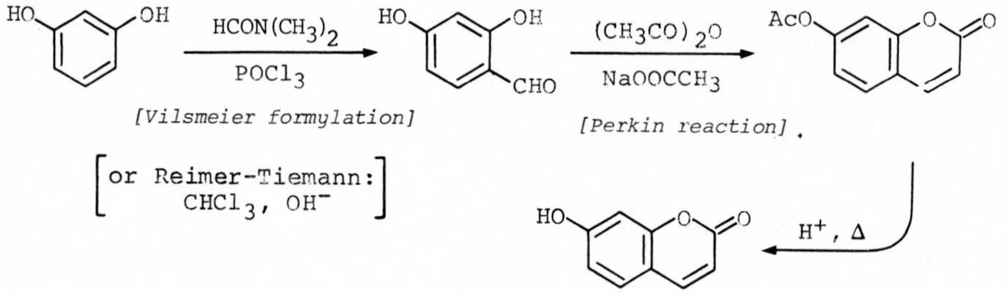

NOTE: the lactone ring hydrolyzes less readily than a normal phenol ester, especially in acid.

20. For $C_{20}H_{21}O_4N$, Zeisel determination gives the partial formula $C_{16}H_9N(OCH_3)_4$. The oxidation to a ketone indicates:

$$C_{15}H_7N(CH_2)(OCH_3)_4 \longrightarrow C_{15}H_7N(CO)(OCH_3)_4$$

The oxidation products from the ketone can be explained on the basis that the initial products undergo further reaction:

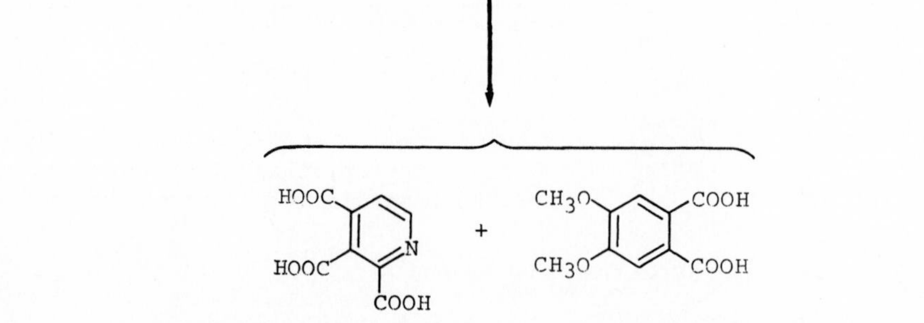

Note that the two primary products are $C_{15} \rightarrow C_9 + C_6$; this is the only combination that can give the molecular formula of the ketone by working backwards. Furthermore, the COOH of both primary product carboxylic acids must come from the same ketone carbonyl. Thus, the structures of the ketone and of papaverine must be:

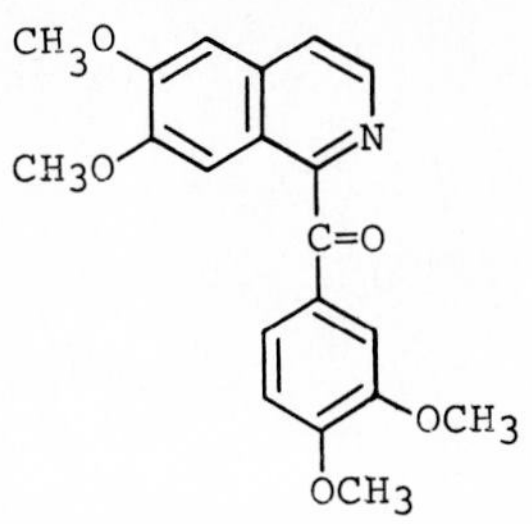

ketone

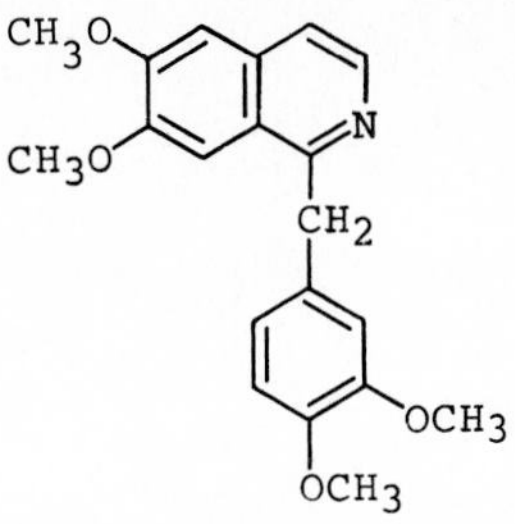

papaverine

21.

$$C_6H_5\overset{\overset{\displaystyle NOH}{\|}}{C}{\diagdown}_H \quad \overset{OH^-}{\rightleftharpoons} \quad \left[C_6H_5\overset{\overset{\displaystyle N-O^-}{\|}}{C}{\diagdown}_H \quad \longleftrightarrow \quad C_6H_5\overset{\overset{\displaystyle N=O}{|}}{\underset{\displaystyle H}{C^-}} \right] \quad \overset{Cl_2}{\searrow}$$

$$C_6H_5\overset{\overset{\displaystyle NOH}{\|}}{C}-Cl \quad \overset{H_2O}{\rightleftharpoons} \quad \left[C_6H_5\overset{\overset{\displaystyle N-O^-}{\|}}{C}-Cl \quad \longleftrightarrow \quad C_6H_5\overset{\overset{\displaystyle N=O}{|}}{\underset{}{C}}-Cl \right] \quad \overset{OH^-}{\rightleftharpoons} \quad C_6H_5\overset{\overset{\displaystyle N=O}{|}}{\underset{\displaystyle H}{C}}-Cl$$

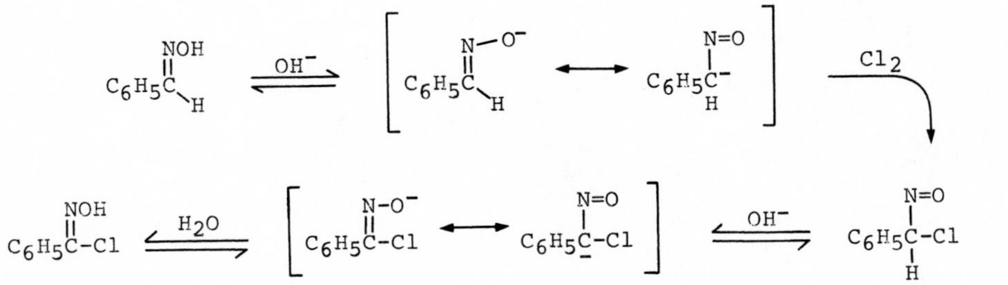

This mechanism is actually similar to the chlorination of phenol in basic solution
(Section 30.6.B.1).

22.

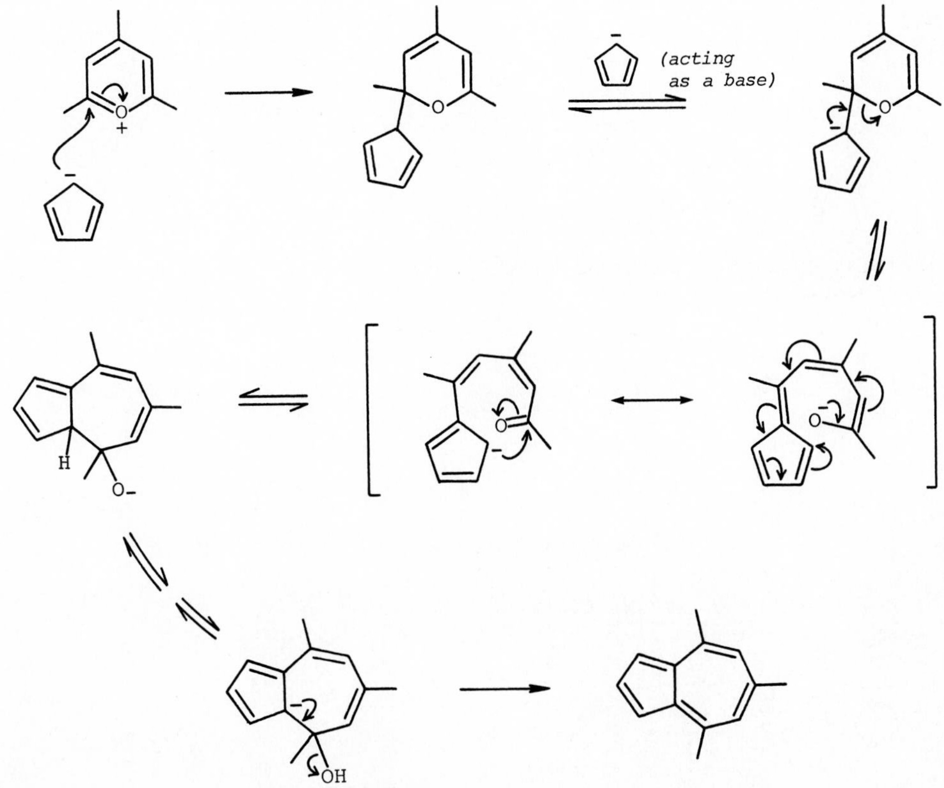

32.E Supplementary Problems

S1. Name each of the following compounds.

(a)

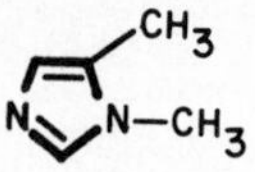

(d)

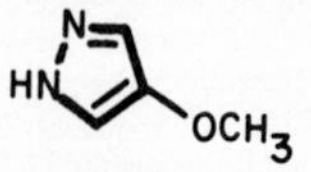

(g)

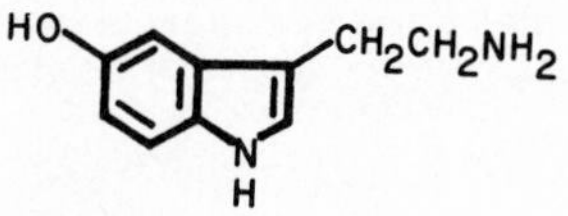

(b)

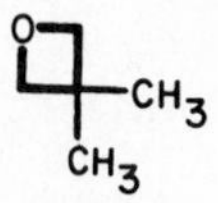

(e)

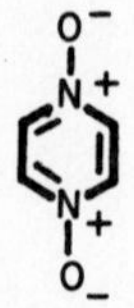

(h)

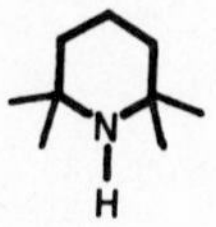

(c)

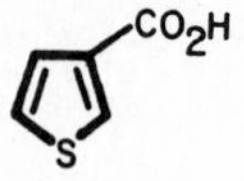

(f)

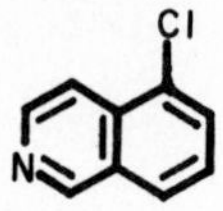

S2. Write the structures of the following compounds.

(a) 4-bromofuran-3-carboxaldehyde
(b) 3-nitrobenzofuran
(c) *trans*-2,3-diphenyloxirane
(d) N-acetylpyrrole

(e) 2-chloroquinoline
(f) 5-nitroisoxazole
(g) 6-methylthiopurine
(h) 2-methylthietane

S3. What is the major product to result from each of the following reaction sequences?

(a)

$$CH_3\overset{O}{\overset{\|}{C}}CH_2CO_2Et \quad H_3O^+ \quad \Delta$$

(b)

$$Cl_2 \quad AcONO_2$$

(c)

$$CH_2O \quad NaCN \quad LiAlH_4$$
$$HCl$$

(d)

$$H_2NOH \quad Br_2$$
$$FeBr_3$$

(e)

$$C_6H_5CH_2MgBr \quad HNO_3$$
$$H_2SO_4$$

(f)

$$H_2O_2 \quad HNO_3 \quad Zn$$
$$H_2SO_4 \quad HCl$$

(g)

$$NaNH_2 \quad CO_2 \quad CH_3OH$$
$$H^+$$

(h)

$$C_6H_5CH_2CH=O \quad \overset{NH_3}{\underset{H_2/Pt}{\longrightarrow}} \qquad \overset{P_2O_5}{\underset{\Delta}{\longrightarrow}}$$

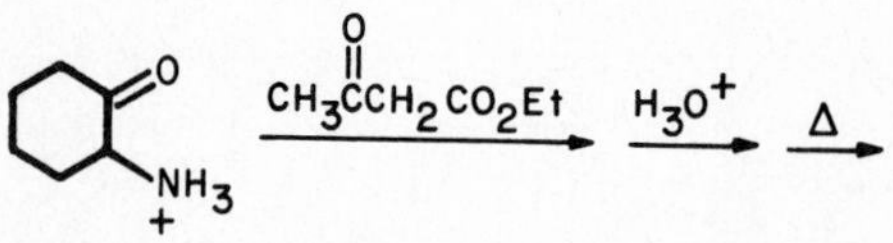

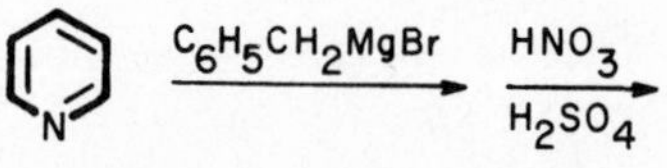

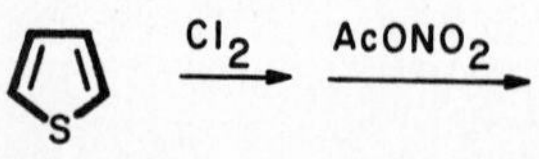

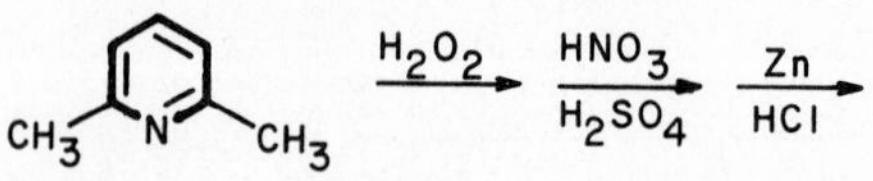

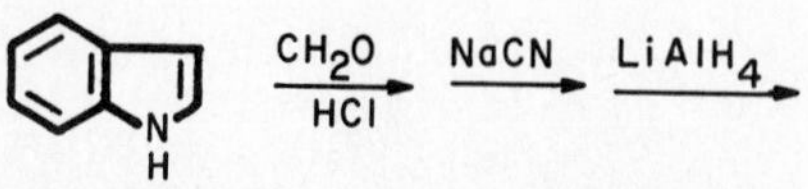

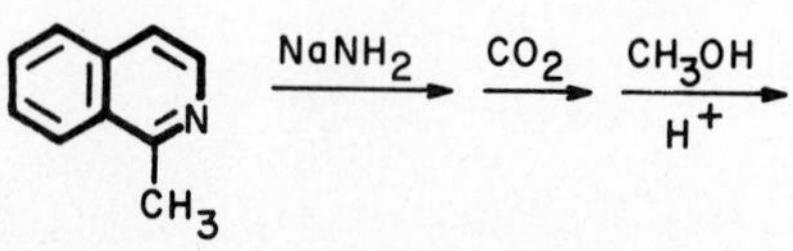

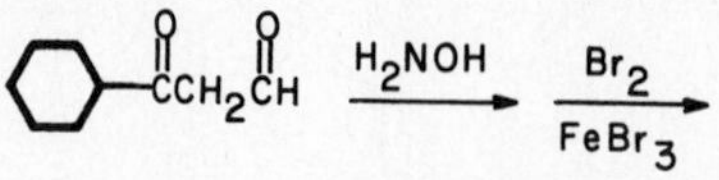

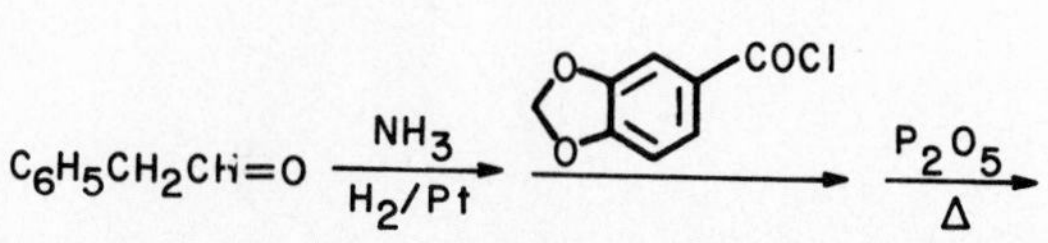

S4. Devise a synthesis of each of the following compounds, using the indicated starting material and any other reagents.

(a)
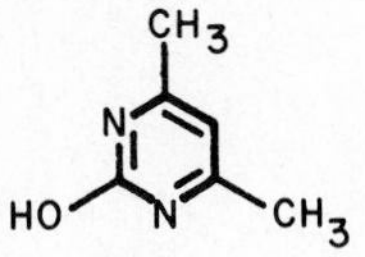
from urea

(b)
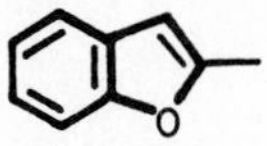
from phenol

(c)
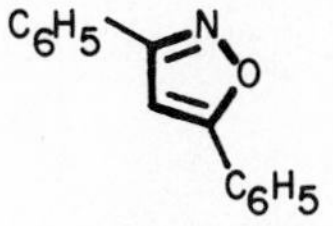
from toluene

(d)
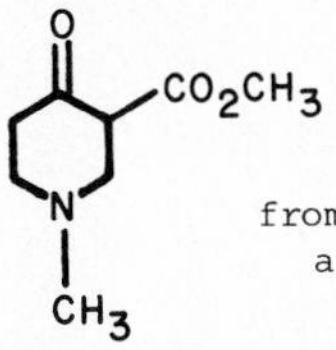
from methyl acrylate

(e)
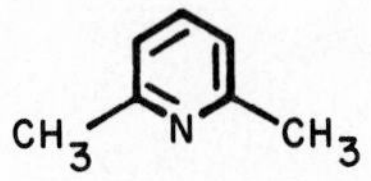
from acetic acid

(f)

from cyclopentane

(g)
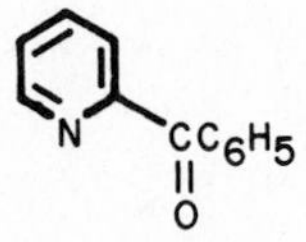
from pyridine

(h)
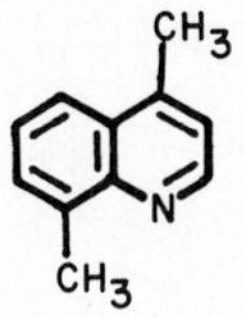
from toluene

S5. Show how to synthesize each of the following compounds from non-heterocyclic precursors.

(a)
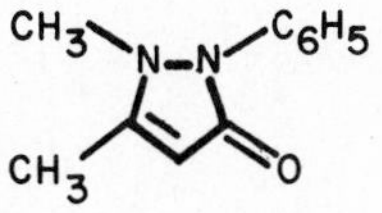
antipyrine, an ingredient in many commercial headache remedies

(b)
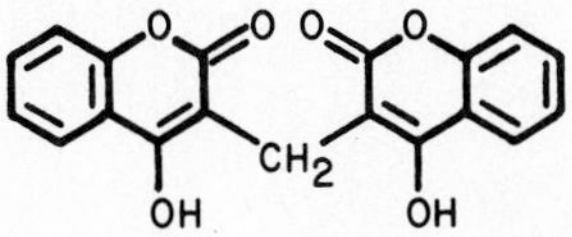
dicumarol, a compound isolated from sweet clover which causes a severe bleeding tendency in cattle

(c)
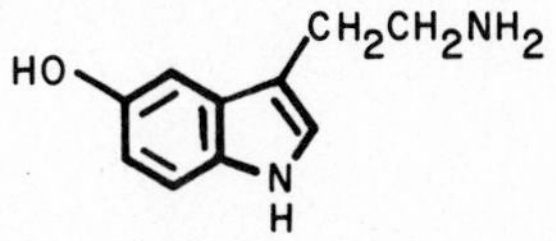
serotonin, one of the molecules involved in the transmission of nerve impulses in the brain

(d)
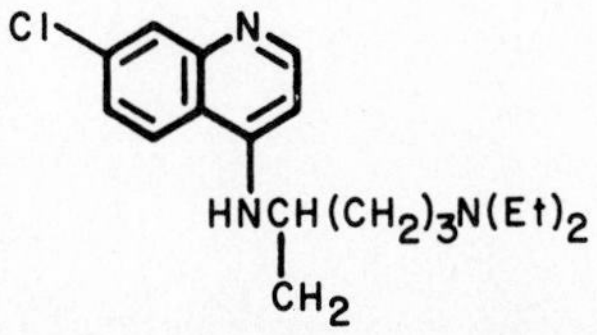
chloroquine, an important antimalarial drug

(e)
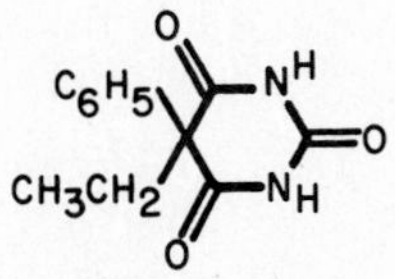
phenobarbital, one of the barbiturates (sedatives, depressants)

S6. Write a reasonable mechanism for each of the following transformations.

(a)

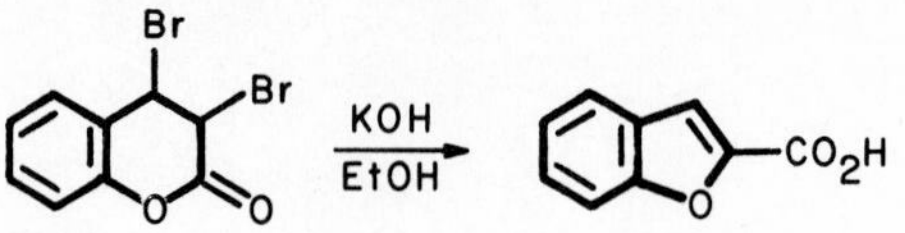

(b)

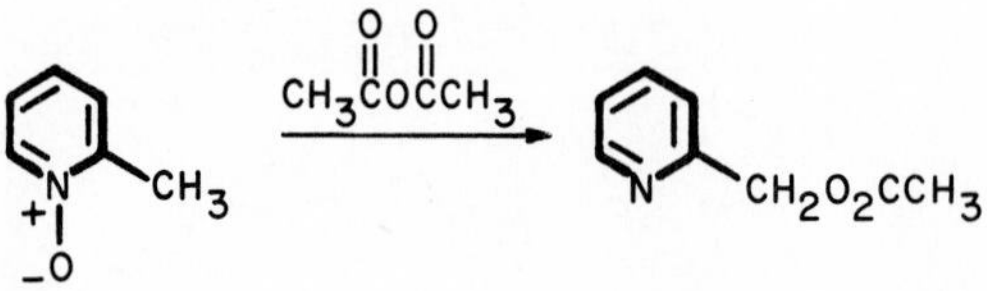

(c)

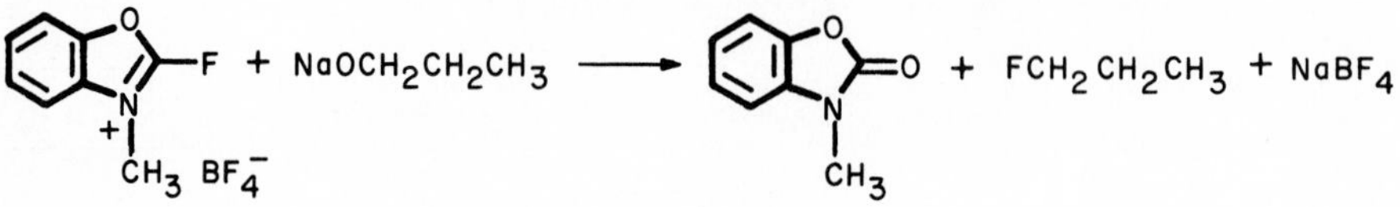

(d)

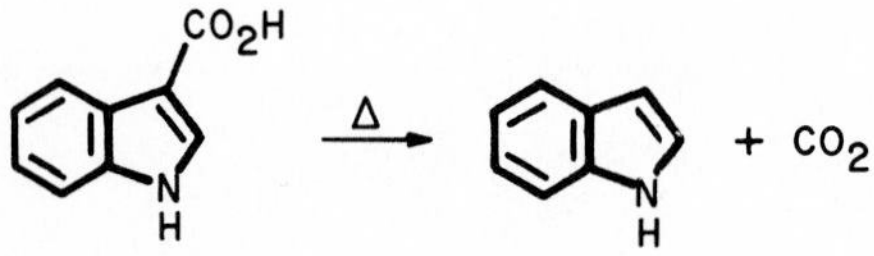

S7. Rank the following compounds in order of basicity.

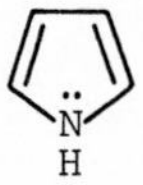

[1]

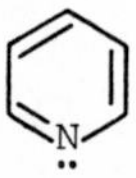

[2]

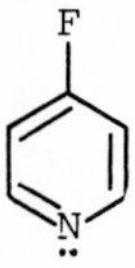

[3]

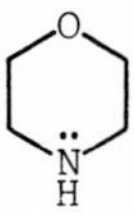

[4]

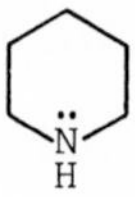

[5]

S8. Explain the differences in reactivity of the three furan derivatives illustrated below:

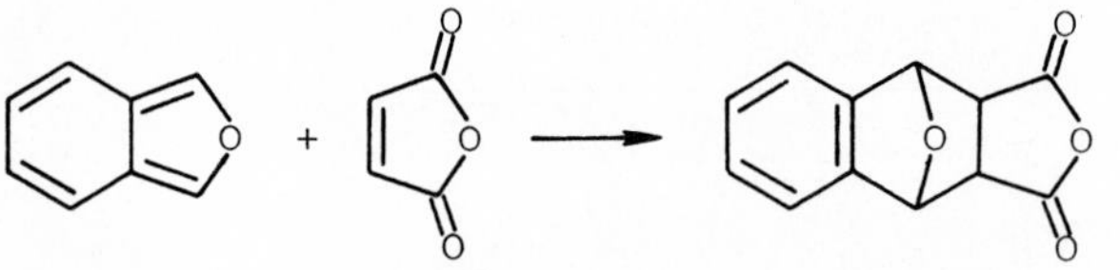

very fast reaction, even at -50 °C

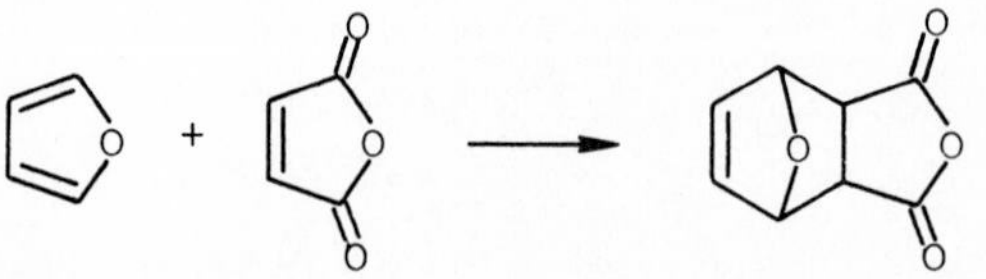

reacts at room temperature

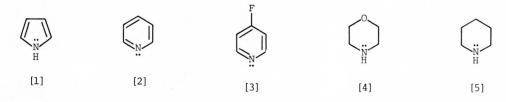

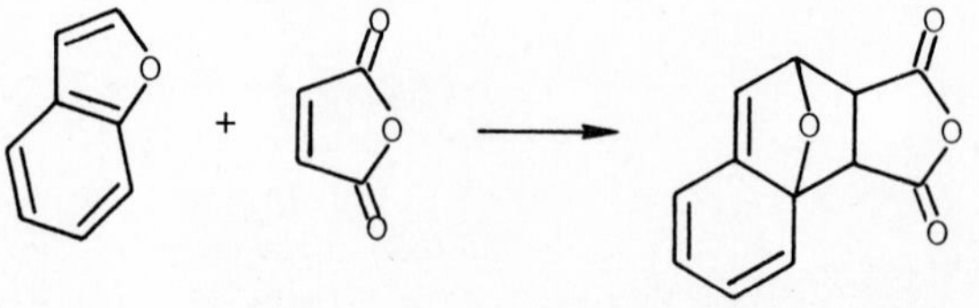

no reaction, even at 100 °C

32.F Answers to Supplementary Problems

S1. (a) 1,5-dimethylimidazole
(b) 3,3-dimethyloxetane
(c) 3-thiophenecarboxylic acid
(d) 4-methoxypyrazole
(e) pyrazine di-N-oxide
(f) 5-chloroisoquinoline
(g) 3-(2-aminoethyl)-5-hydroxyindole (serotonin)
(h) 2,2,6,6-tetramethylpiperidine

S2.

(a) 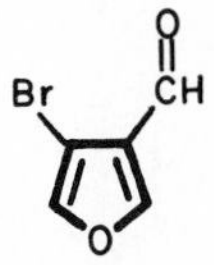

(b) 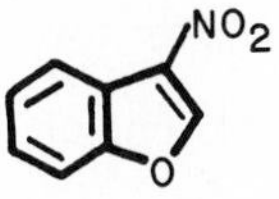

(c) 

(d) 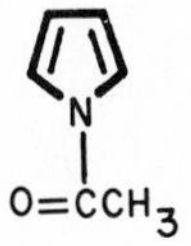

(e) 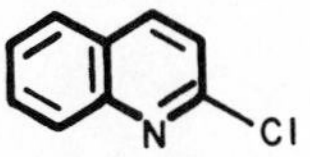

(f) 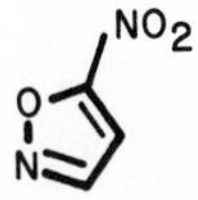

(g) 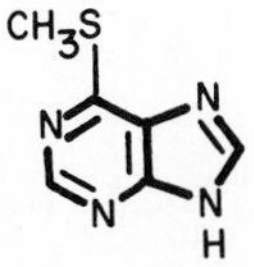

(h) 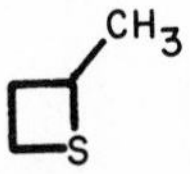

S3.

(a) 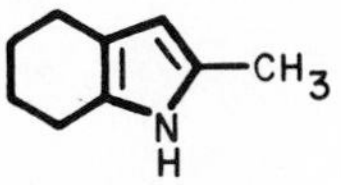

(b) 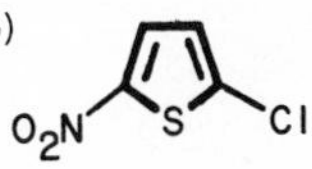

(c) 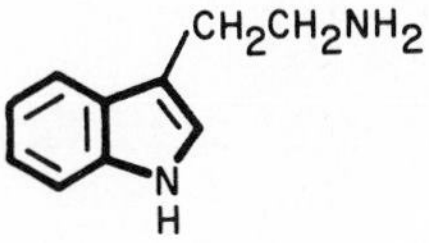

(d) 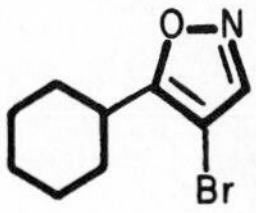

(e) 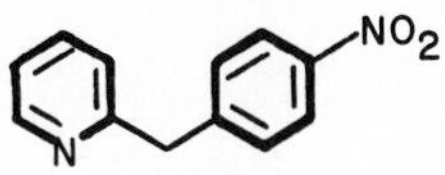

(f) 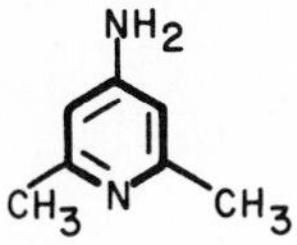

(g) 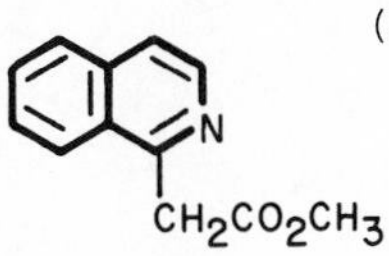

(h) 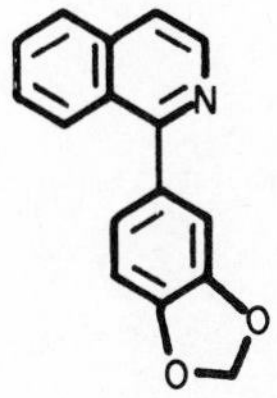

S4.

(a) 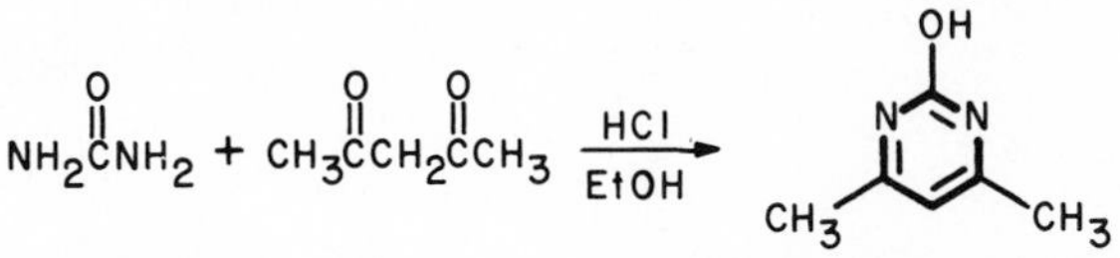

(b) 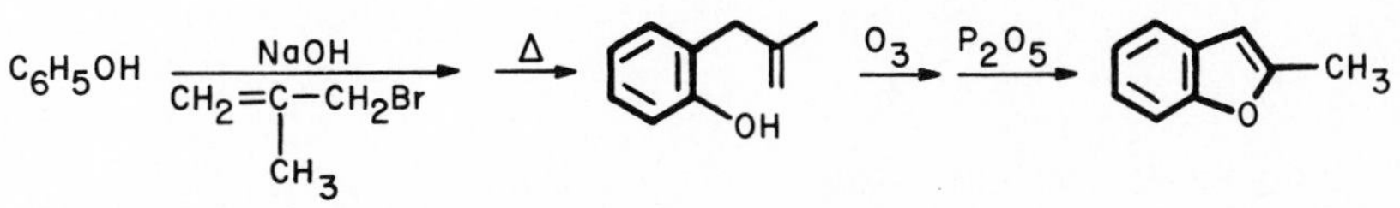

(c) 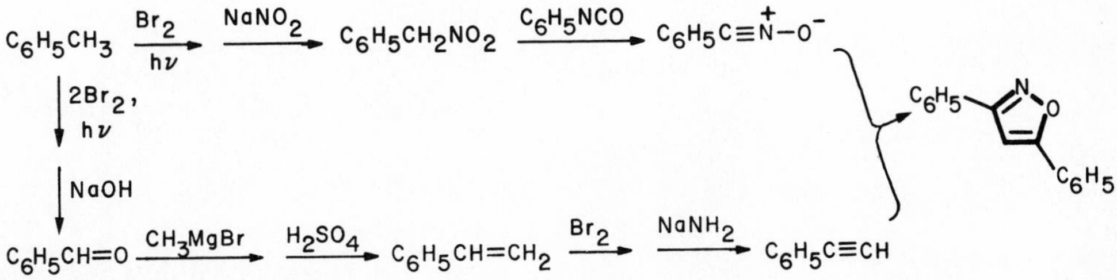

or

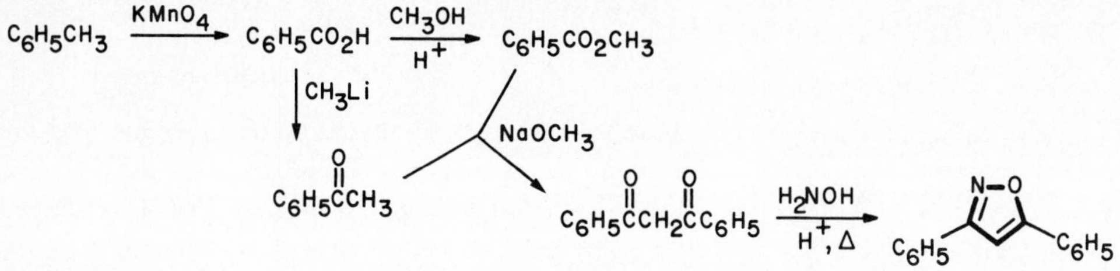

$C_6H_5CH_3$
KMnO4
$C_6H_5CO_2H$
CH3OH
H+
$C_6H_5CO_2CH_3$
CH3Li
NaOCH3
$C_6H_5CCH_3$
$C_6H_5CCH_2CC_6H_5$
H2NOH
H+, Δ
N–O
C6H5
C6H5

(d)

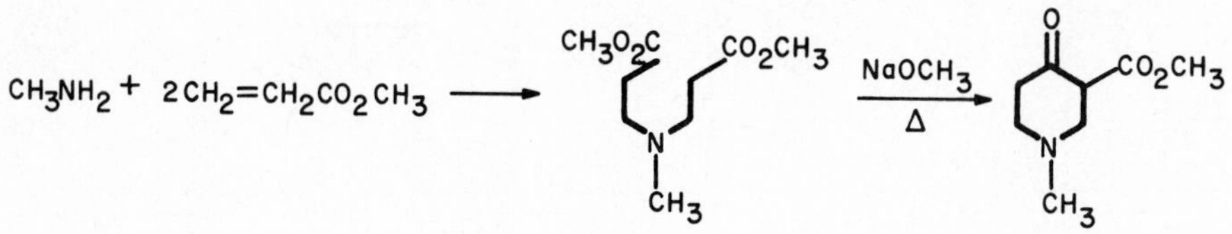

CH3NH2 + 2CH2=CH2CO2CH3
CH3O2C
CO2CH3
N
CH3
NaOCH3
Δ
CO2CH3
N
CH3

(e)

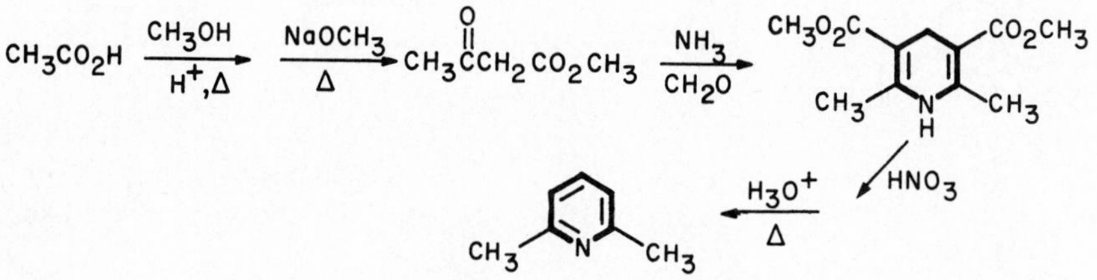

CH3CO2H
CH3OH
H+,Δ
NaOCH3
Δ
CH3CCH2CO2CH3
NH3
CH2O
CH3O2C
CO2CH3
CH3
N
H
CH3
HNO3
H3O+
Δ
CH3
N
CH3

(f)

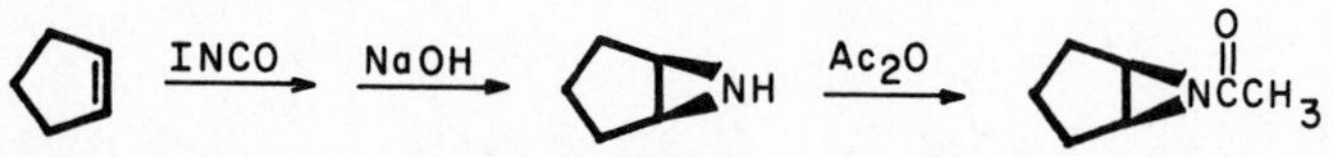

INCO
NaOH
NH
Ac2O
NCCH3

(g)

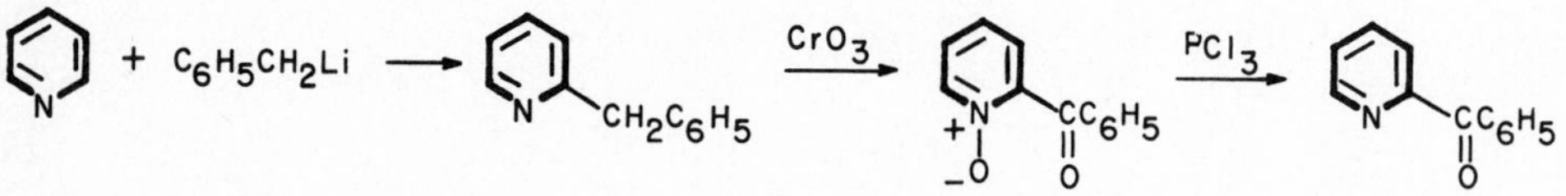

N
+ C6H5CH2Li
N
CH2C6H5
CrO3
N
CC6H5
PCl3
N
CC6H5

(h)

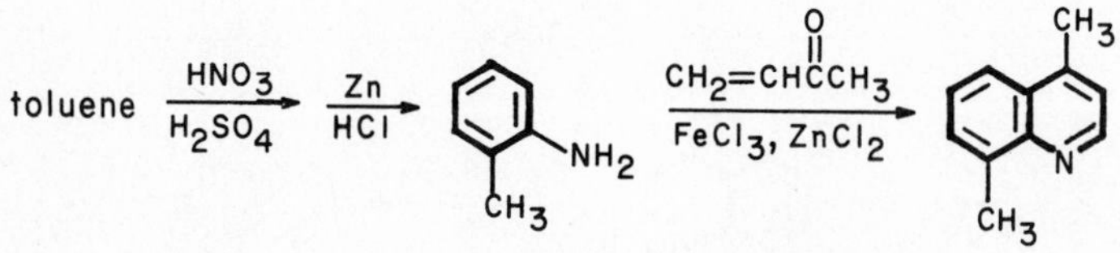

toluene
HNO3
H2SO4
Zn
HCl
NH2
CH3
CH2=CHCCH3
FeCl3, ZnCl2
CH3
N
CH3

S5.

(a)

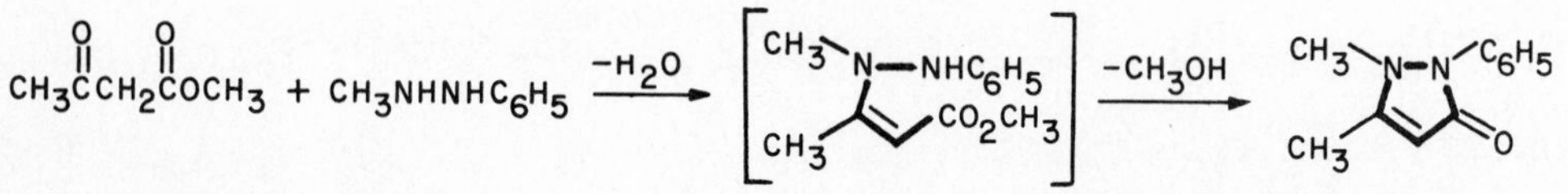

CH3CCH2COCH3 + CH3NHNHC6H5
−H2O
CH3
N
NHC6H5
CO2CH3
CH3
−CH3OH
CH3
N
N
C6H5
CH3
O

(b)

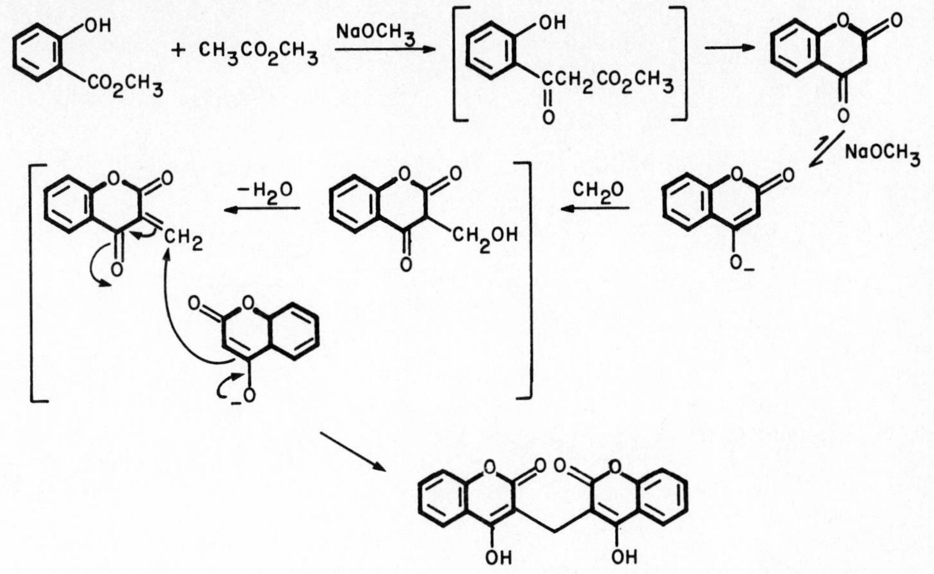

(c)

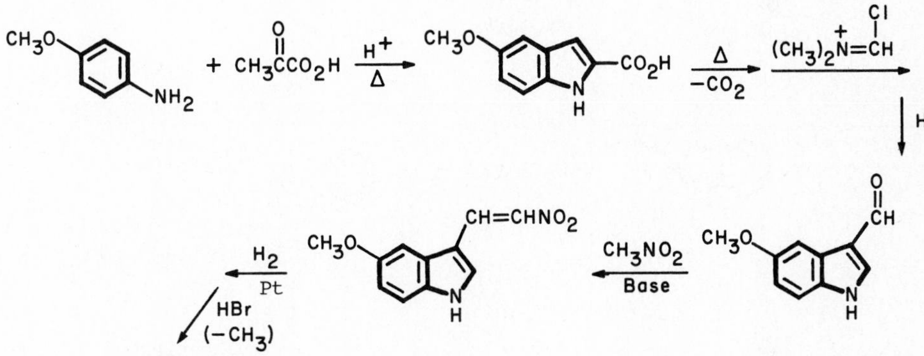

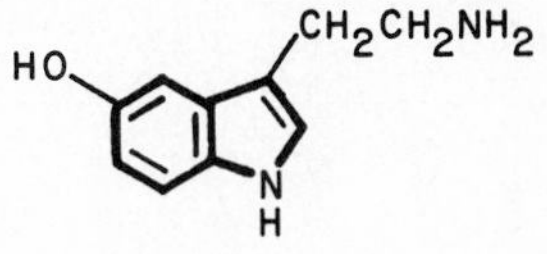

(alternatively, refer to the sequence given in problem #S3 (c) *above)*

(d)

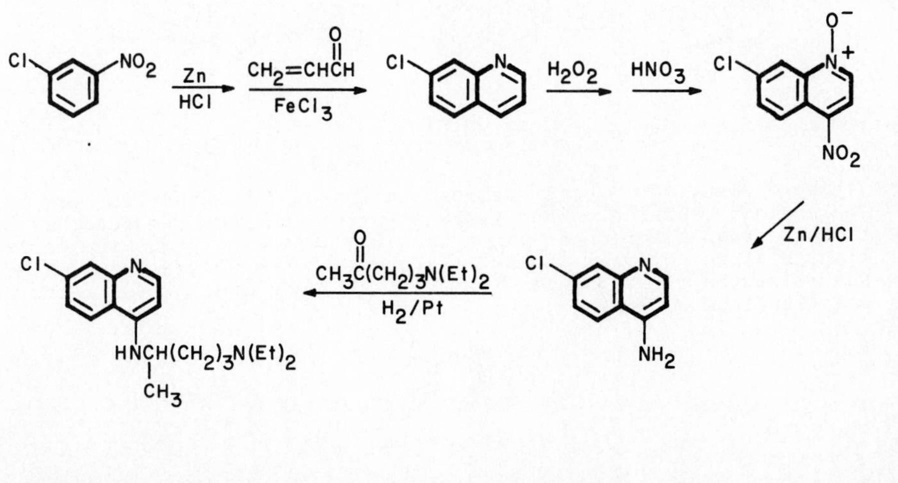

(e)

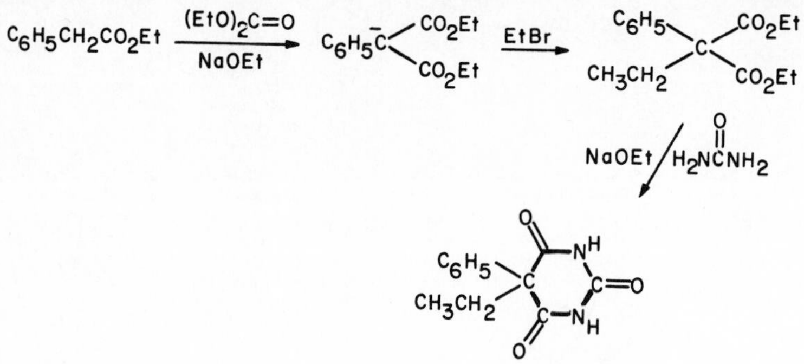

S6. (a)

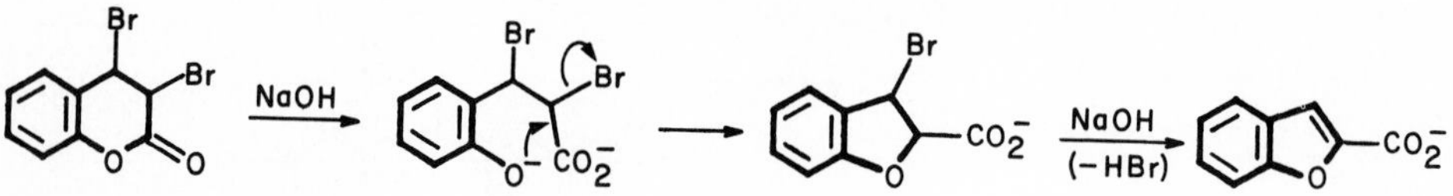

(b)

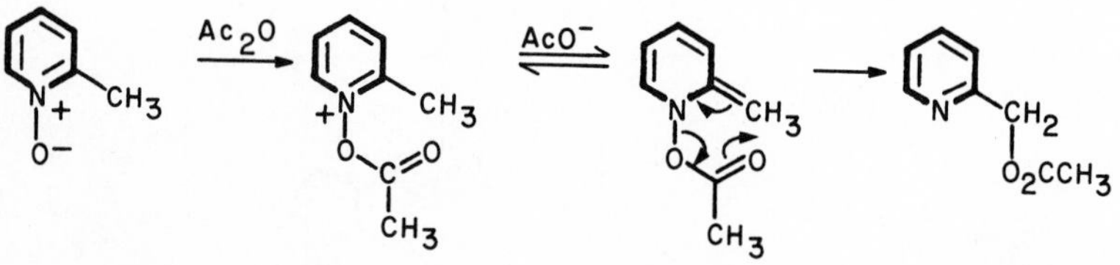

(c)

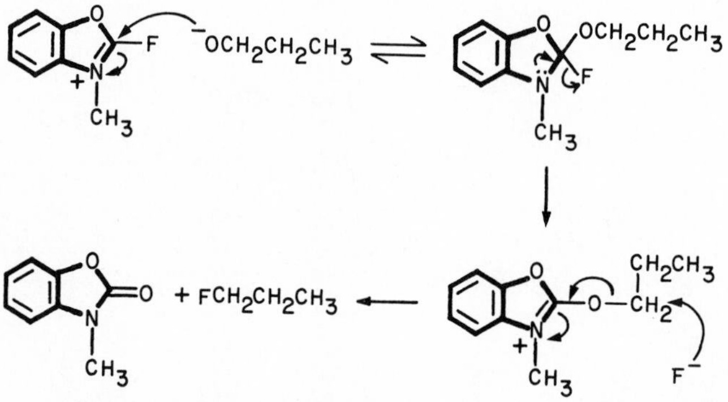

(d)

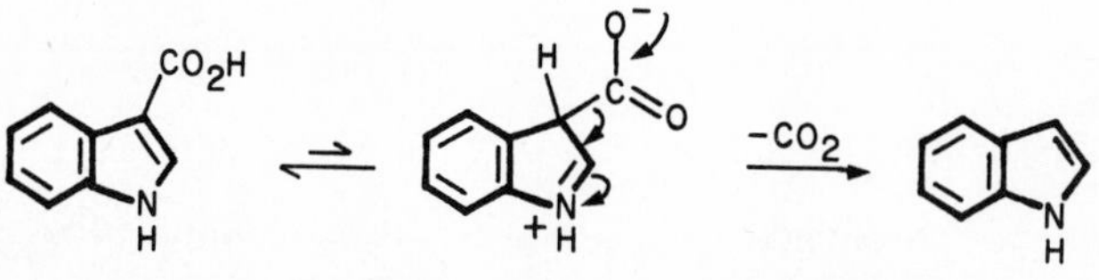

S7. *most basic* [5] > [4] > [2] > [3] > [1] *least basic*

S8. In each case, the aromatic stabilization of the furan ring is lost during the Diels-Alder reaction. When isobenzofuran undergoes the reaction, it gains the aromatic stabilization of a benzene ring, which helps to accelerate the reaction relative to furan itself. In contrast, benzofuran would lose the aromatic stabilization of its benzene ring as well, and that prevents the Diels-Alder reaction from occurring.

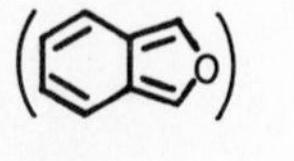

isobenzofuran

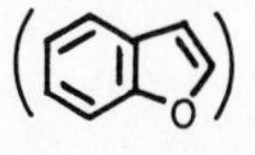

benzofuran

34.C Answers to Exercises

(34.1.B)

(a)

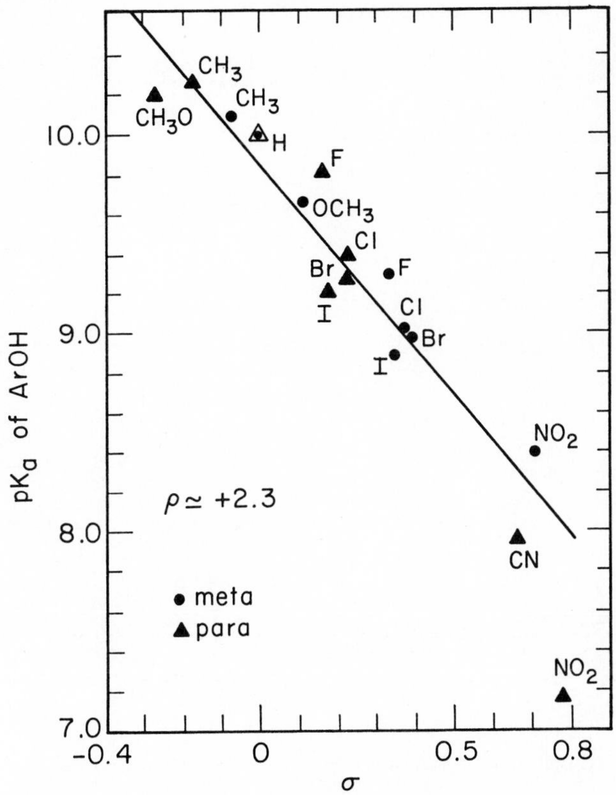

A similar plot for the substituted anilinium ions gives a value for ρ of approx. 3.1. Rho (ρ) is greater for ionization of phenols and anilinium ions than it is for benzoic acids, because in the first two cases, the negative or positive charge is on an atom which is attached directly to the ring. Not only is the inductive influence greater, but the π-system of the ring is directly involved through resonance.

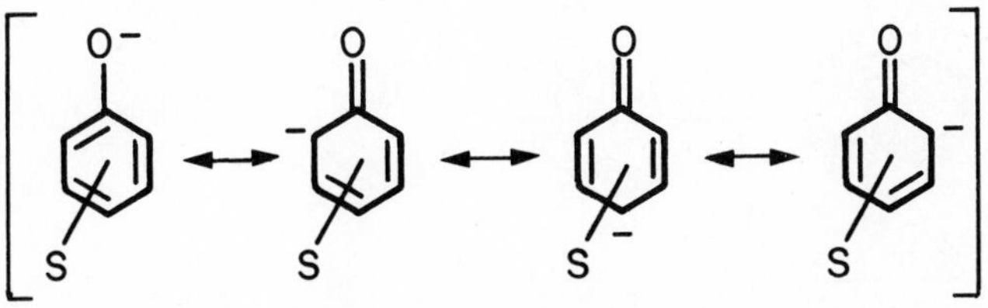

(b) pK of C_6H_5COOH = 4.20 - σ($\underline{m}$-IO_2) = 4.20 - 3.50 = 0.70.
σ($\underline{p}$-IO_2) = 4.20 - 3.44 = 0.76.

The $-IO_2$ group has the Lewis structure which is drawn at the right. If iodine is allowed to expand its octet, we may also write:

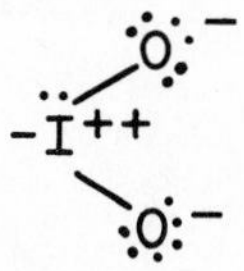

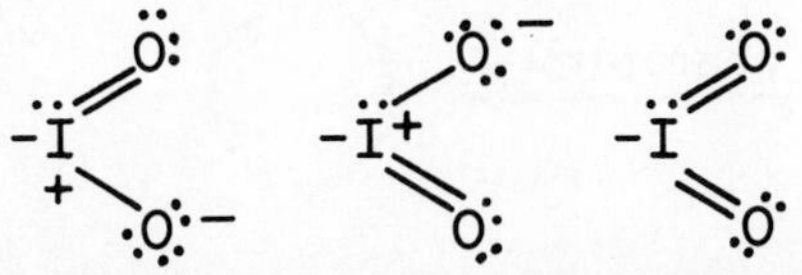

The net result gives the $-IO_2$ group a polar character similar to that of an NO_2 group. Note that $-NO_2$ has similar σ values.

(c) Electron-donating groups should stabilize the partial (+) charge of the transition state, so ρ is negative.

(d) ρ should be greater in magnitude when the positive charge of the transition state is closer to the benzene ring; that is, the magnitude of ρ is greater for (c) than for (d).

(e) The transition state has a negative charge, so ρ is positive.

(f) A plot of log k vs. σ gives a good straight line with $\rho = +1.6$.

m-Trifluoromethyl has $\sigma = 0.43$. From the plot, the calculated log k is -3.91, or $k = 1.23 \times 10^{-4}\ \text{sec}^{-1}$.

(g)

$$K = \frac{[ArN_2O^-][H^+]^2}{[ArN_2^+]}$$

$$\log K = \log \frac{[ArN_2O^-]}{[ArN_2^+]} + 2 \log H^+$$

$$pK = -\log \frac{[ArN_2O^-]}{[ArN_2^+]} + 2\ pH\ , \text{ and } \quad \frac{1}{2} \log \frac{[ArN_2O^-]}{[ArN_2^+]} = pH - \frac{1}{2} pK$$

ρ should be negative (electron-attracting groups, with positive σ, enhance acidity and give lower pK values).

(34.2.A)

(a)

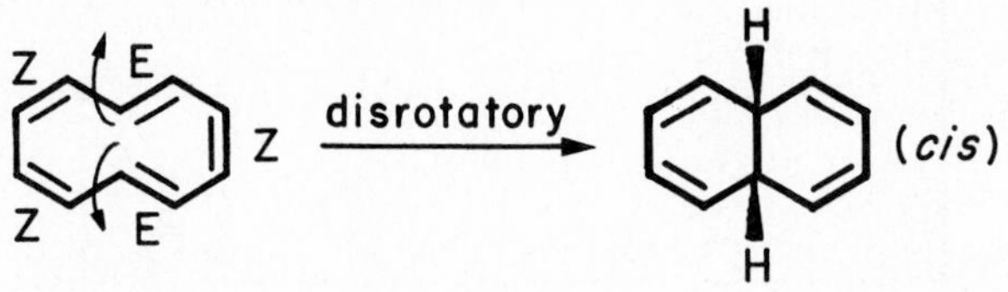

six = 4n + 2 electrons involved in the electrocyclization; therefore disrotatory

(b)

(i)

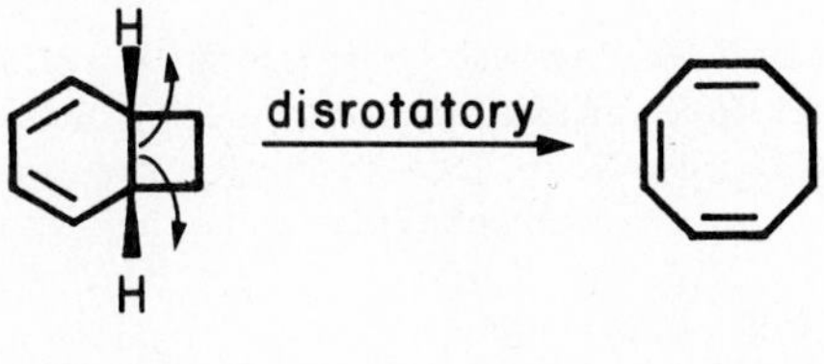

stable and easily formed

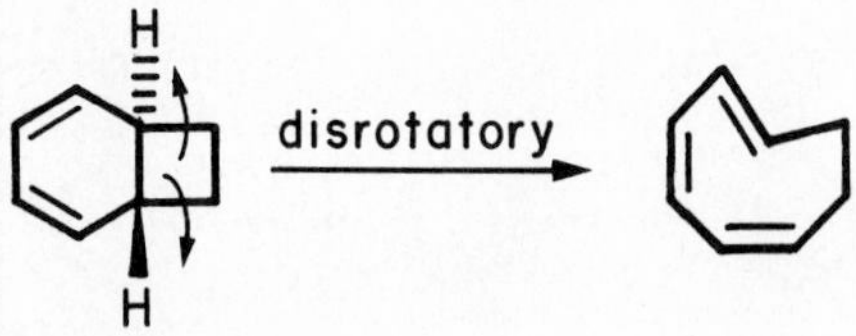

highly strained because of *trans* double bond, therefore formed with greater difficulty

(ii)

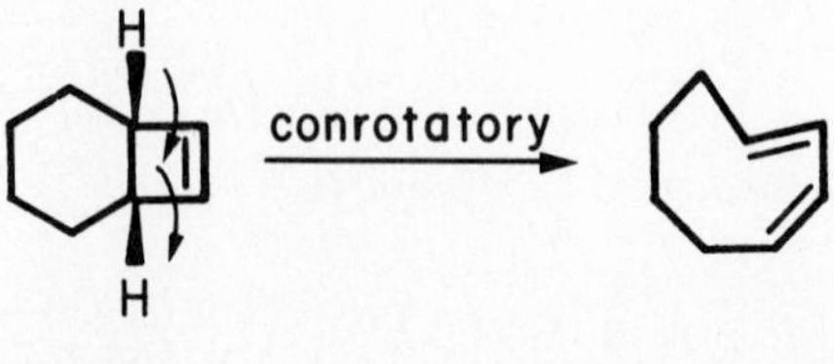

highly strained and hard to form

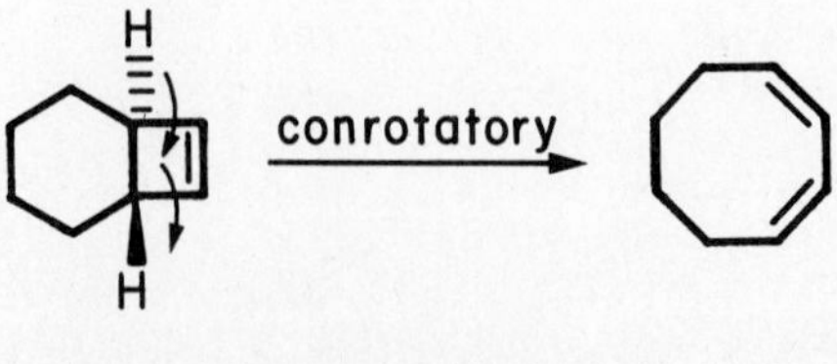

stable and easily formed

(c)

(i)

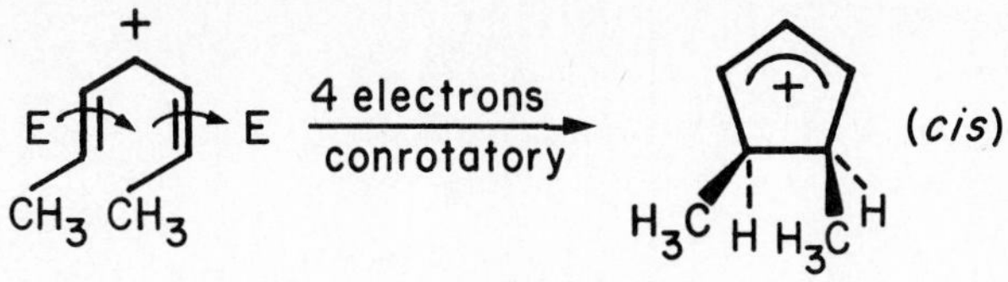

(ii)

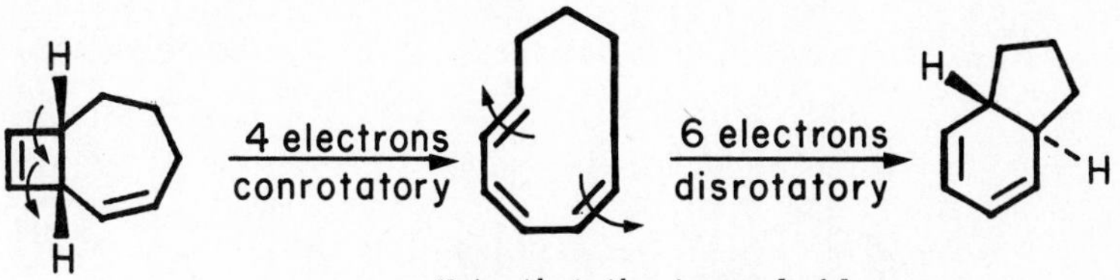

Note that the *trans* double bond is in a nine-membered ring and forms readily

(iii)

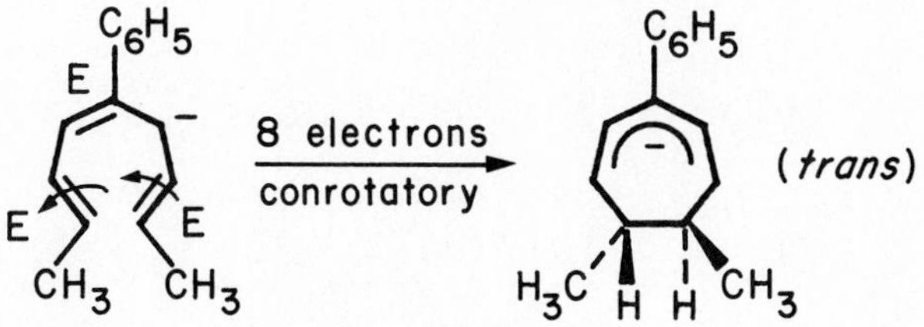

(34.2.B)

(a)

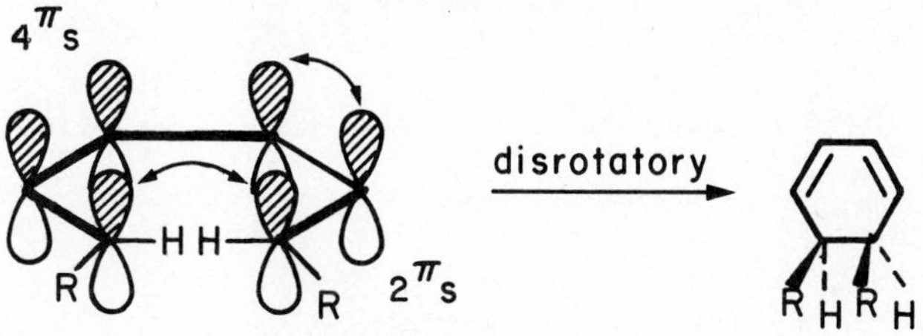

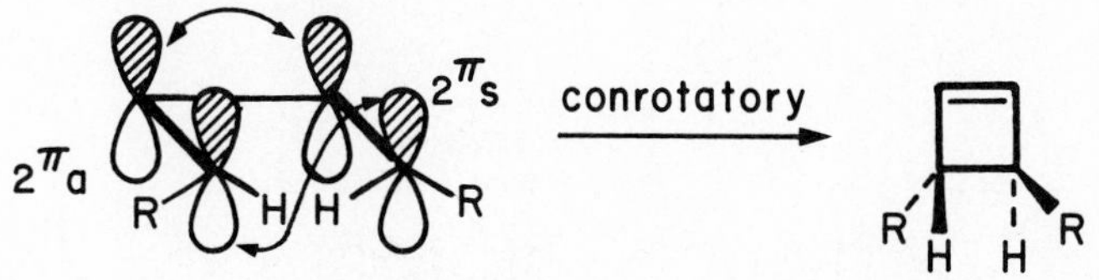

(b) (i) $4^{\pi}s + 4^{\pi}s$: disallowed (the allowed $4^{\pi}s + 4^{\pi}a$ is geometrically unlikely)

(ii) $8^{\pi}s + 2^{\pi}s$: allowed

(iii) $2^{\pi}s + 2^{\pi}s + 2^{\pi}s$: allowed (iv) $12^{\pi}s + 2^{\pi}s$: allowed

(34.2.C)

(a)

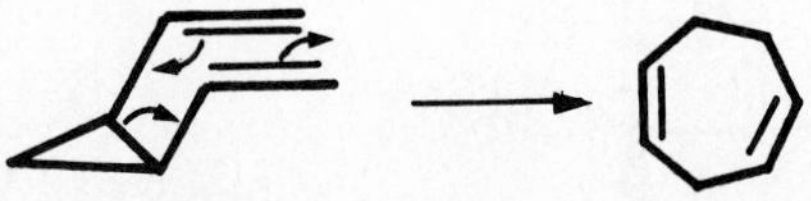

In the *cis* isomer, the vinyl groups are sterically positioned for reaction

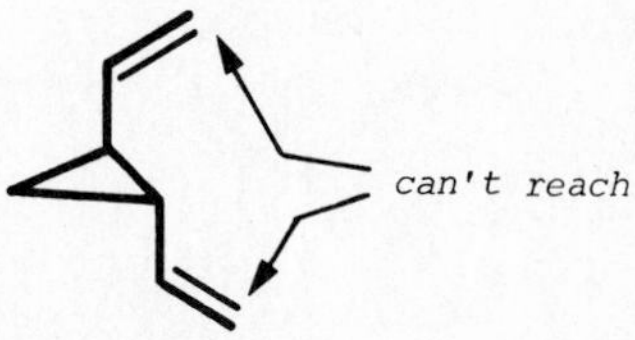

trans isomer

(b)

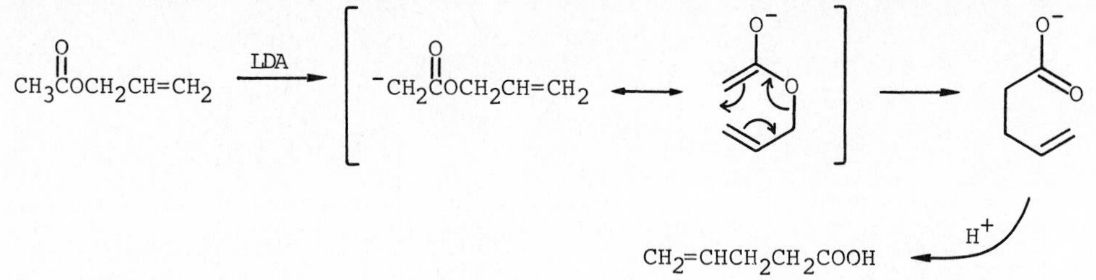

(c) (i)

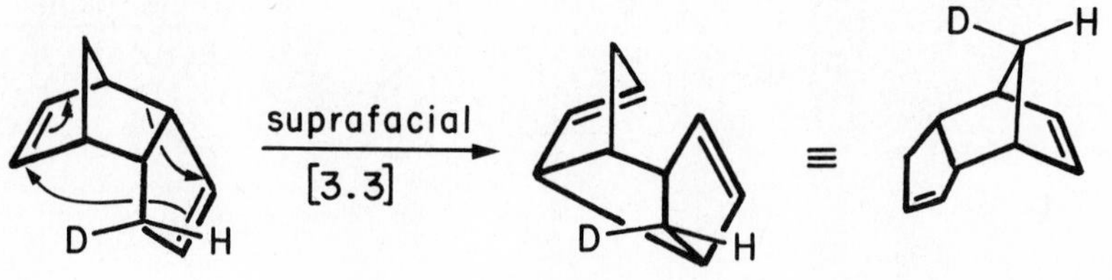

This is the product from an allowed [3.3]sigmatropic rearrangement. The other isomer cannot arise by an allowed pathway.

(ii)

six-electron Möbius, disallowed

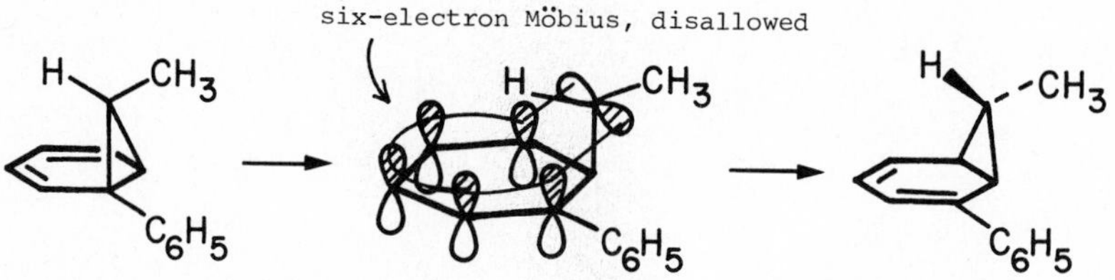

The allowed pathway would give the other stereoisomer via:

six-electron Hückel

(iii)

four-electron Möbius, allowed

(iv)

eight-electron Hückel, disallowed

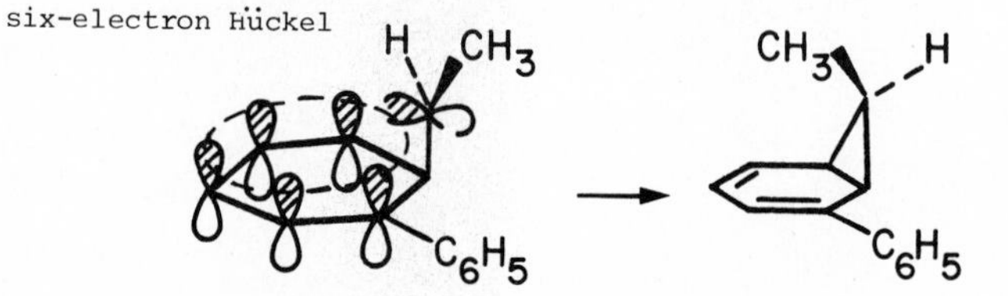

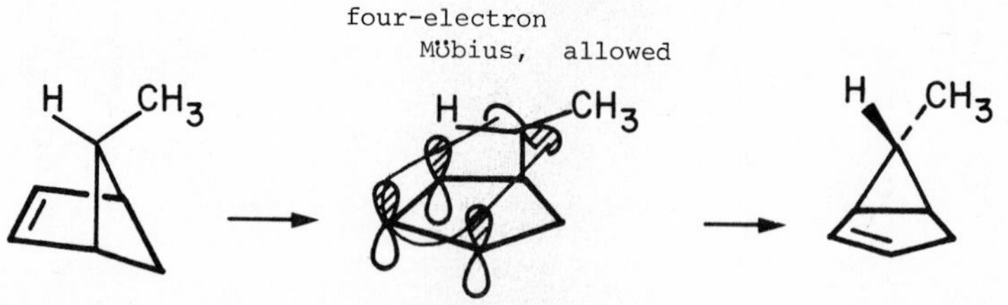

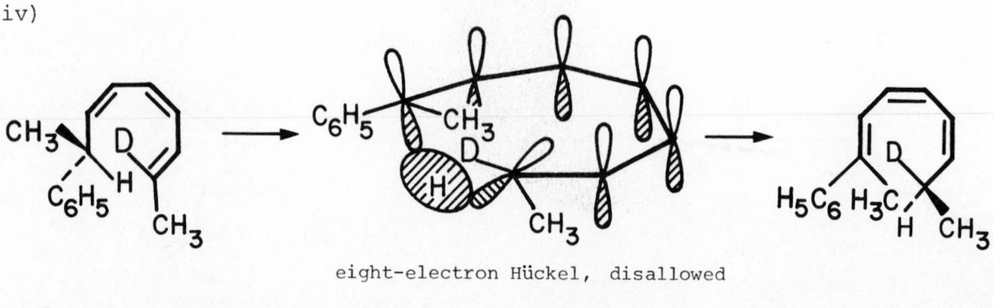

The allowed pathway would be antarafacial:

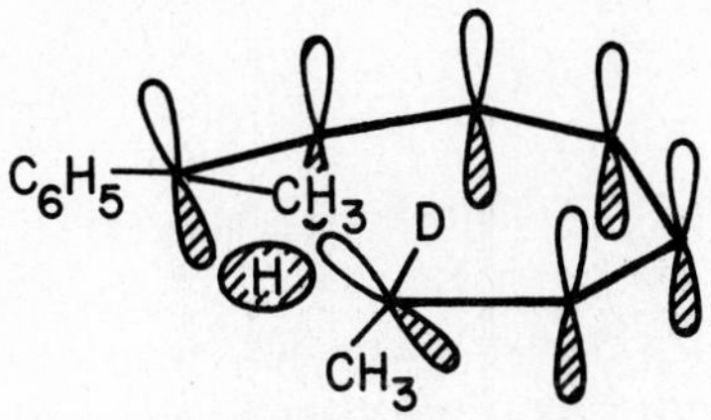

eight-electron Möbius (one negative overlap)

APPENDIX I: GLOSSARY

This glossary provides definitions, brief explanations, and comparisons for many of the words and concepts introduced in the text. After each entry, the section(s) in the text where the term can be found or was first introduced is given in parentheses. Many of the explanations are cross-referenced; terms which appear in italics in the explanations are themselves defined in this glossary.

Most of the bold-faced terms in the text have been included, except those whose definition is given explicitly in the text and which you can find directly through the index. It also does not include name reactions, which again you can find through the index in the text.

Absolute configuration (7.3). Actual 3-dimensional relationship of substituents on a *chiral center*. Named according to the *R-S convention*, or in sugar and amino acid chemistry, by the *"D-L" terminology*.
See *Relative configuration*.

Acene (31.5). Linear, fused polybenzenoid hydrocarbon.

Acetal, ketal (13.7.B).

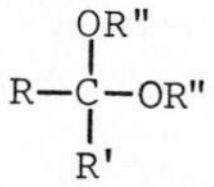

acetal: R,R" = alkyl, R' = H
ketal: R,R',R" = alkyl

These are in equilibrium with the aldehyde or ketone plus alcohol only under acidic conditions. The equilibrium is usually driven to the side of acetal or ketal by removing the water by *azeotropic distillation* using a *Dean-Stark trap*. Ketals and acetals are stable to base and are often used as *protecting groups*.

Achiral (7.1). See *Chiral*.

Activating (23.5). An activating substituent on an aromatic ring increases the electron density of the ring so that it undergoes electrophilic aromatic substitution more readily.

Acylammonium salts (19.7.B). Unstable products of the reaction of acyl halides with tertiary amines:

$$RC(=O)\overset{+}{N}R'_3\ X^-.$$

Acylation (23.4.A). The replacement of a hydrogen (usually) with an acyl group, as in Friedel-Crafts acylation of aromatic rings or acylation of an alcohol with an acid chloride, etc.

Acylium ion (13.7.C.2,19.6, 23.4.A). $[R\text{-}C\equiv O^+ \leftrightarrow R\text{-}\overset{+}{C}{=}O]$

1,2-Additions vs 1,4-additions (conjugate additions) (20.3.A). Additions across one π-bond are called *1,2-additions*, regardless of the actually numbering scheme of the molecule. Addition across a *conjugated* system, for example to the two ends of a *conjugated diene* or across a *conjugated enone*, are called *conjugate* or *1,4-additions*.

Addition-elimination mechanism (19.6). See *Nucleophilic addition-elimination mechanism*.

<u>Aglycon</u> (28.10). The non-sugar part of a *glycoside*.

<u>Alditol</u> (28.5.D). The polyalcohol resulting from reduction of the carbonyl group of a sugar.

<u>Aldonic acid</u> (28.5.E). The monocarboxylic acid resulting from oxidation of the aldehyde carbon of an *aldose*.

<u>Alkylation</u> (23.4.B). The replacement of a hydrogen (usually) with an alkyl group, as in Friedel-Crafts alkylation of an aromatic ring, or alkylation of a ketone via its enolate, etc.

<u>Allyl, allylic</u> (20.1). An *allylic system* is one involving 3 adjacent, overlapping p-orbitals, making 3 molecular orbitals, and filled with 2 (*allyl cation*), 3 (*allyl radical*), or 4 (*allyl anion*) electrons. It is an example of *conjugation*.

An *allylic position* is one next to a double bond.

Reactions of *allyl systems* are said to proceed with *allylic rearrangement* if the double bond moves in the course of the reaction.

<u>Amphoteric</u> (29.1, 29.3). Having both acidic and basic character in the same molecule.

<u>Angle of rotation</u>, α (7.2). The amount by which plane polarized light is rotated by an *optically active* sample.

<u>Anisotropic</u> (7.2, 11.3.A, 12.3.B, 22.3.A). Not having the same effect in all directions. *Diamagnetic anisotropy* refers to the *deshielding* effect of electron density which is adjacent to but not surrounding the nucleus being observed in the nmr. (See *chemical shift* and *ring current*.)

<u>Anomers, anomeric</u> (28.3). *Anomers* are stereoisomeric sugars which differ in configuration only at the *hemiacetal* or *hemiketal* carbon, also known as the *anomeric* carbon.

<u>Antarafacial</u> (34.2.B). On opposite faces of a π system. See *suprafacial*.

<u>Anti, Gauche, Syn, Eclipsed, Staggered</u> (5.2). These are all terms used to describe three-dimensional relationships of substituents at opposite ends of a single bond.

When the substituents at one end are "lined up" with the substituents at the other end, the molecule is said to have an eclipsed conformation of the single bond. The molecules drawn below are shown in eclipsed conformations.

<u>Eclipsed</u>

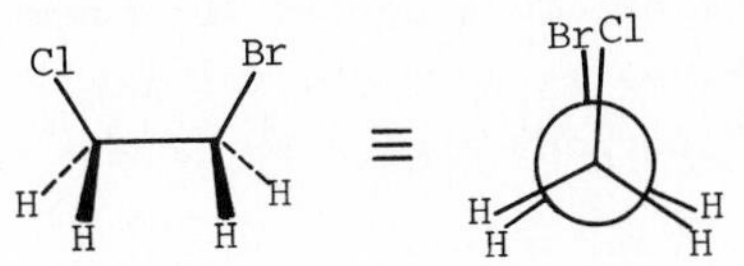

<u>A</u>

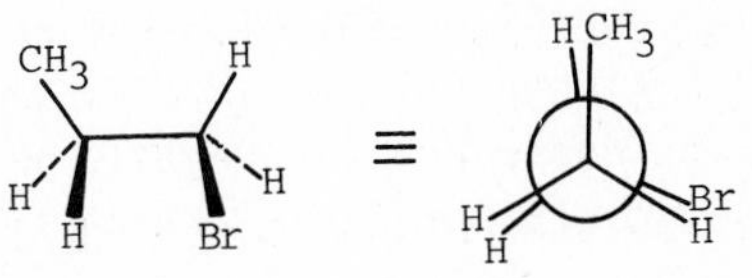

<u>B</u>

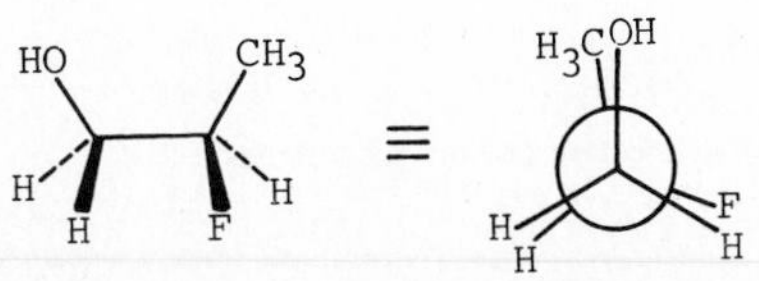

<u>C</u>

In the eclipsed conformation A, the chlorine and bromine atoms are *syn* to each other (coplanar and "pointing in the same direction"). Similarly, the methyl and the hydroxy groups in the eclipsed conformation C are *syn*.

When the substituents at one end are "in between" those at the other end, the molecule is in a *staggered* conformation, as shown for the molecules below. The staggered conformations are more stable (lower potential energy) than the eclipsed.

Staggered

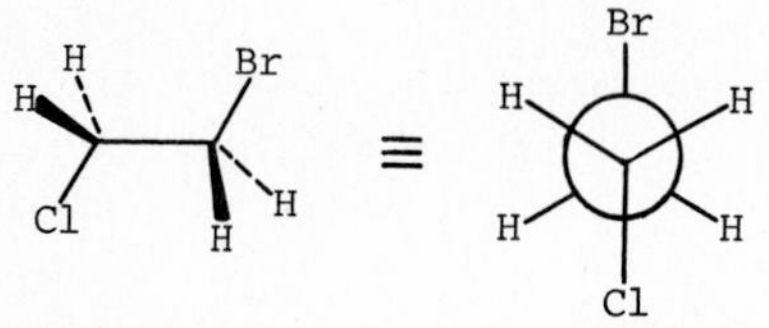

D

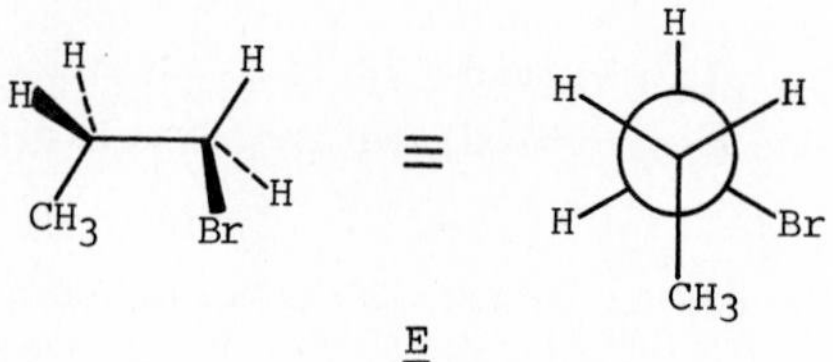

E

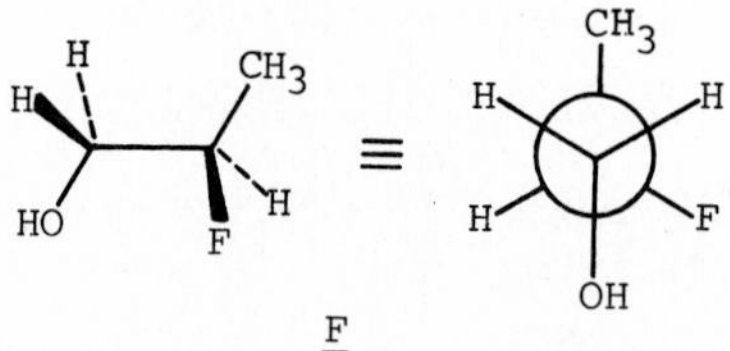

F

In the staggered conformations D and F, the bromine and the chlorine, and the methyl and hydroxy groups, respectively, are *anti* to each other (coplanar, but "pointing in opposite directions"). In the staggered conformation E, the bromine and the methyl group are said to be *gauche* to each other (dihedral angle = 60°); the staggered conformation F has a *gauche* relationship between the fluorine and hydroxy substituents.

Anti-aromatic (22.7.A). See *Aromatic*.

Antibonding orbital (2.7). See *Orbital*.

Anti-Markovnikov (11.6.C). An orientation opposite to that predicted by *Markovnikov's rule* for electrophilic addition to an alkene. It is the overall result of the free-radical addition of HBr to an alkene or alkyne, or of their hydration via hydroboration/oxidation (11.6.D).

Annulene (22.7.E). A monocyclic $(CH)_n$ hydrocarbon, for example cyclodecapentaene is [10]annulene.

Aromatic (22.1.A, 22.1.C, 22.7.A). *Aromatic* compounds are those which have a special stabilization of their electronic systems due to a cyclic arrangement of $4n+2$ π-electrons. See *Hückel 4n + 2 rule*, *resonance energy*, and *delocalization energy*. Compounds with cyclic π-systems of 4n electrons appear to have a special destabilization and are called *antiaromatic*.

An *aromatic* hydrogen or other substituent is one which is directly attached to an aromatic ring.

Aromatization (31.3.B). Formation of an *aromatic* ring from a less unsaturated precursor, usually by a *dehydrogenation* process.

<u>Asymmetric center or carbon</u> (7.1). See *Chiral*.

<u>Asymmetric induction</u> (7.8). *Asymmetric induction* is said to occur when a *chiral center* is formed with a preference for one *absolute configuration* over the other.

<u>Atomic orbital</u> (2.5). See *Orbital*.

<u>Autocatalytic</u> (13.6.D). If one of the products of a reaction is a catalyst for the reaction, the reaction is said to be *autocatalytic*. Often such reactions exhibit an *induction period*.

<u>Autoxidation</u> (10.11.B, 13.8.A). Reaction with atmospheric oxygen, usually to generate an *alkylperoxide*, ROOH, or a peroxycarboxylic acid, $RC(=O)OOH$.

<u>Average bond energy</u> (6.5). The average energy required to dissociate each of the bonds of a certain type in a molecule. It differs from the *bond dissociation energy*, which refers to a specific, instead of an average, bond of a given type.

<u>Axial and equatorial</u> (5.7). In the most stable conformation for the cyclohexane ring, called the "chair" form, there are two orientations which substituents can adopt. If they point up and down, perpendicular to the average plane of the six-membered ring, they are *axial*; if they point out somewhat horizontally relative to the ring, they are called *equatorial*, as drawn below:

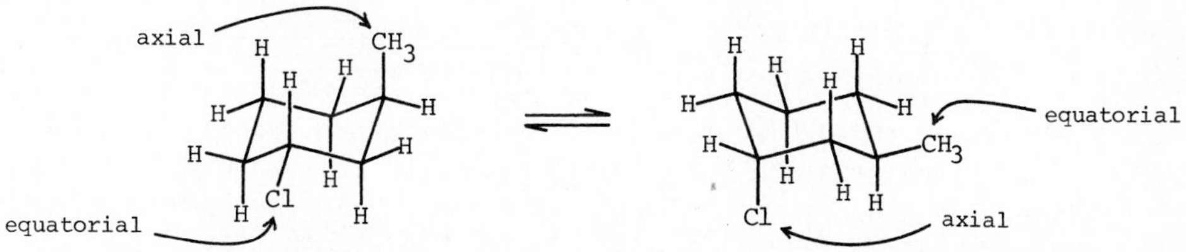

Note that when the cyclohexane ring flips between the two chair conformations, the axial substituents become equatorial and vice versa. You can see this easily if you have a set of models.

The equatorial positions are less *sterically hindered* and therefore more stable for substituents to occupy than the axial ones.

<u>Azeotropic distillation</u> (13.7.B). Two immiscible solvents will distill as a mixture. This forms the basis for removing water from an equilibrium process by distilling with benzene or toluene. The distillate separates into two phases, with the water being drawn off in a *Dean-Stark trap* and the benzene or toluene returned to the reaction flask.

<u>Beer's law</u> (21.3). In ultraviolet *spectroscopy*: $\log \frac{I_o}{I} = \varepsilon cd$, where I_o, I = light intensity before and after passing through cell, ε = *extinction coefficient*, c = concentration in moles $liter^{-1}$, and d = path length in cm.

<u>Bent bonds</u> (5.7, 11.1.A). When geometric constraints force the bond angles around an atom to be smaller than the preferred angles between the orbitals, the single bonds "bend", which is to say the atomic orbitals do not overlap in their normal end-to-end fashion. Instead, the center of electron density in the bonding orbital lies off to the side of the internuclear axis, and the bond is weaker (see Figure 2.9). This behavior is most important for cyclopropane.

Benzyl, benzylic (22.5). A *benzylic* radical is one involving a carbon radical directly attached to an aromatic ring. Benzylic cations and anions are analogous. A benzylic position is one immediately adjacent to an aromatic ring.

Benzyne (30.3.A.2). A highly reactive intermediate in the *elimination-addition mechanism* for nucleophilic aromatic substitution:

Betaine (13.7.G). A *zwitterionic* intermediate. In the Wittig reaction the betaine may be in equilibrium with an *oxaphosphetane.*

Bimolecular (8.5). See *Molecularity.*

Biosynthesis (18.1, 26.7, 34.7). The synthesis of compounds by living organisms.

Bond (2.7). A net attractive interaction between two atoms. The most important bonds in organic chemistry are *covalent bonds,* which result when two atoms share electrons. The behavior of the shared electrons is described by a *molecular orbital,* and the means by which the sharing is accomplished is overlap of the appropriate *atomic orbitals* on each atom. A *single bond* results from the sharing of two electrons in one molecular orbital; a *double* (triple) bond results from the sharing of four (six) electrons in two (three) molecular orbitals.

A *dative* or *donor bond* usually involves the interaction between a vacant metal orbital and a filled orbital (often a *lone pair*) of a *ligand.*

Bond dissociation energy (D or DH°) (6.1). The energy required to break a bond *homolytically* and separate the two radicals.

α- and β-Branching (8.8). Alkyl groups attached to the carbon undergoing a substitution reaction are α-branches; those on an adjacent carbon are β-branches. Both contribute to *steric hindrance* in *S_N2-reactions.*

Bridged, fused, and spiro ring systems.

A *bridged* compound is a bicyclic compound in which the rings share more than two atoms.

A *fused* compound is one in which the two rings share two adjacent atoms.

A *spiro* compound is one in which the rings share only one atom.

Carbenoid (11.6.F). An organometallic complex which behaves like a carbene ($R-\ddot{C}-R'$) source in its reactions.

Carbinol (10.2). A little-used system of alcohol nomenclature.

Carbinolamine (13.7.C). A *hemiaminal:* $R_2C\begin{smallmatrix}\nearrow OH \\ \searrow NR'_2\end{smallmatrix}$

Carbocation (2.4). A molecule in which a carbon atom is only trivalent and has only six *valence* electrons. The simplest carbocation is the methyl cation The positively charged carbon is sp^2 *hybridized,* bonding to the hydrogens via the three sp^2 orbitals and leaving the remaining p orbital vacant. $\begin{matrix} H \\ H:\ddot{C}:H \\ + \end{matrix}$.

Carbocation rearrangement (10.7.B). Migration of a substituent, with its bonding electron pair, from an adjacent carbon to a cationic carbon so as to generate a more stable *carbocation.*

e.g., 2° carbocation → 3° carbocation, or ring-expansion of a four- to a five-membered ring.

<u>Center of symmetry</u> (7.6). An object has a center of symmetry when the exact same environment is encountered at the same distance in both directions along any line through a particular point, which is called the center of symmetry. Such an object is *achiral*.

<u>Chain reaction</u> (6.3). A chain reaction is one in which (relatively few) *initiation* steps take place, followed by (many) *propagation* steps which convert the starting materials to products, and finally, by (again relatively few) *termination* steps. These reactions usually involve radical intermediates, and transformations involving radical intermediates are usually chain reactions. A chain reaction process is shown below:

$$\text{I–I} \xrightarrow[\text{or } \Delta]{h\nu} 2\ \text{I}\cdot$$
$$\text{I}\cdot + \text{A} \longrightarrow \text{I}' + \text{A}\cdot$$
Initiation

$$\text{A}\cdot + \text{B} \rightarrow \text{A}' + \text{B}\cdot$$
$$\text{B}\cdot + \text{A} \rightarrow \text{A}\cdot + \text{B}'$$
Propagation: overall process is $\text{A} + \text{B} \rightarrow \text{A}' + \text{B}'$

$$\text{A}\cdot + \text{B}\cdot,\ \text{A}\cdot + \text{A}\cdot,\ \text{or } \text{B}\cdot + \text{B}\cdot \rightarrow \text{AB, AA, or BB}$$
Termination

<u>Chair conformation</u> (5.7). The most stable conformation of a cyclohexane ring. In the chair conformation (depicted under *axial*, above), all of the C-C bonds have the staggered conformation and each carbon is free to adopt its preferred tetrahedral geometry.

<u>Characterization</u> (3.4). Determination of the physical and chemical properties of a compound. For example, you could characterize 1-tetradecene ($CH_3(CH_2)_{11}CH{=}CH_2$) by determining its melting point (-12 °C), boiling point (232 °C), reaction with $KMnO_4$ (to give a carboxylic acid, $CH_3(CH_2)_{11}COOH$), combustion analysis (85.7% C, 14.3% H), and by recording its nmr, ir, and mass *spectra*, etc.

<u>Charge-transfer complex</u> (30.7.D). A complex formed by the face-to-face interaction of two π-systems, one electron-rich and the other electron-poor. There is a certain amount of electron density transferred from the electron-rich (donor) to the electron-poor (acceptor) system.

<u>Chemical shift</u> (9.4). The difference in resonant frequency in the nmr between a reference compound (usually tetramethylsilane, "*TMS*") and the nucleus of interest, usually measured in parts per million (*ppm*) *downfield* from TMS; this is the so-called "δ scale".

Downfield = *deshielded* = higher frequency = lower field = "to the left" in most spectra.

Upfield = the opposite of downfield.

The presence of electron density *around* a nucleus is *shielding;* the withdrawal of electron density or the presence of electron density *next to* a nucleus (*diamagnetic anisotropy*) is *deshielding.*

Chiral (7.1). A molecule is *chiral* if it is not superimposable on its mirror image; the two mirror image molecules are *enantiomers*. A carbon atom with four different substituents is an *asymmetric* or *chiral center.* A molecule with an odd number of *chiral centers* is *chiral*, but with an even number there exists the possibility of *achiral*, *meso* compounds. *Achiral* molecules are *optically inactive*, but *chiral* molecules are not necessarily *optically active*, they could be present as a *racemic* mixture (equal amounts of both *enantiomers*).

Chloromethylation (23.4.B). Replacement of an aromatic hydrogen (usually) with a chloromethyl group, usually using CH_2O, HCl, and $ZnCl_2$.

Combination bands (14.2). See *Overtone.*

Condensation (13.7.C). Combination of two molecules with elimination of water or other small molecule, as in the formation of an *imine* or in the aldol condensation (13.7.F).

Condensed formulas (3.1). See *Structural formulas.*

Configuration:

Electronic: The specific distribution of electrons in *atomic* or *molecular orbitals.*

Stereochemical: The arrangement in 3 dimensions of the substituents on a *chiral center.* See *Absolute configuration.*

Configurational isomers (11.1.A). See *conformation.*

Conformation (5.2), Conformational isomers (11.1.B) = conformers (5.3). There is a fine distinction between *conformers* and other types of *isomers* which is confusing at times. Anytime there are molecules which have the same formula but are put together differently in three dimensions, we can call them isomers. A distinction is drawn between those isomers which interconvert rapidly at ordinary temperatures (without breaking any bonds usually), and those which interconvert very slowly or not at all (and which usually require bonds to be broken and remade during the process). The former are *conformers*, and the different arrangements in three dimensions which a molecule can adopt easily (without breaking bonds) are called different *conformations.* Those which interconvert slowly are called *configurational isomers.*

Conrotatory and disrotatory (34.2.A). In an *electrocyclic* reaction, if the two ends of the *conjugated* system rotate in the same direction (e.g., clockwise) during ring closure the reaction is said to be *conrotatory;* if they rotate in opposite directions (one clockwise, one counterclockwise) it is a *disrotatory* ring closure. *Electrocyclic* ring opening reactions are evaluated the same way.

Coordinatively saturated (15.6). A transition metal which has achieved the *18-electron configuration* is *coordinatively saturated.*

Conjugate acid, (conjugate base) (4.5). The species which results from loss of a proton from a molecule is its *conjugate base;* the species which results from protonation of a molecule is its *conjugate acid;*

$$HA \rightleftharpoons A^- + H^+$$

$$B: + H^+ \rightleftharpoons \overset{+}{B}H$$

A^- and B: are conjugate bases of HA and $\overset{+}{B}H$, respectively; HA and $\overset{+}{B}H$ are conjugate acids of A^- and B:, respectively.

Conjugation (19.5,20). π-Orbital overlap from more than two atoms. Two double bonds are conjugated if they are adjacent to each other (C = C - C = C, *conjugated diene*) and are *unconjugated* if there is one or more sp^3 hydridized carbons in between (C = C - C - C = C , *unconjugated diene, isolated double bonds).*

A *conjugated enone* is one in which the carbonyl group and double bond are adjacent (C = C - C = O).

Coupling constant, J (8.6). See *Spin-spin splitting.*

Covalent bond (2.7). See *Bond.*

Cracking (6.2). *Pyrolysis* of alkanes to give shorter-chain alkanes and alkenes.

Cross-conjugation (20.3.B). A *conjugated* system which is branched rather than linear is *cross-conjugated.*

Crown ethers (10.12.A). Cyclic polymers of ethylene oxide:

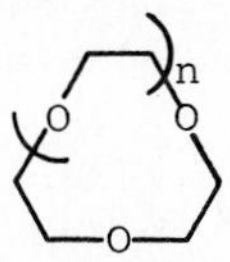

These are important as *phase-transfer catalysts.*

Cumulated double bonds (20.2.C). Double bonds which share the same carbon atom (C = C = C).

Cycloaddition reaction (20.5, 32.5.B, 34.2.B). A reaction in which electron movement occurs in a cyclic manner during the course of addition of a conjugated molecule to a π-bond (for example, the Diels-Alder reaction or *1,3-dipolar cycloaddition).*

D-L convention (28.2). Systems of nomenclature for indicating the *absolute configuration* of *amino acids* and *sugars:* the molecule is drawn in a *Fischer projection* with the carbon chain written vertically and the most oxidized end at the top. If the *hetercatom* substituent (hydroxy or amino group) projects to the right, that *chiral center* is D, if it's on the left it's L. Whether a sugar belongs to the D- or the L-series is determined by the last *chiral center* in the chain. The configuration of an amino acid is determined by the α-carbon.

Deactivating (23.5). The opposite of *Activating.*

Dean-Stark trap (13.7.B). See *Azeotropic distillation.*

Decarbonylation (15.6). Loss of carbon monoxide, usually referring to the reaction catalyzed by transition metal complexes like Wilkinson's catalyst ($[(C_6H_5)_3P]_3RhCl$).

Decarboxylation (27.6.B.3, 27.7.C.2). Loss of carbon dioxide.

Decoupling (9.9). Strong electromagnetic irradiation of a nucleus at its resonant frequency in an nmr spectrometer causes it to change its allowed *quantized* orientation in the magnetic field rapidly on the *nmr time scale*, averaging out its effect on nuclei to which it is *coupled*. This reduces the observed *coupling constant* J to 0. It is useful in proton nmr for interpreting splitting patterns, and is routinely used in cmr to simplify the spectra and improve the signal-to-noise ratio.

Off-resonance decoupling in cmr is a partial decoupling of the protons from the carbons, leaving a small, residual coupling so that the number of hydrogens attached to each carbon can be determined.

Degenerate (14.2, 22.1.C). Energy levels which are different but equal in energy are *degenerate*.

Degree of association (13.6.B). How tightly bound an *ion-pair* is. In non-polar solvents, ions are poorly *solvated* and the *ion pairs* are tightly associated. In polar, particularly hydroxylic solvents, *solvation energies* are high, and ions are only loosely associated. The association of ions usually reduces their reactivity.

Dehydration (11.5.B). Loss of water, usually with reference to an alcohol losing a molecule of water to give an alkene.

Dehydrogenation (22.1.E). The opposite of *hydrogenation*. *Dehydrogenation* involves the removal of hydrogen from a molecule and the introduction of *unsaturation*.

Dehydrohalogenation (11.6.B.3). Removal of HX, usually referring to E2 reaction of an alkyl halide.

Deinsertion (15.6). See *Insertion*.

Delocalization energy (22.1.B). The hypothetical difference in energy between the actual distribution of electrons in a *conjugated* system and a system of identical geometry with electronic isolation of the π-bonds from each other. Contrast with *empirical resonance energy*, (See *Resonance energy*).

Deshielded, shielded (9.4). See *Chemical Shift*.

Deuteration (15.5.A). Introduction of deuterium into a molecule. It can be accomplished by *deuterium exchange* or by reaction of an organometallic reagent with D_2O, among other methods.

Deuterium exchange (13.6.A). Equilibration of ionizable hydrogens with deuterium atoms from the solvent, usually used to determine how many α-hydrogens (enolizable hydrogens) are present in a ketone or other carbonyl compound.

Dextrorotatory, levorotatory (7.2). A sample which rotates plane polarized light clockwise (+ direction) is *dextrorotatory* (counterclockwise is *levorotatory* (-)).

Diagmagnetic shielding and deshielding (9.4). See *Chemical Shift*, *Anisotropy*, and *π-Electron circulation*.

Diastereomer (7.6). *Stereoisomers* which are not *enantiomers*. If more than one *chiral center* is present in a molecule, changing the *configuration* of all of them gives the *enantiomer*. Changing the configuration of *less* than all of them gives a *diastereomer*.

Diastereomeric salts (29.4.f). See *Resolution*.

Diaxial (11.6.B.3). This is equivalent to *anti* in a cyclohexane system.

Diazotization (24.7.B). Conversion of an amino into a diazonium group: $RNH_2 + HONO + H^+ \rightarrow RN_2^+ + 2H_2O$.

Dienophile (20.5). The "monoene" component of the Diels-Alder reaction. Electron-withdrawing groups on a dienophile usually increase its reactivity.

Dilution principle (20.1.C). A *bimolecular* reaction is slowed down more on dilution than a *unimolecular* reaction.

Dimer, trimer (11.6.G). See *Polymer*.

1,3-Dipolar cycloaddition (32.5.B). A *cycloaddition reaction* in which the *conjugated* component is 3 atoms long and is *zwitterionic* or with *zwitterionic* character.

Dipole moment, μ(8.1). A dipole moment arises whenever positive and negative charges are separated. The magnitude depends on separation, d, and charge, q: $\mu = q \cdot d$. The general quality of "polarity" depends in part on the presence of a dipole moment in a molecule.

Dispersion force = London force = Van der Waals attraction (5.1). These three terms all describe a very weak interaction which results in an attraction between two molecules. It is different from the more easily understood ionic or dipole interactions, and depends instead on differences in the "instantaneous" distribution of electrons. The most important features to remember for this force are: (1) it is the primary force of attraction between hydrocarbon molecules; (2) it decreases very rapidly with distance; and (3) its magnitude is proportional to the "surface area" of a molecule.

Disproportionation (6.2). *Disproportionation* involves two of the same or similar molecules reacting with each other to produce different products:

$$2A \rightarrow B + C$$

The *cracking* of alkanes provides an example of this, as does the Cannizzaro-reaction (13.8.D).

Disrotatory (34.2.A). See *Conrotatory*.

Donor-acceptor complex (30.7.D). See *Charge-transfer complex*.

Double bond (2.3). See *Bond*.

Double bond character (19.1). Degree of π-bonding character, often evaluated as amount of contribution from a resonance structure containing a double bond, as in the amide linkage.

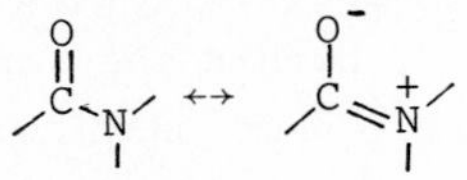

Doublet (9.6). See *Spin-spin splitting*.

Downfield, upfield (9.4). See *Chemical Shift*.

E (entgegen) and Z(zusammen)(11.2). Nomenclature for describing the configuration of a double bond.

E2 (Elimination-bimolecular) (8.11, 11.5.A). The most common mechanism for an elimination reaction. It involves simultaneous removal of the proton by a base, formation of the π-bond, and departure of the leaving group. All the orbitals involved must be coplanar, and this is usually accomplished in an *anti* relationship:

Eclipsed (5.2). See *Anti*.

Edman degradation (29.6.C.2). A method for the stepwise removal and identification of the N-terminal amino acids from a *peptide* chain.

α-, β-, and γ-effects (9.9). Characteristic effects on *chemical shift* in cmr spectroscopy which depend on specific substituents and their position relative to the nucleus observed.

Eighteen electron rule (15.6). Just as a 2nd or 3rd period element tries to achieve an octet of *valence electrons* in its bonding arrangement, so does a transition metal try to achieve a filled shell of 2s + 10d + 6p = 18 electrons through its bonds to *ligands* and other groups. A metal which has achieved an *18-electron configuration* is *coordinatively saturated*.

Electrocyclic reaction (34.2.A). A reaction in which a conjugated molecule undergoes a cyclic rearrangement of electrons to form a molecule with one less π-bond and one more ring, or the reverse of this process.

Electron affinity (2.2). (See definition on page 6 in text.)

Electron-attracting and -donating (10.4). See *Inductive effect*.

π-Electron circulation (11.3.A). The motion of all electrons is altered by a magnetic field. The electron motion induced in the π-electrons of alkenes and aromatic rings causes a *downfield (deshielding)* effect on the *chemical shifts* of nuclei attached to them. This effect is known as *diagmagnetic deshielding*.

Electron count (2.2, 15.6). The number of electrons in the valence shell of an atom or metal is determined for the purpose of evaluating *formal charges*, *filled octets*, *18-electron* configurations, etc.

Electron density (2.5). Refers to the *probability* of finding an electron in a given region of space. This probability is given by the square of the *wavefunction* for the electron; high probability corresponds to high electron density.

Electronegative (2.2). Exerting a strong attraction for electrons. The electronegativity of the elements increases as one goes up and to the right in the periodic table. The more electronegative a group is, the better it is able to stabilize a negative charge, within the constraints of the *octet rule*. Electronegative elements usually need only one or two additional electrons to complete their filled octet.

Electronic transition (21.1). Change of the electronic state of a molecule or atom, usually from the *ground electronic state* to an *excited electronic state* or vice versa. Such transitions are important in ultraviolet *spectroscopy* and photochemistry (34.4).

Electrophile, electrophilic reagent. The opposite of *nucleophile* and *nucleophilic reagent*. An electrophile contributes the vacant *orbital* when a bond is formed by the reverse of a *heterolytic* process:

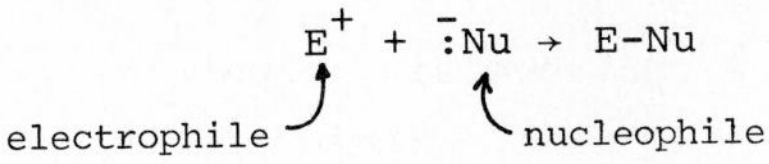

A proton is the simplest electrophile.

Electropositive (2.2). The opposite of *electronegative*. Electropositive elements are those at the left of the periodic table, which achieve a complete *valence* shell (filled *octet*) by losing rather than gaining one or two electrons.

Elimination-addition mechanism (30.3.A.2). For nucleophilic aromatic substitution via a *benzyne* intermediate.

Empirical formula (3.4). The empirical formula of a compound expresses the ratio of elements present. Compare with *molecular formula*.

Enantiomers (7.1). *Stereoisomers* which differ only by being mirror images of each other. See *Chiral* and *Diastereomer*.

Endo and exo (20.5). In a bicyclic compound, the configuration of a non-bridgehead substituent can be specified by its relationship to the other bridges. If it points in the same direction as the longer bridge, it is *endo*; if it points toward the shorter bridge, it is *exo*: e.g.,

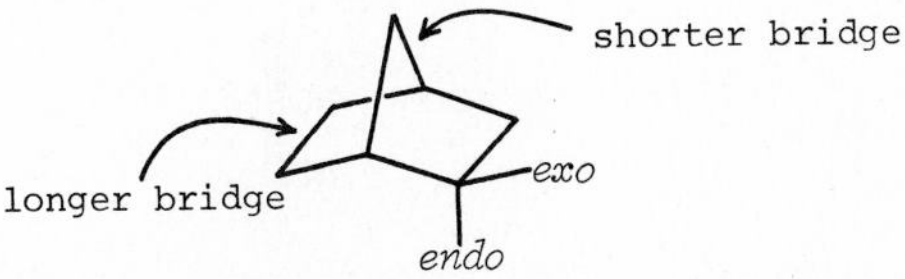

In the transition state of the Diels-Alder reaction, if a substituent on the *dienophile* points toward the diene, it is *endo*; if it points away it is *exo*.

Envelope (5.7). The most favorable conformation for a cyclopentane ring, in which one carbon is pushed out of the plane of the other four in order to minimize the eclipsing interactions of all of the hydrogens. This conformation is not

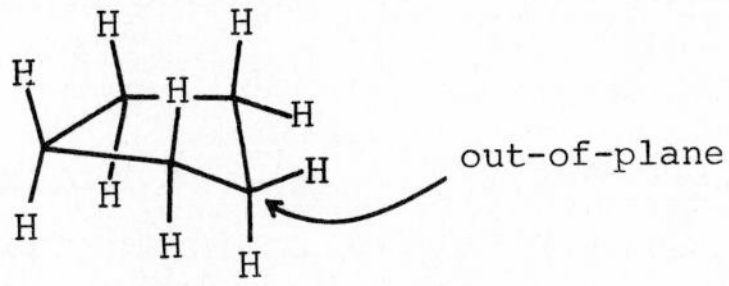

fixed, and all five carbons of a cyclopentane ring can take the out-of-plane position interchangeably. In this instance, the interconversion is referred to as *pseudorotation*.

Enzyme (28.3). A *protein* which catalyzes a chemical reaction.

Equatorial (5.7). See definition under *Axial*.

Equilibrium (4.1). Although the term equilibrium is defined in the text, it is important that you understand the distinction between *equilibrium* and rate. The rate constant for a reaction is a measure of how *fast* it goes; the equilibrium

constant is a measure of how *far* it goes. There is not necessarily any connection
between these two constants: there are many highly *exothermic* reactions which go
very slowly (for instance, the decomposition of TNT in the absence of a detona-
tion), and many reactions which are only slightly exothermic but which go very
rapidly (for instance, the neutralization of a weak acid with a weak base).

  Erythro and threo (27.3.A). A controversial nomenclature system for specify-
ing the relative *configuration* of two *chiral centers*. It arose from carbohydrate
chemistry, and is used with Fischer projections in the following way: with the
carbon chain written vertically, if the "similar" substituents on two chiral
centers are on the same side, the relationship between those carbons is *erythro*.
If they are on opposite sides, the relationship is *threo*. Complications arise in
systems other than carbohydrates and when it is difficult to decide what the
"similar" substituents are.

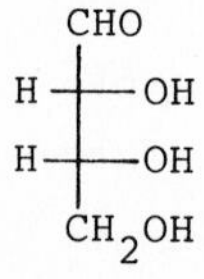

$$
\begin{array}{cc}
\text{CHO} & \text{CHO} \\
\text{H}\!\!-\!\!\text{OH} & \text{HO}\!\!-\!\!\text{H} \\
\text{H}\!\!-\!\!\text{OH} & \text{H}\!\!-\!\!\text{OH} \\
\text{CH}_2\text{OH} & \text{CH}_2\text{OH} \\
\text{D-}erythrose & \text{D-}threose
\end{array}
$$

  Ester, esterification (10.7.C, 10.7.D, 18.7.C.2). The compound formed from
the loss of water (formally) between an alcohol and an acid is an *ester*:

$$\text{ROH} + \text{HO}\overset{\text{O}}{\overset{\|}{\text{C}}}\text{R'} \rightarrow \text{RO}\overset{\text{O}}{\overset{\|}{\text{C}}}\text{R'}, \text{ a carboxylic ester}$$

$$2\,\text{ROH} + \text{HO}\overset{\text{O}}{\underset{\text{O}}{\overset{\|}{\underset{\|}{\text{S}}}}}\text{OH} \rightarrow \text{RO}\overset{\text{O}}{\underset{\text{O}}{\overset{\|}{\underset{\|}{\text{S}}}}}\text{OR}, \text{ a sulfate diester}$$

$$3\,\text{ROH} + \text{HO-}\overset{\text{OH}}{\overset{|}{\underset{..}{\text{P}}}}\text{-OH} \rightarrow \text{(RO)}_3\text{P:}, \text{ a phosphite triester}$$

(but: ROH + HBr → RBr , an alkyl halide)

  Exact mass (17.3). The actual mass, to at least 0.0001 atomic mass unit
accuracy, of a molecule or molecular fragment, which allows you to distinguish
between *molecular formulas* of the same *nominal mass*.

  Excited electronic state (21.1). See *Ground electronic state*.

  Exo (20.5). See *Endo*.

  Extinction coefficient, ε (21.3). A measure of the probability that a
quantum of electromagnetic radiation with the correct energy will be absorbed
and result in an *electronic transition*. It is related to the amount of light
transmitted by a sample by *Beer's law*.

  First-order vs non-first order nmr spectra (9.6, 9.7). When the magnitude of
the difference in *chemical shift* (Δν) of two nuclei is much greater than their
coupling constant, J, the spectrum is said to be *first-order*, and the *spin-spin
splitting* patterns follow the usual rules. When Δν ≈ J, the spectrum is *non-
first-order*, and the splitting patterns are very complex.

  First order reaction (4.3). A true first order reaction is one in which the

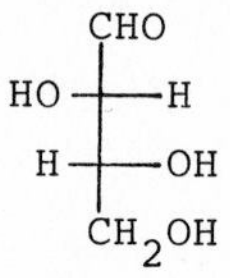

rate-determining step involves only *one* molecule. Therefore the equation for the rate of reaction includes only the concentration of that molecule. If the reaction actually involves two molecules, but one is in large excess (for instance, solvent), the reaction is called a *pseudo first order reaction* because the rate equation still has only one concentration as a variable.

Fischer projection (7.5). A system for indicating 3-dimensional structures in two dimensions. In a Fischer projection the horizontal bonds are understood to project forward and the vertical bonds back:

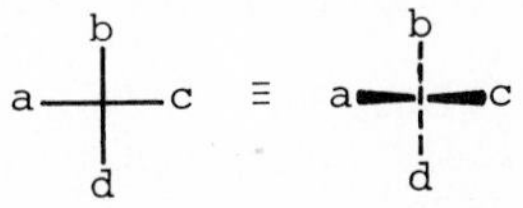

In a Fischer projection, exchange of the positions of any pair of substituents or a 90° rotation of the whole picture leads to a representation of the other *configuration* at that chiral center. Any even combination of these changes (for example 2 pair-wise exchanges, 3 pair-wise exchanges and a 90° rotation, or a 180° rotation) lead to a picture of the same molecule again.

Fluorescence (34.4.A). Loss of a photon from the first *excited singlet state* and transition of the molecule to the *ground state.*

Formal charges (2.2). The difference between the number of *valence* electrons controlled by an atom in the elemental state and in its bonding arrangement in a molecule. The number of valence electrons in the elemental state corresponds to its column in the periodic table; in a molecule, an atom is considered to control all the valence electrons it does not share, and half of those it does share.

Franck-Condon transition (34.4.A). An *electronic transition* which occurs more rapidly than atomic (vibrational) motions in a molecule; also called a *vertical transition.*

Free radical (6.1). Any molecule having unpaired electrons. Usually refers to carbon atoms with only three substituents and seven *valence electrons,* such as

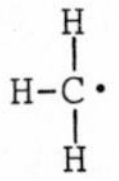

Front side attack (8.5). A possible mode of substitution stereochemistry. It occurs only in rare instances in nucleophilic displacement reactions (which ordinarily involve *inversion* (S_N2) or *racemization* (S_N1)), but is seen in some electrophilic displacement reactions.

Functional groups (3.3). The reactive parts of molecules. These are small, frequently-occurring groups of atoms, such as the hydroxy group or carboxy group, which exhibit a typical reactivity in a wide variety of molecules. For example,

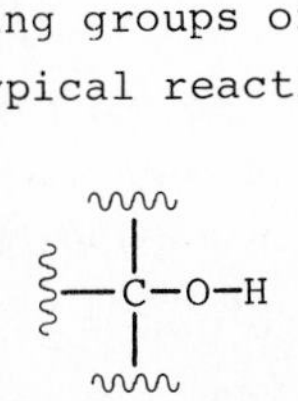

hydroxy group

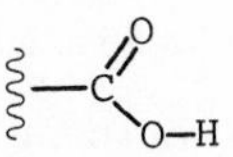

carboxy group

all molecules having a carboxy group are acidic. A list of the most important functional groups is found in Table 3.1 of the text.

Fundamental vibrational modes (14.2). Modes of vibration which involve more than one bond; all vibrational modes of molecules larger than two atoms are fundamental modes.

Fused ring system (31.1). See *Bridged*.

Gauche (5.2). See definition under *Anti*.

Geminal (11.3.A, 12.5.B). Attached to the same carbon. See *Vicinal*.

Gibbs Standard Free Energy Change, $\Delta G°$ (4.2). This represents the amount of energy available for work that would be released during a reaction if all of the starting material in its standard state were converted to all of the product in its standard state. (In solution, the standard state is about 1 M.) Important equations to remember are $\Delta G° = \Delta H° - T\Delta S°$, which indicates how the *enthalpy* ($\Delta H°$) and the *entropy* ($\Delta S°$) of the reaction contribute to the free energy ($\Delta G°$); and $\Delta G° = -RT \ln K$, which relates the free energy change to the equilibrium constant K.

Glycoside (28.3). Cyclic *acetals* or *ketals* of a sugar with another alcohol, called the *aglycon*.

Glyme(s) (10.12.A). Dimethyl ethers of short ethylene oxide polymers: $CH_3\text{-}(OCH_2CH_2)_n\text{-}OCH_3$. See *Crown ethers*.

Ground electronic state (21.1). The lowest energy distribution of electrons in the molecular orbitals of a molecule or the atomic *orbitals* of an atom. Any higher energy distribution, for example, one resulting from promotion of one of the electrons from a *bonding* to an *antibonding orbital*, is an *excited electronic state*.

Halonium ion (11.6.B.3). A divalent halogen cation, usually resulting from the addition of electrophilic halogen to a π-bond.

Harmonic oscillator approximation (14.2). The approximation of a bond vibration as a system which obeys *Hooke's Law*.

Haworth projection (28.3). A semi-perspective drawing of the cyclic acetal or ketal form of a sugar to indicate its stereochemistry. Useful for making the transition from *Fischer projections* to perspective drawings of chair conformations.

Heat of combustion (6.4). Enthalpy released on complete oxidation of a compound: $C_nH_m \rightarrow n\ CO_2 + m/2\ H_2O$. This number provides the same information on the thermodynamic stability of a molecule as the *heat of formation*, but references it to a different standard state. *Heats of combustion* are in fact the values which are obtained experimentally; *heats of formation* are then calculated using the known heats of combustion of the elements.

Heat of formation (5.5). The *enthalpy* released on forming a compound from its elements. Comparison of the heats of formation of isomers is an indication of their relative thermodynamic stability. Compare with *Heat of combustion* (6.4).

Heisenberg Uncertainty Principle (2.5). The only part of chemistry the philosophers really like. It sets a lower limit on the accuracy with which we can know both the position and momentum of a particle. For instance, the more precisely we define how an electron is moving, the less accurately we can know where it is, and vice versa.

Hetero (3.2). Not carbon, hydrogen (or a metal, usually). For example, O, N, F, S etc. are all *heteroatoms*, and cyclic compounds in which there are ring atoms other than carbon are called *heterocycles*.

Heterolysis, heterolytic cleavage (6.3). Cleavage of a bond in which both electrons in the bonding orbital depart with one of the pieces: $A{:}B \rightarrow A^+ + {:}B^-$. There will always be a change in formal charge on the two atoms involved in such a process. Contrast with *Homolysis*.

Hofmann rule (24.7.E). "In the decomposition of quaternary ammonium hydroxides, the hydrogen is lost most easily from CH_3 , next from RCH_2, and least easily from R_2CH."

Homolysis, homolytic cleavage (6.3). Cleavage of a bond in which one of the two bonding electrons departs with each of the pieces: $A{:}B \rightarrow A\cdot + \cdot B$. There is no change in formal charge when this happens. Contrast with *Heterolysis*.

Hooke's law (14.2). States that the force needed to stretch or compress a spring (or bond) is directly proportional to the distance it is stretched or compressed.

Hückel 4n + 2 rule (22.7). "Monocyclic π-systems with 4n + 2 electrons show relative stability compared to acyclic analogs."

Hückel molecular orbital (34.3.A). See *Orbital*.

Hybridization (2.8). A recombination of the *orbitals* of a free atom to enable it to make stronger bonds when it is in a molecule. Whereas the atomic state of carbon has one 2s and three 2p orbitals, carbon is hybridized to provide four $2sp^3$ orbitals (sp^3-hybridized), one 2p and three $2sp^2$ orbitals (sp^2-hybridized), or two 2p and two 2sp orbitals (sp-hybridized) in its molecules. The hybridized orbitals have a characteristic spatial relationship which is reflected in the geometry of the molecule. The exponents correspond to the fractional character of a given atomic orbital in the hybrid; for example, in sp^3, the fractional p-character is 3/(3 + 1), or 0.75. Note that the exponent of s is always unity. (N.B. Don't confuse *hybrid orbitals* with *resonance hybrids.)*

Hydration. The opposite of *dehydration*. Usually the addition of water across the π-bond of an alkene to give an alcohol (11.6.B), of an alkyne to give a ketone (12.6.B), or of a carbonyl compound to give a gem-diol (carbonyl hydrate) (13.7.A).

Hydrogen bond (4.5, 10.3). A dipole-dipole interaction between a hydrogen atom bonded to an electronegative element and an electron *lone pair* on another electronegative atom. It is a weak bond (~5 kcal $mole^{-1}$) but important in the chemistry of alcohols, carboxylic acids, amines and similar compounds.

Hydrogenolysis (13.8.E, 22.6.B). Cleavage of a single bond by hydrogenation: $A{-}B + H_2 \xrightarrow{cat.} AH + HB$. Usually encountered in Raney-nickel desulfurization or in hydrogenation of benzyl alcohols, etc.

Hydrolysis (10.6.A, 19.6). See *Solvolysis reaction*.

Hydrophilic (29.7.B.2). The opposite of *hydrophobic*: polar, often ionic, well-*solvated* by water, water soluble.

Hydrophobic (29.7.B.2). Repelling water. Nonpolar, hydrocarbon chains are poorly *solvated* by water molecules and are excluded from aqueous solution. They form *lipid* bilayers, *micelles*, or separate phases in which they provide their own *solvation*.

Hydroxylation (27.3.A). Usually, addition of two hydroxyl groups to a double bond to give a 1,2-diol.

Hyperconjugation (8.12, 21.4). The mechanism by which a *carbocation* is stabilized by sharing the electron density of bonds to the adjacent carbon. This results in the stability sequence: $3° > 2° > 1° > CH_3^+$ for carbocations.

It is also important in characterizing the interaction of the σ bonds of an alkyl group with an adjacent π-system, as in ultraviolet spectroscopy.

Induction period (13.6.D). If a reaction is *autocatalytic* it will accelerate as it proceeds. The time *before* the reaction gets itself going is called the *induction period*.

Inductive effect (10.4, 18.4.B). The *electron attracting* or *donating* effect of a nearby dipole. For instance, *electronegative* elements like halogens are *electron attracting*; alkyl groups are *electron donating*.

Infrared active, inactive (14.2). See *Selection rule*.

Initiation steps, Initiator (6.3). See *Chain reaction* (6.3).

Inner Salt (29.1). See *Zwitterion*.

Insertion (15.6). A type of reaction in transition metal chemistry in which a donor *ligand* undergoes insertion into a σ-bond between the metal and another group. The reverse of this process is called *deinsertion*. The *electron count* of the metal decreases by two on insertion.

Integrated intensity (9.5). The area under the curve of a given peak in the nmr. For proton nmr, this usually corresponds to the relative number of protons which resonate at that frequency. In *cmr*, *saturation* and *relaxation* effects affect the integrated intensities of different peaks differently, and the area ratios do not correspond to the relative number of carbons.

Interference (2.6). An addition of *wave functions (orbitals)* which results in their cancellation in a certain region of space. This occurs during the formation of an *antibonding* orbital, for instance, in which an atomic orbital wavefunction of one sign overlaps with an orbital of opposite sign on the other atom. The opposite of interference is *reinforcement*.

Intermolecular, intramolecular (8.14). *Intermolecular* = between two or more molecules; *intramolecular* = within the same molecule.

Internal conversion (34.4.A). Relaxation of an *excited* vibrational *state* to a *ground* vibrational *state*.

Intersystem crossing (34.4.A). Conversion of the first *excited singlet state* to the first *excited triplet state*.

Inversion (8.5). The normal stereochemistry seen in displacement reactions occurring by the S_N2 *mechanism*. It arises from simultaneous bonding of the incoming *nucleophile* and the departing *leaving group* to opposite lobes of the *orbital* on carbon:

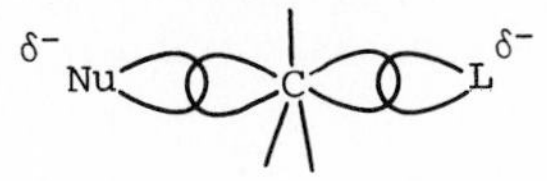

Ionic character (15.1). A highly polarized bond, one between an *electronegative* and an *electropositive* element, has a lot of ionic character. A symmetrical, *covalent* bond between identical groups has no ionic character (e.g., the C-C bond of ethane).

Ionization potential (2.2, 17.4.A). The energy required to remove an electron from an atom or molecule.

Isoelectric point (29.3). The pH at which the average charge on a molecule (usually a *zwitterionic* amino acid or *peptide*) is zero. For the simple amino acids (with only one acidic and one basic group), this is the average of pK_1 and pK_2.

Isolated double bonds (20.2.A). See *Conjugation*.

Isomers (3.6). Compounds which have the same molecular formula but different structures. See *Structural Formulas* and *Stereoisomers*, and the table at the beginning of Chapter 3 in this Study Guide.

J (9.6). See *Coupling Constant*.

Karplus curve (9.6). Figure 9.21. Relates *coupling constant* J to the dihedral angle between two adjacent C-H bonds.

Kekulé structure (2.2). A structure in which a line between two atoms indicates a *covalent single bond* involving two shared electrons. A double bond is represented by a double line, etc. Contrast with *Lewis* structures.

Ketal (13.7.B). See *Acetal*.

Ketyl (27.3.A). A *radical anion* produced by addition of an electron to a ketone; an intermediate in the pinacol reaction.

Kinetic control (20.2.B, 34.8.B). A reaction in which the products are not in equilibrium with the starting materials is under *kinetic control*; that is, the product obtained is the one that is formed the fastest. This is not always the most stable product: see *Thermodynamic control*.

Kinetic order (8.7). Equal to the sum of the exponents of all the concentrations expressed in the rate equation. If the rate depends only on the concentration of one component, the reaction is *first order*; if it depends on the concentration of two components, or on the square of the concentration of one component, the reaction is *second order*.

Leaving Group (8.10). The group which departs in a displacement or an elimination reaction. The nucleophile [or base] attacks and the leaving group leaves, either before (S_N1 [or E1], for example) during (S_N2 [or E2]), or afterwards (reactions of carboxylic acid derivatives, for example). The less basic the departing species is, the better a leaving group it is.

Levorotatory (7.2). See *Dextrorotatory.*

Lewis acid. A Lewis acid is a molecule or ion which furnishes a vacant orbital for bonding. The term is most commonly applied to the strong Lewis acids like $ZnCl_2$, $AlCl_3$, $FeCl_3$, BF_3, etc. However, all *electrophiles* are formally Lewis acids, and the term is used in this sense in the organic chemistry of transition metals (see next entry).

Lewis acid association (15.6). Bond formation between a transition metal with a Lewis acid (*electrophile*). The reverse process is called *Lewis acid dissociation;* the *electron count* of the metal does not change.

Lewis base. A Lewis base is a molecule or ion which has two electrons available for bonding. All nucleophiles are Lewis bases.

Lewis base association (15.6). The formation of a bond to a transition metal in which the two electrons to be shared originate with the *ligand* or other entering group. The reverse process is called *Lewis base dissociation.* The *electron count* of the metal increases by +2 on *Lewis base association.*

Lewis structure (2.2). A structure in which each *valence* electron is indicated by a dot ("electron dot structures"), and a *covalent single bond* is represented by two dots between the atoms. These are helpful in keeping track of electrons, *formal charges*, and valences. Note that "core" electrons are not shown. That is, for carbon only the four valence electrons are depicted; the 1s electrons are understood to be present as well.

Ligand (15.6). Usually refers to a molecule or ion which is stable by itself but which is also capable of forming a *dative* or *donor bond* to a transition metal.

Linear free energy relationship (34.1). An indication that a given structural feature affects related reactions in more or less the same way. Seeing whether such a relationship holds for similar reactions is useful for deciding whether they really are related, with similar mechanisms, etc.

London force (5.1). See *Dispersion force.*

Lone pair electrons (2.2). See *Nonbonded electrons.*

Long range coupling (12.3.B). Coupling observed over longer distances than the usual H-C-C-H system. It occurs with specific conformations of H-C-C-C-H, with alkynes: H-C-C≡C-H, and frequently with nuclei other than protons (for example ^{13}C, ^{19}F, ^{31}P).

Macroscopic isotope (23.2). A non-radioactive isotope used in labeling studies. In contrast to a *tracer isotope*, most of the labeled positions will contain a macroscopic isotope. The presence of a macroscopic isotope is determined by nmr spectroscopy or by mass spectrometry.

Magnetic moment (9.3). Some atomic nuclei behave as if they were spinning (hence the term nuclear "spin") and act like small magnets. Nuclei with a nuclear spin *quantum number* of 1/2 (^{1}H, ^{13}C, ^{19}F, ^{31}P, among others) are allowed only two orientations in the presence of an external magnetic field: aligned with the field or against it (the α- and β-spin states). The difference in energy of these two orientations gives rise to the phenomenon of nuclear magnetic resonance spectroscopy (nmr).

Magnetically equivalent (9.4). Having the same *chemical shift*. Equivalent nuclei (for instance the two hydrogens in acetylene) have to be magnetically equivalent; many others can be equivalent on the *nmr time scale* because of rapid conformational changes (for example the hydrogens of a methyl group); still others may be quite different structurally but coincidentally have the same *chemical shift*.

Markovnikov's rule (11.6.B.1). Original formulation: In the acid-catalyzed addition of water to an alkene (*hydration*), the proton goes on the carbon that already has the greater number of protons (less substituted carbon).

Modern formulation: In electrophilic addition to a π-bond, the *electrophile* adds so as to generate the most stable carbocationic intermediate.

Mass spectrum, mass spectrometer (17.1). A *mass spectrum* is a record of the mass-to-charge ratio of the ions produced when a molecule is subjected to electron impact in a *mass spectrometer*.

McLafferty rearrangement (17.4.A). A two-bond fragmentation mechanism for *radical cations* of carbonyl compounds in the mass spectrometer. It requires the presence of a hydrogen γ to the carbonyl group.

Mechanism (4.1). The details of how the atoms and electrons are reorganized during the course of a reaction.

Meso (7.6). An *achiral* molecule with *chiral* carbon atoms is a *meso* compound. There must be an even number of chiral centers and they must have opposite configurations so that there is a *center* or *plane of symmetry* in the molecule.

Meta-directors (23.5). Substituents on an aromatic ring which destabilize adjacent positive charge and thereby favor electrophilic aromatic substitution at positions *meta* to themselves. See *ortho, para-directors*.

All *m-directors* are *deactivating*.

Metallation (30.3.B). Replacement of a hydrogen or a halogen with a metal cation. Deprotonation is usually accomplished with a strong base like lithium diisopropylamide, alkyllithiums, or sodium amide; halogens are replaced with metals using alkyllithiums (*transmetallation*) or the metals themselves.

Metallocene (22.7.F). "Sandwich" or "half-sandwich" compounds which involve molecules with cyclic π-systems acting as *ligands* to transition metals, the coordination being via the π-systems.

Micelle (18.4.D). Molecules with both *hydrophilic* and *hydrophobic* regions often cluster together in aqueous solution in *micelles*, in which the non-polar chains are in the interior and the polar regions are on the exterior.

Migratory aptitude (13.8.A). Relative ease with which a particular substituent moves over to an adjacent atom in rearrangements like carbocation rearrangements, the Baeyer-Villiger reaction, or acylnitrene or -carbene rearrangements. Usually H > phenyl > 3° > 2° > 1° > CH_3.

Mixed Claisen condensation (27.7.A). Claisen condensation between a ketone and an ester to give a β-diketone.

Möbius molecular orbital (34.2.A). See *Orbital*.

Molecular formula (3.4). The molecular formula of a compound expresses the total number of atoms of each element present. It is always the same as, or a multiple of, the *empirical formula*. For instance, the *empirical formula* for glucose is CH_2O, but the *molecular formula* is $C_6H_{12}O_6$.

Molecular ion, M^+ (17.3). The ion produced in a mass spectrometer on ejection of an electron from a molecule.

Molecular orbital (2.7). See *Orbital*.

Molecularity (8.7). The molecularity of a reaction is the number of molecules involved in the rate-determining transition state. For instance, an S_N2 displacement reaction is a typical *bimolecular* reaction; an S_N1 displacement is a typical *unimolecular* reaction. *Termolecular* reactions, and higher, are very rare.

Monochromator (14.11). An instrument which disperses a beam of light into its component frequencies, so that a narrow range of frequencies can be focused on the sample. It is an important component of infrared and ultraviolet/visible spectrophotometers.

Monomer (11.6.G). See *Polymer*.

Multiplet (9.6). See *Spin-spin splitting*.

Mutarotation (28.3). A change in *optical rotation* which results from the equilibration of *anomers*.

Nitrogen inversion (24.1).

Nmr Time Scale (9.4). About 10^{-3} sec, the length of time a molecule must exist in a discrete state for it to be observable by nmr.

Node (2.5). A surface which separates regions of an *orbital* which have opposite signs. For instance, a 2p orbital aligned with the z axis has as its node the xy plane, and the antibonding orbital of a hydrogen molecule has as its node a plane perpendicular to the internuclear axis.

Nominal mass (17.3). The integral mass of a molecule or molecular fragment. See *Exact mass*.

Non-bonded electrons (2.2, 24.1). *Valence electrons* not involved in a *covalent bond*. Oxygen normally has two pairs of non-bonded electrons and nitrogen one, for example: H$\ddot{\text{O}}$H, H_3N:. These *non-bonded electrons* are also called *lone pair* electrons, and they are the electrons involved when these molecules act as *Lewis bases* or *nucleophiles*.

Nuclear spin (9.3). See *Magnetic moment*.

Nucleophile, nucleophilic reagent (8.7). A *Lewis base*. In its most general sense, any species which can furnish a pair of electrons to make a bond can be considered a nucleophile: any time a bond is formed by the reverse of a heterolytic process, the species that contributes the electrons is a nucleophile. In discussing displacement reactions, those which involve attack by a nucleophile and loss of a *leaving group* are called nucleophilic displacement reactions.

Nucleophilic addition-elimination mechanism (19.6). The most important mechanism whereby substitution reactions occur in carboxylic acid derivatives. In its general form:

$$\underset{\underset{\displaystyle O}{\|}}{R\text{-}C\text{-}X} + H\text{-}Nu \rightleftharpoons \underset{\underset{\displaystyle X}{|}}{\overset{\overset{\displaystyle OH}{|}}{R\text{-}C\text{-}Nu}} \rightleftharpoons \underset{\underset{\displaystyle O}{\|}}{R\text{-}C\text{-}Nu} + H\text{-}X$$

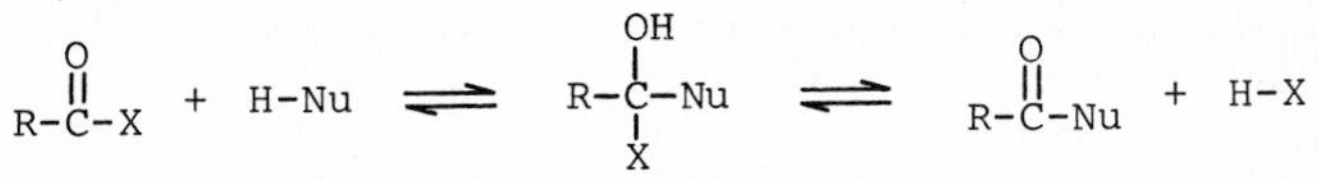

Nucleophilic aromatic substitution can also occur via an addition-elimination mechanism (30.3.A.1).

Octet (2.2). A filled *valence* shell for elements in the second and third periods of the periodic table. This arrangement for electrons is very stable, and achieving a filled octet is the most important factor in determining how many and what kinds of bonds an element will form in its molecules.

Off-resonance decoupling (9.9). See *Decoupling*.

Optically active (7.2). Capable of rotating the *plane of polarization* of *plane polarized light*. An *optically active* sample must be composed of *chiral* molecules, but the converse is not always true: a *racemic* mixture of chiral molecules is *optically inactive*.

Orbital (2.5). An equation which describes the behavior of an electron. The term is synonymous with *wavefunction*, although conceptually chemists envision a volume having specific shape and orientation when they think of orbitals and a mathematical equation when they think of wavefunctions. An *atomic orbital* describes the behavior of an electron in the vicinity of a single nucleus. *Molecular orbitals* result from the combination (overlap) of two or more atomic orbitals and describe an electron which is shared by several nuclei. Molecular orbitals can be further classified into *bonding*, *nonbonding*, and *antibonding* orbitals, depending on whether the electron distribution in the molecular orbital is more favorable (lower energy), unchanged (same energy), or less favorable (higher energy) than in the component atomic orbitals. Bonding orbitals result from the *reinforcing* overlap of atomic orbitals, antibonding orbitals from their *interfering* overlap.

Cyclic molecular orbitals are classified as *Hückel* if they have an even number (or zero) of positive/negative overlaps around the ring; they are classified as *Möbius* if there is an odd number of such interactions.

Ortho, meta, and para (22.2.A). Positions on a benzene ring relative to a substituent.

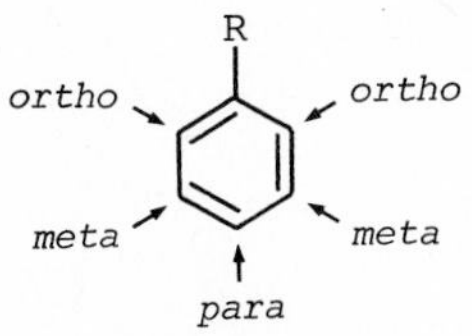

Ortho, para-directors (23.5). Substituents on an aromatic ring which stabilize adjacent positive charge and thereby favor electrophilic aromatic substitution at positions *ortho* and *para* to themselves. See *meta-directors*.

o,p-Directors can be either *activating* or *deactivating*.

Overlap (in nmr spectroscopy)(9.7). When two nuclei resonate at similar *chemical shifts*, their peaks *overlap* (fall on top of one another).

Overtones, combination bands (14.2). Bands which result in an infrared spectrum from simultaneous change of more than one energy level.

Oxaphosphetane (13.7.G). A cyclic intermediate in the Wittig reaction (see *betaine*):

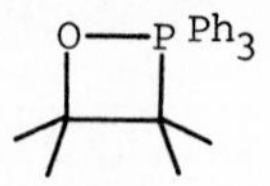

Oxidative addition (15.6). A type of reaction in transition metal chemistry in which the two groups at the ends of a σ-bond both become bonded to the metal. The reverse process is called *reductive elimination*. The *electron count* of the metal decreases by two on oxidative addition.

Oxidative coupling (12.6.F). Joining two molecules with overall loss of two protons and two electrons, as in the oxidative coupling of terminal alkynes to give diynes (Eglinton reaction).

Oxidative deamination (29.5.C). The opposite of *reductive amination*.

Oxonium ion structure (2.4). A structure which involves three bonds to an oxygen atom, and therefore a positive *formal charge* on the oxygen. Although localization of a positive charge on the *electronegative* element oxygen would appear to be difficult, the atom still retains a filled *octet valence* shell (compare with a *carbocation*), and this consideration is the most important. The simplest oxonium ion is the hydronium ion: H:Ö:H with + H below (H:$\ddot{\text{O}}$:H, $^+$H) .

Partial rate factors (23.7). In electrophilic aromatic substitution, the partial rate factor is the reactivity of a given position on a substituted aromatic ring relative to benzene, corrected for the number of equivalent positions.

Peptide (28.1). A *peptide bond* is an amide linkage between two amino acids. *Peptides* are dimers, trimers, oligomers, etc. (*condensation polymers*) of amino acids.

"Peroxides" (116.6). Usually *alkylperoxides* (ROOH) present in impure materials as a result of *autoxidation*. They are often initiators of free radical chain reactions, and are distinctly different from the reagents hydrogen peroxide (HOOH) or peracids (RCO_3H).

Phase-transfer catalysis (24.5.B). Transfer of ionic reagents from an aqueous or solid phase into an organic solvent, where they show enhanced reactivity. Phase-transfer catalysts are either *quaternary ammonium salts* or *crown ethers*.

Phosphorane (13.7.G). See *Ylide*.

pK = -log K (4.5).

$$K_a = \frac{[H^+][A^-]}{[HA]} \qquad pK_a = \text{measure of acidity}$$

$$K_b = \frac{[HA][OH^-]}{[A^-]}$$

$$pK_a + pK_b = 14$$

The lower the pK_a, the more acidic HA is and the less basic the *conjugate base* A^- is. When pH = pK_a, $[A^-] = [HA]$.

<u>Plane of symmetry</u> (7.6). An object has a plane of symmetry if it can be divided into two halves which are mirror images of each other. Such objects are *achiral*.

<u>Plane polarized light</u> (7.2). Light in which the electric field vectors of all the light waves lie in the same plane, called the *plane* of *polarization*. An *optically active* sample rotates the plane of polarization of a beam of plane polarized light which passes through it.

<u>Poisoned catalyst</u> (12.6.A). A catalyst whose efficiency and activity have been reduced with a "poison". Lindlar's catalyst ($Pd/BaSO_4$, poisoned with quinoline) is an example.

<u>Polarimeter</u> (7.2). An instrument for measuring *optical activity*.

<u>Polarizability</u> (8.2). Often thought of in an intuitive sense as "softness" or "mushiness" of an atom, polarizability refers to the ease of deforming the electron density around an atom. It increases as one descends in the periodic table because the valence orbitals lie further from the nucleus and the electrons are held less tightly. In general, in protic solvents, the more polarizable a *nucleophile* is the more potent it is as a nucleophile.

<u>Polarized light</u> (7.2). See *Plane polarized light*.

<u>Polymer, polymerization</u> (11.6.G, 27.6.B.2). The sequential linking of *monomers* to produce *dimers*, *trimers*, molecules of intermediate size (*telomers* or *oligomers*), and then of very large size (polymers) is called *polymerization*.

In the formation of an *addition polymer*, the bonds between the monomers are made by addition to a double bond, by a cationic, radical, or anionic process.

In the formation of *condensation polymers*, the bonds are made with the elimination of a small by-product molecule.

Copolymers incorporate more than one *monomer* in the chain, in contrast to *homopolymers* in which all the units are the same.

<u>PPM = parts per million</u> (9.4). See *Chemical Shift*.

<u>Primary, Secondary, Tertiary, Quaternary</u> (6.1, 24.1, 24.5.A). R is understood to be a carbon substituent, such as an alkyl group.

	Carbon	Hydrogen	Radical	Carbocation	Carbanion	Halide	Alcohol
Primary:	R-CH$_3$	R-C(H)(H)-H	H·C·H, R	H·C$^+$·H, R	H·C$^-$·H, R	RCH$_2$X	RCH$_2$OH
Secondary:	R-C(H)(H)-R'	R-C(H)(H)-R'	R·C·R', H	R·C$^+$·R', H	R·C$^-$·R', H	R-C(X)(H)-R'	R-C(OH)(H)-R'
Tertiary:	R-C(H)(R'')-R'	R-C(H)(R'')-R'	R·C·R', R''	R·C$^+$·R', R''	R·C$^-$·R', R''	R-C(X)(R'')-R'	R-C(OH)(R'')-R'
Quaternary:	R-C(R'')(R'')-R'						

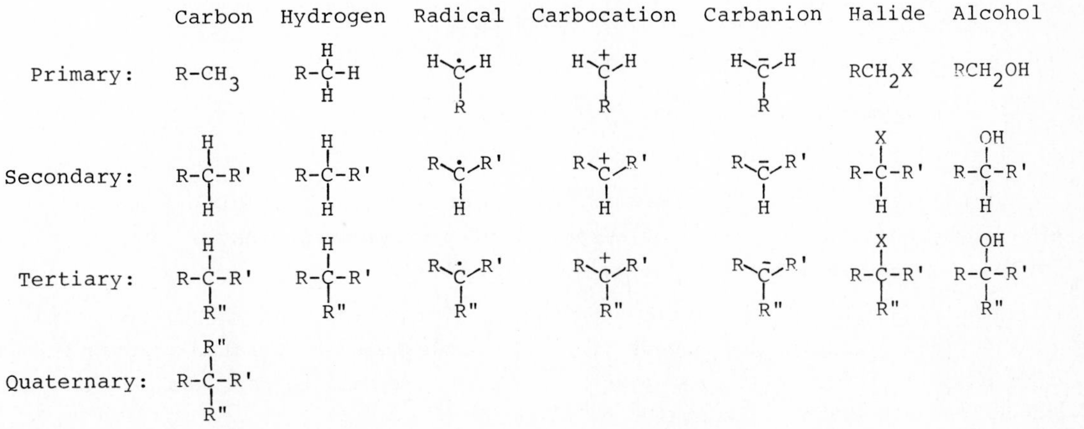

```
                Amine
Primary:        RN̈H2
Secondary:      R-N̈H-R'
Tertiary:       R-N̈-R'
                  |
                  R"

                  R'
                  |
Quaternary:     R-N-R"
                 +|          (ammonium salt)
                  R
```

Principle of microscopic reversibility (6.3). "If the easiest way to get from Yosemite Valley to Mono Lake is through Tuolomne Meadows and over Tioga Pass, then the easiest way from Mono Lake to Yosemite Valley is over Tioga Pass and through Tuolomne Meadows", which is to say that the transition state for the forward reaction will be identical to the transition state for the back reaction.

Probability function (2.5). Squaring the value of a *wave function* at each point in space gives a probability function, which describes the likelihood that the electron will be found at that point. Whereas the wave function itself will have regions in which its value is negative and regions in which it is positive, the probability function is always positive (or zero).

Propagation step (6.3). See *Chain reaction.*

Protecting group (16.4). A functional group introduced to mask the reactivity of another functional group so that an otherwise interfering reaction can be carried out elsewhere on the molecule. It is later removed and the original functional group is regenerated.

functional group	protecting group
alcohols	t-butyl ethers (10.11.A)
	silyl ethers (16.4)
ketones and aldehydes	*acetals* and *ketals* (13.7.B)
anomeric carbons	*glycoside* formation (28.5.A)
amino acids	carbobenzoxy (Cbz) and t-butoxycarbonyl (Boc) groups (29.6.B)

Proteins (29.1). High molecular weight poly*peptides.*

Proton decoupled (9.9). See *Decoupling.*

Pseudo first-order reaction (4.3). See *First-order reaction.*

Pseudorotation (5.7). See *Envelope.*

Pyrolysis (6.2, 19.11). Cleaving a bond or a molecule simply by heating it hot enough, as in the *cracking* of alkanes, the pyrolytic elimination of acetate or xanthate esters (19.11) or of sulfoxides (26.3).

Quantum numbers (2.5), Quantized (9.2). At the atomic level, only certain *electronic configurations*, speeds of rotation, degrees of vibration, orientations in a magnetic field, etc., are possible. These are said to be *quantized*, and their specific value is indicated by the *quantum numbers.*

Quantum yield (34.4.B). The fraction of molecules which proceed to products after absorbing a photon in a photochemical reaction.

Quartet (9.6). See *Spin-spin splitting.*

Quaternary (6.1). See *Primary.*

R-S Convention (7.3). System of nomenclature for indicating the *absolute configuration* of a *chiral center.* Involves use of the "*sequence rule*".

Racemic mixture, racemate, racemic compound (7.4). An equimolar mixture of the two *enantiomers* of a compound. It is *optically inactive.* A *racemic compound* is a *racemic mixture* in which both *enantiomers* are present in the same crystal structure.

Racemization (7.4, 11.6.C). The equilibration of one *enantiomer* with the other, to produce a *racemic, optically inactive* mixture consisting of equal amounts of both *enantiomers.*

Radical (3.5). From the point of view of nomenclature, a *radical* is a piece of a molecule which is considered as a unit. For instance, "methyl", "phenyl", and "1-chloroethyl" are radicals used as part of the name: 2-(1-chloroethyl)-1-methyl-4-phenylcyclohexane.

In chemical reactions, *radicals* are usually present only as short-lived, unstable intermediates, because they have unpaired electrons and lack valence octets. Free radical halogenation (section 6.3) is a good example of a type of reaction which involves radical intermediates.

Radical anion (12.6.A). An anion with an odd number of electrons; usually obtained by addition of an electron to a neutral, even-electron molecule.

Radical cation (17.1). A cation with an odd number of electrons; usually formed from a neutral, even-electron molecule by ejection of an electron, as in the ionization chamber of a *mass spectrometer* on electron impact.

Reducing vs non-reducing sugar (28.5.E). A sugar in equilibrium with its open chain form in alkaline solution (that is, a *hemiacetal* or *hemiketal*) is a *reducing sugar* because it can be oxidized by Fehling's or Tollen's reagents. If the sugar exists as a *glycoside*, it is a *non-reducing sugar.*

Reduction potential: see *Standard reduction potential.*

Reductive amination (24.6.F). Conversion of a carbon-oxygen double bond to a carbon-nitrogen single bond by reduction of an imine or immonium ion intermediate.

Reductive elimination (15.6). The reverse of *oxidative addition.*

Reinforcement (2.7). The overlap of *orbitals* having the same sign, so that their *wavefunctions* add together; the opposite of *interference.* The overlap of two atomic orbitals to make a bonding molecular orbital is an example of reinforcement.

Relative configuration (25.6). This term refers to a *stereochemical* relationship between two or more *chiral centers.* If the *absolute configuration* of two chiral centers is known, so is their *relative configuration*, but it is possible to know their *relative configuration* without knowing their *absolute configuration.*

Terms such as *meso, erythro* and *threo*, and *cis* and *trans* (in cyclic systems) all describe *relative configurations*.

Relaxation (9.5, 13.4). In nmr spectroscopy, *relaxation* is the restoration of the spin distribution of the sample to its equilibrium value. For a ^{13}C nucleus, relaxation is greatly speeded up by hydrogens directly attached to it. If relaxation is slow (for example, for *quaternary* or carbonyl carbons), *saturation* of the nmr signal occurs easily and only a weak resonance peak is seen. See *Integrated intensity*.

Resolution (29.4.F). Separation of the *enantiomers* of a *racemic mixture*, usually by the formation and separation of *diastereomeric salts* or other derivatives, and regeneration of the *enantiomers*.

Resonance (in *Spectroscopy*) (9.4). When the energy of electromagnetic radiation matches the energy difference between two quantum states, as given by the relationship $E = h\nu$, absorption or emission of radiation by the sample is possible, and the system is said to be in *resonance*.

Resonance energy (22.1.B). The special stabilization that an *aromatic* π-system has over a hypothetical system of similar electronic *configuration*, but which lacks the *delocalization* of the aromatic system.

The *empirical resonance energy* is determined by experimental comparison of actual molecules. It differs from *delocalization energy*, which results from comparison with theoretical models.

Resonance hybrid (2.4). A molecule which cannot be adequately represented by a single written structure, but can be understood as a hybrid of two or more *resonance structures*. (N.B. Don't confuse *resonance hybrids* with *hybrid orbitals*.)

Resonance structures (2.4). Structures depicting a molecule which differ only in the distribution of electrons (as opposed to nuclei). They are most important and useful when the alternative bonding arrangements they represent are of similar energy, in which case the molecule they describe is said to be a *resonance hybrid* of the two structures.

Retention (12.6.E). The opposite of *inversion*.

Ring current (22.3.A). The π-*electron circulation* induced in an *aromatic* system by an external magnetic field. See *Anisotropic* and *Chemical shift*.

S_N1 (Substitution Nucleophilic Unimolecular) (8.12). A description of the mechanism of a displacement reaction which involves only one molecule in the rate determining step and which takes place first with ionization of the carbon-leaving group bond (slow) and then attack by the nucleophile (fast). Usually goes with loss of configuration at the carbon atom, i.e., racemization if that is the only chiral center present. The rate depends on carbocation stability (3° > 2° > 1°).

S_N2 (Substitution Nucleophilic Bimolecular) (8.7). A description of a mechanism of a displacement reaction, as the expanded name above implies. These reactions usually occur with inversion of configuration at the carbon undergoing attack. The rate depends on lack of steric hindrance, and is slowed either by α-*branching* (1° > 2° > 3° in rate) or β-*branching* (e.g., "neopentyl" systems *very* slow).

Sanger method (29.6.C.2). A method for determining the N-terminal amino acid of a *peptide* by labeling it with a dinitrophenyl group followed by hydrolysis of the peptide and identification of the derivatized amino acid.

Saponification (19.12.B). Alkaline hydrolysis.

Saturated (3.3). From the point of view of *molecular formula*, *saturated* means that there are 2n + 2 hydrogens for every carbon present in the molecule, after a hydrogen has been added for every halogen atom present and subtracted for every nitrogen atom present. For instance, ethane (C_2H_6), 1,2-dichloroethane ($C_2H_4Cl_2$), and ethylamine (C_2H_7N) are all counted as $C_2H_6 = C_nH_{2n+2}$, where n = 2.

From the point of view of structure, *saturated* means containing only single bonds (σ bonds). It differs from the definition above only for cyclic compounds. For instance, cyclohexane (C_6H_{12}) has only single bonds, but the formula is C_nH_{2n}.

See *Unsaturated*.

Saturation (in the nmr) (9.5, 13.4). Equalization of the populations of the α- and β-spin states, which leads to disappearance of the nmr signal. (The strength of the signal depends on the difference in population of these spin states.) Saturation is opposed by *relaxation*. (See *Integrated intensity*).

Schiff base (13.7.C). A substituted *imine*: 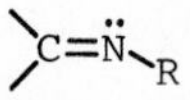

Second-Order kinetics (8.5). See *Kinetic Order*.

Second-Order reaction (4.3). Usually, one in which two molecules are involved in the rate-determining step, and in which, therefore, the rate equation has the concentration of both molecules as variables.

Secondary (6.1). See *Primary*.

Selection rule (14.2). A quantum mechanical requirement for a transition between two energy levels to be allowed. In infrared spectroscopy, the transition must result in a change in the dipole moment of the molecule. Such transitions are *infrared active*.

In ultraviolet-visible spectroscopy, the two electronic states are subject to other quantum mechanical constraints for the *electronic transition* to be allowed.

Sequence rule (7.3). See *R-S convention*.

Sigmatropic rearrangement (34.2.C). See definition in text.

Single bond (2.3). See *Bond*

Singlet state (34.4.A). An electronic state in which all the electron spins are paired, that is, in which the molecule has a net electronic spin of zero.

Solid phase technique (29.6.B). A method for the synthesis of poly*peptides* on a polymer support.

Solvation (4.3), Solvation energy (8.6). Solvation refers to the interactions between solvent molecules and the molecules dissolved in them (the solute). These interactions often influence the mechanism and rate of a reaction in solution, and therefore frequently make it difficult to compare liquid and gas phase data.

The *solvation energy* is an indication of the strength of these interactions, and is the energy released on transferring a solute from the gas phase into a solvent. It reflects how well the solvent molecules stabilize ions, neutral molecules, or transition states, etc. Difficulties in estimating solvation energies often hinders detailed understanding of reaction kinetics.

Solvolysis reaction (8.12). A substitution reaction in which the nucleophile is the solvent. If the nucleophile is water it is called *hydrolysis:*

$$A{-}B + RO{-}H \rightarrow A{-}O{-}R + H{-}B$$

Specific deuteration (15.5.A). See *Deuteration.*

Specific rotation, [α] (7.2). $[\alpha]_D = \frac{\alpha}{l \cdot c}$ ← *angle of rotation*

length in decimeters (l); concentration in g/mL (c)

Spectra, spectroscopy (3.4, 14.1). *Spectroscopy* is the experimental evaluation of the way in which a substance interacts with electromagnetic radiation, usually for the purpose of structure determination. More detailed descriptions can be found in the Chapters on Nuclear Magnetic Resonance Spectroscopy (Chapter 9), Infrared Spectroscopy (Chapter 14), and Ultraviolet Spectroscopy (Chapter 21). Mass spectrometry (Chapter 17) is another experimental technique for structure determination, although it is not truly a "spectroscopy".

Spin-spin splitting (9.6). Results from the *coupling* of the *magnetic moments* of two nearby, *magnetically nonequivalent* nuclei, so that the orientation of one of them (with or against an external magnetic field) affects the *chemical shift* of the other. The amount that nucleus A affects the chemical shift of nucleus B is called the *coupling constant*, J, and is equal to the amount that nucleus B affects nucleus A, too.

In simple cases, n adjacent protons cause a splitting into n+1 peaks (*doublets, triplets, quartets*, etc.), but the situation can easily become complicated by non-equivalent coupling constants (two *different* adjacent nuclei) and *non-first-order* effects (*multiplets*). *Magnetically equivalent* nuclei do not split each other.

Spiro ring system (27.3.B.1). See *Bridged.*

Staggered (5.2). See definition under *Anti.*

Standard reduction potential (15.4.B, 30.7.C). An indication of the *electron affinity* of an atom, ion, or molecule: the more *electropositive* a metal is, for example, the more easily it gives up its electron to form the cation, and the more negative the *standard reduction potential* of the cation is. The more positive the *reduction potential* of a quinone is, the stronger an oxidizing agent it is.

Stationary state (34.4.B). A reaction at equilibrium, usually in reference to a photochemical equilibration.

Stereoisomers (7.1). Molecules which differ only in the three-dimensional relationship between their atoms.

Stereospecific (11.5.A, 16.2, 27.3.A). Requiring or generating a particular three-dimensional relationship. For example, the S_N2 reaction proceeds *stereospecifically* with *inversion*, and the *E2* reaction is usually stereospecific for the *anti* relationship between the *leaving group* and the proton which is being removed.

Stereospecific is also used in the sense of giving only one *stereoisomer*, and is a desirable goal in planning the synthesis of a molecule with chiral centers or *cis* or *trans* double bonds.

Steric hindrance (8.8). An effect which arises when two molecules or parts of molecules try to occupy the same space at the same time. It is responsible for the γ-*effect* in *cmr* (9.9), preference for *equatorial* versus *axial*-substitution on cyclohexane rings (5.7), the slowing of S_N2 displacement reactions by α- or β-*branching* (8.8), the greater stability of *trans* over *cis* alkenes (11.4), etc.

Structural formulas (3.1). Drawings of molecules which show how the atoms are connected. They contrast with *empirical formulas* and *molecular formulas* because a given structural formula represents only one of all possible structural *isomers*. Even more specific are stereo formulas (Chapter 7). Structural formulas may be very detailed

```
              H H H
              | | |
(e.g., H-C-C-C-O-H for 1-propanol)   or condensed   (CH3CH2CH2OH).
              | | |
              H H H
```

As you become more familiar with organic structures, you will use line formulas to save time. In these simple structural formulas, the C's and H's on carbon are omitted, and only the bonds between carbon and non-hydrogen atoms are drawn (for example, 1-propanol is drawn: ).

Suprafacial (34.3.B). On the same face of a π-system. See *Antarafacial*.

Tautomers, tautomerism (13.6.A). *Isomers* which differ only in the placement of the protons; refers only to systems involving heteroatoms and which interconvert fairly rapidly.

Telomer (11.6.C). See *Polymer*.

Termination (6.3). See *Chain reaction*.

Tertiary (6.1). See *Primary*.

Thermodynamic control (20.2.B, 34.8.A). A reaction in which the products are in equilibrium with the starting materials is said to be under thermodynamic control. That is, the product obtained is the most stable one. See *Kinetic control*.

Three center two-electron bond (15.1). A bond between three atoms, incorporating three *orbitals* in the *bonding molecular orbital*, which is occupied by two electrons.

Tracer isotope (23.2). A radioactive isotope (e.g., tritium (^3H) or ^{14}C) used in labeling studies. Only a small fraction of the atoms at a position labeled with a *tracer isotope* are actually the radioactive isotope itself. The presence of a radioactive isotope is determined with an instrument known as a liquid scintillation counter.

Transesterification (19.7.A). Exchange of alkyl groups in an ester by acid- or base-catalysis, usually via the *nucleophilic addition-elimination mechanism*.

Transmetallation (30.3.B). See *Metallation*.

Triplet (in nmr spectroscopy)(9.6). See *Spin-spin splitting.*

Triplet sensitizer (34.4.B). A molecule which is readily converted to its excited *triplet state* photochemically, and which will react with a different *ground state* molecule to convert it to its *triplet state.*

Triplet state (in photochemistry)(34.4.A). A molecule or atom in which two electron spins are unpaired, resulting in a net electronic spin of 1.

Unsaturated (3.5). An *unsaturated* molecule contains double or triple bonds (π-bonds). This is reflected in the formula of a hydrocarbon: for every π bond, there are two fewer hydrogens. Compare ethane (C_2H_6), ethylene (C_2H_4), and acetylene (C_2H_2). From the point of view of *molecular* formula, however, a ring results in the same differences: 1-butene ($CH_3CH_2CH{=}CH_2 = C_4H_8$) or cyclobutane

$$\begin{array}{c} CH_2-CH_2 \\ | \quad\quad | \\ CH_2-CH_2 \end{array} = C_4H_8$$

. Each π-bond *and* ring in a compound, therefore, is considered to be a *degree of unsaturation* from the point of view of molecular formula.

See *Saturated.*

Valence electrons (2.2). Those in the outermost shell of an atom. The attempt by an atom to achieve a filled *octet* of valence electrons is chiefly responsible for the bonding arrangements it undergoes.

Van der Waals forces (5.1). See *Dispersion force.*

Vertical transition (34.4.A). See *Franck-Condon transition.*

Vicinal (11.6.B.3 , 12.5.B). Attached to adjacent carbons. See *Geminal.*

Vinylogy (27.7.B.1). The similarity of a compound in which two functional groups are separated but conjugated by a double bond, with that in which the two groups are directly connected. For example, 4-methoxy-3-penten-2-one is a *vinylogous* ester.

Wavefunction (2.5). The equation which describes the behavior of an electron. The wavefunction can have regions in which its value is positive and regions in which it is negative, just as an ocean wave has peaks and troughs. These signs have no connection with the electron charge, which is always negative. The square of the wavefunction is a *probability function.*

Woodward's rules (21.4). An empirical method for predicting the position of the $\pi \rightarrow \pi^*$ *electronic transition* of *conjugated* systems in ultraviolet *spectroscopy.*

Ylide (13.7.G, 26.8). A neutral molecule which has a formal negative charge on carbon as in *phosphoranes* ($R_3\overset{+}{P}-\bar{C}H_2$) and sulfur ylides ($R_2\overset{+}{S}-\bar{C}H_2$).

Z (11.2). See *E.*

Zero point energy (6.1). The difference between the lowest point on a potential energy diagram for a bond and the lowest energy attainable (at 0°K) within the constraints of the *Heisenberg Uncertainty Principle.*

Zwitterion (29.1). A molecule which is an *inner salt,* that is, in which both the cation and anion are part of the same molecule.

APPENDIX II: SUMMARY OF FUNCTIONAL GROUP PREPARATIONS

Functional group interconversions can be discussed as reactions of one function or preparations of another. The following summary lists the preparations of various functional groups discussed in this textbook with reference to each place the reaction is used. Products shown are those for normal work-up of the reaction. Although examples are included with more than one functional group, specific reactions of polyfunctional compounds are not included. Abbreviations used are

R = alkyl and cycloalkyl; for some cases may also apply to R=H, R=Ar
Ar = aryl
X = halide
Y = leaving group; may be X, sulfonate, and so on
[H] = several reducing agents
[O] = several oxidizing agents

The importance of a given type of reaction or functional group transformation can be gauged roughly from the number of times it is cited.

Acetals and Ketals

Aldehydes or ketones

$$RC(=O)R' + R''OH \xrightarrow{H^+} RC(OR'')_2R'$$ 382-386, 475, 792, 850, 855, 900-901, 905, 906, 923, 926

Ethyl orthoformate

$$ArMgX + CH(OEt)_3 \longrightarrow ArCH(OEt)_2$$ 452

Acid Anhydrides

Acyl Halides

$$RC(=O)Cl + R'CO_2^- \longrightarrow RC(=O)OC(=O)R'$$ 551, 612

Carboxylic acids

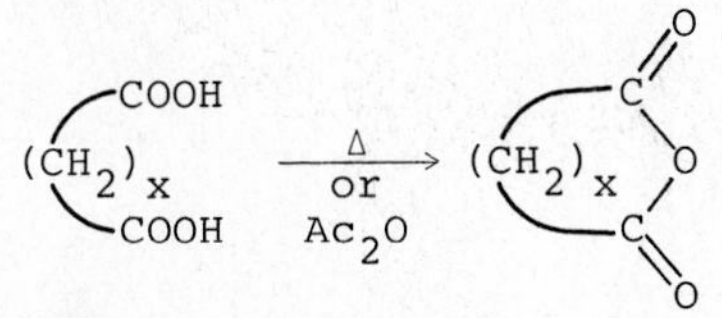

550, 863, 867

x = 2 or 3

Acyl Halides

Carboxylic acids

$$RCOOH + SOCl_2 \longrightarrow RCOCl$$ 519, 547

$$RCOOH + PCl_5 \longrightarrow RCOCl$$ 519

$$RCOOH + PBr_3 \longrightarrow RCOBr$$ 513, 519, 547, 608

Alcohols

Aldehydes and ketones: Carbanion additions

$$\underset{\substack{\| \\ O}}{R\overset{}{C}R'} + R''MgX \text{ or } R''Li \xrightarrow{\hspace{1cm}} \xrightarrow{H^+} RR'R''COH$$

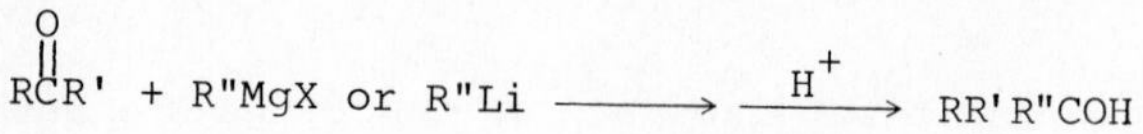

449–451, 453,
465, 470, 475, 602, 603,
855, 990, 1187

$$\underset{\substack{\| \\ O}}{R\overset{}{C}R'} + R''C\equiv C^- \xrightarrow{\hspace{1cm}} \xrightarrow{H^+} RR'\overset{OH}{\underset{|}{C}}C\equiv CR''$$

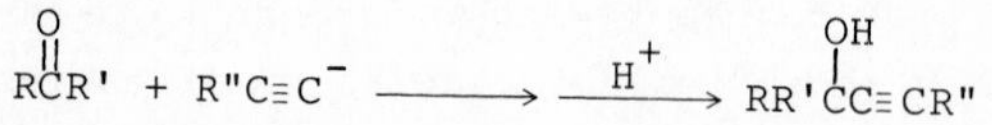

391, 465, 594

Aldehydes and ketones: Reductions

$$\underset{\substack{\| \\ O}}{R\overset{}{C}R'} + LiAlH_4 \text{ or } NaBH_4 \longrightarrow RCHOHR'$$

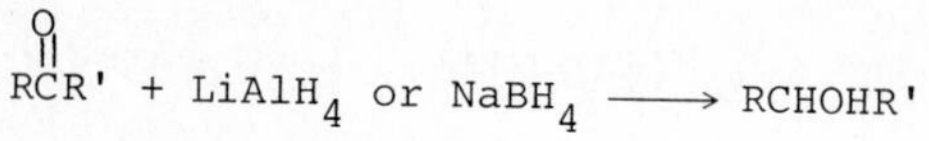

401–403, 466, 604, 639, 847, 855,
895, 908, 1050

$$\underset{\substack{\| \\ O}}{R\overset{}{C}R'} \xrightarrow{H_2/catalyst} RCHOHR'$$

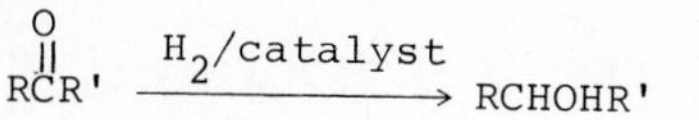

403, 466, 670, 908

$$2RCHO \xrightarrow{OH^-} RCOOH + RCH_2OH$$
(Cannizzaro reaction)

403, 404

$$RCHO + H_2C=O \xrightarrow{OH^-} RCH_2OH$$
(crossed Cannizzaro)

404, 847

$$\underset{\substack{\| \\ O}}{R\overset{}{C}R'} \xrightarrow{Mg} \xrightarrow{H^+} R\overset{R'}{\underset{OH}{C}}\overset{R'}{\underset{OH}{C}}R$$

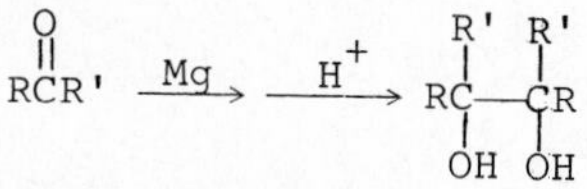

847, 850

Alkanes

$$RH + O_2 \xrightarrow{catalyst} ROOH \xrightarrow{H_2/cat.} ROH$$

246

Alkenes

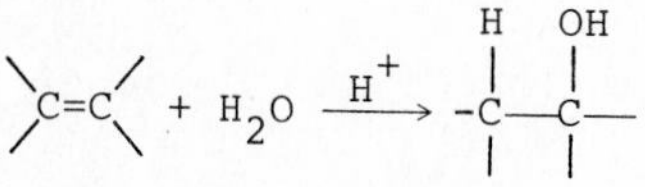

$$\diagup C=C\diagdown + H_2O \xrightarrow{H^+} -\overset{H}{\underset{|}{C}}-\overset{OH}{\underset{|}{C}}-$$

306, 307, 815

$$\diagup C=C\diagdown + HOX \longrightarrow -\overset{X}{\underset{|}{C}}-\overset{OH}{\underset{|}{C}}-$$

265, 321, 467, 1063

$$\diagup C=C\diagdown \xrightarrow[H_2O]{Hg^{2+}} -\overset{Hg^+}{\underset{|}{C}}-\overset{OH}{\underset{|}{C}}- \xrightarrow{NaBH_4} -\overset{H}{\underset{|}{C}}-\overset{OH}{\underset{|}{C}}-$$

311

$$\diagup C=C\diagdown \xrightarrow{B_2H_6} \xrightarrow{H_2O_2} H-\overset{|}{\underset{|}{C}}-\overset{|}{\underset{|}{C}}-OH$$

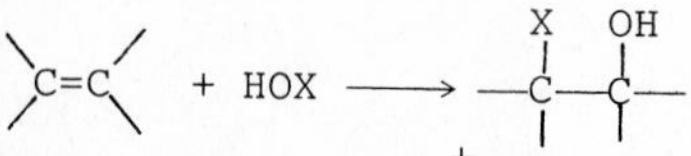

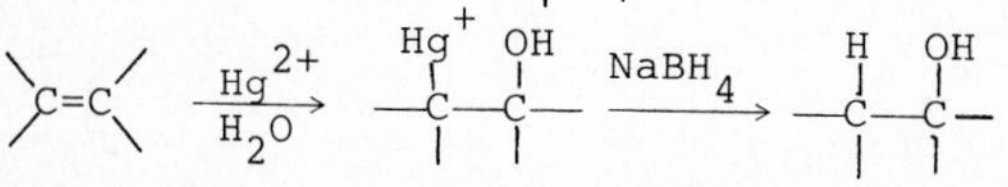

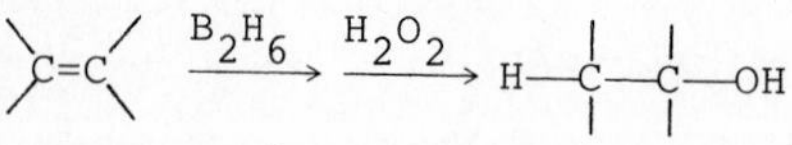

315–317, 467

$$RCH=CHR \xrightarrow{O_3} \xrightarrow{NaBH_4} RCH_2OH$$

320

$$\diagup C=C\diagdown \xrightarrow{KMnO_4} HO-\overset{|}{\underset{|}{C}}-\overset{|}{\underset{|}{C}}-OH$$

318, 845, 846

$$\diagup C=C\diagdown \xrightarrow{OsO_4}{H_2O_2} HO-\overset{|}{\underset{|}{C}}-\overset{|}{\underset{|}{C}}-OH$$

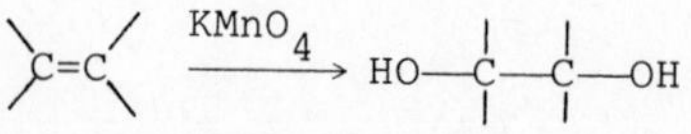

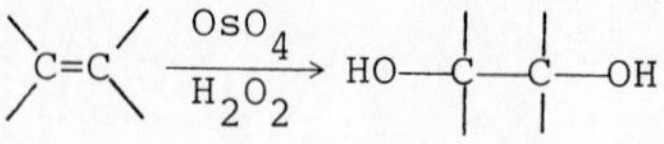

318, 466, 844, 849, 850

Amines

$$RNH_2 \xrightarrow{HNO_2} ROH \qquad\qquad 756$$

Carboxylic acids

$$RCOOH \xrightarrow{LiAlH_4} RCH_2OH \qquad\qquad 403, 515$$

Esters

$$R'COOR \xrightarrow[H_2O]{H+ \text{ or } OH^-} R'COOH + ROH \qquad 58, 59, 245, 540-543, 568, 665,$$
$$858, 917$$

$$R'COOR \xrightarrow[\text{or } LiBH_4]{LiAlH_4} R'CH_2OH + ROH \qquad 403, 555, 847$$

$$RCOOEt + R'MgX \text{ or } R'Li \longrightarrow \xrightarrow{H^+} RR'_2COH \qquad 552, 553$$

$$RMgX + (EtO)_2C=O \longrightarrow R_3COH \qquad\qquad 553$$

$$R'COOR \xrightarrow[\text{or } H_2/cat.]{Na/EtOH} R'CH_2OH + ROH \qquad 555$$

Ethers

$$ROR \xrightarrow{HX} ROH + RX \qquad\qquad 269, 476$$

Halides and sulfonates

$$RY \xrightarrow{H_2O \text{ or } OH^-} ROH \qquad\qquad 51, 56, 154, 160, 177, 244, 466,$$
$$575, 851$$

$$RX \xrightarrow[\text{ether}]{Mg} \xrightarrow{O_2} \xrightarrow{H^+} ROH \qquad 448$$

Oxiranes

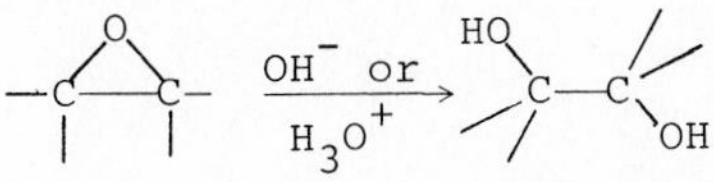

266, 267, 467, 844, 1065

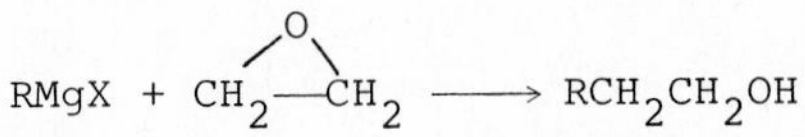

$$RMgX + CH_2\!-\!CH_2 \longrightarrow RCH_2CH_2OH \qquad 452, 453, 470$$

Peroxides

$$ROOR' \xrightarrow{H_2/cat.} ROH + R'OH \qquad\qquad 246$$

Phenols

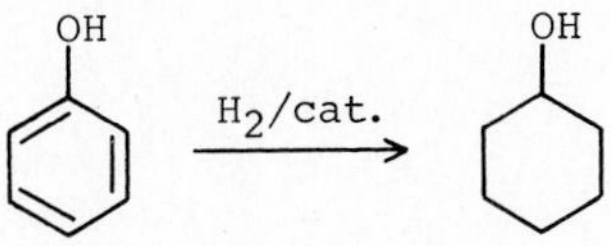

1185

Aldehydes

Acetals

$$RCH(OR')_2 \xrightarrow{H^+} RCHO + R'OH \qquad 893, 901, 902, 904$$

Acid derivatives

$RCOCl \xrightarrow[Li(t\text{-}BuO)_3AlH]{H_2/cat.\ or} RCHO$	554
$RCONR'_2 \xrightarrow{LiAlH(OEt)_3} RCHO$	557
$RCH(OH)COOH \xrightarrow{CaO} \xrightarrow[H_2O_2]{Fe^{+3}} RCHO$	916, 917
$RCN \xrightarrow{DIBAL} RCHO$	557
$ArCN \xrightarrow{SnCl_2,\ HCl} ArCHO$ (Stephen reduction)	558
$ArMgX + CH(OEt)_3 \longrightarrow ArCHO$	452

Alcohols

$RCH_2OH \xrightarrow{[O]} RCHO$	256-259, 367, 466, 600
$RCH(OH)CH(OH)R' \xrightarrow[Pb(OAc)_4]{HIO_4\ or} RCHO + R'CHO$	849, 850, 911-913
$R_2C(OH)CH_2OH \xrightarrow{H^+} R_2CHCHO$	1099

Alkenes

$RCH{=}CHR \xrightarrow{O_3} \xrightarrow[(CH_3)_2S]{Zn/AcOH\ or} RCHO$	320, 367

Alkynes

$RC{\equiv}CH \xrightarrow{B_2H_6} \xrightarrow[OH^-]{H_2O_2} RCH_2CHO$	349-350, 368

Arenes

$ArH + CO + HCl \xrightarrow{AlCl_3} ArCHO$	699
$ArH + HCON(CH_3)_2 \xrightarrow{POCl_3} ArCHO$ (Vilsmeier reaction)	761

Enol ethers

$R_2C{=}CHOR' \xrightarrow[H_2O]{H^+} R_2CHCHO$	386

Phenols

$ArO^- + CHCl_3 \longrightarrow Ar(OH)(CHO)$ (Reimer-Tiemann reaction)	1004

Alkanes and Arenes

Alcohols

$$\underset{\underset{R'}{|}}{\overset{\overset{R}{|}}{Ar\overset{|}{C}}}\text{—OR (or COR)} \xrightarrow[\text{HClO}_4]{\text{H}_2/\text{Pd}} \underset{\underset{R'}{|}}{\overset{\overset{R}{|}}{Ar\overset{|}{C}}}\text{—H}$$ 671, 672, 951, 957

Aldehyde

$$\text{RCHO} \xrightarrow{\text{L}_3\text{RhCl}} \text{RH}$$ 459

Alkenes

$$\underset{}{\overset{}{C}}=\underset{}{\overset{}{C}} \xrightarrow{\text{H}_2/\text{cat.}} \text{H—}\overset{|}{\underset{|}{C}}\text{—}\overset{|}{\underset{|}{C}}\text{—H}$$ 303, 304, 465, 473–475, 605, 670, 839, 1186, 1188

$$\underset{}{\overset{}{C}}=\underset{}{\overset{}{C}} \xrightarrow[\text{or catalyst}]{\overset{\text{initiator}}{\text{radical}}} \left(\text{—}\overset{|}{\underset{|}{C}}\text{—}\overset{|}{\underset{|}{C}}\text{—}\right)_n$$ 323–327, 610

$$\text{Ar—}\overset{|}{\underset{|}{C}}=\overset{|}{C} \xrightarrow{\text{Li/NH}_3} \text{Ar—}\overset{|}{C}\text{H—}\overset{|}{C}\text{H—}$$ 675, 1189

$$\text{—}\overset{\overset{O}{\|}}{C}\text{—}\overset{|}{C}=\overset{|}{\underset{|}{C}}\text{—} \xrightarrow{\underset{\text{NH}_3}{\text{Li}}} \text{—}\overset{\overset{O}{\|}}{C}\text{—}\overset{|}{C}\text{H—}\overset{|}{C}\text{H—}$$ 605

Alkynes

$$\text{RC}\equiv\text{CR}' \xrightarrow{\text{H}_2/\text{cat.}} \text{RCH}_2\text{CH}_2\text{R}'$$ 344, 474

Amines

$$\text{ArNH}_2 \xrightarrow{\text{HONO}} \text{ArN}_2^+ \xrightarrow{\text{H}_3\text{PO}_2} \text{ArH}$$ 796, 797, 802, 1032

Arenes

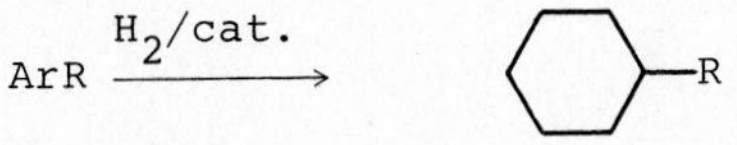

$$\text{ArR} \xrightarrow{\text{H}_2/\text{cat.}} \text{⬡—R}$$ 670, 862

Cyclohexanes

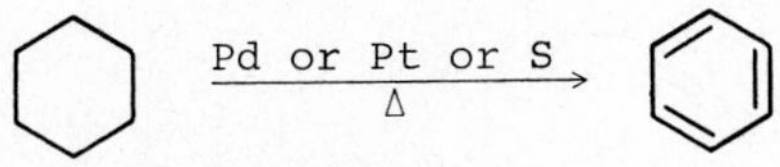

$$\text{⬡} \xrightarrow[\Delta]{\text{Pd or Pt or S}} \text{⬡}$$ 645, 1040, 1050

Halides and sulfonates

$$\text{RX} \xrightarrow{\text{LiAlH}_4} \text{RH}$$ 403, 671

$$\text{RNa + RX} \longrightarrow \text{RR}'$$ 443

$$\text{ArLi or ArMgX + RX} \longrightarrow \text{ArR}$$ 579, 991

$$\text{ArX + RX} \xrightarrow{\text{Na}} \text{ArR}$$ 991
(Wurtz-Fittig reaction)

ArI + Cu ———→ Ar-Ar   991
(Ullmann reaction)

$$\text{ArH + RX} \xrightarrow{\text{AlCl}_3} \text{ArR + ArR'}$$
(Friedel-Crafts alkylation)   699-702, 720

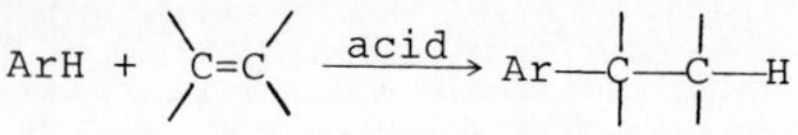

$$\text{ArH} + \begin{array}{c}\backslash\\/\end{array}C=C\begin{array}{c}/\\\backslash\end{array} \xrightarrow[\Delta]{\text{acid}} \text{Ar}-\overset{|}{\underset{|}{C}}-\overset{|}{\underset{|}{C}}-\text{H}$$ 701, 821, 996, 1185

$$\text{Ar}_2\text{CH}_2 \text{ (or Ar}_3\text{CH)} \xrightarrow{\text{NaNH}_2} \xrightarrow{\text{RX}}$$
$$\text{Ar}_2\text{CHR (or Ar}_3\text{CR)}$$ 670

$$\text{ArSO}_3\text{H} \xrightarrow[100°]{\text{H}_3\text{O}^+} \text{ArH}$$ 820

Ketones

$$\underset{\text{RCR'}}{\overset{\overset{O}{\|}}{}} \xrightarrow[\underset{\Delta}{\text{KOH/diethylene glycol}}]{\text{H}_2\text{NNH}_2} \text{RCH}_2\text{R'}$$
(Wolff-Kishner reduction)

$$\underset{\text{RCR'}}{\overset{\overset{O}{\|}}{}} \xrightarrow{\underset{\text{HCl},\Delta}{\text{Zn(Hg)}}} \text{RCH}_2\text{R'}$$
(Clemmensen reduction)

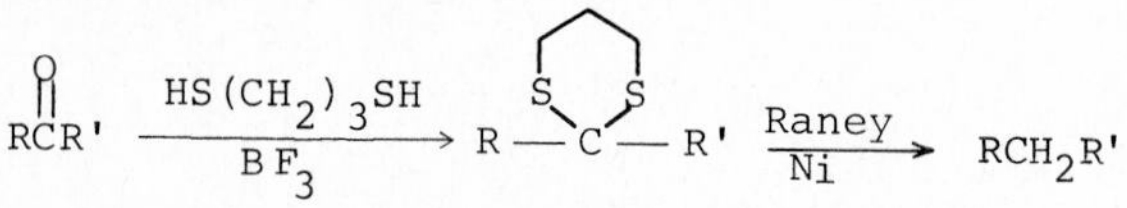

$$\underset{\text{RCR'}}{\overset{\overset{O}{\|}}{}} \xrightarrow[\text{BF}_3]{\text{HS(CH}_2)_3\text{SH}} R-\overset{S\frown S}{\underset{|}{C}}-R' \xrightarrow{\underset{\text{Ni}}{\text{Raney}}} \text{RCH}_2\text{R'} \qquad 406, 813$$

$$\underset{\text{ArCR}}{\overset{\overset{O}{\|}}{}} \xrightarrow{\text{H}_2\text{-Pd}} \text{ArCH}_2\text{R}$$ 671

Organometallics

$$\text{RMgX, RLi, or R}_3\text{Al} \xrightarrow{\text{R'OH}} \text{RH}$$ 447, 474, 476, 671

$$\text{R}_4\text{Si} \xrightarrow{\text{HCl}} \text{RH}$$ 447

Alkenes

Alcohols

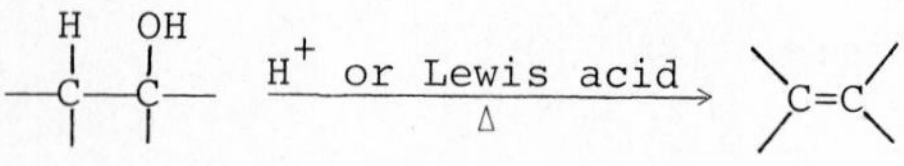

$$-\overset{\text{H}}{\underset{|}{C}}-\overset{\text{OH}}{\underset{|}{C}}- \xrightarrow[\Delta]{\text{H}^+ \text{ or Lewis acid}} \begin{array}{c}\backslash\\/\end{array}C=C\begin{array}{c}/\\\backslash\end{array}$$ 252, 262, 300, 301, 398, 465, 471, 593, 594, 600, 667, 852, 861, 1040, 1050, 1188

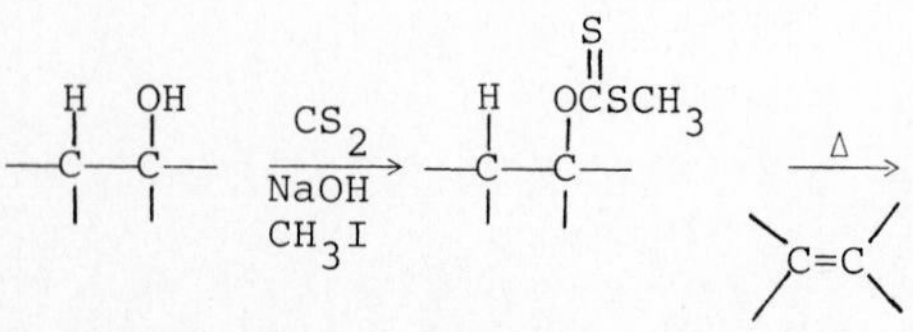

$$-\overset{\text{H}}{\underset{|}{C}}-\overset{\text{OH}}{\underset{|}{C}}- \xrightarrow[\underset{\text{CH}_3\text{I}}{\text{NaOH}}]{\text{CS}_2} -\overset{\text{H}}{\underset{|}{C}}-\overset{\overset{\overset{S}{\|}}{\text{OCSCH}_3}}{\underset{|}{C}}- \xrightarrow{\Delta} \begin{array}{c}\backslash\\/\end{array}C=C\begin{array}{c}/\\\backslash\end{array}$$ 565

Aldehydes and ketones

$$\underset{\text{RCR'}}{\overset{\overset{O}{\|}}{}} + \text{Ph}_3\text{P=CHR''} \qquad \underset{\text{RCR'}}{\overset{\overset{\text{CHR''}}{\|}}{}}$$ 398, 465, 470, 593
(Wittig reaction)

Alkynes

$RC{\equiv}CR' \xrightarrow{H_2/cat.}$ (cis-alkene, R and R' on same side, H and H on same side)

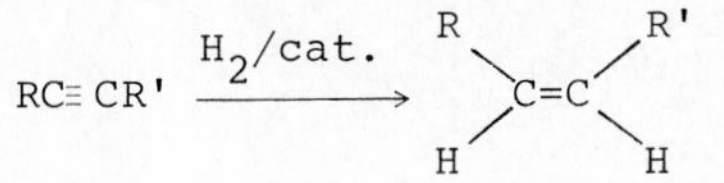

345, 346, 466, 470, 473, 594

$RC{\equiv}CR' \xrightarrow{B_2H_6} \xrightarrow{AcOH}$ (cis-alkene)

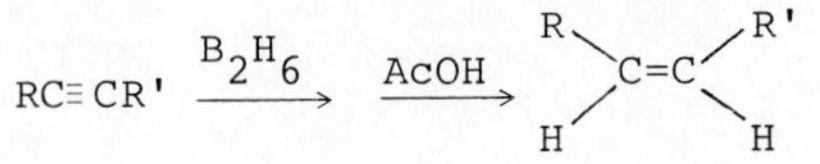

349-350

$RC{\equiv}CR' \xrightarrow[\text{liq. } NH_3]{Na}$ (trans-alkene)

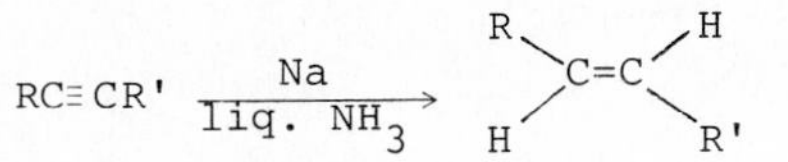

345, 346, 466

Amine oxides

(β-amino oxide with O^-—$N^+(CH_3)_2CH_3$) $\xrightarrow{\Delta}$ $C{=}C$

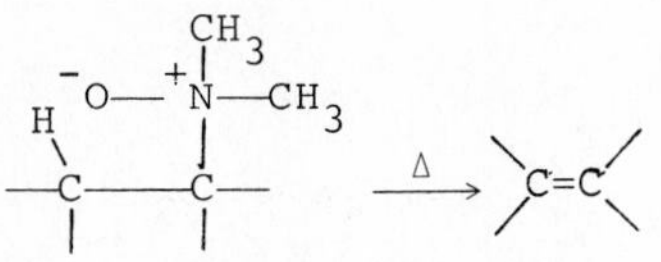

767

Ammonium hydroxides

(compound with $\overset{+}{N}Me_3$ and OH^-) $\xrightarrow{\Delta}$ $C{=}C$

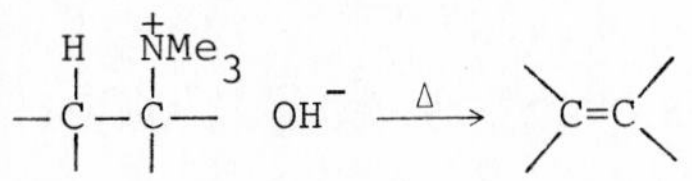

763-766

Arenes

$ArR \xrightarrow[NH_3]{Na}$ (1,4-dihydro product with R)

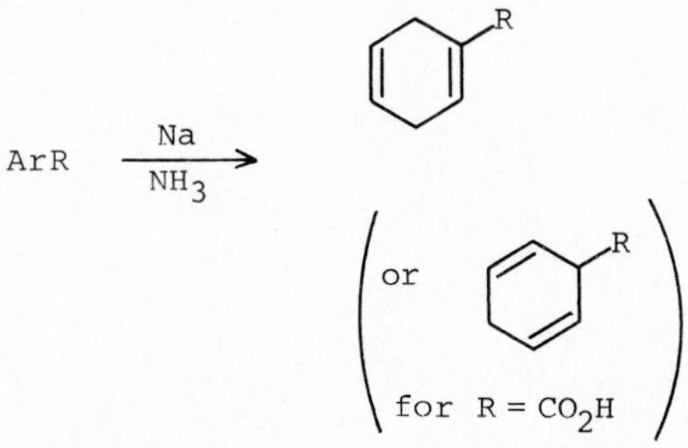

672-674, 1043

(or 2,5-dihydro product with R)

for R = CO$_2$H

674

Esters

(β-acyloxy compound, $OC R$ with $C{=}O$) $\xrightarrow{\Delta}$ $C{=}C$ + RCOOH

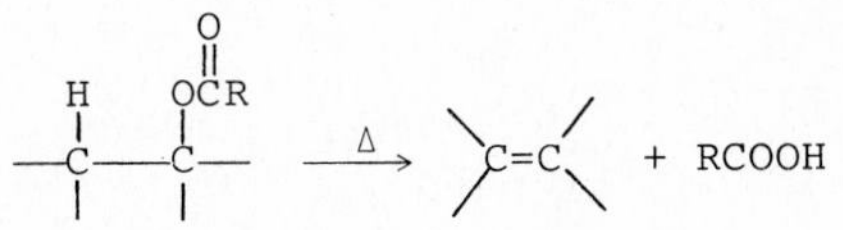

564-566, 593

Ethers

(β-alkoxy compound, OR) $\xrightarrow[\Delta]{H^+}$ $C{=}C$

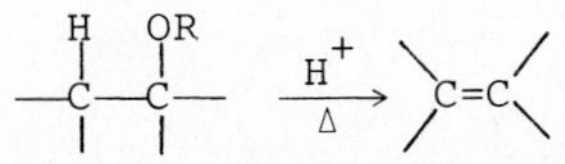

263, 386

Halides and sulfonates (X = halogen; Y = X or sulfonate)

(compound with H and Y) $\xrightarrow{base}$ $C{=}C$

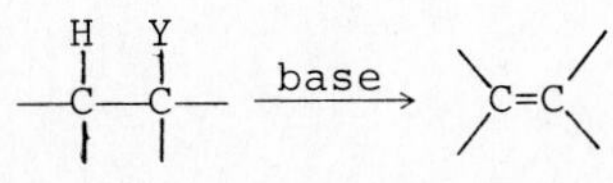

171, 172, 177, 178, 244, 248, 262, 293-299, 311, 352, 466, 470, 608, 611, 764, 857

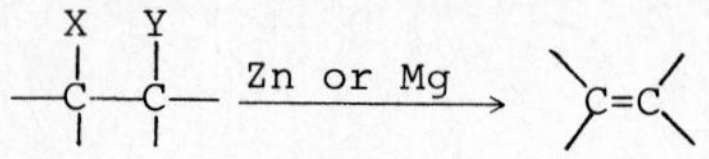

$$\underset{X \quad Y}{\overset{|\ \ \ |}{-\underset{|}{C}-\underset{|}{C}-}} \xrightarrow{\text{Zn or Mg}} \quad \diagdown C = C \diagup \qquad\qquad 444,\ 611$$

$$RMgX + CH_2=CHCH_2Y \longrightarrow RCH_2CH=CH_2 \qquad 579,593$$

$$ArCu + CH_2=CHCH_2Y \longrightarrow ArCH_2CH=CH_2 \qquad 918$$

Sulfoxides

$$\underset{|}{\overset{H}{-\underset{|}{C}-}}\underset{|}{\overset{\overset{O}{\diagdown} \underset{\diagup}{S}-R}{\underset{|}{C}-}} \xrightarrow{\ \Delta\ } \diagdown C = C \diagup \qquad\qquad 812$$

Alkynes

$$RCX_2CH_2R' \xrightarrow{\text{base}} RC\equiv CR' \qquad\qquad 341,\ 342$$

$$RCHXCHXR' \xrightarrow{\text{base}} RC\equiv CR' \qquad\qquad 341,\ 342$$

$$RC\equiv C^- + R'Y \longrightarrow RC\equiv CR' \qquad\qquad 342,\ 343,\ 346,\ 386,\ 464,\ 473$$

Amides

Acid anhydrides

$$(RCO)_2O + R'NH_2 \longrightarrow RCONHR' \qquad 549,\ 550,\ 747,\ 754,\ 944,\ 950,\ 960$$

Acyl halides

$$RCOCl + R'NH_2 \longrightarrow RCONHR' \qquad 548,\ 549,\ 753,\ 949,\ 950,\ 959$$

Carboxylic acids

$$RCOOH + R'NH_2 \xrightarrow{\ \Delta\ } RCONHR' \qquad 575,\ 753,\ 762,\ 802,\ 865,\ 868,\ 954,$$
$$RCOOH + R'NH_2 \xrightarrow{\text{DCC}} RCONHR' \qquad\qquad\qquad\quad 958,\ 959,\ 962$$

Esters

$$RCOOR' + R''NH_2 \longrightarrow RCONHR'' \qquad 949,\ 955$$

Ketones

$$R_2C=O + HN_3 \xrightarrow{H^+} RCONHR \qquad 946,\ 947$$
$$\text{(Schmidt reaction)}$$

Nitriles

$$RCN \xrightarrow[\text{or } H_2O_2/OH^-]{H^+} RCONH_2 \qquad 544,\ 545,\ 610,\ 868$$

Amines

Acyl halides

$$RCOCl + N_3^- \longrightarrow RCON_3 \xrightarrow[H_2O]{\Delta} RNH_2 \qquad 751$$
$$\text{(Curtius reaction)}$$

Aldehydes and ketones

$$\underset{RCR'}{\overset{O}{\overset{\|}{}}} + R''NH_2 \ (\text{or } NH_3) \xrightarrow{H_2/cat.} \underset{RCHR'}{\overset{NHR''}{\overset{|}{}}}$$
$$\underset{(\text{or } RCHR')}{\overset{NH_2}{\overset{|}{}}} \qquad 748,\ 749$$

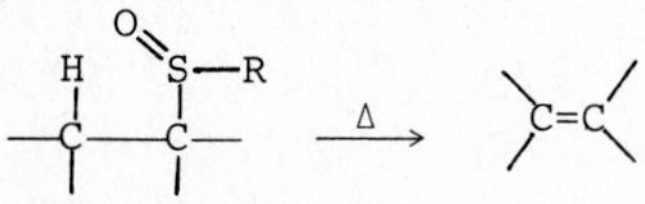

$RC(=O)R' + NH_3 + HCN \longrightarrow RCH(NH_2)CN$	945, 946

Amides

$RCONR'R'' \xrightarrow[OH^-]{H^+ \text{ or}} RCOOH + HNR'R''$	543, 762, 802, 946, 955
$RCONR'R'' \xrightarrow{LiAlH_4} RCH_2NR'R''$	403, 555, 556, 750
$RCONH_2 + Br_2 \xrightarrow{NaOH} RNH_2$ (Hofmann rearrangement)	563, 751, 752

Amines

RNH_2 (or $RR'NH$) $\xrightarrow[\Delta]{CH_2O,\ HCOOH}$ $RN(CH_3)_2$ (or $RR'NCH_3$) (Eschweiler-Clarke reaction)	749

Carboxylic acids

$RCOOH + NaN_3 \xrightarrow[\Delta]{H^+} RNH_2$ (Schmidt reaction)	751, 946, 947

Halides and sulfonates

$RY + R'NH_2$ (or NH_3) $\longrightarrow RR'NH$ (or RNH_2)	160, 179, 180, 742, 743, 785, 943, 946, 1064, 1065, 1068

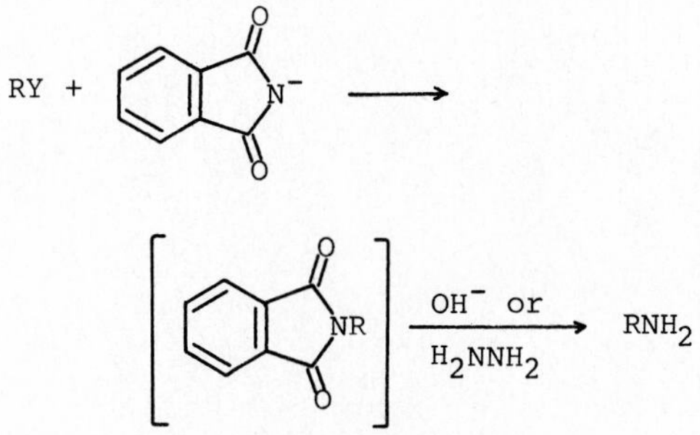

744, 943, 944

$RY + N_3^- \longrightarrow RN_3 \xrightarrow[\text{or } LiAlH_4]{H_2/\text{cat.}} RNH_2$	787
$ArX + NH_3 \longrightarrow ArNH_2$	765, 985
$ArX + NH_2^- \xrightarrow{\text{liq } NH_3} ArNH_2 + Ar'NH_2$	987

Nitriles

$RCN \xrightarrow[H_2/\text{cat.}]{LiAlH_4 \text{ or}} RCH_2NH_2$	403, 557, 746, 747, 865, 866

Nitro compounds

$RNO_2 \xrightarrow{LiAlH_4} RNH_2$	403
$RNO_2 \xrightarrow[H_2/\text{cat.}]{Fe,\ H^+ \text{ or}} RNH_2$	745, 746, 783, 802, 984

Oximes

$$\underset{\text{RCR'}}{\overset{\text{NOH}}{\|}} \xrightarrow[\text{or LiAlH}_4]{\text{H}_2/\text{cat.}} \underset{\text{RCHR'}}{\overset{\text{NH}_2}{\|}}$$ 747, 748, 944

<u>Azides</u>

$$\text{RY} + \text{N}_3^- \longrightarrow \text{RN}_3$$ 160, 751, 787, 839

$$\underset{\text{RCH}\!-\!\!-\!\text{CH}_2}{\overset{\text{O}}{\triangle}} + \text{N}_3^- \longrightarrow \underset{\text{RCHCH}_2\text{N}_3}{\overset{\text{OH}}{|}}$$ 787

<u>Carboxylic Acids</u>

Acid anhydrides

$$(\text{RCO})_2\text{O} + \text{H}_2\text{O} \longrightarrow \text{RCOOH}$$ 544

Acyl halides

$$\text{RCOCl} + \text{H}_2\text{O} \longrightarrow \text{RCOOH}$$ 544

Alcohols

$$\text{RCH}_2\text{OH} \xrightarrow{[\text{O}]} \text{RCOOH}$$ 258, 510, 910, 918-920

$$\text{RCHOHCH}_2\text{R'} \xrightarrow{[\text{O}]} \text{RCOOH} + \text{R'COOH}$$ 258

Aldehydes

$$\text{RCHO} \xrightarrow{[\text{O}]} \text{RCOOH}$$ 399, 473, 510, 511, 600, 909-911,
                 915-920, 923

$$2\text{RCHO} \xrightarrow{\text{OH}^-} \text{RCOOH} + \text{RCH}_2\text{OH}$$ 403
    (Cannizzaro reaction)

Alkenes

$$\text{RCH}=\text{CHR'} \xrightarrow{\text{KMnO}_4} \text{RCOOH} + \text{R'COOH}$$ 319, 742, 862

Alkynes

$$\text{RC}\equiv\text{CR'} \xrightarrow{\text{KMnO}_4} \text{RCOOH} + \text{R'COOH}$$ 350

Amides

$$\text{RCONH}_2 \xrightarrow[\text{or HONO}]{\text{H}^+ \text{ or OH}^-} \text{RCOOH}$$ 543, 955

Arenes

$$\text{ArR} \xrightarrow{[\text{O}]} \text{ArCOOH}$$ 668, 669, 716, 782, 863

Esters

$$\text{RCOOR'} \xrightarrow{\text{H}^+ \text{ or OH}^-} \text{RCOOH} + \text{R'OH}$$ 58, 59, 245, 540-543, 568, 608, 858,
                 879, 883

$$\text{RCOOCH}_2\text{Ar} \xrightarrow[\text{Pt}]{\text{H}_2} \text{RCOOH}$$ 951, 957

Halides

$$\text{RX} \xrightarrow[\text{ether}]{\text{Mg}} [\text{RMgX}] \xrightarrow{\text{CO}_2} \text{RCOOH}$$ 451, 465, 476, 509, 1185

ArX $\xrightarrow{n-\text{BuLi}}$ $\xrightarrow{\text{CO}_2}$ ArCOOH   990

Hydroxy ketones

$\underset{\text{RCCHR'}}{\overset{\text{O OH}}{|| |}}$ $\xrightarrow{\text{HIO}_4}$ RCOOH + $\underset{\text{R'CH}}{\overset{\text{O}}{||}}$ 855

Ketones

$\underset{\text{RCCH}_3}{\overset{\text{O}}{||}}$ $\xrightarrow{X_2, \text{OH}^-}$ RCOOH   377, 608

(Haloform reaction)

$\underset{\text{RCCH}_2\text{R'}}{\overset{\text{O}}{||}}$ $\xrightarrow{[O]}$ RCOOH + R'COOH   401, 862

$\underset{\text{RC-CR}}{\overset{\text{O O}}{|| ||}}$ $\xrightarrow{\text{OH}^-}$ $\underset{R_2\text{CCOOH}}{\overset{\text{OH}}{|}}$ 877, 878

(Benzilic acid rearrangement)

Nitriles

RCN $\xrightarrow{\text{H}^+ \text{ or OH}^-}$ RCOOH   509, 544, 608, 857, 862, 915, 916, 945

Enamines

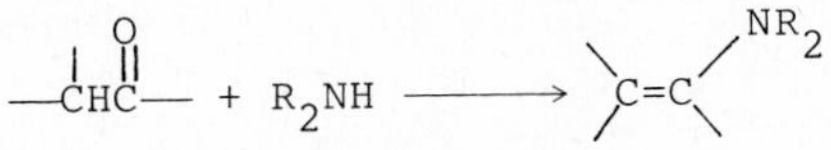

$—\underset{|}{\overset{|}{\text{CH}}}\underset{}{\overset{\text{O}}{||}}\text{C}—$ + $R_2\text{NH}$ $\longrightarrow$ $\text{C=C}\diagup^{NR_2}$ 768-770

$—\underset{|}{\overset{|}{\text{CH}}}—\underset{|}{\overset{|}{\text{CH}}}—NR_2$ $\xrightarrow{\text{Hg(OAc)}_2}$ $\xrightarrow{\text{OH}^-}$ $\text{C=C}\diagup^{NR_2}$ 770

Epoxides (see Oxiranes)

Esters

Acid anhydrides

(RCO)$_2$O + R'OH $\longrightarrow$ RCOOR'   907

Acyl halides

RCOCl + R'OH $\longrightarrow$ RCOOR'   546, 547, 553, 608, 1000

Alcohols and phenols

ROH + R'COOH $\xrightarrow{\text{H}^+}$ R'COOR   516-519, 858-859, 861

ROH + R'COCl $\longrightarrow$ R'COOR   546, 547, 553, 855, 863, 1000

ROH + (R'CO)$_2$O $\longrightarrow$ R'COOR   547, 1000, 1001

Amides

RCONH$_2$ + R'OH $\xrightarrow{\text{H}^+}$ RCOOR'   548, 610

Carboxylic acids

RCOOH + CH$_2$N$_2$ $\longrightarrow$ RCOOCH$_3$ 512

RCOOH + R'OH $\xrightarrow{\text{H}^+}$ RCOOR'   516-519, 858-859, 861, 949, 950, 955, 1185

RCO$_2^-$ + R'Y $\longrightarrow$ RCOOR'   160, 860, 961

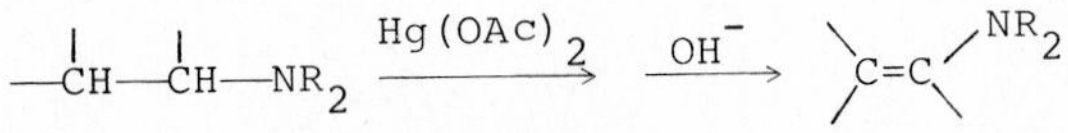

Esters

$$RCOOR' + R''OH \xrightarrow{H^+ \text{ or } R''O^-} RCOOR'' + R'OH \quad 548, 865$$

Halides and sulfonates

$$RY + R'CO_2^- \longrightarrow R'COOR \qquad 160, 172, 245, 512, 665, 817, 860,$$
$$961$$

Ketones

$$\overset{O}{\overset{\|}{R C R}} + R'CO_3H \longrightarrow \overset{O}{\overset{\|}{R C O R}} \qquad 400, 401, 858$$

Ethers

Alcohols

$$ROH \xrightarrow[\Delta]{H^+} ROR \qquad 252, 261\text{-}262, 268, 300, 815, 840, 848$$

Alkenes

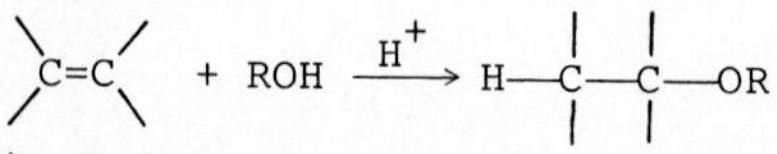

$$\overset{}{C=C} + ROH \xrightarrow{H^+} H-\overset{|}{\underset{|}{C}}-\overset{|}{\underset{|}{C}}-OR \qquad 263, 476$$

$$\overset{}{C=C} + ROH + Hg(OAc)_2 \longrightarrow \xrightarrow{NaBH_4} -\overset{H}{\underset{|}{C}}-\overset{|}{\underset{|}{C}}-OR \qquad 311$$

$$\overset{}{C=C} + ROH + X_2 \longrightarrow -\overset{OR}{\underset{|}{C}}-\overset{|}{\underset{X}{C}}- \qquad 310$$

Alkynes

$$RC\equiv CH \xrightarrow[R'OH]{R'O^-} \overset{OR'}{\underset{|}{RC}}=CH_2 \qquad 349, 386$$

Halides and sulfonates

$$RY + R'O^- \text{ (or } R'OH) \longrightarrow ROR' \qquad 160, 170, 172, 174, 177, 268, 270,$$
$$293, 815, 840, 904, 906, 923, 1066$$

$$RY + ArO^- \longrightarrow ROAr \qquad 999, 1060$$

$$ArX + RO^- \longrightarrow ArOR \qquad 1000$$

Ketals

$$\overset{OCH_3}{\underset{OCH_3}{RCCH_2R}} \xrightarrow{\Delta} \overset{OCH_3}{\underset{}{RC}}=CHR \qquad 386$$

Halogen Compounds

Alcohols

$$ROH + HX \longrightarrow RX \qquad 248\text{-}251, 254\text{-}256, 575, 576, 667$$

$$ROH + SOCl_2 \longrightarrow RCl \qquad 254\text{-}256, 470, 578, 667, 671$$

$$ROH + PCl_5 \longrightarrow RCl \qquad 254$$

$$ROH + PBr_3 \longrightarrow RBr \qquad 254\text{-}256, 466, 1185$$

$$ROH + PI_3 \longrightarrow RI \qquad 254\text{-}255$$

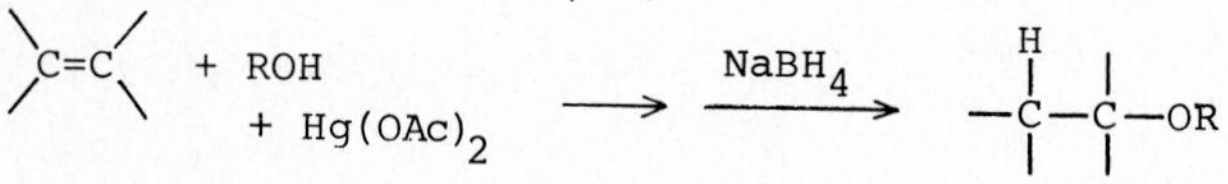

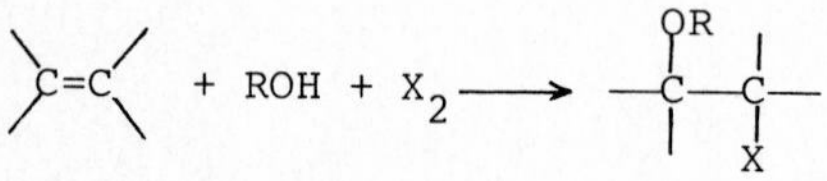

Alkanes

$RH + X_2 \xrightarrow{h\nu} RX$	104-114, 139-143, 465, 513, 583

Alkenes

$>C=C< + HX \longrightarrow H-\overset{\mid}{\underset{\mid}{C}}-\overset{\mid}{\underset{\mid}{C}}-X$	305, 306, 312-314, 347, 589, 866
$>C=C< + X_2 \longrightarrow X-\overset{\mid}{\underset{\mid}{C}}-\overset{\mid}{\underset{\mid}{C}}-X$	307-310, 341, 352, 466, 590, 601, 690
$>C=C< + HOX \longrightarrow HO-\overset{\mid}{\underset{\mid}{C}}-\overset{\mid}{\underset{\mid}{C}}-X$	265, 321, 467, 1063
$>C=C< + CX_4 \longrightarrow X-\overset{\mid}{\underset{\mid}{C}}-\overset{\mid}{\underset{\mid}{C}}-CX_3$	313, 314
$-\overset{\mid}{C}=\overset{\mid}{C}-\overset{H}{\underset{\mid}{C}}- \xrightarrow{NBS} -\overset{\mid}{C}=\overset{\mid}{C}-\overset{Br}{\underset{\mid}{C}}-$	582

Alkynes

$RC\equiv CR' + HX \longrightarrow RCH=CXR'$	347-349, 352
$RC\equiv CR' + X_2 \longrightarrow RCX=CXR'$	347, 348

Amines

$ArNH_2 \xrightarrow{HONO} ArN_2^+ \xrightarrow{I^-} ArI$	793
$ArNH_2 \xrightarrow{HONO} ArN_2^+ \xrightarrow[\Delta]{CuX} ArX\ (X = Cl,\ Br)$	792, 794, 984
$ArNH_2 \xrightarrow[BF_3\,(or\ HPF_6)]{HONO} ArN_2^+\ BF_4^-\ (or\ PF_6^-) \xrightarrow{\Delta} ArF$	795-796, 802

Arenes

$ArH + X_2 \longrightarrow ArX$	690-694, 703, 705, 718, 719, 760, 762, 820, 982, 983, 1001, 1008, 1052, 1053
$ArH + CH_2O + HCl \xrightarrow{ZnCl_2} ArCH_2Cl$	701, 1081
$ArCH_3 + Cl_2 \xrightarrow{h\nu} ArCH_2Cl$	662-664
$ArCHR_2 + Br_2 \xrightarrow{h\nu} ArCBrR_2$	663, 664
$ArH + CCl_4 \xrightarrow{AlCl_3} Ar_3CCl$	702

Carboxylic acids

$RCH_2COOH + Br_2 \xrightarrow{PBr_3} RCHBrCOOH$ (Hell-Volhard-Zelinsky reaction)	513, 514, 608, 859
$RCOOH \xrightarrow{Ag^+} \xrightarrow[CCl_4]{Br_2} RBr$	520
$RCOOH \xrightarrow[Br_2]{HgO} RBr$	520
$RCOOH \xrightarrow[LiCl]{Pb(OAc)_4} RCl$	521

Ethers

$ROR' + HX \longrightarrow ROH + R'X$ or $RX + R'X$	263, 269

Halides and sulfonates

$RY + X^- \longrightarrow RX$ 154-160, 171, 255, 578

Ketones

$-\overset{|}{C}H-\overset{O}{\overset{\|}{C}}- \xrightarrow[H^+ \text{ or } OH^-]{X_2} -\overset{X}{\underset{|}{\overset{|}{C}}}-\overset{O}{\overset{\|}{C}}-$ 375-377

Aldehydes and ketones

$R\overset{O}{\overset{\|}{C}}R' \text{ (or H)} + R''NH_2 \longrightarrow \underset{R' \text{ (or H)}}{\underset{|}{RC}}=NR''$ 388-390, 600, 747, 748, 855, 917, 945

$+ H_2NNHR'' \longrightarrow \underset{R' \text{ (or H)}}{\underset{|}{RC}}=NNHR''$ 389, 914

Amines

$R_2CHNR'_2 \xrightarrow{Hg(OAc)_2} R_2C{=}\overset{+}{N}R'_2$ 770

Ketones

Acyl Halides

$RCOCl + R'MgX \text{ (or } R'_2CuLi \text{ or } R'_2Cd) \longrightarrow R\overset{O}{\overset{\|}{C}}R'$ 551, 552

$RCOCl + CH_2N_2 \longrightarrow R\overset{O}{\overset{\|}{C}}CHN_2 \text{ or } R\overset{O}{\overset{\|}{C}}CH_2Cl$ 788

Alcohols

$RR'CHOH \xrightarrow{[O]} R\overset{O}{\overset{\|}{C}}R'$ 256-259, 366, 466, 667, 871, 907, 1189

$-\underset{|}{\overset{OH}{\overset{|}{C}}}-\underset{|}{\overset{OH}{\overset{|}{C}}}- \xrightarrow[Pb(OAc)_4]{HIO_4 \text{ or}} -\underset{|}{\overset{O}{\overset{\|}{C}}} + \underset{|}{\overset{O}{\overset{\|}{C}}}-$ (849-850, 911-913)

$R_2\overset{OH}{\overset{|}{C}}-\overset{OH}{\overset{|}{C}}R_2 \xrightarrow{H^+} R_3C\overset{O}{\overset{\|}{C}}R$ 847, 848

Alkenes

$\gt C=C \lt \xrightarrow{KMnO_4} \gt C=O + O=C \lt$ 319

$\gt C=C \lt \xrightarrow{O_3} \gt C=O + O=C \lt$ 319, 320 (367)

Alkynes

$RC\equiv CR \xrightarrow[Hg^{2+}]{H^+} R\overset{O}{\overset{\|}{C}}CH_2R$ 348, 368

$RC{\equiv}CR \xrightarrow{B_2H_6} \xrightarrow[OH^-]{H_2O_2} RC(=O)CH_2R$ (350)

$RC{\equiv}CR \xrightarrow{KMnO_4} RC(=O)\text{-}C(=O)R$ 350

Amino Alcohols

$R_2C(OH)CH_2NH_2 \xrightarrow{HONO} RC(=O)CH_2R$ 756, 757

Arenes

$ArH + RCOCl \xrightarrow{\text{Lewis acid}} ArC(=O)R$ 697-698, 720, 761, 869

$ArH + (RCO)_2O \xrightarrow{AlCl_3} ArC(=O)R$ 868, 869, 1009, 1035, 1040

$ArH + RCOOH \xrightarrow[\text{or Lewis acid}]{H^+} ArC(=O)R$ 869, 870, 1009, 1011, 1040, 1050

$ArCH_2R \xrightarrow{[O]} ArC(=O)R$ 1036

Dithioacetals

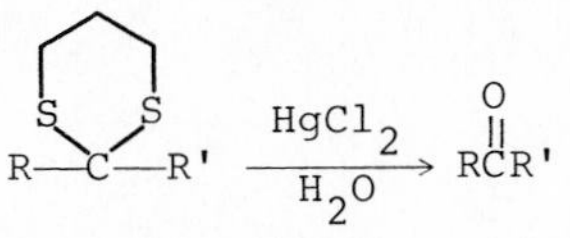

832, 833

Enol ethers

$RC(OR')=CR_2 \xrightarrow[H_2O]{H^+} RC(=O)CHR_2$ 386, 1077

Esters

$RCH_2COOEt + R'COOEt \xrightarrow{\text{base}} RCH(COR')COOEt$ 560-562, 870, 871

Ketones

$RCH_2C(=O)R' \xrightarrow{SeO_2} RC(=O)\text{—}C(=O)R'$ 871

Nitriles

$RCN + R'MgX \text{ (or } R'Li) \longrightarrow RC(=O)R'$ 553

$RCN + ArOH \xrightarrow[HCl]{ZnCl_2} Ar(OH)(COR)$ 1009

(Houben-Hoesch reaction)

Oxiranes

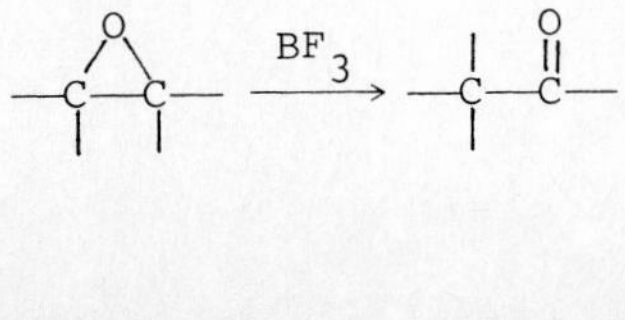

268

Nitriles

Aldehydes and ketones

$$\mathrm{R\overset{O}{\overset{\|}{C}}R'\ (or\ H) + HCN \longrightarrow RR'(or\ H)\overset{OH}{\overset{|}{C}}CN}$$ 391, 392, 464, 610, 857, 915

$$\mathrm{R\overset{O}{\overset{\|}{C}}R'\ (or\ H) + HCN + NH_3 \longrightarrow RR'\overset{NH_2}{\overset{|}{C}}CN}$$ 945, 946

$$\mathrm{-\underset{}{\overset{|}{C}}{=}\overset{|}{C}-\overset{O}{\overset{\|}{C}}- + HCN \longrightarrow -\overset{CN}{\overset{|}{\underset{|}{C}}}-\overset{|}{C}H-\overset{O}{\overset{\|}{C}}-}$$ 601-602, 862

Amides

$$\mathrm{RCONH_2 \xrightarrow[SOCl_2]{P_2O_5\,or} RCN}$$ 563, 865

Amines

$$\mathrm{ArNH_2 \xrightarrow{HONO} ArN_2^+ \xrightarrow{CuCN} ArCN}$$ 794, 802

(Sandmeyer reaction)

Halides and sulfonates

$$\mathrm{RY + CN^- \ (or\ CuCN) \longrightarrow RCN}$$ 160, 168, 311, 464, 509, 608, 665, 862, 866

$$\mathrm{ArX + CuCN \longrightarrow ArCN}$$ 794

Oximes

$$\mathrm{RCH{=}NOH \xrightarrow{Ac_2O} RCN}$$ 917

Sulfonic acids

$$\mathrm{ArSO_3H \xrightarrow[\Delta]{NaCN} ArCN}$$ 922

Nitro Compounds

Amines

$$\mathrm{ArNH_2 \xrightarrow{CF_3CO_3H} ArNO_2}$$ 759

$$\mathrm{ArNH_2 \xrightarrow{HONO} ArN_2^+ \xrightarrow[Cu]{NO_2^-} ArNO_2}$$ 795

Arenes

$$\mathrm{ArH + HNO_3 \longrightarrow ArNO_2}$$ 695, 696, 703, 704, 716, 718, 719, 760, 763, 803, 984, 1006-1008, 1036, 1041

Halides and sulfonates

$$\mathrm{RY + NO_2^- \longrightarrow RNO_2}$$ 168, 781

Ketones

$$\mathrm{R\overset{O}{\overset{\|}{C}}R' + CH_3NO_2 \xrightarrow{base} R\overset{OH}{\overset{|}{\underset{\underset{R'}{|}}{C}}}CH_2NO_2}$$ 781

Organometallics

Aluminum

$RC{\equiv}CH + (\textit{i}Bu)_2AlH \longrightarrow RCH{=}CHAl(\textit{i}Bu)_2$ 446

Boron

$RCH{=}CH_2 + B_2H_6 \longrightarrow (RCH_2CH_2)_3B$ 315, 317, 446

$RC{\equiv}CR + B_2H_6 \longrightarrow (RCH{=}CR)_3B$ 349, 350

Cadmium

$RMgX + CdCl_2 \longrightarrow R_2Cd\,(R = \text{prim.})$ 445

Copper

$RLi + CuI \text{ (or CuBr)} \longrightarrow RCu$ 445

$RLi + CuI \longrightarrow R_2CuLi$ 453, 454, 551, 603, 992

Lithium

$RX + Li \xrightarrow{\text{ether}} RLi$ 339, 442, 990

$ArBr + RLi \longrightarrow ArLi$ 991

Magnesium (Grignard reagents)

$RX + Mg \xrightarrow{\text{ether}} RMgX$ 441, 579, 580, 990

Mercury

$RLi \text{ or } RMgX + HgCl_2 \longrightarrow R_2Hg$ 445

$RCH{=}CH_2 + Hg(OAc)_2 \xrightarrow{R'OH} RCH(OR')CH_2HgOAc$ 311, 440

Silicon

$RMgX + SiCl_4 \longrightarrow R_4Si$ 445

Sodium

$RX + Na \longrightarrow RNa$ 443

$R_2Hg + Na \longrightarrow RNa$ 446

Tin

$RLi + SnCl_4 \longrightarrow R_4Sn$ 445

Oxiranes

Alkenes

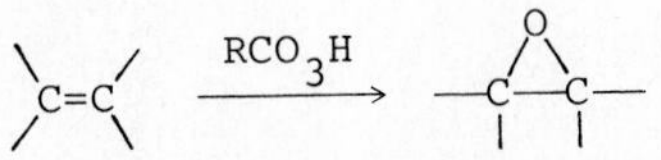

266, 320, 321, 467, 844

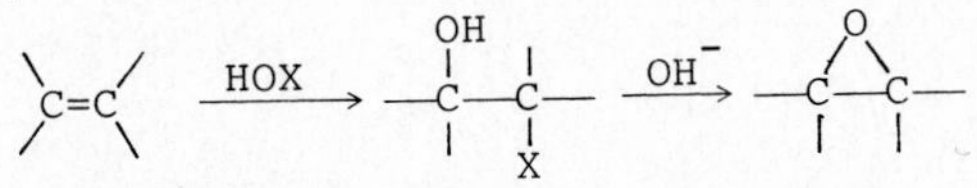

265, 310, 321, 467, 1063

Phenols

Amines

$ArNH_2 \xrightarrow{HONO} ArN_2^+ \xrightarrow[\Delta]{H_2O} ArOH$ — 792, 996

Ethers

$ArOR + HX \longrightarrow ArOH + RX$ — 1011, 1012

Halides

$ArX \xrightarrow{OH^-} ArOH$ — 985, 987, 996

Hydroperoxides

$R_2C(Ar)OOH \xrightarrow{H^+} RC(=O)R + ArOH$ — 997

Sulfonic acids

$ArSO_3H \xrightarrow[\Delta]{NaOH} ArOH$ — 822, 996

Phosphorus Compounds

Phosphate esters

$nROH + Cl_nP(=O)(OR')_{3-n} \longrightarrow (RO)_nP(=O)(OR')_{3-n}$ — 826, 827

Phosphines

$RMgX + PCl_3 \longrightarrow R_3P$ — 824

Phosphite esters

$ROH + PCl_3 \xrightarrow{pyridine} (RO)_3P:$ — 828

Phosphonate esters

$(RO)_3P: + R'X \longrightarrow (RO)_2P(=O)R'$ — 829

Phosphonium salts

$R_3P: + R'X \longrightarrow R_3\overset{+}{P}R'\ X^-$ — 160, 397, 593, 824

Ylides

$R_3\overset{+}{P}CHR'_2 \xrightarrow{n\text{BuLi}} R_3\overset{+}{P}\overset{-}{C}R'_2$ — 397, 472, 593, 830

Polymers

By addition

$>C=C< \longrightarrow -(\overset{|}{\underset{|}{C}}-\overset{|}{\underset{|}{C}})_n-$ — 323-327, 594, 610

By condensation

$HY'-R-YH \longrightarrow -(Y'-R)- + HY$ — 786, 860, 865, 954, 955

Sulfur Compounds

Disulfides

$2\ RSH \xrightarrow{[O]} RSSR$	810

Sulfides

$RS^- + R'Y \longrightarrow RSR'$	808, 809, 1064, 1066, 1068
$>C=C< + RSH \longrightarrow H-\overset{\vert}{\underset{\vert}{C}}-\overset{\vert}{\underset{\vert}{C}}-SR$	313
$R\overset{O}{\overset{\Vert}{C}}CHR' \xrightarrow{R''SSR''} R\overset{O}{\overset{\Vert}{C}}\underset{R'}{\underset{\vert}{C}}HSR''$	810

Sulfinic acids

$ArSO_2Cl \xrightarrow[H_2O]{Zn} ArSO_2H$	823

Sulfonamides

$RSO_2Cl + R'NH_2 \longrightarrow RSO_2NHR'$	823, 956

Sulfonate esters

$ArSO_2Cl + ROH \xrightarrow{pyridine} ArSO_2OR$	255, 578, 817, 822, 1092

Sulfones

$R_2SO \xrightarrow{R'CO_3H} R_2SO_2$	811
$R_2S \xrightarrow[\text{or } R'CO_3H]{KMnO_4} R_2SO_2$	811
$ArSO_2Cl + Ar'H \xrightarrow{AlCl_3} ArSO_2Ar'$	823
$RSO_2^- + R'X \longrightarrow RSO_2R'$	823

Sulfonic acids

$RY + SO_3^{2-} \longrightarrow RSO_3^-$	167, 816
$R\overset{O}{\overset{\Vert}{C}}H + HSO_3^- \longrightarrow R\overset{OH}{\overset{\vert}{C}}HSO_3^-$	817
$ArSO_2Cl \xrightarrow{H_2O} ArSO_3H$	811
$ArH + H_2SO_4 \longrightarrow ArSO_3H$	818-820, 1007, 1008

Sulfonium salts

$R_2S + R'X \longrightarrow R_2\overset{+}{S}R'\ X^-$	160, 812

Sulfonyl halides

$$ArSSAr \xrightarrow[HNO_3]{Cl_2} ArSO_2Cl \qquad\qquad 811$$

$$RSO_3Na \xrightarrow{PCl_5} RSO_2Cl \qquad\qquad 817$$

$$ArH \xrightarrow{ClSO_3H} ArSO_2Cl \qquad\qquad 820, 822$$

Sulfoxides

$$R_2S \xrightarrow{H_2O_2} R_2SO \qquad\qquad 811$$

Thiols

$$RY + HS^- \longrightarrow RSH \qquad\qquad 160, 313, 808$$

$$RY + H_2NCSNH_2 \longrightarrow \xrightarrow{OH^-} RSH \qquad\qquad 809$$

$$RX \longrightarrow RMgX \xrightarrow{S} RSH \qquad\qquad 809$$

$$RSSR \xrightarrow[NH_3]{Li} RSH \qquad\qquad 810$$

$$\text{C=C} + H_2S \xrightarrow{h\nu} HC-C-SH \qquad\qquad 313$$

$$ArNH_2 \xrightarrow{HONO} ArN_2^+ \xrightarrow[EtOCS_2^-;OH^-]{HS^- \ or} ArSH \qquad\qquad 793, 802$$

Thiocyanates

$$RX + {}^-SCN \longrightarrow RSCN \qquad\qquad 160, 165, 742$$

FORMATION OF C—C BONDS

Acyloin and Pinacol Condensations

$$RCOOEt \xrightarrow[ether]{Na} RCOCHOHR \qquad\qquad 851$$

$$\underset{RCR'}{\overset{O}{\|}} \xrightarrow{Mg} \xrightarrow{H^+} R-\underset{R'}{\overset{OH}{\underset{|}{\overset{|}{C}}}}-\underset{R'}{\overset{OH}{\underset{|}{\overset{|}{C}}}}-R \qquad\qquad 847, 848$$

Alkene Addition Reactions

Michael additions

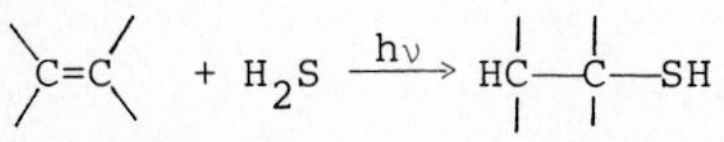

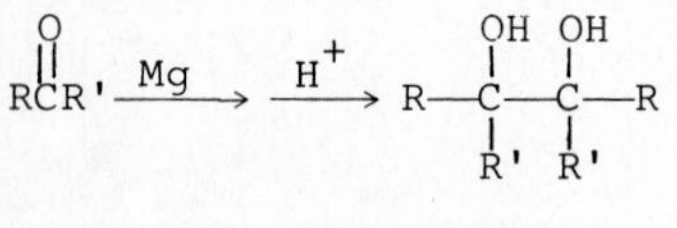

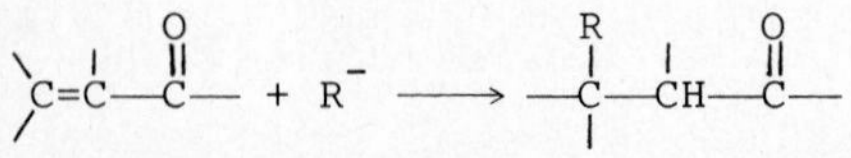

$$\text{C=C}-\overset{O}{\overset{\|}{C}}- + R^- \longrightarrow -\overset{R}{\underset{|}{C}}-CH-\overset{O}{\overset{\|}{C}}- \qquad\qquad 882-883$$

(from ketone, β-dicarbonyl or enamine)

Diels-Alder cycloaddition

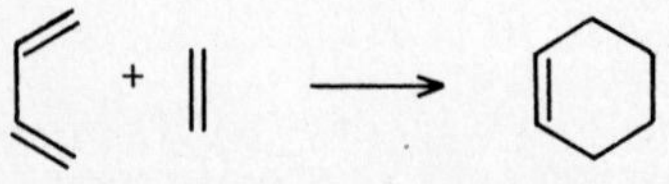

613-617, 679, 862, 1023, 1041, 1051, 1052, 1105, 1141-1143, 1186

Carbenoid additions

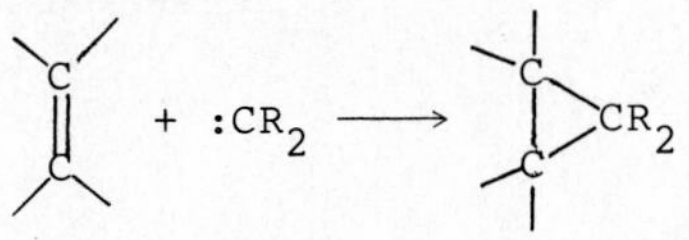

322, 323, 467, 741, 789

Claisen rearrangement

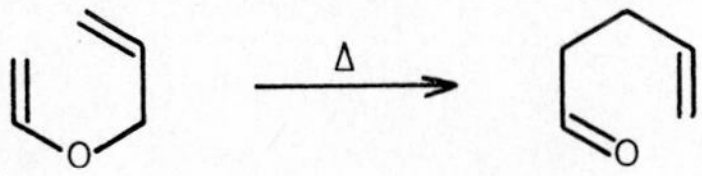

1014

Carbon tetrahalide addition

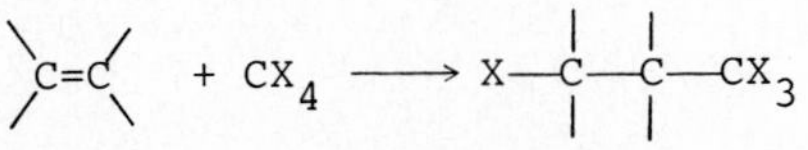

$\rangle C=C \langle \; + \; CX_4 \longrightarrow X-\overset{|}{\underset{|}{C}}-\overset{|}{\underset{|}{C}}-CX_3$ 313, 314

Alkylation of Carbon Acids

Acetylides

$RC{\equiv}C^- \; + \; R'Y \longrightarrow RC{\equiv}CR'$ 342, 343, 346, 386, 464, 473

Benzylic carbanions

Ar_2CH_2 (or Ar_3CH) $\xrightarrow{NaNH_2}$

$\xrightarrow{RX} Ar_2CHR$ (or Ar_3CR) 670

Dithianes

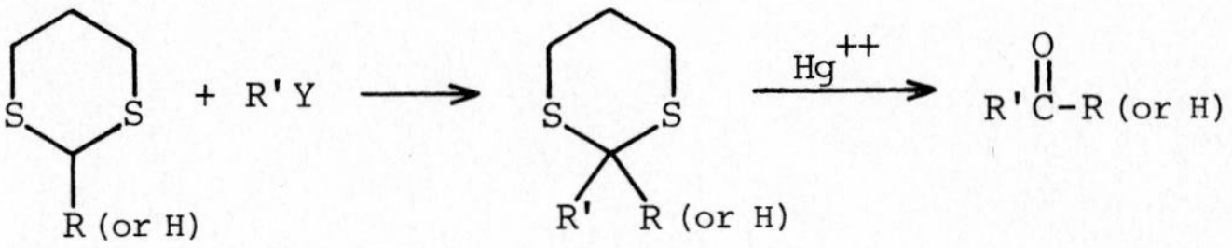

832, 833

Enamines

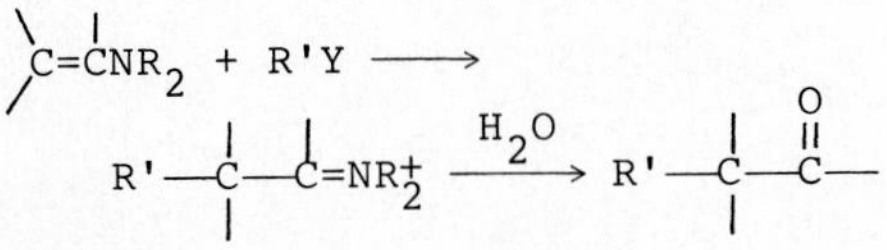

$\rangle C=\overset{|}{C}NR_2 \; + \; R'Y \longrightarrow$

$R'-\overset{|}{\underset{|}{C}}-\overset{|}{C}=NR_2^+ \xrightarrow{H_2O} R'-\overset{|}{\underset{|}{C}}-\overset{O}{\overset{\|}{C}}-$ 769, 884

Enolates

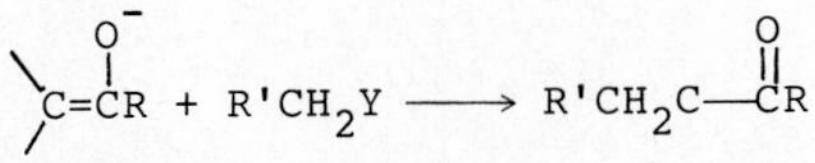

$\rangle C=\overset{O^-}{\overset{|}{C}}R \; + \; R'CH_2Y \longrightarrow R'CH_2C-\overset{O}{\overset{\|}{C}}R$ 374, 465, 559, 674, 742

$(EtOOC)_2CH^- \; + \; RY \longrightarrow (EtOOC)_2CHR$ 877, 878, 944-946

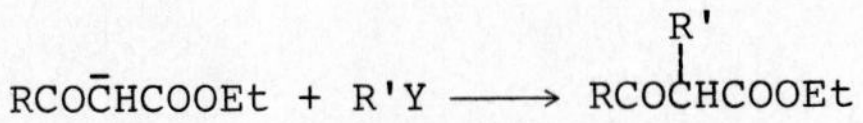

$RCO\bar{C}HCOOEt \; + \; R'Y \longrightarrow RCO\overset{R'}{\overset{|}{C}}HCOOEt$ 878-880, 884

$R\bar{C}HCOOEt \; + \; R'Y \longrightarrow RR'CHCOOEt$ 559

Phosphorus- and sulfur-stabilized carbanions

$RCH_2Z \xrightarrow{nBuLi} \xrightarrow[R'_2C=O]{R'X \text{ or}} RCH(R')Z$ or $RCHZ(COHR_2)$	833-834

($Z = PO_3R''_2$, SR'', $\overset{+}{S}R''_2$, SO_2R'')

Alkyne Coupling

$RC{\equiv}CH \xrightarrow[O_2]{Cu^{+2}} RC{\equiv}C{-}C{\equiv}CR$	351
$RC{\equiv}C^- + BrC{\equiv}CR' \longrightarrow RC{\equiv}C{-}C{\equiv}CR'$	351
$RC{\equiv}CH \xrightarrow[NH_4Cl]{Cu^{+1}} RCH{=}CHC{\equiv}CR$	351, 594

Carbonyl Condensation Reactions

Aldol type

$RCH_2CHO \xrightarrow{base} RCH(CHO)CHOHCH_2R$	392, 393, 396, 465, 847, 852, 944
$RC(=O)CH_2R' + R''CHO \longrightarrow RC(=O)C(R'){=}CHR''$	393-396, 465, 473, 599, 852
$RC(=O)CH_2R' + R''C(=O)R''' \xrightarrow{H^+} RC(=O)C(R'){=}CR''R'''$	395, 599-600, 883, 884
$RCH_2COOEt \xrightarrow{R'_2NLi} R\bar{C}HCOOEt \xrightarrow{R''C(=O)R'''} R''{-}C(OH)(RCHCOOEt){-}R'''$	560, 858
$RC(=O)R' + R''COCH_2COOEt \longrightarrow RC(OH)(R'){-}CH(COOEt)COR''$ or $RR'C{=}C(COOEt)(COR'')$	880, 881

Benzoin

$ArCHO \xrightarrow{CN^-} ArCOCHOHAr$	851, 852

Claisen and Dieckmann condensations

$RCOOEt + R'CH_2COOEt \xrightarrow{EtO^-} R'CH(COOEt)COR$	560-562, 870, 871
$RCOOEt + R'CH_2C(=O)R'' \xrightarrow{EtO^-} RC(=O){-}CH(R')C(=O)R''$	872, 873

Horner-Emmons

$RC(=O)R + R'C(=O)\bar{C}HPO_3Et_2 \longrightarrow R_2C{=}CHC(=O)R'$ 609, 831

Knoevenagel

$RCHO + CH_2(COOH)_2 \xrightarrow{\text{amine}} RCH{=}C(COOH)_2$ 880, 881

Mannich

$RC(=O)CH_2R' + CH_2O + HNR''_2 \xrightarrow{H^+} RC(=O)CH(R')CH_2NR''_2$ 770

Perkin

$ArCHO + (RCH_2CO)_2O \xrightarrow{RCH_2CO_2^-} ArCH{=}C(R)COOH$ 609, 610, 1079

Nitro carbanions

$RCH_2NO_2 + RC(=O)R' \xrightarrow{\text{base}} RCH(NO_2)C(OH)R'_2 \longrightarrow R{-}C(NO_2){=}CR'_2$ 781

Wittig

$Ph_3P{=}CHR + R'_2C{=}O \longrightarrow R'_2C{=}CHR$ 398, 465, 470, 593

Cyanide Reactions

$RY + CN^- \longrightarrow RCN$ 160, 168, 311, 464, 509, 608, 665, 862

$RC(=O)R' + HCN \longrightarrow RR'C(OH)CN$ 391, 392, 464, 610, 857, 915

$RC(=O)R' + HCN + NH_3 \longrightarrow RR'C(NH_2)CN$ 945, 946

$-\overset{|}{C}{=}\overset{|}{C}-C(=O)- + HCN \longrightarrow -\underset{|}{\overset{CN}{\overset{|}{C}}}-\overset{|}{C}H-C(=O)-$ 601-602, 862

$ArX + CuCN \longrightarrow ArCN$ 794

Grignard and Related Organometallic Reactions

Acyl halides

$RCOCl + R'MgX \longrightarrow RC(=O)R'$ 551

$RCOCl + R'_2CuLi \longrightarrow RC(=O)R'$ 552

$RCOCl + CH_2N_2 \longrightarrow RC(=O)CHN_2 \text{ or } RC(=O)CH_2Cl$ 788

Aldehydes and ketones

$RMgX \text{ (or RLi)} + R'C(=O)R'' \text{ (or H)} \longrightarrow RC(OH)(R')R'' \text{ (or H)}$ 449-451, 465, 469, 470, 475, 602, 603, 855, 990

$R\overset{O}{\overset{\|}{C}}R'$ (or H) + $BrCHR''COOEt \xrightarrow{Zn}$ $R\overset{OH}{\overset{|}{\underset{R'\text{ (or H)}}{\underset{|}{C}}}}CHR''COOEt$ 560

$R\overset{O}{\overset{\|}{C}}R'$ (or H) + $RC{\equiv}C^-$ ($\overset{+}{Mg}X$, Li^+, or Na^+) $\longrightarrow$ $R\overset{OH}{\overset{|}{\underset{R'\text{ (or H)}}{\underset{|}{C}}}}C{\equiv}CR''$ 391, 465, 594

Carbon dioxide

$RMgX$ (or RLi) $\xrightarrow{CO_2}$ RCOOH 451, 465, 476, 508, 990, 1185

Esters

$RCOOEt + R'MgX$ (or R'Li) $\longrightarrow RR'_2COH$ 552

$RMgX + (EtO)_2C{=}O \longrightarrow R_3COH$ 553

Halides

$RMgX + CH_2{=}CHCH_2X \longrightarrow RCH_2CH{=}CH_2$ 579, 593

Nitriles

$RCN + R'MgX$ (or R'Li) $\longrightarrow R\overset{O}{\overset{\|}{C}}R'$ 553

Orthoesters

$ArMgX + HC(OEt)_3 \longrightarrow \xrightarrow{H^+} ArCHO$ 452

Oxirane

$RMgX + CH_2\overset{O}{—}CH_2 \longrightarrow RCH_2CH_2OH$ 452, 453, 470

Unsaturated ketones

$RMgX$ (or R_2CuLi) +

$\rangle C{=}\overset{|}{C}—\overset{O}{\overset{\|}{C}}— \longrightarrow —\overset{R}{\overset{|}{\underset{|}{C}}}—\overset{|}{C}H—\overset{O}{\overset{\|}{C}}—$ 603, 604, 839

__Formation of C—C Bonds to Aromatic Rings__

Amine derivatives

$ArNH_2 \xrightarrow{HONO} ArN_2^+ \xrightarrow{CuCN} ArCN$ 794, 802

$ArNH_2 \xrightarrow{HONO} ArN_2^+ \xrightarrow[OH^-]{Ar'H} Ar—Ar'$ 797-798, 802

$ArNHNHAr \xrightarrow{H^+} H_2NAr'Ar'NH_2$ 784, 1032
(benzidine rearrangement)

Arenes

$ArH + RCOCl\ (\text{or}\ (RCO)_2O) \xrightarrow{\text{Lewis acid}} Ar\overset{O}{\overset{\Vert}{C}}R$	697, 698, 720, 761, 868-870, 1009, 1035, 1040
$ArH + RCOOH \xrightarrow[\text{Lewis acid}]{H^+\ \text{or}} Ar\overset{O}{\overset{\Vert}{C}}R$	869, 870, 1009, 1011, 1040, 1050
$ArH + CO + HCl \xrightarrow[AlCl_3]{ZnCl_2\ \text{or}} ArCHO$	699
$ArH + HCON(CH_3)_2 \xrightarrow{POCl_3} ArCHO$	761
$ArH + CH_2O + HCl \xrightarrow{ZnCl_2} ArCH_2Cl$	701
$ArH + RX \xrightarrow{AlCl_3} ArR\ (+ Ar'R + ArR')$	699-702, 720
$ArH + {>}C{=}C{<} \xrightarrow{H^+} Ar{-}\overset{\vert}{\underset{\vert}{C}}{-}\overset{\vert}{\underset{\vert}{C}}{-}H$	701, 821, 996, 1185

Aryl halides and organometallics

$ArX + RX \xrightarrow{Na} ArR$	991
$ArX \xrightarrow{Cu} ArAr$	991
$ArX \xrightarrow{Mg} \xrightarrow{CO_2} ArCOOH$	990
$ArX \xrightarrow{CuCN} ArCN$	794
$ArLi \xrightarrow{CuBr} \xrightarrow{O_2} Ar{-}Ar$	992

Phenols

$ArO^- + RCHO \longrightarrow HOAr'\overset{OH}{\overset{\vert}{C}}HR$	1002
$ArO^- + CO_2 \longrightarrow Ar(OH)(COOH)$	1003
$ArO^- + CHCl_3 \longrightarrow Ar(OH)(CHO)$	1004
$ArOH + RCN \xrightarrow[HCl]{ZnCl_2} Ar(OH)(COR)$	1009
$ArO\overset{O}{\overset{\Vert}{C}}R \xrightarrow{AlCl_3} Ar(OH)(COR)$	1011
$ArOCH_2CH{=}CH_2 \xrightarrow{\Delta} Ar(OH)(CH_2CH{=}CH_2)$ (Claisen rearrangement)	1012, 1013, 1145